Shock Waves

Klaus Hannemann · Friedrich Seiler (Eds.)

Shock Waves

26th International Symposium on Shock Waves, Volume 1

Springer

Editors

Dr. Klaus Hannemann
German Aerospace Center, DLR
Bunsenstraße 10
37073 Göttingen
Germany
Klaus.Hannemann@dlr.de

Prof. Dr. Friedrich Seiler
French-German Research Institute
of Saint Louis, ISL
5 rue du Général Cassagnou
68301 Saint-Louis
France
Seiler@isl.tm.fr

Co-Editors

Daniel Banuti
Martin Grabe
Volker Hannemann
Sebastian Karl
Jan Martinez Schramm
Julio Srulijes
Jeremy Wolfram

ISBN: 978-3-540-85167-7 e-ISBN: 978-3-540-85168-4

DOI 10.1007/978-3-540-85168-4

Library of Congress Control Number: 2008934453

Cover design: Sumitra Sarma

Printed on acid-free paper

9 8 7 6 5 4 3 2 1

springer.com

Preface

The 26th International Symposium on Shock Waves was held from July 15 to July 20, 2007 in the "Deutsches Theater in Göttingen", Germany. It was jointly organised by the German Aerospace Center DLR and the French-German Research Institute of Saint Louis ISL. The year 2007 marks the 50th anniversary of this symposium which was held for the first time in 1957 in Boston - at that time named "Shock Tube Symposium". After 1967 in Freiburg, being the first symposium which was held outside the United States of America, and 1987 in Aachen, it was an honour for us to host the symposium again in Germany and to celebrate this memorable anniversary. Moreover, we were pleased to organise ISSW26 in the same year in which 100 years of institutionalised German aerospace research is commemorated. This event refers back to the long history of basic and applied contributions to applied mechanics, hydro-, aero-, and gas dynamics which was started in Göttingen by Ludwig Prandtl at the beginning of the last century. In 1907, Ludwig Prandtl founded the "Modellversuchsanstalt für Aerodynamik der Motorluftschiff-Studiengesellschaft" (Institute for Testing of Aerodynamic Models of the Powered Airship Society) in Göttingen and he is therefore considered one of the key founding fathers of institutionalised aerospace research. Later the "Modellversuchsanstalt Göttingen" became the "Aerodynamische Versuchsanstalt AVA" (Institute for Aerodynamic Testing), a precursor of the modern day German Aerospace Center DLR.

After the first and second announcement of the symposium was distributed in summer 2006, 464 expressions of interest from 29 countries were received. The call for abstract submission resulted in a total number of 336 abstracts. Each abstract was reviewed by two members of the ISSW26 Scientific Review Committee which consisted of 49 renowned experts in the field of shock wave research. The final program of the symposium contained nine plenary lectures, 161 oral and 87 poster contributions presented in 47 sessions. The posters were presented in two dedicated sessions without overlapping oral presentations.

Two hundred and eighty two participants, of which 71 were students, from 23 countries registered. The nations from which the participants originated were: Germany (66), Japan (48), Russia (26), USA (25), Australia (16), France (14), Canada (11), India (10), Israel (10), Republic of Corea (10), China (9), United Kingdom (9), Netherlands (5), South Africa (5), Belarus (3), Switzerland (3), Taiwan (3), Brasil (2), Norway (2), Singapore (2), Italy (1), Poland (1) and Ukraine (1). In addition, 50 partners enrolled for the companions program of ISSW26.

Following the opening ceremony, the symposium was started with the Paul Vieille Lecture, given by Prof. Hans Hornung, California Institute of Technology, providing an exciting presentation on: "Relaxation Effects in Hypervelocity Flow: Selected Contributions from the T5 Lab".

A new facet for ISSW was the initiation of the Student Award of the International Shock Wave Institute ISWI. The mission of ISWI, which was founded two years ago, is to promote international and interdisciplinary collaboration in all areas of shock wave research through the organisation of conferences, awards and honours and to facilitate liaison with other organisations with similar interests and activities. We hope that with the ISWI Student Award we start a tradition for all the following ISSWs which will further strengthen the links between the ISWI and the shock wave symposia. The award was given to the best two student presentations at ISSW26 in Göttingen. Oral- and poster presentations given by the student participants were considered for the prize and the jury consisted of members of the International Advisory Committee and Session Chairpersons of ISSW26. The ISWI Student Award, which was endowed with $1000 US each, was presented during the opening ceremony of the ISSW26 Dinner Banquet by Prof. Kazuyoshi Takayama, President of the International Shock Wave Institute. The winners are:

- Khadijeh Mohri, Imperial College, London, for her presentation on: “Supersonic Flow Over Axisymmetric Cavitites” (together with R. Hillier) and
- Roger Aure, University of Arizona, for his presentation on: “Particle Image Velocimetry Study of Shock induced Single Mode Richtmyer-Meshkov Instability” (together with J.W. Jacobs).

Something unique was the venue of ISSW26. The international shock wave community met in the “Deutsches Theater in Göttingen” (German Theatre) where, during the season, works of classical and modern theatre are performed. Based on the comments we received from many participants we can conclude that our hope that the combination of artistic atmosphere and science will result in an inspiring symposium was fulfilled. The theatre, which was opened in 1890 with the performance of Friedrich Schiller’s “Wilhelm Tell”, is located right next to the remains of the mediaeval town wall which surrounds the historic city centre of Göttingen. Much of the urban life takes place within the old town walls and all places of interest including hotels and restaurants could be reached by a leisurely ten to fifteen minute walk.

The scientific program was complemented by a number of social events. The lunches were provided in the Old Mensa (refectory) of the University of Göttingen located in the city centre. On Monday evening Göttingen was introduced to the ISSW26 participants during a city tour which finished with a reception of the mayor of Göttingen in the Old Town Hall. The over 1050 years old town of Göttingen is set amid the rolling mountain landscape of southern Lower Saxony in the central part of Germany. The landmark of Göttingen is the Gänseliesel figure on the fountain at the market square in front of the Old Town Hall. This art nouveau statue shows a girl herding geese. It is a tradition that all PhD graduates from the University kiss the statue after passing their examination. This makes her “the most kissed girl in the world”. In addition to the charm of the city centre with its evocative relics of past centuries, Göttingen is well known in the scientific community for its large university, numerous research institutes and 44 Nobel Prize winners who lived or worked in Göttingen. On Wednesday, the symposium excursion led the ISSW26 participants to the city of Wernigerode located at the northern border of the Harz mountains. Wernigerode is a town with gorgeous half-timbered houses - the oldest of which can be traced back to the year 1400. The Town Hall is a highlight of medieval architecture and high above the city the Castle of Wernigerode is located, an originally romanesque castle dating from the 12th century which was extensively altered over the centuries. The symposium excursion ended with a barbecue at the Göttingen site of DLR. On Thursday night, the traditional symposium dinner banquet took place in the main foyer, the glass foyer and the “Keller” (cellar) of the Deutsches Theater. The ceremonial opening of the banquet in the Große Haus of the Deutsches Theater included addresses, the presentation of the ISWI student award, the announcement of the ISSW27 venue and musical entertainment by the "Trio con Brio" of the Göttingen Symphonic Orchestra. On Friday afternoon the ISSW26 was concluded by a facility tour of the Göttingen DLR site.

A companions program was organised during ISSW26, which led the partners and family members of the participants to various locations in the southern Lower Saxony and Göttingen area including the "Sleeping Beauty Castle" Sababurg, the Weser river, Marienburg Castle near Hildesheim, the City of Einbeck, the Plesse Fort, Herzberg Castle, the former German - German border, the Harz mountains and the city of Goslar.

During the meeting of the International Advisory Committee of ISSW26, the venue for ISSW27 which will be held in 2009, was selected. Three excellent proposals were presented to the IAC which shows that there is a continuing interest in ISSW. The 27th International Symposium on Shock Waves will be held in St. Petersburg, Russia and will be chaired by Dr. Irina Krassovskaya of Ioffe Institute, Russian Academy of Sciences. On behalf of the Organising Committee of ISSW26, we congratulate Irina Krassovskaya and wish ISSW27 great success.

ISSW26 could not have been realised without the support of the host institutions, the German Aerospace Center DLR and the French-German Research Institute of Saint Louis ISL, and this is gratefully acknowledged. Further, ISSW26 was generously sponsored by the German Research Foundation DFG and the state of Lower Saxony. The companies LaVision, MF Instruments, Palas, Shimadzu, Springer and VKT could be acquired who supported ISSW26 with a stand at the conference venue.

On behalf of the Organising Committee of ISSW26, we would very much like to thank all participants who came to Göttingen to support the symposium with their attendance and the presentation of a poster or a talk. We would also like to express our gratitude to the members of the International Advisory Committee and the Scientific Review Committee for their continuous support during the preparation and running of the conference. Further, we would like to thank those colleagues who served as session chairperson and who guaranteed an accurate performance of the parallel sessions.

Finally, the support and enthusiasm of all the people who supported the organisation of ISSW26 including the project management of DLR, the logistics section and the secretaries of the DLR Institute of Aerodynamics and Flow Technology, the DLR canteen staff, the ISL team, the students, the team of the Deutsches Theater, the catering team of the Bistro DT Keller and the members of the Spacecraft Section and their families is highly acknowledged. The team spirit of these people was the basis of the realisation of ISSW26 and their friendly appearance was greatly appreciated by the symposium participants.

Göttingen, July 2007

Klaus Hannemann (Chairman)
Friedrich Seiler (Co-Chairman)

The 26th International Symposium on Shock Waves

Hosted by German Aerospace Center, DLR
and French-German Research Institute of Saint-Louis, ISL
Göttingen, Germany
July 15-20, 2007

Chairman

Klaus Hannemann (German Aerospace Center)

Co-Chairman

Friedrich Seiler (French-German Research Institute of Saint-Louis)

Local Organising Committee

Klaus Hannemann
Friedrich Seiler
Sebastian Karl
Jan Martinez Schramm
Sumitra Sarma
Berthold Sauerwein
Julio Srulijes
Susanne Strempel
Alexander Pichler
Jeremy Wolfram

International Advisory Committee

Takashi Abe
Holger Babinsky
Robert Bakos
Gabi Ben-Dor
Salvatore Borrelli
Martin Brouillette
Douglas Fletcher
Nikita Fomin
Sudhir Gai
Boris Gelfand
Victor Golub
Jagadeesh Gopalan
Klaus Hannemann
Ronald Hanson
Richard Hillier
Ozer Igra
Katsuhiro Itoh
In-Seuck Jeung
Zonglin Jiang
Valeriy Kedrinskiy
Shen-Min Liang
Assa Lifshitz
Meng-Sing Liou
Frank Lu
Kazuo Maeno
Richard Morgan
Herbert Olivier
Elaine Oran
Allan Paull
K.P.J. Reddy
Graham Roberts
Akihiro Sasoh
Christof Schulz
Friedrich Seiler
Joseph Shepherd
Beric Skews
Kazuyoshi Takayama
Eleuterio Toro
David Zeitoun
Fan Zhang

Scientific Review Committee

Takashi Abe
Holger Babinsky
Robert Bakos
Gabi Ben-Dor
Martin Brouillette
Douglas Fletcher
Sudhir Gai
Victor Golub
Jean François Haas
Gerald Hagemann
Klaus Hannemann
Marc Havermann
Richard Hillier
Ozer Igra
Katsuhiro Itoh
Mikhail Ivanov
Zonglin Jiang
Mark Kendall
Harald Kleine
Doyle Knight
Wilhelm Kordulla
Assa Lifshitz
Frank Lu
Jan Martinez Schramm
Tim McIntyre
David Mee
Brian Milton
Richard Morgan
Christian Mundt
Herbert Olivier
Allan Paull
Rolf Radespiel
K.P.J. Reddy
Graham Roberts
Akihiro Sasoh
Christof Schulz
Richard Schwane
Friedrich Seiler
Joseph Shepherd
Beric Skews
Julio Srulijes
Johan Steelant
Hideyuki Tanno
Evgeny Timofeev
Euleterio Toro
Zbigniew Walenta
Craig Walton
David Zeitoun
Fan Zhang

Event Coordination and Graphic Design

Sumitra Sarma

Website Programming and Database Support

Jeremy Wolfram

Partners Program

Monika Hannemann

Venue, Deutsches Theater in Göttingen and Deutsches Theater Bistro

Wolfgang Bertram
Carsten Hoffmann
Ulrich Klötzner
Claudia Schmitz

Sponsors

Deutsche Forschungsgemeinschaft
Land Niedersachsen
Deutsches Zentrum für Luft-und Raumfahrt e.V.
Deutsch-Französisches Forschungsinstitut Saint-Louis

Exhibitors

LaVision GmbH
MF Instruments GmbH
Palas GmbH
SHIMADZU
Springer
VKT

Contents

Volume 1

Part I Plenary Lectures

Part II Blast Waves

Part IV Detonation and Combustion

Part VI Facilities

Part VII Flow Visualisation

Part VIII Hypersonic Flow

Part IX Ignition

Contents

Volume 2

Part XII Multiphase Flow

Part XIII Nozzle Flow

Part XIV Numerical Methods

Part I

Plenary Lectures

Relaxation effects in hypervelocity flow: selected contributions from the T5 Lab

H.G. Hornung

Graduate Aeronautical Laboratories, California Institute of Technology

1 Introduction

Experimental access to flows in the total enthalpy range from 5 to 20 MJ/kg was greatly increased by Stalker's [1] introduction of the free-piston shock tunnel in the 1960's. Since then, experiments conducted in the later T series of these (T3 in Canberra, T4 in Queensland, T5 at Caltech) and the larger machines (HEG in Göttingen, HIEST in Japan) have changed the landscape of gasdynamics in this range enormously. Specifically, a number of previously unknown phenomena, some showing dramatic effects, were discovered by such experiments. The scope of this paper is much too constraining to do justice to the whole body of contributions of these research groups. Accordingly I will restrict it to a selection from those more familiar to me and that have come from the T5 laboratory. Much of the work in the total enthalpy range of the free-piston shock tunnels is concerned with the main enemy of this regime: heat flux. For this reason, blunt body flows, boundary-layer transition and shock interactions are of special importance. Even before the introduction of the high enthalpy facilities these problems had been studied in great detail in cold hypersonics, where the important phenomena that are brought about by vibrational excitation and dissociation are, of course, completely missed.

2 Flow over spheres

Because the Mach number independence principle applies with very good accuracy in the case of hypersonic blunt-body flows, and since recombination reactions are not of great importance in the shock layer of a blunt body, binary scaling is very useful for such flows. This enables the number of independent similarity parameters to be reduced to two, one characterizing the rate of energy absorption by relaxation processes such as vibrational excitation and dissociation, and one characterizing the amount of energy absorbed by relaxation processes. When there is only one relaxation rate these parameters are easy to define. However, by considering the flow along the stagnation streamline of a blunt body, Wen [2] was able to define these two parameters for situations in which multiple species and reactions occur:

$$\Omega = \frac{\rho_s d R \Gamma}{\rho_\infty u_\infty^3 C_{ps}} \left(\sum_{i=2}^{n} h_{c_i} \frac{dc_i}{dt} \right)_s$$

$$\Lambda = 1 - \frac{u_e^2}{u_s^2},$$

where ρ, d, R and Γ are density, characteristic length, the universal gas constant and inverse molecular weight, c_i are the species mass fractions, t is time, h is specific enthalpy,

u is flow velocity and C_p is specific heat at constant pressure. The subscripts s and ∞ denote conditions immediately behind the shock and in the free stream respectively. The subscript c_i denotes partial differentiation. The subscript e denotes conditions at equilibrium after a normal shock. All of the variables in these parameters are determined by the free-stream conditions, the properties of the gas mixture and the shock-jump relations.

The parameter Ω may be thought of as the ratio of the rate of energy absorption by the relaxation processes to the rate at which kinetic energy arrives at the bow shock. The parameter Λ may be thought of as the ratio of the energy absorbed by the relaxation to the kinetic energy in the case of a normal-shock flow. These two parameters define the flow field in the shock layer of a blunt body. Specifically, Wen showed that the dependence of the shock stand-off distance in flows over spheres can be obtained explicitly in terms of these two parameters and made extensive comparisons with experiments from T5, which broadly substantiated the theory. An example is shown in Fig. 1.

Wen also obtained the important result that the stagnation-point heat flux formula for dissociating air given by Fay and Riddell [3] predicts the experimental values very well for other gases as well.

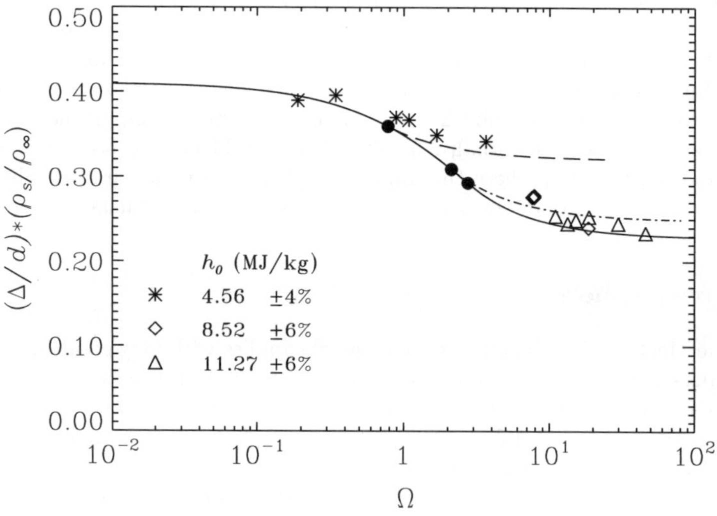

Fig. 1. Shock stand-off distance on a sphere in carbon-dioxide flow. Comparison of theory with experiment

3 Shock interaction

3.1 Shock-shock interaction

The enemy "heat flux" becomes particularly nasty in the situation when (even a weak) oblique shock impinges on the bow shock of a blunt body in the vicinity of the stagnation region. This was first shown by Edney [4] who classified the possible interactions into six types, of which type IV was the most serious. In flow of a perfect diatomic gas, amplification ratios of heat flux over the value without impingement of up to 16 were observed. More particularly, Edney estimated the effect of relaxation by using a perfect gas with reduced ratio of specific heats (γ) and found that the amplification factor with $\gamma = 1.2$ was increased to as much as 60.

An extensive investigation of this problem in high-enthalpy flow of nitrogen was conducted by Sanderson [5], see also [6]. In this study the impinging oblique shock was generated by a wave-rider in order to obtain as nearly two-dimensional a shock as possible at the location where it impinges onto the bow shock of a circular cylinder. Sanderson was able to obtain flow field information in the form of holographic interferograms, see, *e. g.*, Fig. 2, as well as detailed heat flux distributions on the cylinder.

Fig. 2. Holographic interferogram of type IV shock-on-shock interaction in nitrogen flow at $h_0 = 12$ MJ/kg. The locations of the heat flux gauges are shown in the overlayed scale. Note the detailed resolution of the flow field, including the high density in the supersonic jet near the stagnation point, and the flow separation at the top rear of the cylinder.

Sanderson chose three conditions at different total enthalpy, and investigated all six interaction types. Among his many results the most important are, that the large increase of heat flux amplification due to relaxation as estimated by Edney does not occur, and a simple formula for practical use that accurately predicts the heat flux amplification. The former result had actually already been observed in an experiment at a particular condition by Kortz [7] in the HEG. Of course, the wealth of detailed field information provides ample test cases for computations.

3.2 Shock boundary layer interaction

Separation induced by the interaction of a shock with a boundary layer is another area in which the main enemy, heat flux peaks, hits hard. Deflection of a control surface can induce this, and high heat flux occurs in the region where the separated flow reimpinges on the surface. This is also an area that is like a can of worms. Many competing effects make it almost impossible to find simple predictive rules, as was discovered by Davis [8] who made a very detailed experimental and theoretical study of shock boundary layer interaction in a concave corner. Davis used holographic interferometry and surface heat flux and pressure measurements. Correlation of his data did show high-enthalpy real-gas effects, among them one related to free-stream dissociation. Again, his extensive data base provides excellent test material for computation.

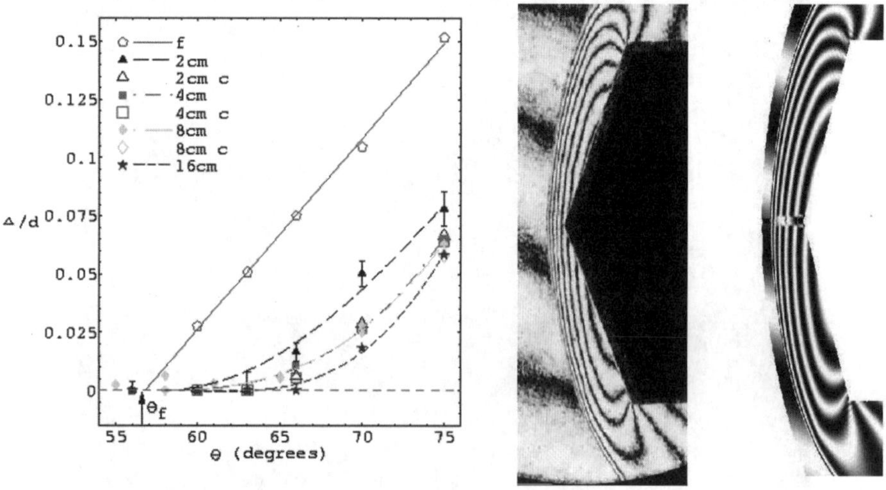

Fig. 3. Carbon dioxide flow over a cone. LEFT: Plot of shock stand-off distance normalized by base diameter against cone half-angle. The straight line emerging at the frozen detachment angle θ_f shows the behavior of frozen flow. The full symbols give experimental data in high-enthalpy relaxing flow for cones of different sizes. The open symbols show corresponding computed values. The scale effect arises from the characteristic length due to relaxation. Carbon-dioxide flow at 3 km/s. RIGHT:Comparison of experimental and computed interferograms of carbon dioxide flow at 4 km/s. The computed interferogram takes into account the finite fringes in the free stream. These are the cause of the asymmetry of the fringe pattern.

4 Shock detachment from cones

As had already been discovered by Smith [9] and Houwing [10] relaxation has an enormous influence on the process of shock detachment from wedges and cones. In order to extend the knowledge base in the case of cones, Leyva [11] made an extensive experimental study using holographic interferometry backed by numerical computations using the software of Candler [12]. The effect is well characterized by the plot of stand-off distance versus cone angle with cone size as a parameter shown on the left of Fig. 3. The right side of the figure shows a comparison of experimental and computed interferograms.

The intimate relation between the phenomena of the detachment process and the behavior of the sonic line in the cone flow with relaxation was brought out by a theoretical and computational study, the latter with fine resolution of the cone angle, in Hornung and Leyva [13].

5 Boundary layer transition

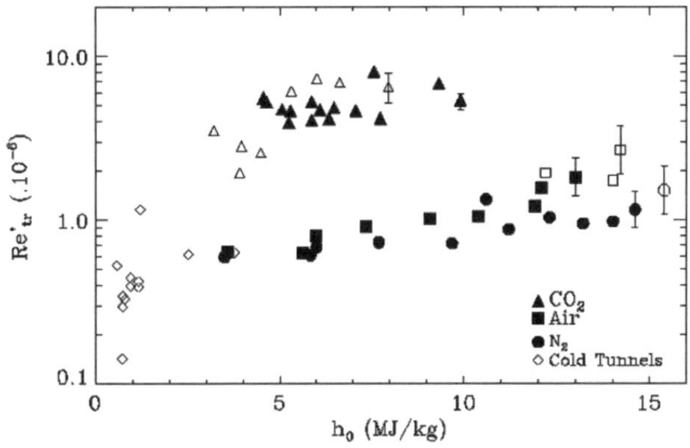

Fig. 4. Plot of the Reynolds number at transition on a 5° half-angle cone. The Reynolds number is based on distance from the tip and is evaluated at the reference conditions in the laminar boundary layer preceding transition. Note the very large transition delay with respect to the data from cold hypersonics, especially in the case of carbon dioxide flows. Data from Germain [14] and Adam [15].

Transition to a turbulent boundary layer is also a cause of heat flux increase. It often amounts to a factor of around four. In hypervelocity flow transition differs in two important ways from low-speed behavior. First, the instability responsible for the path to transition is not the viscous instability but an instability in which acoustic noise is amplified in the boundary layer which acts as a waveguide, amplifying frequencies that are of order U_∞/δ. Secondly, relaxation processes strongly damp acoustic disturbances in a broad range of frequencies around the inverse relaxation time. In consequence the

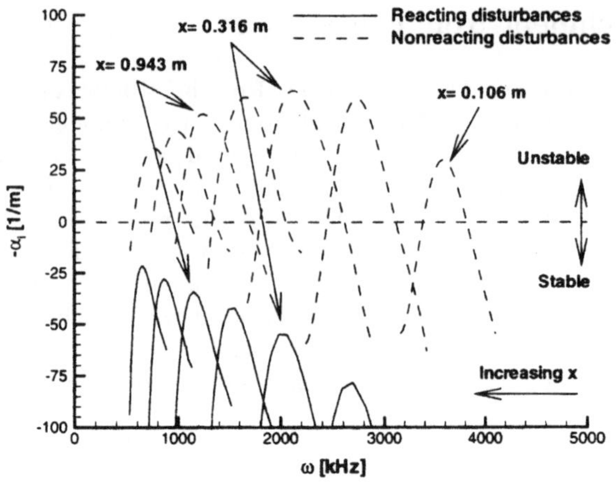

Fig. 5. One of the results of Johnson *et al.* [16] showing the growth rates of disturbances at different distances x from the cone tip with and without relaxation in a carbon-dioxide flow. Note how the presence of relaxation (solid curves) causes the growth rates to be reduced to negative values while the non-reacting values (dashed curves) are positive. Note also the high frequencies involved in high-enthalpy flows.

transition to a turbulent boundary layer can be dramatically delayed by relaxation effects, as may be seen in Fig. 4.

The transition delays observed by Germain and by Adam were dramatically confirmed to be caused by relaxation damping of the acoustic instability by Johnson *et al.* [16] as may be seen in an example of their numerical study of transition at the conditions of the T5 experiments, see Fig. 5.

6 Instability of the shear layer generated by a curved shock

The work in this area began with an exploratory computational study motivated by the expectation that the vorticity generated by a curved bow shock of a long blunt-nosed body would form a shear layer that eventually has to become unstable. Since the vorticity generated is approximately proportional to the density ratio across the shock and relaxation causes the density ratio to increase, we expected to find the shear layer to become unstable particularly early in high-enthalpy flow. The exploratory study showed that this density ratio effect was indeed present. Fig. 6 shows an example in which the shear layer becomes unstable within a nose radius or so from the nose, and forms such strong structures that they push shock waves and perturb the bow shock significantly.

In the computational study it was found that the shear layer remained stable within a region of ten nose radii from the nose of the body, if the normal-shock density ratio was less than 14. These results persuaded us to study the problem experimentally. Unfortunately, the highest normal-shock density ratio achievable in the shock tunnel is around 13. However, a modification of T5 to turn it into a gun that was made by Joe Shepherd and Mike Kaneshige could be used to propel spherical projectiles into a test section containing any gas. By using propane gas, density ratios up to 20 could be achieved across

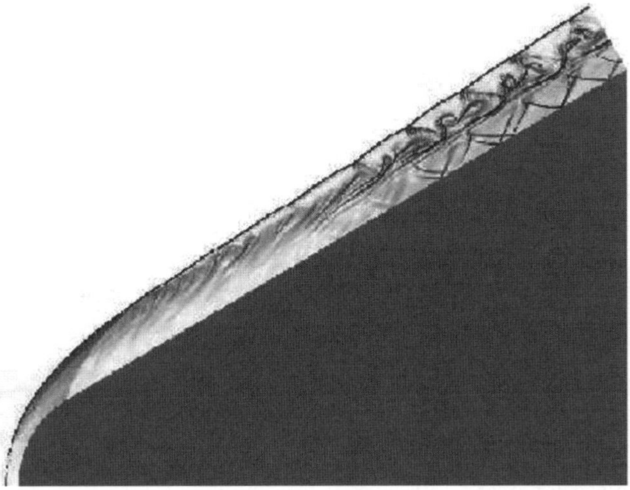

Fig. 6. Computed flow over a blunted wedge at a condition where the normal-shock density ratio is 17. Note the growth of weak disturbances introduced in the free stream grow dramatically in the shock layer to an extent where the bow shock is perturbed.

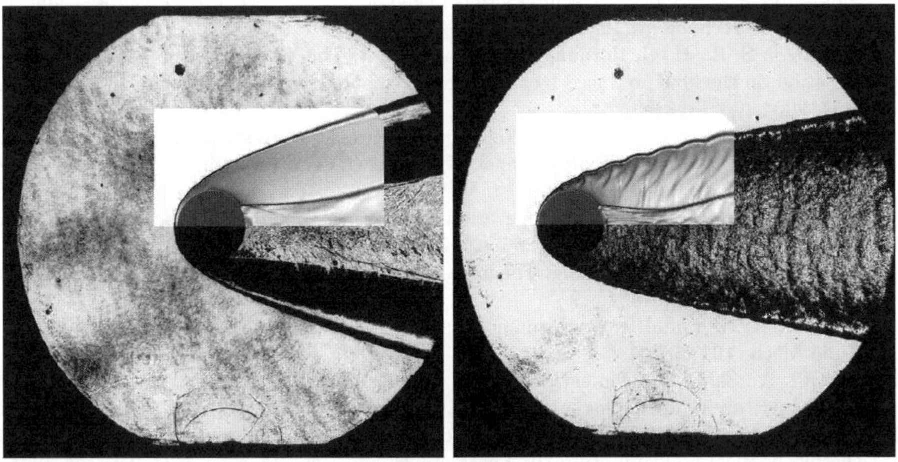

Fig. 7. Spherical projectile in propane LEFT: 2.2 km/s giving a density ratio of 13: smooth flow. RIGHT: 2.7 km/s giving a density ratio of 20: violently unstable flow. Computational result is overlayed.

the bow shock of a spherical projectile. Propane has a large number of vibrational degrees of freedom that are easily excited at the projectile velocities that can be reached in the T5 gun. Two examples of the experimental results are shown in Fig. 7, see also Hornung and Lemieux [17]. This dramatic effect becomes important, *e. g.*, in entry into the carbon dioxide atmosphere of Mars, where density ratios in the 20's occur. The rough flow resulting from the instability is bound to increase heat flux drastically.

7 Conclusions

Only selected examples from only one research group have been presented here, yet the range of important new and sometimes unexpected effects of relaxation that have been discovered by using Stalker's brainchild free-piston shock tunnels is quite amazing. A great deal more can be found by studying the output of the other research groups in Brisbane, Göttingen, Kakuda and New York.

Acknowledgement. The work described in sections 2, 3, 5, and 6 was supported by the United States Air Force Office of Scientific Research.

References

1. Stalker R. J.: A study of the free piston shock tunnel. AIAA J. **5**(12), 2160, 1967.
2. Wen C.-Y. and H. G. Hornung: Non-equilibrium dissociating flows over spheres. J. Fluid Mech. **299** 389-405, 1995.
3. Fay J. A. and F. R. Riddell: Theory of stagnation-point heat transfer in dissociated air. J. Aero. Sci. **25** 73-85, 1958.
4. Edney B. E.: Effects of shock impingement on the heat transfer around blunt bodies. AIAA J. **6**(1) 15-21, 1968.
5. Sanderson S. R.: Shock wave interaction in hypervelocity flow. Ph. D. thesis, California Institute of Technology, 1995.
6. Sanderson S. R., H. G. Hornung and B. Sturtevant: The influence of non-equilibrium dissociation on the flow produced by shock impingement on a blunt body. J. Fluid Mech. **516** 1-37, 2004.
7. Kortz S., T. J. McIntyre and G. Eitelberg: Experimental investigation of shock on shock interaction in the high-enthalpy tunnel Göttingen (HEG). 19th Int. Symp. Shock Waves, Marseille, 1993.
8. Davis J.-P. and B. Sturtevant: Separation length in high-enthalpy shock/boundary layer interaction. Phys. Fluids **12**(10), 2661-2687, 2000.
9. Smith G. H. and H. G. Hornung: The influence of relaxation on shock detachment. J. Fluid Mech. **93** 225-239, 1979.
10. Houwing A. F. P. and H. G. Hornung: Shock detachment from cones in a relaxing gas. J. Fluid Mech. **101** 307-319, 1980.
11. Leyva I. A.: Shock detachment process on cones in hypervelocity flow. Ph. D. thesis, California Institute of Technology, 1999.
12. Candler G. V. and R. W. MacCormack: Computation of weakly ionized hypersonic flow in thermochemical nonequilibrium. J. Thermophys. and Heat Transfer **5**(3), 266-273, 1991.
13. Hornung H. G. and I. A. Leyva: Sonic line and shock detachment in hypervelocity cone flow. Proceedings of IUTAM Symposium Transsonicum IV. H. Sobieczky (Ed.) Kluwer Academic Publishers.
14. Germain P.: The boundary layer on a sharp cone in high-enthalpy flow. Ph. D. thesis, California Institute of Technology, 1994.
15. Adam P. H.: Enthalpy effects on hypervelocity boundary layers. Ph. D. thesis, California Institute of Technology, 1997.
16. Johnson H. B., T. Seipp and G. V. Candler: Numerical study of hypersonic reacting boundary layers on cones. Phys. Fluids **10** 2676-2685, 1998.
17. Hornung H. G. and P. Lemieux: Shock layer instability near the Newtonian limit of hypervelocity flow. Phys. Fluids **13**(8) 2394-2402, 2001.

CFD contributions to high-speed shock-related problems
- examples today and new features tomorrow -

K. Fujii

Institute of Space and Astronautical Sciencc
Japan Aerospace Exploration Agency
Sagamihara, Kanagawa 229-8510, JAPAN

Summary. Computational Fluid Dynamics (CFD) have extensively contributed to high-speed shock-wave research. With study examples by the author's group in the past, effectiveness of CFD both for design of transportation vehicles and for understanding of fluid physics is discussed. Trend of CFD for further use is then discussed based on recent applications and three key features: computer progress, spectral-like high-resolution scheme and LES and LES/RANS hybrid method are focused. Recent CFD research reveals that high speed flows even the ones considered to be steady state have inherently unsteady nature that requires LES-like computations for successful simulations. Such simulations require remarkably higher grid resolution but emerging numerical techniques having spectral-like high-resolution would help reducing the number of grids required for such simulations. The paper is summarized by the address to the issues about future CFD.

1 Introduction

Computational Fluid Dynamics (CFD) study for high-speed flows in aerospace engineering was initiated in early 70's. The embedded shock wave was automatically captured in the computer simulation [1] and the design process of commercial aircraft was drastically changed. CFD study for shock-related problems was initiated much earlier before that . For example, Peter Lax developed a technique for computing high-speed flows including shock waves which represent discontinuities in the flow variables [2] in 1954. At any rate, a lot of CFD studies appeared especially after supercomputers was developed in late 1970's. Some are CFD study for fundamental shock-related problems and some are practical applications on shock-related problems. Theoretical and experimental studies had been two main pillars of research tools and CFD became a third pillar with the progress of computers. Now, CFD simulations became much easier than before [3] and even a small academic group can make important contributions to shock wave study.

Conventional RANS(Reynolds-averaged Navier-Stokes) simulations using turbulence models are widely spread as a useful tool for many research areas. With this as a background, CFD technology in high-speed flow research is considered to be in the first matured stage. Former paper by the present author looked back the CFD history and propose a new direction of CFD in the future [4]. When considering difference between CFD studies in 1980's and 2000's, we surprisingly notice that there is a little difference except that CFD simulations became much easier for more complex geometries. However, with three emerging technologies: computer progress, spectral-like high-resolution schemes and LES and LES/RANS hybrid methods, CFD for high-speed flow problems has started to move to a new stage. In the following sections, some of the author's studies are presented to show how CFD studies would shift its direction under the trend above.

2 Example in the Past - CFD became a good analysis tool

Present author has been engaged in the CFD study for high-speed trains for many years [5]- [7]. Some of the earlier results were used for the design of front cars of Shinkansen (rapid) trains in Japan. Recently, magnetic levitation trains are under development by Japan Railway Central (JRC) and extensive CFD study for the aerodynamic problems are conducted under the collaboration between JRC and the author's laboratory in ISAS/JAXA (Fig. 2).

Initial effort in 1990's was a CFD studyfor a aerodynamic problem of the train entering a tunnel. Figure 1 shows the schematic picture of the flow mechanism [5]. When a train enters a tunnel at high speed, strong pressure wave is created in front and it propagates toward the exit of the tunnel. The pressure wave tends to become steeper and strong booming noise appears outside the exit. The booming noise is one of the important aerodynamic problems for the development of Shinkansen trains as there are many tunnels in Japan. Comparison with the existing theories, experiments and field measurements showed that the CFD approach was useful for understanding the mechanism and for finding a method to alleviate the formation of compression wave inside a tunnel. Based on the simulation results, a new theory called v_{wall} theory was developed.

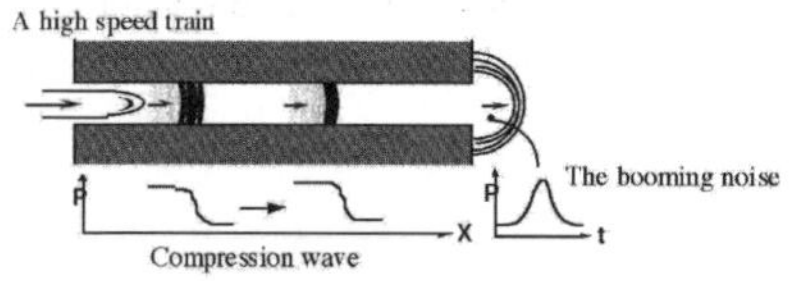

Fig. 1. Mechanism of the booming noise generation

Fig. 2. Example of the simulation for Shinkansen and Maglev trains

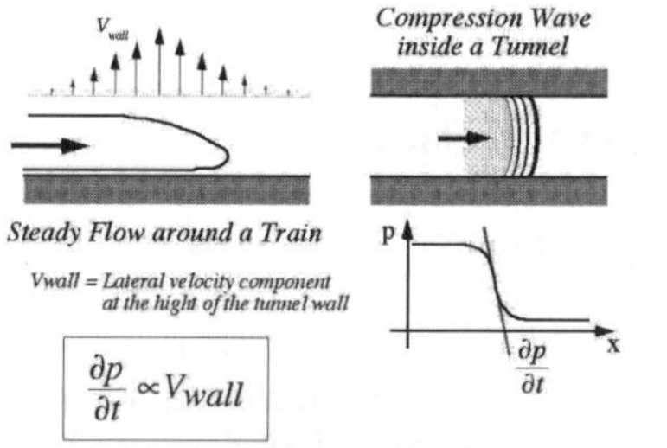

Fig. 3. Pressure gradient and the v_{wall}

Only the essence of the v_{wall} theory is presented [5], [7]. Let's consider change of instantaneous streamlines in time. When a tunnel comes closer to the train, it shrinks the streamtube near the train. Pressure increase inside the tunnel occurs due to this streamtube being narrowed. There exists outward velocity component near the train running outside the tunnel and that is diminished after the train enters the tunnel. Here, we define the integral of the outward velocity component normal to the tunnel wall as v_{wall} as schematically shown in Fig. 3. Some mathematical manipulations show that the gradient of the compression wave inside the tunnel is proportional to the v_{wall}. We call this to be "v_{wall} theory".

In summary, this theory can predict formation of the compression wave accurately only based on the steady-state flow filed. This newly-developed v_{wall} theory became a good engineering tool for the front car design alleviating the booming noise of high-speed trains. CFD gives us large three-dimensional data from which widen our views on flow features and help our understanding of fluid physics in addition to its capability as an analysis tool.

3 Evolutionary effort and New trend in CFD

As shown in the previous section, conventional RANS(Reynolds-averaged Navier-Stokes) simulations are widely spread and CFD in high-speed flow research is considered to be in the first matured stage. Gradual change that moves CFD to on the second matured stage occurs recent years. There are three key features behind it: computer progress, spectral-like high-resolution scheme and LES and/or LES/RANS hybrid method.

First factor: computer progress

In this section, a study example that became possible with the first key feature "computer progress" is presented. Simulations for problems including complex shock waves required large amount of computer time and therefore the discussions have been based on the restricted number of cases. With the progress of computer hardware and related software, studies with number of cases with different flow and/or geometrical parameters became feasible.

The authors' group has been engaged in both experimental and numerical studies of under-expanded jets impinging on an inclined flat plate [8], [9]. Flow fields of the M2.2 jet impinging on an inclined flat plate at various plate angles, nozzle-plate distances and pressure ratios were experimentally investigated. The results suggested that all the flow fields could be classified into three types of flow structure (See Fig. 4) [8]. However, flow structure details were not clarified because analysis was based on the schlieren pictures and the pressure contours over the flat plate obtained from PSP images.

Simulations with conventional method (RANS simulations with a second-order TVD upwind scheme together with a well-known turbulence model) were consequently carried out [9]. Almost fifty cases with different geometrical and physical parameters were simulated and the results were discussed. In Fig. 5, Pressure distributions at the pressure ratio PR=7.4 for four different plate angles and four different plate distances are shown as an example. There observed are several types of pressure peaks. Some cases have single peak and some cases have a few different type of peaks. Figures 6 shows an example of the flow structure details found through the analysis of the simulation results. Flow structure and the associated pressure peaks are much better understood when the flow field is analyzed using whole the three-dimensional data for many cases.

Second factor: LES and LES/RANS hybrid methods

There is an obvious shift in the CFD research from RANS simulations to Large Eddy Simulations (LES). The shift is supported by the rapid progress of computer performance, but more importantly, we start to recognize that the nature of flow physics, even from the engineering viewpoint, require unsteady flow simulations. Separated flows are inherently unsteady and steady recirculating region may be the result of a time average of strongly unsteady flows. Capturing such unsteady flow behavior is inevitable for the analysis and eventual control of the flows.

Requirement of the grid numbers for LES becomes enormous. In addition, fine mesh resolution near the wall boundaries limits the time step size for the computation. Therefore, it still remains difficult to apply LES to complex flows at high Reynolds numbers. To

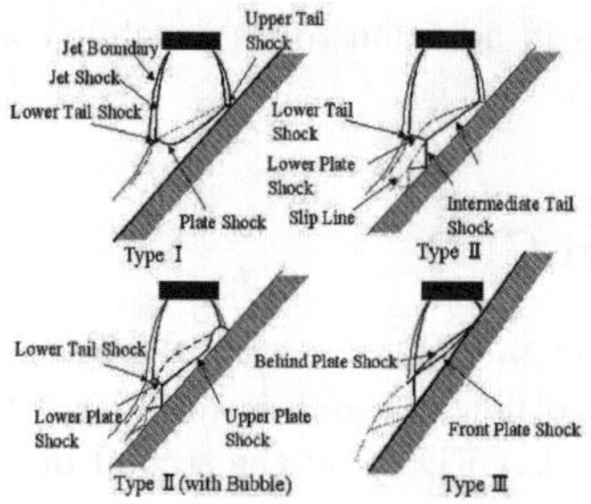

Fig. 4. Schematic pictures of typical flow fields for various plate-angle(PR>4)

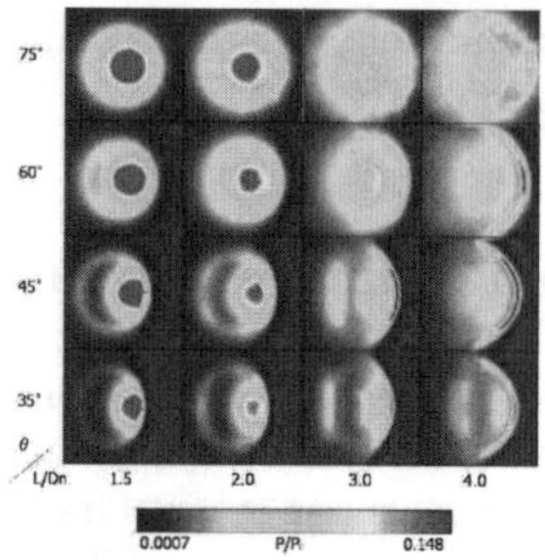

Fig. 5. Pressure contours on the plate surface: PR 7.4

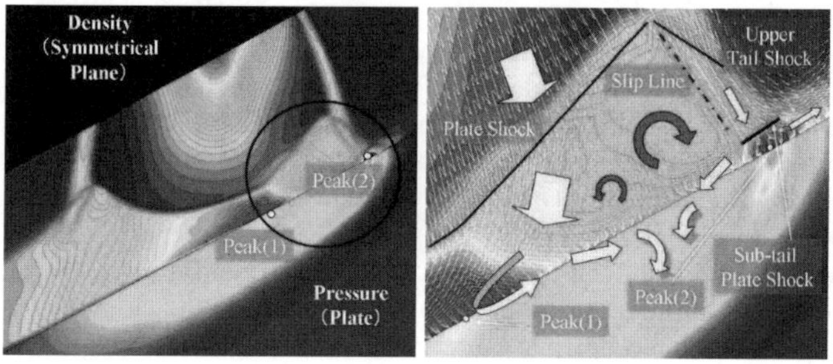

Fig. 6. Mechanism of the pressure peak (2) ($PR = 7.4$, $L/Dn = 3.0$, $\theta = 60°$)

overcome these difficulties, LES/RANS hybrid methodology was proposed, where RANS formulation is applied near the solid surface, while the LES formulation is applied to massively-separated flow regions. The hybrid methodology is considered to require much less computational cost than LES as it alleviates the required mesh resolution near (and along) the walls and the resultant time-step limitation.

As an example of this approach, simulation of supersonic base flow is presented [10]. Accurate simulation of base flows is important as the base drag influences aerodynamic characteristics of the vehicle at certain speed range. As shown in Fig. 7 schematically, supersonic base flow includes a large recirculation region and interaction of shear layer with expansion and compression waves appears. Even with such practical importance, pressure distributions over the base were not well estimated in any simulation using RANS models.

Figures 8(a) and 8(b) show instantaneous views of the computed vorticity magnitude contours computed by the LES/RANS hybrid method and RANS method. There exists strong flow unsteadiness in Fig. 8(a), whereas no unsteadiness is observed in Fig. 8(b). When time-dependent results computed by the LES/RANS hybrid method is averaged in time, vorticity magnitude contours become similar to Fig. 8(b) except the size of recirculating region. This indicates that steady flow field typically known is a time-average of the flow field with strong unsteadiness. The time-averaged base pressure distributions along the base surface are compared with the experiment in Fig. 9. The RANS result shows a strong variation of the pressure distributions due to a strong reverse flow, whereas LES/RANS hybrid method shows a flat distribution which agrees well with the experiment.

The results above show that capturing the unsteady nature of flow field leads to accurate flow simulations and the LES/RANS hybrid method including DES method is appropriate for accurate simulation of complex flows under reasonable computer resources. The LES/RANS hybrid method is a powerful method but at the same time has limitations, which we should keep in mind. One obvious deficiency is flow transition. Developing a method to capture "Scale Effects" is an important but still unresolved issue in CFD [4]. A series of flow simulations at subsonic, transonic to supersonic conditions were later carried out, resulting in better understanding of physics of cylindrical base flows [11].

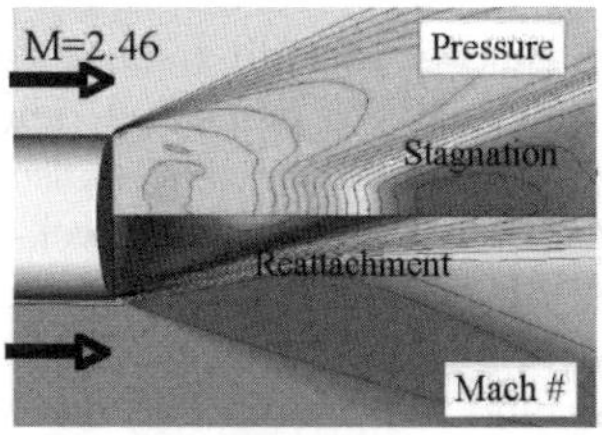

Fig. 7. Feature of supersonic base flows

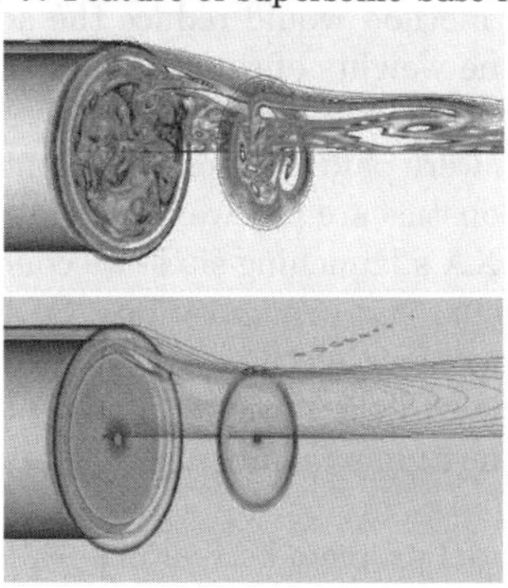

Fig. 8. Instantaneous view of the vorticity magnitude contours: (a) LES/RANS hybrid (top), (b) RANS (bottom)

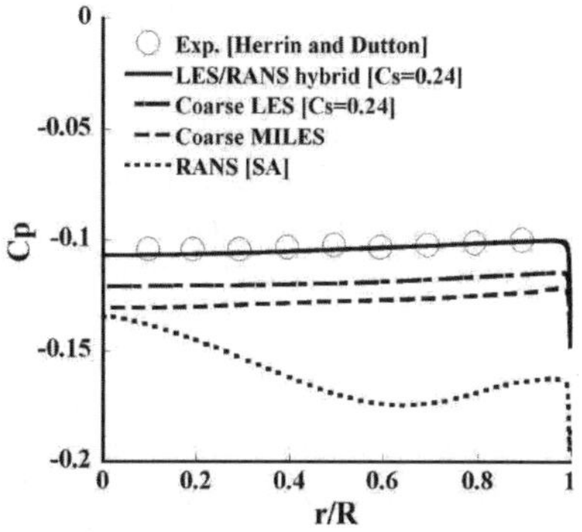

Fig. 9. Pressure distributions in the radial direction at the base

Third factor: Spectral-like schemes

Recently, compact difference scheme receives people's attention. The scheme uses Pade approximation, the idea of which is not new, but with the high-order filtering technique, Lele [12] showed compact difference scheme has a potential for practical simulations. Implementation in the generalized coordinate systems and real applications shown by Gaitonde and Visbal [13] expanded the use. Other compact schemes also appeared as their derivatives. The spectral-like schemes reduce the number of grid points, for instance, necessary for capturing vortex structure. CFD researchers usually consider that 20 to 25 points are necessary inside one vortex but the number is reduced a few times for each direction when the compact scheme is used. In three dimensions, reduction of required computer memory and computer time becomes enormous.

One important shortcoming of the compact scheme is that it cannot be applied to simulations of flows with shock wave. Nonlinear formulation is required to capture discontinuities and various non-linear schemes, such as Weighted Essentially Non Oscillatory (WENO) scheme [14] and Weighted Compact Non-linear Scheme(WCNS) [15] were developed. Also effort to improve filtering techniques in the original compact scheme is pursued [16]. Among them, we consider that an explicit version of WCNS is efficient and still keeping certain level of high resolution of the flow features applicable to practical simulations for shock-related problems [17].

Acoustic noise generated by the impingement of an over-expanded supersonic jet on a flat plate is of practical importance in many engineering applications. Rocket launching is an example. Both satellite and rocket vehicle are exposed to severe structural environment due to acoustics from rocket plumes. Such a plume structure including shock cell and stand off shock interaction with jet shear layers is physically complicated, and such flow structures determine characteristics of emanated acoustic waves. There also exists interaction of plume flows with the complex launcher structures. There is a well-known model of rocket plume acoustics proposed by NASA SP-8072 [18], in which acoustic load over the vehicle is estimated using a semi-empirical method based on the existing numerous experimental data. The model gives us easy and fair estimation method but they usually require empirical factors adjusted to particular launchers, and extension to new launching sites is not easy. Better estimation method would reduce the acoustic and vibration tests of satellites, which may increase the weights of satellites and shorten the period of satellite development. There are many experimental and numerical reports available for plume acoustics but particular attentions were paid to discrete and intensive tones rather than broadband tones. Rocket plume acoustics are not well understood.

As a practical problem of helping the design of JAXA's launching sites, we conducted simulations of plume flows for real rocket and launching site configurations. So far, we only used conventional second-order upwind shock capturing method. As shown in Fig. 10, compression waves reflect over the launcher surface and the tower, and go up toward the fairing of the rocket. Acoustics emit mainly from the plume mainly after the potential core (supersonic region inside the plume).

Simulation using 7th-order WCNS scheme for a model problem is tried in parallel [19]. Primary purpose is to understand the aero-acoustic mechanisms of over-expanded supersonic jet impinging on a flat plate with and without holes. For finding the required grid resolution for capturing sound radiations from the jet and finding important phenomenon of plume acoustics, two-dimensional axisymmetric flow simulations were carried out. Obviously, axisymmetric flow assumption is not valid, but effectiveness of WCNS scheme is discussed as our first step. Figure 11 shows the pressure contour plots of the model problem of rocket plume from a single solid motor impinging on the launcher with a hole. Strong compression waves emanate from certain area downstream of the plume with a strong directivity. Phenomenon is same as in Fig. 10, but reflected compression waves are better captured in this computation by the WCNS scheme.

4 Requirements for New Type of Simulations

Progress of computer hardware has been a major factor that accelerated CFD research since supercomputers appeared in late 1970's. Speed of leading-edge supercomputers

Fig. 10. Pressure contours of rocket plume simulations (RANS simulation)

Fig. 11. Pressure contours: model probem

increases almost twice every year[4]. Now, Japan plans to develop a class of 10 PFLOPS supercomputer in 2011 and U. S. may have a similar plan. Performance of such leading-edge computers would be roughly ten to hundred million times faster than the first commercial supercomputer that first appeared late 1970's. In Table 1, author's personal estimation of required computer time and memory is presented. The estimation is based on the assumption that simulation requires LES with a spectral-like scheme. The number is based on a rough estimation that such a scheme would reduce the required memory and cpu time about 100 times smaller than conventional second-order schemes. The numbers obviously depend on many factors such as computer architecture and others. From the estimation, simulations of flows at 10^5 order of Reynolds number will become feasible soon and even LES simulations at higher Reynolds numbers may become feasible when Petaflops computers become available.

Table 1. Number of grid points and computer time required for spectral-like methods

	Lower than $Re = 10^5$ MAV, UAV, Mars Aircraft	$Re = 10^6$ Wind tunnel level	Higher than $Re = 10^7$ Civil transports
No. Grid points (required memory size)	10^7 (10 GB)	10^9 (1 TB)	10^{11} (100 TB)
Computational time on 1 TFLOPS computer	10-50 hours	1,000-5,000 hours	100,000-500,000 hours

5 Final Remarks

Computational Fluid Dynamics (CFD) contributed to analysis of high-speed shock-related problems extensively. Some of the study examples by the author's group in the past were presented, which showed effectiveness of CFD both for design of transportation vehicles and for understanding of fluid flows. Trend of CFD for further use was discussed based on the recent effort in the CFD community and its applications. There are three key features: computer progress, spectral-like high-resolution scheme, and LES and LES/RANS hybrid method. Recent studies revealed that high-speed flows showing steady state features in the experiment are inherently unsteady and that requires LES-like computations for successful simulations. Such simulations require very high grid resolution, and spectral-like high-resolution scheme would help reducing the number of grids and computer time necessary. With these emerging techniques, CFD will have another prospect and hopefully will contribute much more to high-speed shock-related problems.

References

1. Murman, E. M. and Cole, J. D., Calculation of Plane Steady Transonic Flows, AIAA J., Vol. 9, pp. 114-121, 1971
2. Lax, P. D., Weak Solutions of Nonliner Hyperbolic Equations and their Numerical Computation, Comm. Pure Appl. Math., Vol. 7, pp. 159-93, 1954
3. Shang, J. S., A Glance Back and Outlook of Computational Fluid Dynamics, ASME FEDSM2003-45420, 2003
4. Fujii, K., Progress and Future Prospects of CFD in Aerospace-Wind Tunnel and Beyond, Progress in Aerospace Sciences, on International Review Journal, Vol. 41, pp. 455-470, 2005
5. Ogawa, T. and Fujii, K., Prediction and Alleviation of a Booming Noise Created by a High-speed Train Moving into a Tunnel, The ECCOMAS Computational Fluid Dynamics Conference, 1996
6. Fujii, K. and Ogawa, T., Aerodynamics of High Speed Trains Passing by Each other, Computers & Fluids, Vol. 24, pp. 897-908, 1995
7. Ogawa, T. and Fujii, K., What Have We Learned from CFD Research on Train Aerodynamics, Frontiers of Computational Fluid Dynamics 1998, Ed. by D. A. Caughey and M. M. Hafez, World Scientific, November 1998
8. Nakai, Y., Fujimatsu, N. and Fujii, K., Experimental Study of Underexpanded Supersonic Jet Impingement on an Inclined Flat Plate, AIAA J. Vol. 44, pp. 0001-1452, 2006
9. McIlroy, K. and Fujii, K., Computational Analysis of Supersonic Uinderexpanded Jets Impinging on an Inclined Flat Plate, AIAA Paper 2007-3859, June 2007
10. Kawai S. and Fujii K., Computational Study of a Supersonic Base Flow Using Hybrid Turbulence Methodology, AIAA Journal, Vol. 43, 2005, pp. 1265-1275, 2005
11. Kawai S. and Fujii K., Time-series and Time-Averaged Characteristics of Subsonic and Supersonic Base Flows, , AIAA J., Vol. 45, pp. 289-301, 2007
12. Lele, S. K., Compact Finite Difference Schemes with Spectral-Like Resolution, J. Comp. Phys., Vol. 103, pp. 16-42, 1992
13. Gaitonde, D. V. and Visbal, M. R., Further development of a Navier-Stokes solution procedure based on higher-order formulas, AIAA Paper 99-0557, 1999
14. Liu, X.D., Osher, S. and Chan, T., Weighted essentially non-oscillatory schemes, J. Comp. Phys., Vol. 126, pp. 200-212, 1996
15. Deng, X. G. and Zhang, H., Developing high-order weighted compact nonlinear schemes, J. Comp. Phys., Vol. 165, pp. 22-44, 2000
16. Fiorina, B. and Lele, S. K., Artificial nonlinear diffusivity method for supersonic reacting flows with shocks, J. Comp. Phys., Vol. 222, pp. 246-264, 2007
17. Nonomura, T. and Fujii, K., Increasing order of Accuracy of Weighted Compact Non-Linear Scheme, AIAA Paper 2007-893, Jan. 2007
18. Eldred, K. M., Acoustic Loads Generated by the Propulsion System, NASA SP-8072, June 1971
19. Kawai, S., Tsutsumi, S., Takaki, R. and Fujii, K., Computational Aeroacoustic Analysis of Overexpanded Supersonic Jet Impingement on a Flat Plate with/without Hole, FEDSM2007-3756, 5th Joint ASME/JSME Fluid Engineering Conf., July 2007

Explosive eruptions of volcanos: simulation, shock tube methods and multi-phase mathematical models

V. Kedrinskiy

Lavrentyev Institute of Hydrodynamics, Siberian Branch of the Russian Academy of Sciences Lavrentyev pr.15, 630090 Novosibirsk (Russia)

Summary. The paper[1] presents the short survey both of theoretical and experimental results as well as some approaches to the creation of unsteady multi-phase mathematical models and to the working out of the experimental methods of the simulation of the eruption processes based on the method of hydrodynamic pulse shock tubes. The full system of the equations for the study of the initial stage of explosive eruption as well as the results of numerical studies on the magma state dynamics taking into account the gravity, diffusive processes and dynamically changing viscosity will be considered. The experimental simulation of magma fragmentation carried out within the framework of shock tube methods has shown, that the magma fragmentation can be happened also in a result of flow stratification: the flow division into the system of the vertical jets of the spatial form which then are disintegrated on separate fragments.

1 Introduction

It's well known that magma is generated deep within the Earth, contain enough large amount of SiO_2 and dissolved water. The high pressure keeps the gas in solution in a magma. But when magma rises along a conduit the pressure is decreased, the gas comes out of solution and forms a separate gas phase (bubbles). Bubbles continue to grow in size as pressure is reduced. If the magma has a high viscosity, then the gas will not be able to expand easily, and thus, gas pressure in bubbles will build up. When the bubbly magma arises to the Earth's surface the liquid films around bubbles rupture (explode). Explosive bursting of bubbles will fragment the magma into clots of liquid that will cool as they move through the air. These solid clots become tephra (called blocks and bombs) or volcanic ash (sand-sized). Explosive eruptions are favored by high gas content and high viscosity of andesitic (52-63% silica) and rhyolitic (more 63% silica) magmas. So Rhyolitic magma has the viscosity, ranging between $10^6 \div 10^8$ times more viscous than water (10^{-3}Pa sec). It's considered the magma fragmentation efficiency is controlled such factors as overpressure, temperature, vesicularity, phenocryst and microlite content.

2 The initial state and mechanics of eruptions: classification problems

The analysis of information published to the present time allows one to generalize the data on the explosive eruptions basing on some common signs which characterize the

[1] The work was carried out at the financial support of RFBR Project 06-01-00317, SB RAS Project 4.14.4 and Project SS-8583.2006.1

pre-eruption states of volcanic systems (closed and opened types) as well as the eruption mechanics.

"Classical" structure of pre-eruption state: cooled magma or rock fragments make the solid crust (plugs) blocking a crater and trapping hot gas and magma under the ground between infrequent eruptions. *Mechanics of eruption:* the hot gas builds up until the pressure becomes enough great to blow out these obstructions with strong explosion as a column of ashes, pumice (instantaneously cooled magma containing multitude air pores), and also breadcrust bombs and blocks which fly high up into the air. *Model*: we have classical scheme of gas-dynamics shock tube (GDST).

Practically the same state corresponds to so called Tephra Jets / Blasts eruptions arising in a result of **Magma and groundwater interactions:** lave lake partly filling in the crater destroys its walls. Its level drops below the groundwater table, the lava delta collapses completely and forms a permeable plug blocking a crater. *Mechanics of eruption:* water comes into contact with hot fragments of rock and magma, steam pressure builds up, destroys a plug and ejects along with pre-exiting fragments of rock causes explosions. *Model*: we have explosive expansion of mixture "steam+rock fragments" in shallow lava lake as the analog of shallow underwater explosions.

In the case of Kilauea's pre-1924 explosive eruptions (the data of US Geological Survey) the volcano's summit crater was so deep that its floor was below the water table, letting ground water seep in to form a lake. *Mechanics of eruption:* when magma erupted into the lake water, explosions of steam ejects extremely hot ash-laden steam clouds (pyroclastic surges) out of the crater. *Model*: the typical "shellow underwater explosion" in which "detonation products" are the magma fragmenting into tiny ash particles+steam.

The close processes are observe for Littoral lava fountains forming in a result of **Lava and seawater interactions:** state of this system is determined by the partial collapse of a delta's leading edge (delta front) and significant fracture of a solidified rock around the active lava tubes. *Mechanics of eruption:* this state allows seawater to seep into the tube system and to mix with lava within the confines of a lava tube. The pressure increases and becomes enough high to cause explosions that blast a hole through the roof of the tube. This type of lava-seawater explosion produces fountains of molten lava, steam, bombs and smaller tephra fragments that reach heights of more than 100 m [1]. *Model*: underground steam explosion in a lava tube located close to a ground surface.

Open system of explosive type: Erebus Volcano. *Initial pre-eruption state of systems (dynamic state):* when rhyolitic magma rises to the Earth's surface the pressure is decreased, the gas comes out of solution and forms a separate gas phase that we see as bubbles.In andesitic and rhyolitic magmas the liquid films around the bubbles don't allow the gas to expand rapidly and thus, pressure builds up inside of the gas bubbles. *Mechanics of eruption:* when the magma reaches the Earth's surface the liquid films explode, and a fragmented magma is ejecting through the volcanic vent. This occurs when the volume of bubbles is about 75% of the total volume of the magma column [2]. *Model:* the classical mechanics of magma fragmentation under dynamic decompression through the process of vesiculation, stage of its transformation in foam structure and its disintegration as the "gas-particles" flow.

In [3] the near-vent eruption dynamics of Erebus volcano have studied. According to the data of the experimental (Doppler-radar method) studies the exploding gas bubbles do not oscillate prior to their burst. In all observed cases they were rapidly expanding and exploding in the last phase of their approach to the free surface. This shows that oscillatory eruption models are not applicable to strombolian systems that are similar to Erebus.

3 Shock-tube method: experimental simulation

Doubtless [4], that laboratory experiments modeling the dynamics of volcanic flows and, in particular, shock-tube (ST) experiments are an important component of researches of the mechanics of magma fragmentation under pulse decompression waves. These processes are far from full understanding and main preference is given up practically exclusively to the mechanism of cavitation breakdown. One of examples is the simplest scheme of volcano and eruption model presented in [6], which, at bottom, is the analog of the scheme of classical gas-dynamic shock tube (GDST) where a magma in compressed state is in a chamber of high pressure. Upon the breakdown of the magma plug (diaphragm in scheme), the magma moves upward along the channel (conduit). The pressure drop triggers processes of nucleation, diffusion, and tends a magma to state of a foam type. It is assumed that the breakdown of the foam should result in the inversion of the two-phase state — transition from a cavitating liquid to a dust–gas mixture [5]. Of course, GDST method is well analog of explosive systems of close type but it requires using high hydrostatic pressure for the laboratory simulation. As alternative, we have suggested to use also method of hydrodynamic shock tubes (HDST) equipped by pulse x-ray apparatus for the liquid samples with free surfaces (Fig. 1) [5]. Sample (Fig. 1, HDST, (4)) is loaded by shock wave (SW), initiated by piston impact or pulse magnetic field (Fig. 2), and then (after SW reflection from free surface) is decompressed by rarefaction wave (RW) in real time. Method HDST and a system of consistent "shock wave–rarefaction wave" loading give a possibility to simulate, in a dynamic regime, the sudden decompression of magma compressed statically up to high pressures before eruption. Fig. 1, frames (1)–(6), shows the typical dynamics of cavitation development in liquid sample loaded by SW as consistent series of the x-ray pictures (70 nsec exposure time, for each frame). So, we can note that Dobran's scheme works, at least for some initial time interval.

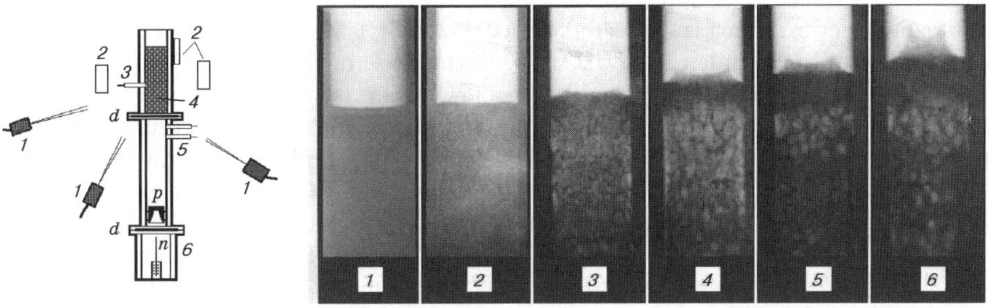

Fig. 1. HDST's scheme and dynamics of cavitation development, time interval 200 μ *sec*, 70 nsec exposure time for each frame

Electromagnetic HDST (Fig. 2) generates short (microsecond-range) shock waves in the sample being examined and, thus, allow one to use liquid drops of tens or hundreds of microliters in volume as the samples for a more precision study of the flow structure. In the drop lying on a conducting diaphragm, a shock wave is produced by a magnetic-field pulse generated under the diaphragm by discharge of a high-voltage capacitor bank onto a helical coil. Fig. 2 shows the cellular structure of a viscous drop ($\mu \approx 0.2$ Pa·sec, 1 cm size) developed to $t = 200$ (a) and 700 (b) μsec after the loading of a shock wave. Later, the drop takes the form of cellular dome. Mechanics of this effect is explained by so called internal "cavitative explosion" as result of fast development of bubbly clusters in the drop center. Two last frames of Fig. 2 demonstrates the late stages of dynamic breakdown of a drop of a gum-resin–acetone solution [7]. This solution can be used as an analog of magma melt on such its properties as crystallization and dynamically increasing viscosity in a result of simultaneous crystallization of the solution during its dynamic loading by rarefaction wave. It turned out that the flow structure resulting from the dynamic breakdown of a gum-resin–acetone drop differs radically from the above-mentioned one: instead of dome cellular structure the system of vertical jets (Fig. 2) arises, which are broken up later into fine droplets.

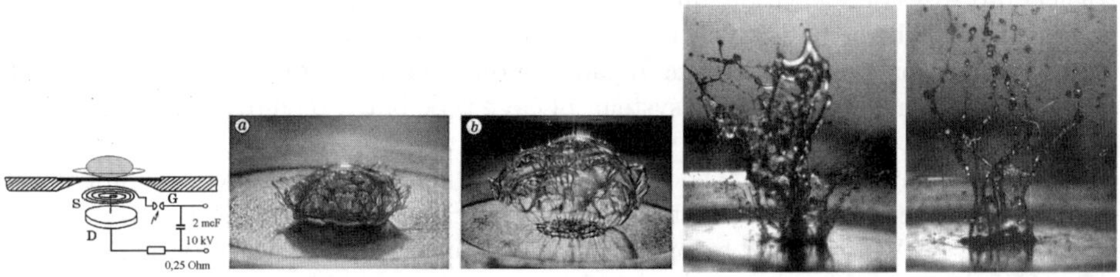

Fig. 2. Electro-magnetic HDST; cellular structure of "boiling" viscous drop (a, b) at 200 and 700 μsec; jet structure of flow for gum-resin-aceton drop broken up by RW loading

Fig. 3 shows the eruption structure features as the enlarged images of the flow segments directly near the exit of the GDST low-pressure channel for two magnitudes of liquid sample viscosity: $\mu \approx 0.2 Pa\ sec$ at T=19^{0}C, and for $\mu \approx 2.6 Pa\ sec$ at T=42^{0}C. It is evident that in the initial stage of the breakdown, transition from a cavitating state to a foam structure which is characteristic for water [5], is also observed for a liquid with a two orders of magnitude higher viscosity. An increase in the scale and a microsecond exposure made it possible to resolve the fine flow structure, which turned out to change significantly with time (Fig. 3a): the flow is stratified into a system of vertical jets having a spatial shape. As the viscosity of the liquid being broken ($\mu \approx 2.6$ Pa·sec and $T = 42°$C) is further increased by an order of magnitude, the flow is almost completely stratified and takes a more distinct jet nature, as it's evident from Fig. 3b.

The probable cause of this effect was found in studies of extrusive eruptions of gas saturated liquids of various viscosities by GDST-method. The experimental statement has corresponded to the extrusive eruption of the liquid saturated by a pressurized gas from a reservoir. The results are given in Fig. 3c-e for ordinary water ($\mu = 0.001$ Pa·sec). It is easy to see that in the structure of the low-velocity (extrusive) eruption, a system of buoyant dense bubble clusters forms by the time $t = 50$ msec (d), a gas slug, as a result of the bubble coalescence, forms by the time $t = 200$ msec (e) in one of the clusters, and at $t = 500$ msec (c), the flow structure transforms in "gas slug–bubble cluster" combination system.

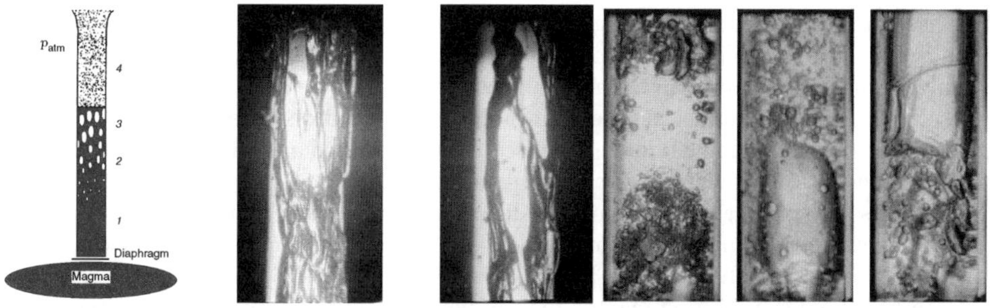

Fig. 3. GDST scheme; the flow stratification for μ $0.2 Pa\ sec$ at T=19^{0}C (1st fr.), and for T=42^{0}C, μ $2.6 Pa\ sec$(b); 3 last frames of slug-flow formation: 50 (c), 100 (d) and 200 (e) msec

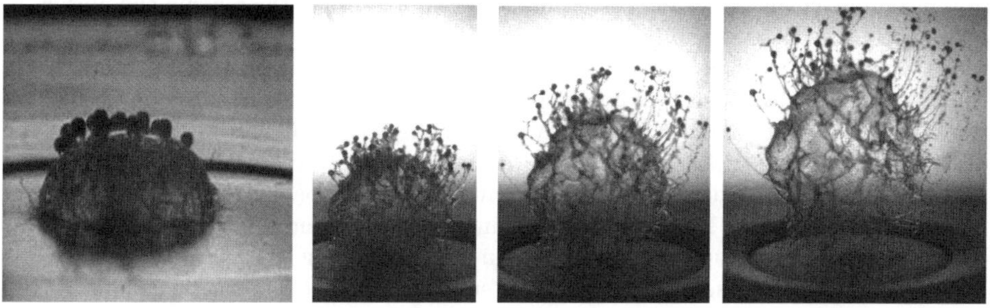

Fig. 4. Breakdown of a drop of a "water-particles" mixture, separation of flows: 0.4 (a), 0.8 (b), 1.2 (c) and 1.600 (d) msec

Note, that some types of volcanoes are characterized by a peculiar combinative structure of eruption, which is accompanied by a powerful ejection of magmatic "bombs" upward to a few kilometers and a simultaneous lava ejection (like to Hawaiian type of eruption). One can assume that in the intervals between eruptions, the magma in the chamber and conduit is a strongly crystallized melt, in which the spontaneous formation of crystalline clusters and zones of glass transition are possible. This state can be considered as a metastable one with a nonuniform distribution of the crystalline phase (clusters). Upon a sudden decompression in a result of intense nucleation the magma flow in a conduit can be treated as a three-phase system consisting of a liquid magma, bubbly regions, and crystalline and glassy clusters (probably, with internal cavitation zones). The dynamics of three-phase magma breakdown was simulated using mixtures of solid particles of arbitrary shape and sizes 1–3 mm with water-glycerol or gum-resin-acetone - drops. According to the experimental data the solid particles are ejected from the main flow, have much higher velocity, and form their own system, which is almost independent on the cavitating liquid (Fig. 4,water-particle sample).

4 Dynamics of magma state under decompression wave

Volcano eruptions are unique nature phenomena which is obviously determined by the dynamics of initial state of magma-melt under decompression waves, by the kinetics of phase transformations changing its properties and cavitation development. Among many publications dealing with this issue, one can note [8–11] where the attempts to simulate eruption dynamics in the general formulation were made. Some of papers involve consideration of particular processes that accompany eruption. These are papers on the growth dynamics of a single bubble in a viscous gas-saturated melt [12–14] and associated thermal effects [15], mechanism of magma solidification under decompression [16,17], etc. Nevertheless, despite significant efforts undertaken there are still many issues that require special consideration and simulations within the framework of multi-phase mechanics.

Multi-phase math-model. The statement analogous GDST is considered. A vertical column of a gas-saturated magma melt of height H in the gravity field is adjacent to the magma chamber at the bottom and is separated by a diaphragm from the ambient medium on the top (atmospheric pressure is denoted by p_0). z axis is directed vertically upward with the origin at the column–chamber interface. The initial pressure in magma in the column–chamber system corresponds to the pressure in the magma chamber with allowance for hydrostatics: $p_i(z) = p_{\mathrm{ch}} - \rho_0 g z$, where ρ_0 is the magma density and p_{ch} is the pressure at $z = 0$. It's assumed that the gas dissolved in magma has initially an equilibrium concentration C^{eq} whose dependence on pressure p is determined by Henry's law $C^{\mathrm{eq}}(p) = K_{\mathrm{H}}\sqrt{p}$ where K_{H} is Henry's constant [18]. At the initial time ($t = 0$), the diaphragm confining the melt is broken, the surface $z = H$ becomes free, and a rarefaction wave starts propagating vertically downward over the magma. The gas dissolved in magma turns out to be supersaturated behind the wave front, which results in spontaneous nucleation and growth of gas bubbles in the melt volume. The process considered is described by one-dimensional system of conservation laws and the kinetic equations [19]:

$$\frac{\partial\rho}{\partial t} + \frac{\partial(\rho v)}{\partial z} = 0; \qquad \frac{\partial v}{\partial t} + v\frac{\partial v}{\partial z} = -\frac{1}{\rho}\frac{\partial p}{\partial z} - g + \frac{1}{\rho}\frac{\partial}{\partial z}\Big(\mu\frac{\partial v}{\partial z}\Big),$$

$$p = p_0 + \frac{\rho c^2}{n}\Big\{\Big(\frac{\rho}{\rho_0(1-k)}\Big)^n - 1\Big\},\ \mu = \mu^* \exp\frac{E_\mu(C)}{(k_{\mathrm{B}}T)},\ J = J^* \exp\frac{(-W_*}{(k_{\mathrm{B}}T}),$$

$$R\ddot{R} + \frac{3}{2}\dot{R}^2 = \frac{(p_{\mathrm{g}} - p)}{\rho_{liq}} + \frac{4\nu\dot{R}}{R}, \qquad \frac{4}{3}p_{\mathrm{g}}R^3 = \frac{m_{\mathrm{g}}k_{\mathrm{B}}T}{M},$$

$$\frac{dm_{\mathrm{g}}}{dt} = 4\pi R^2\rho D\Big(\frac{dC}{dr}\Big), \qquad \frac{dT}{dt} = \mathrm{Ku}\frac{dX}{dt} - \mathrm{Ku}'(4\pi/3)N_b z_0^3\frac{dm_g}{dt}.$$

Here $E_\mu(C) = E_\mu^*(1 - k_\mu C)$ is the activation energy, E_μ^* is the activation energy for the "dry" melt, k_μ is the empirical coefficient, μ^* is the preexponent, k_{B} is the Boltzmann constant, M is the molecular weight of the gas [20]. As was noted above, decompression leads to spontaneous nucleation of gas bubbles in the melt.The frequency of homogeneous nucleation J is determined by the work $W_* = 16\pi\sigma^3/(3\Delta p^2)$ spent on formation of a critical nucleus [14]. The pressure p_{g} is determined by the diffusion gas flow from the supersaturated melt to the bubble, which is found with the use of the quasi-steady solution of the equation of gas diffusion in the melt: $C(r) = C_i - (C_i - C^{\mathrm{eq}}(p_{\mathrm{g}}))R/r$, where r is the radial coordinate and C_i is the gas concentration far from the bubble. It follows from relation for $C(r)$ the diffusion boundary layers (X_D - its relative volume) are formed around the bubbles and the nucleation of bubbles occurs only outside the diffusion layers. Hence, the bubble nucleation will be terminated when the diffusion layers of the growing bubbles completely cover the entire volume ($X_D \to 1$).

Calculation of k, μ and χ distributions. The constructed system of equations completely determines the initial dynamics of magma-melt state in the gravity field. The calculations were performed for the following problem parameters : magma-column height $H = 1$ km, pressure in the volcano chamber located at a depth of approximately 7 km, $p_{\mathrm{ch}} = 1700$ atm, and temperature $T = 1150$ K, which corresponds to the melting point of magma at a pressure $p = p_{\mathrm{ch}}$ [17]. The following values of magma properties are used in the calculations: $\rho_0 = 2300$ kg/m^3, $K_{\mathrm{H}} = 4.33 \cdot 10^{-6}$ Pa$^{-1/2}$, $D = 2 \cdot 10^{-11}$ m^2/sec, $\sigma = 0.076$ J/m^2, $E^*_\mu = 5.1 \cdot 10^{-19}$ J, $k_\mu = 11$, and $\mu^* = 10^{-2.5}$ Pa $\cdot$ sec. The distributions of the viscosity μ, the gas phase concentration k and the crystalline phase concentration χ in the magma flow are presented at three time instants: $t = 2(1), 3.8(2)$, and $6.1s(3)$. Note that, due to the rise of the bubble cavitation, the column length increases and, according to the calculation data, by the end of the process, it attains a value of 1.5 km (Fig. 5). The calculations demonstrate that, by the instant $t = 6.1s$,the pressure within the magma column approaches some steady-state. The character of its distribution along column turns out to be close to the distributions of viscosity (it has increased in magnitude by six orders of magnitude), the concentrations of gas- and crystalline-phases. The calculation results presented suggest that, approximately by the sixth second after the beginning of the decompression, the glass transition process in the three-phase magma with "frozen-in" 0.3-mm-thick bubbles almost ends on a considerable portion of the column. Then the magma structure is close (in parameters) to the structure of real pumice [18].

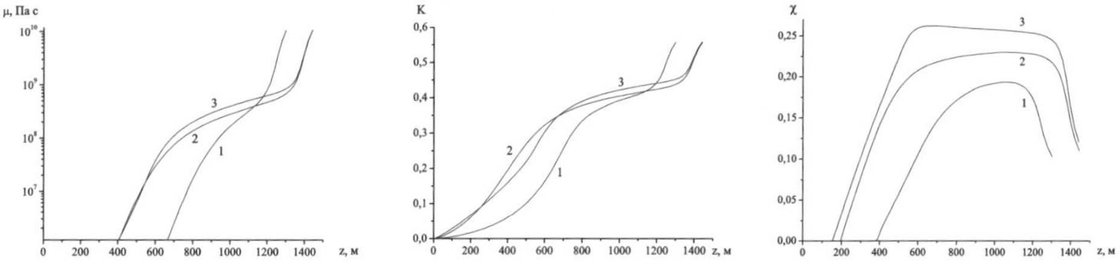

Fig. 5. Distributions of viscosity ν, volumetric concentrations of gas k and crystallin phases χ for $t = 2(1), 3.8(2)$, and $6.1s(3)$.

5 Conclusion

The results of the experimental simulation presented have shown that the magma fragmentation mechanism can be determined by flow stratification into a system of vertical jets of a spatial shape. The model of three-phase pre-eruption magma state was simulated using a "liquid–solid particles" mixture. Lab-experiments have shown that the particles are ejected from the main flow and form separate flow moving with significantly more high velocity. The full system of equations was proposed to study the problems of dynamics of magma state under rarefaction-wave passing over the magma-melt column in the gravity field. Numerical studies have shown that the magma viscosity in the degassing processes grows up by 6-7 orders and its state is characterized by the process of the glass transition of flow which includes "frozen-in" 0.3-mm-thick bubbles.

References

1. Mattox T. N. and Mangan M. T.: Littoral hydrovolcanic explosions: a case study of lava – seawater interaction at Kilauea Volcano, Journal of Volcanology and Geothermal Research, (1997), v. **75**, p. 1–17.
2. Woods Andrew W.: The dynamics of explosive volcanic eruptions, Reviews of geophysics, (1995), V. **33**, No. 4, p. 495–530.
3. Gerst, A, et al: The First Second of a Strombolian Eruption: Velocity Observations at Erebus Volcano, Antarctica. EOS Trans. AGU, 87(52), Fall Meet. Suppl., (2006), Abstract V31G-04.
4. Gilbert J.S. and Sparks R.S.J.(eds): The Physics of Explosive Volcanic Eruptions, Geological Society, London, Special Publications, V. **145**, (1998).
5. Kedrinskii V. K. : Nonlinear problems of cavitation breakdown of liquids under explosive loading, J. Appl. Mech. Tech. Phys., No. 3, p. 361–366, (1993).
6. Dobran F. : Non-equilibrium flow in volcanic conduits and application of the eruption of Mt. St. Helens on May 18 1980 and Vesuvius in Ad. 79, J. Volcanol. Geotherm. Res., V. **49**, 285–311 (1992).
7. Kedrinskii V.K., Makarov A.I., Stebnovsky S.V., and Takayama K.: Explosive Volcanic Eruption: Some Approaches to Simulation. Combustion, Explosion and Shock Waves, (2005), V. **41**, No. 6, p. 193–201, Kluwer Academic/Plenum Publisher.
8. Wilson L. : Relationships between pressure, volatile content, and eject in three types of volcanic explosion, J. Volcanol. Geotherm. Res., V. **8**, p. 297–313, (1980).
9. Slezin Yu. B. : *Mechanism of Volcanic Eruptions (Steady Model)* [in Russian], Nauchnyi Mir, Moscow, (1998).
10. Papale P., Neri A., and Macedorio G.: The role of magma composition and water content in explosive eruptions. 1. Conduit ascent dynamics, J. Volcanol. Geotherm. Res., V. **87**, p. 75–93, (1998).
11. Barmin A. A. and Mel'nik O. E.: Hydrodynamics of volcanic eruptions, Usp. Mekh., No. 1, p. 32–60, (2002).
12. Lyakhovsky V., Hurwitz S., and Navon O.: Bubble growth in rhyolitic melts: Experimental and numerical investigation, Bull. Volcanol., V. **58**, No. 1, p. 19–32, (1996).
13. Navon O., Chekhmir A., and Lyakhovsky V.: Bubble growth in highly viscous melts: Theory, experiments, and autoexplosivity of dome lavas, Earth Planet. Sci. Lett., V. **160**, p. 763–776, (1998).
14. Proussevitch A.A. and Sahagian D.L.: Dynamics of coupled diffusive and decompressive bubble growth in magmatic systems, J. Geophys. Res., V. **101**, No. 8, p. 17 447–17 456, (1996).
15. Proussevitch A.A. and Sahagian D.L.: Dynamics and energetics of bubble growth in magmas: Analytical formulation and numerical modeling, J. Geophys. Res., V. **103**, No. B8, p. 18 223–18 251, (1998).
16. Hort M.: Abrupt change in magma liquidus temperature because of volatile loss or magma mixing: Effects on nucleation, crystal growth and thermal history of the magma, J. Petrology, V. **39**, No. 5, p. 1063–1076, (1998).
17. Chernov A.A.: A model of magma solidification during explosive volcanic eruptions, J. Appl. Mech. Tech. Phys., V. **44**, No. 5, p. 667–675, (2003).
18. Stolper E.: Water in silicate glasses: An infrared spectroscopic study, Contrib. Mineral. Petrol., V. **81**, p. 1–17, (1982).
19. Kedrinskii V.K., Davydov M.N., Chernov A.A., and Takayama K: The Initial Stage of Explosive Volcanic Eruption: The Dynamics of the Magma State in Depression Waves, Doklady Physics, V. **51**, No 3, p. 140–143, (2006), Pleiades Publishing, Inc.
20. Persikov E.S.: The viscosity of magmatic liquids: Experiment, generalized patterns. A model for calculation and prediction. Applications,Physical Chemistry of Magmas V. **9** Advances in Physical Geochemistry, Springer-Verlag, New York (1991), p. 1–40.

Ignition delay time measurements at practical conditions using a shock tube

E.L. Petersen

Mechanical, Materials and Aerospace Engineering, University of Central Florida, Orlando, FL, 32816, USA

1 Introduction

Experiments are ongoing in our laboratory to study the ignition kinetics of hydrocarbon- and hydrogen-based mixtures behind reflected shock waves at conditions representative of those seen in gas turbine and other engine applications. Engine conditions herein imply elevated pressures (10 atm and higher) and undiluted fuel-air mixtures. Lower temperatures (less than 1100 K) are also of interest in recent years for several reasons, including the possibility of autoignition in premixed systems prior to entering the main combustor and in the validation of chemical kinetics models at low-to-intermediate temperatures. Several different blends of methane-hydrocarbon and hydrogen-carbon monoxide fuels have been studied in recent years, and provided herein are sample results as well as details on performing measurements at higher pressures and undiluted fuel-air mixtures.

2 Shock-Tube Techniques

2.1 Facilities

To date, the bulk of our ignition tests have been conducted in a high-pressure shock tube with the capability of achieving reflected-shock pressures as high as 100 atm, described in more detail by Petersen et al. [1]. This facility has a 16.2-cm-diameter driven section that is 10.7 m long. The relatively large inner diameter helps to minimize reflected-shock boundary layer effects [2]. Ignition delay times are obtained behind the reflected shock using a combination of OH* or CH* chemiluminescence and pressure measurements, as discussed in more detail below.

A new, second shock-tube facility targeting not just higher pressures but also lower temperatures and longer test times has recently been built, as discussed by de Vries et al. [3]. The single-pulse shock tube uses either lexan diaphragms or die-scored aluminum disks of up to 4 mm in thickness. The modular design of the tube allows for optimum operation over a large range of thermodynamic conditions from 1 to 100 atm and between 600-4000 K behind the reflected shock wave. Test times up to 20 ms can be obtained using the proper driver-driven configuration featuring a longer driver section. The system includes a smart gas handling and vacuum system; high temporal and spatial resolution, multi-channel data acquisition boards; and the capability to apply several optical diagnostics. Details on the layout are presented in Fig. 1.

2.2 Endwall Versus Sidewall Measurements

Chemiluminescence emission from exited species such as OH* or CH* can be a convenient and effective diagnostic for monitoring ignition delay times in shock-heated mixtures.

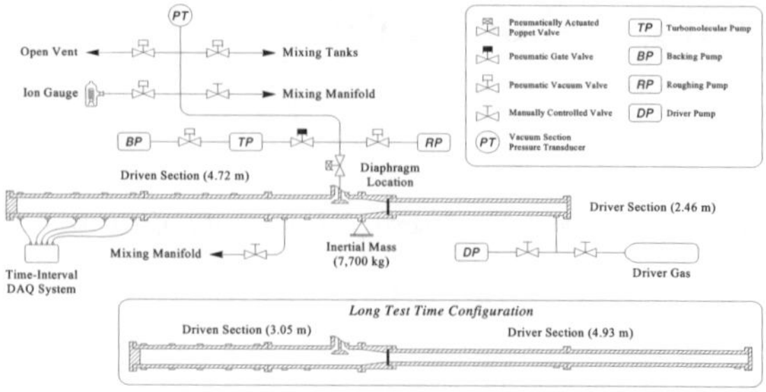

Fig. 1. Shock-tube facility, showing two configurations, depending on desired test time; from de Vries et al. [3]

Ideally, the ignition delay time obtained from the radical-species emission signal should agree with ignition delay time as obtained from the pressure trace. Under ideal shock-tube conditions, ignition behind the reflected shock wave occurs first at the endwall, so the measurement of endwall pressure is often considered the best way to determine ignition delay time when such an increase in pressure is available, which is generally the case for the experiments of interest herein. We have shown in recent experiments in conjunction with a simple optical model that endwall emission can indeed be employed reliably to measure ignition delay times when the ignition event is abrupt, as in undiluted fuel-air mixtures. Figure 2 shows an example for a methane-hydrocarbon mixture at 38.4 atm; while the ignition inferred from the endwall emission and pressure traces is similar for both, the ignition inferred from the sidewall appears accelerated due to gas dynamic effects after ignition occurs at the endwall.

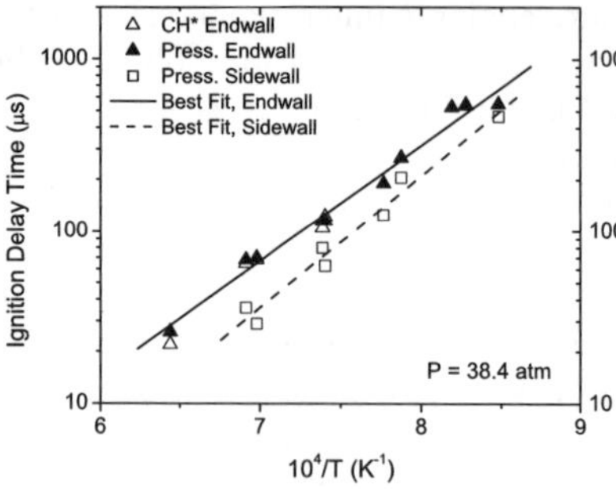

Fig. 2. Ignition delay times derived from three diagnostics: sidewall and endwall pressure traces and endwall CH* for an undiluted fuel-air mixture of 0.0784 CH_4 + 0.0058 C_2H_6 + 0.0029 C_3H_8 + 0.1917 O_2 + 0.7211 N_2 ($\phi = 1.0$); average pressure is 38.4 atm. Ignition appears faster in the sidewall pressure diagnostic, while the endwall CH* results agree with the endwall pressure ignition times

2.3 Extended Test Times

To study combustion chemistry at low temperatures in a shock tube, it is of great importance to increase experimental test times, and this can be done by tailoring the interface between the driver and driven gases. While driver-gas tailoring techniques are certainly not new, we have recently used unconventional driver-gas tailoring with the assistance of tailoring curves to increase shock-tube test times from 1 to 15 ms for reflected-shock temperatures below about 1000 K [4]. As we have seen, He/CO_2 and He/C_3H_8 driver mixtures provide a unique way to produce a tailored interface and, hence, longer test times, when facility modification is not an option. The tailoring curves can be used to guide future applications of this technique to other configurations, sa shown in Fig. 3.

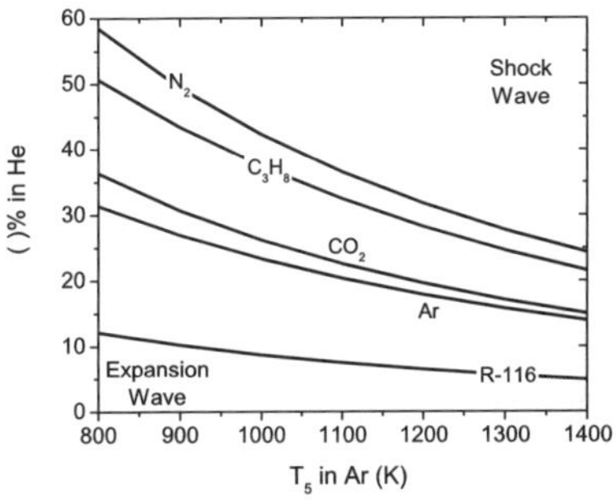

Fig. 3. Gas-specific tailoring curves for the driver-gas combinations of N_2, C_3H_8, CO_2, Ar, and R-116 mixed in helium (molar percent) as a function of the test temperature in Ar ($T_1 = T_4 = 298K$) under ideal shock-tube conditions; from Amadio et al. [4]

2.4 Wall Heat Transfer

In most shock-tube experiments, the test times are short enough so that it can be safely assumed that the temperature conditions remain constant since heat losses are minimal. However, there is some concern that higher pressures and longer test times may allow significant heat loss to the walls of a shock tube, thus creating observable deviations from the isothermal assumption. In a recent study, we applied a 2-D, axisymmetric heat transfer model to estimate the heat loss to the shock-tube walls for extreme reflected-shock conditions approaching 20 ms for pressures as high as 20 atm [5]. The resulting solution allows for the calculation of an average gas temperature over the entire hot gas region, and a typical result is shown in Fig. 4, which displays the change in average gas temperature relative to the initial T_g for a range of temperatures at 20 atm. Note that at a test condition of 800 K and 20 atm, the average gas temperature only decreases by about 5 K. Also, for a typical shock-tube experiment that ends within 3 ms, the average gas temperature for a 1400-K experiment at 20 atm decreases by about 6 K. The general conclusion is that heat losses to the walls should not be a problem for most shock tubes, even at longer test times, and increasing test pressure actually reduces the conduction heat loss from the gas to the walls.

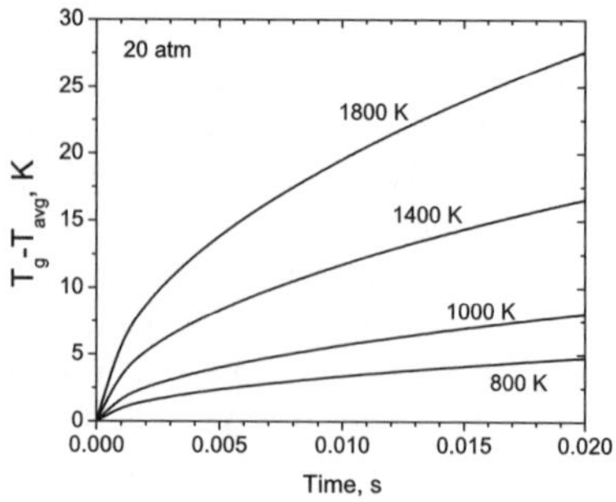

Fig. 4. Decrease in average Ar gas temperature from the radial heat transfer model for a pressure of 20 atm and 4 different initial gas temperatures and a stainless steel wall; from Frazier et al. [5]

3 Methane-Based Fuel Blends

Because of their importance for the power generation industry, the ignition chemistry of methane-based fuel blends continues to be a topic of active research. Although much work has been done on methane oxidation and natural gas oxidation through the years, few data existed prior to the recent studies for significant levels of hydrocarbon-fuel addition to methane, namely for volumetric percentages of the heavier hydrocarbons on the order of 10 to 50 percent. Several studies have been conducted recently in our laboratory for undiluted fuel-air mixtures at pressures up to 40 atm.

In most of these fuel-blend studies, particularly for temperatures about about 1100 K, the ignition that occurs is considered a strong ignition due to the abrupt pressure rise. Figure 5 from de Vries and Petersen [6] shows a typical result for a 50/50 CH_4/C_2H_6 mixture at 19 atm, 803 K. Note that the endwall pressure and CH* emission traces provide the same ignition time, as mentioned above.

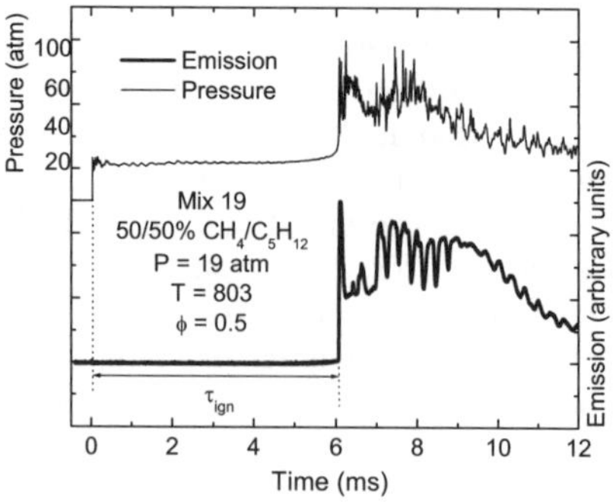

Fig. 5. Endwall pressure and CH* emission for a 50/50 CH_4/C_2H_6 mixture at 19 atm, 803 K; from de Vries and Petersen [6]

3.1 Lean Fuel Blends

Shock-tube experiments and chemical kinetics modeling were performed to further understand the ignition and oxidation kinetics of lean methane-based fuel blends at gas turbine pressures. Such data are required because the likelihood of gas turbine engines operating on CH_4-based fuel blends with significant (10% or more) levels of hydrogen, ethane, and other hydrocarbons is very high. Ignition delay times were obtained for fuel mixtures consisting of CH_4, CH_4/H_2, CH_4/C_2H_6, and CH_4/C_3H_8 in ratios ranging from 90/10 to 60/40% [7]. Lean fuel/air equivalence ratios ($\phi = 0.5$) were utilized, and the test pressures and temperatures ranged from 0.54 to 30.0 atm and from 1090 to 2001 K, respectively. Major reductions in ignition delay time were seen with the fuel blends relative to the CH_4-only mixtures at all conditions. A methane/hydrocarbon-oxidation chemical kinetics mechanism developed in recent studies (including the present data set) [8-13] was able to reproduce the high-pressure, fuel-lean data for the fuel/air mixtures very well, as seen in Fig. 6 for two methane-hydrogen mixtures.

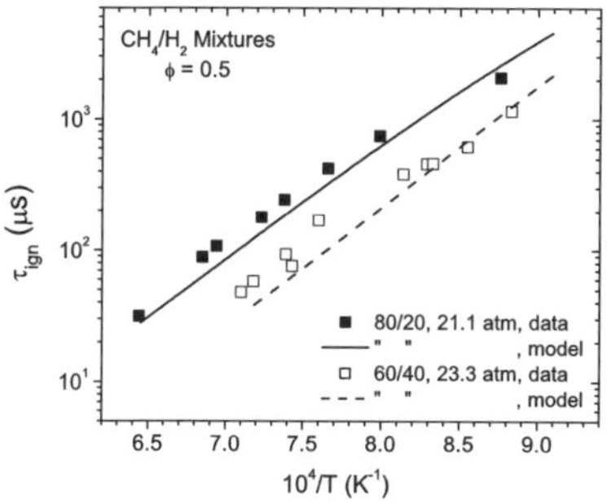

Fig. 6. Comparison between model [13] and experiment [7] for mixtures of methane and hydrogen in air ($\phi = 0.5$); from Petersen et al. [7]

3.2 Gas Turbine Autoignition Study

To determine gross autoignition behavior for lean, pre-mixed power generation gas turbine engines over a wide range of possible blends, undiluted natural-gas-based mixtures combining CH_4 with C_2H_6, C_3H_8, C_4H_{10}, C_5H_{12}, and H_2 were tested at engine-relevant conditions, i.e. at an average pressure of 20 atm, target temperatures near 800 K, and an equivalence ratio of $\phi = 0.5$ [6]. A statistical matrix approach was employed to cover as wide a range of multiple-fuel blends as possible, resulting in 21 binary and ternary mixtures with methane content in the blends as low as 50% by volume in some mixtures. The mean ignition time for all mixtures around 800 K was found to be 7.9 ms with a standard deviation of 1.9 ms, or less than 25%, which is small compared to the relatively high reduction effect at higher temperatures. While most of the fuel blends exhibited strong ignition behavior (as in Fig. 5), the pure methane tests exhibited only a weak ignition at 800 K. Assuming either type of ignition to be worthy of concern for autoignition in gas turbines, it was found that pure methane at low temperatures (800 K) and gas turbine pressures (20 atm) has about the same ignition delay time as other mixtures.

3.3 Methane-Propane Study

Shock-tube experiments and chemical kinetics modeling were performed to further understand the ignition and oxidation kinetics of various methane-propane fuel blends at gas turbine pressures. Ignition delay times were obtained behind reflected shock waves for fuel mixtures consisting of CH_4/C_3H_8 in ratios ranging from 90/10 to 60/40% [13]. Equivalence ratios varied from $\phi = 0.5$ to 3.0 at test pressures from 5.3 to 31.4 atm and with temperatures as low as 1042 K. Figure 7 shows typical results with a comparison between the data and the new model. It was found that the reactions involving $CH_3\dot{O}$, $CH_3\dot{O}_2$, and $\dot{C}H_3 + O_2/H\dot{O}_2$ chemistry were very important in reproducing the correct kinetic behavior.

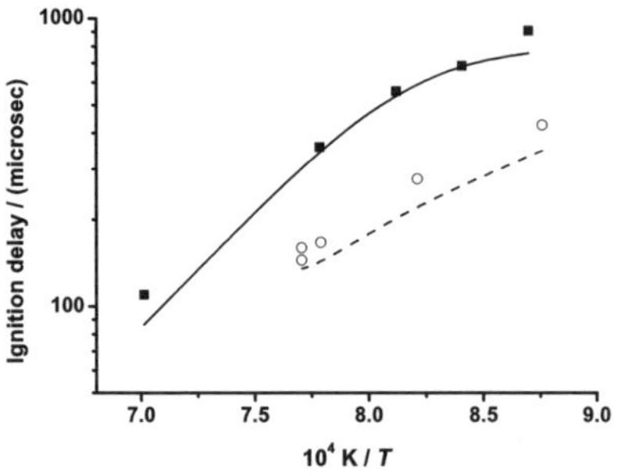

Fig. 7. Experimental results versus model predictions (lines) from the methane-propane study of Petersen et al. [13] at $\phi = 2.0$, 27.0 atm, black squares: 90% CH_4/10% C_3H_8, circles: 70% CH_4/30% C_3H_8. Dashed line corresponds with open symbols

4 Hydrogen-Based Mixtures

Hydrogen-based ignition experiments at practical conditions are also being conducted. These experiments utilize fuel blends that are based on syngas mixtures containing mostly carbon monoxide and hydrogen. Since the composition of coal-derived fuels such as syngas can vary substantially across the industry, there is the need to explore the ignition of blends containing a wide range of hydrogen content and also fuel blends containing water vapor.

Some recent results are provided in Fig. 8 for a syngas mixture in air at $\phi = 0.4$ at two different pressures (9.6, 30.8 atm) in comparison with two modern CO/H_2 chemical kinetics models [14,15]. At the higher temperatures, there is excellent agreement between the models and the data, but at temperatures below about 1100 K (right figure of Fig. 8), the measured ignition times are considerably faster than those predicted by the mechanisms. This problem has been seen in shock tubes for pure hydrogen fuels and may be connected to non-homogeneous ignition of the mixture. The cause of this early ignition is important for practical systems and is currently under debate, and further studies in this regard are being conducted in our laboratory.

Relatively early ignition in hydrocarbon systems is also seen in some shock-tube experiments, which may be related to what is seen in hydrogen-based systems. Some further discussion with respect to hydrocarbon ignition at high pressure and lower temperature are discussed briefly in the following section.

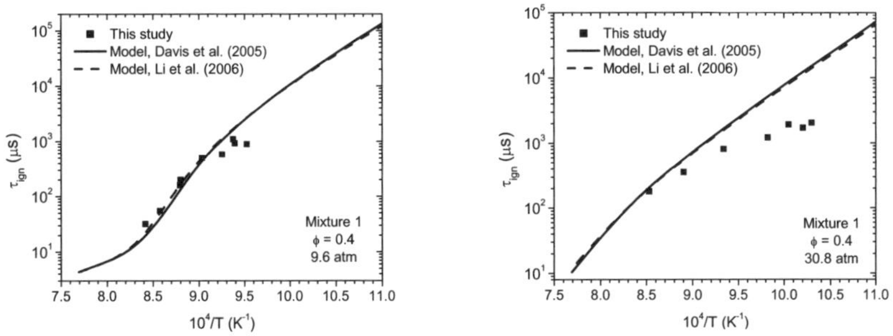

Fig. 8. Comparison of results for a syngas-air mixture (0.1143 H_2 + 0.0065 CO + 0.0056 CH_4 + 0.0150 CO_2 + 0.0014 Ar + 0.1791 O_2 + 0.6781 N_2) with chemical kinetics models [14,15] for two different pressures and a fuel/air equivalence ratio of 0.4

5 Ignition at Lower Temperatures

At higher pressures and temperatures below about 1100 K, ignition delay times obtained from shock-tube experiments of methane and other lower-order hydrocarbons are faster than those obtained from rapid compression machines under overlapping conditions. Similarly, the shock-tube ignition data are also faster than the predictions of detailed kinetics models, sometimes by an order or magnitude or more. However, shock-tube measurements from various laboratories tend to agree amongst themselves when the same pressure and emission features are employed for the definition of ignition time (Fig. 9).

Several examples of these discrepancies are presented by Petersen et al. [16]. One common feature of the faster shock-tube data is the appearance of mild ignition accompanied by a significant pressure increase prior to the main ignition event; such effects were documented by Fieweger et al. [17]. It is shown that when the data are corrected for the corresponding increase in temperature due to this gas compression, good agreement is seen between the shock-tube and the model/RCM data. Two examples of this adjustment for increased temperatures due to the early ignition are shown in the colored symbols in Fig. 9. The further investigation of the faster ignition seen at lower temperatures in shock tubes is a topic of continuing study in our laboratory, particularly into the origins of the early, perhaps non-homogeneous ignition events.

Acknowledgement. Several individuals at the University of Central Florida participated in all aspects of the research, as cited in the referenced articles, including Danielle Kalitan, Joel Hall, Jaap de Vries, Christopher Aul, Anthony Amadio, Viktorio Antonovski, Mouna Lamnaouer, Stefanie Simmons, Shatra Reehal, Schuyler Smith, Corey Frazier, and Dr. Alain Kassab.

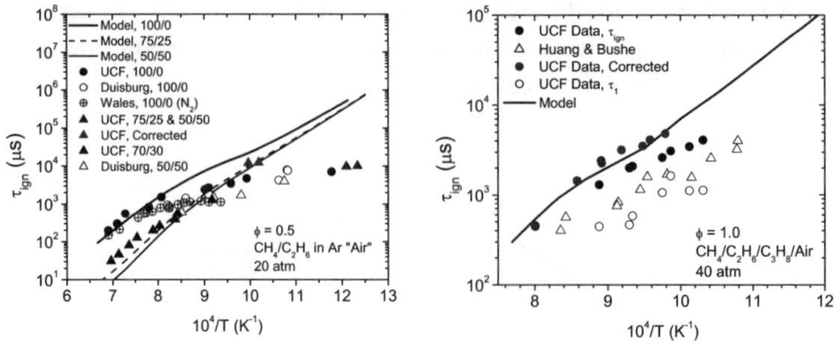

Fig. 9. Ignition for methane fuel blends. Left plot: CH_4 and CH_4/C_2H_6 fuels. Right plot: 0.0867 CH_4 + 0.0034 C_2H_6 + 0.0011 C_3H_8 in air at $\phi = 1.0$; from Petersen et al. [16] Corrections are in the plotted temperature to account for temperature rise due to early, combustion-induced pressure increase

References

1. Petersen E.L., Rickard M.J.A., Crofton M.W., Abbey E.D., Traum M.J., Kalitan D.M.: Meas. Sci. Tech. **16**, 1716 (2005)
2. Petersen E.L., Hanson R.K.: Shock Waves **10**, 405 (2001)
3. de Vries J., Aul C.J., Barrett, A., Lambe, D., Petersen, E.: Paper 0913, these proceedings
4. Amadio A.D., Crofton M.W., Petersen E.L.: Shock Waves **16**, 157 (2006)
5. Frazier C., Kassab A., Petersen E.: Paper 0910, these proceedings
6. de Vries J., Petersen E.L.: Proc. Combust. Inst. **31**, 3163 (2007)
7. Petersen E.L., Smith S.D., Hall J.M., de Vries J., Amadio A., Crofton M.W.: J. Eng. Gas Tubines Power, in press (2007)
8. Ó Conaire M., Curran H.J., Simmie J.M., Pitz W.J., Westbrook C.K.: Int. J. Chem. Kinet. **36**, 603 (2004)
9. Fischer S.L., Dryer F.L., Curran H.J.: Int. J. Chem. Kinet. **32**, 713 (2000)
10. Curran H.J., Fischer S.L., Dryer F.L.: Int. J. Chem. Kinet. **32**, 741 (2000)
11. Curran H.J., Gaffuri P., Pitz W.J., Westbrook C.K.: Combust. Flame **114**, 149 (1998)
12. Curran H.J., Gaffuri P., Pitz W.J., Westbrook C.K.: Combust. Flame **129**, 253 (2002)
13. Petersen E.L., Kalitan D.M., Simmons S., Bourque G., Curran H.J., Simmie J.M.: Proc. Combust. Inst. **31**, 447 (2007)
14. Davis, S.G, Joshi, A.V., Wang, H., Egolfopolous, F.: Proc. Combust. Inst. **30**, 1283 (2005)
15. Li, J. Zhao, Z., Kazakov, A., Chaos, M., Dryer, F.L., Scire, J.J.: Int. J. Chem. Kin. **39** 109 (2007)
16. Petersen E., Lamnaouer M., de Vries J., Curran H., Simmie J., Fikri M., Schulz C., Bourque G.: Paper 0911, these proceedings
17. Fieweger K., Blumenthal R., Adomeit G.: Combust. Flame **109**, 599 (1997)

Material processing and surface reaction studies in free piston driven shock tube

K.P.J. Reddy[1], M.S. Hegde[2], and V. Jayaram[2]

[1] *Department of Aerospace Engineering, Indian Institute of Science, Bangalore, India*
[2] *Solid State and Structural Chemistry Unit, Indian Institute of Science, Bangalore, India*

Summary. Recently established moderate size free piston driven hypersonic shock tunnel HST3 along with its calibration is described here. The extreme thermodynamic conditions prevalent behind the reflected shock wave have been utilized to study the catalytic and non-catalytic reactions of shock heated test gases like Ar, N_2 or O_2 with different material like C_{60} carbon, zirconia and ceria substituted zirconia. The exposed test samples are investigated using different experimental methods. These studies show the formation of carbon nitride due to the non-catalytic interaction of shock heated nitrogen gas with C_{60} carbon film. On the other hand, the ZrO_2 undergoes only phase transformation from cubic to monoclinic structure and $Ce_{0.5}Zr_{0.5}O_2$ in fluorite cubic phase changes to pyrochlore ($Ce_2Zr_2O_{7\pm\delta}$) phase by releasing oxygen from the lattice due to heterogeneous catalytic surface reaction.

1 Introduction

The performance capabilities of shock tubes have increased progressively and they have become important laboratory tools with applications in many areas of research. The strength of the shock wave traveling inside the driven section dictates the thermodynamic conditions of the test gas. The shock wave strength can be increased by using lighter driver gas such as helium at either very large driver pressure or temperature. In the free piston driven shock tube (FPST) the helium driver gas at both high temperature and pressure is produced by adiabatic compression using a heavy piston traveling at high speed inside a compression tube [1]. Using highly compressed and heated helium driver gas, test gases with specific enthalpies exceeding 25 MJ/kg are routinely produced in a FPST. We have recently constructed a moderate size free piston driven shock tunnel, HST3 with piston mass of 20 kg operating at free stream Mach number of 8. While the tunnel is used for the research work in hypersonic aerodynamics, the shock tube portion is also used for synthesizing new materials as well for studying the surface chemical reactions using various substrates and test gases.

We describe the design features and the performance parameters of the HST3 tunnel in this paper. Then we describe the application of the shock tube portion of HST3 for studying some of the interesting surface reactions on materials like C_{60} carbon, zirconia and ceria substituted zirconia. These high temperature materials are exposed to different shock heated test gases (Ar, N_2 or O_2) behind the reflected shock wave for about 2 to 4 ms and then examined for heterogeneous surface reactions using different experimental methods. The details of the techniques used for examining the shock treated samples and the results obtained are also presented in this paper.

2 HST3 Free Piston Driven Hypersonic Shock Tunnel

The basic analysis of piston motion in the compression tube given in ref. [2] is used to design the HST3 tunnel [3]. The HST3 tunnel shown schematically in Fig. 1 consists of a 10m long 0.165 m diameter compression tube (CT) inside of which a 20 kg piston is driven by high pressure nitrogen gas in a 1.3m long gas reservoir of 0.45m diameter. The CT is connected to a 4.5m long shock tube of 39mm diameter using a lock-nut arrangement shown in Fig. 1b. An inertial mass of about 2tons is mounted at the CT and shock tube junction to dampen the amplitude of oscillation of the tunnel due to the shifting of CG during the piston motion inside the compression tube. The shock tube is connected to a convergent divergent conical nozzle of 10^0 cone angle and 0.3 m exit diameter through a thin paper diaphragm. The Mach 8 flow from the conical nozzle exits into the 2.5m long 1.0m diameter cylindrical dump tank as a free jet.

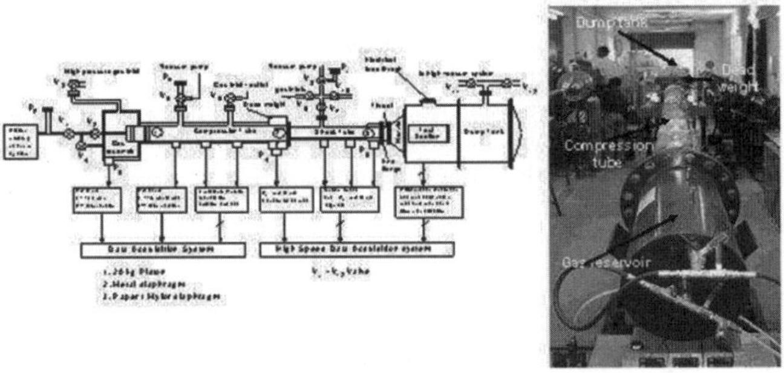

(a) (b)

Fig. 1. (a) Schematic diagram of HST3 with details of instrumentation and (b) photograph of the HST3 tunnel.

The HST3 tunnel is calibrated for a typical flow enthalpy of about 5 MJ/kg using reservoir pressure of 18 bar and the helium gas pressure of 1 bar in the CT. The light reflected from the black piston rings on the piston is used to measure the piston speed and acceleration at four locations along the CT and the pressure jump at the end of the compression stroke is recorded using a pressure transducer mounted at the end of the CT. The aluminium diaphragm with rupture pressure predetermined using a 39mm diameter shock tube ruptured at about 56 bar creating a normal shock wave traveling into the shock tube filled with test gas at a typical pressure of 0.1 bar. The shock speed is measured using two pressure transducers mounted a known distance apart on the shock tube and the pressure behind the primary and the reflected shock waves is measured using a pressure transducer mounted at the end of the shock tube. The free stream Mach number in the test section (dump tank) is estimated by the total pressure from the pitot probe mounted at the middle of the flow. The pressure signals recorded in a typical run are shown in Fig. 2. The measured values of the diaphragm rupture pressure, stagnation pressure at the end of the shock tube and the pitot pressure are used to numerically estimate the performance parameters of the HST3 tunnel. It is found that the piston landing is very smooth with minimum damage to the polyurethane piston stopper mounted on the pressure plate, as expected in a tuned operation.

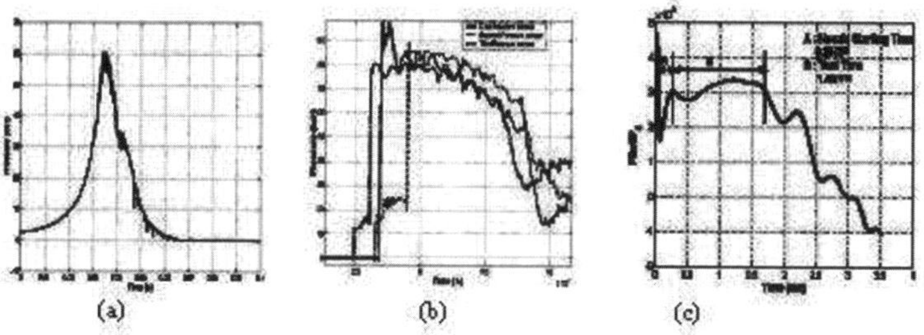

Fig. 2. Typical pressure signals at the (a) end of the compression (P_4), (b) from the pressure transducers used for shock speed measurement and the transducer mounted at the end of the shock tube (P_5) and (c) the normalized pressure signal from the pitot mounted in the test section.

3 Experimental Details

The high temperature materials required for present experimental work are prepared in the laboratory. The C_{60} carbon film was deposited on a Macor substrate by direct heating of C_{60} using tungsten filament in a vacuum chamber to perform non-catalytic surface reaction studies. High temperature materials like, nano ZrO_2 and nano $Ce_{0.5}Zr_{0.5}O_2$ in powder form are synthesized by single step combustion method in the laboratory to perform catalytic surface reaction studies [4]. The pellets of 10mm diameter were made using 100mg of each compound and compressed with 40 kN load using pellet pressing unit to use in the shock tube experiments.

Shock tube portion of free piston driven shock tunnel is used to conduct the experiments by isolating the nozzle-test section/dump tank portion by a flange at the end of the shock tube. Experiments are carried out by filling the compression tube with 1bar helium gas, while the shock tube is filled with any one of the test gases like Ar, N_2 and O_2 at 0.1 bar pressure. The test samples were mounted on the inner surface of the end flange. The deposited C_{60} film was subjected to shock in presence of nitrogen to synthesize carbon nitride films. The pellets of ZrO_2 and $Ce_{0.5}Zr_{0.5}O_2$ samples are mounted on the end flange of the shock tube to perform catalytic reaction in presence of test gas. For every run, the shock speed and the pressure signal at the end of the shock tube are recorded using piezoelectric pressure transducers. The temperature behind the primary (T_2) and reflected (T_5) shock waves have been calculated using 1-d normal shock theory assuming the specific heat ratio γ of 1.4. However, including the real gas effects due to the high temperature and assuming that the molecule goes from non-equilibrium state to equilibrium state as per Light Hill model [5] the effective value of specific heat ratio γ will be smaller which in turn reduces the temperature substantially. The experimental and calculated data for γ of 1.4 are listed in Table 1.

The characterization of samples was carried out before and after exposure to shock waves using different experimental methods. For example, XRD of samples were recorded using Philips X-ray powder diffractometer (X-pert Philips) at room temperature and Photoelectron spectra (XPS) were recorded in an ESCA-3 Mark II spectrometer (VG Scientific Ltd., England) using AlK_α radiation (1486.6 eV). The morphology and microstructure of samples were analyzed using SEM (Philips SIRION). Raman spectra

(Dilor XY Micro Raman) were recorded at room temperature and FTIR spectra were recorded using Perkin Elmer FTIR Spectrometer (SPECTRUM-1000).

		Experimental values				Calculated values		
Run	Sample	P_1 bar/test gas	ΔT	P_2 bar	P_5 bar	Vs (m/s)	Ms	T_5
R74	C_{60}	$0.05/N_2$	201	17.9	107	2487	7.0	6850
R36	ZrO_2 pallet	$0.1/O_2$	181	9.44	72	2762	8.4	9550
R42	ZrO_2 pallet	$0.1/N_2$	189	9.51	67	2645	7.5	7720
R75	$Ce_{0.5}Zr_{0.5}O_2$	0.1/Ar	206	9.48	50	2427	7.5	12810

Table 1. Experimental and calculated values for $\gamma = 1.4$

4 Results and discussion

4.1 Investigation of synthesized carbon nitride films (non-catalytic reaction)

After exposing C_{60} to shock heated nitrogen gas, investigation for carbon nitride film was taken up using different experimental methods. XPS study is used to investigate the chemical bonding of carbon nitride formed after the shock tube experiment. The deconvoluted spectra of C(1s) and N(1s) of carbon nitride film after exposure to shock wave are shown in Figs. 3a and 3b, respectively. The C(1s) photoelectron peak could be resolved into four peaks. The main peak detected at 285.5 eV is attributed to sp^2 type C=N bond and a smaller peak at 287.0 eV corresponds to sp^3 type C-N bond. A

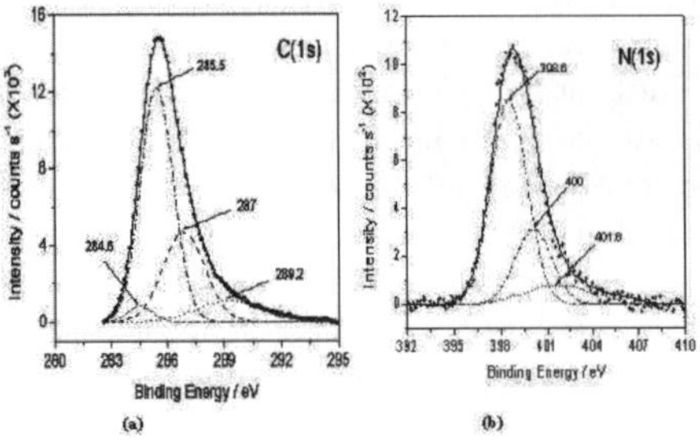

Fig. 3. Deconvoluted XPS spectra of CNx film synthesized using C_{60} film, (a) C(1s) peak and (b) N(1s) peak.

peak at 289.2 eV is assigned to C-O bond formed at the surface of the CNx film due to oxygen present in the atmosphere. Additional small peak at binding energy of 284.6 eV is attributed to C-C bond in carbon nitride as shown in Fig. 3a. XPS peak of N(1s) was deconvoluted to three peak being located at 398.6 eV, 400 eV and 401.6 eV which are attributed to sp^3-C-N, sp^2-C=N and N-O bonds, respectively shown in Fig. 3b. Binding energies of C(1s) and N(1s) presented here are in agreement with the reported values in the literature for CNx film prepared by other methods. It is confirmed that the CNx film synthesized by exposing C_{60} film to shock heated nitrogen gas is due to non-catalytic surface reaction. SEM micrograph of as-deposited C_{60} film shows a flat smooth surface as shown in Fig. 4a. After exposing C_{60} film to strong shock heated nitrogen, a total change in morphology shows carbon nitride crystals. The microstructure reveals features like near spherical and spherical crystallites of different sizes as shown in Figs. 4b, c. The

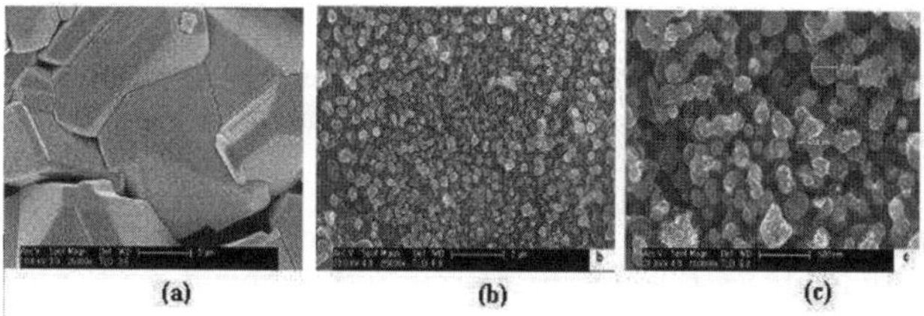

Fig. 4. SEM micrograph of as-prepared C_{60} sample and synthesized carbon nitride films, (a) pure C_{60} film, (b) CNx film and (c) CNx film at higher magnification

CNx films synthesized by shock tube method were further analyzed using Raman and FTIR spectroscopy. A Raman spectrum taken for carbon nitride film is shown in Fig. 5a. Deconvoluted D and G peaks of Raman spectrum at 1260, 1350, 1503 and 1593 cm^{-1} represent the presence of $\beta - C_3N_4$ and CNx like carbon nitride. FTIR spectra shows broad absorption band between 1000 and 2000 cm^{-1} due to carbon nitride film as shown

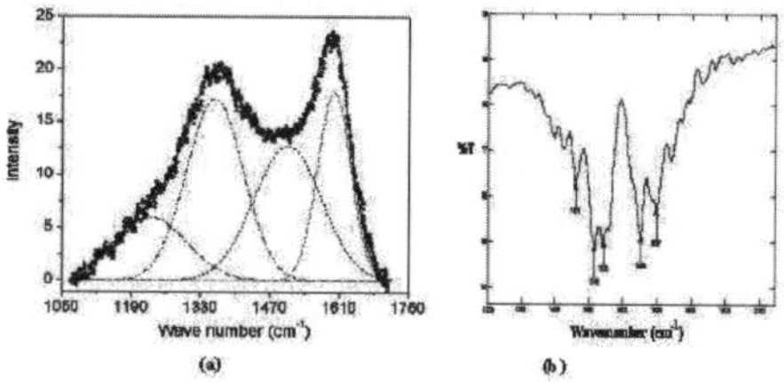

Fig. 5. (a) Micro Raman spectra of CNx and (b) FTIR spectra of synthesized CNx film

in Fig. 5b. This absorption band covers the frequency range of the stretching vibration of both C-N and C=N bonds between carbon atoms. The C-N and C=N absorption at 1500-1600 cm^{-1} and 1300-1500 cm^{-1} respectively, is well resolved, and the absorption at 1500-1600 cm^{-1} correlate to Raman D and G peaks.

It is clear from the above data that non-catalytic reactions have occurred when strong shock heated nitrogen gas interacted with the surface of carbon film. In addition, many physical and chemical processes can take place, like the surface melting of C_{60} carbon material, non-catalytic surface reaction and re-crystallization of carbon nitride compound. The non-catalytic chemical reaction can be written as 3C + $2N_2 \rightarrow C_3N_4$. Due to this reaction a stable C_3N_4 in solid form has indeed formed on the surface of carbon film.

4.2 Catalytic surface reaction with ZrO_2

XRD studies of pure ZrO_2 (as prepared sample) show cubic structure as shown in Fig. 6a. Figure 6b. shows the XRD of shock treated samples in presence of oxygen gas and a

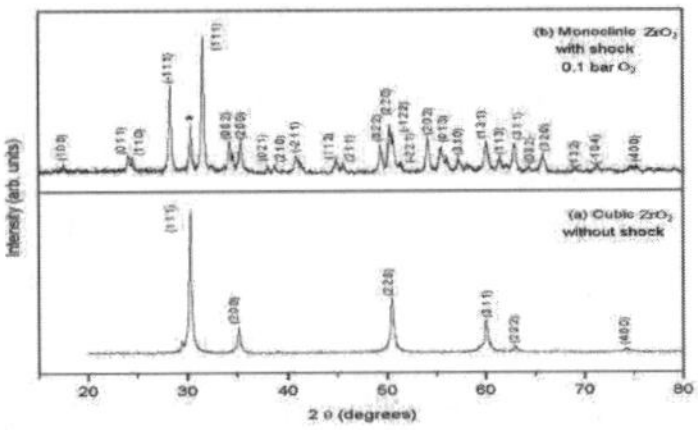

Fig. 6. Fig.6. XRD of powder pellet sample, (a) as prepared pellet sample shows cubic ZrO_2 phase, (b) after exposure to shock heated 0.1 bar O_2 the sample shows monoclinic phase.

monoclinic ZrO_2 structure along with the characteristic diffraction lines are assigned with hkl values. The study shows the stabilization of metastable c-ZrO_2 phase transformed to stable monoclinic ZrO_2 phase. The deconvoluted XPS spectrum of Zr (3d) core level peaks at 182.7 and 184.7 eV due to Zr $3d_{5/2}$ and $3d_{3/2}$ states show no change in the binding energy after the shock, which confirms that ZrO_2 does not change its chemical nature. Fig. 7a shows SEM micrographs of pure c-ZrO_2 and Fig. 7b shows shock-exposed

Fig. 7. SEM micrograph of ZrO_2 powder sample, (a) as-prepared sample, (b) after exposure to shock in 0.1 bar O_2 and (c) after exposure to shock in 0.1 bar N_2).

sample with needle type growth of monoclinic-ZrO_2 crystal. We can observe the growth of monoclinic needles type crystals with aspect ratio (Length/Diameter) of about 15 to 20. These micrographs show the huge melt of ZrO_2 and the growth of needles at the tip of the melt as shown in Fig 7c may be due to instantaneous cooling process.

The catalytic reaction process occurs when ZrO_2 undergoes 3-body catalytic surface recombination reaction with non-dissociated gas as: $O_2 + M_1 \rightarrow O_2 + M_2$ and reaction with dissociated oxygen is; $O + O + M_1 \rightarrow O_2 + M_2$ (where M_1 is c- ZrO_2 and M_2 is m- ZrO_2). In both the cases the third body (ZrO_2) retains the same composition but heat transfer takes place from gas molecules to third body (M_1).

4.3 Catalytic surface reaction with $Ce_{0.5}Zr_{0.5}O_2$

When shock heated argon at 12800K at a pressure of 50 bar interacted with the sample, the compound became black in colour suggesting the reduced $Ce_{0.5}Zr_{0.5}O_2$phase. Figures 8a, b show the powder XRD pattern of cubic $Ce_{0.5}Zr_{0.5}O_2$ and exposed to shock heated Ar, gas at initial pressure at 0.1 bar shows pyrochlore phase. With Ar the surface of the pellet turned completely black due to reduction, but with N_2 there was only a slight change in colour without turning into black. The compound remained pale yellow when

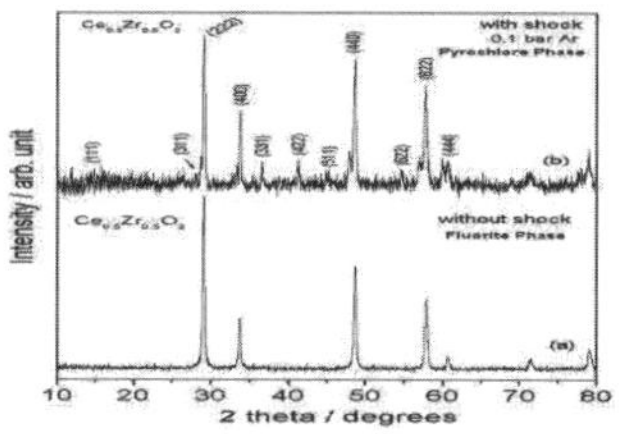

Fig. 8. XRD of $Ce_{0.5}Zr_{0.5}O_2$ pellets, (a) as prepared sample with Fluorite Structure and (b) after exposure to shock wave in Argon test gas, showing pyrochlore structure.

exposed to shock heated O_2 gas. The XRD pattern is indexed to a pyrochlore structure with lattice parameter a=10.64 $\mathring{A}$. Indexed pattern clearly demonstrates the formation of reduced $Ce_2Zr_2O_{6.3}$ pyrochlore phase. Thus, significant amount of lattice oxygen is released and the new phase is obtained with in 2-3ms, the test time of the shock tube. In contrast, recently Tinku et al. have shown that nano $Ce_{0.5}Zr_{0.5}O_2$ ($Ce_2Zr_2O_8$) can be reduced to $Ce_2Zr_2O_{6.3}$ pyrochlore phase by heating the compound to 1100 K for 7 days in presence of hydrogen atmosphere [6].

The oxidation states of the $Ce_{0.5}Zr_{0.5}O_2$ films were investigated by core-level XPS study. Before exposure to shock, XPS shows Ce and Zr in +4 states, but after exposure to shock in presence of Ar the sample shows Ce in +3 states and Zr(3d) in Zr +2 states. The XPS of Ce(3d) and Zr(3d) confirm reduction of both Ce and Zr to lower valent states. Figs. 9a, b show the SEM images of sample before and after exposure to shock

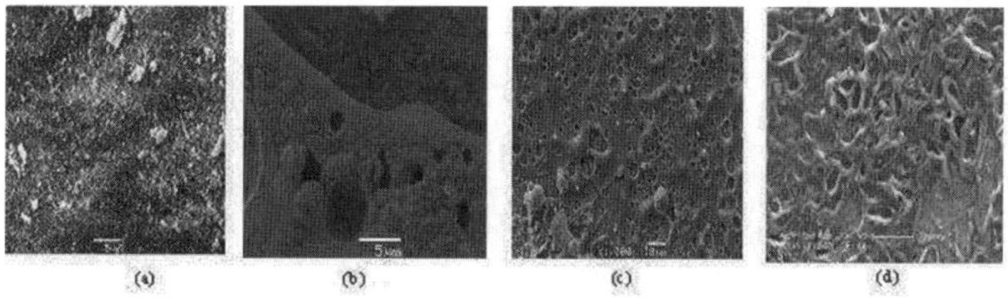

Fig. 9. (a) SEM Micrograph of $Ce_{0.5}Zr_{0.5}O_2$ sample before and after exposure to shock in presence of 0.1 bar test gas (b) With Ar colour changed to black, (c) with N_2 gas, colour changed but not completely black (d) with O_2 gas no change in colour.

heated Ar test gas. Clearly one can see changes in the morphology and the observed colour change from pale yellow to black is due to reduction of compound from $Ce_2Zr_2O_8$ to $Ce_2Zr_2O_{7-\delta}$. Fig. 9c shows SEM micrograph of shock treated compound in presence of 0.1 bar N_2 gas. Here also the colour of the compound changed, but not to complete black. Fig. 9d shows SEM micrograph of shock heated compound with 0.1 bar oxygen, but there was no change in colour. In presence of different test gases the micrograph clearly shows the melting and re-crystallization of the sample. Catalytic reaction and Oxygen Storage Capacity (OSC) of $Ce_{0.5}Zr_{0.5}O_2$ is presented in the following reaction. Under shock, it is clear that $Ce_{0.5}Zr_{0.5}O_2$ releases oxygen as follows:

$(Ce_{0.5}Zr_{0.5}O_2) \times 4 \rightarrow Ce_2Zr_2O_8$

$Ce_2Zr_2O_8 \rightarrow Ce_2Zr_2O_{7\pm\delta} + [O_{1\pm\delta}]_L$ (after exposed to Ar)

Lattice Oxygen (O_L) released from $Ce_2Zr_2O_8$ can react with high temperature air molecules present in the shock layer of the re-entry space vehicles. The three-way-catalyst involves removal of all the three gases like NO_x, CO and hydrocarbon. These reactions occur due to lattice oxygen released from $Ce_2Zr_2O_8$ compound which will reduce the temperature of the body of the reentry space vehicles and simultaneously remove NO_x, CO and hydrocarbons present in shock layer due to ablation of heat shield materials.

5 Conclusion

We have established and calibrated a moderate size free piston driven hypersonic shock tunnel. The tunnel has been used for basic aerodynamic research on large blunt cones at freestream Mach number of 8 with 5.0 MJ/kg specific enthalpy. The shock tube portion of the FPST has been utilized to investigate non-catalytic and catalytic reactions of different test gases at elevated thermodynamic conditions behind the reflected shock wave on the surface of different substrates. This study shows that the carbon C_{60} undergoes non-catalytic surface reaction by forming carbon nitride film (harder than diamond) with shock heated N_2 test gas. The ZrO_2 and $Ce_{0.5}Zr_{0.5}O_2$ compounds support catalytic surface reactions while c-ZrO_2 changes to m-ZrO_2 and fluorite $Ce_{0.5}Zr_{0.5}O_2$ changes to pyrochlore $Ce_{0.5}Zr_{0.5}O_2$ structure due to transfer of heat from the test gas.

Acknowledgement. We thank ISRO, DST, DRDL, AR&DB and The Director, IISc for the financial support. We thank HEA Lab team for their support during the experimental work and operation of FPST.

References

1. Stalker R J: *Preliminary results with a free piston shock tunnel*, AIAA Jl, Vol. 3, pp. 1170-1171, 1965.
2. Hornung H G: *Piston motion in free piston driver for shock tubes and tunnels*, FM 88-1, GALCIT, Pasadena, California,1988
3. V. Menezes, S. Kumar, Maruta, K. P. J. Reddy, and K. Takayama,: *Hypersonic flow over a multi-step afterbody*,Shock Waves Vol. 14, pp 421-424,2005.
4. Reddy K P J, Keshavamurthy K S and Reddy N M: *Design calculation for the proposed IISc free piston driven shock tunnel*, Detailed Results, Report No-94, HEA-1,1994.
5. Aruna S T and Patil K C: *Combustion synthesis and properties of nanostructured ceria-zirconia solid solutions*, Nano Structured Materials, 10, pp. 955-964,1998. .
6. Lighthill M J. : *Dynamics of a dissociating gas Part I: Equilibrium flow*, Journal of Fluid Mechanics, 2,pp.1-32,1957
7. Tinku Baidya, Hegde M S and Gopalakrishnan J : *Oxygen-release/storage property of* $Ce_{0.5}Zr_{0.5}O_2$ *(M=Zr, Hf) oxides: interplay of crystal chemistry and electronic structure*,J. Phys. Chem. B, (2007) (in print)

Molecular dynamics of shock waves in dense fluids

S. Schlamp[1] **and B.C. Hathorn**[2]

[1] *ETH Zürich, Institute of Fluid Dynamics*
Sonneggstr. 3, 8092 Zürich (Switzerland)
[2] *Oak Ridge National Laboratories, Division of Computer*
Oak Ridge, TN 37831 (USA)

Summary. The shock structure problem is one of the classical problems of fluid mechanics and at least for non–reacting dilute gases it has been considered essentially solved. Here we present a few recent findings, to show that this is not the case. There are still new physical effects to be discovered provided that the numerical technique is general enough to not rule them out a priori.

1 Introduction

Like Direct Simulation Monte Carlo (DSMC), Molecular Dynamics (MD) is particle–based and does not directly solve a macroscopic governing equation (e.g. Navier–Stokes). But unlike DSMC, MD simulates the movement of each real particle, i.e., there is no distinction between real and computational particles.

While DSMC can be shown to solve the Boltzmann equation, MD solves no macro- or microscopic governing equation for a fluid; only the Newtonian equations of motion are solved. It hence does not require any a priori assumptions or knowledge about the fluid, e.g. an equation of state and the transport coefficients (as required by macroscopic governing equations) or a collision model (as required by DSMC).

The forces between pairs of particles are calculated based on some potential function, the choice of which is the only form of modeling required. These range from hard–sphere models over generic potentials to many–parameter functions, which reproduce the thermodynamic and physical properties of fluids and solids quantitatively over a wide range of densities and temperatures. The computational cost of calculating forces between all possible particle pairs in the domain scales as N^2, but this growth can be made linear by a finite cut–off radius beyond which molecules are not considered for the force calculation.

MD was first proposed by Alder and Wainwright in the 1950s [1]. It does not make the assumptions underlying the Boltzmann equation and is thus well–suited even for high densities and can even be used for simulations of solids. Hoover [2] and Holian et al. [3], for example, use MD to simulate shock waves in dense monatomic gases. They find that the discrepancies between the Navier–Stokes and the MD results is smaller than for dilute gases. Tsai and Trevino [4] perform MD simulations of shock waves in liquids.

MD has been applied repeatedly for this fluid mechanical problem (e.g. [3, 5, 6]). These works focus on the steady–state profile. They reproduce the overshoot of the translational temperature perpendicular to the plane of the shock wave, which has been predicted by Yen [7]. A setup similar to the one used here, namely the creation of the shock by an impulsively accelerated piston, has also been studied in Refs. [8–10]. These authors obtain temporally resolved data for the formation of the steady state profile from an initially quiescent fluid. All of the above only consider monatomic gases, i.e., do not consider rotational degrees of freedom. Steady–state profiles for a shock in dilute nitrogen has been obtained using a hybrid method (MD + Direct Simulation Monte Carlo) by Tokumaso and Matsumoto [11]. In the present work, for the first time, the shock structure of a diatomic dense fluid is considered.

Sec. 2 first gives a review of the macroscopic shock structure. Each of Secs. 3 through 5 will then discuss one phenomenon which has only recently come to light (see Refs. [12–14], respectively, for more details). Please also note a second paper by the same authors presented at this conference, which gives an additional example, namely the existence of long–range correlations of the molecular velocities (or see Ref. [15] for more details). These are all aspect of a broader research project carried out over the last several years at ETH Zurich [16]. Here we only report on some of the results for the shock wave in nitrogen, but the same results (where applicable) have been observed in argon.

2 Macroscopic shock structure

The shock thickness Λ (e.g. measured by the maximum density gradient) for dilute gases tends to infinity for weak shocks, but is in the range between 2 to 4 mean–free paths for $M > 1.5$ with a slightly increasing tendency for large Mach numbers. For a given Mach number, it increases slightly (relative to the mean–free path) with the density. Over this short distance, the fluid state changes significantly so that it is immediately obvious that the continuum hypothesis is not applicable within the shock. Fig. 1 shows a particular example for a shock in dense nitrogen. Yet the dilute shock structure is very similar when properly scaling the horizontal axis. All quantities have been nondimensionalized by the pre– and post–shock properties, i.e., zero corresponds to the pre–shock state and unity is the post–shock value.

The breakdown of the LTE (local thermodynamic equilibrium) hypothesis is evident from the fact that $T_{rot} \neq T$ and even more so from the fact that the directional temperatures are not equal[3]. Note in particular the well–known temperature overshoot of the shock–normal temperature component, whose magnitude depends on the Mach number and the ratio of specific heats, and which has been confirmed experimentally.

The local Knudsen number based on the local density gradient and the local mean–free path reaches a peak value of ~ 0.2 in Fig. 1. Our results for dense fluids indicate that the asymmetry parameter (or shape factor) is less than unity whereas is is found to be greater unity in dilute gases for comparable Mach numbers.

3 Anisotropic molecular orientations

Let the orientation angle θ be the angle between the x–axis and the inter–nuclear axis. Because the nitrogen molecule is symmetric, θ is mapped to the range $0 \ldots \pi/2$ without loss of generality. Similarly, ϕ $(0 \ldots \pi/2)$ denotes the angle between the x–axis and the angular momentum vector. Histograms of the molecules in shock–parallel slices based on θ and on ϕ are then calculated, transformed into probability density functions, and compared to distribution functions of purely random orientations.

[3] The directional temperature only considers the molecular velocity fluctuations in one spatial direction.

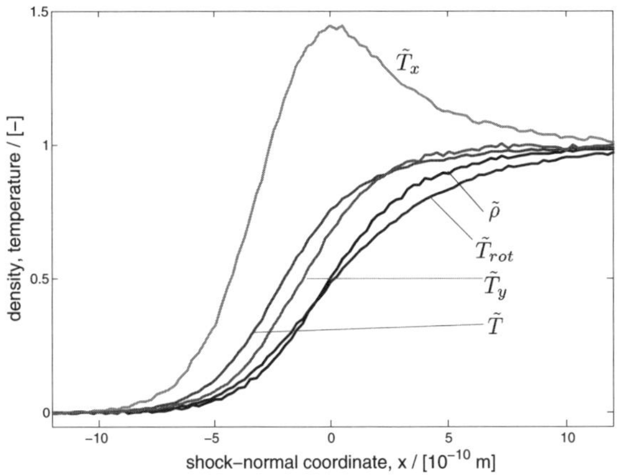

Fig. 1. Shock structure ($M = 3.56$) in dense nitrogen ($T_1 = 300$ K, $\rho_1 = 370.9$ kg/m^3, $T_2 = 978$ K, $\rho_2 = 741$ kg/m^3, $\Delta u = 985$ m/s). The velocity profile (not shown for clarity) is similar to T_y.

Fig. 2(a) & (b) shows the result. The color coding refers to the over– (red) or under–population (blue) of certain angles. It can be seen that θ and ϕ are distributed randomly upstream and downstream of the shock. Within the shock wave ($|x/\Lambda| \leq 1$), large θ are overpopulated, while small θ are underpopulated. The trend is weak (at least for these fluid conditions), but clearly visible above the background noise. The opposite behavior is observed for the angular momenta. Within the shock, small values of ϕ are overpopulated, large ϕ are underpopulated. But the effect for the angular momentum orientation is weaker.

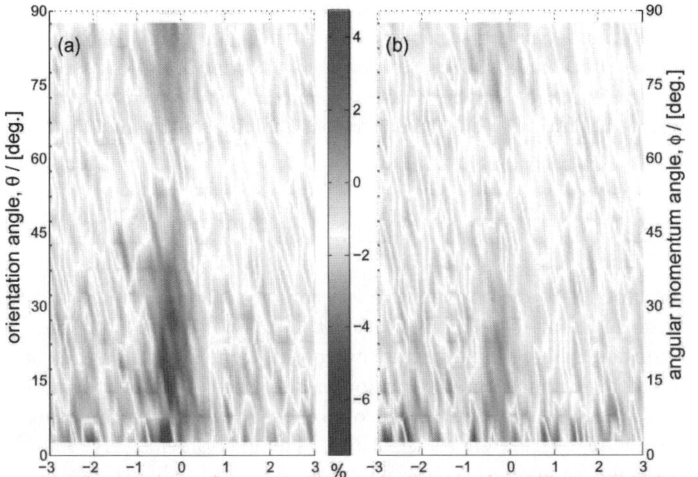

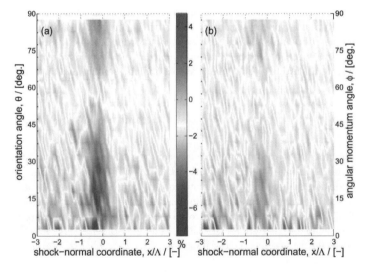

Fig. 2. Orientation angle distributions across shock wave; (a, left) population enhancement of orientation angles relative to a purely random distribution; (b, right) ditto for the orientation of the angular momentum vector. Flow is from left to right. The horizontal scale has been nondimensionalized by the shock wave thickness $\Lambda = 7.5$ Å.

We have proposed a nondimensional parameter governing the magnitude of the anisotropy, which includes the elongation of the molecule, the magnitude of the density gradient, the density, and the curvature of the potential function between molecules. Additional simulations with different flow conditions are required to confirm this scaling.

A posteriori (see Ref. [12] for a proposed mechanism), it seems obvious that such an alignment effect should exist and one wonders why it has not been observed before. One must note that all of the following conditions must be met to observe the phenomenon: a) the molecular model must be non–spherical; b) it is a dense gas effect. Simulations in dilute gases will not produce an alignment; c) realistic (i.e., smooth) interaction potentials are required. Simulations of hard–sphere molecules will not produce an alignment; d) the effect is weak. Large sample sizes are required to observe it above the statistical fluctuations.

4 Higher moments of the velocity distribution function

All common macroscopic governing equations of fluid mechanics are derived from the Boltzmann equation. The Euler and the Navier–Stokes equations, for example, are the zeroth and first order series expansions with respect to the Knudsen number (Chapman–Enskog expansion). The closure problem consists of expressing the heat flux and the stress tensor as a function of the other quantities. This implicitly results in making certain assumptions about the moments of the velocity distribution function above some order. The Euler equations, for example, follow from the Boltzmann equation in the high collision rate limit. In this limit, the molecular velocities follow an equilibrium, i.e., Maxwell–Boltzmann distribution, for which all odd moments are zero. The Navier–Stokes equations account for non–zero skewness, but similar assumption for the moments of order four and up are required in their derivation from the Boltzmann equation. Higher–order terms of the Chapman–Enskog series are the Burnett and super–Burnett equations, for which the closure problem is shifted to moments of order 5 and 6, respectively [17].

Let $\boldsymbol{\xi}^i = (\xi, \eta, \zeta)$ and $\mathbf{c}^i$ be the location and the velocity vector of molecule i in the shock–fixed reference frame, respectively. We define the central moments as follows:

$$\mu_0 = N = \sum_{i=1}^{N} 1 \qquad \mu_{1,\alpha} = \mathbf{u}_\alpha = \frac{1}{N}\sum_{i=1}^{N} \mathbf{c}_\alpha^i \qquad \mu_{k>2,\alpha} = \frac{1}{N}\sum_{i=1}^{N} \left(\mathbf{c}_\alpha^i - \mathbf{u}_\alpha(\xi^i)\right)^k \tag{1}$$

Greek subscripts denote components of vectors or tensors. Roman subscripts indicate the order of the moment. The moments for $k > 2$ are tensors, but only the diagonal elements are considered here. The sum is over all molecules within a slice $|x - \xi^i| \leq \delta x/2$.

The third and higher moments are normalized by the respective power of the standard deviation $\sqrt{\mu_{2,\alpha}}$. Just as the temperature can be related to the second moment, the heat flux vector can be related to the third central moment. The fourth and higher even moments are expressed as excess moments, i.e., the value of the moment which a Maxwell–Boltzmann distribution would have is subtracted ($\mu_{4,MB} = 3$, $\mu_{6,MB} = 15$, $\mu_{8,MB} = 105$, $\mu_{10,MB} = 945$). An equilibrium distribution would thus correspond to an excess moments of zero.

Fig. 3 shows the higher central moments of the velocity distribution function across the shock wave. The even moments are plotted in Fig. 3(a), the odd moments in Fig. 3(b). The solid lines are for the direction along the direction of the main flow. The dotted lines are for one of the in–plane velocity components. Upstream and downstream, all excess moments are zero, consistent with a Maxwell–Boltzmann distribution of a fluid in equilibrium.

The even moments of order four and higher of the velocity distribution function across the shock wave exhibit a sign reversal. They are positive on the cold side of the shock, but slightly negative on the hot side of the shock. This means that the velocity distribution function changes from having fat tails to having slim tails, at least with respect to the molecular velocities along the shock–normal direction. The distribution function for the in–plane velocity components does not have a sign reversal. We do not expect that this is a dense gas effect. Experimental and numerical data for dilute gases, from which the higher moments can be extracted, is available in the literature, but to the authors' knowledge, the effect has not been reported previously. The location where the higher out–of plane moments first deviate from zero does not depend on he order of the moment, i.e., the trend for the lower moments that the temperature (second moment) changes upstream of the flow velocity (first moment) and the density (zeroth moment) is not continued or it approaches a limit asymptotically (alse see Fig. 1).

The magnitude of the higher moments does not decrease with the order of the moment. The opposite is observed. This is significant when considering appropriate closure relations for the atomistic governing equations when deriving macroscopic governing equations from them. The influence on the macroscopic quantities will, for most practical purposes, be negligible because the higher moments are predominantly affected by the (few) particles in the tails of the distribution function. The effect could, however, be large for flows in which high kinetic energy collisions play a significant role, such as for chemically reacting flows.

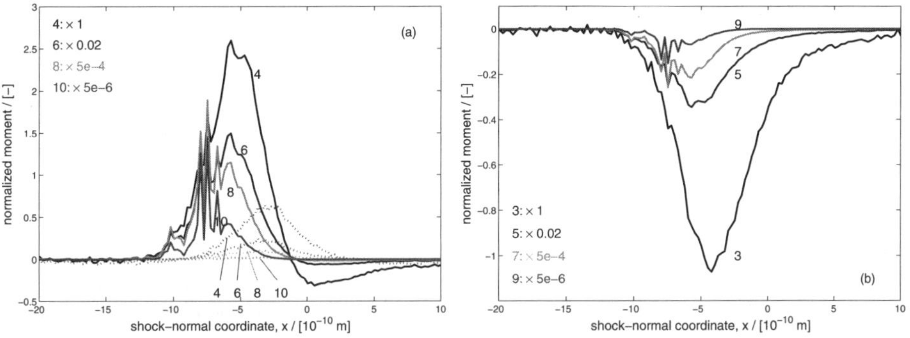

Fig. 3. Even (a, left) and odd (b, right) moments of the velocity distribution function for a shock wave in dense nitrogen. The labels indicate the order of the moment. The solid lines are for the molecular velocities in shock–normal direction, the dotted lines for velocities within the shock plane. The odd moments for the shock–parallel velocities are zero (within the measurement uncertainty) and are not shown. The curves are scaled to fit in the same axes. The scaling factors, which have been applied for each order, are shown in each panel.

5 Three–dimensional structure of a plane shock wave

The term "shock structure" usually refers to the variation of the various thermodynamic state variables along the shock–normal direction. When measuring, simulating, or theoretically considering the shock structure, data is either averaged along the direction parallel to the shock wave, or it is implicitly assumed that the shock structure is one–dimensional, i.e., that a plane shock wave is truly plane.

Suppose that the shock wave is not plane on microscopic scales. Then, in a Gedankenexperiment, the averaged shock structure can be decomposed into two parts: First, the shock structure one would obtain if the shock wave was indeed truly plane (Fig. 4a). In this case, the local shock structure would be identical to the averaged shock structure everywhere. Second, a broadening effect due to deflections up– or downstream of the shock location from its mean. Consider, for example, the (purely theoretical) case where the shock is a discontinuity locally, but that the plane connecting these discontinuities is wavy (Fig. 4b). Spatial averaging along the in–plane directions would then produce a smooth shock profile with a thickness governed by the amplitude of the plane's deflections. These two limiting cases are shown in Fig. 4(a) & (b), respectively. The averaged profile (shown at the bottom of Fig. 4) could be identical, even though the local profiles (e.g. a step function in Fig. 4b) are qualitatively different. In our Gedankenexperiment, the averaged shock structure is a superposition of these two cases. Fig. 4(c) & (d) shows two limiting cases with respect to the co–movement of different iso–lines. While the local shock structure is the same everywhere in Fig. 4(c) (yet different from the macroscopic structure), even the local structure varies spatially.

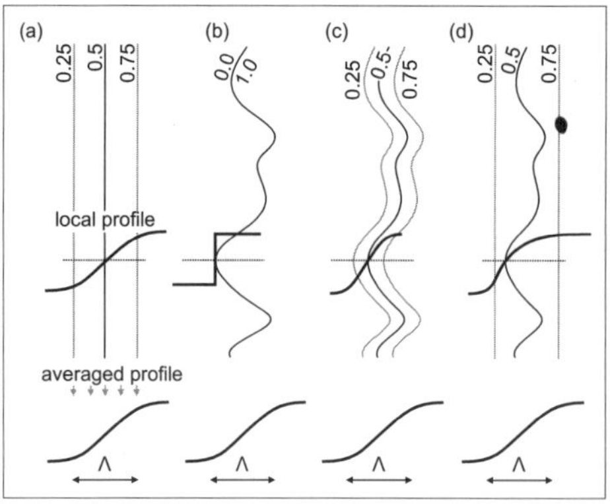

Fig. 4. Schematic of some limiting cases for the three–dimensional shock wave structure; (a) truly plane shock wave; (b) wavy discontinuity; (c) strong correlations between iso–density planes; The local shock structure is the same everywhere, but shifted up– or downstream. (d) no correlation between iso–density planes.

We determine the local density and velocity shock locations $x_{s,\rho}$ and $x_{s,u}$ by finding the location x for given y and z, where the density and the mean velocity within a spherical neighborhood of radius $R = 20$ Å is the mean between the pre– and the post–shock state.

We find local shock deflection comparable to the mean–free path. The velocity shock is smoother than the density shock and there is no significant correlation between these deflections. In order to quantify the effect of the three–dimensional structure on the overall shock structure, consider the limiting case shown in Fig. 4(b). Assume that the local shock structure is discontinuous at the local shock location x_s. Then, the cumulative

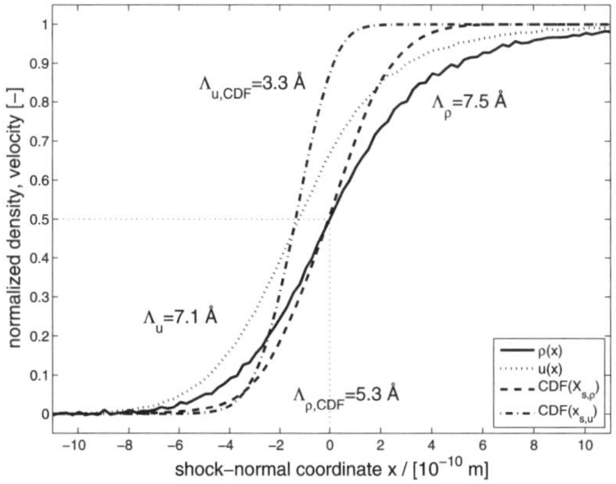

Fig. 5. Density and velocity profiles across shock wave in nitrogen; solid line: actual structure of density shock; dashed line: cumulative distribution function for $x_{s,\rho}$; dotted line: actual structure of velocity shock; dash–dotted line: cumulative distribution function for $x_{s,u}$.

distribution function (CDF) for x_s would be the corresponding averaged shock structure. This is shown in Fig. 5. The solid line and the dotted line are the actual averaged profiles for the density and the velocity across the shock, respectively (as shown in Fig. 1). The velocity shock leads the density shock. The dashed curves are the CDFs for $x_{s,\rho}$ and $x_{s,u}$ (the latter leading the former). Based on these curves, a shock thickness Λ_{CDF} can be calculated for this hypothetical case. The result is also shown in Fig. 5. Λ_{CDF} represents 71% (density) and 46% (velocity) of Λ. One could say that half or more of the macroscopic shock thickness can be explained by the three–dimensional structure of the shock wave.

The three–dimensional structure is a necessary consequence of the breakdown of the continuum hypothesis on the length scale of the shock thickness. On this microscopic level, the shock is no longer propagating into a homogeneous medium, but into one with local fluctuations of the density, velocity, and temperature. The local speed of propagation for the shock wave will hence vary spatially. The shock remains mostly planar for two reasons: through information exchange in the in–plane directions (molecules have velocity components not just in the shock–normal direction) and because the density, velocity, and temperature average out to their macroscopic values along each shock–normal trajectory.

The re–interpretation of the shock thickness to be partially due to a “surface roughness” of the shock wave might be considered merely semantics, but we believe that this point of view provides an interesting and intuitive alternate interpretation.

6 Conclusions

The effects described owe their discovery to the ever increasing computational resources. The computations require large sample sizes to reveal subtle effects and provide good statistics and had to employ the most expensive (yet most general) computational technique to allow their existence in the first place. But if such measures are taken, then there is still “new physics” left to be discovered even in such a classical problem.

Acknowledgement. This work was sponsored by the Division of Computer Science and Mathematics and used resources of the Center for Computational Sciences at Oak Ridge National Laboratory, which is supported by the Office of Science of the U.S. Department of Energy under Contract No. DE–AC05–00OR22725. Thanks to Mitchio Okumura at the California Institute of Technology for useful discussions.

References

1. B. J. Alder and T. E. Wainwright. Phase transition for a hard sphere system. *Journal of Chemical Physics*, 27:1208–1209, 1957.
2. W. G. Hoover. Structure of shock wave front in a liquid. *Physical Review Letters*, 42(23):1531–1534, 1979.
3. B. L. Holian, W. G. Hoover, B. Moran, and G. K. Straub. Shock–wave structure via non–equilibrium molecular–dynamics and Navier–Stokes continuum mechanics. *Physical Review A*, 22(6):2798–2808, 1980.
4. D. H. Tsai and S. F. Trevino. Thermal relaxation in a dense liquid under shock compression. *Physical Review A*, 24(5):2743–2757, 1981.
5. E. Salomons and M. Mareschal. Usefulness of the Burnett description of strong shock waves. *Physical Review Letters*, 69(2):269–272, 1992.
6. O. Kum, W. G. Hoover, and C. G. Hoover. Temperature maxima in stable two–dimensional shock waves. *Physical Review E*, 56(1):462–465, 1997.
7. S. M. Yen. Temperature overshoot in shock waves. *Physics of Fluids*, 9(7):1417–1418, 1966.
8. A. K. Macpherson. Formation of shock waves in a dense gas using a molecular–dynamics type technique. *Journal of Fluid Mechanics*, 45:601–621, 1971.
9. J. Horowitz, M. Woo, and I. Greber. Molecular dynamics simulation of a piston–driven shock wave. *Physics of Fluids (Gallery of Fluid Motion)*, 7(9):S6, 1995.
10. M. Woo and I. Greber. Molecular dynamics simulation of piston-driven shock wave in hard sphere gas. *AIAA Journal*, 37(2):215–221, 1999.
11. T. Tokumasu and Y. Matsumoto. Dynamic molecular collision (DMC) model for rarefied gas flow simulations by the DSMC method. *Physics of Fluids*, 11(7):1907–1920, 1999.
12. S. Schlamp and B. C. Hathorn. Molecular alignment in a shock wave. *Physics of Fluids*, 18(9):096101, 2006.
13. S. Schlamp and B. C. Hathorn. Higher moments of the velocity distribution function in dense–gas shocks. *Journal of Computational Physics*, 223(1):305–315, 2007.
14. S. Schlamp and B. C. Hathorn. Three–dimensional structure of a plane shock wave. *submitted to Physics of Fluids*, 2007.
15. S. Schlamp and B. C. Hathorn. Incomplete molecular chaos within dense–fluid shock waves. *submitted to Physical Review E*, 2007.
16. S. Schlamp. *Shock Wave Structure in Dense Argon and Nitrogen – Molecular Dynamics Simulations of Moving Shock Waves.* Habilitation thesis, ETH Zurich, Institute of Fluid Dynamics, Sonneggstr. 3, 8092 Zurich, Switzerland, 2007.
17. C. Cercignani. *The Boltzmann Equation and its Applications.* Springer–Verlag, New York, 1988.

SBLI control for wings and inlets

H. Babinsky and H. Ogawa

Department of Engineering, University of Cambridge
Trumpington St., Cambridge CB2 1PZ, United Kingdom

1 Introduction

The application of flow control to shock wave / boundary layer interactions (SBLIs) has been widely studied in recent decades. Two important practical applications for shock control technology are transonic wings (Fig. 1) and supersonic engine inlets (Fig. 2). In both cases the overall performance is significantly affected by shock wave / boundary layer interactions and it is hoped that flow control can provide considerable benefits.

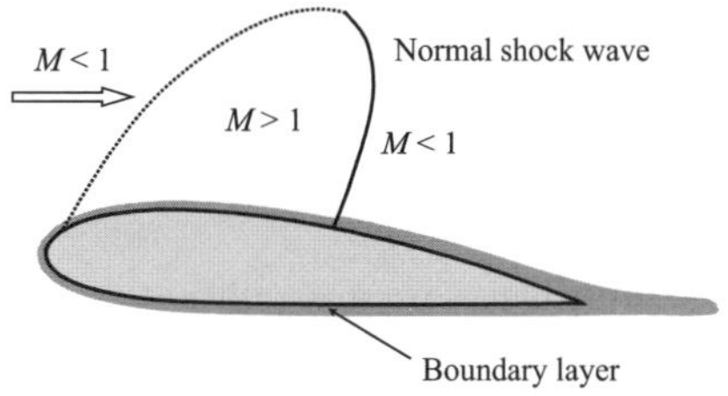

Fig. 1. Transonic wing

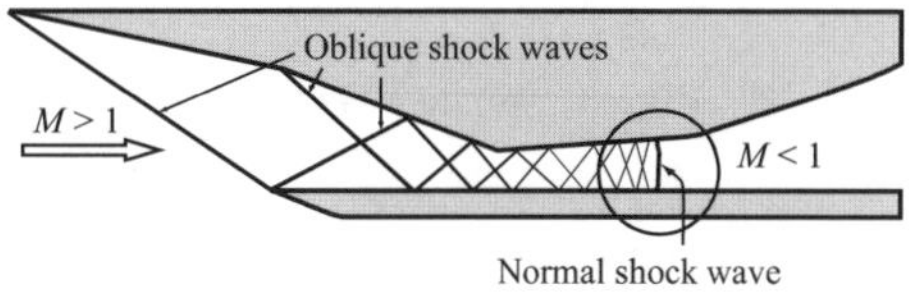

Fig. 2. Supersonic engine inlet

Current methods of SBLI control can be classed into two groups: Shock control and boundary layer control. The latter type modifies the boundary layer ahead of a shock interaction to prevent or delay shock-induced separation. This class of control is particularly suitable in applications where strong shock waves are encountered and flow separation is the prime performance degradation mechanism. Supersonic engine inlets fall into this category. Shock control, on the other hand, attempts to modify the structure of a SBLI with the aim to reduce the overall stagnation pressure loss. This type of control is more relevant to transonic airfoils which are generally operated far from shock-induced separation. To some extent, the two types of control have contradictory effects; boundary layer control often tends to increase stagnation pressure losses through the shock, while shock control often causes a degradation of the boundary layer. This paper presents a brief overview of the current state of the art of flow control for SBLIs.

2 Boundary Layer Control

In situations where the presence of SBLI can cause flow separation and performance degradation flow control can change the boundary layer properties to improve the resistance to adverse pressure gradients and delay or prevent shock induced separation.

Pearcey [1] reported on a number of suitable boundary layer control methods, such as blowing, suction and vortex generators (including air-jet vortex generators). Of these techniques, vortex generators (VGs) have proved to be the most successful and continue to be the most widely used today. Fig. 3 shows examples of traditional vortex generators whose height is comparable to the boundary layer thickness. Most of these devices have successfully demonstrated a delay or prevention of shock-induced separations [1–4].

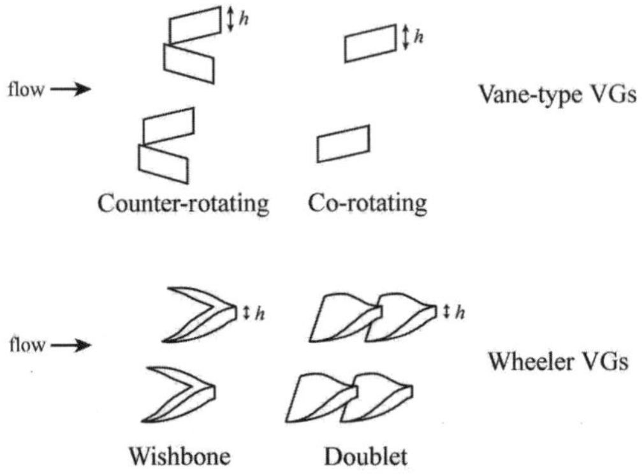

Fig. 3. δ-scale VGs [5] ($h \sim \delta$)

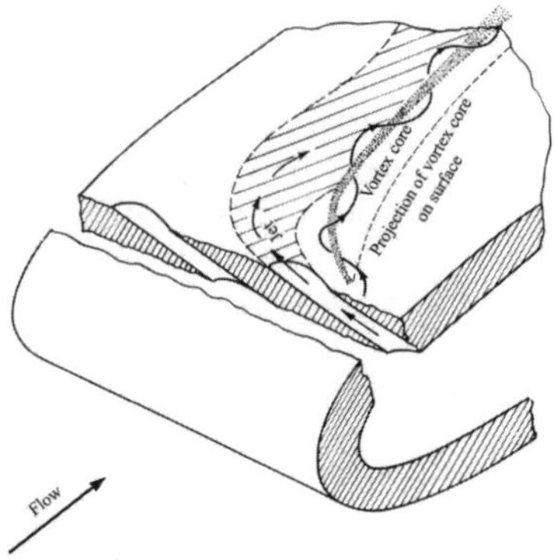

Fig. 4. Air-jet VGs [1]

While undoubtedly highly successful, vortex generators have several significant disadvantages. Primarily, they incur considerable device drag which is a particular drawback if they are only required for a small part of the operating envelope. Secondly, their use in inlets is problematic because of concerns about their mechanical robustness given the catastrophic consequences in case of parts breaking off and entering the engine.

Air-jet vortex generators (Fig. 4) can overcome some of the problems associated with traditional VGs [1,6,7]. While air-jet VGs have been shown to be capable of reducing or eliminating shock induced separation, they require significant mass flow rates and this in turn makes their application on aircraft heavy and costly.

More recently, pulsed jets have been used to achieve boundary layer control in many low speed applications. Zero mass flux pulsed jets (or virtual jets) have the distinct advantage that they do not need a supply of high pressure air and this significantly reduces their installation weight and complexity. For virtual jets and pulsed jets in general it has been demonstrated that the unsteady nature of the control can be more effective than steady air jets. However, while widely investigated in many low speed experiments (under effectively incompressible conditions), their use to control SBLI flows is hampered by the need for very powerful devices which so far are only at an experimental stage. Nevertheless, a number of researchers [8–10] have already demonstrated the ability of pulsed jets to control SBLI induced separation and a comprehensive investigation of pulsed jets for unsteady SBLIs is included in a current European research programme (UFAST).

Somewhat more mature than pulsed or virtual jets are micro-vortex generators (or sub-boundary layer VGs). Micro-VGs are similar to traditional VGs, but their sizes are much smaller. Typical heights of successful designs lie between 10% and 40% of boundary layer thickness. This small size incurs much reduced device drag, giving low off-design

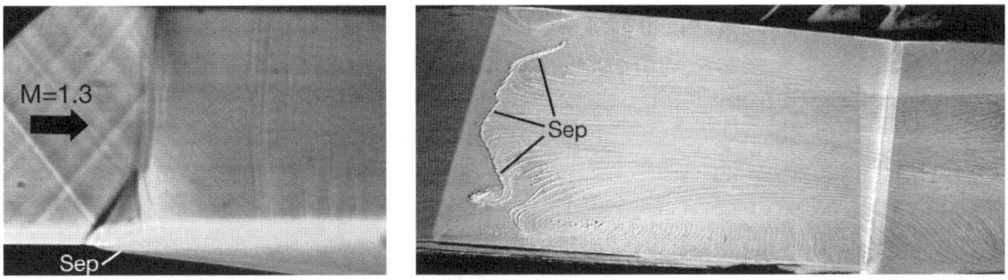

(a) without control

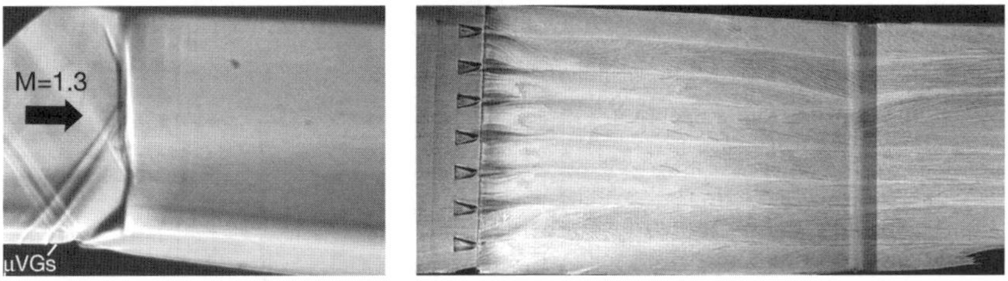

(b) with control

Fig. 5. Schlieren images and surface oil flow patterns of simplified transonic inlet flow with/without micro-VG flow control placed ahead of the shock wave

penalties. However, current research suggests that micro-VGs need to be placed closer to the point of separation in order to be effective. In a study of normal shock / boundary layer interactions at $M = 1.5$ [11], micro-ramps have considerably reduced separations while micro-vanes were able to completely eliminate separation. On transonic airfoils, micro-ramps have been shown experimentally to delay shock-induced separations with significantly reduced device drag when compared to traditional VGs [12]. More recently, RANS computations performed by NASA Glenn Research Center on inlet configurations [13], have demonstrated that micro-ramps have the ability to produce benefits comparable to traditional boundary layer bleed. Fig. 5 illustrates the potential of boundary layer control [14]. Here the flow in a simplified inlet duct, where a normal shock wave is followed by a subsonic diffuser, is shown with and without micro-vortex generators. It can be seen that these devices are capable of keeping the flow attached throughout, which would lead to significantly enhanced inlet performance.

However, while boundary layer flow control can prevent shock induced separation, it has little effect on the strength of the shock wave and cannot therefore reduce the shock-induced stagnation pressure loss. In fact, often the shock wave is strengthened by the application of boundary layer control (because the interaction length is reduced) and this leads to slightly enhanced stagnation pressure losses.

3 Shock control

Much of the research on shock control has focused on its application to transonic aircraft wings, where wave drag is a prime concern. As seen in Fig. 6, the aim of shock control is to reduce wave drag by smearing the shock wave at the rear of the supersonic domain. This changes the flow structure at the foot of the shock, replacing the (predominantly normal) shock wave with a λ-type structure. Such an effect can be achieved by deflecting the flow outside the boundary layer away from the airfoil surface. The streamline deflection causes the formation of an oblique shock wave, which eventually merges with the main shock. Due to the fact that the pressure rise across this leading oblique shock wave is less than the pressure jump across the normal shock a second shock (the rear leg) is necessary to achieve the required overall pressure jump, completing the λ-shape. Since multiple, but weaker, shock waves incur less stagnation pressure drop for the same pressure jump, the effect of control is to reduce the shock losses in the λ region and this has the overall effect of reducing wave drag. Simple theoretical considerations suggest that wave drag reductions of up to 2/3rds can be achieved on typical transonic airfoils [19].

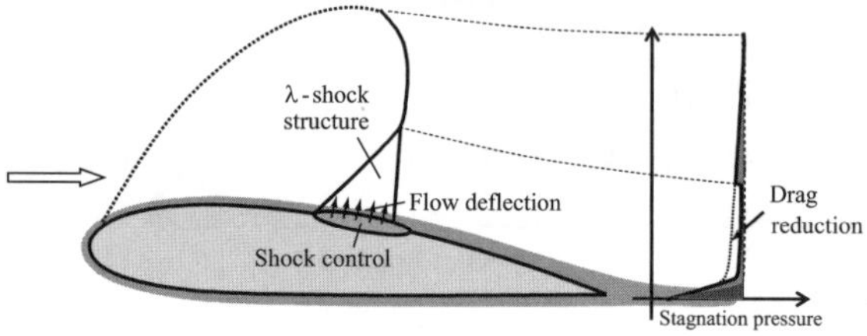

Fig. 6. Transonic wing with shock control

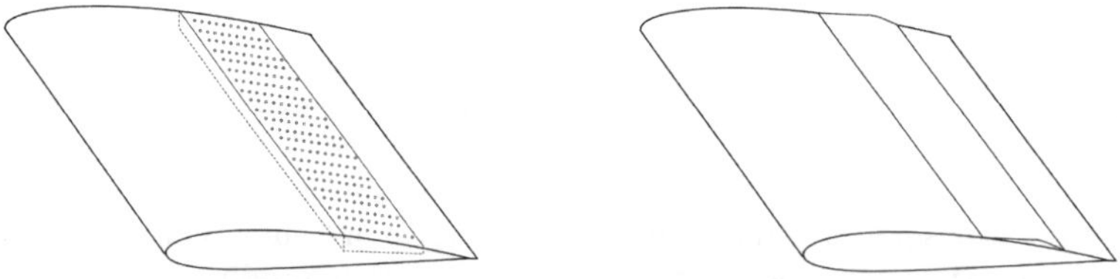

(a) passive control (b) 2D bump control

Fig. 7. 2-dimensional shock control techniques

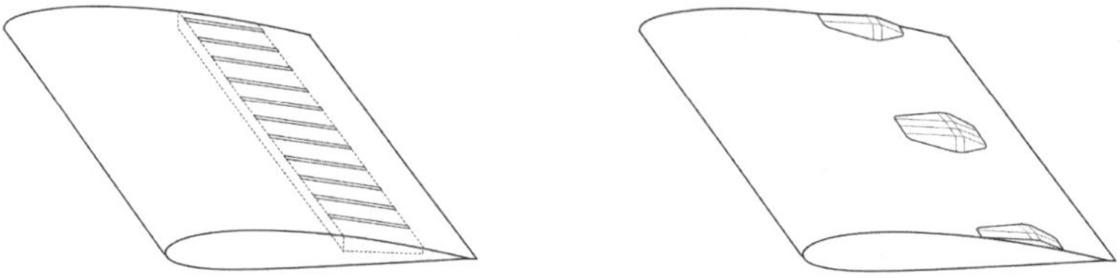

(a) slot/groove control (b) 3D bump control

Fig. 8. 3-dimensional shock control techniques

Shock control methods can be divided into 2-dimensional and 3-dimensional techniques as shown in Figs. 7 and 8. In each category the simplest methods are those that achieve a flow displacement by adding a bump to the surface contour [15,16]. However, the same effect can be achieved through wall transpiration by so-called passive control which was first suggested by Bushnell and Whitcomb at NASA Langley (as reported

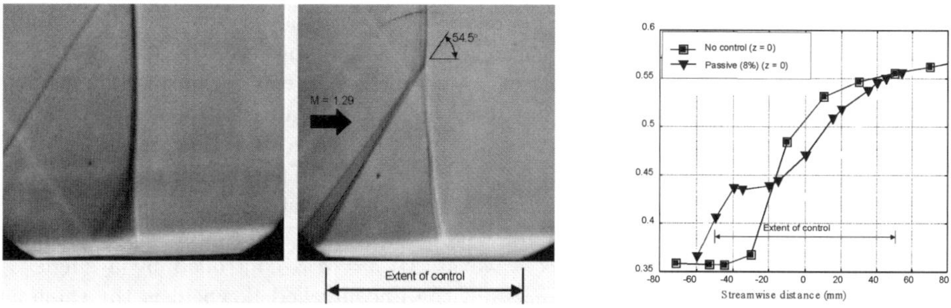

(a) schlieren photos (b) surface pressure distributions

Fig. 9. Normal SBLI with/without passive control [18]

by Bahi et al. [17]). The change in shock structure due to control can be seen clearly in Fig. 9 (a) when comparing schlieren images of a normal shock flow with and without control [18]. Fig. 9 (b) shows the corresponding surface pressure distributions measured along the floor of the wind tunnel. It can be seen that the leading shock leg introduces an initial pressure rise which is followed by a near constant pressure plateau. This plateau is located underneath the λ-shock region and indicates that the Mach number ahead of the rear shock leg is significantly reduced. The strength of control can be measured by the level of plateau pressure achieved in the λ region. Greater flow deflection at the start of the control region produces stronger leading shock legs and gives rise to greater plateau pressures. This in turn lowers the flow Mach number in this domain, which reduces the strength of the trailing shock. In general there exists an optimum flow deflection for maximum reduction of stagnation pressure losses across the shock system, which can be determined theoretically [19].

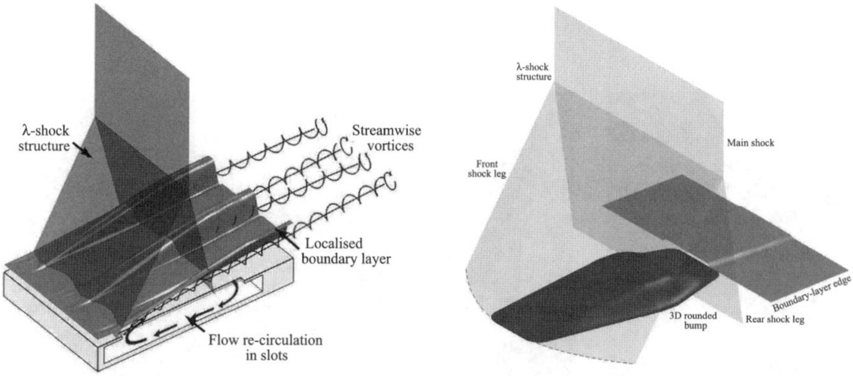

(a) slots (b) 3D bump (no separation)

Fig. 10. Normal shock interaction controlled by 3D devices

While most traditional shock control methods can be classed as 2-dimensional (the control applies uniformly along the span of a wing), 3-dimensional devices are modifications of these basic ideas, such as mesoflaps or slots /grooves which are derivations of passive control [23] and three-dimensional contour bumps (see Fig. 8). It has been shown [20–24] that such 3-dimensional methods which only cause a localised deflection of the flow can, if spaced correctly, bring about a global alteration of the shock structure in the spanwise direction, as shown schematically in Fig. 10 (a). In other words, an array of 3-dimensional devices is capable of generating a region of near-uniform pressure underneath a λ-shock system. The success of 3-dimensional controls can be judged by the extent and uniformity of this pressure plateau in streamwise and spanwise direction.

Figure 11 shows a normal shock wave at $M_\infty = 1.3$ controlled by a single 3-dimensional contour bump. For comparison, the uncontrolled flowfield is the same as in Fig. 9 (a). Here, the shape of the shock control bump has been carefully optimised to avoid flow separation and as a result there is almost no additional viscous drag due to the presence of control. An integration of the stagnation pressure downstream of flow control demonstrates that stagnation pressure losses are reduced by as much as 30% in this flow. Similar savings on wings and airfoils have so far only been demonstrated for 2D devices [15, 16] but research into 3D controls is currently under way.

While shock control bumps are generally aimed at weak shock waves in the absence of shock induced separation, there are also indications that the presence of such devices can be beneficial at higher Mach numbers (for stronger shocks). Figure 12 compares surface oil flow visualisations for the interaction of a turbulent boundary layer with a normal shock wave at $M_\infty = 1.5$ both, with and without bump control. Here the baseline flowfield features a quasi-2D separation. In contrast, the flow in the presence of control exhibits significantly reduced separation, which is broken up into discrete regions surrounded by attached flow. Not only does this reduce the detrimental effects of separation, the highly 3-dimensional flow downstream of the bump also introduces streamwise vortices similar to those encountered in slot control (Fig. 10 (a)). Since strong shock waves on airfoils are commonly seen under off-design conditions near the onset of large flow separation around the trailing edge, it is thought that a well-designed 3-dimensional shock control bump may also offer advantages under such conditions, where more traditional 2-dimensional devices generally perform poorly.

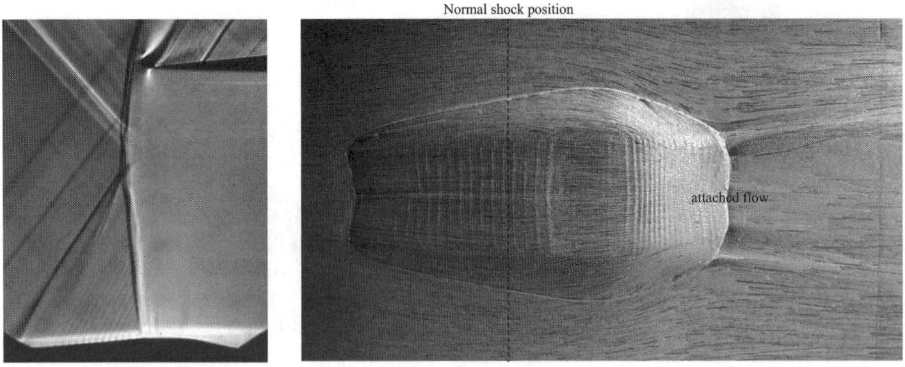

(a) schlieren photo (b) surface oil flow visualisation

Fig. 11. 3D-bump-controlled flow field at $M_\infty = 1.3$

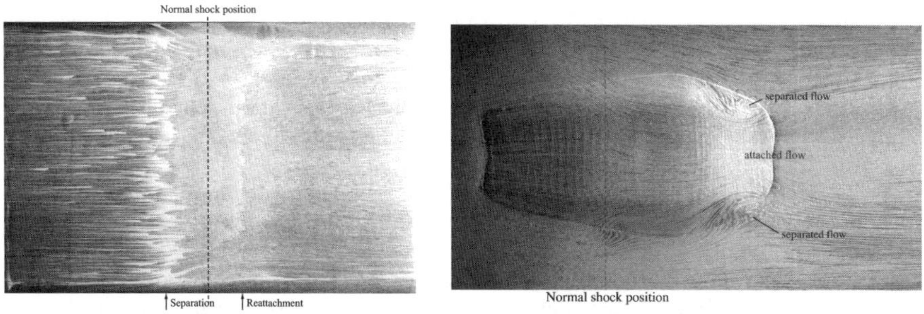

(a) without control (b) with 3D bump control

Fig. 12. SBLIs with/without 3D bump control at $M_\infty = 1.5$

4 Conclusion

Flow control for shock boundary layer interactions can be used successfully to either delay shock-induced separation (boundary layer control) or reduce shock-induced stagnation pressure losses (shock control). To date, the most mature and successful boundary layer control method is the vortex generator and more advanced versions of this in the form of sub-boundary layer VGs (or micro-VGs) promise at least the same effect at reduced device drag. Wall transpiration techniques (eg virtual jets), although promising, remain in an experimental phase and pose significant installation problems. Among the shock control methods, the most successful device to date is the contour bump which has demonstrated its potential of reducing stagnation pressure losses without incurring additional viscous drag. Three-dimensional bumps may offer significant additional benefits under off-design conditions where there are indications that they may be able to combine the functions of shock control and boundary layer control.

References

1. Pearcey H.H.: Shock-Induced Separation and Its Prevention by Design and Boundary Layer Control, In: *Boundary Layer and Flow Control*, vol 2, Pergamon Press, 1961, pp 1166-1344
2. Mitchell G.A. and Davis R.W.: Performance of Centerbody Vortex Generators in an Axisymmetric Mixed-Compression Inlet at Mach numbers from 2.0 to 3.0, NASA TN D-4675, 1968
3. Neumann H.E., Wasserbauer J.F. and Shaw R.J.: Performance of Vortex Generators in a Mach 2.5 Low-Bleed Full-Scale 45-Percent-Internal-Contraction Axisymmetric Inlet, NASA TM X-3195, 1975
4. Reichert B.A. and Wendt B.J.: Improving Curved Subsonic Diffuser Performance with Vortex Generators, AIAA Journal, vol 34, No 1, 1996, pp 65-72
5. Lin J.C.: Review of research on low-profile vortex generators to control boundary-layer separation, In: *Progress in Aerospace Sciences*, vol 38, Elsevier, 2002, pp 389-420
6. Doerffer P. and Szwaba R.: Shock Wave-Boundary Layer Interaction Control by Streamwise Vortices, In: *Proceedings of the 21st International Congress of Theoretical and Applied Mechanics*, Warsaw, Poland, Springer Verlag, 2005
7. Szumowski A. and Wojciechowski J.: Use of Vortex Generators to Control Internal Supersonic Flow Separation, AIAA Journal, vol 43, No 1, 2005, pp 216-218

8. Amitay M., Smith D.R., Kibens V., Parekh D.E. and Glezer A.: Aerodynamic Flow Control over an Unconventional Airfoil Using Synthetic Jet Actuators, AIAA Journal, vol 39, No 3, 2001, pp 361-370
9. Seifert A., Greenblatt D. and Wygnanski I.J.: Active separation control: an overview of Reynolds and Mach numbers effects, Aerospace Science and Technology, vol 8, No 3, 2002, pp 569-582
10. Lee C., Hong G., Ha Q.P. and Mallinson S.G.: A piezoelectrically actuated micro synthetic jet for active flow control, Sensors and Actuators A: Physical, vol 108, No 1, 2003, pp 168-174
11. Holden H. and Babinsky H.: Effect of Microvortex Generators on Separated Normal Shock / Boundary Layer Interactions, Journal of Aircraft, vol 44, No 1, 2007, pp 170-174
12. Ashill P., Fulker J. and Hackett K.: Research at DERA on Sub Boundary Layer Vortex Generators (SBVGs), 39th AIAA Aerospace Sciences Meeting and Exhibit, Reno, NV, AIAA Paper 2001-0887, 2001
13. Anderson B.H., Tinapple J. and Surber L.: Optimal Control of Shock Wave Turbulent Boundary Layer Interactions Using Micro-Array Actuation, 3rd AIAA Flow Control Conference, San Francisco, CA, AIAA Paper 2006-3197, 2006
14. Babinsky H., Makinson N.J. and Morgan C.E.: Micro-Vortex Generator Flow Control for Supersonic Engine Inlets, 45th AIAA Aerospace Sciences Meeting and Exhibit, Reno, NV, AIAA Paper 2007-0521, 2007
15. Milholen II W.E. and Owens L.R.: On the Application of Contour Bumps for Transonic Drag Reduction, 43rd AIAA Aerospace Sciences Meeting and Exhibit, Reno, NV, AIAA Paper 2005-0462, 2005
16. Pätzold M., Lutz T., Krämar E. and Wagner S.: Numerical Optimization of Finite Shock Control Bumps, 44th AIAA Aerospace Sciences Meeting and Exhibit, Reno, NV, AIAA Paper 2006-1054, 2006
17. Bahi L, Ross J.M. and Nagamatsu H.T.: Passive shock wave/boundary layer control for transonic airfoil drag reduction, 21st AIAA Aerospace Sciences Meeting and Exhibit, Reno, NV, AIAA Paper 1983-137, 1983
18. Gibson T., Babinsky H. and Squire L.C.: Passive Control of Shock Wave / Boundary-Layer Interactions, Aeronautical Journal, vol 104, No 1033, 2000, pp 129-140
19. Ogawa H. and Babinsky H.: Evaluation of Wave Drag Reduction by Flow Control, Aerospace Science and Technology, vol 10, No 1, 2006, pp 1-8
20. Smith A.N., Babinsky H., Fulker J. and Ashill P.R.: Shock-Wave / Boundary-Layer Interaction Control Using Streamwise Slots in Transonic Flows, Journal of Aircraft, vol 41, No 3, 2004, pp 540-546
21. Holden H.A. and Babinsky H.: Separated Shock-Boundary-Layer Interaction Control Using Streamwise Slots, Journal of Aircraft, vol 42, No 1, 2005, pp 166-171
22. Wong W.S., Qin N. and Sellars N.: A Numerical Study of Transonic Flow In A Wind Tunnel Over 3D bumps, 43rd AIAA Aerospace Sciences Meeting and Exhibit, Reno, NV, AIAA Paper 2005-1057, 2005
23. Srinivasen K., Loth E. and Dutton J.C.: Aerodynamics of Recirculating Flow Control Devices for Normal Shock / Boundary Layer Interactions, AIAA Journal, vol 44, No 4, 2006, pp 751-763
24. Ogawa H. and Babinsky H.: Shock / Boundary-Layer Interaction Control Using Three-dimensional Bumps for Transonic Wings, 45th AIAA Aerospace Sciences Meeting and Exhibit, Reno, NV, AIAA Paper 2007-0324, 2007

Shock pattern in the plume of rocket nozzles - needs for design consideration

G. Hagemann

EADS Astrium Space Transportation GmbH, Propulsion & Equipments, Munich 81663, Germany

Summary. For ideal nozzles, basically two different types of shock structures in the plume may appear for overexpanded flow conditions, a regular shock reflection or a Mach reflection at the nozzle centreline. Especially for rocket propulsion, other nozzle types besides the ideal nozzles are often used, including simple conical, thrust-optimized or parabolic contoured nozzles. Depending on the contour type, another shock structure may appear: the so-called cap-shock pattern. The exact knowledge of the plume pattern is of importance for mastering the engine operation featuring uncontrolled flow separation inside the nozzle, appearing during engine start-up and shut-down operation. As consequence of uncontrolled flow separation, lateral loads may be induced. The sideload character strongly depends on the nozzle design, and is a key dimensional load for the nozzle's mechanical structure. It is shown especially for the VULCAIN and VULCAIN 2 nozzle, how specific shock pattern evolve during transients, and how - by the nozzle design - undesired flow phenomena can be avoided.

1 Introduction - Rocket Nozzle Contours and Plume Pattern

Rocket engine thrust chamber design requirements typically include high performance demands at minimum weight in a specified, limited geometric envelop. Thus, a derived nozzle requirement asks for maximum thrust contribution at a given nozzle length. Different types of rocket nozzle contours are adopted by space propulsion industry to master the design challenges. The classical design approach is based on ideal contours (designed with method of characteristics, delivering a homogeneous 1-D exit profile), but being significantly truncated to limit overall thrust chamber length. As of today, practically all Eastern world rocket engines are designed with truncated ideal contours; examples are the Japanese LE-7 and the Russian RD-0120. Hoffman [1] proposed in 1961 a design approach for further shortening truncated ideal nozzles without significantly altering performance, by linear compression. This approach was recently adopted for the successor of the LE-7 engine, the LE-7A engine.

In 1959, Rao [2] has formulated a direct optimisation approach to maximise performance for a pre-scribed nozzle length, based on the method of characteristics. Later, Rao simplified this design approach by showing that thrust-optimised contours are very similar to rather simple parabolic contours [3]. This approach has been adopted practically by all modern Western world rocket engines, including the American SSME and RS-68, and the European VULCAIN and VULCAIN 2 [1].

Table 1 summarises characteristic data of the above referred engines. These boosterand main stage engines feature rather high area ratio nozzles to maximise engine performance during high altitude and in-vacuum flight phases. Except for the RS-68, the

[1] The VULCAIN and VULCAIN 2 engines have been developed by Snecma SAFRAN in France, which shared the development with partners in Europe.

nozzle area ratios have been maximised, but facing the constraint of sufficient margin against uncontrolled flow separation. For nominal load point operation at sea-level, this leads to an effectively full-flowing nozzle extension. The supersonic exhaust gases are highly overexpanded, and are re-compressed behind the nozzle exit by means of shocks to the ambient pressure condition.

Engine	Launcher	Chamber pressure	Nozzle area ratio	Vacuum thrust	Vacuum I_{sp}
VULCAIN	ARIANE 5 G	110 bar	45	1145 kN	431 sec
VULCAIN 2	ARIANE 5 ECA	115.5 bar	59, with TEG-injector at 32	1350 kN	434 sec
LE-7	H-II	127 bar	54	1078 kN	446 sec.
LE-7A	H-II A	121 bar	52 (original), 39 (modified), with H_2-injection at 36	1098 kN	436 sec.
RD-0120	Energia	218 bar	85.7	1961 kN	455 sec.
SSME (Bl. II)	Space Shuttle	204 bar	77.5	2278 kN	453 sec.
RS-68	Delta IV	97 bar	21,5	3313 kN	410 sec.

Table 1. Characteristic rocket engine data.

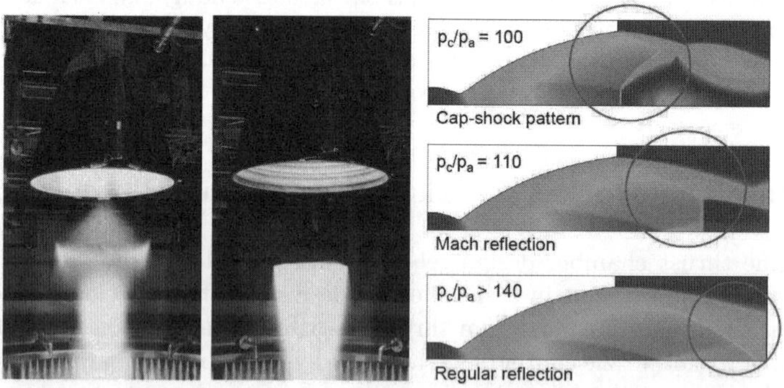

Fig. 1. Nozzle plume pattern observed in the VULCAIN engine: Cap-shock pattern (left) and classical Mach disk (center). Numerical simulations (right), also with regular reflection[2] (bottom right) [Mach number distribution, increasing from blue to red]

It is obvious that the divergent nozzle contour directly influences the nozzle plume pattern. The plumes of ideal plane and axisymmetric supersonic expansion nozzles are frequently discussed in the literature, e.g. Ref. [4]- [6]. Pending on the contour type as first order influencing factor and the operational condition as second order influencing factor, three different plume patterns may be observed in the highly overexpanded plume: the regular shock reflection, the Mach reflection, and / or the so-called cap-shock pattern. Figure 1 displays two photos of the VULCAIN engine during ground operation. At nominal chamber pressure operation, the nozzle extension generates a highly overexpanded flowfield, with an average 1-D exit pressure of approx. 0,18 bar. The exhaust gases are recompressed to ambient pressure, the cap-shock pattern (left photo) and the classical Mach disk (right photo) are observed in the plume of VULCAIN.

[2] For axisymmetric flowfields, in principle the pure regular reflection does not exist. A small Mach [4] stem appears at the centreline. This apparent regular reflection is referred to as the regular reflection in this paper.

Figure 1 also illustrates the VULCAIN engine plume pattern simulated by CFD analysis for different pressure ratios from chamber pressure to ambient pressure, featuring all three plume pattern. For the lower pressure ratio, the cap-shock pattern appears in the plume; being transferred for increasing pressure ratios to the Mach reflection with its characteristic disk. The numerical simulations compare well for the two photos from engine test with capshock pattern and Mach disk. For even higher pressure ratios (that cannot be reached on ground operation due to fixed ambient pressure and limitations on max. chamber pressure), the regular reflection will appear, only being observed during flight.

The regular reflection and the Mach reflection may be observed in the plume of all nozzle types, while the cap-shock pattern can only be observed for nozzles featuring an internal shock, such as thrust-optimized, parabolic or compressed truncated ideal nozzles. The internal shock stems from the - from a pure aerodynamic point of view - not perfect contour design, and is induced shortly downstream of the nozzle throat. With the Mach number distribution shown in Figure 1, it can be easily traced inside the nozzle.

Based on analysis performed in the late 90's, it has been shown that the cap-shock pattern results from the interaction of the inverse Mach reflection of the internal shock with the recompression shock. The cap-shock pattern and its transition to the Mach reflection are influenced by the contour shape, and by the operational conditions of the thrust chamber. Further, a significant hysteresis is observed for the pressure ratio (chamber to ambient) at the instant of transition between engine start-up (or up-ramping) and shut-down (or downramping), see Ref. [5] for a detailed discussion.

Fig. 2. VULCAIN thrust chamber during start-up (at approx. 40 bar chamber pressure), with free shock separation (FSS, left) and restricted shock separation (RSS, right) [Mach number distribution as inlay, increasing from blue to red]

2 Separation and side-load characteristics of rocket nozzles

The exact knowledge of the plume pattern is of special importance for mastering the engine operation featuring uncontrolled flow separation inside the nozzle, occurring normally only during transient flow conditions, i.e. engine start-up and shut-down operation. As consequence of uncontrolled flow separation, lateral loads may be induced. The side-load character strongly depends on the nozzle design, and is a key dimensional load case for the nozzle's mechanical structure. Principally, two separation patterns are observed, the free shock separation (abbreviated as FSS in the following) and the restricted shock separation (abbreviated as RSS in the following). Both patterns are illustrated in Figure 2. With FSS,the flow downstream the separation remains separated from the wall. This pattern is observed in any type of overexpanded nozzle, and has been subject to intensive analytical and numerical analyses [9]. With RSS, the separated flow re-attaches again to the nozzle wall, and further expands along the wall. Figure 3 depicts the characteristics of restricted shock separation. This separation pattern is driven by the cap-shock pattern, and thus is likely to occur only inside nozzles featuring an internal shock, such as compressed truncated ideal contours, thrust-optimised and parabolic contours. Key driver for the re-attachment is the momentum balance across the cap-shock pattern, with the radial momentum towards the wall delivered by the reflected internal shock as illustrated in Figure 4. The relative distance of the separation location and the normal Mach stem may be used to characterize the instant of transition.

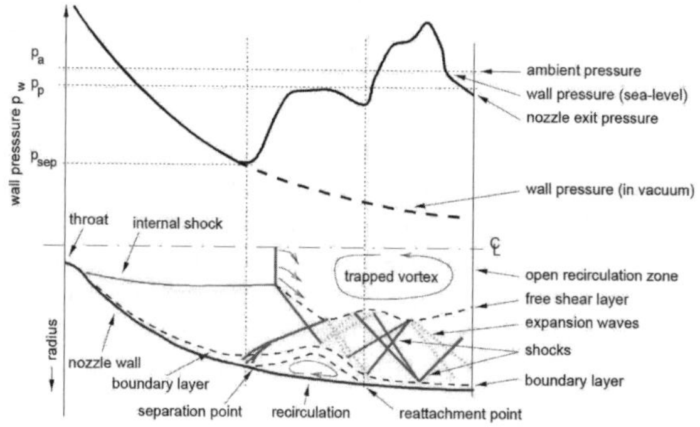

Fig. 3. Characteristics for restricted shock separation pattern, RSS

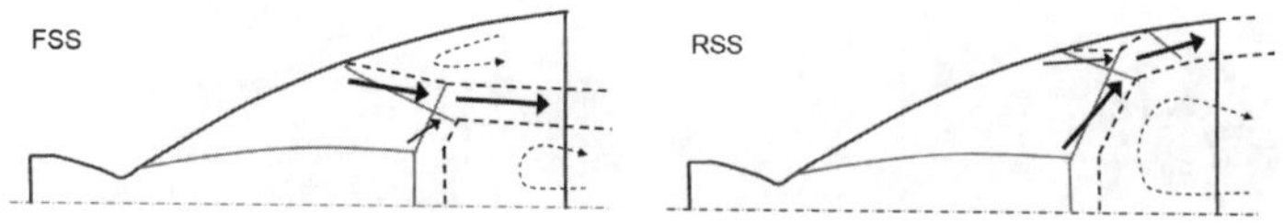

Fig. 4. Momentum balance across cap-shock pattern, as key driver for re-attachment

The criticality of RSS is twofold: First, the transition from free- to restricted shock separation imposes significant impulsive side-loads to be handled by the lightweight nozzle structure. The re-transition process from RSS to FSS shortly occurs prior to the fullflowing, and is of oscillative character, associated again with side-loads. And second, the re-attached supersonic flow remaining attached to the wall introduces a significant heat load towards the wall with the risk of local overheating due to continuous boundary layer disturbances by means of shock- and expansion waves (see also Figure 3, downstream of reattachment point). Figure 5 illustrates both critical elements. The excessive heat load for the re-attached condition with RSS is clearly visible inside the radiation-cooled 40 kN ceramic nozzle demonstrator (shown in the left part of the Figure). The re-attached flow region is visualized by the much stronger radiation due to significant higher temperatures.

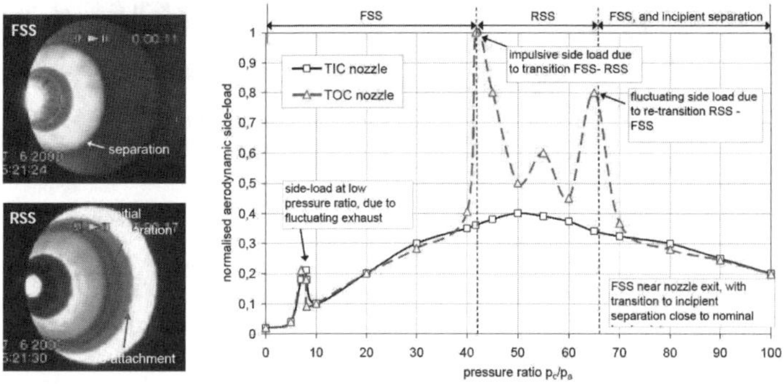

Fig. 5. 40 kN thrust chamber in FSS- (top left) and RSS condition (bottom left); and typical side-load profile for truncated ideal (TIC) and thrust-optimized (TOC) nozzle

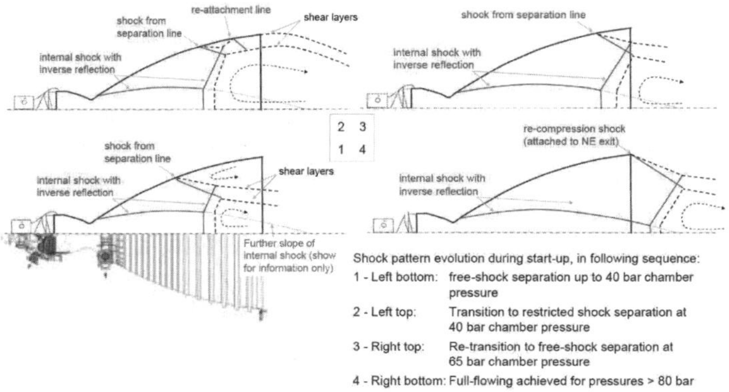

Fig. 6. Nozzle plume pattern observed in the VULCAIN rocket engine: shock pattern evolution during start-up, with free-shock and restricted shock separation[3]

2.1 Characteristics for truncated ideal nozzles: RD-0120, LE-7

In truncated ideal nozzles, only free shock separation occurs. This has also been demonstrated within the European Calo-Programme by means of an extensive test programme with a 40 kN thrust chamber being operated with LOX/H_2 [10]. Key-side load driver are random pressure pulsation across the separation zone. For rocket engines equipped with truncated ideal nozzles such as RD-0120 and LE-7, no problems with sideloads have been reported. It can be stated in principle, that for this nozzle design side-loads are not regarded as critical issue [9], [11].

2.2 Characteristics for thrust-optimized / parabolic nozzles: VULCAIN, SSME

As discussed before, this nozzle type introduces the risk of a change in separation pattern from free- to restricted shock separation. The plume instability shortly before and at its

[3] Shock and expansion wave structures not further depicted beyond cap-shock pattern or Mach disk, for purpose of simplification (also valid for Figure 8).

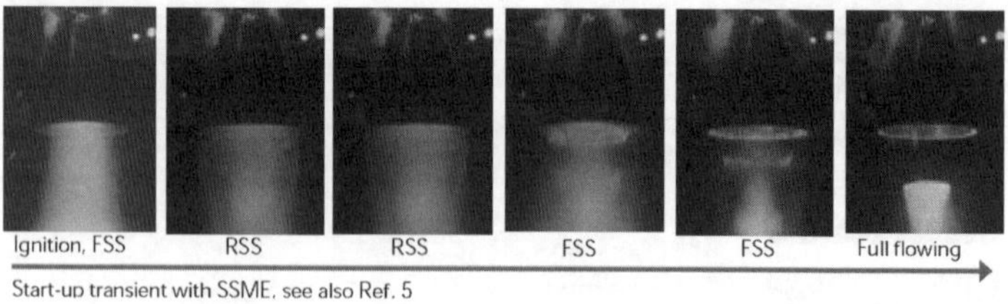

Fig. 7. Nozzle plume pattern observed in the SSME rocket engine: shock pattern evolution during start-up, with free-shock and restricted shock separation

sudden FSS to RSS change is key side-load driver. Due to its impulsive character, this load case has to be carefully handled within the design. Figure 6 sketches the shock pattern evolution inside the VULCAIN nozzle extension during start-up. Maximum side-loads being induced by the transition process close to 40 bars are assessed to approx. 3% of the vacuum thrust level. The transition process, and the re-transition process from restricted to free shock separation are also the key side-load driver inside the SSME engine, based on the analysis of the author. This finding is in line with a reported SSME nozzle failure to side-loads, likely to be caused by the oscillative re-transition from restricted shock separation to free shock separation.4 The picture sequence shown in Figure 7 summarises the transition and re-transition inside the SSME nozzle during start-up.

2.3 Thrust-optimized / parabolic nozzle without RSS: VULCAIN 2

The transition from FSS to RSS is driven by the momentum balance across the cap-shock pattern. Consequently, the nozzle designer has the chance to avoid the transition process by design in such way, that the momentum balance yields always a balance in favour of FSS. This may be achieved by stabilizing the separation point movement during the critical pressure ratios. For example, in VULCAIN 2 the transition from FSS to RSS does not occur due to separation stabilization at the turbine exhaust gas (TEG) injector. With the TEG injector, turbine exhaust gases are introduced into the nozzle for performance and cooling reasons, being a design element available with open cycle engines. It is noted that RSS would occur for a clean nozzle wall without TEG-injector. The following sequencing of the flowfield occurs for increasing chamber pressure, see also Figure 8:

- During start-up, FSS occurs for lowest chamber pressures until separation reaches TEG injector.
- While FSS separation line is stabilized at the TEG-injector, cap-shock structure moves further downstream towards the nozzle exit, with further increasing chamber pressure.
- Due to specific shape of internal shock (being reflected at centreline inside the nozzle), cap-shock disappears while separation starts to jump across the TEG-injector. Maximum side-loads are generated by an asymmetric detachment of the flow from the TEG-injector.
- Further separation appears in FSS condition until incipient separation is reached at nozzle exit, with classical Mach disk in plume.

Figure 9 includes three images taken during the early development of the VULCAIN 2 engine. As shown in the picture to the right, the classical Mach disk is visible at nominal operation under sea-level condition with the full length nozzle extension. Early within the development programme, specific tests with shortened skirt were performed to validate the engine performance at rather low combustion chamber pressure. Within the transient startup, high speed photographs were taken to visualise the plume pattern. The left and center picture are extracted, and proof the existence of the cap-shock pattern for low-level off-design points.

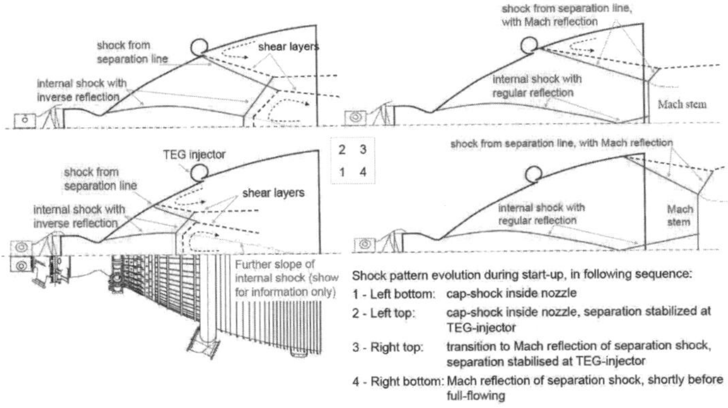

Fig. 8. Nozzle plume pattern in the VULCAIN 2 rocket engine: shock pattern evolution during start-up, from cap-shock to classical Mach disk (right), only featuring FSS

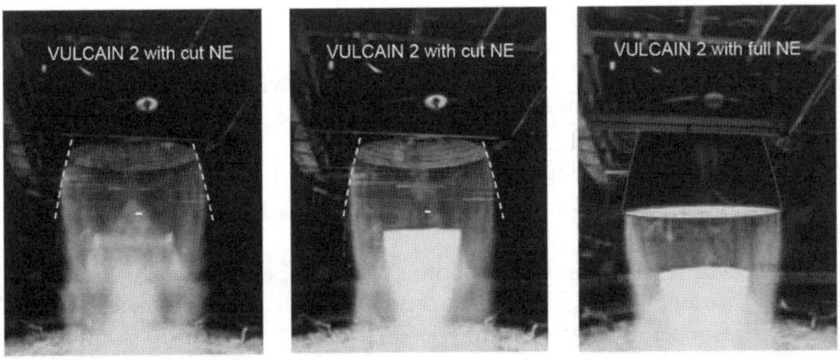

Fig. 9. Nozzle plume pattern observed in the VULCAIN 2 rocket engine: classical Mach disk (with nominal skirt, right, and cut skirt, center) and cap-shock pattern (with cut skirt, left) [Courtesy Snecma SAFRAN, DLR]

2.4 Characteristics for compressed truncated ideal nozzles with RSS: LE-7A

Due to the internal shock induced in compressed truncated ideal nozzles, this design introduces also the risk of a change in separation pattern from free to restricted shock separation. Thanks to the documentation of the problematic encountered with the LE-7A, it is shown how significant the RSS problematic can be, not being adequately considered within the design [11]. Major problems occurred during engine development due to the appearance of RSS causing local overheating of wall. Further problems were encountered by large side-loads, due to an asymmetric detachment of the separation point from a rearmounted H_2-film-injector. The film-injector is closely mounted to the nozzle exit to enable a dismounting of the skirt for low chamber pressure operation. Profound analyses, also by means of subscale tests, were performed to understand and successfully master the design problem. As first countermeasure, the nozzle was significantly cut from area ratio 59 down to 39 to avoid the origin of the large side-loads, the asymmetric detachment.

3 Conclusion

Rocket nozzles belong to the critical components of a rocket engine, being exposed to extreme aerodynamic, thermal and mechanical loads during both transients and steadystate. When changing the design from classical truncated ideal design to other contour types allowing for further length reduction, the correct knowledge on separation pattern and associated side-loads are of significant importance. Especially for restricted shock separation, the excessive overhead-load and the side-load character has to be mastered with the thermo-mechanical design. It has also been show that by design, the risk of separation pattern transition may be diminished.

References

1. Hoffman, J., D., “Design of Compressed Truncated Perfect Nozzles”, AIAA-85-1172, July 1985.
2. Rao, G.V.R., “Exhaust Nozzle Contours for Optimum Thrust”, Jet Propulsion, June 1958, pp. 377–382.
3. Rao, G.V.R., “Approximation of Optimum Thrust Nozzle Contours”, ARS Journal, June 1960, p. 561.
4. Moelder, S., Gulamhussein. A., Timofeev, E., and Voinovich, P., “Focusing of Conical Shocks at the Centreline of Symmetry”, Proceedings of the 21st International Symposium on Shock Waves, July 1997.
5. Frey, M., “Behandlung von Strömungsproblemen in Raketendüsen bei Überexpansion”, Ph.D. Dissertation, Institute of Aerodynamics and Gasdynamics, University of Stuttgart, July 2001.
6. Shapiro, A.H. Compressile Fluid Flow, John Wiley & Sons, New York, 1953.
7. Hagemann, G., Immich, H., Nguyen, T., and Dumnov, G., “Rocket Engine Nozzle Concepts”, “Liquid Rocket Thrust Chambers: Aspects of Modeling, Analysis, and Design”, Progress in Astronautics and Aeronautics, Volume 200, 2004.
8. Terhardt, M., Hagemann, G., and Frey, M., “Flow Separation and Side-Load Behaviour of the VULCAIN Engine”, AIAA 99-2762, 1999.
9. Hagemann, G., Frey, M., Koschel, W., “On the Appearance of Restricted Shock Separation in Rocket Nozzles”, Journal of Power and Propulsion, Vol.18, No.3, 2002.
10. Hagemann, G., Preuss, A., Grauer, F., Frey, M., Kretschmer, J., Ryden, R., Jensen, K., Stark, R., Zerjeski, D., “Flow Separation and Heat Transfer in High Area Ratio Rocket Nozzles”, AIAA-2004-3684, July 2004.
11. Watanabe, Y., Sakazume, N., Yonezawa, K., Tsujimoto, Y., “LE-7A Engine Nozzle Separation Phenomenon and the Possibility of RSS Suppression by the Step inside the Nozzle”, AIAA-2004-4014, July 2004.
12. Goetz, O., and Monk, J., “Combustion Device Failures during Space Shuttle Main Engine Development”, Proceedings of the 5th International Symposium on Long Life Combustion Device Technologies, Chattanooga, Tennessee, October 27-30, 2003.

Part II

Blast Waves

A quick estimate of explosion-induced stress field in various media

A. Sakurai[1] **and K. Tanaka**[2]

[1] *Tokyo Denki University, 101-8457 Tokyo, (JAPAN)*
[2] *National Institute of Mathematical and Chemical Research, Tsukuba 305-8565, (JAPAN)*

Summary. This is to provide to estimate stress field directly induced by an explosion in varieties of non-gaseous media. We utilize the point-source blast wave solution for this, in which the difference between the stress fields in different media is represented by two parameters; i.ie. the density and the index to the stress-strain relation of medium. This fact combined with an existing peak stress P and duration T data from, for example, free-air explosion is to provide the estimation of P, T values expected from a burst in a material of a certain density and index. The approach is extended also to cases of surface and near surface bursts with use of an appropriate energy-partitioning ratio.

1 Introduction

We are concerned with the procedure to estimate the magnitude of stress field induced directory by an explosion to varieties of non-gaseous media such as silt, clay, sand and gravel ([1]). We utilize the point source blast wave solution of ideal gas.

Basis of this approach is as follows: Most of materials under high-pressure such as produced by explosion behaves as a fluid. So that the basic feature of the stress field due to an explosion can be similar to the one for the blast wave in gas, while salient feature of the point source blast wave solution in an ideal gas is given by its two parameters, namely, the density ρ_0 of the uniform atmosphere and the adiabatic index γ. Accordingly, an explosion generated stress-field in a certain density and polytropic index can be estimated from the blast wave solution of these density and index.

Furthermore, a modification of the formula combined with a proper energy-partitioning ratio can provide the necessary data for surface or near surface explosion at an air-medium interface.

2 Point source blast wave solution in ideal gas

Consider the blast wave generated by an instantaneous release of a finite amount of energy E at a point o in a uniform ideal gas of pressure p_0, density ρ_0 and the adiabatic index γ. The dynamic behavior of this wave is described by the point source blast wave solution of compressible fluid dynamic equations (Sakurai [2]). In particular, the velocity U of its front shock wave is given for large U values as,

$$\left(\frac{U}{c}\right)^2 \left(\frac{R_0}{R}\right)^3 = J_0(\gamma) + O\left(\left(\frac{U}{c}\right)^{-2}\right), \quad c = \sqrt{\gamma p_0/\rho_0}, \; R_0 = \left(\frac{E}{4\pi p_0}\right)^{1/3} \quad (1)$$

where $J_0(\gamma)$ is a constant related to the energy integral of the blast wave equation and its value is given by the function of γ, as it is shown graphically in Fig. (1) ([2]).

We have also, $U = dR/dt$, $R = R(t)$ and this is combined with eq. (1) in integration to give

$$\frac{ct}{R_0} = \frac{2}{5}\left(\frac{R}{R_0}\right)^{5/2}\sqrt{J_0(\gamma)} + \cdots. \tag{2}$$

Furthermore, the pressure of the front shock $P(R,\gamma)$ and the duration of the blast wave $T(R,\gamma)$ are given as

$$\begin{aligned} P(R,\gamma) &= \frac{2\gamma}{\gamma+1}\cdot\frac{E}{4\pi J_0(\gamma)}R^{-3} + \cdots, \\ T(R,\gamma) &= A\cdot\frac{2}{3}\sqrt{\frac{4\pi}{E}}\cdot\sqrt{\frac{\rho_0 J_0(\gamma)}{\gamma}}R^{5/2} + \cdots, \end{aligned} \tag{3}$$

where A is a constant depending on how the duration time being defined.

While, corresponding P, T values in real case such as in free-air ($\gamma = 1.4$) have been well documented in the entire range of R in a standard form as seen in Fig. (2).

3 Blast wave in non-gaseous media

First, we consider point source blast waves in different ideal gases of different γ, ρ_0 but from a same amount of energy E, to which we can see in eq. (3) that P, T values depend on two parameters γ, ρ_0 only. Accordingly, the peak pressure (or stress) P and the duration T at the same range R from the same point source energy E in different media of different γ and the density ρ_0 are given roughly as,

$$P \propto \frac{1}{J_0(\gamma)}\frac{2\gamma}{\gamma+1},\ T \propto \sqrt{\frac{\rho_0}{\gamma}}\sqrt{J_0(\gamma)}. \tag{4}$$

We can use this equation (4) to have P, T values for any γ, ρ_0 case from known $\bar{P}, \bar{T}$ values for $\bar{\gamma}, \bar{\rho}_0$ case as;

$$P = F_1\bar{P},\ T = sF_2\bar{T} \tag{5}$$

where

$$F_1 = \frac{J_0(\bar{\gamma})}{J_0(\gamma)}\frac{\gamma}{\gamma+1}\frac{\bar{\gamma}+1}{\bar{\gamma}},\ F_2 = \sqrt{\frac{\bar{\gamma}}{\gamma}}\sqrt{\frac{J_0(\gamma)}{J_0(\bar{\gamma})}},\ s = \sqrt{\frac{\rho_0}{\bar{\rho}_0}}. \tag{6}$$

Most conveniently we can utilize the air-blast data for $\bar{P}, \bar{T}$ values such as given in Fig. 2, which give the peak pressure and the positive duration values at reduced range λ ($RW^{-1/3}$) from the explosion source of W-TNT equivalent and the portion of the pressure curve near $R = 0$ being added by the universal $P \propto R^{-3}$ from the point source theory since the close-in characteristics are different to each individual medium. We have $\bar{\gamma} = 1.4$, $\bar{\rho}_0 = 1.293 \times 10^{-3}\ g/cm^3$ there. F_1, F_2 values for various γ values are computed from the $J_0(\gamma)$ values as given in Fig. 1 and they are shown graphically in Fig. 3.

We utilize the above approach not only to a case of different gas but also for a non-gaseous material of density ρ_0 and having a property of compression phase represented by an effective value of $\gamma > 0$. Suppose the (dynamic) strain (σ)-strain (ε) curve of a material [3], [4] is concave up-ward, its stress wave tends to propagate with a steep shock front. The γ value is determined from the stress-strain test data as above from a best fit of the curve at the stress level of the present interest, so that we have approximately $\sigma \propto \varepsilon^\gamma$

there for the material being in rest. The formulae can be simplified further for large p, ρ to a polytropic gas type relation as, $p \propto \rho^{\gamma}$. With use of these ρ_0, γ as determined to represent effectively the property of a medium, we may follow the procedure above to find the needed P, T values from the existing $\bar{P}, \bar{T}$ data as given in Fig. 2 for free-air through eq. (2) with F_1, F_2 values as given in Fig. 3, Figs. 4,5 show the comparison of the results so obtained with some test data as well as code calculations for the corresponding cases, where Fig. 4 is for a kind of clay with $\gamma = 1.4$, $\rho_0 = 1.52g \cdot cm^{-3}$ so that we have $F_1, F_2 = 1$ and we use the same P value as $\bar{P}$ as given in Fig. 2, while T is given by $\bar{T}$ data in Fig. 3 multiplied by the factor $s = 34$; Fig. 5 is for a kind of sand with $\gamma = 2.2 \sim 2.6$, $\rho_0 = 1.49g/cm^3$ so that we have $F_1 = 2.3 \sim 2.6$, $F_2 = 0.7 \sim 0.6$ and $s = 33$ from eq. (6) and we get P, T values from $\bar{P}, \bar{T}$ values multiplied by these factors above. In the computation, shock compaction model for porous brittle material is applied at density higher than voidless density material [4]. Porosity is assumed to be 50%.

Further we extend the procedure above to the cases of surface and near surface explosions to have the P, T values in non-gaseous medium under the surface. There in particular we see that the feature of shock propagation directly under the explosion source is similar to that for a contained burst from the reduced charge weight W_e [5], [6], $W_e = aW$, $a < 1$. Results for an example of a surface burst for kind of clay with $\gamma = 1.4$, $\rho_0 = 1.52g/cm^3$ and $a = 0.2$ are compared with an available test data [1] in Fig. 6.

4 Conclusion

The use of point-source blast wave solution is found to be effective to have the quick estimate of salient features of the stress field by an explosion to varieties of non-gaseous media such as silt, clay, sand and gravel.

References

1. A. Sakurai and T.Tanaka, A method to estimate explosion-induced stress field in various media based on point source blast wave solution, Shock-Wave and High-Strain-Rate Phenomena,Elsevier(2001) 655-659
2. A. Sakurai, Blast Wave Theory, Basic Developments in Fluid Dynamics, ed. Holt, Academic Press, (1965) 309-375
3. K. Tanaka, Proc. Eighth Symposium on Detonation, NSWC MP 86-194(1985) 548.
4. K. Tanaka, Shock waves in Condensed Matter-1983,Elsevier(1983) 203.
5. A. Sakurai, A realistic approach for describing the explosion-generated axi-symmetric wave propagating in a half space, Fifth Smposium on Detonation, ONR, ACR-184 (1970) 379-384
6. A. Sakurai, Blast wave from a plane source at an interface, J. Phys.Soc. Japan,36 (1974) 610

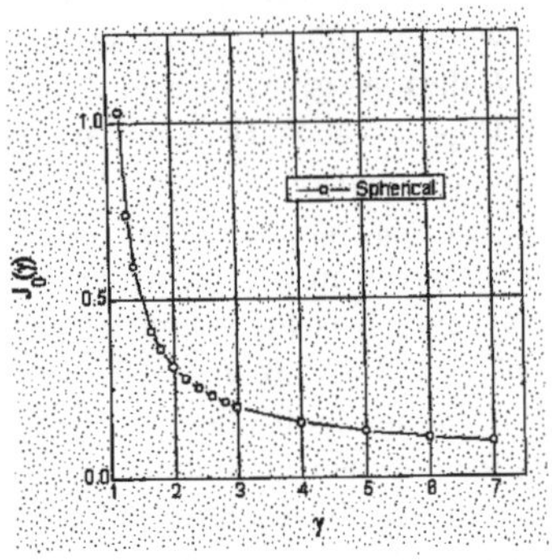

Figure 1 $J_0(\gamma)$, γ: adiabatic index

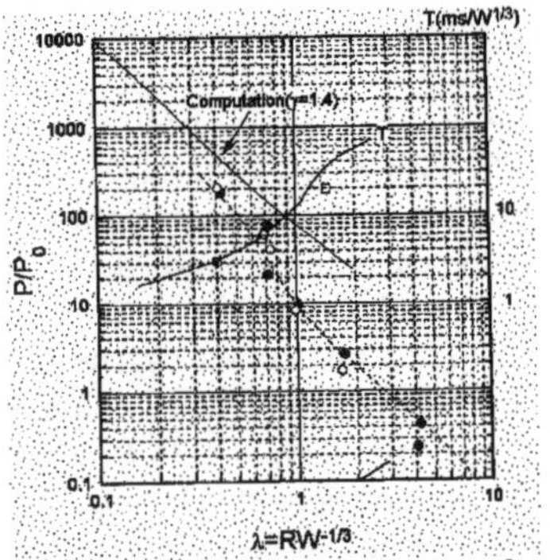

Figure 4. P, T **values for clay (** $\gamma=1.4, s=34; F_1=F_2=1$ **) compared with experimental and computational data.**

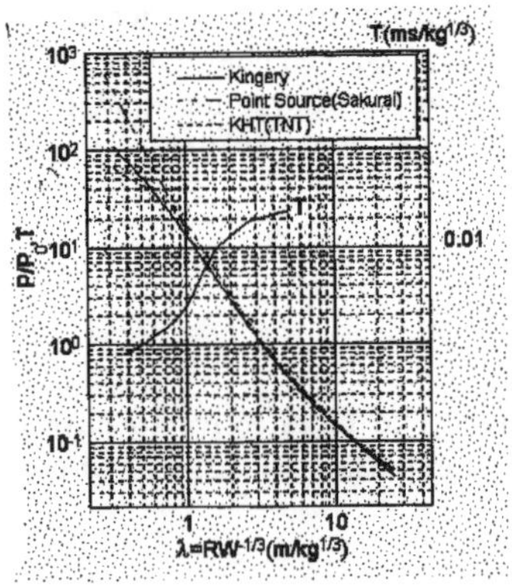

Figure 2. TNT-equivalent free-air P, T **values in scaled range.**

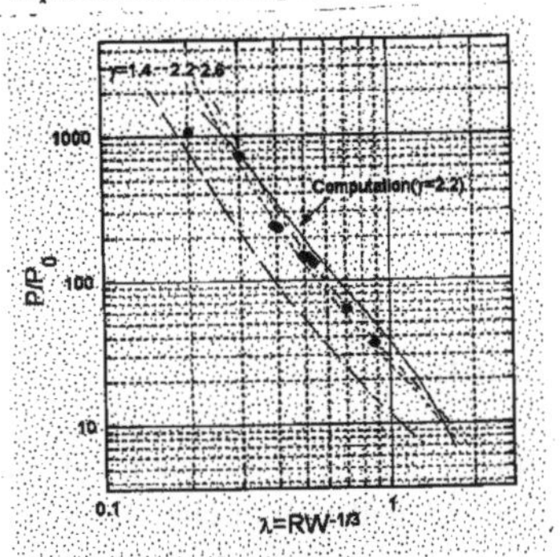

Figure 5. P values for sand $(\gamma=2.2\sim2.6, s=33. F_1=2.3\sim2.7, F_2=0.7\sim0.6)$ **compared with experimental and computational data**

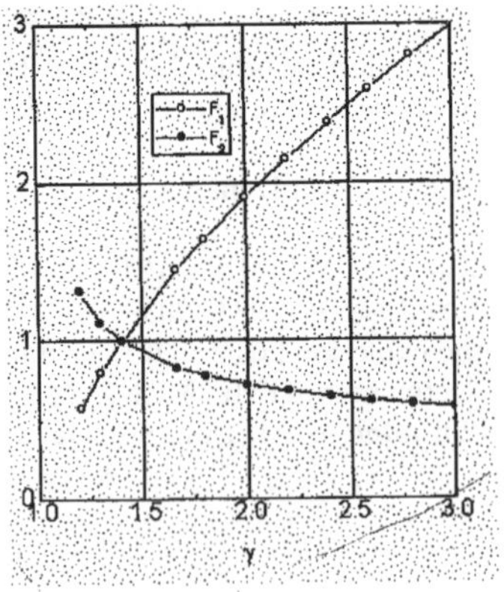

Figure 3. $F_1(\gamma,\bar{\gamma}), F_2(\gamma,\bar{\gamma})$, $\bar{\gamma}=1.4$.

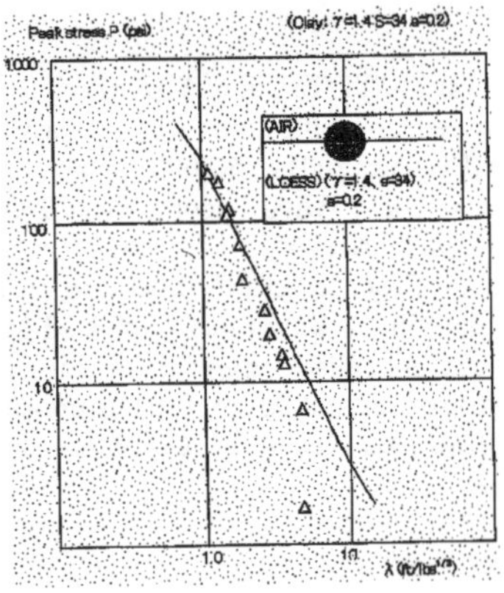

Figure 6. P **values for surface burst on clay** $(\gamma=1.4, s=34)$ **with a=0.2 compared with test data**

Attenuation properties of blast wave through porous layer

K. Kitagawa[1], S. Yamashita[2], K. Takayama[3], and M. Yasuhara[1]

[1] *Dept. of Mechanical Engng., Aichi Institute of Technology, Toyota, JAPAN.*
[2] *Chugoku Kayaku Co., LTD., Hiroshima, JAPAN*
[3] *Tohoku University Biomedical Engineering Research Organization, Sendai, JAPAN*

Summary. The propagation and attenuation of blast waves through porous layers is one of the important research topics related to safety assessment and prevention from explosion hazards. This study is a part of series of research on shock in complex media. Blast waves passing through porous layer such as polyurethane foam with high-porosities($\approx$ 98%) and low densities($\approx$ 28kg/m^3) ; and sand layer having low-porosities($\approx$ 45%) and high densities ($\approx$ 1500kg/m^3) are tested. Peak overpressures in polyurethane foam and sand layer decrease quickly than that in air, as the blast wave is degenerated into compression waves due to its interaction with three-dimensional porosity distribution. Pressure attenuation caused in present complex media are in between 10 to 40% of the peak overpressure value in air.

1 Introduction

Industrial, physical, medical and environmental mechanism to attenuate strong impulse against a structure by some mediums, have attracted the interest of many investigators [2–10]. Blast wave attacking human body may give damages to hearing, lungs and or to life itself. The blast waves in air, water, and complex media are investigated to predict their behavior as functions of both magnitudes of explosion (weight of explosive) and distance from explosives. Blast wave propagation in porous layers or porous media yet has not been completed research field. The introduction of scaled blast wave parameters normalized by enables the result of a small scale explosion in porous layers to convert into large-scale one. In order to experimentally investigate this problem, detonating micro explosives are traditionally used, which provide an insight to establish analytical modeling of the shock/porous media interaction and also to the performance of attenuation and protection to the dynamic process. Mechanical properties of cellular material are investigated by Gibson and Ashby [5]. The reflection and propagation of shock wave into compressible foam, are investigated experimentally by Skews et al. [13], and numerically by Baer [1], Levy et al. [11] , Olim et al [12] etc.. Kitagawa et al. [7] investigated effects of unsteady drag on the dynamic interaction of the shock wave with elastic foam. Kitagawa et al. [8,9] investigated effect of pressure attenuation of shock wave by porous materials, while Britain et al. [3] investigated effect of shock wave attenuation by granular filters. The aim of the present research is to investigate the effect of pressure attenuation of blast wave by porous mediums. Blast wave experiments for porous layer are conducted by detonating micro-explosives (silver azide and penthrite pellets). It is deduced that the dynamic behavior of pressure attenuation of blast wave corresponds to the momentum loss of the blast wave and shock wave by complex medias.

2 Porous material and porous layer

The important property of complex media is its relative density ρ_c/ρ_s, where $\rho_c(=\phi_s\rho_s)$ denotes its density and ϕ_s the volume fraction of solid phase. ρ_s the density of solid material from which the cellular material is composed. It is assumed that the density of the solid material ρ_s is constant and incompressible. The porosity ϕ_g (volume fraction of gas phase) of porous material is defined as,

$$\phi_g = 1 - \frac{\rho_c}{\rho_s}. \tag{1}$$

Two different kinds of porous mediums are used in these experiments that are polyurethane foam and sand. Static and dynamic properties of polyurethane foam are investigated by Gibson and Ashby [5]. The open-cell type polyurethane slab shaped foams are produced by Inoac Co., Japan. Foam xx denotes nominal (statistical mean) numbers of open cells per inch in x, y and z directions.

In the present paper, each foams having high-porosities with different cell number per unit length, are investigated; foam 13 is of density ρ_c = 28.7kg/m^3 with porosity ϕ_g = 0.977,and foam 48 is of density ρ_c = 22.0kg/m^3 with porosity ϕ_g = 0.982, respectively. The material density of polyurethane is ρ_s=1200kg/m^3.

Two kinds of sand layers have lower porosity and higher density compared with polyurethane foams. The solid particle diameters of sand layer are below 0.2mm. Sand 1 is of density ρ_c = 1458kg/m^3 with porosity ϕ_g = 0.450; while sand 2 is of high density ρ_c = 1500kg/m^3 with low-porosity ϕ_g = 0.419. The material density of solid particle in sand is referred as ρ_s=2650kg/m^3. It is assumed that sand layers are incompressible and chemically inert. Solid particles are of identical size and have spherical shape, which is maintained during the shock impact.

3 Blast Wave Experiment

Propagation and attenuation of blast waves in porous layers is experimentally investigated by detonating micro-explosives (silver azide and penthrite pellets), as shown schematically in Figs. 1a and 1b. The silver azide(AgN_3) pellets are manufactured by Showa Kinzoku Kogyo Co. Ltd., Japan, and delivered as cylindrical charges each with its mass of approximately 10mg, since the material density is 3770kg/m^3, each cylinder with 1.5mm in diameter has aspect ratio (length over diameter) of unity. The penthrite(PETN) pellets are manufactured by Nippon Kayaku Co. Ltd., Japan, and the material density is 1770kg/m^3, so that for a diameter of 5.5mm with a mass of approximately 100mg, the cylinder has with aspect ratio of 1.8. In the blast wave experiment, the charge is glued to optical fiber and ignited by the irradiation of pulsed Nd:YAG laser (1064nm, 7ns pulse duration, 25mJ/pulse) or Ar Ion CW laser(514.5nm, 4W) fed through this fiber. The face-on overpressure is measured dynamic pressure in complex media and is conducted by piezo-electric type pressure transducers (Kistler 603B and PCB HM113A) at positions from P_1 to P_4, as shown in Figs. 1a and 1b.

The scaling law of blast property comes from the cube root relation that occurs regularly in the analysis of explosive effect. The scaled distance of blast wave is defined as [2,6],

$$Z = \frac{R}{W^{\frac{1}{3}}}. \tag{2}$$

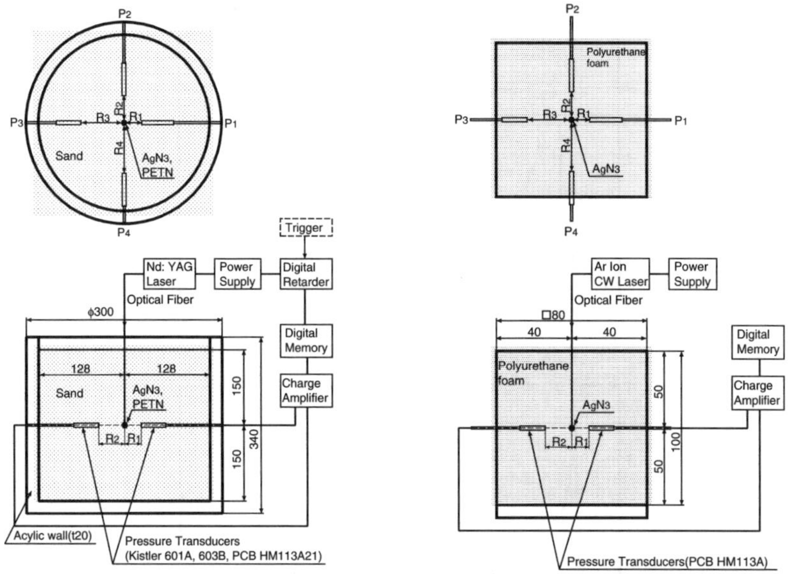

Fig. 1. Schematic description of setup for overpressure measurement. **a** sand layer and **b** polyurethane foam.

where R denotes the distance from the charge to measuring device and W the charge mass. The scaling law of explosives extends to evaluate differences in ambient conditions. Eq. (2) can be written as [4, 14],

$$Z_s = \frac{R}{(WT_a)^{\frac{1}{3}}} P_a^{\frac{1}{3}} = \frac{R}{W^{\frac{1}{3}}} \left(\frac{P_a}{P_0}\right)^{\frac{1}{3}} \left(\frac{T_0}{T_a}\right)^{\frac{1}{3}}. \tag{3}$$

P_a denotes ambient pressure, T_a the ambient temperature, and P_0 and T_0 are the reference conditions (0.101325MPa and 288.15K), respectively.

4 Results and Discussion

Figures 2a, 2b, 3a and 3b show typical time variations of experimental face-on overpressure for the sand 1, the sand 2 and polyurethane foam, respectively. Figures 2a and 2b show pressure histories at distances 20, 30, 40 and 50mm from the center of the charge. Figures. 3a and 3b show at distances 30, 40, 60 and 85mm from the center of the charge, respectively. In Figs. 2a and 2b, pressure rise at R=20 or 30mm show a first peak pressure by the impingement of blast wave, and then pressure decrease quasi-exponentially until the pressure reaches atmospheric (zero pressure). Peak overpressure decreases as distance R from the center of the charge increases, due to the momentum loss of blast wave in sand layer. Peak overpressures in sand layers decrease much more quickly than that in air, as the blast wave is degenerated into compression wave by its interaction with dense and heavy particle distributions. Here, differences of up to about 20 times exist in overpressure both silver azide(AgN_3) and penthrite(PETN) for present experimental

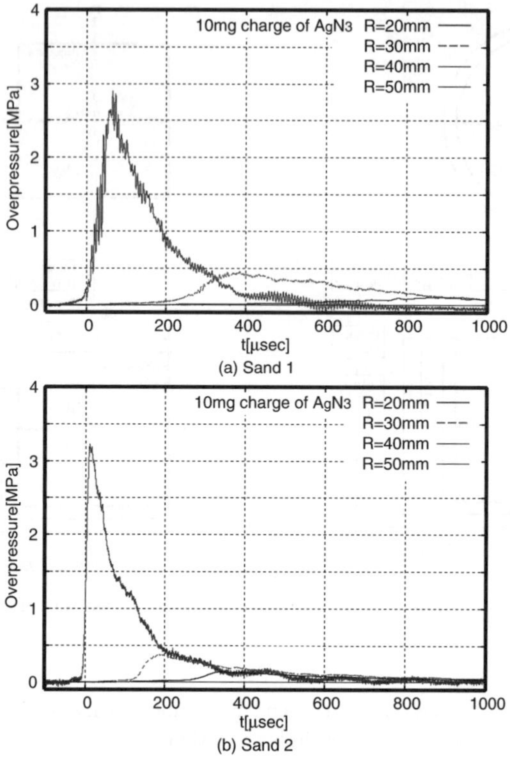

Fig. 2. Measured overpressure histories at positions R=20, 30, 40 and 50mm from the center of the charge for (**a**) the dry sand 1 and (**b**) the dry sand 2(10mg charge of silver azide)

condition. These differences of overpressure have different effects of explosion, because the mass, energy and impulse of explosion in PETN is greater than AgN_3. An explosive can be considered to follow the scaling laws if all overpressure measurements for various charge masses and ambient conditions can be represented by one curve when plotted with the scaled distance given in Eq.(3). Figure 4 shows the experimentally obtained face-on overpressure versus scaled distance in comparison with air. It is seen from Fig. 4 that the pressure attenuation caused in dry sands are about 40% of the peak overpressure value in air for $Z < 0.8$, while its attenuation are about 10% of the peak overpressure for $1.5 < Z < 2.8$ and about 5% of the peak overpressure for $Z > 3$. Effect of momentum loss of blast wave comes from unsteady drag interaction with three-dimensional porosity distribution in sand layer. Peak overpressures in foam 48 are compared with foam 13, which show the decrease to about $50 \sim 10\%$ of the first pressure pulse. The momentum loss comes from caused by the structure and cell number of foams. The pressure attenuation caused in foam 48 are about 10% of the peak overpressure value in dry sand for $Z < 1$. Effect of pressure attenuation for foam 48 is the highest compared to sand layer. The present complex medias can dominantly affect the pressure attenuation of the blast wave.

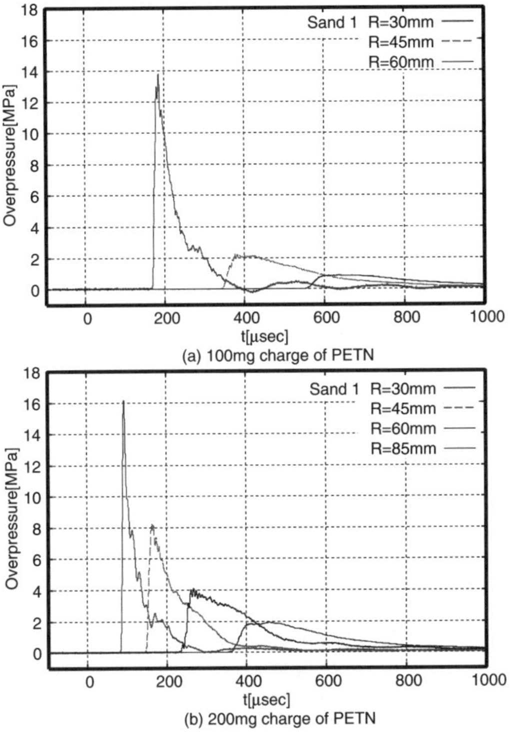

Fig. 3. Measured overpressure histories at distance of R=30, 45, 60 and 85mm from the center of the charge for the dry sand 1, (**a**) 100mg and (**b**) 200mg charge of penthrite)

5 Conclusions

Effects of pressure attenuation of blast wave by porous layers are investigated experimentally. It is deduced that the dynamic behavior of pressure attenuation corresponds to the momentum loss of blast wave by complex medias. (a) Peak overpressures in sand layers decrease much more quickly than that in air, as the blast wave is degenerated into compression wave by its heavy interaction with dry sand layer having three-dimensional porosity distribution. (b) Pressure attenuation caused in dry sands are about 5~40% of peak overpressure value in air. The sand layer can dominantly affect to the pressure attenuation of the blast wave. (c) Peak overpressures in fine cell structure foam decrease to about 50 ~ 10% of the first pressure pulse for coarse cell structured foam, which show that the decrease of momentum is caused from the complex three-dimensional structure and cell number of foams. (d) Pressure attenuation caused in polyurethane foams are about 10% of peak overpressure value in sand layer.

Acknowledgement. This work is supported in part by Grant-in-Aid for Scientific Research (C) No. 19560184 offered by Japan Society for the Promotion of Science, Japan.

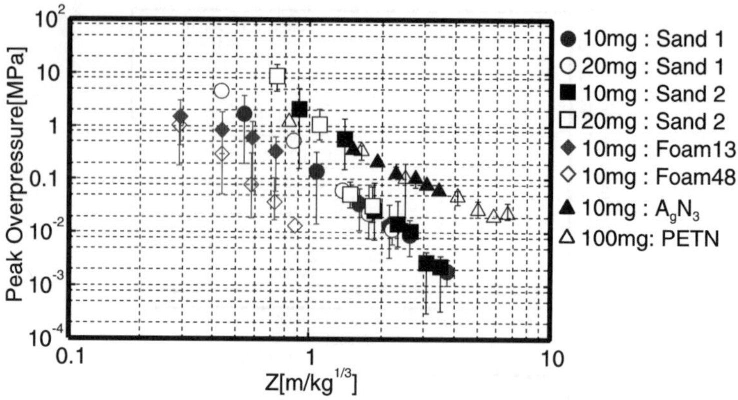

Fig. 4. Peak overpressure versus scaled distance

References

1. Baer, M.R.: A numerical study of shock wave reflections on low density foam, Shock Waves, **2**, 121–124(1992)
2. Baker, W. E.: *Explosions in Air.* University of Texas Press(1973)
3. Britan, A., Ben-Dor, G., Igra, O., Shapiro, H.: Shock waves attenuation by granular filters, Int. J. Multiphase Flow, **27**, 617–634(2001)
4. Cooper, W. P.: *Explosive engineering.* Wiley-VCH(1996)
5. Gibson, L. J., Ashby, M.F.: *Cellular solids: structure and properties.* Pergamon Press(1988)
6. Kinney, G. F., Graham K. J.: *Explosive Shocks in Air, 2nd edn.* Springer-Verlag(1985)
7. Kitagawa, K., Jyonouchi, T., Yasuhara, M.: Drag difference between steady and shocked gas flows passing through a porous body, Shock Waves, **11**, 133–139(2001)
8. Kitagawa, K., Yamashita, S., Ojima, H., Takayama, K., Yasuhara, M.: Propagation and attenuation of blast wave and shock wave in porous layer, In: Takayama., K. (eds.) *22nd International Symposium of Interdisciplinary Shock Wave Research.*(2005)
9. Kitagawa, K., Takayama. K, Yasuhara, M.: Attenuation of shock waves propagating in polyurethane foams, Shock Waves,(2006)
10. Kleine, H., Dewey, J. M., Ohashi, K., Mizugaki, T., Takayama, K.: Studies of the TNT equivalence of silver azide charges, Shock Waves, **13**, 123–138(2003)
11. Levy, A., Ben-Dor, G., Sorek, S.: Numerical investigation of the propagation of shock waves in rigid porous materials: flow field behavior and parametric study, Shock Waves, **8**, 127–137(1998)
12. Olim, M., van Dongen, M.E.H., Kitamura, T., Takayama, K.: Numerical simulation of the propagation of shock waves in compressible open-cell porous foams, Int. J. Multiphase. Flow., **20-3**, 557–568(1994)
13. Skews, B.W., Atkins, M.D., Seitz, M.W.: The impact of a shock wave on porous compressible foams, J. Fluid. Mech., **253**, 245–265(1993)
14. Zukas, J. A., Walters P. W.: *Explosive Effects and Applications.* Springer-Verlag(2003)

Blast wave reflection from lightly destructible wall

V.V. Golub, T.V. Bazhenova, O.A. Mirova, Y.L. Sharov, and V.V. Volodin

Associated Institute for High Temperatures, Russian Academy of Sciences
Izhorskaya st. 13/19, 125412 Moscow, Russia

Summary. The paper presents the results of experimental study of the action of blast waves on the obstacle made of different materials. The pressure in the front of reflected blast wave is compared in the cases of its interaction with a rigid metal wall and the destructible wall made of weakly cemented sand.

1 Introduction

The shields made of various materials for the blast wave attenuation are proposed. In [1] the behavior of concrete under blast wave loading was investigated and offered to design refuges made of concrete plates with use of special armoring constructions. Interaction of blast wave with the protected obstacle depends substantially on the material of shield and on the relationship of its thickness with the length of the positive phase of blast wave. For example, in [2] it is shown that the polyurethane foam screen is effective for attenuation of shock waves when the thickness comparable with the length of the positive phase of blast wave. In [3] it was established that at the interaction of shock wave with the rigid wall, covered with the layer of polyurethane, the pressure on the wall is considerably higher than that in the absence of coating.
Among the number of proposals on the attenuation of blast effect the use of granular materials [4] and walls made of damp sand [5] as the enclosure of dangerously explosive volume are examined. In [5], [6] and [7] the process of destroying the obstacle made of weakly cemented sand under the action of blast wave is investigated. It was established that the material is destroyed without the formation of large fragments. After this the particles of sand are pulverizing, taking away the substantial part of the explosion energy and attenuating the transmitted shock wave. The investigated material was proposed for creating the explosion-resistant constructions [8].
When the explosion occurs indoors destructive blast effect increases due to the blast wave reflection from the walls. In this case it is necessary to take measures for attenuate of the reflected wave action. The purpose of this work is a study of the methods of the reflected blast wave attenuation.
One-dimensional case - blast effect on the obstacles from different material in the channel was investigated.
Action on a shield of two types of shock waves was investigated: short wave in which the peak of pressure is accompanied by rapid pressure drop in a rarefaction wave of underpressure, and long wave in which the pressure behind front drops slowly. The first type characterizes action of explosion in close distance from the source, the second - in far distance in the scale carried to radius of a source. The short waves are attenuated by the rarefaction wave and rapidly changed velocity and pressure profile underpressures

easied by a wave, change in the speed and the structure of pressure profile. Long waves are close to ones with constant parameters.

2 Experimental setup

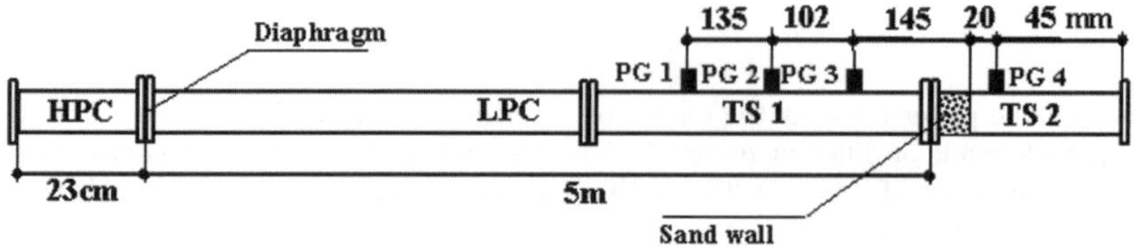

Fig. 1. Shock tube: HPC - high pressure chamber, LPC - low pressure chamber, TS1, TS2 - test sections, PG1-PG4 - pressure gauges

In the experiments the $70 \times 70mm^2$ cross section shock tube was used (fig. 1). The low pressure chamber(LPC) was $6m$ long. For the simulation of short blast wave the shock tube with the short high-pressure chamber(HPC) was used. HPC length was $23cm$. After the rapture of diaphragm the shock wave begin to propagate into the low pressure chamber. In the high pressure chamber during the pushing gas outflow the wave of rarefaction is generated. It reflects from the closed end and propagates into the low pressure chamber, overtaking the shock wave. With the low Mach numbers of shock wave and the short high pressure chamber the head of rarefaction wave rapidly overtook the shock wave and the profile of the falling pressure, similar to pressure profile behind the blast wave, creates after it. Blast wave was simulated analogously in [9].

For the simulation of long blast wave the shock tube with the high-pressure chamber with a length of 1 m was used. In this case the rarefaction wave, reflected from the end of high pressure chamber, did not overtake the shock wave in the low pressure chamber at the observation point. Pressure gauge showed a constant pressure profile after the incident shock wave.

Two test sections ($TS1$, $TS2$) were installed at a distance of $5m$ from the diaphragm. In the wall of the first section pressure gauges $PG2$ and $PG3$ for measuring the speeds of incident and reflected waves, and for registration of pressure changing behind the incident and reflected waves were mounted.

In the beginning of the second section ($TS2$) different obstacles were installed: the rigid metallic wall, glass wall and wall made of weakly cemented sand.

The second research section was supplied with transparent walls made from plexiglas for the visualization of the destruction of sand wall after the blast wave loading. Distance from the wall to the closed end of shock tube was equal to $52cm$. Measuring of the shock wave velocity was carried out by base method with the accuracy of 1%.

The propagation of sand front after the decomposition of wall was recorded with the digital high-speed camera Cordin-530 in the regime of frame survey. Pressure on the side wall of shock tube was measured by piezoelectric gauges PCB 113A36. Mach number of shock wave in the different experiments measured at a distance of $14.5mm$ from the obstacle was $1.6 - 1.73$. Overpressure at the shock wave front was about $2.2bar$. This pressure corresponds, for example, to the explosion of $20kg$ TNT at a distance of $5m$ from the measuring point [10].

3 Experimental results

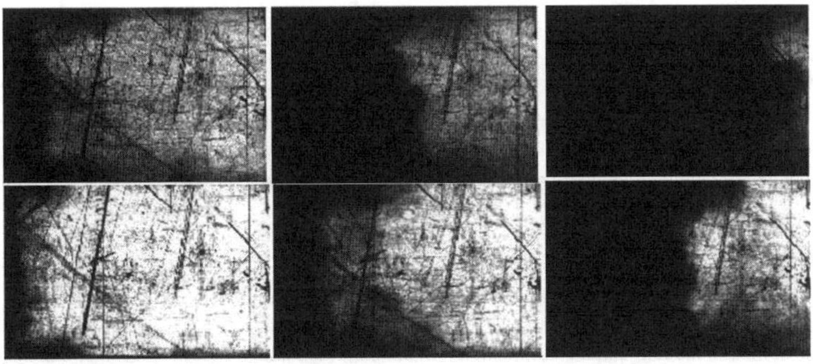

Fig. 2. Successions of sand boundary images for the case of interaction with the compressed sand wall (thickness h = 12mm - top line, 16 mm - bottom line) for time intervals 35, 105 and 175 μs

The cinegrams of the process of sand wall destruction are presented in the fig.2. Obtained on the basis of the cinegrams sand front trajectory after interaction with the blast wave are presented in fig.3 (left). The differentiation of trajectories obtained the speed of sand front depending on the thickness of sand wall fig.3 (right).
The speed of sand front V decreases with an increase of the thickness of the sand layer h and, respectively its mass according to the exponential law:

$$V = 233 * exp(-\frac{h}{16.33}) + 8.92$$

The oscillograms of pressure on the side wall of shock tube showed that the pressure amplitude in the front of the wave reflected from the wall depends on the material of the wall (fig.4). At the reflection from the rigid wall the pressure sharply increases and falls in the rarefaction wave. At the reflection from the glass wall before its destruction the pressure at the front is close to its values, obtained with the reflection from the rigid wall and with the calculation for the shock wave. Behind the front the sharp drop in pressure, caused by the rarefaction wave, which appears after the destruction of glass arise. At the reflection from the wall made of the sand behind the front a less sharp drop in the pressure, than with the reflection from the glass wall is observed. The pressure in the blast wave front at the reflection from the sand wall is less than the pressure at the reflection from the rigid wall.

In the fig.5 (left) the measured values of pressure in the shock wave front, reflected from the sand walls of different thickness are presented. Pressure is normalized by the value of pressure in the front of the incident shock wave with the appropriate Mach number. The comparison of the pressure at the front of the reflected blast wave after

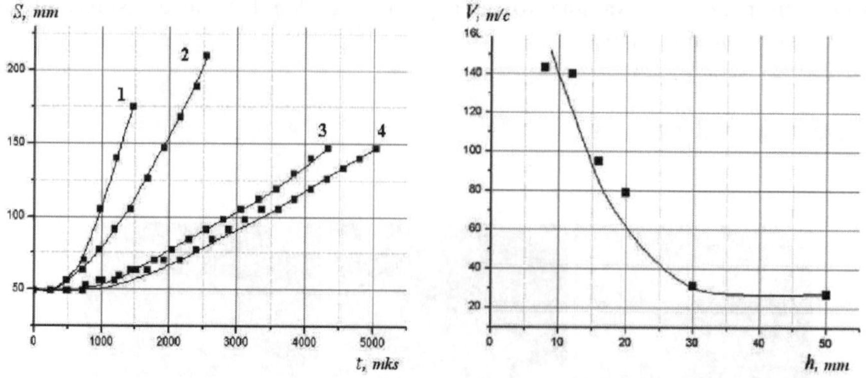

Fig. 3. Trajectory of sand front as result of shock wave impact. 1,2,3,4 - sand wall thicknesses 9, 16, 30, 50 mm accordingly (left), the velocity of sand front depending on sand wall thickness (right)

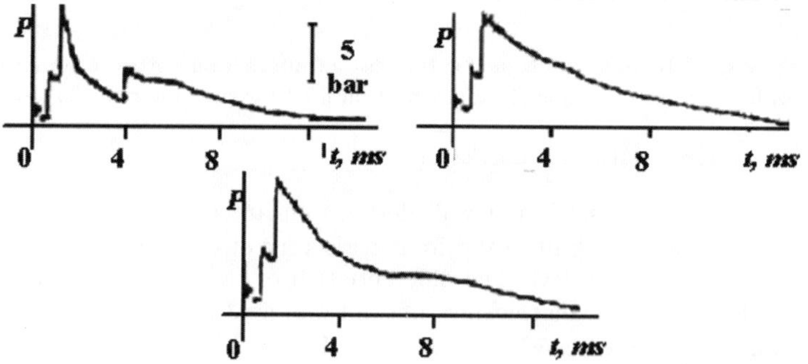

Fig. 4. Pressure histories of blast wave reflection from glass(top line, left), from steel(top line, right) and from sand wall(bottom line)

its interaction with the rigid metallic wall and the destructible wall made of weakly cemented sand is carried out. Attenuation coefficient is higher, the thinner the layer of sand (fig.5(right)).

4 Discussion

The attenuation coefficient of the long shock wave reflected by the wall made of the sand, observed in the experiment is less than the short one, because the short wave is unstable and additionally loses intensity at the propagation from the point of speed measuring to the wall and vice versa.
Attenuation of reflected shock wave can depend on two factors: attenuation at the shock propagation through the sand and creation of rarefaction wave at the destruction of the

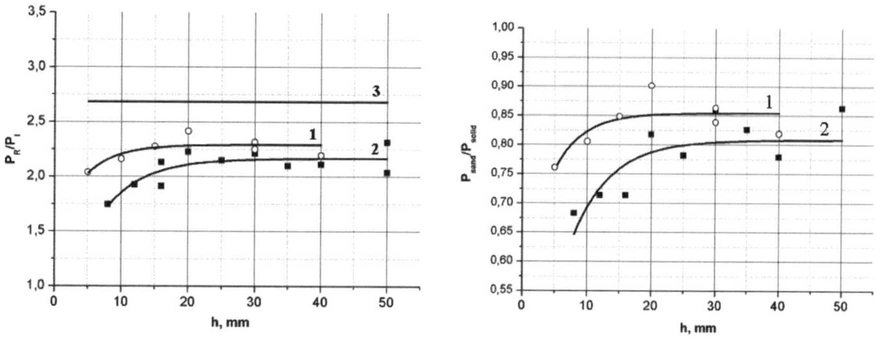

Fig. 5. Pressure on reflected shock wave front(left). 1-long wave, 2-blast wave, 3-shock wave reflected from rigid wall. Attenuation factor(right). 1-long wave, 2-blast wave

wall. For determining the role of attenuation of shock wave at the transition through the layer of sand the tests on the reflection of shock wave from the rigid wall, protected by the layer of the sand of a thicknesses in the range from 10 to 30 mm were carried out. These experiments were conducted for both the long and the short shock waves. The cases, when the end was not protected, or protected by the layer of sand were compared. In the investigated range of the parameters, the pressure amplitude at the interaction of shock wave with the layer of the sand was substituted with the pressure after the interaction with the rigid wall. It indicates that the attenuation of the reflected shock wave is determined by the action of rarefaction wave that is forming at the destruction of the wall. The role of the shock wave attenuation by the rarefaction wave, which appears at the destruction of the sand layer, is demonstrated by cinegramms of the sand shield destruction process fig.2. Reflected shock wave attenuation coefficient is higher when the shock interacts with the thinner wall, which is destroying more rapid and the rarefaction wave is stronger (fig.3).

5 Conclusion

1. Pressure at the front of the reflected blast wave after its interaction with the destructible wall made of granular material is less than that after the interaction with the rigid wall.
2. After the destruction of the wall made of granular material under the action of shock wave the speed of the destructible material front falls the exponentially with an increase of the wall thickness.
3. Reflected shock wave attenuation coefficient is higher when the shock interacts with the thinner wall, which is destroying more rapid.
4. The established regularities offer the possibility to attenuate both passed and reflected shock waves by the destructible walls.

References

1. Krauthammer T., Frye M., Schoedel T.R., Selter M. Advanced SDOF approach for structural concrete systems under blast and impact loads. Procceedings of the 11th International Symposium of Interaction of Minitions Effects with Structures, Manheim, Germany, May 5-9, 2003.
2. Gelfand B. E., Gubanov A.B.,Timofeev E.I. Vzaimodeystvie vozdushnih udarnih voln s poristim ekranom. Izvestija AS USSR. MZG. 1983. V 4. P. 79-84
3. Gvozdeva L.G., Faresov U. M. Issledovanie osobennostey rasprostranenija i otrazenija voln davlenija v poristoy srede. PGTF. 1984. Vipusk 19. P. 153-156.
4. 4. Bakken J., Slungaard T., Engebretsen T., Cristensen S O. Attenuation of shock waves by granular filters. Shock waves. 2003. V. 13. N. 1. P. 33-40.
5. Pokrovskiy G. I. Vzriv. Moscow. Nedra.1980.190 p.
6. Golub V.V., Lu F. K., Medin S.A., Mirova O. A., Parshikov A.N., Petukhov V.A., Volodin V.V. Blast wave attenuation by lightly destructible granular materials. Proceedings of 24th International Symposium on Shock Waves, Beijing, China, July 11-16, paper 1891, P. 1-6, 2004.
7. Mirova O.A.Vzaimodeystvie vzrivnoy volni s pregradoy iz slabosvjazanogo granulirovanogo materiala.Sbornik: Nauchnie trudi Instituta teplofiziki extremalnih sostojanij OIVT RAN./ pod redakziey V. E. Fortova i A.P. Lihacheva. Moscow : OIVTRAN Vipusk 6. 2005.P.54
8. Golub B.B., Medin S.A, Petuhov V.A. and all. Vzrivozazhitniy ekran. Patent 2206062 from 10.06.2003
9. John A.G., Gardner K.D., Lu F.K., Volodin V.V., Golovastov S.V., Golub V.V. Shock wave impact on weak concrete. Procceedings of the 25th International Symposium on Shock Waves, Bangalore, India, July 11-16, paper 1817, P. 1-6, 2005.
10. Jakovlev U.S. Gidrodinamika vsriva. Leningrad:Sudpromgiz. 1961. P. 149.

Gram-range explosive blast scaling and associated materials response

M.J. Hargather[1], G.S. Settles[1], and J.A. Gatto[2]

[1] *Gas Dynamics Laboratory, Mechanical and Nuclear Engineering Department, The Pennsylvania State University, University Park PA 16802 (USA)*
[2] *Transportation Security Laboratory, US Department of Homeland Security, William J. Hughes Technical Center, Atlantic City NJ 08405 (USA)*

Summary. Laboratory-scale gram-range explosive blast testing of materials is shown to be feasible. Blast loading from different explosive compounds is coupled to a witness plate through the air by way of a shock wave of known strength, measured optically. The resulting witness-plate deflection is also measured by a high-speed optical method. An attempt is made to relate the material response to blast loading parameters, especially impulse, by scaling arguments. More work is needed on this topic, and a discussion of future research directions is included. The promise of gram-range testing is to take on at least some of the burden now carried by expensive, dangerous, time-consuming full-scale explosive testing.

1 Introduction

Blast resistant material development requires understanding the explosive blast parameters and their effects on a material sample. Explosive blasts have been studied extensively on a large scale [1] [2] and recently on the laboratory scale as well [3] [4]. Material responses to blasts, however, are comparatively poorly understood.

Typical material blast-response tests are conducted at full-scale, with material samples larger than $1m^2$ [5] [6]. These full-scale tests can require $10-100kg$ explosive charges placed at distances up to $100m$ away, which forces the tests outdoors into relatively-uncontrolled settings. At this scale, instrumentation becomes difficult and expensive, often only yielding piezoelectric overpressure profiles at limited locations and a visual record of material survival or failure. Optical methods to reveal shock waves in the field (e.g. background distortion and sunlight shadowgraphy [7]) are often crude and weather-dependent. Overall, the researcher may get relatively little data for his time and expense.

In contrast, laboratory experiments provide a controlled environment with much-better instrumentation. Unfortunately few laboratory-scale blast experiments have been performed. An exception is Nurick's group, who have performed extensive lab blast tests on panels of various geometries, materials, and levels of reinforcement [8] [9] [10]. However, these tests are conducted with the explosive charge placed in physical contact with the material sample under test. Coupling blast energy to materials through the air via a shock wave is the more practical case, and it also allows precise optical shock-loading measurements to be made.

The research presented here applies our previous experience with gram-scale explosions to improving the understanding of how blast parameters such as overpressure, duration, and impulse affect the deformation response of a clamped aluminum plate, and how the process might be scaled.

2 Experimental procedure

Two explosive compounds are used here to examine a range of blast parameters: Pentaerythritol tetranitrate (PETN) is a common and well-documented secondary explosive [11] and triacetone triperoxide (TATP) is a primary explosive recently linked to terrorist activities [12]. These explosives were previously studied using shadowgraph techniques and high speed digital videography to measure the shock radius as a function of time [4]. Example results are shown in Figure 1a, where the horizontal axis is the shock wave radius scaled to represent a $1g$ charge. The shock wave profile, from Rankine-Hugoniot theory and piezoelectic pressure-gage data, provides the peak overpressure magnitude and duration as functions of radius from the charge. The positive explosive impulse is calculated as the integral of pressure from the time of shock-wave arrival to the end of the positive pressure duration [13]. Figure 1b shows the explosive impulse versus radius due to $1g$ charges of PETN and TATP and a $3g$ charge of TATP.

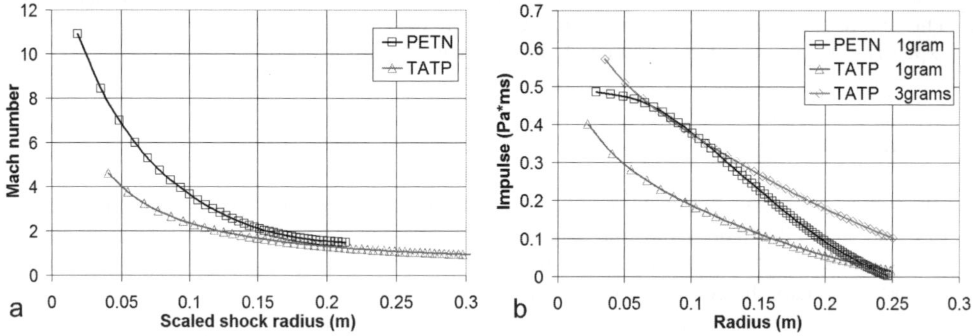

Fig. 1. a shock wave Mach number as a function of scaled radius for PETN and TATP charges **b** explosive impulse as a function of radius for various PETN and TATP charges

In this research so far, aluminum witness plates are deformed by the gram-range explosions. Aluminum alloy 3003, $0.406mm$ thick, was selected because it is readily available and its static and low-speed properties are well-documented, making it an ideal material for initial high-speed blast deformation research. The witness plates are $0.3m$ square, and are bolted into a rigid mounting fixture with a $0.25m$ circular opening as diagramed in Figure 2a. The plate fixture is then oriented so that the side of the plate opposite the explosive can be optically recorded to determine its real-time 3-D motion in response to the explosive loading, Figure 2b. Some refer to this as a "shock-hole" test geometry.

3-D witness-plate motion is determined from the stereo camera records using commercial software by Correlated Solutions. The rear of the test panel is spray-painted with a random black dot pattern on a white ground using automotive primer paint. The two high-speed cameras then record images of the plate from different view angles. With a calibration, the software can determine the 3-D location of the dots on the plate surface at every image interval. To achieve the optimum spatial and temporal resolution possible with our twin Photron APX-RS high-speed digital cameras, every test is recorded at 36000 frames/s with a frame resolution of 512x128 pixels. This resolution allows the entire width of the plate and a $0.06m$-high "slice" to be tracked throughout the test.

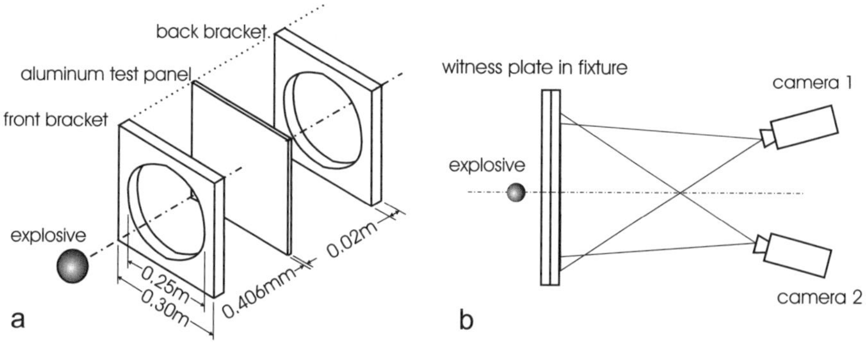

Fig. 2. **a** schematic of explosive charge and witness plate setup **b** top view schematic of witness plate and high-speed cameras for 3-D plate deformation measurement

In this manner one can document witness-plate deformation profiles for different explosive blast input parameters. These parameters are varied by changing the explosive compound, the charge mass, and the charge stand-off distance from the witness plate.

3 Experimental results

The explosive shock-loading input parameters are determined from previous work [4] and are known to have small variability from charge-to-charge. The recorded data are the dynamic motion and permanent deformation of the witness plate. The input parameters are selected to compare similar PETN and TATP charges, and to cause permanent deformation but not rupture of the witness plate. Table 1 gives a summary of tests performed to date. (Dynamic deflection data are missing from some tests due to data acquisition problems.) The tests with 1.71 and 1.75g TATP charges were performed to evaluate repeatability; resulting plate deflections show 7% and 2% differences in dynamic and static deflections, respectively, which are considered to be within the repeatability error expected in materials testing.

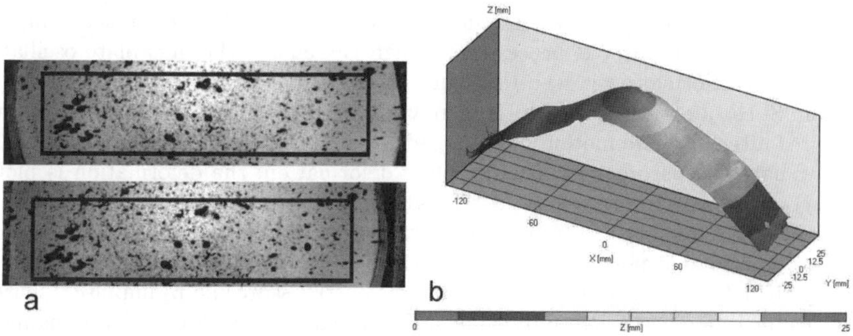

Fig. 3. **a** sample stereo image pair with area of witness plate deflection measurement boxed **b** xyz plot of the witness plate deflection as extracted from the stereo images

Table 1. Summary of witness-plate tests to date

explosive	mass g	standoff m	overpressure atm	duration ms	impulse $atm * ms$	final defl. mm	dynamic defl. mm
PETN	0.80	0.057	37	0.024	0.443	10.9	14.4
	0.87	0.046	50	0.019	0.471	15.1	-
	0.88	0.040	60	0.016	0.489	17.7	-
	0.88	0.035	70	0.014	0.502	19.4	-
	0.94	0.032	80	0.013	0.528	21.8	23.1
	1.00	0.055	43	0.023	0.494	17.3	18.0
TATP	1.36	0.052	20	0.035	0.349	9.5	15.0
	1.71	0.042	28	0.028	0.396	14.8	16.9
	1.75	0.042	28	0.028	0.396	14.5	18.2
	1.77	0.032	42	0.022	0.464	20.1	22.2

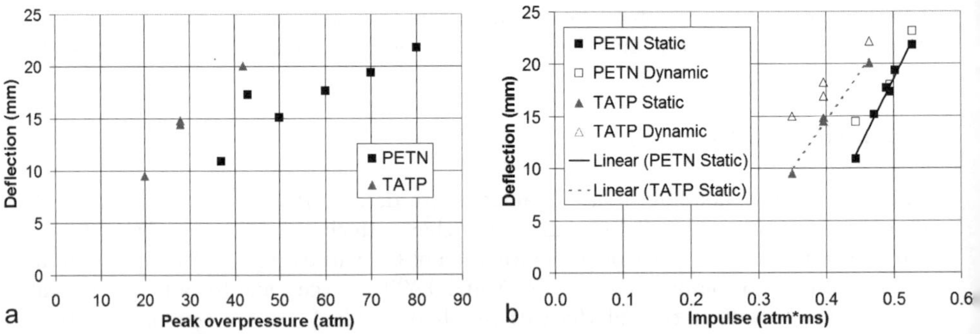

Fig. 4. **a** maximum witness plate deflection as a function of peak overpressure **b** maximum witness plate deflection as a function of impulse

Each test results in a series of stereo image pairs that is processed by the Correlated Solutions software. A sample pair of images and the resulting 3-D surface of the witness plate at that instant are given in Figure 3. The approximate area of analysis is boxed in Figure 3a. The circular edge of the mounting bracket can also be seen in the images.

Each image-pair dataset is processed through the end of the first plate oscillation. The maximum deformation is seen to occur within 0.4ms of the beginning of plate deformation. Subsequent plate motion is merely elastic oscillation. The final deformation is measured after all plate motion has stopped.

To evaluate the parameters affecting plate deformation, the deformation is plotted as a function of peak overpressure, overpressure duration, and explosive impulse. For purposes of this calculation, the pressure decay is approximated to be linear from the peak overpressure to atmospheric pressure over the duration interval. Figure 4 shows the witness-plate deformation as a function of a) peak overpressure and b) impulse. Figure 4b also shows the difference between dynamic and static deflection: Each plate dynamically deforms about 3mm beyond its eventual final deformation. This could be due to elastic deformation at the edges of the plate, which adds to dynamic deformation but contributes nothing to the permanent deformation profile.

The graph of deflection versus peak overpressure, Figure 4a, shows that most of the data for a given explosive compound lie along a single line, with the exception of one PETN test. This outlying point, however, collapses with the other data when deflection is plotted as a function of impulse, Figure 4b. This single test, conducted with a 15% larger charge mass than the others, reveals the importance of mass scaling and the use of explosive impulse as the controlling blast parameter.

4 Conclusions and future work

Laboratory-scale gram-range blast testing of materials can be safely and successfully performed. The experiments are highly repeatable and are amenable to instrumentation yielding blast overpressure loading and resulting witness-plate deflection. The dynamic and permanent deformation of a witness plate can be estimated, based on these results, if the explosive blast parameters are known. Before performing materials tests, however, the explosives being used require optical characterization in terms of their shock wave Mach number versus radius profiles in order to understand the variation of overpressure, duration, and impulse that they produce.

The most important parameter for gram-range blast testing appears to be the explosive impulse. The assumption of a triangular overpressure profile is a first approximation, but it is insufficient to accurately estimate and model the explosive blast loading of different explosive compounds.

Future work needs to reconcile the deflection vs. impulse curves of PETN and TATP, which currently do not collapse upon a single line. The difference between these curves could potentially be eliminated by computing the impulse from the actual pressure decay profile instead of the linear-decay assumption [14]. The decay parameter varies with distance from the charge center and initial attempts to scale using it have failed.

Once a better scaling approach has been identified, more testing is planned using the same explosive compounds over a wider range of charge masses. Future work will also include other explosive compounds in order to improve the understanding of blast parameters and to confirm scaling laws. The ultimate goal is to fully understand the deformation of simple aluminum witness plates in order to predict the final and dynamic deflections for a given explosive over a broad range of impulse and overpressure loading.

Another research direction should examine how the blast scaling is affected by witness-plate thickness and dimensions. The strain rate experienced by the witness plates can also be examined in order to estimate material properties which could then be used to validate numerical models of simple materials like aluminum subjected to blast loading. Ultimately, the techniques being developed in this research should be extended to other, more exotic blast-resistant materials and to the development of a comprehensive understanding of material blast response.

Acknowledgement. This research was supported by the Transportation Security Laboratory, US Department of Homeland Security, via subcontract to the Battelle Corporation. The authors thank J. D. Miller and L. J. Dodson for their assistance.

References

1. Dewey, J. M., "Air velocity in blast waves from TNT explosions," *Proc. Roy. Soc. A*, **279** 1378:366-385 (1964)
2. Dewey, J. M., "Expanding spherical shocks (blast waves)" *Handbook of Shock Waves* **2**:441-481, ed. G. Ben-Dor, O. Igra, and E. Elperin (Academic Press, 2001)
3. Kleine, H., Dewey, J. M., Ohashi, K., Mizukaki, T., and Takayama, K., "Studies of the TNT equivalence of silver azide charges," *Shock Waves* **13** 2-123-138 (2003)
4. Hargather, M. J., Settles, G. S., and Gatto, J. A., "Optical measurement, characterization, and scaling of blasts from gram-range explosive charges, *Proc. 4th International Aviation Security Technology Symp.*, (Washington DC, Nov. 2006). This paper is downloadable in PDF form from `http://www.mne.psu.edu/PSGDL/publicationswebpage.html`.
5. Houlston, R., Slater, J. E., Pegg, N., and DesRochers, C. G., "On analysis of structural response of ship panels subjected to air blast loading," *Computers and Structures* **21** 1-2:273-289 (1985)
6. Jacinto, A. C., Ambrosini, R. D., and Danesi, R. F., "Experimental and computational analysis of plates under air blast loading," *Intl. J. Impact Engineering* **25** 10:927-947 (2001)
7. Settles, G. S. *Schlieren and Shadowgraph Techniques* (Springer-Verlag, 2001)
8. Nurick, G. N. and Martin, J. B., "Deformation of thin plates subjected to impulsive loading – A review 2. Experimental studies," *Intl. J. Impact Engineering* **8** 2:171-186 (1989)
9. Nurick, G. N. and Shave, G. C., "The deformation and tearing of thin square plates subjected to impulsive loads – An experimental study," *Intl. J. Impact Engineering* **18** 1:99-116 (1996)
10. Yuen, S. C. K. and Nurick, G. N., "Experimental and numerical studies on the response of quadrangular stiffened plates. Part I: Subjected to uniform blast load," *Intl. J. Impact Engineering* **31** 1:55-83 (2005)
11. Cooper, P. W., *Explosives engineering* (Wiley-VCH, 1996)
12. McKay, G. J., "Forensic characteristics of organic peroxide explosives (TATP, DADP and HMTD)," *Kayaku Gakkaishi - J. Japan Explosives Soc.* **63** 6:323-329 (2002)
13. Held, M., "Blast waves in free air," *Propellants Explosives Pyrotechnics* **8** 1:1-7 (1983)
14. Kinney, G. F. and Graham, K. J., *Explosive shocks in air* (Springer-Verlag, 1985)

High-speed digital shadowgraphy of shock waves from explosions and gunshots

M.M. Biss, G.S. Settles, M.J. Hargather, L.J. Dodson, and J.D. Miller

Gas Dynamics Laboratory, Mechanical and Nuclear Engineering Department, The Pennsylvania State University, University Park PA 16802 (USA)

1 Introduction

Shadowgraph and schlieren methods [1] have served to image shock waves for almost 150 years, but the traditional field-of-view of these instruments is often too small for large-scale experiments. On the other hand, optical methods to reveal shock waves in the field (e.g. background distortion and sunlight shadowgraphy) are often crude and weather-dependent. Between these extremes lie some useful but little-used optical approaches for large fields-of-view, two of which are exemplified here.

Shadowgraphy was invented by Robert Hooke around 1672, though centuries went by before it was first applied to ballistics. Toepler, Mach, and C. V. Boys used open electric sparks to illuminate high-speed physics over 100 years ago, and Boys published the first spark-shadowgram photo of a bullet in flight [2]. Cranz and Schardin introduced "focused" shadowgraphy as part of their famous multi-spark high-speed camera [3], and H. E. Edgerton of strobe-lamp fame demonstrated in 1958 a simple and elegant direct-shadow technique for large-scale explosions using a retroreflective screen [4] [5]. The intervening years saw many applications of "focused" shadowgraphy but few of Edgerton's retroreflective-screen technique – henceforth called *Edgerton shadowgraphy* – except in ballistics and helicopter-rotor testing [6] [7].

The "focused" shadowgraph technique differs from direct shadowgraphy in that it produces an optical image by way of a focusing lens, rather than simply projecting a mere shadow. The focusing lens brings the test field into more-or-less sharp focus, for example, on the sensor plane of a camera. This technique is often used in ballistics, where the sensitivity of schlieren is not needed while the robustness of shadowgraphy is required. However, its name has caused controversy since, by definition, the shadowgraph effect disappears in a sharply-focused image [1]. In fact, "focused" shadowgrams are seldom sharply focused, though we will explore that topic further here. "Almost-focused" or "poorly-focused" shadowgraphy better describes the actual case.

Why continue to develop a technique that is so simple and so old as shadowgraphy? Because it is ubiquitous in its utility and, amazingly, it has yet to reach its full potential. The simplest of all the optical flow diagnostics, shadowgraphy is also the best for imaging shock waves. It reveals shock waves clearly while de-emphasizing other, less-abrupt flow features [1]. Yet there is still a need for robust large-scale shadowgraph methods for outdoor testing and explosive events. The most recent development in shadowgraphy is the replacement of the rather-painful traditional high-speed photographic methods by digital imaging, which lends simplicity to high-speed shadowgraphy, promotes quantitative as well as the traditional qualitative analyses, and yields both high-resolution still images and high-speed videography at lower (but still usable) resolution.

The goal of this paper is to combine modern high-speed videography with Edgerton and "focused" shadowgraphy in order to explore the shock waves generated by laboratory-scale explosions and the discharge of firearms. Allotted space permits only a fraction of these results to be shown, but see also [5], [8], and [9], some or all of which are downloadable from `http://www.mne.psu.edu/PSGDL/publicationswebpage.html`.

2 Experimental methods

2.1 Z-type focused shadowgraph system

Our z-type focused shadowgraph setup is identical to a z-type schlieren system without a knife edge cutoff, as shown in Figure 1. It consists of twin 0.76m-diameter f/5 schlieren-quality parabolic mirrors, a 1kW Oriel xenon arc lamp, and a high-speed digital camera. The focusing ability of these optics is put to good use in what follows. However, this setup requires two expensive, heavy, fragile glass parabolas that make it unsuitable for large-scale outdoor applications.

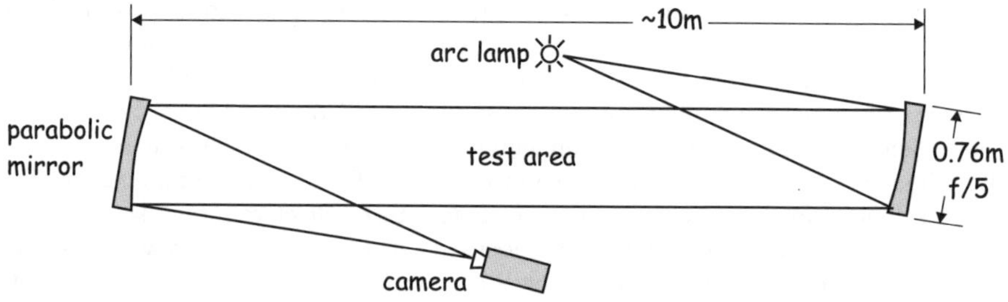

Fig. 1. Z-type focused shadowgraph system, top view

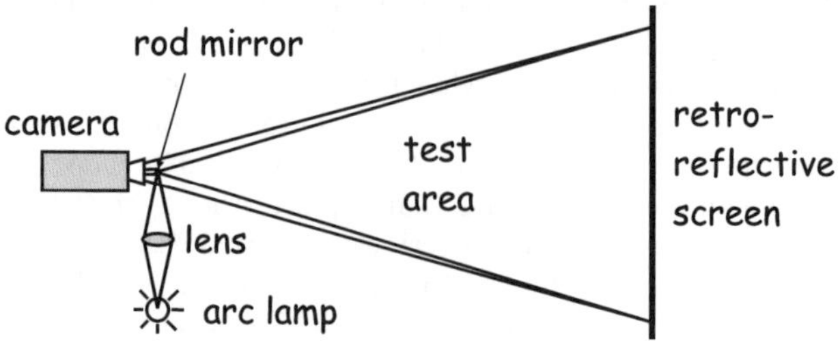

Fig. 2. Edgerton retroreflective shadowgraph system, top view

2.2 Edgerton retroreflective shadowgraph system

Our Edgerton shadowgraph setup, Figure 2, is distinct in several ways from the z-type approach. Its simplicity, robustness, and ease of use make it appealing for large-scale and even outdoor applications, e.g. [9]. It consists only of a retroreflective screen, high-speed digital camera, and $1kW$ Oriel xenon arc lamp, the latter two components being conveniently mounted on a small optical breadboard. However, the Edgerton shadowgraph cannot be focused and is not capable of the previous system's razor-sharp results.

The slight double-imaging inherent in Edgerton's original setup [4] was avoided by fitting the camera lens with a clear filter having a centered 45° rod mirror to direct the illuminating beam toward the screen without seriously blocking the coincident reflected light. The retroreflective screen is made of 3M Scotchlite^{TM} 7610, a high gain, industrial grade, exposed-lens, plastic-based material pre-coated with a pressure-sensitive adhesive. Screens made of this material can be provided by an industrial supplier [10]. The present screen is $2.4m$ square and cost about \$4000US, but we also have a $4.9m$ square screen that provides a $2.4m$ square test area [9].

More detail and setup images of the Edgerton shadowgraph are provided in [5], and are thus omitted here due to space limitations.

2.3 Photron APX-RS high-speed digital camera

Serving both shadowgraph systems is a Photron APX-RS digital video camera. Its CMOS image sensor provides 1024x1024 (i.e. 1 Mb) frame resolution at frame rates up to 3000/s. It can record at 10,000 frames/s with 512x512 pixel resolution and is capable of 250,000 frames/s at further-reduced image size. Frame exposure is independently controllable down to 1 μs, the value used here. A fiber-optic link connects the APX-RS to its controlling laptop computer, upon which the results are viewed. The camera acquires 18 Gb of image data in 6 real-time seconds of memory. Rapid events thus require no triggering, since the camera records continuously and overwrites its memory until stopped. Results are immediately available for viewing. The camera is rated for a 100g shock and has survived powerful explosions at close range [9]. Similar cameras are available from other manufacturers, e.g. the Vision Research Phantom camera. Compared to the now-obsolete high-speed photographic cameras, these modern digital wonders can be operated by anyone after a few minutes training.

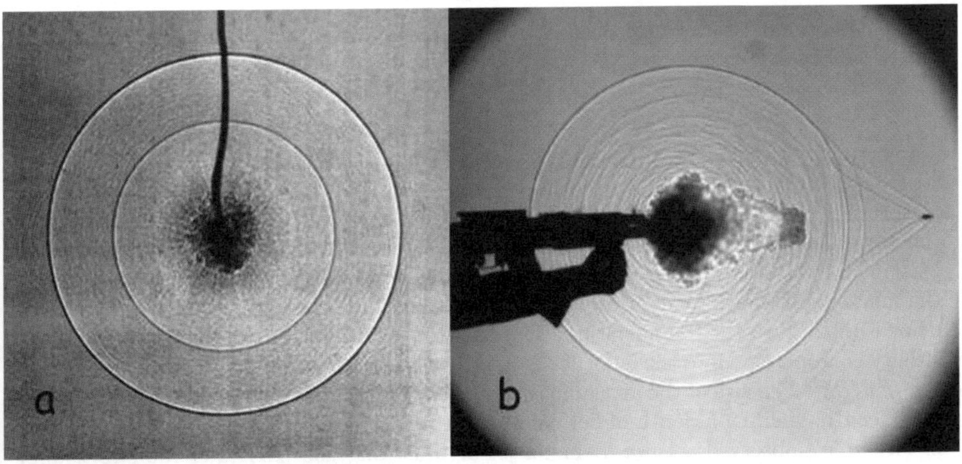

Fig. 3. Edgerton shadowgrams from [5], **a** the explosion of $1g$ of TATP, revealing primary and secondary shock waves, **b** firing the infamous AK-47 submachine gun in single-shot mode

3 Results and discussion

In our previous work along similar lines [5], gram-range explosives and the discharge of several firearms were imaged by Edgerton shadowgraphy. Two examples from this earlier work are reproduced in Figure 3.

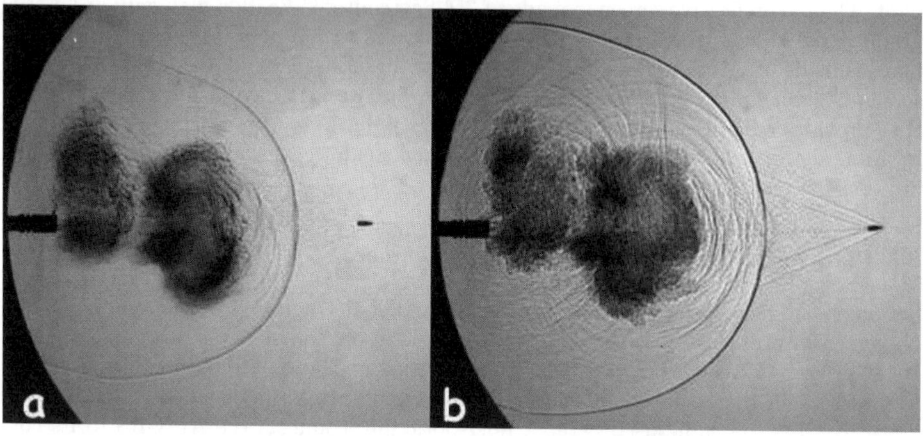

Fig. 4. "Focused" shadowgrams of .223 automatic rifle fire, **a** sharply focused, **b** defocused $1m$

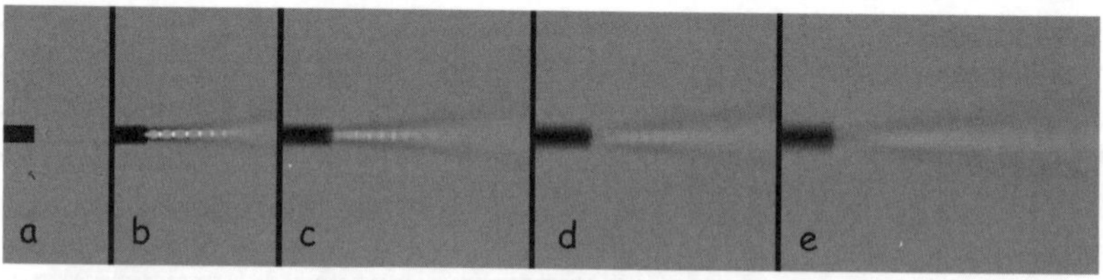

Fig. 5. Effect of defocusing on a small jet of 1,1,1,2-tetrafluoroethane in air, **a** sharply focused, **b** defocused $0.25m$, **c** defocused $0.50m$, **d** defocused $0.75m$, **e** defocused $1.0m$

3.1 Z-type focused shadowgraph results

Theoretical considerations notwithstanding, strong shock wave shadows are visible with this optical system even when sharply focused, as illustrated in Figure 4a. The muzzle blast of the automatic rifle clearly has sufficient lateral extent to be visible even when the muzzle is in sharp focus. However, many weaker flowfield features only appear upon defocusing, Figure 4b. To understand this, refer to Sec. 6.2.2 of [1]: The shadowgraph sensitivity is proportional to the defocus distance times the Laplacian of the refractive index, e.g. $\partial^2 n/\partial y^2$. The size of the arc-lamp source is about $3mm$ compared to the $3.8m$ mirror focal length, yielding a depth-of-field of about $1m$ within which to resolve features $1mm$ or larger. We observed a $1mm$-diameter jet of "dust-off" gas (1,1,1,2-tetrafluoroethane) across the optical axis in order to focus the shadowgram image. As shown in Figure 5, this jet is almost invisible when sharply focused, but its shock diamonds are clear when defocused $0.25m$, where there is finite sensitivity but the blur is only $0.25mm$. However, when defocused $1m$ the blur defeats the shadowgram sensitivity and the shock diamonds are no longer seen. This compromise among sensitivity, blur, and feature size is always an issue in shadowgraphy.

Figure 6 shows 7 frames of Colt 9mm submachine gun fire (124-grain American Eagle ammunition) observed at 75,000 frames/s. These frames are not sequential, but rather cover an interval of 300 μs between Figs. 6b and 6g. This closeup view reveals shock wave emergence from the muzzle brake attached to the weapon prior to the exit of the bullet. After firing five previous rounds, unburned powder is seen emanating from the muzzle of the gun. Some of these powder grains are hurled at supersonic speed as revealed in Figs. 6e-6g. Shock wave diffraction through the multiple orifices of the muzzle brake is also revealed in Figs. 6b-6d. As discussed previously in [5], the gas-dynamic understanding of muzzle brakes, suppressors, and other issues of firearm design could benefit from the sort of high-speed imaging shown here.

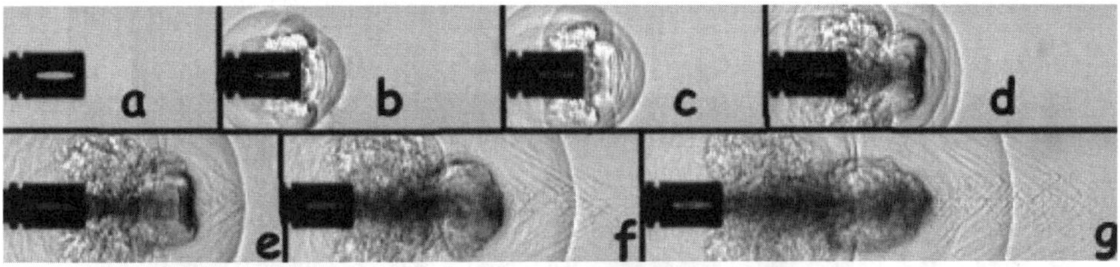

Fig. 6. Sequence of frames from $9mm$ submachine gun fire, muzzle closeup view observed by focused shadowgraphy

3.2 Edgerton retroreflective shadowgraph results

Gram-range explosive charges of TATP were observed using Edgerton shadowgraphy at 10,000 frames/s, e.g. Figure 7. The shock wave radius vs. time can be measured from such a frame sequence by way of a length calibration, yielding shock Mach number vs. radius. Such quantitative results are very useful in gram-range explosive testing, as described in a companion paper [11].

Acknowledgement. We thank E.M. Freemesser of the Monroe County NY Dept. of Public Safety, and Pennsylvania State Police Trooper E. F. Spencer, Jr. for providing the firearms and assisting in the experiments.

References

1. Settles, G. S. *Schlieren and Shadowgraph Techniques*, (Springer-Verlag, 2001)
2. Boys, C. V., "On electric spark photographs; or, photography of flying bullets, etc., by the light of the electric spark," *Nature* **47** 1219:440-446 (1893)
3. Cranz, C. and H. Schardin, "Kinematographie auf ruhendem Film und mit extrem hoher Bildfrequenz," *Zeitschrift für Physik* **56**, 147-183 (1929)
4. Edgerton, H. E., "Shockwave photography of large subjects in daylight" *Review of Scientific Instruments* **29** 2:171-172 (1958)

5. Settles, G. S., T. P. Grumstrup, J. D. Miller, M. J. Hargather, L. J. Dodson, and J. A. Gatto, "Full-scale high-speed 'Edgerton' retroreflective shadowgraphy of explosions and gunshots," *Proc. 5th Pacific Symp. on Flow Visualisation and Image Processing, PSFVIP5*, paper 251 (Australia, 27-29 September 2005)
6. Biele, J. K., "Point-source spark shadowgraphy at the historic birthplace of supersonic transportation - A historical note," *Shock Waves* **13** 3:167-177 (2003)
7. Parthasarathy S. P., Y. I. Cho, and L. H. Back, "Wide-field shadowgraphy of tip vortices from a helicopter rotor," *AIAA Journal* **25** 1:64-70 (1987)
8. Settles, G. S, "High-speed imaging of shock waves, explosions and gunshots," *American Scientist* **94** 1:22-31 (2006)
9. Hargather, M. J., G. S. Settles, J. A. Gatto, T. P. Grumstrup, and J. D.Miller, "Full-scale optical experiments on the explosive failure of a ULD-3 air cargo container," *Proc. 4th International Aviation Security Technology Symp.*, (Washington DC, Nov. 2006)
10. Virtual Backgrounds, 101 Uhland Road, Ste. 106, San Marcos, TX 78666 (USA), 1-800-831-0474, `www.virtualbackgrounds.net`
11. Hargather, M. J., G. S. Settles, and J. A. Gatto, "Gram-range explosive blast scaling and associated materials response," *Proc. ISSW26*, paper 3131 (July 2007)

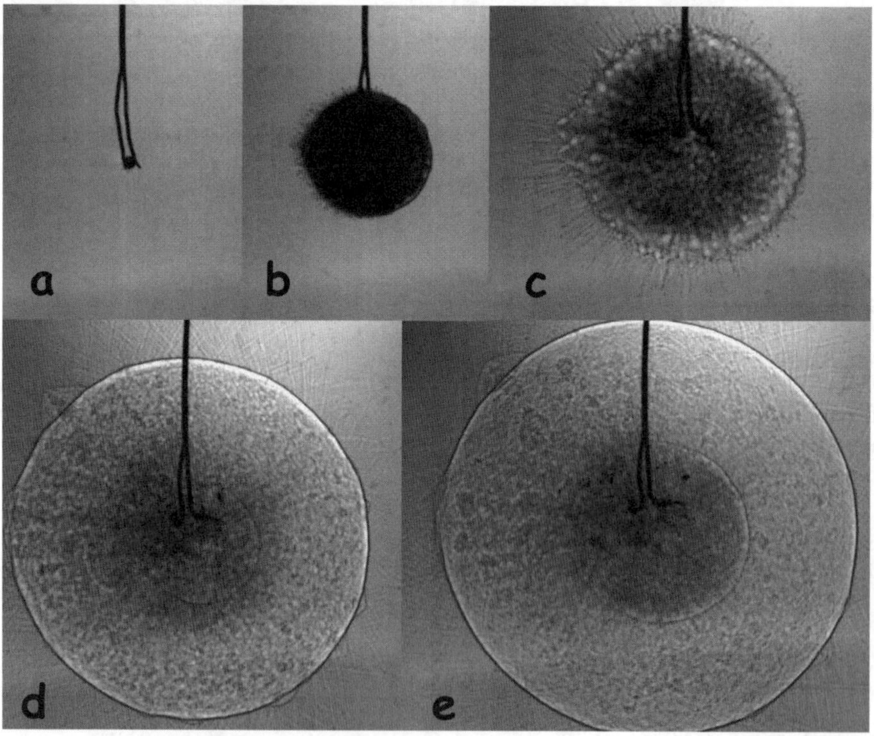

Fig. 7. Selected frames from the explosion of a 1*g* spherical TATP charge observed by Edgerton shadowgraphy

Modelling high explosives (HE) using smoothed particle hydrodynamics

M. Omang[1,3], S. Børve[2,3], and J. Trulsen[3]

[1] *Norwegian Defence Estates Agency, Postbox 405 Sentrum 0103 Oslo (Norway)*
[2] *Norwegian Defence Research Establishment, Postbox 25, 2027 Kjeller (Norway)*
[3] *Institute of Theoretical Astrophysics, University of Oslo, Postbox 1029 Blindern, 0315 Oslo (Norway)*

Summary. In this paper we present results from numerical simulations of high explosives, using a constant volume method and an axis-symmetric Regularized Smoothed Particle Hydrodynamics method. Empirical and numerical results show satisfactory agreement for 1 kg of detonating TNT charge. The method is further challenged with the study of shock propagation and shock reflection in complex geometries. The shock reflection pattern is altered by introducing barriers of different shapes. The effect of such barrier structures are studied.

1 Introduction

The assumption of axis-symmetry or spherical symmetry has a number of applications in the shock wave discipline. In the case of a free-field detonation of a spherical high explosive, for instance, the choice of spherical symmetry allows a 3D problem to be reduced to a 1D computation, thereby saving computational resources and time. In this work we take advantage of symmetry assumptions to investigate blast wave propagation in an axis-symmetric chamber-tunnel system. In the case of an unexpected detonation in such a configuration we want to evaluate the effect of barrier structures positioned in front of the tunnel entrance. For the latter problem, we use an axis-symmetric description of the numerical method Smoothed Particle Hydrodynamics, which is further described in section 2. The modelling of high explosives with a constant volume model is discussed in section 3 for a 1 kg charge of the high explosive (HE) TNT. The constant volume model is implemented in the chamber-tunnel configuration presented in section 4. The results are further discussed and summarised in section 5.

2 Symmetry assumptions in SPH

In [5,6] an approach to implementing symmetry assumptions in SPH was presented. The description takes advantage of SPH interpolation theory, that a field function can be expressed in terms of an interpolation integral of itself with a suitable kernel function. Depending on the symmetry assumptions made, spherical, cylindrical or planar, the interpolation integral is computed for the chosen generic kernel. Lagrangian formalism further provides a new set of equations of motion, specific for the chosen symmetry.

Results from benchmark tests show satisfactory performance with low noise level, also in regions close to the symmetry axis. The new approach is further challenged in [7] were we look at shock collisions and complex shock reflection patterns. Here we also present a test which shows that the convergence rates are comparable to those reported with various finite difference methods.

In the current work we use an extension to the original SPH method called Regularized Smoothed Particle Hydrodynamics (RSPH) [1–3]. In RSPH we allow an optimization of the resolution, by introducing smoothing lengths of stepwise smaller or larger size, corresponding to grid refinements in finite difference methods. The particles are distributed in rectangular boxes together with particles of their own kind (same smoothing length). The smoothing length difference for neighbouring boxes should vary by a factor of two. The particle interactions across boxes are preformed via auxiliary particles ensuring that particles of one smoothing length only interact with particles of the identical smoothing length. At regular time intervals the particle distribution is redefined, via a particle regularization, the new particle set given from the particle properties of the old set. Based on fex pressure, density or velocity gradients, the new particle set may obtain a new and better suited resolution.

3 Modelling of high explosives

There is a number of suggested approaches as to how high explosives can be implemented for numerical purposes. In the current work we use the constant volume approach. The constant volume approach implies that the detonation front is assumed to propagate through the HE at an essentially infinite velocity with no loss of energy. Consequently, when the simulations are initiated, the high explosive is assumed to be completely converted into gas, homogeneously distributed throughout the original volume of the high explosive. The simulations are started at this point in time, thereby allowing the assumption of an ideal gas law. In practical experiments one way of defining the initialisation of the high explosive, is by mounting a wire on the outside of the charge. When the wire is broken, a signal is recorded, which is defined to be time zero.

In our numerical simulations, we use the parameters given in Table 1 for 1 kg of TNT. If the high explosive is assumed to be spherical, we can use our 1D spherical symmetric code for the initial evolution. In Figure 1 we have plotted our numerical results together with the results from different empirical databases [4, 8].

In the upper part of Figure 1, the overpressure and shock time-of arrival are plotted as functions of distance from the charge centre. The agreement between the experimental and numerical results is satisfying, as the stars and plus signs representing experimental and numerical data are plotted almost on top of each other. For the representation of the positive shock duration t_+, the match with the empirical data is less accurate, whereas the pressure impulse plot shows satisfactory agreement. Both positive duration and pressure impulse measurements are sensitive to fragment hits, consequently the reported results may be less accurate. In the literature there is also less empirical data available on pressure impulse and positive shock duration.

E (MJ/kg)	Q(kg)	ρ(kg/m^3	R_0(mm)
4.29	1.0	1630	52.7
4.29	314	1630	358.0

Table 1. TNT parameters used in the constant volume model (2nd row). Data used in the complex structures study are give in the 3rd row.

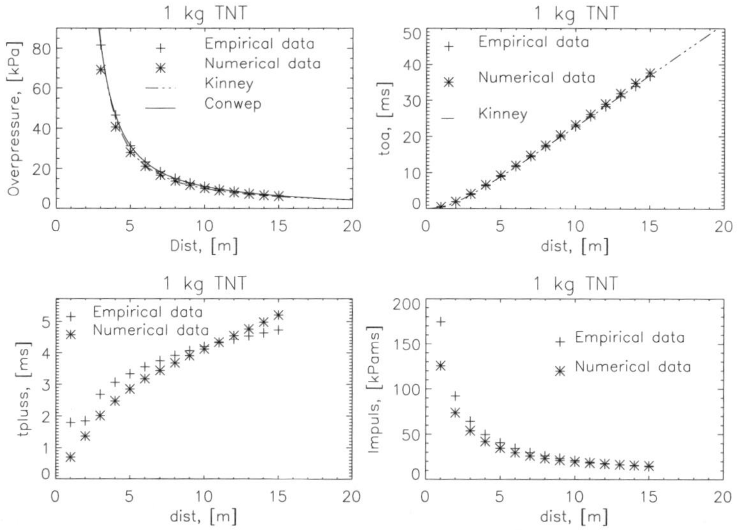

Fig. 1. Numerical and experimental results of overpressure, shock time of arrival, positive shock duration, and pressure impulse from the detonation of 1 kg of the high-explosive TNT.

4 Internal detonations

The study of internal detonations was initiated to investigate shock propagation and shock reflections from a high explosive detonated inside a chamber. The chamber has an opening to a straight tunnel with exit to free-field. A sketch of the construction is given in Figure 2. We have also assumed the tunnel exit to be part of a vertical wall. In front of the tunnel entrance, two different barriers, a t-shaped and a u-shaped, are tested separately, and compared to a similar open system. In the study we use a spherically symmetric code to simulate the detonation and initial state of the explosion. The initial data are given in the 3rd row of table 1, for a charge density of 5 kg/m^3. At $t = 4$ ms the results are saved and used as input in an axis-symmetric RSPH code. In Figure 4 we have plotted the initial conditions for density. Due to symmetry assumption, the center of the chamber is located on the z-axis. A complex shock reflection pattern is obtained as the shock reaches the chamber walls and are reflected. The left panel in Figure 3 illustrates this situation. As the second panel illustrates, multiple reflection occurs as the initial shock propagates along the tunnel. This is also the case for the third panel, where the shock is about to reach the exit. In the forth panel of the shock has expanded into free-field, and a vortex structure has been formed on the outside. In the last panel, the shock is reflected from the u-shaped barrier, and a Mach reflection is about to form on the wall to the right of the barrier.

We use numerical sensors positioned at different angles and radii's from the tunnel entrance. In the right panel of Figure 4 we have plotted the overpressure for three of these positions in the flow-field. The results from the open, u-shaped, and t-shaped barrier are plotted with solid, dotted and dashed lines, respectively. In Figure 4a the numerical sensor is positioned inside the tunnel, and as expected the overpressure of the different

Fig. 2. Chamber-tunnel configuration with a HE charge (the sphere) detonated inside. A vertical wall (not shown) is assumed to surround the exit. The two barrier structures, t-shaped and u-shaped are shown to the right. Both tunnel, chamber and barriers are cylindrically symmetric.

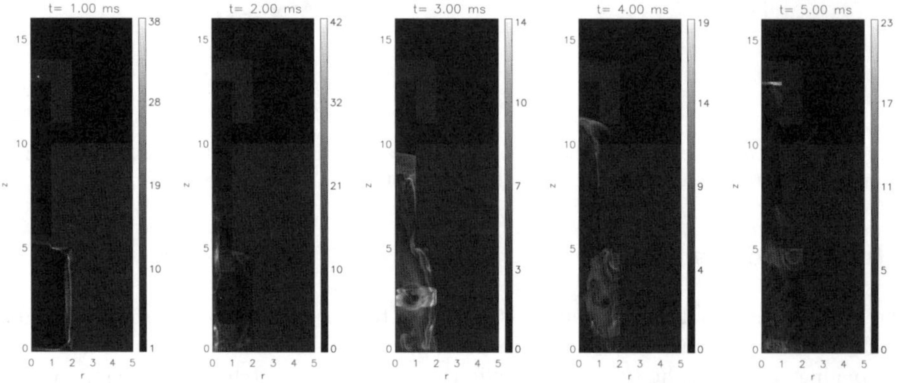

Fig. 3. Numerical results at t = 1.0, 2.0, 3.0, 4 and 5 ms from simulations of HE detonation inside a chamber-tunnel configuration. A u-shaped barrier is positioned in front of the tunnel exit.

configurations are similar in shape. In Figure 4b the sensor is positioned outside the tunnel exit, in this case the t-shaped barrier shows a much higher overpressure peak than the other sensors. This is to be expected, since the position chosen is on the lower barrier wall. The sensor position is moved to the right in Figure 4c. In this case we observe that both barrier structures have served to reduce the peak pressure measured. In the open tunnel case, the peak pressure is four times higher.

5 Conclusion and discussion

In this paper we have demonstrated the ability of RSPH to simulate shock and shock reflections from high explosives in an axis-symmetric geometry. A constant volume model is tested on a 1 kg charge of TNT. The agreement with empirical data is found to be

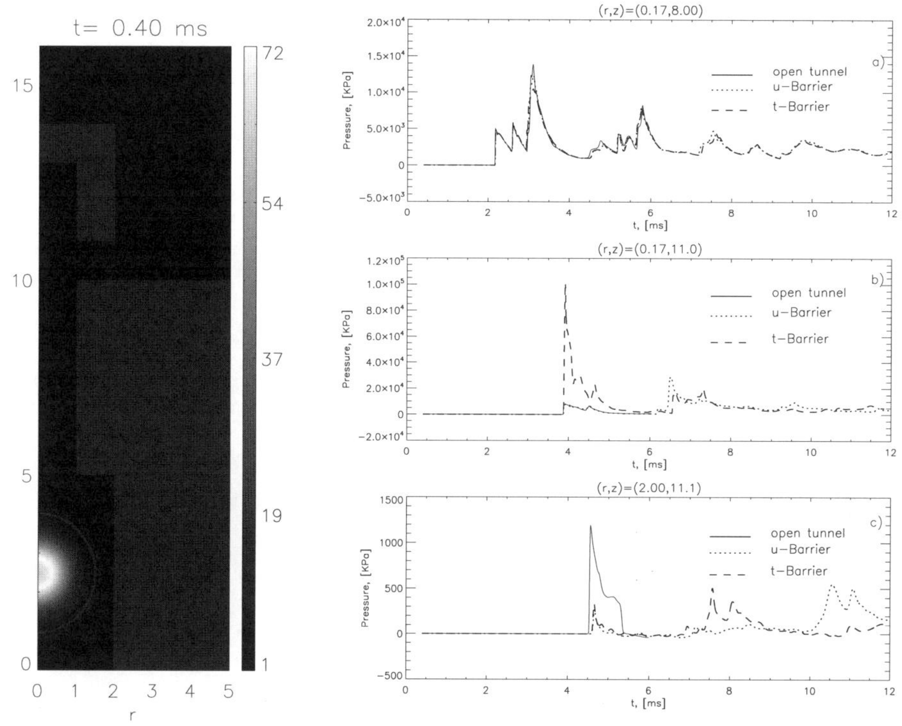

Fig. 4. Initial conditions for the u-shaped barrier simulation, left panel. In the right panel, overpressure as a function of time for three different positions are given.

satisfactory for overpressure, time of arrival and pressure impulse. The fit is less accurate for the positive duration, which may be due to the explosive model used. Results from explosions in complex chamber-tunnel systems are also presented. The results illustrate the complex shock reflection pattern obtained, and the effect of introducing barrier structures in the flow field outside the entrance.

References

1. Børve, S., Omang, M., & Trulsen, J. (2001). *Regularized Smoothed Particle Hydrodynamics: A new approach to simulating magnetohydrodynamic shocks.* Astrophys. J., **561**, 82–93.
2. Børve, S., Omang, M., & Trulsen, J. (2005). Regularized Smoothed Particle Hydrodynamics with improved multi-resolution handling. J. Comput. Phys., **208**(1), 345–367.
3. Børve, S., Omang, M., & Trulsen, J. (2006). Multidimensional MHD Shock Tests of Regularized Smoothed Particle Hydrodynamics. Astrophys. J. , **652**, 1306–1317.
4. Kinney, G. F. & Graham, K. J. (1985). *Explosive shocks in air (2nd edition). Shock and Vibration.*
5. Omang, M., Børve, S., & Trulsen, J. (2005). Alternative kernel functions for Smoothed Particle Hydrodynamics in cylindrical symmetry. Shock Waves, **14**(4), 293–298.
6. Omang, M., Børve, S., & Trulsen, J. (2006). SPH in spherical and cylindrical coordinates. J. Comput. Phys., **213**(1), 391–412.

7. Omang, M., Børve, S., & Trulsen, J. (2007). Shock collisions in 3d using an axi-symmetric Regularized Smoothed Particle Hydrodynamics RSPH code. Shock Waves.
8. US Army (1991). CONWEP- conventional weapons effects program, version 2.00. US Army Engineer Waterways Experimantel Station, Vicksburg, MS USA.

Modification of air blast loading transmission by foams and high density materials

B.E. Gelfand[1], M.V. Silnikov[2], and M.V. Chernyshov[2]

[1] *Semenov Institute of Chemical Physics RAS, Kosygina Str. 4, 119991 Moscow (Russia)*
[2] *Special Materials Research Institute, Bolshoy Sampsonievskiy Ave. 28A, 194044 Saint Petersburg (Russia)*

1 Introduction

Studies of the impact of blast shock waves on porous media were executed to find the methods and materials for the blast loading attenuation on different targets. The results of sound waves attenuation due to propagation through porous flexible media (for example, foams with open or closed cells) served as a background for that approach. It was shown that the transmitted pressure wave reduced its magnitude if a slab of such material is placed at some distance ahead of a protected surface [1]- [3]. But no attenuation of transmitted wave occurs if the gap does not exist (foam slab is placed right against a solid surface). The early results on amplification effects with open-cell and closed-cell flexible foams were published in [4], [5]. A comprehensive description of such phenomena in foam blocks exposed to the shock wave loading is presented in [6]. The unique compressive strain-stress relation for the flexible cellular materials leads to several interesting observations on the characteristics of one-dimensional stress wave transmission by a cellular media which are important for understanding the blast and impact mitigation or attenuation using such materials. Discovered baric effects represent the fundamental background for the construction of personal protection systems using flexible and compressible materials. The materials of such types sometimes were rested against the thoracic walls of experimental bio-objects exposed to blast overpressure to investigate the transmission of stress waves into the thorax. It was shown that the foams acted as acoustic couplers and resulted in significant augmentation of visceral injury [7], [8]. Direct tests [7] also displayed that decoupling and elimination of injury may be achieved by application of a high acoustic impedance sublayer on the top of the foamy slab.

2 Methods, materials, and blast wave parameters

Protecting single and multi-layer screens consisting of various structures containing porous media were considered in [7]- [12] where the damping or amplifying properties of such interlayer were studied. The protecting screens transforming the impulse loading were placed at the steel wall [10]- [12] or water surface [7]. The ultimate aim of that tests was the optimization of the porous filler parameters for its maximum ability to damp the loading. The screens made of foam rubber (density 40 kg/m^3) with different thickness as well as alternating multilayer structures made of steel and foam rubber were investigated [10]- [12]. The screens made of foamy pearlite (density 60 kg/m^3) were also used at some tests [10]- [12]. The loading was carried out by the blast of spherical RDX charge (weight 0.15 kg, radius $r \sim 35$ mm located at the distance of 0.5 m from the wall. Initial Mach number of the blast wave (BW) was $M = 3.27$ [13]. The length ΔL of the blast wave was determined in the accordance to information in [14]. The screens were mounted at the wall and the pressure gauge was set into the wall under the screen. The scheme of the tests is presented at Figure 1.

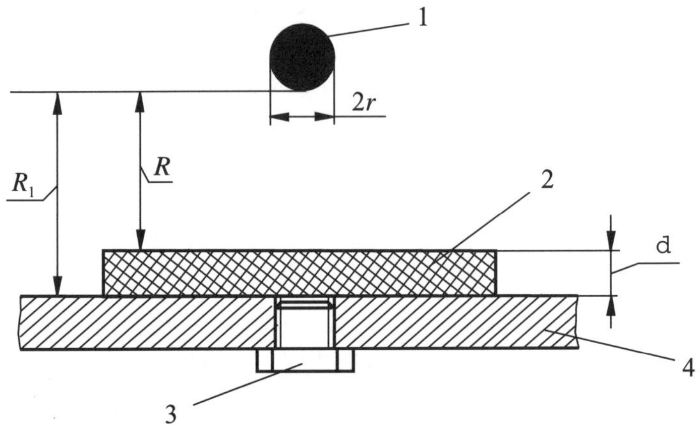

Fig. 1. Schematic overview of experimental setup, Afanasenko et at. (1990): 1 – HE charge, 2 – flexible screen, 3 – pressure gauge, 4 – steel plate

It is of great interest to determine the optimal porous material for maximal pressure mitigation and to find the relationship between the damping degree, blast magnitude, and blast impulse duration. The spherical charges of RDX with the mass of 0.034 kg (radius of the charge was $r = 22$ mm) and 0.215 kg ($r = 40$ mm) were used for this purpose [10]- [12]. The magnitude of the overpressure ($\Delta P_i = P_i - P_0$ for incident BW and $\Delta P_r = P_r - P_0$ for reflected one) and BW-impulse duration τ were measured and calculated [14], [13] at each test run with the changing of the charge weight and the distance R from the charge to the screen. The similar testing procedure was described at [9], [15], [16]. Sometimes porous interlayer was filled by sand, rubber crumb, or steel shots [10]- [12]. Such materials as copper (thickness $\delta = 0.58$ mm) or resin-bonded Kevlar ($\delta = 15$ mm) incorporated as a facing of the foam screen were also used for the analysis of stress transmission at blast wave loading [7]. The experimental results of [10]- [12] are presented at Table 1 and, only for foam-rubber screen, at Figure 2.

It is evident from the Table 1 and Figure 2 that the imposition of foam rubber screens with the thickness of the single layer $\delta < 80$ mm leads to the increase of blast wave overpressure at the wall behind the screen (the coefficient of the amplification is $\kappa > 1$). Here the coefficient of the amplification κ is the ratio of the reflected BW overpressures with and without the protective screen. Thereof it can be seen from tests with $\delta = 80$ mm, 60 mm, and 70 mm that the reflected BW can be amplified. Screens with $\delta = 100 - 150$ mm result in the overpressure decrease comparing with the walls loaded by explosion without the screen ($\delta = 0$). One can see the increase of BW duration and transformation of blast wave into compression wave (CW in the tests with $\delta > 80$ mm). The typical wall overpressure-time histories obtained in the tests are presented in Figure 3. Initial parameters for the tests with cited numbers are listed at the Tables 2 and 3.

BW length (ΔL)	Thickness of the screen	Foam-rubber (40 kg/m^3)			Pearlite (60 kg/m^3)		
mm	mm	ΔP, MPa	τ, μsec	κ	ΔP, MPa	τ, μsec	κ
250	0	5.8	230	1	5.8	230	1
250	30	13	130	2.24	7.0	120	1.2
250	50	9.3	200	1.6	2.87	210	0.49
250	60	8.5	240	1.46	-	-	-
250	70	7.6	630	1.31	-	-	-
250	80	-	-	-	2.2	400	0.38
250	100	1.92	1060	0.33	1.35	420	0.23
250	150	1.12	1260	0.19	0.66	660	0.11

Table 1. The dependence of loading parameters on the flexible screen structure and thickness. HE weight – 0.15 kg RDX, distance $R = 0.5$ m ($R/r = 14$), $M = 3.27$.

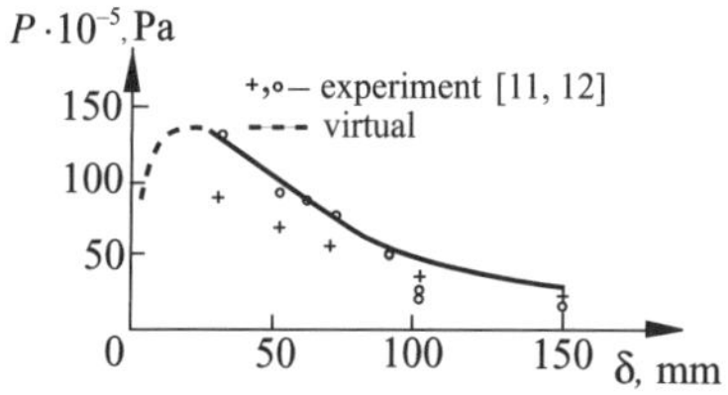

Fig. 2. The maximal overpressure loading at the wall for foam-rubber screens with different thickness at constant blast wave parameters.

Tect No.	δ_{fr}, mm	δ_{st}, mm	δ, mm	$\delta/\Delta L$
105	70	0	70	0.25
106	100	0	100	0.40
111	30	0	30	0.12
113	150	0	150	0.60
116	2×30	2×2.5	65	0.26
117	60	5	65	0.26
124	3×30	3×2.5	97.5	0.39

Table 2. Numbers of the tests and geometrical parameters of the screens. Here δ_{fr} is the thickness of foam-rubber layer, δ_{st} is the thickness of steel, δ is the total thickness of the screen.

The analysis of the stratified multilayer screens of the equal thicknesses and masses (tests 116 and 117) shows that two-layer screen is the most effective for BW-overpressure mitigation. Also two-layer screen is better than three-layer screen with the larger thickness and mass because of the pressure disturbance stretching and magnitude reducing (test 124).

The impulse received by the back plate due to the interaction with BW, $I = \int_0^{\tau} \Delta P(t)dt$, does not clearly depend on the screen structure and remained almost constant under the same loading conditions.

The test results [10]- [12] collected in [17] for two-layer screens of the similar structure but different type of porous material are presented in Table 4. The loading overpressure amplitude and the characteristic duration times behind the screens are listed there. Mach

BW length, mm	Thickness and structure of the screen, mm	ΔP, MPa	τ, μsec	κ
250	2.5 mm (steel) + 30 mm (foam-rubber, 40 kg/m^3) + 2.5 mm (steel) + 30 mm (foam-rubber, 40 kg/m^3); $\delta = 65$ mm	0.6	5000	0.1
250	5 mm (steel) + 60 mm (foam-rubber, 40 kg/m^3); $\delta = 65$ mm	0.28	2160	0.05
250	3 layers, each consisting of 2.5 mm (steel plate) + 30 mm (foam-rubber, 40 kg/m^3); $\delta = 97.5$ mm	0.65	2690	0.11

Table 3. The dependence of loading parameters on the screen structure and thickness. Multilayer screens. HE weight – 0.15 kg RDX, distance $R = 0.5$ m.

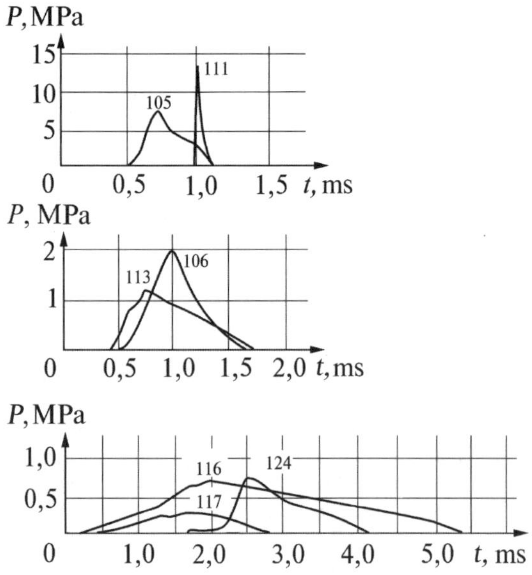

Fig. 3. Typical overpressure – time records at hard wall surface under protected screen.

numbers and the lengths of impacting BW are also presented basing on [14] and [13]. The relative distances R/r from the charge to the screen are also given at the tables.

As it is seen from the Tables 1 and 3-4, the foam rubber, rubber comb, and foamy pearlite allow sufficient reducing of BW overpressure amplitude at protected walls with simultaneous increase of impulse duration. The screens filled by sand and steel shot are less effective. Such screens cause the increase of overpressure at protected walls in a number of cases concerned.

The dependence of the damping properties on BW impulse duration and amplitude are also of great importance. Pressure amplitude reduction slightly differs for various screens for BW pulse generated by 0.215 kg RDX charge at the distance of 1.4 m. On the contrary, considerable difference of damping properties of the distance of 0.5 m was observed.

Distance to the screen	Parameters of BW loads	Without screen	Steel-sand screen	Steel-rubber screen	Steel-pearlite screen	Steel – rubber crumb screen
$R = 1.4$ m	ΔP, MPa	0.46	0.16 ($\kappa = 0.34$)	0.12 ($\kappa = 0.26$)	0.16 ($\kappa = 0.34$)	0.19 ($\kappa = 0.4$)
$R/r = 35$	τ, ms	1.31	4.1	4.2	3.6	4.1
	ΔL, mm	430	430	430	430	430
	M	1.35	1.35	1.35	1.35	1.35
$R = 1$ m	ΔP, MPa	1.22	0.83 ($\kappa = 0.68$)	0.21 ($\kappa = 0.17$)	0.30 ($\kappa = 0.24$)	0.33 ($\kappa = 0.27$)
$R/r = 25$	τ, ms	1.04	2.6	3.7	2.6	4.0
	ΔL, mm	360	360	360	360	360
	M	1.91	1.91	1.91	1.91	1.91
$R = 0.5$ m	ΔP, MPa	16.6	6.40 ($\kappa = 0.38$)	1.17 ($\kappa = 0.07$)	3.5 ($\kappa = 0.21$)	1.24 ($\kappa = 0.075$)
$R/r = 13$	τ, ms	0.38	0.50	3.60	1.40	3.90
	ΔL, mm	300	300	300	300	300
	M	4.60	4.60	4.60	4.60	4.60

Table 4. Loads at walls caused by BW generated by 0.215 kg HE charge through the following screen steel ($\delta = 2$ mm) + porous material ($\delta = 40$ mm).

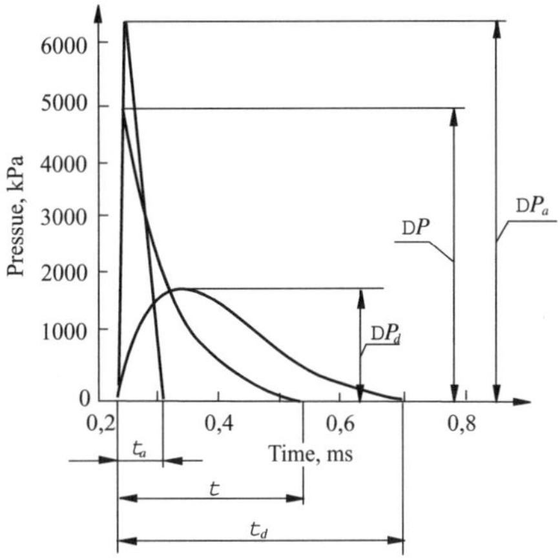

Fig. 4. Three cases of overpressure – time history due to blast wave loading. Here ΔP_a and τ_a are amplitude and duration of amplified blast loading (short screen), ΔP and τ are the same for basic loading without screen, ΔP_d and τ_d are the same for damped blast loading (long screen).

3 Conclusions

The critical thickness δ^* of the porous layer exists for the BW loading with triangular overpressure-time profile. Three possible results in relation to the critical thickness are presented at Figure 4.

If the thickness of porous layer $\delta < \delta^*$, the overpressure amplitude of BW reflected from the rigid wall under the porous layer increases comparing with the overpressure of reflected wave without the layer. If $\delta > \delta^*$, the overpressure level at the wall behind the screen is lesser than in the case of loading without protection.

The value δ^* of the critical thickness depends of the amplitude and the duration of the BW.

Two-layer protective screen is most effective among investigated structures. The denser sublayer should be located from the side perceiving loading. The pressure amplitude mitigation is proportional to the thickness of the porous sublayer.

Foam-rubber, rubber crumb, foamy pearlite and polyurethane are the most effective among used porous materials of two-layer protective screens.

References

1. Gelfand B. E., Gubanov A. V., Timofeev E. I.: Interaction of shock waves in air with porous barrier. Izv. Acad. Nauk. Mech. zhidkosti i gaza, no. 4, 1983 (in Russian)
2. Seitz M. W., Skews B. W.: Shock impact of porous plug with fixed gap between plug and wall. Proceedings of the 20th ISSW, ed. by Sturtevart B., Shepherd J. E., Hornung H. G., World Sciectific, vol. 2, 1996.
3. Kitagawa K., Takayama K., Yasuhara M.: Attenuation of shock waves propagation in polyurethane foams. Shock Waves, vol. 15, no. 6, 2006.
4. Gelfand B. E., Gubin S. A., Kogarko S. M.: Investigation of special characteristics of propagation of pressure wavs in porous medium. Sov. Physics. Appl. mech. thechn. phys., vol. 15, no. 6, 1975 (in Russian).
5. Gvozdeva L. G., Faresov Yu. M.: Calcilating the parameters of steady shock waves in porous compressible media. Zh. Techn. Phys., vol. 55, 1985 (in Russian).
6. Seitz M. W., Skews B. W.: Effect of compressible foam properties on pressure amplification during shock wave impact. Shock Waves, vol. 15, no. 3-4, 2006.
7. Cooper G. J., Townend D. J., Cater S. R., Pearce B. P.: The role of stress waves in thoracic visceral injury from blast loading. J. of Biomechanics, vol. 24, no. 5, 1991.
8. Li Q. M., Meng H.: Attenuation or enhancemnt — a one-dimensional analysis of shock transmission in solid phase of cellular material. Int. J. of Impact Eng., vol. 27, no. 12, 2002.
9. Nerenberg J., Makris A., Klein H., Chichester C.: The effectiveness of different personal protective ensembles in preventing blast injury to the thorax. Proc. of the 4th Intern. Symp. on Technology and Mine Problems. Mounteray, California, 2000.
10. Afanasenko S. I., Nesterenko V. F.: Application of the condensed porous materials for the maximal stress reduction in walls of spherical explosive chamber. Proc. of VII-th Intern. Symp. on Use of the Energy of Explosion, Pardubice, 1988.
11. Afanasenko S. I., Grigoriev G. S., Klapovsky V. E.: The influence of the thickness and material of porous screens on explosive impulse transmission. In: The Processing of Materials by Impulsive Loadings, Sib. Branch of Sov. Acad. Sci., Novosibirsk, 1990 (in Russian).
12. Afanasenko S. I.: The application of porous media for the attenuation of shock waves near walls of explosion chambers. In: The Processing of Materials by Impulsive Loadings, Sib. Branch of Sov. Acad. Sci., Novosibirsk, 1990 (in Russian).
13. Hyde D. W. Conventional weapon effects (ConWep), USAEWES/SS-R, 1992.
14. Adushkin V. V.: Parameters of shock wave in vicinity of HE charge at explosion in air. Sov. Physics. Appl. mech. thechn. phys., vol. 1, no. 5, 1961 (in Russian).
15. Makris A., Nerenberg J., Lee L. H. S.: Attenuation of blast wave with cellular material. Proceedings of the 20th ISSW, ed. by Sturtevart B., Shepherd J. E., Hornung H. G., World Sciectific, vol. 2, 1996.
16. Lee J. J., Frost D., Lee J. H. S. The transmission of a blast wave through a deformable layer. Proc. of the 19th International Symposium on Shock Waves, ed. by Brun R. and Dumitrescu L. Z., vol. 3, 1995.
17. Gelfand B. E., Silnikov M. V. Explosions and Blast Control. Asterion, St. Petersburg, 2004, 296 p.

Multifunctional protective devices for localizing the blast impact when conducting the ground and underwater operations of explosion cutting

B.I. Palamarchuk, A.T. Malakhov, A.N. Manchenko, A.V. Cherkashin, and N.V. Korpan

Paton Welding Institute, 11 Bozhenko St., 03680, Kiev, Ukraine

Summary. The paper describes the practical implementation of the results of studying the shock waves in heterogeneous media of foamy and bubble structure. Proceeding from the revealed features of attenuation of blast wave parameters and their interaction with structures and biological objects, heterogeneous protective devices "Mach" have been developed for localization of the impact of condensed explosives at performance of ground and underwater operations on explosion cutting. The developed devices have a wide sphere of application during performance of blasting operations, when it is necessary to lower the parameters of the shock waves generated by the blast in the environment, in order to prevent damage to the adjacent objects, structures and live organisms. The paper gives the examples of their application at elimination of accidents in oil- and gas pipeline transportation and cutting of piping in NPP, as well as during performance of underwater explosion cutting of pile foundations of offshore stationary platforms. The devices have a protective sheath from heterogeneous materials in combination with anti-fragment shields. The optimum combination of the characteristics of the protective sheath and shield materials al-lowed development of effective, compact and mobile protective devices of modular type.

Interest to study of explosion waves in gas-liquid media arises from the possibility of their effective use to lower the blast loads on engineering and biological objects. The basic idea of their application for this purpose consists in using the high compressibility of dispersed gas-liquid media (GLM) in blast waves (BW), leading to considerable dissipative losses at propagation of blast waves in them. This effect is related to the fact that in dispersed GLM the heat capacity depends on the condensed phase, and compressibility - on the gas phase. It should be noted that use of two-phase media for blast wave damping started as far back as 1960ties. However, this work mostly involved development of bubble screens in front of the protected object, or air-water curtains in the path of blast wave propagation. In this presentation attention is focused on the issues of localizing the impact of the blast and development of approaches, aimed at understanding the mechanism of blast wave damping at the stage of their formation and evolution in gas-liquid media.

The purpose of this presentation is to describe the results of studies of the blast-wave processes occurring at a blast in gas-liquid media, aimed at development of elective means of projection of biological objects from blast waves.

For investigation of the structure and parameters of blast waves in dispersed media we used both piezoelectric and electret pressure transducers, also those of piezotron type for field conditions. In addition, in order to study the blast waves various blade-type transducers were developed, as well as electret transducers to study the BW reflection from a rigid wall, where the natural frequency was up to 1.5 MHz. Parameters of quasi-acoustic BW were measured, using Brule-Kier microphones and Kistler pressure transducers, which allowed studying BW parameters in the region of high pressures up to several kilobars.

It is characteristic that the dependencies obtained by numerical methods for BW attenuation with distance, both at frozen relaxation processes and at thermal equilibrium of the phases, fall above the experimental data. This is related to the fact that an adiabatic dispersion of detonation products is assumed in calculations. However, in reality at explosion of condensed explosives in GLM the detonation products are filtered through GLM, and already at the dispersion stage the detonation products interact with the liquid particles, this leading to additional energy take-off from detonation products, not allowed for in numerical simulation. The dependencies indicate that the effectiveness of BW suppression by foams for explosive charges, for which experiments were performed (explosive weight from several grams up to several kilograms), is above the design value. Therefore, if we increase the weight of the charge, the filtration effects will have a less and less significant contribution to the overall energy balance, and eventually the effectiveness of BW damping by two-phase media will be lower. This should be taken into account when using two-phase media for BW suppression at firing of explosive charges of more than 100 kg weight. The correctness of these conclusions is indicated that formation and separation of BW from detonation products in the foam proceed much closer to the blast center than in air. Using one reference point it is possible to determine the characteristic time of thermal relaxation of GLM, and obtain a satisfactory agreement of the experimental and design data, taking into account the depth of running of relaxation processes, allowing for the properties of the medium and thermophysical properties of gas.

Now let us see how the volume fraction and degree of completion of heat transfer processes in two-phase media (air, water, foam, bubble media) affect the pressure on BW front at fixed distance from the charge equal to 30 charge radii (fig. 1). It is seen that pressure dependence in the wave on liquid concentration at freezing of the relaxation processes of heat transfer, unlike a similar dependence at complete heating of the liquid, does not have a minimum lying in the area of liquid concentrations corresponding to foams. That is, thermal losses are maximum in the range of liquid concentration corresponding to foams, and, therefore, these media are the most effective for BW suppression. Now in bubble media at gas concentrations equal to several percent, the degree of completion of heat transfer processes no longer is the determinant factor for BW damping. The determinant factor is this case is gas concentration in a bubble medium, which is what determines the degree of BW attenuation. It should also be taken into account, that in order to preserve the effectiveness of protection with greater depth, it is necessary to maintain a constant concentration of gas in the bubble medium, i.e. increase the gas flow rate. This is exactly why at great depths for BW suppression it is better to use two-phase media with a rigid frame, for instance, foam metal.

If we generalize all the experimental data on BW attenuation at a blast in energy coordinates, using the results of application of the strong blast theory and the obtained energy equivalents, which depend on the degree of expansion of the detonation products in LGM, then we will see, that in energy coordinates (where E is the internal energy covered by BW front and referred to the blast energy W^*), irrespective of the medium (water, air, foam, bubbled sand) all the data are in one echelon (fig. 2). That is, we have established a criterion of similarity, which enables us considering BW attenuation in energy coordinates in different media and predicting their parameters.

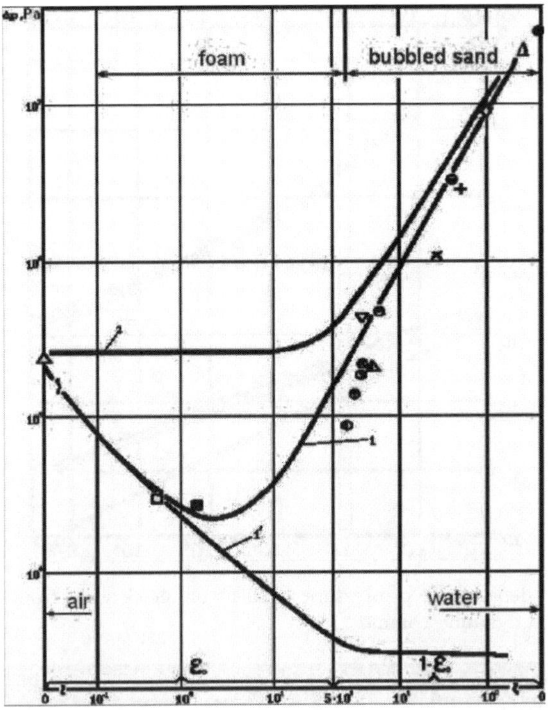

Fig. 1. Dependence of excess pressure on SW front on volume content of k-phase at a blast in gas-liquid media of water-air: 1 - for condensed explosives with W^*; 2 - at "frozen" relaxation processes; 1' - for a point source

Obtained results were the basis to develop various protection scenarios, in particular, for localizing underwater explosion of shaped charges. These are different modifications of protective devices, allowing engineering problems of underwater cutting to be solved. Here are specific examples: this is a multilayered structure - a sandwich (metal - concrete - metal - concrete). This structure is severed by firing one explosive charge of 1 kg weight (fig. 3). Effectiveness of protection from BW is such that fish is swimming 3 m from the blast center. This is dismantling of a platform, using underwater shaped charges. The fig. 4 shows profile of superfluous pressure BW in water.

Given below is retrofitting of systems of localizing the impact of the blast to solve the problem of severing metal radioactive scrap. Use of gas-liquid foams provides not only BW damping, but also dust-depression so as to prevent transfer of radioactive particles during the blast. Therefore, foam application in this case solves two problems simultaneously. The figs. 5, 6 show fire-blast-proof cutting systems for conducting technological operations in an explosive atmosphere. The protective device should not only localize the BW, but also prevent inflammation of gas mixtures, if work is performed in an explosive atmosphere.

Effectiveness and reliability of localizing was checked many times at explosions of fragmentation charges of strong explosives with TNT equivalent of up to 2 kg. "Mach" localizers of blast impact have been tried out for many years, and are recommended for designing industrial samples.

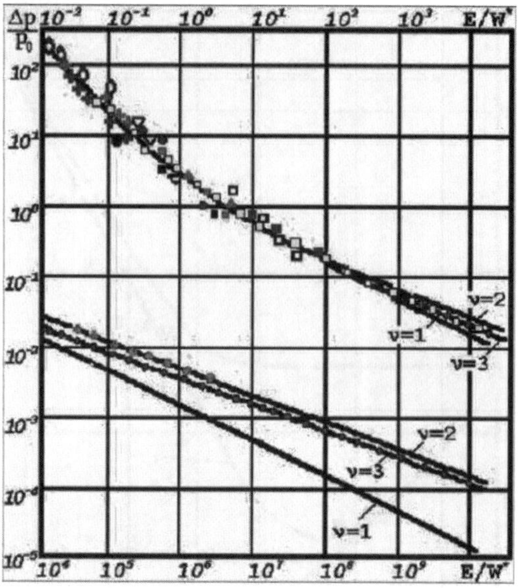

Fig. 2. Generalized dependence of pressure gradient on shock wave front in single-phase and multi-phase media on dynamic volume

Fig. 3. Cross-section of an explosion-cut column consisting of two concrete-filled steel pipes ($325 \times 11\,\mathrm{mm}$, $219 \times 8\,\mathrm{mm}$)

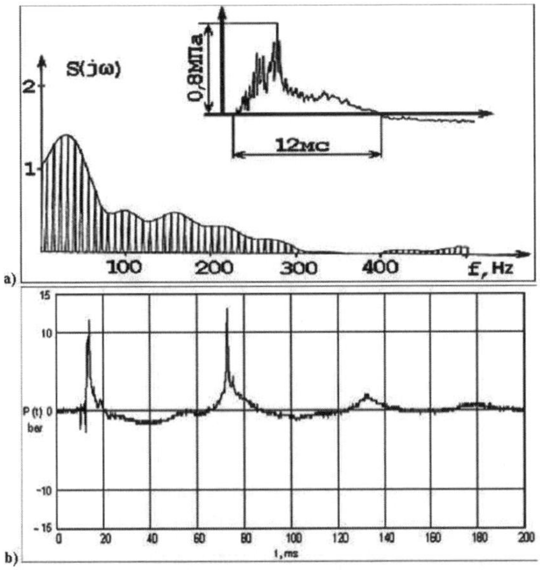

Fig. 4. Profile of superfluous pressure BW in water at explosion of a charge in weight of 1 kg RDX in combined bubble protection and its peak spectrum wave - (a), and also profile BW and its reflexion from lines of demarcation between bubble medium - water at explosion of a cylindrical charge in medium ($\beta = 1\%$), radius 1.2 m, running mass a charge - 12 g/m and distance from a charge to the gauge - 0.4 m - (b)

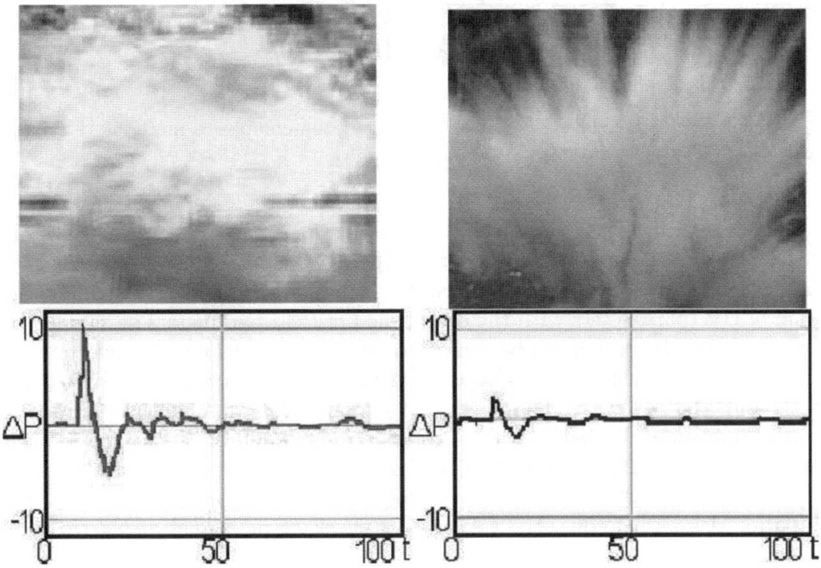

Fig. 5. The records show: open charge blast on the left; blast using "Mach" localizer on the right

Fig. 6. "Mach" protective device based on heterogeneous composite material "Palmakor" of 7 kg mass

In the end we will notice, that such safe systems for structure severing have already been developed and are applied together with the developed means of control of process operations, using the energy of explosion, which is mandatory. The developed methods of prediction of BW parameters with the established critical parameters of damage to biological objects and structures can be used with success also to solve the inverse problems of blast-engineering examination. This concerns primarily examination of unauthorized explosions and acts of terror. The developed applied programs allow determination of what has happened and what was the explosion energy. We have created information modules of support of explosion technologies in the higher risk zone. Such modules ensure not only control of the technological process, but also, for instance when working in a radiation-dangerous zone, the telemedical check of personnel, using satellite and cellular communication means. In other words, both the technological process, and the operators conducting the blasting operations under extreme or increased risk conditions are under control.

Propagation of the shock wave generated by two-dimensional beam focusing of a CO_2 pulsed laser

S. Udagawa[1], Y. Yamamoto[2], S. Nakajima[3], K. Takei[3], K. Ohmura[2], and K. Maeno[3]

[1] *Graduate School of Science and Technology, Chiba University, 1-33 Yayoi, Inage, Chiba, 263-8522, Japan*
[2] *Graduate School of Engineering, Chiba University, 1-33 Yayoi, Inage, Chiba, 263-8522, Japan*
[3] *Faculty of Engineering, Chiba University, 1-33 Yayoi, Inage, Chiba, 263-8522, Japan*

1 Introduction

To simulate the field for the actual scale explosion experiments is sometimes very difficult. From the reasons of safety-space, schedules, and human factor the number of actual field explosion experiments has been strictly limited. Therefore a small scale shock wave generator is considered to be very useful for experimentally simulating the explosion phenomena[1].

This study deals with the blast effects by using laser -induced micro shock wave. A plasma and a shock wave, which generated by a focusing pulsed CO_2 laser beam into a two-dimensional acrylic resin structure model by a cylindrical lens, have been visualized by the schlieren method. Moreover, the numerical simulation for compressible and inviscid flow with thermal equilibrium has been conducted to clarify the propagation characteristics of the laser induced shock wave.

2 Experimental set up

2.1 Measurement of the laser energy

As the first step, we have performed a measurement of the energy of an incoming pulsed CO_2 laser and the focusing plused laser after transmitting the plasma to obtain the energy into the plasma. Figure 1 shows a schematic drawing of the measurement system of the energy of a pulsed laser. The left side of Fig. 1 shows the measurement system of the energy of an incoming pulsed laser. The laser beam enters a joule meter (OPHIR OPTRONICS, F150A-SH) after forming by passing through a beam cutter. The right side of Fig. 1 shows the measurement system of the energy of a focusing pulsed laser after passing through a plasma. A pulsed laser after forming by the beam cutter, generates a plasma at the focal point after passing through a cylindrical lens (f=190.5mm). A pulsed laser after transmitting a plasma, enters the joule meter to estimate the plasma energy.

2.2 Visualization experiment of the shock waves

Figure 2 shows the schematic drawing of the experimental apparatus and the arrangement of the optical system. A pulsed laser emitted from CO_2 laser head (LUMONICS, TEA 103-2) generates a two-dimensional plasma at the focal line by a cylinder lens after

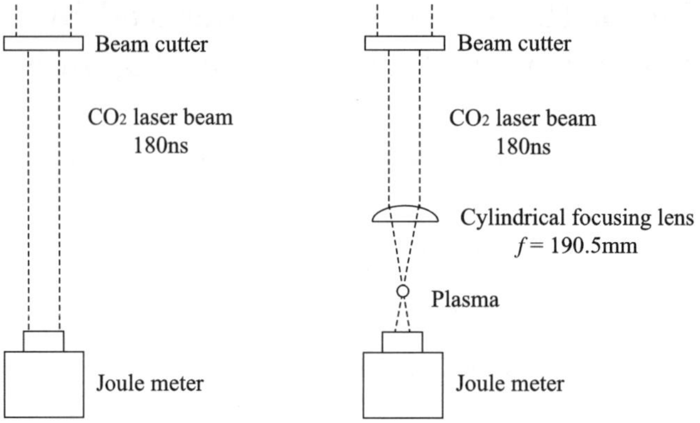

Fig. 1. Schematic drawing of the measurement system of the energy of a pulsed laser

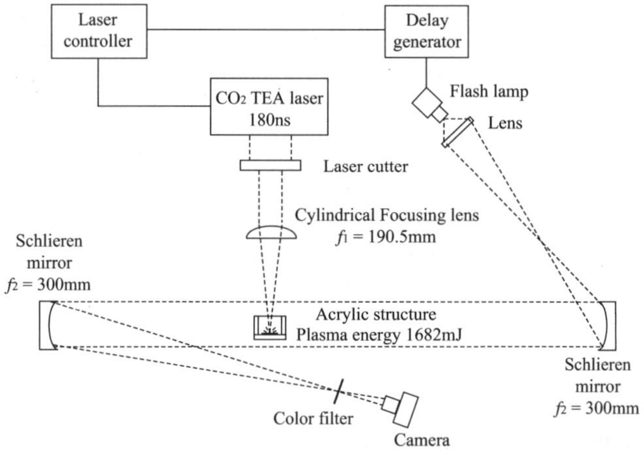

Fig. 2. Schematic diagram of laser focusing apparatus and the arrangement of optical system

forming by passing through the beam cutter. The inner wall of an acrylic structure and tpointocal line are placed on the same distance.

The shock wave generated by a focusing laser beam near the inner wall of a structure model is visualized in a structure model and after discharging from the open end with different delay times. Figure 3 shows the schematic drawing of an acrylic rectangular structure model.

The color schlieren method is used for visualizing the shock waves. The signal of the laser emission is put into a delay generator (Sugawara Laboratories, Digital retardar RE-306) and make 180ns light pulse of a flash lamp (Sugawara Laboratories, NPL-5) after making a designated delay time by the delay generator. The flash light refrects at the first parabolic mirror (f2=300mm, d=130mm) after passing through a converging lens as a small light source of the schlieren optical system. The flash light after reflecting the first parabolic mirror passes through the observation section as a parallel light beam and is reflected and collected by a second parabolic mirror (f2=300mm, d=130mm). A color

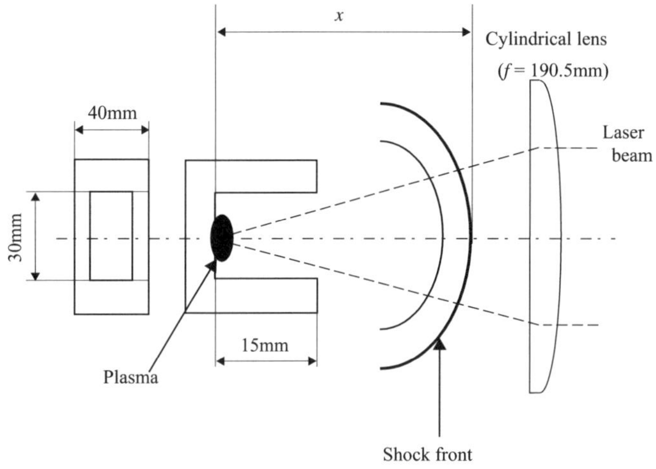

Fig. 3. Schematic drawing of an acrylic rectangular structure model

filter is settled at the focal point of a second mirror and the density gradient change at the observation section is recorded as a color schlieren image by a digital still camera (Canon, EOS Kiss Digital N).

3 Numerical simulation

In this study the numerical simulation has been also performed by using the calculation code for a plasma-induced shock wave[1].

3.1 Real gas effect

A plasma generated by focusing a pulsed laser is considered that a part of the energy of a pulsed laser has been absorbed by the air and the energy ionizes and dissociates of the gas. Consequently, it is necessary that the numerical simulation should include the real gas effects to analyze the acutual phenomena with higher accuracy. In this study the real gas effects of the high temperature equilibrium state is considered under the assumption that the velocities of the chemical reactions such as the ionization and the dissociation of air are negligible small compared with the propagation velocity of the shock wave. The real gas effects are calculated by Thermochemistry and Normal-Shock Computer Program (JPL). The JPL program is the calculation program of the state equation which can calculate the gas properties of the high temperature region and changes in the chemical composition.

3.2 Computational scheme

A two-dimensional axi-simmetric Euler equation system is used as the primitive equations. AUSM-DV scheme is used for the discretization of the convection term. MUSCL scheme is used to obtain the high order accuracy for the space to increase the resolution. A second ordered Runge-Kutta scheme is used for the time integration. The state equation of the high temperature equilibrium is used for considering the real gas effects.

3.3 Computational grid, initial conditions and boundary conditions

The flow field of the two-dimensional wave propagation is assumed as Figure 4 (a) to compare with the results of the visualization using a rectangular structure model.

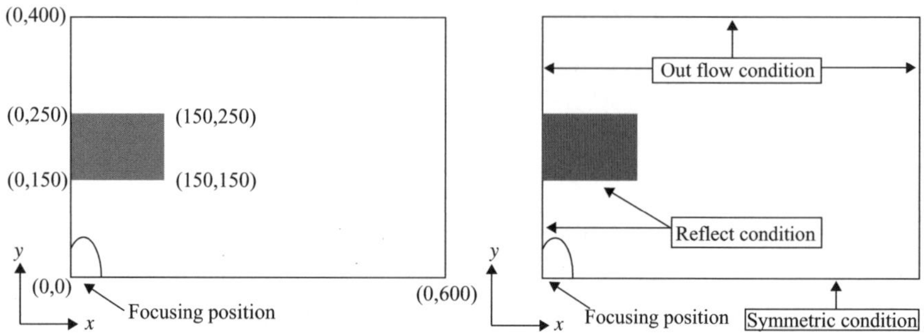

Fig. 4. (a) Schematic drawing of the flow field for the numerical simulation (left), (b) Schematic drawing of the boundary condition for the numerical simulation (right)

The real grid size is 0.1mm for x and y directions and the boundary conditions are shown in Fig.4 (b). A high pressure region is assumed as the plasma near the focusing position and othor low pressure region is settled for the surrounding air. The plasma energy, which was obtained measurement of a pulsed laser energy, is converted into the high pressure region.

4 Results and discussion

Figure 5 shows the results from the experiments and the numerical simulation. The time underwritten indicates the delay time from the laser irradiation. The upper figures show the isopycnic diagram from the numerical results and the lower figures show the color schlieren images from the experiments. A plasma is indicated by a white part near the center of the structure model in the lower side of Fig. 5, and the two-dimensional shock wave propagates from the center of a plasma to the open exit with the reflections from the inner walls of the structure.

Figure 6 shows the time variation of the shock wave propagation (x-t diagram). The horizontal and the vertical axes are the time from the generation of a plasma and the distance between the inner wall of the structure model and shock wave front on the center axis. The lozenge points are obtained from color schlieren images and the error bar is represented by 3σ. The dashed line shows the result from the numerical simulation. In the figure discrepancy is still presented, where the detailed comparison and the improved check for our CFD should be continued.

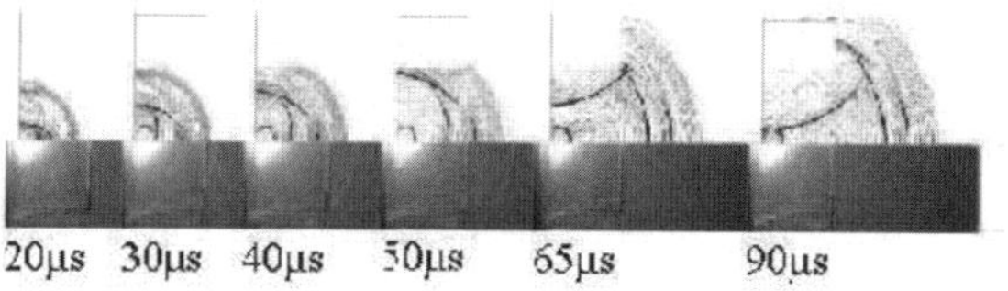

Fig. 5. Results form the experiments and the numerical simulation

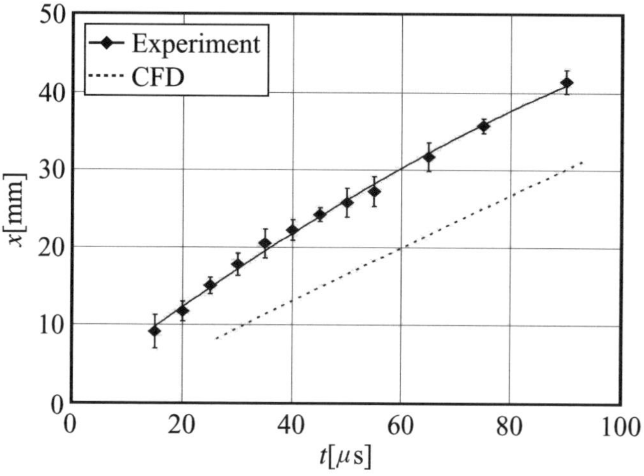

Fig. 6. The time variation of the shock wave propagation

5 Conclusion

In this study we have performed the visualization experiment for a focusing pulsed laser induced shock waves in a rectangular structure model by using the color schlieren method. We have also performed the numerical simulation for two-dimensional Euler equation. From our results the following points can be concluded.

1. The experimental set up with an acrylic rectangular structure model is very convenient to visualize the propagation of laser induced shock wave discharged from the open side, and the interaction between the inner wall of the structure model and the shock wave.
2. We confirm that the numerical and the experimental results show similar propagation characteristics. The propagation distance, however, of the shock wave by the numerical simulation is shorter than that by the experimental results. The given parameters seem to be insufficient, which should be checked further.
3. In our preliminary measurement of the energy of a plasma, we sometimes measured the energy of a generated plasma when the energy of a focused laser exceeds the critical energy of air breakdown before laser focuses completely. The result of the numerical simulation using the data from preliminary measurement as the parameter does not indicate the agreement with the experimental results. Therefore, the energy of a plasma obtained from our preliminary measurement is not enough.

4. The accuracy of the numerical simulation can be improved by obtaining the precise plasma energy which accords with the experimental results. The accurate prediction of the plasma energy should be performed.

References

1. T. Hashimoto, K.Maeno, T. Matsumura, Y. Nakayama, K. Okada, and M. Yoshida: Propagation of laser-induced blast wave discharged from open end of plass structures. In: *23rd ISSW Abstracts* (2001) pp 202

Simulation of strong blast waves by simultaneous detonation of small charges - a conceptual study

A. Klomfass, G. Heilig, and H. Klein

Fraunhofer Ernst-Mach-Institut, Eckerstr.4, 79104 Freiburg, Germany

1 Introduction

Terrorist bombings are considered a major threat to government and corporate buildings and to sensitive industrial structures. A blast resistant design of such buildings and structures is in certain cases mandatory. The same holds for military vehicles, where attacks with so-called Improvised Explosive Devices (IEDs) are in many of today's operations the most frequent threat. The masses of high explosives used in terrorist attacks typically range up to 1000 kg, a mass that can be stored and transported in a small truck.

Full scale tests with such large masses of explosives can rarely be realized for research issues or as part of a design process. Thus numerical simulations and model scale experiments are employed whenever possible. However, full-scale tests are eventually demanded for prototype qualification.

In the present paper a concept is analyzed which offers the potential to experimentally simulate strong blast waves by the simultaneous detonation of a large number of small charges in a long tunnel. As the charges are arranged in a plane perpendicular to the tunnel axis the concept is in the following called "blast panel". It is characterized by a number of parameters: the distance between the charges, the mass and the shape of the charges, the dimensions of the panel and the distance to the test object within the tunnel. The concept can be considered effective if the blast wave parameters of the simulated single charge event are reproduced well and if the total amount of HE is well below the mass of the simulated single charge.

The response of blast-loaded structures depends on two major parameters: the peak pressure and the positive impulse (overpressure-time integral) of the incident blast wave. The dependency of these parameters on charge mass and distance is well known for the detonation of a single spherical charge. The blast waves generated by the blast panel are thus to be evaluated and assessed in terms of these quantities.

The performance of the blast panel concept is here analyzed by numerical simulations. A number of simplified (2D) computations have been carried out in order to evaluate the effects of the various design parameters on the resultant blast parameters. Relations for peak pressures, positive impulse and decay time (positive overpressure duration) are derived from these simulations. As a major result, a quantitative comparison between the detonation of a single spherical charge and the blast panel in terms of blast parameters is obtained. This comparison shows in which range of masses and distances the concept can be used for practical blast simulation.

2 Ideal Blast Waves and Failure Diagrams

The detonation of a spherical charge in a free field (or a semi-spherical charge on rigid planar ground) causes at a distance R a static-overpressure time-curve which can be analytically approximated by the Friedlander-function

$$p(t) = P\left(1 - \frac{(t - t_a)}{\tau}\right) e^{-\alpha(t-t_a)/\tau} \tag{1}$$

where P and τ denote the peak overpressure and the positive time duration. The wave form parameter α determines the shape of the decaying part of the wave. Time integration of this curve in the interval $[t_a, t_a+\tau]$ renders the so-called positive overpressure impulse I. The dependencies of the blast-wave parameters P, I, τ, α on charge mass m and distance R have been examined thoroughly for spherical TNT charges. Correlations are found e.g. in [1].

The response of thin-walled structures such as building walls, windows and other shell-like or beam-like objects to blast-loading can be calculated on a simplified basis from 1DOF-models [2]. Such models assume that the response is dominated by a single mode, which typically is a characteristic displacement or deflection of the structure. Further it is assumed, that failure of the structure occurs at a critical amplitude in the considered mode. For a certain class of load functions it can be shown that the maximum amplitude, and thus the occurance of failure, depends on two load parameters only: the peak load and the total supplied impulse. Alternatively the same dependency can be expressed in terms of the peak load and the characteristic loading time (the latter in relation to the natural frequency of the loaded structure). This fact allows the definition of "PI"-diagrams or failure diagrams for structures. An example is given in figure 1. Parameter combinations that lie below the failure curve are considered safe, while such values above the curve are considered hazardous. Given a combination of peak load and impulse (or time duration) the criticality of the loading to the structure can thus be easily assessed.

In accordance with this concept of failure curves it can be stated, that, in order to simulate the effects of a detonation on a certain structure, the two parameters peak load and transferred impulse must be replicated by the applied testing method.

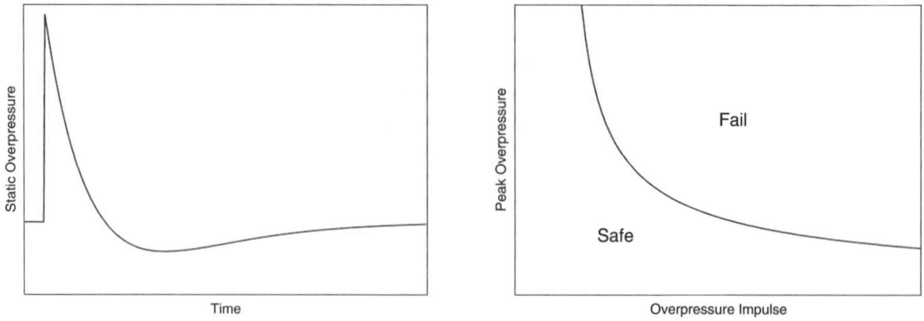

Fig. 1. Pressure-time curve for an ideal blast wave (left) and example of an PI diagram (right)

3 The Concept of a Blast Panel

Figure 2 illustrates the concept of a blast panel as considered in this work. The arrangement consists of a regular planar matrix of identical high explosive charges, which shall be ignited simultaneously. At the position of the loaded object in a distance x to the panel, it is desired to obtain a planar wave with a spatially homogeneous flow field behind the shock front. Obviously, these conditions can be satisfied in any distances only if the blast panel extends over the entire cross section of the tunnel as indicated in the figure. The shape of the tunnel cross section does not affect the generated waves, if the characteristic tunnel size is large compared to the average distance between charges.

In this study we do not consider the effects of wave reflections at the tunnel ends. Thus the findings of this study hold only, if the panel is placed somewhere in mid tunnel position or immediately at a solid closed end of a sufficiently long tunnel.

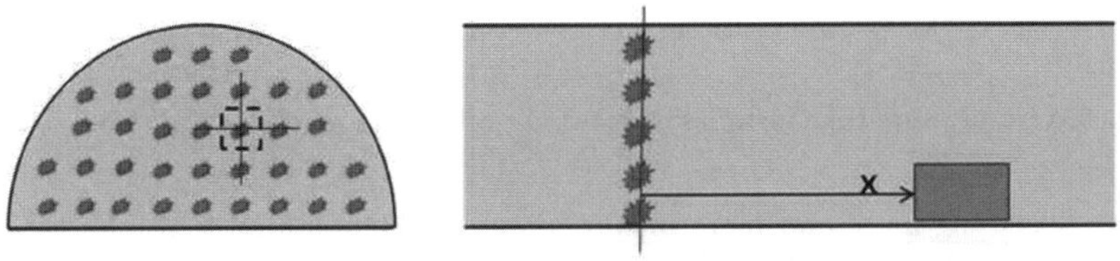

Fig. 2. Arrangement of charges in a tunnel cross section (here semi-circular) and illustration of an elementary cross section (left), tunnel with blast panel and test object (right)

4 Numerical Analysis

The regular arrangement of charges within the panel allows a simplified analysis, where only a single elementary cross section as shown in figure 2 is considered. This elementary cross section has rectangular shape. For the analysis we approximate this by a circular cross section with equal area, which allows a reduction to an axially symmetric 2D problem. Using this approximation we carried out a parameter study, where the charge shapes (cylindrical charges with varying L/D in the range L/D=0.3 to 10), and the charge masses were varied. As the influence of the charge shape was found to be small all further computations were carried out with L/D=1 charges.

The computations were carried out with the AUTODYN code, which solves the conservation equations for inviscid gases by explicit time integration. Air was modelled as an ideal gas, the TNT detonation products where described by a JWL equation of state according to the code standard material library. All computations where carried out in two steps. The detonation process and the initial blast wave propagation was in the first step calculated with a high resolution within a small computational domain. Before the blast front leaves this domain, a re-mapping to the final computational domain was carried out and the computation continued. An elementary cross section of 1 m^2 was considered in the study. As can be seen in figure 3 the pressure time curves exhibit strong oscillations at close distances (for the considered cross section up to distances of about 6m), while at larger distances the pressure curves become rather smooth due to

Table 1. Dimensions and Resolution of Computational Grids used in the study. R: outer radius of axially symmetric grid, L total length of grid, ΔR and Δx: resolution in radial and axial directions.

Step 1	R=0.50m L=0.5m ΔR=0.25cm Δx=0.25cm
Step 2	R=0.56m L=15m ΔR=1.2cm Δx=1.2cm - 5cm

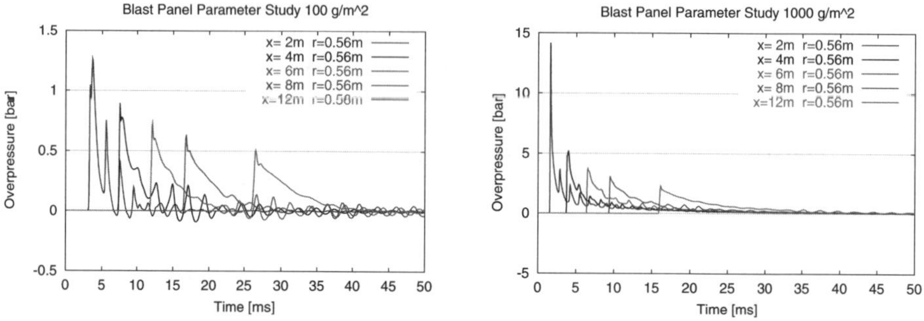

Fig. 3. Overpressure-time histories at selected positions in an elementary cross section

the wave becoming more and more planar. From the computational results the charts in figure 4 were obtained, which show the dependency of peak static overpressure and positive impulse on charge mass per area and distance. The decay time was here determined as the time difference between arrival time and the time when 90 % of the positive impulse are accumulated. For further evaluation the analytical correlations given below were derived from the discrete data. In these equations m_A denotes the charge mass per area in kg/m^2. Further units used are bar, ms and m.

5 Comparison with single Charge Blast Waves

Using the above correlations and the corresponding correlations for blast parameters of single spherical charges from [1], the following PI diagram was constructed. Shown in grey are the parametric curves for the single spherical charge in free field and shown in red are the parametric curves for the blast panel in mid position within a long tunnel. The diagram can be used as follows: for a given combination of mass and distance of a single spherical charge in free field the values of peak overpressure and positive impulse can be determined through interpolation between the grey curves; referring to the red curves the values of mass per area and distance can be obtained for a blast panel which renders the same blast parameters (peak static overpressure and positive static overpressure impulse). Alternatively the given correlations can be inverted to obtain the mass per area and distance values for given blast parameter values. Some selected matches are summarized in the table below. As the listed examples show, the decay times of the blast waves in the tunnel and those of the ideal blast waves do not match in all cases, although the impulses and peak pressures are the same. This is due to different wave shapes. The latter result from of the fact, that the ideal blast wave is spherical and the wave in the tunnel is planar. Furthermore it can be seen, that some listed cases may be unrealistic from a practical point of view as large distances to the blast panel (50m and more) and thus very long tunnels are required. Also those cases were very close distances

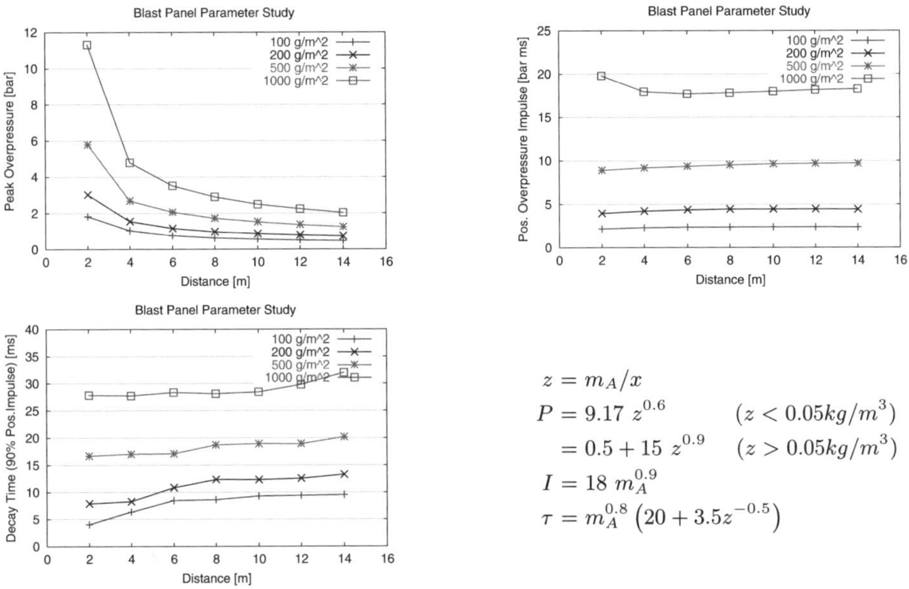

$$
\begin{aligned}
z &= m_A/x \\
P &= 9.17\ z^{0.6} && (z < 0.05kg/m^3) \\
&= 0.5 + 15\ z^{0.9} && (z > 0.05kg/m^3) \\
I &= 18\ m_A^{0.9} \\
\tau &= m_A^{0.8}\left(20 + 3.5z^{-0.5}\right)
\end{aligned}
$$

Fig. 4. Dependencies of peak overpressure P, positive impulse I and decay time τ on specific charge mass m_A and distance x. Correlations in units of bar, ms, kg, m as derived from displayed data. Range of validity: $z < 0.5kg/m^3$

Table 2. Blast Panel parameters matching ideal spherical blast waves. Left part: Mass and Distance of free field spherical charge, peak static overpressure, pos. static overpressure impulse and duration for ideal blast wave, right part: required charge mass per area and distance to blast panel for matching peak overpressures and impulses. Due to different wave shapes the decay times do not match exactly

	Ideal Blast Wave					Blast Panel		
no.	m[kg]	R[m]	P[bar]	I[bar ms]	τ[ms]	m_A[kg/m^2]	x[m]	τ[ms]
1	100	10	1.70	3.23	5.8	0.150	2.5	7.5
2	100	20	0.37	1.77	10.3	0.075	15.5	9.0
3	100	50	0.09	0.71	16.4	0.028	64.0	10.5
4	200	20	0.61	2.77	10.9	0.120	11.6	10.2
5	200	50	0.12	1.13	19.2	0.045	63.0	12.8
6	500	20	1.18	4.90	11.5	0.230	7.1	12.3
7	500	50	0.19	2.09	22.5	0.090	58.0	16.0
8	1000	20	2.03	7.35	11.7	0.370	4.7	14.5
9	1000	50	0.28	3.30	24.7	0.152	49.5	18.5

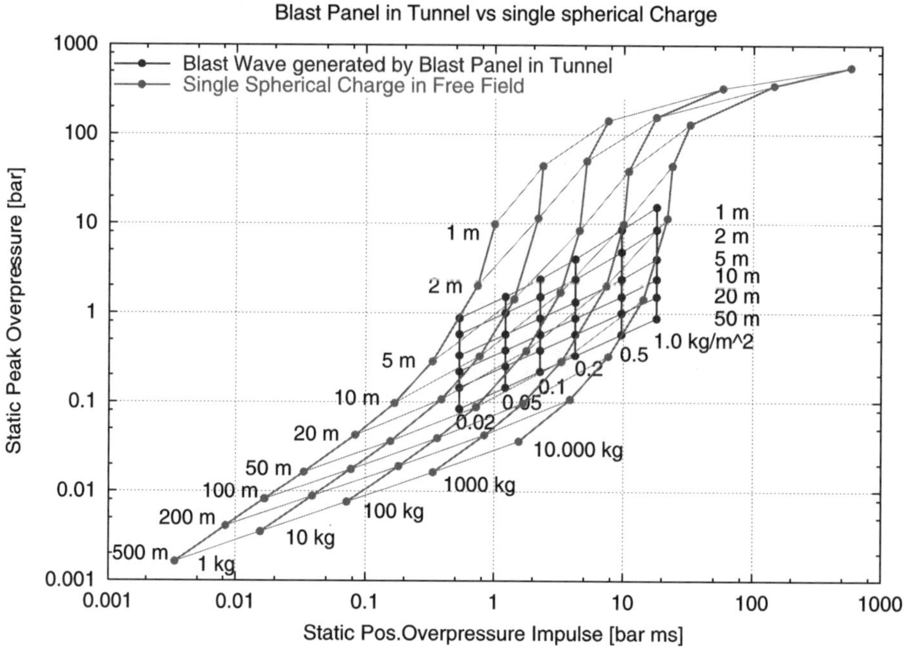

Fig. 5. PI chart of ideal free field blast waves (large array of curves) and blast waves generated by a blast panel in a long tunnel (smaller array)

to the blast panel are required (no.1, no.9) seem unrealistic. This is due to the fact, that, in order to obtain a planar and sufficiently homogeneous wave at such close distances, the charges must be very densely distributed within the panel. For comparison we refer to figure 3: with one charge per square meter the resulting wave becomes sufficiently homogeneous only at a distance of about 8 m. From this we can conclude, that in order to obtain a sufficiently homogeneous wave at e.g. 4 m distance the charges must be as close as 0.5 m from each other within the panel. This may require an exceedingly large number of very small charges. Nevertheless are there cases (e.g. no.2, no.5, no.6) which appear to be of practical use. This holds for both the required specific charge mass and the distance to the panel.

The presented results were derived on the basis of a simplified analysis, i.e. with respect to an elementary cross section of circular shape. For selected cases we also carried out full 3D simulations of an entire tunnel with 75 m^2 and up to 66 evenly distributed charges. The results of these computations were in a reasonable agreement with those of the simplified analysis.

References

1. G. F. Kinney, K. J. Graham: *Explosive Shocks in Air*, Springer-Verlag 1985
2. W. E. Baker, P. A. Cox, P. Westine: *Explosion Hazards and Evaluation*, Elsevier Scientific Publishing, 1983

Small-scale installation for research of blast wave dynamics

N.P. Mende, A.B. Podlaskin, and A.M. Studenkov

A. F. Ioffe Physical Technical Institute RAS, St.Petersburg, Russia

Summary. A compact, safe and low-cost blast tube has been developed at the Ioffe Institute. The tube allows one to solve a variety of tasks in physical modeling blast wave dynamics. The diagnostic complex of the installation includes the means of local measurements of pulse pressure and the optic means of image recording.

1 Introduction

Experimental investigations of blast waves in air have a long history due to the obvious practical importance of the problem. In connection with the known features of today's life the experimental tests in this field find financial and organizational obstacles. The use of classical shock tubes is possible but costly, and the use of explosive materials is limited by safety rules and normally require travels to faraway proving grounds, that is also costly. The present paper introduces a new blast tube created in the Ioffe Institute for investigation of blast waves in their interaction with various objects, which tube is characterized by low maintenance costs and provides full safety when working indoors.

2 Working principle and design of the installation

The tube models one-dimensional propagation of the blast wave. It consists of a blast chamber at one end, of a channel, screwed to the chamber, and of a test section/nozzle, screwed to the channel at the opposite end. (see Fig.1). The channel of the tube consists of one or more sections (400 to 700 mm in length), put together with the help of threaded junctions. For the tasks modeling flat symmetry of the wave a rectangular channel 30×30 mm in cross-section is used. For the tasks modeling axial symmetry of the wave the channel with round cross-section of 12 to 35 mm in diameter is used. The latter can be ended by one of a set of conical nozzles, which can be screwed to its end. The key feature of the design of the installation is the use of electric igniters as the source of the blast wave. The factory-made igniters provide highly synchronous operation (without any additional explosive). The "EKV-2M" devices (Fig.2) have a threaded shell, an exploding wire inside it (connected between the side wall of the shell and the central contact point at the end wall) and a portion of the TNRS explosive 0.9 grams under the protective foil. The principle of electric pulse ignition provides better (in comparison with mechanical ignition) synchronization of several explosions in one experiment among themselves (guaranteed accuracy is $1 \mu s$) and with other processes controlled in the experiment. Threaded shell of the EKV provides easy loading of the breech block of the tube. Electric circuitry and commutative fittings distribute equal pulses of current among all charged igniters in the experiment. The igniters are mounted in radial direction to send the force of flame towards the axis. This prevents the products of explosion from flying fast along the direction of propagation of the blast wave which should be formed evenly in the tube channel as a result of this explosion. The blast wave is normally formed by simultaneous explosion of several igniters (3 to 12) and propagate along the tube to the test section. The strength of the wave can be varied by quantity of the simultaneously exploded igniters and by length of the channel, and usually counts some tens of atmospheres of pressure at the front of the wave.

The test section of the installation, geometrically corresponding to the square 30 × 30 mm channel, is flat, formed by two parallel glass windows. Its internal geometry can be varied with the help of a set of removable inserts, which form different channels or nozzles between the windows. Those windows are designed for the use of optical methods of diagnostics of the flows. In the cases of study of interaction between blast wave and solid bodies, the tested bodies also can be easily placed in the test section. For the round channel, some conical nozzles are screwed instead of the flat test section.

The installation is supplied with a system of diagnostics. For local pressure measurements, piezo ceramic pressure transducers are used in 2, 4 or 6 signal transmission lines, with the appropriate electronics for static pressure recording. For integral registering of flow pattern in the test area (e.g., blast wave – solid body interaction), a schlieren device IAB-451 (with the view field 200 mm in diameter) and high-speed photography means are used.

3 The use of the installation

The blast tube has been put into operation in 2003 and has been used in several experiments. As examples of its work, some results are presented below.

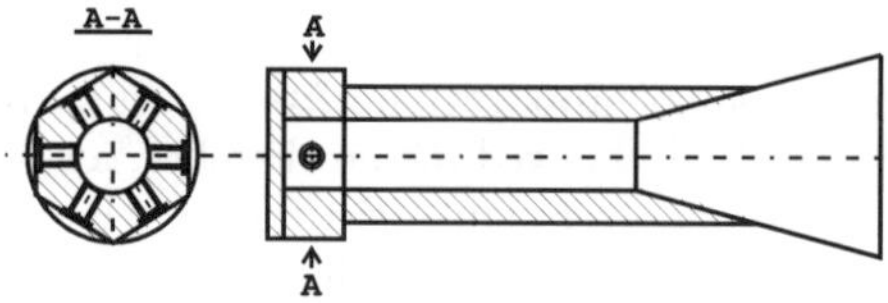

Fig. 1. The general scheme of the blast tube. The cross-section A-A demonstrates the blast chamber with the threaded openings for the igniters

Fig. 2. Electric igniters EKV-2M. Overall view

First case: the blast tube with a round channel 34 mm in diameter and with a conical nozzle (full angle is 30°) has been used for investigation of the flow field in the area of interaction between a supersonic flying body and a blast wave propagating from a side of the body's trajectory. The strength of the blast wave in the moment of interaction, was evaluated by measurements of axisymmetric interferogram. The jump of gas density at the front of the blast wave was estimated as approximately 1.5 (when exploding 3 igniters). The task has been fulfilled at the ballistic range of the Ioffe Institute. The scheme of the experiment can be seen at Fig.3. The body, accelerated by a powder gun flied at the velocity approximately 1250 m/s. The synchronization of explosion is provided by a system of contactless sensors (light barriers). The flowfield visualization is performed by an optical interferometer, constructed on the base of a schlieren device IAB-451. Photographic recording was implemented onto a wide (320 mm) film with the use of ruby laser OGM-20. An example of experiment result is shown in Fig.4. This series of results has been used for comparison with correspondent results of CFD calculation of the similar problem in the form of "computer visualization".

Second case: the investigation of pressure history on a wall at the end of the cannel of the blast tube, when this wall is covered by samples of porous material. The square channel has been used, the pressure signal from the piezo transducers recorded by electronic oscilloscopes. Typical examples of results are demonstrated in Fig.5.

Polyurethane foam samples were tested under blast wave loading generated from a single detonator explosion over a 1100 mm long channel.

In the case presented in Fig.5a, there was no air gap between the foam sample and the end wall. The first curve in Fig.5a (graphs are obtained by digitalization of experimental oscilloscopic curves) corresponds to the test without any porous sample, the second one - to the test of 90 mm long sample pressed to the end wall. This graph demonstrates a known effect of pressure jump amplification in such conditions. A different shape of

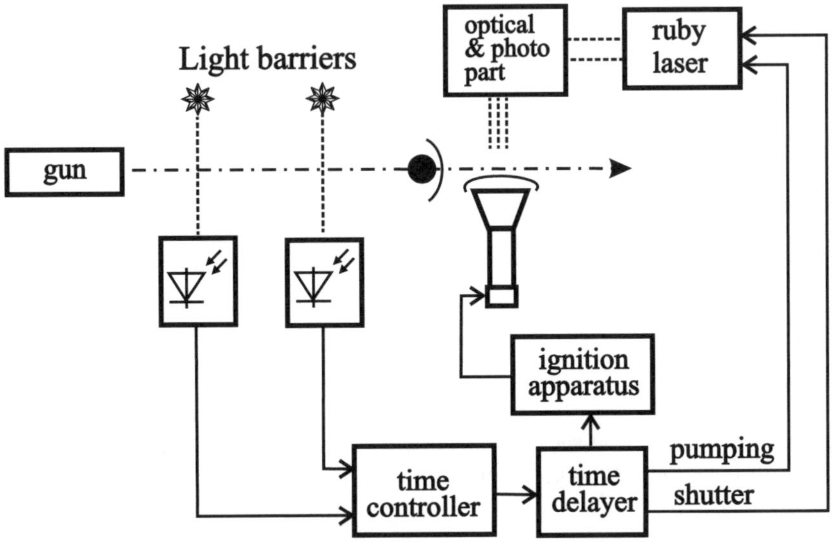

Fig. 3. Schematic of an experiment at a ballistic range

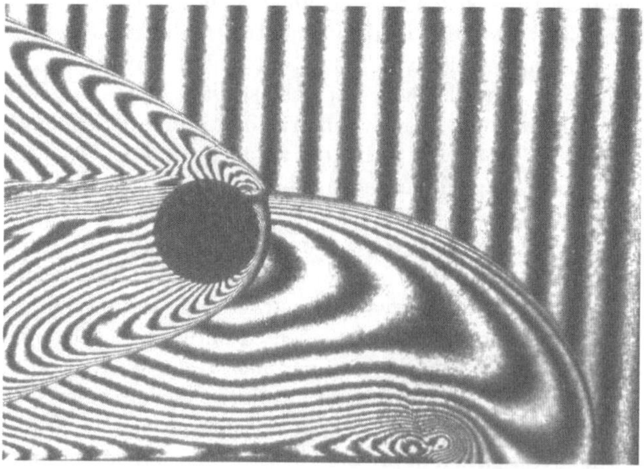

Fig. 4. Result of experiment from Fig.3. Interferogram of a lateral interaction of blast wave and moving body

curves can be seen in the condition of air gap present between porous material and the end wall. In Fig.5b a graph made of three test results is shown. The first curve, as before, presents wall pressure history with no foam. Curves 2 and 3 correspond to the tests with polyurethane samples 15 and 45 mm long. Air gap width in those tests was 5 mm. The analysis of the results give a qualitative correspondence to the effects reported in [1,2]. It is not possible to analyze the quantitative correspondence because of important difference in physical modeling methods. In the noted papers classical shock waves had been generated in shock tubes. They are principally different from blast waves in our case. Short positive phase of the pressure pulse (and even with negative phase) provides decompression waves in the sample under test, which influence substantially the dynamics of interaction process. In [2] such effects used to be seen in cases of very long samples of foams (in classical shock tube the gas parameters behind the shock wave keep constant 2 orders of magnitude longer than in blast tube). Nevertheless, the fact of qualitative correspondence of the results proves correctness of the method of physical modeling and proper interpretation of the results of this series of experiments, which are being conducted for investigation of blast protective properties of different materials and structures.

4 Conclusion

So it may be concluded that the experimental installation presented in this paper combine a number of favorable handling features, like the following:

- Low maintenance costs (no high pressure gas holders, compressors, mechanisms);
- Easy handling and safety in use (factory-made devices are used for explosion, with very small amounts of explosive enclosed inside a reliable shell);
- Compact design (can be arranged on a standard workbench or desk);

– Quick recharging (duration of a working cycle is a few minutes, can be operated by one man).

Meanwhile, as any tool operated with explosive materials, the installation can be handled only by qualified and certified work staff and only if the organization operating it has a license for work with explosive materials.

References

1. L.G.Gvozdeva, J.M.Faresov, V.P.Fokeev: 'Interaction of air blast waves with porous and compressible media' JPMTF # 3 111–115 (1985)
2. B.E.Gelfand, A.V.Gubanov, E.I.Timofeev: 'Interaction of air blast waves with porous screen' MJG # 4 79–84 (1983)

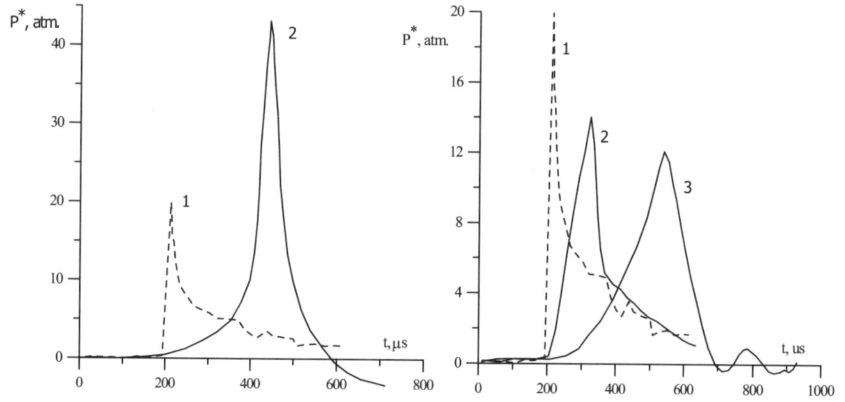

Fig. 5. Pressure history at the end wall of the blast tube behind a porous screen. (**a**) Case of no air gap between the porous screen and the wall. (**b**) Case of 5mm air gap between the porous screen and the wall

Strong blast in a heterogeneous medium

B.I. Palamarchuk

Paton Welding Institute, 11, Bozhenko St., 03680, Kiev, Ukraine

Summary. The paper is devoted to the issues of the theory of Strong Blast in a heterogeneous medium, having a wide range of inner degrees of freedom. Unlike the classical definition of the Sedov-Taylor problem, in which the shock wave is simulated by mathematical rupture, in this paper the shock wave is considered in terms of synergy, as a partially ordered localized structure, in which the range of inner degrees of freedom determines the dissipative properties of the medium. An important feature of the derived family of solutions of the problem of Strong Blast with power non-linear dependencies is presence of physical indeterminateness of development of irreversible shock-wave processes, characteristic for the processes of self-organizing in thermodynamically open systems with different physical nature. Special attention in the paper is given to the central solution of the problem of Strong Blast (PSB), derived in the form of special logarithm-hyperbolic functions, involving Fibonacci structural invariants. Comparative analysis of the theoretical results and earlier obtained transformations and automodel solutions of the problem of Strong Blast in a heterogeneous medium in the classical definition has been performed. It is shown that the earlier obtained for the heterogeneous medium transformations of classical equations and solutions of PSB are approximate in the general case.

An important feature of dispersed gas-liquid media, by which they differ from gas, is the fact that the dissipativeness of the medium is related not so much to gas molecule collisions, as to complex effects of interaction of the gas and condensed phases. This generates a wide range of inner degrees of freedom in the shock wave, creating a diverse spectrum of shock and detonation wave structures, not realizable in homogeneous media. Earlier obtained solutions (2-6) of the problem of Strong Blast in heterogeneous media containing an incompressible phase, revealed important features of the influence of incompressible phase ε volume fraction depending on spatial symmetry. Equations (4, 6) correlating the coordinate systems have the following form:

$$\begin{aligned} \bar{\rho} &= a^{2(\nu-1)}(1-\varepsilon)^{-1}\rho, \\ \bar{u} &= a^{1-\nu}u, \\ d\bar{r} &= a(1-\varepsilon)dr + a^{\nu}\varepsilon u dt, \\ d\bar{t} &= a^{\nu}dt, \end{aligned} \tag{1}$$

$$\left(\frac{\partial \bar{t}}{\partial r}\right)_t = 0, \quad \bar{p} = p, \quad \left(\frac{\partial}{\partial t} + u\frac{\partial}{\partial r}\right)a^{\nu-1} = 0, \tag{2}$$

where $a = r/\bar{r}$.

In the general case of arbitrary ν, coefficient a is the function of spatial symmetry and expresses the non-linearity of transformation of spatial coordinates at the change of medium volume. However, complete analogy is preserved only for plane symmetry at $\nu = 1$, when $a = 1$ can be assumed. For spherical and cylindrical symmetries the transformation does not yield an accurate solution. The analogy of mathematical recording of the laws of the impulse and full load is preserved, but the analogy of recording for the equation of continuity is disturbed, so that the full mass balance is not preserved. In particular, for spherical symmetry at $\Gamma = 1.4$ and $\varepsilon = 0.5$ the error is up to 10%, becoming greater at increase ε.

In dimensionless values the variables of the flow in cross-hatched coordinates are expressed as:

$$\bar{R} = \bar{\rho}/\bar{\rho}_0^1, \quad \bar{V} = \bar{u}/\bar{D}, \quad \bar{P} = \bar{p}/\bar{\rho}_0^1(\bar{D})^2,$$
$$\bar{\eta} = \frac{\bar{r}(t)}{\bar{r}_f(t)},$$
$$\bar{\chi} = \left(\frac{r_f}{\bar{r}_f}\right)^{\nu} \cdot \frac{\bar{r}_f}{\bar{D}}, \tag{3}$$

here:

$$d\bar{t}_1 = \left(\frac{r_f}{\bar{r}_f}\right)^{\nu} d\bar{t}_f = dt, \quad \bar{t}_1 = t, \quad \rho_0 = \bar{\rho}_0\left(1-\varepsilon_0\right)\left(\frac{r_f}{\bar{r}_f}\right)^{2(\nu-1)}. \tag{4}$$

At transition to $\bar{t}_1 = t$ system, equations of constraint become:

$$(1-\varepsilon_0)\,\bar{V} = \left(\frac{\eta}{\bar{\eta}}\right)^{\nu-1} V, \quad \bar{P} = \frac{P}{1-\varepsilon_0}, \quad \varepsilon = \varepsilon_0 R,$$
$$\bar{R} = R\left(\frac{\bar{\eta}}{\eta}\right)^{2(\nu-1)} \cdot \frac{1-\varepsilon}{1-\varepsilon_0}, \quad \bar{t}_1 = t, \tag{5}$$
$$\left[\bar{\eta} - \left(\frac{\eta}{\bar{\eta}}\right)^{\nu-1} \cdot \frac{1-\varepsilon}{1-\varepsilon_0}\eta - \varepsilon\bar{V}\right] dt = \bar{\chi}\left[d\bar{\eta} - \left(\frac{\eta}{\bar{\eta}}\right)^{\nu-1} \frac{1-\varepsilon}{1-\varepsilon_0} d\eta\right]$$

Thus, solving a similar problem of Strong Blast in gas, we will find cross-hatched variable flows behind the shock front $\bar{R}$, $\bar{V}$, $\bar{P}$ and will determine approximate values R, V, P, using (6).

The causes for violation of the analogy of movement of the heterogeneous medium and gas at $a = 2, 3$ can be readily established. For this purpose it is sufficient to write the above transformations (6) in the following form:

$$V_1 = \mu\frac{V}{1-\varepsilon_0}, \quad P_1 = \frac{P}{1-\varepsilon_0}, \quad R_1 = \frac{1-\varepsilon_0}{\mu^2} \cdot \frac{R}{1-\varepsilon_0 R}, \quad \mu = \left(\frac{\eta}{\eta_1}\right)^{\nu-1},$$
$$\eta\mu = \eta_1\left(1+\varepsilon_0\left(\mu^2 R_1 - 1\right)\right) - \varepsilon_0\mu^2 V_1 R_1, \tag{6}$$

$\eta_1 = \frac{\mu}{1-\varepsilon}\left[(1-\varepsilon_0 R)\,\eta + \varepsilon_0 V R\right]$ or $\eta_1 = (V-\eta)\,\frac{\mu}{1-\varepsilon}\,(1-\varepsilon R)$. Then, for the equation of continuity:

$$(V-\eta)\frac{dR}{d\eta} + \frac{R}{\eta^{\nu-1}}\frac{d}{d\eta}\left(\eta^{\nu-1}V\right) = (V-\eta)\,R' + RV' + RV\frac{\eta-1}{\eta} = 0 \tag{7}$$

for a two-phase medium, and

$$(V_1-\eta_1)\frac{dR_1}{d\eta_1} + R_1\left(\frac{dV_1}{d\eta_1} + V_1\frac{\nu-1}{\eta_1}\right) = 0 \tag{8}$$

for an ideal gas.

Using transformations, we obtain (6) from (7),(8). For this purpose, we will write equation (8) in derivatives of η, and will calculate the derivatives:

$$(V_1 - \eta_1)\frac{R_1'}{R_1} + V_1' + V_1\frac{\nu-1}{\eta_1}\cdot\eta_1' = 0, \tag{9}$$

prime sign is derivative of $\eta - \frac{d}{d\eta}$;

$$\frac{R_1'}{R_1} = -2\frac{\mu'}{\mu} + \frac{R'}{R(1-\varepsilon_0 R)}, \quad V_1' = \frac{\mu}{1-\varepsilon_0}\left(V' + V\frac{\mu'}{\mu}\right), \quad (\nu-1)\frac{\eta_1'}{\eta_1} = \frac{\nu-1}{\eta} - \frac{\mu'}{\mu},$$

where $\mu = \left(\frac{\eta}{\eta_1}\right)^{\nu-1}$.

From the transformations it is seen that $\mu = \left(\frac{\eta}{\eta_1}\right)^{\nu-1}$ can be presented as the function of η: $\mu = \mu(\eta)$, therefore $\mu' = \frac{d\mu}{d\eta}$ makes sense.

As a result, we have from (8):

$$(V-\eta)\frac{R'}{R} - 2(V-\eta)(1-\varepsilon R)\frac{\mu'}{\mu} + V' + V\frac{\mu'}{\mu} + \frac{\nu-1}{\eta}V - V\frac{\mu'}{\mu} = 0, \tag{10}$$

whence in view of (7): $\frac{\mu'}{\mu}(V-\eta)(1-\varepsilon_0 R) = 0$, which is satisfied only at $\nu = \mu = 1$.

Note that the derived solution is valid in the general case and for relaxing media. Influence of the volume fraction of the condensed phase on the laws of shock wave motion and on parameter value on the shock wave can be established analytically, without finding distributions of R, V, P. With this purpose, let us use expression for E_0 into the following form:

$$E_0 = \sigma(\nu)\rho_0 r_f^{\nu} D^2 \frac{(1-\varepsilon_0)^2}{\Gamma - 1}\psi,$$
$$\psi = \int_0^1 \left(\bar{P} + \frac{\Gamma-1}{2}\bar{R}\left(\bar{V}\right)^2\right)\bar{\eta}^{\nu-1}d\bar{\eta}. \tag{11}$$

Using the dimensional methods we will obtain the equation of shock wave trajectory:

$$r_f = \left(\frac{E_0}{\alpha\bar{\rho}_0}\right)^{\frac{1}{\nu+2}} \cdot \left(\frac{t}{1-\varepsilon_0}\right)^{\frac{2}{\nu+2}},$$

where

$$\bar{D} = \frac{2}{(\nu+2)(1-\varepsilon_0)} \cdot \left(\frac{E_0}{\alpha\rho_0}\right)^{\frac{1}{2}} \cdot r_f^{-\frac{\nu}{2}}, \tag{12}$$
$$\alpha = \frac{4\sigma\psi}{(\nu+2)^2(\Gamma-1)}.$$

As for discontinuities in the shock wave front the movement analogy is complete, let us establish analytical dependence $\alpha(\Gamma)$, considering the impact transition in a heterogeneous medium, having a wide range of inner degrees of freedom. In [4] Fibonacci structural invariants were used for this purpose, which are effectively applied in solving the dynamic chaos problems, and describing the processes of structural genesis in irreversible processes of different nature. In the case of the strong blast problem the main hypothesis of applicability of Fibonacci invariants is based on that at a distance from thermodynamic equilibrium the partially ordered structures, alongside the property of scale-time invariance, also have the property of self-consistent development of system parts as a whole. As is known, the first property is initially assumed, and follows from the self-simulating solution of the strong blast problem. The second property requires the presence of a large number of degrees of freedom in the shock-wave transition zone, which, however, cannot be allowed for in terms of the classical models. The derived non-linear equations, correlating the entropy production with thermodynamic state of the particle in the shock-wave transition zone allow introducing this lacking property into the mathematical model of the strong blast in a heterogeneous medium with $0 < u* < 1$ and determining the required and sufficient conditions, providing the self-consistent movement of continuum particles, compressed in the shock wave. Conditions leading to the central solution of non-linear equations using Fibonacci invariants and the solution proper have the following form:

$$\frac{\ln\left(\frac{4E_0}{muD}\right)}{\ln(u*)_{u*\to 1}} \to \mathrm{phi}, \qquad \frac{\ln u^*}{\ln\left(\frac{muD}{4W}\right)_{u*\to 0}} \to \mathrm{phi}, \tag{13}$$
$$\mathrm{Phi}\cdot\ln(v^* i^*)\,\mathrm{Phicosl}(s^*) - \ln(u^*)\cdot\mathrm{Phisinl}(s^*) = 0$$

where

$$u* = \frac{u}{D}, \quad v* = \frac{v}{v_0}, \quad m = \frac{4}{3}\pi r^3,$$
$$i* = \frac{\rho u}{\rho_0 u_0}, \quad s* = \frac{i*^2}{p*}, \quad p* = \frac{p}{p_0}, \quad \mathrm{Phi} = \frac{1}{phi} = \frac{\sqrt{5}+1}{2},$$
$$m = \frac{4}{3}\pi\rho(1-\varepsilon)r^3,$$

and Phi-functions are defined as follows:

$$\mathrm{Phisinl}(s*) = \frac{\mathrm{Phi}^{\ln s*} - \mathrm{Phi}^{-\ln s*}}{2}, \qquad \mathrm{Phicosl}(s*) = \frac{\mathrm{Phi}^{\ln s*} + \mathrm{Phi}^{-\ln s*}}{2}. \tag{14}$$

Here, index 0 denotes the parameters of the compressed medium state “optimum” in terms of structural genesis.

Equation correlating the medium parameters in a shock wave with relative velocity of movement of the compressed medium particles, does not contain the blast energy as a parameter, and, thus, expresses the dynamic properties of the continuum compressed in an irreversible shock-wave process.

Analytical expression for α has the following form:

$$\alpha = \frac{8\pi}{75}\cdot\frac{1}{\Gamma-1}\cdot\left(\frac{2}{\Gamma+1}\right)^{\mathrm{phi}\cdot\mathrm{Phitanl}\left[(\Gamma-1)\cdot e^{-\mathrm{phi}}\right]} \tag{15}$$

and closes the system of equations of the strong blast problem, thus allowing the singularity in the blast center to be avoided. Theoretical dependence and presence of earlier unknown symmetry of the strong blast problem is confirmed by the results of direct numerical calculations of Sedov – Taylor problem for specific values Γ in the range of $\Gamma - 1$ variation from 10^{-3} up to 10.

The derived solution is versatile and allows analyzing in the explicit form the parameters of shock waves both in gas and in dispersed gas-liquid media with an arbitrary volume fraction of the condensed phase. Comparison with Zeldovich model shows (fig. 1) that if the final compression of the medium is not allowed for, the solution of the strong blast problem lacks the symmetry, inherent to the derived analytical solution, and confirmed by the results of numerical calculations of the full system of gas-dynamic equations in Sedov-Taylor solution. The above solution allows more precise determination of variable

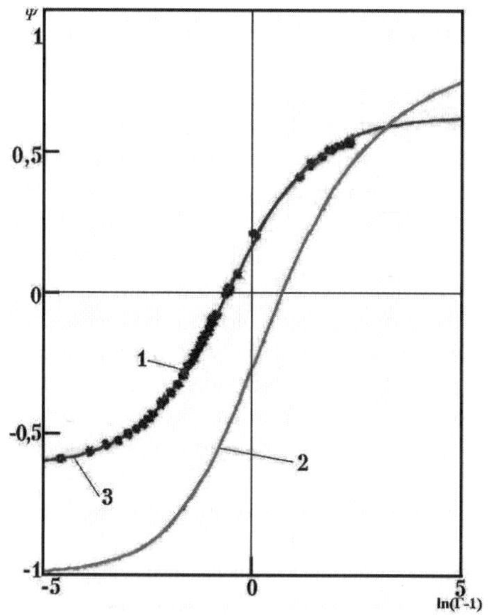

Fig. 1. Dependence of a dimensionless functional of the solution of the strong blast problem in limit case ($\varepsilon = 0$): 1 - Sedov solution, 2 - Zeldovich solution, 3 - solution, using Phi-invariants

distribution behind the strong shock wave front and finding the condition at which a complete analogy is observed between the motion of gas and gas-liquid medium. Investigation of Phi- functions let us to establish the following properties, useful for studying of self-organising of irreversible processes:

Property 1. Property of a reversibility:

$$\mathrm{Phisinl} = \frac{\mathrm{Phi}^{\ln x} - \mathrm{Phi}^{-\ln x}}{2} = \frac{x^{\ln \mathrm{Phi}} - x^{-\ln \mathrm{Phi}}}{2}. \tag{16}$$

Property 2. At $k \to \infty$:

$$\frac{\mathrm{Phisinl}(2k)}{\mathrm{Phicosl}(2k-1)} \to \mathrm{Phi}. \tag{17}$$

Statement 1.

$$\mathrm{Phicosl}^2(x) - \mathrm{Phisinl}^2(x) = 1. \tag{18}$$

Statement 2. In a five dimensional Euclidean space:

$$F(2k-1) = 2\cos(\alpha)\mathrm{Phicosl}\left(\mathrm{e}^{2k-1}\right), \tag{19}$$

where cos(α) is guiding the hypervector, carried out from a coordinates origin in a point (1,1,1,1,1).

At values $x = \mathrm{e}^k$, where $k \in Z$, the Phisinl(x) are generate even terms of a series of the Fibonacci, while Phicosl(x) generate odd terms of a sequence of the Fibonacci. The analytical formula for any term $F(k)$ of a sequence of the Fibonacci appears to be like this:

$$F(k) = \frac{2}{\sqrt{5}}\text{Phisinl}\left(\mathrm{e}^0\right), \frac{2}{\sqrt{5}}\text{Phicosl}\left(\mathrm{e}^1\right), \frac{2}{\sqrt{5}}\text{Phisinl}\left(\mathrm{e}^2\right), \frac{2}{\sqrt{5}}\text{Phicosl}\left(\mathrm{e}^3\right), \ldots, \text{Phisinl}\left(\mathrm{e}^{2k}\right), \text{Phicosl}\left(\mathrm{e}^{2k+1}\right) \ldots \qquad (20)$$

Thus entered Phi-functions allow to connect continuous and discrete properties of functional space of the conditions realised in natural systems of the various nature. Thus entered Phi-functions allow to connect continuous and discrete properties of functional space of the conditions realised in natural systems various both physical, and the biological nature.

As Fibonacci functions and invariants define growth, formation and stability of functioning of biological systems [5, 6], it allows to speak about unity and universality of proof self-organization of complex systems of the physical and biological nature far from thermodynamic equilibrium state. That is why to have a more profound understanding of the processes of self organization and self regulation in the thermodynamically opened systems it is essential to develop mathematics systems based on Fibonacci numbers together with Phi number as a structural invariant including the special functions based on them.

Let's notice, that the presented Phi-solution problem of Strong Blast is not the only thing. The family of solutions of a problem received reflects natural variability of processes self-organization. Determination of rules of selection of quasi-stable conditions and border of uncertainty of macroparameters in heterogeneous medium represent a challenge, important for deeper understanding of interaction of shock waves with biological objects and substance. Discussion of methods and results of the solution of this problem is beyond this short article.

References

1. Pai, S.I.., Menon S., Fan Z.Q.: Similarity solutions of a strong shok wave propagation in a mixture of gas and dusty paticles, Int.J. Eng. Sci., **18**, 12,(1980)
2. Kudinov, V. M., Palamarchuk, B. I., Vakhnenko, V. A.: Attenuation of a strong shock wave in a two-phase medium, Soviet Physics Doklady, **28**, (1983).
3. Vakhnenko, V. O., Kudinov, V. M., Palamarchuk, B. I.: An analogy of the motion of a two-phase medium consisting of an incompressible phase and a gas phase with the motion of a gas, Akademiia Nauk Ukrains'koi RSR, Dopovidi, Seriia A, (1983)
4. Palamarchuk, B. I.: Dynamics of shock-wave processes at explosion in gas-liquid media: thesis for the degree of Dr. of Sci. (Phys.- Math.) , RNTs "Kurchatov Institute", Moscow, (2003)
5. Bodnar, O. Ya.: Geometry of phyllotaxis, NASci of Ukraine, Dopovidi, **9**, (1992), pp.9-14.
6. Knyshov G.V., Nastenko E. A., Palamarchuk B.I., Palets B. L., Burkot A. N., Lukoshkina T. A., Rysin S. V.: Golden section invariants in proportions of arterial pressure values in normal and pathological conditions of bloodflow, Cardiovascular surgery, Annual Transactions of Ukrainian Association of Cardiovascular Surgeons, Part 10, Kyiv, (2002), pp. 121-125.

Part III

Chemically Reacting Flows

Atomized fuel combustion in the reflected-shock region

B. Rotavera and E.L. Petersen

Mechanical, Materials and Aerospace Engineering, University of Central Florida, Orlando, FL, 32816, USA

Summary. A liquid-spray injection system was implemented to conduct measurements of fundamental, heterogeneous combustion phenomena including activation energy, ignition delay time, and soot production of hydrocarbon fuels. The study of liquid-spray combustion holds significance in a wide range of industries and applications including combustor design and analysis in power generation gas turbines, in addition to efficiency and emission studies of developmental automotive and aviation fuels. Details on the design and implementation of the technique are provided. Repeatability of the autonomous liquid-spray system has been qualified, and measurements of the ignition delay times of 1.0-μm kerosene droplets at low pressures have been obtained.

1 Introduction

While shock tubes are traditionally used to study the combustion of gas-phase mixtures, the ability to use a shock tube to study the combustion of liquid fuels broadens their capability. However, the combustion of liquid sprays in shock tubes is a multifaceted problem comprised of complex chemical reactions, heterogeneous processes, and timing issues. Chief concerns to address in heterogeneous shock-tube studies are those of spatial uniformity and system repeatability. Although under utilized, the spray technique has been employed by others in past shock-tube combustion research.

A number of approaches have been attempted and successfully achieved over the past several decades. Early work in performing two-phase experiments involved injecting small quantities of fuel through the orifice of an automotive injector directly into the reflected-shock region where the fuel meets the high-temperature, high-pressure environment to study autoignition behavior and properties [1]. Suspending a premixed spray column prepared by an ultrasonic atomizer has also been employed. Using this approach, ignition delay time measurements and activation energies for cetane and its dependence on distance from the endwall were obtained [2], highlighting the importance of droplet breakup and subsequent micromist formation in the combustion process. Droplet breakup and combustion have been studied in years past using a variable-position, free-falling stream of relatively large, monodisperse droplets using the Rayleigh capillary instability technique [3] to examine aerodynamic shattering and subsequent combustion as a function of oxygen concentration in an otherwise inert medium for mean droplet diameters ranging from 950 μm to 1480 μm. Octane-painted quartz plates were employed to study the effect of pre-flame reactions with the incident shock wave on the combustion of octane and iso-octane [4]. Fuel quantities from 1 mg to 10 mg were applied to the plate, which provided measurement of temperature coefficients. It was inferred from this technique that pre-flame reactions controlled the course of flame propagation.

More recently, the effect of nitrate additives on the combustion properties of aviation fuels JP-7 and JP-8 has been studied using a single-stroke, positive-displacement fuel pump to pressurize and transfer liquid fuel through a diesel injector into the endwall region shortly after formation of the reflected-shock wave [5]. Nitrate and nitrate/water mixtures were added to ethanol to examine respective effects on soot formation at varying locations within the endwall region [6] by injecting 40 mg of the blended fuel into a 10 percent O_2/Ar environment. Additional fuel additive work using shock-tube endwall-injection includes the effect of water and alcohols on ignition delay time, soot production, and the nature of the combustion process [7] of propanol and tetradecane. An automotive injector producing 30-μm droplets was used to introduce the fuel and respective additives to the endwall region in 65-mg quantities, and it was concluded that the main effect of water addition to fuels influences the type and strength of the combustion leading to a proposed correlation with the amount of soot produced.

Environmental concerns have given rise to an interest in the combustion of vegetable oils and their esters as a replacement for hydrocarbon fuels in compression ignition engines. The incident shock wave was used to replicate the compression mechanism within a diesel engine and to combust small quantities of fuel [8]; it was concluded that the ignition delay time of the ester is longer at any temperature than the corresponding hydrocarbon. To account for residual gases in an actual diesel engine, endwall injection was extended to include the use of burnt gas [9], and driver-gas tailoring was employed to extend the test times past the time required for the entire range of the diesel combustion process (10 ms). Additionally, measurements of evaporation rates of n-dodecane and water using endwall injection of micron-sized droplets have been performed to determine rate constants behind shock waves [10].

With the limited amount of experimental data existing for the combustion of liquid fuels in shock tubes, the present work is directed toward obtaining fundamental measurements and arriving at empirical correlations of activation energies and other related phenomena. Altering various conditions such as spray concentration, mass flow rate, and injection time over a broad range of reflected-shock temperature and pressure are examples of possible interests. Furthermore, efforts contributing to the previous efforts on the effects of fuel additives on combustion and ignition properties of liquid fuels are being extended to synthetic and other alternative fuels.

Outlined in the present paper is the experimental approach taken in performing combustion studies with atomized hydrocarbon fuels. A description of the shock-tube facility is presented, including the process taken to atomize the fuel, after which a discussion of the experiments performed to date ensues.

2 Experimental Approach

Among the difficulties in performing two-phase shock-tube experiments are the need for precise timing capabilities and accurate control over the amount of fuel being studied. Since timing is of much concern in liquid-spray experiments, the approach in the present work focuses on dynamic temporal control over an automated electronic control system from which the fuel is atomized and injected. Due to the relatively short period of time existing for a given shock-tube experiment, emphasis was placed on producing droplets approaching the sub-micron level to minimize the time for vaporization and mixing. Utilizing the shock-tube technique in the study of liquid sprays holds the advantage that volatile liquid fuel blends can be studied without distilling any constituents due to varying vapor pressures. A secondary motivation in attaining small droplet sizes is the fact that micron-sized droplets, due to their low weight and small size, have high drag and tend to follow the motion of the bulk carrier gas, namely the shock-tube driven gas.

The University of Central Florida (UCF) liquid-spray combustion facility hosts a 6.1-m long shock tube with the ability of using a windowed endwall region for optical diagnostics and high-speed photography of ignition and combustion processes (Fig. 1). The internal dimensions for the circular driver and square driven sections are 7.6 cm and 10.8 cm, respectively. The driver section (L/D = 23.6) is supplied helium through an electronically enhanced solenoid valve employed to control the time taken to rupture the diaphragm and generate the shock wave. Shock-front velocity is precisely measured through the use of four, high-frequency piezoelectric pressure transducers (PCB 113A) mounted atop the driven section in conjunction with 120-MHz counters/timers (Phillips P6666) using respective distances between pressure sensors and the recorded time intervals. Computation of shock velocity serves as one of two input variables, the other of which being the initial pressure of the driven gas (P_1), to arrive at a characteristic solution of the Rankine-Hugoniot shock relations with thermodynamic properties from the Sandia database to define conditions within the reflected-shock region.

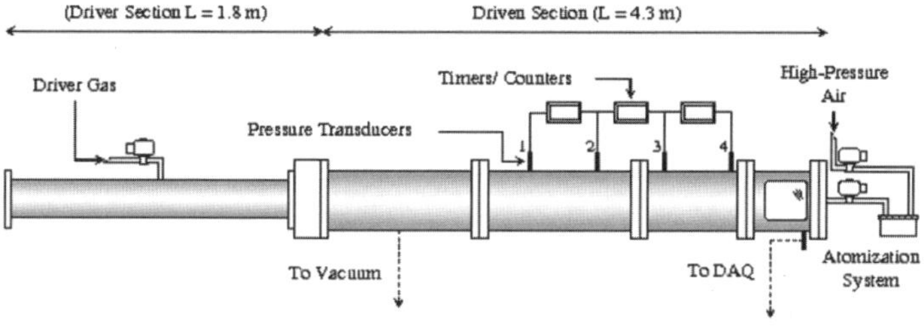

Fig. 1. Liquid-spray shock-tube facility

Sidewall pressure measurements are made using a 500-kHz quartz pressure transducer (Kistler 603B1) and recorded using two, 16-bit, 10-MHz Gage Applied Sciences data acquisition boards, and sidewall emission during the combustion event is tracked through the measurement of CH* chemiluminescence via a photo-multiplier tube (Hamamatsu Type 1P21) in a custom-made housing, the signal of which is amplified using a low-noise pre-amplifier (SRS SR560). A narrowband filter (10-nm FWHM) centered at 430 nm was utilized to capture the visible CH* chemiluminescence. Test times for the above mentioned shock-tube geometry are within 1-2 ms with time extensions up to 5 ms obtainable through the use of hydrocarbon-helium tailored driver-gas mixtures [Amadio (2006)]. Reflected-shock pressures up to 10 atm are obtainable.

Atomized fuel is introduced into the test region via a particle generator which is appended directly to the endplate such that fuel is sprayed along the longitudinal axis of the shock tube and is capable of producing droplets on the order of 5 μm or less in mean diameter depending on inlet pressure. The particle generator holds 72.6 mL of fuel in its reservoir, and micron-sized droplets are created by the expansion of compressed air through a 0.5-mm diameter orifice. The compressed air acts as a high-velocity jet creating an area of low pressure near the orifice which, in turn, causes the liquid in the reservoir to be drawn upward through the orifice and into the jet. Resulting from the high pressure of the air stream, the liquid fuel is then shattered into micron-sized droplets, and the compressed air then carries the droplets through the outlet tube and into the test region.

The automated liquid-spray system is comprised of the particle generator and two electronically controlled solenoid valves (Fig. 2) both of which are governed by a programmable, 4-channel pulse generator (BNC 565). The first channel on the pulse generator controls the driver-appended solenoid valve, while the second and third channels control the high-pressure air supply and a normally open valve, respectively. The normally open solenoid valve is programmed to close after expiration of the set delay period, precisely defining, for a given mass flow rate, the quantity of fuel being studied. The fourth channel enables control over recording intervals with a high-resolution camera (PCO 1200hs) capable of practical recording speeds up to 5000 frames per second and nanosecond exposure time.

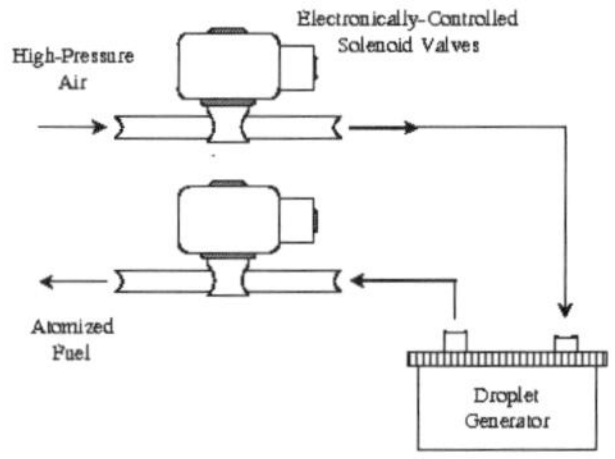

Fig. 2. Electronically controlled atomization system

The programming capability of the pulse generator along with the responsiveness of the solenoid valves provides for precise temporal control over the injection process enabling accurate quantification of the amount of fuel entering the test region. The particle generator utilizes a pressure regulator which allows for variation of the mass flow rate and spray concentration. Ranges of mass-flow rate from 1.83 mg/s to 10.80 mg/s and spray concentrations of 1.1×10^4 particles/cm^3 to 4.3×10^4 particles/cm^3 are obtainable for respective inlet pressures. With inlet pressure variation from 482 kPa to 34 kPa, the capability of a 1 μm to 5 μm mean droplet diameter range is available. Shock generation and atomization system initiation are governed by the pulse generator which regulates the supply voltage width and delay to each of the valves comprising the spray system, ensuring repeatable and controllable fuel injection.

Kerosene droplets produced by the atomizer form a cone-shaped structure and take on an arithmetic mean diameter of 1.0 μm using air at a supply inlet pressure of 482 kPa, resulting in a concentration on the order of 2×10^4 particles/cm^3. The spray cone produced by the particle generator measures approximately 7.6 cm in peak diameter and 15.2 cm in length at a mass flow rate of 10.8 mg/s.

3 Results

Due to the level of complexity inherent to heterogeneous shock-tube studies, repeatability within the test region is paramount to the study of liquid-spray combustion. A series of experiments was therefore conducted under similar test conditions to characterize the repeatability of the spray injection process. Pulsation of the atomizer for a period of 4 s for an inlet pressure of 482 kPa introduces 43 mg of kerosene and a pressure contribution to the initial pressure of the driven gas of 13 torr from the addition of the carrier air. An additional 1 torr of air was added to the shock tube prior to initiating the control system providing for an initial test-section pressure (P_1) of 14 torr.

Reflected-shock conditions (T_5, P_5) computed from the shock velocity and initial pressure are representative of the vapor component (air) since the liquid spray resides primarily near the endwall, and the composition of the medium through which the shock traverses is that of air. Measurements of the maximum spray cone length and the length of the driven section over which the shock wave develops reveal that the path is composed of air throughout 96 percent of its travel.

Ignition measurements using kerosene were performed over a narrow reflected-shock temperature range of 1330 - 1410 K and a pressure of 1.1 atm to demonstrate the technique and to check for repeatability. Figure 3 shows a typical plot of sidewall emission collected by the photomultiplier tube and the corresponding sidewall pressure measurement. Since the driven gas is air and therefore composed of diatomic gases, bifurcation of the reflected shock is evident, as in Petersen and Hanson [12]. Utilizing an endwall pressure measurement in future experiments rather than a sidewall measurement allows for removal of bifurcation effects and a clearer indication of time zero (i.e., the time of passage of the main normal shock wave).

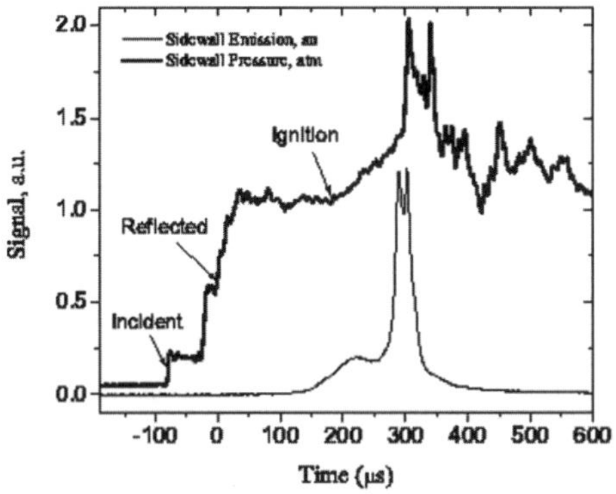

Fig. 3. Sidewall emission and pressure trace of kerosene droplets 1 micron in mean diameter. T_5 = 1408 K, P_5 = 1.14 atm; ignition time = 160 μs

CH* emission due to the combustion process behind the reflected wave has been repeatable, and although the behavior and duration of the combustion has varied slightly, the time at which the event initiates has been consistent. The ignition appears to occur in two stages, with the initial pressure and CH* increase due to ignition as defined in

Fig. 3, followed by a steeper pressure and CH* increase. Further characterization of the ignition behavior in the present system is underway for the baseline kerosene fuel.

Spatial uniformity has been ensured from the repeatability of emission and pressure traces and is being quantified in two different manners. The first technique involves laser scattering with a HeNe laser at various locations within a detachable endwall section, separate from the main facility, with exact internal dimensions as that of the shock-tube test section. The effect of time on both particle agglomeration and identification of void formation within the spray region are being examined. The time simulated in the detachable section is synonymous with that in the actual shock-tube experiment, where the spray is suspended for a finite amount of time prior to shock wave initiation.

Additionally, a Malvern spray particle analyzer is being used to examine droplet distribution. To visualize the injection and ignition process, the high-speed camera system is being utilized near the endwall of the shock tube to visualize the combustion process and to ensure that flame growth does not occur on the walls of the test region. A wide range of temperatures and pressures is being explored to examine the effect of the various conditions on the combustion type and progression of the liquid spray. Further studies include spray combustion experiments at elevated pressures to replicate practical gas turbine conditions in the new heated, high-pressure shock tube described by de Vries et al. (2007). With capabilities of reflected-shock pressures up to 100 atm and test times up to 20 ms, a broadened range of experimental data for the combustion of liquid hydrocarbon fuels is obtainable.

References

1. Mullaney, G.J.: Ind. Eng. Chem. **50**, 53 (1958)
2. Miyasaka, K., Mizutani, Y.: *Ignition of Sprays Behind a Reflected Shock*, Modern Developments in Shock Tube Research, Proc. Tenth Int. Shock Tube Symp., Kyoto (1975) pp. 429-436.
3. Wierzba, A.S., Kauffman, C.W., Nicholls, J.A.: Comb. Sci. Tech. **9**, 233 (1974)
4. Nettleton M.A.: Fuel **53**, 99 (1974)
5. Sidhu S.S., Graham J.L., Kirk D.C.: *Investigation of Effect of Additives on Ignition Characteristics of Jet Fuels: JP-7 and JP-8*, 22nd Int. Symp. Shock Waves, Paper No. 3810, (1999)
6. Cadman P.: *Shock-Tube Combustion of Liquid Sprays of Ethanol: Effect of Additives on Ignition Delays and Combustion Characteristics*, Proc. 20th Int. Symp. Shock Waves, Vol. 2 (1995) pp. 1039-1044
7. Kunz A., Wang R., Cadman P.: *Liquid Spray Combustion of Propanol/Tetradecane/Water Mixtures*, 21st Int. Symp. Shock Waves, Paper No. 1691 (1997)
8. Williams R.L., Cadman P.: *The Combustion and Pyrolysis of Droplet Sprays of Vegetable Oils at High Temperatures and Pressures in a Shock Tube*, 22nd Int. Symp. Shock Waves, Paper No. 3330 (1999)
9. Tsuboi T., Hozumi T., Hayata K., Ishii K.: Combust. Sci. Tech. **177**, 513 (2005)
10. Hanson T.C., Davidson D.F., Hanson R.K.: *Shock Tube Measurements of Water and n-Dodecane Droplet Evaporation behind Shock Waves*, AIAA Paper 2005-350 (1997)
11. Amadio A.D., Crofton M.W., Petersen E.L.: Shock Waves **16**, 157 (2006)
12. Petersen E.L., Hanson R.K.: Shock Waves **15**, 333 (2006)
13. de Vries J., Aul C.J., Barrett A., Lambe D., Petersen E.L.: *Shock tube development for high-pressure and low-temperature chemical kinetics experiments*, These Proceedings, Paper No. 0913 (2007)

Coupling CFD and chemical kinetics: examples Fire II and TITAN aerocapture

P. Leyland[1], **S. Heyne**[1], **and J.B. Vos**[2]

[1] *Institute of Energy Sciences, Ecole Polytechnique Fédérale de Lausanne, CH-1015 Lausanne*
[2] *CFS Engineering, CH-1015 Lausanne, Switzerland*

Summary. The investigation of the thermochemical field around planetary reentry capsules, and their experimental scaled models is made by coupling aerothermodynamic simulation of the corresponding flow field to a library of chemical kinetics of the specific atmosphere. As examples, the FIRE II reentry and TITAN aerocapture entry which have been studied in collaboration with the Hypersonics centre of the University Queensland, where scaled geometries are tested in their expansion tube facilities. Examples are given over equivalent sphere geometries to test the models of the kinetics, and complete capsule configurations for Titan are presented with simplified models.

1 Introduction

Investigation of the chemical kinetics of the shock layer of hypersonic vehicles can be estimated either by experimental means: especially using shock tubes, expansion tubes, and non reflected shock tubes facilities, or by performing flight experiments, or by estimating with the use of chemical kinetic databases and analysis tools, and of course theoretical studies. During the last 20 years, the use of computational fluid dynamics for reacting flows (CFD), can be added to this list, where the flow field equations are solved coupled with the species conservation equations. The chemical source terms and reactions are taken from models derived by the former methods. For Earth reentry, the gas is composed of air, and for reentry, the number of species involved is limited (5 to 9 for instance) and fully coupled aerothermodynamic-chemical kinetics can be made. For other planetary atmospheres, as for Mars, Venus and the recent experiment of the aerocapture in Titan's atmosphere, give a challenging task.

In this paper, a test bed for the study of analysing the chemical kinetics and aerodynamic flow field is taken from experiments made at the University of Queensland, Australia in two of their expansion tubes. Reentry of the successful flight campaign of Earth reentry of FIRE II was studied in the small X1, and the aerocapture of the Huygen's capsule in Titan atmosphere in the larger tunnel, X3, [4–6].

The scaling in experimental facilities tends the chemistry to have a very short activity time. For the Fire II experiment in X1 the flow becomes close to chemical equilibrium quickly and the whole shock layer can be considered in equilibrium. Also radiation heat flux is very small compared to flight due to binary scaling.

In situations such as Titan's aerocapture, the non-equilibrium solution remains much longer and the post shock flow is in non-equilibrium upto and well over the body and behind, [3]. The conditions considered are those of the X3 expansion tube shots and equivalent flight point. The model was 1:40.8 scaled. Radiation heat flux is thus maintained at a reasonable level, see table 1.

2 Coupling CFD and Chemical Kinetics

The speeds at which these planetary capsules enter the atmospheres can produce a highly reacting flow behind the outer bow shock that forms in front of the capsule. This is particularly the case in the actual flight experiment. In the case of experimental conditions, equivalent flight conditions can be realised for the dynamics applying binary scaling. However the chemical kinetics are influenced by the reduced size. Dissociation, re-combination and ionisation are molecular collisional procedures with characteristic time scales, and hence depend on the dimension of the shock layer. Consequently, the radiative contribution of the gas to heat fluxes are significantly lower than in flight. For Earth reentry (air) consisting of 5 species (plus eventually ionisation species) and 17 reactions, it is usual to adopt either Park's model [1] for forward reaction rates and equilibrium coefficients, or the data of Hanson and Salimian [2]. For Titan aerocapture entry, the atmosphere is composed of 95 % N2 and 5% CH4. The Gokcen's models [7] have updated Nelson's models [8, 9] and include the significant reactions for the methane decomposition.

The modelisation of the aerothermodynamics of the flow field is covered by the generalised Navier Stokes equations for the fluid dynamics augmented by the chemical species conservation equations and the appropriate source terms with the reaction rates:

$$\frac{\partial \rho Y_s}{\partial t} + div(\rho Y_s(\mathbf{u} + \mathbf{v_s})) = \dot{\omega_s}$$

The chemical production term ω_s represents the net production or destruction of the mass of species s by chemical reactions. The forward reaction rates are in general given by an Arrhenius equation,

$$\mathrm{k}_{f,k} = A_k T^{\alpha_k} exp\left(-E_k/T\right) \tag{1}$$

where E_k is the activation energy, and A_k and α_k are constants. The values of these constants comes from experimental data and sensitivity analysis and are part of chemistry data banks. Extensive datasets are available in public domain databases, particularly on combustion.

The choice of the chemical kinetic model depends on the choice of dominant species of the gas mixture, to consider a set of consistent reactions. These, together with the values of the constants for the reaction rates form the computational mechanism. This mechanism has to be complemented by the thermodynamic properties for the species given as polynomials that have to be extended beyond 6000K for non-equilibrium chemistry situations. The number of species and reactions, the coefficients of these polynomials, the chemical kinetic mechanism, constitute the chemical kinetic model that has to be coupled to the Navier-Stokes equations.

For air chemistry it is usual to implement the chemical reaction model directly within the code. The number of species are relatively low, and the characteristic time scales of the chemical kinetics of air are compatible with such an approach.

2.1 Coupling Method

The CFD solvers considered are NSMB and CFD++. The NSMB flow solver is a multi-block structured code developed by a consortium including EPFL since 1990, and now is monitored by CFS Engineering. The advantages of NSMB is that being "in-house" code-development and modification is done directly in the sources. CFD++ is a commercial software of MetaComp Technologies. The finite rate chemistry can be implemented "by hand" within the code interface as long as the species in question are in the internal database. One of the great advantages of the chemistry modules, is that information on the transport coefficients can be extracted and evaluated as curve fits. This is particularly interesting for Titan atmosphere where the transport coefficient modeling remains a difficulty. In NSMB, for gas mixtures other than air, the strategy is to link the CFD calculation of pressure, internal energy and temperature with a chemical kinetics model. The Chemical model is setup by the means of CHEMKIN. The SENKIN II module (in fortran) has even be implemented directly in the NSMB code, [10].

The detailed model of Gokcen includes 28 species and 74 reactions. This was then simplified to his reduced model of 21 species and 35 reactions based on sensitivity analysis using CHEMKIN, maintaining the important species contribution to radiating exchanges. Here, we have further reduced this model, maintaining the main radiating species, to 16 species: C2, C, CH4, CH3, CH2, CH, CN, CN+, H2, H, N2, N, N+, HCN, NH, e-.

3 Results

The first results were using air chemistry directly implemented in the NSMB code for the Fire Capsule in the X1 expansion tube conditions. The model was 1:27.7 and the dimensions of the nozzle small. This led to a post shock flow that was in chemical equilibrium. Park's 5 species equilibrium model was used. The CFD quality was assured by a very fine regular multi-block structured grid, (see Figure 2). The main feature to be compared was the heat flux on the heat shield, and the comparison between CFD and X1 experimental results gave a discrepancy of 13 %. (CFD 12.7kW/cm2, X1 $\approx$ 14 KW/cm2). The measurements were at the conditions made in Table 1. Both NSMB and CFD++ were used and were in good agreement.

Then the Titan aerocapture chemical kinetics and flow field was considered in both the X3 expansion tube and flight. In X3 a good representation of Titan atmosphere was obtained, [3]. The complete kinetic model of the methane and nitrogen dissociation behind the bow shock wave is unfeasible, so then Gokcen's reduced model was implemented as a chemical mechanism file for CHEMKIN and the internal SENKIN model of NSMB. The model allows for several ionised species and hence is particularly interesting for observing the radiating gas field in the shock layer. Two 53 block structured grid were made over the capsule, with 4.1 and 12 million grid points, as the domain covers the whole capsule and the back part in order to capture the non-equilibrium phenomena extensively. (As the flow is in non-equilibrium even well past the body). However this then led to memory problems. A further reduced Gokcen model was obtained by observing the solutions over equivalent cylinders are the relative importance of the species. Then sensitivity analysis was performed on the corresponding reactions to test compatibility. At the present time the calculations were impossible in a coupled way over the complete Titan capsule, a weak coupled approach was used.

The calculations of the flow field of the Titan aerocapture capsule are shown in figures 1 and the flow solution is then used at several steps and lines to evaluate the chemistry using a 1D approach in CHEMKIN. In this way an estimation of the heat flux on measured points could be made. The measurements were at the conditions made in Table 1. The total heat fluxes on 2 points were compared (CFD: 0.78 kW/cm2, 0.6 KW/cm2 and X3: $\approx$ 0.7 KW/cm2, 0.5 KW/cm2). However, the CFD does not treat directly the radiation component which is present in the field particularly in flight, but also in the X3 experiment. (The binary scaling reduces considerably the proportion of radiative heat flux, $\dot{q}_{r\,X3} \approx 6\%$, flight $\dot{q}_{r\,X3} \approx 90\%$).

The calculations over equivalent cylinders allow a fully coupled approach - using the call to SENKIN within the NSMB code itself. The results are shown in the figures 3 for critical species. The CH4 dissociates completely very quickly behind the shock, giving rise to species of which the highly radiating CN, but also significant quantities of N+ in the near wall layer, CN+ and also a small but non zero contribution of electrons (10^{-6}). These species all contribute to the shock layer plasma. N2 very partially dissociates in both the flight and X3 conditions. The main differences in the chemical kinetics of flight and X3 conditions are in the relaxation of the species behind the shock, but the similarity is very close [11].

4 Figures

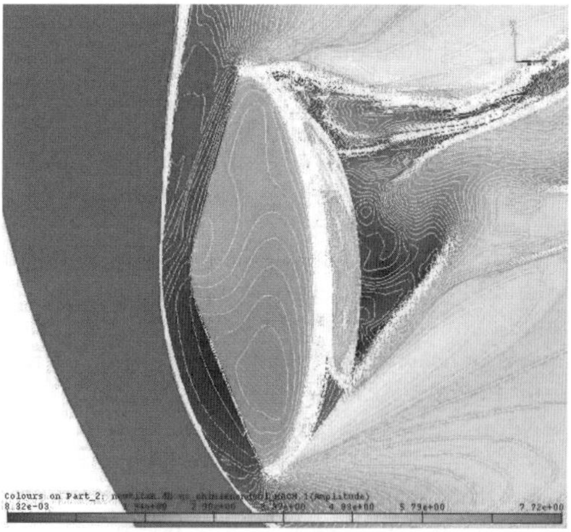

Fig. 1. *Temperature on the body and iso-Mach for the TITAN aerocapture capsule in X3 conditions. The solution field is continued pas the body due to capture the whole field.*

Table 1. Experimental conditions for FIRE II and Titan aerocapture.

Facility	conditions	flight
X1	$\rho L = 4.410^{-3}$	2.810^{-4}
X1	$V_\infty = 9.3 - 10.17 km/s$	
X1	$\rho\ kg/m3 = 2.4310^{-3}$	$8.1710^{-3} - 1.310^{-3}$
X1	p_{st} KPa = 18.8 - 25	0.021

Facility	conditions	flight
X3	$\rho L = 5.210^{-4}$	5.210^{-4}
X3	V_∞= 5.83 km/s	5.76
X3	$\rho\ kg/m3 = 2.4310^{-3}$	1.4910^{-4}
X3	p Pa = 1130	6.9

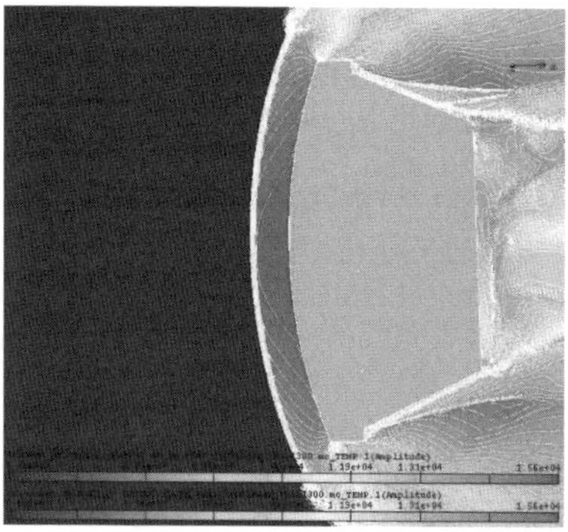

Fig. 2. *Temperature field for Fire II Capsule in X1 conditions. The flow field is in chemical equilibrium.*

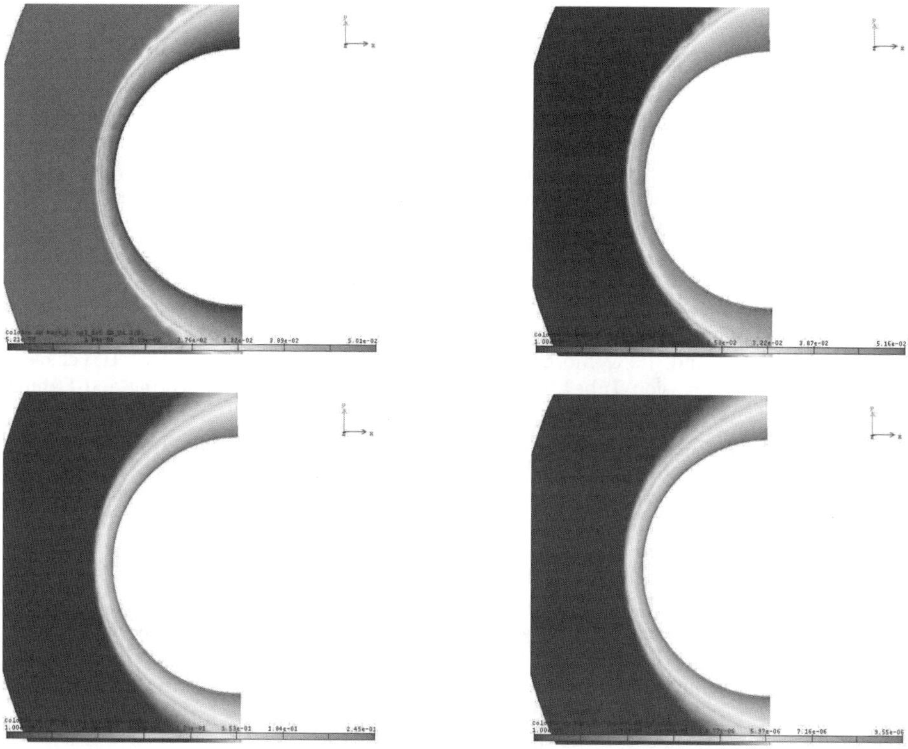

Fig. 3. *Testing chemical kinetics of the flight and X3 conditions for the Titan aerocapture shows evolution over a cylinder showing complete dissociation of CH4 (top left) and the creation of highly radiating CN in the shock layer (top right), and a significant quantity of N+ (bottom low), and a small amount of electrons, (bottom right).*

5 Conclusions

The diagnosis of the chemical kinetic modelling in shock layers in hypersonic flight and experimental conditions is studied using a coupling between a large scale Navier Stokes solvers with chemical kinetics under the form of chemistry data bases and CHEMKIN type. This allows use of the functionalities of sensitivity analysis for the different species and reactions, and the use of the chemical kinetic mechanisms to evaluate the source terms of the kinetics in the flow field solver. The study of the flow fields over FIRE II and Titan where investigated in this way and apart from the CN production that is well known, importance to other species, especially the ionised nitrogen N+, CN+ and electrons are present. A further reduced model of 16 species was devised which proved successful.

Acknowledgement. Acknowledgments are made to Tahir Gokçen for his help with his reduced model, and to the group of R. Morgan at the University of Queensland, Australia, for their extensive collaboration.

References

1. Park, C. "On Convergence of Chemical Reacting Flows". AIAA 85 - 0427
2. Hanson, R.K. and Salimian S. "Survey of Rate Constants in the N/H/O System". In: Combustion Chemistry, Edited by Gardiner, W.C. Springer Verlag, 1984
3. B. Capra. Ph D Thesis. University Queensland, Australia. 2006
4. Capra, B. R.,Leyland, P., & Morgan, R. G. "New gauge design to measure radiative heat transfer to a titan aerocapture vehicle in expansion tubes", in 5th European Symposium on Aerothermodynamics for Space Vehicles, Cologne, Germany, ESA Publications 2005
5. Capra, B. R.,Leyland, P., & Morgan, R. G. , "Subscale testing of the FIRE II Vehicle in a Superorbital Expansion Tube"; in 42nd AIAA Aerospace Sciences Meeting and Exhibit, Reno, Nevada, 2005
6. Capra, B. R., Morgan, R. G., & Leyland, P., "Heat Transfer Measurements of the First Experimental Layer of the FIRE II Reentry Vehicle in Expansion Tubes"; in 5th European Symposium on Aerothermodynamics for Space Vehicles, Cologne, Germany, ESA Publications 2005
7. T. Gökçen : "N_2-CH_4-Ar chemical kinetic model for simulations of atmospheric entry to Titan", AIAA Paper 2004-2469, 2004.
8. J. Olejniczak, M. Wright, D. Prabhu, N. Takashima, B. Hollis, E. Zoby, and K. Sutton : An analysis of the radiative heating environment for aerocapture at Titan, AIAA Paper 2003-4953, 2003.
9. Takashima, N., Hollis, B.R., Zoby, E. V., et al. 2003, in 39th AIAA/ASME/SAE/ASEE Joint Propulsion Conference and Exhibit, Huntsville, Alabama
10. A. Roubaud, V. Mahulkar, P. Leyland, D. Favrat, J. Vos.
In-cylinder combustion modeling with simplified chemistry; In proceedings of the International Multidimensional Engine Modeling Conference, Detroit, MI, USA; March 7, 2004
11. R. J. Gollan : `poshax`: Computing post-shock thermochemically relaxing flow behind a steady normal shock, Division of Mechanical Engineering Report 2007/09, The University of Queensland, May, 2007.

Experimental investigation of catalytic and non-catalytic surface reactions on SiO_2 thin films with shock heated oxygen gas

V. Jayaram[1], G.M. Hegde[2], M.S. Hegde[1], and K.P.J. Reddy[2]

[1] *Solid State and Structural Chemistry Unit, Indian Institute of Science, Bangalore, India*
[2] *Department of Aerospace Engineering, Indian Institute of Science, Bangalore, India*

Summary. The present study is to investigate the interaction of strong shock heated oxygen on the surface of SiO_2 thin film. The thermally excited oxygen undergoes a three-body recombination reaction on the surface of silicon dioxide film. The different oxidation states of silicon species on the surface of the shock-exposed SiO_2 film are discussed based on X-ray Photoelectron Spectroscopy (XPS) results. The surface morphology of the shock wave induced damage at the cross section of SiO_2 film and structure modification of these materials are analyzed using scanning electron microscopy and ion microscopy. Whether the surface reaction of oxygen on SiO_2 film is catalytic or non-catalytic is discussed in this paper.

1 Introduction

The Thermal Protection Systems (TPS), are composite materials with major constituents of SiO_2 .When a re-entry vehicle flies at very high Mach number, a very strong bow shock forms in front of the body and dissociation of air takes place inside the shock layer. If the flow is in non-equilibrium, finally dissociated gas recombines at the wall, which causes the increase in aerodynamic heating. To prevent recombination of dissociated atoms, uses of TPS with low catalytic properties are required. A new model describing heterogeneous catalysis on the surface of SiO_2 materials with mixture of gases is presented in the literature [1]. TPS materials having low catalytic property will prevent from recombination of dissociating oxygen atoms on the surface and release of massive dissociation energy [2]. SiO_2 based materials are widely used for their low catalytic properties. However, even slight catalysis of these materials increases considerable amount of heat flux on the body compared to non-catalytic surface [3]. Surface reaction on SiO_2 film in presence of oxygen gas is not reported in literature. We are presenting here the surface reaction on SiO_2 film with shock heated O_2 test gas where the Focused Ion Beam (FIB) technique [4] has been used to understand the damage occurring due to strong shock wave interaction.

2 Experimental Details

2.1 Shock tube experiment

The shock tube portion of the recently established free piston driven hypersonic shock tunnel (HST3) is used for this purpose. To conduct shock tube experiment, gas reservoir chamber is filled with 18 bar nitrogen gas before launching a 20kg piston, the compression tube is filled with helium gas of 1 bar pressure and the shock tube is filled with high purity O_2 (99.995 percent) at 0.1 bar pressure. Sudden supply of the high-pressure gas by opening a valve behind the piston sets its motion in the compression tube and as a result, the piston gets maximum acceleration. Motion of the heavy piston in compression tube adiabatically compresses the helium gas and there by bursts the primary diaphragm. It produces strong primary and reflected shock waves in the shock tube. The photograph of shock tube portion of FPST and shock tube end flange mounted with SiO_2 sample is shown in Fig.1. The reflected shock wave produces high pressure and temperature oxygen gas which interacts with the sample mounted on the end flange of the shock tube.

Fig. 1. (a) Photographs of the shock tube and position of pressure sensors to measure shock speed and P_5 pressure and (b) SiO_2 film samples mounted on end flange of the shock tube

The typical pressure signals acquired to measure the shock speed and reflected shock pressure (P_5) are shown in Fig.2a and 2b respectively. The temperature behind the reflected shock wave calculated using 1-D normal shock theory with γ= 1.4 is about 8500 K. However, at this temperature, the internal degrees of freedom of O_2 get excited leading to non-equilibrium state which reduces the γ value. Hence the calculations are also presented for γ=1.285 assuming the molecules under non-equilibrium condition. Thus the actual reflected shock temperature is about 6500 K lasting for about 2-3 ms at shock Mach number 8. The dissociated oxygen at this temperature interacts with the sample mounted at the end of the shock tube.

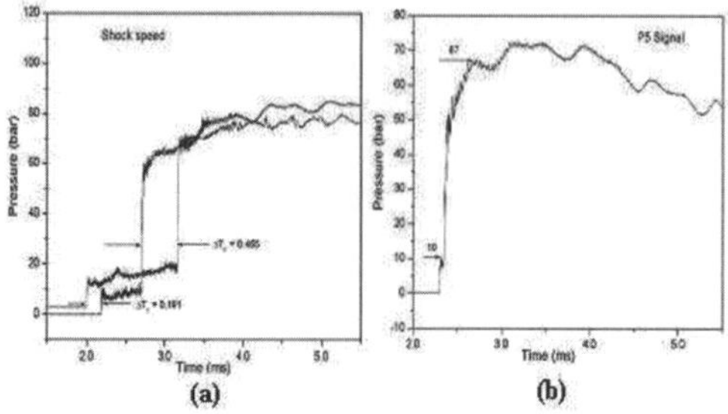

Fig. 2. Data acquired from pressure sensors at the end of the shock tube. (a) Time history for shock speed measurement. (b) Reflected shock pressure (P_5) data.

2.2 Characterization

Thin film SiO_2 was deposited on silicon substrate by electron beam reactive evaporation technique. The morphology of the SiO_2 films was analyzed using SEM (Philips SIRION with EDX). XPS of pure and shock exposed SiO_2 were recorded in an ESCA-3 Mark II spectrometer (VG Scientific Ltd., England) using AlK_α radiation (1486.6 eV). FIB milling and ion microscopy of SiO_2 films were carried out using instrument manufactured by FEI Company (model PN 21021-B).

3 Results and discussion

3.1 XPS and SEM studies

XPS of pure SiO_2 film coated on Si(100) substrate before exposure to shock is shown in Fig. 3a and shock exposed SiO_2 film in presence of O_2 gas is shown in Fig. 3b. XPS of Si(2p) in pure SiO_2 film shows well resolved two peaks; one due to Si at 99.9 eV and another due to SiO_2 at 103.4 eV. Even after deconvolution the resolved peaks show same binding energy for Si and SiO_2 species before exposing to shock, as shown in Fig.3a. After exposure to shock the intensity of the peak at binding energy 103.3 eV increased compared with that of 99.5 eV peak. Deconvoluted spectra of Si(2p) shows three peaks at different binding energies 99.3 eV, 101.2 eV and 103.3 eV, corresponding to Si(2p) of pure Si, SiO (Si^{2+}) and SiO_2

(Si^{4+}) respectively, which are in good agreement with binding energies reported by Zhu et al. [5] as shown in Fig. 3b. Intensity of Si(2p) peak at 103.3 eV due to SiO_2 increased in addition to new peak developed due to SiO species at 101.2 eV. This indicates

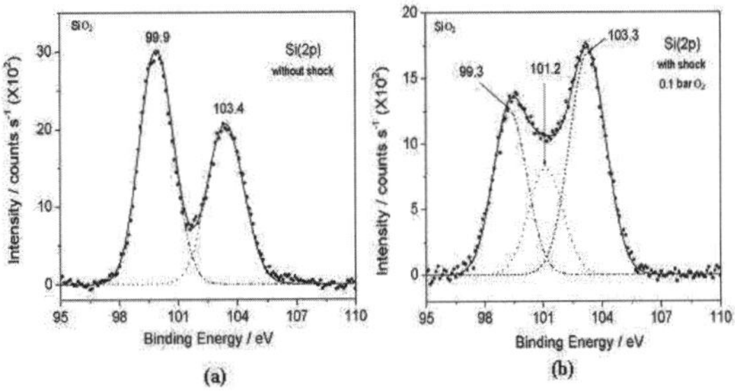

Fig. 3. (a) XPS of SiO_2 film on Si substrate before exposure to shock wave (b) Deconvoluted spectra of Si (2P) of SiO_2 film after subjected to shock in presence of O_2.

that the shock heated O_2 molecule dissociated into oxygen atoms which reacted and diffused into the junction of SiO_2/Si interface. Formation of SiO suboxide species in solid phase at the interface is an interesting phenomenon. Integrated area under the Si(2p) peak at 101.2 eV is proportional to the total yield of SiO compound due to shock. Increase in concentration of oxygen and high stress developed due to instantaneous rise and fall of shock temperature with in 2-3ms could have initiated the reaction between Si and O at the interface of SiO_2 and Si. The SEM micrograph and EDX spectra of pure SiO_2 film deposited on Si substrate was taken and is shown in Fig. 4a. The EDX spectra show the O_2 and Si presence on the surface. After exposure to shock the micrograph of the ablated SiO_2 material is shown in Fig. 4b. The micrograph shows damage along with ablated SiO_2 film. It is difficult to estimate the thickness of the ablated material and the nature of damage due to shock wave.

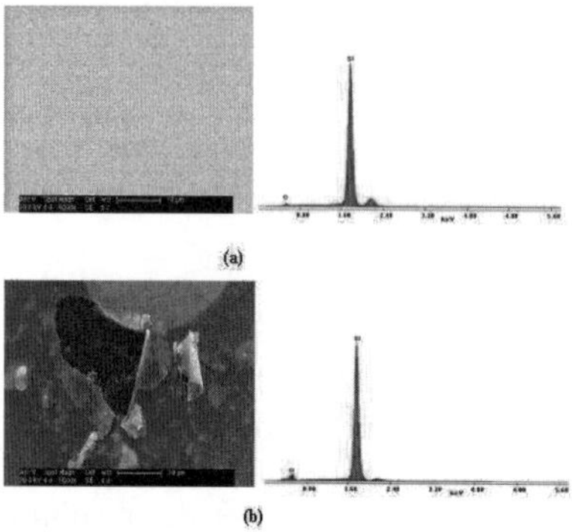

Fig. 4. SEM Micrograph of SiO_2 film. (a) As deposited SiO_2 film on Si substrate with EDX (b) After subjected to shock, with ablation of material along with EDX spectra.

3.2 Focused ion beam studies

FIB can produce an image of a sample in a manner analogous to a scanning electron microscope. The clear channeling contrast is unique to images generated by a scanning ion beam microscopy. The change in microstructure in the cross-section of SiO_2 film and substrate layer can be seen in the ion micrograph. We are introducing this method for the first time to analyze ablation of thin film materials and the shockwave induced damage in the cross section of the film. Scanning ion beam microscopy is used to observe the change in morphology at the cross-section of the film, and also to analyze the thickness of the ablated material due to shock.

Figure 5(a) shows the micrograph of as deposited SiO_2 thin film on silicon substrate with clear FIB image. After exposure to shock in presence of O_2 gas, the damage on the surface of pure SiO_2 film and micrograph of the ablated materials from the SiO_2 surface is shown in Fig. 5(b). Ion milling for the area 25 μm X 15 μm with a depth of about 2 μm was performed on unexposed film as shown in Fig. 5(c). Micrograph of the polished portion of the milled area was taken after tilting the specimen to 45.2^0. The two parallel lines in the micrograph show the thickness of the deposited SiO_2 film (about 0.3 μm). The line marked as (A) is the clear bifurcation between the SiO_2 and Si substrate layer.

A milling for 25μm X 15 m area to a depth of 2.5 μm was performed on shock exposed sample to study microstructure at the cross section of SiO_2 film. Specimen was tilted to 46.2^0 before taking micrographs. Investigation of the microstructure at the cross sectional area of SiO_2 film shows no bifurcation line between SiO_2 and Si substrate after exposure to shock as shown in Fig.5(d). The letter marked (X) shows that the material might have ablated or reacted due to shock and we cannot see the interface line. In Fig. 5(c) we can observe uniform microstructure but in the shock exposed SiO_2 film we cannot see the same line. The letter marked Y show change in morphology developed due to impact of

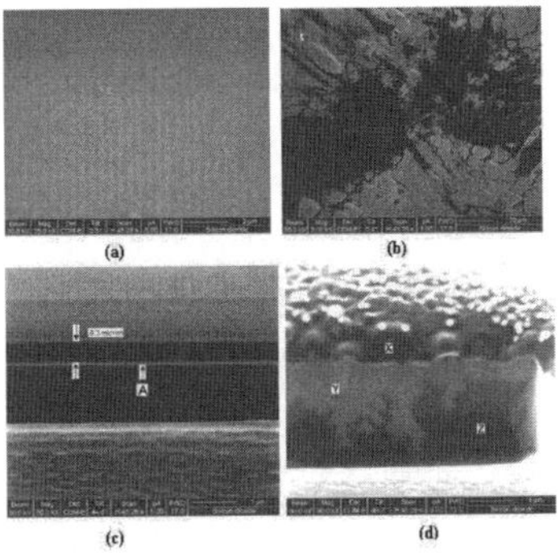

Fig. 5. Fig. 5. FIB images from ion microscopy of (a) pure SiO_2 film deposited on Si substrate (b) ablated SiO_2 film (0^0 tilt) after exposed to shock with oxygen (c) the cross section of the film after milling and 45.2^0 tilting, without shock (d) the cross section of the film after milling and 46.2^0 tilting, with shock

high temperature, high pressure shock waves (thermal stress). At position marked as Z we observe nearly the same morphology as unexposed shock wave without much change. FIB milling and ion microscopy gives information about the change in microstructure at the cross section of the material due to surface reaction of shock heated oxygen gas. A comparison of only the surface morphological changes in Fig.5(a) and (b) with that of 5(c) and (d) suggest that FIB is a very useful technique to identify the reaction occurring at different depths of the sample due to shock wave interaction.

3.3 Catalytic studies

The surface reaction with SiO_2 thin films with shock heated oxygen is as follows. SiO_2(s)+O(g) $\rightarrow$ SiO(s)+O_2 and SiO+O$\rightarrow$$SiO_2$ At high substrate temperature and low oxygen pressures an "active" oxidation reaction can occur as 2Si(s)+O_2(g)$\rightarrow$2SiO(g). The formation of volatile SiO can also take place as a result of high temperature SiO_2/Si decomposition when oxygen partial pressure is low: Si(s)+SiO_2(s)$\rightarrow$ 2SiO(g). In the presence of gaseous oxygen the re-oxidation reaction of 2SiO(g)[or 2SiO surface] can have following reaction:2SiO(g)[or 2SiO surface]+O_2(g)$\rightarrow$ 2SiO_2(s). When SiO_2 is exposed to shock heated oxygen gas, if SiO (s) is formed then the reaction is non-catalytic and if SiO_2 is formed then it undergoes catalytic reaction.

4 Conclusion

XPS studies confirm the development of an additional peak due to SiO sub-oxide species in the interface between SiO_2 / Si which is due to surface non-catalytic reaction. Both SEM and FIB micrographs show the ablation of SiO_2 films due to shock. The micrographs taken using FIB at the cross section of the SiO_2 film requires careful analysis to understand the damage due to residual stress induced inside the SiO_2 film due to strong shockwave interaction. Surface catalytic/non-catalytic reaction and ablation of material take place when strong shock heated oxygen gas reacts with SiO_2 film.

Acknowledgement. We thank ISRO, DST, DRDL, AR&DB and The Director, IISc for the financial support. We thank HEA Lab team for their support during the experimental work and operation of FPST.

References

1. Scott C D: *Catalytic recombination of nitrogen and oxygen on high temperature reusable surface insulation*, AIAA 15th Thermophysics Conference, Snowmass, Colo.,July 14-16,pp192-212,1980.
2. Jumper E J and Seward W A: *Model for oxygen recombination on silicon dioxide surfaces Part 2, implications toward reentry heating*, 30th Aerospace Science Meeting & Exhibit, Reno, NV, AIAA-92-0811,January 6-9,1992.
3. Kurotaki T: *Construction of catalytic model on* SiO_2 *- based surface and application to real trajectory*, 34th AIAA Thermo physics Conference, Denver, CO, AIAA A00-33689 June,pp19-22,2000.
4. Orloft J: *High-resolution focused ion beams*, Rev. Sci. Instrum.,64,5,pp1105-1130,1993.
5. Zhu X P, Tomiyuki Yukawa, Makoto Hirai, Hisayuke Suematsu, Weihua Jiang, Kiyoshi Yatsui, Nishiyama H and InoueY: *X-ray photoelectron spectroscopy characterization of oxidated Si particles formed by pulsed ion-beam ablation*,Applied Surface Science, 252,pp5776-5782,2006

Experimental investigation of interaction of strong shock heated oxygen gas on the surface of ZrO_2 - a novel method to understand re-combination heating

V. Jayaram[1], M.S. Hegde[1], and K.P.J. Reddy[2]

[1] *Solid State and Structural Chemistry Unit, Indian Institute of Science, Bangalore, India*
[2] *Department of Aerospace Engineering, Indian Institute of Science, Bangalore, India*

Summary. Interaction of shock heated test gas in the free piston driven shock tube with bulk and thin film of cubic zirconium dioxide (ZrO_2) prepared by combustion method is investigated. The test samples before and after exposure to the shock wave are analyzed by X-ray diffraction (XRD), X-Ray Photoelectron Spectroscopy (XPS) and Scanning Electron Microscope (SEM). The study shows transformation of metastable cubic ZrO_2 to stable monoclinic ZrO_2 phase after interacting with shock heated oxygen gas due to the heterogeneous catalytic recombination surface reaction.

1 Introduction

Zirconium dioxide (zirconia) is an extremely good refractory material. It offers chemical and corrosion inertness to temperatures well above the melting point of alumina. At very high temperature (>2650 K) the material has a cubic (c), and at intermediate temperature (1450 to 2650 K) it has a tetragonal (t), and below 1450 K it has a stable monoclinic (m) structure. During the tetragonal to monoclinic transformation all structural parameters show linear temperature dependence. Strong nonlinearity was found in the transformation from m to t direction [1]. Both the tetragonal (t-ZrO_2) and monoclinic (m-ZrO_2) structure can be considered as distortions of the cubic structure. Transformation from tetragonal to monoclinic is rapid and is accompanied by an increase in volume by 3-5 percent that causes extensive cracking in the material. The controlled, stress induced volume expansion of the tetragonal to monoclinic inversion is used to produce very high strength, hardness and tough varieties of zirconia available for mechanical and structural applications [2]. ZrO_2 has bulk permittivity of approximately 25 (high dielectric constant) and low thermal conductivity suitable for aerospace applications as high temperature materials. In this paper we discuss the transformation of metastable cubic ZrO_2 to stable m-ZrO_2 phase in presence of high temperature and high pressure shock heated oxygen gas due to catalytic surface recombination reaction.

2 Experimental Details

2.1 Synthesis of ZrO_2 powder and film on Macor substrate

The test material zirconia is synthesized in the laboratory using the following procedure. 23.3 mmoles of $Zr(NO_3)_4$ and 56.8 mmoles of glycine were dissolved in 35 ml of water to get a clear solution. The solution was kept in the furnace at 450^0C in a 300 ml crystallizing dish. After dehydration, the solution ignited into a flame, rising the temperature to about 1000oC dwelling for about 30-40 seconds which leads to the chemical reaction

$9Zr(NO_3)_4 + 20C_2H_5NO_2 \rightarrow 9ZrO_2 + 40CO_2 + 10N_2 + 50H_2O$.

The product was left for 1 hour at 500^0C to remove carbonaceous impurities, if any and ZrO_2 was obtained as white powder [3]. Synthesis of ZrO_2 films on Macor substrate was carried out by the following procedure.10 g of $Zr(NO_3)_4$ was dissolved in water to give a clear solution. Few Macor pieces of 1x1cm^2 were dipped into the solution and taken out and kept in the furnace at 500^0C. After combustion a thin film coating of ZrO_2 remains on the Macor substrate. As prepared ZrO_2 pellets and film samples were taken for experimental investigation before exposing to the shock heated gas in the shock tube.

2.2 Shock tube experiment

The schematic diagram of a free piston driven shock tube is shown in Fig. 1. The shock

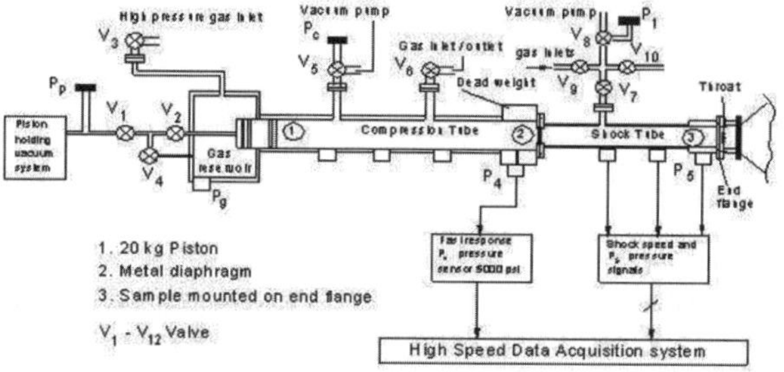

Fig. 1. Schematic diagram of fully instrumented free piston driven shock tube

tube was operated by filling the high pressure gas reservoir with 18 bar N_2, compression tube with 1 bar He and shock tube with 0.1 bar ultra pure oxygen (99.995 percent). All the electronic systems are connected to the data acquisition system. Samples were mounted on the end flange of shock tube as shown in Fig. 2a. To launch 20kg piston the valves V_2 and V_4 were opened which sets the heavy piston in to motion in compression

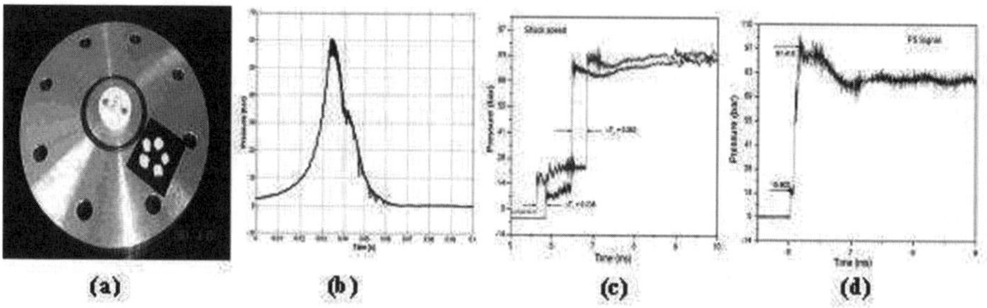

Fig. 2. (a) Photograph of ZrO_2 samples mounted on the end flange of the shock tube, (b) pressure (P_4) signal recorded at the end of the compression tube (c) data acquired from the pressure transducers used for shock speed measurement, and (d) pressure behind the shock wave at the end of the shock tube(P_5).

tube which adiabatically compresses the helium gas and there by increases its pressure and temperature. This high pressure and high temperature helium gas bursts the primary diaphragm, which produces strong shock wave in the shock tube. The typical pressure signals recorded at the end of the compression tube, shock speed signal and reflected shock pressure signals are shown in Figs. 2b, 2c and 2d, respectively. This strong shock wave interacts with the ZrO2 sample mounted on the end flange of the shock tube. Different experimental and calculated values are listed in Table 1. At high temperature dissociated oxygen gas considered to be at non-equilibrium state reacts with the sample

		Experimental values				Calculated values		
Run	Sample	P_1 bar/test gas	$\triangle T$	P_2 bar	P_5 bar	Vs (m/s)	Ms	T_5
R36	ZrO_2 pallet	$0.1/O_2$	181	9.4	72.0	2762	8.4	9550
R35	ZrO_2 film	$0.3/O_2$	238	15.8	97.0	2100	6.4	5620
R42	ZrO_2 pallet	$0.1/N_2$	189	9.5	67	2645	7.5	7720
R76	ZrO_2 pallet	$0.3/N_2$	268	14.1	89	1865	5.3	3950

Table 1. Experimental and calculated values for γ=1.4

mounted on the end flange of the shock tube. The above phenomenon can be observed in the pressure signal shooting up to 97 bar for a small duration of test time, as shown in Fig. 2d. Non-equilibrium to equilibrium temperature of the oxygen gas molecule may vary between 9500 K and 7300 K in one of the experiments listed in Table 1. Similar high pressure and high temperature shock heated test gases are produced to study the interaction of different gases with high temperature materials.

2.3 Characterization

Shock exposed ZrO_2 samples were taken for analysis using different experimental methods. Diffraction data for all samples were recorded using CuK_α radiation from automated Philips X-ray powder diffractometer (X-pert Philips). The diffraction peaks of crystalline phases were compared with those of standard data reported in JCPDS data file. The morphology of the ZrO_2 films was analyzed using SEM (Philips SIRION with EDX). Photoelectron spectra of pure and shock exposed ZrO_2 were recorded in an ESCA-3 Mark II spectrometer (VG Scientific Ltd., England) using AlK_α radiation (1486.6 eV).

3 Results and discussion

3.1 XRD and XPS studies

Both ZrO_2 pellets and film samples after exposure to shock wave were taken for XRD studies. The characteristic indexed diffraction lines due to ZrO_2 as-prepared sample is shown in Fig. 3(a). The shock treated sample in presence of oxygen gas shows a monoclinic ZrO_2 structure as shown in Fig. 3(b). The cubic phase zirconia with space group Fm3m changes to monoclinic ZrO_2 with space group P21/a (14). The powder XRD contain cubic ZrO_2 form deeper side of the pellet and the lines corresponding to cubic ZrO_2 are indicated by (*). XRD of ZrO_2 powder sample exposed to shock heated

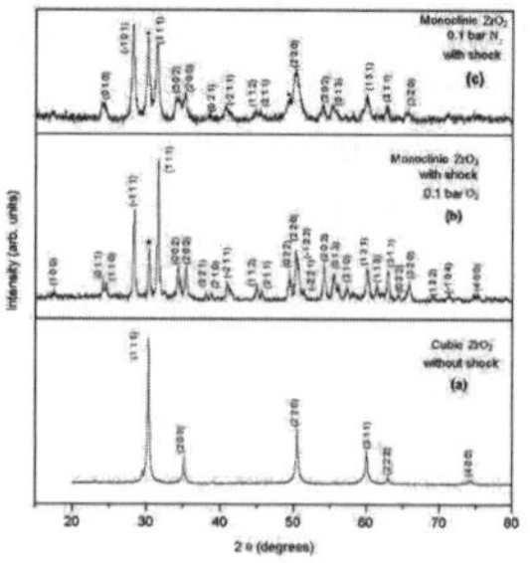

Fig. 3. XRD of ZrO_2 pellet sample, (a) before exposure to shock wave showing cubic ZrO_2 phase, (b) after subjected to shock heated 0.1 bar O_2 (monoclinic phase) and (c) after subjected to shock heated 0.1 bar N_2 shows monoclinic phase.

0.1 bar nitrogen gas in the shock tube is shown in Fig. 3(c). The cubic ZrO_2 changed to monoclinic structure with impurity lines of cubic structure. With different test gas pressure in the shock tube we observed that the metastable c-ZrO_2 phase changes to stable m-ZrO_2 phase. XPS of Zr (3d) core level spectra were recorded before and after exposure to shock and the deconvoluted spectrum are shown in Fig. 4. The spectra of Zr(3d) core level shows two peaks at 182.7 and 184.7 eV corresponding to the binding energies of Zr $3d_{5/2}$ and $3d_{3/2}$states. These energies correspond to Zr in +4 states. In the shock exposed ZrO_2, intensity increases due to densification of ZrO_2 after melting. There is no change in the binding energy of Zr(3d) after the shock, which confirms that ZrO_2 does not change its chemical nature.

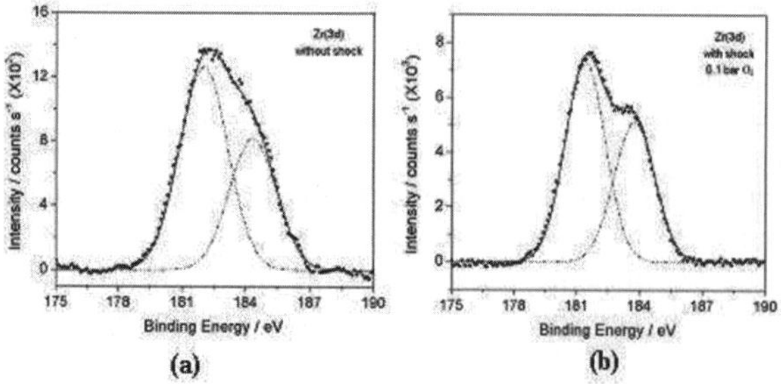

Fig. 4. Deconvoluted XPS spectra of Zr(3d) peak of ZrO_2 sample, (a) pure ZrO_2 sample and (b) after exposure to shock heated O_2 gas at 0.1 bar.

3.2 SEM studies

The transformation of metastable (c and t phase) ZrO_2 changes to the monoclinic phase due to decrease in surface energy in case of nano-crystalline samples. In the present investigation, XRD study shows the nano-crystalline ZrO_2 prepared by combustion methods stabilizes to cubic phase. The cubic-ZrO_2 pellets and thin films on Macor substrate when exposed to shock heated oxygen gas at different pressures and temperature transform to stable m-ZrO_2 phase. SEM micrographs of such samples with EDX are shown in Fig. 5 along with the unexposed ZrO_2 sample shown in Fig.5a. Needle type growth of monoclinic-ZrO_2 crystal can be clearly seen in Fig. 5(b). Length of the needles varies

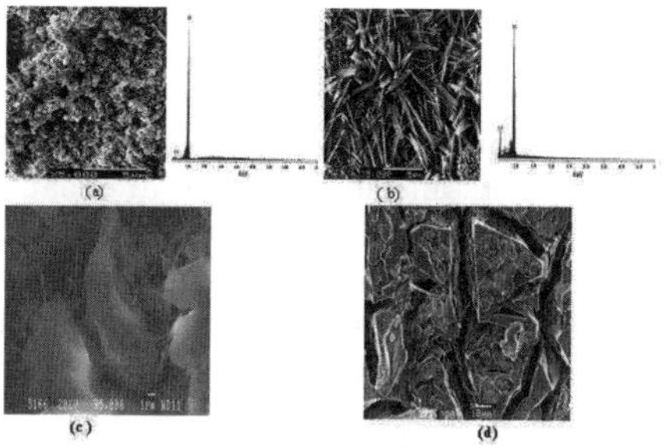

Fig. 5. SEM micrograph of ZrO_2 powder sample, (a) pure sample with EDX spectra, (b) after exposure to shock at 0.1 bar O_2 with EDX spectra, (c) after exposure to shock at 0.1 bar N_2 and (d) ZrO_2 film after exposure to shock at 0.1 bar O_2.

from 2 to 10 μm with variable thicknesses. Micrographs show the growth of monoclinic crystals with aspect ratio (Length/Diameter) of about 15 to 20 and further, ZrO_2 compound melted in the form of globules and needles growing on the surface of the globules is visible. For the sample exposed to shock at 0.1 bar nitrogen shown in Fig. 5(c), many needle-like crystals of small and big sizes along with the melted sample were seen. The ZrO_2 sample had melted and needle-like crystals grew from the melt. It is important to know that, in the shock tube experiment the temperature of the gas will be more than 6000 K and it remains for only 2-3 ms, after which the temperature falls to room temperature in another 2-3 ms. In shock tube experiments, the rise and fall in temperature is instantaneous and spontaneous. The growth of needles from the melt occurs during the instantaneous cooling process. Micrographs in Fig. 5(d) show shock exposed ZrO_2 film deposited on Macor substrate. Surface morphology of the film shows the cracking of ZrO_2 film due to transformation from c-ZrO_2 to m-ZrO_2. There was increase in volume and the change in volume was due to phase transformation which led to cracks of 5-10 μm width in the film.

3.3 Catalytic studies (Surface re-combination heating)

Most of the ablative heat shield materials used as TPS in re-entry space vehicles undergo catalytic wall reaction with ablation of materials. Surface heating is very high in case of catalytic wall reaction compared to non-catalytic wall reaction. Present experimental investigations of shock heated gas interaction on the surface of the ceramic material are important in aerospace applications. High temperature oxygen with vibrational excitation (below 3000 K) or dissociation (above 3000 K) undergoes 3-body catalytic surface recombination reaction with ZrO_2. Reaction with high temperature oxygen molecule (not-dissociated gas) is given by $O_2 + M_1 \rightarrow O_2 + M_2$, reaction with dissociated oxygen is $O + O + M_1 \rightarrow O_2 + M_2$, and reaction with high temperature nitrogen molecule (not-dissociated gas) is like $N_2 + M_1 \rightarrow N_2 + M_2$, where the third body M_1 is cubic ZrO_2 and M_2 is monoclinic ZrO_2 in all these reactions. The catalytic surface recombination reaction has taken place on ZrO_2 sample in presence of nitrogen and oxygen test gases.

Due to high temperature reaction, densification and oxygen diffusion [4, 5] could have happened on the surface of transformed monoclinic ZrO_2 samples.

3.4 Conclusion

Shock treated ZrO_2 in presence of O_2 and N_2 gas in the free piston driven shock tube were examined by different experimental techniques to confirm surface catalytic reaction. XRD study shows that both pellets and film samples are changed from cubic structure to monoclinic ZrO_2 structure. SEM micrograph reveals the growth of needle type monoclinic crystals with aspect ratio more than 20 and in the case of film sample the micro cracks appeared due to increase in volume. XPS study definitely shows that there is no change in the chemical composition of ZrO_2 sample before and after exposure to shock in presence of test gases.

Acknowledgement. We thank ISRO, DST, DRDL, AR&DB and The Director, IISc for the financial support. We thank HEA Lab team for their support during the experimental work and operation of FPST.

References

1. Boysen H, and Frey F: *Neutron powder investigation of the tetragonal to monoclinic phase transformation in undoped zirconia*, Acta. Cryst., B47, (1991) pp881-886.
2. Subba Rao E C, Maiti H S, and Srivatava H S: *Martensitic transformation in zirconia*, Phys. Status solidi A, 21 (1984) pp9-40.
3. Purohit R D, Saha S and Tyagi A K: *Combustion synthesis of nanocrystalline ZrO_2 Powder: XRD, Raman spectroscopy and TEM studies*,Materials Science and Engineering B,130,pp57-60,2006.
4. Brossnann U, Wurschum R, Sodervall U, and Schaefer H E: *Oxygen diffusion in ultrafine grained monoclinic ZrO_2*, J. Appl. Phys.,85,pp7646-7654,1999.
5. Theunissen G S A M, Winnubst A J A, and Burggraaf A J: *Surface and grain boundary analysis of doped zirconia ceramics studied by AES and XPS*, J. Mater. Sci.,27,pp5057-5066,1992.

Further studies on initial stages in the shock initiated H_2 - O_2 reaction

K. Yasunaga[1], **D. Takigawa**[1], **H. Yamada**[1], **T. Koike**[1], **and Y. Hidaka**[2]

[1] *Department of Chemistry, National Defense Academy, Hashirimizu, Yokosuka 239-8686, Japan*

[2] *Chemistry and Biology, Graduate School of Science and Engineering, Ehime University, Bunkyo-cho, Matsuyama 790-8577, Japan*

Summary. Induction times τ obtained in the shock initiated H_2-O_2 reaction were studied by observing the growth of O-atoms and were discussed by comparing formerly reported τ based on OH-radicals from a different laboratory. Conclusion is that the dependences of τ on reactants as well as buffer gas are not simple, as was expected, and the reaction proceed in a very complex way from very initial stages of reaction.

1 Introduction

The H_2-O_2 reaction mechanism has almost been established and used to simulate many combustion systems of hydrocarbon fuels. Shock tube techniques with optical observing methods played important roles in the H_2-O_2 reaction studies. Induction time τ is primary parameter to study the reaction steps and the usefulness have been demonstrated by many researchers. Rate determining step in the H_2-O_2 reaction is $H + O_2 = OH + O$ (Reaction 1) due to the endothermicity, $\Delta H_R = 70.7$ kJ mol^{-1}, where ΔH_R is heat of reaction at 298 K, and parameter, $\tau[O_2]_0$, has been used in actual simulation studies, where $[O_2]_0$ is initial concentration of O_2. It was expected that the relation between log $\tau[O_2]_0$ and $1/T$ would have a linearity of Arrhenius plots independent of compositions of test gases investigated. Although various compositions of $H_2/O_2/Ar$ were studied and reported to get the H_2-O_2 reaction mechanism, concentration dependences of initial H_2 and O_2 as well as Ar on τ have not been reported yet. Gardiner et al. [1] did extensive and consecutive studies of the reaction using their shock tube and UV-absorption technique for OH-radicals, and published a lot of papers. Their research was done in the 12-14W laboratory at the department of chemistry of the University of Texas at Austin, USA. They made several 12-14W reports mainly expressing raw data obtained in their shock tube work, and one of them concerns the relation between log $\tau[O_2]_0$ and $1/T$ observed for 13 test gas mixtures of $H_2/O_2/Ar$ for many years and by many researchers.

Figure 1 shows the collection of their results on τ which were well computer simulated by assuming [OH] at τ, threshold value of [OH], to be 2.5×10^{-10} mol cm^{-3}. Below this threshold value, no information on the reaction could be obtained. We see that $\tau[O_2]_0$ depend on the test gas compositions as well as T on the contrary to our expectation. We, therefore, proceeded multiple linear regression analysis for their data by assuming $\tau = \text{Const.}[O_2]_0{}^{\alpha}[H_2]_0{}^{\beta}[Ar]^{\gamma}\exp(E/RT)$, where Const., α, β, γ and E are constants to be evaluated and R is the gas constant. Figure 2 shows the data points re-evaluated and the regression line was obtained as Eq. 1: $\tau/\text{s} = 10^{-13.65}\ [O_2]_0{}^{-0.399}\ [H_2]_0{}^{-0.183}\ [Ar]^{-0.617}\exp(64.1\ \text{kJ mol}^{-1}/RT)$, where concentrations are in mol and cm units. Overall activation energy, E, is 5 kJ mol^{-1} less than ΔH_R. The power of $[O_2]_0$ is not -1 and the power of $[H_2]_0$ is about half of $[O_2]_0$, but is not zero. The most interesting point is that τ depends on [Ar] as well. If we neglect the influence of [Ar] on τ, τ/s $= 10^{-10.124}[O_2]_0{}^{-0.398}[H_2]_0{}^{-0.180}\exp(65.7\text{kJ mol}^{-1}/RT)$ can be obtained. The powers of $[O_2]_0$ and $[H_2]_0$ are almost equal in both cases. The parameter, $\tau[O_2]_0$, seems to be inadequate to investigate the progress of H_2-O_2 reaction by simply assuming the rate determining step to be Reaction 1. Such a mixture dependence on τ must be due to a late time of the reaction when the reaction progressed and a chain reaction system among Reaction 1, O + H_2 = OH + H (Reaction 2), and OH + H_2 = H_2O + H (Reaction 3), has been predominant.

In this study, we adopted shock tube techniques combined with atomic resonance absorption spectroscopy (ARAS) for O-atoms. Since ARAS has a high sensitivity for growth of O-atoms, it is expected to be capable of observing the reaction progress from the commencement of the reaction through O-atoms and to get another images on the reaction through τ.

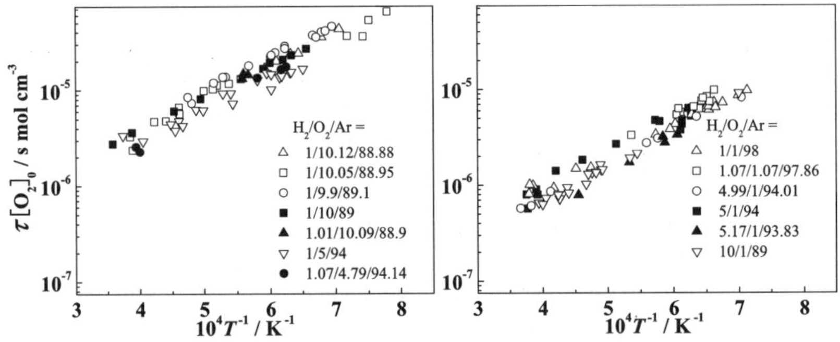

Fig. 1. $\tau[O_2]_0$ from OH-radicals vs. $10^4/T$ obtained by Gardiner's group [1] for 13 $H_2/O_2/Ar$ mixtures. Due to a great number of test gas mixtures studied, data points are divided and shown in the two fugures. Experiments were done using incident shock tube at starting pressure 10 Torr. OH-radicals were observed with UV-absorption at 306.7 nm using microwave excited Bi emission.

2 Experimental

Details of diaphragm-less incident shock tube with test gas flowing system and ARAS used in this study were reported before [2]. Since this shock tube system has been proved to work like a conventional shock tube under very clean conditions without contamination by air leaking, only present experimental conditions are described here briefly. This system would have also a merit of no necessity of equipping extra gas storing systems.

Test gases, whose compositions were determined through differences of each element gas flow rate using a mixing gas generator (GM-4B, Kofloc Co. Ltd), were continuously poured into the test section of the shock tube and extruded away at end part of the test

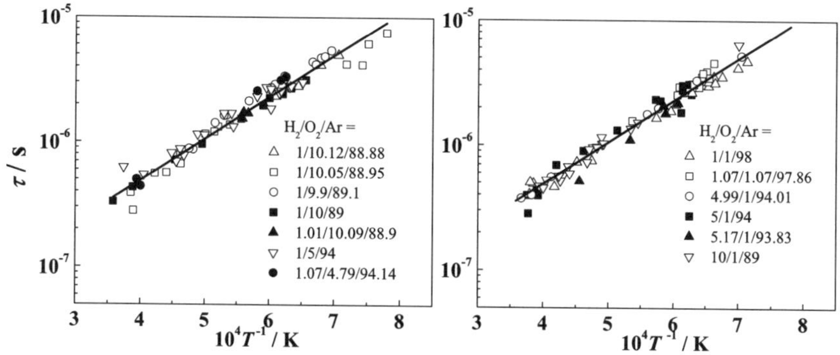

Fig. 2. Re-evaluation of τ using Eq. 1 for 13 $H_2/O_2/Ar$ mixtures. The solid line shows the result of least squares analysis for all the re-evaluated data.

section with a rotary pump. The purities of H_2, O_2 and Ar were higher than 99.999%. Incident shock waves were actuated during the continuous test gas flow at 20 Torr.

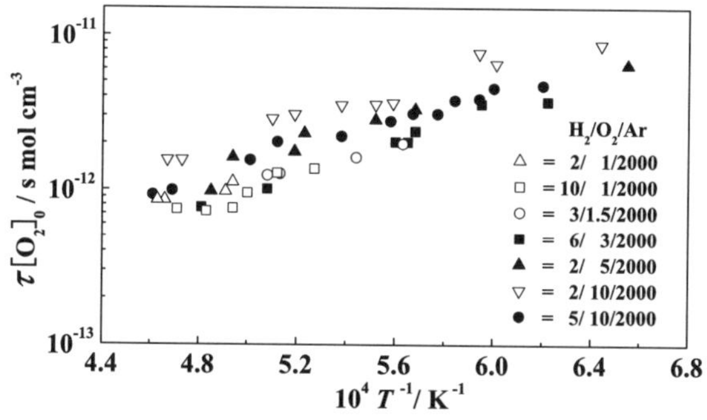

Fig. 3. $\tau[O_2]_0$ from O-atoms vs. $10^4/T$ evaluated for 7 $H_2/O_2/Ar$ mixtures. Test gas compositions are shown with flow rate(ml $\min^{-1}$) of each gas. O-atoms were observed with VUV-absorption at 130.5 nm using mirowave excited O-atoms emission.

3 Results and discussion

Figure 3 shows the relation between the products, $\tau[O_2]_0$, and $1/T$ for 7 $H_2/O_2/Ar$ mixtures. Although the value range of $\tau[O_2]_0$ between the 7 test gas compositions at same temperature is narrower than that in Fig. 1, the products did not merge on one

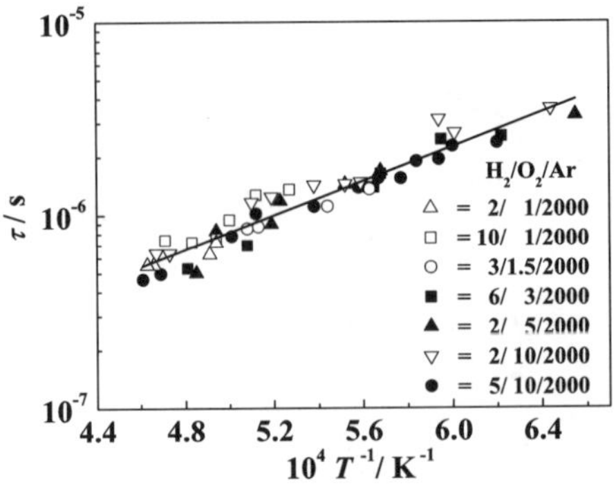

Fig. 4. Re-evaluation of τ using Eq. 2 for 7 $H_2/O_2/Ar$ mixtures. The solid line shows the result of least squares analysis for all the re-evaluated data.

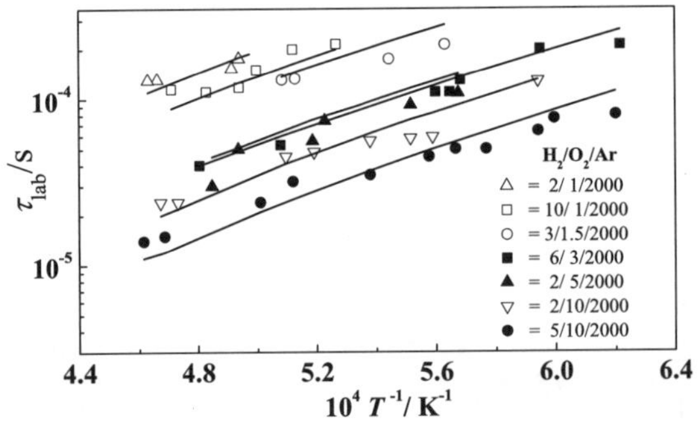

Fig. 5. Computer simulation results using the reaction mechanism in Ref. 3 for laboratory time τ based on O-atoms.

straight line either. Figure 4 shows the data points evaluated as above and the regression line was determined as Eq. 2: $\tau/\mathrm{s} = 10^{-14.32}[O_2]_0{}^{-0.772}[H_2]_0{}^{-0.227}[Ar]^{-0.044}\exp(85.1\ \mathrm{kJ\ mol^{-1}}/RT)$. The expression without [Ar] is: $\tau/\mathrm{s} = 10^{-14.118}[O_2]_0{}^{-0.782}[H_2]_0{}^{-0.225}\exp(84.5\ \mathrm{kJ\ mol^{-1}}/RT)$. In comparison to the above OH-radical case, the τ value depends on $[O_2]_0$ very much, whereas the influence of [Ar] is negligible. The power dependence of τ on $[H_2]_0$ is about one fourth of $[O_2]_0$ but is not zero under the present experimental conditions. The features of H_2-O_2 reaction can be said that the reaction progress through strong coupling between the chains, H, O, and OH.

As is shown in Fig. 5, computer simulation values for τ using recent reaction mechanism of the H_2-O_2 reaction [3] were almost accordance with the measurements when threshold of $[O] = 7.0 \times 10^{-13}\ \mathrm{mol\ cm^{-3}}$.

For OH-radicals, as stated above, the threshold value of $[OH] = 2.5 \times 10^{-10}\ \mathrm{mol\ cm^{-3}}$. Figure 6 shows how the reaction progress around the threshold values of [O] and [OH].

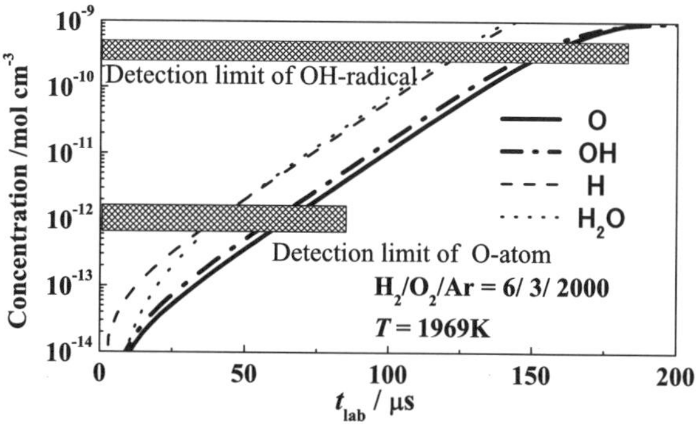

Fig. 6. Initial growth profiles of O-atoms, OH-radicals and other products obtained by computer simulation as in Fig.5. Each bar shows about ranges of detection limits of OH-radicals and O-atoms depending on the light source conditions used.

For [O] profile, an exponential growth with a time constant is seen and mutual effects by Reactions 1, 2, and 3 on the growth of [O] seem not to change. For [OH] profile, the reaction progressed very much and a simple exponential growth is not appropriate any more around τ.

Lastly, we have to mention about the buffer gas effect on the reaction after studying the Gardiner's laboratory report [1]. We did not find large differences between the two data of τ with buffer gas Ar and Kr/He. Therefore, we did not proceed the ARAS experiment for the test gases with Kr/He furthermore.

References

1. Gardiner, W. C., Jr., Pre-1970 induction time data for the hydrogen-oxygen reaction. *UT CB12W Laboratory Reports*, 11, 1970
2. Koike, T. Ichino, H., Yasunaga, K., Hidaka, Y., Reactions of C_2 oxy-hydrocarbons at high temperatures. *Proceedings of the 24th International Symposium on Shock Waves*:639-643, 2004
3. O'Conaire, M., Curran, H. J., Simmie, J. M., William, J. P., Westbrook, C. K., A comprehensive modeling study of hydrogen oxidation, *Int. J. Chem. Kinet.*, 36:603-622, 2004

Shock-tube development for high-pressure and low-temperature chemical kinetics experiments

J. de Vries, C. Aul, A. Barrett, D. Lambe, and E. Petersen

Mechanical, Materials and Aerospace Engineering, University of Central Florida, Orlando, FL, 32816, USA

Summary. A new shock-tube facility targeting lower temperatures and long test times is described. The single-pulse shock tube uses either lexan diaphragms or die-scored aluminum disks of up to 4 mm in thickness. The modular design of the tube allows for optimum operation over a large range of thermodynamic conditions from 1 to 100 atm and between 600-4000 K behind the reflected shock wave. Test times up to 20 ms can be obtained using the proper driver-driven configuration featuring a longer driver section. The system includes a smart gas handling and vacuum system; high temporal and spatial resolution, multi-channel data acquisition boards; and the capability to apply several optical diagnostics. The new facility allows for ignition delay time, chemical kinetics, high-temperature spectroscopy, vaporization, atomization, and solid particulate studies. Details on the layout are presented, and the largest potential contributors to the overall uncertainty are identified.

1 Introduction

Fundamental data such as characteristic times and species time histories at practical conditions are invaluable for the improvement and extension of chemical kinetics models to the region of interest for practical applications. The sharply risen interest in fuel flexibly issues concerning land-based power generation gas turbines over the past decade confirms the need for an apparatus capable of testing fundamental combustion properties of a large variety of fuels. Practical concerns among power generation gas turbines include autoignition in premixed systems (de Vries et al. 2007 [1]), flashback, blowoff, and combustion instability (Lieuwen et al. 2006 [2]). Shock tubes are ideal for such measurements and have been utilized extensively in providing measurements of rate coefficients for specific reactions, ignition delay times, and for the validation and improvement of entire mechanisms. Shock-tube ignition data at higher pressures and low-to-intermediate temperatures are scarce but are required for the validation of chemical kinetics models which are tuned primarily with higher-temperature and lower-pressure data. Ignition data at lower temperatures however require longer test times since the chemistry occurs more slowly.

Amadio et al. [3] have shown that shock-tube test times can be extended by the use of unconventional driver gases, such as CO_2/H_2 mixtures. Similar techniques were used for the investigation of automotive fuel blends such as the work by Ciezki and Adomeit [4], Fieweger et al. [5], Herzler et al. [6,7], and Zhukov et al. [8]. From these studies, it is evident that in the lower-temperature regime, the ignition behavior often deviates away from linearity when presented on an Arrhenius plot. Such behavior can even lead to negative temperature coefficients (NTC) as found by Fieweger et al. for n-heptane mixtures [5]. Lower-temperature (1000 K or less), longer test time shock-tube experimental data are relatively sparse, especially for gas turbine fuel blends at practical fuel-air mixture ratios.

A common way of measuring the autoignition time in this regime is with rapid compression machines (RCMs). It has been noticed recently that shock-tube experiments can in some cases disagree significantly with RCM data, especially for methane-based fuel blends [9]. Numerous suggestions have been given for this disagreement including heat transfer effects, reflected-shock bifurcation with the boundary layer, wall effects, diaphragm particle contaminants, or incident-shock chemical priming prior to reflected shock arrival [9, 10]. Several experiments have been performed including the usage of schlieren optics and/or high-speed photography [6, 10]. The facility described herein allows the study of these phenomena with a large (15 cm), polished inside diameter which has been specifically designed for these conditions. Optical access throughout the driven section allows for absorption experiments to investigate the reaction chemistry at locations other than near the endwall.

In addition to ignition delay time measurements, a shock tube can be utilized for heterogeneous combustion processes and for shock and detonation waves through aerosol-laden mixtures. The near instantly obtained test conditions of temperatures between 600 and 4000 K and pressures between 1 and 100 atm are accomplished within a controlled environment. Extended test time conditions allow for lower-temperature experiments and liquid-spray or atomization studies.

This paper presents an overview of the present shock-tube facility located at University of Central Floridas Gas Dynamics Laboratory in Orlando, Florida. Described herein are specific hardware and vacuum details, the velocity detection technique, the gas handling methodology, and the data acquisition system. An uncertainty analysis for the test temperature is performed in a similar fashion to that described by Petersen et al. [11].

2 Facility design

The total facility consists of the shock-tube hardware, control system, data acquisition system, vacuum section, and the velocity detection system. A schematic of the gas handling and the shock tube in both the conventional and long test time configuration is given in Fig. 1. A description of each key component is given as follows.

2.1 Hardware

Both the driver and driven sections of the shock tube are made of 304 stainless steel. The driver section has an ID of 7.62 cm and 1.27-cm wall thickness. The driven section has an ID of 15.24 cm, also with 1.27-cm wall thickness. The inside of the driven section is polished to a surface finish of 1 μm RMS or better. In the conventional configuration, the driver length is 2.46 m and the driven length is 4.72 m. When long test times are needed for lower-temperature experiments, the shock tube can be reconfigured to have a 4.93-m driver section and a 3.05-m driven length. All driven connections are weldless and designed for high pressure, easy removal, and minimal flow/shock perturbations between sections. The design for the driven connection is similar to that described by Petersen et al. [11] and is shown in Fig. 2. A 7700-kg inertial mass is permanently attached to the driven section to minimize shock-induced vibration of the complete assembly, particularly any displacement in the axial direction.

Pressure transducer and viewing window access is provided through 25 ports located along the tube. The protrusions on the ports are given curvature to match the inside

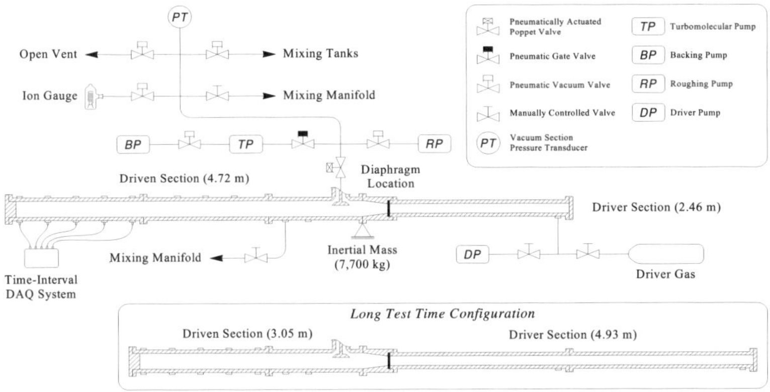

Fig. 1. Shock-tube facility, showing both configurations

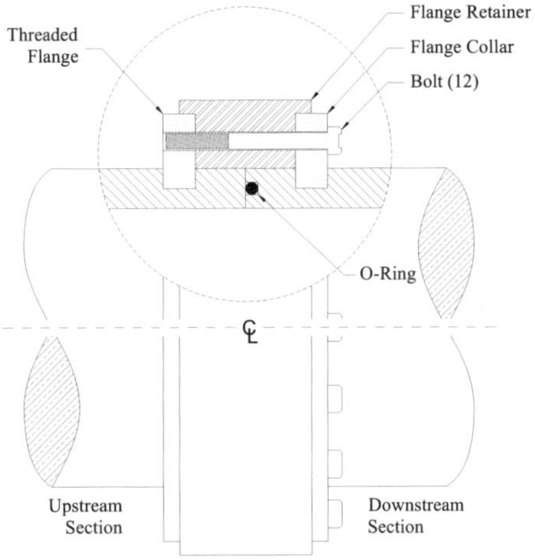

Fig. 2. Weldless driven connection

diameter of the tube as seen in Fig. 3 to minimize flow and shock obstructions in the test section. The pressure in the tube is constantly monitored by a Setra GCT-225 0-3000 psi pressure transducer. Wave speed and test pressure conditions are measured through five PCB P113A piezoelectric pressure transducers alongside the tube and one PCB 134A located at the endwall. Post reflected-shock conditions are obtained by using the incident wave speed and the initial condition in the driven tube. Five equally spaced pressure transducers offer four velocities that are then curve fitted to give the incident wave speed at the endwall location. It is shown by Petersen at al. that this technique can maintain the uncertainty below 10 K [11]. The breech-loaded assembly allows for both lexan and aluminum diaphragms (see Fig. 4). Lexan diaphragms are used for test pressures up to about 10 atm, and pre-scored aluminum diaphragms are used for pressures up to 100 atm. When lexan diaphragms are used, a special cutter is utilized to facilitate breakage of the diaphragm and prevent diaphragm fragments from tearing off.

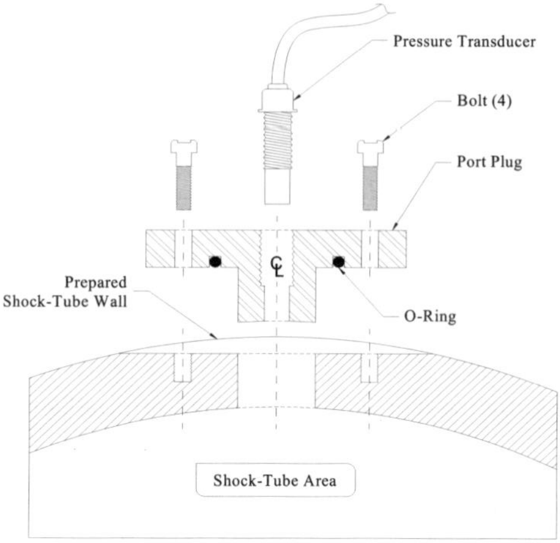

Fig. 3. Pressure transducer port

2.2 Mixing tanks

Test mixtures are created in three different mixing tanks of 1.22 m, 1.83 m, and 3.05 m length made from 304 stainless steel tubing with 15.24-cm ID and 1.27-cm wall thickness. The 1.8-m length mixing tank can be heated for low vapor pressure experiments. The pressure in the mixing tanks is measured using three Setra GCT-225 pressure transducers (2 x 0-250 psi and 1 x 0-500 psi). All mixing tanks are connected to the vacuum system and can be pumped down to pressures below 10^{-6} Torr. The heated mixing tank is directly connected to the driven section so mixed gases do not have to pass through uninsulated plumbing. Different gases are passed through a perforated stinger in the center of each mixing tank to allow for turbulent mixing.

2.3 Vacuum system

A high-vacuum system has been designed to create high-purity mixtures. The driven section is pumped down to about 50 mTorr using a Varian DS402 (410 L/min) roughing pump. At approximately 50 mTorr, a Varian 551 (450 L/sec with He) Turbo-molecular pump with a Varian DS302 (285 L/min) backing pump takes over which can pump the system down to 10^{-6} Torr. The pressure is measured using two MKS Baratron model 626A capacitance manometers (0-1000 Torr and 0-10 Torr) and an ion gauge for high vacuum. The driver tube is evacuated by a separate Varian DS102 vacuum pump (114 L/min). A pneumatically driven poppet valve matching the inside diameter of the driven

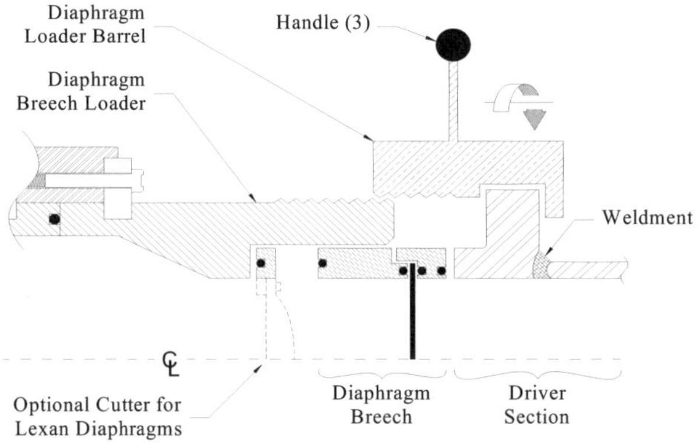

Fig. 4. Breech diaphragm assembly

section is used to separate the tube from the vacuum system. This poppet-valve design allows for a 7.62 cm passage between the vacuum cross section and the driven tube.

3 Conclusion

A new shock tube facility for chemical kinetics and heterogeneous combustion experiments has been described. The weldless flanges of the driven section allow for the shock tube to be configured in favor of the test conditions needed. The thermodynamic conditions behind the reflected shock wave can be 1-100 atm and 600-4000 K. The test pressure can be set by using either lexan or pre-scored aluminum diaphragms. The facility is ideal for gas phase as well as heterogeneous mixtures at high to intermediate temperatures.

Acknowledgement. This work was supported by the National Science Foundation, contract number CTS-0547159, and in part by a Provost Fellowship from the University of Central Florida.

References

1. de Vries J., and Petersen E.L.: Proc. Combust. Inst. **31**, 3163 (2007)
2. Lieuwen T., McDonell V., Petersen E.L., Santavicca D.: Fuel Flexibility Influences on Pre-Mixed Combustor Blowout, Flashback, Autoignition, and Stability. ASME Paper GT2006-90770 (2006)
3. Amadio A.R., Crofton M.W., Petersen E.L.: Shock Waves **16**, 157 (2006)
4. Ciezki H.K., and Adomeit G.: Combust. Flame **93**, 421 (1993)
5. Fieweger K., Blumenthal R., Adomeit G.: Combust. Flame **109**, 599 (1997)
6. Herzler J., Jerig L., Roth P.: Combust. Sci. Technol. **176**, 1627 (2004)
7. Herzler J., Jerig L., Roth P.: Proc. Combust. Inst. **30**, 147 (2005)
8. Zhukov V.P., Sechenov V.A., Starikovskii A.Yu.: Combust. Flame **140**, (2005)

9. Petersen E.L., Lamnaouer M., de Vries J., Curran H., Simmie J., Fikri M., Schulz C., Bourque G.: Discrepancies Between Shock Tube and Rapid Compression Machine Ignition at Low Temperatures and High Pressures. Proceedings of the 26th International Symposium on Shock Waves, Springer-Verlag (2007)
10. Goy C.J., Moran A.J., Thomas G.O.: Autoignition Characteristics of Gaseous Fuels at Representative Gas Turbine Conditions. Proceedings of TURBO EXPO AMSE GT-0051 (2001)
11. Petersen. E.L., Rickard M.J.A., Crofton M.W., Abbey E.D., Traum M.J., Kalitan D.M.: Meas. Sci. Technol. **16**, 1716 (2005)

Shock-tube study of tert-butyl methyl ether pyrolysis

K. Yasunaga[1], **Y. Hidaka**[1], **A. Akitomo**[1], **and T. Koike**[2]

[1] *Chemistry and Biology, Graduate School of Science and Engineering, Ehime University, Bunkyo-cho, Matsuyama 790-8577, Japan*
[2] *Department of Chemistry, National Defense Academy, Hashirimizu, Yokosuka 239-8686, Japan*

Summary. The high temperature pyrolysis of tert-butyl methyl ether (TBME) was studied behind reflected shock waves using single-pulse (heating time between 1.5ms and 2.2ms) and time-resolved IR (3.39μm) and UV-absorption (195nm) methods. The studies were done using mixtures, 1.00%, 0.05%, 0.025%, 0.017% and 0.013% TBME diluted in Ar, in the temperature range 1000-1500K at total pressures between 1.0 and 9.0atm. From a computer-simulation study, a 56-reaction mechanism that could explain all our data was constructed. The reaction TBME $\rightarrow$ $iC_4H_8+CH_3OH$ and its limiting high-pressure rate constant 9.6×10^{13}exp(-59.9 kcal mol$^{-1}/RT$) s^{-1} were found to be important for reproducing experimental results.

1 Introduction

Ethers are expected to be an additive in gasoline, an octane improver or a diesel fuel. Tert-butyl ethyl ether as an additive in gasoline, which is synthesized from biomass ethanol and iso-butene, has merit of reducing CO_2 emission into atmosphere from the concept of carbon neutral. In the combustion of dimethyl ether as diesel fuel, relatively clean exhausted gas is emitted. As for TBME, it is used as an octane improver for the replacement of alkyl lead compounds. The modeling of the combustion of ethers is important for a better understandings of predicting the emissions of pollutants, as well as for effective use of energy. The gas-phase oxidation of pure TBME has been investigated using several techniques, such as a static reactor [1], a flow reactor [2], shock tubes [3–5] and jet-stirred reactors [6–8]. In the present study, we propose a pyrolysis mechanism and rate constants of TBME, which is an essential part of its oxidation mechanism, using shock tube above 1000K and compare with the former studies.

2 Experimental

Three shock tubes were used in this study. The first was a magic-hole-type with 4.1 cm i.d. [9]. The reacted gas mixtures were extracted into a pre-evacuated vessel (50cm^3) through the valve near the end plate, analyzed by three serially connected gas-chromatographs each having a TCD. The gas-chromatographic analysis which was similar to that in Hidaka et al. [10] was used to determine the concentrations of reactant and products within an accuracy of $\pm$4%. An effective heating time was determined using the same method as that in Hidaka et al. [10] with an accuracy of $\pm$5% under our experimental conditions. The concentrations of observed compounds were compared with those simulated. The modeling methodology for this comparison was shown in detail in Hidaka et al. [9]. The second shock-tube was a standard-type shock tube with 4.1cm i.d. connected to laser absorption system which was the same as that described in Hidaka et al. [11]. The third shock-tube was a standard-type shock tube connected to UV absorption system. UV light obtained by deuterium lamp passing through shock tube with 7.1cm i.d. at observation station, dispersed into 195nm by monochromator (half width=1.6nm) was detected by photomultiplier (R208; Hamamatsu photonics). The detail of this shock tube was described in Koike et al. [12]. Test gas mixtures employed were 1.00%, 0.05%, 0.025%, 0.017% and 0.013% TBME diluted in Ar. The initial pressure (P_1) was fixed to 50 torr in IR absorption and gas analyzing experiments, while it was changed from 100 to 300 torr in UV absorption experiments in order to investigate the pressure dependence of the rate constant for TBME = iC_4H_8 + CH_3OH. The Ar from Teisan Co. and Iwatani Co. specified to be 99.999% pure was used without further purification. The TBME (Sigma-Aldrich) specified to be 99.8% pure was frozen, degassed a number of times and purified by trap-to-trap distillation before use.

The computer calculations used in this study were essentially the same as those described in Hidaka et al. [10]. The computer routine used was Gear-type integration of the set of differential equations describing the chemical kinetics under constant density conditions for reflected shock waves. Reverse reactions were automatically included in the computer program through equilibrium constants computed from thermochemical data. The basic thermochemical data source was the JANAF table [13]. The heat of formation of TBME was adopted from reported table [14]. The thermochemical data of TBME decomposition products of tC_4H_9O, aC_4H_9O, raTBME and rdTBME were estimated using a group additivity method reported by Benson [15]. The structures of above four radicals are described in Fig.1. The heats of formation used for the species, TBME, tC_4H_9O, aC_4H_9O, raTBME and rdTBME were -60.9 kcal mol^{-1}, -16.6 kcal mol^{-1}, -14.4 kcal mol^{-1} , -19.0 kcal mol^{-1} and -19.0 kcal mol^{-1} at 0 K, respectively. Thermochemical data other than the JANAF table and above were adopted from Table 2 shown in Hidaka et al. [10, 16].

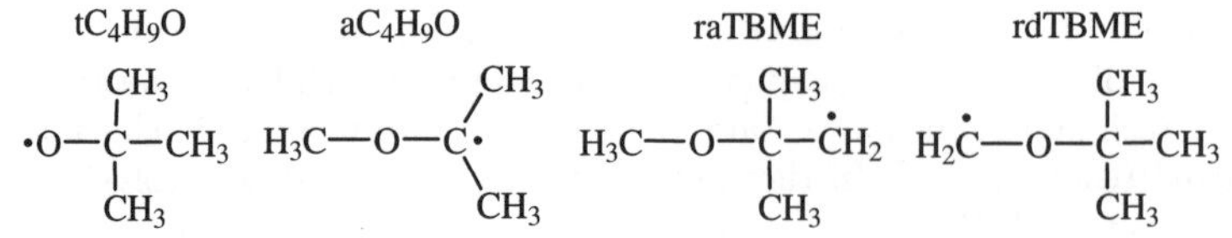

Fig. 1. Structures of four radicals produced in TBME pyrolysis.

3 Results and discussion

The mixture of 1.0% TBME diluted in Ar was heated in the temperature range 1000-1500K using magic-hole-type shock-tube. The reacted gas mixture was analyzed by gas chromatographs. The temperature dependence of distribution for main products is shown in Fig.2. Under our experimental conditions, the main products detected were CH_3OH, iC_4H_8, CO, CH_4, C_2H_6, CH_2O, C_3H_6, acetone, C_2H_4, aC_3H_4(allene), pC_3H_4(propyne) and C_2H_2.

The 1.0% TBME diluted in Ar was heated in the temperature range 1000-1500K and the profiles of IR-laser absorption in the TBME pyrolysis were measured at 3.39μm. A_t was defined as the following equation. $A_t = \log(I_f/I_t)/\log(I_f/I_0)$, where I_f was the

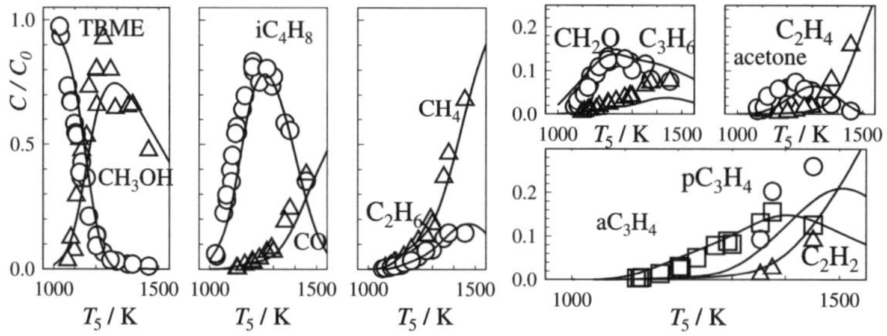

Fig. 2. Comparison of the observed main products distribution (symbols) with those calculated (curves), where C_0 is the initial concentration of TBME and C is the concentration of reactant and each main product. The effective heating time at 1000, 1200, 1400 and 1600K were 2.20, 2.00, 1.75 and 1.55ms, respectively.

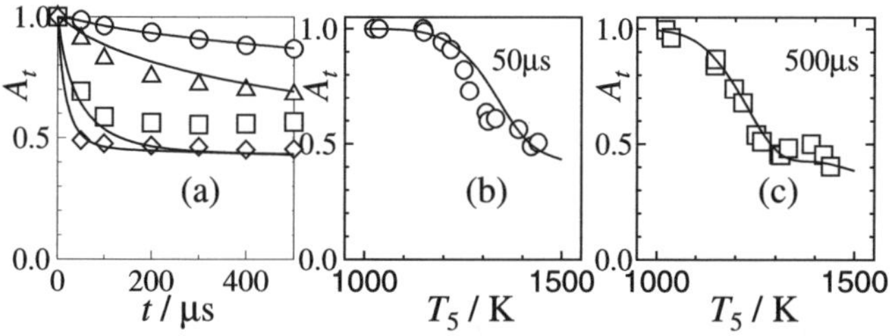

Fig. 3. (a)Comparison of the observed time-profiles A_t(symbols) with those calculated (curves) using 1.0% TBME in Ar:open circles, 1150K :open triangles, 1218K :open squares, 1357K :open diamonds, 1422K.(b)Comparison of the observed A_t (symbols) with those calculated (curves) using 1.0% TBME in Ar at 50μs. (c)Comparison of the observed A_t (symbols) with those calculated (curves) using 1.0% TBME in Ar at 500μs.

signal voltage corresponding to the full intensity and I_0 and I_t were the signal voltages corresponding to the absorption intensity at the reflected shock front at (t=0) and at time t, respectively. Four typical A_t at 3.39μm vs. time profiles for 1.0% TBME diluted in Ar are shown in Fig. 3(a). The relationships between absorption data A_t at 50 and 500μs and the shock heated temperature using mixture 1.0% TBME diluted in Ar are shown in Fig.3(b) and (c). In order to determine the calculated A_t of the reactant and each product at 3.39μm, their extinction coefficients were measured. The equations for the extinction coefficient of TBME and iC_4H_8 are $\epsilon_{IR}(\text{TBME}) = 1.53 \times 10^5 - 48.2 \times T$ cm^2mol^{-1} and $\epsilon_{IR}(iC_4H_8) = 5.87 \times 10^4 - 20.6 \times T$ cm^2mol^{-1}, respectively. Those for the other species detected were shown in previous reports, Hidaka et al. [10, 17].

The mixtures of 0.05%, 0.025%, 0.017% and 0.013% TBME diluted in Ar were heated in the temperature range 1170-1330K using standard-type shock-tube connected to UV absorption system. In order to reproduce the observed absorbance, $\log(I_f/I_t)$, the extinction coefficients of TBME and iC_4H_8 were measured. The equations for extinction coefficient were $\epsilon_{UV}(iC_4H_8) = 3.44 \times 10^6$ cm^2mol^{-1} and $\epsilon_{UV}(\text{TBME}) = -1.08 \times 10^6 + 1080 \times T$ cm^2mol^{-1}, respectively. Typical absorbance profile in TBME pyrolysis at 195nm is shown in Fig.4(a).

From the reaction mechanism of TBME pyrolysis including the reactions in Table 1, 93% of TBME is consumed by reactions (1),(3) and (4) and 6% by reactions (6) and (8) at 400 μs under the condition in Fig.4(a). Though the influence of secondary reactions was not neglected completely, TBME pyrolysis proceeds mainly via unimolecular reactions. Considering only the reactions (1),(3) and (4), the time dependent concentrations of iC_4H_8 and TBME were described as $[iC_4H_8]_t = \{k_4/(k_1+k_3+k_4)\}[\mathrm{TBME}]_0[1 - \exp\{-(k_1+k_3+k_4)t\}]$ and $[\mathrm{TBME}]_t = [\mathrm{TBME}]_0\exp\{-(k_1+k_3+k_4)t\}$, respectively, where $[\mathrm{TBME}]_0$, $[iC_4H_8]_t$ and $[\mathrm{TBME}]_t$ are the initial concentration of TBME and the concentration of iC_4H_8 and TBME at time t, respectively. Above equations were used as a model function combined with Lambert-Beer law to reproduce the absorbance profiles obtained by UV absorption experiments. The solid curve in Fig.4(a) shows the best-fit (k_1+k_3=650 s^{-1},k_4=2470 s^{-1}) curve for sum of absorbance from TBME and iC_4H_8 obtained by non-linear least squares method. Dotted and chain curves show the individual absorbance of TBME and iC_4H_8, respectively. The reliabilities of estimated k_1+k_3 and k_4 were examined by changing them independently. The solid curve with filled circles and squares show the calculated curves obtained by changing k_1+k_3 to $2\times(k_1+k_3)$ and k_4 to $2\times k_4$, respectively. It was found that the calculated profile was not sensitive to the variation of k_1+k_3 value, so the accuracy of estimated k_1+k_3 values is low.

Arrhenius plots of k_1+k_3 and k_4 is shown in Fig.4(b). The estimated rate constant $k_4 = 9.6 \times 10^{13} \exp(-59.9\ \mathrm{kcal\ mol^{-1}}/RT)\ s^{-1}$ is shown, compared with those of Brocard et al. [1] at 0.25-1.3atm and 573-773K and Goldaniga et al. [7] at 10atm and 800-1150K. The difference of k_4 values between Brocard et al. and Goldaniga et al. is about 6 times. It infers the pressure dependence of k_4. On the other hand, k_4 values estimated by our UV absorption experiments did not show the noticeable pressure dependence at the pressure range 2.5-8.9 atm. Our k_4 value, moreover, could explain the products distribution described in Fig.2 and A_t proflies described in Fig.3(a)(b) and (c) around 1 atm. We concluded that k_4 reached its limitting high-pressure value.

Table 1. Elementary reactions and rate constant expressions.

No.	Reaction	A	E	ΔH_0
1	TBME = $tC_4H_9O + CH_3$	1.5×10^{15}	76,500	80,100
2	TBME = $tC_4H_9 + CH_3O$	1.0×10^{15}	80,000	85,000
3	TBME = $aC_4H_9O + CH_3$	5.0×10^{15}	76,500	82,300
4	TBME = $iC_4H_8 + CH_3OH$	9.6×10^{13}	59,900	15,900
5	TBME + H = raTBME + H_2	3.0×10^{14}	13,000	-9,500
6	TBME + H = rdTBME + H_2	1.0×10^{14}	6,000	-9,500
7	TBME + CH_3 = raTBME + CH_4	6.0×10^{12}	12,000	-9,500
8	TBME + CH_3 = rdTBME + CH_4	2.0×10^{12}	7,000	-9,500
9	raTBME = $iC_4H_8 + CH_3O$	5.0×10^{14}	24,800	24,700
10	rdTBME = $iC_4H_9 + CH_2O$	2.5×10^{14}	19,800	10,600
11	$tC_4H_9O = C_2H_6CO + CH_3$	1.0×10^{14}	7,000	4,000
12	$aC_4H_9O = C_2H_6CO + CH_3$	1.0×10^{14}	7,000	1,800

Rate constants in the form, $A\exp(-E/RT)$, in cm, mol, cal and K units. Heat of reactions, ΔH_0, in cal $\mathrm{mol^{-1}}$ unit.

The k_4 estimated by above method was used in simulation to reproduce our other experimental results as described in Table 1, while that of k_1+k_3 was not adopted because

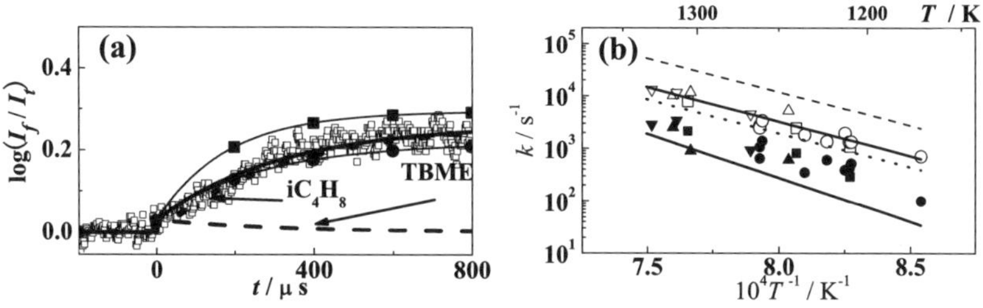

Fig. 4. (a)Typical absorbance profile in TBME pyrolysis at 195nm. Shock conditions: 0.05% TBME in Ar, T_5=1262K, P_5=2.8 atm, $\rho_5 = 2.7\times10^{-5}$ mol cm^{-3}. solid line, the best fit curve: dotted line, absorbance from iC_4H_8: chain line, absorbance from TBME. (b) Arrhenius plots of k_4 and k_1+k_3. Upper solid line, k_4 estimated by UV absorption method: lower solid line, sum of k_1 and k_3 described in Table 1: dotted line, k_4 value of Brocard at 0.25-1.3atm and 573-773K: chain line, k_4 value of Goldaniga at 10atm and 800-1150K: open symbols, k_4 value estimated: filled symbols, k_1+k_3 value estimated: circles, P_5=2.5-2.8atm, 0.05%TBME in Ar: triangle, P_5=5.6-6.1atm, 0.013%TBME in Ar :inverted triangles, P_5=5.5-6.0atm, 0.025%TBME in Ar :squares, P_5=7.9-8.9atm, 0.017%TBME in Ar.

of its low accuracy. A TBME pyrolysis mechanism including the most recent pyrolysis sub-mechanisms for formaldehyde, methane, ethane, ethylene, acetylene, propene, allene and propyne reported (See Hidaka et al. [9, 11, 16–20]) was constructed because these might be produced in TBME pyrolysis at temperatures above 1000K. To interpret our data, computer modeling was performed. The possible initial steps in the high temperature pyrolysis of TBME might be the molecular elimination reaction to iC_4H_8 and CH_3OH, which was already discussed above, C-C bond fission and C-O bond fission. The rate constants of reactions (1) and (3) were estimated to explain the distribution of acetone shown in Fig.2 , because two kinds of radicals tC_4H_9O and aC_4H_9O produced via reactions (1) and (3) decompose producing acetone and CH_3 via reactions (11) and (12). The same activation energy was assumed and pre-exponential factors were adopted based on the number of CH_3 group for reactions (1) and (3). The rate constant of reaction (2) was assumed. The produced H atom may abstract H atoms from two different positions in TBME by the reactions (5) and (6). These reactions produce two different radicals raTBME and rdTBME. The CH_3 radical may also abstract the H atoms from TBME by reactions (7) and (8) producing raTBME and rdTBME. Both of the radicals decompose producing iC_4H_8, CH_2O and H via reactions (9), (10) and unimolecular decomposition of iC_4H_9 and CH_3O. At 1350K using 1.0% TBME mixture, reactions (1)-(8) consumed about 20% of TBME at 10μs and about 80% at 95μs, especially about 65% and 80% of consumed TBME was due to reaction (4) at 10 and 95 μs, respectively. The shapes of IR absorption profiles at 3.39μm are also sensitive to reaction (4), because the main absorbers are TBME and iC_4H_8 . As are seen in Fig.3(a),(b) and (c), simulation results could explain experimental results well. The consumption of TBME and formation rate of main products, CH_3OH, CO, iC_4H_8 and acetone were very sensitive to the variation of the rate constants k_1, k_3 and k_4. The formation rate of CH_2O was sensitive to (5)-(8), especially (6) and (8). Thus reaction mechanism proposed in this study could explain all of our data, especially the rate constant for reaction (4) estimated by UV absorption experiments was found to be very important for reproducing experimental results.

References

1. Brocard, J. C., Baronnet, F., and O'Neal, H. E. Chemical kinetics of the oxidation of methyl tert-butyl ether (MTBE). *Combust. Flame*, 52:25-35, 1983
2. Norton, T. A., Dryer, F. L. The flow reactor oxidation of C_1-C_4 alcohols and MTBE. *Twenty-Third Symposium (International) on Combustion* :179-185, 1990
3. Dunphy, M. P., Simmie, J. M. Combustion of methyl tert butyl ether. Part I:Ignition in shock waves. *Combust. Flame*, 85:489-498, 1991
4. Curran, H. J., Dunphy, M. P., Simmie, J. M., Westbrook, C. K., Pitz, W. J. Shock tube ignition of ethanol, isobutene and MTBE: experiments and modeling. *Twenty-Fourth Symposium (International) on Combustion*:769-776, 1992
5. Fieweger, K., Blumenthal, R., Adomeit, G. Shock-tube investigations on the self-ignition of hydrocarbon-air mixtures at high pressures *Twenty-Fourth Symposium (International) on Combustion* :769-776, 1992
6. Ciajolo, A., D'anna, A., Kurz, M. Low-temperature oxidation of MTBE in a high-pressure jet-stirred flow reactor *Combust. Sci. and Tech.* 123:49-61, 1996
7. Goldaniga, A., Faravelli, T., Ranzi, E., Dagaut, P., Cathonnet, M. Oxidation of oxygenated octane improvers: MTBE, ETBE, DIPE, AND TAME *Twenty-Seventh Symposium (International) on Combustion* :353-360, 1998
8. Glaude, P. A., Battin-leclerc, F., Judenherc, B., Warth, V., Fouenet, R., Come, G. M., Scacchi, G., Dagaut, P., Cathonnet, M. Experimental and modeling study of the gas-phase oxidation of methyl and ethyl tertiary butyl ethers *Combust. Flame* 121:345-355, 2000
9. Hidaka, Y., Nakamura, T., Miyauchi, A., Shiraishi, T., Kawano, H., Thermal decomposition of propyne and allene in shock waves *Int. J. Chem. Kinet.* 21:643-666, 1989
10. Hidaka, Y., Higashihara, T., Nimomiya, N., Oshita, H., Kawano, H., Thermal isomerization and decomposition of 2-butyne in shock waves *J. Phys. Chem.* 97:10977-10983, 1993
11. Hidaka, Y., Taniguchi, T., Tanaka, H., Kamesawa, T., Inami, K., Kawano, H. Shock-tube study of CH_2O pyrolysis and oxidation *Combust. Flame* 92:365-376, 1993
12. Koike, T. Kudo, M. Maeda, I. Yamada, H. Rate constants of $CH_4+M=CH_3+H+M$ and $CH_3OH+M=CH_3+OH+M$ over 1400-2500K *Int. J. Chem. Kinet.* 32:1-6, 2000
13. M.W. Chase et al in JANAF thermochemical tables. 3rd Ed *J. Phys. Chem. Ref. Data* 1, 1985
14. Sharon, G. L. et al. Gas-phase ion and neutral thermochemistry *J. Phys. Chem. Ref. data* 17, 1988
15. Benson, SW. Thermochemical kineics. 2nd Ed *Wiley New York*, 1976
16. Hidaka, Y., Hattori, K., Okuno, T., Inami, K., Abe, T., Koike, T. Shock-tube and modeling study of acetylene pyrolysis and oxidation. *Combust. Flame* 107:401-417, 1996
17. Hidaka, Y., Nakamura, T., Tanaka, H., Jinno, A., Kawano, H., Higashihara, T., Shock-tube and modeling study of propene pyrolysis. *Int. J. Chem. Kinet.* 24:761-780, 1992
18. Hidaka, Y., Sato, K., Henmi, Y., Tanaka, H., Inami, K., Shock-tube and modeling study of methane pyrolysis and oxidation. *Combust. Flame* 118:340-358, 1999
19. Hidaka, Y., Sato,K., Hoshikawa, H., Nishimori, T., Takahashi, R., Tanaka, H., Imai, K., Ito, N., Shock-tube and modeling study of ethane pyrolysis and oxidation. *Combust. Flame* 120:245-264, 2000
20. Hidaka, Y., Nishimori, T., Sato, K., Henmi, Y., Okuda, R., Inami, K., Higashihara, T., Shock-tube and modeling study of ethylene pyrolysis and oxidation. *Combust. Flame* 117:755-776, 1999

Temperature dependence of the soot yield in shock wave pyrolysis of carbon-containing precursors

A. Drakon[1], **A. Emelianov**[1], **A. Eremin**[1], **A. Makeich**[1], **H. Jander**[2], **H.G. Wagner**[2], **C. Schulz**[3], **and R. Starke**[3]

[1] *IHED AIHT Russian Academy of Sciences, Moscow 127412, Russia*
[2] *IPC, Göttingen University, Gottingen 37077, Germany*
[3] *IVG, University of Duisburg-Essen, Duisburg 47048, Germany*

Summary. In this work a study of the actual gas-phase temperatures during carbon particle formation from different carbon precursors behind shock waves has been carried out. A pyrolysis of C_3O_2, CCl_4, C_2Cl_4 and C_6H_6 at initial ("frozen") temperatures behind the reflected shock wave in the range 1500 - 3000 K have been studied. A significant difference between frozen and real temperatures behind shock waves was found, which was related to the heat consumption of the precursor decomposition and the heat release during particles formation. The obtained results have shown that the real temperature dependence of particle yield in all measured mixtures does not depend on the kind and initial concentration of precursors and close to that measured in premixed flames.

1 Introduction

Soot formation in pyrolysis of carbon-containing gas-phase materials (carbon-particle precursors) is important for many practical applications. The underlying processes have been studied in a number of works. One of the most remarkable results of these studies is the temperature dependence of the yield of final condensed carbon that was determined via extinction measurements. The observed optical density of the particle-laden mixture shows a bell shape when plotted as a function of temperature of shock-heated gas mixtures. However the position of the bells differs for different precursors [1], [2]. It was concluded in [3], [4] that the actual temperature can differ significantly from the calculated post-shock values due to the thermal effect of precursor decomposition and particle formation. Therefore, the temperature dependency of the optical density and of the particles size in the various mixtures must be reconsidered in view of the actual temperatures of the reacting mixtures. To this end, in the present work the actual gas-phase temperature during carbon particle formation from different precursors has been measured in a time resolved manner after the reflected shock wave using absorption/emission spectroscopy at 2.7 μm.

2 Measurements

The experiments were carried out behind reflected shock waves in a stainless-steel shock tube with an inner diameter of 80 mm with a 2.5 m long driver section and a 6.3 m long driven section at IVG Duisburg-Essen. The process of carbon particle formation was studied in mixtures of Ar with small amount of different carbon precursors (C_3O_2, CCl_4, C_2Cl_4, C_6H_6) and addition of several percents of CO_2. CO_2 was added as a target molecule for IR thermometry to enable temperature measurements before the onset of particle formation as well. The initial pressures of the test mixture were varied between 30 and 130 mbar. To determine the initial post-shock gas properties velocities of the shock waves were measured from the signals of piezoelectric pressure gauges. The resulting "frozen" post-shock gas temperatures T_5 were determined based on one-dimensional gasdynamic theory and the assumption of ro-vibrational equilibrium and frozen chemical reactions. In our experiments temperatures and pressures were varied in the ranges 1400 K $< T_5 <$ 3000 K and 2.5 $< p_5 <$ 3.50 bar. The maximum test time behind the reflected shock wave varied depending on experimental conditions in the range 800 - 1000 μs. The IR emission-absorption measurements are based on two identical optical arrangements. For detection a dense region of the CO_2 absorption/emission spectrum at $\lambda = 2.7$ μm was selected by band pass filters. While one channel measured emission only, the second channel detected the combination of absorption and emission in an identical volume within the probe region. As a reference light source an electrically-heated tungsten tube (inner diameter 3 mm, tube length 40 mm) was used. Internal reflection inside the tube ensured "near-black-body" conditions (emissivity ≈ 0.95 at 2.7 μm). Light from the open end of the incandescent tube was collimated with a CaF_2 lens (f = 150 mm, d = 25 mm) and send through the probe volume. The effective brightness temperature of the radiation at $\lambda = 2.7$ μm inside the probe volume was measured with a pyrometer and set to $T_0 = 1700 \pm 20$ K in all experiments. The time resolution of the detection system was about 10 μs. To record the particle yield, light attenuation was detected with the beam of a HeNe laser at 633 nm.

3 Results

In order to check the performance of the temperature measurement experiments were performed in a test mixture, that contained a small amount of CO_2 in argon only. At temperatures below 2500 K this mixture is non reactive, and can therefore be used for testing the applied technique. In each experiment, emission and absorption-emission signals at $\lambda = 2.7$ μm and laser-attenuation signal at 633 nm were measured. From emission and emission-absorption signals at $\lambda = 2.7$ μm measured during experiment the post-shock temperature was determined as a function of time. A good agreement between measured temperature and calculated "frozen" temperature T_5 was observed (see Figure 1A). Note, that the temperature peak visible on the plot immediately after the reflected shock is due to schlieren effects in the absorption-emission signal. Therefore the temperature measurements during the first 10 - 15 μs after the reflected shock were neglected. The estimated error of the temperature measurements depended on the difference between T_0 and T_5 and varied between $\pm$ 20 K at $T_5 \approx T_0$ to $\pm$ 50 K at $T_5 \approx T_0$ + 300 K. Measurements in reactive mixtures with 3%C_3O_2+5%CO_2 in Ar were performed in the range of "frozen" post-shock temperatures of T_5 = 1400 - 3000 K. According to [1] this temperature range included the whole region of the observed bell-shaped dependence of optical density and particle size that has been observed before. The extinction profile at 633 nm and the evaluated temperature obtained at T_5 = 1743 K and p_5 = 3.2 bar are displayed in Figure 1B. One can see, that during the first 150 μs the temperature is stable and it is in a good agreement with the "frozen" post-shock temperature T_5 = 1743 K. The measured temperature indicates that the C_3O_2 decomposition is essentially temperature-neutral. The extinction signal indicates a fast particle growth without induction time. At t $\approx$ 150 μs after the reflected shock the temperature rises by $\approx$ 100 K at the same time as the onset of particle formation that was observed from a significant increase of extinction at 633 nm. Note, that at high "frozen" temperatures $T_5 > 1850$ K no induction time was observed and measured temperatures immediately show values significant higher than the "frozen" post-shock temperature.

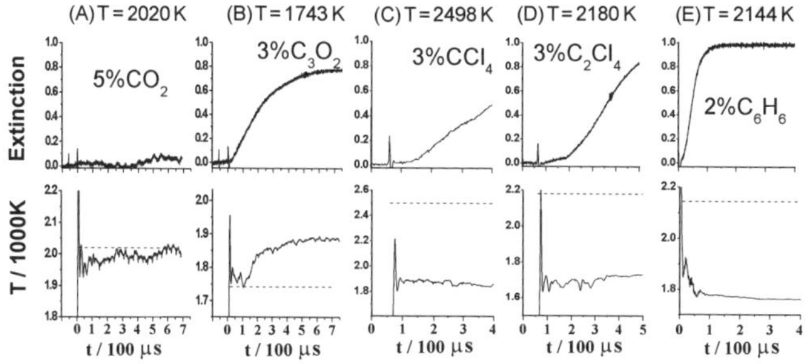

Fig. 1. Examples of the evaluated temperatures and extinctions at λ = 633 nm in the mixtures: (A) 5%CO_2+Ar; (B) 3%C_3O_2+5%CO_2+Ar; (C) 3%CCl_4+5%CO_2+Ar; (D) 3%C_2Cl_4+5%CO_2+Ar; (E) 2%C_6H_6+5%CO_2+Ar.

In the following experiments the carbon particle formation during CCl_4 decomposition was investigated. In Figure 1C extinction and measured temperature signals in mixture 3%CCl_4+5%CO_2+Ar at T_5 = 2498 K and p_5 = 3.3 bar are presented. The measured temperatures after the reflected shock wave immediately show values that are significant lower than the "frozen" post-shock temperatures due to decomposition of CCl_4. The process of CCl_4 decomposition at this temperature is very fast, therefore, the temperature decrease can not be resolved temporally and one can see the net result and schlieren effect. The extinction signal rises slowly after an induction time of approximately 100 μs which indicates the formation of particles. The temperature drop in this case is about 600 K and the temperature value is stable during the experiment. In the mixture of 3%C_2Cl_4+5%CO_2+Ar at T_5 = 2180 K and p_5 = 3.1 bar (Fig. 1D) the behavior of presented signals looks like in previous case, and the measured temperature shows values about 500 K below T_5. The last experimental series was devoted to study of C_6H_6 decomposition. At low temperatures $T_5 < 2100$ K pyrolysis of C_6H_6 proceeds rather slow and gradual decrease of temperature during the experiment is observed (Fig. 1E). The performed measurements allow to evaluate the difference between the real gas-phase temperature T_{exp} (established due to the pyrolysis of initial precursors and particles formation) and the "frozen" T_5 temperatures.

In Figure 2 the dependence of this difference $\Delta T = T_{exp} - T_5$ is shown for all investigated mixtures. The measured temperature value T_{exp} was taken at the time of maximum extinction signal value. In C_3O_2 containing mixtures the temperature difference increases with increasing post-shock temperature T_5 (Fig. 2A). This means that the heat release during carbon particle formation is higher than the heat consumption of the C_3O_2 decomposition. That is typical only for the experiments with C_3O_2 mixtures. In all other mixtures the temperature difference has a negative value (Fig. 2B - D), i.e. the real temperature T_{exp} was lower than the frozen temperature T_5. One can see that the absolute difference between the real and frozen temperatures increase with the increase of post-shock temperatures. As it can be seen from the plots, in all cases the dependencies $T = f(T_5)$ are linear. The main reason for that is the increase of the degree of decomposition of initial species with the temperature rise. On the other hand, the observed temperature difference ΔT has a nonlinear dependence on the concentration of reacting molecules. In the mixtures with CCl_4, a decrease of its concentration by a factor of 2 causes the ΔT to decrease by just 20-30% (Fig. 2B) and in case of mixtures containing C_6H_6 a decrease of initial concentration of C_6H_6 to 8 times leads to reduction of ΔT by a factor of 3 (Fig. 2D). This fact can be explained by incomplete precursor decomposition in rich mixtures and by the influence of exothermic reactions of products of partial precursors decomposition (by example, C_xH_y formation from C_6H_6 fragments during the pyrolysis of C_6H_6 in a case of high C_6H_6 concentration).

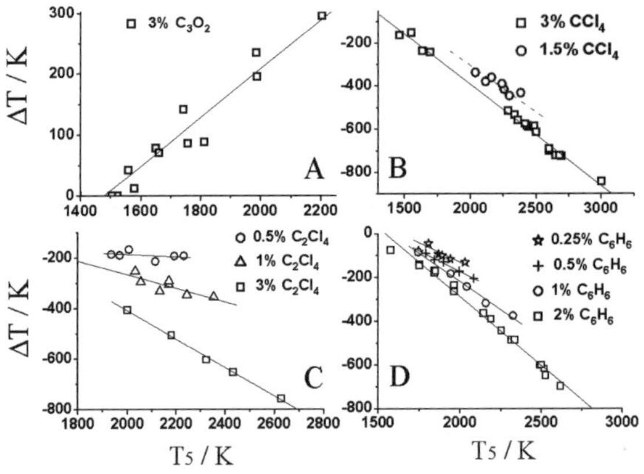

Fig. 2. Difference between measured and frozen temperatures for all investigated mixtures.

The next stage of data analysis was devoted to a correlation between the measured temperatures and particle yield. The particle yield Y was determined from the relation:

$$Y = -\frac{\ln (I_1/I_0))}{\alpha \cdot [C] \cdot l} \tag{1}$$

Here, I_0 and I_1 are the incident and transmitted light intensities, $[C]$ and l are the total available carbon concentration and the shock tube diameter, respectively, α is the extinction coefficient for the given wavelength. Since α can depend on the size and the structure of carbon particles [5], only the values of normalized optical density $D = Y \cdot \alpha$ were considered:

In Figure 3 the dependencies of the final normalized optical density on frozen T_5 (upper plots) and measured T_{exp} (lower plots) temperatures in all investigated mixtures are presented. One can see in the upper plots of Figure 3 the typical bell-shaped curve of optical density, which show the variable behavior for various mixtures and even for various concentrations of reacting molecules. On the other hand, the lower plots demonstrate a good coincidence of the positions of the bells at different concentration precursors in each mixtures once they are plotted as a function of T_{exp}. Note, that the right part of the bell measured in C_3O_2 is shifted towards higher temperatures (Fig. 3A), while on all other plots the position of the bells not only shifted toward lower temperatures, but also become much narrower. These tendency can be explained by the fact that in C_3O_2 at temperatures above 1600 K the heat release of particle formation prevails over the heat consumption of the precursor decomposition [4], in contrast to the reverse situation in all other mixtures.

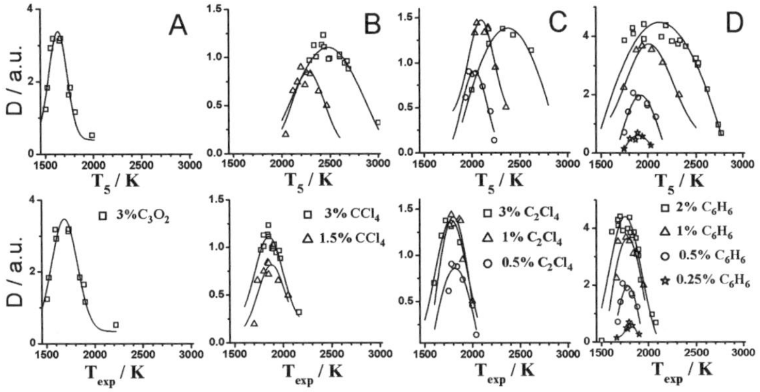

Fig. 3. Dependencies of the final normalized optical density D on "frozen" T_5 (upper plots) and measured T_{exp} (lower plots) in all investigated mixtures

In Figure 4A the resulting plots for all investigated mixtures are given. All the data show the similar temperature dependence of soot yield independently of the kind of precursors molecules (Fig. 4A). For comparison, at the same plot the data of soot volume fraction measured in premixed C_6H_6-air flames [6] is shown. One can see that for all precursors the maximum of particle yield are found in the rather narrow temperature range 1600 - 1850 K. In Figure 4B the temperature corresponding to maximum particle yield T_{max} depending on the relative initial concentration of carbon atoms for all investigated species are given. The most non-isothermal effect of precursor pyrolysis in the mixtures containing CCl_4 is observed, while in C_3O_2 mixtures no noticeable shift of T_{max} at any concentration could be observed. This plot testifies, that the real temperature corresponding to maximum particle yield in all measured mixtures does not depend on initial concentration of precursors. The minor difference of T_{max} for various carbon precursors in pyrolysis processes can be explained by the different exothermic reactions of products of initial molecule decomposition. Note, that in all combustion processes, where soot formation is controlled by the actual flame temperature, T_{max} is always observed in the narrow range 1600 - 1700 K, which is in a good coincidence with the values of T_{max}, observed in C_3O_2 pyrolysis, where no secondary molecular exothermic reactions takes place. Thus, the presented results give a serious hint that the last stage of carbon particle growth proceeds via the same mechanism in hydrocarbon pyrolysis and combustion as well as in hydrogen-free conditions.

4 Conclusion

The direct temperature measurements during pyrolysis of C_3O_2, CCl_4, C_2Cl_4 and C_6H_6 behind reflected shock waves at initial ("frozen") temperatures behind the shock wave

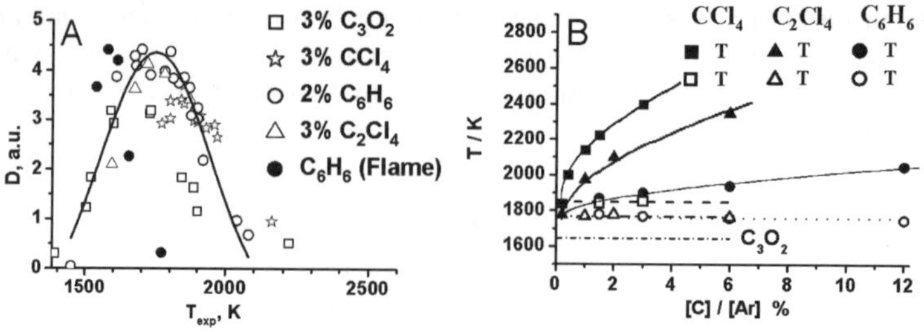

Fig. 4. Temperature dependence of the final soot optical density at shock wave pyrolysis of various carbon bearing molecules (A), and temperatures corresponding to the position of the maximum of the bell-shaped curves in dependence of the initial carbon concentration in mixture (B)

front in the range 1500 - 3000 K have been performed via absorption/emission spectroscopy at 2.7 μm.

A significant difference between frozen and real temperatures behind shock waves was found. The observed difference was related to the heat consumption of the precursor decomposition and the heat release during particles formation.

The obtained results have shown that the real temperature dependence of particle yield in all measured mixtures does not depend on the kind and initial concentration of precursors. For all mixtures these dependencies are close to each others and close to those measured in premixed flames [6]. Based on these results it can be assumed that the final stage of carbon particle formation in all pyrolysis and combustion processes proceeds via on identical mechanism.

This work was supported by Russian Foundation for Basic Research and Göttingen Academy of Sciences.

References

1. Bauerle S., Karasevich Y., Slavov S., Tanke D., Tappe M., Thienel T., Wagner H. Gg.: Soot formation at elevated pressures and carbon concentrations in hydrocarbon pyrolysis. In: *Proc. Combust. Inst.*, vol.25, 1994 ,P. 627-634.
2. Kellerer H., Wittig S.: Growth and coagulation of soot particles at high pressures. In: *Proc. of ISSW*, vol. 21, 1997, P. 177-182
3. Drakon A., Eremin A., Makeich A.: Time-resolved temperature measurements during pyrolysis and nanoparticle growth processes behind shock waves. In: *Proc. of ISSW*, vol. 25, 2005, Paper 1033 (on CD).
4. Emelianov A, Eremin A, Makeich A, Jander H, Wagner H. Gg., Starke R, Schulz C.: Heat release of carbon particle formation from hydrogen-free precursors behind shock waves. In: *Proc. Combust. Inst.*, vol. 31, 2006, P. 649-656.
5. Emelianov A, Eremin A, Jander H. and Wagner H. Spectral and Structural Properties of Carbon Nanoparticle Forming in C3O2 and C2H2 Pyrolysis behind Shock Waves// In: *Proc. Combust. Inst.*, vol. 29, 2002, P. 2351-2358.
6. Boenig M., Feldermann C., Jander H., Luers B., Rudolph G., Wagner H. Gg. Soot formation in premixed C_2H_4 flat flames at elevated pressures. In: *Proc. Combust. Inst.*, vol. 23, 1990, P. 1581-1587.

Thermal reactions of o-dichlorobenzene. Single pulse shock tube investigation

A. Lifshitz, A. Suslensky, and C. Tamburu

Department of Physical Chemistry, The Hebrew University, Jerusalem 91904, Israel
corresponding author: A. Lifshitz assa@vms.huji.ac.il

Summary. The thermal reactions of o-dichlorobenzene were studied behind reflected shock waves in a single pulse shock tube covering the temperature range 1150–1600 K at overall densities of $\sim 3 \times 10^{-5}$ mol/cm^3. Products were determined by gas chromatography using both flame ionization and mass selective detectors with Porapak-N and Megabore PLOT CP-A 1203/KCL columns. There are three reaction channels that characterize the thermal reactions of o-dichlorobenzene: fragmentation, isomerization and formation of the three isomers of trichlorobenzene. Fragmentation is the dominant channel. It is followed by isomerization and formation of the trichlorobenzenes. A detailed product distribution in the fragmentation of o-dichlorobenzene, and the relative yields of the three isomers of trichlorobenzene are reported.

1 Introduction

Chlorinated hydrocarbons, both aliphatic and aromatic are of extreme importance in the balance of ozone in the atmosphere. The reactions of chlorinated hydrocarbons fall into two categories: 1. Decomposition and combustion reactions at high temperatures. Here, in addition to the basic research the practical application is associated with attempts to dispose of these compounds using normal means without environmental hazards and 2. Heterogeneous reactions on microsurfaces simulating the processes that take place in the stratosphere.

When fuel molecules are subjected to the conditions that prevail in combustion, in parallel to the oxidation process, they undergo thermal decomposition and fragmentation that are independent of the presence of air. The study of these decompositions in the absence of air is essential to reveal the mechanism of the overall combustion process. The shock tube is the ideal tool for studying combustion processes and high temperature degradations since it perfectly imitate the conditions that prevail in combustion. That is, pressure, temperature and reaction dwell times that are typical to combustion reactions.

Whereas numerous studies have been published by chemical kineticists in the past and at present, describing catalytic and photochemical reactions of chlorinated aromatics we are not aware of studies on their homogeneous thermal decompositions at high temperatures. (Since there is a huge number of references describing the catalytic and photochemical reactions of dichlorobenzenes, citations will not be given.)

In this article we present a study of the homogeneous decomposition of o-dichlorobenzene using the single pulse shock tube technique.

2 Experimental

The thermal reactions of o-dichlorobenzene were studied behind reflected shock waves in a pressurized driver single pulse shock tube covering the temperature range of approximately 1150–1600 K at overall densities of $\sim 3\times10^{-5}$ mol/cm^3. The driven section of the shock tube was 4 m long and was divided in the middle by a ball valve. The driver had a variable length up to a maximum of 2.7 m and could be varied in small steps in order to tune for the best cooling conditions. A 36-L dump tank was connected to the driven section near the diaphragm location in order to prevent reflection of transmitted shocks. "Mylar" polyester films of various thicknesses were used as diaphragms to separate the driver and the driven sections.

After pumping down the tube to approximately 10^{-5} Torr, the reaction mixture was introduced into the section between the ball valve and the end plate of the driven section, and pure argon into the section between the diaphragm and the valve, including the dump tank. To increase the vapor pressure of o-dichlorobenzene, the tube, the storage bulb and the gas manifold were maintained at 170 °C with an accuracy of ±2 °C.

After firing a shock, gas samples were transferred directly from the tube through a heated outlet in the end plate of the driven section into gas the chromatographs.

Reflected shock temperatures and density ratios were calculated from the measured incident shock velocities using the three conservation equations and the ideal gas equation of state. In view of the high temperatures covered in this study a chemical thermometer could not be used. Dwell times of approximately 2 ms were measured with an accuracy of ~5%. Cooling rates were approximately 0.5–1×10^6 K/s.

Figure 1 shows a typical pressure record obtained as an output of a miniature high frequency - high temperature pressure transducer, showing the reflected shock heating and the cooling processes.

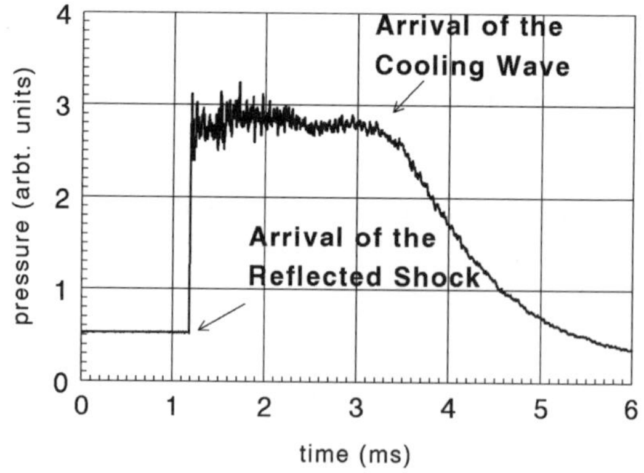

Fig. 1. Pressure record showing the reflected shock heating and the cooling process

Products were determined by gas chromatography using both flame ionization and mass selective detectors with Porapak-N and megabore PLOT CP-A 1203/KCL columns. They were used for both, identification and quantitative yield determination. Figures 2(l) and 2(r) show typical gas chromatograms of products obtained in the decomposition of o-dichlorobenzene run at $T_5 = 1580$ K. The products in 2(l)are the low molecular weight products with a shorter elution time whereas 2(r) gives the aromatic higher molecular weight products that contain several chlorine atoms each.

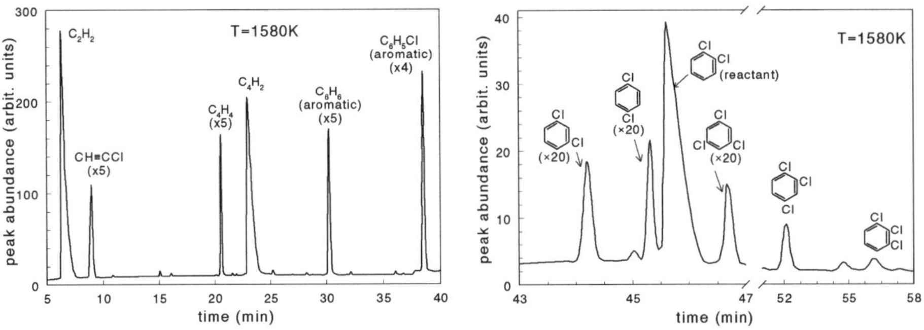

Fig. 2. o-dichlorobenzene chromatograms. Left - low molecular weight products using Porapak-N column, Right - High molecular weight products using Megabore PLOT CP-A 1203/KCL column. The numbers by the peaks are multiplication factors by which the original peaks have been multiplied

3 Determination of product concentrations

To avoid errors resulting from irreproducibility of the detectors from one run to another, concentrations of reaction products $C_5(\mathrm{pr})_i$ were calculated relative to the concentration of the reactant. Evaluation of the concentration of an individual product $C_5(\mathrm{pr})_i$ from its GC peak area was done using the following relations that are based on carbon atom balance:

$$C_5(\mathrm{pr})_i = A(\mathrm{pr}_i)_t / S(\mathrm{pr}_i) \times \{C_5(\mathrm{dichlorobenzene})_0 / A(\mathrm{dichlorobenzene})_0\}$$

$$C_5(\mathrm{dichlorobenzene})_0 = \{p_1 \times \%(\mathrm{dichlorobenzene}) \times \rho_5/\rho_1\}/100RT_1)\}$$

$$A(\mathrm{dichlorobenzene})_0 = A(\mathrm{dichlorobenzene})_t + \frac{1}{N}\sum_i \{n(\mathrm{pr}_i) \times A(\mathrm{pr}_i)_t / S(\mathrm{pr}_i)\}$$

$C_5(\mathrm{dichlorobenzene})_0$ is the concentration of dichlorobenzene behind the reflected shock prior to decomposition, $A(\mathrm{dichlorobenzene})_0$ is its calculated GC peak area prior to decomposition and N is the number if its carbon atoms, 6 in dichlorobenzene. $A(\mathrm{pr}_i)_t$ is the peak area of a product i in the shocked sample, $S(\mathrm{pr}_i)$ is its sensitivity relative to that of the reactant and $n(\mathrm{pr}_i)$ is the number of its carbon atoms. ρ_5/ρ_1 is the compression behind the reflected shock wave and T_1 is the temperature of the reaction mixture prior to shock heating, i.e. the temperature of the shock tube, 170 °C in the present study.

4 Results

In order to determine the distribution of reaction products, some 40 tests were run with mixtures containing 0.3% o-dichlorobenzene in argon, covering the temperature range 1150-1650 K. Extents of pyrolysis starting from approximately one hundredth of one percent were determined for many of the products. Details of the distribution of reaction products are given graphically in Fig's 3(l) and 3(r). The percent of a given product in the figures, corresponds to its mole fraction in the post-shock mixture (not including HCl, Cl_2, Ar and H_2) irrespective of the number of its carbon atoms. Products of extremely small yields are not shown in the figures.

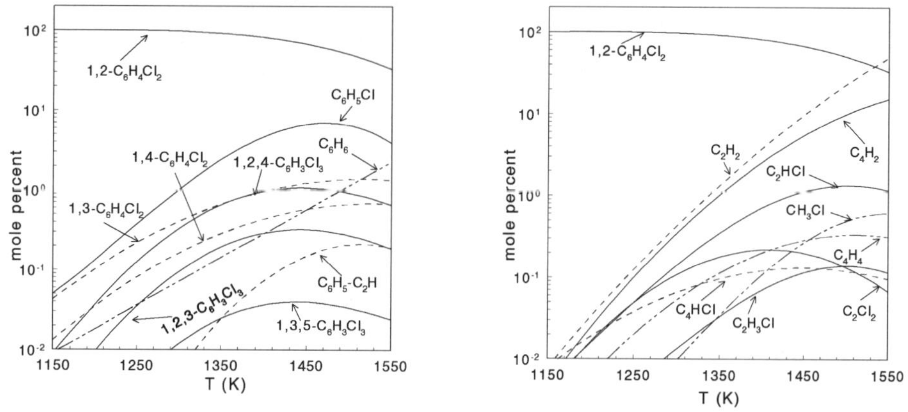

Fig. 3. Product distribution in the decomposition of o-dichlorobenzene. Left - aromatic products, Right - aliphatic products
Errata: The line for C_2Cl_2 was introduced by mistake and should be omitted from the figure.

5 Discussion

There are three reaction channels that characterize the thermal reactions of dichlorobenzene: fragmentation, isomerization and formation of the three isomers of trichlorobenzene. Fragmentation is the dominant channel. It is followed by isomerization and formation of trichlorobenzene. The fragmentation mechanism of o-dichlorobenzene is very similar to that of benzene, except for the lower temperatures where fragmentation begins, owing to the weaker C–Cl bond in chlorobenzene relative to the C–H bond in benzene. At the low temperature end of the study, yields of similar products with and without a chlorine such as C_2HCl and C_2H_2 *ect.* are practically the same. As the temperature increases and further decomposition takes place, the ejection of chlorine atom is much faster than that of a hydrogen atom so that the concentrations of the chlorinated products decrease much faster than those of the unchlorinated compounds. This can be seen in Fig's 4(1) and 4(1), as an example, where C_4HCl is compared to C_4H_2 and C_2HCl is compared to C_2H_2.

It is interesting to examine the relative yields of the three isomers of trichlorobenzene. In the 1,2,4- and 1,2,3-isomers one chlorine atom replaces a hydrogen atom without a shift of another chlorine atom in the ring. In the 1,3,5- isomer, there is a chlorine atom shift in addition to a Cl atom dissociative recombination. As can be seen in Fig's 2 and 3, this additional process drastically reduces the yield of the 1,3,5-isomer of trichlorobenzene, relative to the other two.

We have also performed a series of experiments with both o-and p-dichlorobenzene to examine the relative yields of the trichlorobenzene isomers that are obtained from these two dichloroenzene isomers. The results are shown in Fig 5. The production of

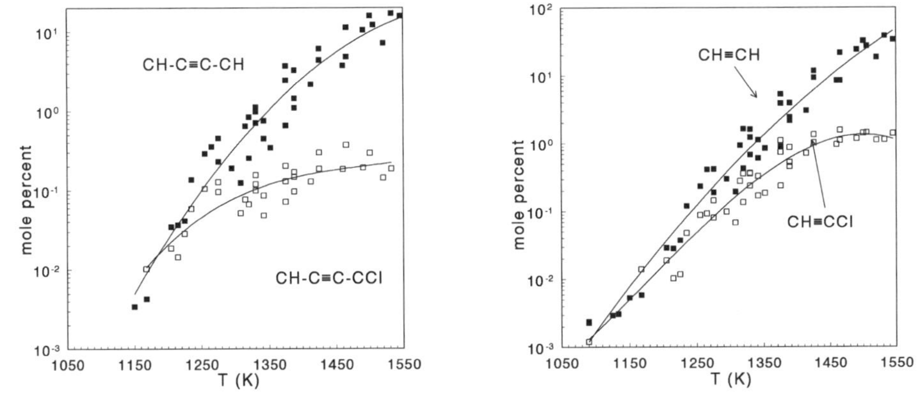

Fig. 4. Data points showing a comparison between the yields of C_4HCl and C_4H_2 and the yields of C_2HCl and C_2H_2

1,2,4-trichlorobenzene does not require a chlorine atom shift in both the ortho and the para isomers. The production rates of 1,2,4-trichlorobenzene from both isomers are thus equal. The production 1,2,3-trichlorobenzene requires, on the other hand, a chlorine atom shift in the para isomer, but not in the ortho isomer. Its yield from o-dichlorobenzene is considerably higher than that from the para isomer.

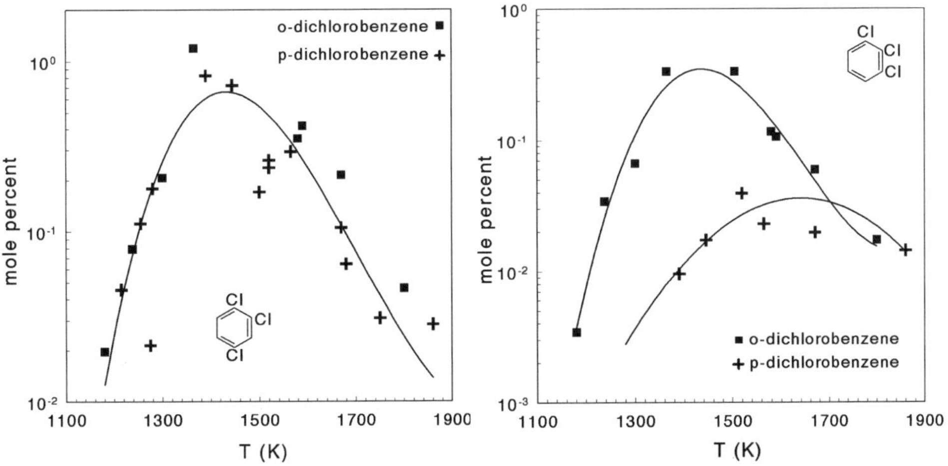

Fig. 5. 1,2,3 and 1,2,4-trichlorobenzene production from ortho and para dichlorobenzene

6 Conclusion

When o-dichlorobenzene diluted in argon is subjected to shock heating and is elevated to high temperatures (above ~1100 K) it fragmentizes, isomerizes to 1,3 and 1,4-dichlorobenzene, and produces the three isomers of trichlorobenzene: 1,2,3 , 1,2,4 and 1,3,5. Fragmentation to aliphatic compounds is the major reaction channel and occurs at much lower temperature than the equivalent fragmentation of benzene owing to the relatively weak C–Cl bond.

Acknowledgement. This work was supported by a grant from the ISF, The Israel Science Foundation.

Wall heat transfer in shock tubes at long test times

C. Frazier, A. Kassab, and E.L. Petersen

Mechanical, Materials and Aerospace Engineering, University of Central Florida, Orlando, FL, 32816, USA

1 Introduction

In recent years in our laboratory, relatively long chemical kinetics experiments are being employed behind reflected shock waves [1,2]. These experiments are performed in upwards of 15-ms test times with the common assumption that the region behind the reflected shock wave can be considered isothermal prior to the main chemical reaction. While the isothermal assumption is applicable to short test times as heat transfer between the hot gas and the cold walls can be neglected, there is some concern that longer test times may allow significant heat loss to the walls of the shock tube, thus creating observable deviations from the isothermal assumption.

Heat loss to the walls of the shock tube has a well-known solution when considering the flow behind the incident shock wave [3–6]. Solution of this incident-shock problem involves convection from the moving gas behind the incident shock wave as transferred to the wall via the moving gas in the growing boundary layer. The focus of these classic problems however was on the boundary layer temperature profile and not on the resulting average hot gas temperature. On the other hand, a closed-form analytical treatment of the wall heat transfer behind the reflected shock wave is difficult because the motion of the reflected shock wave through the gas that has been previously conditioned by the incident wave brings the gas to zero velocity. The induced gas flow that created the growth of the boundary layer behind the incident shock wave is then no longer present, but the fluid in the boundary layer does not necessarily stop moving [7, 8].

The approach taken herein as a first step is to treat the gas/wall interaction behind the reflected shock wave as one of pure conduction, keeping with the assumption that the gas behind the reflected shock is stationary. In this way, an analytical model can be developed for parametric studies designed to gauge the effect of temperature, pressure, geometry, and test gas on the heat transfer at long times. The more complex treatment that includes the the shock-boundary layer interaction will be treated in a later paper.

Although the problem of wall heat transfer in shock tubes has been studied reasonably well, no comprehensive evaluations of the effect of heat transfer on the test gas uniformity at long times and in conditions of elevated test pressure were found. The present study for the first time also evaluates the impact of heat loss to the shock-tube walls in the endwall region using a fully 2-dimensional solution in $T(r, x, t)$, with emphasis on the temperature distribution in the entire hot-gas region as a function of test time. Provided in this paper are details on the analytical solution for three different models: 1) the endwall, or $T(x, t)$ solution; 2) the radial heat transfer problem to the sidewall, or $T(r, t)$; and, 3) the heat transfer in the entire endwall region behind the reflected shock wave, or $T(r, x, t)$. Calculations of temperature uniformity in the hot gas region, thermal thickness, and wall temperature are provided for the shock-tube geometry of primary interest herein.

2 Heat Transfer Model

As aforementioned, the primary mode of heat transfer that is assumed to occur in the present study is conduction, which applies to both the hot test gas medium and the cold shock-tube walls. The basic problem is one where the post-shocked, hot test gas is immediately exposed to the cold walls of the shock tube. Hence, at time $t = 0^-$, the gas is assumed to be at temperature T_g, and the walls are assumed to be at the initial temperature T_a. At time $t = 0^+$, the wall and hot gas are in contact and transfer heat via conduction per Fourier's law of heat conduction.

Other assumptions include ideal gas; constant specific heat and thermal conductivity at a given initial test-gas temperature, T_g; no chemical reactions; and the motion of the reflected shock wave is neglected, so the initial conditions in the endwall region under consideration are assumed to be after the shock has passed. This latter assumption is valid when one considers the overall time scales of ms after shock passage that are of interest to the present study relative to the time it takes the shock to pass through the gas in the endwall region, which is on the order of hundreds of μs.

2.1 $T(x,t)$ Solution

For the endwall region of the shock tube, the solution of the $T(x,t)$ problem assumes a semi-infinite solid wall is instantaneously put into contact with a semi-infinite, hot gas. The basic problem is well known for shock tubes [9–12]. Figure 0910fig1 presents the basic problem and nomenclature. Because the duration of the test is very brief, there

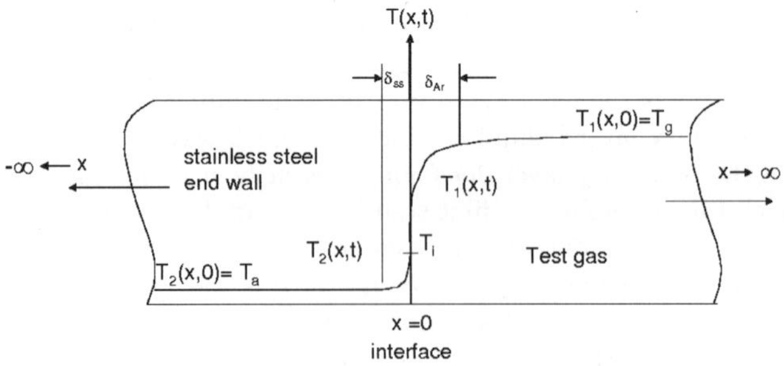

Fig. 1. Endwall region heat conduction model. The hot gas is to the right of the interface, and the cold wall is to the left

is not enough time for the thermal penetration depth in the solid, $\delta_{ss}(t)$, to reach the outside wall of the shock tube. Consequently, the endwall/gas region of the shock tube is modeled herein as two conducting, semi-infinite media each initially at uniform but different temperatures suddenly joined at $x = 0$ in perfect thermal contact at $t = 0^+$. The gaseous (right-most) medium 1, for $x > 0$, is taken as the hot test gas with thermal conductivity k_1 and thermal diffusivity α_1, initially at temperature T_g, while medium 2, for $x < 0$, is taken as the solid endwall with thermal conductivity k_2 and thermal diffusivity α_2, initially at temperature T_a, as illustrated in Fig. 1

Enforcing continuity of temperature and heat flux at the interface, the temperatures in each medium are, respectively [13,14]

$$for\, x > 0: \; T_1(x,t) = T_a + \frac{T_o}{1+\frac{k_2\sqrt{\alpha_1}}{k_1\sqrt{\alpha_2}}}\left[1+\frac{k_2\sqrt{\alpha_1}}{k_1\sqrt{\alpha_2}}erf\Big(\frac{x}{2\sqrt{\alpha_1 t}}\Big)\right]$$
$$for\, x < 0: T_2(x,t) = T_a + \frac{T_o}{1+\frac{k_2\sqrt{\alpha_1}}{k_1\sqrt{\alpha_2}}}\left[1-erf\Big(\frac{|x|}{2\sqrt{\alpha_2 t}}\Big)\right] \tag{1}$$

where $T_o = T_g - T_a$. Evaluating the interface temperature, T_i, at $x = 0$ reveals that it remains constant until the thermal penetration depth reaches the left, outside wall of the solid endwall and is given by [13,14],

$$T_i = T_a + T_o\Big(\frac{k_1\sqrt{\alpha_2}}{k_1\sqrt{\alpha_2}+k_2\sqrt{\alpha_1}}\Big) \tag{2}$$

Results from computations under the test conditions reveal that the thermal penetration depth in the endwall material, $\delta_{ss}(t)$, is much smaller than the endwall thickness at all times under consideration.

2.2 $T(r,t)$ Solution

This problem can be considered as the radial analog of the endwall model, where the region of concern is the shock-tube lateral wall/gas region as illustrated in Fig. 2. The radial temperature distributions corresponding to the model in Fig. 2 can be obtained by means of Laplace transforms, and they are given as functions of integrals on $[0,\infty]$ of Bessel functions of the first kind, J_ν, and second kind, Y_ν, of order ν as well as the conductivity and thermal diffusivity ratios [14] as

$$T_1(r,t) = T_g + (T_g - T_a)\frac{4\rho_k}{\pi^2}\int_0^\infty \frac{J_1(\mu)J_o(\frac{\mu r}{R_o})}{\mu^2[\phi(\mu)^2+\psi(\mu)^2]}e^{-\mu^2 Fo}d\mu \tag{3}$$

$$T_2(r,t) = T_g + (T_g - T_a)\frac{2\rho_k}{\pi}\int_0^\infty \frac{J_1(\mu)[J_o(\sqrt{\rho_\alpha}\frac{\mu r}{R_o})\phi(\mu) - Y_o(\sqrt{\rho_\alpha}\frac{\mu r}{R_o})\psi(\mu)]}{\mu[\phi(\mu)^2+\psi(\mu)^2]}e^{-\mu^2 Fo}d\mu \tag{4}$$

where $\rho_k = \frac{k_1}{k_2}$, $\rho_\alpha = \frac{\alpha_1}{\alpha_2}$, $Fo = \frac{\alpha_1 t}{R_o^2}$, $\phi(\mu) = \rho_k J_1(\mu)Y_o(\sqrt{\rho_\alpha}\mu) - \sqrt{\rho_\alpha}J_o(\mu)Y_1(\sqrt{\rho_\alpha}\mu)$, and $\psi(\mu) = \rho_k J_1(\mu)J_o(\sqrt{\rho_\alpha}\mu) - \sqrt{\rho_\alpha}J_o(\mu)J_1(\sqrt{\rho_\alpha}\mu)$. Evaluation of the interface temperature, $T_i = T_1(R_o,t) = T_2(R_o,t)$, using these expressions reveals that, in the temperature and pressure ranges of interest and for the shock-tube geometry under consideration, the value does not differ from that given by the 1D endwall interface temperature given by Eq. 2 for times up to at least 10 ms. Moreover, the thermal penetration depth in the solid sidewall obtained from Eq. 4 never reaches the outer sidewall radius within the duration of the tests, i.e. up to 20 ms, for the conditions and wall thickness of interest herein.

As such, a simpler model for the radial temperature distribution can be adopted for computations, namely, that of a cylinder imposed with constant wall temperature, T_i, whose value is given by Eq. 2. The radial temperature distribution of the test gas is then,

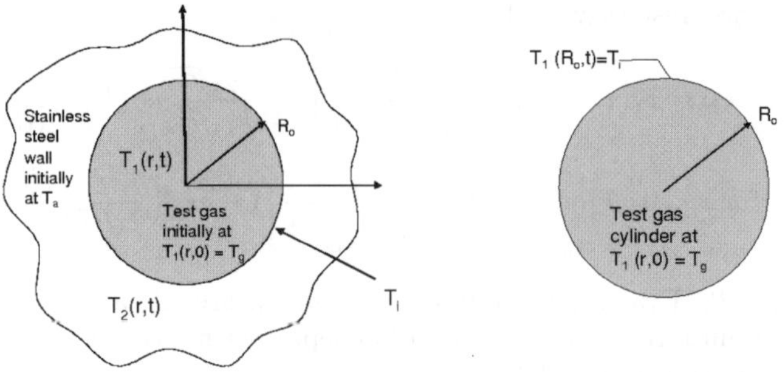

Fig. 2. Radial model of wall/gas interface. Left figure is a gas cylinder in contact with a solid medium; right figure is a gas cylinder imposed with a constant wall temperature, T_i

$$T_1(r,t) = T_i + 2(T_g - T_i)\sum_{n=1}^{\infty} \frac{J_o(\lambda_n r)}{(\lambda_n R_o) J_1(\lambda_n R_o)} e^{-\alpha_1 \lambda_n^2 t} \tag{5}$$

where the eigenvalues λ_n are the zeroes of J_o divided by R_o.

Using the resulting $T(r,t)$ distribution in the test gas, an average gas temperature can be determined at the desired test time. Treating the problem as one of only heat conduction provides a simple estimate of the effects of wall heat transfer at longer test times. To the authors' knowledge, the consideration of the radial heat conduction problem for the conditions behind the reflected shock wave as outlined here has not been previously considered in the shock-tube literature. For relatively large-diameter shock tubes, the $T(r,t)$ solution for the thermal layer thickness and temperature distribution approaches that of the 1D, $T(x,t)$ case considered above.

2.3 $T(r,x,t)$ Solution

By combining the radial solution with the endwall solution, the heat transfer in the entire endwall region of the shock tube can be modeled analytically. The time-dependent axial and radial temperature distribution in the test gas can be found by use of product solutions as

$$T_1(r,x,t) = T_i + (T_g - T_i)\, S(x,t)\, C(r,t) \tag{6}$$

where

$$S(x,t) = \frac{T_1(x,t) - T_i}{T_g - T_i} \quad \text{and} \quad C(x,t) = \frac{T_1(r,t) - T_i}{T_g - T_i} \tag{7}$$

and $S(x,t)$ is the 1D semi-infinite medium solution of model 1, Eq. 1, while $C(r,t)$ is the 1D radial solution of model 2, Eq. 5.

3 Results and Discussion

Each of the models presented above was applied to a range of temperatures from 800 to 1800 K for pressures of 1, 20, and 50 atm for a test gas of Argon. One specific geometry was chosen for the calculations to coincide with the recent experiments at elevated pressures and intermediate temperatures [1,2] using the facility described by Petersen et al. [15]. In summary, the shock-tube wall material was stainless steel, and the internal diameter of the tube was 16.2 cm with a sidewall thickness of 9.5 mm.

From the $T(x,t)$ (endwall) solution, the general result is that the thermal boundary layer is thinnest at the higher pressures and for the lower T_g. Therefore, the results of the calculations favor thinner thermal layers (and hence more uniform, high-temperature core regions) at the conditions where longer test times are more likely for kinetics experiments, that is for higher pressures and lower temperatures. For example, the thickness at the worst-case temperature for this study (1800 K) is only 4.6 mm at a time of 20 ms. Additionally, the wall surface temperature experiences a "jump" condition at $t = 0^+$, but this jump is only at most a few degrees for the conditions herein; the wall temperature remains at this value until the thermal layer in the solid reaches the outer wall (which does not happen within the time frame of interest herein). For instance, the wall temperature only increases by 3.1 K for the test conditions of 50 atm and 1400 K.

For the $T(r,t)$ solution, δ_{Ar} is identical to the solution from the $T(x,t)$ problem because of the relatively large inner diameter. More importantly, the radial solution allows for the calculation of an average gas temperature over the entire hot gas region. A typical result is shown in Fig. 3, which displays the change in average gas temperature relative to the initial T_g for a range of temperatures at 20 atm. Note that at a test condition of 800 K and 20 atm, the average gas temperature only decreases by about 5 K. Also, for a typical shock-tube experiment that ends within 3 ms, the average gas temperature for a 1400-K experiment at 20 atm decreases by about 6 K.

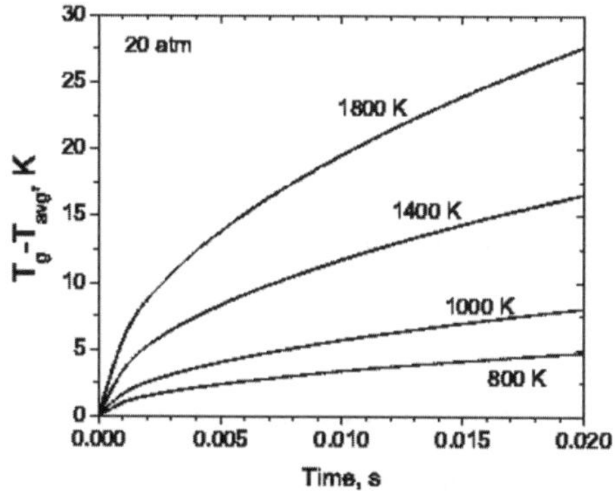

Fig. 3. Decrease in average Ar gas temperature from the radial model for a pressure of 20 atm and 4 different initial gas temperatures.

Finally, the results of the entire endwall region are modeled by the $T(r,x,t)$ solution. A typical result is shown in Fig. 4 for a case with an initial gas condition of 1800 K, 1 atm at 20 ms. The regions of the thermal layers are evident from this lower-pressure case. At a suitable distance outside of the endwall thermal layer, the temperature distribution is the same as that predicted by the $T(r,t)$ model. Of interest to the present study is a location 16 mm from the endwall where the sidewall optical access port is located for chemical kinetics experiments. Figure 4 shows that even in the worst-case conditions of higher temperature and lower pressure, the hot gas region is effectively outside of the endwall thermal layer even at 20 ms.

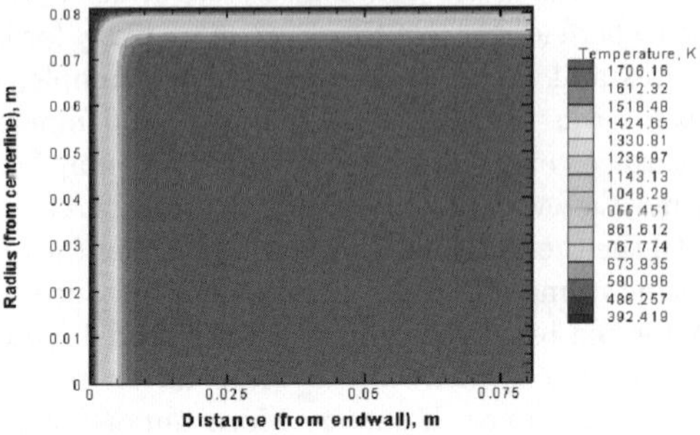

Fig. 4. Complete $T(r, x, t)$ solution for the entire endwall region at $t = 20$ ms for an initial test condition of 1800 K, 1 atm in Ar.

Calculations for other shock-tube diameters and test gases will be presented in a later study. Since any convective effects at the sidewall behind the reflected shock wave are neglected, we expect that their inclusion would increase the heat transfer between the hot gas and the solid wall material.

References

1. Amadio A.D., Crofton M.W., Petersen E.L.: Shock Waves **16**, 157 (2006)
2. de Vries J., Petersen E.L.: Proc. Combust. Inst. **31**, 3163 (2007)
3. Bromberg R.: Jet Propulsion **26**, 737 (1956)
4. Hartunian R.A., Russo A.L., Marrone P.V.: J. Aerospace Sciences **27**, 587 (1960)
5. Mirels H.: *Laminar Boundary Layer behind Shock Advancing into Stationary Fluid*, NACA TN 3401 (1955)
6. Mirels H.: *Boundary Layer behind Shock or Thin Expansion Wave Moving into Stationary Fluid*, NACA TN 3712 (1956)
7. Wilson G.J., Sharma S.P., Gillespie, W.D.: *Time-Dependent Simulation of Reflected-Shock/Boundary Layer Interaction in Shock Tubes*, Shock Waves @Marseille I - Proc. 19th Int. Symp. on Shock Waves, Springer-Verlag (1995) pp 439-444
8. Nishida M., Lee M.G.: *Reflected Shock Side Boundary Layer Interaction in a Shock Tube*, Proc. 20th Int. Symp. on Shock Waves, Vol. I, World Scientific (1996) pp 705-710
9. Goldsworthy F.A.: J. Fluid Mech. **5**, 164 (1959)
10. Sturtevant B., Slachmuylders E.: Phys. Fluids **7**, 1201 (1964)
11. Baganoff D.: J. Fluid Mech. **23**, 209 (1965)
12. Hanson R.K.: *Study of Gas-Solid Interaction Using Shock-Wave Reflection*, Shock Tube Research-Proc. 8th Int. Shock Tube Symp., Chapman and Hall (1971) pp 58
13. Eckert E.M., Drake R.: *Analysis of Heat and Mass Transfer*, Hemisphere (1972)
14. Luikov A.V.: *Analytical Heat Diffusion Theory*, Academic Press (1968)
15. Petersen E.L., Rickard M.J.A., Crofton M.W., Abbey E.D., Traum M.J., Kalitan D.M.: Meas. Sci. Tech. **16**, 1716 (2005)

Part IV

Detonation and Combustion

A study on DDT processes in a narrow channel

K. Nagai, T. Okabe, K. Kim, T. Yoshihashi, T. Obara, and S. Ohyagi

Saitama University, Graduate School of Science and Engineering
255 Shimo-Ohkubo, Sakura-ku, Saitama, 338-8570 (Japan)

Summary. One of the fundamental problems to be studied on a Pulse Detonation Engine (PDE) is the deflagration to detonation transition (DDT). For the development of the PDE, it is essential to shorten a distance of detonation transition that is called a detonation induction distance (DID). We carried out an experimental study of DDT in a narrow channel with height of 1-5mm by using pressure and soot track records in oxyhydrogen mixtures. Detonation limits was discussed according to height of tube and equivalence ratio. According to pressure history and soot track record, detonation velocity, DDT process and DID was discussed. Over driven detonation and attenuated detonation was observed in the narrow channel. DID that measured by soot track record applied to experimental formula for oxygen and hydrogen system.

1 Introduction

A Pulse Detonation Engine (PDE) is expected to be applied to the next-generation system and the various fields. The features are high heat efficiency, high specific impulse and simple structure. One of the fundamental problems to be studied on the PDE is the deflagration to detonation transition (DDT). For the Development of the PDE, it is essential to shorten a distance of detonation transition that is called a detonation induction distance (DID). In general, it is known that DID will be short, in case tube diameter is smaller, high combustion velocity and high pressure. However, it is not yet understood DDT process and other phenomenon in narrow channel. Because of relation to detonation limit, critical propagation diameter, cell width and velocity deficit for initial pressure and equivalent ratio. We carried out an experimental study of DDT in a narrow channel with height of 1-5mm by using pressure and soot track records in oxyhydrogen mixtures.

2 Experimental

Experiment was carried out in a narrow channel by using pressure transducers, ionization probes and soot track records. propagation velocity and pressure were measured with pressure histories and ionization current. DID and cell width were searched with the soot track records. Figure 1 shows the schematic of experimental apparatus. A detonation tube is fabricated by two aluminum blocks, which has 1000mm in length, 126mm in width and 56mm in height. A rectangular detonation tube is formed by one aluminum block that has narrow channel with length of 808mm, width of 8mm and depth of 1 to 5mm. Pressure transducer (PCB113A21, Piezotronics Co.Ltd.) are embedded on the aluminum block at five positions(P1, P2, P3, P4, P5), which are placed at a distance of 180 mm each. The ignition plug is placed at 50 mm upstream from P1, three ionization probes (I2, I3, I4) is placed in face of P2, P3, P4. The experiments were carried out under the conditions in Table 1. The tube is filled up to initial pressure with oxyhydrogen mixture. When it ignites with a spark plug, pressure wave and combustion wave are generated. Then the signals of the pressure transducers and the ionization current measured. The propagation velocity and pressure obtained from the signal data. In order to observe DID and cell width, the visualization by the soot film method was performed. The sooted thin plate was a length of 830mm, a width of 20mm, a thickness of 0.5mm. The plate was sooted by kerosene flames.

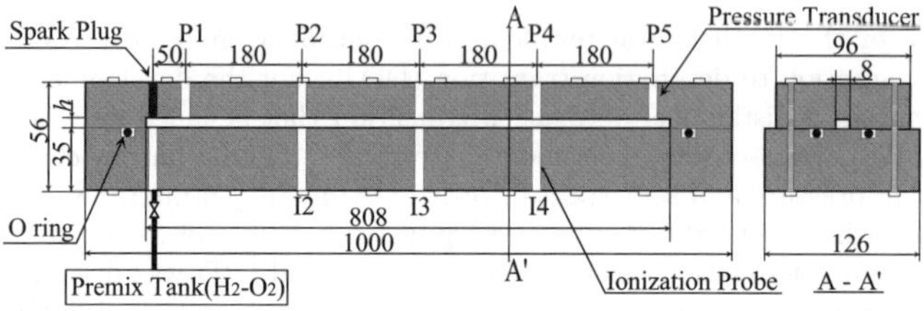

Fig. 1. Schematic of experimental apparatus

Table 1. Experimental Conditions

Mixture	H_2 - O_2
Equivalence Ratio, ϕ	0.5, 1.0, 1.5
Initial Pressure, P_0	20 - 150 [kPa]
Height of Channel, h	1, 2, 5 [mm]

3 Results and discussion

3.1 Pressure and propagation velocity

The pressure wave and following combustion wave propagate by ignition, after that DDT occurs. Figure 2(a),(b) shows typical two pressure histories measured in the present experiments. Experimental conditions are the cases of height (h) = 5mm, equivalence ratio $\phi = 1.0$. A horizontal axis is measured the time from ignition, the vertical axis is the nondimensional pressure.

In the case of Fig. 2 (a), a combustion wave instantly catches up with the pressure wave generated by ignition then detonation initiate(type a). This pressure histories is the case of initial pressure (P_0) = 150 kPa. At P2 to P4, reaction fronts of ionization current observed simultaneously the steep pressure so that it is concluded that they are detonation. The nondimensional peak pressure of P2 and P3 is 29 and 22 each. The average propagation velocity between P2 - P3 is 2951 m/s. The Chapman-Jouguet (CJ)

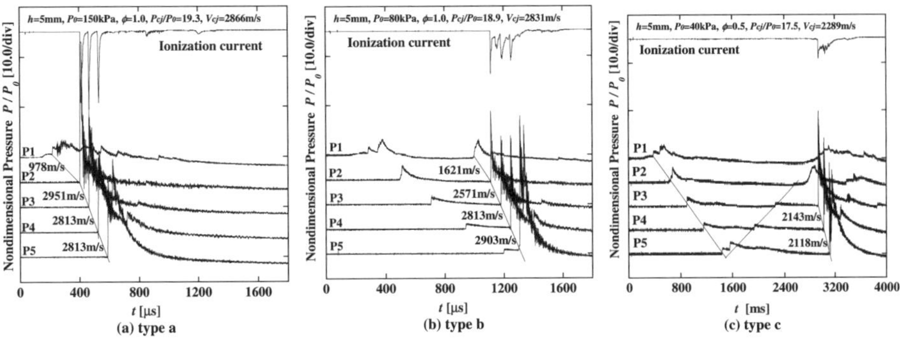

Fig. 2. Pressure histories:(a)type a(left), (b)type b(center), (c)type c(right)

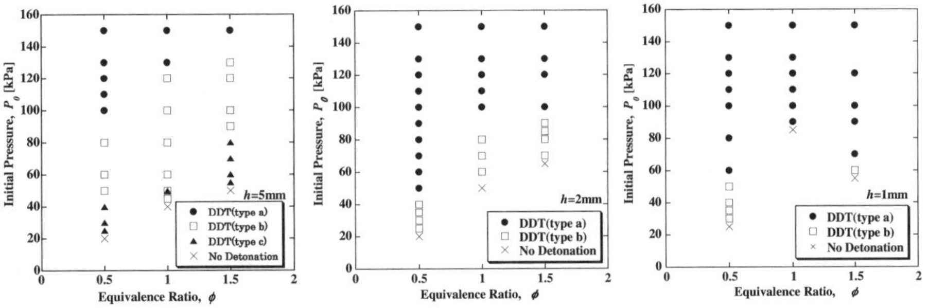

Fig. 3. Detonability Limit:h=5mm(left), h=2mm(center),h=1mm(right)

nondimensional pressure is 19.3, the CJ detonation velocity is 2866 m/s by chemical equilibrium calculation. It turns out that it is overdriven detonation between P2-P3.

In the case of Fig.2(b), where the initial pressure is 80 kPa and the equivalence ratio is 1.0, a blast wave propagates at first due to combustion in the induction port which has a considerable volume compared with the combustion tube itself. Ignition of the mixture in the tube has occurred at about 1 millisecond and it develops to detonation. The average propagation velocity of this preceding blast wave is in the range of 700-860 m/s and it decays considerably at P5. The ignition of the main mixture should be occurred due to a hot burned gas which is injected from the induction port. It should be mentioned that the main mixture had been ignited by the spark plug at first. But the created flame kernel propagates slowly through the main tube and should be overtaken by the burned gas injected from the port. The combustion wave developed to detonation between P2 and P3. It is still accelerate through P4 and P5 by the effects of the preceding blast wave. At P5, it may be affected by the reflected shock wave at the downstream end of the tube. For such a case is called as type b.

Figure 2(c) shows the case where initiation of detonation is retarded so far after the arrival of reflected shock wave at the upstream end. This case is called as type c. The equivalence ratio is 0.5 and the initial pressure is 40 kPa. The initiation may be affected by the reflected shock wave and the second ignition has been occurred at P2. It should be concluded that the effect of the tube length is also important because the reflected shock wave may affect the ignition.

The experimental result are shown in Figure 3 about the relation between equivalence ratio (ϕ) and initial pressure (P_0). Detonation limit is totally wide for larger P_0, lower height (h) of channel in h=5, 2mm. For type a, detonation range will become widely when height(h) becomes low (e.g. detonation limit is P_0=150 for h=5, that is 70kP for h=1). For h=1mm, the range of type a is wide. However detonation range is narrow and it was not observed type b at ϕ=1.0. For equivalence ratio, detonation limit is the most wide at ϕ=0.5.

3.2 Soot track record

Figure 4 shows pressure histories and soot track records measured simultaneously in the experiments. Experimental conditions are height (h) =1 mm, initial pressure (P_0) =100 kPa, and equivalence ratio (ϕ)= 0.5(type a). The propagation direction is from left to right. Fig. 4 (a), (b) and (c) are at position of 130mm, 230mm, and 700mm each from an ignition point (spark plug). Fig. 4(a) is between P1 and P2. Invisible fine cells 1mm less appear on the side wall from the position of the ignition point to 140mm. And visible regular cells are observed from 145mm in position. DDT occurred on the side wall in the position 140mm from ignition point, so that DID is 140mm. After that, cells became large, its size was 0.5-2mm at position 230mm (P2) from ignition point. This size corresponds to the cell width of other experiments results. Moreover, Cell width were 8mm or more(Fig. 4(c)), cell disappeared at the right end of tube. In pressure histories, nondimensional peak pressure is 16.7 and the average of propagation velocity is 2195 m/s between P2 to P3. Nondimensional CJ pressure is 18.8, CJ detonation velocity is 2324 m/s by the chemical equilibrium calculation. It is overdriven detonation between P1 to P2, it has already decreased between P2 to P3. It agrees with soot track record. Cell width depends on mixture, equivalence ratio, initial pressure etc. In the present study, cell width varied from 1mm less to 8mm in narrow channel. In the narrow channel, DDT early occurs, detonation is not able to keep steady condition, and early decreased because of velocity deficit for friction of side wall and heat loss.

Figure 5 is the pressure histories obtained simultaneously with the soot track record of type c. Experimental condition is the case of h =5 mm, P_0 =50 kPa and ϕ =1.0. CJ detonation velocity is 2804 m/s and the nondimensional CJ pressure is 18.5 by chemical equilibrium computation. The soot track record with the position (P1-P2) of 190mm from the ignition point (spark plug) is shown in Fig. 5 (a). Fig. 5 (b) and (c) are at position of 220mm and 250mm each from an ignition point. The combustion wave propagates from left-hand side to right-hand side, the reflected pressure wave propagates from right-hand side to left-hand side. Under 1mm fine cells appear in the whole from the position of about 210mm from the ignition point. This position will be collision point. Unlike Fig. 4(a), cells appear not on the a side wall but in the whole. From the pressure histories, the reflected pressure wave interact with combustion wave, the nondimensional peak pressure is 20.9 at P2. DDT occurs between P1-P2. That position of DDT for the soot track record corresponds with collision.

3.3 DID and Cell width

Figure 6 shows the relation of the initial pressure and DID at ϕ=1.0. It is known that DID become shorter for smaller tube diameter and higher initial pressure. It shows the exper-

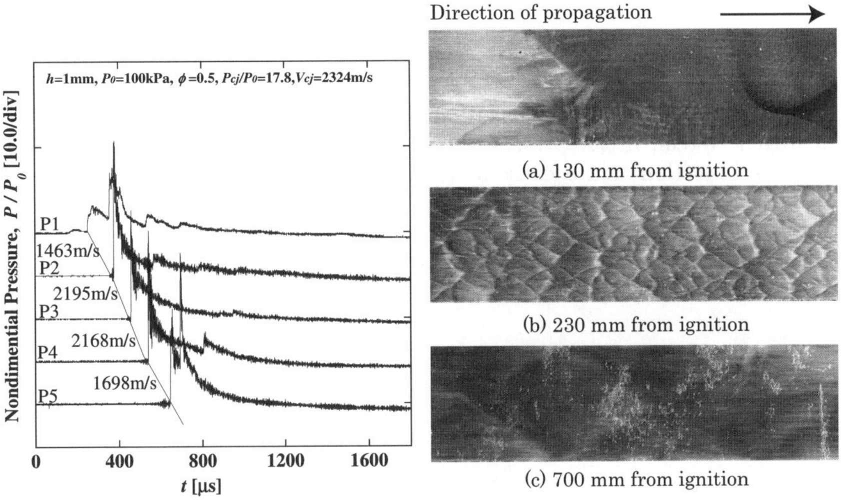

Fig. 4. Pressure histories (Left) and soot track record (Right) (h=1mm, P_0=100kPa, ϕ=0.5) :type a

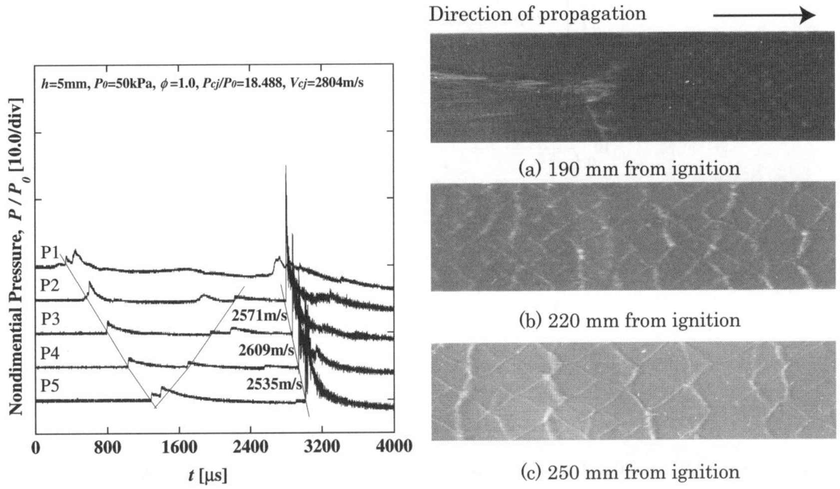

Fig. 5. Pressure histories (Left) and soot track record (Right) (h=5mm, P_0=50kPa, ϕ=1.0) :type c

imental formula of DID that proposed from various experimental results in oxyhydrogen mixture.

DID that measured by soot track record aplly to following experimental formula for oxygen and hydrogen system:

$$\frac{l}{d} = \alpha \frac{P_{atm}}{P_0}, (\alpha \approx 30\ to\ 70) \quad (1)$$

where l is DID, d is tube diameter, P_{atm} is atmospheric pressure, P_0 is initial pressure. DID that measured by soot track record applied to experimental formula (1) for oxygen and hydrogen system. In the present study, tube diameter (d) was used equivalent diameter as cross-section area because it was used rectangular tube. α varied from 32 to 70 for various ϕ and h. It was approximated by α= 50 at h =1 mm, α= 40 at h = 2, 5 mm in ϕ=1.0. Though the measured DID of type c is 210mm, DID from formula (1) is 720mm. DID became short of 500mm, because it was interaction the combustion wave with the pressure wave.

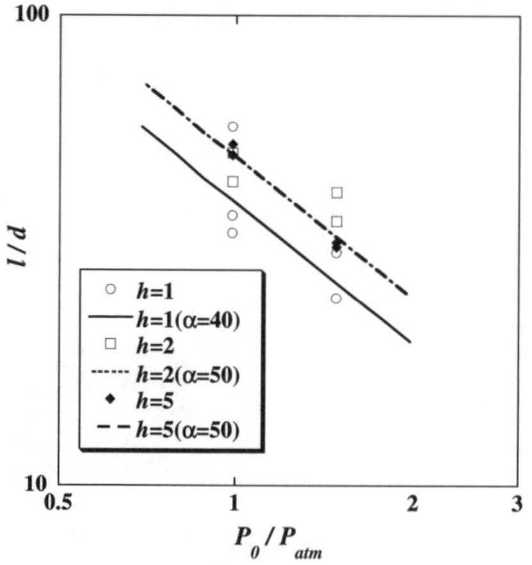

Fig. 6. Variation of DID with an initial pressure (ϕ=1.0)

4 Conclusion

DDT processes in a narrow rectangular channel were studied for oxygen and hydrogen system. In the narrow channel, DDT early occurs, detonation is not able to keep steady condition, and early decreased. The reason is for friction and heat loss of side wall. DID that measured by soot track record applied to experimental formula for oxygen and hydrogen system.

References

1. Nettleton M.A., Gaseous Detonations : their nature, effects and control, Chapmanand Hall, 1987

Combustion in a horizontal channel partially filled with porous media

C. Johansen and G. Ciccarelli

Queen's University, 130 Stuart St, Kingston ON K7L3N6, Canada

Summary. Experiments were carried out to investigate the combustion phenomenon in a horizontal channel partially filled with 12.7 mm diameter ceramic oxide beads. A 1.22 m long, nominally 45 mm thick layer of ceramic oxide beads is located at the ignition end of a 2.44 m long, 76 mm square channel. The channel is filled with a 31 percent nitrogen diluted stoichiometric methane-oxygen mixture and an automotive spark ignition system is used to ignite a flame at the end of the channel. Instrumentation includes ionization probes and piezoelectric pressure transducers. High-speed Schlieren video was used to capture the structure of the explosion front. Experiments were performed at room temperature and at an initial pressure in the range of 14 kPa to 50 kPa. At low initial pressures, a mode of combustion was observed where the reaction in the gap above the bead layer dominates the combustion within the porous media. As the initial pressure of the mixture was increased, combustion within the porous media became more dominant and began to drive the combustion in the gap above the layer. Pressure tme-history and flame time-of-arrival data indicate that transition to detonation occurs within the bead layer section of the channel for mixtures at an initial pressure of 30 kPa and higher. As the initial pressure is increased, transition to detonation occurs closer to the ignition source in the gap above the bead layer.

1 Introduction

Flame propagation in porous media has been investigated for many years in the context of explosion safety. Porous media is often used as a flame arrester in the chemical industry but it can also be used as an integral part of the process line. Babkin [1] proposed that different steady-state flame propagation regimes exist in porous media that are characterized by the propagation mechanism and the front velocity. The regimes of interest to this study include high velocity flames (0.1-10 m/s), sound velocity flames (100-300 m/s), low velocity detonation (500-1000) and normal detonations (1500-2000 m/s). The flame quenching limit in a porous media consisting of spherical beads has been defined by a critical Peclet number of roughly 65, where the characteristic length scale and velocity is taken to be the bead diameter and laminar flame speed [2, 3]. Pinaev [4] studied the behavior of flames propagating in porous media over the complete propagation regime. Pinaev noted that for front speeds greater than 5 m/s there is a pressure rise that coincides with the reaction zone, and that the pressure front steepens with propagation velocity. In order to achieve a CJ detonation the pore size must accommodate at least thirteen detonation cells widths [5].

The propagation regimes observed in porous media are analogous to the phenomena observed in obstacle laden tubes [6]. For a given thermodynamic condition, the geometry governs the flame acceleration and ultimately the steady-state explosion front propagation velocity. In these two geometries the flow path is characterized by the bead diameter and the obstacle blockage ratio. Both of these flow geometries introduce turbulence into the unburned gas flow ahead of the flame that enhances the reaction rate, and conversely adds wall surface area that extracts energy from the reaction front as a result of enhanced momentum and heat losses. These characteristics lead to enhanced flame acceleration but ultimately limit the peak velocity.

Experiments investigating flame propagation in porous media are typically carried out in a vertically oriented tube where the cross section is filled with the media. In this study a layer of beads is located at the bottom of a horizontal channel, such that a long empty cavity exists in the upper part of the channel. Flame propagation in this geometry is the result of a combination of influences associated with the porous media and the cavity above. The combustion front in this unique geometry is tracked by ionization probes and pressure transducers. High speed photography is used to capture the structure of the combustion front.

2 Experimental Apparatus

Experiments were carried out in a 2.44 m long horizontal channel partially filled with a 45 mm thick layer of ceramic oxide beads. The 1.22 m length bead layer is located at the ignition end of 76 mm wide square channel (Fig. 1). The channel is filled with a 31 percent nitrogen diluted stoichiometric methane-oxygen mixture and an automotive spark ignition system is used to ignite a flame.

Fig. 1. Experimental apparatus

Each non-optical module is equipped with two instrumentation ports spaced 30.48 cm apart and 15.24 cm from the end flanges on both the top and bottom surfaces. The optical module is equipped with a total of eight instrumentation ports spaced 15.24 cm apart and 7.62 cm from each end flange. The instrumentation configuration for all of the flame acceleration tests involved four piezoelectric pressure sensors mounted in instrumentation ports on the top surface. Ionization probes were mounted in each remaining instrumentation port on the bottom and top channel surface and protruded 15 mm into the test section. In some configurations, ionization probes were mounted in both the top and bottom channel surfaces at the same axial position allowing for a comparison of flame speed through the porous media and above it in the gap. In the optical module, 19 mm thick glass panels are integrated into the channel front and back sides to facilitate 445 mm of unobstructed optical access. Visualization of flame propagation in the gap above the porous media is obtained through a high-speed z-Schlieren photography system. Although the Schlieren system cannot detect flame propagation in the porous media due to the non-transparency of the bead layer, fast rates of combustion in the porous media were detected in the form of light emitted between the beads. The size of each Schlieren image was restricted to a height of 7.62 cm and a length of 25.4 cm, which correspond to the aluminum side cut-out length and diameter of the parabolic mirror, respectively. Experiments were performed with 12.7 mm diameter beads at room temperature at initial pressures of 14 kPa to 50 kPa. Experiments were also undertaken with the channel completely filled with beads to understand the nature of combustion through porous media alone.

3 Results and Discussion

The flame velocity obtained from the ionization probes distributed along the length of the partially filled channel is shown in Fig. 2. Magnitudes of velocity in Figs 2a and 2b were calculated from ionization probes in the gap above the bead layer and from probes imbedded inside the bead layer on the bottom surface of the apparatus, respectively. A large range of flame propagation velocities are observed over the total initial pressure range (p_i = 15 - 50 kPa). At low initial pressures (p_i = 15 kPa), which correspond to relatively insensitive mixtures, the flame velocity accelerates to a maximum velocity of 900 m/s at the end of the bead layer. For the remaining length of the unobstructed channel, the flame propagated at this maximum velocity. From a comparison of flame time-of-arrival data obtained from pairs of ionization probes located at the same position along the channel length, it was found that the flame position in the gap above the bead layer was slightly ahead of the flame position in the porous media at these low pressures. By the end of the bead layer (x= 1.22 m) flame-time-of arrivals in both the gap and within the porous media are the same.

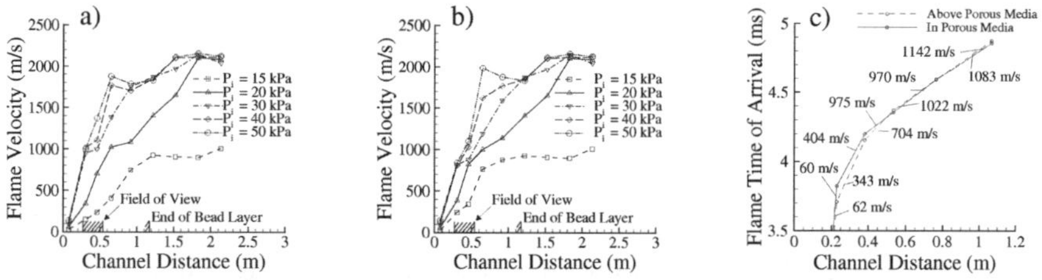

Fig. 2. Axial flame velocity distribution along partially filled channel of 1/2 inch beads a) propagation above beads b) propagation through beads c) x-t diagram at p_i = 20 kPa

At higher initial pressures (p_i = 20 kPa), a similar initial lag in flame position within the porous media is observed relative to the flame front above the layer. However, flame acceleration within the porous media increases faster than that in the gap above the bead layer. Before the end of the bead layer, the flame propagating within the porous media reaches the axial position of the flame front in the gap (see the x-t diagram in Fig. 2c). At this point, a transition occurs where the driving mode of combustion occurs within the bead layer as opposed to in the gap above the porous media as seen at lower pressures.

The flame velocity in both the porous media and in the gap approach 1100 m/s at the end of the bead layer. Transition from deflagration-to-detonation (DDT) is observed as the flame propagates past the bead layer into the remaining length of the channel. At higher initial pressures (p_i = 30 - 50 kPa), the flame position within the porous media is ahead of the flame position in the gap above the bead layer as combustion in the porous media dominates the acceleration process. The flame accelerates to a maximum of 1850 m/s by the end of the bead layer and then up to the Chapman Jouguet velocity of V_{CJ} = 2148 m/s. Combustion in both the gap and the bead layer is driven by turbulence produced ahead of the flame. In the gap, turbulence is produced predominantly at the bead surface and in the beads it is produced as jets in the bead cavities. In the bead layer severe heat loss limits flame propagation for less reactive mixtures to less than 1 m/s [1]. As the mixture reactivity increases (e.g., higher initial pressure) a limit is reached where supersonic propagation is possible [4].

The failure of DDT to occur within the bead layer can be better understood by observing flame acceleration with the channel filled with beads. As shown in Fig. 3, when the apparatus is completely filled with porous media, flame velocities reach a maximum of 800-1000 m/s in the 15-50 kPa initial pressure range.

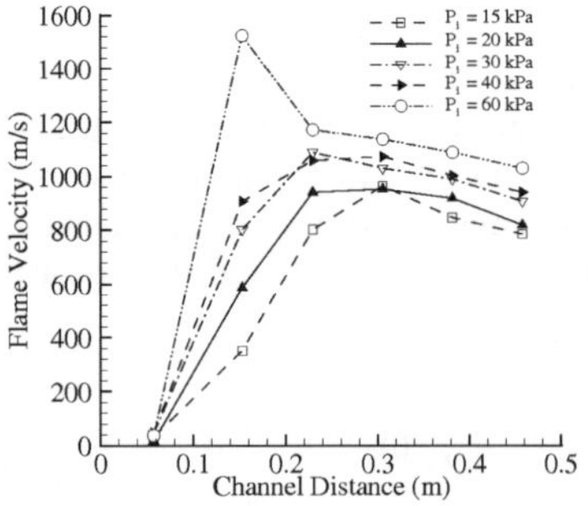

Fig. 3. Distribution of flame velocity along channel length through porous media (1/2 inch beads)

Although rates of flame acceleration increase with increasing initial pressure, the flame velocity does not reach detonation velocities in the pressure range tested. DDT within the porous media is not possible because the detonation cell size is larger than the average pore size. However, for tests in the partially filled channel, detonation is possible as the detonation cell size becomes less than the gap height above the porous media. A sooted foil placed at the end of the bead section in the partially filled channel at an initial pressure of 50 kPa confirms that the transition to detonation occurred in the gap above the beads. The distributions of pressure along the channel length for the partially filled configuration corresponding to experiments with an initial pressure of 20 kPa are shown in Fig. 4.

Due to the limit in number of available pressure sensors, Fig. 4 is a compilation of two experiments at similar conditions. All ionization probes and pressure sensors located at 38, 53, 76, and 107 cm downstream of the ignition source correspond to one experiment

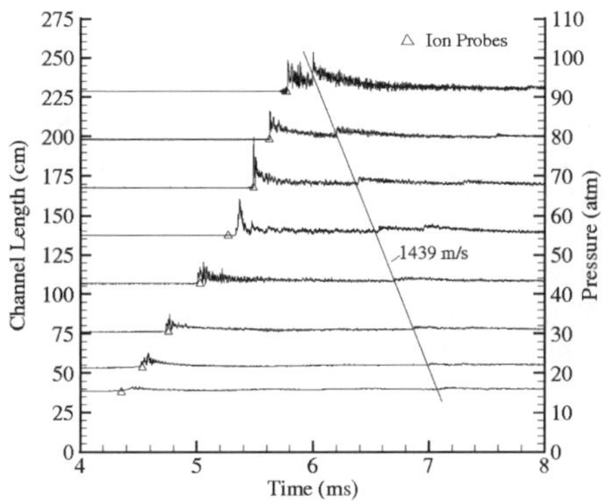

Fig. 4. Distributions of pressure along channel length with a partially filled bead layer (p_i = 20 kPa)

at an initial pressure of 20 kPa. The remaining pressure sensors correspond to a test at similar conditions and is synchronized to the first experiment. At axial positions near the ignition source, the pressure rises slowly ahead of the flame due to the expansion of the combustion products. As the flame accelerates, the pressure traces become steeper until a shockwave forms further downstream. During transition to detonation, the shock compression of the unburned gas causes autoignition to occur and the flame and shock become coupled, separated by a small induction time. The discrepancy between the ionization probe and pressure sensor located at a streamwise location of 137 cm shown in the Fig. is an error due to the synchronization of the two experiments. The reflected shock wave can be tracked as it propagates back towards the ignition end of the channel at a speed of 1439 m/s.

Visualization of the transition process, observed in Fig. 2c as a rapid rise in velocity, is needed to further understand the different modes of combustion in a partially filled channel with porous media. Fig. 5 shows two types of combustion modes corresponding to mixtures with initial pressures of 15 kPa and 30 kPa, respectively. At the lower pressure, the Schlieren image shows the propagation of an oblique flame, with the leading edge located at the bead surface where the highest turbulence intensity in the gap is expected. The absence of light emission from the bead layer would indicate slow burning predominately in the downward direction. In this mode of combustion, the turbulent flame in the gap above the bead layer drives combustion within the porous media. At higher pressures (see Fig. 5b) flame acceleration is greatly enhanced due to the intense burning in the bead layer visualized in the Schlieren images as light emanates from between the beads along the glass surface. The leading edge of the oblique flame in the gap is at the same location as the explosion front propagating in the bead layer. In this mode of combustion, propagation of the flame front in the bead layer drives burning in the gap above the layer with a precursor shock. At 20 kPa, a transition from one combustion mode to the other is observed in the videos, while for 15 kPa, a similar transition occurs just after the video field of view (see Figs 2a and 2b for the location of the camera field of view).

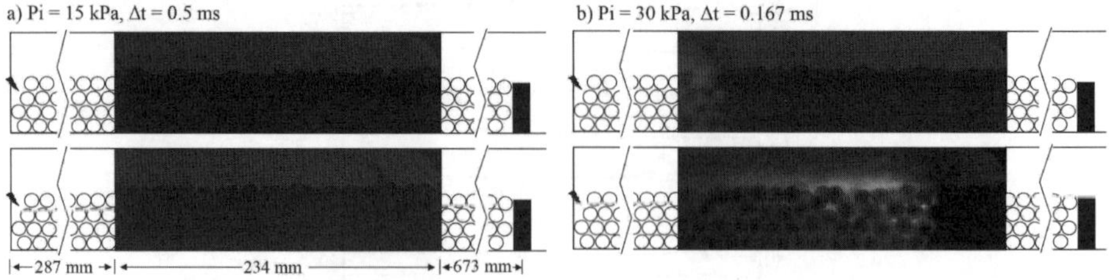

Fig. 5. Propagation of flame front along partially filled channel for two different initial pressures

4 Conclusions

Experiments with nitrogen diluted stoichiometric methane-oxygen revealed two different modes of combustion in the initial stages of flame acceleration in a horizontal channel partially filled with porous media. High-speed Schlieren photography reveal the turbulent structure of the oblique flame front above the porous layer at low initial pressures. At higher pressures, light emanating from the bead layer indicates intense burning in the porous media that dominates combustion in the gap. Ionization probes embedded within the bead layer confirm this transition as the flame front in the bead layer leads the flame in the gap. A configuration with a completely filled channel with porous media reveals an upper limit to the flame velocity below the Chapman Jouguet detonation velocity. At low initial pressures in the partially filled channel, transition to detonation was not observed in the porous media but occurred after the bead layer. At higher initial pressures, transition to detonation occurred above the gap as the cell size is smaller than the gap height. At higher initial pressures in the partially filled channel, the transition occurs closer to the ignition source in the gap above the bead layer.

References

1. Babkin V.S., Korzhavin A.A., and Bunev V.A., Propagation of premixed gaseous explosion flames in porous media, Combustion and Flame, vol. 87, 182-190, 1991.
2. Babkin V.S., The problems of porous flame-arresters, Prevention of Hazardous Fires and Explosions, V.E. Zarko et al. (eds.), Kluwer Academic Publisher, Netherlands, 199-213, 1990.
3. Joo P., Duncan K., and Ciccarelli G., Flame Quenching Performance of Ceramic Foam, Combustion Science and Technology, 178(10-11):1755, 2006.
4. Pinaev A.V., Combustion modes and flame propagation criteria for an encumbered space, Combustion, Explosion, and Shock Waves 30(4), 454-461, 1994.
5. Makris A., Shafique H., Lee J.H., and Knystautas R., Influence of mixture sensitivity and pore size on detonation velocities in porous media, Shock Waves, vol. 5, pp. 589-95, 1995.
6. Peraldi O., Knystautas R., and Lee J.H., Criteria for transition to detonation in tubes, Proceedings of the Combustion Institute, vol 21, 1629-1637, 1986.

Continuum/particle interlocked simulation of gas detonation

A. Kawano and K. Kusano

The Earth Simulator Center, Japan Agency for Marine-Earth Science and Technology, 3173-25 Showa-machi Kanazawa-ku Yokohama Kanagawa (Japan)

Summary. A multiscale simulation method for gas detonations is proposed. This method is performed by the interlocking of a continuum fluid model and a particle based molecular model. The simulation system is spatially divided into the two kinds of domains, in one of which the near-thermal equilibrium is satisfied but not in the other. The dynamics of the near-thermal equilibrium domains is modeled by the continuum fluid dynamic equations. In the nonthermal equilibrium domain, which corresponds to the denotation front, the dynamics is modeled by the Boltzmann equation. This method has been tested on a model detonation system.

1 Introduction

The simulation of a combustion system is usually performed using the fluid dynamic method based on the continuum and the local thermal equilibrium approximations. In this method the chemical reactions are assumed to be described by a chemical kinetic model with reaction rates written as a function of temperature by the Arrhenius equation. However, this method is likely to be broke down in detonation simulations. The local Knudsen number within the shock front of a gas detonation becomes large because of the strong steepness of state variables. This leads that the detailed structure of the shock front cannot be reproduced accurately by the usual continuum method. In addition, chemical reactions with large heat release may generate a strong non-thermal equilibrium state in the detonation reaction zone behind the shock front.

The effect of non-thermal velocity distributions on gas phase reactions, disregarded in the usual continuum fluid dynamic models, has been investigated on the basis of the molecular kinetic theory using the Boltzmann equation for several decades. Prigogine and Mahieu [1] have studied the effect of the heat of reaction on the velocity distribution using the Chapman-Enskog (CE) method. Although the CE solution is limited to slow reactions with small departures from equilibrium, it was pointed out that the heat of reaction can parturb the reaction rate from the equilibrium rate to an appreciable extent. Koura [2] has used a Monte Carlo method to describe a highly exothermic reaction with low activation energy. He found that the reaction rate can be increased by more than 100%. Anderson and Long [3] have indicated that the assumptions required for the Chapman-Jouguet and Zeldovich-von Neumann-Döring theories may not be valid for a one-dimensional detonation with very fast reactions. Bruno and coworkers [4] have investigated a one-dimensional model detonation system and showed that the non-thermal equilibrium character of the velocity distribution function in the shock front modifies substantially the kinetics of the reactions and the overall profile of the reaction zone.

Despite the success of the molecular kinetic theory, it allows for the analysis of smaller systems than continuum fluid dynamic methods due to its high computational cost. The dynamics of a detonation is considered as a multiscale phenomenon, in which the microscopic reactive/unreactive collision processes and the macroscopic fluid dynamics are mutually interacted. In a detonation, fortunately, most spatial areas except the shock front and the reaction zone are in near-thermal equilibrium in which a continuum fluid model is applicable. In this study we propose a new multiscale simulation model for gas detonations, which is performed by the interlocking of a continuum fluid model and a particle-based molecular kinetic model. The simulation system is spatially divided into the two kinds of domains. One occupies most of the system in which the near-thermal equilibrium is satisfied. We call this domain the *continuum domain*. The other domain, which we call the *particle domain*, lies in the vicinity of the shock front, where strong non-thermal equilibrium may take place. The dynamics of the continuum domain is treated by a macro-scale continuum fluid model based on the Navier-Stokes equation. The particle domain is treated by a molecular kinetic model based on the Boltzmann equation.

2 Computational Model

In this paper we employed a model chemical reaction system with three molecular species R (reactant molecule), Y (radical), and P (product molecule). For simplicity, all of the molecules are regarded as hard spheres with the same molecular mass m and diameter d without internal degrees of freedom, i.e., they are calorically perfect and have constant heat capacities as for monoatomic molecules. We suppose that the following bimolecular reactive collision processes occur in the system:

$$\text{Chain initiation:} \qquad \mathrm{R}+\mathrm{M}' \overset{k^{\mathrm{Y,M}'}_{\mathrm{R,M}'}}{\longrightarrow} \mathrm{Y}+\mathrm{M}', \tag{1}$$

$$\text{Chain branching:} \qquad \mathrm{R}+\mathrm{Y} \overset{k^{\mathrm{Y,Y}}_{\mathrm{R,Y}}}{\longrightarrow} \mathrm{Y}+\mathrm{Y}, \tag{2}$$

$$\text{Chain termination:} \qquad \mathrm{Y}+\mathrm{M}'' \overset{k^{\mathrm{P,M}''}_{\mathrm{Y,M}''}}{\longrightarrow} \mathrm{P}+\mathrm{M}'', \tag{3}$$

where M' and M'' are the proxies for R or P and that for Y or P, respectively, and $k^{\mathrm{Y,M}'}_{\mathrm{R,M}'}$, $k^{\mathrm{Y,Y}}_{\mathrm{R,Y}}$, and $k^{\mathrm{P,M}''}_{\mathrm{Y,M}''}$ are the rate coefficients of the reactions. The reverse reactions are neglected.

The reactive collision processes (1), (2), and (3) are described by the reactive hard sphere model [5]. In this model the reaction probability $P^{\gamma,\delta}_{\alpha,\beta}$ that the reaction $\alpha+\beta \to \gamma+\delta$ occurs for a collision of α and β is given by

$$P^{\gamma,\delta}_{\alpha,\beta} = \begin{cases} 0 & \text{if } E^{\gamma,\delta}_{\alpha,\beta} \geq E_{\mathrm{col}}, \\ \zeta\left(1-\dfrac{E^{\gamma,\delta}_{\alpha,\beta}}{E_{\mathrm{col}}}\right) & \text{if } E^{\gamma,\delta}_{\alpha,\beta} < E_{\mathrm{col}}, \end{cases} \tag{4}$$

where $E^{\gamma,\delta}_{\alpha,\beta}$ is the activation energy of the reaction, ζ is the steric factor, and E_{col} is the collision energy given by $\mu_{\mathrm{r}}g^2/2$, where $\mu_{\mathrm{r}} = m/2$ is the reduced mass of the pair, and g is the relative speed of the pair. This model results in Arrhenius behavior of the reaction rate coefficient

$$k^{\gamma,\delta}_{\alpha,\beta} = \zeta\sigma\sqrt{\frac{8k_{\mathrm{B}}T}{\pi\mu_{\mathrm{r}}}}\exp\left(-\frac{E^{\gamma,\delta}_{\alpha,\beta}}{k_{\mathrm{B}}T}\right), \tag{5}$$

where σ is the total cross section for the pair, and T is the temperature.

The molecular parameters used in this study are shown in Table 1.

We describe the time evolution of the distribution function $f_\alpha(\mathbf{r}, \mathbf{c}, t)$ of chemical species α for position $\mathbf{r}$ and velocity $\mathbf{c}$ by the Boltzmann equation

$$\begin{aligned}\frac{\partial f_\alpha}{\partial t} &+ \mathbf{c}\cdot\nabla f_\alpha \\ &= \sum_\beta \int [f_\alpha(\mathbf{r},\mathbf{c}',t)f_\beta(\mathbf{r},\mathbf{c}_1',t) - f_\alpha(\mathbf{r},\mathbf{c},t)f_\beta(\mathbf{r},\mathbf{c}_1,t)]\, g\,\sigma_{\alpha,\beta}^{\alpha,\beta}(g)\,\mathrm{d}\Omega\,\mathrm{d}\mathbf{c}_1 \\ &\quad + \sum_{\beta,\gamma,\delta} \int f_\gamma(\mathbf{r},\mathbf{c}',t)f_\delta(\mathbf{r},\mathbf{c}_1',t)\, g'\,\sigma_{\gamma,\delta}^{\alpha,\beta}(g')\,\mathrm{d}\Omega\,\mathrm{d}\mathbf{c}_1 \\ &\quad - \sum_{\beta,\gamma,\delta} \int f_\alpha(\mathbf{r},\mathbf{c},t)f_\beta(\mathbf{r},\mathbf{c}_1,t)\, g\,\sigma_{\alpha,\beta}^{\gamma,\delta}(g)\,\mathrm{d}\Omega\,\mathrm{d}\mathbf{c}_1,\end{aligned} \tag{6}$$

where $g = |\mathbf{c} - \mathbf{c}_1|$ and $g' = |\mathbf{c}' - \mathbf{c}_1'|$ are the relative speeds, $\sigma_{\gamma\delta}^{\alpha\beta}(g)$ is the cross section for the collision process $\alpha + \beta \to \gamma + \beta$ with the relative speed g, Ω is a solid angle, and the primes denote postcollisional velocities. The reactive cress section for process $\alpha + \beta \to \gamma + \beta$, is given by $\sigma_{\alpha\beta}^{\gamma\delta}(g) = \sigma P_{\alpha\beta}^{\gamma\delta}(g)$. In this study we solve Eq. (6) by use of a particle-based scheme called the direct simulation Monte Carlo method [7].

The dynamics of the continuum domain is modeled by the compressible fluid dynamic equations as follows:

$$\frac{\partial \rho}{\partial t} + \nabla\cdot(\rho\mathbf{u}) = 0, \tag{7}$$

$$\rho\frac{\partial \mathbf{u}}{\partial t} + \rho\mathbf{u}\cdot(\nabla\mathbf{u}) = -\nabla\cdot\mathbf{P}, \tag{8}$$

$$\frac{\partial e}{\partial t} + \nabla\cdot(e\mathbf{u}) = -\nabla\cdot\mathbf{q} - \mathbf{P} : (\nabla\mathbf{u}), \tag{9}$$

$$\frac{\partial \rho_\alpha}{\partial t} + \nabla\cdot(\rho_\alpha\mathbf{u}) = -\nabla\cdot(\rho_\alpha\mathbf{V}_\alpha) + w_\alpha, \tag{10}$$

with

$$\mathbf{P} = p\mathbf{I} + \frac{2}{3}\mu(\nabla\cdot\mathbf{u}) - \mu[(\nabla\mathbf{u}) + (\nabla\mathbf{u})^T], \tag{11}$$

Table 1. Molecular parameters for the model detonation system

Particle mass	$m = 30$ g/mol
Total cross section for all pair	$\sigma = 0.5027$ nm^2
Steric factor	$\zeta = 1$
Activation energies	$E_{\mathrm{R,M'}}^{\mathrm{Y,M'}} = 125.5$ kJ/mol
	$E_{\mathrm{R,Y}}^{\mathrm{Y,Y}} = 41.84$ kJ/mol
	$E_{\mathrm{Y,M''}}^{\mathrm{P,M''}} = 62.76$ kJ/mol
Formation enthalpies	$\Delta H_{\mathrm{f}}(\mathrm{R}) = 0$ kJ/mol
	$\Delta H_{\mathrm{f}}(\mathrm{Y}) = 20.92$ kJ/mol
	$\Delta H_{\mathrm{f}}(\mathrm{P}) = -20.92$ kJ/mol

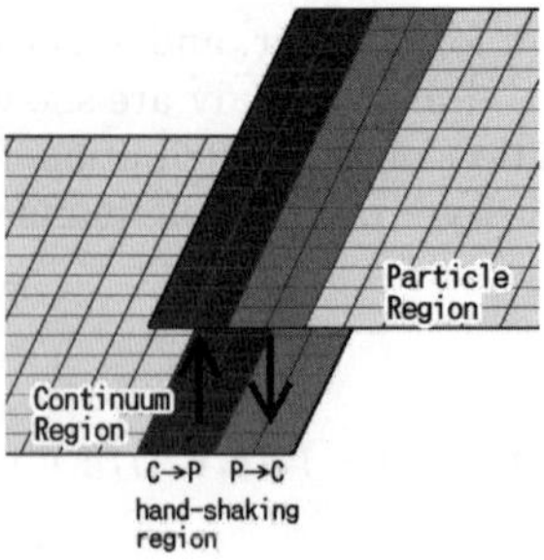

Fig. 1. Illustration of the handshaking between particle and continuum models.

$$\mathbf{q} = -\lambda \nabla T + \sum_{\alpha} h_\alpha \rho_\alpha \mathbf{V}_\alpha, \tag{12}$$

$$\omega_{\mathrm{R}} = \frac{1}{m^2}(-\rho_{\mathrm{R}}^2 k_{\mathrm{R,R}}^{Y,R} - \rho_{\mathrm{R}}\rho_{\mathrm{P}} k_{\mathrm{R,P}}^{Y,P} - \rho_{\mathrm{R}}\rho_{\mathrm{Y}} k_{\mathrm{R,Y}}^{Y,Y}), \tag{13}$$

$$\omega_{\mathrm{Y}} = \frac{1}{m^2}(\rho_{\mathrm{R}}^2 k_{\mathrm{R,R}}^{Y,R} + \rho_{\mathrm{R}}\rho_{\mathrm{P}} k_{\mathrm{R,P}}^{Y,P} + \rho_{\mathrm{R}}\rho_{\mathrm{Y}} k_{\mathrm{R,Y}}^{Y,Y} - \rho_{\mathrm{Y}}^2 k_{\mathrm{Y,Y}}^{\mathrm{P,Y}} - \rho_{\mathrm{Y}}\rho_{\mathrm{P}} k_{\mathrm{Y,P}}^{\mathrm{P,P}}), \tag{14}$$

$$\omega_{\mathrm{P}} = \frac{1}{m^2}(\rho_{\mathrm{Y}}^2 k_{\mathrm{Y,Y}}^{\mathrm{P,Y}} + \rho_{\mathrm{Y}}\rho_{\mathrm{P}} k_{\mathrm{Y,P}}^{\mathrm{P,P}}), \tag{15}$$

where ρ_α, ρ, $\mathbf{u}$, e, $\mathbf{P}$, $\mathbf{q}$, $\mathbf{V}_\alpha$, ω_α, p, $\mathbf{I}$, μ, λ, and h_α are the mass density of species α, the total mass density defined as $\sum_\alpha \rho_\alpha$, the velocity, the specific total energy, the stress tensor, the heat flux, the diffusion rate of α, the reaction term for α, the pressure, the unit tensor, the viscosity coefficient, the thermal conductivity, and the enthalpy of α, respectively. The transport coefficients μ, λ and $\mathbf{V}_\alpha$ is obtained from the molecular kinetic theory. Body forces, the Soret and Dufour effects, pressure gradient diffusion, bulk viscosity, and radiation fluxes are ignored. All species are assumed as perfect gases which obey the gas equation of state

$$p = T \sum_{\alpha} \rho_\alpha R_\alpha, \tag{16}$$

where R_α is the specific gas constant of α. Equations (7), (8), (9), and (10) are solved by an approximate Riemann solver called the HLLC method [6].

As illustrated in Fig. 1, the particle and continuum domains are interlocked via handshaking regions. The hand-shaking region consists of C → P and P → C connections, which pass information from the continuum domain to the particle domain and vice versa. The C → P connection generates simulation particles in the particle domain drawn from the thermal-equilibrium distribution corresponding to the macroscopic quantities received from the corresponding part of continuum domain. The P → C connection obtains macroscopic quantities of each particle cell by particle averaging, and then it passes these quantities to the corresponding part of the continuum domain.

3 Test calculations

We have applied our method for a 2-dimensional detonation system to test its validity.

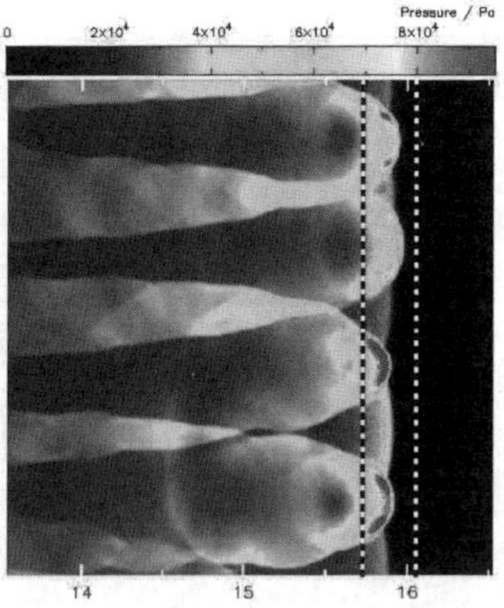

Fig. 2. The pressure distribution around the detonation front. The region bounded by dotted line corresponds to the domain for the particle-based model.

Figure 2 shows a typical result of the interlocked simulation for detonation propagation. The narrow part enclosed by dotted lines corresponds to the particle domain. In this simulation, the particle domain can automatically track the propagating shock front. The result shows that the characteristic structure of detonation front with triple points is clearly reproduced by the particle model, and the flow field runs properly toward downwind through the hand-shaking region.

It is well-known that two-dimensional detonation propagates with successive collisions of triplet points, which form a scale-like cellular structure. Maximum pressure history shown in Fig. 3 reveals that the result of the the interlocked model is in good agreement with the particle-based simulation, in which all molecular kinetics is taken into account in the whole domain. This indicates the advantage of the interlocked simulation from the point of view of computational efficiency.

Further work has been undertaken to compare the results obtained by each method in detail and to evaluate the kinetic effects qualitatively.

4 Conclusions

We propose a multiscale simulation model for gas detonations, which is performed by the interlocking of a continuum fluid model and a particle based molecular model. Our model has the potential to perform accurate simulation of detonation phenomena which takes account of the effect of the molecular kinetics as well as that of the continuum fluid dynamics with the Arrhenius-type chemical kinetics.

Acknowledgement. The calculations presented in this paper were done on the Earth Simulator under support of Japan Agency for Marine-Earth Science and Technology.

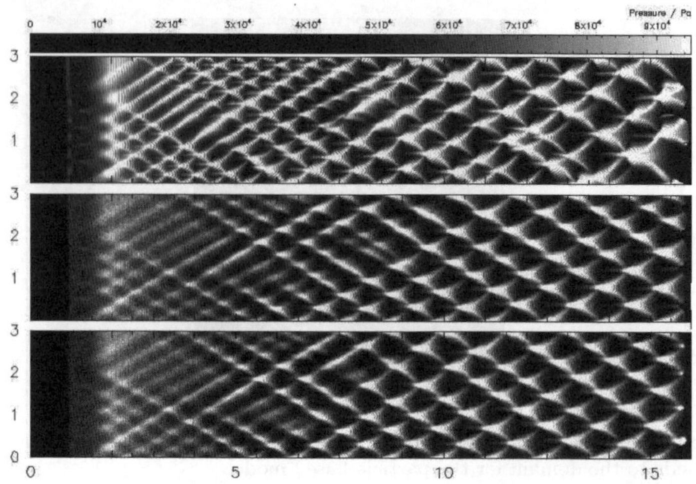

Fig. 3. The distribution of maximum pressure history. Top, middle, and bottom represent the results of continuum model, particle model and the interlocked model, respectively.

References

1. I. Prigogine and M. Mahieu: Sur la perturbation de la distribution de Maxwell par des réactions chimiques en phase gazeuse. Physica **16**, 51 (1950).
2. K. Koura: Nonequilibrium velocity distributions and reaction rates in fast highly exothermic reactions. J. Chem. Phys. **59**, 691 (1973).
3. J. B. Anderson and L. N. Long: Direct Monte Carlo simulation of chemical reaction systems: Prediction of ultrafast detonations. J. Chem. Phys. textbf118 15 (2003).
4. D. Bruno, M. Capitelli, and S. Longo: Effect of translational kinetics on chemical rates in a direct simulation Monte Carlo model gas phase detonation. Chem. Phys. Lett. **380**, 383 (2003).
5. J. I. Steinfeld, J. S. Francisco, W. L. Hase: *Chemical Kinetics and Dynamics.* 2nd edition (Prentice Hall, 1998).
6. P. Batten, N. Clarke, C. Lambert, and D.M. Causon: SIAM J. Sci. Statist. Comput. **18** 1553 (1997).
7. G. A. Bird: *Molecular Gas Dynamics and the Direct Simulation of Gas Flows.* (Clarendon Press, Oxford, 1994).

Dependence of PDE performance on divergent nozzle and partial fuel filling

Z.X. Liang, Y.J. Zhu, and J.M. Yang

Department of Modern Mechanics, University of Science and Technology of China, Hefei, 230027, China

Summary. A grid adaptive finite volume method combined with kinetic chemical reaction model was adopted to simulate the flow field of detonation wave propagating through a divergent channel. A series of ballistic pendulum experiments were also carried out for the qualitative comparison. The investigation were focused on the performance of pulse detonation engine (PDE) depending on the nozzle shape and the partially fuel fill in the chamber. It was found that the impulse of the straight detonation tube with a hydrogen-oxygen mixture achieves its maximum at a certain filling ratio. On the other hand, as to the nozzles with a larger divergent angle, the impulse increases with increasing fuel filling, although in most cases the specific impulse increases with decreased fuel filling ratio, which shows that the PDE performance can be obviously improved when the nozzle shape and fuel filling are designed properly.

1 Introduction

A pulse detonation engine (PDE) has many advantages such as simple structure, high efficiency and specific impulse, which is expected to be one of the propulsion machineries of new generation. Because the strong irreversibility during the process of detonation wave propagating through nozzle results in the energy loss, a proper design of nozzle will be helpful to convert more chemical energy into mechanical energy of propulsion. Therefore the behaviors of unstationary detonation nozzle flow need to be investigated, since the traditional theories and concepts of nozzle design are not able to predict the PDE performance correctly.

Furthermore, the partial fuel filling effect in nozzle is seldom considered, although the impact on the PDE performance is not negligible. A typical PDE operation cycle might be mainly divided into three stages of filling, detonation and exhaust. People usually require intuitively that the detonation wave catches up with the combustible gas mixture-air interface at the exit of thrust tube for the optimal performance. Whether it is true, or when this condition is not satisfied, how it would influence the PDE performance is still a problem that needs to be investigated. M. Cooper and J. E. Shepherd [1] have conducted a series of experiments about the partial fill effect in thrust tube using stoichiometric ethylene-oxygen as combustible. Li et al. [2] have simulated numerically the partial fill effect and analyzed the PDE performance, also took ethylene as fuel. Li et al. [3] [4] have used hydrogen-oxygen as fuel to study the wave interaction and the structure of flow field when detonation wave is propagating through a divergent nozzle of PDE. One of their results is that the thrust of a fully fuel filled nozzle is larger than the filling air one significantly.

Based on the work of Li et al. [3] [4], focusing on the factors of nozzle-shape and fuel filling ratio, this paper studies the flow field of PDE to conduct performance analysis numerically and experimentally. We hope it could provide some supports for the economical and efficient use of PDE.

2 Numerical Methods

A grid adaptive finite volume method combined with kinetic chemical reaction model was adopted to simulate axisymmetric or 2-D flow of a single PDE cycle. The flow governing equations was discretized using cell-centered finite volume method based on unstructured quadrangle grid. The program uses MUSCL-Hancock algorithm to produce a solution of the Euler equations which is second order accurate in space and time [5]. A time-split method was used to handle the flow governing equations and chemical reaction governing equations, which involves a chemical model of 11-species and 23-reaction to simulate the hydrogen-oxygen reaction, respectively [6]. The credibility of the computational method mentioned above has been validated by Li et al. [3] [4].

The mixture-based specific impulse is defined as $I_{sp} = \frac{I}{M}$, where M is the combustible mixture mass in detonation tube (including nozzle). Three different nozzle shapes, straight tube, straight tube plus a 15°nozzle and straight tube plus a 45°nozzle, are chosen for the numerical simulation. As to the first type of shape, say straight tube case, when it is not fully filled with fuel in the tube, the rest filled with air could be considered as a 0°nozzle. The straight detonation tube is $6.1cm$ long with a length-to-diameter ratio of 15.25. There are 4 cases of the filling conditions, 4.5, 5.0, 5.5 and 6.1 cm, representing 74%, 82%, 90% and 100% of filling length, respectively. A 0.08-cm section next to the end wall of tube is set as initiation region, in which the pressure and the temperature are $3.3MPa$ and $3000K$, respectively. Those in the rest part are $0.1MPa$ and $300K$. As to the other two nozzle shapes, the length of straight tube is $5.0cm$, and the nozzle is a 1.1cm-long taper of constant angle. There are three kinds of filling conditions for both nozzles. The first one is fully filled with combustible mixture, the second one is half-filled($0.55cm$ fuel length in nozzle), and the last one is no-filled (namely the nozzle filled with non-combustible air and the combustible mixture is only filled in straight tube). The mixture is stoichiometric hydrogen-oxygen and the ambient air is composed of 22% oxygen and 78% nitrogen.

3 Experimental setup and procedure

The detonation tube is $30mm$ inner diameter and $1000mm$ long. An additional straight tube (or 0° nozzle) of $90mm$ long is connected to the main one. One end of the tube is sealed with a spark plug, a pressure transducer, and a gas-inlet valve. A $10\mu m$ thick Mylar diaphragm was used to separate the hydrogen-oxygen mixture from atmospheric air. It was placed between the tube and the nozzle (91.7% filling ratio of mixture length to the total tube length) or at the end of the nozzle (fuel fully filled in the tube and nozzle). The tube was hung from the ceiling by two steel wires in a ballistic pendulum arrangement. The masses of the wires for hanging were neglected in the momentum analysis. The total mass of ballistic pendulum is $8.73kg$. At the beginning of experiment the tube was evacuated. The hydrogen and oxygen composing the fuel mixture, which was stoichiometric or at an equivalence ratio ϕ 3:2, were added to the tube to an initial fill pressure of $101kPa$ one after the other with several steps. Then the ballistic pendulum was placed still for about 10 minutes for the mixing. The spark plug ignited the combustible mixture at the tube's end wall. After a DDT process detonation wave was formed, run and broke the diaphragm. Combustion products were exhausted into environment air at last. A digital vidicon was used to capture the displacement of the tube. The credibility of the experiment method has been validated by Li hui-huang et al. [4].

4 Numerical Results

4.1 Straight nozzle

The simulation of straight tube is an axisymmetric flow. The histories of thrust, impulse and specific impulse are showed in Fig. 1. When the right-travelling detonation wave reaches the interface of mixture-air, a transmitted shock enters into the air medium, and at the same time, a weak reflected shock propagates upstream toward the thrust wall of tube because the speed of sound in the fuel mixture is higher than that in air, which differs from the case of ethylene-oxygen [1] [2]. When the left-travelling shock reaches the end wall, it produces a small jump on the plot of thrust (presented in Fig. 1, the symbol 6.1cm_f4.5cm, for example, in the figure represents that the length of tube is 6.1cm and that the length of fuel mixture is 4.5cm), followed by a reflected expansion wave that is produced by the interaction of Taylor wave with the interface that decreases the thrust gradually. As the increase of the fuel filling, or the interface further downstream, the delay time of the trust jump increases accordingly. On the other hand, the exhaust expansion waves generated at the tube exit speed the descent of thrust. It also means that the descent of thrust consists of two parts, the first part is resulted from the expansion wave caused by the interaction of Taylor wave with the interface, and the second part is the expansion wave produced when the detonation or shock wave reach the open end of the tube.

If the detonation wave meets the interface at the exit, the combination of reflected shock and expansion wave weakens the reflected shock. The thrust jump is only 4% larger than the force plateau. On the contrary, it is 24% in the case of partial fill and the descent is much more gently. It can be seen in Fig. 1, that the impulse of fully filled case is less than that of partial fill case, which is different from the results of other references [1] [2] that the impulse decrease as the fuel filling ratio increases. This is because of that the Molecular weight of Hydrogen is much less than that of hydrocarbon fuel so that the sound speed of hydrogen-oxygen is higher than that of air on the other side of the interface, which causes the reflected shock wave to increase the thrust.

In present work, the case of 82% fuel filling gets the highest impulse, followed by the cases of 90%, 74% and 100% in turn. The impulse of 82% fuel filling is slightly 4% higher than that of 100%. If we consider the consumption of fuel, the benefit is much better. Decreasing the fuel filling, the mixture-based specific impulse I_{sp} goes up. The specific impulse of 74% fuel filling is 37% larger than that of 100% case. E. Wintenberger [7] gets a full-filled specific impulse result of 172.9sec by their theory model. The result of Zitoun and Desbordes [8] is 226sec. Our result is 195.5sec.

4.2 15°nozzle

Simulations of axisymmetric flow and 2-D planar flow are conducted respectively to analyze the similarities and differences. In order to compare the results, a width is given to the model of 2-D planar flow so that the end wall areas of straight tube are same.

From basic PDE configuration we know that the thrust mainly consists of two parts, the nozzle thrust and the end wall thrust. As to 2-D planar 15°nozzle, it is necessary to point out that the thrust peak of fully filled is higher than that of half-filled but its action is short and decreases quickly. For the later stage, the thrust of half-filled is higher than that of fully filled case, accordingly it generates higher impulse. As for the reason why with more fuel in the tube but obtaining less impulse, it might be explained with

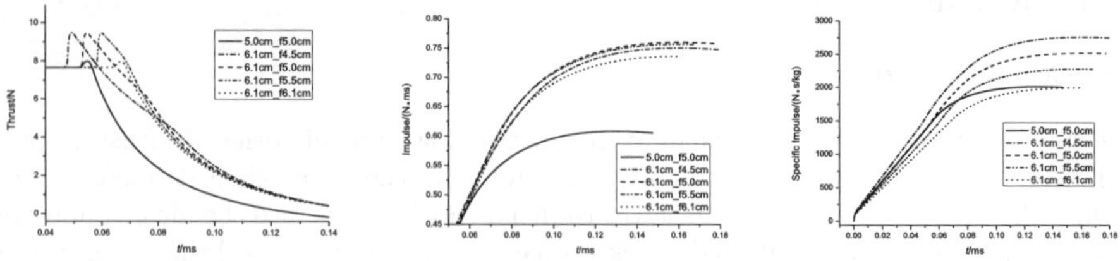

Fig. 1. Pressure time history at end wall (left), impulse (center) and specific impulse (right) for straight detonation tubes

help of the similarity to the case of straight tube in which the best performance achieved by a partially fuel filling.

Fig. 2 shows the impulse from both two parts, namely the thrust wall and the nozzle wall under the three filling conditions. It can be found that the impulse provided by end wall is dominant. So that the efficiency of 15° nozzle is analogous to that of straight tube. It also means if the half angle is too small, or the size of nozzle exit is not large enough, it will not be able to improve the PDE performance obviously.

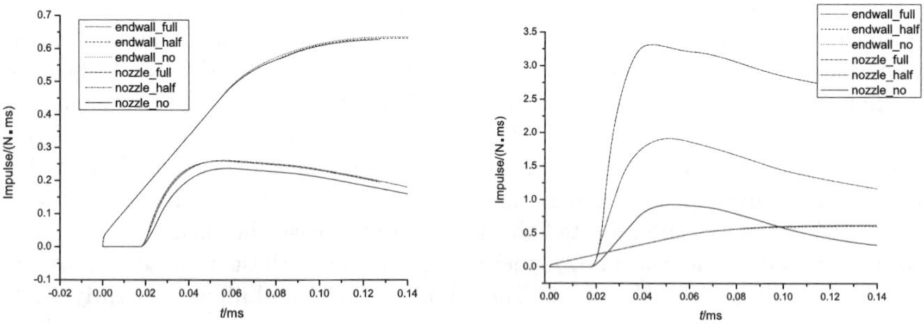

Fig. 2. Impulse on nozzle and thrust wall : 15° 2-D nozzles (left) and axisymmetric 45° (right)

4.3 45°nozzle

In the case of axisymmetric, the contribution of fuel filling length in a 45° nozzle is similar to that of 15°. The results in Fig. 2 show that the impulse goes up with the increase of fuel filling ratio and the impulse on nozzle is comparative to that on end wall even in the case of no fuel filled in nozzle. The change in fuel filling makes little difference on the impulse at the thrust wall. The three curves generated at thrust wall are almost of superposition and the difference is less than 2%. It could draw a conclusion that the total impulse difference is contributed from the nozzle.

In order to compare the effect of different nozzles and filling conditions, Fig. 3 shows the specific impulse of axisymmetric 15° and 45°nozzles. It can be found that the specific impulse of 45°nozzle is much larger than that of 15° one, especially when there is no filled or half filled in the nozzle. So the effect of nozzle is considerable. However, a bigger half angle nozzle has larger mass and bigger drag force during high speed flight, which is not included in the present work. Besides, as it has been mentioned above, that the case of full filling gets higher thrust and impulse, but from Fig. 3 it can be found that it also has a lower specific impulse, which means that the increase in impulse due to the additional detonable mixture may not adequately compensate for the increase in mass. Comparing to that, the case of half filled gets a benefit of close I_{sp} and larger impulse. So that the fuel filling in nozzle is an important factor worthy to pay attention to.

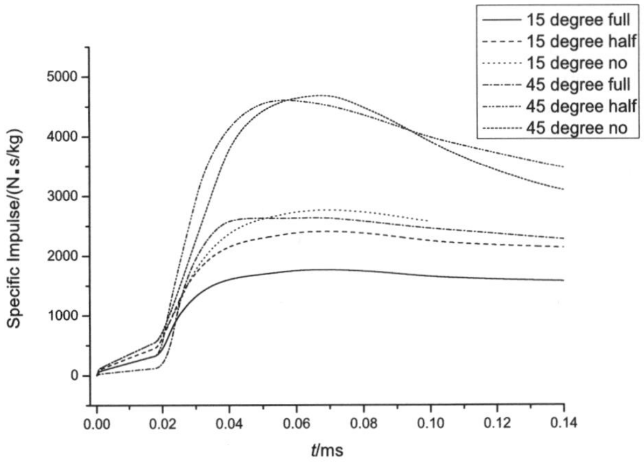

Fig. 3. Specific impulse of different nozzles and filling conditions (axisymmetric)

5 Experimental Results

The pendulum experiments, with careful arrangement and operation, were capable of telling the difference between the fully filled case and 91.7% length ratio case, which offers an obvious qualitative support of the above conclusion for straight tube case(see table 1). The impulse decreases as the fuel filling ratio increases. There are two control factors in the experiments, mixture filling length and mixture component molar ratio. With the same component molar ratio, the 91.7% cases got a higher impulse than that of 100% cases. When the molar ratio of the hydrogen to oxygen increases from 2:1 to 3:1, in which the sound speed in the fuel mixture increases accordingly, so that the reflected shock is strengthened. As a result, the experimental data shows that the impulse of 91.7% cases at a equivalence ratio 1.5 increase slightly in comparison with that of stoichiometric ones.

Fuel Filling	Equivalence Ratio ϕ	Impulse $[N \cdot s]$
91.7%	1.0	$0.460 \pm 2\%$
91.7%	1.5	$0.425 \pm 2\%$
100%	1.0	$0.428 \pm 2\%$
100%	1.5	$0.393 \pm 3\%$

Table 1. Ballistic pendulum experiment results

6 Summary

Present work provided some descriptions of the dependence of PDE performance on divergent nozzle and partial fuel filling ratio. It could be concluded that:

1. The impulse of the straight detonation tube with a hydrogen-oxygen mixture does not always increase with increasing fuel filling. It achieves its maximum at a certain filling ratio, although the specific impulse increases with the decreased fuel filling ratio in most case.

2. Without the consideration of the drag force from outer flow, a 45° axisymmetric nozzle usually enhance the PDE performance and its I_{sp} increases too. If half angle of a nozzle is too small, it has a characteristics similar to the case of straight tube.

3. As to larger half angle nozzle, its impulse increases with the increase in fuel filling usually. However, high fuel filling ratio may decrease specific impulse.

Acknowledgement. The authors gratefully acknowledges the support of K. C. Wong Education Foundation, Hong Kong.

References

1. Cooper and J. E. Shepherd: The Effect of Transient Nozzle Flow on Detonation Tube Impulse, AIAA Paper 2004-3914, July 2004.
2. Li C, Kailasanath K, and Patnaik G: A Numerical Study of Flow Field Evolution in a Pulse Detonation Engine. AIAA Paper 2000-0314, Jan. 2000.
3. LI Hui-huang, YANG Ji-ming, XU Li-gong: A numerical simulation on the nozzle flow of pulse detonation engine. Journal of Propulsion Technology, 25(6), 2004.
4. LI Hui-huang, ZHU Yu-jian, YANG Ji-ming: Experimental and numerical investigation of gaseous detonation diffraction through a divergent nozzle based on double exposure holography. Explosion and Shock Waves, 25(5), 2005
5. Mingyu Sun: Numerical and experimental studies of shock wave interaction with bodies. Ph.D thesis of Tohoku University, 1998.
6. Kee R J, Rupley FM, Miller J A: Chemkin-III: A fortran chemical kinetics package for the analysis of gas-phase chemical and plasma kinetics. SAND 96-8216, 1996.
7. Wintenberger E, Austin J M, Cooper M, et al: An analytical model for the impulse of single-cycle pulse detonation tube, Journal of Propulsion and Power, 19(1), 2003.
8. Zitoun R and Desbordes D: Propulsive Performances of Pulsed Detonations, Combustion Science and Technology, 144:93-114, 1999

Direct Monte-Carlo simulation of developing detonation in gas

Z.A. Walenta and K. Lener

PWSZ Jaroslaw,
Czarnieckiego 16, 37-500 Jaroslaw (Poland)

Summary. The research on gaseous detonation has recently become a very important issue, mainly due to safety reasons in connection with increasing use of gaseous fuels. To simulate detonation, the Direct Monte-Carlo Simulation technique has been proposed, together with simple model of molecular collisions, making it possible to heat the gas in a way similar to the processes in the flame. Such model is capable of producing waves, having the features characteristic for detonation (Dremin [2]). In the present work the influence of finite reaction time and the inverse reaction upon formation and extinguishing detonation, in the framework of this model, has been investigated.

1 Introduction

The Direct Monte-Carlo Simulation (DMCS) technique (Bird [1]) has proven to be a very powerful tool for solving flow problems, particularly connected with presence of solid walls of complex geometry. It offers also a possibility of taking into account relaxation phenomena and chemical reactions (Bird [1], Larsen et al [3]). This however increases substantially complexity of the computer programs and the necessary computing times (although other methods of computation may be even more demanding). Fortunately, in gaseous detonation the state of the medium is far from equilibrium, combustion proceeds very fast, therefore majority of complex relaxation processes at the molecular level may, perhaps, be disregarded. The most important remaining factor seems to be the produced thermal energy. In the present work we investigate how the two additional factors: the inverse chemical reaction (important at high temperatures) and finite reaction time (i.e. delay in energy release) influence the detonation - particularly its formation and decay. In our first simulations (Walenta et al [5]), when investigating the mechanism of extinguishing detonation by cooling and friction at the walls, these factors were disregarded.

2 Model of a combustible gas

As always in the Direct Monte-Carlo Simulations the gas is treated as an ensemble of molecules colliding with each other and with walls and moving along straight lines with constant speed between collisions. For simplicity of calculations all molecules are modeled as identical hard spheres. It is assumed, that some molecules, uniformly distributed in space, carry certain amount of "internal" energy of unspecified character. This may be released and transformed into kinetic energy during collision, if the two colliding molecules approach each other with sufficiently high velocity and only one of them is carrying energy (see Figure 1a for definition of "velocity of approach"). The energy may be released either in the first collision of this kind (no delay), or during some later collisions (delayed energy release). The relative velocity of the molecules after collision is then increased suitably (Figure 1b). A molecule which had lost its "internal" energy

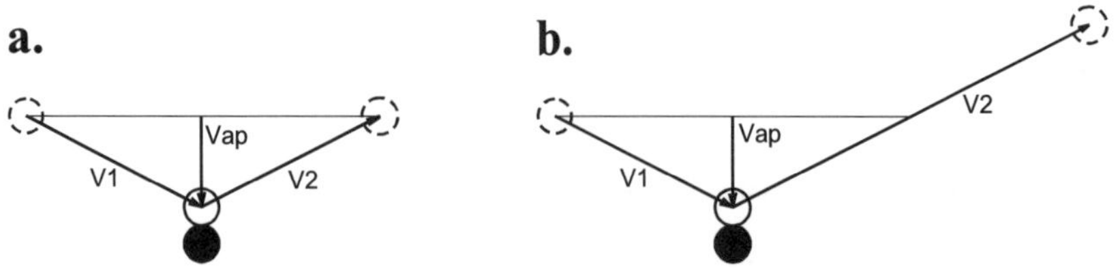

Fig. 1. Collision of two molecules in reference frame connected to one of them: a - elastic, b - with energy release. V1 - relative velocity of the molecules before collision, V2 - relative velocity after collision, Vap - "velocity of approach" of the molecules.

may regain it if it collides with a molecule, which cannot carry this kind of energy, and if the "velocity of approach" is higher than velocity corresponding to this "internal" energy (the "inverse reaction"). The relative velocity of the molecules after collision is then decreased suitably.

3 Details of calculation

In our calculations the standard DMCS procedure, (Bird, [1]) was employed: only binary collisions were taken into account - the pairs of molecules for collisions were selected with the use of the ballot-box scheme, as proposed by Yanitskiy [6]. One-dimensional geometry was used to investigate the detonation initiation and development, and three-dimensional geometry to investigate its decay in a narrow channel. In 1-D geometry (plane wave, no walls) the number of molecules in most cases was equal to about 1 million (in some cases 3 million), in 3-D geometry (flow in a pipe) it was equal to about 13 million.
Contrary to the usual DSMC procedure, it was not possible to perform a number of calculations for the same initial conditions and different starting values for the random number generator, and take the average to make the results less noisy. The outcome was unphysical - the front part of the averaged shape of the detonation wave was not steep - obviously because of instability of detonation, resulting in different positions of the wave in different calculation runs.
The calculation area was divided into 1000 cells (in some cases 3000 cells) in 1-D geometry and about 1.3 or 2.6 million cells in 3-D geometry. The axial dimension of a cell was equal to 1 unit of length, (equal to 1 mean free path - lambda - of the gas molecules in the undisturbed area). The diameter of the pipe in 3-D geometry was equal to 100 mean free paths.
The wave was initiated by removal of a "diaphragm", placed at $X = 100$ units of length from the origin of coordinates. The gas in front of the diaphragm contained 10, 20 or 30 per cent of the molecules carrying the "internal" energy. The energy released in a single collision increased the relative velocity of the colliding molecules by the value equal to 10 times the most probable molecular speed. The "threshold velocity" of approach of the colliding molecules, necessary to release the "internal energy", was taken 5.48 times the most probable molecular speed.
The parameters of the driver gas (behind the diaphragm) were: case 1 - temperature 10 times higher than that of the driven gas, pressure such, that after diaphragm removal

the shock wave of Mach number $Ms = 2$ was produced; case 2 - the same as in case 1, only behind the diaphragm a "buffer" section, filled with cold, inert gas was placed; case 3 - pressure equal to that of the driven gas, temperature 100 times higher.
For pipe flows (investigation of the influence of walls) only case 1 was employed. At X less than X0 (400 or 600 units) the molecules reflected from the walls specularly, i.e. without exchange of tangential momentum and energy. Such flow without losses was necessary for the detonation to develop.
For larger values of X, the "diffuse reflection" was assumed (Maxwell [4]). The molecules hitting the wall stuck to it and were subsequently re-emitted, with kinetic energies corresponding in average to the temperature of the wall, in directions selected at random. Such reflection corresponds to maximum possible friction and heat exchange and is appropriate for most technologically prepared surfaces.

4 Results

Figures 2 to 5 show selected results of the simulations.
Figure 2 shows a plane, perpendicular waves, moving along X - axis in a positive half-space, without influence of walls. Distributions of gas temperature in terms of distance along the X - axis for several instants evenly spread in time (every 20 mean collision times) are presented. In the left part of Figure 2 gas contains 20 per cent of the "energetic" component, in the right part - 10 per cent. Energy is released in the first collision with sufficient "velocity of approach" (no delay). The parameters of the driver gas are as in case 1, described in the previous section. In Figure 2 - left the detonation wave forms

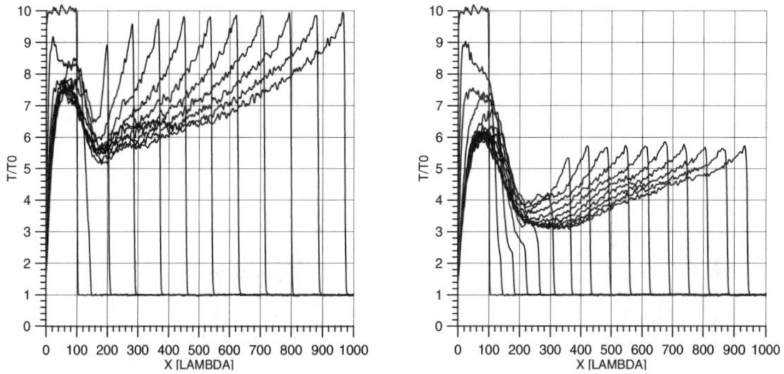

Fig. 2. Temperature distributions inside a plane detonation wave, initiated by shock and contact with hot driver gas. Driven gas contains 20 per cent (left) or 10 per cent (right) of energetic component. T0 - initial temperature of the driven gas; lambda - mean free path.

very quickly and then moves with constant speed and constant intensity, as expected. In Figure 2 - right, formation of the detonation wave is much slower. Its details are therefore clearly visible. First, the primary shock wave forms, following the diaphragm removal. The shock then gradually speeds up and increases its intensity, transforming into a detonation wave, which then moves with constant speed and constant intensity, lower than that shown in Figure 2 - left.

Figure 3 shows the waves in the same configuration, only the parameters of the driver are as described in case 2 - shock wave passing initially through a buffer of cold, inert gas. In the left part of Figure 3 (20 per cent of the "energetic" component) detonation

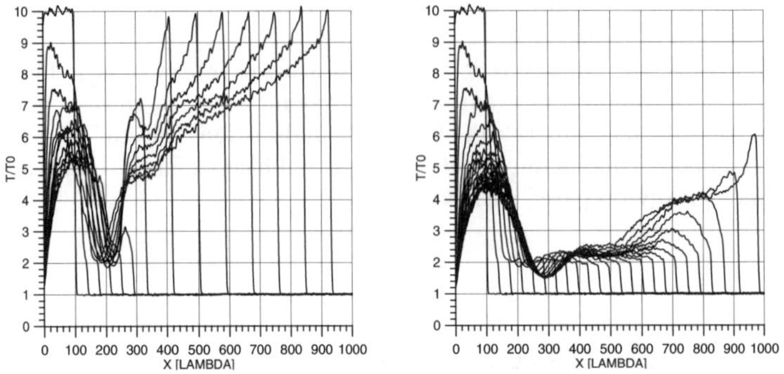

Fig. 3. Temperature distributions inside a plane detonation wave initiated by shock wave only. Driven gas contains 20 per cent (left) or 10 per cent (right) of energetic component.

develops fast, although not as fast as in Figure 2 - left, and then moves with the same speed as in this figure. For gas containing only 10 per cent of the "energetic" component (Figure 3 - right) the time necessary for detonation to develop is very long (much longer than in all previous cases) - detonation appears only at the end of the simulation region. Figure 4 shows the detonation waves transformed from a flame, ignited by temperature difference only (case 3). In the left part of the figure, gas contains 30 per cent of the "energetic" component, in the right - 20 per cent. Figure 4 - left shows detonation developing

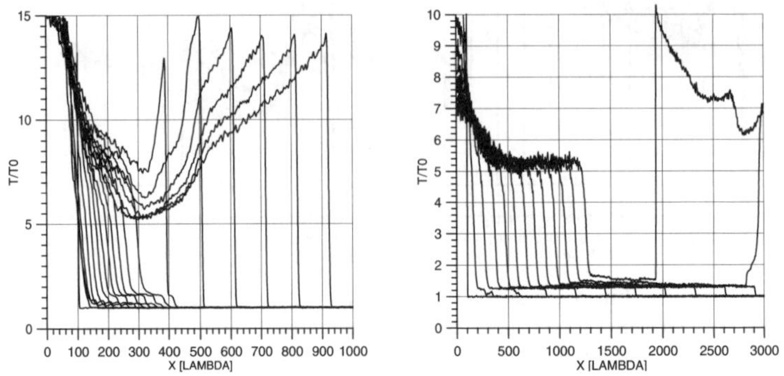

Fig. 4. Temperature distributions inside a plane detonation wave initiated by contact with hot driver gas of the same pressure as driven gas. Driven gas contains 30 per cent (left) or 20 per cent (right) of energetic component.

slower than those in Figures 2 - left, and 3 - left (in spite of higher percentage of "energetic" component), but later moving with the expected, high speed and intensity. In

Figure 4 - right the primary wave does not transform into detonation at all. (The time period between subsequent wave positions in this simulation is equal to 250 mean collision times, i.e. it is 12.5 times longer than that in the previous pictures). Detonation wave appears only after reflection of the primary shock from the solid wall at the end of the computation domain.
In a gas containing only 10 per cent of "energetic" component, temperature difference without an initial shock wave does not produce detonation at all.
All computations reported till now were performed with the inverse reaction taken into account. Additional computations, without inverse reaction (not shown in this text) gave the same results (within scatter of the DSMC technique).
Figure 5 shows temperature diagrams for waves in gas containing 20 per cent of "energetic" component, moving in a cylindrical pipe 100 mean free paths in diameter. The values shown, have been averaged over the cross-section of the pipe. The wave was initiated, as before, by removal of a "diaphragm", placed at $X = 100$ units of length from the origin. The parameters of the driver gas were as in case 1 - temperature 10 times higher than that of the driven gas, pressure such, that the shock wave of Mach number $Ms = 2$ was produced. In Figure 5 - left the energy was released in the first collision

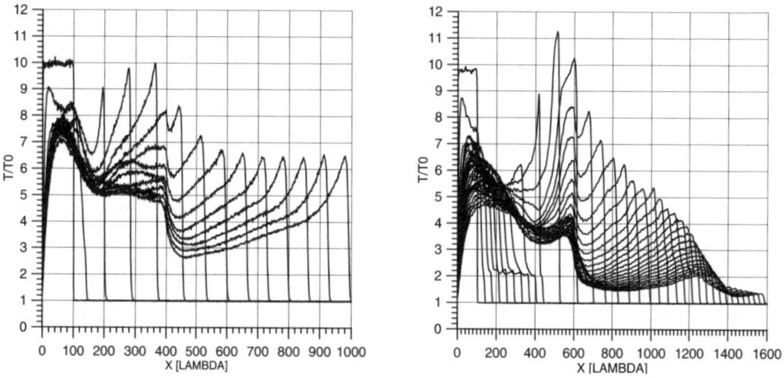

Fig. 5. Temperature distributions in a detonation wave, moving along a narrow pipe. Gas contains 20 per cent of energetic component. Energy release in the 1-st collision; X0 = 400 units (left). Energy release in the 6-th collision; X0 = 600 units (right).

with the velocity of approach above the threshold value (no delay); in Figure 5 - right the energy was released in the 6-th collision (in this case the threshold value of the velocity of approach was decreased to 4.47 times the most probable molecular speed - otherwise detonation did not develop.). The initial part of Figure 5 - left, up to $X = X0$ (area of no friction and no heat exchange), looks similar to that in Figure 2 - left; formation of shock and detonation wave is evident. For $X > X0$ the walls cool the gas down and decrease its velocity, however the temperature decrease is not sufficient to extinguish the flame. The detonation wave slows down and gets weaker, still it eventually moves ahead with constant speed and intensity. In Figure 5 - right the development of detonation takes more time. Later, for $X > X0$ (friction and heat exchange at the walls present) the gas temperature falls down below the ignition point; gas stops burning, the shock still moves ahead, however unsupported by the flame front gets weaker and weaker. Similar picture of development and decay of detonation was obtained for gas containing 10 per cent of "energetic" component, releasing its energy in the first collision (no delay).

5 Conclusions

A method of simulation of the detonation phenomenon, based on the Direct Monte-Carlo Simulation technique has been proposed. One of the advantages of this method is, that it enables studying detonation in narrow channels and its decay under the influence of wall friction and heat exchange. Thanks to that, this method may be used for simulation and design of the devices extinguishing detonation in pipelines.
The proposed model of a "detonating gas" makes it possible to "heat" the gas in a simple way, simulating combustion. It has been shown, that flame fronts and shock waves, propagating in such "detonating gas", may transform into detonation.
The inverse reaction, introduced to the model at the present stage of work, seems to have little or no influence on formation of detonation. This may be due to the fact, that at initial stage of detonation the temperature of the medium is not high enough to promote this reaction.
Similarly - the inverse reaction has no influence upon stationary detonation in the medium insufficiently energetic - as before the temperature of the medium is insufficient.
For more energetic media the velocity of the detonation wave and the maximum temperature in the combustion zone may be decreased by the presence of the inverse reaction only slightly.
The performed simulations indicate in addition, that the influence of the inverse reaction upon extinguishing detonation is negligible too. The factors, which seem to be important, are: energy released in a single collision, fraction of molecules carrying energy, threshold velocity for initiation of the energy release, delay of the energy release (in mean collision times) and Knudsen number (ratio of the mean free path to the diameter of the channel). Influence of these factors has been demonstrated qualitatively; quantitative description requires much more simulations and also a comparison with experiment.

References

1. Bird G. A.: *Molecular gas dynamics and the direct simulation of gas flows*, Clarendon Press, Oxford, 1994.
2. Dremin A. N.: *Toward detonation theory*, Springer, 1999.
3. Larsen P. S., Borgnakke C., In: *Rarefied Gas Dynamics*, (ed. M. Becker and M. Fiebig), 1, Paper A7, DFVLR Press, Porz-Wahn, Germany, 1974.
4. Maxwell J. C.: The scientific papers of James Clerk Maxwell (W.D. Niven, ed.) New York: Dover, 2, 706, 1952.
5. Walenta Z. A., Teodorczyk A., Witkowski W.: Simple model of a detonating gas for use with the Direct Monte-Carlo Simulation technique, In: *Mechanics of the 21st Century* (ed. W. Gutkowski and T. A. Kowalewski), Paper 12911, Springer, 2005.
6. Yanitskiy V. E., Belotserkovskiy O. M.: The statistical method of particles in cells for the solution of problems of the dynamics of a rarefied gas. Part I, Zh. Vychisl. Mat. Mat. Fiz., 15, 1195-1208; Part II, Zh. Vychisl. Mat. Mat. Fiz., 15, 1553-1567, 1975.

Effects of detailed chemical reaction model on detonation simulations

N. Tsuboi[1], M. Asahara[2], A.K. Hayashi[2], and M. Koshi[3]

[1] *Institute of Space and Astronautical Science, Japan Aerospace Exploration Agency, 3-1-1 Yoshinodai, Sagamihara, Kanagawa 229-8510 (Japan)*
[2] *Depertment of Mechanical Engineering, Aoyama Gakuin University, 5-10-1 Fuchinobe, Sagamihara, Kanagawa 229-8558 (Japan)*
[3] *Department of Chemical System Engineering, The University of Tokyo, 7-3-1 Hongo, Bunkyo-ku, Tokyo 113-8656 (Japan)*

Summary. Unsteady two-dimensional simulations were performed for hydrogen/air mixtures in order to evaluate the effects of the detailed chemical reaction model on detonations. The reaction models in the present study are Petersen and Hanson model, and Koshi model. The pressure, specific heat release, and OH mass fraction are not affected by the reaction models, however, HO_2 and H_2O_2 mass fractions are dependent on them. These features also appear in the results of the zero-dimensional ignition simulation for the high pressure gas mixture. Therefore the chemical kinetic models for the high pressure affects the results of detonation simulations because the chemical properties in high pressure are significant different between them.

1 Introduction

Detonation is a shock-induced combustion wave propagating through a reactive mixture or pure exothermic compound, and has been studied from the safety engineering point of view such as for coal mine explosions or from the scientific point of view of astrophysics as in star explosions. Recent numerical simualtions on gasgeous detonations can be carried out in not only two-dimension but also three-dimension with a detailed chemical reaction model [1], [2], [3], [4]. For hydrogen/air gas mixture in atmospheric pressure, most of the detailed reaction model have nine species and about 18 elemental reactions to obtain reasonable results. Instantaneous high pressure more than 1 MPa appears near the detonation front, however, there are few reaction models to include such a high-pressure effects because most of the reaction models are constructed by using experimental data on atmospheric pressure. One of the important high-pressure effects is a pressure dependence on reaction coefficients such as HO_2 and H_2O_2 chemistries near the second and third explosion limits. However, their effects on the detonation cell size, local heat release, and instantaneous chemical species are unclear now.

One of the reaction models including the pressure dependence is the Petersen and Hanson model ([5]). A feature of this model (PH model) is the pressure dependence on a forward reaction coefficient including the collision reaction with a third body; in other words, HO_2 and H_2O_2 chemistries near the second and third explosion limits are considered. Koshi model [6], which has been developed for the high pressure combustion of hydrogen, includes important reactions at high pressure above the second explosion limit such as $H_2 + HO_2 \rightleftharpoons H + H_2O_2$, $H + H_2O_2 \rightleftharpoons H_2O$, and $H + O_2 + M \rightleftharpoons HO_2 + M$ as well as $OH + OH + M \rightleftharpoons H_2O_2 + M$.

Some of authors have simulated the detonations by using PH model for the two-dimensional and also three-dimensional problems. We also studied about the effects of the chemical reaction models on the detonation simulations [7] to conclude that PH model is valid to apply the detonation simulations, however, more research should be nessecery. In this paper, we discuss the effects of detailed chemical reaction model including pressure dependence on the detonation structure.

Table 1. Hydrogen-Air detailed reaction model by Koshi [6]

Reaction Considered	$A_k^{a)}$	n_k	$E_{a_k}^{)b}$	Note
(1) $OH + H_2 \rightleftharpoons H_2O + H$	2.14×10^8	1.5	3449.0	
(2) $H + O_2 \rightleftharpoons OH + O$	1.00×10^{14}	0.0	14850.0	
(3) $O + H_2 \rightleftharpoons OH + H$	5.00×10^4	2.7	6290.0	
(4) $OH + HO_2 \rightleftharpoons H_2O + O_2$	2.89×10^{13}	0.0	-497.0	
(5) $H + HO_2 \rightleftharpoons OH + OH$	1.69×10^{14}	0.0	874.0	
(6) $H + HO_2 \rightleftharpoons H_2 + O_2$	4.28×10^{13}	0.0	1411.0	
(7) $H + HO_2 \rightleftharpoons H_2O + O$	3.01×10^{13}	0.0	1721.0	
(8) $O + HO_2 \rightleftharpoons O_2 + OH$	3.25×10^{13}	0.0	0.0	
(9) $HO_2 + HO_2 \rightleftharpoons H_2O_2 + O_2$	4.20×10^{14}	0.0	12000.0	
(10) $HO_2 + HO_2 \rightleftharpoons H_2O_2 + O_2$	1.30×10^{11}	0.0	-1630.0	
(11) $OH + OH \rightleftharpoons O + H_2O$	4.33×10^3	2.7	-2485.0	
(12) $HO_2 + H_2 \rightleftharpoons H + H_2O_2$	1.97×10^{12}	0.0	21510.0	
(13) $H_2O_2 + H \rightleftharpoons H_2O + OH$	8.00×10^{14}	0.0	9110.0	
(14) $H_2O_2 + OH \rightleftharpoons H_2O + HO_2$	1.70×10^{18}	0.0	29407.0	
(15) $H_2O_2 + OH \rightleftharpoons H_2O + HO_2$	2.00×10^{12}	0.0	427.0	
(16) $H_2O_2 + O \rightleftharpoons HO_2 + OH$	6.62×10^{11}	0.0	3974.0	
(17) $H + O_2 + M \rightleftharpoons HO_2 + M$	5.90×10^{17}	-0.8	0.0	c
(18) $H + O_2 + N_2 \rightleftharpoons HO_2 + N_2$	1.05×10^{15}	0.0	-1640.0	
(19) $H + H + M \rightleftharpoons H_2 + M$	1.00×10^{18}	-1.0	0.0	d
(20) $H + H + H_2 \rightleftharpoons H_2 + H_2$	9.20×10^{16}	-0.6	0.0	
(21) $H + H + H_2O \rightleftharpoons H_2 + H_2O$	6.00×10^{19}	-1.3	0.0	
(22) $H + OH + M \rightleftharpoons H_2O + M$	1.60×10^{22}	-2.0	0.0	e
(23) $H + O + M \rightleftharpoons OH + M$	6.20×10^{16}	-0.6	0.0	e
(24) $O + O + M \rightleftharpoons O_2 + M$	1.89×10^{13}	0.0	-1788.0	f
(25) $H_2O_2(+M) \rightleftharpoons OH + OH(+M)$	3.00×10^{14}	0.0	45500.0 k_∞	g
	1.80×10^{17}	0.0	42924.0 k_0	g

Forward reaction rates: $k_{f,k} = A_k T^{n_k} \exp(-E_{a_k}/RT)$
a) units of mole-cm-sec
b) unit of cal/mole
c) Listed values are for Ar. Third body efficiencies are 3.5(H_2O), 1.4(O_2), 3.0(H_2), 1.0(all others). M does not include N_2.
d) Third body efficiencies are 1.0. M does not include H_2 and H_2O.
e) Third body efficiencies are 5.0(H_2O), 1.0(all others).
f) Third body efficiencies are 15.4(H_2O), 2.4(H_2), 1.0(all others).
g) Third body efficiencies are 6.0(H_2O), 2.0(H_2), 1.0(all others). Troe formulation [5].

2 Numerical Method

The governing equations are the two-dimensional Euler equations with 9 species (H_2, O_2, H, O, OH, HO_2, H_2O_2, H_2O, and N_2). These equations are explicitly integrated by the

Strang type fractional step method. The chemical reaction source terms are treated in a linearly point-implicit manner in order to avoid a stiff problem. A second-order Harten-Yee non-MUSCL type TVD scheme is used for the numerical flux in the convective terms. In the present simulation, PH model and Koshi model are used for chemical kinetics to solve detonation problems. PH model has a pressure-dependence rate coefficients for $OH + OH + M \rightleftharpoons H_2O_2 + M$, which varies between low-pressure and high-pressure limits. The rate constants of important reactions of $H_2{+}HO_2 \rightleftharpoons H{+}H_2O_2$ and $H{+}H_2O_2 \rightleftharpoons H_2O$ in Koshi model were evaluated by means of the transition state theory combined with the molecular orbital calculations of the transient states. The reaction mechanism and rate constants of Koshi model are listed in Table.1.

The computational grid is orthogonal system with 1601x401. The grid size of 5 μm gives a resolution of 32 grid points in the half reaction length which equals 160 μm for H_2 at atmospheric pressure. Therefore the channel width is 2 mm.

3 Results and Discussions

The maximum histries of pressure, OH, HO_2, and H_2O_2 and the time histories of local specific energy release contours are shown in Fig. 1. The detonation cellular structures for both chemical reaction model are irregular because of the effects of N_2 as a inert gas and the local explosions of unburned gas behind the detonation front. The difference between these reaction models is small in Figs. 1(a)$\sim$(c), however, maximum HO_2 and H_2O_2 massfraction contours are significantly dependent on the reaction model. PH model generates HO_2 and H_2O_2 more than Koshi model. This reason is considered to be the chemical mechanism on both reaction models in high pressure condition, therefore we estimated the unsteady behavior of chemical species in a constant volume by using 0-thmensional simulation.

Figure 2 shows the comparison of time-dependent chemical species histries for $p = 1$ atm and T=1000K. When temperature decreases with pressure of 1 atm, the second explosion limit appears which is governed by balance of chain branching and termination of $H + O_2 \rightleftharpoons OH + O$ and $H + O_2 + M \rightleftharpoons HO_2 + M$. Chemical behavior in the second explosion limit is well known by the experimental data and most of the chemical reaction models include its limit. Therefore the chemical behavior at p=1atm for both models is expected to be similar results. Figure 2 shows that H and OH exponentially increases and HO_2 becomes a constant value close to the ignition time at t=0.2 msec for PH model and t = 0.15 msec for Koshi model. H_2O_2 for both models also increase exponentially until ignition time, however, Koshi model delays rather than PH model.

Figure 3 shows the comparison of time-dependent chemical species histries for $p = 10$ atm and T=1000K. Third explosion limit appears over the pressure of appriximately 10 atm. Third explosion limit is governed by the formation and decomposition of H_2O_2 such as $HO_2 + HO_2 \rightleftharpoons H_2O_2 + O_2$, $H + H_2O_2 \rightleftharpoons HO_2 + H_2$, and $H_2O_2 + M \rightleftharpoons OH + OH + M$. Both results show that H_2O_2 is decomposited and H,O, and OH increase exponentially near the ignition time. The amount of H_2O_2 for Koshi model is smaller than that for PH model. The amount of H_2O_2 is equal to that of HO_2 for PH model. Therefore the chemical behaviors between PH model and Koshi model are significantly different.

CJ pressure behind the detonation is approximately 18 atm and instantaneous pressure near the detonation front is larger than CJ pressure. Pressure and OH massfraction in Fig. 1 for both models are similar, however, HO_2 and H_2O_2 are significantly different. The zero-dimensional simulations in a constant volume show that amount of HO_2 and

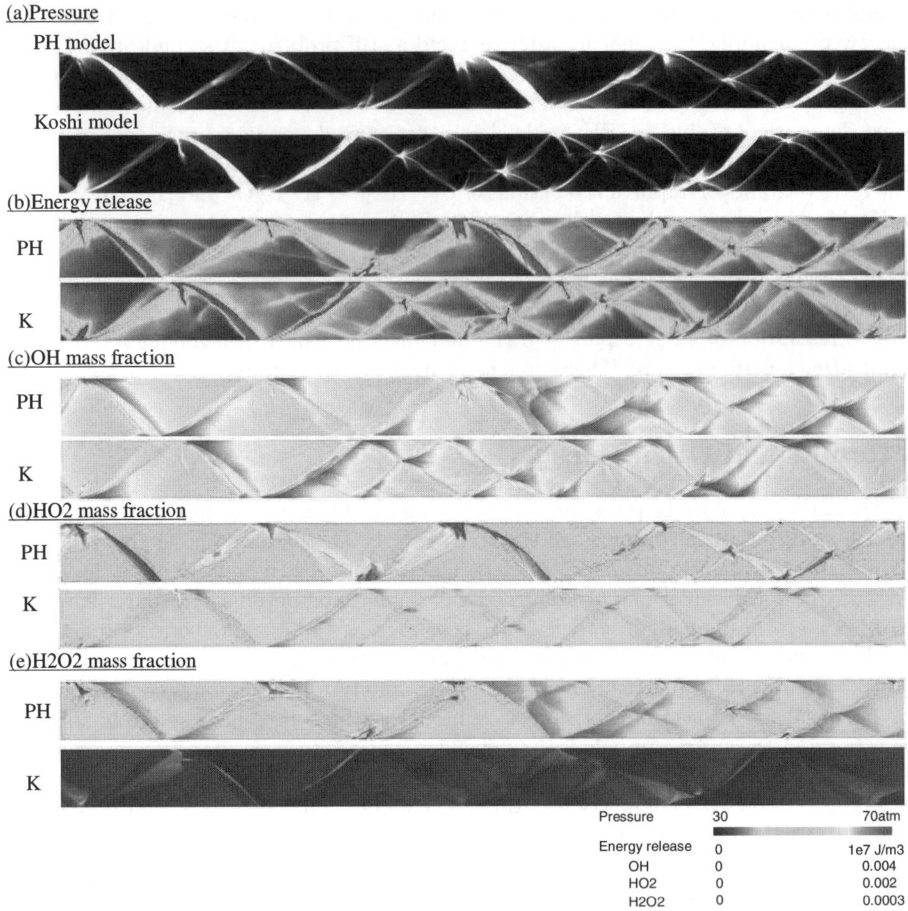

Fig. 1. Effects of reaction models on maximum histories of (a)pressure, (c)OH mass fraction, (d)HO_2 mass fraction, (e)H_2O_2 and on (b) time history of the local specific energy release contours. Detonation propagates from left to right

H_2O_2 for PH model is larger than those for Koshi model in high pressure condition. The trend appeared in the zero-dimensional simulations for HO_2 and H_2O_2 coinsides the two-dimensional results and the chemical kinetic model in such a high pressure affects on these results.

Chemical kinetics in high pressure situation are now being researched by many researchers in order to determin the chemical reaction model including high pressure effects. The detonations are also affected by the chemical reaction model as shown in the present results. The chemical reaction model affects on the cell size as well as cell irreguarality therefore vaild chemical reaction model should be constructed for combustion data with high pressure.

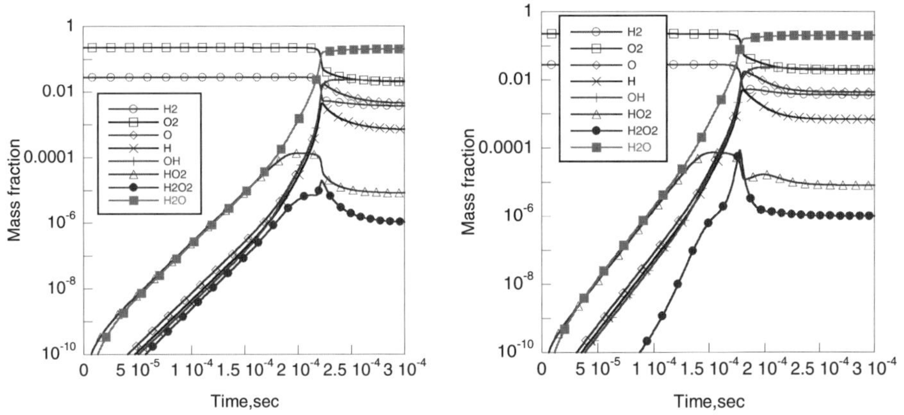

Fig. 2. Comparison of time-dependent chemical species histories for p=1atm, T=1000K. PH model (left), Koshi model(right)

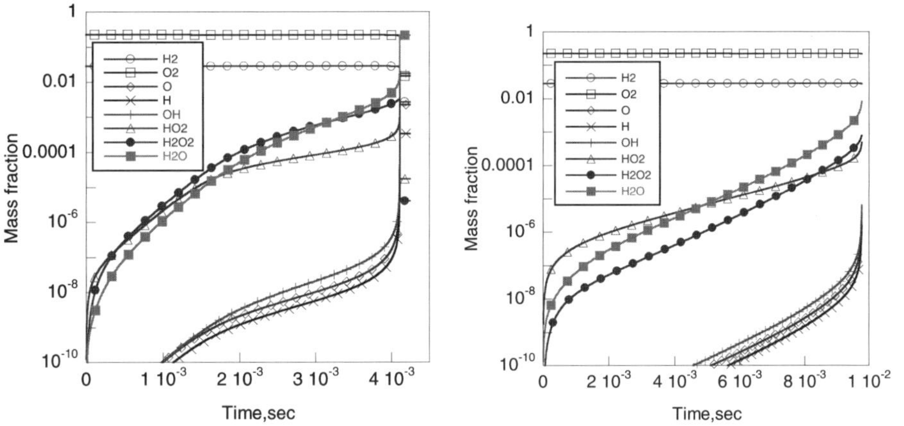

Fig. 3. Comparison of time-dependent chemical species histories for p=10atm, T=1000K. PH model (left), Koshi model(right)

4 Conclusions

Unsteady two-dimensional simulations were performed for hydrogen/air mixtures in order to evaluate the effects of the detailed chemical reaction model on detonations. The pressure, specific heat release, and OH mass fraction are not affected by the reaction models, however, HO_2 and H_2O_2 mass fractions are dependent on them. HO_2 and H_2O_2 mass fractions calculated by PH model are more than those by Kochi model. These features also appear in the results of the zero-dimensional ignition simulation for the high pressure gas mixtures. Therefore the chemical kinetic models for the high pressure affects the results of detonation simulations because the chemical properties in high pressure are significant different between them.

Acknowledgement. This research was collaborated with Center for Planning and Information Systems in ISAS/JAXA. This study was also partially supported by Industrial Technology Research Grant Program in 2006 from New Energy, Industrial Technology Development Organization(NEDO) of Japan, and Grant-in-Aid for JSPS fellows by Ministry of Education, Culture, Sports, Science and Technology of Government of Japan, subject no. 19360092.

References

1. N. Tsuboi, S. Katoh, A.K. Hayashi, Proc. Comb. Inst. 29 (2002) 2783-2788.
2. K. Eto, N. Tsuboi, A.K. Hayashi, Proc. Comb. Inst. 30 (2005) 1907-1913.
3. N. Tsuboi, A.K. Hayashi, Proc. Comb. Inst. 31 (2006) 2389-2396.
4. N. Tsuboi, K. Eto and A.K. Hayashi, Combustion and Flame 149, No.1/2(2007), 144-161.
5. E.L. Petersen, R.K. Hanson, J. Propulsion Power 15 (1999) 591-600).
6. M. Koshi, Combustion Science and Technology, Vol.7 (2000) 153-162(in Japanese).
7. K.Eto, N. Tsuboi, A.K. Hayashi, 19th International Colloquium on the Dynamics of Explosions and Reactive Systems (2003) No.178.

Effects of flame jet configurations on detonation initiation

K. Ishii[1], **T. Akiyoshi**[1], **M. Gonda**[1], **and M. Murayama**[2]

[1] *Department of Mechanical Engineering, Yokohama National University, 79-5, Tokiwadai, Hodogaya-ku, Yokohama, 240-8501, Japan*
[2] *Aero-Engine & Space Operations, Ishikawajima-Harima Heavy Industries Co., Ltd, 229 Tonogaya, Mizuho-machi, Nishitama-gun, Tokyo 190-1297 Japan*

Summary. Deflagration-to-detonation transition (DDT) using flame jets was experimentally studied to improve performance of pulse detonation engines. The detonation tube has a total length of 7350 mm and an inner diameter of 50 mm. Three types of flame jets were arranged by changing configuration of sub-chamber: two flame jets colliding at the tube center (counter type), two flame jets forming a swirl inside the detonation tube (offset type), and one flame jet supplied in the axis direction of the detonation tube (conventional type). Test gas was a stoichiometric acetylene-air mixture. The experimental results show that the shortest DDT time is provided by a swirling flame jet emanating in the oblique direction to the tube axis. The DDT time decreases with increase in the orifice diameter and length of the sub-chamber. It was also found that DDT distance is not significantly affected by types of the flame jets and slightly decreases with decrease in length of the sub-chamber.

1 Introduction

Initiation of detonation in a short time and a short distance is practically important to realize a pulse detonation engine (PDE) studied intensively because of its simple construction and its theoretical high thermal efficiency. Since direct initiation by using a high energy source is not realistic, detonation in PDE tubes is generally initiated through deflagration to detonation transition (DDT) process. However a method of detonation initiation is still one of main tasks for fuel-air mixtures with low detonability.

To enhance DDT process insertion of an obstacle such as a Shchelkin spiral is widely used [1]. One of the authors reported significant reduction of DDT length by placing small tubes bundled densely near the ignition point [2]. Jet initiation [3]- [5], in which an accelerating flame is introduced to a main chamber through an orifice, makes it possible to initiate detonation without DDT process under appropriate initial conditions. In this case significant enhancement of mixing between an unburned mixture and combustion products causes local explosion leading finally to onset of detonation. However, successful jet initiation is dependent on detonability of mixtures, orifice size, and jet velocity.

In the present work, effects of configuration of flame jets on DDT characteristics have been studied experimentally to reduce time and length needed for detonation initiation.

2 Experimental apparatus

2.1 Flame jet

Three types of flame jets were tested as shown in Fig. 1: (a) counter type, in which two opposed flame jets collide with each other at the tube center; (b) offset type, in which two flame jets are ejected to form a swirling flame; (c) conventional type, in which flame jetting along the axis of the detonation tube. Spark plugs for automobile-use and conventional pressure transducers were placed at the end of the sub-chamber in each configuration. An inner diameter of the sub-chamber was 32 mm and its length, L was 59 mm and 89 mm. An orifice plate for providing flame jet was inserted between the main chamber and the sub-chamber and diameter of the orifice, d was varied from 4 mm to 12 mm. In addition three orientation angles of flame jet to the axis of the detonation tube, α of 45°, 60 °, and 90° were arranged in the offset type so that the two flame jets formed double spirals in the detonation tube, as shown in Fig. 2. In the case of offset 45° and 60°, only sub-chamber length of 89 mm was tested because of construction requirement.

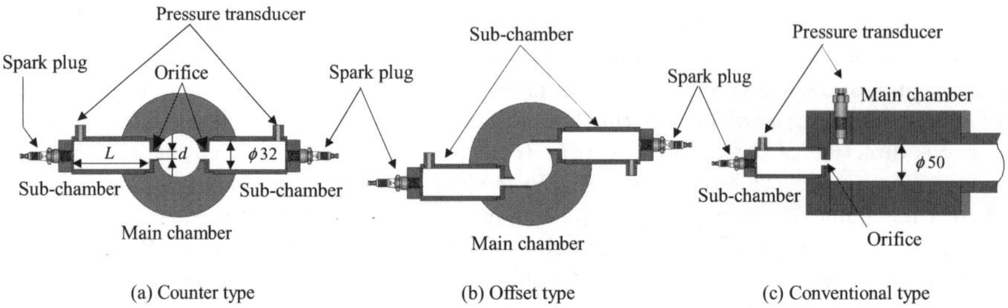

Fig. 1. Configuration of sub-chamber.

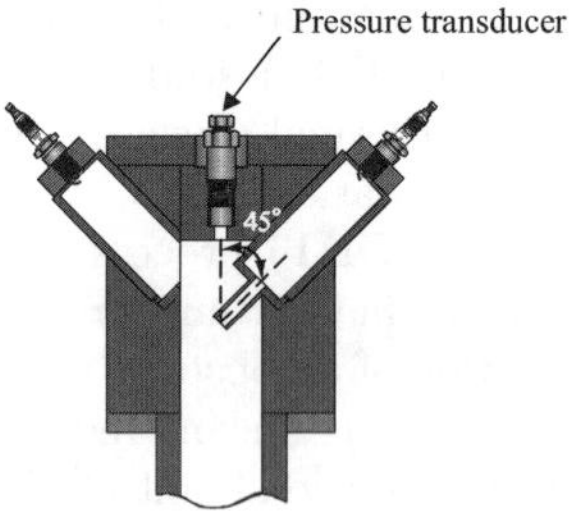

Fig. 2. Schematic of construction of offset type (offset 45°).

2.2 Detonation tube

The detonation tube used in the present work was a stainless tube of total length 7350 mm having inner diameter of 50 mm as shown in Figure 3. The detonation tube was composed of the main chamber equipped with the sub-chamber and a rest part of a tube. Three pressure probes of P4 to P6 and five ion probes of I1 to I5 were fixed to the tube to detect arrival of the pressure and reaction front. At the end of the tube a photodiode was mounted for estimation of DDT time [5]- [6], since self-emission from flame front increases rapidly at onset of detonation. As a test gas a stoichiometric acetylene-air mixture was charged into the detonation tube under a room temperature and 0.1 MPa. Ignition system, which is essentially the same as used in automobile engines, gives spark discharge to two spark plugs at the same time for the counter type and the offset type. More than 4 tests were repeated under the same condition and results were evaluated using average of measurement value.

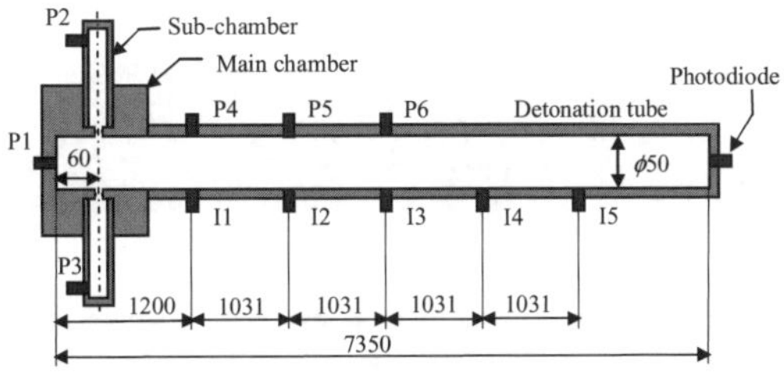

Fig. 3. Schematic of detonation tube. P and I denote pressure and ion probe, respectively. Dimensions in mm.

3 Results and Discussion

3.1 High Speed Image of Flame Jets

High speed images of flame jets ejected to the main chamber were taken through an observation window placed at the end flange. In Figure 4 (a) two flame jets generated in the counter type are observed at 3.0 ms after spark discharge. They collide with each other at the tube center and then turn to the horizontal direction at 3.1 ms. Afterwards the flame spread the whole cross-sectional area of the tube. With reference to the offset type, two flame jets are ejected from the orifice at 3.6 ms in Figure 4 (b) and at 3.0 ms in Figure 4 (c). In the subsequent images the flame jets travel rotating clockwise. Since the flame jets form double spirals, time needed for spreading out in whole cross-sectional area of the tube seems to be longer than in the counter type.

3.2 DDT

Figure 5 shows effects of types of flame jets on DDT time and distance. It is found that in all conditions DDT time decreases with increase in the orifice diameter. Among these three types the offset type shows shorter DDT time and in particular offset 60° gives the shortest. Advantage of the counter type to the conventional type is dependent on the sub-chamber length and the orifice diameter and thus no general tendency can be found. Figure 5 (a) also indicates that DDT time decreases with decrease in the sub-chamber length for $d = 4$ mm and 6 mm, although the opposite tendency is found for $d = 8$ mm and 12 mm.

In the present work it was confirmed that DDT distance, which was estimated using DDT time and x-t diagram of reaction front, was almost coincide with that measured from smoked foil records. Without any flame jet DDT distance was found to be 4 m to 5 m, while use of flame jet reduces DDT distance to 3.5 ± 0.3 m. Figure 5 (b) shows that the shorter sub-chamber gives shorter DDT distance in almost all conditions and configuration of sub-chamber does not significantly affect DDT distance, although it slightly decreases with the sub-chamber length.

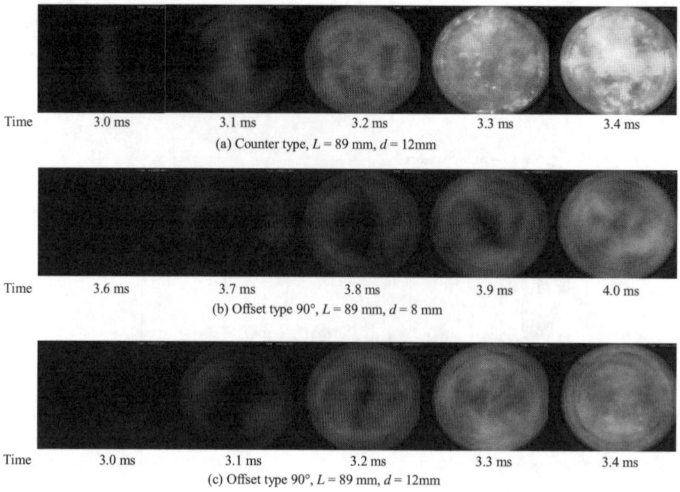

Fig. 4. Direct photographs of flame jet ejected into the main chamber as a function of elapsed time from spark discharge. *L*: sub-chamber length; *d*: orifice diameter

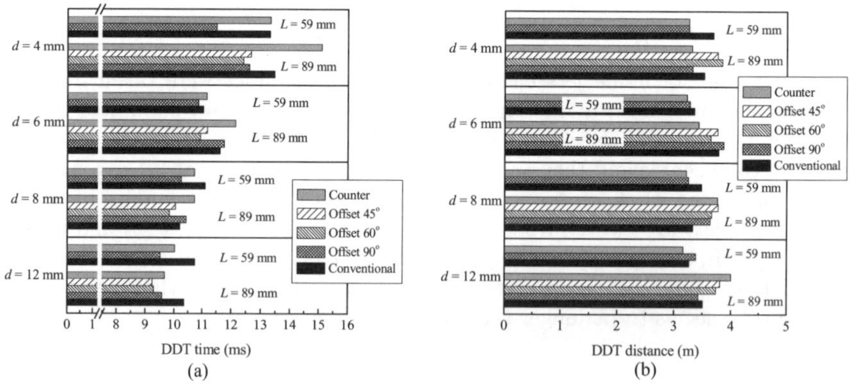

Fig. 5. DDT time and distance for various configuration of sub-chamber. *L*: sub-chamber length; *d*: orifice diameter.

3.3 Pressure Profiles

Figure 6 shows pressure profiles in the sub-chamber and in the detonation tube for offset 90°. Time of 0 corresponds to beginning of spark discharge. In Figure 6 (a) P1, pressure in the main chamber increases slightly after spark discharge, which indicates ejection of an unburned mixture. At 3 ms P1 rapidly rises and reaches 0.4 MPa. This peak pressure is due to a shock wave generated by intensive combustion through rapid mixing between an unburned gas and combustion products, not due to detonation initiation. This shock wave is accompanied with rarefaction waves and then P1 decreases gradually until onset of detonation. P2, pressure in the sub-chamber continues to rise just after spark discharge and pressure ratio of P2 to P1 suggests that flame jet is choked at the orifice. P4, P5 and P6 detect the shock wave in order of the probe location, where P5 and P6 were used only for detection of pressure waves, not for pressure measurement. Pressure rise in P4 at 12 ms is owing to a retonation wave, because DDT distance in this condition is about 3 m, which lies downstream from the position of P4.

In Figure 6 (b) the larger orifice diameter of 12 mm causes earlier rise of P1 and earlier generation of the shock wave than in Figure 6 (a). From pressure ratio of P2 to P1 it is found that the orifice does not choke the flame jet. As shown in Figure 6 (c) the longer sub-chamber gives higher peak pressure of 0.7 MPa than in Figure 6 (b), although time of pressure rise of P1 becomes later. This peak pressure is related to increase in heat release rate caused by larger orifice diameter and high velocity of flame jet in the choke condition. However, jet initiation found in Refs [3]- [5] did not occur in the present work and detonation was always initiated through DDT process.

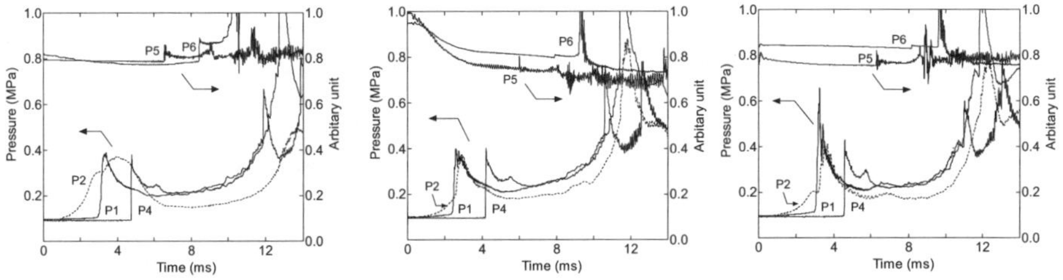

(a) Offset type 90°, L = 59mm, d = 6 mm (b) Offset type 90°, L = 59mm, d = 12 mm (c) Offset type 90°, L = 89mm, d = 12 mm

Fig. 6. Pressure profiles in the sub-chamber and the detonation tube. P1 $\sim$ P6 denotes location of pressure probes in Fig. 3. L: sub-chamber length; d: orifice diameter.

3.4 Ejection time of flame jet

DDT time needed for detonation initiation can be divided into two parts: One is period from the beginning of spark discharge to ejection of flame jet, τ_1 and the other is one from ejection of flame jet to onset of detonation, τ_2. In the present work the ejection time was assumed to be the same as that of pressure rise of P1 as shown in Figure 6 and it was confirmed that τ_1 obtained in such a manner was coincide with that in the high speed images. Figure 7 shows again that larger orifice diameter and shorter sub-chamber contribute to early ejection of flame jet. It is also found that longer sub-chamber has a tendency to reduce τ_2, which is shown in Figure 8. In the case of longer sub-chamber, choke condition at the orifice leads to much supply of flame jet with high velocity. Although larger orifice reduces τ_2 slightly, effects of the orifice diameter on τ_2 is not as significant as DDT time shown in Figure 5. With reference to configuration of flame jet, the offset type shows the shortest τ_2, in particular offset 60° demonstrates the beset performance. In Figure 8 the conventional type having one nozzle gives shorter τ_2 in some cases than the counter type having two nozzle. This result suggests that ejection of flame jet in wider space is more effective to promote rapid detonation initiation than generation of turbulence by mutual collision of flame jets.

From the above results it is found that for reduction in DDT time early ejection of flame jet provided by shorter sub-chamber and use of the offset type with larger orifice diameter are effective as compared to use of long sub-chamber generating a strong shock wave.

4 Conclusion

Promotion of deflagration to detonation transition was experimentally studied by varying orifice diameter, sub-chamber length, and configuration of flame jets. The experimental results show that the shortest DDT time is provided by the offset type, in particular flame jet emanating in the oblique direction to the tube axis. The DDT time decreases with increase in the orifice diameter and the sub-chamber length. It was also found that DDT distance is not significantly affected by configuration of sub-chambers and slightly decreases with decrease in length of the sub-chamber.

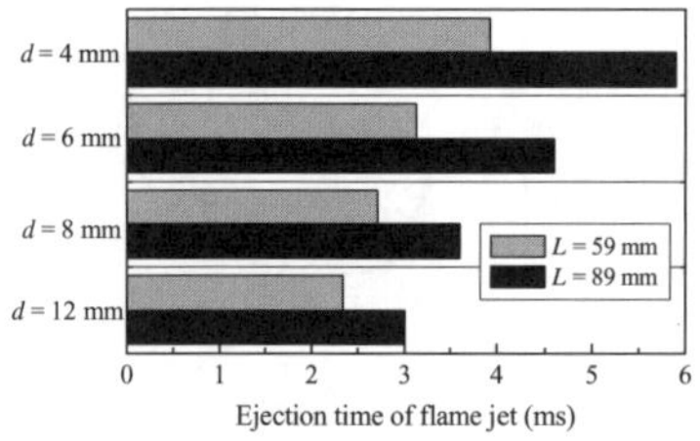

Fig. 7. Ejection time of flame jet into main chamber, τ_1. L: sub-chamber length; d: orifice diameter.

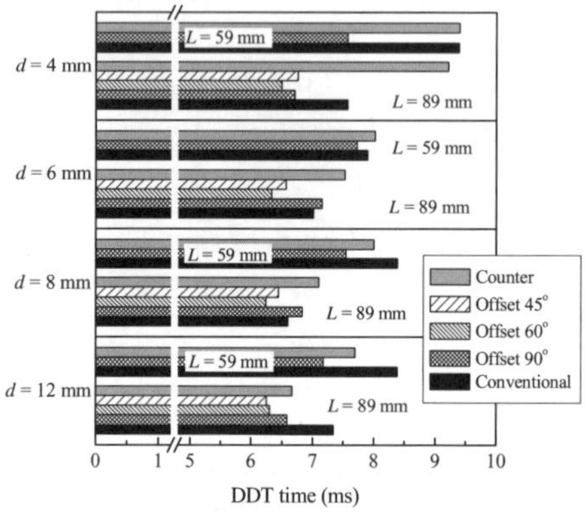

Fig. 8. Time from flame jet ejection to DDT for various configuration of sub-chamber, τ_2. L: sub-chamber length; d: orifice diameter.

References

1. Lindstedt, R. P., Michels, H. J.: Deflagration to Detonation Transitions and Strong Deflagrations in Alkane and Alkene Air Mixtures, Combustion and Flame, vol.76, pp. 169-181, 1989.
2. Ishii, K., Tanaka, T: A Study on Jet Initiation of Detonation Using Multiple Tubes, Shock Waves, vol. 14, pp. 273-281, 2005.
3. Knystautas, R., Lee, J. H., Moen, I., Wagner, H. Gg.: Direct Initiation of Spherical Detonation by a Hot Turbulent Gas Jet, Proceedings of the. 17th Symposium (International) on Combustion, pp. 1235- 1245, 1979.
4. Carnasciali, F., Lee, J. H. S., Knystautas, R.: Turbulent Jet Initiation of Detonation, Combustion and Flame, vol.84, pp. 170-180, 1991.
5. Thomas, G. O., Jones, A.: Some Observations of the Jet Initiation of Detonation, Combustion and Flame, vol.120, pp. 392-398, 2000.
6. Ishii, K., Grönig, H.: Re-initiation Process of Detonation Waves at Low Pressures, Transactions of the Japan Society of Mechanical Engineers, Series B, vol. 67, pp. 1244-1249, 2001.

Experimental and theoretical investigation of detonation and shock waves action on phase state of hydrocarbon mixture in porous media

D.I. Baklanov, L.B. Director, S.V. Golovastov, V.V. Golub, I.L. Maikov, V.M. Torchinsky, V.V. Volodin, and V.M. Zaichenko

Joint Institute for High Temperatures, Izhorskaya st. 13/19, 125412, Moscow, Russia

Summary. Experiments were carried out in filtration setup that allow to simulate physical processes in real gaseous condensate and oil stratums. Detonation combustion chamber was attached as shock wave generator.

Liquid phase deposition in the process of model binary hydrocarbon mixture filtration and system behavior dynamics after the action of detonation waves sets of different intensity and frequency were investigated experimentally and numerically.

1 Introduction

The coefficient of the extraction of resources from the depths of gas-condensate bore holes several times less than from the oil wells. Besides the complex geological conditions characteristic for the gas-condensate stratum this difference is explained by the special feature of the thermodynamic properties of gas condensate. Gas condensate is the complex mixture of methane and higher derivatives of a methane number. The phase diagram of this mixture includes the region so-called "retrograde region", in which the formation of the retrograde liquid is possible with lowering in the pressure. This liquid evaporates with further decrease of pressure. A pressure drop and a change in the temperature near the face of bore hole occurs with the selection of condensate from the productive layer. Gas condensate in the critical zone partially is condensed with the formation of retrograde liquid, which fills pore space and it prevents the output of gas phase. The filtration of multicomponent two-phase mixture to the face of bore hole causes an increase in the saturation by the condensate of pore space in comparison with the process of differential static condensation up to the formation of "condensate plug". In this case the quality of the obtained raw material deteriorates, because its most valuable part is concentrated in the difficult extractive liquid fraction [1].

The calculation of the influence of all factors, which determine the behavior of gas-condensate system, under the conditions for full-scale experiment represents practically insolvable task. Therefore for predicting the development of real layers and study of laws governing the intra-stratified filtration is required the creation of the models, based on the theory of multicomponent filtration taking into account the thermophysical properties of hydrocarbon mixtures, and also the thermodynamic and hydrodynamic conditions, under which the process of filtration occurs.

In this paper the appearance of gas-condensate plug and destruction it by a series of shock waves were investigated to restor the debit of the gas condensate.

2 Gas-condensate plug

2.1 Experimental investigation

The experimental base have been created to study the thermophysical properties of stratified fluids, the processes of filtration with the thermobaric conditions of real it is stratified and the simulation of the processes of action on the gas-condensate medium. Experimental setup "PLAST" can ensure pressure to 40 MPa and temperature to 673 K. Stratified conditions were simulated during these parameters and experiments with liquids and gases of different fractional composition were conducted. As the one-dimensional model of stratum the thermostatically controlled pipe prepared from the stainless steel and filled with quartz sand with a length of 2.2 m and by an inside diameter of 10 mm was used.

As the model the mixture methane-n-butane with the molar concentration of methane is more than 0.65 was selected. In this case the pressure of the beginning of condensation comprises a little larger 100 bar, and mixture in the two-phase state is located in the retrograde region.

Mixture methane-n-butane was prepared in the high-pressure cylinder at a pressure 13 MPa, the mass concentration of methane comprised on the average for the different experiments of about 45%, the n-butane - 55%. The temperature of thermostating changed in the range 290-350 K. These parameters ensure the presence of the mixture in the retrograde region of phase diagram with conducting of experiment. Experimental section (ES) was filled up with the washed quartz sand of fraction 0.09 - 0.125 mm. The measured coefficient of the permeability of section equaled $3.8*10^{-11}m^2$. After the filling ES with mixture at a pressure 12 MPa a constant pressure differential on was supported during the experiment. Pressure at the entrance was 12 MPa, at the output - from 4 to 9 MPa in the different experiments.

The dependence of the expenditure of gas phase per output of ES on the time, data in composition of initial mixture are represented in Fig. 1. The expenditure of mixture sharply falls after 250th second since the beginning of the experiment, and its composition changes to the side of the decrease of the content of n-butane. This testifies about the formation in the pore space of gas-condensate plug and the redistribution of the content of components in the liquid and vapor phases. The mass concentration of methane from initial 42.4% increases to 66.4%.

2.2 Numerical investigation of gas-condensate plug

The nonstationary processes of filtering the hydrocarbon mixture were simulated in the one-dimensional approximation under the isothermal conditions. Model consisted of the hydrodynamic part, which describes the process of two-phase filtration in the porous medium in the approximation of Darcy's law, and thermodynamic, which describes the parameters of the phase equilibrium of system [2] . The equations of hydrodynamics are supplemented with the equation of states of liquid and vapor phases and with thermodynamic equilibrium conditions. Equation of states for the mixtures take the same form as for the clean components. In this case the coefficients of the equation of state of multicomponent system are determined by the properties of the components, which form mixture, and with the portion of each of them in the mixture.

The generalized cubic four-coefficient equation of state Van-der-Waals type, developed specially for the natural oil-gas-condensate mixtures for the pressures to 100 MPa

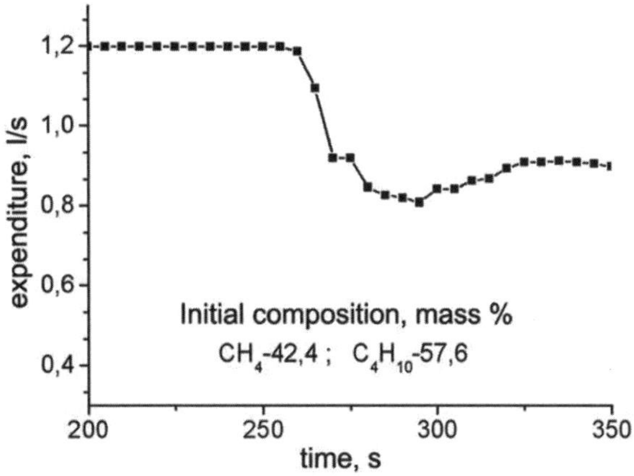

Fig. 1. Consumption of a gas-phase at the outlet of ES. Pressure at the point of entry - 12 MPa, at the point of outlet - 9 MPa, Temperature of an incoming mixture - 303 K, temperature of a thermostate - 310 K.

and temperatures to 473 K, was used for calculating the properties of binary mixture in the steam and liquid phases. In this case the accuracy of design characteristics of the components of hydrocarbon systems is substantially higher than with the use of classical two-coefficient equations (Penga-Robinson, Redlich-Kvonga, etc.). The concentrations of components in the equilibrium phases were calculated on the basis of the fundamental position of thermodynamics about the equality of the volatilities of the components of mixture in the coexisting phases. In this case the forces of surface tension and capillary effects were considered. The value of the phase permeability was determined experimentally.

Finite-difference method was used with the use of a uniform grid on the coordinate and multistage method with the automatic calling sequence on the time, the algorithm of Geer [3].

For increasing the velocity of convergence and decrease of error of finite-difference equation the method of one-dimensional correction additionally is used.

During the first stage was simulated the process of the accumulation of condensate in the critical zone on the model binary mixture methane-n-butane.

The thermostatically controlled experimental section (tube with an inside diameter of 10 mm, with a length of 2 m), filled with quartz sand was simulated. The characteristics of porous medium were determined in the preliminary experiments. The pressure at the entrance into the experimental section and at the output from it was selected as the boundary conditions. In the calculations the following parameters of experimental section and operating conditions were used: pressure at the entrance - 11.7 MPa, pressure at the output - 4.5, 6.5, 8.0 MPa, the initial state of installation - P = 11.7 MPa, working mixture - methane-n-butane, the molar concentration of methane - 0.75, temperature - 330 K, porosity - 0.35, the coefficient of absolute permeability - 10-12 m^2.

The pressure profiles along the length of section at different pressures at the output are shown on Fig. 2a. In accordance with the phase diagram the pressure of maximum condensation for the regime in question is about 80 bar. At smaller pressures the zone of an abrupt change in the gas saturation displaces to the entrance of experimental section.

The distributions of condensate along the length of the model of layer are shown in Fig. 2b.

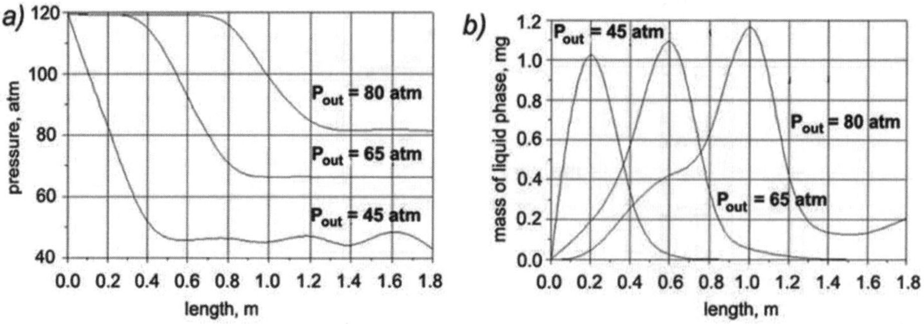

Fig. 2. Steady pressure profiles along the length of the section (a) and distribution of condensate along the length of section in the steady-state regime (b)

From the results of calculations with the initial data, which correspond to the conditions for experiment, it follows that the formation of gas-condensate plug at a temperature 330 K to does not occur. At a temperature 290 K in the time interval between 25th and the 30th by seconds a sharp drop in the expenditure of mixture occurs, the mass concentration of methane in this case composes 70%. Thus, is a qualitative correspondence of results of experiments and calculated dependencies with the observed under the natural conditions special features of gas-condensate systems.

3 Destruction of the condensate plugs by detonation and shock waves

Use for the destruction of the condensate plugs by detonation and shock waves is one of the promising methods of technological action on the gas-condensate layers. A pressure drop from the the intra-stratified up to a pressure of on the collector occurs in the sufficiently narrow region (several meters). With reduction in the debit for the destruction of condensate plug in the zone of the producing well it is proposed to achieve the short-term effect on hydrocarbon intra-stratified system by the detonation waves, which are generated by chemical means in the mouth of bore hole.

3.1 Complex experimental setup and detonation combustion chamber

For the experimental check of the proposed method was produced the installation of complex installation “detonation combustion chamber - PLAST”, and preliminary experiments on the pulse effect on gas-condensate system on the stand were carried out. The schematic of experimental setup is presented in Fig. 3.

As the device, which generates powerful shock waves, steel cylindrical detonation combustion chamber (4) with an inside diameter of 18 mm and a length of 3000 mm used, closed from one end. In the tail end of the combustion chamber diaphragm knot

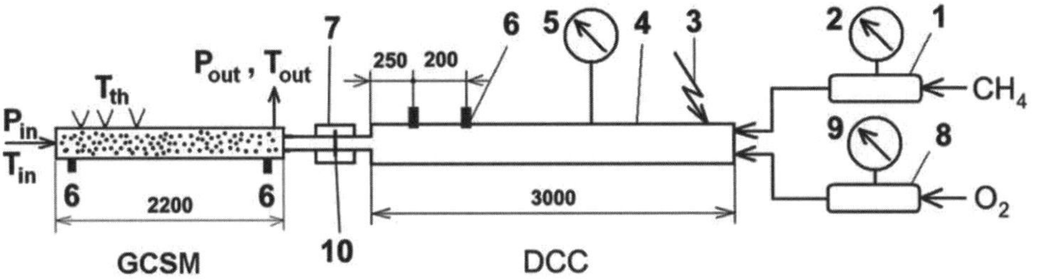

Fig. 3. Schematic of complex experimental setup with the detonation combustion chamber. GCSM - gaseous condensate stratum model, DCC - detonation combustion chamber, PT - piezoelectric pressure transducer. 1,8 - buffer capacities, 2,5,9 - manometers, 3 - spark-gap, 4 - combustion chamber, 6 - piezoelectric pressure transducer PCB, 7 - diaphragm knot, 10 - copper diaphragm.

(7) with copper diaphragm was located (10). Combustion chamber preliminarily was blown through, and then was filled up simultaneously with oxygen and methane from buffer capacities (1,8) in the stoichiometric ratio up to a pressure of 10 bar. In this case throughout the entire length of combustion chamber was formed stoichiometric detonable mixture.

At initial pressure of detonable mixture 1, 5 and 10 bar the detonation wave velocity equaled 1500-1600 m/s, pressure in the shock wave equaled 40, 150 and 250-300 bar accordingly. Energy of combustion of methane in detonation combustion chamber equaled 4, 20 and 40 kJ accordingly.

3.2 Numerical calculation of detonation wave interaction with gas-condensate mixture

It was assumed that after a sharp drop in the expenditure in the section of pipe, which imitates the mouth of bore hole, the detonation wave (sequence of waves) is generated, which is extended along the semi-infinite pipe. Basic assumptions of the model: interaction of shock wave is considered only with the gas phase; there is no mechanism of the attenuation of detonation wave; the final kinetics of phase transitions is not considered.

With the passage of shock wave the properties only of gas phase of collector change. In each calculated cell the properties of gas, phase equilibrium recount in accordance with the pressure. In the calculations it entered relationship for the intensity of detonation waves in the form $P_2/P_1 = 10$.

Figure presents action by the packet of waves during 1 second and results of calculating the debit of model bore hole (Fig. 4).

In the approximation of one-dimensional task and model binary hydrocarbon mixture methane-n-butane preliminary estimations showed that under the short-term influence by the pulse packets of detonation waves the hydrocarbon system from the state with the practically zero debit of gas again returns to the initial single-phase state (Fig. 5); moreover the characteristic times of action several orders less than the time of the formation of condensate plug.

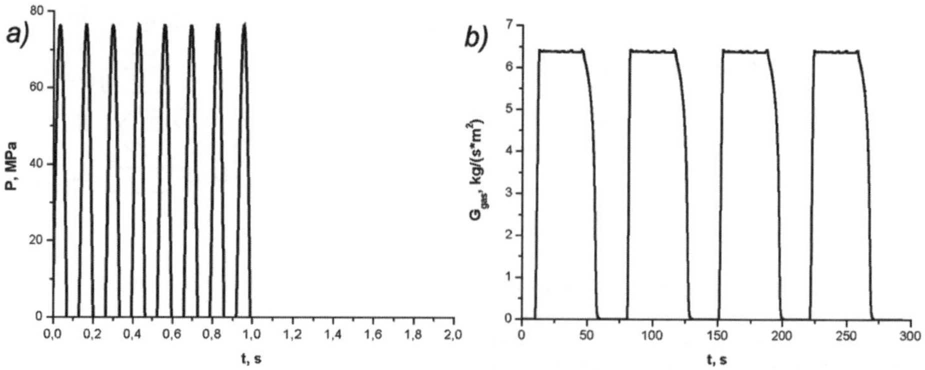

Fig. 4. Packet of the detonation waves (a) and dependence of the yield of gas on the time (b). The packet of the waves on (b) imposes just after each plugging. The frequency of the pocket was equal to 0.014 Hz

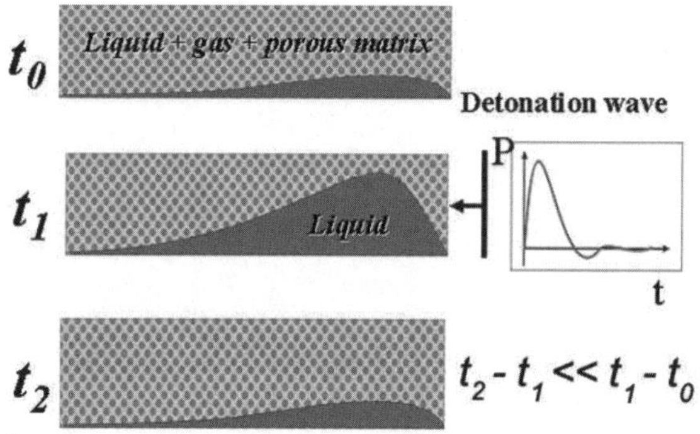

Fig. 5. Schematic of action on well bottom zone of gaseous condensate fields.

4 Conclusion

1. It was shown both experimentally and numerically that appearance of gas-condensate plug is possible using model binary methane-n-butane mixture.
2. It was shown numerically that the condensate plug can be destroyed by a series of strong shock wave with intensity of 80 bar and with frequency of 10 Hz. So that the time of affection by this series is substantially smaller than the period of plug condensation.

References

1. A.Mirzadzanzade et.al.: *Phoundation of gas production tecnology*, Moscow: Nedra, 2003. 880pp
2. A. Brusilovsky: Oil and Gas Geology. 1997, N7, pp.31-38
3. A. Tihonov, A. Samarskiy: *Equations om mathematical physics.* Moscow: Nayka, 1977

Experimental and theoretical study of valveless fuel supply system for PDE

D.I. Baklanov, S.V. Golovastov, N.W. Tarusova, and L.G. Gvozdeva

Associated Institute for High Temperatures, Izhorskaya st. 13/19, 125412, Moscow, Russia

Summary. Recently great interest is manifested in the development of the rocket systems, which use high-frequency detonations. Reaching a maximally possible repetition frequency of cycle, i.e., the smallest possible cycle time is the most important technical task. In IHED of JIHT RAS the model of PDD with the valveless fuel supply system is created, which makes it possible to noticeably reduce the cycle time. Furthermore, the system possesses another number of advantages. It allows to create the concentration of fuel variable along the length of combustion chamber, what make possible to govern the formation of detonation and to reduce the transition time from deflagration to detonation. This article presents the results of experimental and theoretical studies of the valveless fuel supply system.

1 Introduction

In IHED of JIHT RAS the detonation combustion chamber (DCC), is developed and acts with the valveless fuel supply system. It is possible to decrease the duration of cycle due to the use of nonstationary processes with the filling of combustion chamber [1]. Maximum repetition frequencies reached 95 Hz with a constant cooling by water the walls DCC. The time of installation operation is limited only by fuel stock. Furthermore the unit is developed, which simulates one cycle of work DCC [2].

For reconstructing a full wave pattern of PDE operation, pressure and luminosity were simultaneously registered by piezosensors and photodiodes in control sections of the DCC. The processing of signals from piezosensors and photodiodes made it possible to: differentiate fronts of different nature; plot the distance-time diagram for flames, shock, and detonation waves in the DCC; determine the location of detonation onset and parameters behind the detonation wave.

On the basis of these studies the hypothesis has been advanced about the fact that the process of the "starting" of gas-dynamic valves can be considered as the decoupling of arbitrary break in the shock tube, counting combustion chamber – as a high-pressure chamber, and supply mains – as a low-pressure chamber [3].

For the approval of this idea it was necessary to conduct experiences with the pressure sensors and the photodiodes in the fuel supply system. This cannot be made at the installation, which works in the frequency mode, since the feeder tubes are too narrow. Therefore for this work the modification of combustion chamber was used, which simulates one cycle of DCC operation. The experiment can help to: reconstruct the wave dynamics of shocks and contact surfaces in the supply system; determine the instant when fuel components started filling the DC and estimate pressure distribution in the feed manifolds and in the DCC.

2 Experimental studies on the modified combustion chamber

The special modification of the camera was used, which simulates one cycle of DCC operation (Fig. 1). Modification consists in the fact that, the collectors of supply system were extended at which pressure sensors and photodiodes were established.

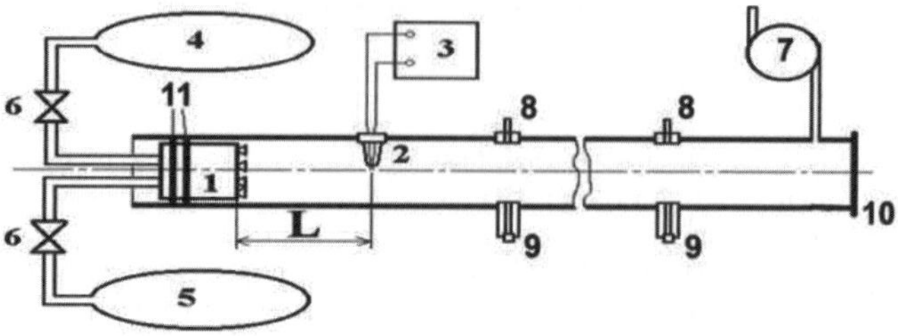

Fig. 1. The schematic of detonation combustion chamber (DCC). 1 – injection block (IB), 2 – spark-gap, 3 – ignition system, 4 – chamber oxygen chamber, 5 – hydrogen chamber, 6 – pneumovalves, 7 – vacuum pump, 8 – pressure transducers PCB 113A34, 9 – light sensors FD 256, 10 – flange, 11 – O-rings.

The separate supply of propellant components in DCC through IB was accomplished, which has six injectors – three injectors for each gas, oxidizer and fuel. The mixing of components occurred directly in the combustion chamber. As the means of diagnostics piezoelectric pressure transducer PCB-113A34 were served for measuring the velocity of shock waves propagation in DCC (error 0.5 %), located along the axis of combustion chamber. In one cross-section with them the photodiodes FD-256, which fix flame front were located (error 1 %). Signals were written on the four-beam digital memory oscillograph Tektronix 3014B (frequency of channel – 100 MHz).

The injector block is presented on (Fig. 2). Components of a detonable mixture is feeded through collectors (2) to injectors (3), separated at the block of injectors.

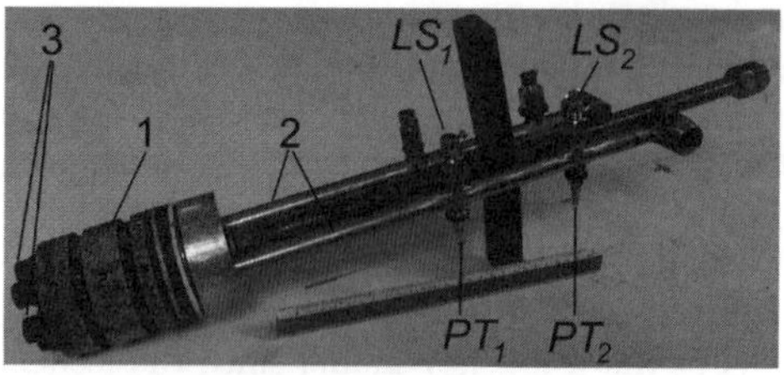

Fig. 2. The block of injectors. 1 – block of injectors, 2 – fitting collectors, 3 – injectors. PT_1, PT_2 – pressure transducers PCB 113A34, LS_1, LS_2 – light sensors FD 256.

Experimental studies were conducted in the stoichiometric mixtures of hydrogen with oxygen in the DCC with a diameter of 83 mm. Hydrogen and oxygen were fed through the collectors with a diameter of 10 mm, and length 3 m. Collectors were specially extended in order to have the capability to investigate the wave picture of processes not only in the combustion chamber, but also in the collectors. For registering the wave pattern in the collectors, two pressure sensors and photodiode were set into them. The oscillograms of pressure and luminous flux were obtained in the collectors (Fig. 3), what make possible to determine the shock wave appearance and contact surfaces velocities. In the collectors of N_2 supply the pressure was 14.5 bar, while in the collectors of O_2 – 9.5 bar, that secured entering in DCC in 30 ms 2.32 dm^3 of hydrogen and 116 dm^3 of oxygen. Pressure in DCC was equaled 1 bar. The ignition of mixture occurred 30 ms after the start of the set-up. Electrical discharge causes induction signal simultaneously in all measuring lines.

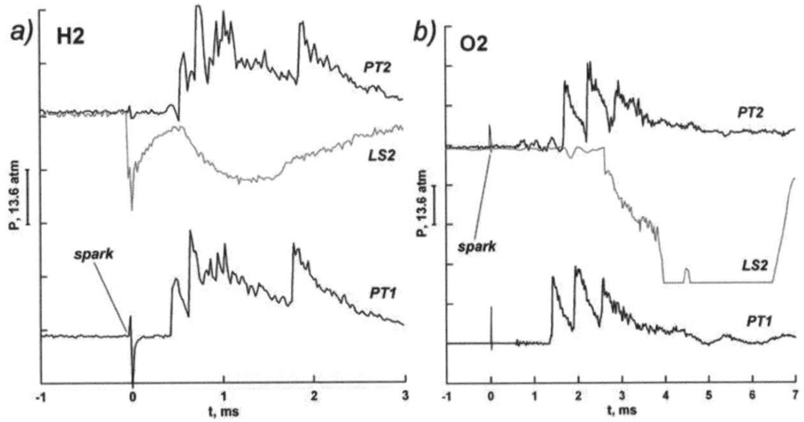

Fig. 3. Oscillograms of pressure (PT_1, PT_2) and luminous flux (LS_2) in the collectors of the supply of hydrogen and oxygen. PT_1 and PT_2 – pressure sensors, LS_2 – the light sensor

In the combustion chamber the transition of deflagration to detonation occurred. Retonation wave is propagated to the side of collectors, while detonation – to the opposite side. Retonation wave falls into the injector block, which is a complex device. Retonation wave experiences a number of sequential reflections. The resulting lift of pressure causes the outflow of the burned gases from the combustion chamber into the collectors and the appearance of shock waves in them. The arrival of shock wave into the sensor stations is at first noticed in the hydrogen main, since the speed of sound in hydrogen considerably (about 4 times) exceeds the speed of sound in oxygen. Time between the ignition of mixture and the arrival of shock wave at PT_1 (Fig. 3, a) is 423 s, in 106 s the shock wave is fixed of PT_2. Taking into account the speed of the motion of hydrogen (900 m/s) in the collector, Mach number of shock wave is equaled 1,47. Contact surface lags behind the shock wave on 105 s. The observed second shock waves are caused by the reflection of retonation and detonation waves from the end walls of DCC. In the oxygen main shock wave (Fig. 3, b) is fixed considerably later, in 1350 s after ignition with the sensor PT_1, and by sensor PT_2 in 1660 s. Taking into account the speed of the motion (400 m/s) of oxygen in the collector Mach number of shock wave is equaled 2,26. Contact surface lags behind the shock wave on 987 s. If we assume that the speed of contact surface is constant, then from these data it is possible to obtain the speed of the motion of the escaping burned gases.

Thus, experiments showed that actually "closing" of gas-dynamic valves occurs in accordance with our assumptions about the disintegration of arbitrary break in the combustion chamber. In the moving system shock waves clearly are fixed. Their speed is measured and it is evident that the speeds in the mains of supply are different in the main of hydrogen and oxygen. This gives grounds for the control capability of knocking process in the combustion chamber due to the creation of variable concentration along the length of combustion chamber.

3 Analytical calculation of the valveless fuel supply system for PDE, which operates in the frequency mode

Analytical calculation of the valveless fuel supply system have been developed for the PDE installation, operating in a frequency mode. The operation of the detonation combustion chamber was examined with a diameter of 36 mm and length 7 m. The frequency of detonation waves was 2 Hz. Investigation were made in the stoichiometric mixture of methane with oxygen. The mathematical model of the calculation of the stage of gas-dynamic valves "closing" of is depicted in fig. 4.

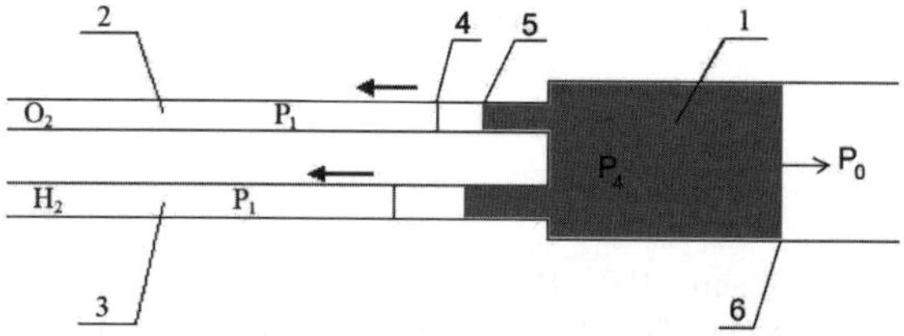

Fig. 4. Mathematical model of valveless supplied supply system. 1 – detonation combustion chamber; 2 – oxygen supply manifold; 3 – methane supply manifold, 4 – shock front, 5 – contact surface, boundaries of outflowing burned gases, 6 – detonation wave.

The following known equations for calculating of shock waves and rarefaction waves in the combustion chamber were used.

3.1 Equations for shock wave and rarefaction wave

$$u_2 = \left(\frac{p_2}{p_1}\right)\sqrt{\frac{2/\gamma_1}{(\gamma_1+1)p_2/p_1+(\gamma_1-1)}},$$

$$u_3 = \frac{2a_4}{\gamma_4-1}\left[1-\left(\frac{p_3}{p_4}\right)^{(\gamma_4-1)/2\gamma_4}\right]$$

The first formula shows the dependence between the speed of the outflowing from the combustion chamber u_2 gases behind the shock waves in the low-pressure chamber (i.e., in the fuel supply system) and a pressure ratio of the appearing shock waves p_2/p_1. The second formula describes the connection between lowering of the pressure in the high-pressure chamber p_3/p_4 (in the combustion chamber) and the speed of the gases escaping into the supply system u_3. It was assumed that in the combustion chamber the detonation of stoichiometric hydrogen-oxygen mixture at an initial pressure 1 bar and an initial temperature T = 291 K occurs. It is supposed that the transition of deflagration to detonation occurs instantly [4]. At the thrust wall the time-constant parameters are obtained, as this was shown in the work of Zeldovich [5].

Calculations according to these formulas give us the discharge velocities of the burned gases into the supply system. Then it is possible: to determine at what depth the penetration of the hot burned gases will occur inside the conduits and to find the mass of the burned gases in the conduits. This can be made, after determining the times of the outflow of the burned gases through the open end of the combustion chamber.

The results of calculations are given in Table 1.

Table 1. Unsteady flow parameters of combustion products flowing out of detonation combustion chamber into supply manifolds

	Oxygen manifold, O_2	Methane manifold, CH_4
Contact surface velocity, w, m/s	244	318
Pressure behind shock wave, P, bar	7.9	7.4
Mach number of shock wave, M	1.55	1.5
Shock wave velocity, U_1, m/s	502	687
Time of combustion product outflowing into supply manifold, t, s	0.003956	0.003956
Filling distance of penetration of contact surface in manifold, l, m	0.965	1.258
Feeling distance of penetration of shock wave in manifold , L, m/s	1.986	2.718

The following step is the calculation of the outflow of the cooled burned gases from the conduits back into combustion chamber. These times will be different for the conduits of fuel and oxidizer. Because of this difference the "opening" of gas-dynamic valves will begin in the different time for the fuel and the oxidizer. Thus, changing the diameters of supply lines and pressure in them, it is possible to assign different laws of filling of detonation combustion chamber.

The calculations of the reverse discharge of the burned gases from the supply mains to the combustion chamber were given according to the formulas for the steady flow (Table 2).

Table 2. Reverse flow parameters of combustion products out of supply manifold towards detonation combustion chamber

	Oxygen manifold, O_2	Methane manifold, CH_4
Mass flow of combustion products, Q_M, kg/s	0.0282968	0.019609
Reverse flow time of combustion products out of supply manifold into detonation combustion chamber, t, s	$0.84 * 10^{-3}$	$1.58 * 10^{-3}$

$\Delta t = 0.74 * 10^{-3}$ s.

We see that as a result of the wave processes there is a difference in the time of the "opening" of gas-dynamic valves for the fuel and the oxidizer.

4 Concluding Remarks

The work presents the results of experimental and theoretical studies of DCC operation with the valveless fuel supply system. Experiments were conducted on the special combustion chamber, which simulates one cycle of work DCC in the hydrogen-oxygen mixture [$2H_2 + O_2$]. Pressure sensors and photodiodes were located not only inside the combustion chamber, but also in the feed system. The phase of "closing" of gas-dynamic valves is investigated. It is shown that the shock waves before the emerging burned gases from DCC appear. Experiments confirmed the lawfulness of the method of calculation of the phase of "closing developed by the authors", using an equation for the decoupling of arbitrary break in the shock tube.

The system of DCC filling is analytically calculated, which operates in the frequency mode in the stoichiometric methane-oxygen mixture. For the calculation, both the steady-state equations and the nonstationary methods were used. It is shown, that as a result of the wavelike nature of the operation of valveless fuel supply system, it is possible to attain the different time of "opening" gas-dynamic valves. Thus, the specified distribution of fuel concentration along the length DCC can be obtained, that will make possible to reduce the process of transition from deflagration to detonation.

Acknowledgement. This work was partly supported by the Russian Foundation for Basic Research (grant No. 05-08-33614).

References

1. D. Baklanov, L. Gvozdeva, N. Scherbak: Formation of high-speed gas flow at combustion in the regime of multi-step detonation. In: *Gaseous and heterogeneous detonations: Science to applications.* Eds. G. Roy, S. Frolov, K. Kailasanath, and N. Smirnov. Moscow, 1999. ENAS Publ. 141-52.
2. S. Golovastov, V. Golub, V. Volodin, D. Baklanov: Effect of chamber closed end and mixture velocity on deflagration-to-detonation transition. In: *Nonequilibrium Processes. Combustion and Detonation.* Eds. G. Roy, S. Frolov, A. Starik. Moscow, 2005. TORUS PRESS, V.1, pp. 310-316.
3. D. Baklanov, L. Gvozdeva, N. Tarusova: Application of valveless fuel supply system for detonation control by creating variable mixture composition along the combustion chamber. In.: *Pulsed and continuous detonations.* Eds. G. Roy, S. Frolov, J. Sinibaldi. Moscow, 2006. TORUS PRESS Ltd., 376 pp.
4. K. Stanukovich: *Unsteady flows of continuous medium*, State publishing house of technical theoretical literature. Moscow, 1955.
5. J. Zeldovich, A. Kompaneec: *The detonation theory*, State publishing house of technical theoretical literature. Moscow, 1955.

Experimental investigation of the ignition spark shock waves influence on detonation formation in hydroxygen mixtures

D.I. Baklanov, S.V. Golovastov, V.V. Golub, and V.V. Volodin

Associated Institute for High Temperatures, Izhorskaya st. 13/19, 125412, Moscow, Russia

Summary. The paper is devoted to the investigation of the influence of shock waves reflected from back and side walls of cylindrical tube on length of detonation formation. The experiments were conducted at cylindrical detonation tube of 83 mm in diameter and 2510 mm in length. Detonation tube is equipped with system of feeding with detonable mixture, and evacuation, ignition system. Combustion diagnostics was conducted with pressure transducers PCB 113A34 and photodetectors FD-256 mounted in four consequent stations along the tube. Results are presented as oscillograms, x-t diagrams and DDT length dependencies on the varied parameters. Experiments have shown that in the case of weak spark the location of ignition point essentially affects the length of detonation onset.

1 Introduction

Since second half of the last century scientists had noted significant influence of shock waves on flame front that leads to detonation onset in several cases. It is well known that shock waves reflections from detonation tube walls and interactions of shock waves with flame front affects combustion development and in several cases lead to detonation formation. K.I. Schelkin [1] have shown the influence of distance between detonation tube closed end and ignition location on detonation onset length is noticed but quantitative data are not presented. In the paper of Ya.B. Zeldovich [2] correlations of detonation direct initiation critical energy are presented but there is no information about investigations of detonation formation at subcritical initiation energies.

The aim of this work is investigation of detonation formation process at subcritical initiation energies depending on spark energy value and spark plug location relatively the detonation tube closed end and sidewall.

The process of transition the combustion into the detonation in the gases is characterized by the distance, which will pass flame front, before will arise detonation. Predetonation distance is determined by fuel-air ratio, by its pressure and temperature, and also by geometry of detonation combustion chamber (DCC): by the position of the closed end relative to the system of ignition, by arrangement and by the form of obstacles and precombustion chambers in the channel DCC. Turbulence influences the acceleration of the flame, which leads to the detonation. In the quiescent mixtures the flame is accelerated due to auto-turbulence.

In the paper the shock-wave effect, which appear as a result of the acceleration of flame front, on the process of transition the combustion into the detonation in the moving detonable mixture was investigated. The influence of the speed of the flow of detonable mixture and position of the place of initiation on detonation onset with the weak initiation and with the energies of initiation, close to the critical was investigated.

2 Experimental procedure

Experiments were conducted in the cylindrical detonation combustion chamber (DCC), which length was $2500mm$ and inside diameter corresponded to $83mm$, that is in 80 times more than the size of the detonation cell of stoichiometric hydrogen-oxygen mixture ($\lambda = 1mm$). Setup made it possible to work both with the quiescent mixture and with that moving. The spark discharge, which occurred in the combustion chamber, served as initiator. Distance from it to the combustion chamber closed end L changed from 0.3 tube diameters to 7 tube diameters. The velocity of propagation of shock waves in DCC were measured with the aid of the pressure sensors (PCB-113A34) and speed of the motion of flame front with the aid of the photosensitive devices (FD-256). The schematic of experimental setup is presented in fig. 1.

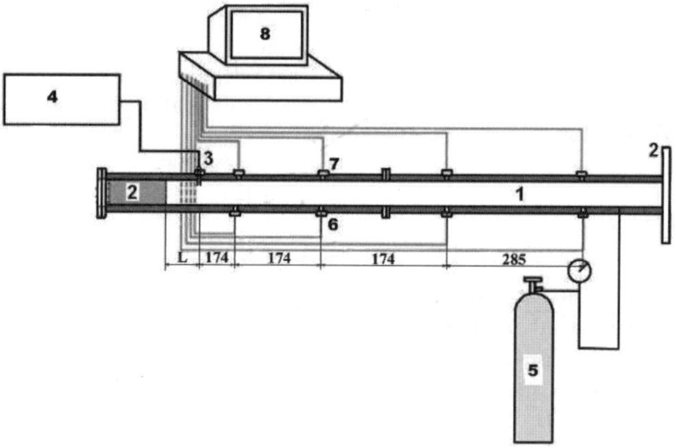

Fig. 1. The schematic of experimental setup. 1 - detonation combustion chamber (DCC), 2 - flange, 3 - spark plug, 4 - ignition system, 5 - mixture chamber, 6 - pressure transducers PCB 113A34, 7 - light gauges FD 256, 8 - data acquisition system

Before the experiment detonation chamber was pumped out up to a pressure of $1Torr$. After this, it was filled by the preliminarily prepared detonable mixture up to a pressure of $1atm$. The combustion, which passes into the detonation, was initiated by the spark, location and energy of which changed from one experiment to the next. In the process of experiment the indications of the pressure and light sensors were recorded.

3 Results processing

Typical signals time histories and x-t diagrams of the combustion development, which are based on the experimental data, are represented in Fig. 2. Trajectories of detonation and retonation waves were extrapolated to the intersection and location of detonation onset was found. The distances of the detonation onset, obtained from the x-t-diagrams, were substituted to the plots depending on the changing parameter. One of the plots is presented in fig. 3. One can see the characteristic U-shape curves that were described in [1]. This makes it possible to consider the results of experiments reliable.

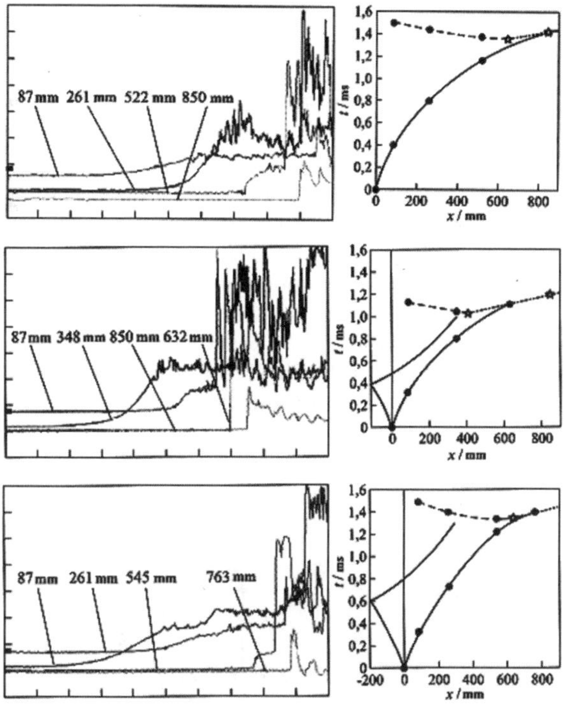

Fig. 2. Oscillogramms (left) and x-t diagrams (right) of combustion development at $P_0 = 1bar$, $E = 0.1E_{cr}$, $d = 32mm$ and $L = 26mm$, $113mm$ and $200mm$ top to bottom

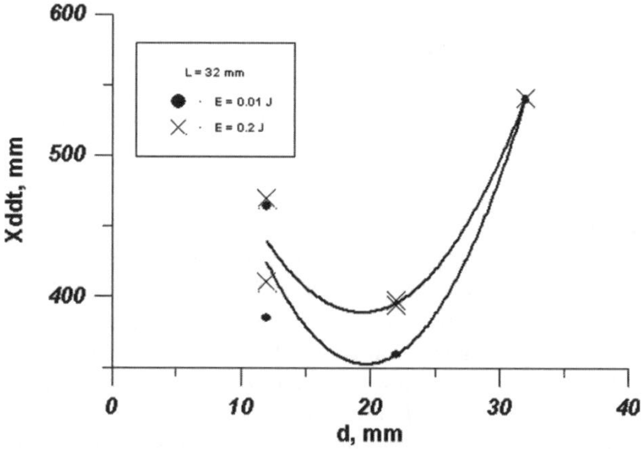

Fig. 3. Dependences of detonation onset length on the distance d between the discharge gap and sidewall at two igniting spark energy values

4 Discussion

When the initiated shock wave, which is followed by the chemical reaction zone, propagates in the confined volume filled with detonable mixture, the shock wave decelerates and weakens due to the energy dissipation in spherical shock and rarefaction fan overlapping. Intensive heat release accelerates the shock wave but at the subcritical ignition energy values this acceleration is insufficient for detonation formation close to the ignition point. Shock wave reflection from the detonation chamber walls leads to the pressure and temperature increase in the point of reflection and the further shock compression of the flame region. This can cause detonation onset or additional acceleration of flame. Reflected shock wave, when it interacting with the flame front, can: or to accelerate flame front, without leading to the detonation; or to form detonation; or to overtake the already formed detonation wave. Fig. 4 shows the DDT length dependence on the distance from back wall to the spark location at various values of ignition energy.

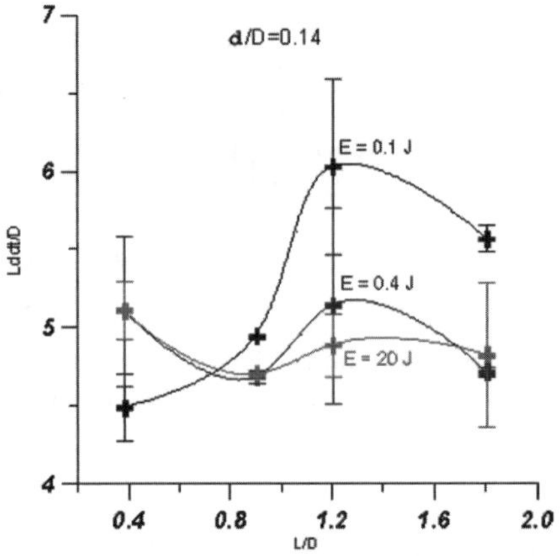

Fig. 4. Dependences of detonation onset length on the distance between the discharge gap and back wall

The study of the influence of the position of the ignition of detonable mixture on detonation onset was conducted. From the obtained results it follows that minimum L_{DDT} corresponds $L/D = 1.2$ (D - inside diameter DCC). When $L = 1.2D$ expanding products of combustion create compression waves, which, being reflected from the closed end DCC, overtake the compression waves, which are extended to the opposite from the end side, additionally accelerating that not reacted gas this leads to its additional compression, to the acceleration of flame front and finally to the decrease of predetonation distance. In such a case, when discharger is located sufficiently close to the closed end DCC ($L = 0.2 - 0.3D$), compression waves are created only in one direction. The waves reflected be negligible, that it does not lead to the additional compression. With $L = 0.2D$ the compression waves reflected do not manage to overtake the compression waves, which are moved in opposite from the end direction to the moment of the appearance of detonation. It is also discovered, that with the energy densities of the initiation of less than $0.08E_{cr}$ (E_{cr} - planar detonation ignition critical energy calculated by [2]) on the appearance of detonation is substantial the influence of the position of ignition relative to the closed end of DCC. Moreover, the less the energy density, the more expressed this influence. At energy densities higher than $0.33E_{cr}$ influences of initiator on the medium become so great with the energy densities of large that the position of initiation does not play role. With the intermediate values of energy the displacement of the position of discharger affects predetonation period only.

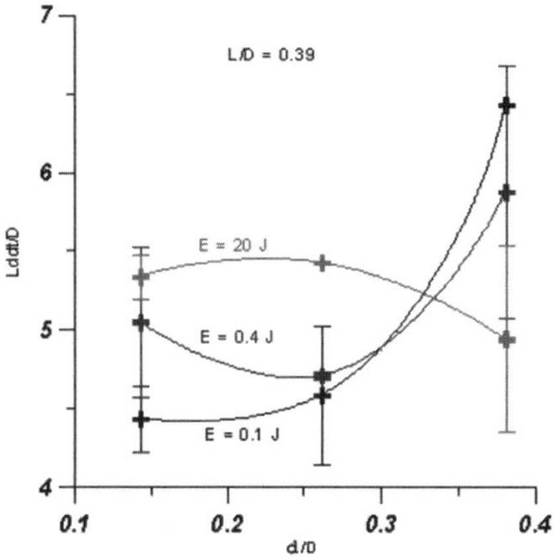

Fig. 5. Dependences of detonation onset length on the distance between the discharge gap and sidewall

Fig. 5 shows the DDT length dependence on the distance from sidewall to the spark location at various values of ignition energy. It is possible to note that with the decrease of distance from the spark to the wall of tube, an increase in the energy of spark leads to the larger length of the deflagration to detonation transition. With the initiation of combustion it is nearer to the tube axis, an increase in the energy reduces predetonation distance. This phenomenon can be explained by the mutual influence of the focusing of shock waves in the tube and the reduction in the density, temperature and volumetric release of energy in the rarefaction wave, which follows behind the shock wave. Powerful spark in the vicinity of the tube wall ignites the region of gas, compared in the dimensions with the discharge gap, in this case the powerful shock wave is followed by the wave of strong rarefaction, which cools mixture to the temperature lower than initial, and considerably decreases density. Accordingly [2] reduction in gas density noticeably increases the length of predetonation distance.

5 Conclusions

An experimental study of shock-wave, generated by the initiating spark and reflected from the tube walls, effect on the length of the deflagration to detonation transition is carried out.

It is shown that the shock wave reflected from the back wall of tube can substantially accelerate detonation onset. Predetonation distance proved to be smallest at $L = 1.2D$. The location of spark discharger relative to the axis of tube in conjunction with the spark energy value influence the initial stage of the detonation formation. In the case of the spark discharger location in the vicinity of tube sidewall an increase in the spark energy leads to an increase in the predetonation distance.

References

1. Shchelkin KI, Troshin YaK: *Gasdynamics of Combustion*, edition (Mono Book Corp. Baltimore; 1965) 255 pages
2. Zeldovich YaB, Kogarko SM, Simonov NI: Journal of technical physics **26**, 8 (1957), pp. 1744-1752

Experimental study on the nonideal detonation for JB-9014 rate sticks

L. Zou, D. Tan, S. Wen, J. Zhao, and C. Liu

Laboratory for Shock Wave and Detonation Physics Research, Institute of Fluid Physics, CAEP P.O.Box 919-103, Mianyang 621900 (P.R.China)

Summary. JB-9014 is a kind of insensitive explosives with high energy and high safety which consists of 95 percent wt. TATB and 5 percent wt. binder. Its detonation performance is closely relative to diameter, temperature and confinement. Its local normal detonation velocity Dn depends primarily on the local total curvature k. In this paper, series experiments of the JB-9014 rate sticks with air, aluminum and brass boundary were carried out at the temperatures -30 C and 24 C. At each temperature four diameters were fired: 10, 12.5, 15 and 30 mm respectively. The objective of this study is to investigate the diameter effect, temperature effect and confinement effect of JB-9014 rate sticks. By means of optical and electrical test methods, steady detonation velocities and wave shapes were measured precisely in above experiments. Wave shape data were fitted with an analytic form, by which $Dn(k)$ curves were calculated.

1 Introduction

The perfect explosives need to combine high energy with high safety characteristics together. Most of those explosives are based on TATB composition, and known as LX-17 and PBX 9502 (USA), PTC (Russia), EDC-35 (England), T1 and T2 (France), JB-9014 (China). The basic detonation performance and parameter of those similar TATB-based compositions are very useful in both fundamental research and engineering application. Campell measured the diameter effect curve and failure diameter at -55 C, 24 C and 75 C for PBX 9502 [1]. Hill et al. further investigated the wave shape and the relation of normal detonation velocity and curvature in above three temperatures [2]. Hutchinson et al. studied the change tendency of steady detonation velocity of rate stick vs. diameter and density for EDC-35 [3]. In this study, the joint optical and electrical techniques are used to measure the steady detonation velocity and wave shape of JB-9014 rate sticks. Some evolutive tendencies are revealed for different diameter, different confinement and different temperature.

2 Experimental assembly

The rate stick design and the experimental assembly are illustrated schematically in Figure 1. The experimental assembly is composed of rate stick design, shot box, optical system and electrical system. The explosive was pressed to the desired density, and then machined into 80 or 100 mm long cylinders as it is rather difficult to press and machine the entire charge in a single one. Charge segments are linked together with a kind of glue. The aspect ratio of the charge for different diameter ranges from 10 to 20. The charge assembly was loaded between two aluminum plates. The detonator was held on one aluminum plate. The other plate is the optical end-plate and pushed against the main charge. The detonation velocity can be calculated by the time signals of electrical pins on the alloy connector. The shot box is made of foam and wood. For cold shots, the temperature was controlled by dry nitrogen operated by a mass flow controller.

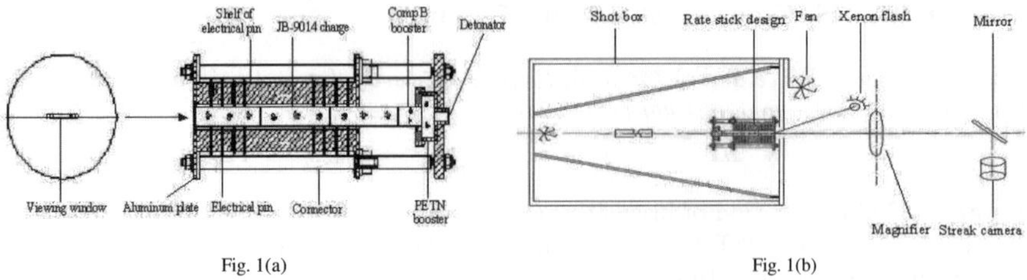

Fig. 1. Rate stick design (left) and sketch of experimental assembly (right)

3 Detonation velocity

The experimental data are tabulated in Table 1 and plotted in Figure 2(a). In Table 1, the uncertainty of experimental temperature (T) is 0.05 C. Compared with the radius of 24 C, the corrected coefficient of JB-9014 charge radius (R) at -30 C is 3.9x10^{-5} C^{-1}. Detonation velocities (D_0) are corrected to those at the density of 1.890 g/cm^3 with a coefficient of 0.003. Figure 2(a) shows the comparison of the diameter-effect curves at -30 C and 24 C. As R varies from 5 to 15 mm, D_0 increases monotonically and tends to ideal detonation velocity D_{CJ} for both temperatures. Another feature is that the two diameter-effect curves have a point of intersection. When R is less than 11.21 mm, D_0 at 24 C is larger because of the effect of temperature on reaction zone. While R is larger than 11.21 mm, D_0 at 24 C is smaller because of the effect of temperature on density. The crossover denotes the balance case of two temperature effects.

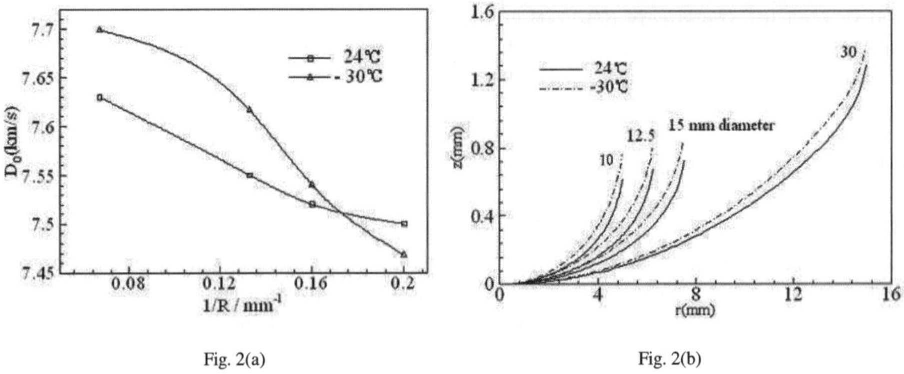

Fig. 2. diameter-effect curves (left) and wave front curves (right)

shot run	Exp. tem. (T, C)	R observed (mm)	R^* corrected (mm)	Charge Length (L,mm)	Charge Density (g/cm^3)	D_0 observed (km/s)	$D_0{}^*$ corrected (km/s)
21-2	24	4.99	4.99	200.16	1.890	7.502	7.502
21-5	24	6.25	6.25	200.03	1.890	7.521	7.521
21-8	23.5	7.50	7.50	239.90	1.890	7.555	7.555
21-9	23.5	15.00	15.00	299.95	1.895	7.632	7.632
30-1	-29.9	4.98	4.97	200.15	1.886	7.481	7.468
30-5	-30.3	6.23	6.22	200.06	1.890	7.538	7.540
30-6	-30.3	7.48	7.46	240.18	1.892	7.612	7.616
30-8	-30.4	14.97	14.94	300.25	1.893	7.690	7.699

Table 1. Steady-state detonation velocity

4 Steady wave shape

Detonation wave shapes are recorded by SJZ-15 high speed streak camer. The scan speed of camer is 24 km/s. The uncertainty of scan speed is 0.2 percent. The wave shape data was fitted using the following series,

$$z(r) = -\sum_{i=1}^{n} \alpha_i \{\ln[\cos(\eta \frac{\pi r}{2R})]\}^i \tag{1}$$

where $z(r)$ is wave front curve, R is the charge radius, α_i and η are fitting parameters [4]. Table 2 gives the fitting parameters. As is shown in Figure 2(b), wave shape is slightly flatter in hotter charges than colder ones for same diameter. Detonation wave propagates more ideally as charge diameter increases for same temperature.

shot run	Exp. tem. (T, C)	R^* corrected (mm)	$D_0{}^*$ corrected (km/s)	α_i/mm	α_2/mm	α_3/mm	η
21-2	24	4.99	7.502	0.3104	0.0443	0.0125	0.9330
21-5	24	6.25	7.521	0.3235	0.0467	0.0088	0.9529
21-8	23.5	7.50	7.555	0.3643	0.0748	0.0147	0.9573
21-9	23.5	15.00	7.632	0.8268	0.2397	0.0329	0.9757
30-1	-29.9	4.97	7.468	0.3983	0.0607	0.0131	0.9380
30-5	-30.3	6.22	7.540	0.4496	0.0932	0.0154	0.9541
30-6	-30.3	7.46	7.616	0.4558	0.0985	0.0158	0.9609
30-8	-30.4	14.94	7.699	0.9198	0.2591	0.0320	0.9780

Table 2. Fitting parameters of wave front curves

5 $D_n(k)$ curves

$Dn(k)$ relation can be calculated by D_0 and $z(r)$. The normal detonation velocity is,

$$D_n(r) = D_0 \cos[\theta(r)] = \frac{D_0}{\sqrt{1 + z'(r)^2}} \tag{2}$$

And the local total curvature is,

$$\kappa(r) = \sqrt{\frac{(z''(r))^2}{[\sqrt{1 + z'(r)^2}]^3}} + \sqrt{\frac{(z'(r))^2}{r^2[\sqrt{1 + z'(r)^2}]}} \tag{3}$$

The $Dn(k)$ curves for Table 2 are exhibited in Figure 3(a). When k is smaller than 0.05, $Dn(k)$ curves are concave upward and trace nearly a common path for any diameter and any temperature. When k is larger than 0.05, Dn varies with curvature, diameter, temperature and $Dn(k)$ curves tend to diverge.

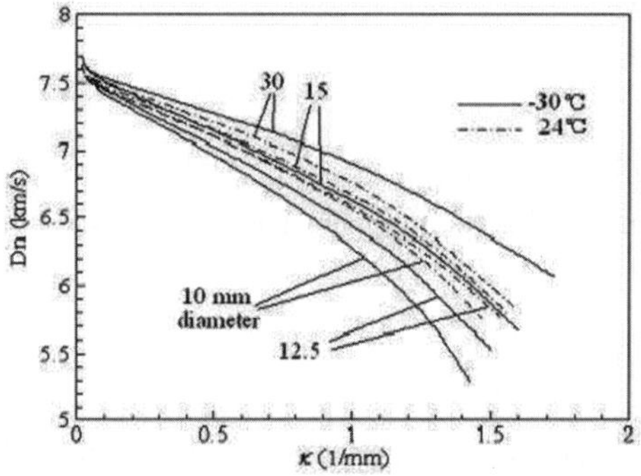

Fig. 3(a)

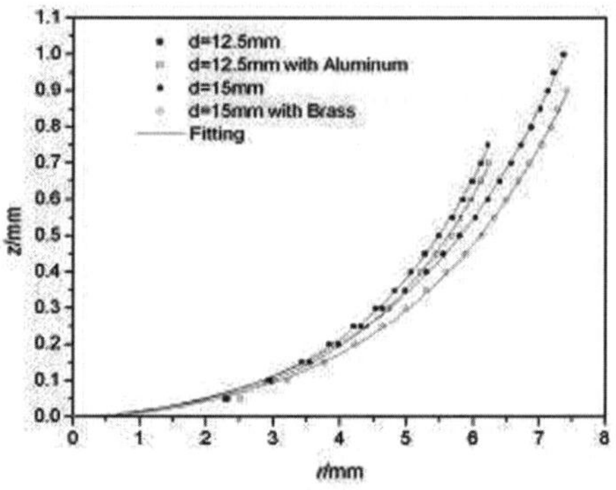

Fig. 3(b)

Fig. 3. $Dn(k)$ curves for different temperature (left) and different boundary (right)

6 Confinement boundary

Figure 3(b) shows the wave front curves of JB-9014 rate sticks with air, aluminium and brass boundary at 12.5 and 15 mm diameter, respectively. It can be seen that when the confinement is stronger, wave front becomes flatter and normal detonation wave propagates faster.

7 Conclusion

Series experiments are performed to investigate the nonideal detonation of JB-9014 rate sticks by use of the joint optical and electrical test techniques. Based on present results, for each diameter, wave shape is slightly flatter in hotter charges than colder ones. For each temperature, detonation wave propagate more ideally as charge diameter increases. When k is smaller than 0.05,$Dn(k)$ curves are concave upward and trace nearly a common path for any diameter and any temperature. When k is larger than 0.05, Dn varies with curvature, diameter, temperature and $Dn(k)$ curves tend to diverge. For each confinement, when the confinement is stronger, wave front becomes flatter and normal detonation wave propagates faster. Impedance of boundary plays an important role in the confinement effect.

Acknowledgement. The authors wish to thank Qing Fang for some helpful discussions, Guang-Sheng Zhang and Zhi He for some assistance to experiments reproted here.

References

1. Campell A W. Diameter effect and failure diameter of a TATB-based explosive. Propellants, Explosive, Pyrotechnics, 9: 183-187, 1984
2. Hill L G, Bdzil J B and Aslam T D. Front curvature rate stick measurements and detonation shock dynamics calibration for PBX 9502 over a wide temperature range. Proceeding of 11th International Symposium on Detonation, Office of Naval Research, Colorado, 1029-1037, 1997
3. Hutchinson C D, Foam G C W. Initiation and detonation properties of the insensitive high explosive TA-TB/Kel-F 800 95/5. Proceeding of 9th International Symposium on Detonation, Office of Naval Research, Oregon, 123-132, 1989
4. Bdzil J B. Steady-state two-dimensional detonation. Journal of Fluid Mechanics, 108: 195-226, 1981

References

Experimental study on transmission of an overdriven detonation wave across a mixture

J. Li[1], K. Chung[2], W.H. Lai[1], and F.K. Lu[3]

[1] *National Cheng Kung University, Institute of Aeronautics and Astronautics, Tainan, Taiwan*
[2] *National Cheng Kung University, Aerospace Science and Technology Research Center, Tainan, Taiwan*
[3] *University of Texas at Arlington, Mechanical and Aerospace Engineering Department, Aerodynamics Research Center, Arlington, Texas, USA*

Summary. Two sets of experiments were performed to achieve a strong overdriven state in a weaker mixture by propagating an overdriven detonation wave via a deflagration-to-detonation transition (DDT) process. First, preliminary experiments with a propane/oxygen mixture were used to evaluate the attenuation of the overdriven detonation wave in the DDT process. Next, experiments were performed wherein a propane/oxygen mixture was separated from a propane/air mixture by a thin diaphragm to observe the transmission of an overdriven detonation wave. A simple wave intersection model showed that the rarefaction effect must be included to ensure that the post-transmission wave properties are not overestimated. The experimental results showed that the strength of the incident overdriven detonation plays an important role in the wave transmission process. After the wave transmission process, the propagation of the detonation directly correlates with detonability limits.

1 Introduction

The phenomenon of the transmission of detonation from one mixture to another of different sensitivity is of interest in the development of pulse detonation engines (PDEs). To ensure the success of a PDE, a predetonator concept has been proposed [1]. A detonation wave can then be generated with relative ease in the predetonator filled with an energetic material and transmitted into a larger diameter, main combustor filled with a gaseous or liquid hydrocarbon-air mixture. Nevertheless, Schultz and Shepherd [2] indicated that a detonation wave with a $C_3H_8 + 5O_2$ mixture failed to transmit into a larger tube with a $C_3H_8 + 5(O_2 + \beta N_2)$ mixture ($\beta \geq 0.76$). On the other hand, Desbordes and Lannoy [3] showed that a Chapman-Jouguet (CJ) detonation wave propagating into a less sensitive mixture can result in an overdriven detonation condition, in which the cell width decreases with a higher degree of overdrive D^* and in turn a decrease in the diffraction critical diameter. Here the diffraction critical diameter is the minimum tube diameter for the successful transmission of a detonation wave through an abrupt area expansion. This result can be applied in the predetonator concept to let the detonation wave propagate across a mixture change prior to an area change. Furthermore, in some cases, an overdriven detonation wave was observed in the deflagration-to-detonation transition (DDT) process [4]. It can therefore be expected that an overdriven detonation wave propagating into a less sensitive mixture may sustain a detonation wave with some degree of overdrive than the propagation of a CJ detonation wave. Transmission of detonation waves across a mixture change has been studied from the 1950s. These studies, however, did not examine the transmission of an overdriven detonation wave in detail. For such reasons, the present work examines the transmission of an overdriven detonation from a C_3H_8/O_2 mixture to a C_3H_8/air mixture.

2 Experimental Setup

A preliminary test concerning the relation between the degree of overdrive and the length of the detonation wave decay was carried out. Four smooth, aluminum 6061-T6 tubes, 914.4 mm long with inner tube diameters $d = 25.4$, 50.8, 101.6, and 152.4 mm were used. It must be noted that the tube diameter used in the next set of experiments was $d =$ 50.8 mm because this size is near the detonability limits of a stoichiometric C_3H_8/air mixture. Here the detonability limit is the propagation limit of a stable, self-sustained detonation. For the smooth-wall tube, detonability limit is $d \geq \lambda_{CJ}$ [5].

A weak ignition system was used at the closed end. After evacuating the tube to 20–30 Pa, propane and O_2 were pumped into the tube. To ensure the homogeneity of the mixture, a circulation pump was used. In addition, the concentration of the mixture was calibrated by gas chromatography (CHINA 9800 GC/TCD) for each run. The uncertainty of the equivalence ratio was ± 0.05 for 95% confidence. Signals from the sensors were digitized and stored in LeCroy 6810 high-speed data acquisition modules, typically at 500 ksamples/s. Five piezoelectric pressure transducers (PCB 112A) were mounted along the streamwise direction to estimate the propagation speed of the detonation wave. A nonstationary cross-correlation technique (NCCF) [6] was adopted to compute the uncertainty of the propagation speed (estimated to be 8.4–34.2%). A photodiode (Hamamatsu S6468-10) was installed at the closed end to detect the onset time of the detonation waves. The DDT run-up distance X_{ddt} and its uncertainty were estimated based on this onset time of detonation and the trajectory of the pressure wave [6].

In the subsequent set of experiments, a 38 μm thick diaphragm separated the C_3H_8/O_2 and C_3H_8/air mixtures to examine transmission of an overdriven detonation wave across a mixture. A schematic of this experimental setup is shown in Fig. 1. Based on the preliminary test, the diaphragm location from the closed end L_D was set to 152.4, 203.2 and 254 mm in an attempt to vary the strength of the overdriven detonation wave. The C_3H_8/O_2 mixture compositions were nearly stoichiometric ($\phi = 0.9$, 1.0, 1) while the C_3H_8/air mixture compositions varied from $\phi = 0.6$–1.8. The uncertainty of the equivalence ratio of C_3H_8/O_2 mixture and C_3H_8/air mixture were ± 0.05 and ± 0.03 for 95% confidence, respectively. Two circulation pumps, one for each section, were used for about 10 minutes. Ten piezoelectric pressure transducers (PCB 113A22) were mounted along the streamwise direction. To examine the success or failure in propagating a detonation wave, a pressure transducer was mounted next to the open end at $X = 1137.9$ mm.

3 Transmission of Overdriven Detonation

3.1 Transmitted detonation wave

An idealized wave intersection model was used to calculate the states of the transmitted detonation wave D_t. With the pressure and particle velocity remaining continuous across the contact surface, the state of D_t was determined. First, the rarefaction is neglected. The intersection I_1 of the two curves DH and IE is the state of D_t, Fig. 2. Here E represents the burned products just behind the incident overdriven detonation wave. However, for the overdriven detonation wave, because the velocity of the detonation wave front relative to the burned products is subsonic, a rarefaction will follow and overtake the detonation front and further decreases the particle velocity of the burned products. Thus, the effect of the rarefaction should be considered. From the characteristic relations, the Riemann invariant is conserved along a backward characteristic at each instant of the overdriven detonation state. Thus, the particle velocity of the burned products u_r just behind the incident overdriven detonation wave D_i is

$$u_r/c_u = [2/(\gamma+1)](U_i/c_u) - 2/(\gamma+1) \quad (1)$$

Here, U_i is velocity of D_i, and c_u is the sound speed in uniform region. The pressure just behind the incident overdriven detonation wave can be derived from the isentropic relations. As a result, the state of the burned products just behind the incident overdriven detonation wave changes from E to E'. Assume that E' is the state of the burned products at the moment of wave impact at the mixture interface. The state across the contact surface I_2 can be calculated by intersecting DH and IER. Table 1 shows the calculation for an incident overdriven detonation wave with degree of overdrive ${D_i}^* = 1.0$, 1.1, 1.2, 1.3. The table shows that the stronger incident overdriven detonation wave, the stronger is the transmitted detonation.

Table 1. Calculation results of the transmitted wave state, donor: stoichiometric C_3H_8/O_2 at 1 atm and 298 K; acceptor: stoichiometric C_3H_8/air at 1 atm and 298 K

${D_i}^*$	$U_t(I_1)$ (m/s)	${D_t}^*(I_1)$	$U_t(I_2)$ (m/s)	${D_t}^*(I_2)$	$U_t(I_3)$ (m/s)	${S_t}^*(I_3)$
1.0	1996	1.11	1996	1.11	1669	0.93
1.1	2493	1.39	2114	1.17	1839	1.02
1.2	2813	1.56	2268	1.26	2043	1.14
1.3	3105	1.73	2429	1.35	2242	1.25

3.2 Transmitted shock wave

Previous studies show that the transmitted detonation wave is not initiated instantaneously [7]. Thus, the transmission of the incident detonation wave can be reasonably thought to be equivalent to the transmission of a shock wave. A similar calculation as in section 3.1 was made for the transmitted shock wave S_t in which the effect of the rarefaction was also included. The results are also shown in Table 1. The velocity $U_t(I_3)$ of S_t is lower than the velocity $U_t(I_2)$ of D_t. In addition, a steady, one-dimensional, shock heating, explosion model [8] was used to calculate the induction time after the shock heating of reactive gas mixtures. The calculations show the Mach number and postshock temperature increase while τ decreases with a higher ${D_i}^*$. To instantaneously initiate a transmitted detonation wave, a high ${D_i}^*$ is required.

4 Results and Discussion

The results of the preliminary test on the effect of the degree of overdrive D^* on the DDT run-up distance for different tube diameters are shown in Fig. 3 (left). The decay of D^* is correlated with X/X_{ddt}. We can see that D^* decreases with increasing X/X_{ddt}. The length of decay of the overdriven detonation wave as it approaches the CJ detonation is estimated to be about $2X/X_{ddt}$. From these observations, we can conclude that in order to obtain a high D^* for the incident overdriven detonation wave, the interface of mixture change should be placed close to the DDT location.

The experimental results of the wave transmission process reveal that the wave transmission can be classified into three modes:

A. Decay from an overdriven state to a lower overdrive, and decay to a near-CJ state.
B. Decay from an overdriven state to a lower overdrive, and accelerate to a higher overdrive, and then decay to a near-CJ state.
C. Decay from an overdriven state to a sub-CJ state, and transition to an overdriven state, and then decay to a near-CJ state.

Examples of these three modes of wave transmission are shown in Fig. 3 (right). For mode A, two distinct modes are observed. The displacement-time diagram for mode A1 is shown in Fig. 4 (left). The highly overdriven detonation wave decays abruptly at the diaphragm and then maintains an overdriven state ($D_t{}^* = 1.43$). This overdriven detonation wave gradually decays to a near-CJ state ($D_t{}^* = 0.9$). In this mode, a transmitted overdriven detonation wave was obtained because of the high $D_i{}^*$. The rarefaction led to the continuous decay of the transmitted overdriven detonation wave. Because the cell width λ depends on the degree of overdrive, the decay process will increase λ. Once λ becomes so large that it lies outside the detonability limit $d > \lambda$, the detonation wave fails to be sustained. For $\phi_t = 0.94$, $\lambda \approx 60$ mm at the CJ detonation state, which is larger than the tube diameter. This is the most likely interpretation that a detonation wave is not found.

Figure 4 (right) shows the displacement-time diagram for mode A2. The transmitted overdriven detonation wave after diaphragm rupture was obtained with $D_t{}^* = 1.32$ and gradually decayed to a near-CJ state. Then an overdriven detonation wave with $D_t{}^* = 1.17$ was found after the attenuation process. Further downstream the attenuation process occurred again. This repeated process of decay, transition, and decay was regarded as an unstable detonation wave [9].

In mode B, the slightly overdriven state of $D_t{}^* = 1.07$ is obtained after the diaphragm rupture. However, a gradually acceleration took place to develop a higher degree of overdrive at $D_t{}^* = 1.25$. Then the overdriven detonation wave decayed to a near-CJ state of $D_t{}^* = 0.93$ just as for modes A1 and A2. For the acceleration process to occur instead of an attenuation process after diaphragm rupture, it is thought that a transmitted shock wave was formed. Because of the less reactive mixture in this case ($\phi_t = 0.76$), the induction time to react was too long and the detonation wave was not able to be initiated instantaneously, as discussed in section 3.2. In mode C, a sub-CJ state where $D_t{}^* = 0.77$ was observed after the diaphragm rupture. The strength of the incident overdriven detonation wave of this mode ($D_i{}^* = 1.08$) is apparently lower than the other modes. Thus, a transmitted shock wave is formed. Then the transition to an overdriven detonation wave takes place.

Among these modes, an unstable detonation wave appears in modes A2 and C, which was found in most of the tests. Mooradian and Gordon [9] noted the unstable detonation wave near the detonability limit as the overdriven detonation wave decays approaching the CJ state. Otherwise, it can be noticed that there is no detonation wave near the open end in some modes, Fig. 5. The figure shows that the survival of the detonation wave is mainly related to the mixture composition and the detonation wave propagating velocity. The critical conditions are $\phi_t \geq 0.87$ and $D_t{}^* \geq 0.82$, which are directly connected to λ. This implies that the detonability limit is the dominant factor in the propagation of the detonation wave after the process of wave transmission.

5 Conclusions

In wave transmission process, a transmitted overdriven detonation wave takes place instantaneously when a strong incident overdriven detonation wave is used. A near-CJ state of the incident wave leads to a transmitted shock wave, and then the transition to an overdriven detonation occurs downstream. Whenever the transmitted overdriven detonation wave is attained instantaneously or by transition, the process of the transmitted overdriven detonation wave decaying to a near-CJ state occurs in all tests. In most tests, there is an unstable detonation wave observed after the attenuation process approaching to a near-CJ state. This may be attributed to that the increase of the cell width in the attenuation process lies outside the detonability limits.

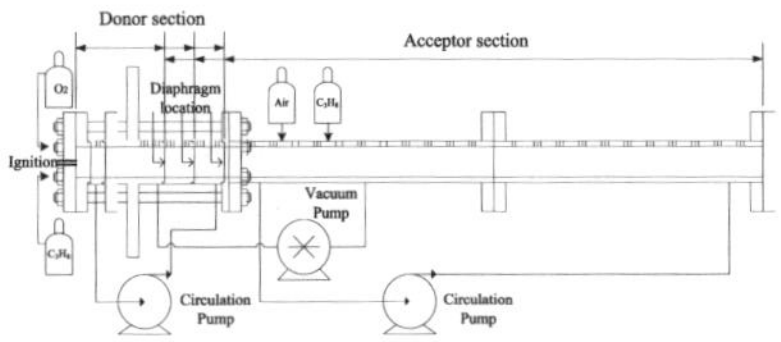

Fig. 1. Experimental facility and instrumentation

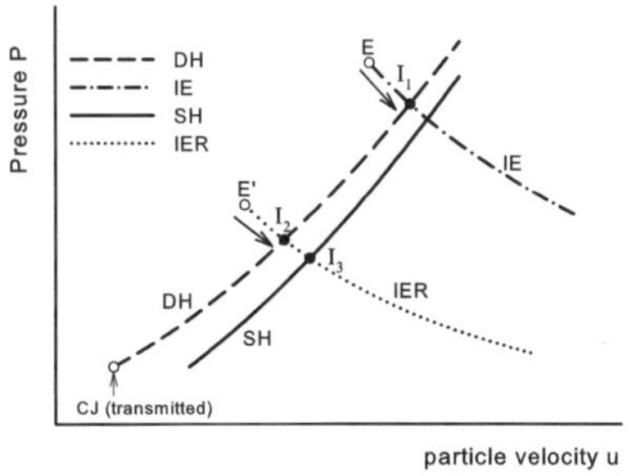

Fig. 2. *P–u* diagram; *IE*: isentropic-expansion curve passing D_i product; *IER*: *IE* curve considering rarefaction; *DH*: the locus of all possible final states for the D_t; *SH*: the locus of all possible postshock states

References

1. J.O. Sinibaldi, C.M. Brophy, C. Li, K. Kailasanath: AIAA paper 2001-3466 (2001)
2. E. Schultz, J.E. Shepherd: *Detonation Diffraction through a Mixture Gradient*, Technical Report FM00-1, GALCIT (2000)
3. D. Desbordes, A. Lannoy: Effects of a Negative Step of Fuel Concentration on Critical Diameter of Diffraction of a Detonation. In: *Dynamics of Detonations and Explosions: Detonations*, Progress in Astronautics and Aeronautics, vol 133, ed by A.L. Kuhl et al. (AIAA, 1991) pp 170-186
4. P.A. Urtiew, A.K. Oppenheim: *Proc. R. Soc. Lond. A* **295** (1966) 13-28
5. R. Knystautas, J.H. Lee, O. Peraldi, C.K. Chan: Transmission of a Flame from a Rough to a Smooth-Walled Tube. In: *Dynamics of Explosions*, Progress in Astronautics and Aeronautics, vol 106, ed by J.R. Brown et al. (AIAA, 1986) pp 37-52
6. J. Li, W.H. Lai, K. Chung, F.K. Lu: *Shock Waves* **14** (5-6) (2005) 413-420
7. M.S. Kuznetsov, V.I. Alekseev, S.B. Dorofeev, I.D. Matsukov, J.L. Boccio: *Proc. Combust. Inst.* **27** (1998) 2241-2247
8. R.E. Mitchell, R.J. Kee, SHOCK: A General Purpose Computer Code for Predicting Chemical Kinetic Behavior behind Incident and Reflected Shocks, SAND82-8205 (1982)
9. A.J. Mooradian, W.E. Gordon: *J. Chem. Phys.* **19** (9) (1951) 1166-1172

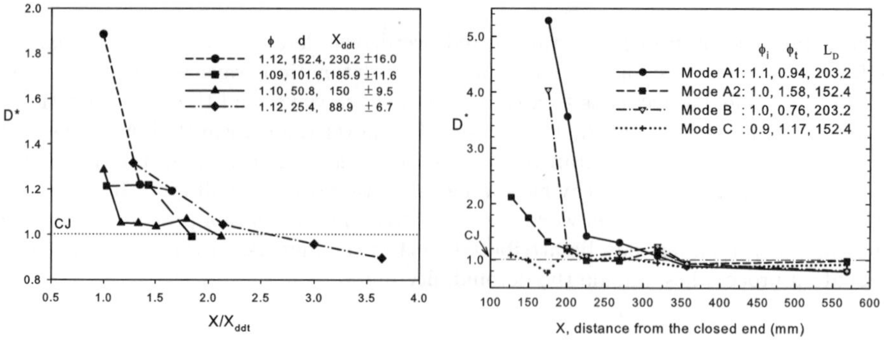

Fig. 3. Degree of overdrive with different tube diameters (left); modes of wave transmission process (right)

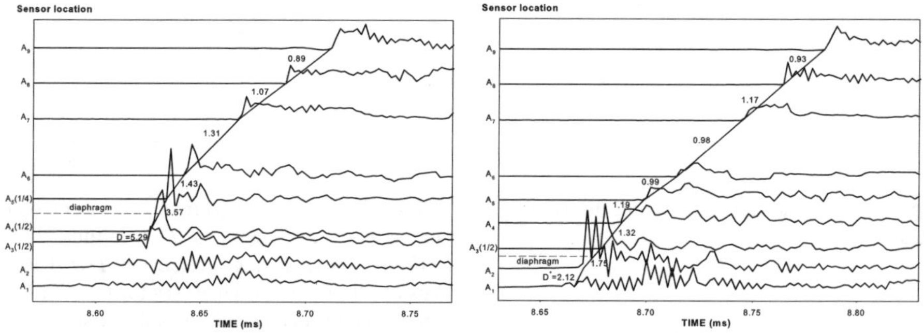

Fig. 4. The displacement-time diagram for mode A1 (left); the displacement-time diagram for mode A2 (right)

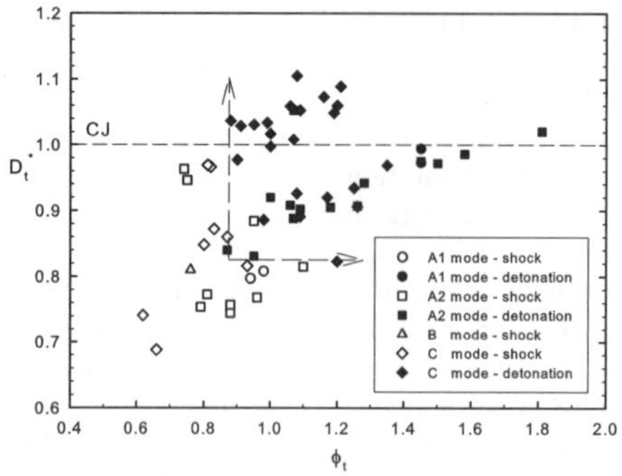

Fig. 5. The wave behavior near the open end for all tests

Flow vorticity behavior in inhomogeneous supersonic flow past shock and detonation waves

V.A. Levin and G.A. Skopina

Russian Academy of Sciences, Institute of Automation and Control Processes, 5 Radio Street, Vladivostok 690041 (Russia)

Summary. The study dwells on behavior of velocity vortex vector on surface of discontinuity in inhomogeneous supersonic flow of flammable gas past a body with formation of shock or detonation wave. In general case, free stream comes as vortex flow with a preset distribution of parameters. Expressions for the vortex vector components in special wave-based coordinate system have been found. It has been shown that in this case wave-normal component of the vortex remains continuous in going through the surface of discontinuity, and for axial-symmetric flows, a value that equals to relation of tangential vortex component to density also remains continuous, though these values individually are discontinuanced.

1 Introduction

Truesdell [1] was the first to get expression for the vortex past curved stationary shock wave at constant values of free stream. Other authors derived the same expression at a later time. In general case formulae for vector components of the vortex post arbitrary curved wave were derived by Lighthill [2] under the assumption that shock wave has infinite intensity. Formulae for vector components of the vortex post shock wave of any intensity at constant values of free stream parameters were derived in [3–5]. In the given paper the vorticity post curvilinear stationary detonation wave in vortical super sonic flow is determined.

2 Calculation of vorticity post stationary detonation wave

Let detonation wave is in stationary vortical flow of flammable gas. Detonation wave is considered as the surface of strong discontinuity where heat value Q is released in the combustion of gas mass unit. In general case Q can depend on the parameters of free stream (at Q=0 we have usual shock wave). In this case, gas motion is described by the following equation set:

$$\rho\frac{\partial v_i}{\partial x_i} + v_i\frac{\partial \rho_i}{\partial x_i} = 0, \quad \frac{\partial p}{\partial x_k} + \rho v_i\frac{\partial v_k}{\partial x_i} = 0, \quad v_i\frac{\partial}{\partial x_i}\left(\frac{p}{\rho^{\gamma}}\right) = 0. \tag{1}$$

Here v_1, v_2, v_3 - corresponding velocity components in Cartesian coordinate system (x_1, x_2, x_3), ρ - density, p - gas pressure, γ - isentropic exponent, indexes i, k take value 1,2,3, summation over i.

Relations (1) are correct in all zone of gas flow, and on the surface of discontinuity laws of conservation of mass, impulse and energy are fulfilled.

$$\rho v_\nu = \rho_0 v_\nu, \quad p + \rho v_\nu^2 = p_0 + \rho_0 v_{0\nu}^2, \quad v_\beta = v_{0\beta}, \quad v_\tau = v_{0\tau}, \\ \frac{1}{2} v_\nu^2 + \frac{\gamma}{\gamma - 1}\frac{p}{\rho} = \frac{1}{2} v_{0\nu}^2 + \frac{\gamma}{\gamma - 1}\frac{p_0}{\rho_0} + Q. \tag{2}$$

Here the values with index "0" denote the parameters of gas in front of wave, and values without indexes - post wave, $v_\nu = v_i \nu_i$ - normal component of velocity, $v_\beta = v_i \beta_i$ and $v_\tau = v_i \tau_i$ - tangent component of velocity, ν_i,β_i,τ_i - components of normal vector and tangent vectors to the surface of discontinuity.

To find the components of vortex vector $2\boldsymbol{\omega} = rot\boldsymbol{v}$ let us introduce on the surface of discontinuity the curvilinear orthogonal coordinate system connected with wave.

Let Σ be the surface of detonation wave determined by the equation $x_i = x_i \left(y^1, y^2\right)$, where y^1, y^2 - curvilinear coordinates on the surface, $x_1,\ x_2,\ x_3$ - coordinates in Cartesian system.

The vectors with coordinates $\partial x_i / \partial y^\alpha = x_{i,\alpha}$ (α=1,2) are tangent to the surface vectors. Here comma denotes derivative.

Let in every point of surface Σ be possible to build unit vector of normal $\boldsymbol{\nu}$, to this surface directed towards flow post front. Let us introduce such local Cartesian coordinate system (q_1, q_2, q_3) that axis $q_3$is directed in the line of normal to Σ, and axes q_1, $q_2$coincide with the main directions in tangent surface. Let Λ_1, Λ_2 be the lines of intersection of Σ with the surfaces q_1=0 and q_2=0, correspondingly, $\boldsymbol{\beta}$ and $\boldsymbol{\tau}$ - unit vectors tangent to these lines, l,s, - lengths of lines' arcs Λ_1, Λ_2, counted in the direction of $\boldsymbol{\beta}$, $\boldsymbol{\tau}$. Thus, we can take parameters l and s as parameters y^1, y^2, correspondingly. Then the surface will be determined by equation $x_i = x_i\,(l, s)$.

Vectors $\boldsymbol{\beta}$, $\boldsymbol{\tau}$, $\boldsymbol{\nu}$ are the unit vectors along axes q_1, q_2, q_3, and v_β, $\boldsymbol{v}_\tau$, v_ν corresponding components of velocity $\boldsymbol{v}$. Vector tangents are determined as $\boldsymbol{\beta} = r_{,l}$, $\boldsymbol{\tau} = r_{,s}$, $\mathbf{r} = \{x_1, x_2, x_3\}$. Vector of normal $\boldsymbol{\nu}$ satisfies the following equalities

$$\nu_i \nu_i = 1, \quad x_{i,\alpha} \nu_i = 0. \tag{3}$$

Hereinafter it is conventional to sum over recurring indexes, the Latin indexes take value 1,2,3, Greek indexes take value l and s.

From differential geometry it is known that

$$\nu_{i,\alpha} = -\kappa_\alpha x_{i,\alpha}, \quad \alpha = (l, s)\,, \tag{4}$$

where k_l, k_s are the principal curvatures of surface Σ. Curvature κ_α will be positive, if curve Λ_α is turned by its convexity in the line of free stream.

Let some function $f\,(x_i)$ be determined in coordinate system x_i. Then on surface Σ the function $f\,(x_i\,(l, s)) = f\,(l, s)$ is determined.

The partial derivatives of function f with respect to coordinates of surface x_i are connected with the derivatives with respect to curvilinear coordinates l and s by relations

$$f_{,i} = f_{,n}\nu_i + f_{,\alpha} x_{i,\alpha} = f_{,n}\nu_i + f_{,l}\beta_i + f_{,s}\tau_i, \tag{5}$$

where $f_{,n} = f_{,i}\nu_i$ - derivative with respect to normal.

If we go on from the derivatives with respect to Cartesian coordinates to derivatives with respect to surface ones using relation (5), and take into account that tangent components of velocity remain unchanged in going through detonation wave, then, using relations (1)-(5), let us calculate the vorticity components in coordinate system connected with the wave:

before wave

$$\omega_{0\nu} = (v_{0\tau,l} - v_{0\beta,s})/2,$$
$$2\omega_{0\beta} = p_{0,s}/\rho_0 v_{0\nu} + (v_{0\beta}v_{0\tau,l} + v_{0\tau}v_{0\tau,s})/v_{0\nu} + v_{0\nu,s},$$
$$2\omega_{0\tau} = -p_{0,l}/\rho_0 v_{0\nu} - (v_{0\beta}v_{0\beta,l} + v_{0\tau}v_{0\beta,s})/v_{0\nu} - v_{0\nu,l};$$

past wave

$$\omega_\nu = (v_{0\tau,l} - v_{0\beta,s})/2,$$
$$\omega_\beta = \frac{\rho}{\rho_0}\omega_{0\beta} - \frac{(\rho_0-\rho)^2}{2\rho\rho_0}v_{0\nu,s} + \left(1-\frac{\rho_0}{\rho}\right)\frac{v_{0\nu}\rho_{0,s}}{2\rho_0} + \left(1-\frac{\rho}{\rho_0}\right)\frac{p_{0,s}}{2v_{0\nu}\rho_0},$$
$$\omega_\tau = \frac{\rho}{\rho_0}\omega_{0\tau} + \frac{(\rho_0-\rho)^2}{2\rho\rho_0}v_{0\nu,l} - \left(1-\frac{\rho_0}{\rho}\right)\frac{v_{0\nu}\rho_{0,l}}{2\rho_0} - \left(1-\frac{\rho}{\rho_0}\right)\frac{p_{0,l}}{2v_{0\nu}\rho_0}.$$

Thus, normal component of vorticity remains continuous function $\omega_\nu = \omega_{0\nu}$ in going through surface of discontinuity, and vortex component past detonation wave in either of the two directions of the lines of principal curvatures depends on initial vorticity in the same direction, gas parameter derivatives in perpendicular direction, gas parameters and relation of densities ρ/ρ_0.

In the case of axial-symmetric flow, when derivative along one of the principal directions is equal to zero $(\partial/\partial s)$, for vortex component in perpendicular direction a law of conservation of value $\omega_\beta/\rho = \omega_{0\beta}/\rho_0$ is fulfilled, as for nonstationary flows [6].

3 Vorticity past stationary detonation wave for flows with constant parameters.

If gas parameters before wave are constant, then vorticity before wave also is equal to zero, correspondingly normal component of vortex vector will be equal to zero, and tangential vortex components will be depend on the product of curvature and tangential velocity component in perpendicular direction and multiplier depending on density ratio, which in its turn is a function of parameters, determining medium state, and value of heat added to gas unit mass

$$\omega_\beta = \frac{1}{2}\kappa_s v_{0\tau}\left(1-\frac{\rho}{\rho_0}\right)^2\Big/\frac{\rho}{\rho_0}, \quad \omega_\tau = -\frac{1}{2}\kappa_l v_{0\beta}\left(1-\frac{\rho}{\rho_0}\right)^2\Big/\frac{\rho}{\rho_0}.$$

The obtained results fully coincide with the formulae for calculating vortex value past curvilinear stationary shock wave at constant values of gas initial parameters [1–5]. It follows from these formulae that with Mach number growing, vorticity grows as well, and under other factors being the same the value of vortex past detonation wave will be less than one past shock wave, because value ρ/ρ_0 decreases as the energy-release in wave increases.

Let us consider the plane axial-symmetric flow past detonation wave with the constant gas parameters before detonation wave. In this case only normal to velocity vector component of vortex is different from zero

$$\omega_\tau = -\frac{1}{2}\kappa u_0 cos\alpha\left(1-\frac{\rho}{\rho_0}\right)^2\Big/\frac{\rho}{\rho_0},$$

where u_0 is the velocity of undisturbed flow, α is slope angle of discontinuity surface with regard to direction of undisturbed flow, κ is curvature of discontinuity surface. Curvature of discontinuity surface is a decreasing function, and will be equal to zero at the point of transition of detonation wave to CJ one for axial-symmetric flows, or it will tend to zero for the flows with the plane waves [7]. At the point of transition of over-compressed to CJ regime the wave has a contact of the third order.

Density ratio ρ/ρ_0 is determined from the laws of conservation on the surface of discontinuity (2)

$$\frac{\rho}{\rho_0} = \frac{1+\gamma M_{0n}^2 + \sqrt{\left(M_{0n}^2-1\right)^2 - qM_{0n}^2}}{\left(\gamma-1\right)M_{0n}^2 + q/\left(\gamma+1\right)+2}. \tag{6}$$

Here $M_{0n} = M_0 sin\alpha$, $M_0 = u_0/a_0$ is Mach number, a_0 is speed of sound, $q = 2Q\left(\gamma^2-1\right)/a_0^2$ is immeasurable value of heat added to gas unit mass, for shock wave q=0.

In ratio (6) subradical expression must be non-negative that in its turn gives the limits within which the streamline $\alpha_J \le \alpha \le \pi/2$ is possible. Here α_J is slope angle of tangent to the wave at the point of transition to CJ regime.

$$\alpha_J = arcsin\left(\frac{1}{M_0}\sqrt{\frac{2+q+\sqrt{q\left(4+q\right)}}{2}}\right). \tag{7}$$

Ratio (7) in its turn permits to find the restrictions for parameters of problem: $0 \le q \le q_*$, where $q_* = \left(M_0 - \frac{1}{M_0}\right)^2$, $M_{0*} \le M_0 \le \infty$, where $M_{0*} = \sqrt{\left(2+q+\sqrt{q\left(4+q\right)}\right)/2}$.

Tangent of slope angle of tangent to wave is equal

$$tg\alpha_J = \sqrt{\frac{2+q+\sqrt{q\left(4+q\right)}}{2\left(M_0^2-1\right)-q-\sqrt{q\left(4+q\right)}}}.$$

For the shock wave $tg\alpha_J = \sqrt{\frac{1}{M_0^2-1}}$. At critical values of problem parameters $q = q_*$ or $M_0 = M_{0*}$, $tg\alpha_J = \infty$, $\alpha = \pi/2$. With increasing of M_0 the slope angle of tangent of wave decreases, energy-release enlarges the slope angle of tangent.

Let us consider flow past the section of detonation wave preceding the onset of CJ regime when the gas parameters are constant. As an example let's assume that when flowing around some solid there is a wave of type $R^2 = \frac{1}{a-x} - b$, which at the point x_J transfers to CJ regime. Function R is equal to zero at the point $x_0 = a - 1/b$, beyond point $x = x_J$ the wave is linear.

The conditions at the point of transition to CJ regime are fulfilled at the following values of function parameters

$$a = 3tg^{-\frac{2}{3}}\alpha_J\Big/2, \quad b = 3tg^{\frac{2}{3}}\alpha_J\Big/4, \quad x_J = tg^{-\frac{2}{3}}\alpha_J\Big/2.$$

Then over all region of flow the wave can be presented as the following function

$$R\left(x\right) = \begin{cases} \pm\frac{tg^{\frac{1}{3}}\alpha_J}{2}\sqrt{\frac{6tg^{\frac{2}{3}}\alpha_J x-1}{3-2tg^{\frac{2}{3}}\alpha_J x}}, & x_0 \le x \le x_J \\ \pm tg\alpha_J x, & x \ge x_J \end{cases}$$

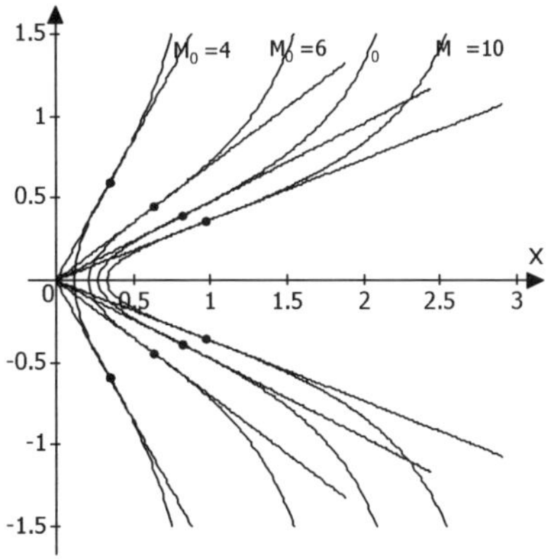

Fig. 1. Detonation wave $R(x)$ and tangent to wave at different values of Mach number for q=10, γ=1, 4.

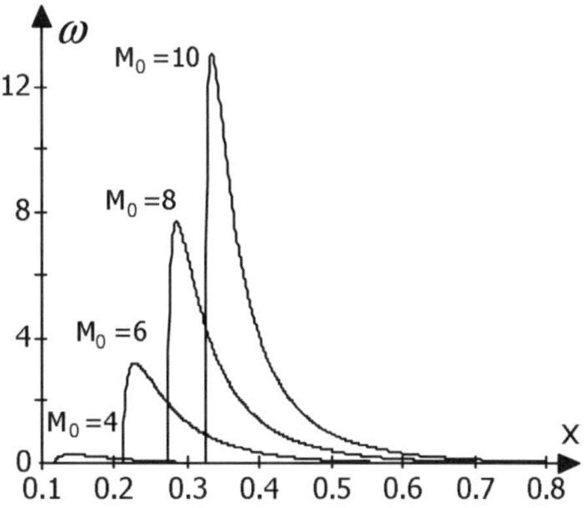

Fig. 2. Diagram of vorticity ω=$-2\omega_\tau/u_0$ at different values of Mach number for q=10, γ=1, 4.

Fig. 1 shows diagram of detonation wave $R(x)$ and tangent to wave at different values of Mach number. The points of wave transition to CJ regime are shown by the points. Beyond these points the wave is a straight line. Fig. 1 shows that the greater is Mach number, the less is slope angle of tangent to wave and the closer is wave to the surface of solid.

Figure 2 shows the diagram of vorticity $\omega = -2\omega_\tau/u_0$ at different values of Mach number.

Vorticity is equal to zero at the point x_0, where the wave is linear ($\alpha = \pi/2$), and at the point of transition to CJ regime x_J. With Mach number growth, vorticity also increases. Energy-release significantly reduces vorticity beyond detonation wave.

4 Conclusion

Thus, the study was made into dependence of value of vorticity flow immediately past stationary detonation wave. The formulae for the components of vortex vector in the attendant coordinate system have been obtained. It has been shown that in this case the normal vortex component remains continuous function in going through the surface of discontinuity.

Also it has been shown that only for axial-symmetric flows the law of conservation of value ω_β/ρ is fulfilled at any distributions of gas parameters in free stream, whether it be shock or detonation wave.

References

1. Truesdell C.: On curved shocks in steady plane flow of an ideal fluid. J. Aeronaut Sci. 1952, No. 19. Paper 826-828.
2. Lighthill M.: Dynamics of Dissociating Gas, Rocket Technology Review. 1957. No. 6. Paper 41-60.

3. Hayes W.D.: The vorticity jump across a gasdynamic discontinuities. J. Fluid Mech. 1957. No. 2. Paper 595-600.
4. Maikapar G.I.: Vortexes Past Head Shock Wave, Fluid and Gas Mechanics. 1968. No. 4. Paper 162-165.
5. Rusanov V.V.: Derivatives of Gas Dynamic Functions past Curved Shock Wave. Moscow, 1973. (Prepr. / IAM USSR AS; No. 18).
6. Levin V.A., Skopina G.A.: Propagation of Detonation Wave in Rotated Gas Flow. Journal of Applied Mathematics and Theoretical Physics. 2004. Vol. 45, No. 4. Paper 3-6.
7. Levin V.A., Cherny G.G.: Asymptomatic Laws of Detonation Waves Behavior, Applied Mathematics and Mechanics. 1967. Vol. 31, No. 3. Paper 393-405.

Ground reflection interaction with height-of-burst metalized explosions

R.C. Ripley[1], L. Donahue[1], T.E. Dunbar[1], S.B. Murray[2], C.J. Anderson[2], F. Zhang[2], and D.V. Ritzel[3]

[1] *Martec Limited, 1888 Brunswick Street, Suite 400, Halifax, Nova Scotia, B3J 3J8, Canada*
[2] *DRDC Suffield, P.O. Box 4000, Station Main, Medicine Hat, Alberta, T1A 8K6, Canada*
[3] *Dyn-FX Consulting Limited, 19 Laird Avenue N., Amherstburg, Ontario, N9V 2T5, Canada*

Summary. Near-field impulse from metalized explosives is dominated by afterburning energy release and is influenced by a small confinement effect that is caused by ground reflected shocks in height-of-burst (HOB) scenarios. The ground reflected shock interacts with the fireball, reheating the detonation products and distorting the fireball interface which promotes afterburning chemistry. The air behind the primary reflected shock is doubly shock heated, which improves the local heating rate of the dispersed metal particles to their ignition temperature. Shock reverberation between the ground and expanding fireball, interaction with the emerging secondary shock, arrival of the incident particle front and localized enhanced chemistry have a compounding effect and lead to increased loading observed on the ground. These effects are of particular interest for structural response resulting from near-field explosions.

1 Height-of-burst scenarios

The experimental configuration for the height-of-burst scenario (Figure 1), including diagnostic locations and mixture formulations is described in the companion paper by Murray et al. [1]. Two in-service weapons-grade thermobaric explosive (TBX) mixtures and two high-metal-content improvised explosive formulations (IEF) were modeled and compared to experiment as well as high-explosive C4 (91% RDX) reference charges. Nominal 20-kg spherical charges cased in thin polyethylene at 3 m HOB are the focus of this paper. All metalized explosives have a nominal 10% C4 central booster by mass. Since the explosive mass and compositions vary in addition to the particle morphology and metal content, direct comparison was not always possible, however phenomenology is discussed. Numerical predictions were compared with high-quality experimental data to augment experimental observations and help elucidate flow physics.

2 Numerical explosion afterburning model

Numerical modeling of metalized explosives remains a challenge in the simulation community due to the dense chemically-reacting multiphase flow, with the degree of confinement increasing the complexity of the simulations. Although even the best models may need to be adjusted based on experimental data, they offer additional insight into the flow field beyond the capability of experimental diagnostics, such as the evolution of afterburning reactivity and the internal temperature distribution. Unlike simple afterburning models, e.g., [2], the present work employs chemically reacting multiphase flow models.

The Chinook code, a parallel 3D unstructured mesh CFD code developed by Martec and DRDC, was used to model the NLT2 trials in [1]. Details of the model are given

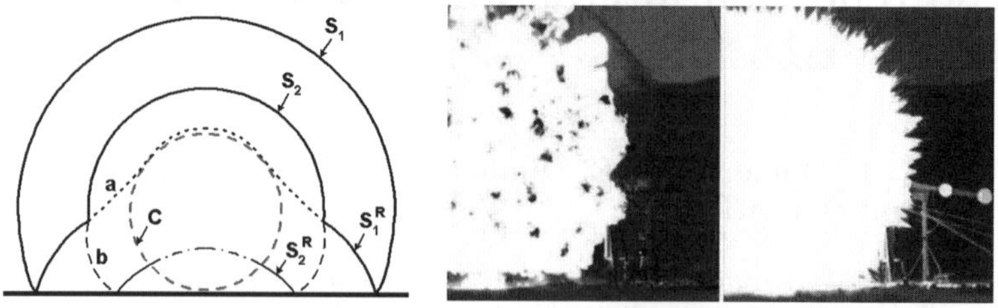

Fig. 1. Illustration of HOB scenario (left to right: schematic, C4 trial, IEF1 trial). Figure nomenclature: S_1, incident primary shock; S_2, incident secondary shock; S_1^R, reflected primary shock; S_2^R, reflected secondary shock; C, contact surface (fireball); a, acceleration of S_1^R through shock-heated zone and fireball; and, b, acceleration of S_2 through double-shock-heated zone.

in Ripley et al. [3]. The multiphase fluid-dynamic equations governing the flow include conservation of mass, momentum, energy, material concentrations, and particle number density. Chinook features a second-order HLLC approximate Riemann solver for both the fluid and particle phases, each of which has independent velocity and temperature fields. Gas/particle interactions are simulated using empirical correlations for drag, convective heat transfer and mass transfer (metal particle combustion). Equations of state (EOS) for each material close the system of equations including: shock-Hugoniot EOS for condensed reactants; JWL for gaseous detonation products; ideal gas law for surrounding atmosphere and air blast; and, a phenomenological EOS for solid particles. High resolution meshes are used to resolve the details of the initiation region (size and location), booster configuration, charge shape, detonation wave propagation through the heterogeneous mixture, expansion and afterburning of gaseous detonation products (fireball), transmission of the blast wave into ambient air, and dispersal, heating and afterburning of metal particle additives with oxidizing gas species. Detonation of the condensed explosive (powdered explosive, solid oxidizer granules and/or liquid monopropellant) is modeled using a global Arrhenius reaction rate. Explosive products fitting parameters and gas species from the booster and bulk explosive are modeled using Cheetah chemical equilibrium for the CJ state. Afterburning reactions are simulated for C, CO, H_2 and C_2H_2 with O_2 from the surrounding air. Metal particle reaction with O_2, CO, CO_2, and H_2O oxidising gas species are computed according to a diffusion-limited d^2 rate law.

3 Near-field TBX phenomena

Fundamental TBX physics are covered in detail in the companion paper [4] and in [5–7]. These are: delayed incident time of arrival, lower primary peak overpressure in the near field, noisy shock front due to particle and fireball jetting, and an advanced secondary shock that diminishes the negative phase and increases positive-phase impulse.

In general, the maximum TBX fireball size is larger than for HE and occurs later in time. Figure 2 compares numerical results for C4 and IEF fireballs. For the C4 reference charge, the maximum fireball radius was 2.41 m, while the IEF fireball radius surpassed that and is still expanding when the reflected primary shock returns and alters the fireball flow field. The C4 detonation products are cold ($\sim$ 500 K) except at the fireball/air interface where gas-phase afterburning occurs. The IEF fireball is hotter due to late-time afterburning of the detonation products and metal particles which accelerates the

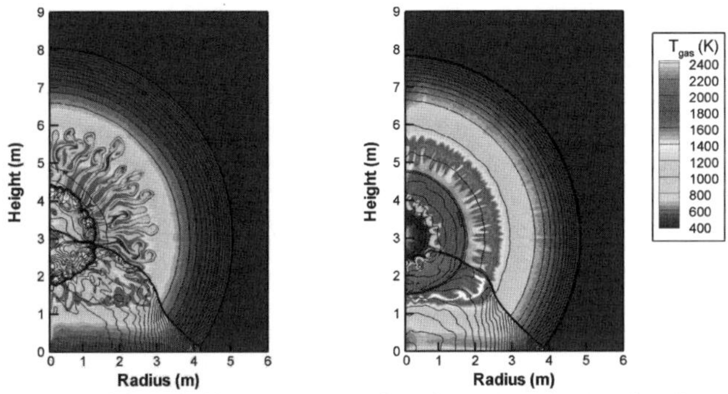

Fig. 2. Comparison of flow-field temperature distribution at 4 ms using 5-mm resolution mesh. C4 reference explosive (left) and IEF1 (right).

ground reflection through the fireball, while the cold booster gasses remain visible at the HOB location. The IEF fireball at maximum expansion is larger and less dense, having expanded at a higher velocity. Measurement of the fireball speed showed that within the first 1.2 ms the C4 fireball expands faster, after which the IEF1 fireball expands more quickly (traveling up to 10% faster at 2 ms). This is significant as it translates into a piston effect that the target observes as impulse. Experimental video exaggerates the IEF fireball size due to reactive particle dispersal outside the detonation products. The Rayleigh-Taylor (RT) instability seen in C4 is less prominent in the IEF explosion due to the smaller density discontinuity, owing to the hot detonation products, and diffusive effects as the particles cross the contact interface. This is consistent with the simple afterburning results presented in [7]. Rather than producing billowing RT instabilities, the IEF contact interface forms coherent thermal streams (Figure 2).

Figure 3 illustrates ground reflection results at gauge P2 (located on the ground at 1.375 m radius). This gauge illustrates the explosive characteristics and relative performance of the formulations (note that the actual charge masses vary from 18.5 kg to 21.3 kg). Delayed S_1 time of arrival (TOA) and lower peak pressure are due to the reduced explosive fraction of the charge mass and the momentum transfer required as the detonation accelerates the metal particles. The TBX formulations feature high VOD explosives with small metal flakes that react promptly in the detonation products.

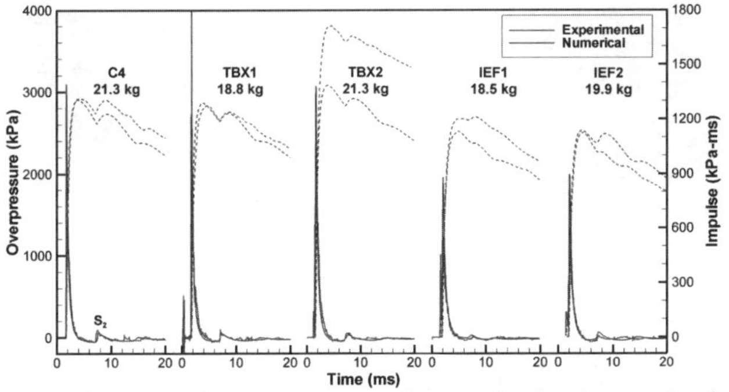

Fig. 3. Comparison of numerical and experimental ground reflection results (gauge P2) for all explosives. Pressure in solid lines; impulse in broken lines. The TOA is 1.81 ms for C4, 1.83 ms for TBX1, 1.69 ms for TBX2, 2.05 ms for IEF1 and 1.80 for IEF2.

The secondary shock (S_2) forms after the collapse and reflection of the recompression shock and must travel through the detonation products, fireball interface, dispersed particle cloud and upwards-traveling ground-reflected shock. Accurate modeling of the secondary shock is therefore dependent on correct EOSs, afterburning reactions and explosive modeling parameters. Advancement of the secondary shock increases impulse by diminishing the negative phase in both TBX and IEF, for example S_2 forms at 2.65 ms for IEF1 compared to 2.95 ms for C4. Figure 3 demonstrates good agreement in both primary and secondary shock arrival times indicating correct physical and numerical models.

4 Multiphase explosion flow field

Even simple models, e.g., [2], can demonstrate non-ideal waveforms in the free field, however near-field and semi-confined explosions require multiphase fluid dynamic solutions [3,5] to capture the thermochemical behaviour. Prediction of peak impulse depends on the positive phase of the pressure history and is particularly sensitive to the secondary shock arrival time. Detonation products afterburning and metal particle combustion are dependent on mixing and spatial distribution of fuels/oxidisers, which significantly affects the secondary shock speed. In the HOB scenario, the ground-reflected shock influences both the distribution and temperature of the reacting flow field.

Approximately 0.5 ms after the incident shock arrival at the ground, the reflected shock interacts with the expanding fireball. For IEF1, the particle front reaches the ground ahead of the incident primary shock. Figure 4 illustrates the spatial distribution of particles relative to the IEF1 fireball. At 2 ms, the dispersed particle front has passed the shock, although considerable particles remain within the fireball and shock-heated air. At 4 ms, the particles reside almost entirely within the shock-heated air, where the drag forces cause the particles to lag the primary shock and aerobic particle combustion occurs. Because of the initially monodisperse particle size distribution and lack of a particle agglomeration model, the particle cloud dispersal is spherical compared to the large-scale jetting observed in the experiment.

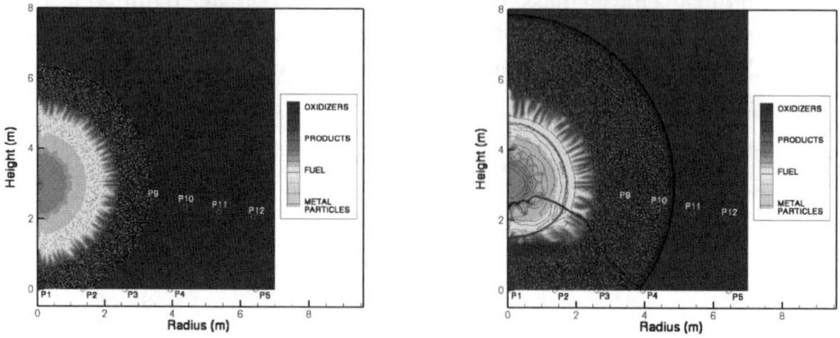

Fig. 4. Composite illustration of IEF1 chemistry at 2 ms (left) and 4 ms (right). Spatial distribution of afterburning gas species (fuels), metal particles and mixing with air oxidizer.

5 Metal particle afterburning

Metal particle reaction depends on spatial distribution, particle temperature and availability of oxidizing gas species. Figure 5 illustrates the particle concentration relative to the shock structure. At this time after the maximum fireball expansion, the particle front lags the incident shock and Mach stem. The highest particle density is concentrated in a layer on the ground, where considerable energy release may increase the ground impulse, depending on the particle temperature and oxidiser availability. Close behind the primary shock, the particle concentration is banded and will contribute to blast strength while the particle ignition temperature is maintained. Afterburning energy release coupled behind the shock was shown by Zarei et al. [2] to provide the optimal blast impulse.

The ground-reflected shock forms a Mach stem and triple point which pass below the elevated gauge array. Figure 5 illustrates pressure histories at five gauge locations ranging from 3.45 m (P9) to 8.45 m (P14) and shows the primary shock (S_1), ground reflected shock (S_1^R), secondary shock (S_2), reflected secondary shock (S_2^R), and tertiary (S_3) and quaternary (S_4) shocks. Agreement is excellent for all flow features up to the tertiary shock. Shockwaves S_3 and S_4 are resolved in the numerical results but are noticeably late as compared to experiment. The incident shock arrival time measured using high-speed video against a zebra-board backdrop showed agreement to within ±0.1 ms for all formulations simulated. Triple-point trajectories were also compared and match triple-point height rankings (see Murray et al. [1]). Peak impulse was compared to experimental results from two orthogonal elevated arrays and far-field ground gauges with typical agreement within the error band of the experimental results.

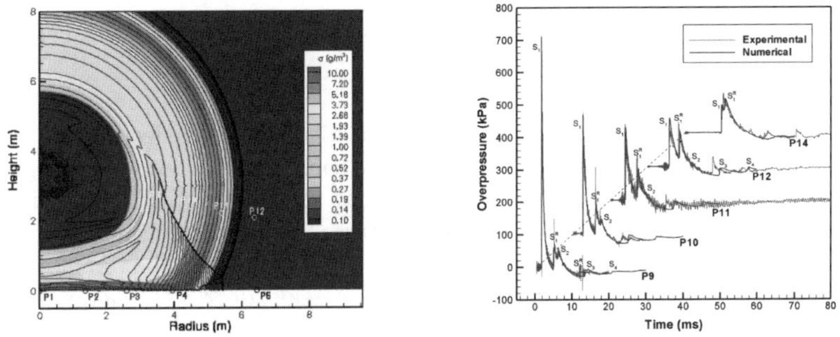

Fig. 5. Particle concentration at 6 ms (left) and comparison of numerical and experimental results along elevated gauge array (right).

6 Heat release and energy partitioning

The small particle size used in the TBX formulations leads to rapid heating to ignition temperature and early-time reaction almost entirely in the detonation products. Reaction of 95% of the initial metal mass occured after 0.11 ms for TBX1 and 0.81 ms for TBX2, with residual unreacted metal less than 1% in both TBXs. In contrast, the large granule size in IEF1 results in a slower heating time and reaction rate allowing substantial aerobic combustion in the shocked air. Expansion cooling occurs before the IEF particles react completely leaving residual metal amounts of 31.2% in IEF1 and 13.1 % in IEF2. For IEF1, the metal particle consumption rate measured using volumetric integration showed that the maximum reaction occurred at about 3 ms, corresponding to substantial aerobic combustion and the triple-point formation time. At 4 ms, particle quenching begins 1 m behind the shock, except in the ground-reflection region where reaction continues until 6 ms. Since substantial particle reaction occurred within the fireball expansion time (< 4 ms), the energy release timescale contributes to the far-field impulse [6].

Table 1 divides the energy release into detonation energy, gas-phase afterburning, metal reactions with detonation products (anaerobic) and metal reactions with air (aerobic). The TBXs showed dominant anaerobic metal reactions due to the rapid heating time and availability of oxidizers. The IEF1 metal particles were slowest to heat, reaching ignition temperature mostly in the air. The particle size effect and afterburning timescale influence on energy partitioning indicates that a cube-root-type mass scaling is not appropriate for these metalized explosives. This has been more extensively studied by Frost et al. [7] and a multi-energy scaling has further been proposed by Leadbetter et al. [6].

Table 1. Energy partitioning (explosives listed in order of increasing metal content)

Explosive Formulation	Detonation Energy (%)	Gas-Phase Afterburning (%)	Anaerobic Metal Reactions (%)	Aerobic Metal Reactions (%)	Total Energy (MJ/kg)
C4	78.4	21.6	0	0	6.934
TBX1	49.8	19.0	22.4	8.8	8.080
TBX2	34.3	9.8	55.3	0.6	9.567
IEF1	23.3	3.6	18.9	54.2	8.876
IEF2	11.3	7.2	45.8	36.9	12.26

7 Conclusion

Understanding the metalized explosion interaction with a single reflecting surface in the HOB scenario is a prerequisite to interpreting more complex situations involving multiple reflecting surfaces and high confinement. Given that the TBX and IEF explosive classes performed as well or better than the C4 reference charge, afterburning of detonation products and metal particle additives contributed substantially to the impulse measured in the near field. Numerical modeling allowed the timescale, degree of reaction and spatial distribution of afterburning components to be studied. Mixing and heating caused by the ground-reflected shock promotes afterburning, and dispersal of larger particles into the shock-heated air contributes to near-field impulse and blast pressure.

References

1. Murray, S.B., Anderson, C.J., Gerrard, K.B., Smithson, T., Williams, K. and Ritzel, D.V.: Overview of the 2005 Northern Lights Trials, Proc. ISSW26, Gottingen, Germany, 2007.
2. Zarei, Z., Frost, D.L., Donahue, L., Whitehouse, D.: Simplified modeling of non-ideal blast waves from metallized heterogeneous explosives, 20th ICDERS, Montreal, Canada, 2005.
3. Ripley, R.C., Donahue, L., Dunbar, T.E., Zhang, F.: Explosion performance of aluminized TNT in a chamber, Proc. 19th MABS, Calgary, Canada, 2006.
4. Ritzel, D.V., Ripley, R.C., Murray, S.B., Anderson, C.J.: Near-field blast phenomenology of thermobaric explosions, Proc. ISSW26, Gottingen, Germany, 2007.
5. Zhang, F., Yoshinaka, A., Anderson, C.J., Ripley, R.C.: Confined heterogeneous blast, Proc. 19th Military Aspects of Blast and Shock, Calgary, Canada, 2006.
6. Leadbetter, J., Ripley, R.C., Zhang, F., Frost, D.L.: Multiple energy scaling of blast waves from heterogeneous explosives, Proc. 21st ICDERS, Poitiers, France, 2007.
7. Frost, D.L., Zhang, F.: The nature of heterogeneous blast explosives, Proc. 19th Military Aspects of Blast and Shock, Calgary, Canada, 2006.

High-fidelity numerical study on the on-set condition of oblique detonation wave cell structures

J.-Y. Choi[1], **E.J.-R. Shin**[1], **D.-R. Cho**[1] **and I.-S. Jeung**[2]

[1] *Department of Aerospace Engineering, Pusan National Univesity, Busan, 609-735, (Korea)*
[2] *Department of Aerospace Engineering, Seoul National Univesity, Seoul, 151-742, (Korea)*

Summary. A comprehensive numerical study was carried out to identify the on-set condition of the cell structures of oblique detonation waves (ODWs). Mach 7 incoming flow was considered with all other flow variables were fixed except the flow turning angles varying from 35 ° to 38 ° . For a given flow conditions theoretical maximum turning angle is 38.2 ° where the oblique detonation wave may be stabilized. The effects of grid resolution were tested using grids from 500×250 to 4,005×1,800. The numerical smoked foil records exhibits the detonation cell structures with dual triple points running opposite directions for the 36 ° to 38 ° turning angles. As the turning angle get closer to the maximum angle the cell structures gets finer and the oscillatory behavior of the primary triple point was observed. The thermal occlusion behind the oblique detonation wave was observed for the 38 ° turning angle.

1 Introduction

Oblique detonation waves (ODWs) stabilized over inclined walls have been considered as a promising combustion means for hypersonic propulsion systems such as ODW engines and ram accelerators. A number of studies were carried out to examine the fundamental characteristics of an ODW and its implementation for propulsion systems.[1-14] Among the various issues of ODW studies, the on-set condition self sustaining ODW was one of the key issues, and the presence of cell structure was considered as a proof of that condition. However, little has been known about the ODW cell structure either by experimentally or by numerically. In numerical aspect, the great difference in the length scales of the chemical induction behind the oblique shock wave and the ODW was the major obstacle for resolving that problem. Choi et al. [15] captured the unsteadiness ODW frontal structures unveiling the various source instability mechanisms from pressure wave interactions generated from vortex dynamics originating from the primary triple points. They could capture the detailed wave structures with help of the systematic numerical approach of grid refinement study and chemical induction length estimations behind oblique shock wave(OSW) and ODW. They showed that the ODW instability has a strong dependency on the chemical the activation energy in same way to the normal detonation waves. However, only single-sided triple point was found from their study since they did the studies only for a fixed flow parameters except the activation energy. Recently, Daimon et al. [16] did a variety numerical study of ODW structures over a hemi-spherical blunt body by changing flow conditions. They showed the cellular structures of ODW wave front and discussed about the flow structures about the ODW detonation cells. They also discussed about the onset condition ODW cell structures from the parametric studies on flow speed, body size and initial pressures. However, it is considered that they failed to identify the onset condition of oblique detonation waves since they could not consider the fundamental thermo-fluidic parameters. The focus of present study is to identify the on-set condition of ODW by resolving the ODW cell structures numerically. Since there were various parameters affecting the ODW characteristics, careful attention should be give to isolate the effect of the thermo-chemical parameters. In the present study a comprehensive numerical study was carried out to investigate the unsteady cell structures of oblique detonation waves (ODWs) for a fixed Mach 7 incoming flow as an extension of the previous studies [15].

2 Theoretical formulation and numerical method

Since the grid requirement for the present study would be significant, a simplest possible formulation is used in this study. Euler equations for compressible inviscid flows and the conservation equation of reaction progress variable are summarized as a following vector form in a two-dimensional coordinate.

$$\frac{\partial}{\partial t}\begin{bmatrix}\rho\\ \rho u\\ \rho v\\ \rho e\\ \rho Z\end{bmatrix}+\frac{\partial}{\partial x}\begin{bmatrix}\rho u\\ \rho u^2+p\\ \rho uv\\ (\rho e+p)u\\ \rho Zu\end{bmatrix}+\frac{\partial}{\partial y}\begin{bmatrix}\rho v\\ \rho uv\\ \rho v^2+p\\ (\rho e+p)v\\ \rho Zv\end{bmatrix}=\begin{bmatrix}0\\ 0\\ 0\\ 0\\ \rho w\end{bmatrix} \tag{1}$$

where, pressure is defined as,

$$p=(\gamma-1)\,\rho\left\{e-\frac{1}{2}(u^2+v^2)+ZQ\right\} \tag{2}$$

Here, Z is reaction progress variable that simulates the product mass fraction and varies from 0 to 1. Q is the dimensionless heat addition by combustion. As a combustion mechanism, one-step Arrhenius reaction model is used to simulate the various regimes of detonation phenomena without the complexity and large computing time for dealing with many reaction steps and detailed properties of reacting species. Thus, the reaction rate in Eq. 1 that depends on mixture concentration is defined as follows.

$$w=(1-Z)k\exp(-E\rho/p) \tag{3}$$

Here, k is reaction constant and E is activation energy. The fluid dynamics equations are discretized by finite volume formulation, and numerical fluxes are calculated by Roe's approximate Riemann solver with interpolated primitives variables by third-order accurate MUSCL-type TVD scheme. The discretized equations were integrated in time by 4th order accurate 4-stage Runge-Kutta scheme (RK4). Since the details about the theoretical formulation and numerical algorithms for the compressible reactive flows has been discussed thoroughly in the previous studies, [17, 18] those will not be discussed further.

3 Thermo-chemical parameters for the morphology of ODW cell structures

Among the various thermo-fluidic parameters for the modeling of ODW, heat release and specific heat ratio is determined by the mixture composition and there is no clue of evidence that they have direct influence on the existence of cell structures. The activation energy is shown to have a strong influence of the degree of instability, but only a single-sided structures were found from the parametric studies of the activation energies. [15] Therefore the only variables that can affect the ODW structures is the flow turning angles regarding from the basic analysis on the morphology of ODW, [1] and only the flow turning angle was considered fro parametric studies for the onset condition of ODW cell structures with all other variables were fixed. The selected thermo-fluidic parameters are dimensionless heat release, Q of 10.0, specific heat ration γ of 1.3 and dimensionless activation energy E of 30.0 from the previous study. [15] These flow parameters were selected to weaken the restriction of grid resolution requirements from the chemical induction length scale analyses. The effects of grid resolution have been studied by using grids from 500×250 to 4,005×1,800. The calculation of half reaction length scale behind ZND structure exhibit that the 10 grid points could be included in the half reaction length behind the ODW and more than 80 point behind the oblique shock wave. The comprehensive study on the grid resolutions for the reaction zone is found in the previous study. [15] Fig. 1 is polar diagram for ODW and OSW with parameters of present study. At C-J point θ is 12.3 $^\circ$ and β is 34.5 $^\circ$. At maximum θ condition of attached ODW, θ is 38.2 $^\circ$ and β is 67.1 $^\circ$. In the previous study for flow turning angle greater than θ_{max} [14], it has been shown that the detached detonation is observed for a wedge of infinite length, but periodic oscillation of the primary triple point could be observed over a wedge of finite length due to the repeated thermal occlusion and relaxation. Therefore it is expected that a dual-headed ODW structures of triple points running opposite directions could be existed where the flow is near choked condition since the transverse waves behind the triple points has a propagation speed close to the speed of sound. The reason why the previous study exhibited only the single headed structures is that flow turning angle of 30 $^\circ$ is has flow speed quite greater than the speed of sound behind the ODW. Therefore flow turning angles from 35 $^\circ$ to 38 $^\circ$ were considered for parametric study.

4 Results and Discussion

Figure 4 is the instantaneous density contours from the ODW simulations with different flow turning angles. Similarly to the previous results [15], present results show the irregular ODW front structures are observed with various wave interaction behind. At 35 $^\circ$ most of the transverse waves move downstream, but a part of wave front shows dual headed wave front. The primary triple point is stationary for this condition. As the turning angle get larger, existence of the dual-headed triple points structure gets clear. At 38 $^\circ$ transverse waves more evenly spaced. Differently from the results of 35 $^\circ$ and 36 $^\circ$, the results of 37 $^\circ$ and 38 $^\circ$ shows a separated or disturbed flow regions along the wall, which reflects the presence of the subsonic flow region. The locations of the primary triple points of 36 $^\circ$, 37 $^\circ$ and 38 $^\circ$ are not stationary but moves back and forward repeatedly, due to the thermal occlusion behind the ODW. Figure 2 is the numerical smoked foil record of ODWs that is produced by the tracking of maximum pressure locations along the ODW wave fronts. The maximum pressure locations were traced along the flow directions at each time and piled up along the flow direction. At 35 $^\circ$, the cell structure is not observed at earlier stage, but begins to appear at middle of the computing time. The presence of the cellular structure is getting clear as the turning angle get larger. Another important observation is that the periodic motion of the primary triple point. Since the primary triple point is stationary at 35 $^\circ$, the trace of it is plotted as a straight line along the bottom, but shows periodic traces for 36° to 38°. The periodicity gets frequent as the turning angle gets larger. Fig. 3 is the magnified plots of the smoked foil records for each cases at the end of the computations. The presence of the cell structure is more clearly shown. It is also understood that the cell size gets smaller as the turning angle gets larger.

5 Conclusion

A comprehensive numerical study was carried out to identify the onset condition of the cell structures of ODWs. From the investigation of the morphology of ODW cell structures, flow turning angle was presumed being a key parameter for the onset condition of the dual-headed triple point structures. Since the transverse waves propagate at near sonic speed and the flow speed becomes subsonic beyond the θ_{max} condition, a parametric study was carried out for the flow turning angles from 35° to 38° those are close to the θ_{max}. Differently from the previous result of single headed structures at 30°, present results exhibits the dual-headed triple point structures with cellular structures in smoked foil records. The cell structure gets finer as the flown turning angle gets larger. At larger turning angles, thermal occlusion is observed behind the ODW that results in a repeated motion of the primary triple point moving forward and backward, contrary to the results at smaller turning angles having stationary position.

References

1. D.T. Pratt, J.W. Humphrey, D.E. Glenn, *J. Prop. Pow.* 7(5) (1991) 837-845.
2. J. E. Shepherd, in: J. Buckmaster, T.L. Jackson, A. Kumar (Eds.), *Combustion in High-Speed Flows*, Kluwer, Dordrecht, (1994) 373.
3. E.K. Dabora, D. Desbordes, C. Guerraud, H.G. Wagner, *Prog. Aero. Astro* 133 (1991) 187-204.
4. C. Viguier, A. Gourara, D. Desbordes, B. Deshaies, *Proc. Combust. Inst.* 27 (1998) 2207-2214.
5. H.F. Lehr, *Astronautica Acta* 17 (1972) 589-597.
6. M.J. Kaneshige, J.E. Shepherd, *Proc. Combust. Inst.* 26 (1996) 3015-3022.
7. J. Kasahara, T. Fujiwara, T. Endo, T. Arai, *AIAA J.* 39 (2001) 1553-1561.
8. C.I. Morris, M.R. Kamel, R.K. Hanson, *Proc. Combust. Inst.* 27 (1998) 2157-2164.
9. C. Li, K. Kailasanath, E.S. Oran, *Phys. Fluids* 6(4) (1994) 1600-1611.
10. A.A. Thaker, H.K. Chelliah, *Combust. Theory Modeling* 1 (1997) 347-376.
11. M.V. Papalexandris, *Combust. Flame* 120 (2000) 526-538.
12. L.F. Figueira da Silva, B. Deshaies, *Combust. Flame* 121 (2000) 152-166.
13. G. Fusina, J.P. Sislian, B. Parent, *AIAA J.* 43(7) (2005) 1591-1604.
14. J.-Y. Choi, I.-S Jeung, Y. Yoon, Y., *Proc. Intl. Symp. Shock Waves* 22 (1999) 333-337.
15. J.-Y. Choi, D.-W. Kim, I.-S. Jeung, I.-S., F. Ma., V. Yang. V., *Proc. Combust. Inst.* 31 (2007) 2473-2480.
16. Y. Daimon, A. Matsuo, J. Kasahara, AIAA-2007-1171 (2007)
17. J.-Y. Choi, F.H. Ma, V. Yang, AIAA Paper 2005-1174 (2005).
18. J.-Y. Choi, I.-S Jeung, Y. Yoon, Y., *AIAA J.* 38(7) (2000) 1179-1187.

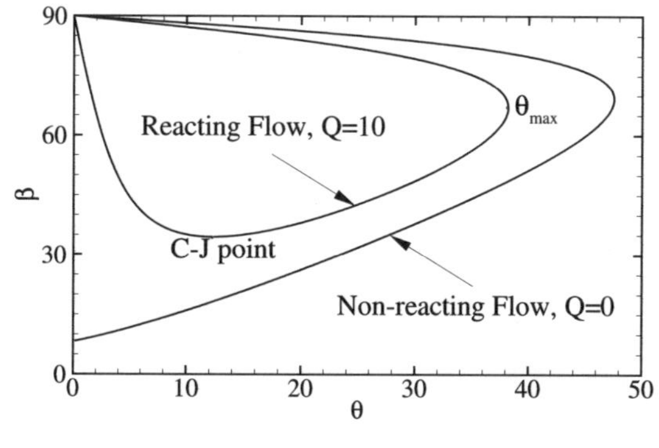

Fig. 1. Detonation polar diagram for Mach number 7 with γ=1.3

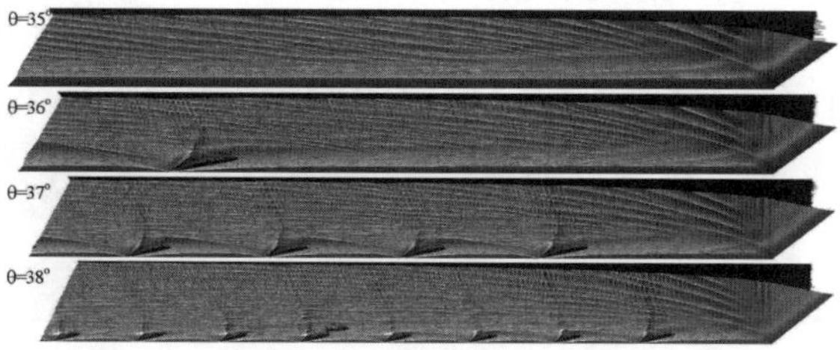

Fig. 2. Numerical smoked-foil records of oblique detonation waves for flow turning angles from 35 ° to 38 °

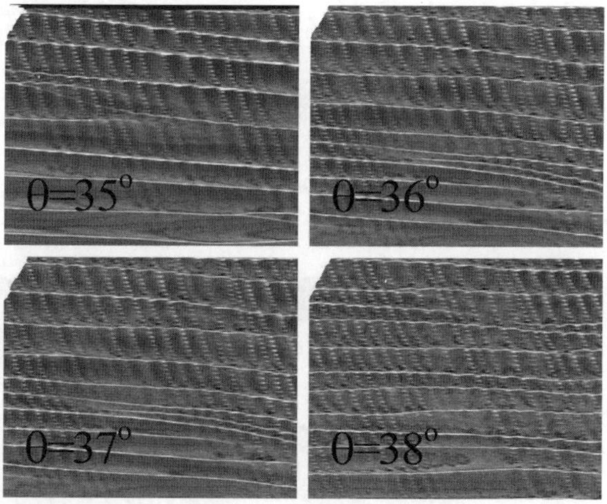

Fig. 3. Magnified plots of numerical smoked-foil records showing cell structures of oblique detonation waves.

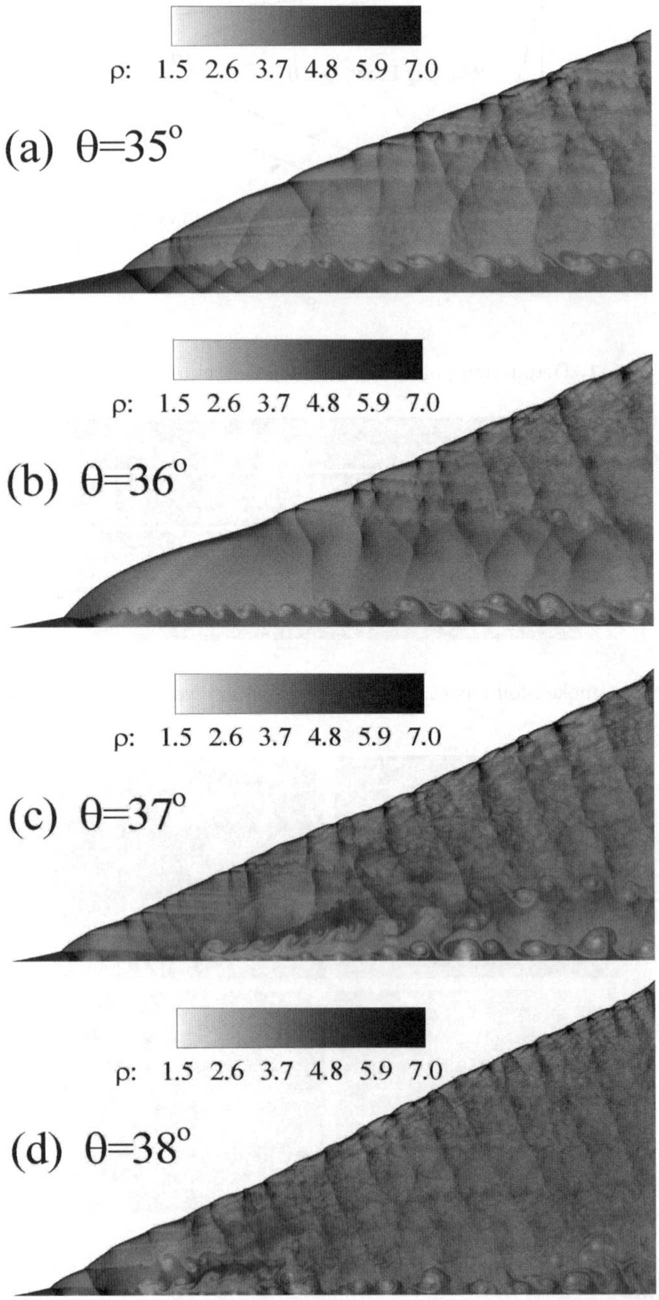

Fig. 4. Instantaneous density contours for flow turning angles from 35° to 38°

High-pressure shock tube experiments and modeling of n-dodecane/air ignition

S.S. Vasu, D.F. Davidson, and R.K. Hanson

Mechanical Engineering Department,
Stanford University, Stanford, CA, 94305 (USA)

Summary. We have measured ignition delay times of n-dodecane/air mixtures over a range of conditions including pressures of 18-31 atm, temperatures of 943-1177 K, and equivalence ratios (ϕ) of 0.5-1.0, utilizing the heated, high pressure shock tube (HPST) at Stanford University. The shock tube and mixing assembly were heated to 120 and 170 C, respectively, to prevent condensation of n-dodecane fuel (because of its low vapor pressure). Ignition delay times behind reflected shocks were measured using side−wall pressure and OH^* emission diagnostics. Shock tube ignition time measurements can provide excellent validation targets for refinement of jet fuel kinetic modeling, and n-dodecane is widely used as the principal representative for n-alkanes in jet fuel surrogates. To the best of our knowledge, we report the first gas-phase shock tube ignition time data for n-dodecane. We also provide comparisons with the ignition delay time predictions of two detailed JP-8 mechanisms that include n-dodecane as an important JP-8 surrogate component.

1 Introduction

Understanding the combustion in practical aero-combustors is necessary to characterize and improve combustor efficiency and to identify and eliminate those processes leading to pollutant formation. Practical jet fuels such as Jet-A, JP-8 and other kerosene-based fuels contain thousands of compounds, and in most cases, n-alkanes constitute more than 50 % by volume of jet fuels (Edwards [1]). Composition of jet fuels varies from shipment to shipment, thus complicating fundamental studies. Detailed chemical kinetic mechanisms of even the individual components of jet fuels have not been understood well, although the development of experimental kinetic databases for surrogate jet fuels, which are mixtures of few hydrocarbon compounds, have been a focus of much research attention recently (Colket et al. [2]). There exist numerous jet fuel surrogates in the literature, which are usually developed to approximate some of the physical and chemical properties related to the combustion of jet fuels; such surrogates are desirable for reproducibility and tractability in experimental and computational studies. Most surrogates utilize a pure hydrocarbon to represent each key chemical class found in jet fuels, and n-dodecane is widely used as the representative for n-alkanes in jet fuel surrogates (Violi et al. [3]). Hence, there is an urgent need for understanding the combustion characteristics of n-dodecane, and yet the lack of ignition delay time data alone, has been a major factor impeding the development of an accurate chemical kinetic mechanism for n-dodecane combustion (Ranzi et al. [4]). Dependence of ignition delay time (τ_{ign}) on temperature, pressure and composition is critical in describing the combustion of liquid fuels in engines and combustion chambers of various types, as well as in optimizing external combustion. Shock tubes are nearly ideal devices for studying ignition phenomena as they provide well-controlled step changes in temperature and pressure, well-defined time zero and ignition delay times, and are not affected by surface or transport problems. Also, shock tubes can reproduce nearly identical pressures and temperatures from shock experiment to experiment. Due to the high boiling points of large hydrocarbon components of Jet-A and JP-8 fuels, heating of the shock tube and associated mixing facilities is necessary to prepare a homogeneous gas phase mixture for ignition time studies, and this partially explains why there are so few data on this topic. Shock tube ignition time measurements of n-dodecane mixtures thus can provide critical validation targets for refinement of n-dodecane and jet fuel surrogate kinetic mechanisms, and is the focus of this paper. We also provide comparisons with the predictions of following two detailed JP-8 (where n-dodecane is used as a representative of the n-alkane fraction of jet fuels) mechanisms: Ranzi et al. [5](280 species, 7800 rxns.), and Zhang et al. [6](208species, 1087 rxns.).

2 Experimental Method

All high-pressure ignition delay times were measured for n-dodecane/air mixtures behind reflected shock waves using Stanford high-purity, high-pressure shock tube (HPST), which was designed to attain pressures as high as 600 atm. The shock tube driver section is 3 m long with a 7.5 cm internal diameter. Helium was used as the driver gas. The stainless steel driven section of the HPST is 5 m long with a 5 cm internal diameter. A fuller description of the shock tube can be found in Petersen [7]. Mixtures of research grade n-dodecane (Sigma-Aldrich) and synthetic air (Praxair 79% N_2, 21% O_2) were prepared in a 12.8 liter, magnetically-stirred, thermally insulated, stainless-steel mixing tank. In the present study, the mixing tank and connecting gas lines to the shock tube were heated to 170 C and 125 C respectively, and the shock tube was heated to 120 C. Mixtures were prepared in a similar manner to that described by Vasu et al. [8] for Jet-A/air. In this study, the ignition delay time is defined as the time interval between the arrival of the reflected shock wave and the onset of ignition at the sidewall observation location. The arrival of the reflected shock wave was marked by the step rise in pressure, and the onset of ignition was observed in both the pressure history and the OH^* emission history. The onset of ignition from the pressure history and OH^* emission were defined by locating the time of steepest rise and linearly extrapolating back in time to the pre-ignition baseline. The two methods give very similar results (±1%) and ignition delay times are readily identifiable with both diagnostics; example data are shown in Fig. 1. In comparing experimental data with model predictions, the model calculations are done for homogeneous, adiabatic conditions behind reflected shock waves, with a constant-volume constraint using the CHEMKIN 4.0 software package (refer Vasu et al. [8] for details). Details about the implications of this assumption for the ignition process are described in Davidson and Hanson [9].

3 Results and Discussion

We performed several series of n-dodecane/air ignition delay time measurements; n-dodecane/air experiments with equivalence ratios of ϕ=1.0 and ϕ=0.5 are compared in Fig. 2. The data for τ_{ign} are characterized by small scatter and show distinct variation with ϕ for the temperature range studied near 20 atm.

To the best of our knowledge, we report the first gas-phase shock tube ignition time data for n-dodecane. A clear trend of longer ignition delay times for the lean (ϕ=0.5)

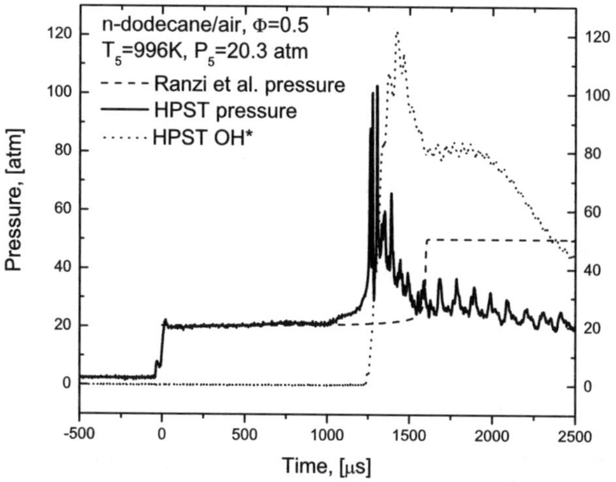

Fig. 1. Sample n-dodecane/air experimental pressure and OH^* emission data. Modeled pressure trace using the Ranzi et al. [5] mechanism is also shown.

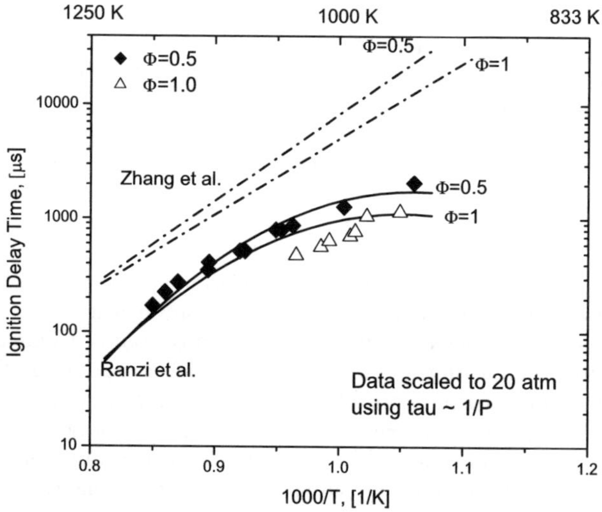

Fig. 2. n-dodecane/air τ_{ign} results. τ_{ign} predictions of the Ranzi et al. [5] and the Zhang et al. [6] mechanisms are also shown.

as compared to the stoichiometric case (ϕ=1.0) is noticed in Fig. 2. Past ignition time measurements from our laboratory conducted by both Davidson et al. [10] and Vasu et al. [11] reported similar trends in iso-octane/air mixtures for the same equivalence ratios. The reason for this trend is explained in detail by Curran et al. [12] in a comprehensive modeling study of oxidation using iso-octane. In effect, the higher the concentration of fuel in the mixture, the faster the ignition process becomes. Curran et al. [12] found that at temperatures below approximately 1150K, increasing ϕ, i.e. increasing the fuel concentrations, increases the alkyl-hydroperoxide radical pool production and results in shorter ignition delay times. After the initial fuel decomposition steps, the increase in the radical pool concentration is responsible for the rapid reactions associated with ignition.

We have used the $\tau_{ign} \sim 1/P$ relation to scale the ignition delay time results in the small pressure range of these experiments. This is a good first approximation for jet fuels and surrogate components and has been experimentally verified for the case of jet fuels in previous studies by Vasu et al. [8]. However, the actual pressure dependence of the ignition delay time is expected to vary with temperature, and studies are needed to more completely characterize this dependence. The temperature dependence of the ignition delay time also clearly varies with temperature. For the data in Fig. 2, ignition delay times at high temperatures (above $\sim$ 1000 K) were observed to decrease gradually as the temperature is increased. However, at low temperatures, ignition delay times show NTC-type behavior, in which the ignition delay time remains nearly constant or decrease as the temperature decreases. Interestingly, this NTC behavior is similar to that reported for n-heptane (a diesel surrogate) ignition delay times in air by Gauthier et al. [13] from our laboratory. In the case of n-dodecane, there is a slight evidence of NTC type ignition time roll-off below 1000K (Fig. 2.). NTC behavior is due to the formation of peroxy radicals, and large n-alkanes such as n-dodecane have a stronger tendency to show NTC than small or branched alkanes. Clearly, measurements need to be extended to the low temperature region to characterize the NTC roll-off.

The current data set provides critical validation targets for jet fuel surrogate mechanisms. Ignition delay time predictions using the JP-8 mechanisms of Ranzi et al. [5] and Zhang et al. [6] are presented in Fig. 2. The Ranzi et al. [5] mechanism gives very close agreement in magnitude with data (especially at ϕ=0.5), and is able to capture the roll-off and τ_{ign} variation with ϕ as seen in data. The Ranzi et al. [5] model slightly overpredicts the ignition delay times at ϕ=1 (above 980 K). Recently, we reported (Vasu et al. [8]) that the Ranzi et al. [5] mechanism applied to the Violi et al. [3] JP-8 surrogate mixture (73.5% by volume of this surrogate is n-dodecane) predicted a roll-off trend similar to that seen in our measured ignition delay times near 20 atm for Jet-A/air mixtures (though the mechanism predicted stronger roll-off than the data). Comparison of an experimental pressure profile with the Ranzi et al. [5] modeled pressure profile is presented in Fig. 1 for a 20.3 atm test condition. The comparison serves to confirm that up to the point of ignition, the constant volume assumption is a good representation of the shock tube behavior (Davidson and Hanson [9]). However, simple constant volume calculations do not capture the experimentally measured pressure oscillations caused by the blast wave after ignition. By contrast, the Zhang et al. [6] mechanism predicts higher ignition times for the temperature range studied at 20 atm for both equivalence ratios. However, this mechanism was not designed for very low temperature ignition applications. The inability of the n-dodecane sub-mechanism of the Zhang et al. [6] JP-8 mechanism to predict low temperature n-dodecane ignition delay times is likely due to the absence of peroxy chemistry in their mechanism. Also note that both the mechanisms show no difference in ignition delay times between the two equivalence ratios at high temperatures (near 1200 K) for the 20 atm pressure condition (Fig. 2). Fig. 3 presents a comparison

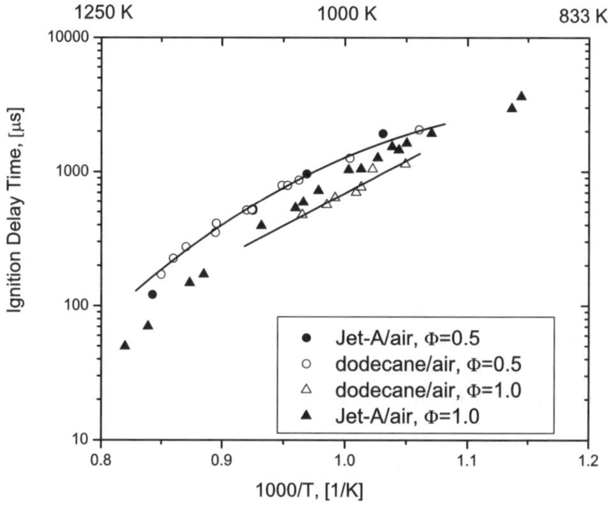

Fig. 3. τ_{ign} comparison of n-dodecane/air and Jet-A/air. Jet-A/air results are from Vasu et al. [8]. Data scaled to 20 atm using $\tau_{ign} \sim 1/P$. Lines are solid fit through n-dodecane data.

of ignition delay time measurements for n-dodecane/air and Jet-A/air (Vasu et al. [8] from the same facility). At ϕ=0.5, ignition delay times of both these mixture do not show any significant difference and at ϕ=1, there is only a slight difference (25%) in τ_{ign} of these fuels. This suggests that using n-dodecane as a single component surrogate to jet fuels may be adequate for studying certain combustion characteristics of jet fuels such as gas-phase ignition delay time. This similarity between n-dodecane and jet fuels has been observed earlier in other type of experimental studies. For example, Eigenbrod et al. [14] observed the similarity between n-dodecane and kerosene droplet induction times, and Holley et al. [15] found that extinction strain rates of n-dodecane and jet fuel non-premixed flames are even closer. However, it should be noted that pollutant or soot emissions often depend on trace fuel species and other additives present in the real fuel, and will not, in general, be reproduced by a simple chemical surrogate. These comparison studies further affirm the important role of n-dodecane in jet fuel surrogate development.

4 Conclusions

Ignition delay times of n-dodecane/air mixtures were measured over a range of conditions including pressures of 18-31 atm, temperatures of 943-1177 K, and equivalence ratios of 0.5-1.0, utilizing the heated, high pressure shock tube (HPST) at Stanford University. To the best of our knowledge, we report the first gas-phase shock tube ignition time data for n-dodecane. The measured ignition delay data showed the following: low scatter; distinct variation with equivalence ratios; and NTC-type roll-off behavior at low temperatures. A clear trend of longer ignition delay times for the lean (ϕ=0.5) as compared to the stoichiometric case (ϕ=1.0) is noticed. Comparison of ignition delay time measurements for n-dodecane/air and Jet-A/air (measurements of Vasu et al. [8] from the same facility) indicated that at ϕ=0.5, ignition delay times of both these mixtures are same and at ϕ=1, there is only a slight difference in τ_{ign} of these fuels, for the pressures studied. This could suggest that using n-dodecane as a single component surrogate to jet fuels may be adequate for studying certain combustion characteristics of jet fuels such as ignition delay time. Performance studies of two JP-8 mechanisms (Ranzi et al. [5] and Zhang et al. [6]), where n-dodecane is an important surrogate component for jet fuels, yielded the following: the Ranzi et al. [5] mechanism gave the closest agreement in magnitude with data and was able to capture the roll-off trend seen in data; the Zhang et al. [6] mechanism predicted very long ignition delay times compared to the data for the temperature range studied near 20 atm and was not able to predict any NTC roll-off. The inability of the n-dodecane sub-mechanism of the Zhang et al. [6] JP-8 mechanism to predict low- temperature n-dodecane ignition delay times is likely due to the absence of peroxy chemistry in their mechanism. The present data will be incorporated into the Fundamental Shock tube Kinetics Database, currently being developed at Stanford University [16].

Acknowledgement. This research was sponsored by the Army Research Office (ARO), with Dr. Ralph Anthenian as technical monitor (Contract No. DAAD19-01-1-0597).

References

1. Edwards T.: J. Prop. Power, vol. 19, no. 6, pp. 1089-1107, 2003
2. Colket M., et al.: AIAA Paper 2007-770, 2007
3. Violi A., et al.: Comb. Sci. Tech., vol. 174, nos. 11 & 12, pp. 399-417, 2002
4. Ranzi E., et al.: Ind. Eng. Chem. Res., vol. 44, pp. 5170-83, 2005
5. Ranzi E.: http://www.chem.polimi.it/CRECKModeling/kinetic.html, 2006
6. Zhang R.H., et al.: Proc. Combust. Inst., vol. 31, pp. 401-409, 2007;
7. Petersen E.L.: Ph.D. thesis, Stanford University, CA, 1998
8. Vasu S.S., Davidson D.F., Hanson R.K.: Jet fuel ignition delay times: Shock tube experiments over wide conditions and surrogate model predicitons . Combustion and Flame, in press, 2007
9. Davidson D.F., Hanson R.K.: International J. of Chemical Kinetics, vol. 36, no. 9, pp. 510-23, 2004
10. Davidson D.F., et al.: Proc. Combust. Inst., vol. 30, pp. 1175-82, 2005
11. Vasu S.S., Davidson D.F., Hanson R.K.: Shock Tube Measurements and Modeling of Ignition Delay Time in Lean Iso-octane/Air, Proc. of the 25th ISSW, Bangalore, India, 2005
12. Curran H.J., et al.: Combust. and Flame, vol. 129, no.3, pp. 253-80, 2002
13. Gauthier B.M., et al.: Combust. and Flame, vol. 139, no. 4, pp. 300-311, 2004
14. Eigenbrod C., Moriue O., Weilmunster P., Rath H.J.: Development of a Simple Model Fuel for Kerosene Droplet Ignition, 28th Int. Annual Conf. of Fraunhofer Institut Chemische Technologie, Karlsruhe, Germany, pp. 42.1-42.14, 1997
15. Holley A.T., et al.: Proc. of the Fall Meet., WSS/CI, Stanford, CA, Paper 05F-56, 2005
16. Davidson D.F., Hanson R.K.: available at: http://hanson.stanford.edu/news.htm, 2006

Implicit-explicit Runge-Kutta methods for stiff combustion problems

E. Lindblad[1], D.M. Valiev[2], B. Müller[3], J. Rantakokko[1,4], P. Lötstedt[1], and M.A. Liberman[5]

[1] *Department of Information Technology, Uppsala University, Box 337, 751 05 Uppsala, Sweden*
[2] *Materials Science and Engineering, KTH, 100 44 Stockholm, Sweden*
[3] *Department of Energy and Process Engineering, Norwegian University of Science and Technology, Kolbjørn Hejes vei 2, 7491 Trondheim, Norway*
[4] *UPPMAX, Uppsala University, Box 337, 751 05 Uppsala, Sweden*
[5] *Department of Physics, Uppsala University, Box 530, 751 21 Uppsala, Sweden*

Summary. New high order implicit-explicit Runge-Kutta methods have been developed and implemented into a finite volume code to solve the Navier-Stokes equations for reacting gas mixtures. If only the stiff chemistry is treated implicitly, the linear systems in each Newton iteration are simple and solved directly. Numerical simulations of deflagration-to-detonation transition (DDT) show the potential of the new time integration for computational combustion.

1 Introduction

The present work is aimed at gaining further understanding of the basic mechanisms controlling deflagration-to-detonation transition (DDT). The understanding of DDT is not only a major challenge of combustion theory, but also important for safety problems and for detonation propulsion engines. The Landau-Darrieus hydrodynamic flame instability plays an important role. Friction and roughness of the tube walls make the flow ahead of the flame nonuniform and result in bending the flame front, which leads to flame acceleration. The accelerating reaction front acts as semitransparent piston, generating a pressure wave ahead. However, the acceleration is too weak to generate strong enough shock waves for triggering detonation. For a flame propagating from the closed end of a semi-infinite tube, it was recently shown that the formation of a preheat zone ahead of the flame is the basic physical mechanism of the deflagration-to-detonation transition, if the preheat zone is extended enough to provide a positive feedback for considerable enhancement of the flame acceleration [1].

Recent 2D studies [2, 3] of a flame propagating from the closed to the open end of a tube show that DDT occurs for a fast flame either due to preheat zone formed within the flame fold developed due to the Landau-Darrieus instability, when the influx of heat from the folded reaction zone increases temperature inside the fold or near the tube wall due to hydraulic resistance caused by the friction or roughness at the tube walls. The simulations [2] were performed using a parallel version of a general code developed by L.-E. Eriksson [4]. This code solves the Navier-Stokes equations for reacting gas mixtures using a third-order upwind-biased finite volume method for the inviscid fluxes and a second-order central discretization of the viscous fluxes with an explicit second order Runge-Kutta time integrator. Future additions include simulations in 3D, complex chemistry and the influence of turbulence. These additions increase the stiffness of the governing equations and therefore the time stepping method must be improved.

New time integrators implement second-, third- and fourth-order implicit-explicit Runge-Kutta methods [5] to use the more expensive implicit method only for the stiff part of the problem to overcome the severe stability restriction of the economical explicit method, e.g. for treating complex chemistry. The resulting nonlinear systems are local and solved efficiently by Newton's method directly. For if only the chemistry is treated implicitly, the nonlinear systems can be solved independently in each cell by a few Newton iterations. Tests show that the increased stability of the implicit methods and the efficiency of the explicit methods provide efficient time integrators for the intended stiff combustion problems.

2 Finite volume method for the Navier-Stokes equations

With the finite volume method, the 2D compressible Navier-Stokes equations in integral form are discretized for each cell in a computational grid by the approximation

$$\frac{d\mathbf{U}_{i,j}}{dt} Vol_{i,j} + \sum_{s=1}^{N} [(\mathbf{F} - \mathbf{F}_v) n_x A + (\mathbf{G} - \mathbf{G}_v) n_y A]_s = \mathbf{S}_{i,j} Vol_{i,j}. \tag{1}$$

The volume averaged vector of the conservative variables in the cell $\Omega_{i,j}$ is $\mathbf{U}_{i,j}$, $\mathbf{S}_{i,j}$ is the volume averaged source vector, and $Vol_{i,j}$ is the area of the cell. Since we consider structured grids with quadrilaterals as control volumes, the cells have $N = 4$ faces. A_s is the length of the cell interface s. The cell averages $\mathbf{U}_{i,j}$ are the unknowns in the cell-centered finite volume method.

The inviscid flux vectors for the x- and y-directions in (1) are $\mathbf{F} = \mathbf{F}_1$ and $\mathbf{G} = \mathbf{F}_2$ and $\mathbf{F}_v = \mathbf{F}_{v1}$ and $\mathbf{G_v} = \mathbf{F}_{v2}$ are the viscous flux vectors for the x- and y-directions. $(x_1, x_2)^T = (x, y)^T$ and $(u_1, u_2)^T = (u, v)^T$. The density, pressure, temperature, total energy per unit mass, total enthalpy, enthalpy of species k, mass fraction of species k, diffusion coefficient of species k, production rate of species k, viscosity, and heat conduction coefficient, are denoted by ρ, p, T, E, H, h_k, Y_k, D_k, ω_k, μ, and κ, respectively. Then the conservative variables, the inviscid and viscous flux vectors of a thermally perfect gas mixture of n species read

$$\begin{aligned}
\mathbf{U} &= (\rho, \rho u_1, \rho u_2, \rho E, \rho Y_1, \ldots, \rho Y_{n-1})^T, \\
\mathbf{F}_j &= (\rho u_j, \rho u_1 u_j + p\delta_{1j}, \rho u_2 u_j + p\delta_{2j}, \rho H u_j, \rho_1 u_j, \ldots, \rho_{n-1} u_j)^T, \\
\mathbf{F}_{vj} &= (0, \tau_{1j}, \tau_{2j}, \sum_{l=1}^{2} u_l \tau_{lj} + \kappa \frac{\partial T}{\partial x_j} + \rho \sum_{k=1}^{n} \left(D_k h_k \frac{\partial Y_k}{\partial x_j} \right), D_1 \rho \frac{\partial Y_1}{\partial x_j}, \ldots, D_{n-1} \rho \frac{\partial Y_{n-1}}{\partial x_j})^T
\end{aligned} \tag{2}$$

and $\mathbf{S} = (0, 0, 0, 0, \omega_1, \ldots, \omega_{n-1})^T$ is the source term. For a Newtonian fluid, the components of the shear stress tensor are $\tau_{ij} = \mu \left(\frac{\partial u_i}{\partial x_j} + \frac{\partial u_j}{\partial x_i} \right) - \frac{2}{3} \mu \left(\sum_{k=1}^{2} \frac{\partial u_k}{\partial x_k} \right) \delta_{ij}$ where $\delta_{ij} = 1$ if $i = j$ and $\delta_{ij} = 0$ if $i \neq j$.

We have to approximate the flux vectors at the cell interfaces. The inviscid flux vectors are discretized by a third-order upwind-biased approximation of the characteristic variables and using a total variation diminishing (TVD) limiter [4] [6]. Central discretizations are employed for the viscous fluxes at the cell interfaces. The volume averaged nonlinear source term is approximated by

$$\mathbf{S}_{i,j} \approx \mathbf{S}(\mathbf{U}_{i,j}). \tag{3}$$

After the finite volume discretization of equation (1), we have a system of ordinary differential equations (ODEs) for the time dependent cell averages $\mathbf{U}_{i,j}$

$$y' = f(y) + g(y)\,, \tag{4}$$

where y denotes the vector of all $\mathbf{U}_{i,j}$, $f(y)$ the vector of all inviscid and viscous flux contributions, and $g(y)$ the vector of all source terms $\mathbf{S}_{i,j}$. Thereby, we classify the right hand side of the ODE (4) into a non-stiff part f and a stiff part g. Other classifications are possible and discussed in [5].

3 Implicit-explicit Runge-Kutta methods

We have developed high order implicit-explicit Runge-Kutta (IERK) methods to efficiently solve separable stiff problems (4) [5, 7]. While first-order IERK methods have frequently been used to treat the stiff source terms from chemistry implicitly in hypersonic flow and combustion simulations [8, 9], two of our new IERK methods are second order accurate and one is fourth order accurate. Similar high order IERK methods have only recently been available [10–14].

An explicit Runge-Kutta (ERK) method is used to solve the non-stiff part f and a diagonally implicit Runge-Kutta (DIRK) method is employed to solve the stiff part g. A general s-stage implicit-explicit Runge-Kutta (IERK) method consists of an s-stage ERK and an s-stage DIRK method with common weighting coefficients $b_i, i = 1, ..., s$. The following tableaus define the ERK and DIRK methods of an IERK method [5, 7]:

$$\begin{array}{c|cccc}
 & 0 & & & \\
 & \varepsilon_{21} & 0 & & \\
 & | & & 0 & \\
 & \varepsilon_{s1} & - & \varepsilon_{s,s-1} & 0 \\
\hline
 & b_1 & - & - & b_s
\end{array}
\qquad
\begin{array}{c|cccc}
 & a_{11} & & & \\
 & a_{21} & a_{22} & & \\
 & | & & & \\
 & a_{s1} & - & - & a_{ss} \\
\hline
 & b_1 & - & - & b_s
\end{array} \tag{5}$$

The approximate solution y^{n+1} at $t = (n+1)\Delta t$ is defined by

$$\begin{aligned}
y^{n+1} &= y^n + \Delta t \sum_{i=1}^{s} b_i k_i \,, \text{ where} \\
k_i &= f\left(y^n + \Delta t \sum_{j=1}^{i-1} \varepsilon_{ij} k_j\right) + g\left(y^n + \Delta t \sum_{j=1}^{i} a_{ij} k_j\right), \\
i &= 1, \ldots, s.
\end{aligned} \tag{6}$$

The coefficients of our 4th order accurate 5-stage IERK method denoted as IERK45 method are given in [5].

4 Results

4.1 Deflagration-to-detonation transition (DDT) due to flame folding

One of the typical numerical pictures of DDT due to formation of the appropriate flame fold at the tube wall is shown in Fig. 1, where the sequence of tonal images depicting the square of pressure gradient gives a cinematic impression of the dynamics of the flame front during the incipient stage of the transition from deflagration to detonation [2]. These images resemble the Schlieren photographs of laboratory experiments, though the latter visualize gradients of the density rather than of the pressure.

A detailed investigation shows [2]: (i) formation of the large-scale preheat zone (preconditioning) in the unreacted gas trapped within the fold interior at the tube wall, (ii) acceleration of the fold-tip, and (iii) the pressure elevation and formation of a high pressure peak. The transition occurs when the pressure peak becomes high enough to produce a shock capable of supporting detonation. This requires the fold to be narrow and deep enough. Otherwise one ends up with a moderately strong pressure wave insufficient for triggering the transition.

4.2 Explicit and implicit-explicit Runge-Kutta methods

To show the advantages of using IERK methods for combustion a series of DDT simulations have been performed. The flux is calculated using the combustion code of [4]. The original code uses a second order explicit Runge-Kutta method known as Gary's method [15] for the time stepping. For our tests it has been replaced by implicit-explicit Runge-Kutta methods of order 2, 3 and 4. The third order method is derived in [12], and the second and fourth order methods are derived in [5, 7]. Simulations showing order of accuracy and applications of IERK methods to various model problems are also presented in the cited source papers.

Since the chemical source term (3) only depends on cell data, the implicit iteration can be performed by solving a $(3+n) \times (3+n)$ system per cell where n is the number of species. In the present simulations a mixture of two species representing burned and unburned fuel is used. The production rate of the fuel is given by an Arrhenius expression, cf. [2]. This reduces the implicit treatment of the source term to a scalar equation which can be solved using scalar Newton iteration without having to solve any linear systems or to compute Jacobian matrices. The overhead for solving the chemistry implicitly is therefore very small. Fig. 2 shows the temperature contours obtained by using the first order implicit Euler method for the chemistry treatment. Whereas the explicit Runge-Kutta method becomes unstable, the present semi-implicit method remains stable during the transition [5]. The higher order IERK methods have not yet been fully implemented in the code, but preliminary results are promising. The new high order methods have comparable efficiency as Gary's explicit method for the current setup. For more realistic chemistry the stiffness is greatly increased and the IERK methods are then expected to surpass the explicit methods regarding both stability and efficiency.

5 Conclusions

New semi-implicit time discretization methods have been developed for the conservation laws governing combustion. While the stiff source terms due to chemistry are treated implicitly, the inviscid and viscous fluxes are discretized explicitly. As the implicit part is restricted to the the local solution of the stiff chemistry ordinary differential equations in each grid cell, the high order implicit-explicit Runge-Kutta methods are expected to be more efficient than purely explicit or purely implicit methods. The advantages of the implicit treatment of the stiff chemistry source terms have been demonstrated for deflagration-to-detonation transition (DDT).

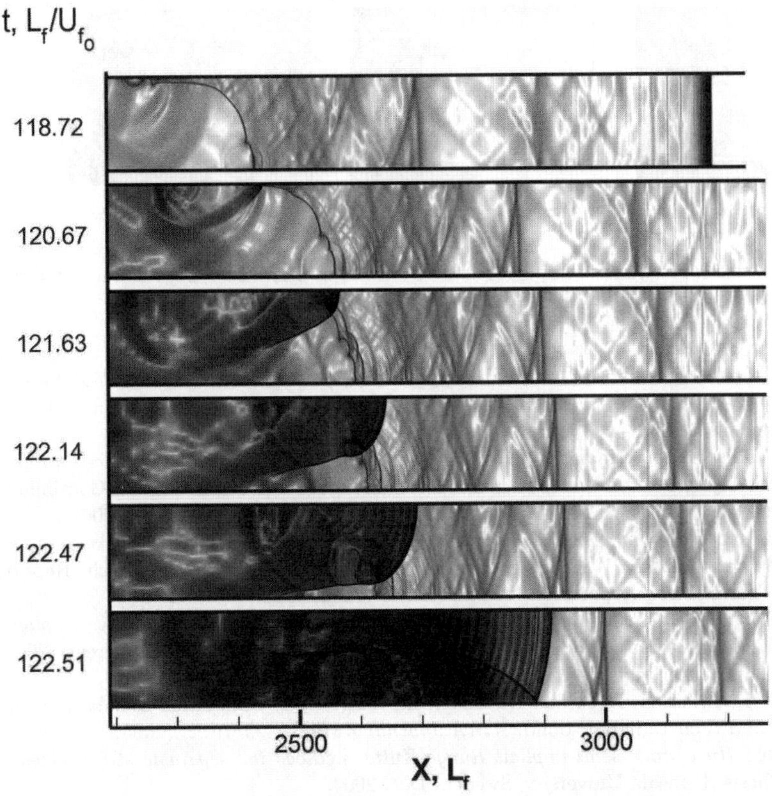

Fig. 1. Time sequence of images for the flame/shock dynamics near the transition point at the wall. Stronger shading corresponds to higher pressure gradient. The time and distance are referred to as L_f/U_{f0} and L_f, respectively. $L_f = \mu_u/(Pr\rho_u U_{f0})$ is the flame width, where U_{f0} is the incipient velocity of the planar flame, ρ_u and μ_u the density and dynamic viscosity of the unburned fuel, respectively. The incipient velocity of the planar flame U_{f0} corresponds to the Mach number $M_{f0} = U_{f0}/c_u = 0.06$, where c_u is the speed of sound in the unburned fuel. Reaction order $m = 2$, tube width $D = 140L_f$, dimensionless activation energy $\epsilon = E^a/(R_u T_b) = 4$, expansion ratio $\Theta = T_b/T_u = 10$. T_u and T_b are the temperatures of the unburned and burned gases, respectively [2]

Acknowledgement. Fellowships for E. Lindblad by EU and Göran Gustafsson Foundation are gratefully acknowledged by M. Liberman and B. Müller, respectively.

References

1. Liberman M. A., Kagan L., Sivashinsky G., Valiev D., *Flame acceleration due to preheating within the flame fold and hydraulic resistance as a mechanism of Deflagration-to-Detonation Transition*, 21st International Colloquium on the Dynamics of Explosions and Reactive Systems (ICDERS-2007), Poitiers (2007).

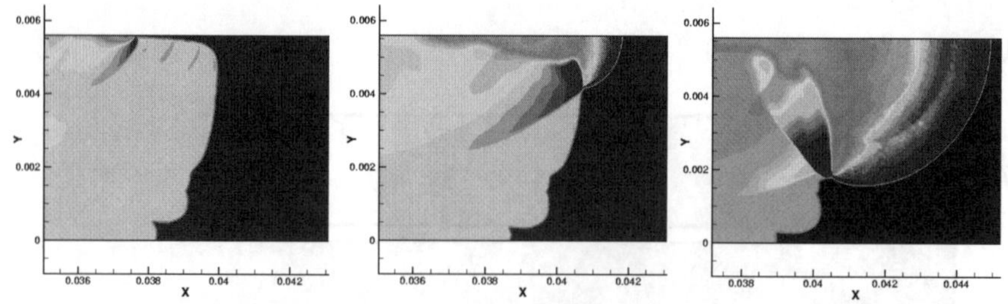

Fig. 2. Temperature contours with implicit chemistry treatment. When using the implicit Euler method for chemistry, the DDT evolves with retained stability

2. Liberman M. A., Sivashinsky G., Valiev D., Eriksson L.-E., *Numerical simulation of deflagration-to-detonation transition: Role of hydrodynamic flame instability*, Int J. Transport Phenomena, **8**, 253 - 277 (2006).
3. Kagan L., Valiev D., Liberman M., Sivashinsky G., *Effects of hydraulic resistance and heat losses on the deflagration-to detonation transition*, Proceedings of Pulsed and Continuous Detonations, Eds. G.Roy, S. Frolov, J. Sinibaldy, Torus Press LTd. 119-123 (2006).
4. Eriksson, L.-E., *Development and validation of highly modular flow solver versions in G2DFLOW and G3DFLOW series for compressible viscous reacting flow*, Tech. Report, 9970-1162, Volvo Aero Corporation, Trollhättan, Sweden, 1995.
5. Lindblad E., Valiev D.M., Müller B., Rantakokko J., Lötstedt P., Liberman M.A., *Implicit-explicit Runge-Kutta method for combustion simulation*, ECCOMAS CFD Conference 2006, TU Delft, The Netherlands, 5-8 September 2006, 20 pages (2006).
6. Andersson, N., Eriksson, L.-E. and Davidson, L., Large-Eddy Simulation of subsonic turbulent jets and their radiated sound. *AIAA Journal*, **43**(9), 1899–1912, (2005).
7. Lindblad, E., *High order semi-implicit Runge-Kutta methods for separable stiff problems*, Master's Thesis, Uppsala University, Sweden, Dec. 2004.
8. Bussing, T.R.A. and Murman, E.M., Finite-volume methods for the calculation of compressible chemically reactive flows. *AIAA J.*, **26** (9), 1070–1078, (1988).
9. Eriksson, L.-E., *Time-marching Euler and Navier-Stokes solution for nozzle flows with finite rate chemistry.* Tech. Report, 9370-794, Volvo Flygmotor AB, Trollhättan, Sweden, 1993.
10. Ascher, U.M., Ruuth, S.J. and Spiteri, R.J., Implict-explicit Runge-Kutta methods for time-dependent partial differential equations, *Appl. Numer. Math.*, **25**, 151–167, (1997).
11. Pareschi, L. and Russo, G., Implicit-explicit Runge-Kutta schemes and applications to hyperbolic systems with relaxation, *J. Scientific Comp.*, **25** (1/2), 129–155, (2005).
12. Yoh, J.J. and Zhong, X., New hybrid Runge-Kutta methods for unsteady reactive flow simulation, *AIAA J.*, **42** (8), 1593–1600, (2004).
13. Yoh, J.J. and Zhong, X., New hybrid Runge-Kutta methods for unsteady reactive flow simulation: Applications, *AIAA J.*, **42** (8), 1601–1611, (2004).
14. Kennedy, C.A. and Carpenter, M.H., Additive Runge-Kutta schemes for convection-diffusion-reaction equations, *Appl. Numer. Math.*, **44**, 139–181, (2003).
15. Gary, J., On certain finite difference schemes for hyperbolic systems, *Math. Comp.*, **18**, 1–18, (1964).

Near-field blast phenomenology of thermobaric explosions

D.V. Ritzel[1], **R.C. Ripley**[2], **S.B. Murray**[3], **and J. Anderson**[3]

[1] *Dyn-FX Consulting Ltd., 19 Laird Avenue North, Amherstburg, Ontario N9V 2T5, Canada*
[2] *Martec Limited, 1888 Brunswick Street, Suite 400, Halifax, Nova Scotia B3J 3J8, Canada*
[3] *DRDC Suffield, P.O. Box 4000, Station Main, Medicine Hat, Alberta T1A 8K6, Canada*

Summary. Depending on charge parameters, the near-field blast flow from multi-phase explosives can exhibit a wide range of wave-dynamic and hydrodynamic behaviour, including the development of jetting and previously unidentified transient interfaces. Jetting from the fireball is consistently observed in many experiments and may be a key mechanism controlling particle combustion, yet is not reproduced by CFD modelling when particle/particle interactions are neglected. Although not explicitly including the necessary physics for jetting, multi-phase explosion modelling was used to investigate the early blast-wave flow for a range of charge types to identify what controls particle distributions, including the conditions which may lead to jetting observed in reality.

1 Introduction

A thermobaric explosive (TBX) refers herein to any multi-phase formulation intended to augment blast strength by post-detonation combustion. This effect is typically achieved through burning of reactive metal particulate co-dispersed during the expansion of detonation products. In this regard, standard munition fills such as tritonal ($\sim$ 80/20 TNT/Al powder) may be considered TBX, although research beginning in the mid-90s [1] shows that the blast from cased munitions with this fill is often greatly diminished due to quenching of the afterburn phase. Successful TBX formulations must be carefully tuned to exploit the hydrodynamics of material flows and their interfaces, as well as the wave-dynamics of the early multi-phase blast flow to ensure the reactive afterburning materials are at the necessary location, at the necessary time, and in the necessary condition to react at the correct rate to promote the blast propagation. As well as requiring astute design of the formulation with regard to the particulate and CHNO matrix, the charge geometry, scale, casing, initiation, and surrounding environment also affect performance and must be factored. The analysis and understanding of TBX charge performance has challenging requirements for new theoretical models, computational solvers, and experimental diagnostics, none of which can be said to be perfected for this task at this time. The following article is a companion paper to the series describing the 'Northern Lights II' trial series [2] and related CFD modelling [3] investigating how particular mix formulations and charge parameters affect TBX blasts. The current paper discusses some of the key phenomenology appearing to control how particles become distributed, including perspectives from both experiments and computational modelling.

2 Observations of Jetting

Whereas theory and CFD modelling for blast-wave flows from conventional homogeneous explosive charges show that a gasdynamic contact front of detonation products drives

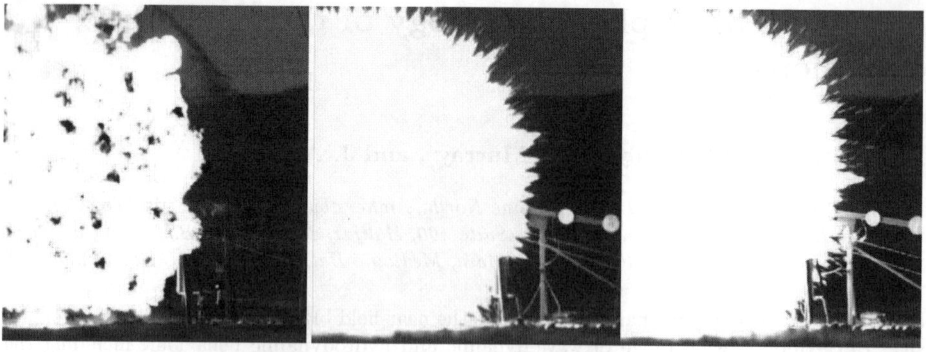

Fig. 1. Comparison of fireball interface features for: (left) classical case of high-explosive blast (C4) where the main shock is clearly distinguished ahead of a billowing RT contact front; (middle) multi-phase mixture IEF1 with reactive particulate six-fold the size of those for IEF2; and (right) multi-phase mixture IEF2. The charges are all spherical, centrally initiated, having roughly 20 kg of fill with very light polyethylene casing; the IEF descriptions are given in [2].

a shock-wave into the surrounding atmosphere, observations from TBX blast suggests a possibly important variant of this paradigm. A classical Rayleigh-Taylor (RT) front, characterized by a 'billowing' interface, develops from the detonation-product expansion of standard condensed-phase high-explosives such as TNT or C4; a smooth and 'stable' contact front develops from a typical fuel-air-explosive (FAE) fireball. These characteristic differences of fireball contact fronts are well-known features of spherically or cylindrically expanding blasts due to the respective differences in density across the decelerating interfaces. Isolated jetting has indeed been observed from the fireball of conventional explosives, however, these are true anomalies invariably attributable to some irregularity of the charge. There are also distinctive end-jets that develop from cylindrical explosive charges. However, as shown in Fig. 1, and as distinguished from these, the fireball for many or most TBX blasts is characterized by prolific and regular jetting giving a uniform 'spiny' texture to the entire fireball surface. A distinct shock front is not discernable separating from the TBX fireball during this stage as it is from a typical high-explosive RT interface, but instead the main blast front appears actually formed from the 'ballistic' shocks of the multitudinous jetting protrusions. That is, although there is an annular zone of shock-heated air between an RT fireball and the shock front, the equivalent TBX zone is entirely filled by jetting streams of dense particle/detonation-product gases with their associated ballistic shocks. The jetting protrusions do not penetrate a primary shock front, rather they seem to *develop* the primary shock front. Clearly, jetting is a uniquely 3D hydrodynamic phenomenon and must involve some manner of wake-dynamics and likely agglomeration of particles; the physics modelling required for these are not included in most CFD codes currently being applied to resolve TBX blast flows. Jetting must also involve different mechanisms for particle transport, mixing, and combustion compared to theories neglecting particle/particle interactions.

Important observations from Fig. 1 are that the two TBX fireballs are indistinguishable despite the greatly different particle sizes, and secondly, no distinct shock front leads the jetting fireball. CFD modelling neglecting particle/particle interactions indicates the main particle fronts from either improvised explosive formulation (IEF) should not overtake the 'primary' shock. Therefore, jetting may be an 'equalizer' for particle distribution mechanisms, inhibiting optimal dispersals for some cases and augmenting dispersal for others in comparison to results presuming no particle interactions. Further evidence of jetting from multi-phase charges has been observed in the case of cylindrical charges with light metal casing, as shown in Fig. 2. The photograph shows distinct streaks of metal-oxide deposits on a metal 'witness plate' placed end-on to a 2-kg cylindrical charge. The multi-phase charge components, casing, and size in this case were entirely different than for Fig. 1, but for this very reason suggest that jetting is a phenomenon which crosses broad domains of multi-phase charge types. The streaks in this case may be artifacts from jetting between splits of the casing during its early break-up, or possibly from the wake-flow behind fragments. Regardless of the particular mechanism or hydrodynamics, the streak deposits clearly show that coherent streams of possibly agglomerated particulates will develop from an initially uniform distribution within the explosive matrix.

3 Computational Modelling

The computational studies described in [3,4] were reviewed and extended with the particular interest to track and understand how particle distributions develop in multi-phase blast flows. The Chinook code has been described in a companion paper in these proceedings [3]. As with most codes currently being applied to multi-phase blast, Chinook has not been currently focused to resolve jetting, such as accounting for particle/particle interactions in 3D, including possible agglomeration or wake dynamics. Notwithstanding this limitation and other simplifications for the physics and combustion solvers, numerical experiments with such codes can be very instructive to show indicators, or precursors, to what might develop should such additional flow solvers or higher resolutions be applied.

Fig. 2. Streak deposits of metal oxides from a cased cylindrical multi-phase charge. (photograph courtesy of DSTO, Australia)

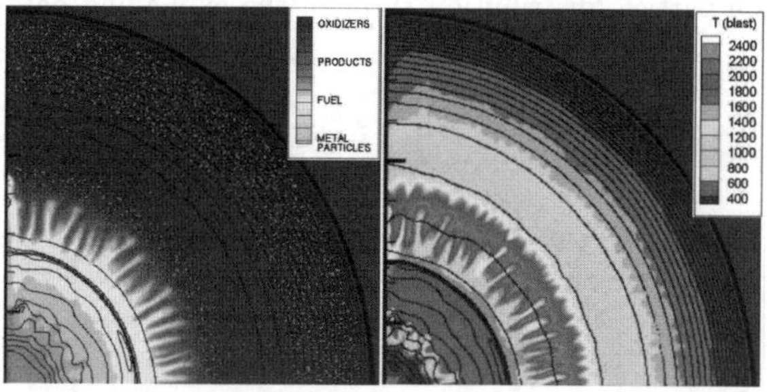

Fig. 3. Computational model from [3] for the IEF1 blast at 4 ms showing possible precursors to jet formation being developed.

3.1 Interface Features

Realistic RT instabilities will develop in inviscid CFD modelling of conventional blast interfaces due to numerical viscosity and are triggered by grid perturbations. In the modelling of reacting multi-phase blast, that interface has reason to be further affected and possibly exaggerated since light and hot secondary-combustion product species are being evolved at this interface. Furthermore, momentum transfer by particles across such gasdynamic interfaces may stretch these perturbations outwards.

Distinctive radial protuberances beneath the outer surface of the fireball can be seen in the modelling results for the IEF1 blast as shown in Fig. 3. Although initiated by numerical RT instabilities, coherent 'thermal streams' develop within the edge of the fireball whereby hot product gases surround tongues of cooler fuel materials projecting out from deeper within the fireball. It is not clear whether turbulence would in fact destroy these features or whether there are forces causing them to be maintained or enhanced as distinct streams. Any such features, although only evident here in these thermal and product/fuel maps, would have associated variances in viscosity, density, and flow velocity between the protrusions and the surrounding gas. If in fact particle/particle interactions do occur, such as agglomeration in the hotter shrouds around these protrusions and especially at the leading edge or in wake-flows behind the leading edge, it could be speculated that such features may be precursors to the jet formations seen in reality.

3.2 Development of Particle Distributions

The detonation-phase and early shock interactions with particles are described elsewhere [5]. However, the matter of how entire particle distributions subsequently develop during the early blast flow is a complex interplay between particles and gas flow. Three case-study examples were used to illustrate how the booster (if it exists), the outswept but rearward-facing secondary shock, the fireball contact surface, and the primary shock-front all influence the shaping of particle distributions and their conditioning (the examples in Figs. 5 and 6 are from calculations in Reference [4]). These inherent gasdynamic features of near-field blast will have different degrees of influence dependent on the charge and particle parameters. Figure 4 shows the particle distribution development for the IEF1 blast. The role of an 'oversized' booster in this case (10% of the charge mass, well beyond that needed to initiate the mixture), is shown here to have the key role to launch the particle distribution through the expanding detonation gases and into the zone of

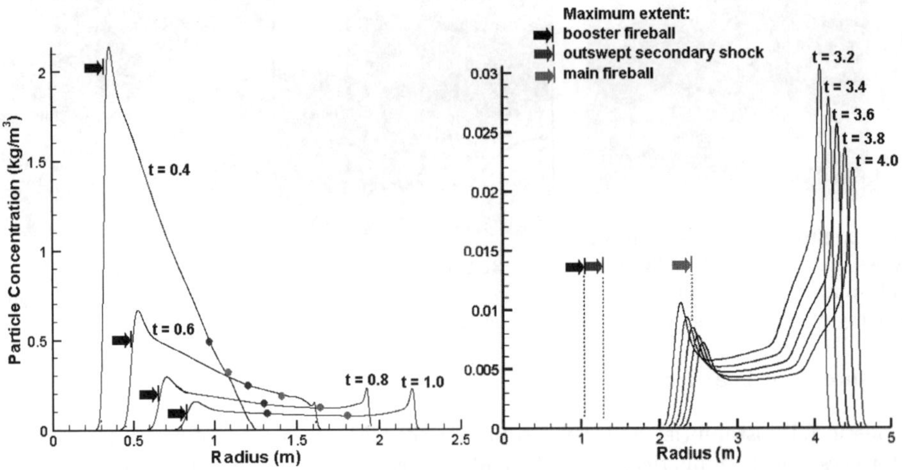

Fig. 4. Development of the particle distribution for the IEF1 blast (large particles) showing how the contact front for the booster (black arrow) launches the distribution into the critical zone of shock-heated air just beyond the main fireball. In Figs. 4 and 5, the blue and red dots on the profiles track the positions for the outswept secondary shock and gasdynamic fireball front respectively. Time is shown in milliseconds.

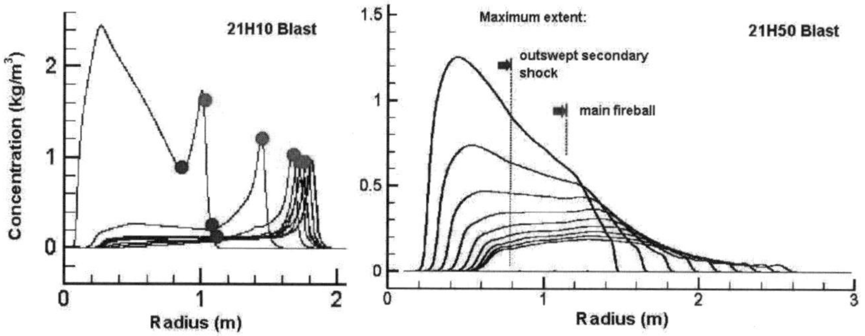

Fig. 5. Development of particle distributions for 21.3-cm diameter nitromethane charges heavily loaded with aluminum particles, nominally 13 μm (left) and 54 μm (right)

shock-heated air beyond the fireball edge. This projection is achieved in the necessary timescale to possibly boost the blast front, although the particles in this case are too large to burn rapidly enough once exiting the fireball. The particle momentum transfer across the gasdynamic fireball interface greatly smoothes the density discontinuity there to a ripple. The central portion of the particle distribution becomes quite uniform in concentration and velocity throughout this outward projection. In reality, the presence of high concentrations of particles at similar relative velocity may lead to wake effects whereby particles preferentially fall into streams.

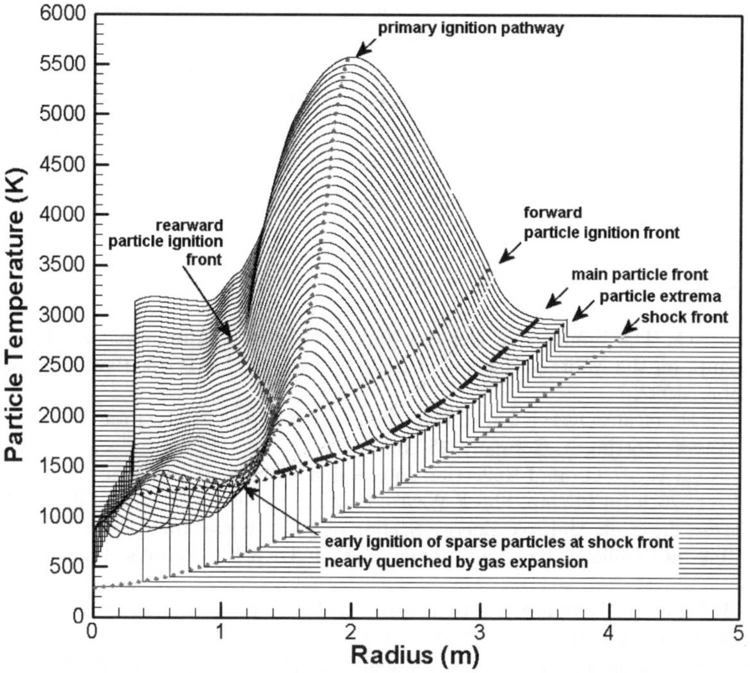

Fig. 6. Wave diagram showing the delayed ignition of the particle cloud for the 21H50 blast, initiated by early shock heating of a narrow band of particles behind the front of the particle distribution.

The outswept secondary shock is the principal decelerator for the early gasdynamic outflow and also provides substantial shock-heating for material passing through it. This outswept shock and the gasdynamic contact surface for the fireball have negligible effect on the above distribution due to the overwhelming momentum of the particles. However, as shown in Fig. 5 for charges without boosters and having smaller particles, these features can dominate the shaping of the particle distributions. In the case of the 21H10 charge having nominally 13-μm particles, the secondary shock nearly stops the outflowing particles, causing accumulation beyond it, yet (except for a few extrema) these particles do not have sufficient momentum to penetrate the shock front. The distribution becomes frozen in that earlier shape as the detonation product flow comes to rest. The 21H50 particles, having the higher momentum of their nominal 54-μm size, develop a banded distribution whereby a clearly defined rearward edge is shaped by the outswept secondary shock while the leading edge is drawn out by the primary shock flow. However, in both cases these edges lag the gasdynamic features that caused them. The 21H50 blast provides a remarkable case where the shell of shock-heated air beyond the fireball provides just enough heating of a small band of particles to initiate a 'thermal explosion' of the particle distribution at delayed time, as shown in Fig. 6.

4 Conclusions

Observations of the fireballs and resulting blast waveforms from multi-phase explosive charges suggest that jetting of particulate/gaseous streams may be a process which is affecting the secondary combustion, yet is not currently included in blast CFD modelling which neglects particle/particle interactions. Although not explicitly including modelling to account for jetting, computational results for various TBX blasts were analyzed to assess what factors control the development of particle distributions and their conditioning. The roles of booster, outswept secondary shock, gasdynamic fireball interface, and primary shock have all been demonstrated to have strong influences on the shaping of distributions, including banding, as well as ignition of particle clouds.

References

1. Needham, C.: The Influence of Delayed Energy Release in Aluminized Explosives, Proc. 15th Symp. on the Military Aspects of Blast and Shock, MABS15, Banff, Canada, 14-19 September, 1997.
2. Murray, S.B., et al.: Overview of the 2005 Northern Lights Trials, Proc. 26th Int'l Symp. Shock Waves, Gottingen, Germany, 15-20 July, 2007.
3. Ripley, R.C., et al.: Ground Reflection Interaction with Height-of-Burst Metalized Explosions, Proc. 26th Int'l Symp. Shock Waves, Gottingen, Germany, 15-20 July, 2007.
4. Leadbetter, J., et al: Multiple Energy Scaling of Blast Waves from Heterogeneous Explosives, 21st Int'l Colloq. Dyn. Expl. React. Sys., ICDERS21, Poitiers, France, 23-27 July, 2007.
5. Ripley R.C., et al.: Detonation Interaction with Metal Particles in Explosives, Proc. 13th Int'l Det. Symp., pp. 214-223, Norfolk, U.S.A, 23-28 July, 2006.
6. Ritzel, D.V., Smithson, T., Murray, S.B., and Gerrard, K.B.: Experimental Study of Afterburning Explosive Charges, 19th ICDERS, Hakone, Japan, July 27 - August 1, 2003.

Numerical and theoretical analysis of the precursor shock wave formation at high-explosive channel detonation

P. Vu, H.W. Leung, V. Tanguay, R. Tahir, E. Timofeev, and A. Higgins

Department of Mechanical Engineering, McGill University, 817 Sherbrooke Street West, Montreal, Quebec H3A2K6, Canada

1 Introduction

When a detonation propagates in an explosive layer that only partially fills a channel, the rapidly expanding detonation products can form a piston and drive a precursor shock wave (PSW) in the air gap between the explosive layer and the channel confinement, ahead of the detonation as illustrated in Fig. 1a (see [1] and references there). The experiments with nitromethane [2] demonstrated that the phenomena is sensitive to the initial air pressure in the channel: at high enough initial pressure a precursor will not form. Taking into account that the detonation products pressure is hundreds of thousands of atmospheres, it is surprising that the initial air pressure as low as 10 to 20 atmospheres may be high enough to prevent the formation of PSW. This was also confirmed by the numerical simulations [2] for a single value of air gap width.

In the present paper we conduct a systematic numerical study to determine the influence of the initial air pressure on the formation of PSW for various sizes of air gap in the channel. We also include the case of an axisymmetrical channel (Fig. 1b). A simple analytical model was developed in [2] where the criterion for the formation of a precursor shock is Mach reflection of the oblique shock at the channel upper wall (see Fig. 2a). Although results from this simple model agreed quite well with experiments and CFD, it is noted in [2] that the model seems to be in contradiction with some experimental and numerical findings and it cannot predict the effect of width of the air gap. In the present paper we examine the unsteady numerical flowfields in detail to reveal the actual mechanism of the precursor shock formation, which is then subjected to an analytical treatment (both for the planar and axisymmetrical cases). The new analytical model is compared with the CFD results and experimental data for channels with various air gaps.

2 Problem setup and numerical code

We attach the frame of reference to the detonation front which then represents a stationary boundary of the computational domain (see Fig. 2). This approach is considerably

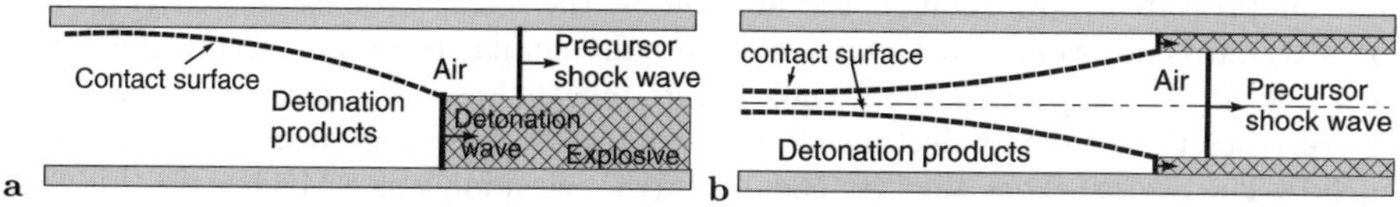

Fig. 1. Schematic of the precursor shock wave formation in **a** planar and **b** axisymmetrical channels.

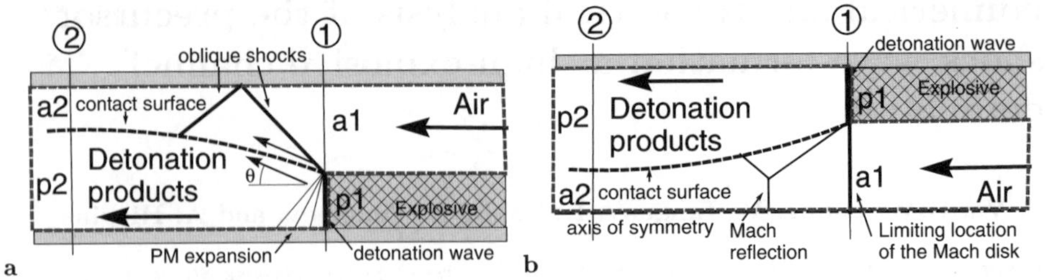

Fig. 2. Schematic of the computational domains (red dashed lines) and flow patterns in **a** planar and **b** axisymmetrical channels.

simpler than moving boundary techniques or the simulation of the detonation wave and energy release using a shock-capturing scheme. The thickness of the explosive layer (nitromethane) is kept constant at 5 mm while the air gap is varied in size from 5 mm to 30 mm. The flow parameters at the detonation wave boundary correspond to the CJ condition for detonation products, as determined using Cheetah 2.0 [3]: $M = 1$, $V = V_{CJ} = 4482$ m/s, $p = p_{CJ} = 118,900$ bar, $\rho = \rho_{CJ} = 1591$ kg/m^3. In the reference frame attached to the detonation wave the air stream enters from the right (Fig. 2) at $V = V_{\text{det}} = 6149$ m/s, $T = 300$ K, and selected initial pressure. The air and detonation products are treated as ideal gases with $\gamma = 1.4$ and $\gamma = 2.69$ (from Cheetah), respectively. The numerical simulations are performed using a two-dimensional, locally adaptive, unstructured Euler solver [4].

3 Numerical flowfield analysis

In the numerical simulations, the detonation is initiated inside the channel (Fig. 2). As soon as the computation starts the detonation products enter the domain and begin to expand. An oblique shock wave then forms, deflecting the air stream along the contact surface which separates the channel air from the expanding products. The oblique shock reflects from the upper wall as schematically shown in Fig. 2a. For all air gaps and initial air pressures considered in the present study, the reflection is always of regular type, not exhibiting transition to Mach reflection in the course of flow development. It appears that the expanding detonation products and the upper channel wall form a converging duct through which the channel air flows. The primary oblique shock undergoes multiple interactions with the wall and the contact surface as shown in Fig. 3a. The air flow is not choked at that moment (see the sonic line in the right image). The subsequent flow development depends on the initial air pressure. If the pressure is sufficiently high, it prevents further expansion of the detonation products, the air flow in the duct remains non-choked, the flowfield similar to Fig. 3a stabilizes and PSW is not formed. If the initial air pressure is low enough, the air flow in the duct becomes choked (Fig. 3b). Soon afterwards a normal shock wave, conceptually similar to an unstarting shock in a supersonic air inlet, appears near the choked duct throat (Fig. 3c). It propagates upstream and overtakes the regular reflection of the primary oblique shock from the upper wall (Fig. 3d). The normal shock then propagates further seemingly as a Mach stem (Fig. 3e). Eventually it goes past the detonation front (Fig. 3f) and becomes a PSW.

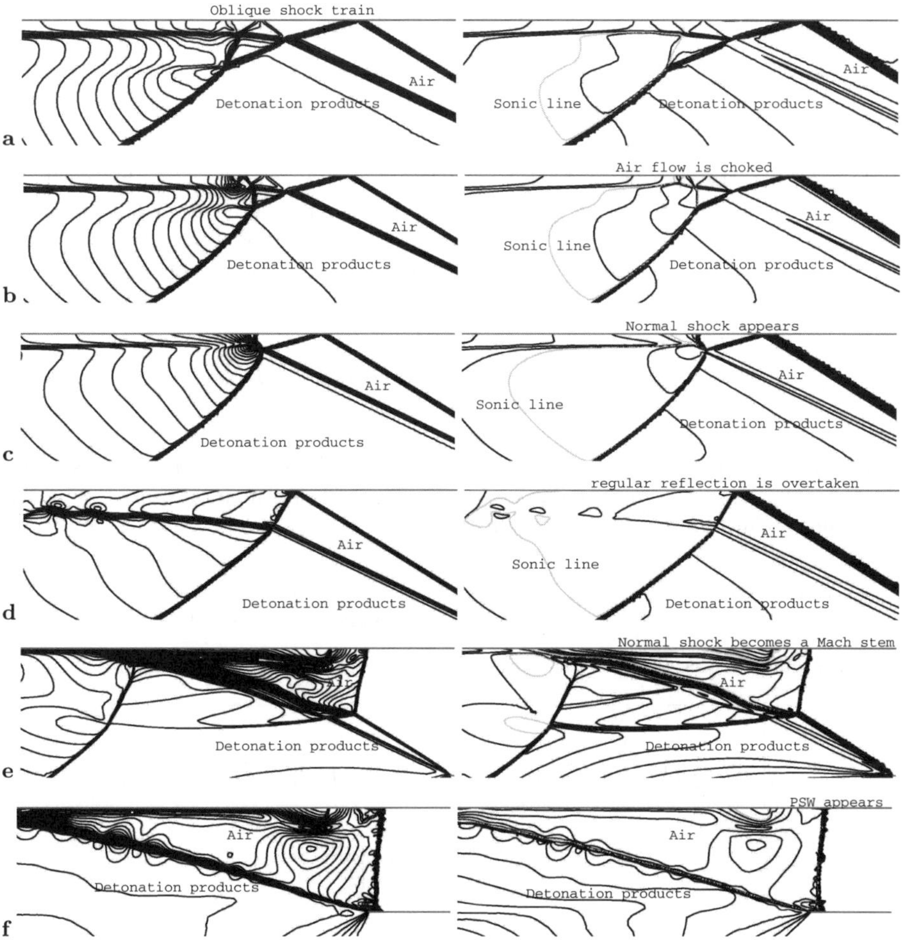

Fig. 3. Flow evolution (**a-f**) near the upper wall of the planar channel (Fig. 2). Right images – Mach number contours ($M = 1$ shown as green line); left images – velocity contours. The air gap is 5 mm; the initial air pressure is 65 atm.

4 Quasi-one-dimensional analytical model

The numerical simulations have shown that a precursor shock wave forms when the flow in the air gap is choked. This is analogous to the unstarting of a supersonic diffuser. The mechanism of PSW formation can be subjected to an analytical treatment similar to [5], assuming that the two flows of air and detonation products, on either side of the contact surface, can each be considered as a quasi-one-dimensional one. Let us consider two cross-sections of the quasi-one-dimensional ducts: entry cross-section 1 at which all parameters are known, and cross-section 2 at which the air gap is choked ($M_{a2} = 1$), see Fig. 2 for notations.

Two more criteria are needed to solve the problem. The first, which is a purely geometrical one, is that the total channel cross-sectional area remains constant:

$$A_{a1} + A_{p1} = A_{a2} + A_{p2} \ . \tag{1}$$

The second criterion is that the pressures at sections $a2$ and $p2$ be equal:

$$p_{a2} = p_{p2} \ . \tag{2}$$

Strictly speaking, the pressure in both layers should match everywhere at the interface. However, this cannot be achieved in our one-dimensional treatment. Therefore, it is assumed that the pressure matches only at the throat, which hopefully, is sufficiently far downstream of the detonation for multi-dimensional effects to be negligible.

Using pressure ratios and assuming fully isentropic flow of the detonation products, Eqn. (2) becomes:

$$\left(\frac{p_{a2}}{p_{0a2}}\right)_{M_{a2}=1} \left(\frac{p_{0a2}}{p_{0a1}}\right)\left(\frac{p_{0a1}}{p_{a1}}\right)_{M_{a1}} p_{a1} = \left(\frac{p_{p2}}{p_{0p2}}\right)_{M_{p2}} \left(\frac{p_{0p1}}{p_{p1}}\right)_{M_{p1}=1} p_{p1} \tag{3}$$

Here the decrease in stagnation pressure for air due to the presence of shocks is to be evaluated later; M_{p2} is an unknown. Introducing cross-sectional area ratios and using the $p_o A^* = const$ relation, we have instead of Eqn. (1):

$$A_{a1} + A_{p1} = \left(\frac{A_{a2}}{A^*_{a2}}\right)_{M_{a2}=1} \left(\frac{p_{0a1}}{p_{0a2}}\right)\left(\frac{A^*_{a1}}{A_{a1}}\right)_{M_{a1}} A_{a1} + \left(\frac{A_{p2}}{A^*_{a1}}\right)_{M_{p2}} A_{p1} \ . \tag{4}$$

The system of Eqns. (3) and (4) can be rewritten in the following manner:

$$M_{p2} = \sqrt{\frac{\gamma_p+1}{\gamma_p-1}\left[\left(\frac{2+(\gamma_a-1)M_{a1}^2}{\gamma_a+1}\right)^{\frac{\gamma_a}{\gamma_a-1}}\left(\frac{p_{0a2}}{p_{0a1}}\right)\left(\frac{p_{a1}}{p_{p1}}\right)\right]^{\frac{1-\gamma_p}{\gamma_p}} - \frac{2}{\gamma_p-1}} \ , \tag{5}$$

$$\frac{A_{a1}}{A_{p1}} = \frac{1 - \frac{1}{M_{p2}}\left[\frac{2+(\gamma_p-1)M_{p2}^2}{\gamma_p+1}\right]^{\frac{\gamma_p+1}{2(\gamma_p-1)}}}{M_{a1}\frac{p_{0a1}}{p_{0a2}}\left[\frac{\gamma_a+1}{2+(\gamma_a-1)M_{a1}^2}\right]^{\frac{\gamma_a+1}{2(\gamma_a-1)}} - 1} \ . \tag{6}$$

Substituting Eqn. (5) into Eqn. (6), one obtains the critical area ratio at a given air pressure required to choke the air gap. Inverting, one can plot the critical pressure for given channel dimensions.

To solve the above equations, one must model the stagnation pressure drop across the train of oblique shocks in the air gap. Let us consider the primary oblique shock. This shock is curved because the Prandtl-Meyer expansion (Fig. 2a) reflects from the bottom of the channel and interacts with the shock. As a first approximation, let us consider the explosive thickness to be infinite and therefore, the shock to be straight. To find its strength (angle) we need to solve the well-known self-similar problem of the interaction of two supersonic streams. First, note that both the flow turning angle and pressure behind the oblique shock must match the flow turning angle and pressure behind the Prandtl-Meyer expansion:

$$\theta_{pC} = \theta_{aC} \ \ , \ \ p_{pC} = p_{aC} \ . \tag{7}$$

The subscript C refers to the contact surface between the air and the detonation products.

For the detonation products undergoing a Prandtl-Meyer expansion the turning angle and the pressure can be related to the flow Mach number:

$$\theta_{pC} = \nu(M_{pC}) = \sqrt{\frac{\gamma_p + 1}{\gamma_p - 1}} \arctan \sqrt{\frac{\gamma_p - 1}{\gamma_p + 1}(M_{pC}^2 - 1)} - \arctan \sqrt{M_{pC}^2 - 1} \ , \qquad (8)$$

$$p_{pC} = p_{p1} \left[\frac{\gamma_p + 1}{2 + (\gamma_p - 1) M_{pC}^2} \right]^{\frac{\gamma_p}{\gamma_p - 1}} \ . \qquad (9)$$

For the flow across the oblique shock wave we have:

$$\tan \theta_{aC} = \frac{2(M_N^2 - 1)\cot \sigma}{(\gamma_a + 1) M_{a1}^2 - 2M_N^2 + 2} \ , \quad p_{aC} = p_{a1} \frac{2\gamma_a M_N^2 - (\gamma_a - 1)}{\gamma_a + 1} \ , \qquad (10)$$

where $M_N^2 = M_{a1}^2 \sin^2 \sigma$, σ is the shock angle. The above system can be solved simultaneously for the shock angle. With the shock angle known, it is straightforward to calculate the stagnation pressure drop across the oblique shock. Similarly, one can calculate the reflected shock angle and include the stagnation pressure drop due to the reflection.

The above model is valid for rectangular channels. In the axisymmetrical case, the criterion for precursor shock wave formation is slightly different. In the axisymmetrical case the oblique shock reflection from the axis of symmetry is always of Mach type as schematically shown in Fig. 2b. The subsonic flow behind the Mach disk (essentially a normal shock) accelerates and is choked at the throat formed by the expanding detonation products. Our numerical simulations show that the size of the Mach disk increases and it moves to the right (Fig. 2b) when the initial air pressure decreases. Therefore, for formation of a precursor shock wave, the normal shock (Mach disk) must be located at the "lip" of the inlet, corresponding to the location of the detonation wave (the limiting location is shown in Fig. 2b). This is analogous to the unstarting of an axisymmetric supersonic inlet diffuser. This criterion is even simpler than in the rectangular geometry. In this case, Eqns. (1)-(6) still hold while the stagnation pressure drop term corresponds to that across a normal shock wave with Mach number M_{a1}.

5 Comparison and discussion

The numerical and analytical results are compared with the experiments [1,2] in Fig. 4. It turns out that all three sets of data are qualitatively similar, predicting the same trend and even the same order of magnitude for the critical pressure in spite of gross oversimplifications built into the numerical and analytical models. Below we address some factors which might influence our models in an attempt to achieve better quantitative agreement and/or explain the reasons behind the existing discrepancies.

As seen from Fig. 4b, the value of γ for the detonation products significantly influences the results. Thus, one of the future research directions is to repeat the simulations using the JWL equation of state for the detonation products. The temperature of the air behind PSW reaches approximately 10,000 K and high-temperature gasdynamics effects are to be taken into account as well.

In the experiments, nitromethane was initiated outside the channel and afterwards the resulting detonation wave and expanded detonation products entered the channel. The possible influence of this entry stage on the precursor formation is evaluated for 30 mm air gap using numerical simulations. The reference frame is still attached to the detonation wave. Initially a steady state without the upper wall is reached. At the next stage of the simulation the upper wall moves and interacts with the oblique shock and detonation products. It turns out that the critical pressure remains essentially the same.

Another observation is that at the initial air pressures close to the critical one the formation of the precursor is relatively slow and it is not clear whether or not a PSW has been formed when the detonation front reaches the end of the 2m long channel used in the experiments. It takes 325 μs for the detonation wave to cover the distance. In the computation illustrated in Fig. 3 (5 mm air gap, 65 atm initial air pressure as compared to the critical value of 76.5 atm) the PSW forms (i.e., it is observed ahead of the detonation wave) after approximately 30 μs. Thus, it seems to be possible that PSW formation may take longer than 325 μs at the initial air pressures close to the critical one. Experimentally, such a case would be considered as "no precursor" one. It is clear from the above values that the experimental error due to the effect would not be large enough to explain the apparent discrepancies in Fig. 4. However, it is to be noted that in the experiments [1] the precursor is additionally slowed down due to the presence of boundary layers. This effect is not taken into account in our numerical simulations.

The numerical and analytical models also neglect viscous and turbulent effects. However, from the considerations on the boundary layer influence given in [1] as well as from simple Fanno flow with variable area analysis it follows that boundary layer effects should lead to higher critical pressures, not lower ones.

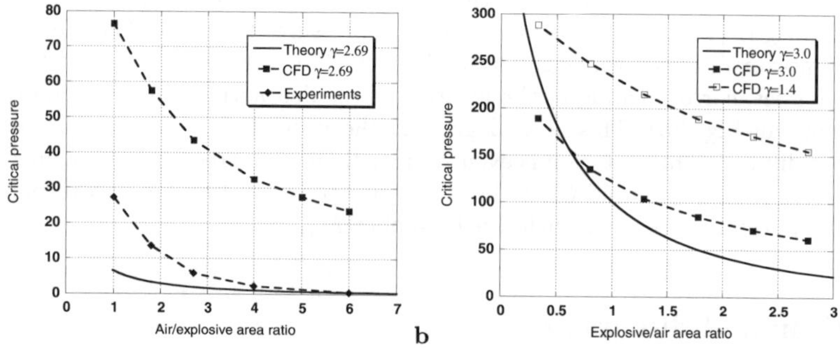

Fig. 4. Comparison of the theoretical, numerical and experimental results for the critical air pressure (in atm) above which PSW is not formed, for **a** planar and **b** axisymmetrical channels.

Acknowledgement. The study was partially supported by NSERC Discovery grant 298232-2004.

References

1. Tanguay V., Higgins A.J.: J. Applied Physics **95**(11):6159-6166 (2004)
2. Tanguay V., Vu P., Oliver P.J., Timofeev E., Higgins, A.J.: AIAA Paper 2005-0278 (2005)
3. Fried L., Howard W.M., Souers P.C.: Cheetah 2.0 User's Manual, Lawrence Livermore National Lab, UCLR-MA-117541 Rev. 5 (1998)
4. Tahir R., Voinovich P., Timofeev E. (1993-2007) `Solver II` – Two-dimensional, multi block, multi-gas, adaptive, unstructured, time-dependent CFD application for MS Windows. Email: rabi.tahir@rogers.com, timofeev@sympatico.ca
5. Mitrofanov V.V.: Acta Astron. **3**, 995-1004 (1976)

Numerical study on shockwave structure of superdetonative ram accelerator

K. Sung[1], **I.-S. Jeung**[1], **F. Seiler**[2], **G. Patz**[2], **G. Smeets**[2], **and J. Srulijes**[2]

[1] *Institute of Advanced Aerospace Technology, School of Mechanical and Aerospace Engineering, Seoul National University, Seoul, Republic of Korea*
[2] *Shock Tube Department, French-German Research Institute of Saint-Louis(ISL), P.O.Box34, F68301, Saint Louis, France*

Summary. A numerical study was conducted to investigate shockwave and detonation structure of superdetonative mode Ram Accelerator, based on ISL's RAMAC30 II Shot225 experiment. Govern equation is discretized by Roe's FDS and integrated by LU-SGS implicit time integration. $H_2/O_2/CO_2$ detailed chemical reaction for high pressure was considered. By computation result, shockwave structure is determined by two separation bubble. Detonation wave is sustained by second separation bubble and second separation bubble is growing by pressure difference of detonation wave.

1 Introduction

The concept of ram accelerator is that projectile is flying with synchronized combustion in tube filled with pre-mixed combustible gas mixture. Ram Accelerator facilities have major benefits for hypersonic research. The gas dynamics in ram accelerator is very similar to those in scramjet and oblique detonation wave engines. Therefore, understanding of ram accelerator operation will enhance understanding of supersonic combustion and hypersonic propulsion system.

Despite the number of research programs around the world, a maximum speed was 2.7 km/s only available at the UW (University of Washington) ram accelerator facility starting with subdetonative speeds [1]. Thus, to obtain higher velocity, combustion mode must be transferred to the superdetonative mode, where ignition and combustion occur at supersonic flow speeds. Based on this motivation, ISL (French-German Research Institute of Saint-Louis) has developed a rail tube version of a ram accelerator facility named RAMAC 30 version II that bypassed the gasdynamic transition from subdetonative to superdetonative ignition. In this facility, a cylindrical projectile having conical fore- and after-bodies was launched by a powder gun and accelerated in a ram accelerator tube having four or five guide rails [2]. Although the initial launching speed of a powder gun was only about 1.8 km/s, the superdetonative launch was possible by using the $H_2/O_2/CO_2$ mixture having a lower C-J (Chapman-Jouguet) detonation wave speed than the launching speed.

In the previous experiments, ISL's RAMAC 30 demonstrated that ignition was successful with an aluminium projectile but no combustion was observed in the case of a steel projectile. After the experiment, the aluminium projectile was significantly ablated [3]. These show that there is an important ignition mechanism which is strongly related to the aluminium projectile surface's friction, heat conduction, and combustion.

At the earlier stage of ram accelerator studies, the superdetonative mode operation had been considered to be sustained by an oblique detonation wave in the combustor. Computational studies by Yungster and Bruckner [4] and Li et al. [5] using inviscid flow model and chemical kinetics showed that ram acceleration was possible through this concept at very high velocity ranges. However, more recent viscous analyses by Yungster [6] and Choi et al. [7] showed that combustion could be initiated in the boundary layer due to the aerodynamic heating associated with shock wave/ boundary layer interaction at intermediate velocity ranges where shock-heating was insufficient for mixture ignition. The ISL's ram accelerator experiments, however, revealed that the previous studies on superdetonative combustion characteristics were not applicable to the experimental case and the other combustion characteristics might be more important for the low speed superdetonative mode of operation. Therefore, to understand the combustion mechanism, numerical simulations was conducted for ISL's RAMAC 30 experiments using the experimental configuration and conditions in this study.

2 ISL's RAMAC Test Facility

Based on the need for a hypersonic launching facility, ISL has built two ram accelerators: a 30-mm-caliber-tube, called RAMAC 30, and 90-mm-one, RAMAC 90. The superdetonative combustion mode has been mainly tested in RAMAC 30 which was implemented with rail tube version II since 1997. Figure 1 shows the RAMAC 30 test facility, which consisted of a pre-accelerator tube, a ram tube containing a combustible gas mixture with both ends sealed by diaphragms which were hit and destroyed by the moving projectile forebody nose tip, and a decelerator tube. Two tubes with a length of 2.4m each were used forming a total ram tube length of 4.8m. Projectiles had an inner magnesium core which was fully covered by an aluminum (or sometimes steel was used for different experimental shots) in the combustor as well as at the fore- and afterbodies. Cylindrical projectiles of 130g-150g, 3.0 cm caliber, and 16.1 cm long, could be accelerated to a speed of around 1800 m/s at the exit of the pre-accelerator tube before propagating through the mixture in the rail tube version II.

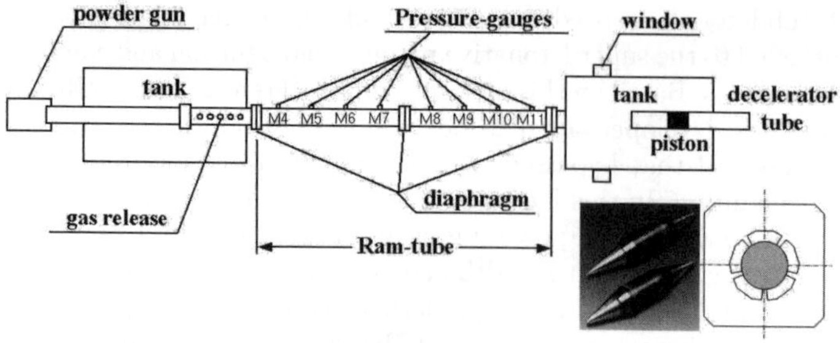

Fig. 1. Schematic of ISL's RAMAC 30 Version II

3 Numerical Methods

For computational study of ram accelerator, a fully coupled form of multi-species conservation equations and Reynolds averaged Navier-Stokes equations coupled with Baldwin-Lomax turbulence modeling was used for axisymmetric geometry. Typical operational pressure of ram accelerator is higher than 50atm and reduced kinetic mechanism for low pressure can not applicable. Therefore finite-rate chemistry model of high pressure should be considered. Petersen and Hanson [8] developed reduced mechanisms to model the combustion characteristics of typical ram accelerator mixtures at pressures approaching 300atm based on the GRI-Mech. In this study, fully detailed 10 species (H, H_2, O, O_2, OH, H_2O, HO_2, H_2O_2, CO, CO_2) and 29 step reaction model for combustion of $H_2/O_2/CO_2$ mixture was considered. Govern equation is discretized by finite volume cell vertex approach. Viscous flux is discretized by central differencing and convective flux is obtained by Roe's flux difference splitting method [9]. Primitive variables are extrapolated at cell interface by MUSCL (Monotonic Upstream method for Scalar Conservation Law) scheme [10]. Discretized equation is integrated by LU-SGS scheme [11]

4 Computational Modelling

Figure 2 shows computational conditions and domain. Diameters of a projectile and a tube are 3.0 cm and 4.2 cm, respectively. In the experiment, the tube wall had a decagonal cross-section with five rails but in this study assumed circular for the axisymmetric simulations. Initial launching speed was 1800m/s from the experiment. Acceleration and speed of projectile is computed and inflow speed is updated by every step. Mixtures for shot 225 had a pressure and temperature of 40 bar and 300K, respectively. The gas composition used for shot 225 was $2H_2+O_2+5CO_2$: a stoichiometric H_2/O_2 mixture diluted with 5 moles of CO_2. The computational domain for the simulation was extended by 1cm before and after the projectile, and it was covered by the 380x100 computational grids that was uniformly distributed in the axial direction and clustered to both walls in the radial direction. Since these simulations had to cover overall flow features of the scale of the entire projectile, the computational resolutions used were limited. Therefore, details of the detonation structure, such as the induction region, might not be well resolved. But our purpose was exploring the overall development and so, current grid resolution was enough to represent the shock and detonation position because the local transit time was sufficiently smaller than the induction time of $2H_2+O_2+5CO_2$ mixture behind the reflected shock where mixture ignition occurred.

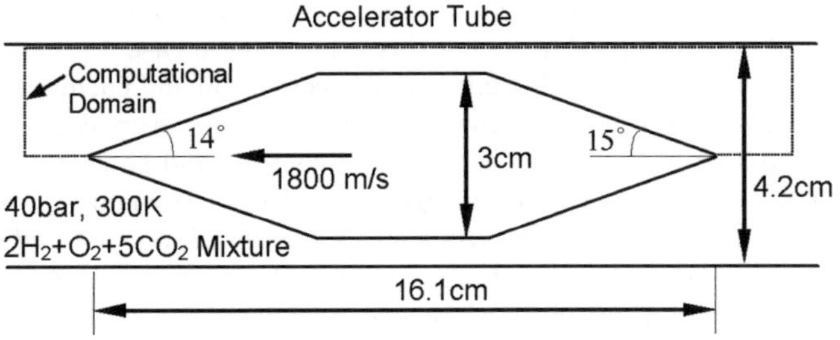

Fig. 2. Computational Domain/Condition

5 Result and Analysis

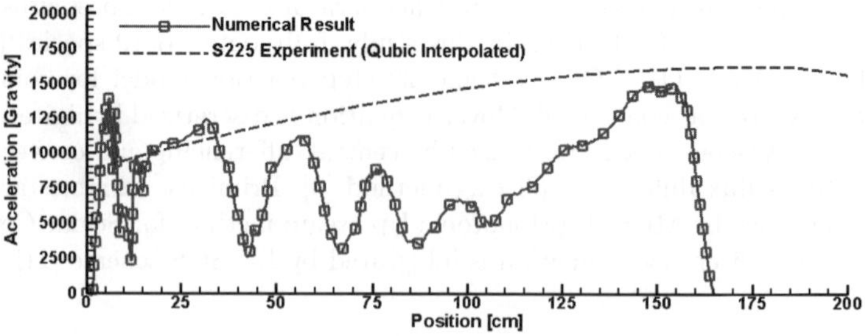

Fig. 3. Acceleration of Projectile

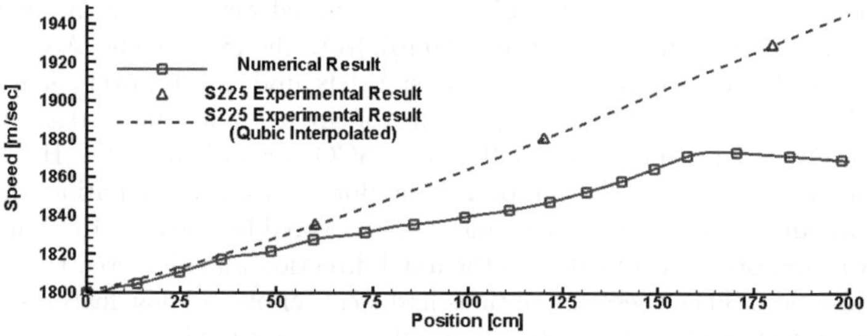

Fig. 4. Speed of Projectile

Figure 3 shows acceleration of projectile. For numerical result, acceleration is fluctuating and peak is decreasing in range 13cm 100cm. Acceleration is maximun at 150cm and after 165cm, acceleration is negative and speed of projectile is decreasing. In experimental case, acceleration is slightly increasing. Figure 4 shows speed of projectile. Initial trend of speed increasing is similar with experimental case. But due to fluctuating of acceleration speed is not increasing as much as experimental case.

Projectile position 55cm is local maximum of acceleration and 65cm is local minimum of acceleration in Figure 3. Figure 5 shows pressure and temperature contour at projectile position 56cm and Figure 6 shows Mach number contour and shockwave structure. In pressure contour, we can find detonation region near 10cm. At Figure 6 shockwave from nose cone is reflected at tube wall and meet a shockwave which is generated by separation bubble in combustor. Behind 8.2cm, there is expansion wave and this is recompressed around 9cm. Strong recompression shockwave is reflected at 10cm of tube wall. Detonation wave is located behind this point. Behind detonation wave, high temperature and pressure gas is expanded on separtion bubble. Expansion wave is reflected at projectile. Pressure of detonation is very high, and enough to acceleration projectile. Pressure on expansion nozzle is much higher than one of nose cone.

Figure 7 shows pressure and temperature contour at projectile position 66cm and Figure 8 shows Mach number contour and shockwave structure. Shockwave structure

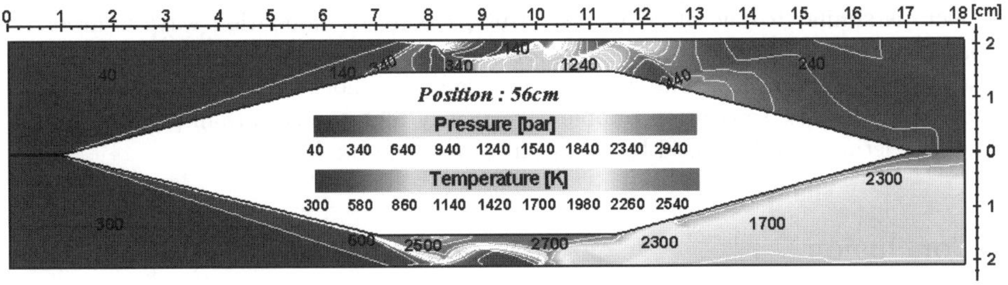

Fig. 5. Pressure and Temperature Contour at 56cm

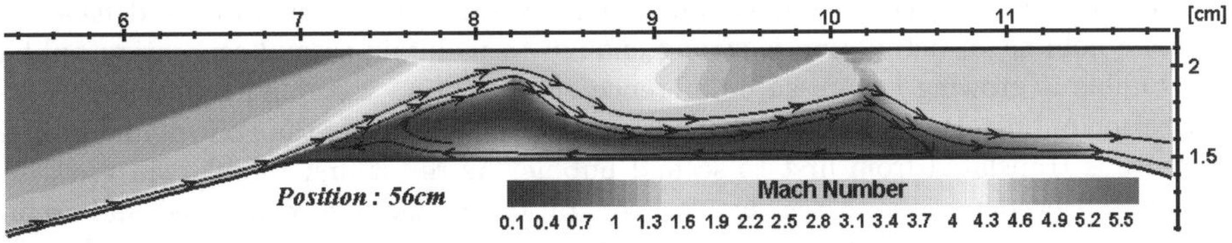

Fig. 6. Mach Number Contour at 56cm

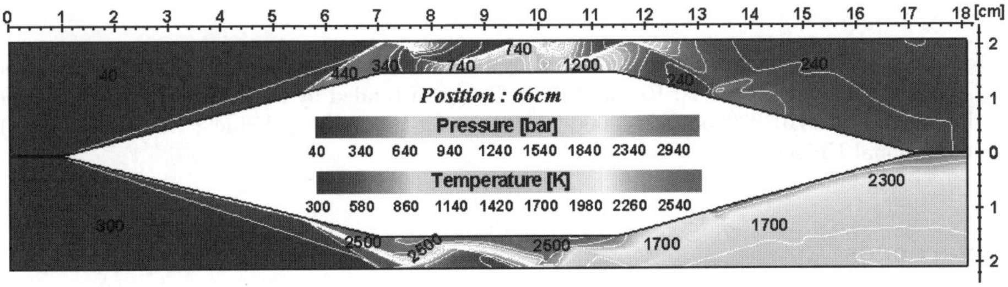

Fig. 7. Pressure and Temperature Contour at 66cm

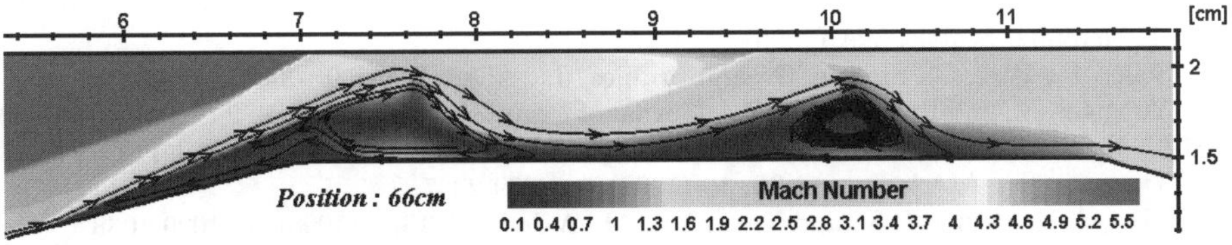

Fig. 8. Mach Number Contour at 66cm

of 66cm is similar to one of 56cm case. But pressure of detonation wave is lower and separation bubble is located more front of combustor. In Figure 8, we can see two separation bubble. Front first bubble located at 6cm and shockwave induced more front of combustor. Because separation bubble is relatively high pressure than pressure on nose cone, acceleration is decreased by this front separation bubble.

6 Conclusion

Shockwave structure is determined by two separation bubble on projectile surface. Figure 5-8 show role of separation bubble. First separation bubble make induced shockwave and pre-heat the inflow. Second separation bubble sustain detonation wave and have role of flame holding. If detonation is strong enough to maintain second separation bubble, this bubble is growing by pressure difference of detonation wave and connected to first separation bubble. Pressure in second separation bubble is higher than one of first bubble, pressure is translated from first to second bubble. As result first bubble will grow and second will be depressed. After this two separation bubble is disconnected and second bubble is grow by detonation wave. This sequence is repeated in fluctuating acceleration region. But if second bubble is in end of combustor(nozzle front) detonation is weaken and separation bubble cannot be maintained.

Acknowledgement. This work was supported by the Korea Science and Engineering Foundation (KOSEF) through the National Research Lab. Program funded by the Ministry of Science and Technology (No. M10500000072-06J0000-07210) and FVRC(Flight Vehicle Research Center), Seoul National University.

References

1. Elvander, J. E., Knowlen, C., and Bruckner, A. P. : *Ram Accelerators* (Springer-Verlag, 1988) pp 55-64
2. Seiler, F., Patz, G., Smeets, G., and Srulijes, J. : *Second International Workshop on Ram Accelerators*, Seattle, WA (1995)
3. Seiler, F., Patz, G., Smeets, G., and Srulijes, J. : AIAA Paper **98-3445** (1998)
4. Yungster, S., and Bruckner, A: P. : Journal of Propulsion and Power **8**, 2 (1992) pp 457-463
5. Li, C., Kailasanath, K., and Oran, E. S. : Combustion and Flame **108**, 1 (1997) pp 173-186
6. Yungster, S. : AIAA Journal **30**, 10 (1992) pp 2379-2387
7. Choi, J.-Y., Jeung, I.-S., and Yoon, Y. : AIAA Journal **36**, 6 (1998) pp 1029-1038
8. Petersen, E.L. and Hanson, R.K. : Journal of Propulsion and Power **15**, 4 (1999) pp 591-600
9. Roe, P. L. : Journal of Computational Physics **43** (1981) pp 357-372
10. Hirsch, C. : *Numerical Computation of Internal and External Flows, Vol 2* (John Wiley & Sons, 1990)
11. Shuen, S. and Yoon, S. : AIAA Journal **27**, 12 (1989) pp 1752-1760

Numerical study on the self-organized regeneration of transverse waves in cylindrical detonation propagations

C. Wang and Z. Jiang

Key Laboratory of High Temperature Gas Dynamics, Institute of Mechanics, China Academy of Science, 15 Beisihuanxi Road, Beijing, 100080, China

Summary. Cylindrical detonation propagation was numerically investigated by solving the two-dimensional multi-component Euler equations implemented with a one-step chemical reaction model. The numerical results demonstrate the evolution of cellular cell bifurcation of cylindrical detonation, and indicate that new cellular cells are generated from the self-organized transverse waves. The local curvature of the cylindrical cellular detonation is found to be a critical issue in the propagation. Originating from curvature variations, the concave front on the Mach stem between two triple points is developed from the flow expansion induced by both transverse wave motion and detonation front diverging. The concave front will focus later and result in the self-organization of transverse waves from which the cellular cell bifurcation takes place. The self-organization of transverse waves is dominated by shock diverging, flow expansion and chemical reactions.

1 Introduction

It is well known that as the cylindrical or spherical detonation waves expand, there is an inherent regeneration of new transverse waves forming new cells asymptotically toward a roughly constant cell size pattern in a continuously enlarging detonation front. The phenomenon is often termed as the cellular cell bifurcation. The work, conducted by Lee et al. had shown that, with the increasing of the surface area of the diverging detonation, the formation of more than one localized explosion at the end of the cycle of a decaying blast wavelet is required (see Ref.([1, 2])).

Current theories of detonation instability concern the instability of planar detonations under conditions of small disturbances. For example, with Fourier decomposition of the linearized reactive hydrodynamics equations, the detonation instability problem can be reduced to a one-dimensional, time dependent problem given transverse wave number $\varepsilon/2\pi$ (see Ref.([3])). The "square-wave" model of a detonation is also used to study the instability of one-dimensional detonation wave (see Ref. [4, 5]). For diverging cylindrical or spherical detonations, the generation of new transverse waves is due to the cellular instability, that is, the self-organization of transverse waves in detonation front. Obviously, the cellular structure instability of diverging cylindrical or spherical detonation can not be modelled as the planar detonation instability. The essential mechanism of the self-organized regeneration of transverse waves is still not well understood.

In this paper, the instability of cylindrical cellular detonation is considered, and the mechanisms of the self-organized regeneration of transverse wave are investigated. The two-dimensional Euler equation implemented with one-step reversible chemical reaction model is solved by using the NND finite difference scheme. The efforts were made to approach some fundamental mechanisms underlying the self-organized regeneration of transverse waves.

2 Numerical method

2.1 Governing equations

By neglecting the effects of viscosity, heat conductivity, and diffusion, the cylindrically diverging detonation is governed by the two-dimensional Euler equation. For perfect gases the equations can be expressed in Cartesian coordinates in the following form:

$$\frac{\partial U}{\partial t}+\frac{\partial F}{\partial x}+\frac{\partial G}{\partial y}=S \tag{1}$$

where $U=[\rho,\rho u,\rho v,\rho E,\rho\lambda]^T$,$F=[\rho u,\rho u^2+p,\rho uv,(\rho E+p)u,\rho\lambda u]^T$,$G=[\rho v,\rho uv,\rho v^2+p,(\rho E+p)v,\rho\lambda v]^T$, $S=[0,0,0,0,\rho W]^T$. ρ, p, and E are the density, pressure and specific total energy of the fluid, respectively. u and v are the fluid velocities in the x and y direction. λ represents the progress variable measuring chemical reaction and W is the chemical reaction source term. The total energy E is expressed with the internal energy $e=p/[\rho(\gamma-1)]+(1-\lambda)Q$ and the kinetic energy $(u^2+v^2)/2$:

$$E=p/[\rho(\gamma-1)]+(u^2+v^2)/2+(1-\lambda)Q \tag{2}$$

where Q is the heat of chemical reaction per unit mass and γ is the polytropic exponent.
Considering one-step reversible chemical reaction,

$$A\Longleftrightarrow B, \tag{3}$$

W is defined by a simplified Arrhenius law:

$$W=[k_f(1-\lambda)-k_r\lambda]exp(E_a/R_oT) \tag{4}$$

where k_f and k_r represent the forward and reverse reaction rates, respectively. E_a, R_o, and T in Eqn. 4 are the active energy, general gas constant, and temperature. The temperature is given as:

$$p/\rho=[(1-\lambda)R_A+\lambda R_B]T \tag{5}$$

where R_A and R_B are the gas constants of reactant and product. For the reversible chemical reaction, k_r is related to k_f through the equilibrium constants π_e by:

$$k_r=k_f/\pi_e \tag{6}$$

Without taking into account the effects of viscosity, heat conductivity, and diffusion, the reactive fluid is determined by the constant parameters k_f, π_e, E_a, Q, γ, R_A, and R_B, as well as the initial state parameters p_0, ρ_0, and T_0.

2.2 Computational scheme

The convective terms of the governing equation are discretized with NND finite difference algorithm with 2th order spatial precision:

$$\begin{aligned}\triangle F&=(F^+_{i+\frac{1}{2},L}-F^+_{i-\frac{1}{2},L}+F^+_{i+\frac{1}{2},R}-F^+_{i-\frac{1}{2},R})\\ \triangle G&=(G^+_{j+\frac{1}{2},L}-G^+_{j-\frac{1}{2},L}+G^+_{j+\frac{1}{2},R}-G^+_{j-\frac{1}{2},R})\end{aligned} \tag{7}$$

where $i\pm 1/2$ and $j\pm 1/2$ denote the position of half grid nodes, and

$$F^{+}_{i+\frac{1}{2},L} = F_i + \frac{1}{2} minmod(\triangle F^{+}_{i-\frac{1}{2}}, \triangle F^{+}_{i+\frac{1}{2}}),\ F^{-}_{i+\frac{1}{2},R} = F_{i+1} + \frac{1}{2} minmod(\triangle F^{-}_{i+\frac{1}{2}}, \triangle F^{-}_{i+\frac{3}{2}}),$$

$$G^{+}_{j+\frac{1}{2},L} = G_j + \frac{1}{2} minmod(\triangle G^{+}_{j-\frac{1}{2}}, \triangle G^{+}_{j+\frac{1}{2}}),\ G^{-}_{j+\frac{1}{2},R} = G_{j+1} + \frac{1}{2} minmod(\triangle G^{-}_{j+\frac{1}{2}}, \triangle G^{-}_{j+\frac{3}{2}})$$

where *minmod* function is defined as

$$minmod(x,y) = sgn(x) \cdot \max\{0, \min[|x|, y \cdot sgn(x)]\} \tag{8}$$

3 Problem specification

The cylindrical detonation propagation in the present study is schematically shown in Fig. 1a. The cylindrical chamber is initially charged with a certain detonable gas. Detonation is directly initiated by an ignition source in the center of the chamber, and the cylindrical detonation will develop and propagate outward.

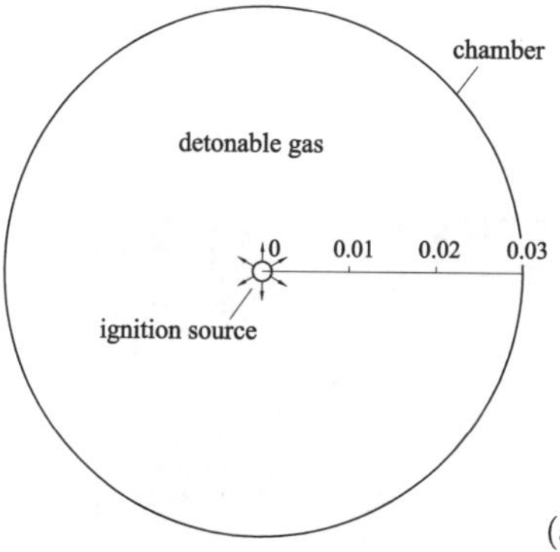

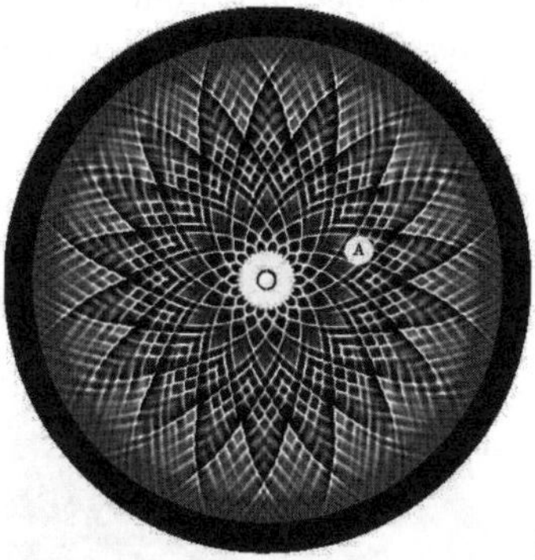

Fig. 1. The model and numerical results of cylindrical detonation propagation

In our numerical simulations, the initial temperature and pressure of the detonable gas are specified at $300K$ and $1atm$. The gas properties of detonable gas and detonation product are selected to be $R_A = R_B = 287.096J/(mol \cdot K)$. The polytropic exponent is taken as $\gamma \equiv 1.4$. The control parameters of chemical reaction are selected as:$k_f = 8.5E8$, $\pi_e = 20$, $E_a = 24kJ$, $Q = 1.58E6J$.

For saving computational time, 1/4 geometry was computed. It should be noticed that in the center of the cylindrical camber, a small circular region is cut to avoid the possible odd points. The inner and outer boundaries are defined as free-slip solid walls, and the radial boundaries are defined as symmetrical boundaries.

The initial ignition source is numerically given as a small region of high pressure and high temperature. After the detonation is initiated, small disturbances on the gas energy are induced to excite the instability of detonation, which helps the rapid detonation of cylindrical cellular structure development.

4 Results and discussion

Fig. 1b shows the numerical cellular structure evolution of the cylindrically diverging detonation waves, which looks similar to the experimental result reported in Ref.([6]). Such the cellular cell evolution indicates a similar process, that is, the self-organization of transverse waves from which the cellular cells remain a roughly similar size. Actually, the cells grow bigger as the detonation expands, but cell bifurcations take place when the cells become sufficiently large. Such cell growing/bifurcating cycles will repeat periodically as the wave front expands, but the periodical time will get longer and longer as the curvature of the cylindrical wave front decreases.

Fig. 2 shows pressure distributions of four successive instants during the period marked with an **A** in Fig. 1b. It is observed that a pair of transverse waves are generated from a triple-points collision and propagates in the directions as indicated with a two-way arrow(Fig. 2a). As the transverse waves propagate the Mach stem, as referred to as in many papers, appears to become a concave surface(Fig. 2b). The concave surface, being a combustion front, focuses itself and a hot spot develops(Fig. 2c). Finally, a detonation bubble develops from the hot spot explosion, and a convex front and a new pair of transverse waves are developed and observable(Fig. 2d).

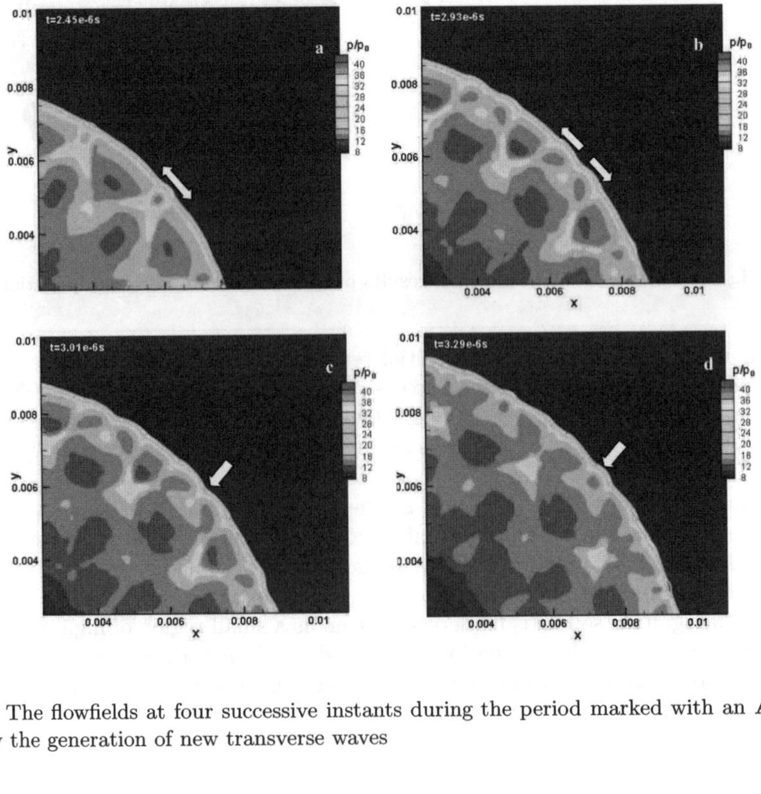

Fig. 2. The flowfields at four successive instants during the period marked with an **A** in Fig. 1b show the generation of new transverse waves

Fig. 2 indicates an important mechanism for transverse wave generation during the cylindrical detonation propagation. The present pair of the transverse waves are generated not from a triple-point collision but from a hot spot explosion. As is well known, the spherical or cylindrical wave front will get much weaker as it expands than the planar detonation wave. Furthermore, two transverse waves propagate away as indicated in Fig. 2b and an additional flow expansion is induced behind them. This expansion acts to further weaken the Mach stem, therefore, the concave front surface occurs when its curvature becomes critical. It is understandable that two kinds of the flow expansion are closely related to the curvature of the detonation front. In the self-organized regeneration process, a certain distance is also necessary to produce enough flow expansion to generate a sufficiently concave front for an effective reaction zone focusing from which a detonation bubble develops, and this distance scale will vary with the detonation front curvature on which the flow expansion depends.

The local speed of detonation varies dramatically from one collision of transverse waves to the next, whereas the average of detonation speed maintains a constant value at $\bar{u} = 2160m/s$, as shown in Fig. 3a. There are three stages in the propagation of the cylindrical detonation. During the detonation initiation stage, the high temperature and pressure in the ignition region induce an over-driven detonation, the speed of which quickly decays to a self-sustained cylindrically diverging detonation. In the first splitting stage, the generation of new transverse waves results in the cellular cell bifurcation, and wave speed oscillation varies in the range of $0.7 \sim 1.4\bar{u}$ due to flow expansion as discussed above. At the time about $6.3\mu s$, the generation of new transverse waves reaches to the secondary splitting stage, which may results from wave front instability. Before the secondary splitting stage, the wave speed undergoes a deeper drop which may come from a possible local decoupling and re-initiation of the cylindrical detonation.

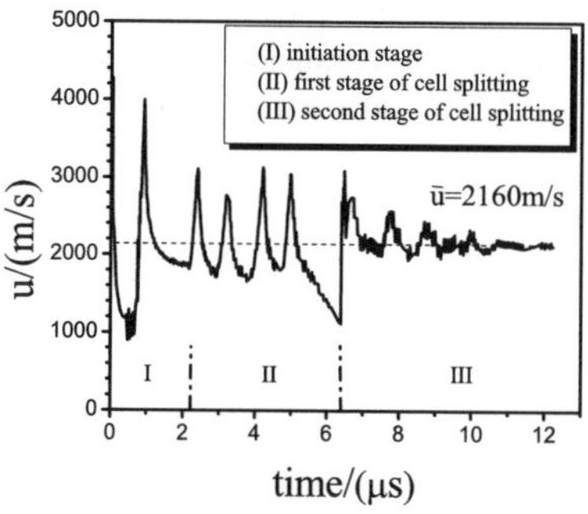

(a)Local detonation speed

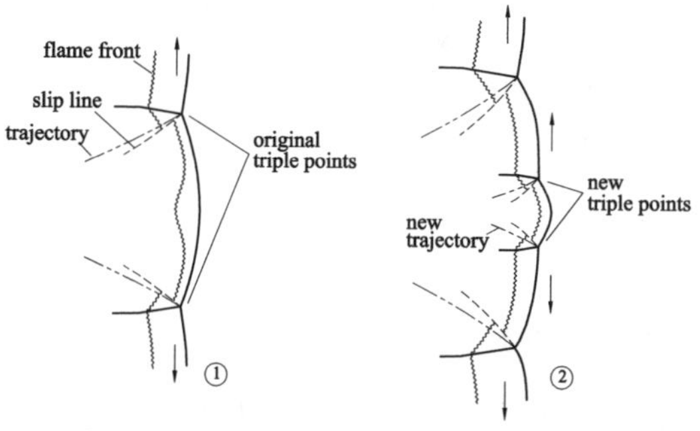

(b)New transverse waves generation

Fig. 3. The local detonation speed and the mechanism of new transverse waves generation

A schematic mechanism of new transverse waves generation can be modelled in Fig. 3b. In the propagation of diverging cylindrical detonation, the Mach stem expands much faster than that of a planar detonation. In certain position between triple points, the bend of Mach stem results in a concave curvature of flame front. With the changes in curvature of leading shock and flame front, and the interaction of chemical reaction and flow expansion, a local explosion is induced and a pair of new transverse waves are generated. Finally, By the self-adjustment of transverse waves, the transverse waves reach to uniform distribution in the circumferential direction.

The cellular structures of detonation can be stable in slightly expansion duct, that is, the cell will not split instantaneously when the cell size depart from its proper size in an unconfined space. So there is a regular cell zone in the flowfield, where the cell maintains a fixed number within a certain distance. When the area of duct expands further, the enlarged cellular cells prepare for their next splitting. In the radial direction, the cellular structures undergo a process consists of splitting, stabilization, and further splitting, which is a typical phenomenon of diverging detonation waves.

The cellular structure instability of detonation involves many factors which may cause various modes of cellular structure splitting or merging. The self-organized regeneration of transverse waves in this paper is a typical characteristic of cylindrically diverging detonations, which reflects the basic mechanism of detonation instability in different manner from planar detonation. The generation of the new transverse waves comes from the bend of detonation front. This can also be concluded by analyzing the experimental cellular structures. Because the influence of the viscosity of fluid and turbulence is not considered in numerical simulation, the self-organization of transverse waves is dominated by shock diverging, flow expansion and chemical reactions.

5 Concluding remarks

In the self-organized regeneration of cylindrically diverging detonation, the flow expansion is induced between two transverse waves which propagate away. This expansion weakens the Mach stem and further leads to the concave shock front development. The concave reaction zone focusing forms a local explosion (hot spot), which results in the generation of new transverse waves. The critical curvature is a key parameter for cellular cell bifurcation and the interaction of shock diverging, flow expansion and chemical reactions is the primary mechanism underlying the cylindrical detonation propagation.

Acknowledgement. This paper was supported by the National Natural Science Foundation of China (90205027, 10602059) and China Postdoctoral Science Foundation(2005037444)

References

1. Lee, J. and Lee, B., Cylindrical Imploding Shock Waves, Physics of Fluids 1965:2148-2152.
2. John H.S. Lee, Initiation of Gaseous Detonation, Ann. Rev. Phys. Chem. 1977.28:75-104.
3. Jerome J. Erpenbeck, Detonation stability for disturbance of small transverse wavelength, The Physics of Fluids, vol.9, 1966:1293-1306.
4. Zaidel, R.M., and Zeldovich, Ya. B., One-dimensional instability and attenuation of detonation. Zh. prikl. Mekh. Tekh. Fiz. 1963(6):59-65. (English Translation: WrightPatterson Aire Force Base, Dayton, Ohio, FTD-MT-64-66, p.85)
5. Alpert, R.L. and Toong, T.Y., Periodicity in exothermic hypersonic flows about blunt projectiles. Astronaut. Acta 1962(17):539-560.
6. Sun C W,Wei Y Z,Zhou Z K., Applied Detonation Physics, Beijing:Defense Industry Press,2000.3(in Chinese).

On the mechanism of detonation initiations

Z. Jiang[1], H. Teng[1], D. Zhang[1], S.V. Khomik[2], and S.P. Medvedev[2]

[1] *LHD of Chinese Academy of Sciences, Institute of Mechanics, Beijing 100080, China*
[2] *Institute of Chemical Physics, Russia Academy of Sciences, 119991 Moscow, Russia*

Summary. Two essential issues in detonation initiations were discussed in this paper, that is, hotspots and reaction zone acceleration. By defining the reaction zone acceleration, the hotspot and the reaction zone acceleration are identified as the primary mechanisms underlying the deflagration to detonation transition. The dominant physical process for two mechanisms is the interaction between gasdynamic non-linearity and chemical reactions that are very sensitive to temperature gradient. In the process, spontaneous waves are generated in the reaction zone, propagate with the local sound speed, decelerate suddenly in the front of the reaction zone because of the sharp temperature gradient, and develop into a pressure pulse acting to elevate the gas temperature that will further induce more intensive chemical reactions. Such a positive feedback loop supports detonation initiation and propagation.

1 Introduction

There are two modes of detonation initiations as classified by Lee [1]: a slow mode where the detonation is developed via an accelerating flame front and a fast mode where the detonation is driven out "instantaneously" by a strong blast that is created by a sufficiently powerful igniter. The slow mode is usually referred to as the self-initiation via the deflagration to detonation transition (short for, DDT). The fast mode is referred to as a direct initiation. However, even when the ignition energy is substantially high above the critical value for the direct initiation, the blast is always observed to decay first below the C-J value, and then, re-accelerates itself back to the C-J value at large radius [2]. This excursion is now known as the quasi-steady period before the final onset of detonation, and also considered to depend on DDT. So, it is believed that exploring DDT is not only important for the self-initiation, but also for the direct initiation.

DDT is a primary combustion phenomenon and has been called one of the major unsolved problems in combustion theory [3] for years because the mechanisms underlying DDT in gaseous energetic mixtures are still not well understood. many experiments on DDT were carried out in the past several decades and the pioneer work showed that DDT is an extremely complex process involving deflagrations, shock interactions, shear layers, turbulence, chemical reactions, flow instability, and all of their interactions with each other. Two issues had been reported to be important for DDT: hotspots and flame acceleration. The hotspots seem to be well understood than the flame acceleration because more complex processes are included in it. To develop an universal theory for describing the mechanisms underlying detonation initiation, more elementary phenomena need to be defined, which must be simple and not involved with each other.

This paper is dedicated to exploring primary mechanisms in detonation initiations by examining two essential issues, i.e. hotspot and reaction zone acceleration. Numerical simulations are carried out by solving two-dimensional multi-component Navier-Stokes equations implemented with an elementary chemical reaction model. Numerical results of two test cases are reported and discussed in detail.

2 Problem specifications

Figure 1(a) shows the sketch of toroidal shock focusing of the first case and its geometric domain consists of a shock tube and solid cylinder, which are co-axially installed with a given scale ratio of $d : D$. The whole domain is filled with the reactive gas of $2H_2 + O_2 + 4N_2$ at the 1.0 atm pressure and the 300 K temperature. The incident shock wave is initially posited near the left boundary, and diffracts along the end of the solid cylinder, and then implodes toward the axis of symmetry. The diffracting shock wave will decay first, but get stronger and stronger as its radius decreases. The converging process deposits some energy around the focal point and provides hotspots with the required thermodynamic environment. The energy thus deposited can be adjusted by changing either the incident Mach number or the domain configuration scale, so that it can be convenient to generate a hotspot and trigger a detonation bubble at a expected location.

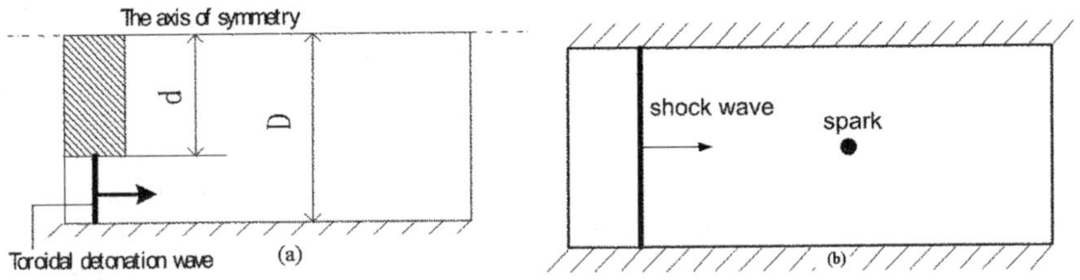

Fig. 1. Schematic of toroidal shock focusing and shock/spark-induced-deflagration interaction

The computational domain for the second test case is shown in Fig.1(b). This is the incident shock/spark-induced-deflagration interaction in a two-dimensional channel and was examined in many papers, for example, in Oran's work [3]. The deflagration front is generated from an electric spark and interacts with the coming incident shock. The detonation initiation will be induced from reaction zone instability.

3 The role of hotspots in detonation initiations

The"hotspot" has been recognized to be universal to all modes of detonation initiations since it was reported as "explosion in the explosion" by Oppenheim [4]. In his paper, a sequence of typical schlieren photographs demonstrated the hotspot was generated from a turbulent flame brush and developed into a detonation bubble. The experiments following his work indicated that the hotspot can originate either at flame fronts, boundary layers, disturbed contact surfaces and locations at which shock waves merged, and its appearance may vary from event to event. The exact physical nature of the process prior and subsequent to the hotspot explosion that leads to a detonation establishment has attracted many researchers for decades, but still remain to be not well understood.

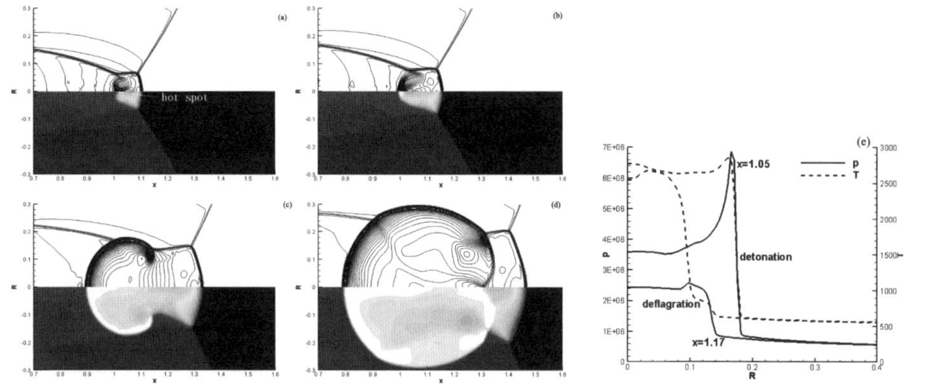

Fig. 2. Sequential pressure and temperature distributions during hotspot development with the pressure and the temperature profiles across detonation bubble at a given moment

Setting the incident Mach number to be 2.70 and 2.72 for the first case, two numerical experiments were carried out. Numerical results showed that the physical phenomena occurring in two tests are dramatically different with each other. For $M_i = 2.70$, the shock reflection pattern is the same to the shock wave focusing in non-reactive gases and no significant chemical reactions are observable. For $M_i = 2.72$, the results in the region around the focal point at four instants are plotted in Fig. 2 and showed that a hotspot was generated and developed in a detonation bubble later. At the beginning of shock focusing as shown in Fig.2(a), there is a hotspot occurring on the axis of symmetry. Inside the hotspot, the explosion-like chemical reactions generate intensive spontaneous waves as shown in Fig.2(b). These waves spread successively behind the reaction zone, but are accumulated in the front of the reaction zone because of sharp temperature gradient, and elevate gas thermo-states that can speed up chemical reactions there. A feedback loop develops quickly, and a complex of a shock and reaction zone appears, and looks like a mushroom referred as to detonation bubble, as shown in Fig. 2(c). The detonation bubble spreads quickly, and engulfs the reaction zone, as shown in Fig.2(d). Finally a closed detonation bubble generates and becomes self-sustained later.

Figure 2(e) shows the pressure and temperature profiles along two lines of $r = 1.05$ and 1.17 at the same time instant of Fig.2(c). The profiles along the line of $r = 1.17$ that passes the downstream part (right) of the mushroom-shaped reaction zone show that the shock wave and the reaction zone are decoupled. The post-shock temperature is about 800K, and reaches to 2700K in the following reaction zone where a pressure drops slightly. Hence, the combustion in the downstream part is a typical deflagration. Examining the profiles along the line of $r = 1.05$ that passes through the upstream part (left) of the mushroom-shaped reaction zone, one can see that the shock and the reaction zone couple tightly and the post-shock parameters reach to the C-J values. This typical profile indicates that the detonation bubble had developed.

To further investigate into detonation bubble developing, a time sequence of pressure, temperature, and OH and H_2O profiles along the axis of symmetry at five instants are shown in Fig. 3. The time interval between the first and the last profile is about 0.53 microseconds, and the presented flow field is limited around the focal point. Examining the

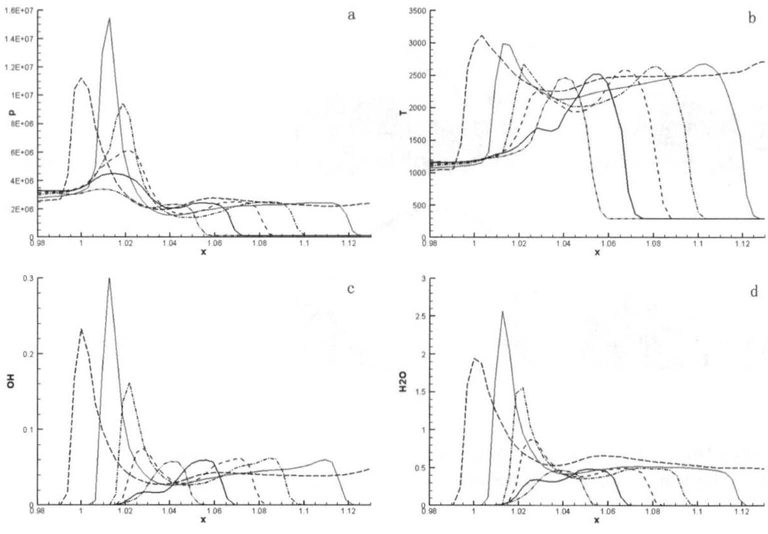

Fig. 3. Sequential profiles of pressure, temperature, OH and H_2O during hotspot explosion

first profile, one can see that the peak temperature is about 2500K, and the corresponding pressure, OH and H_2O are much low, therefore, the combustion at the moment is still in deflagration stage. The reaction zone travels both upstream and downstream. The downstream-travelling zone remains to be deflagration, but in the upstream-travelling zone, the post-shock parameters arise higher and higher with time evolution. Finally, a hotspot develops, as observed in Fig.3(a). The highest peak pressure appearing in front of the reaction zone at the fifth profile indicated with a solid line is about 15.5 MPa, the corresponding temperature is around 3000K, and OH and H_2O also reach to their maximum values, as shown in Fig.3(b)-3(d). This implies that intensive chemical reactions result in a hotspot explosion from which a detonation bubble forms. Shortly after the instant, the leading shock wave decays as it propagates up-stream, and the temperature reaches and stays around 3000K for a longer time, as shown at the sixth profile in Fig.3(b). This demonstrated that the detonation is overdriven at the beginning of detonation bubble developing and decays to the C-J values later. So, it is understood that the hotspot-induced detonation initiation behaves itself like the direct initiation.

4 Reaction zone acceleration

The flame acceleration had been reported and its role in detonation initiation has been recognized for many years. Flow visualization data show that the flame front gets thicker suddenly, and then the onset of detonation develops. In most of literatures, the flame acceleration was taken as the whole DDT process, in which many physical issues were included, even the hotspot explosion. In this paper, the reaction zone acceleration will be investigated and defined as one of the elementary phenomena for detonation initiation. It is expected that the thus-defined reaction zone acceleration can serve as a primary phenomenon like the hotspot for describing detonation initiation and propagation.

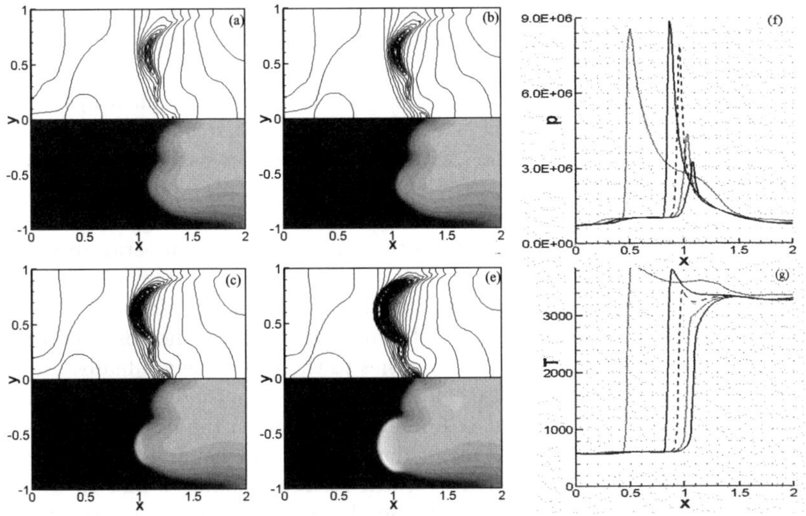

Fig. 4. Sequential pressure (upper) and temperature (lower) distributions during the reaction zone acceleration with sequential pressure and temperature profiles across the reaction zone

Detonation initiation developed from Richtmyer-Meshkov instability of the reaction zone in the second case is presented in Fig.4, and the upstream-travelling reaction zone is shown with pressure distributions in the upper half and temperature variations in the lower half of each picture. A series of spontaneous waves resulting from the disturbed reaction zone are also observable in Fig.4(a). These compression waves are expected to propagate in all directions, but their speeds are affected dramatically by the local thermo-dynamic environment. The downstream-propagating waves leave the reaction zone successively since products temperature behind the reaction zone is almost uniform. The upstream-propagating waves pile up in the front of the reaction zone because of the sudden temperature change from products to reactants. The reaction zone and its temperature front, as shown in Fig.4(a), are convex, so, the post-shock pressure and temperature should fall because of the front diverging effect. However, the isobars in Fig.4(b) becomes denser than those in Fig.4(a), which indicates the more spontaneous waves are generated in the reaction zone and accumulated in its front. Although the convex surface divergence is likely to make the post-shock pressure and temperature fall, the more and more intensive reactions generate the stronger and stronger compressive waves which not only compensate the front diverging effect, but also increase temperature further. The temperature peak forms at the tip of the convex reaction front, and meanwhile the corresponding pressure also becomes higher, as shown in Fig.4(c) where the isobars get more denser. And then, a arc detonation wavelet forms in Fig.4(d) and engulfs the flame front below it. Finally a cellular detonation develops. Detonation initiation occurs on the convex reaction front, and the coupling of pressure waves with successively increasing reactions are fundamentally important.

Time sequence of pressure and temperature profiles along the line of $y = 0.6$ during reaction zone acceleration are plotted in Fig.4(e) and (f). The profiles can be ordered from right to left according to their successive time instants. The line of $y = 0.6$ passes exactly the middle of the convex reaction zone, the parameters along which are useful to demonstrate the reaction zone acceleration. The first pressure profile shows a pressure pulse which indicates the beginning of spontaneous wave accumulation. The peak pressure is getting higher and higher as the reaction zone propagates upstream, and the corresponding temperature front is getting steeper and steeper. This phenomenon indicates the leading shock generation and reaction rate increasing. The peak pressure reaches to its maximum at the fifth profile as shown in Fig.4(e) and the value is about 9.0 MPa. The peak pressure decreases slightly at later instants and remains almost constant since then. Meanwhile a little pulse occurs in the fourth temperature profile and the temperature peak reaches about 3800K in the fifth profile, as shown in Fig.4(f). The maximum temperature finally reaches to 4000K and maintains almost constant since then, and a detonation forms. It is observed that the overshot of the post-shock pressure is slight and the post-shock temperature reaches to the C-J value gradually.

The positive feedback loop is also underlying the hotspot-induced DDT, but there are two primary physical issues that tell differences between the hotspot-induced DDT and the reaction zone acceleration. One of them is that the hotspot explosion happens in an unburned gas pocket, more intensive heat release is induced, and an over-driven detonation develops. For the reaction zone acceleration, chemical reactions take place only in a narrow zone, and no obviously over-driven phenomenon is observed. The other issue is that external perturbations are in need to trigger the hotspot explosion when the unburned gas pocket reaches to a critical thermo-state, otherwise the hotspot may die out without such perturbations.

Being similar to detonation initiation, cellular detonation propagation also depends on these mechanisms. Hotspots are generated from triple-point collisions, and the reaction zone follows incident shocks. The complex of the incident shock and reaction zone is alternately accelerating and decelerating around the C-J state. The acceleration is supported by the hotspot explosion and the deceleration results from the convex front diverging. Their interaction makes the cellular detonation be self-sustained, and there would not be any practical detonation if without either of them. Therefore, the detonation propagation and initiation go on the same primary mechanisms.

5 Conclusion

The hotspot and the reaction zone acceleration are two primary physical mechanisms underlying the deflagration to detonation transition. The first mechanism may be dominate in some cases and the second in other cases, but in most cases, they are observed together. The two mechanisms are also essential issues for cellular detonations, their interaction supports the steady propagation of practical detonations.

References

1. John H. S. Lee, Initiation of Gaseous Detonation, Ann. Rev. Phys. Chem, 28:75-104, 1977
2. John. H. S. Lee and A. J. Higgins, Comments on Criteria for Direct Initiation of Detonation. Phil. Trans. R. Soc. Lond. A, 357: 3503-3521, 1999
3. Elaine S. Oran, Vadim N. Gamezo: Origins of the Deflagration to detonation Transition in Gas-phase Combustion. Combust. Flame 2006.
4. A. K.Oppenheim, R. I. Soloukhirt, Experiments in Gas Dynamics of Explosions. Ann. Rev. Fluid Mech., 5:31-58, 1973.

Overview of the 2005 Northern Lights Trials

S.B. Murray[1], C.J. Anderson[1], K.B. Gerrard[1], T. Smithson[2], K. Williams[2], and D.V. Ritzel[3]

[1] *DRDC Suffield, P.O. Box 4000, Station Main, Medicine Hat, Alberta T1A 8K6, Canada*
[2] *DRDC Valcartier, 2459 Pie-XI Blvd. North, Val-Belair, Quebec G3J 1X5, Canada*
[3] *Dyn-FX Consulting Ltd., 19 Laird Avenue North, Amherstburg, Ontario N9V 2T5, Canada*

1 Introduction

Owing to the success of the Northern Lights Trials in 2002 [1], a second round of trials, dubbed 'Northern Lights Trials 2 (NLT2),' was conducted at DRDC Suffield in July of 2005. The main goal of the program was to develop a database of fireball characteristics that would allow an observer to identify the nature of an arbitrary distant explosion. A secondary objective was to assess the blast performance of military and improvised explosive formulations of interest. A tertiary goal was to compare the blast injury predictions from a number of human surrogates.

2 Experimental Setup and Instrumentation

A series of 22 condensed-phase charges were detonated, including seventeen nominal 20-kg charges and five nominal 150-kg charges. The mixtures included two weapons grade 'thermobaric' explosives (TBXs), two improvised explosive formulations (IEFs), and C4 (91% RDX) reference charges. The influence of charge casing material (polyethylene, aluminum or steel), height-of-burst (3 m and 6 m for the small and large charges, respectively), and charge shape (spherical or cylindrical, L/D $\simeq$ 1.6) were investigated. The booster charges consisted of C4 and were nominally 10% of the total charge mass.

The first thermobaric mixture (TBX1) contained a blend of monopropellant, explosive dust, and metallic flake, while the second mixture (TBX2) comprised monopropellant, metallic flake and solid oxidizer particles. Both mixtures had a putty-like texture. The improvised formulations were composed of oxidizer, metallic particles and a non-energetic fuel, the only difference being the size and morphology of the metallic particles (small spherical for IEF2 versus large granular for IEF1). Both formulations had the consistency of moist powders. Table 1 shows the trial plan and charge details.

The instrumentation employed was extensive. Free-field and reflected blast histories were measured using 24 pressure gauges and four suites of cantilever gauges positioned in three orthogonal arrays. Five high-speed video cameras were used to obtain images of the explosions and to track the incident and reflected shocks as they propagated along an illuminated zebra-board backdrop. Figures 1 and 2 show the experimental setup and a typical video sequence. DRDC Valcartier's Airborne Infrared Imaging Spectroscopy (AIRIS) technology demonstrator was also employed. AIRIS, an air-borne nadir-viewing system, was operated aboard the National Research Council's Convair 580 aircraft. Most measurements were obtained at 9000 feet AGL with the system's 3x telescope to permit sufficient time over the target. The system configuration included two 8x8 element detector array modules capable of simultaneous acquisition in the 2-12 μm region, and three video cameras; one operating in the visible range, one in the 3-5 μm range, and one in the 8-12 μm range. Ground truthing was accomplished by teams from Wright Patterson

Table 1. Summary of trials conducted in Northern Lights Trials 2

Charge Details				Explosive Masses			Casing Masses	
Trial No.	Charge Shape	Casing Material	Mixture	C4 Booster (kg)	Mixture (kg)	NEQ (kg)	Booster (kg)	Charge (kg)
0	sphere	poly	C4	-	20.2	20.2	n/a	0.75
1	sphere	poly	C4	-	21.3	21.3	n/a	0.79
2	cylinder	poly	C4	-	28.4	28.4	n/a	1.48
3	sphere	poly	TBX1	1.7	17.0	18.8	0.11	1.02
4	cylinder	poly	TBX1	1.8	20.9	22.6	0.23	1.52
5	sphere	poly	TBX2	1.7	19.6	21.3	0.11	1.03
6	cylinder	poly	TBX2	1.8	25.2	27.0	0.23	1.44
7	cylinder	poly	IEF1	1.8	23.8	25.5	0.23	1.47
8	sphere	poly	IEF1	1.7	16.7	18.5	0.12	1.01
9	sphere	poly	IEF2	1.7	18.1	19.9	0.12	0.99
10	cylinder	poly	IEF2	1.8	26.3	28.1	0.23	1.49
11	cylinder	steel	TBX1	1.8	23.6	25.4	0.23	6.35
12	sphere	aluminum	TBX1	1.7	16.5	18.3	0.12	2.28
13	sphere	steel	TBX1	1.7	17.7	19.5	0.12	3.62
14	sphere	aluminum	TBX2	1.7	19.4	21.1	0.11	1.95
15	sphere	steel	TBX2	1.7	20.3	22.0	0.11	3.62
16	cylinder	steel	TBX2	1.8	27.3	29.1	0.23	6.55
17	sphere	poly	TBX1	12.9	134.8	147.7	0.47	3.35
18	cylinder	poly	TBX1	12.2	143.5	155.7	1.24	7.02
19	sphere	aluminum	TBX1	13.0	141.2	154.2	0.47	9.94
20	sphere	steel	TBX1	13.0	141.0	154.0	0.47	17.8
21	cylinder	steel	TBX1	12.2	145.8	158.0	1.25	15.3

Air Force Base and DRDC Valcartier using multiple spectrometers and hyper-spectral imagers positioned 1.15 km from Ground Zero. The Canadian team employed a Pirates IV hyper-spectral imaging system and a MR300 spectrometer. Pirates IV was equipped with an 8x8 InSb detector array (2-5 μm) coupled with a 5-cm telescope. The MR300 was configured with a 2.5-cm telescope and two detectors covering the 2-12 μm region.

Personal vulnerability (PV) rigs included the United Kingdom's Thoracic Rig, Australia's AUSman, the United States' GelMan and Blast Test Device (BTD), DRDC Valcartiers MABIL thoracic and head surrogates, and Med-Eng Systems' Hybrid II and III anthropomorphic test devices. These surrogates represent a variety of approaches taken to meet the challenge of assessing human vulnerability to blast. Injuries associated with exposure to blast include burns, displacement and acceleration injuries, eardrum rupture, and injuries to the internal organs; in particular, gas-filled organs such as the lungs and gastrointestinal tract. This was the first time that all rigs had been tested side-by-side.

3 Experimental Results

Only selected representative results can be presented in this brief paper. The shock and fireball dynamics [2] and ground reflection phenomena [3] for a TBX height-of-burst explosion are discussed in companion papers.

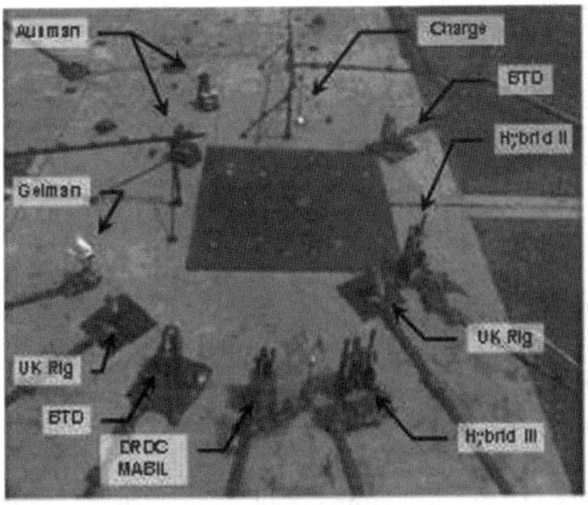

Fig. 1. Trial set-up showing the elevated gauge arrays (top and left), ground gauge array (right), and arrangement of human vulnerability assessment rigs positioned around a 20-kg spherical charge

Fig. 2. Video sequence showing the detonation of a 150-kg, steel-cased, spherical charge. The zebra-board backdrop is visible in the background.

3.1 Pressure Measurements

The pressure data were initially scaled to the average ambient pressure and temperature conditions for the series, and subsequently cube-root scaled to produce plots of incident overpressure ratio $\Delta P/P_o$, positive-phase impulse, $I/W^{1/3}$ (kPa-ms/kg$^{1/3}$), and time of arrival, $t/W^{1/3}$ (ms/kg$^{1/3}$), versus scaled range, $R/W^{1/3}$ (m/kg$^{1/3}$). The casing material had very little influence on the peak pressures and positive-phase impulses for all charges,

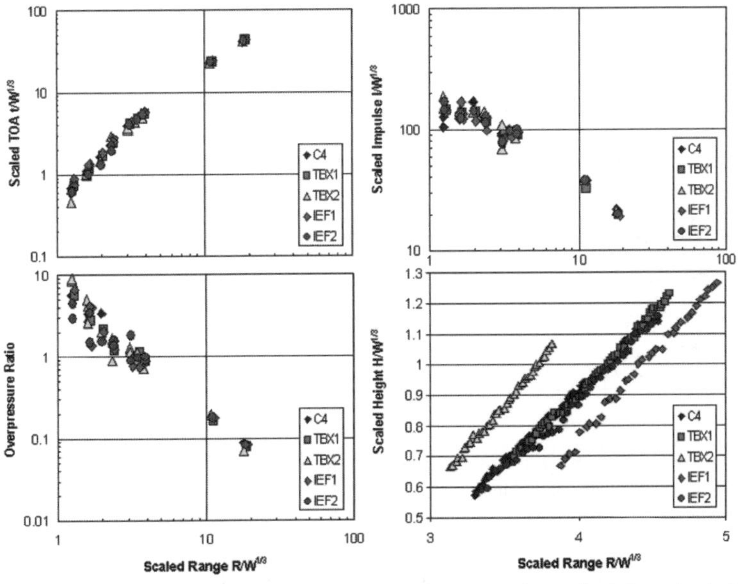

Fig. 3. Scaled time-of arrival, overpressure, positive-phase impulse and triple-point height versus scaled range for small polyethylene-cased spherical charges (from lollipop gauges)

a result that is not unexpected given that the casings were very thin. These parameters also scaled very well for small and large charges.

The performance of the five explosive formulations for small, polyethylene-cased, spherical charges is compared in Fig. 3. The performance is practically identical for scaled ranges greater than 3. TBX2 and IEF2 give the shortest times of arrival and highest impulses for scaled ranges less than 3. However, TBX2 and IEF1 give the highest overpressures for the same range. The triple-point trajectory measurements show that the highest and lowest triple-point heights for a given scaled range are for TBX2 and IEF1, respectively. TBX2 and IEF2 give trajectories nearly identical to C4. Higher heights indicate stronger reflections and earlier transition to Mach reflection.

3.2 Spectroscopy Data

Only representative examples can be described here. Figure 4 shows that the apparent radiant intensity (ARI - watts/sr/cm^{-1}) is higher for a large versus a small charge at all times throughout the burn. The peak intensity ratio is 1.2, while the total intensity ratio is 2.37. The peak gray body apparent temperatures are calculated to be ~1200 C for both charge sizes, as expected. The effect of charge geometry on ARI is shown in Fig. 5. Clearly, the charge shape has a dramatic effect on the observed IR emissions. The peak intensity for the cylindrical charge is 1.7 times higher than for the spherical charge and occurs ~100 ms into the burn versus ~550 ms for the spherical charge. This is likely due to more intense mixing between the fireball and air for the cylindrical geometry.

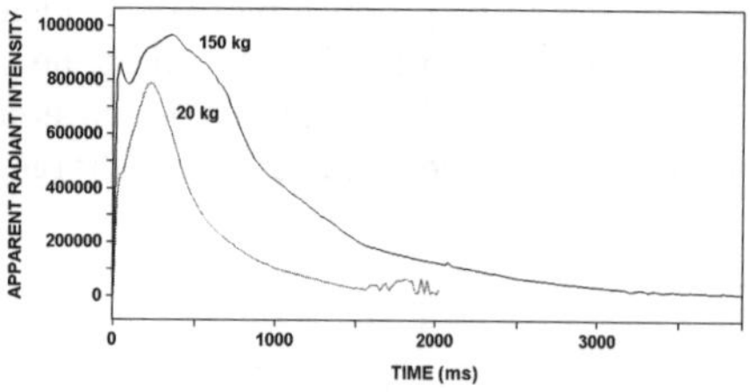

Fig. 4. Apparent radiant intensity versus time into burn for spherical, Al-cased charges containing TBX1: Effect of mass

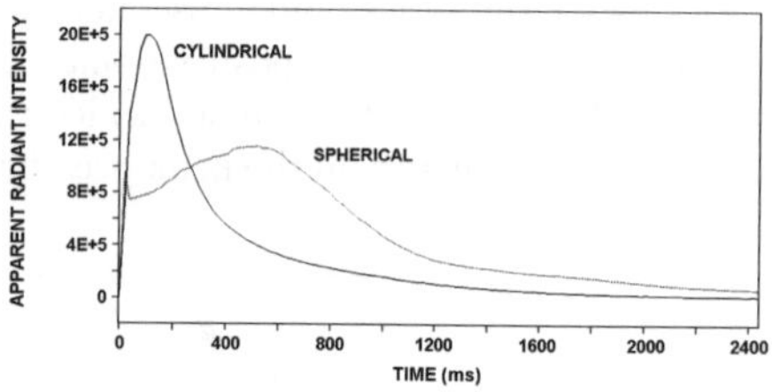

Fig. 5. Apparent radiant intensity versus time into burn for large, steel-cased charges containing TBX1: Effect of charge geometry

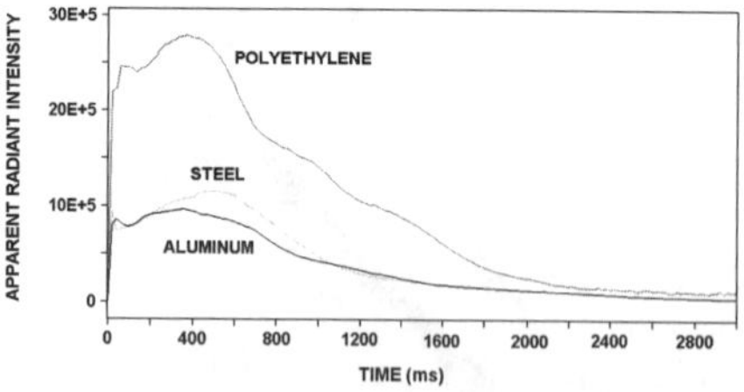

Fig. 6. Apparent radiant intensity versus time into burn for large, spherical charges containing TBX1: Effect of casing material

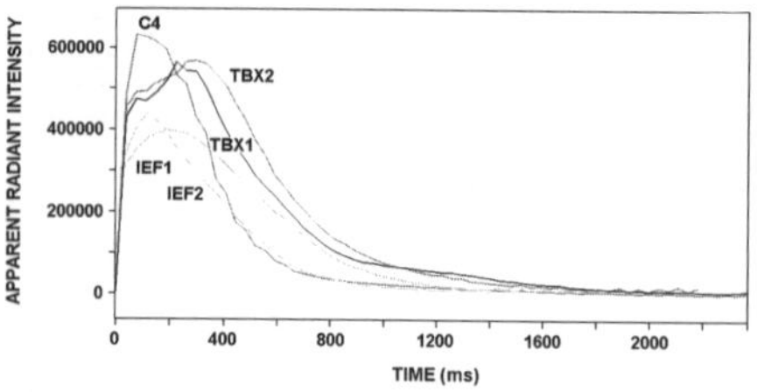

Fig. 7. Apparent radiant intensity versus time into burn for small, polyethylene-cased, spherical charges: Effect of mixture type

Figure 6 shows that the charge casing material has a dramatic effect on the radiant intensity. The peak intensity is 2.9 times that for aluminum and 2.4 times that for steel.

In addition, the polyethylene emissions persist at a higher value and for a longer time. Aluminum- and steel-cased charges exhibit similar behaviours. The corresponding peak temperatures have been calculated to be ~1400 C for polyethylene and ~1200 C for the other two casing materials. This 200 C difference persists throughout the burn. Finally, the influence of charge formulation on the radiant intensity profiles is shown in Fig. 7 for all five mixtures. C4 produces the highest peak emission (0.63). Next comes TBX1 and TBX2 with values of 0.56 and 0.57. The lowest peak emissions are produced by IEF1 and IEF2 (values of 0.40 and 0.44, respectively). C4 and IEF2 emissions peak at ~100 ms into the burn, whereas the peak occurs at 250–350 ms for the other explosives. The ratios of vapour to gray body emissions have also been calculated and show an order of magnitude difference. The peak ratios for C4, TBX1, TBX2, IEF1 and IEF2 are 0.79, 0.33, 1.24, 1.63 and 3.3, respectively. These peaks occur at ~700 ms for C4 and IEF2, and at times of approximately 1, 1.2 and 1.4 seconds for IEF1, TBX2 and TBX1, respectively. This ratio generally increases with increasing metal content and increasing metal particle size for the same metal content.

3.3 Personal Vulnerability Assessments

The rigs were arranged in two concentric circles around the charges (one circle shown in Fig. 1) such that the target peak incident overpressure (based on a C4 spherical reference charge) was approximately 2 bars for the larger radius and 4 bars for the inner radius. For the 20-kg charges, these corresponded to 7 m and 4.5 m radii, respectively. For the 150-kg charges, the radii used were 14 m and 7.5 m, respectively. With the exception of the Hybrid II, the rigs were generally deployed for the lightly cased charges only. The discussions that follow will highlight some results from the rigs deployed by DRDC.

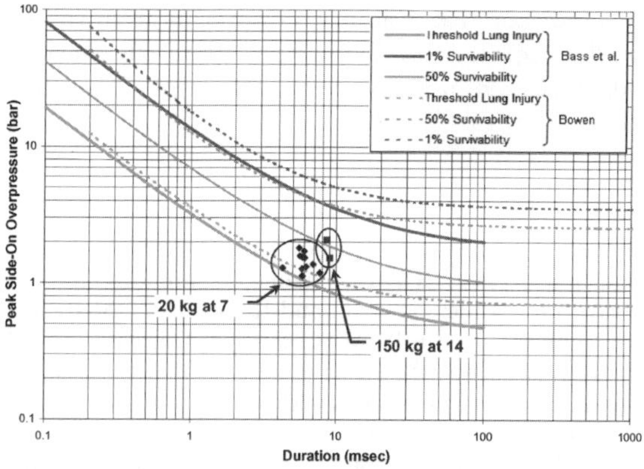

Fig. 8. Primary blast injury assessment based on Bowen [4] and Bass et al. [5] injury risk curves

Figure 8 shows the measured reference pressures and durations for the 20-kg charges at 7 m and 150-kg charges at 14 m plotted against primary blast injury risk curves. Note that the results for the 20-kg charges indicate threshold to minor lung injuries according to the Bowen curves [4], while the curves from Bass et al. [5] would predict generally more severe injuries with a few points bordering on 50% lethality. The loading from the two 150-kg charges are predicted to be more severe with slightly higher overpressures and longer durations. In all cases, the overpressures are above the criteria for a greater than 50% probability of severe ear drum rupture.

The results from the DRDC MABIL torso are shown in Fig. 9. Note the influence of the charge geometry on the response. This trend was common to most of the injury risks investigated with the loading from the cylindrical charges presenting a greater risk of injury than for the spherical charges. Results for the BTD [6] also support this trend.

An example of Hybrid III advanced technology demonstrator (ATD) results is shown in Fig. 10 where the Head Injury Criteria (HIC), a function of the resultant linear acceleration of the head centre of gravity, is plotted for each charge. Data for the cylindrical 20-kg TBX1 charge was lost when gauge mounts in the Hybrid III head were damaged. The HIC associated with a 50% probability of AIS 2 injuries (moderate but not life threatening) is 700, far above the measured values. However, rotational accelerations induced by the blast were sufficient to be associated with a risk of concussion and diffuse axonal injuries. This is consistent with previous test results which showed injury risk for the head to be based on the rotational rather than linear accelerations.

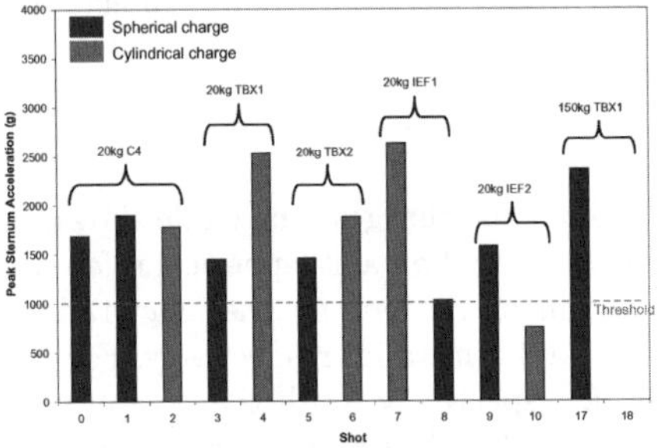

Fig. 9. Peak chest-wall accelerations measured in the DRDC MABIL torso

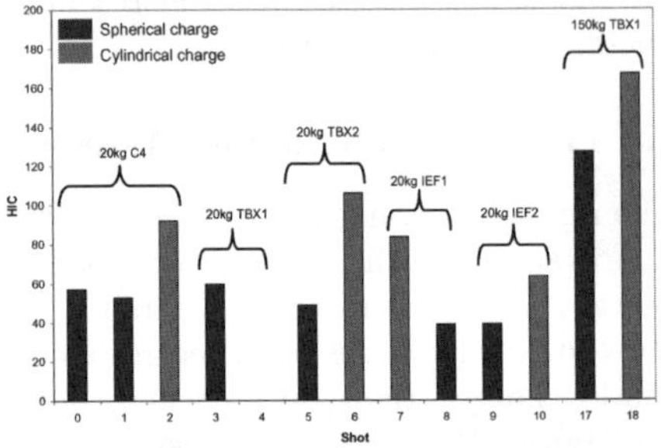

Fig. 10. HIC measured with the Hybrid III ATD

Where there is a significant risk of injury, however, is from the second phase of the Hybrid III response, the impact with the ground. Here, the head accelerations, neck loads, and pelvis loads measured in many of the tests were all associated with high risks of critical injuries [7].

References

1. Ritzel, D.V., Smithson, T., Murray, S.B., and Gerrard, K.B.: Experimental study of afterburning explosive charges, 19th ICDERS, Hakone, Japan, July 27 - August 1, 2003.
2. Ritzel, D.V., Murray, S.B., Anderson, C.J., and Ripley, R.C.: Near-field blast phenomenology of thermobaric explosions, 26th ISSW, Gottingen, Germany, July 15-20, 2007.
3. Ripley, R.C., Donahue, L., Dunbar, T.E., Murray, S.B., Anderson, C.J., Zhang, F., and Ritzel, D.V.: Ground reflection interactions with height-of-burst metallized explosions, 26th ISSW, Gottingen, Germany, July 15-20, 2007.
4. Bowen, I.G., Fletcher, E.R., and Richmond, D.R.: Estimate of man's tolerance to the direct effects of air blast, Lovelace Foundation for Medical Education and Research, Albuquerque, New Mexico, DASA 2113, DA-49-146-XZ-372, 1968.
5. Bass, C.R., Rafaels, K., and Slazar, R.: Pulmonary injury risk assessment for short-duration blasts, Proceedings of the Personal Armour Systems Symposium (PASS 2006). Royal Armouries Museum, Leeds, UK, September 18-22, 2006.
6. Chan, P. and Ho, K.: Northern Lights Trials II blast injury analysis quick-look report, Titan Corporation, Personal Communication, December 14, 2005.
7. Manseau, J., Williams, K., Dionne, J.-P., and Levine, J.: Response of the Hybrid III dummy subjected to free-field blasts – Focusing on tertiary blast injuries, proceedings of 19th Int'l Symosium on Military Aspects of Blast and Shock, Calgary, Canada, October 1-6, 2006.

Physics of detonation wave propagation in 3D numerical simulations

H.-S. Dou[1], B.C. Khoo[2], and H.M. Tsai[1]

[1] *Temasek Laboratories, National University of Singapore, Singapore 117508*
[2] *Department of Mechanical Engineering, National University of Singapore, Singapore 119260*

Summary. Based on the three-dimensional simulation results, the mechanism of detonation waves sustenance is analyzed. The detonation wave of perfect gas with premixed gaseous in a rectangular duct is considered. The governing system composed of the Euler equation and a one-step Arrhenius chemical reaction model is solved with a fifth-order WENO (Weighted Essentially NonOscillatory) discretization scheme and a 3rd order TVD Rouge-Kutta method. The simulation shows that under an initial disturbance, the detonation front finally develops to an unsteady three-dimensional pattern. For the given duct, the flow front displays a periodic quasi-steady "rectangular mode," with the front assuming variation of "convex" and "concave" shapes continuously. It is shown that the detonation propagating depends on the strength and the fraction of the hot spots on the detonation front.

1 Introduction

The common features of gases detonation are there-dimensionality and the characteristic of time-dependent patterns of front evolution. In the past years, a lot of work involving both numerical and experimental can be found in the literature. Numerical simulation as a tool can contribute much towards exploring the physics of detonation and the study of the structure of detonation to increase the overall understanding of the detonation phenomenon. Although there are many reported studies, most of these are confined to two-dimensions, and very few are for three-dimensions [1]. For 3D studies, Williams et al. [2] probably made the first detailed simulation of detonation structure in a duct using the Euler equations and one-step reaction law. The simulation results showed a "rectangular" structure for the front and there is a phase shift for the front motions between two neighbouring walls. The structure of transverse waves observed is much more complicated than that for theepsfbox two-dimensional case. Tsuboi et al. [3] carried out the 3D simulations of detonation in a duct using a more detailed chemical reaction model. They found that there are two types of 3D modes: a rectangular mode and a diagonal mode. On the pattern of maximum pressure history, the cell length in the rectangular mode is about the same as that for the 2D detonation simulations. The cell length of the diagonal mode is only about three-quarters of that for the two-dimensional simulations. Deiterding and Bader [4] made the simulations for 3D detonation using a detailed chemical reaction model. The results showed that there is no phase shift between both transverse directions. Only a "rectangular mode" is observed. The cell length in 2D simulation and that in 3D simulation is about the same, but the maximum velocity in 3D simulation is higher than that in 2D simulation. He et al. [5] made simulations for 1D, 2D, and 3D detonations. The results showed that the size of the cell decreases with the increase of the activation energy and increases with the increase of the energy release. For a two-step induction model, the averaged cell length in 2D case is only about 65%

of that found from 3D simulations. There appears to be significant discrepancies and differences among 3D detonation and especially their relation to the 2D detonations in the literature. It is seen that much more work is needed to clarify these problems.

In the present study, numerical simulation is performed for the propagation of 3D detonation waves in a duct using finite difference methods. Based on the simulation results, the physics of the detonation sustenance is discussed.

2 Governing Equations

The governing equations to describe the fluid flow and the detonation propagation are the three-dimensional Euler equations with a source term which represents the chemical reaction process and in conservative form, these are written as

$$\frac{\partial \mathbf{U}}{\partial t}+\frac{\partial \mathbf{F}}{\partial x}+\frac{\partial \mathbf{G}}{\partial y}+\frac{\partial \mathbf{H}}{\partial z}=\mathbf{S} \tag{1}$$

where the conserved variable vector $\mathbf{U}$, the flux vectors $\mathbf{F}$, $\mathbf{G}$, and $\mathbf{H}$ as well as the source vector $\mathbf{S}$ are given, respectively, as

$$\mathbf{U}=(\rho,\rho u,\rho v,\rho w,E,\rho Y)^T,$$

$$\mathbf{F}=(\rho u,\rho u^2+p,\rho uv,\rho uw,(E+p)u,\rho uY)^T,$$

$$\mathbf{G}=(\rho v,\rho vu,\rho v^2+p,\rho vw,(E+p)v,\rho vY)^T,$$

$$\mathbf{H}=(\rho w,\rho wu,\rho wv,\rho w^2+p,(E+p)w,\rho wY)^T,$$

$$\mathbf{S}=(0,0,0,0,0,\omega)^T.$$

Here u, v, and w are the components of the fluid velocity in the x, y, and z directions, respectively, in Cartesian coordinates system, ρ is the density, p is the pressure, E is the total energy per unit volume, and Y is the mass fraction of the reactant. The total energy E is defined as

$$E=\frac{p}{\gamma-1}+\frac{1}{2}\rho(u^2+v^2+w^2)+\rho qY. \tag{2}$$

Here q is the heat production of reaction, and γ is the ratio of specific heats. The source term ω is assumed to be in an Arrhenius form

$$\omega=-K\rho Ye^{-(T_i/T)} \tag{3}$$

where T is the temperature, T_i is the activation temperature, and K is a constant rate coefficient. For a perfect gas, the state equation is

$$p=\rho RT. \tag{4}$$

As such, equations (1) to (4) constitute a closed system of equations. The above mentioned equations are made dimensionless based on the state of the unburned gas, and the reference length which is chosen as the half-reaction length (L) and is defined as the distance between the detonation front and the point where half of the reactant is consumed by chemical reaction in ZND model.

3 Numerical Methods

The conservation-law form of the governing equations can be written in quasi-linear form as

$$\frac{\partial \mathbf{U}}{\partial t} + A_x \frac{\partial \mathbf{U}}{\partial x} + A_y \frac{\partial \mathbf{U}}{\partial y} + A_z \frac{\partial \mathbf{U}}{\partial z} = \mathbf{S} \tag{5}$$

where $A_x = \frac{\partial \mathbf{F}(\mathbf{U})}{\partial U}$, $A_y = \frac{\partial G(\mathbf{U})}{\partial U}$, and $A_z = \frac{\partial H(\mathbf{U})}{\partial U}$, are the Jacobian matrices. By mathematical derivation and treatment, each entry of the above matrices can be found. Then, the eigenvalues and eigenvectors of the Jacobian matrices A_x ,A_y, and A_z are obtained. The final form of the discretization of the system can be written as:

$$\frac{dU_{i,j,k}(t)}{dt} = -\frac{1}{\Delta x}(\widehat{F}_{i+1/2,j,k} - \widehat{F}_{i-1/2,j,k}) - \frac{1}{\Delta y}(\widehat{G}_{i,j+1/2,k} - \widehat{G}_{i,j-1/2,k})$$

$$-\frac{1}{\Delta z}(\widehat{H}_{i,j,k+1/2} - \widehat{H}_{i,j,k-1/2}) + S_{i,j,k} \tag{6}$$

where, the $\widehat{F}_{i+1/2,j,k}$, $\widehat{G}_{i,j+1/2,k}$, $\widehat{H}_{i,j,k+1/2}$ and are the numerical fluxes which can be calculated using upwinding WENO scheme [6]. Finally, the ordinary differential equation (6) can be integrated by the third order TVD Runge-Kutta method [6].

4 Results and Discussion

We study the case for $f = 1.0$ (overdriven factor), $q = 50$, $T_i = 20$, and $\gamma = 1.20$. The one-dimensional simulation result in natural coordinates is shown in Fig. 1. Firstly, 1D result can be used to validate the 3D code for the detonation problem. Secondly, 1D result can be used to check the mesh convergence and the resolution needing for the detonation front, and to determine the CFL (Courant-Friedrichs-Levy) condition applicable. Thirdly, it can be used as the initial condition for 2D or 3D simulations. The computational geometry for 3D case is a rectangular duct. A detonation wave is propagating from the left to right. for 2D simulation, the boundary condition at the left side is set as reflection type and that at the right side is quiescent. It can be seen that the peak pressure at the detonation front tends towards the CJ pressure.

For the 3D calculation, at time t=0, the flow in both transverse directions is given and set as uniform. The inertial frame of reference moving at the steady CJ detonation velocity is fixed at the shock front. The boundary conditions consist of the unburnt mixture entering the domain at the CJ detonation velocity, a non-reflecting outflow, and the walls modeled by reflection boundary conditions. We added a random 3D perturbation in the form of a localized explosion (energy pulse) just behind the leading shock at the first time step. For the 3D simulation, we use a grid of 241x151x151 for a rectangular domain, in which the normalized dimension of the duct by the half ZND reaction length is 16x20x20, corresponding to a resolution of 15 points in the ZND half reaction length.

Results show that a few small humps are generated at the front with non-uniform amplitudes, and these humps become larger with time. With the increase of hump size, the pressure and the velocity behind these humps gradually increase to values much higher than those of CJ values, while in the areas among these humps the pressure and the velocity gradually becomes lower than those of CJ values. As a result, there is no detonation in these latter areas among the humps, but only chemical reaction (i.e.,

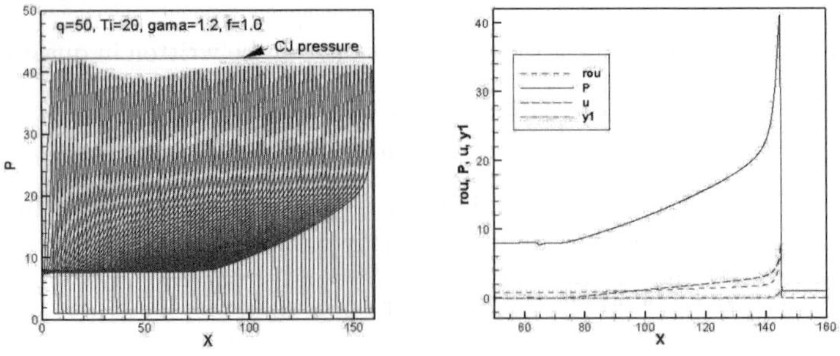

Fig. 1. One-dimensional detonation with resolution of 16 points per half-reaction length, $f = 1.0$, $q = 50$, $T_i = 20$, and $\gamma = 1.20$. Pressure profile with the time (left); Final detonation profile (right).

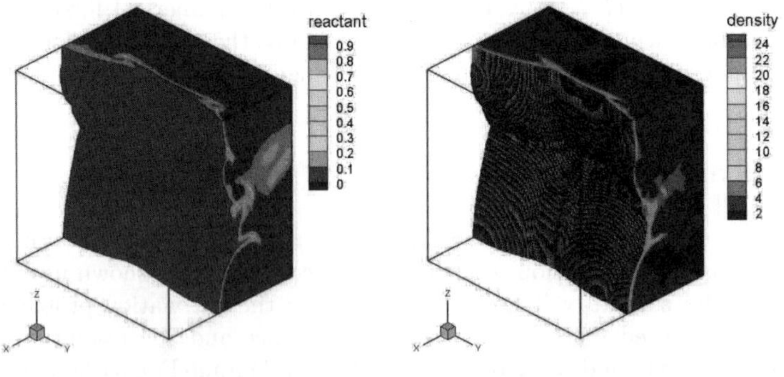

Fig. 2. Contours of various parameters in the wide duct at t=120/140 for a period of time 140.

deflagration) takes place behind the front. From the CJ principle of detonation, these humps are equivalent to the Mach stems for 2D detonation in a plane, while the flat area among the humps are the incident shock wave plane. The fronts of these humps which overtake the incident shock wave, therefore, play a similar role as the Mach stems in two-dimensional detonation. With further time evolution, these small humps wander within the domain and then connect to form larger humps. Simultaneously, transverse waves are formed on the front. After a long time evolution, the detonation front develops as a quasi-steady periodic motion, displaying a "rectangular mode," as shown in Figs. 2 and 3. The flow fronts on the four walls are not in the same phase. The phases of the front motion in z and y directions are different. The flow front shows alternatively "convex" and "concave" shapes. The motion of the front is gradually switched between the mode approximately like "#" and the mode approximately like "+." This phenomenon is similar to that found by Williams et al. [2] and Tsuboi et al. [3], and this mode is referred to as "rectangular" structure or "rectangular mode."

We also calculated the two-dimensional detonation with the same grid and at the same flow and reactant parameters by setting the grid number as unity in the z direction. For

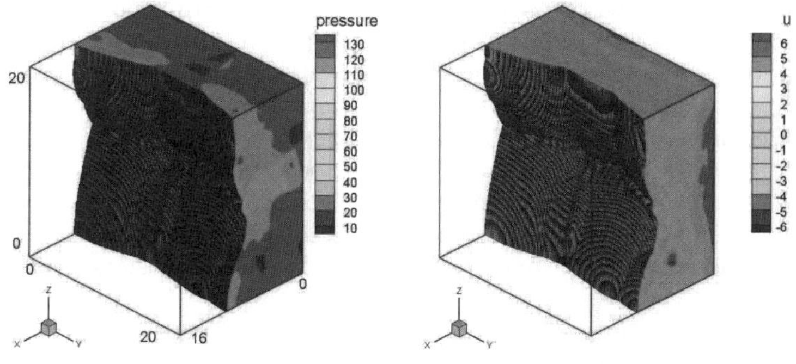

Fig. 3. Contours of various parameters in the wide duct at t=120/140 for a period of time 140.

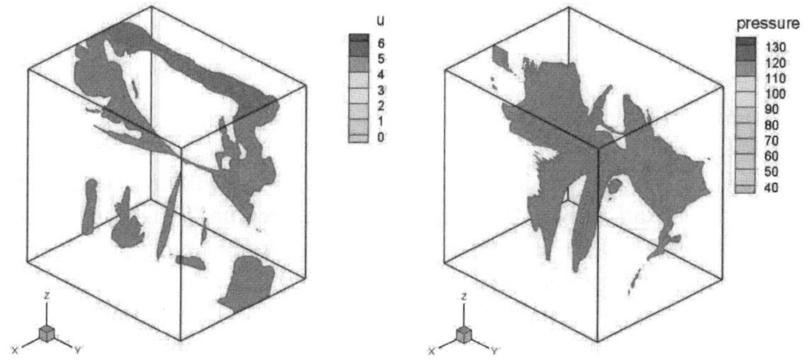

Fig. 4. Areas of velocity larger than the CJ value (left); Areas of pressure larger than the CJ value (right); at t=120/140 for a period of time 140.

3D detonations, the detonations are much more spatially distributed as compared to the 2D detonations. The variables on the front are more non-uniform in 3D case. From 1D, 2D, to 3D, the maximum pressure and the maximum velocity is increased. It is also found that there are more unreacted fuel pockets in 3D detonations. It is observed that the length of the cell in steamwise direction is about 40 in 3D and is about the same as that in 2D simulations.

In 2D and 3D, the detonation does not occur along the whole front, but it only occurs around some areas with high pressure and high velocity behind the convex Mach stem and near the triple points. Thus, hot "spot" are generated in the cores of Mach stems and at the triple points. The detonation is strongest at triple points when the transverse waves collide. At the triple point, the maximum pressure and the maximum velocity is well coupled which is like the 1D detonation, and their values are larger than the CJ values. In other areas along the front (incident shock), the flow front is only a fluid and fuel interface; there is no detonation generation there. This is why there are some unreacted fuel pockets in 2D and 3D detonations as found in experiments, see [6].

We found that the detonation occurrence depends upon the closed coupling of the (maximum) pressure and the (maximum) velocity [6]. The role of transverse waves is that the collision of transverse waves leads to the formation of triple points which provides a closed coupling of the (maximum) pressure and the (maximum) velocity. This strong coupling in the area of triple points makes the detonation overdriven locally and sustains the detonation by the transverse motion of these zones along the front. It can be observed that the pressure in the second half cell is generally less than its CJ value (area passed by incident shock wave), and the velocity in the second half cell is also less than its CJ value. As a result, in these areas, detonation would not be produced and only deflagration reaction occurs. The reaction in the second half cell is heated by the "hot spot" at the triple points and in the Mach stem in neighbour cells via the transverse motion with the front moving.

Figure 4 shows the areas of $p > p_{cj}$ and $u > u_{cj}$ in the domain, respectively, in which detonation can only occur. In other areas along the front, the reactant is heated and ignited by the transverse motion of these hot spots. The strength and the fraction of the hot spots determine the strength of the reaction. If the fraction of hot spots is small, the detonation is weak and unstable. This may be the reason why detonation with large cells is unstable.

5 Conclusion

Detonation wave in a duct is simulated with a 3D finite difference method in which the time integration and the spatial approximation are the TVD Runge-Kutta scheme and the fifth-order WENO scheme, respectively. The simulation result reveals that there are certainly differences between the three-dimensional and two-dimensional detonations, although there are certain common features among them. The detonation structure and the reaction process are more complex for the three-dimensional case. It is shown that the detonation mechanism depends on the closed coupling between the pressure and the velocity. The detonation sustenance depends on the strength and the fraction of the hot spots (zone with high pressure and high velocity, i.e., triple points and the core of Mach stem) on the front. The role of transverse waves is to help the formation of the hot spots.

References

1. Mitrofanov, V. V., Modern view of gaseous detonation mechanism, Progress in Astronautics and Aeronautics, Vol. 137 (Washington, DC: AIAA),1996, pp.327–339
2. Williams, D.N., Bauwens, L., and Oran, E.S., Detailed structure and propagation of three-dimensional detonations, Proceedings of the Combustion Institute, Vol. 26, 1997, pp. 2991-2998
3. Tsuboi, N., Katoh, S., and Hayashi, A.K., Three-dimensional numerical simulation for hydrogen/air detonation: rectangular and diagonal structures, Proceedings of the Combustion Institute, Vol. 29, 2002, pp. 2783-2788
4. Deiterding, R. and G. Bader., High- resolution simulation of detonations with detailed chemistry, In G. Warnecke, editor, Analysis and Numerics for Conservation Laws, Springer, Berlin, 2005, pp.69-91
5. He, H., Yu, S.T.J., and Zhang, Z-C., Direct calculations of one-, two-, and three-dimensional detonations by the CESE method, AIAA Paper 2005-0229, 2005
6. H-S. Dou, H.M.Tsai, B. C. Khoo, J. Qiu., Three-dimensional simulation of detonation waves using WENO schemes, AIAA Paper 2007-1177, 2007

Propagation of cellular detonation in the plane channels with obstacles

V. Levin[1], V. Markov[2], T. Zhuravskaya[3], and S. Osinkin[3]

[1] *Far Eastern Branch of the Russian Academy of Sciences, Institute for Automation and Control Processes, 5 Radio Street, 690041 Vladivostok (Russia)*
[2] *Russian Academy of Sciences, V. A. Steklov Mathematical Institute, 8 Gubkina Street, 119991 Moscow (Russia)*
[3] *M. V. Lomonosov Moscow State University, Institute of Mechanics, 1 Michurinsky pr., 119192 Moscow (Russia)*

Summary. The numerical investigation of propagation of formed cellular detonation wave in the encumbered plane channels filled with the stoichiometric hydrogen-air mixture under normal conditions has been carried out. The set of gas dynamic equations jointly with the set of chemical kinetic equations, which takes into consideration principal features of chemical interaction of hydrogen with oxygen, is solved by a finite-difference method based on the Godunov's scheme. The influence of rigid obstacle (barrier) placed in the channel on the propagation of cellular detonation has been studied. The behaviour of the cellular detonation in the case of abrupt widening of the channel cross section has been considered too.

1 Introduction

The interest to the detonation wave examination has been increased visibly at last time because of the practical necessity. One of the main examination objects is a study of detonation propagation in particular a determination of channel features influence on the process of detonation propagation. The importance of this problem is connected with a prospect to use detonation in reactive devices and generators of powerful pressure impulses.

In the present research the numerical investigation of obstacles influence on the cellular detonation propagation in the plane channels filled with stoichiometric hydrogen-air mixture under normal conditions is carried out. The behaviour of the cellular detonation in the case of abrupt widening of the channel cross section has been considered too.

2 Mathematical model

The propagation of detonation in the plane channel filled with the stoichiometric hydrogen-air mixture under normal conditions (p_0=1atm, T_0=298°K) was examined. The instantaneous electrical discharge of narrow layer form placed near the closed end of the channel was used for the detonation initiation. It was supposed the electrical energy is transformed instantaneously into internal energy of gas mixture.

The system of equations describing plane two-dimensional flows of non-viscous gas mixture is as follows:

$$0 = \frac{\partial \rho}{\partial t} + \frac{\partial (u\rho)}{\partial x} + \frac{\partial (v\rho)}{\partial y}$$

$$0 = \frac{\partial(\rho u)}{\partial t} + \frac{\partial(\rho u^2 + p)}{\partial x} + \frac{\partial(\rho uv)}{\partial y}$$

$$0 = \frac{\partial(\rho v)}{\partial t} + \frac{\partial(\rho vu)}{\partial x} + \frac{\partial(\rho v^2 + p)}{\partial y}$$

$$0 = \frac{\partial(\rho(u^2 + v^2)/2 + \rho h - p)}{\partial t} + \frac{\partial(u\rho((u^2 + v^2)/2 + h))}{\partial x} + \frac{\partial(v\rho((u^2 + v^2)/2 + h))}{\partial y}$$

$$\rho\omega_i = \frac{\partial(\rho n_i)}{\partial t} + \frac{\partial(u\rho n_i)}{\partial x} + \frac{\partial(v\rho n_i)}{\partial y},$$

were x and y – Cartesian coordinates, u and v are the corresponding components of velocity, t is time, ρ, p and h are density, pressure and enthalpy, respectively, n_i is the molar concentration of the ith component of mixture; ω_i is the rate of formation/depletion of the ith component.

The following elementary stages of the hydrogen oxidation are taken into account in the reaction mechanism [1,2]:

$$\begin{array}{ll}
O_2 + H \leftrightarrow OH + O, & O + O + M \leftrightarrow O_2 + M, \\
HO_2 + OH \leftrightarrow H_2O + O_2, & O + H_2 \leftrightarrow OH + H, \\
H + O_2 + M \leftrightarrow HO_2 + M, & OH + OH + M \leftrightarrow H_2O_2 + M, \\
H_2 + OH \leftrightarrow H_2O + H, & HO_2 + H \leftrightarrow OH + OH, \\
H_2O_2 + H \leftrightarrow H_2 + HO_2, & OH + OH \leftrightarrow H_2O + O, \\
HO_2 + H \leftrightarrow H_2 + O_2, & H_2O_2 + H \leftrightarrow H_2O + OH, \\
H + H + M \leftrightarrow H_2 + M, & HO_2 + H \leftrightarrow H_2O + O, \\
H_2O_2 + O \leftrightarrow OH + HO_2, & H + OH + M \leftrightarrow H_2O + M, \\
HO_2 + O \leftrightarrow OH + O_2, & H_2O_2 + OH \leftrightarrow H_2O + HO_2, \\
HO_2 + HO_2 \rightarrow H_2O_2 + O_2, &
\end{array}$$

where M denotes a third particle.

The equations of state for the hydrogen-air mixture have the usual form

$$p = \rho R_0 T \sum_{i=1}^{9} n_i, \quad h = \sum_{i=1}^{9} n_i h_i(T).$$

Here T is the temperature, R_0 is the universal gas constant. The values of the partial enthalpies $h_i(T)$ are borrowed from [3].

Boundary condition on the channel walls and obstacles is equality to zero of normal gas velocity.

The set of gas dynamics equations jointly with the set of chemical reaction rate equations, which takes into account principal features of chemical interaction of hydrogen with oxygen, was solved by a finite-difference method based on the Godunov's scheme [4].

3 Numerical results

The numerical calculations of the detonation propagation in the plane channels with the parallel walls have shown that in the case of energy input by the instantaneous electrical discharge of narrow layer form near the closed end of the channel the formed plane detonation front is curved with time due to instability of the combustion zone and the cellular detonation structure is formed. This fact is conformed to results of early realized

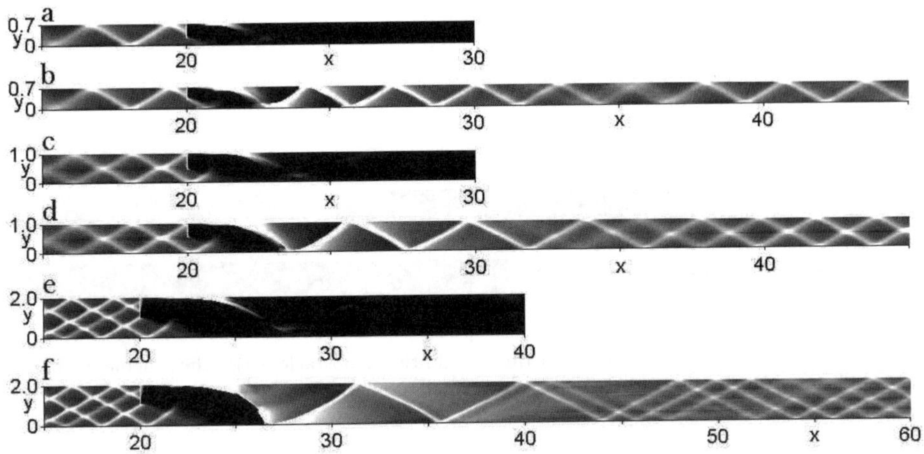

Fig. 1. Propagation of cellular detonation in the channel with undestroyable transversal obstacle for L=20: (**a**) – l=0.7, l_w=0.4; (**b**) – l=0.7, l_w=0.3; (**c**) – l=1.0, l_w=0.5; (**d**) – l=1.0, l_w=0.4; (**e**) – l=2.0, l_w=1.0; (**f**) – l=2.0, l_w=0.9

examination of detonation propagation in the plane channel filled with the stoichiometric hydrogen-air mixture [5] with use less detailed kinetic model of chemical reactions [6].

The influence of undestroyable transversal rigid obstacle (barrier) on the propagation of formed cellular detonation wave in the channel with parallel wall has been investigated under different values of obstacle height l_w and width of channel l. The obstacle is placed at the distance L from the closed end of channel. It has been established the detonation regime of combustion is restored after the interaction with the barrier if the barrier height l_w does not exceed some critical value depending on the channel width, see Fig. 1. Here and then all linear dimensions are divisible by the mean transversal size of detonation cell that obtained in the numerical calculations. Let us note that in the case of conservation of detonation after the passing of barrier the cellular detonation structure is restored only after some time.

The case of detonation destruction after the interaction with the obstacle has been studied. The effect of detonation reestablishment due to additional transversal obstacle placed in the channel has been detected, see Fig. 2. It has been obtained that the possibility of detonation reestablishment depends on such factors as additional obstacle height l_2 and its position L_2.

The propagation of detonation in the channel with one destroyable rigid obstacle (barrier) that is placed at the distance L from the closed end of channel was examined. The influence of obstacle existence time t_w on the conservation of detonation regime in the channel was analysed. Here t_w is the time interval starting from the moment of interaction of leading shock front with transversal obstacle. It has been obtained that in the case of destroyable barrier height l_w being greater than critical one the detonation is conserved if t_w does not exceed some critical value (Fig. 3).

Thereupon it is interesting to consider the propagation of detonation in the channel with destroyable transversal rigid obstacle when its height l_w is equal to width of the channel l. In this case the detonation is conserved if the time of barrier existence t_w does not exceed some critical value. Let us note that in the case when detonation wave

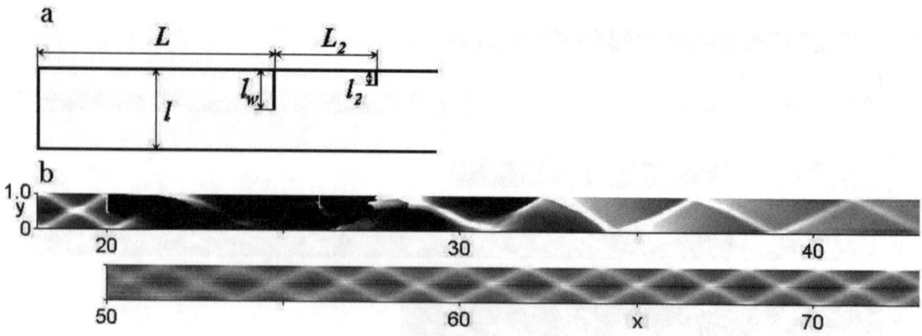

Fig. 2. The detonation reestablishment due to additional obstacle: (**a**) – the scheme of channel with additional obstacle; (**b**) – trajectories of triple points in the case of detonation reestablishment under L=20, l=1.0, l_w=0.5, L_2=6, l_2=0.3

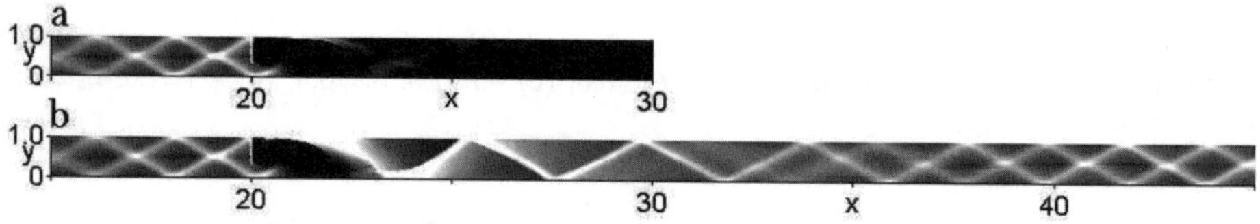

Fig. 3. Propagation of detonation in the channel (l=1.0) with destroyable transversal rigid obstacle of height that is greater than critical (l_w=0.6) for L=20: (**a**) – destruction of detonation under t_w=1mks, (**b**) – conservation of detonation under t_w=0.5mks

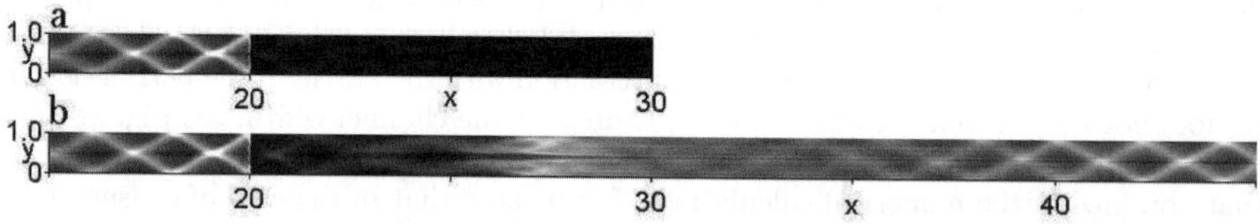

Fig. 4. Propagation of detonation in the channel with destroyable transversal rigid obstacle of height that is equal to width of the channel (l_w=1.0) for L=20: (**a**) – destruction of detonation under t_w=0.3mks, (**b**) – conservation of detonation under t_w=0.2mks

is conserved the restoration character of the cellular detonation structure differs qualitatively from the earlier observed one because of the barrier does not make additional two-dimensional distortion (see Fig. 4).

The behaviour of the cellular detonation in the case of sudden widening of the channel cross section was studied too. It has been established that in the case when channel width l is smaller than the half of critical channel width for the detonation transition into the unconfined space the detonation wave transits into the wide part of channel without failure if the widening h does not exceed some critical value (Fig. 5).

Acknowledgement. This work has been supported by the Russian Foundation for Basic Research (Grants 05-01-00004 and 05-08-33391), by the President Grant for Support of Leading Science Schools of Russian Federation (No. NSh-6791.2006.1).

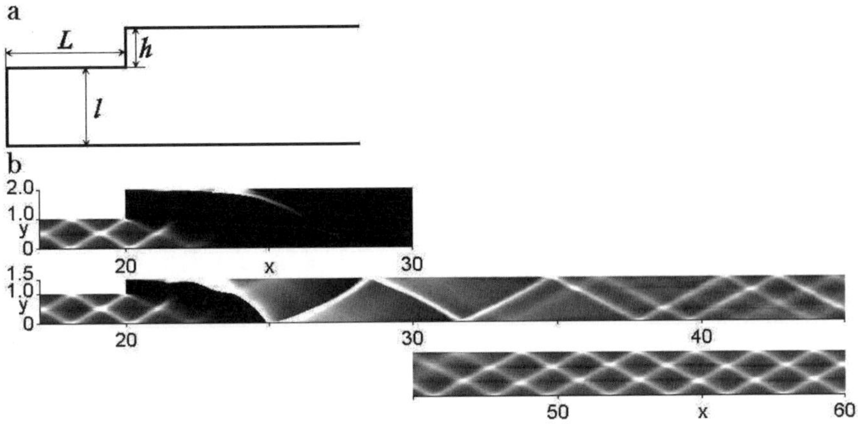

Fig. 5. Propagation of detonation in the case of sudden widening of the channel cross section: (**a**) – the scheme of abrupt widening channel; (**b**) – trajectories of triple points in the case of detonation destruction under h=1.0 and detonation conservation under h=0.5

References

1. Maas U., Warnatz J.: Combustion and Flame **74**, 1(1988)
2. Singh S., Rastigejev Ye., Paolucci S., Powers J.: Combustion Theory and Modelling **5**, 2(2001)
3. Glushko V.P. ed.: *Thermodynamic properties of individual substances*, Moscow (Nauka, 1978) 327p
4. Godunov S.K., Zabrodin A.V., Ivanov M.Ya., Kraiko A.N., Prokopov G.P.: *Numerical solution of multidimensional problems in gas dynamics*, Moscow (Nauka, 1976) 400p
5. Levin V.A., Markov V.V., Zhuravskaya T.A., Osinkin S.F.: Nonlinear wave processes that occur during the initiation and propagation of gaseous detonation. In: *Nonlinear Dynamics, Proc. V.A. Steklov Inst. Math.*, vol 251, Moscow (Nauka, 2005) pp 200–214
6. Takai R., Yoneda K., Hikita T.: Study of detonation wave structure. Proceedings of the 15th International symposium on combustion, Pittsburgh (1974)

Re-initiation of detonation wave behind slit-plate

J. Sentanuhady, Y. Tsukada, T. Obara, and S. Ohyagi

Graduate School of Science and Engineering, Saitama University, 255 Shimo-Ohkubo, Sakura-ku, Saitama-shi, Saitama, 338-8570 Japan

1 Introduction

The study to investigate a quenching mechanism of detonation wave utilizing a slit is of particular importance by considering safety devices to suppress the detonation wave in industries where flammable gases are handled [1,2]. The detonation wave propagated through the slit is disintegrated into a shock wave and a reaction front, since expansion waves generated at a corner of the slit have effects to decrease a temperature and reaction rate behind the shock wave. However, it is understood that the shock wave diffracted from the slit causes re-initiation and transited to detonation wave at downstream region, even though a diameter of open-area is smaller than critical tube diameter [1,2]. It is also well known that reaction front can accelerate rapidly to supersonic velocity when propagating over obstacles. Mitrovanov and Soloukhin [3] reported and was also confirmed by Edward et al. [4] that the critical value to distinguish the propagation of detonation wave is about 13λ for circular tube and about 10λ for rectangular channel, where λ is a cell size of stable detonation wave.

A fundamental observation carried out by Moen et al. [5] clarified if the turbulence intensity is maintained by placing obstacles, the reaction rate and degree of turbulence become highly coupled. Furthermore, experiment and numerical simulation of decoupling and re-coupling processes behind sudden expansion of a tube were conducted by Pantow et al. [6] and Ohyagi et al. [7] to show re-initiation processes of detonation wave after decoupled by diffraction process. These results showed that reflected shock wave and Mach reflection could be a source to re-initiate a detonation wave. However, fundamental mechanisms of re-initiation processes of detonation wave by the interaction of shock wave with another shock wave or tube wall are still open questions.

In this study, experiments are carried out in order to elucidate the re-initiation mechanisms of detonation wave by installing the slit-plate into a detonation tube filled with premixed gas of hydrogen and oxygen. A width of slit w, a distance between two slits x and initial pressure of test gas p_0 are varied and re-initiation processes are visualized using high-speed image converter camera with schlieren optical system.

2 Experimental Setup

Figure 1(a) shows schematic diagram of experimental setup of detonation tube used in this study. The detonation tube has square cross section of 50 mm and 4,100 mm in total length. The tube was divided into three sections, i.e. driver section of 1,000 mm, driven section of 2,700 mm including observation section, and dump tank of 400 mm to decay the detonation wave. In observation section, measuring stations named from

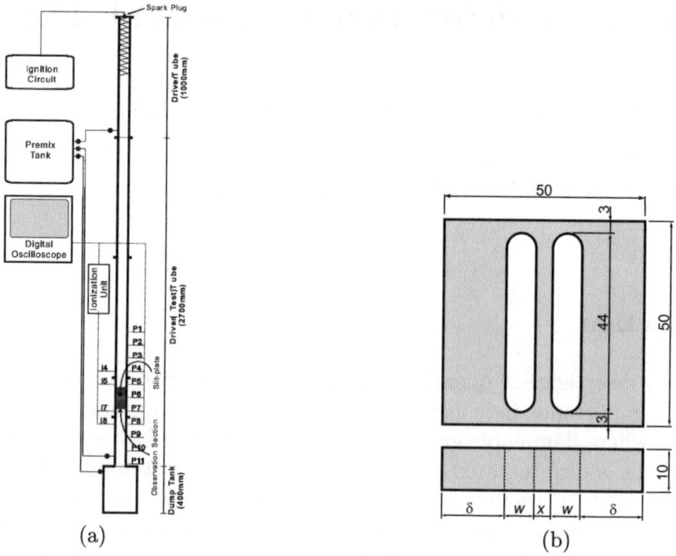

Fig. 1. (a) Schematic diagram of experimental setup and (b) configuration of slit-plate

P1 to P11 were mounted with an interval of 100 mm. Mylar film of 25 μm thickness was inserted between driver section and driven one to separate driver gas and test gas, which were filled with different initial pressure of gas mixture. The Shchelkin spiral coil of 500 mm length, 38 mm pitch was inserted at driver section to decrease detonation induction distance. An automobile ignition plug was also installed at top of the driver section to ignite driver gas.

Figure 1(b) shows configuration of slit-plate used as model in this experiment. A slit-plate having dimension of 50 mm square and 10 mm thickness was inserted at P6 of observation section. A width of slit w and a distance between two slits x were varied as shown in table 1(a).

To record pressure profile of shock wave, four pressure transducers (PCB, model 113A24) were installed at position P4, P5, P7 and P8 as shown in Fig. 1(a). The arrival time of reaction fronts were detected by four ionization probes I4, I5, I6 and I7, oppositely installed to pressure transducers. The outputs of these probes were stored by eight channels digital oscilloscope (Yokogawa, model DL750) to record a profile of shock

Table 1. (a) Slit-plate configuration and (b) initial condition of gas mixture

(a)

Slit-Plate Code	w (mm)	x (mm)
8-10-8	8	10
8-5-8	8	5
8-2-8	8	2
5-10-5	5	10
5-5-5	5	5
5-2-5	5	2

(b)

	Driver Section	Driven Section
Fuel	H_2	H_2
Oxidizer	O_2	O_2
Equivalence ratio	1	1
Initial pressure (kPa)	100	10 ~ 100

wave and reaction front. The phenomenon of re-initiation processes of detonation wave was visualized by using high-speed schlieren photographs with image converter camera (IMACON 792, Hadland Photonics). Soot track records of detonation wave at upstream and downstream region of slit plate were also taken in order to evaluate a re-initiation distance of detonation wave from the end of slit-plate.

A stoichiometric mixture of hydrogen and oxygen with initial pressure of 100 kPa was used as driver gas and initial pressure p_0 ranged from 10 $\sim$ 100 kPa was used as test gas. An initial condition of gas mixture is shown in Table 1(b).

3 Results

Based on a photograph taken by high-speed image converter camera, soot track record and pressure histories, re-initiation of detonation wave behind slit-plate could be classified into mainly two types, i.e. the detonation wave is re-initiated by the interaction of two shock waves (SSI) and by interaction of shock wave with wall (SWI). Furthermore, based on repetition of interaction, re-initiation mechanisms are classified into four types, where first time of shock-shock interaction (SSI1), second times of shock-shock interaction (SSI2), first time of shock-wall interaction (SWI1) and second times of shock-wall interaction (SWI2).

3.1 The first shock-shock interaction (SSI1)

Figure 2(a) is sequential photograph showing diffraction and re-initiation process of detonation wave behind a slit-plate for case of $w = 8$ mm, $x = 10$ mm and initial pressure of $p_0 = 40$ kPa. This result corresponds to soot track record as shown in Fig. 3(a-I), which is obtained with same experimental condition. A detonation wave is propagated to downstream direction. In this case, the cell size of detonation wave is about 5.5 mm, it means that cell number of detonation wave inside the slit is more than unity. A diffracted shock waves (IS) emerged from the slit are decoupled from a reaction front (RF), which is clearly identified in frame number of 1 to 4. The incident shock waves propagated ahead of reaction front has a propagation velocity of about 2,300 m/s interact each other at a center of slit-plate. This interaction produces high-energy enough to generate local explosion just behind the slit-plate. Thereafter, this local explosion generates shock wave indicated as ES, propagated spherically as clearly visible in a frame number 1 to 3. Since the local explosion shock wave is also propagated upstream direction (burned gas region), it is not followed by reaction front and as a result the velocity of the shock wave is decreased to about 1,500 m/s. However, explosion shock wave propagated downstream direction (unburned gas mixture region) directly followed by reaction front with small separation distance, which is indicated as detonation wave. According to frame number 2 and 3, the propagation velocity of initial detonation wave is estimated as 3,200 m/s, which is greater than Chapman-Jouguet (C-J) detonation velocity of 2,794 m/s.

According to a measurement of velocity and soot track record as shown in Fig. 3(a-I), this detonation wave is classified as overdriven detonation wave. The overdriven detonation wave is quickly decelerated to the propagation velocity of C-J velocity as shown in soot track record image, where cell size becomes relatively larger. These results indicate that local explosion shock wave generated by the interaction of the diffracted shock wave plays key role on the re-initiation of the detonation wave and this is classified as type SSI1 (re-initiated by first interaction of shock wave with shock wave).

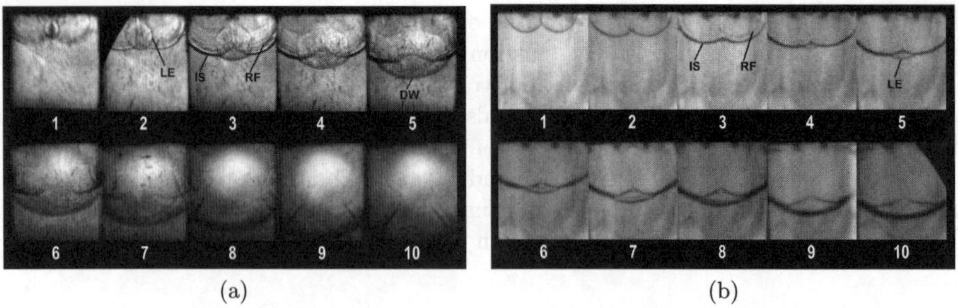

Fig. 2. Sequential schlieren photograph showing (a) re-initiation process of detonation wave by the first shock-shock interaction (IFT = 2 μs, po = 40 kPa, w = 8 mm, x = 5 mm) and (b) re-initiation process of detonation wave by the second shock-shock interaction (IFT = 2 μs, po = 20 kPa, w = 5 mm, x = 10 mm)

3.2 The second shock-shock interaction (SSI2)

Figure 2(b) is sequential schlieren photograph of high-speed video camera showing local explosion due to shock-shock interaction which could not re-initiate detonation wave just behind the slit-plate. A frame interval and exposure time is same as Fig. 2(a), while slit-plate of $w = 5$ mm, $x = 10$ mm is used with initial pressure $p_0 = 20$ kPa. This condition corresponds to soot track record as shown in Fig. 3(a-III). The cell size of detonation wave for this condition is estimated as 9 mm, it means that cell number inside the slit is less than unity. Incident shock waves emerged from the slit are propagated with a velocity of 921 m/s. A local explosion shock wave (ES) is also generated in this case which could be observed at frame number 2 and 3 and propagated with a velocity of 2,177 m/s, lower than C-J velocity of 2,750 m/s. However detonation wave could not be re-initiated by this local explosion because of lower sensitivity of gas mixture and lower velocity of incident shock wave emerged from the slit compared with a case of with Fig. 2(a). The local explosion shock wave (ES) overtakes the incident shock wave (IS) and transits to a strong shock wave shown as black layer in frame number 8 to 10.

The phenomena after frame number of 10 could not be shown in this figure because of limitation of active coverage of camera. However by analyzing soot track record as shown as Fig. 3(a-III), it is confirmed that detonation wave is failed to re-initiate after that the coupled shock wave interact with surface of wall. Later, reflected shock waves generated by interaction of coupled shock wave with surface of wall and propagated transversally collide each other at the center of tube. This collision produces hot-spot region and strong Mach stem at center of tube, which is identified as white region at position of shock-shock interaction. An overdriven detonation wave is re-initiated at downstream of white region as shown as fine detonation cells. As shown in Fig. 3(a-III), the overdriven detonation wave is transited to stable detonation at downstream region, where cell size becomes relatively larger. This behavior of re-initiation process is classified as type SSI2 (re-initiated by second interaction of shock wave with shock wave).

3.3 Re-initiation observed by soot track record

Since coverage area of an image converter camera on the observation section is limited, the re-initiation generated at downstream position is understood by using soot track record.

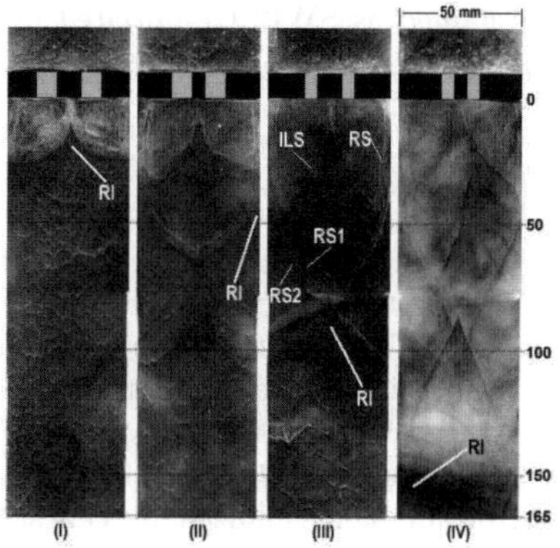

(a)

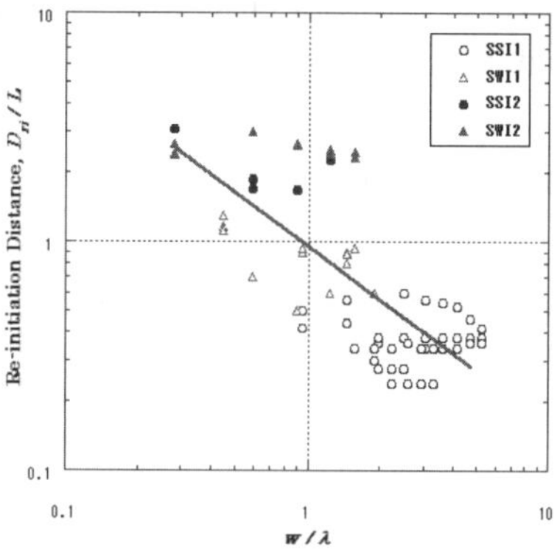

(b)

Fig. 3. (a) Soot track record of detonation wave, (I) SSI1, slit-plate 8-10-8, $p_0 = 40$ kPa, (II) SWI1, slit-plate 8-5-8, $p_0 = 20$ kPa, (III) SSI2, slit-plate 5-10-5, $p_0 = 20$ kPa, (IV) SWI2, slit-plate 5-5-5, $p_0 = 20$ kPa and (b) relationship of non-dimensional re-initiation distance D_{ri}/L and non-dimensional width of slit w/λ

Figure 3(a) shows the images of soot track record showing four types of re-initiation processes of detonation wave behind the slit-plate up to 165 mm.

Figure 3(a-I) is a case of the first shock-shock interaction, named as type SSI1 where detonation is re-initiated by direct interaction of diffracted shock waves at center of the tube. This interaction induces overdriven detonation wave at a region indicated as RI, where fine cellular pattern is observed. However the overdriven detonation wave is quickly attenuated to stable detonation wave, which is shown as relatively larger cell size of detonation wave at downstream of re-initiation point.

Figure 3(a-II) is soot track record obtained from a slit-plate of $w = 8$ mm, $x = 5$ mm and initial pressure of driven section of $p_0 = 20$ kPa. In this case, re-initiation of detonation wave is failed at position of shock-shock interaction, incident shock wave and local explosion shock wave are coupled to be strong shock wave and it will interact with surface of wall. The shock wave is reflected from the wall and temperature behind reflected shock wave becomes high value, where compressed unburned gas mixture and turbulent boundary layer have formed. If temperature is high enough, local explosion will be generated and will induce overdriven detonation wave near the surface of wall as indicated as RI in Fig 3(a-II). This type of re-initiation is characterized by type SWI1 (re-initiated by the first interaction of shock wave with surface of wall).

As explained in section 3.2, Fig. 3(a-III) is the case, detonation wave is re-initiation by the second interaction of two shock waves reflected from both wall and propagated transversally. This type of re-initiation would be occurred if re-initiation by both types SSI1 and SWI1 has failed and this type of re-initiation is characterized as SSI2.

Figure 3(a-IV) is a case where distance between the slit x is changed to 5 mm. The initial pressure of test gas is same as Fig. 3(a-III). In this case, interaction of incident shock wave, emerged from slits could not produce strong local explosion below the slit-plate. After four times of interaction, the diffracted shock wave is reflected from the wall secondly to re-initiate the detonation wave at position as indicated as RI in Fig. 3(a-IV). Same as previous re-initiation phenomena, overdriven detonation wave is also observed in this case. This type of re-initiation is named as SWI2 (re-initiated by the second interaction of shock wave with surface of wall).

4 Discussion

As is described above, the re-initiation distance D_{ri} from the end of the slit-plate is changed by the width of slit w, distance between two slits x and initial pressure of test gases p_0. The re-initiation distance of detonation wave D_{ri} might be correlated using width of slit w and cell size λ, which is inversely proportional to a reaction rate of the test gas mixture. Figure 3(b) shows a relationship between re-initiation distance and width of the slit, where vertical axis is re-initiation distance D_{ri} normalized by the width of square detonation tube L, horizontal is width of slit w normalized by cell size λ. Non-dimensional re-initiation distance is decrease as the w/λ is increased. Furthermore, most of the re-initiation type indicated SSI1 is occurred for the condition w/λ is greater than unity. Therefore, the detonation wave directly re-initiated by the first interaction of diffracted shock waves is found for experimental condition that at least one cell is emerged from the slit. When non-dimensional distance w/λ is less than unity, detonation wave has tendency to be re-initiated by the type SWI1, SSI2 or SWI2.

5 Conclusions

Experiments were conducted in order to investigate the re-initiation processes of detonation wave utilizing slit-plate and schlieren photograph was obtained to show re-initiation processes of detonation wave. When a detonation wave is propagated through the slit, it is consistent that the detonation wave is once failed to transmit, even though relatively high-initial pressure of test gas. Shock waves diffracted from the slit interact each other at center of the tube and local explosion is observed producing explosion shock waves. This explosion shock wave is enough to re-initiate the detonation wave for the case that w/λ is greater than unity. The detonation wave is re-initiated by shock-wall interaction for the case that w/λ is less than unity.

References

1. Ciccarelli G., Boccio J. L., Detonation wave propagation through single orifice plate in a circular tube, Proc. of 27th Symp. (Int.) on Combustion, 2233-2239, 1998
2. Jayan S., Tsukada Y., Yoshihashi T., Obara T., Ohyagi S., Re-initiation of detonation wave behind a perforated plate, Proc. of 20th Int. Coll. on the Dynamics of Explosions and Reactive Systems, CD-ROM, 2005
3. Mitrovanov V. V., Soloukhin R. I., The diffraction of multifront detonation waves, Soviet Physics-Doclady, vol. 9:12, 1055-1058, 1965
4. Edwards D. H., Thomas G. O., Nettleton M. A., The diffraction of a planar detonation wave at an abrupt area change, J. Fluid Mech., vol. 95, 79-96, 1979
5. Moen I. O., Donato M., Knystautas R., Lee J. H., Flame acceleration due to turbulence produced by obstacles, Comb. and Flame, vol. 39, 21-32, 1980
6. Pantow E. G., Fischer M., Kratzel Th., Decoupling and recoupling of detonation waves associated with sudden expansion, J. Shock Waves, vol. 6, 131-137, 1996
7. Ohyagi S., Obara T., Hoshi S., Cai P., Yoshihashi T., Diffraction and re-initiation of detonations behind a backward-facing step, J. Shock Waves, vol. 12, 221-226, 2002

Shock-to-detonation transition due to shock interaction with prechamber-jet cloud

S.M. Frolov, V.S. Aksenov, and V.Y. Basevich

N.N. Semenov Institute of Chemical Physics, Russian Academy of Sciences, Moscow 119991, Russia

Summary. A new principle of detonation initiation due to controlled interaction of a propagating shock wave (SW) with the cloud of hot explosive gas formed by prechamber flame jets in the stoichiometric propane–air mixture was demonstrated experimentally. Detonation initiation was shown to be conditioned by synchronization of cloud autoignition with its shock-induced compression.

1 Introduction

There exist two classical approaches for gaseous detonation initiation, namely, direct initiation with a strong source [1] and deflagration-to-detonation transition (DDT) [2,3]. In the course of direct initiation, a strong primary SW is generated. The temperature, pressure and compression phase duration in such an SW are sufficient for triggering fast exothermic chemical reactions in the close vicinity to the lead shock front. In this case, a detonation forms after a certain relatively short transition period. For the DDT, there is no need in the strong primary SW. The flame arising from a weak ignition source changes shape due to various instabilities and nonuniformities, thus leading to progressive thermal expansion of the reactive mixture and formation of an SW. After a certain relatively long transition period, autoignition of the mixture occurs in the region between the SW and the accelerating flame, leading to a detonation.

The other approach, different from those mentioned above, was suggested in [4,5]. A possibility to initiate detonation due to acceleration of an initially weak SW by a travelling ignition pulse was demonstrated experimentally. In this case, fast exothermic reactions behind a lead shock front were triggered by the external ignition source rather than by the SW itself. The external ignition source, travelling with the SW, triggered chemical reactions in the explosive gas, thus promoting fast shock-to-detonation transition (SDT). In [4,5], successive triggering of seven electric discharges, mounted equidistantly along the tube, was sufficient for initiating a detonation in the stoichiometric propane–air mixture at normal initial conditions at a distance of 12 to 14 tube diameters. A necessary condition for detonation initiation was careful synchronization of the triggering time of each electric discharge with the SW arrival to its position. The research outlined in this paper continues the studies reported in [4,5]. Contrary to [4,5], forced ignition of the reactive mixture behind the propagating SW was achieved using a classical prechamber [6] rather than a series of electric discharges.

2 Experimental Setup

Figure 1 shows the schematic of the experimental setup. The main elements of the setup are an SW-generator, straight detonation tube 60 mm in diameter and 3.5 m long, and a prechamber. The SW-generator (solid-propellant gas generator) was mounted at one end of the detonation tube. The other end of the tube was closed. The prechamber was screwed to the lateral wall of the tube at a certain distance from the SW-generator.

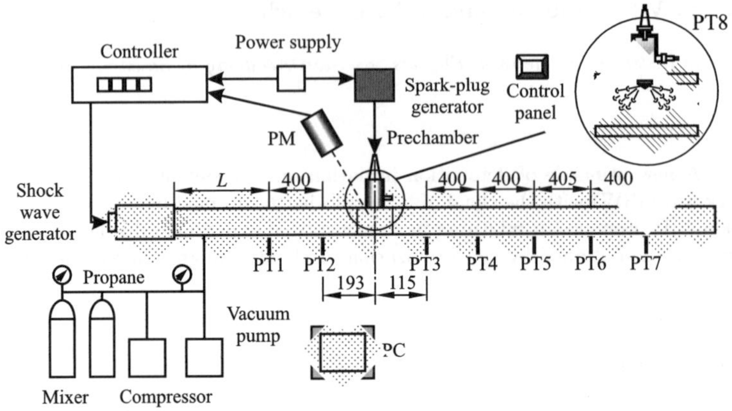

Fig. 1. Schematic of the experimental setup. Dimensions are in millimeters

The prechamber was a steel cylindrical chamber 31 mm in diameter and 50 mm long. A standard spark plug was fixed at one end of the prechamber, while the other end was equipped with a changeable conical nozzle with the apex angle of 120° (see the insert in Fig. 1) and two round orifices 5 mm in diameter connecting the prechamber with the volume of the detonation tube. The thickness of the nozzle wall was 1 mm. The axes of the nozzle orifices and the tube were positioned in the same plane.

The measuring system included 8 high-frequency piezoelectric pressure transducers PT1 to PT8 and a photomultiplier (PM). The pressure transducers were used to register the SW arrival time to the corresponding tube cross section and for the analysis of wave processes in the prechamber and in the SW traversing the cloud of prechamber gases. From now on, the cloud of prechamber gases will be referred to as the prechamber cloud. For detecting the luminosity of combustion products by the PM, a provision was made for an optical window. The data acquisition system comprised the analog-to-digital converter and a PC.

3 Experimental Procedure

Before each run, the tube and prechamber were evacuated and filled with the stoichiometric propane–air mixture at normal initial conditions (a temperature of 293 ± 2 K and pressure of 0.1 MPa). In the preliminary experiments, the SW-generator and prechambers were tested and the measuring procedure was established. The tests with the SW-generator were aimed at measuring the parameters of the shock waves propagating along the detonation tube without mixture ignition in the prechamber. The error in determining the mean SW velocity was estimated as 2.5%. The preliminary tests with the prechambers were aimed at providing ignition of the prechamber cloud with a significant delay after the prechamber jets issued from the nozzles. These tests were conducted by trying different nozzle units without activating the SW-generator. With the nozzle unit containing two round orifices 5 mm in diameter, the delay time between mixture ignition in the prechamber and prechamber cloud autoignition in the tube was about 35–40 ms.

In the experiments on SW – prechamber cloud interaction, all measuring units were activated by a digital controller upon spark plug triggering in the prechamber. The SW-generator was activated after a certain preset time delay resulting in the formation of a primary SW. The SW propagated along the tube and arrived at the prechamber nozzle position either before or after prechamber cloud autoignition. For the quantitative description of the SW – prechamber cloud interaction, a concept of SW arrival delay time with respect to the autoignition event registered by the PM was introduced. In the following, this delay time will be referred to as the SW arrival delay τ. The activation time of the SW-generator was chosen based on the preliminary tests described above.

4 Experimental Results

The primary SW was generated by a solid-propellant SW-generator, a cylindrical combustion chamber equipped with the changeable nozzle and a bursting diaphragm. Before the test, a small amount (up to 1.50–2.25 g) of porous cotton propellant was placed in the chamber. The propellant was ignited by a primer. The total time of propellant combustion was 1–2 ms. The compression phase duration of the resultant SW was determined by the nozzle diameter. The thickness and material of the bursting diaphragm determined the initial SW velocity. The initial SW velocity was also varied by changing the distance between the SW-generator and the prechamber. This distance was $L = 1.563$ m in experimental Series 1 and $L = 1.293$ m in Series 2 (see Fig. 1). The tests in Series 1 and 2 were performed mainly with the shock waves of initial velocity $V = 700$–850 and 1080–1120 m/s, respectively. In addition, the tests with the elevated initial SW velocities, $1000 \leq V \leq 1250$ m/s, were made within Series 1.

Figure 2 shows the measured dependencies of the mean SW velocity on the distance travelled along the tube in Series 1 (solid curves) and 2 (dashed curves). Section $X = 0$ in Fig. 2 corresponds to the prechamber axis position, and the horizontal dotted line corresponds to the theoretical CJ detonation velocity in the stoichiometric propane–air mixture.

Curves *1* and *9* were plotted based on the results of preliminary tests without mixture ignition in the prechamber. It is seen that in Series 1, the mean velocity of the primary SW (curve *1*) dropped monotonically from about 710 m/s at $X = -0.35$ m to 650 m/s at $X = 0.7$ m. Contrary to curve *1*, the baseline curve *9* in Series 2 shows gradual acceleration of the SW along the tube from 1100 m/s at $X = 0.35$ m to 1650 m/s at $X = 1.7$ m. Similar behavior of primary shock waves in Series 2 was registered at large "negative" SW arrival delays $\tau < -3$ ms.

In the tests of Series 1 with the elevated initial SW velocities (curves *7* and *8* in Fig. 2), secondary explosions were detected when the SW was traversing the prechamber cloud. Figure 3 shows the pressure records corresponding to $\tau = 12.2$ ms (curve *7* in Fig. 2). The SW velocity at the measuring segment PT1–PT2 was 1057 ± 30 m/s. At the segment PT2–PT3, the primary SW accelerated to 1481 ± 40 m/s, and then decelerated to 1282 ± 30 m/s at the segment PT3–PT4 and to 770 ± 15 m/s at the segment PT4–PT5. The pressure record of PT3 indicates that a secondary explosion occurred behind the primary shock wave. The secondary SW propagated at a velocity of 2200–2300 m/s rapidly approaching the primary one (see pressure records of PT4 and PT5).

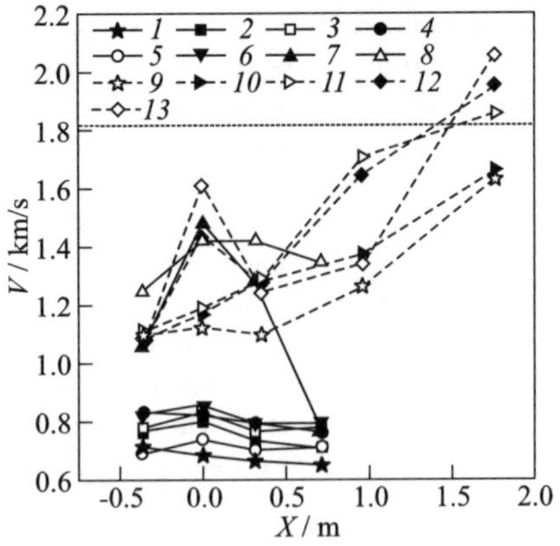

Fig. 2. Measured dependencies of the mean velocity of the primary SW on the distance travelled along the tube at different SW arrival delay to the prechamber cloud. Series 1 (solid curves): *1* — without prechamber, *2* and *3* — $\tau = -0.6$ ms, *4* — -0.4, *5* — 0.0, *6* — 2.3, *7* — 12.2, and *8* — $\tau = 31.0$ ms; Series 2 (dashed curves): *9* — without prechamber, *10* — $\tau = -2.0$ ms, *11* — 2.6, *12* — 4.6, and *13* — $\tau = 5.6$ ms. Section $X = 0$ corresponds to the prechamber axis location. Horizontal line corresponds to the theoretical CJ detonation velocity in the stoichiometric propane–air mixture

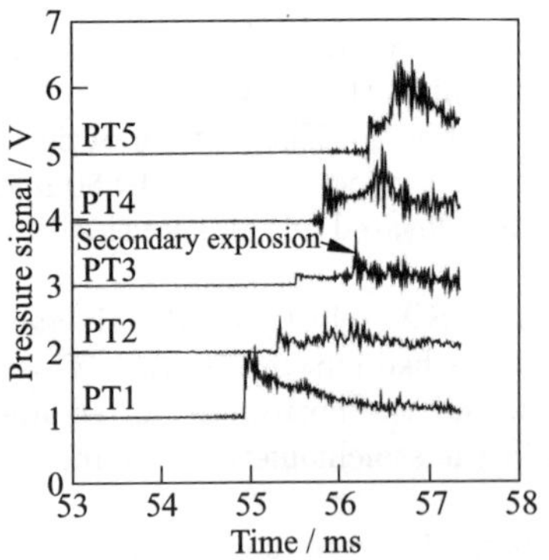

Fig. 3. Records of pressure transducers PT1 to PT5 in the test of Series 1 with the SW arrival delay time $\tau = 12.2$ ms (see curve *7* in Fig. 2)

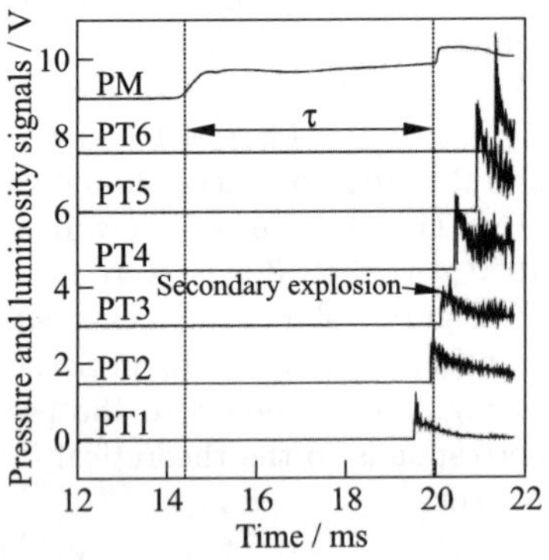

Fig. 4. Records of pressure transducers PT1 to PT6 and photomultiplier (PM) in the test of Series 2 with the SW arrival delay time $\tau = 5.6$ ms (see curve *13* in Fig. 2)

The most interesting were the tests with mixture ignition in the prechamber in Series 2. At SW arrival delays τ from 2.6 to 5.6 ms (curves *11–13* in Fig. 2), acceleration of a primary SW to the mean velocity of 1850–2100 m/s, exceeding the CJ detonation velocity, was detected. Such high values of the mean SW velocities indicate that downstream the prechamber cloud, the SDT occurred via the stage of overdriven detonation formation. The shortest predetonation distance ($X \approx 1.3$–1.5 m) was attained at short positive SW arrival delays ($\tau = 2.6$ and 4.6 ms, curves *11* and *12* in Fig. 2). At the delay of $\tau = 5.6$ ms (curve *13* in Fig. 2 and Fig. 4), the primary SW, after passing the cloud, propagated initially at the velocity typical for the "negative" delays τ ($\tau = -2$ ms, curve *10* in Fig. 2). However, at a distance of $X \approx 1.0$–1.7 m, a sharp increase in the mean SW velocity up to 2100 m/s occurred due to collision of the secondary SW formed in the prechamber cloud with the primary SW.

Discussion and conclusion

Thus, synchronization of SW arrival to the prechamber cloud with cloud autoignition made it possible to accelerate the SDT phenomenon. In the tests of Series 2, for the transition of the SW of Mach number 3.2 to a detonation, a distance of about 1.3–1.5 m (22–25 tube diameters) was required. This distance should be compared with that needed for detonation onset in the tube without mixture ignition in the prechamber. Note that in the tests without mixture ignition in the prechamber, a detonation was not observed (see curve *9* in Fig. 2). If curve *9* is extrapolated to the line $V = 1804$ m/s, the estimate of 2.2 m (37 tube diameters) for the predetonation distance could be obtained. This means that SW – prechamber cloud interaction resulted in the reduction of the predetonation distance by a factor of 1.5. The possibility of such a reduction was determined by the initial SW velocity and compression phase duration, as well as by the SW arrival time to the prechamber cloud. Early or late SW arrival to the cloud did not lead to a significant change in SW evolution. The effect of predetonation distance reduction was detected only at careful synchronization of SW arrival to the prechamber cloud with cloud autoignition.

The physical mechanism of such a resonant interaction of the SW with the prechamber cloud is most probably connected with the enhanced sensitivity of the explosive mixture in the cloud, preconditioned to autoignition. In such conditions, mixture compression and heating in the SW with the Mach number of about 3.2 appeared to be sufficient for triggering fast exothermic reactions in the close vicinity to the lead shock front and detonation onset. This mechanism has much in common with the SWACER-mechanism suggested in [7].

The interpretation of the present experimental findings is also in line with the early experimental data of Shchelkin and Sokolik [8] who studied the effect of preliminary cool-flame oxidation of *n*-pentane–oxygen mixtures on the predetonation distance. Shchelkin and Sokolik discovered a sharp reduction (up to a factor 1.5 to 2) of the predetonation distance depending on the ignition timing of the preconditioned explosive mixture.

In the phenomena observed herein, a proper role could be also played by the classical mechanism of flame acceleration due to SW – flame interaction leading to a multiple increase in the flame surface area [2,3]. However, the fact that the considerable reduction of the predetonation distance was obtained only at a certain (resonant) delay time of SW arrival to the prechamber cloud indicates the minor role of this mechanism. Nevertheless, at long SW arrival delays, when the effects under discussion were less pronounced, the pressure transducers located downstream the prechamber registered secondary explosions behind the primary SW. These explosions might promote the detonation onset in tubes longer than that used in the present study.

Acknowledgement. This work was partly supported by the International Science and Technology Center, project No. 2740.

References

1. Zel'dovich, Ya. B., S. M. Kogarko, and N. I. Simonov. 1957. *Sov. J. Technical Physics* 86(8):1744.
2. Sokolik, A.S. 1960. *Self-ignition, flame, and detonation in gases*. Moscow: USSR Acad. Sci. Publ.
3. Shchelkin, K. I., and Ya. K. Troshin. 1963. *Gas dynamics of combustion*. Moscow: USSR Acad. Sci. Publ.
4. Frolov, S. M., V. Ya. Basevich, V. S. Aksenov, and S. A. Polikhov. 2004. *Doklady Physical Chemistry* 394(2):222.
5. Frolov, S. M., V. Ya. Basevich, V. S. Aksenov, and S. A. Polikhov. 2003. *J. Propulsion Power* 19(4):573.
6. Sokolik, A. S., and V. P. Karpov. 1958. In *Combustion and mixture formation in diesel engines*. Moscow: USSR Acad. Sci. Publ. Part 1. 483.
7. Lee, J. H. S., I. O. Moen. 1980. *Progress Energy Combust. Sci.* 6(4):359.
8. Shchelkin, K. I., and A. S. Sokolik. 1937. *Sov. J. Physical Chemistry* 10:484.

Shock-to-detonation transition in tube coils

S.M. Frolov[1], **I.V. Semenov**[2], **I.F. Ahmedyanov**[2], **and V.V. Markov**[3]

[1] *N. N. Semenov Institute of Chemical Physics, Russian Academy of Sciences, Moscow 119991, Russia*
[2] *Institute for Computer Aided Design, 2nd Brestskaya Str. 19/18, Moscow, 123056, Russia*
[3] *Steklov Mathematical Institute, Gubkin Str. 8, Moscow, 119991, Russia*

Summary. Experimental and computational studies of reactive shock wave propagation through tube coils with different curvature radii demonstrate that tube coils promote shock-to-detonation transition.

1 Introduction

There are several ways to decrease the predetonation distance in gaseous explosive mixtures. In 1920s, Laffitte [1, 2] experimentally demonstrated that a decrease in the tube diameter and an increase in the initial pressure of an explosive mixture decrease the predetonation distance. Shchelkin and Sokolik [3] detected a decrease in the predetonation distance after preliminary thermal treatment of fuel. Shchelkin [4] revealed that the aerodynamic conditions in the channel play a leading role in the deflagration-to-detonation transition (DDT). He showed that, if an obstacle in the form of a wire spiral (a Shchelkin spiral) is placed in the channel, the predetonation distance decreases considerably. Since then there were many studies of DDT in straight tubes with obstacles (McGill University, Caltech, CNRS, Russian schools, etc.)

Recently the topic of DDT and shock-to-detonation transition (SDT) have attracted much attention in view of possible application of detonation to propulsion. Brophy et al. [5] showed that the predetonation distance in the case of ignition of a mixture by a nanosecond corona discharge in combination with a Shchelkin spiral is smaller than that in the case of ignition of a mixture by an arc discharge. Frolov et al. [6] proved experimentally that the predetonation distance can be significantly decreased by accelerating a weak primary shock wave by a traveling forced ignition pulse. For this purpose, several (up to 7) electric dischargers mounted along the tube were triggered synchronously with the primary shock wave arrival at their position. Frolov et al. [7–11] discovered and studied experimentally a significant effect of tube coils and U-bends on reduction of the predetonation distance at SDT. It was found that the shock waves propagating at velocities above 800–900 m/s in the stoichiometric gaseous propane–air mixture (51-mm U-bent tube) and above 900–1000 m/s in the heterogeneous n-hexane–air and n-heptane–air mixtures (36-mm and 28-mm tubes with coils) were capable of transitioning to detonation after passing 180-degree U-bends or tube coils. For obtaining such shock waves we used a weak electric igniter and a Shchelkin spiral.

The objective of the experimental and computational research outlined in this paper was to better understand the phenomena accompanying SDT in the course of shock propagation through the tube coils represented by circular 360-degree loops of tubes with different curvature radii.

2 Experimental Studies

According to Frolov et al. [7–11] curved surfaces are capable of promoting SDT due to the onset of exothermic centers in the core flow behind the propagating shock wave. In view of it, SDT could be expected to arise more readily in curved rather than straight tubes. Thus, tube coils and bends should promote SDT efficiently. As a matter of fact, the experimental studies described in this Section demonstrate that SDT can be facilitated by placing a tube coil on the way of the propagating shock wave.

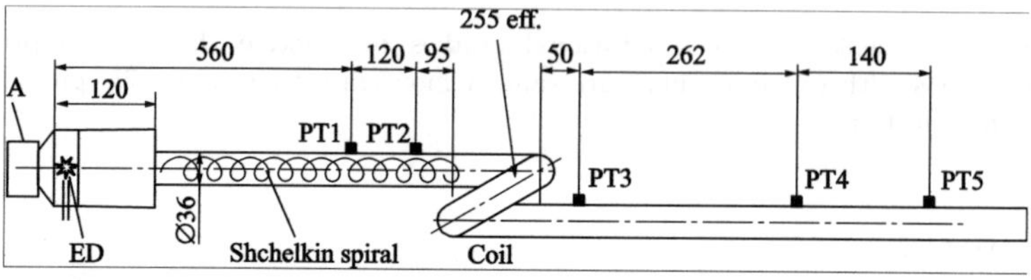

Fig. 1. Schematic of the 36-mm i.d. tube with a combination of Shchelkin spiral and tube coil

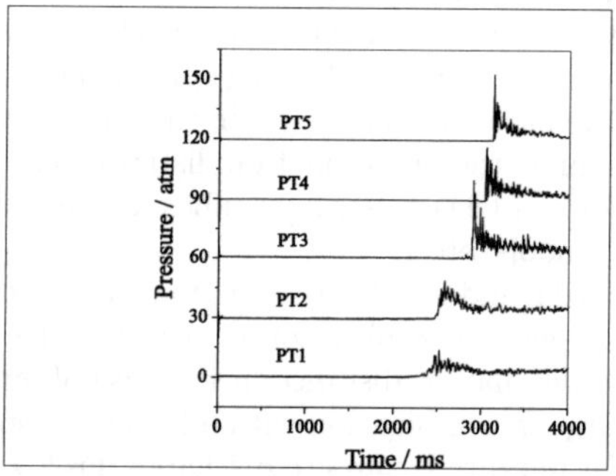

Fig. 2. Pressure records in the run with the ignition energy of 60 J

Figure 1 shows the schematic of the experimental setup comprising a smooth-walled tube 36 mm in diameter with liquid-fuel atomizer A, electric igniter ED, Shchelkin spiral, and a removable section with a 360-degree tube coil. The air-assist atomizer described in detail by Frolov et al. [9] provided the entire mass flow rate of two-phase n-hexane–air mixture through the tube. The mean droplet size at a distance of 60 mm downstream from the atomizer nozzle, measured by a soot sampling method, was 5–6 μm. The three-electrode electric igniter was mounted at a distance of 60 mm downstream from the nozzle exit and was used for two-phase mixture ignition. Ignition pulse duration and energy were varied within the following ranges: 20–40 μs and 20–240 J. Shchelkin spiral 600 mm long was wounded from the steel wire 4 mm in diameter with the pitch of 18 mm and was used for flame acceleration and generation of the shock wave ahead of the flame. The coil was fabricated from the standard segments of 36-millimeter tube by electrical welding. Internal walls of the tube inside the coil were smooth. The length of the coil along the tube axis was 255 mm. The tube was mounted on the experimental test rig equipped with fuel and air supply systems, as well as control and diagnostic systems described by Frolov [10]. Piezoelectric pressure transducers PT1 to PT5 were used to register the shock wave velocity and pressure profiles. Two sets of experiments were made: one without coil section and the other with coil section.

In the experiments without coil section, the maximal shock wave velocity at the exit from Shchelkin spiral was 900–1000 m/s and no DDT was observed. Variation of the spiral wire diameter, as well as spiral length and pitch did not result in any significant change in the maximal shock wave velocity up to ignition energies of 240 J.

Figure 2 shows the records of pressure transducers PT1 to PT5 in the run with the coil at the ignition energy of 60 J. In this run, the detonation wave was registered at the exit from the coil (transducer PT3). Detonation arises inside the coil at a distance of about 1 m from the igniter (about 28 tube diameters). The detonation wave propagates with the constant mean velocity of 1750 ± 20 m/s till the end of the tube. The minimal ignition energy required for detonation initiation in this setup was about 55 J. At lower ignition energies, the flame kernel was blowing-off by the high-speed two-phase flow issuing from the atomizer nozzle. Similar experiments were made in the 28-millimeter tube equipped with the same air-assist atomizer, igniter, and similar Shchelkin spiral and removable coil. The maximal shock wave velocity at the exit from Shchelkin spiral in the experiments with the 28-mm tube was 900–1000 m/s as well.

The experiments with the 36-mm and 28-mm tubes demonstrated that shock waves propagating at the velocity on the order of 900–1000 m/s were capable of transitioning to a detonation once they passed a curved section in the form of tube coil. For better understanding of the coil effect on the SDT phenomenon, extensive computational studies were performed.

3 Computational Studies

In the computational studies, for resolving spatial effects inherent in shock wave propagation through a tube coil, the statement of the problem was based on three-dimensional (3D) compressible Euler equations coupled with the single-step kinetic equation for a single progress variable calibrated to represent the stoichiometric propane–air mixture. For solving a set of governing equations, a special parallel LU-SGS algoritm was developed based on the implicit finite-difference scheme applying finite-volume spatial discretization with first-order Harten – Lax – van Leer or Godunov approximation of fluxes. The unstructured computational grids of tube coils contained up to 7,800,000 cells with the minimal and maximal cell sizes of 0.25 and 0.7 mm, respectively. The time integration step was less than 100 ns. The calculations were performed using up to 100 processors on the multiprocessors system MVS-15000BM of the Joint Supercomputer Center of the Russian Academy of Sciences and took up to 15 CPU hours per each run.

Shock-to-detonation transition in two configurations of tube coils filled with the stoichiometric propane–air mixture at normal conditions was considered. The first configuration was a circular 360-degree loop of the 28-millimeter tube 365 mm long (measured along the tube axis). The second configuration was a circular 360-degree loop of 36-millimeter tube 255 mm long (also measured along the tube axis). These configurations corresponded to those tested experimentally. The inlet and outlet of each coil were positioned in the same normal plane and were in direct contact with each other (Fig. 3). The primary shock wave at the coil inlet was modeled using Rankine–Hugoniot relationships at the shock discontinuity.

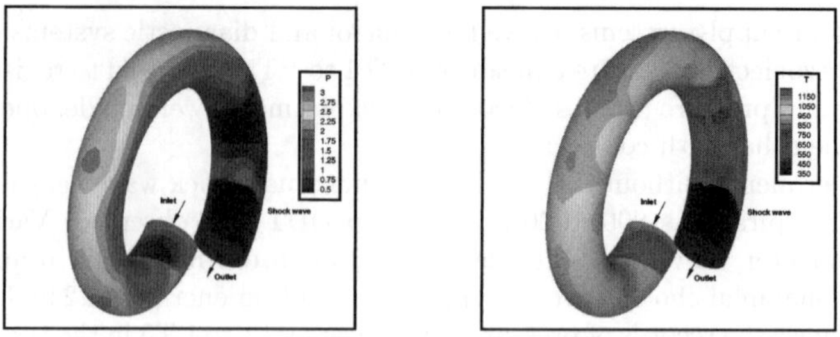

a *b*

Fig. 3. Snapshots of pressure (*a*) (in MPa) and temperature (*b*) (in K) in the 28-millimeter tube coil with the primary shock Mach number of 3.5. Time: 275 μs. No ignition

Figures 3 and 4 show the snapshots of pressure, temperature, and fuel (propane) mass fraction in the 28-millimeter tube coil in the situation with no ignition inside the coil when shock waves of Mach number 3.5 (Fig. 3) and 3.7 (Fig. 4) pass through it. The snapshots clearly indicate the onset of regular hot spots on the compressive wall inside the coil. To show the instantaneous position of the lead shock front in Fig. 4, we have plotted the isobar of 5 atm (thick black curves). The hot spots with elevated temperature and pressure are generated due to successive reflections of the propagating shock wave. This phenomenon is very similar to that observed in the course of shock wave propagation in a straight tube with regular obstacles in the form of orifice plates or Shchelkin spiral. Contrary to the coil, the use of obstructed tubes for obtaining detonations results in considerable momentum loss in the shock induced flow.

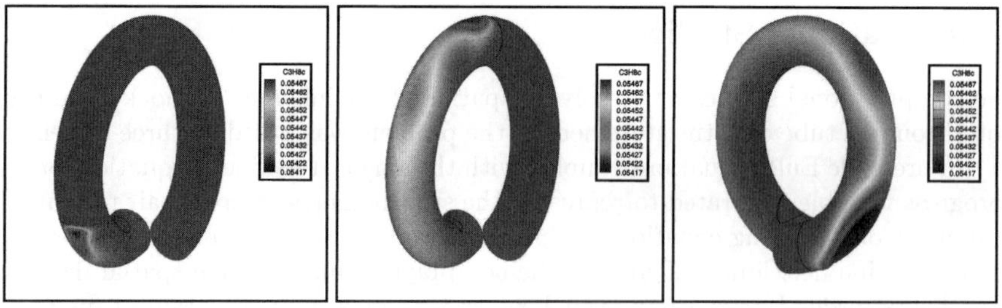

a *b* *c*

Fig. 4. Snapshots of propane mass fraction distributions in the 28-millimeter tube coil with the primary shock Mach number of 3.7. Black curves correspond to the isobar of 5 atm. Time: (*a*) 60, (*b*) 170, and (*c*) 290 μs. No ignition

At the incident shock Mach number of 3.9, the 28-millimeter tube coil promotes detonation initiation, as demonstrated in Fig. 5. The detonation arises after the third reflection of the incident shock wave from the compressive wall of the coil. The critical incident shock Mach number required for detonation initiation in the 36-mm tube coil (Fig. 6) appeared to be somewhat lower than for the 28-millimeter tube coil, which is in line with our experimental findings for U-bent tubes of different diameter [11]. In Fig. 6, the detonation originated from the hot spot arising after the first reflection of the incident shock wave from the compressive wall.

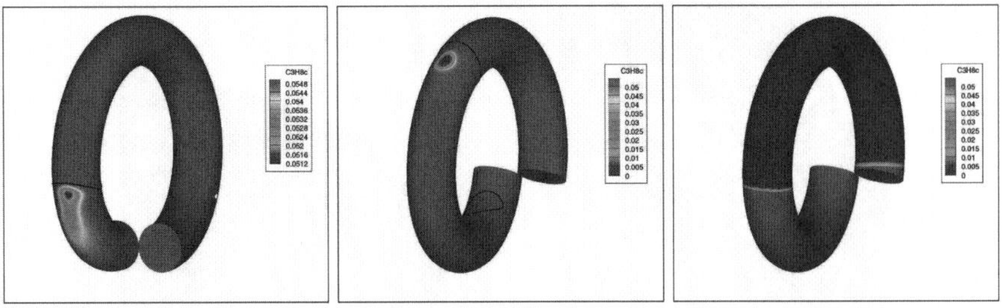

a *b* *c*

Fig. 5. Snapshots of propane mass fraction distributions in the 28-millimeter tube coil with the primary shock Mach number of 3.9 Time: (*a*) 60, (*b*) 170, and (*c*) 240 μs. Detonation

The results of calculations for the stoichiometric propane–air mixture reveal salient features of the SDT phenomena and are in qualitative agreement with the experimental observations for the two-phase *n*-hexane–air mixture. The minimal (critical) Mach numbers of the primary shock wave required for SDT in the coils of first and second configurations were 3.9 and 3.6, respectively. Several discrete locations of SDT were observed in the calculations in the vicinity to the collision sites of shock waves reflected from expansive and compressive tube walls. At near-critical conditions, the SDT occurred in the second half-loop of the coil closer to the outlet. With increasing the shock wave Mach number above the critical value the transition region moved to the first half-loop of the coil closer to the inlet. The 3D effects were more pronounced for the tube coil of the second configuration with a smaller curvature radius and manifested themselves in the considerable deviation of the lead shock front from the planar shape (see Fig. 6).

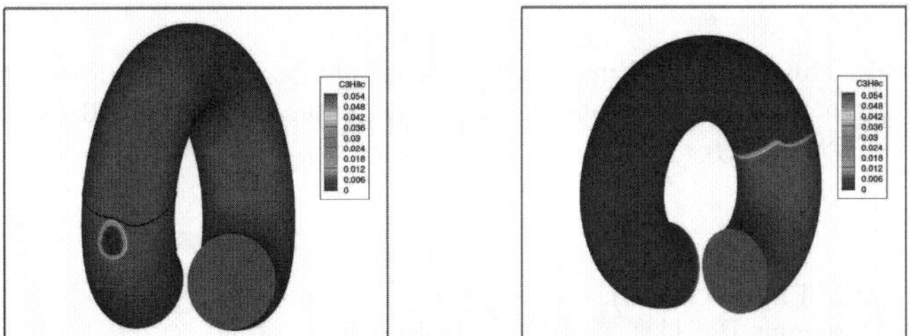

a *b*

Fig. 6. Snapshots of propane mass fraction distributions in the 36-mm tube coil with the primary shock Mach number of 3.7 Time: (*a*) 40 and (*b*) 110 μs. Detonation

For better understanding the SDT phenomenon in the coil, we studied carefully the spatial temperature distributions in the vicinity to the ignition site in Fig. 5*b*. Figure 7*a* shows the exploded view of the propane mass-fraction snapshot of Fig. 5*b* with instantaneous temperature isolines (thick black curves) superimposed. The numbers at the isolines correspond to the static temperature in Kelvin. In addition to the distribution of propane mass fraction on the compressive wall surface (Fig. 7*a*), Fig. 7*b* shows the propane mass-fraction distribution in the coil section normal to the surface which is cut

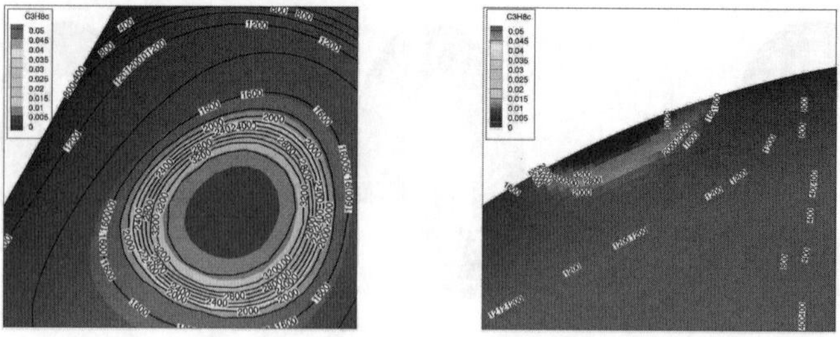

a *b*

Fig. 7. Predicted temperature isolines (black curves with temperature values in K) and propane mass fraction contours in the vicinity of the initial detonation bubble in the 28-millimeter tube coil with the primary shock Mach number of 3.9 at 170 μs. Detonation

through the center of the hot spot. One can see an extended zone around the spot which is thermally preconditioned for selfignition: the temperature of the fresh propane–air mixture is at the level of 1600–2000 K, whereas the temperature behind the incident shock wave is at the level of 1100–1200 K. The temperature distribution in this zone is nonuniform and anisotropic: temperature gradients appear to be steeper in the direction normal to shock wave propagation. Thus the spontaneous ignition wave thus originated tends to propagate predominantly along the direction of shock wave propagation giving rise to an asymmetric detonation bubble. The bubble then grows predominantly in the direction of the lead shock front and in the opposite direction giving rise to detonation and reactive retonation waves. These observations testify that the detonation forms due to spontaneous generation of strong blast waves in the exothermic centers. The mechanism of blast wave formation due to propagation of fast spontaneous flames in the compressible medium was studied elsewhere [12].

Acknowledgement. This work was supported by Russian Foundation for Basic Research grant 05-08-50115a, International Science and Technology Center project N2740 and President RF Grant (No. MK-5068.2007.1).

References

1. Laffitte P.: Ann. Phys. **4** (1925).
2. Laffitte P., Dumanois P. : C.R. Acad. sci. **184** (1926).
3. Shchelkin K. I., Sokolik A. S.: Sov. J. Phys. Chemistry **10** (1937).
4. Shchelkin K. I.: Sov. J. Exp. Theor. Phys. **10** (1940).
5. Brophy C. M., Sinibaldi J. O., Wang F., et al.: AIAA Paper No.2004-0834 (2004).
6. Frolov S. M., Basevich V. Ya., Aksenov V. S., Polikhov S. A.: J. Propulsion and Power **19**, 4 (2003).
7. Frolov S. M., Basevich V. Ya., Aksenov V. S. In: *Application of detonation to propulsion*, ed. by Roy G. D., Frolov S. M., Shepherd J. (Moscow: TORUS PRESS, 2004) pp. 240–249.
8. Frolov S. M., Aksenov V. S., Basevich V. Ya.: Doklady Physical Chemistry **401**, 1 (2005).
9. Frolov S. M., Basevich V. Ya., Aksenov V. S., Polikhov S. A.: J. Shock Waves **14**, 3 (2005).
10. Frolov S. M.: J. Propulsion and Power **22**, 6 (2006).
11. Frolov S. M., Aksenov V. S., Shamshin I. O. Proceedings of the Combustion Institute **31** (2007).
12. Zel'dovich Ya. B., Gelfand B. E., Tsyganov S. A., Frolov S. M., Polenov A. N. In: Progr. Astr. Aeron., Dynamics of Explosions **114** (1988).

Simulation of hydrogen detonation following an accidental release in an enclosure

L. Fang[1], **L. Bédard-Tremblay**[1], **L. Bauwens**[1], **Z. Cheng**[2], **and A.V. Tchouvelev**[2]

[1] *University of Calgary, Mechanical & Manufacturing Engineering, Calgary, Alberta, Canada*
[2] *A. V. Tchouvelev & Associates, Mississauga, Ontario, Canada*

Summary. The risk of detonation in an electrolyzer following a hydrogen leak is analyzed numerically. The scenario under investigation assumes a leak leading to the presence of a detonable mixture. Numerical simulation is performed using an ENO scheme. The dispersion pattern used as an initial condition for the detonation simulation is obtained from a previous simulation of the release process. The leak occurs in the horizontal direction, leading to a horizontally elongated cloud. Detonation is triggered by adding energy into a narrow region within the detonable cloud. The detonation fails at the top and bottom soon after ignition, but it progresses on the left and right side, in the direction of release. As the detonation front moves into leaner mixture, it eventually fails everywhere, but the weaker shock wave is still strong enough to cause some damage as it reaches the enclosure walls.

1 Introduction

The purpose of this study was to evaluate the value of computational fluid dynamics to assess the risk of detonation following a hydrogen leak in an enclosed environment. A scenario is considered in which a high pressure pipe rupture results in a leak of 42 g of hydrogen inside the electrolyzer enclosure. Dispersion data obtained from a separate numerical simulation were used as initial conditions [1]. Detonative ignition was simulated by adding energy into a narrow region. The code used in the detonation simulation reported here is based upon an ENO scheme; it has been parallelized using the MPI protocol.

Detonation is a concern from a safety standpoint because hydrogen is known to detonate over a much wider concentration range than usual hydrocarbon fuels, and the high pressure associated with the shock can be very mechanically damaging [2,3]. Safety measures used to prevent detonation with other fuels are of very limited use with hydrogen in the type of situation under consideration. For instance, insuring that the hydrogen concentration cannot reach the lower detonation limit is impossible since in the current scenario, the concentration varies from 100% at the location of the leak, all the way across the detonability range, down to 0% where no hydrogen has yet reached. Moreover, hydrogen is known for requiring very little energy to ignite; it has been found that static electricity, shock compression and even viscous effects may possibly lead to ignition [4–6]. As a result, ensuring that ignition will not occur is difficult.

A detonation wave consists of a shock coupled with a reaction zone; the temperature increase across the shock leads to much faster chemistry, which is extremely temperature-sensitive; finally the expansion due to the heat released supports the shock. Freely propagating waves travel supersonically at the so-called Chapman-Jouguet (CJ) speed, which depends on the heat release but is independent of the details of the chemistry. Here, it ultimately depends upon hydrogen concentration. Failure occurs when the leading shock and the reaction zone are no longer coupled.

In the next section, the formulation used is described in detail. This is followed by an overview of the numerical method that was used. Finally, results are shown and discussed.

2 Formulation

The flow is described by the inviscid, nonconducting, reactive Euler's equations. The electrolyzer is modeled as a rectangular box, with a length of 4 m and a height of 2.54 m. The simulation was performed in two dimensions; therefore the computational domain represents a vertical slice taken halfway across the width. The leak occurs on the left of the enclosure, where the pipe failure is taken as having occurred. The grid used is made up of 2000 cells in the enclosure length, and 1778 cells in its height.

Hydrogen dispersion data following the pipe rupture were obtained from a separate numerical simulation, [1], in which the equipment in the enclosure was taken into account, as well as a fan which circulates air at a rate of 4000 SCFM. The concentration cloud is then used as the initial condition for the current simulation. Fig. 1 shows the result of the dispersion simulation, 500 ms after the beginning of the release, which is the assumed ignition time. In the current simulation, the equipment contained in the enclosure is neglected. Shock reflections on obstacles potentially increase the risk of deflagration-to-detonation-transition and detonation propagation; thus neglecting the equipment inside the enclosure may lead to underestimating the risks. On the other hand, the current simulation was only performed in two dimensions, which is conservative. To some extent these two simplifications balance each other out. Finally, further dispersion after ignition is also neglected, which is reasonable since the dispersion velocities are orders of magnitude smaller than the wave velocity.

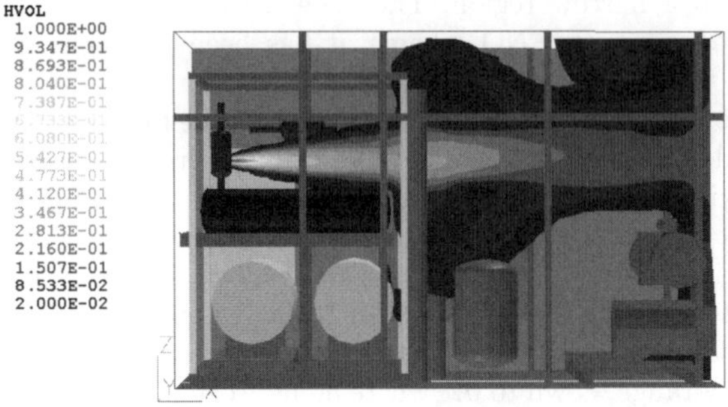

Fig. 1. Dispersion Pattern 500 ms after beginning of release

A single step Arrhenius scheme is used to model the chemistry, recently proposed by Gamezo et al. [7]. More complex schemes could easily be used but the main effect of the concentration gradient due to dispersion, on the CJ speed, is covered by the single step model. Also, using more complex models would not give better results. Indeed, on the grid used, covering the entire domain without being prohibitive, the wave thickness is not well resolved, with only one or two grid points across the wave thickness. Intermediate species included in more complex schemes would occur on even thinner zones, poorly resolved to the point of being useless.

The enclosure walls are taken to be rigid. Because of inertia, the time scale characterizing wall motion is orders of magnitude longer than for wave motion.

Finally, ignition was simulated by adding energy to a narrow region that represents the assumed ignition point, located approximately in the center of the detonable cloud.

3 Numerical Solution

The code used evolved from a code first developed by Xu. et al. [8], that has been extensively modified over the years. It is based upon an ENO scheme and it has been parallelized using the MPI protocol. It is well adapted for parallel computer architecture. The code has been used extensively to study the effect of chemical kinetics models on detonation cells [9–12]. It is second order accurate in time and in space.

4 Results

Results are shown in Fig. 2. All results are presented as Schlieren-like pictures, for mass fraction, pressure and temperature gradients, respectively, with the gray scale associated with gradients in the quantity being shown. Because Schlieren-like pictures show the first derivative of the primitive variable, results shown are computed with one order of accuracy less than the primitive variables, and well-resolved Schlieren-like results require a fine mesh. The faint traces observed on the plots are the result of numerical noise resulting from the concentration gradients.

In Fig. 2, the outer dark circular line is the shock, while the inner one represents the reaction front. On pressure gradient plots, the reaction front is not obvious, but it is very clear on temperature gradient plots. Indeed, across the reaction zone, there is a large temperature jump but a small pressure difference, in contrast with shocks, across which both variables jump. On the results, a detonation is present when the reaction front and the shock are indistinguishable from one another because they travel as a single front.

The results show that ignition results in a detonation, which quickly fails at the top and bottom, but progresses for a while toward the right and left. Around 704 μs after ignition, the detonation also fails towards the right wall, and shortly after, it fails everywhere. In all cases, failure occurs because the reaction front encounters leaner mixture, therefore the chemical reaction slows down. This is also observed on the mass fraction plots, where the combustion products are represented as a white zone. The interface between burnt and unburnt mixture nearly stops propagating. Because the shock is no longer supported by chemistry, it slows down, and the temperature behind it drops, further slowing the chemistry.

The interface between burnt and unburnt mixture is subject to hydrodynamic, Richtmyer-Meshkov and Rayleigh-Taylor instabilities, which can be observed on the upper side. These result in a convoluted reaction front, which is observed at a later point in the computation. Physically, the interface propagates as a flame, under diffusive mechanisms not included in the current physical model and requiring a much finer resolution. At the current resolution, numerical diffusion leads to an artificial propagation speed likely larger than the actual one.

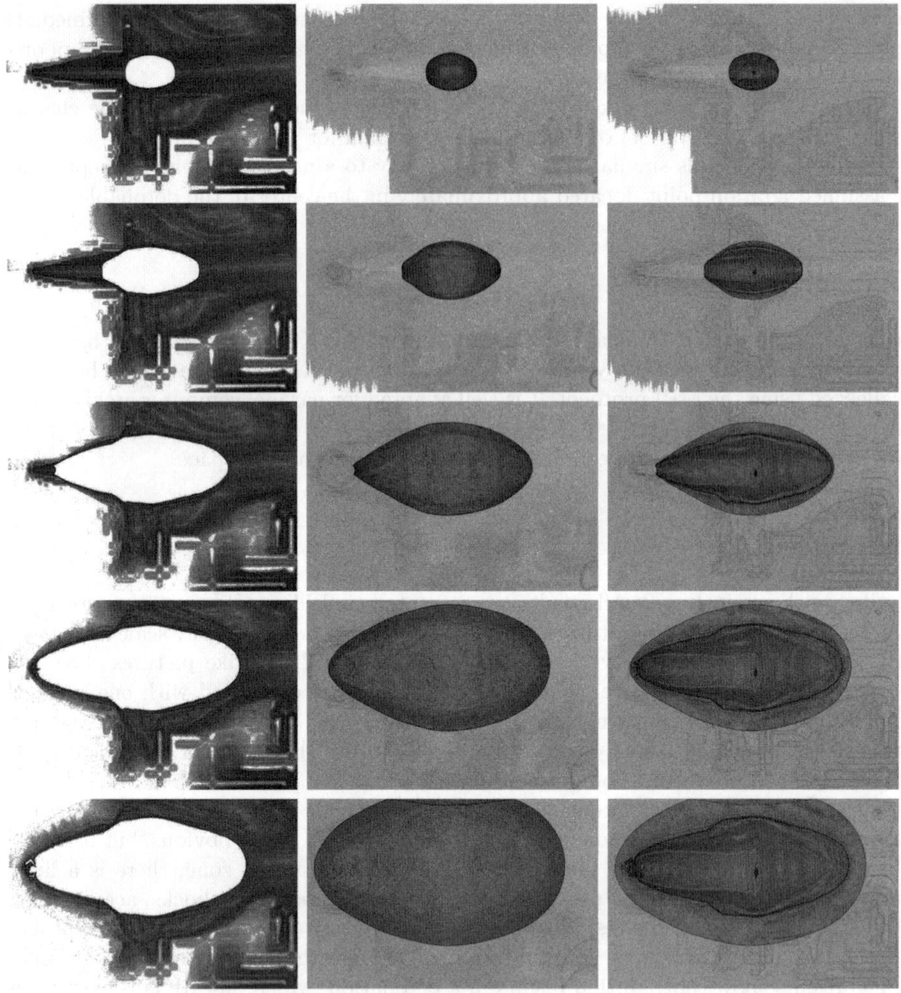

Fig. 2. Mass fraction (left), pressure (center) and temperature gradients (right), from top to bottom, 176, 352, 704, 1060 and 1410 μs after ignition

When reaching the top wall the shock is still strong, since the ignition location is close to the top wall. This occurs about 1100 μs after ignition. In contrast, it has weakened more when reaching the side and bottom walls. Results for impulse are shown in Fig. 3, for the left, right and top wall. The peak pressure at the top wall is relatively high, but since the wall is subjected to that pressure for a very short period of time, the impulse gives a better indication of possible damage. On average, the impulse is 500 Ns/m^2. Considering a ceiling panel with weight per unit area of 10 kg/m^2, such an impulse would result in an initial velocity of 50 m/s, if the panels were not attached to the side walls.

The results obtained from the simulation are conservative, since the hydrogen concentration that was considered is the highest. One would expect that the impulse values would decrease to the left and right of the cross-section that was simulated.

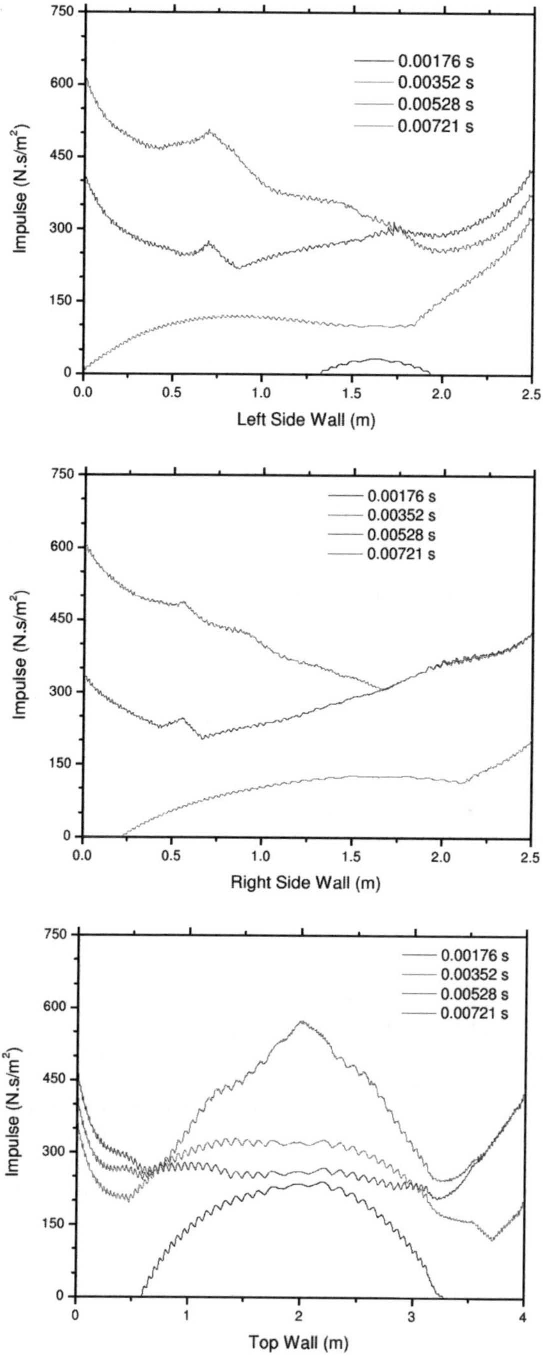

Fig. 3. Impulse on left, right and top wall

5 Conclusion

Results show that numerical simulation can be used to assess the damages resulting from a hydrogen detonation in an enclosure. In the scenario under consideration, even though the detonation failed in all directions, impulse on the walls is still of some significance. Values obtained should be taken into account for enclosure design.

There are however clear limitations. First, the simulation was performed in two dimensions, and even then, the wave thickness is not well resolved. Three-dimensional simulations and/or resolving the wave thickness properly would entail very large computations.

Acknowledgement. Work supported by the Natural Sciences and Engineering Research Council of Canada, by Natural Resources Canada and by the Auto21 Network of Centres of Excellence.

References

1. Tchouvelev, A.V., Hay D.R., Bénard, P. et al., *Quantitative Risk Comparison of Hydrogen and CNG Refuelling Options.* Final Technical Report to Natural Resources Canada, March 2006.
2. Fickett, W. and Davis, W.C., *Detonation*, University of California Press, 1979.
3. Strehlow, R.A., *Combustion Fundamentals*, McGraw-Hill Series in Energy, Combustion and Environment, 1984.
4. Astbury, G.R. & Hawksworth, S.J., Spontaneous Ignition of Hydrogen Leaks: A Review of Postulated Mechanisms, *Proc. First Int. Conference on Hydrogen Safety*, Pisa, 2005.
5. Dryer, F.L., Chaos, M., Zhao, Z., Stein, J.N., Alpert, J.Y. & Homer, C.J., Spontaneous Ignition of Pressurized Releases of Hydrogen and Natural Gas into Air. *Combust. Sci. Technology*, 179:663-694, 2007.
6. Wolanski, P. & Wojcicki, S., Investigation into the Mechanism of the Diffusion Ignition of a Combustible Gas Flowing into an Oxidizing Atmosphere, *Proc. Combust. Inst.*, 14:1217-1223, 1972.
7. Gamezo, V.N., Ogawa, T. & Oran, E.S., Numerical Simulations of Flame Propagation and DDT in Obstructed Channels Filled with Hydrogen-Air Mixtures, *Proc. Combust. Inst.*, 31, 2007.
8. Xu, S.J., Aslam, T., & Stewart, D.S., High Resolution Numerical Simulation of Ideal and Non-ideal Compressible Reacting Flows with Embedded Internal Boundaries, *Combustion Theory Modelling*, 1:113-142, 1997.
9. Liang, Z. & Bauwens, L., Detonation Structure with Pressure Dependent Chain-Branching Kinetics, *Proc. Combust. Inst.* 30:1879-1887, 2005.
10. Liang, Z. & Bauwens, L., Z. Liang and L. Bauwens. Cell Structure and Stability of Detonations with a Pressure Dependent Chain-Branching Reaction Rate Model. *Combust. Theory Modelling* 9:93-112, 2005.
11. Liang, Z. & Bauwens, L., Liang, Z. and Bauwens, L. Detonation Structure under Chain Branching Kinetics, *Shock Waves* 15:247-257, 2006.
12. Bauwens, C.R.L., Bauwens, L. & Wierzba, I., Accelerating Flames in Tubes - An Analysis. *Proc. Combust. Inst.* 31:2381–2388, 2007.

Spectroscopic studies of micro-explosions

G. Hegde[1], A. Pathak[2], G. Jagadeesh[1], C. Oommen[1], E. Arunan[2], and K.P.J. Reddy[1]

[1] *Department of Aerospace Engineering, Indian Institute of Science, C. V. Raman Avenue, Bangalore, 560012, India*
[2] *Inorganic and Physical Chemistry Department, Indian Institute of Science, C. V. Raman Avenue, Bangalore, 560012, India*

Summary. NONEL tube finds vast applications in civil and military because of its safe and confined explosion technique. Spectroscopic and chemical analysis of a NONEL tube with an uniform mixture of HMX and Al is reported here. Peak temperature obtained at the open end of the NONEL tube due to the detonation of the explosive has been calculated using Planck's radiation law. The products of the chemical reaction taking place due to the ignition of HMX + Al are characterized using FTIR spectroscopy.

1 Introduction

Study of controlled micro-explosions in the laboratory [1] is very essential for a better understanding of the near field explosion dynamics. Understanding explosion dynamics assumes significance especially in the backdrop of increasing frequency of premeditated explosions (terrorist attacks) or accidental explosions in industries around the globe. These kind of bench mark experiments are useful for validating many complex CFD codes that are increasingly used in disaster management studies.

The present report is an effort to get a detailed spectroscopic information and the chemistry involved in micro-explosion generated at the open end of a non-electrical (NONEL) tube. A typical NONEL tube (M/s Dyno Nobel, Sweden) explosive transfer system (ETS) consists of a plastic tube of approximately 1 mm inner diameter and 3 mm outer diameter with a thin layer of explosive material coating deposited on its inner wall during the tube extrusion process. In this case, the coating comprises a small amount of combustible explosive octahydro-1,3,5,7-tetranitro-1,3,5,7-tetrazocine (High-Molecular-weight rdX, HMX) and aluminium (Al) adhered to the inner walls of the tube. The amount of explosive is of the order of 18 mg/m. Explosion inside the tube is initiated using an electric spark. A shock wave is established by the ignition of the explosive material that propagates at a supersonic speed of 2,000 m/s. The shock wave and the products of the explosion come out only through the open end. The NONEL tube is an ingenious device because it has a high propagation speed and is very safe to handle. It finds application in mining blasts, quarrying, constructions, crew escape systems in military aircrafts; ordnance systems in launch vehicles and missiles and launch vehicle flight termination systems, which requires the highest functional reliability.

The chemistry of micro explosions of HMX provides crucial information about the energy liberated during such blasts. The main aim of the work is to get an accurate estimate of the temperature at the open end of the NONEL tube using spectroscopic measurements. Further, we intend to identify the major products formed during the ignition of HMX and Al to get a better idea of the chemistry involved.

2 Structure of HMX

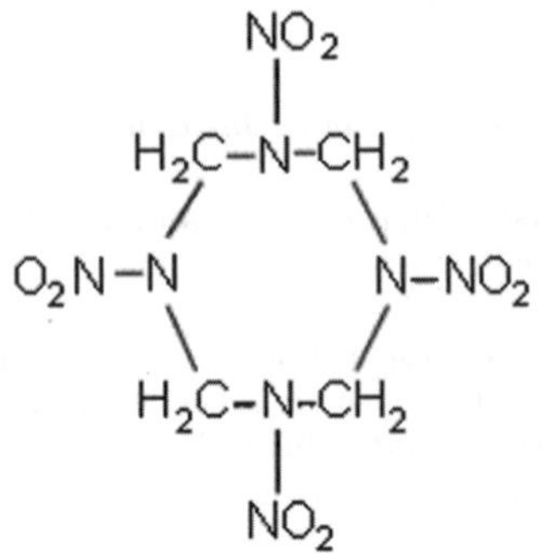

Fig. 1. Structure of octahydro-1,3,5,7-tetranitro-1,3,5,7-tetrazocine (HMX)

The structure of HMX is shown in Fig. 1. Four solid phases of HMX are known to exist: β - HMX, α - HMX, γ - HMX, and δ - HMX. γ - HMX is just a hydrated version of HMX [2, 3]. The thermal stabilities of the four crystalline forms are $\beta > \alpha > \gamma > \delta$. The most stable structure is of β - HMX that has a density of 1.9 g/cm^3. The most stable structure at high temperature is the δ - HMX having a density of 1.8 g/cm^3 [4]. Transformation of β - HMX to δ - HMX involves a significant increase in volume and surface area. β - HMX has a chair like structure while δ - HMX has a boat like structure. Boat like structure results in bringing the highly reactive $> N - NO_2$ groups in close proximity of each other and this subsequently increases the reactivity of HMX at high temperatures. The initial step in HMX decomposition involves breaking of intermolecular attraction between HMX molecules. This leads to dissociation of single molecules with the formation of various radicals such as CH, NO, CH_2, etc. These radicals are short lived and may recombine or react with other species / wall to form more stable products.

3 Experimental methods

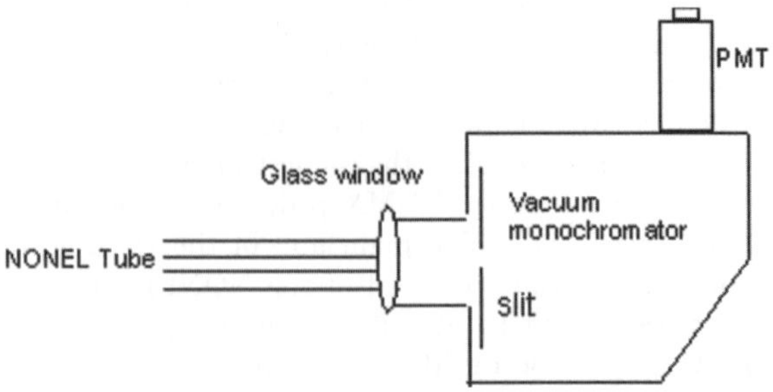

Fig. 2. Schematic of the experimental set up for temperature measurement

The schematic of the experimental set up involved in the measurement of temperature of the NONEL tube is shown in Fig. 2. A monochromator with an attached photo-multiplier tube (PMT) has been used to measure the emission signal at the open end at different wavelengths. Short lengths (0.5 m) of the NONEL shock tube, inserted inside a brass rod of 10 mm outer diameter, are placed in the vicinity of the monochromator slit, which is separated by a 1 mm thick BK 7 glass window. The emission signal of the explosion is recorded using a digital oscilloscope interfaced with the photomultiplier tube. A typical emission signal at 341 *nm* is shown in Fig. 3. The emission measurements are done at intervals of 20 *nm* covering 340 *nm* to 550 *nm* range.

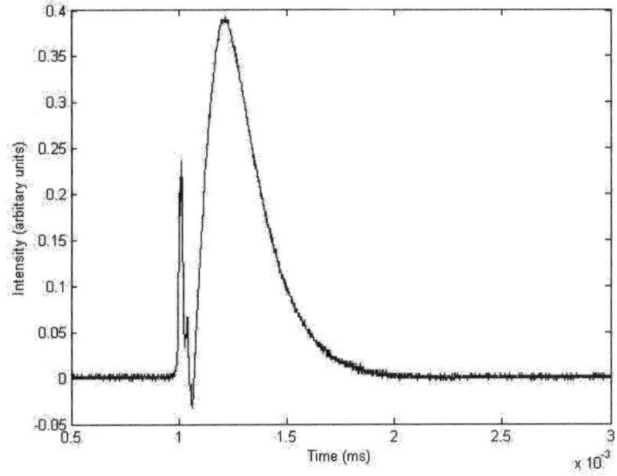

Fig. 3. Emission signal obtained in the digital oscilloscope interfaced with the photo-multiplier tube

For analyzing the products formed after the ignition of HMX, a vacuum cell is designed with KBr windows fixed at both ends of the tube (Fig. 4). The NONEL tube is inserted in the cell through a valve controlled tubular opening at the centre. The products of ignition coming out of the NONEL tube are collected in the cell. A NICOLET Fourier Transform Infrared (FTIR) spectrometer model NEXUS 870 FT-IR is used to record the mid-infrared spectra of the products in the range 500 cm^{-1} to 4000 cm^{-1} at a resolution of 4 cm^{-1}.

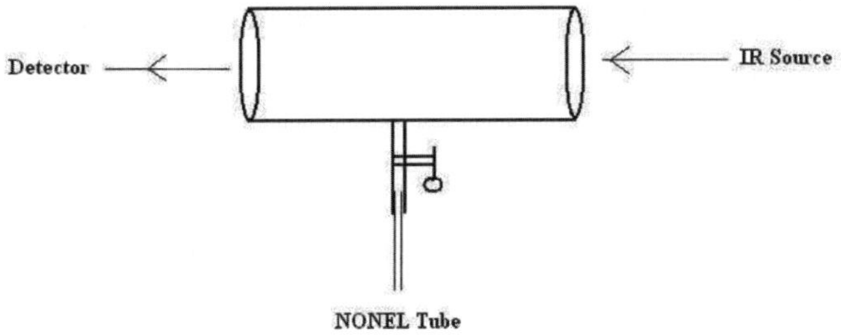

Fig. 4. Schematic of the set–up used for the measurement of FTIR spectra of the products.

4 Results and discussion

4.1 Temperature measurement of the detonating explosive

The detonation inside a NONEL tube generates a shock wave followed by chemical reaction zone through which the explosive is converted into combustion products. The chemical energy released in this process is responsible for the detonation. The explosive in this case is HMX + Al. Since, particles of metal oxides (Al_2O_3) are formed, the spectrum is always mainly continuous and may be assumed to be close to that of a black body [5]. In such cases, the temperature measurement may be done using the Planck's radiation law that provides an accurate measure of temperature of a black body. According to the Wien's displacement law derived from Planck's law, $\lambda_{max} = b/T$; where, λ_{max} is the wavelength at which maximum energy is being emitted and the constant $b = 2.8978\ mK$. T is the temperature corresponding to the maximum energy emitted at λ_{max}. The Planck's black body radiation law has been widely used to measure detonation temperature and is also employed in techniques like optical pyrometry [6, 7].

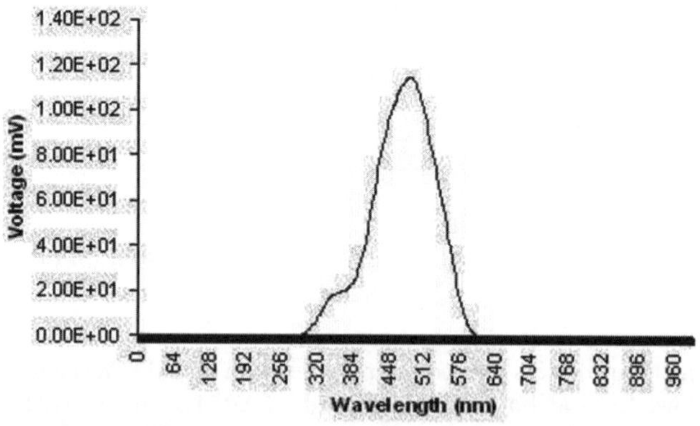

Fig. 5. Curve showing the amplitude of emission signals plotted against corresponding wavelengths. The curve follows Planck's black body radiation law and λ_{max} is obtained from this plot (see text for details).

The peak temperature at the open end of the NONEL tube, corresponding to the maximum energy emitted, is measured using Wien's displacement law. The energy emitted for wavelengths in the range 340 nm to 550 nm is recorded in terms of the amplitude (voltage) of the signal obtained in the digital oscilloscope. The emissions are measured at intervals of 20 nm. The amplitudes of each signal is plotted against corresponding wavelengths (Fig. 5). λ_{max} is obtained from this plot. The maximum energy emitted corresponds to the wavelength, $\lambda_{max} = 487\ nm$. The peak temperature T corresponding to this λ_{max} is 5950.72 K. This is the peak temperature attained by the detonation heated system at the open end of the NONEL tube.

4.2 Dissociation of HMX

Owing to the various important applications, thermal dissociation of HMX and the chemistry involved has been extensively discussed in the past [8–10]. At high temperatures, HMX dissociates to form volatile radicals that combine to form stable species. The formation of different radicals depends on the bond energies associated with each bond in the HMX molecule. The $> N - NO_2$ bond fission has been found to be the fastest [8,11]. This has been suggested to be the first step in the dissociation of HMX.

Detailed chemistry of the dissociation is not taken up here because of space restrictions. We concentrate on the formation of stable products and their characterization.

Two different channels have been suggested for the dissociation of HMX. Initially HMX yields N_2O at low temperatures or NO_2 at higher temperatures; former is quickly followed by CH_2O and the latter by HCN. Subsequently, highly exothermic reaction between CH_2O and NO_2 leads to CO, NO and H_2O and constitutes the main source of heat [12].

$$5CH_2O + 7NO_2 - - > 7NO + 3CO + 2CO_2 + 5H_2O \tag{1}$$

4.3 FTIR spectra

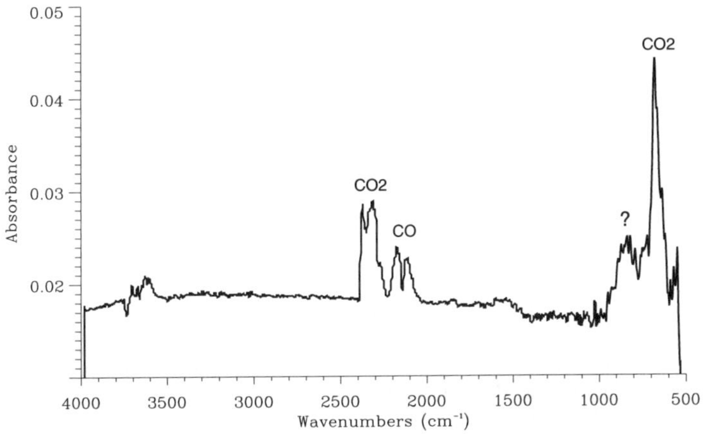

Fig. 6. FTIR spectrum of the detonation products of HMX + Al.

The FTIR spectrum of products of HMX detonation is shown in Fig. 6. The major peaks identified correspond to CO_2, and CO. Peaks at around 680, 2314 and 2370 cm^{-1} are attributed to bending and asymmetric stretching vibrational modes in CO_2 molecule. The weak feature near 3675 cm^{-1} is also attributed to CO_2. Features at 2118 and 2176 cm^{-1} are due to CO molecule. The broad feature at around 840 cm^{-1} remains unidentified. Future experiments at higher resolutions will reveal much better information and may help in identification of this and other features.

In the present NONEL tube experiment, the dissociation of HMX is taking place at temperatures higher than 5900 K. This temperature is a realistic approach to detonation conditions while all previous reported studies have been done at temperatures much below this. The exact chemistry at such temperatures and in the presence of Al is still unknown. Features due to other expected product molecules like N_2O, NH_3, H_2CO and H_2O are completely missing. Al plays a major role in enhancing the ignition of explosives. This increases the temperature at which combustion is taking place. This further decides the final products that appear. Formation of AlO, Al_2O_3 and $Al(OH)_3$ may be possible along with aluminium-nitrogen compounds.

5 Conclusions

We have done spectroscopic measurements on a NONEL tube detonation system having very small amount of the explosive HMX + Al. The experiments have been done at temperatures close to real detonation conditions. Such experiments are crucial in modelling studies of explosions.

Using Planck's radiation law for black bodies, we have calculated the temperature at the open end of the NONEL tube. This temperature comes out to be 5950.72 K.

Chemistry of dissociation of HMX is a major contributor to the energy released. Detailed chemistry has not been dealt here but identification of major products has been done using FTIR spectra of the products. CO_2 and CO have been identified as major products. A broad feature at 840 cm^{-1} remains unidentified. It is planned to carry out more experiments at higher resolutions. This will provide more information on the low concentration products formed. Detailed chemical kinetics study is proposed to be done in the near future.

Acknowledgement. Financial assistance received from Armament Research Board, Department of Science and Technology and IISc-ISRO Space Technology Cell and technical help from High Energy Materials Research Laboratory are acknowledged.

References

1. L.C. Yang, H.P. Do: Nonelectrical Tube Explosive Transfer System. In: *AIAA Journal* vol 38 (12), 2000, 2260.
2. F. Goetz, T.B. Brill, J.R. Ferraro: J. Phys. Chem. **82**, 1912 (1978).
3. P. Main, R.E. Cobbledick, R.W.H. Small: Acta Cryst. **41**, 1351 (1985).
4. T.R. Gibbs, A. Popolato (Eds.): *LASL Explosive Property Data*, (University of California Press, Berkley, CA, 1980).
5. A.G. Gaydon, H.G. Wolfhard: *Flames: Their structure, radiation and temperature*, IIIrd edition (Chapman and Hall Ltd., London, 1970).
6. F.C. Gibson, M.L. Bowser, C.R. Summers, F.H. Scott, C.M. Mason: J. App. Phys. **29**, 628 (1958).
7. B. Leal-Crouzet1, R. Bouriannes, G. Baudin1, J.C. Goutelle: Eur. Phys. J. App. Phys. **8**, 189 (1999).
8. R. Shaw, F.E. Walker: J. Phys. Chem. **81**, 2572 (1977)
9. M.L. Hobbs: Thermochimica Acta **384**, 291 (2002)
10. C.J. Cobos: Journal of Molecular Structure: THEOCHEM **714**, 147 (2005)
11. S. Zhang, T.N. Truong: J. Phys. Chem. A **104**, 7304 (2000)
12. T.B. Brill, P.J. Brush: Phil. Trans. Royal Soc. London A **339**, 377 (1992).

Structural response to detonation loading in 90-degree bend

Z. Liang, T. Curran, and J.E. Shepherd

California Institute of Technology, Pasadena, USA 91125

Summary. The structural response due to detonation propagation through a 90-degree bend in a circular tube was experimentally examined. Hoop strain measurements were obtained at key locations along the tube to measure elastic deformation. Dynamic pressure signals at the same locations were also recorded to track the detonation wave and record the peak pressure. Of particular interest are the effects of the bend on the magnitude of the pressure and strain in the material when compared to the straight tubes. These geometrical effects are due to the excitation of multiple modes: a short period detonation driven mode and longer period bending modes within the structure not seen in the straight tubes. The excitation of these bending modes serves to increase the maximum strain observed, which translates to greater hazards for industrial piping systems.

1 Introduction

Process plant piping is characterized by straight runs of pipe connected by elbows, tees, valves, pumps, reactor vessels, holding tanks and other features, including detonation and flame arrestors. In addition, the piping system is suspended or supported from a framework that provides reaction forces and limits the motion of the piping [1]. If detonations are possible within the piping, then a comprehensive analysis of the structural response requires consideration of how the detonation will interact with these features and what structural loads will be created.

When a detonation wave propagates through a tube, flexural waves are created that may result in strains that are significantly higher, up to 4 times greater, than strains that would be observed under simple static loading with the same internal pressure. Several conditions can create these higher strain conditions: resonant excitation, interaction of direct and reflected flexural waves and detonation pressure oscillations coupling with flexural waves [2].

When the detonation reaches a closed end, the peak pressure of the reflected shock wave is about 2.5 times the Chapman-Jouguet (CJ) pressure [3] and the pressure decays as the wave moves away from the reflecting surface. The reflected shock wave will induce flexural waves in the pipe which will interfere with the waves that were created by the incident wave. Constructive interference of these waves leads to the maximum strain values being observed at times corresponding to the passing of the reflected wave [1].

Detonation propagation through an elbow or tee is an example of detonation diffraction [4] which, depending on the direction of curvature, may cause the detonation to intensify or weaken [5]. Thomas [6] has carried out experiments on a plastic piping network, measuring both the forces on the supports and the strains on the pipes. Substantial motion of the pipe supports was observed in these tests, raising the possibility of piping containing the explosion at early times but failing due to excessive distortion of the supports.

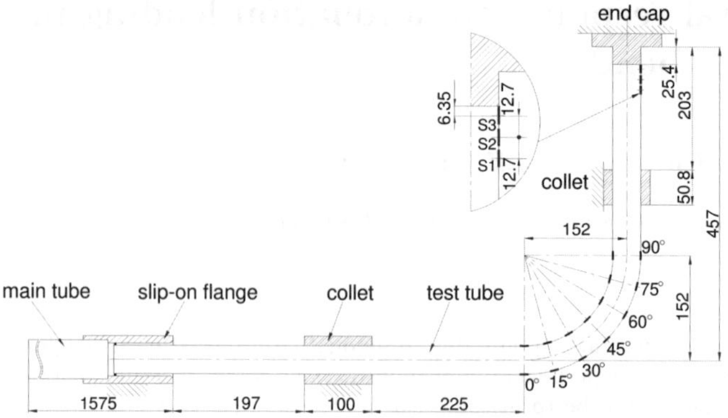

Fig. 1. Schematic of the experimental setup. The length unit is mm.

Previous experiments have been performed in our laboratory to investigate the effects of detonation waves on straight tubes [2,7]. The focus of the present paper is on detonation propagation through a 90 degree bend, which is part of a larger study [9] in our laboratory examining a range of structural elements including bends and tees.

2 Experimental Setup

The experimental setup consists of a main detonation tube (76 mm ID, 1.5 m long and 6.4 mm thick) connected to a carbon-steel test tube (41.3 mm OD and 0.15 mm thick) by a slip-on flange with an O-ring seal, as shown in Fig. 1. The period of oscillation for the fundamental hoop mode of the test tube is 24 μs, corresponding to a frequency of 41.4 kHz. This is the highest characteristic frequency with which we expect to observe oscillations in the strain signals [1,2].

The test tube consisted of a 0.61 m (24 in) straight section, a 90-deg bend with a radius of 0.152 m (6 in), and a straight section of 0.305 m (12 in). Two collets were used to support the test tube, one located 0.225 m (8.9 in) before the bend and 0.025 m (1 in) after the bend. The terminating end of the test tube was closed.

Three types of test tubes were used. Case 1: plain tubes with strain gauges were placed every fifteen degrees along the bent portion (15-90°), six on the intrados, six on the extrados, and three on the reflecting end. Cases 2 and 3: the tubes were modified with welded pressure transducer adapters at the locations corresponding to the strain gauges in Case 1. Pressure transducers were located at 15° increments along the bend on the extrados for Case 2 and on the intrados for Case 3. A stoichiometric ethylene-oxygen mixture was ignited by an electrical spark at the end of the main detonation tube, and a Schelkin spiral was used to accelerate the flame to a detonation before entering the test section.

3 Results and Discussion

All the tests were carried out at a room temperature 21-23°C and an initial pressure of 1 bar. The ideal detonation [8] for the test mixture has a CJ velocity $U_{CJ} = 2376$ m/s,

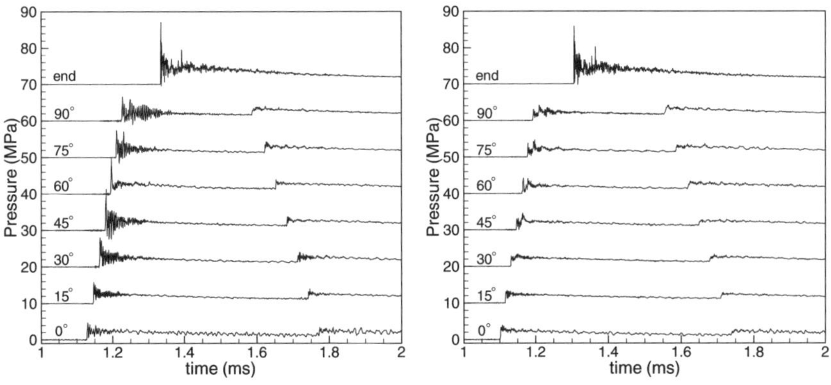

Fig. 2. Pressure histories at locations on the a) extrados and b) intrados. Traces are offset by 10 MPa for visibility.

CJ pressure $P_{CJ} = 3.33$ MPa, and reflected CJ pressure $P_{CJref} = 8.34$ MPa. Replica tests showed good reproducibility of the near-CJ detonation conditions before entering the test tube.

Detonation diffraction through the bend [4] results in the generation of compression waves on the extrados and expansion waves on the intrados, visible in Fig. 2. As a consequence, the peak pressures (Fig. 3a) on the extrados are all larger than those on the intrados. On the intrados of the bend, the peak pressure decreased to a minimum of P_{CJ} at 30°, then increased to $\approx 1.3\ P_{CJ}$ at 90°. On the extrados, the peak pressure increased to a maximum of 3-3.5 P_{CJ} at 45°–60° and decreased to about 2 P_{CJ} at 90°. We believe that this is due to the transverse shock waves generated by the diffraction within the bend and the peak value will be smaller when the length of the extended section after the bend is larger. At the reflecting end, the peak pressure was about twice P_{CJref}, the nominal value for straight tube.

Figure 3b demonstrates the change of the average wave speed, computed using the pressure wave arrival times, along the bend. The wave propagated at near-CJ velocity at the beginning of the bend (0°). The velocity decreased to $\approx 0.8\ U_{CJ}$ at 60° on the intrados, and increased to $\approx 1.4\ U_{CJ}$ at 75° on the extrados. After the peak, the wave speed on the extrados decreased to $\approx 0.8\ U_{CJ}$, the wave speed on the intrados was nearly U_{CJ}, and the wave at 90° was slightly overdriven ($\approx 1.1\ U_{CJ}$), consistent with the pressure profile and the wave being tilted at the exit of the bend. The measured hoop strains (Fig. 4) show the effects of incident and reflected waves as structural oscillations over a range of time scales. The radial oscillations induced by the incident detonation result in peak hoop strains on both the extrados and intrados of about 300 μstrain and 350– 575 μstrain due to the reflected shock wave.

The strain gauges near the end (Fig. 4c) show traces that are nearly identical to those obtained with straight samples (not shown in this paper). This suggests that although the peak reflected pressure is higher than for a straight tube, the next effect of the bend on radial structural motion is negligible after a propagation distance of two bend radii. The maximum hoop strain magnitude near the end was approximately 600 μstrain.

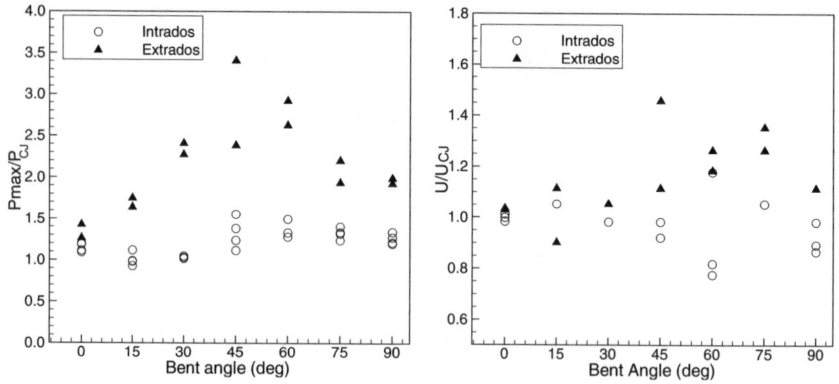

a) b)

Fig. 3. a) Peak pressure (normalized by P_{CJ}) and b) detonation speed (normalized by U_{CJ}) vs. the bend angle.

In addition to radial oscillations, a bending mode is excited due to the forces created by the change in direction of the detonation wave as it passes through the bend. This results in a long-period (≈4 ms) oscillation (Fig. 4d) observed in the hoop strain. This period corresponds to the flexure of the tube in a beam-like mode between the two collets (Fig. 1) holding the tube fixed to the support structure.

The bend in the pipe results in time-dependent forces on the pipe due to both the internal pressure acting on the bend and the change in flow direction. As shown in Fig. 5, forces F_x and F_y will be generated in the plane of the elbow in the direction of the pipe segments upstream and downstream of the bend. During the passage of the detonation wave through the bend, these forces will have a complex time dependence due to the waves created by the diffraction processes. Later, once these waves have died down, the forces can be estimated from the momentum balance for a steady flow. In terms of the average properties, the forces will be $F_x = A_1(P_1 + \rho_1 u_1^2)$ and $F_y = A_2(P_2 + \rho_2 u_2^2)$. Immediately behind the detonation front, the dynamic pressure ρu^2 is of the same order as the static pressure P so that the force will be a factor of two higher than computed on the basis of pressure alone. As the Taylor wave propagates through the bend, the flow will eventually come to rest and only the pressure will contribute to the forces.

We have analyzed the measured strain signals by computing the dynamic load factor (DLF), the ratio of the measured peak strain to the peak strain expected in the case of quasi-static loading $DLF = \epsilon_{max}/\epsilon_{static}$ where $\epsilon_{static} = \Delta PR/Eh$. For our test tubes, the parameters are E=210 GPa, R=20 mm (mean of inner and outer radius), h=1.5 mm (wall thickness), ϵ_{max} is the measured peak strain. The DLF values shown in Table 1 were computed in two ways. The values for DLF_{exp} were based on the static strain that would be expected from the experimentally measured peak pressure (ΔP=P_{max}) and the values for DLF_{CJ} were based on the calculated CJ pressure (ΔP=P_{CJ}-P_a).

The average peak pressures (Table 1) are always smaller on the intrados than on the extrados, but the maximum strains are larger on the intrados than the extrados between 0–60°. Using the experimental peak pressures, we find values of the DLF that are between 1 and 2, indicated the loading is in the regime intermediate to impulsive and "sudden loading" [1]. For the purposes of estimating peak deformations the dynamic load factor

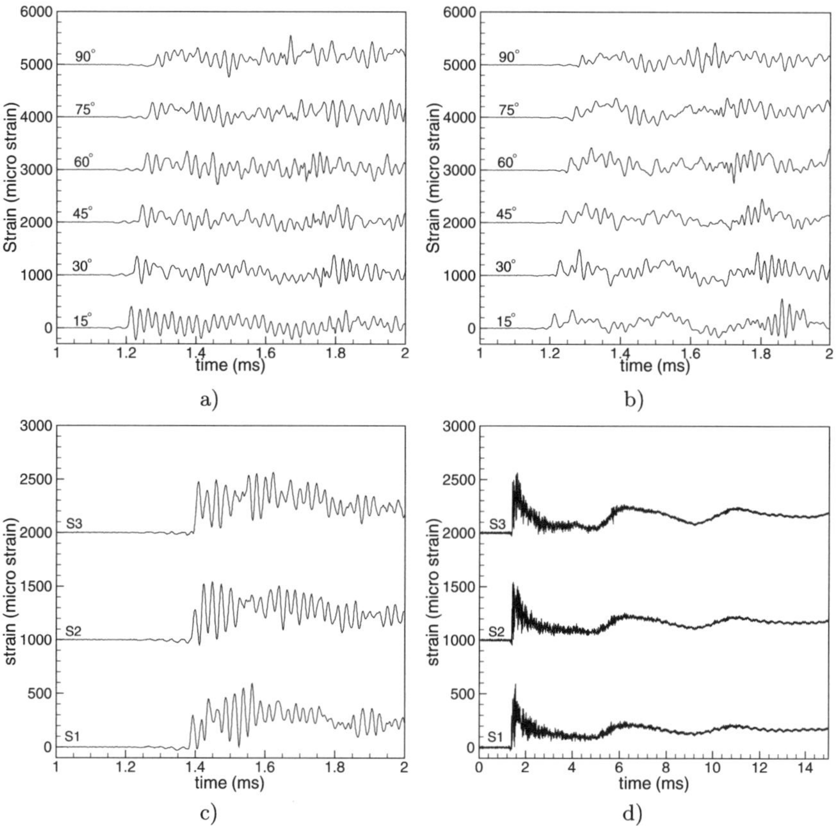

Fig. 4. Strain histories at locations along the bend on the a) extrados, b) intrados, c) reflecting end, and d) over a longer time scale for S_1-S_3. Traces are plotted with a vertical offset of 1000 μstrain.

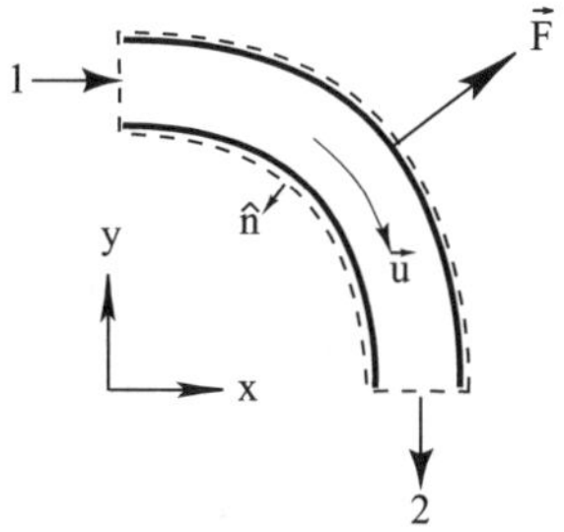

Fig. 5. Forces on pipe bend created by pressure and flow.

DLF_{CJ} based on the CJ pressure is most useful since it does not require experimental measurements. From the present data, we see that $1.7 < DLF_{CJ} < 2.4$ depending on the location of the measurement. This is valid for both the short duration hoop oscillations and the longer-duration beam oscillations.

Table 1. Average values of measured peak pressure P_{max}, peak strains ϵ_{max} and the corresponding dynamic load factors.

	extrados				intrados			
angle	ϵ (μstrain)	P (MPa)	DLF_{exp}	DLF_{CJ}	ϵ (μstrain)	P (MPa)	DLF_{exp}	DLF_{CJ}
15°	423	5.63	1.18	1.99	513	3.34	1.43	2.42
30°	366	7.79	0.74	1.72	500	3.43	1.01	2.36
45°	390	9.63	0.63	1.84	448	4.39	0.73	2.11
60°	382	9.23	0.65	1.80	485	4.55	0.82	2.29
75°	406	6.88	0.93	1.91	396	4.40	0.90	1.87
90°	479	6.51	1.15	2.26	366	4.18	0.88	1.83

4 Conclusion

The structural response due to detonation propagation through a 90° bend was examined. Due to the detonation diffraction process, the peak pressure on the extrados was substantially larger than on the intrados of the bend. The peak values of strain on the extrados and intrados were comparable. Oscillatory hoop strains were observed with short periods similar to the flexural modes observed with straight tubes and a longer period mode corresponding to beam bending excited by the change in wave direction through the bend. A maximum dynamic load factor (based on P_{CJ}) of 2.4 was measured. The peak hoop strains associated with the bending mode were comparable to those created by detonation wave reflection from a closed end.

References

1. J.E. Shepherd: 'ASME Structural Response of Piping to Internal Gas Detonation'. PVP2006-ICPVT11-93670, (2006).
2. W. M. Beltman and J.E. Shepherd: 'Linear Elastic Rsponse of Tubes to Internal Detonation Loading'. Journal of Sound and Vibration, 252(4), (2002) pp. 617.
3. J.E. Shepherd, A. Teodorcyzk, R. Knystautas, and J.H. Lee: 'Shock waves produced by reflected detonations'. Progress in Astronautics and Aeronautics, 134, (1991), pp. 244-264.
4. J.E. Shepherd, E. Schultz, and R. Akbar: 'Detonation diffraction'. Proceedings of the 22nd International Symposium on Shock Waves, 1, (2000), pp. 4148.
5. G. O. Thomas and R. L. Williams: 'Detonation interactions with wedges and bends'. Shock Waves, 114, (2002) pp. 481-491.
6. G. O. Thomas: 'The response of pipes and supports generated by gaseous detonations'. Journal of Pressure Vessel Technology, 124, (2002) pp. 66-73.
7. T. Chao and J.E. Shepherd: 'Detonation loading of tubes in the modified shear wave regime'. Proceedings of the 24nd International Symposium on Shock Waves, 2, (2005) pp. 865–870.
8. W.C. Reynolds: 'The element potential method for chemical equilibrium analysis: implementation in the interactive program STANJAN'. Stanford University, (1986).
9. Z. Liang, T. Curran and J. E. Shepherd: 'Structural Response to Detonation Loading in Piping Components', California Institute of Technology Report FM2006.008 (2006).

Study on perforated plate induced deflagration waves in a smooth tube

Y.J. Zhu[1], **Z.X. Liang**[1], **J.H.S. Lee**[2], **and J.M. Yang**[1]

[1] *Dept. of Modern Mechanics, Univ. of Sci.and Tec.of China, Hefei,230027, P.R.China*
[2] *Dept. of Mechanical Engineering, McGill University, Montreal, H2V3W1, Canada*

Summary. A deflagration wave was generated by putting a perforated plate in the way of a self-sustained detonation wave and interrupting it. We investigated the averaged one dimensional structure of this deflagration wave and its development through streak schlieren photography and wave speed measurement. With the increase of initial pressure, the deflagration wave varies from a laminar structure to a turbulent one. The former usually can not stand the attenuation of background rarefaction, whereas the later is capable of running up followed by an abrupt transition to detonation waves. There is a very unstable critical case between above-mentioned two situations, where the deflagration wave can usually propagate for a relatively long distance in $50-60\%$ CJ detonation speed. Theoretical analysis also reveals that the state of its products in this critical case perfectly matches the constant-volume combustion solution in same mixture.

1 Introduction

When effectively disturbed, a detonation wave tends to fail and switch to a so-called deflagration wave. This disturbance can be generated by a set of acoustic-absorbing walls, barriers, obstacles and so on [2–6]. As the deflagration wave reentrants an ideal situation, re-ignition of detonation may occur under certain circumstance. It was noted that, in this process, there is usually a very rapid transition from detonation to deflagration or the reverse, where the deflagration wave is much faster than an ordinary diffusion flame but in another hand with a typical maximum velocity of about half CJ detonation speed. Naturally, people would like to figure out what mechanism drives this deflagration wave and what kind of characteristics it carries.

Current research concerns mostly the one-dimensional structure and the behavior of this high speed deflagration wave. We put a piece of perforated plate in the way of a self-sustained detonation wave to generate the deflagration wave and let it propagate in the downstream smooth tube. The advantage of this method is that the initially transmitted deflagration does not include large three-dimensional disturbances which are significant and even dominant in other ignition methods. The problem hence can be more reasonably approached by a one-dimensional theoretical model, and this will benefit the understanding of the phenomena.

Experimentally, streak schlieren photos were taken to illustrate the averaged structure of the transmitted deflagration right downstream the plate, and pressure transducers were used to trace the further development of the wave front. We compared and discussed the structures and behaviors of the deflagration waves under different initial pressures. Then the one-dimensional theoretical analysis was carried out to examining the possible categories of this deflagration wave.

2 Experimental Setup

A stoichiometric acetylene-oxygen mixture was tested in present study.

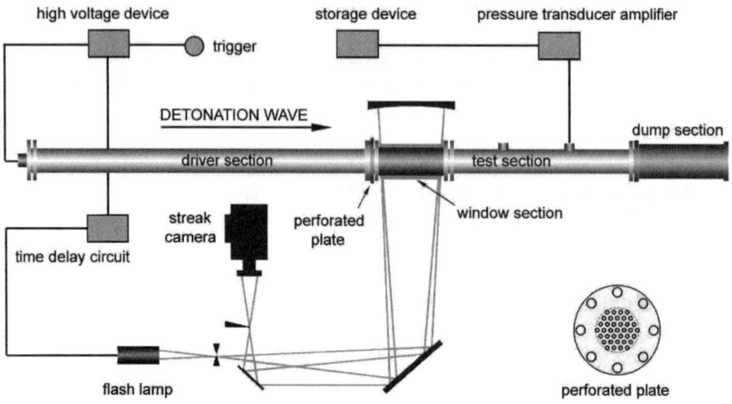

Fig. 1. Schematic of experimental setup

Fig. 1 shows the detailed experimental setup. The experiment was carried out in a set of aligned tubes of 65mm diameter. The whole tube length adds up to 4.5 m, and a perforated plate of 5mm thick separates it into a 2.5 m long driver section and a 2.0 m long test section. The blockage ratio of the perforated plate is about 80%. ϕ5 mm holes uniformly spread on the plate and neighbor holes space about 10 mm apart. Right downstream the perforated plate, there are a pair of 300 mm long 2 mm wide slice windows for streak schlieren photography. The averaged structure and its evolution of the initially transmitted deflagration waves thus could be recorded in the film as an $x-t$ diagram. Meanwhile, pressure transducers and ion probes mounted on the downstream tube traced the further development of the flame and wave front.

3 Experimental Results and Discussion

3.1 Structure of Deflagration Waves in Streak Schlieren Photos

Six typical streak schlieren photos under different initial pressure conditions are exhibited in Fig. 2. The horizontal direction represents the length along window section and vertical the time.

Fig. 2(a),(b) and (c) are streak schlieren photos for initial pressure of 0.5, 1.0 and 2.0 kPa respectively. Fig. 2 (a) shows the clearest structure of a sharp leading shock and a smooth flame trace as well as a peaceful induction zone. This represents a sort of steady laminar deflagration wave. As initial pressure increases, in Fig. 2(b), the flame now become getting thicker by time with a large bifurcation at the end of observing window. This indicates that the flame starts to be out of stable and large scale winkles are formed on the flame surface. Then in Fig. 2(c) with even higher pressure, many spike-like traces appear in the wake of the flame region, and more and stronger pressure waves

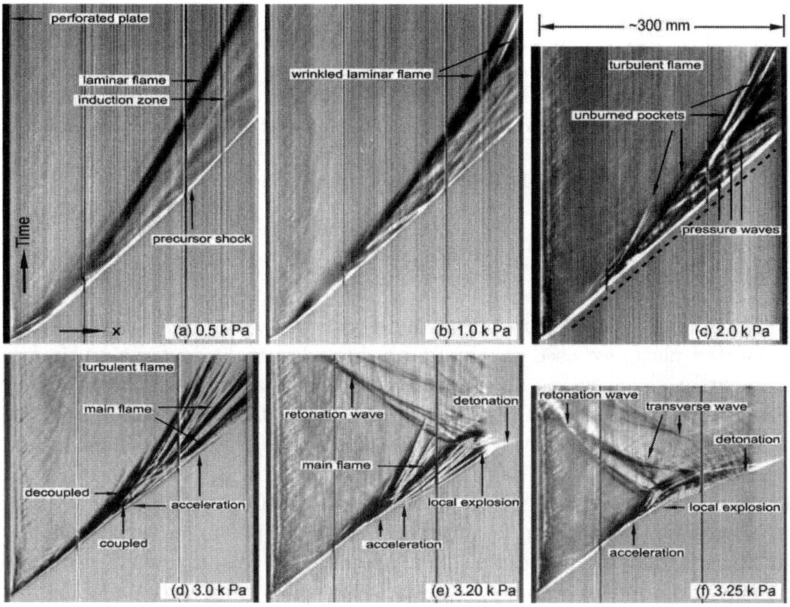

Fig. 2. x-t streak schlieren photos showing the deflagration waves formed under initial pressures of 0.5, 1.0, 2.0, 3.0, 3.2 and 3.25 kPa

come up overtaking the leading shock. In our schlieren photos, the spiky traces represent little high-density packages lagged behind the main flame and disappearing slowly. We believe that they are so-called unburned pockets burning out, which implies the existence of turbulent transport within flame. Comparing the speeds of these three deflagration waves, we found that both flame and precursor shock propagates faster under higher initial pressure. Looking at each of them separately, the deflagration waves in Fig. 2(a) and (b) are slowing down by time, while in Fig. 2(c) the precursor shock maintains a constant speed of about 1100 m/s.

Fig. 2(d) is taken under the initial pressure of 3.0 kPa. Now the deflagration wave structure seems totally different from the former three cases. The flame and shock are coupled with each other without any distinguishable induction zone. In the beginning, the coupled complex propagates at a steady speed of about 1200 m/s. Since higher initial pressure results in a faster reaction speed, the mixture becomes very sensitive to the disturbances and breaks up to a violent turbulent flame quickly [7]. Then a slight increase in initial pressure makes it possible for us to directly observe the transition to detonation, as can be seen in Figs. 2(e) and (f). The location of transition moves toward perforated plate with the increase of initial pressure. A new detonation usually starts from a local explosion inside an accelerating turbulent flame.

According to above-mentioned experimental results and discussion, the deflagration waves right downstream the perforated plate now could be categorized into 4 typical cases (Table 1).

No.	Examples	Initial pressure	Cell width	Flame
A	Fig. 2(a)	0.5 kPa	60.5 mm	Weak laminar flame
B	Fig. 2(b)	1.0 kPa	25.4 mm	Wrinkled laminar flame
C	Fig. 2(c)	2.0 kPa	10.6 mm	Local turbulent flame
D	Fig. 2(d)(e)	3.0,3.2 kPa	6.4,6.0 mm	Turbulent flame

Table 1. Four typical deflagration waves

3.2 Further Development

To find out the further development of the deflagration waves in a longer distance away from perforated plate, we used pressure transducers to make an arriving time schedule for the precursor shock waves at some fixed positions along test section. The data is plotted as an $x-t$ diagram in Fig. 3(a), where $x=0$ is the position of perforated plate. And the shock velocities measured through the last two transducers ($x = 1.183$ m and 1.432 m) versus initial pressure are also plotted out in Fig. 3(b).

It is clear that the further development of deflagration waves differs. Under lower initial pressures, no transition to detonation occurs. The precursor shocks of deflagration have a maximum velocity of about 1200 m/s, and in the lowest pressure that we tested, 1kPa namely, a minimum velocity of about 800 m/s is recorded. Under higher initial pressures, the deflagration will undergo an abrupt transition to detonation. There is a critical pressure range from 1.6 kPa to 1.8 kPa between above two situations.

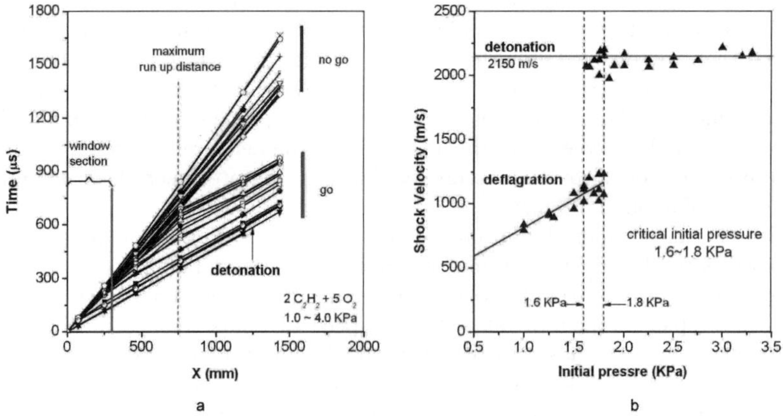

Fig. 3. Experimental results showing further development of the transmitted precursor shocks under different initial pressures

Since rarefaction waves behind incident detonation wave are capable of penetrating into the downstream flow field and attenuating the deflagration wave gradually, the propagating deflagration wave is always in a competition between back-boundary rarefaction and flame acceleration. Combining with the former streak schlieren observations, we found that the deflagration waves of laminar type (Table 1, flame A and B) usually can not overcome the attenuation of rarefaction waves. Therefore, they won't be able to accelerate or even sustain themselves if no external disturbances introduced. But for the turbulent deflagration waves (Table 1, flame D), the overall reaction rate of the mixture is greatly enhanced by turbulent transportation, they hence can defeat the back rarefaction. Due to the feedback and amplification mechanism, this kind of deflagration usually accelerates and DDT is often inevitable. Between these two cases there is a very unstable critical situation (Table 1, flame C) where the deflagration wave can propagate for a relatively long distance at about half CJ detonation speed.

3.3 Rankine-Hugoniot Analysis

Theoretically, present work simplifies the deflagration wave as a one-dimensional structure with two continuities (Fig. 4(a)). The assumptions are as followed:

a. The system consists of two discontinuities, i.e. a precursor shock and a flame. States across these two discontinuities are confined by Rankine-Hugoniot relations.

b. Two gaseous species are considered, i.e. the reactant and product. The thermal properties of these two species are composed from JANAF table.

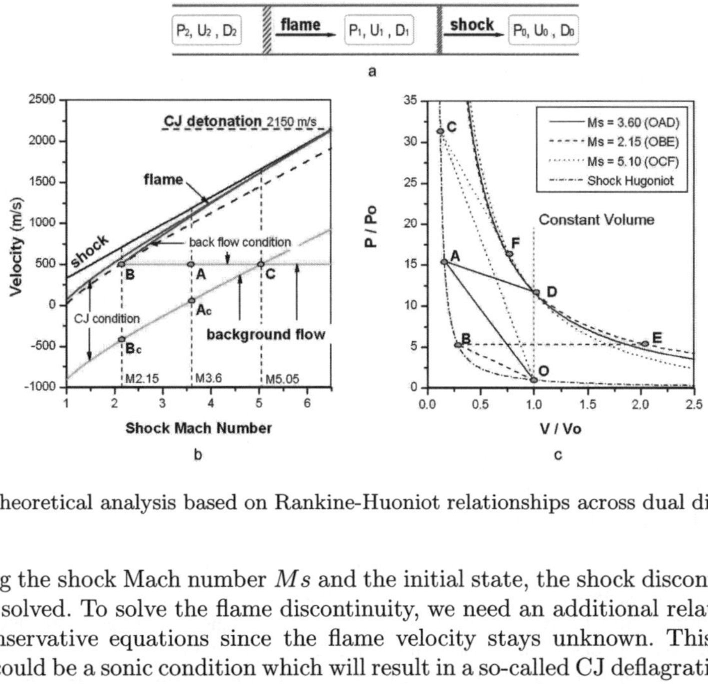

Fig. 4. Theoretical analysis based on Rankine-Huoniot relationships across dual discontinuities

Giving the shock Mach number Ms and the initial state, the shock discontinuity then could be solved. To solve the flame discontinuity, we need an additional relation beyond three conservative equations since the flame velocity stays unknown. This additional relation could be a sonic condition which will result in a so-called CJ deflagration solution. We have it tested in current calculation and the flame and back-boundary particle velocity varying by Ms are plotted in Figure 5*b*. It's obvious that when Ms increases the flame velocity is approaching the leading shock velocity, till in CJ detonation point they meet the same value. If we look at the corresponding particle velocity, we can see that, at the critical deflagration velocity observed in experiment ($Ms = 3.6$), the back-boundary should be nearly at rest. This does not match the fact since the real back-boundary flow velocity is measured to be around 500 m/s, while the possible CJ solution satisfying this velocity gives a precursor shock of Mach 5.05 (Vs = 1700 m/s, Fig. 4(b) point C) which has never been observed in our test. Therefore, the CJ deflagration is out of reason. Rejecting the CJ condition and directly setting back boundary velocity to be 500 m/s as the additional condition. This leads to a minimum shock velocity when the exothermic reaction rate becomes zero (Fig. 4(b) point B, $Ms = 2.15$, $Vs = 710$ m/s). From this point increasing Ms, we get another curve of flame speed varying by Ms. Three typical situations are plotted in $p - v$ plane as Fig. 4(c). In the experimentally observed critical case $Ms = 3.6$ ($Vs = 1200$ m/s), the final state of product locates right around that of constant volume combustion, which matches Thomas's observation [7].

4 Conclusion

We studied the structure and behavior of the perforated plate induced deflagration waves in stoichiometric Acetylene-Oxygen mixture experimentally. A one-dimensional theoretical analysis was also performed to get a better understanding on the phenomena. It was found that:

1. The typical deflagration wave generated by perforated plate was a complex of shock wave and flame. With the increase of initial pressure, it varies from a laminar structure where shock and flame are decoupled, to a turbulent one where shock and flame are strongly coupled.

2. Deflagrations with laminar structure usually can not stand the attenuation of background rarefaction and keeps on slowing down, whereas the turbulent one is capable of accelerate run up to a detonation wave. There is a very unstable critical case between above-mentioned two situations, where the deflagration can usually propagate for a relatively long distance in $50 - 60\%$ CJ detonation speed.

3. The flow behind the critical deflagration wave can not be in CJ state. Instead, it seems to be consistent with that of a constant-volume combustion in same mixture.

Acknowledgement. The author gracefully acknowledges the support of K.C.Wong education foundation, Hong Kong.

References

1. J.H.S. Lee, On the Transition from Deflagration to Detonation, of Progress in Astronautics and Aeronautics series, 1986, Vol.106,pp.644
2. G.Dupre, O.Peraldi, J.H.S.Lee, et al. Propagation of Detonation Waves in an Acoustic Absorbing Walled Tube, Progress in Astronautics and Aeronautics, 1988, Vol.114, pp.248-263
3. A.Teodorcyzk, and J.H.S.Lee, Detonation Attenuation by Foams and Wire Meshes Lining the Walls, Shock Waves (1995), Vol.4, pp.225-236
4. M.I.Radulescu and J.H.S.Lee, The Failure Mechanism of Gaseous Detonations: Experiments in Porous Wall Tubes, Combustion and Flame,2002,131:29-46
5. C. Guo, G.Thomas,J.Li and D.Zhang, et al, Experimental Study of Gaseous Detonation Propagation over Acoustically Absorbing Walls, Shock Waves (2002),11:353-359
6. R.S. Chue, J.H.Lee, T.Scarinci, et al. Transition from Fast Deflagration to Detonation Under the Influence of Wall Obstacles, Progress in Astronautics and Aeronautics, 1992, Vol.153
7. G.O.Thomas, R.J.Bambrey, and C,Brawn, Experimental observations of flame acceleration an transition to detonation following shock-flame interaction. Combust. Theory Modelling, 5(2001)573-594

Unconfined aluminum particles-air detonation

F. Zhang[1], K.B. Gerrard[1], R.C. Ripley[2], and V. Tanguay[3]

[1] *DRDC Suffield, PO Box 4000, Stn Main, Medicine Hat, AB, T1A 8K6 Canada*
[2] *Martec Ltd., 1888 Brunswick Street, Halifax, NS, B3J 3J8 Canada*
[3] *DRDC Valcartier, 2459 Pie-XI Blvd. N, Val-Belair QB G3J 1X5 Canada*

1 Introduction

The detonability of aluminum (Al) particles suspended in air is an important fundamental problem in understanding the limits of multiphase explosion in a metal particles-gas flow. Al particles possess an oxide coating that has a high melting temperature, thus further increasing the difficulty in understanding their detonation mechanism. Micrometric Al-air detonation at atmospheric conditions is feasible in large tubes 0.12-0.3 m in diameter [1,2]. Experiments in tubes also indicated a strong dependence of the Al-air deflagration-to-detonation transition (DDT) and detonation on initial pressure, thus suggesting a dependence on chemical kinetics [3]. These tube-confined experiments mostly showed a spinning detonation or marginal cellular structure, suggesting a characteristic detonation cell size of 0.4-0.6 m for flaked and 1-2 μm atomized Al-air stoichiometric or slightly rich mixtures. In order to remove tube wall influence, unconfined detonation tests are necessary but have not yet been successful or conclusive [4]- [5]. In this paper, unconfined Al-air detonation is further investigated and numerical modeling based on a kinetics-diffusion hybrid reaction model is conducted to explore kinetic dependence of the Al reaction in the process of detonation initiation and DDT.

2 Unconfined Al-air detonability

The charge configuration was arranged in an 18 m long, 90-deg V-shaped steel trough line, in which a PETN cord (21.3 g/m, 6.1 mm in diameter) was located at the bottom vertex and covered by a layer of aluminum powder. Two types of Al particles were used: 47.6 kg atomized Al with a mean diameter of 2-3 μm known as H-2 by Valimet Inc., and 31.5 kg flaked Al. Detonation of the PETN cord dispersed the Al powder in air to a 3 m × 3 m cross section and 18 m long suspension with an average Al concentration of 670 g/m^3 for H-2 or 290 g/m^3 for flakes at a given dispersal time. The Al-air cloud was then initiated at the one end using 8 kg C4 explosive located 1.5 m from the end as well as 1 m from the ground and the steel line. The resultant wave phenomena were recorded using PCB pressure transducers located on the ground along the wave propagation direction, 3 m from the C4 explosive for the first transducer and 2 m intervals for the rest. Detonation cell sizes were registered using a 1.52 m × 1.22 m Al smoke foil installed on the ground between the last two transducers. Two Phantom high-speed video cameras were employed to record the flame propagation, one facing the propagation (6504 fps) and the other in the direction perpendicular to the propagation (10,000 fps).

Figure 1 shows an example of high-speed photographs for a flaked Al-air cloud experiment and Figures 2 and 3 display the pressure histories along the propagation distance for an H-2 Al and a flaked Al test respectively. For H-2 within a 4 m ground distance from C4, the shock velocity decays to 1300 m/s. The shock velocity further decays to 1050 m/s

Fig. 1. Detonation in flaked Al-air (#07115*B*). Upper left: Al suspension, upper right: initiation, lower left and right: detonation propagation

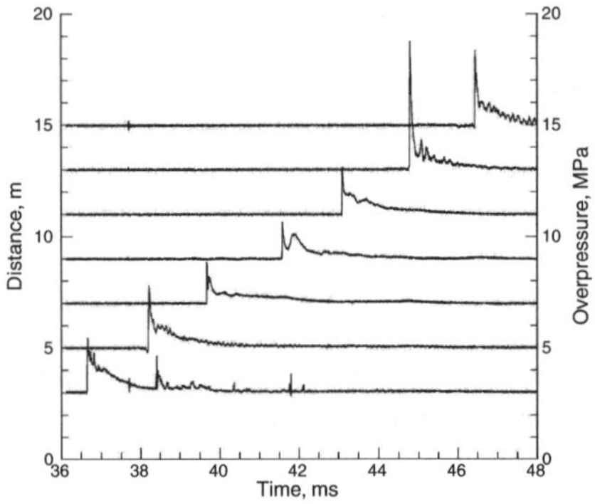

Fig. 2. H-2 Al-air DDT (#07107*B*)

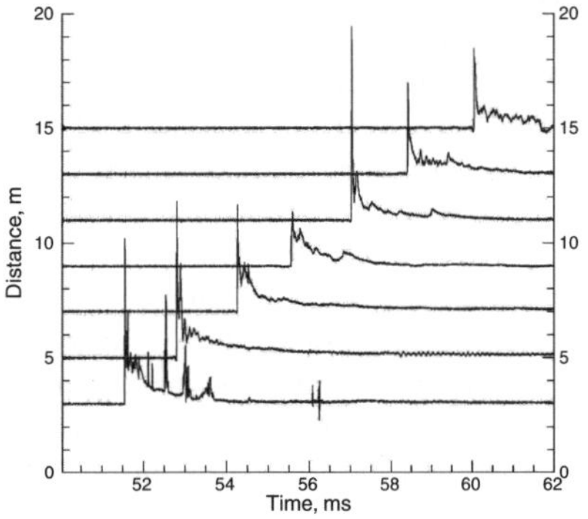

Fig. 3. Flaked Al-air DDT (#07115*B*)

at 9 m, while Al combustion generates multiple compression waves behind the shock front during the propagation. At 13 m, DDT-like phenomena take place with a 5.5 MPa peak pressure and a pressure history featured with oscillations, a typical phenomenon related to the transverse waves. However the propagation velocity remains 1300-1200 m/s to the end of the cloud. For flaked Al shown in Figure 3, the shock velocity is 1600 m/s at 4 m from C4 and decays to 1380 m/s at 7 m. At 11 m, the onset of detonation occurs with a peak pressure of 8.4 MPa. The 2-m averaged detonation velocity reaches a maximum value of 1533 m/s and sustains 1460 m/s in the remainder of propagation. In the oscillating pressure history at 15 m, a transverse wave structure is distinguishable with a primary period of about 350 μs. The smoke foil registers a detonation cell width of 0.55 m, consistent with the pressure oscillation period. These experiments indicate that the 8 kg C4 is not sufficient to directly initiate H-2 Al-air detonation but is close to the critical charge for direct initiation of flaked Al-air detonation. While the Al cloud dimension in the cross section is not sufficiently large, the fact that the critical energy for direct initiation of detonation is near or larger than 8 kg C4 is consistent with the estimate made from the cell size. Note that without Al particles, the air blast overpressure from the 8 kg C4 explosion decays rapidly to 0.186, 0.046 and 0.023 MPa at 5, 10 and 15 m respectively.

3 Aluminum reaction model

Khasainov et al. applied a diffusion reaction model to Al-gas detonations [6]:

$$J_p = n_p \pi d_p^2 \rho_s \frac{\mathrm{d}r_p}{\mathrm{d}t} = \frac{3\sigma_p}{t_b}(1 + 0.276 Re^{1/2} Pr^{1/3}), \quad \text{if } T_p \geq T_{ign}; \tag{1}$$

else $J_p = 0$. In equation (1), the particle burning time is

$$t_b = K d_{p0}^n / X_{oxi}^{\alpha} \tag{2}$$

where J_p is the Al mass depletion rate in kg/m^3/s and n_p, T_p, d_p, r_p, σ_p and ρ_s are the particle number density, temperature, diameter, radius, mass concentration and material density respectively. Re is the Reynolds number based on the velocity difference between the two phases and Pr is the gas-phase Prandtl number. The parameters K, d_{p0}, X_{oxi} and T_{ign} are the rate coefficient, initial particle diameter, mole fraction of oxidizing gases and particle ignition temperature respectively. $n = 2$ has been often used that essentially assumes infinite chemical kinetics. This model is independent of temperature and pressure and assumes a particle ignition temperature above which aluminum reacts. The model has been widely used in CFD codes to model the propagation of detonation assuming an ignition temperature (933-1350 K) much below the oxide melting point ([1, 5, 7]). Apart from diffusion models, Arrhenius kinetics models have also been applied recently to Al-gas detonation [8].

Al reaction can depend on chemical kinetics, as evidenced by the Al-air DDT reported in this paper as well as the previous papers using weak initiation sources in tubes showing strong dependence of detonability on initial pressure [2, 3]. Hence, a kinetics-diffusion hybrid reaction model is proposed in the form:

$$J_p = n_p \pi d_p^2 k_p = n_p \pi d_p^2 \frac{\nu_p W_p}{\nu_{oxi} W_{oxi}} k, \tag{3}$$

$$k = k_d (C_{oxi} - C_{oxi,s}) \quad \text{or} \quad k = k_s C_{oxi,s} = C_{oxi,s} k_0 e^{-E/RT_s} \tag{4}$$

where k_p and k are the mass depletion rate (equaling the mass flux) of the particle and oxidizing gas at the particle surface in the r direction respectively. k_d and k_s are the rate coefficient for diffusion and surface reaction respectively. W, ν, T_s, C_{oxi} and $C_{oxi,s}$ denote the molecular weight, stoichiometric coefficient, particle surface temperature, oxidizing gas mole concentration and its particle surface value respectively. From equation (4), one obtains the hybrid model by making the diffusion and surface chemical reaction expressions equal:

$$k = \frac{k_d k_s}{k_d + k_s} C_{oxi} \tag{5}$$

The diffusion rate coefficient k_d can be obtained by inserting the diffusion model (1-2) ($\alpha = 1$ is assumed for simplicity) into equations (3) and (5):

$$k_d = \frac{\nu_{oxi} W_{oxi}}{\nu_p W_p} \frac{\rho_s d_p}{2 C_{total} K d_{p0}^2} (1 + 0.276 Re^{1/2} Pr^{1/3}) \tag{6}$$

where C_{total} is the mole fraction of total gases. From equation (4) the surface reaction rate coefficient is:

$$k_s = k_0 e^{-E/RT_s}, \quad \text{with } T_s = (T + T_p)/2 \tag{7}$$

Thus, equations (3) and (5-7) form the hybrid reaction model which depends on temperature and pressure via oxidizing gas concentration, and does not need a particle ignition temperature. The model becomes surface kinetics-limited ($k \rightarrow k_s$) when $k_s/k_d \ll 1$, and diffusion-limited ($k \rightarrow k_d$) if $k_s/k_d \gg 1$.

4 Numerical results and discussion

The hybrid model is implemented in a second-order unsteady Eulerian multiphase fluid dynamics code, where the solid phase and gas phase have been treated as two separate flows (different velocity and temperature) and their mass, momentum and energy are exchanged through the source terms. Wall loss terms are not present and the reaction of aluminum and oxygen is assumed to form condensed Al_2O_3. In all calculations in this paper, $K = 4 \times 10^6$ s/m^2, $k_0 = 9 \times 10^5$ s/m^2 and $E = 95.5$ kJ/mole were used.

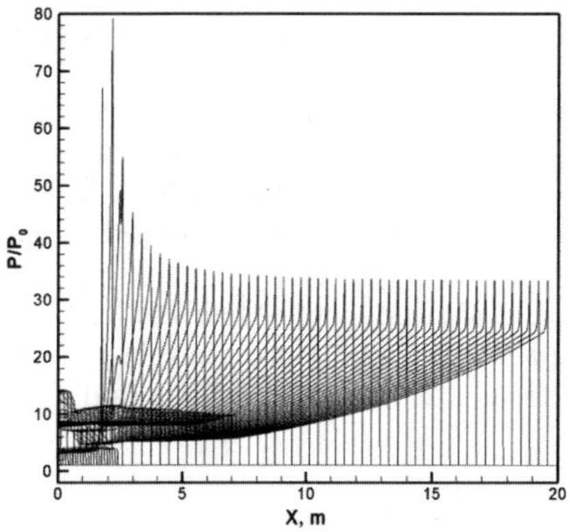

Fig. 4. DDT pressures using the hybrid reaction model

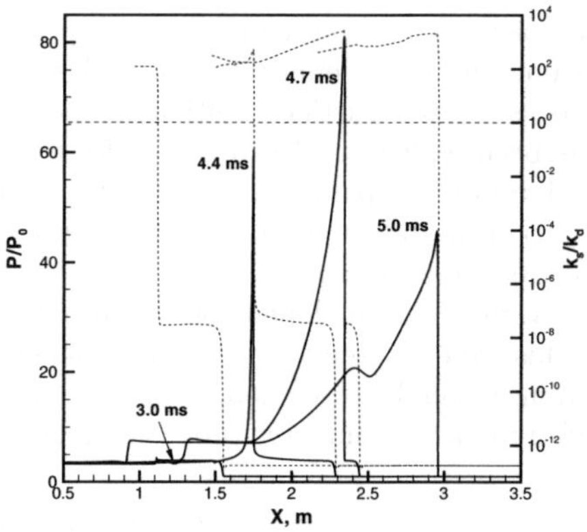

Fig. 5. The early DDT process in Fig. 4. Solid line: p/p_0, dashed line: k_s/k_d.

Figure 4 shows 1D simulation results of DDT in a rich H-2 Al-air mixture at $\sigma_p = 1250$ g/m^3 and an initial pressure of $p_0 = 2.5$ atm. The calculation used a 1 mm numerical cell size and a hot spot zone (2 mm long with $p = 24$ bar, $T = 2843$ K and $Tp = 300$ K). The results clearly show that after an ignition delay behind the shock front, a local Al explosion leads to the onset of detonation. Thereafter the overdriven detonation relaxes towards a detonation, with a 1745 m/s detonation velocity and 33.5 p_0 front pressure at x = 20 m. During the early DDT process, $k_s/k_d < 1$ holds behind the shock front in an induction stage that leads to local explosion, thus indicating a kinetics-limited reaction (Figure 5). As the local explosion develops, the particle temperature rapidly increases and hence results in a rapid increase in k_s/k_d. Once the detonation forms, $k_s/k_d > 1$ holds after a very short kinetics induction time behind the shock, thus showing a diffusion-limited reaction for most of the Al mass. The process is in agreement with the experimental results reported in [3] and therefore the hybrid model properly describes the detonation initiation and DDT. Under the same initial conditions, abrupt DDT via an auto-explosion center cannot be obtained using the diffusion model (1-2) as shown in Figure 6. Figure 8 shows the particle mass concentration field from a 2D detonation simulation for the same Al-air mixture using the hybrid reaction model with a 4 mm numerical cell size. A dense Al layer is developed behind the shock front due to drag compression. The reaction then significantly reduces Al concentration over a length of less than 100 mm, consistent with the reaction length from the 1D simulation. The 2D simulation shows a transverse wave structure with a detonation cell width of 0.33 m. Since most of the Al reaction is diffusion-limited during the detonation propagation, the cell size mainly depends on the diffusion reaction parameters.

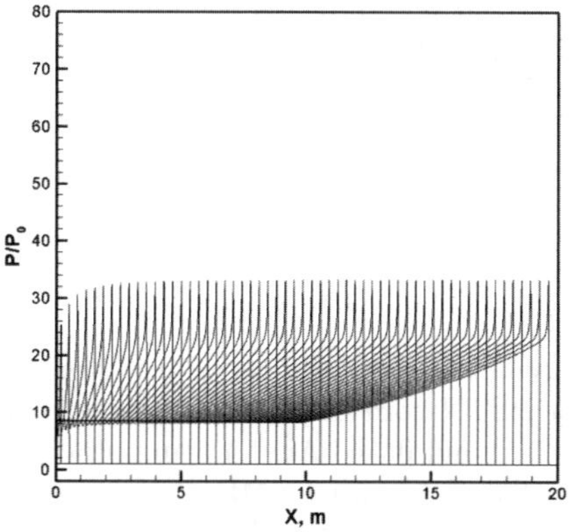

Fig. 6. Detonation using the diffusion model

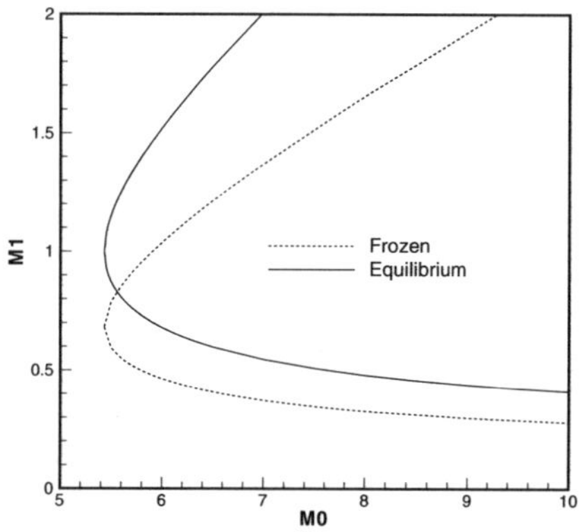

Fig. 7. $M_1 - M_0$ relation

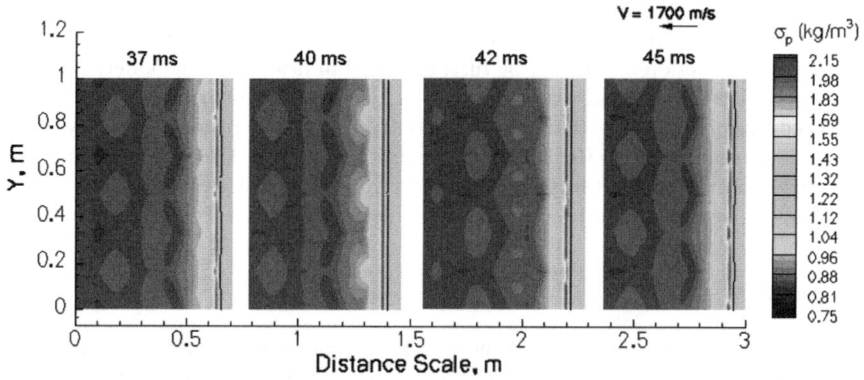

Fig. 8. Particle concentration fields and shock fronts (black lines) from 2D detonation simulation

The above simulations assume the detonation products including condensed Al and Al_2O_3 that are incompressible without contribution to gas pressure. When Al_2O_3 is changed to gas phase in the detonation products, the simulation shows that the detonation velocity and front pressure at 20 m are increased to 2290 m/s and 61 p_0 respectively. This is obvious since the work supporting the shock propagation is done by the expansion of the gaseous detonation products directly. Furthermore, in Figure 4 where the condensed Al_2O_3 is used, the detonation wave becomes steady after a long propagation distance of nearly 19 m, from which the local flow velocity relative to the shock reaches the equilibrium sound speed at a constant reaction length of 160 mm for 99.99% oxygen consumption (the two phase velocity and temperature equilibrium has reached much before). To further examine possible steady solutions, a steady control volume analysis in the shock frame has also been conducted based on mass, momentum and energy conservation. The control volume has the initial condensed Al and air at the entry and the detonation products including nitrogen and condensed Al and Al_2O_3 (incompressible and without contribution to the gas pressure) at the exit. The velocity and temperature equilibrium between the two phases is also assumed at the exit. Figure 7 presents the solution for the local Mach number at the exit, M_1, as a function of detonation Mach number, M_0, with respect to gas-phase sound speed. It is shown that the minimum detonation Mach velocity as the unique solution is attainable at $M_1 = 1$ with respect to equilibrium sound speed while it is subsonic relative to the frozen gas-phase sound speed. This steady solution analysis further indicates that the steady CJ detonation solution may not exist above a limit amount of the condensed detonation products that do not make contribution to gas pressure directly.

5 Conclusion

Unconfined Al-air detonation has been investigated experimentally in a 3 m × 3 m cross section and 18 m long cloud generated using a PETN cord dispersal technique. DDT phenomena were observed for both atomized H-2 Al and flaked Al at 1 atm initial pressure, while detonation was established only in the flaked Al with 4-8 MPa peak pressures, 1460-1530 m/s shock velocities and a 0.55 m detonation cell width. A surface oxidation-diffusion hybrid reaction model was suggested to simulate the kinetics-dependent phenomena of the Al detonation initiation and DDT process observed in experiments. The 1D and 2D unsteady two-phase fluid dynamics modeling without wall loss terms showed the success of the hybrid model, capable of catching both the kinetics-limited transient processes of Al reaction and the propagation of detonation limited by the diffusion reaction. The Al hybrid reaction model will be further tested in describing the kinetics-dependent processes of detonation failure and quenching. Using the detonation products including condensed aluminum oxide that does not contribute to gas pressure directly, a steady solution is attainable only after a long propagation distance. A steady solution based on the control volume analysis further indicated that there exists a limit in the amount of condensed phase detonation products above which the steady CJ solution may not exist.

References

1. Borisov, A.A., Khasainov, B.A., Saneev, E.L., Formin, I.B., Khomik, S.V., Veyssiere, B.: On the detonation of aluminum suspensions in air and in oxygen. In: *Dynamic structure of detonation in gaseous and dispersed media*, ed by Borisov, A.A. (Kluwer, 1991) pp 215-253
2. Zhang, F., Grönig, H., van de Ven, A., Shock Waves **11**, 53-71 (2001)
3. Zhang, F., Murray, S.B., Gerrard, K.B., Shock Waves **15**, 313-324 (2006)
4. Tulis, A.J.: On unconfined detonation of aluminum powder-air clouds. In: *1st Intl. Colloq. on Explosibility of Industrial Dusts*, ed by Wolanski, P. (Warsaw, 1984) pp 178-186
5. Ingignoli, W., Veyssiere, B., Khasainov, B.A.: Study of detonation initiation in unconfined aluminum dust clouds. In: *Gaseous and Heterogeneous Detonations*, ed by Roy, G. Frolov, S., Kailasanath, K., Smirnov, N. (ENAS Publishers, Moscow, 1999) pp 337-350
6. Khasainov, B.A., Veyssiere, B., Archivum Combustionis **7**, 333-352 (1987)
7. Benkiewicz, K., Hayashi, A.K., AIAA Journal **44**, 608-619 (2006)
8. Fedorov, A.V., Khmel, T.A., Combustion, Explosion, and Shock Waves **41**, 435-448 (2005)

Viscous attenuation of a detonation wave propagating in a channel

P. Ravindran[1], **R. Bellini**[1], **T.-H. Yi**[2], **and F.K. Lu**[1]

[1] *Aerodynamics Research Center, University of Texas at Arlington, Arlington, Texas, USA*
[2] *Institute of High Performance Computing, Singapore*

Summary. The initiation and propagation of a detonation wave in a two-dimensional channel is simulated by an Euler and a Navier-Stokes solver. Transport processes were found to play a role in the wave propagation, resulting in a lower wave propagation speed arising from viscous drag.

1 Introduction

The propagation of a detonation wave remains one of the challenging problems in physics. While it is possible to predict the average propagation velocity almost correctly using classical theory [1, 2], there is no theory to explain the internal flow structure satisfactorily. The famous Zel'dovich-von Neumann-Döring theory [3–5] is widely considered to describe the one-dimensional detonation structure correctly. Early experiments uncovered that the reduction to one space dimension is not entirely justified in long tubes. The detonation waves are non-planar and exhibit multi-dimensional substructures [6]. Numerical simulations have lately been used to examine these transient sub-structures but most of these studies thus have far neglected transport phenomena partly to simplify the approach. Moreover, neglecting transport phenomena appeared to be feasible in regions where such phenomena can be neglected. Nonetheless, the presence of walls requires careful consideration of transport phenomena [7].

Recent progress in computing power has enabled Navier-Stokes solutions of detonation waves [8–10]. A preliminary study on the effect of transport processes is reported here. The propagation of a detonation wave is examined by a comparison between an inviscid and a viscous solution.

2 Governing Equations

The governing equations for a high-temperature, chemically reacting viscous flow are modeled with thermodynamic processes and chemical reactions as well as the fluid dynamics. The gas considered is assumed to be thermally perfect and chemically non-equilibrium with finite-rate chemistry. For brevity, the derivations and relations are only summarized here; details can be found in [11]. The equations to be solved are the two-dimensional, unsteady conservation equations extended for multi-species with chemical reactions. The species are assumed to be thermally perfect gas. Thus, the specific heats, enthalpy, entropy and internal energy of each species are functions of temperature only and are expressed as approximations by a least-squares fit method [12]. The expressions and coefficients are valid for 200–6000 K that exceeds the temperature range of interest. Dalton's law is used to obtain the pressure of the mixture from the partial pressures of the individual species. The overall reaction is that for a stoichiometric hydrogen-air mixture. The elementary reaction mechanisms for the mixture are extracted from the GRI-Mech 3.0 [13] database that contains 325 reactions and 53 species. From this database, the 9-species 28-reaction model was chosen. The transport properties and third body reaction data were also extracted from the GRI-Mech 3.0 database.

The viscosity coefficient, diffusion coefficient and thermal conductivity were obtained from classical kinetic theory [14]. The viscosity coefficient and thermal conductivity of species are independent of the pressure whereas the diffusion coefficient is inversely proportional to the pressure. It is noted that the mixture-averaged formulations for the thermal conductivity and diffusion coefficient are only approximations in order to reduce computational time [14].

3 Numerical Formulation

The two-dimensional Navier-Stokes equations were discretized by a finite volume formulation and retaining the integral form with the source term. The convective and viscous fluxes were evaluated at the faces of a cell in the x- and y-direction. The convective fluxes were approximated by flow quantities extrapolated to the left and right sides of the cell face while the viscous fluxes were computed by the averaged variables at a face and then were approximated using a second-order Roe scheme which was extended for a multi-species, thermally perfect and non-equilibrium gas [15,16]. High-order accuracy in space is achieved by using the MUSCL (Monotone Upstream-Centered Schemes for Conservation Laws) approach. By employing an appropriate minmod limiter, the MUSCL scheme becomes first-order accurate in the vicinity of the discontinuities and second-order accurate in smooth regions. For temporal discretization, the governing equations were discretized in time so that they become a system of ordinary differential equations. A two-step Runge-Kutta scheme was utilized to achieve second-order accuracy. Since this scheme is fully explicit, it can be started with known values at time level n. The transport terms were evaluated using Green's theorem [17], but the evaluation of the source terms was much more complex due to the chemical reactions having much shorter time scales than those associated with the flow, resulting in stiffness. The approach used to overcome this difficulty is highlighted in [11]. The temperature must be evaluated at each grid cell whenever the species density was updated for a thermally perfect gas. Likewise, the boundary conditions were evaluated for the wall, supersonic inflow into the channel and outflow.

4 Results

A 50 cm long and 10 cm high channel was used for the detonation channel and the computational domain was discretized into 100×150 grid cells. The grids were clustered near the walls to capture the boundary layer. Adiabatic wall boundary conditions were imposed on the upper and lower sides of the computational domain while supersonic inflow and outflow were imposed on the left and right side respectively, with an ambient pressure of 1 atm imposed at the outlet of the channel. Pre-mixed hydrogen-air gas at an ambient pressure of 2 atm and temperature of 500 K was set to flow into the channel at an incoming Mach number of 2. The mixture was ignited by a localized hot spot at

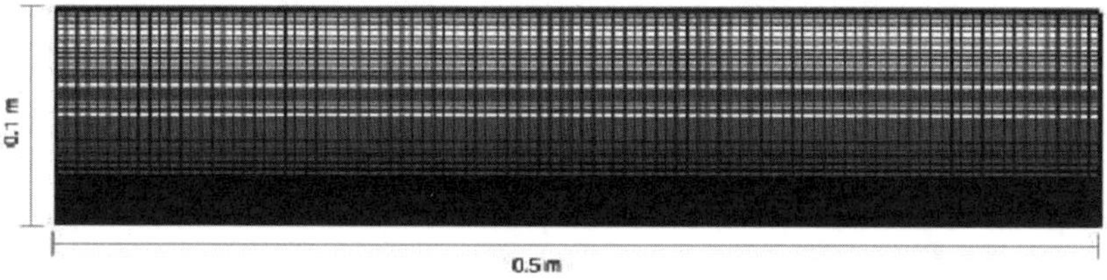

Fig. 1. Schematic of the computational domain

approximately 0.1 m into the channel from the left end. The hot spot was located at the bottom of the channel. It was 1 mm wide and 5 mm high. The pressure and temperature in the hot spot were 30 atm and 3000 K respectively. Detonation waves were immediately generated and the initiation and propagation are shown in the results.

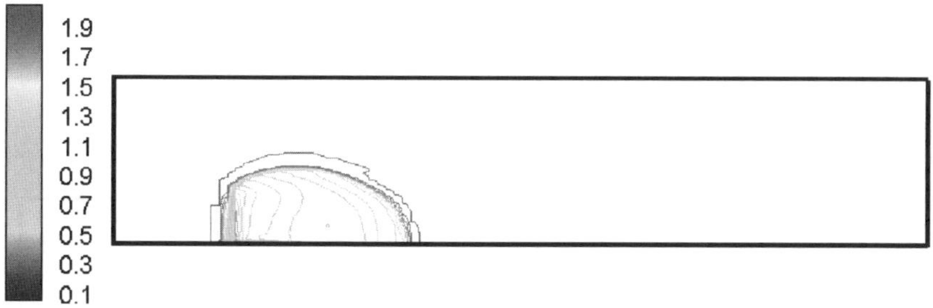

a. Inviscid Flow

b. Viscous Flow

Fig. 2. Mach number distribution: detonation initiation snapshot at $t = 1.8090 \times 10^{-5}$ s

The results of the numerical solution were compared with that of an inviscid algorithm to characterize the effect of viscosity on the flow. Figure 2 shows the initiation of the detonation wavefront at time $t = 1.8090 \times 10^{-5}$ s. Figure 3 indicates that the propagation of the wavefront at $t = 5.9636 \times 10^{-5}$ s is almost half the length of the channel and Fig. 4 shows the wavefront exiting from the channel at $t = 1.0107 \times 10^{-4}$ s. Results in Figs. 5and 6 for the pressure and temperature, respectively, taken at $t = 5.9636 \times 10^{-5}$ s, show a similar pattern in wavefront propagation.

A comparison of the viscous and inviscid results shows that the wavefront propagation is delayed or attenuated due to the presence of viscosity. This can be attributed to the drag arising from the formation of a laminar boundary layer behind the shock wave [18,19]. Also, the pressure and temperature at the wall are found to be higher for viscous solutions, due to the presence of shear layers. The simulation also shows that the wave

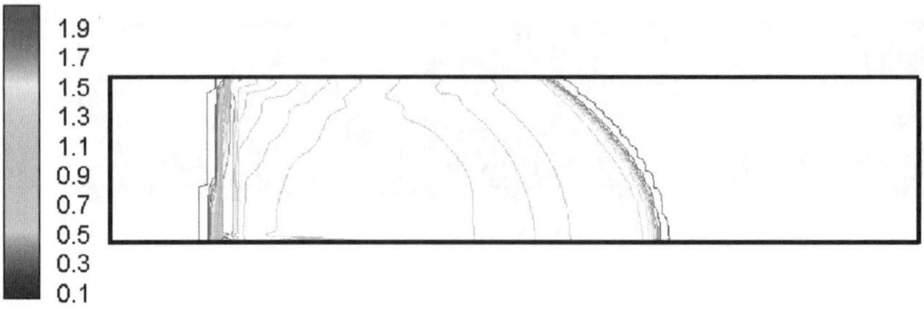

a. Inviscid Flow

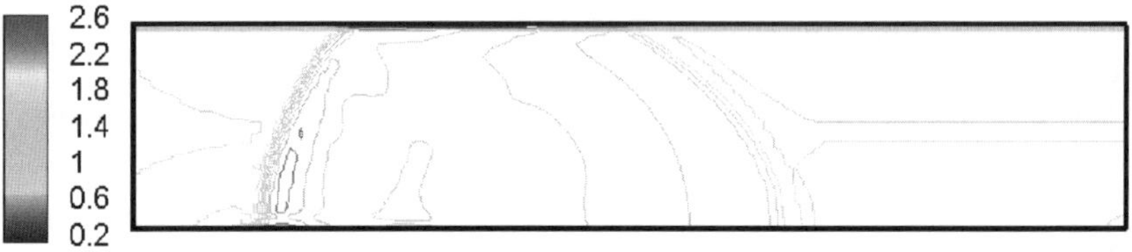

b. Viscous Flow

Fig. 3. Mach number distribution: detonation propagation snapshot at $t = 5.9636 \times 10^{-5}$ s

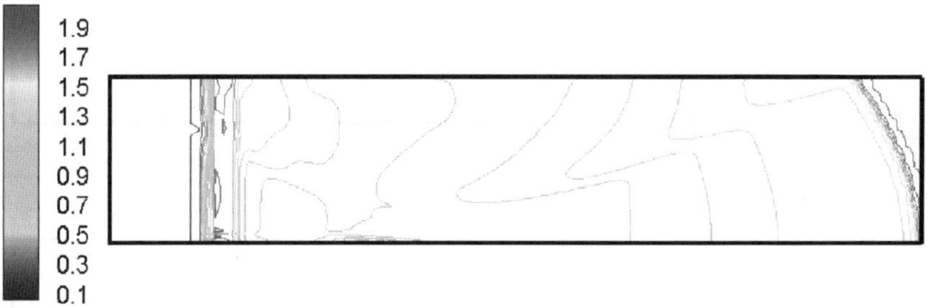

a. Inviscid Flow

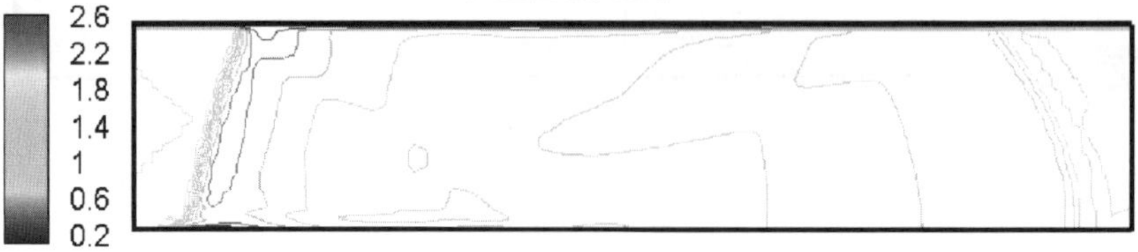

b. Viscous Flow

Fig. 4. Mach number distribution: detonation propagation snapshot at $t = 1.0107 \times 10^{-4}$ s

strength of the downstream moving detonation is significantly stronger than that of the upstream-moving detonation due to the non-zero Mach inflow condition. Another significant difference between the viscous and inviscid flows is observed in the downstream moving wavefront. For an inviscid simulation it was found to be a straight shock but the effect of transport processes is evident in the viscous simulation, wherein the downstream shock in more curved due to the velocity gradient from the walls to freestream.

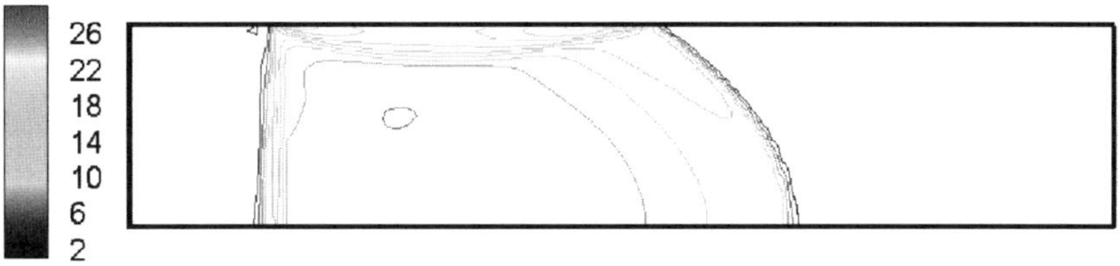

a. Inviscid Flow

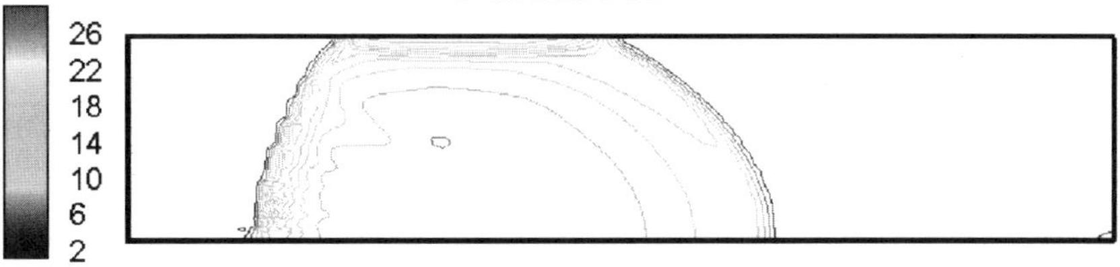

b. Viscous Flow

Fig. 5. Pressure distribution: detonation propagation snapshot at $t = 1.8090 \times 10^{-05}$ s

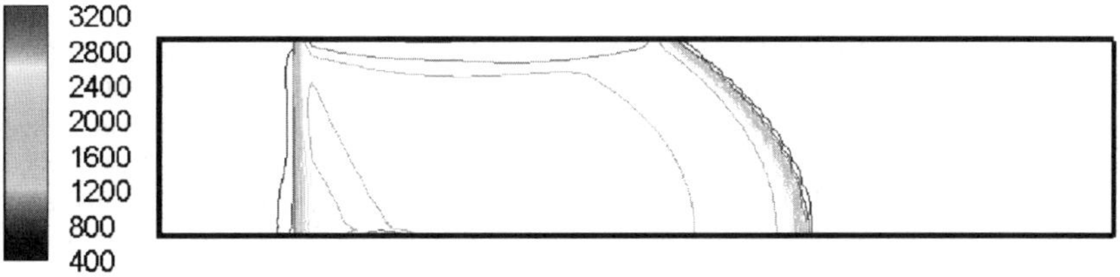

a. Inviscid Flow

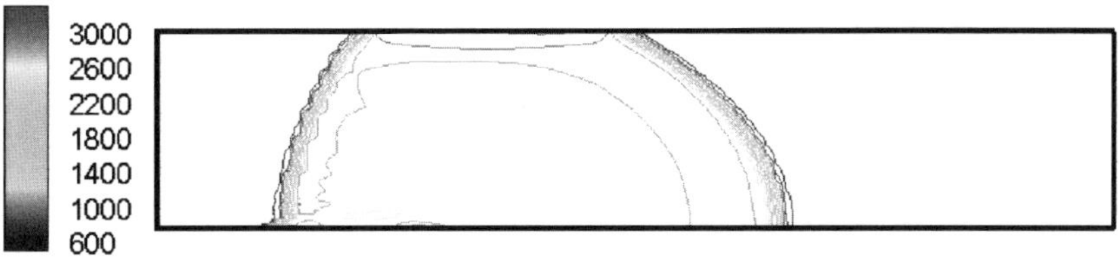

b. Viscous Flow

Fig. 6. Temperature distribution: detonation propagation snapshot at $t = 1.8090 \times 10^{-5}$ s

References

1. Chapman, D.L.: On the rate of explosion in gases. In: *Philos. Mag.*, vol 47, pp 90-104, 1899
2. Jouguet E.: On the propagation of chemical reactions in gases. In: *J. de Mathematiques Pures et Appliquees*, vol. 1, pp. 347-425, 1906
3. Zel'dovich, Ya.B.: On the theory of the propagation of detonations on gaseous system. In: *Zh. Eksp. Teor. Fiz.*, vol. 10, pp. 542-568, 1940
4. von Neumann, J.: Theory of detonation waves. Progress Report to the National Defense Research Committee Div. B, OSRD-549, (April 1, 1942. PB 31090); In: Taub, A. H. (ed). *John von Neumann: Collected Works, 1903-1957*, Vol. 6, Pergamon Press, New York, 1963
5. Döring, W.: On detonation processes in gases. In: *Ann. Phys.*, vol. 43, pp. 421-436, 1943
6. Pintgen, F., Eckett, C.A., Austin, J.M., Shepard, J.E.: Direct observations of reaction zone structure in propagating detonations. In: *Combust. Flame*, vol. 133(3), pp. 221-229, 2003

7. Vitello, P., Souers, P.C.: Stability effects of artificial viscosity in detonation modelling. In: *12th Int. Detonation Symp.*, San Diego, California, Aug 11-16, 2002
8. Oran, E.S., Gamezo, V.N.: Origins of the deflagration-to-detonation transition in gas phase combustion. In: *Combust. Flame*, vol. 148(1-2), pp. 4-47, 2007
9. Akkerman, V., Bychkov, V., Petchenko, A., Eriksson, L.-E.: Accelerating flames in cylindrical tubes with nonslip at the walls. In: *Combust. Flame*, vol. 145(1-2), pp. 206-219, 2007
10. Singh S., Rastigejev Y., Paolucci S., Powers J.M.: Viscous detonation in H_2-O_2-Ar using intrinsic low-dimensional manifolds and wavelet adaptive multilevel representation. In: *Combust. Theory Modelling*, vol. 5(2), pp. 163-184, 2000
11. Yi, T.-H.: Numerical study of chemically reacting viscous flow relevant to pulsed detonation engines. Ph.D. dissertation, Univ. Texas at Arlington, 2005
12. Gordon, S. and McBride, B.J.: Computer program for calculation of complex chemical equilibrium compositions and application I. Analysis. Tech. Rep. NASA RP-1311, 1976
13. Smith, G.P., Golden, D.M., Frenklach, M., Moriarty, N.W., Eiteneer, B.: *GRI-Mech 3.0*, [Online] Available: http://www.me.berkeley.edu/gri-mech
14. Kee, R.J. , Coltrin, M., and Glarborg, P.: *Chemically Reacting Flow*, Wiley Interscience, 2003
15. Grossmann, B., and Cinella, P.: Flux-split algorithms for flows with nonequilibrium chemistry and vibrational relaxation. In: *J. Comp. Phys.*, vol. 88, pp. 131-168, 1990
16. Deiterding, R.: Parallel adaptive simulation of multi-dimensional detonation structures. Ph.D. dissertation, Brandenburgische Technische University, Cottbus, Germany, 2003
17. Hirsch, C.: *Numerical Computation of Internal and External Flows*, Wiley, 1990
18. Liu, W.S., Du, X.X., and Glass, I.I.: Laminar boundary layers behind detonation waves. In: *Proc. Roy. Soc. London. Ser. A, Math. Phys. Sci.*, vol. 387(1793), pp. 331-349, 1983
19. Zhang, F., Chue, R.S., Frost, D.L., Lee, J.H.S., Thibault, P., and Yee, Y.: Effects of area change and friction on detonation stability in supersonic ducts. In: *Proc. Roy. Soc.: Math. Phys. Sci.*, vol. 449(1935) pp. 31-49, 1995

Part V

Diagnostics

A diode laser absorption sensor for rapid measurements of temperature and water vapor in a shock tube

H. Li, A. Farooq, R.D. Cook, D.F. Davidson, J.B. Jeffries, and R.K. Hanson

High Temperature Gasdynamics Laboratory, Department of Mechanical Engineering, Stanford University, Stanford, CA 94305-3032, USA

Summary. Time-resolved measurements of gas temperature and water vapor concentration are made in a shock tube using a novel diode laser absorption sensor (100 kHz bandwidth). Gas temperature is determined from the ratio of fixed-wavelength laser absorption of two water vapor rovibrational transitions near 1.4 μm, and H_2O concentration is determined from the inferred temperature and the absorption for one of the transitions. Wavelength modulation spectroscopy is employed with second-harmonic detection to improve the sensor sensitivity. The sensor is validated in a static cell and shock tests with H_2O-Ar mixtures, yielding an overall accuracy of better than 1.9% for temperature and 1.4% for H_2O concentration measurements over the range of 500-1700 K. The sensor is then demonstrated in a preliminary study of combustion in $H_2/O_2/Ar$ and heptane/O_2/Ar mixtures in the shock tube.

1 Introduction

The design and optimization of many combustion systems rely on accurate chemical reaction mechanisms [1]. For example, kinetic modeling has been applied to study advanced combustion systems such as homogenous charge compression ignition (HCCI) engines. Shock tube measurements have been used extensively in the development and validation of combustion mechanisms, e.g. see [2]. Here we develop a tunable diode laser (TDL) sensor for nonintrusive measurements of gas temperature and H_2O concentration behind reflected shock waves, thereby providing a diagnostic tool useful for studying the combustion mechanisms of hydrocarbon fuels.

TDL absorption sensors have been demonstrated previously for measurements of temperature, pressure, species concentration, and flow velocity in various applications [3] (and references therein). The need to scan the laser wavelength to recover a zero-absorption baseline limits the speed and sensitivity of direct-absorption sensors. Wavelength modulation spectroscopy with second-harmonic detection (WMS-2f), as an extension of absorption spectroscopy, is a well-known technique for improving the signal-to-noise ratio (SNR) [3]. Normalization of the WMS-2f signal with the 1f signal can remove the need for calibration and baseline, making WMS-2f an attractive technique for shock tube experiments. Thus, a fixed-wavelength WMS-2f technique is used in our sensor design to achieve high bandwidth and accuracy.

Water vapor is a major combustion product of hydrocarbon fuels. The strong rovibrational spectrum of water vapor in the near-infrared overlaps with the well-developed telecom laser technology. Therefore, H_2O is chosen as the target species, and gas temperature is inferred from the ratio of fixed-wavelength laser absorption of two selected H_2O transitions near 1.4 μm. The WMS-2f sensor is first validated in a heated static cell and shock tests with H_2O-Ar mixtures, and then is demonstrated in a preliminary study of the combustion of $H_2/O_2/Ar$ and heptane/O_2/Ar mixtures in a shock tube.

2 Fundamental spectroscopy

The fractional transmission of monochromatic radiation through a uniform gas medium of length L [cm] is described by the Beer-Lambert relation:

$$\tau(\nu) = (I_t/I_0)_\nu = exp[-PX_iS(T)\phi_\nu L] \tag{1}$$

where I_t and I_0 are the transmitted and incident laser intensities, P [atm] the total pressure, X_i the mole fraction of the absorbing species, and S [cm^{-2}/atm] and ϕ_ν [cm] the line strength and line shape function of the absorption feature.

For fixed-wavelength WMS-2f detection of absorption, the laser frequency and intensity are sinusoidally modulated:

$$\nu(t) = \bar{\nu} + acos(2\pi ft) \tag{2}$$

$$I_0(t) = \bar{I}_0[1 - i_0cos(2\pi ft)] \tag{3}$$

Here, $\bar{\nu}$ [cm-1] is the center laser frequency, a [cm^{-1}] the modulation depth, $\bar{I}_0$ the average laser intensity at $\bar{\nu}$, and i_0 the intensity modulation amplitude. The transmission coefficient is a periodic even function and can be expanded in a Fourier cosine series:

$$\tau[\bar{\nu} + acos(2\pi ft)] = \Sigma_{k=0}^{\infty} H_k(\bar{\nu}, a)cos(k2\pi ft) \tag{4}$$

For optically thin samples ($PX_iS\phi_\nu L$<0.1), the Fourier components are given by

$$H_0 = 1 - \frac{PX_iSL}{2\pi}\int_{-\pi}^{\pi}\phi[\bar{\nu} + acos\theta]d\theta, H_k = -\frac{PX_iSL}{\pi}\int_{-\pi}^{\pi}\phi[\bar{\nu} + acos\theta]cosk\theta d\theta. \tag{5}$$

A lock-in is used to extract the 1f and 2f signals by multiplying the detector signal by sinusoidal reference signals. The absorption-based WMS-2f and -1f signal are given by

$$S_{2f}(\bar{\nu}) = \frac{G\bar{I}_0}{2}|H_2 - i_0(H_1 + H_3)/2|, S_{1f}(\bar{\nu}) = \frac{G\bar{I}_0}{2}|H_1 - i_0(H_0 + H_2/2)|. \tag{6}$$

where G accounts for the optical-electrical gain of the detection system and the laser transmission variation. The WMS-2f signal is directly proportional to species concentration and line strength, while the 1f signal at line center is close to $G\bar{I}_0 i_0/2$. By normalizing the WMS-2f signal with the 1f signal, common terms such as the laser intensity, detector sensitivity, and laser transmission variations are eliminated. The laser parameters can be determined before the measurements. Therefore, no calibration is needed to scale the simulations to the measurements. Gas temperature can be obtained from the measured ratio of the 1f-normalized WMS-2f signal near the line center of two selected transitions [3]. Species concentration can then be determined from either of the 1f-normalized WMS-2f signals. Additional details of the sensor development can be found in [4].

3 Sensor design and architecture

The H_2O transitions in the 1.3-1.5 μm region are systematically analyzed to select the optimum line pair for the target temperature range of 1000-2000 K and pressure range of 1-2 atm [4]. Two H_2O transitions near 1392 and 1398 nm are selected for the WMS-2f sensor. The spectroscopic parameters for the two H_2O transitions have been measured

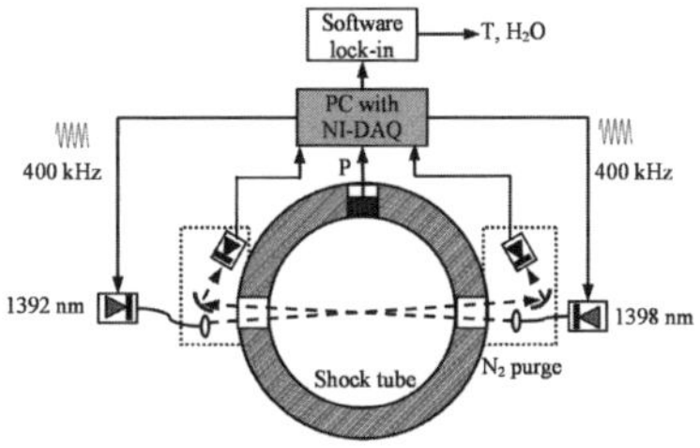

Fig. 1. Schematic of the WMS-2f sensor for temperature and H_2O concentration in shock tube

in a heated static cell [5]. Figure 1 provides a schematic of the WMS-2f sensor. The two diode lasers are sinusoidally modulated by 400 kHz digital waveforms generated by a PC running a 10 MHz National Instruments data acquisition (NI-DAQ) system. The modulation depths are adjusted to maximize the WMS-2f signal. The light from each laser is collimated, transmitted through the shock tube (opposite directions), and focused onto an InGaAs detector. The optics and detectors are enclosed in plastic bags purged by dry N_2 to avoid absorption interference from ambient water vapor. The detector signals are recorded at 10 MHz and demodulated by a digital lock-in program on LabVIEW with a low-pass filter bandwidth of 100 kHz. A Kistler transducer is used to record the pressure time-history at the same location during shock tests.

4 Sensor validations

4.1 Sensor validation in heated cell

The WMS-2f sensor is first validated in a heated static cell. The cell is first evacuated and the background 1f and 2f signals are taken for each laser. The cell is then filled with H_2O-Ar mixture to P=1 atm, and the 1f and 2f signals with absorption are recorded. A detailed description of the experimental setup and procedure can be found in [4].

The top panel of Fig. 2 compares the thermocouple measurements with the temperatures from the sensor measurements. They are in good agreement (standard deviation=1.9%) over the tested temperature range. The bottom panel of Fig. 2 shows the ratio of the H_2O mole fraction measured by the sensor ($X_{Measured}$) and the mole fraction independently measured by direct absorption (X_{Actual}). The standard deviation between the measured and actual H_2O mole fraction is 1.4%. The excellent agreement between measured and actual values confirms the accuracy of the WMS-2f sensor. The errors in Fig. 2 primarily come from the uncertainties in the measured spectroscopic data, thermocouple measurements, and the direct absorption measurements.

4.2 Sensor validation in H_2O/Ar shocks

Shock tests with H_2O-Ar mixtures are conducted to validate the sensor at combustion temperatures. Experiments are performed behind reflected shock waves in a shock tube with an inner diameter of 15.24 cm. The reflected shock temperature is calculated from the measured shock velocities and 1-D shock wave theory, assuming vibrational equilibrium and frozen chemistry. TDL measurements are made at 2 cm from the endwall.

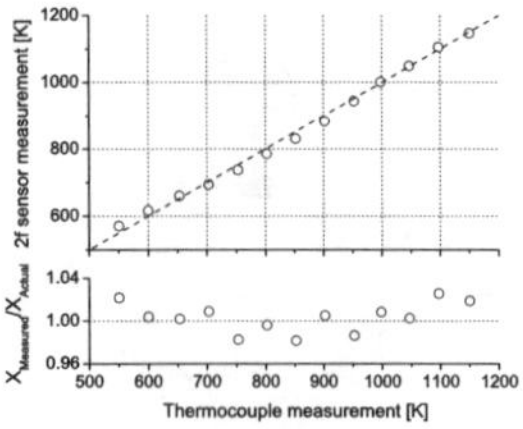

Fig. 2. Validation measurements of the sensor in a static cell. P=1atm, 1.0%H_2O in Ar, L=76cm

Figure 3 shows the measured time-history of temperature and H_2O concentration during a shock in an H_2O-Ar mixture. The sensor has fast response (<10 μs) and a good SNR for temperature measurements. The average measured temperature over the time interval 0.1-1 ms (where T and P are expected to be virtually constant) is 1226±14 K and is in excellent agreement (within 1.2%) with the calculated T_5=1211 K. The average H_2O concentration measured by the sensor is 0.702% (±0.014%) and is in good agreement (within 1.7%) with the direct absorption measurement (0.690%) before the shock. The scatter on the H_2O concentration comes from the noise in individual laser signal as well as in the temperature and pressure measurements.

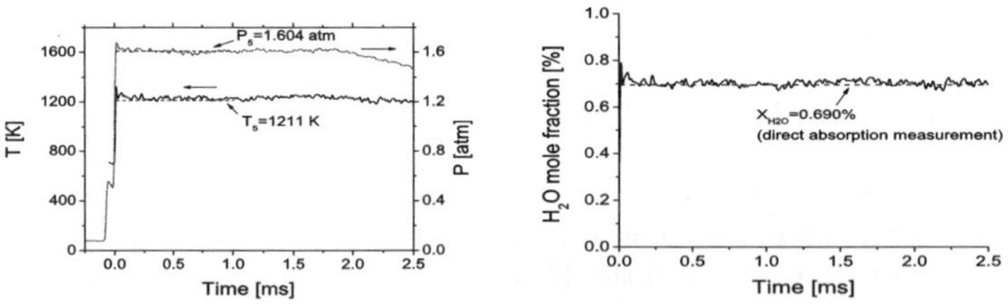

Fig. 3. Measured T and water concentration during a shock with H_2O-Ar mixture. Initial conditions: P_1=59.3 Torr, T_1=295 K; incident shock conditions (calculated): P_2=0.461 atm, T_2=696 K; reflected shock conditions (calculated): P_5=1.604 atm, T_5=1211 K

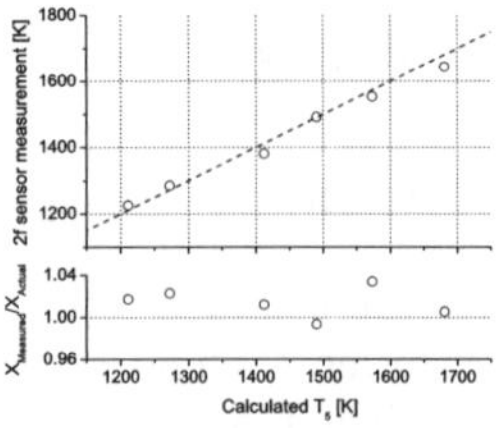

Fig. 4. Validation measurements of the WMS-2f sensor in the shock tube with H_2O-Ar mixtures

Similar tests are performed at different temperatures. The top panel of Fig. 4 provides a comparison of the temperature measured by the sensor (averaged over the 0.1-1 ms period) with the calculated T_5. They are in good agreement (within 1.5%) over the tested temperature range. The bottom panel shows the ratio of the H_2O mole fraction measured by the sensor ($X_{Measured}$) and the one measured by direct absorption (X_{Actual}). They agree within 1.4% over the tested temperature range.These results validate the sensor accuracy for temperature and H_2O concentration measurements at combustion temperatures.

5 Sensor demonstrations

5.1 $H_2/O_2/Ar$ shock

The WMS-2f sensor has been applied in a preliminary study of combustion in $H_2/O_2/Ar$ mixtures in a shock tube. Figure 5 shows the measured temperature and H_2O concentration during a shock with 1.0% H_2/0.625% O_2/Ar as the initial mixture. Here, the sensor bandwidth has been reduced to 25 kHz to improve the SNR. The measurement results are compared with simulations using: 1) the hydrogen mechanism by Conaire et al. [6], and 2) the modified GRI-Mech 3.0 [7] with the measured reaction rate for $H+O_2+M \rightarrow HO_2+M$ from [2]. The simulation is based on a recently developed kinetic model combining CHEMKIN and isentropic expansion (or compression) to the measured pressure in infinitesimal time steps [8]. The simulation results have been shifted by 35 μs to match the measured H_2O rise. This early rise is thought to be due to impurities in the shock tube. The measured temperature is in good agreement (within 1%) with the simulations using both mechanisms. Simulated temperatures using these two mechanisms are within 1%, while simulated H_2O concentrations differ by less than 9%. Simulation with the modified GRI mechanism provides better agreement with the measured H_2O profile. Further work is ongoing to reduce the noise and uncertainty in these measurements.

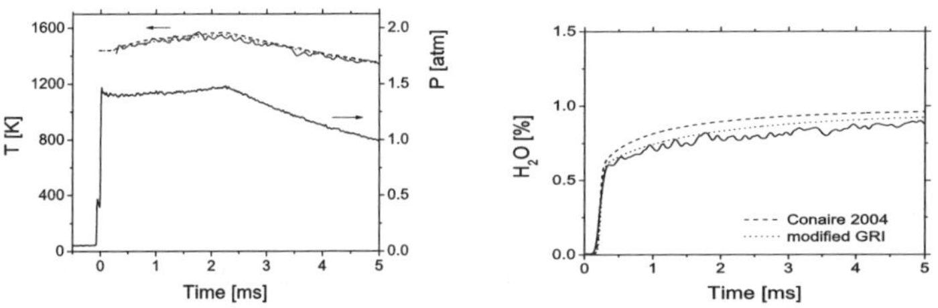

Fig. 5. Measured T and H_2O during a shock with initial mixture: 1.0%H_2/0.625%O_2/Ar (ϕ=0.8). P_1=39.0 Torr, T_1=294 K; P_2=0.370 atm, T_2=793 K; P_5=1.401 atm, T_5=1440 K

5.2 Heptane/O_2/Ar shock

Figure 6 shows the measured temperature and H_2O concentration during a shock with 0.2%heptane/1.85%O_2/Ar as the initial mixture. The measurement results are compared

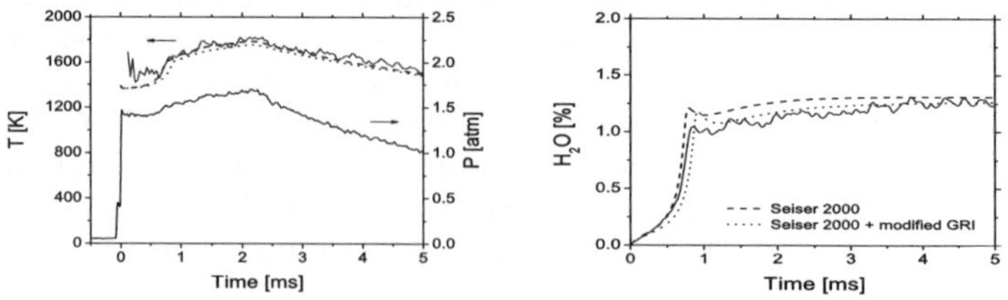

Fig. 6. Measured T and H_2O during a shock with initial mixture: 0.2%heptane/1.85%O_2/Ar (ϕ=1.2). P_1=39.4 Torr, T_1=294 K; P_2=0.372 atm, T_2=776 K; P_5=1.415 atm, T_5=1385 K

with simulations using: 1) the reduced heptane mechanism from Seiser et al. [9], and 2) the reduced heptane mechanism from Seiser et al. combined with the hydrogen mechanism from the modified GRI mechanism. The measured temperature is in good agreement (within 2%) with the simulations using both mechanisms, while the measured H_2O concentration agrees better with simulation using the hybrid mechanism (Seiser 2000+modified GRI) for t>1.2 ms. This is consistent with the observation in the H_2/O_2/Ar shock described above. Further work is in progress to refine these data and simulations.

6 Conclusions

A TDL absorption sensor based on a WMS-2f technique is developed for rapid measurements of temperature and H_2O concentration in a shock tube. The fast response of the sensor is achieved by fixing the laser wavelengths near the line centers of two H_2O rovibrational transitions near 1392 and 1398 nm. The sensor is first validated in a heated static cell and then in shock wave tests with H_2O-Ar mixtures. The overall accuracy is better than 1.9% for temperature and 1.4% for H_2O concentration measurements over the range of tested conditions. The sensor is finally applied in feasibility studies of combustion in H_2/O_2/Ar and heptane/O_2/Ar mixtures in a shock tube. The measured gas temperature and H_2O concentration are compared with simulations using different mechanisms, illustrating the potential of this rapid sensor for testing combustion mechanisms.

References

1. Glassman I.: *Combustion*, 3rd edition (Academic Press, San Diego, CA 1996)
2. Bates R.W., Golden D.M., Hanson R.K., Bowman C.T.: *Phys. Chem. Chem. Phys.*, 3:2337-2342, 2001.
3. Li H., Rieker G.B., Liu X., Jeffries J.B., Hanson R.K.: *Appl. Opt.*, 45(2):1052-1061, 2006.
4. Li H., Farooq A., Jeffries J.B., Hanson R.K.: In preparation.
5. Li H., Farooq A., Jeffries J.B., Hanson R.K.: submitted to *J. Quant. Spectrosc. Radiat. Transfer.*
6. Conaire M.O., Curran H.J., Simmie J.M., Pitz W.J., Westbrook C.K.: *Int. J. Chem. Kinet.*, 36:603-622, 2004.
7. GRI-Mech 3.0 http://www.me.berkeley.edu/gri-mech/, 1999.
8. Li H., Owens Z., Davidson D.F., Hanson R.K.: In preparation.
9. Seiser H., Pitsch H., Seshadri K., Pitz W. J., Curran H. J.: *Proc. Combust. Inst.* 28:2029-2037, 2000.

A novel fast-response heat-flux sensor for measuring transition to turbulence in the boundary layer behind a moving shock wave

T. Roediger[1], H. Knauss[1], J. Srulijes[2], F. Seiler[2], and E. Kraemer[1]

[1] *Institute of Aerodynamics and Gas Dynamics (IAG), Universität Stuttgart, Pfaffenwaldring 21, 70569 Stuttgart(Germany), email: roediger@iag.uni-stuttgart.de*
[2] *French-German Research Institute of Saint-Louis (ISL), 5 rue du Général Cassagnou, 68301 Saint Louis Cedex (France)*

Summary. To study the local heat transfer rate to the shock tube wall and to observe the transition mechanisms from laminar to turbulent in the shock-induced flow, an investigation is performed by use of a new fast response heat flux gauge. The initial shock tube conditions establish a unit Reynolds number range of more than one order of magnitude, reaching from $0.5 \times 10^6 < Re_{unit}/m < 11 \times 10^6$. The new sensor possesses a high spatial and temporal resolution with a time constant down to 1 μs. The temporal resolution allows the detection of the boundary layer transition not only by a rise of the mean heat flux density but also by the detection of the increase of the heat flux fluctuation level.

1 Introduction

In a shock fixed coordinate system, the boundary layer (BL) which grows from the foot of the shock can be regarded as quasi-steady. It differs from the usual stationary BL because here the wall moves with the velocity of the shock wave. The shear stress and heat transfer vary with distance from the "leading edge", here the shock wave, in the same manner as in the stationary case.
Mirels [1] found a solution for the laminar BL equations behind a shock advancing into a stationary fluid. The solution shows that the heat flux at a fixed location x into the wall decreases with $1/\sqrt{t}$, where t is the time after the passing of the shock wave. After BL transition from laminar to turbulent, the unsteady heat flux into the wall increases significantly. Hartunian et al. [2] presents correlation data of transition Reynolds numbers for a wide range.
A new heat flux gauge allows the experimental study of the laminar BL development and of the transition scenario in a new quality and extended unit Reynolds number range with high temporal and spatial resolution.

2 Facility and Instrumentation

The experiments have been carried out in the ISL shock tube ST70A. The driver section is 2 m long and separated by a MYLAR diaphragm from the driven section. For Run 1&2 the driven section is 7 m long and for Run 3, 6.5 m in length. The tube has an inner diameter of 70 mm with honed surface. Connecting elements in between the steel tube segments allow the installation of 6 sensors at a defined cross-section of the tube as well as some pressure transducers along the driven tube to determine the shock velocity. Comparative simultaneous measurements between commercial thin-film gauges produced by SWL [3] and the heat flux gauge called Atomic layer Thermopile (ALTP) have been carried out at a position 6 m downstream of the MYLAR diaphragm. The sensitivity of the ALTP and the specification of the thin-film gauge are listed in Table 1. A detailed

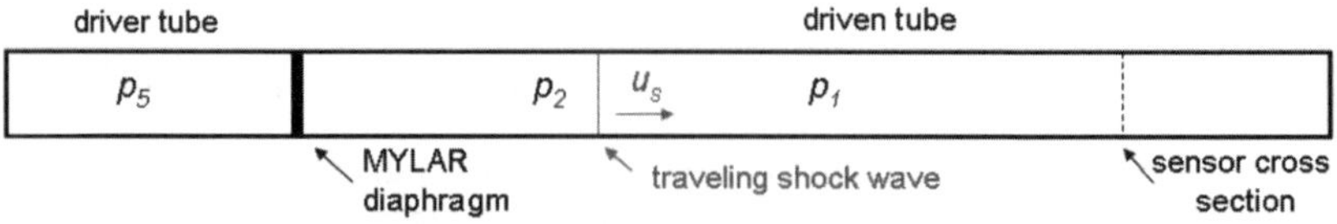

Fig. 1. Schematic of the shock tube ST70A at ISL

Sensor (Serial number)	active area $[mm^2]$	Sensitivity s (GAIN=1)	$\sqrt{\rho c k}$ of substrate $[J/(cm^2 K\sqrt{s})]$	Resistance $[\Omega]$
ALTP (798)	0.4×2	$182\,\mu V/(W/cm^2)$	0.5246	125
Thin Film Gauge (P-657)	0.3×0.9	$3852\,\mu V/(W s^{1/2}/cm^2)$	0.3223	28.67

Table 1. Specification of installed ALTP and thin film gauges

description of the working principle, structure and calibration procedure of the ALTP sensor is found in [4]. The working principle of this sensor is based on the Transverse Seebeck effect. ALTP sensors are isothermal active thin film gauges with a film thickness below 1 μm and a minimum active surface area of approx. 0.4×2 mm^2. Because of its small size and fast response, high spatial and temporal resolution down to 1 μs is attained. Furthermore, the ALTP has a linear characteristic over more than 10 orders of magnitude, ranging from mW/m^2 to MW/m^2.

Low-noise amplifiers (nominal Gain 100) were used for the amplification of the ALTP sensor signals. The signals have been captured with a sampling rate between $6 \div 40$ MHz and 14 bit resolution depending on the experimental condition of the run.

3 Definitions

The common definition of the shock-tube flow Reynolds number, based on laboratory reference coordinates, is

$$Re = \frac{\rho_2 u_2 x}{\mu_2} = \frac{\rho_2 u_2^2 t}{\mu_2} \tag{1}$$

where μ_2 is obtained from Sutherland's law and the characteristic length x is the distance that a particle travels in the free stream relative to the wall in the time t. The definition of the transition Reynolds number differs from the Reynolds number defined above. The relevant characteristic length x_p which describes the distance traveled by a shock induced particle since it was initially set in motion by the shock and reaches the transition point [2], is used as characteristic distance, here indicated by subscript t:

$$Re_t = \frac{\rho_2 u_2 x_p}{\mu_2} = \frac{\rho_2 u_2^2}{\mu_2} \frac{u_s t}{(u_s - u_2)} \tag{2}$$

The Stanton number is defined by

$$St = \frac{q}{\rho_2 u_2 c_{p,2}(T_r - T_w)} \tag{3}$$

where q is the measured heat flux density, T_r the recovery temperature and c_p the specific heat at constant pressure.

4 Experimental Results

The characteristics of the instationary BL and the laminar-turbulent transition behind a moving shock are studied and discussed for a unit Reynolds number range between 0.5 and 11.3×10^6 /m. Three sample runs within the Reynolds number regime mentioned are investigated in detail. Their experimental conditions are displayed in Table 2. The

RUN	$M_s[-]$	u_s [m/s]	$M_2[-]$	u_2 [m/s]	T_2 [K]	$\Gamma[-]$	$Re_{unit} \times 10^{-6}$ [1/m]
1	1.78	618	0.82	352	443	2.32	0.5
2	2.23	776	1.08	517	547	2.99	0.82
3	3.28	1156	1.43	880	881	4.25	11.3

Table 2. Experimental conditions of sample runs

determined shock velocity and Mach number are indicated by subscript s and the conditions behind the shock outside of the BL are indicated by subscript 2. The density ratio across the shock is given by $\Gamma = \rho_2/\rho_1 = u_s/(u_s - u_2)$.
Figures 2 to 4 show the signal response of the ALTP gauge to the shock passage for 3 different unit Reynolds numbers.

The measuring time of the 3 runs varies between 1.5 ÷ 7 ms due the different shock velocities and arrival of the reflected shock wave. The laminar and turbulent region can be clearly identified by a significant rise of measured heat flux density and an increase of the fluctuation level at the same time. The laminar-turbulent transition of Run 3 occurs within the first 100 μs after the shock passage. Fig. 5(left) displays an expansion of this time interval revealing the transition process. A thin film gauge mounted at the same cross section of the tube was used for comparison and its time signal is also plotted in Fig. 4. The time signal of the thin film-gauge does not resolve the transition region due to its larger time constant. The readings in the turbulent region, however, show a very good agreement. The scattering of the measured mean heat fluxes lies within 15%.
The heat flux density postulated by the power law for laminar BL development behind a moving shock wave is plotted for comparison in Fig. 2, 3 and 5(left). The time traces of Run 1 and 2 deviate from the expected decay. The discrepancies are possibly caused by disturbances created by casing effects and conduction errors or non-uniformities in the shock front and the preceding flow due to a slight asymmetric burst of the diaphragm. For Run 3, a laminar BL state can be defined starting immediately behind the shock. It exists only for a very short time interval t_l= 11.4 μs which corresponds to an extent of $x_l = u_2 t_l$ = 10 mm. The existence of a laminar boundary has never been detected for such a high unit Reynolds number due to the confined time response of the gauges. In addition, the time traces of all runs show an increase of the fluctuation level superimposed on the

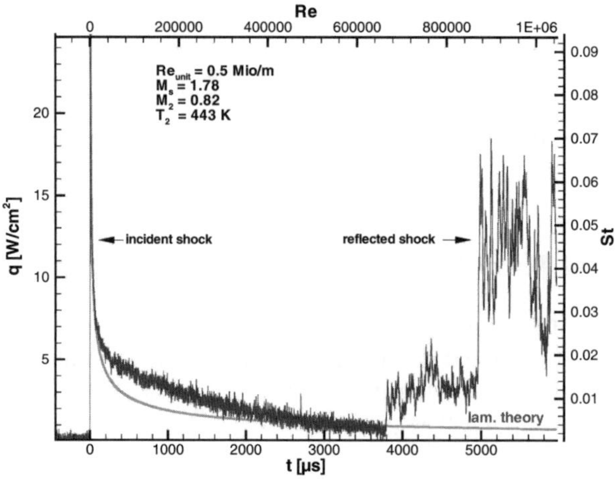

Fig. 2. ALTP time History of Run 1

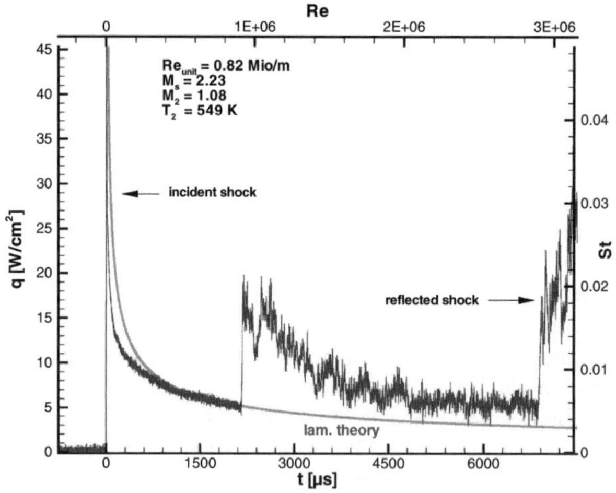

Fig. 3. ALTP time History of Run 2

mean value already within the progression of the laminar BL. The heat flux fluctuations seem to rise to a certain level in the laminar region before it reaches the turbulent state after transition.

The transition from laminar to turbulent in the time traces of Run 1 and 2 occurs fairly abrupt in a nearly step-like rise of the mean value. A transition Reynolds number of $Re_t \approx 1.55 \times 10^6$ for Run 1 and $Re_t \approx 2.7 \times 10^6$ for Run 2 can be defined. The transition scenario clearly differs from the typical slower rise of a steady flat plate BL. The heat flux rise seems to produce a weak compression and the formation of a shock wave that could destabilize the instationary BL and accelerate its transition to turbulence. The explanation for the scenario is however speculative and not clearly understood. In addition, the scenario differs from the observed time history of Run 3. Fig. 5(left) shows that the heat flux does not increase immediately from the point where the trace deviates

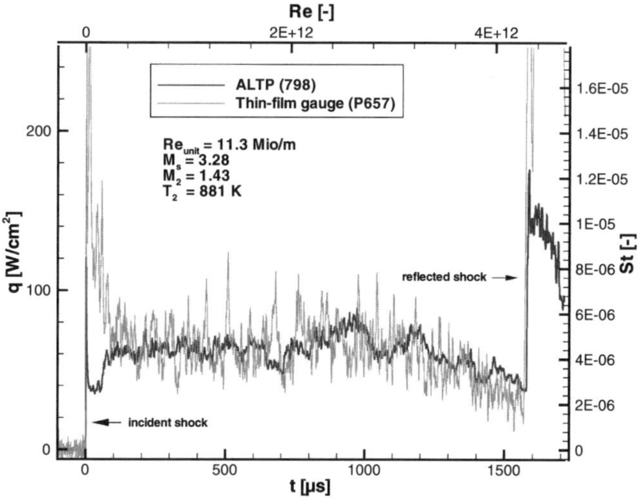

Fig. 4. ALTP and thin film time history of Run 3

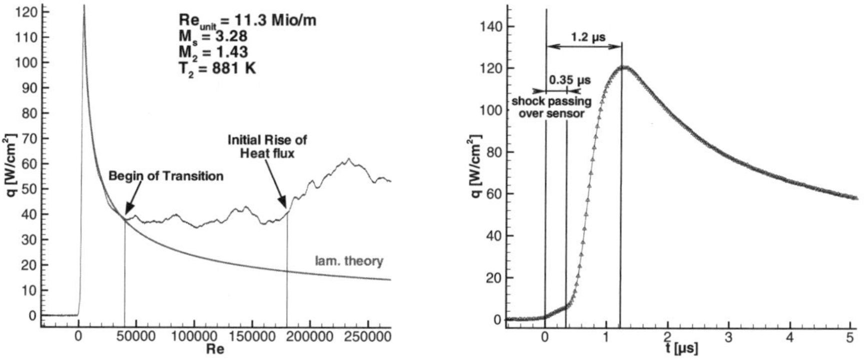

Fig. 5. Details of ALTP signal (Run 3): (left) laminar BL and transition interval; (right) response to incident shock wave.

from laminar theory. The heat flux signal remains constant for another 50 μs before it starts to rise. A transition interval can be defined between this onset and end point of transition corresponding to a transition Reynolds number interval of $5.3 \times 10^5 < Re_t < 2.6 \times 10^6$. A similar time history was observed by Davies and Bernstein [5] on a semi-infinite flat plate. These authors noted that the change in heat flux during BL transition depends on the relative effectiveness of two opposing tendencies. The local thickening of the BL reduces the wall heat flux while the enhanced mixing tends to increase it. It was stated that there might exist a region in which heat flux density remains steady before it finally rises. This statement seems to be in good agreement with the time history of Run 3 and a plausible explanation of the observed transition scenario. A final comparison of the observed transition Reynolds number intervals of all three runs with correlation data presented by Hartunian et al. (Figure 2 in [2]) show a good match for the present shock

strengths of Run 1÷3 equivalent to $T_w/T_2 = 0.34$, 0.53 and 0.66, respectively.
Fig.5(right) shows a magnification of the ALTP response to the incident shock wave. The scale is plotted in microseconds and the data points are marked for 40 MHz sampling rate (no filtering has been applied!). The initial slope of the signal shows the shock passing over the sensor ($t < 0.35$ μs) before it significantly increases and the heat flux reaches a maximum at $t = 1.2$ μs a peak values of 120 W/cm^2. The measured time of the shock passing corresponds exactly to the time, that can be calculated from the shock speed u_s and the width of the sensor normal to the shock front (0.4 mm).
A passing shock wave can also be used for the investigation of the dynamic response of wall mounted gauges. Thus, a time constant of the sensor can be estimated from the time history. Different definitions for this quantity can be found in literature. An approximation of the time history by the error function leads to a time constant of $\tau \approx 0.22\,\mu$s. Sometimes the time interval between the shock arrival and the maximum heat flux is simply given as time constant, which results presently in $\tau \approx 1\,\mu$s. The ALTP time constants are in the range as predicted from theory and dynamic laser calibration [4].

5 Conclusions

Highly time-resolved heat transfer measurements in the instationary BL behind an incident shock wave were carried out by a novel ALTP heat flux gauge. The existence of a laminar BL was found for the first time in a unit Reynolds number range up to 11×10^6. The time histories even resolve the progression of the fluctuation level superimposed on the mean heat flux of the laminar BL. The observed transition scenarios seem to differ within investigated unit Reynolds number regime. Finally, the temporal and spatial resolution of the ALTP resolve the passing shock wave in an time interval of 0.35 μs. The dynamic response confirms the ALTP time constant $\tau \approx 1\,\mu$s.

Acknowledgement. This research was supported by the German Research Foundation (DFG) within the project KN 490/1-2. The assistance of B. Sauerwein/ISL is gratefully acknowledged.

References

1. Mirels, H.: Laminar Boundary Layer behind Shock Advancing into Stationary Fluid. NACA TN 3401, 1955.
2. Hartunian, R.A.; Russo, A.L.; Marrone, P.V.: Boundary-Layer Transition and Heat Transfer in Shock Tubes. Jour. Aeron. Sci., Vol. 27, pp. 587-594 , 1960.
3. Olivier, H.: Thin Film Gauges and Coaxial Thermocouples for Measuring Transient Temperatures. Documentation, Shock Wave Laboratory Aachen, Germany, 2003.
4. Knauss, H., Roediger, T., Gaisbauer, U., Kraemer, E. (IAG, Germany), Buntin D., Smorodsky B., Maslov A. (ITAM, Russia), Srulijes, J., Seiler, F. (ISL, France): A Novel Sensor for Fast Heat Flux Measurements, AIAA Paper 2006-3637, 25th AIAA Aerodynamic Measurement Technology and Ground Testing Conference, San Francisco, 2006.
5. Davies, W.R; Bernstein, L.: Heat Transfer and Transition to Turbulence in the Shock-Induced Boundary Layer on a Semi-infinite flat plate. Journal of Fluid Mechanics, Vol. 36, No. 1, pp. 87-112, 1969.

Application of HEG static pressure probe in HIEST

T. Hashimoto[1], S. Rowan[1], T. Komuro[1], K. Sato[1], K. Itoh[1], M. Robinson[2], J. Martinez Schramm[2], and K. Hannemann[2]

[1] *Japan Aerospace Exploration Agency, Kakuda Space Center, 1 Koganesawa, Kimigaya, Kakuda-shi, Miyagi-ken, Japan 981-1525*

[2] *German Aerospace Center (DLR), Institute of Aerodynamics and Flow Technology, Goettingen, Germany 37073*

Summary. A survey of the test flow in the HIEST free-piston shock tunnel was conducted by obtaining Pitot and static pressure rake measurements. Experiments were performed at stagnation ethalpies of 4, 8, 12, 16 and 20 MJ/kg and stagnation pressures of 20 and 40 MPa. The free-stream Mach number was approximately 7 for all test conditions. A core flow region of approximately 600 mm diameter was observed. Good repeatability of the Pitot and static pressure was obtained at all test conditions.

1 Introduction

This study was performed as one of the cooperative activities between the High Enthalpy Shock Tunnel (HIEST) section of the Japan Aerospace Exploration Agency (JAXA) and the Spacecraft section of the German Aerospace Center (DLR) using the large free-piston driven shock tunnels, HIEST [1] and HEG [2]. Both institutions are carrying out aerothermodynamics research for future space-transportation systems using these high enthalpy facilities. The cooperative activities are based on a mutual interest in exchanging experience and performing common research on the operation, calibration, test technique development and computational fluid dynamics (CFD) support for these unique short-duration ground-based test facilities. Successful cooperative activities during recent years have included mutual visits, the exchange of research personnel and test techniques and the publication of cooperative articles.

Static pressure measurements are very important to the operation of reflected shock tunnels for the calibration of the free-stream flow and the determination of the arrival of the driver gas in the test section. A static pressure probe is used in HEG as part of its permanent probe setup. This paper describes the calibration and initial testing of a static pressure probe rake system in HIEST.

2 Experiment

Calibration tests of a static pressure probe were carried out in the HIEST shock tunnel. Pitot pressure measurements were also obtained. Figure 1 shows the rake probe configuration used in the experiments. The rake was equipped with 16 Kulite XCL-100 pressure transducers to measure the Pitot pressure and 4 Kulite XCQ-093 pressure transducers to measure the static pressure. The Pitot pressure was measured between -400 to 400 mm at 50 mm intervals and the static pressure was measured between -150 to 150 mm at 100 mm intervals.

A detailed schematic of the static probe used in the experiments is shown in Figure 2. The probe is of the same design as those used in the HEG shock tunnel. The sensor

(ϕ2.5 mm) is mounted inside the probe (ϕ5 mm) which produces a fast response time due to the close proximity of the sensor to the flow sampling location.

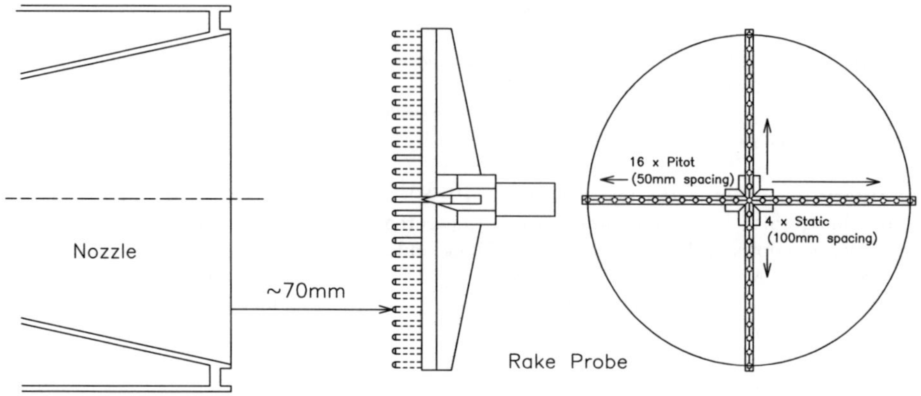

Fig. 1. Rake probe configuration used for the flow survey experiments in HIEST. The rake probe was mounted with 16 Pitot pressure probes and 4 static pressure probes

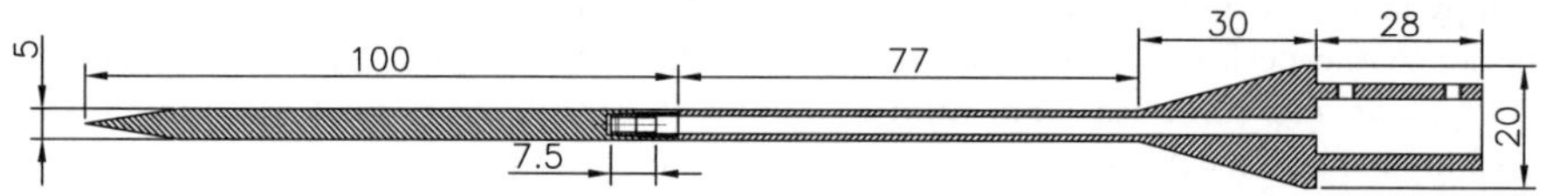

Fig. 2. Schematic of the static pressure probe design

Experiments were performed at stagnation enthalpies of 4, 8, 12, 16 and 20 MJ/kg and stagnation pressures of 20 and 40 MPa. The free-stream Mach number for all test conditions was approximately 7. The flow properties for each test condition calculated using the ESTC [3] and NENZF [4] numerical codes are shown in Table 1. Two shots were performed at each test condition using a test gas of air.

3 Results

Pitot and static pressure traces from an example calibration test are shown in Figure 3. Data is shown for the -50 mm and 50 mm probe locations. The static pressure predicted by the NENZF code is also included for reference.

The test was conducted at a stagnation enthalpy of 4 MJ/kg and a stagnation pressure of 16 MPa. At this test condition, there is approximately 3 ms of steady test time. The close agreement in the rise time of the static and pitot pressure traces indicates that

Table 1. Average flow properties for each test condition produced by ESTC and NENZF numerical codes

Condition	Stag. Pressure (MPa)	Stag. Enthalpy (MJ/kg)	Pressure (kPa)	Temperature (K)	Density (kg/m^3)	Velocity (m/s)	Mach Number
P20H4	16.0	4.3	1.4	339	0.015	2740	7.4
P20H8	22.0	7.5	2.8	745	0.013	3576	6.6
P20H12	19.1	12.0	2.8	1231	0.008	4330	6.2
P20H16	18.8	14.9	2.8	1427	0.006	4729	6.2
P20H20	17.4	19.5	2.6	1679	0.005	5310	6.2
P40H4	31.3	3.8	2.8	298	0.033	2609	7.5
P40H8	37.0	6.6	4.3	627	0.024	3385	6.8
P40H12	30.6	11.0	4.6	1213	0.013	4211	6.1
P40H16	33.0	16.1	5.3	1694	0.010	4930	6.0

the response of both gauges is very good. The static pressure trace also illustrates the difference in the flow establishment process for each gauge. In addition, there is good agreement between the two Pitot pressure traces and two static pressure traces obtained at different locations.

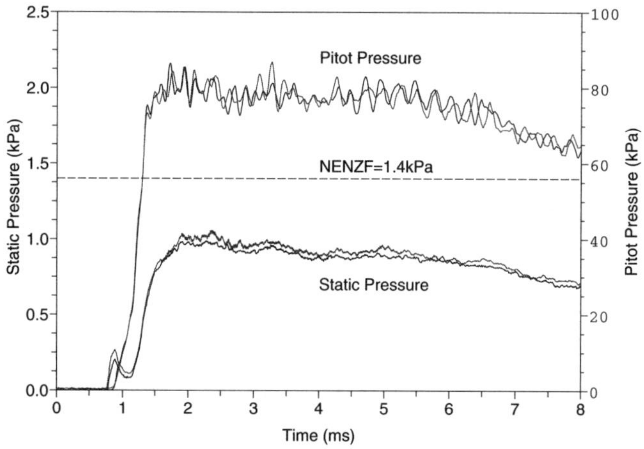

Fig. 3. Pitot and static pressure histories for stagnation enthalpy of 4 MJ/kg and stagnation pressure of 16 MPa at -50 mm and 50 mm locations

3.1 Pitot Pressure

Pitot pressure measurements obtained with the rake probe at a stagnation enthalpy of 8 MJ/kg and stagnation pressures of 20, 40 and 60 MPa are shown in Figure 4. Results are shown for two tests conducted at each test condition. Time histories for the Pitot pressure probe located at 50 mm are shown on the left. The Pitot pressure time history at each probe location was averaged over the test time to obtain the Pitot pressure distributions shown on the right. At a stagnation enthalpy of 8 MJ/kg, there is approximately 2 ms of steady test time available.

The Pitot pressure distributions indicate that a core flow approximately 600 mm in diameter is produced by the HIEST Mach 7 nozzle. In addition, the close agreement in the data obtained for each test conditon confirms the good repeatability of the test flow conditions produced by the HIEST shock tunnel.

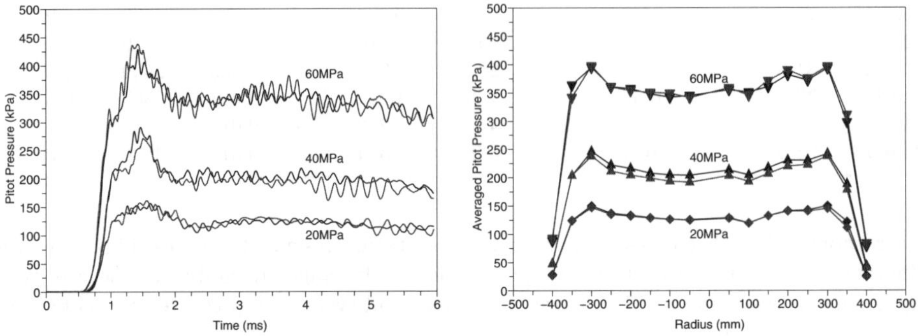

Fig. 4. Pitot pressure time histories (left) and average Pitot pressure distributions (right) for stagnation pressures of 20, 40 and 60 MPa at a stagnation enthalpy of 8 MJ/kg

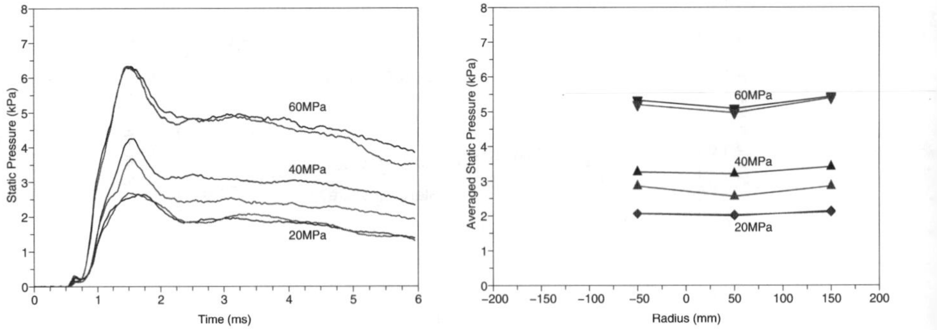

Fig. 5. Static pressure time histories (left) and average static pressure (right) for stagnation pressures of 20, 40 and 60 MPa at a stagnation enthalpy of 8 MJ/kg

3.2 Static Pressure

Static pressure measurements obtained at a stagnation enthalpy of 8 MJ/kg and stagnation pressures of 20, 40 and 60 MPa are shown in Figure 5. As for the Pitot pressure, results are shown for two tests conducted at each test condition. Time histories for the static pressure probe located at 50 mm are shown on the left and average static pressure measurements for all probe locations are shown on the right. The results indicate that the static pressure can be measured with good repeatability at the 20 and 60 MPa test conditions. At the 40 MPa test condition, the static pressure level for the repeat test is lower than expected. It is likely that this is related to the operation of the static pressure probe or the test facility and is under further investigation.

4 Future Work

In the near future, CFD of the nozzle flows in HIEST and HEG will be performed and compared with the experimental data. The investigation of the influence of the choice of test gas on the static pressure determination will be given special emphasis. Tests will be performed with air and nitrogen test gases as both test gases behave differently with respect to, for example, the relaxation process of excited internal degrees of freedom. Due to the sensitivity of the static pressure to the modelling of these relaxation processes in CFD tools, these investigations will provide a valuable method to validate the applied models and improve the determination of the free-stream properties in HIEST and HEG.

In addition, during the current experimental program in HIEST, it was noted that the static pressure probe did not perform well at high enthalpy test conditions. An example of the probe output at a stagnation enthalpy of 16 MJ/kg is shown in Figure 6. The data indicates that the static pressure decreases rapidly during the test time in comparison to the Pitot pressure. It is thought that this phenomena may be related to the probe design and it will be the subject of further experimental investigation.

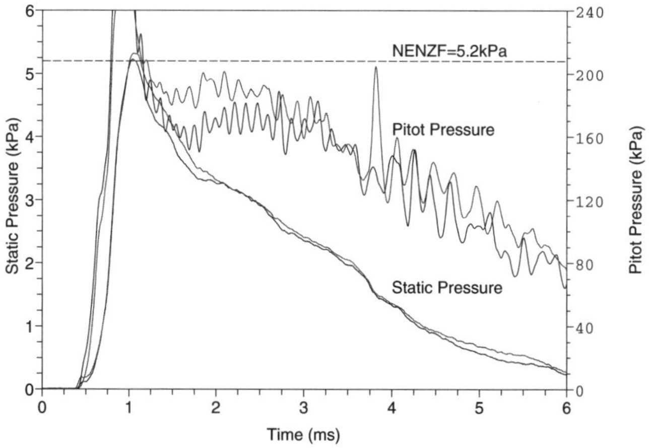

Fig. 6. Pitot and static pressure histories for a total enthalpy of 16 MJ/kg and a total pressure of 40 MPa at -50mm and 50mm (Shot 1373)

5 Conclusions

A static pressure survey of the test flow was conducted for the first time in the HIEST shock tunnel using a static probe design provided by DLR. Pitot pressure rake measurements were also obtained. The results indicated that the HIEST shock tunnel produces a core flow of approximately 600 mm in diameter. Good repeatability of the Pitot pressure was observed at all test conditions. The repeatability of the static pressure was good at all test conditions except the 40 MPa condition.

References

1. Itoh K, Ueda S, Komuro T, Sato K, Takahashi M, Miyajima H, Koga S: Design and Construction of HIEST (High Enthalpy Shock Tunnel). Proc. International Conference on Fluid Engineering, I:353-358 (1997)
2. Hannemann K: High Enthalpy Flows in the HEG Shock Tunnel: Experiment and Numerical Rebuilding. AIAA Paper 2003-0978 (2003)
3. McIntosh MK: Computer program for the numerical calculation of frozen and equilibrium conditions in shock tunnels. Technical Note CPD 169, Department of Supply, Australian Defence Scientific Service, Weapons Research Establishment (1970)
4. Lordi JA, Mates RE, Moselle JR: Computer program for the numerical simulation of nonequilibrium expansions of reacting gas mixtures. NASA Contractor Report CR-472 (1966)

Application of laser-induced thermal acoustics to temperature measurement of the air behind shock waves

T. Mizukaki

Dept. of Aeronautics and Astronautics, Tokai University
1117, Kitakaname, Hiratsuka, Kanagawa, 259-1292 Japan

Summary. Laser-induced thermal acoustics (LITA) were applied to measure the temperature profiles induced behind spherical shock waves generated by high-voltage discharge with energy of 6 J in air. A pulse Nd:YAG laser (wavelength: 532 nm, energy: 300 mJ/pulse, pulse duration (FWHM): 10 ns, line width: 0.005 cm^{-1}) and an CW argon-ion laser (wavelength: 488 nm, power: 4 W) were used for pump and probe laser for LITA measurement, respectively. The peak temperatures were well agreed with the results calculated with the Euler equation. The temperature profiles behind shock waves, however, differed in decay rates. Also the peak temperatures behind shock were determined by the reflected overpressure, and agreed with those by the LITA measurement to maximum error of 5 %.

1 Introduction

The temperature changes induced by propagating shock waves are unsteady phenomena with the duration of several hundreds microsecond. Then, the temperature measurement technique having the time resolution with less than one micro second is required to obtain precise history of the temperature changes. Thermocouples are not suitable to measure the temperature changes because of their response time being around one millisecond if we have the fastest one.

The high pressure generated by underwater shock wave focusing has been successfully applied to noninvasive disintegration of human calculi, which is called extracorporeal shock wave lithotripsy (ESWL) [1]. The basic research has successfully been extended to clinical applications. It is found, however, that, despite clinical success, tissue damage, although not very serious, actually occurs during the ESWL treatment [2]. To clarify the mechanism of the tissue damage by heat induced by underwater shock wave focusing, the temperature profile near focusing point should be investigated by experimental method.

The laser-induced thermal acoustics (LITA) [3] havs been applied for an advanced velocimetry for wind tunnel [4] [5]. LITA allows us to determine the speed of sound in medium by noninvasive detection method with laser beam. The advantages of LITA measurement are, noninvasive measurement, measurement duration of sub-microseconds, test volume with the diameter of sub-millimeters, and needless of tracer particles.

The purpose of the present work is to develope the thermometry for shock-wave-related phenomena by LITA. In the present paper, the results of temperature measurement behind spherical shock waves propagating in air are described.

2 Experimental setup

The experimental setup used here is shown in Fig. 1. A flash Q-switched Nd:YAG laser with a wavelength of 532 nm, maximum output energy of 300 mJ, pulse width of 10 ns, and beam diameter of approximately 10 mm was used for the pump laser. The laser had an injection seeder that narrows the line width of the beam to less than $0.005\,\mathrm{cm}^{-1}$. The pump beams were divided into two paths at a half mirror, and were made parallel by plane mirrors inside optical components **Optics**. The beam reached at the focusing lens L with a focal length of 800 mm and a diameter of 100 mm. At focusing point, the half-crossing angle of the pump beams θ was 0.426 degree. After focusing point, the beams become parallel again so that one is captured at the beam-dumper **BD**, while the other went into the photo detector **PD**, and acted as a trigger. A CW argon-ion laser with a wavelength of 488 nm, maximum output power of 2.5 W, and with beam diameter of approximately 2 mm, was used for probe beam. The probe beam was set in the phase-match condition at the focusing point. A half-crossing angle of the probe beams ψ was 0.391 degree. The beam was captured at the **BD**. A narrowband interference filter with FWHM of 10 nm for 488 nm and a spatial filter reduced noise light in the signal beam. The photomultiplier **PMT** (Hamamatsu H6780) detected signal beam through optical fiber. The beam intensity histories were recorded with a high-speed digital oscilloscope **DPO** (20 GSa, 2.5 GHz, Tektronix TDS7254). Activation signal was generated by a switch box **SW**. Trigger generator **TG** generated main trigger signal right after detecting both activation signal and Q-switch-synchronize signal. Delay generator **DG** added time-delay to main trigger signal to adjust the moment of shock wave generation to laser beam irradiation. Shock generator **SH** produced semi-spherical shock waves by high-voltage discharges eith 6 J. Figure 2 shows the shock generator and the semi-spherical shock waves, at 325 μs after discharge, taken by dissection color schlieren method. Overpressure caused by the generated shock waves, were obtained with a PZT-pressure transducer (PCB Model HM103) **PG** through a signal conditioner **SC** to estimate temperatures.

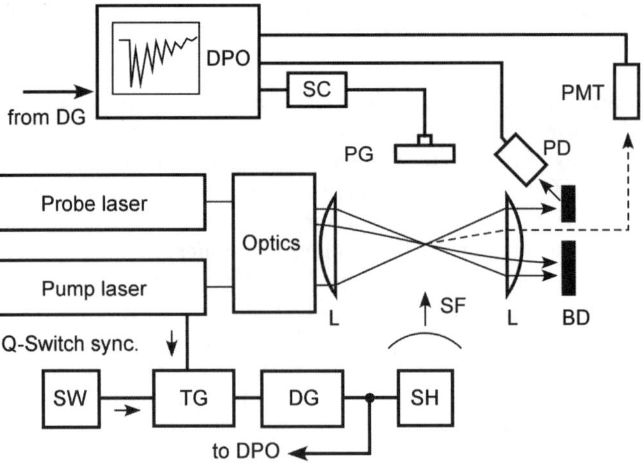

Fig. 1. Experimental setup for LITA measurement behind shock waves.

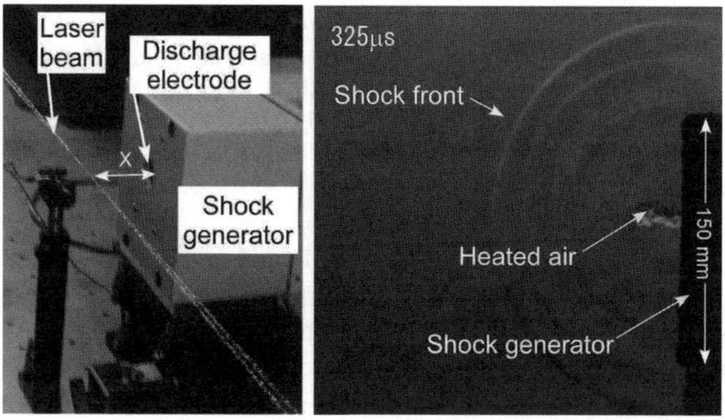

Fig. 2. Overview of shock generator and propagating shock waves.

3 Results and Discussion

A typical history of signal intensity I plotted against the past time after detecting LITA signal is shown in Fig. 3. Temperature measurements were made at the distance from 30 to 60 mm with 10 mm interval. Due to lens effect, the light path of signal beam was refracted by small angle at the moment when shock front past at the measured point. The signal beam intensity detected was decreased due to the displacement of signal beam from center of the detector. Increasing the distance x decreased the intensity changes. With decreasing curvature of the shock front due to distance, the refraction angle of signal beam decreased. The intensity changes caused by the lens effect disappeared within one microsecond. The lens effect was not observed after passing of shock front. In the present paper, we assume the arrival time of shock front at the measured point is at the moment when the peak of the intensity change by lens effect occurs.

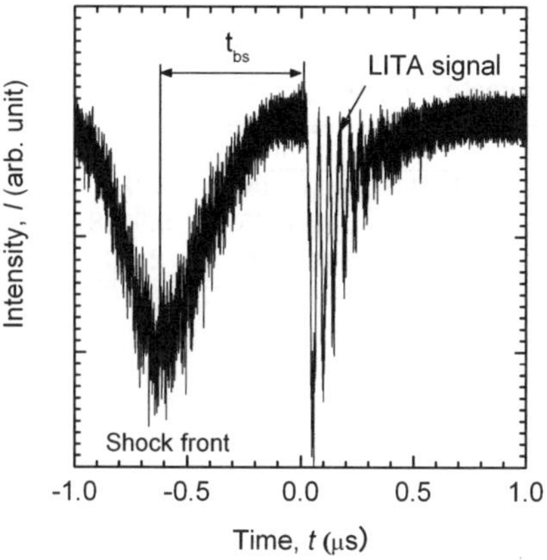

Fig. 3. Typical LITA signal after shock front passing.

Figure 4 shows the dependence of temperature ratio T/T_0 obtained from LITA signal on the past time after shock front passing the measured point. Here, T indicates the absolute temperature determined by LITA signal, T_0 does initial temperature. In Fig. 4, the open circles refer to experimental results, while the solid line does to the fitted curves for temperature decay. The temperature decay was fitted to the exponential function shown as equation (1).

$$T/T_0 = 1 + A \cdot \exp(-t/B) \tag{1}$$

Due to the lens effect, we could not obtain the temperature right behind shock front. Then, the maximum temperature ratios T^0/T_0 induced at shock front were determined by substituting $t = 0$ into the fitted functions for temperature decay. Table 3 summarizes the determined parameters A, B, and the calculated values T^0/T_0. The broken lines represent the results of the numerical calculation by the one-dimensional Euler equation. The shock energy source was the high pressure air with the internal energy of equivalent to discharge energy. We assumed no energy loss was occurred in shock wave generation. Numerical analysis calculated spherical shock waves by energy release in open air, while the discharges in experiments generated semi-spherical shock waves due to the structure of the shock generator here. The internal energy of high pressure air for numerical analysis was set to twice as discharge energy, 12 J. The initial conditions for the numerical analysis are listed in Table 3.

We found that the maximum temperature ratios T^0/T_0 were in rough agreement with numerical results while the decay rates of the temperature behind shock front were inconsistent.

The peak overpressure allows to estimate the temperature ratio behind shock front by simple theory [6]. Figure 5 (left) shows measured overpressure history along the axis at each point where LITA signals were measured. We employed the head-on method providing reflected overpressure here.

Figure 5 (right) shows the temperature ratio with the reflected overpressure. The open circles, the solid line, and the solid circles indicate the values obtained by LITA signal, the values by simple theory, and the differences between them, respectively. We found that the differences between both value were within 5%.

Table 1. Parameters for temperature decay behind shock front

X (mm)	30	40	50	60
T^0/T_0	1.284	1.145	1.091	1.092
A	0.2836	0.1445	0.9088	0.09169
B	1.865	2.211	2.223	2.313

4 Conclusions

To develop the non-intrusive thermometry for significant unsteady phenomena such as shock waves, we applied laser-induced thermal acoustics measurement technique and tried to obtain the temperature distributions behind spherical shock waves generated by high-voltage discharges. The results are as follows:

Table 2. Initial conditions of numerical analysis

Fluid equation:		Euler equation
Analyzed distance:		0 to 120 mm
Mesh period:		0.1 mm
Atmosphere	Equation od state:	Ideal gas
	Reference density:	1.255×10^{-3}kg/cm^3
	γ:	1.400
	Reference temp:	288.2 K
High-pressure air (Initial condition)	Total energy:	12 J
	Area:	0 to 1 mm
	Density:	1.225 kg/m^3
	Specific energy:	2.3386×10^{-9} J/kg

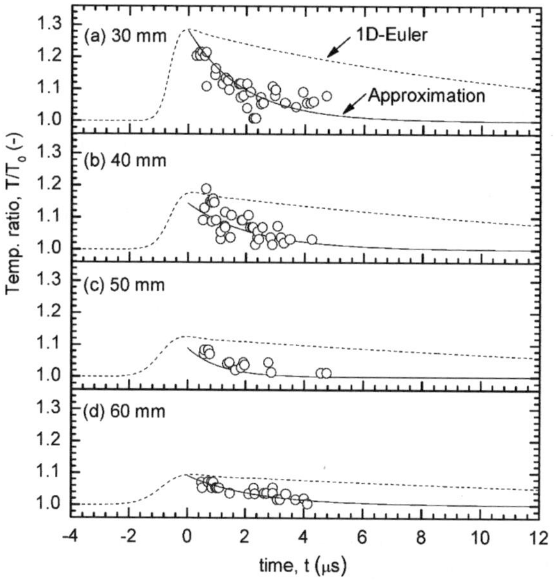

Fig. 4. The temperature profiles behind shock front.

- The temperature distributions behind shock front were successfully obtained at the every 10 mm distance ranging from 30 mm to 60 mm from shock generator.
- The reflection of the signal beam caused by shock front passing at test volume gives limited affection to LITA signal acquisition.
- The peak temperature ratios determined by the LITA signals obtained agreed with the numerical results calculated by the one-dimensional Euler equation.
- The decay constants of the temperature estimated by LITA signals behind shock front were significantly grater than those by the one-dimensional Euler equation.

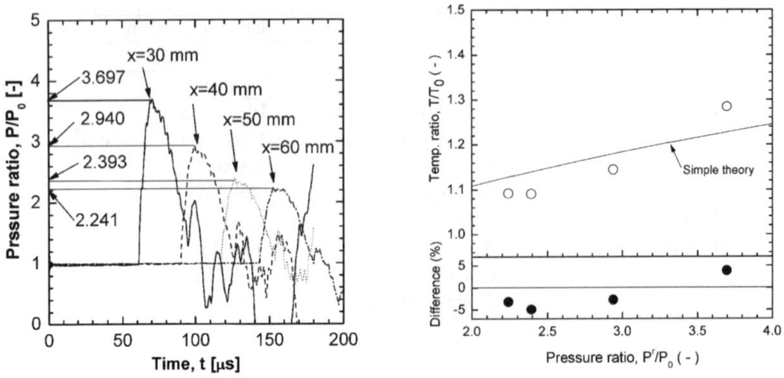

Fig. 5. Overpressure histories (left), comparison of peak temperatures estimated by LITA with by overpressure (right)

– The estimated temperature behind shock front, by reflected overpressure measurements, indicated 5% differences, at maximum, with those by LITA signal measurements.

In the present paper, we examined relatively weak shock waves generated by high-voltage discharges in air while the medical applications of shock waves are usually carried out underwater and with relatively strong shock waves. We will investigate the accuracy of temperature measurement by LITA for strong shock waves and underwater shock waves.

References

1. Takayama K.: Application of underwater shock wave focusing to the development of extracorporeal shock wave lithotripsy. Jpn. J. Appl. Phys., Vol. 32, pp. 2192-2198, 1993.
2. Takayama K., Obara T., Saito K., Kameshima N.: Fourcusing of underwater shock waves and the mechanism of high pressure generation. Transactions of the Japan Society of Mechanical Engineers, Series B, Vol. 56, No. 526, pp. 1579-1582, 1990.
3. Cummings E. B.: Laser-induced thermal acoustics: simple accurate gas measurements. Optics Letter, Vol. 19, No. 17, pp.1361-1363, 1994.
4. Hart R. C., Balla R. J., Herring G. C.: Simultaneous velocimetry and thermometry of air using nonresonant heterodyned laser-induced thermal acoustics, ICASE Report No. 2000-22, 2000.
5. Hart R. C., Balla R. J., Herring G. C., Jenkins L. N.: Seedless laser velocimetry using heterodyne laser-induced thermal acoustics, ICASE Report No. 2001-19, 2001.
6. Kinnery G. F., Graham K. J., *Exprosive Shocks in Air*, Chap. 5, Springer, 1962.

Assessment of rotational and vibrational temperatures behind strong shock waves derived from CARS method

A. Matsuda[1], M. Ota[2], K. Arimura[2], S. Bater[2], K. Maeno[2], and T. Abe[1]

[1] *Japan Aerospace Exploration Agency, The Institute of Space and Astronautical Science 3-1-1, Yoshinodai, Sagamihara 229-8510 Kanagawa (Japan)*
[2] *Chiba University, Faculty of Engineering 1-33, Yayoicho, Inageku, Chiba 263-8520 Chiba (Japan)*

Summary. For the assessment of the representative CARS result, the emission spectroscopy was conducted in order to obtain the rotational and vibrational temperature distribution behind the shock wave. Based on the comparison of temperatures by these method, it was found that the temperature determination method from the CARS spectrum could not be sufficiently accurate. This is caused by the simplified calculation of the theoretical CARS spectrum which is used for the temperature determination. Therefore, for the more accurate temperature determination, the pressure effect should be involved for the theoretical CARS spectrum calculation.

1 Introduction

Recently, the research field of the hypersonic nonequilibrium flow has been reactivated. This is caused by the missions such as HAYABUSA, STARDUST, and so on, in which missions the reentry capsules are planed to enter into the earth atmosphere from the inter-planet orbit. The reentry velocity is predicted to be more than 12 km/s. Up to now, in order to comprehend the reentry flight environment, many numerical and experimental researches have been conducted [1–4]. As a result, we found that the rotational temperature was clearly nonequilibrium with the translational temperature and suggests the discrepancy with the conventional thermochemical model, as shown in Figs.1 and 2.

In order to comprehend the rotational nonequilibrium, Fujita, et.al., theoretically investigated the rotational relaxation by using QCT method [5]. Contrary to the experimental results, his result suggested that the rotational relaxation did not become slower as increasing the translational temperature. Here, it should be noted that in the QCT analysis the electron excitation energy level of N_2 in which the rotational and vibrational temperatures are defined is different from that in the experiment. In the QCT analysis, the rotational and vibrational temperatures are assumed to be for the X state of N_2. On the other hand, in the experiment, the rotational and vibrational temperatures reflect the C state of N_2 which corresponds to the upper energy state of the $N_2(2+)$ spectra. The C state of the N_2 is a highly excited state. Therefore, it becomes questionable whether the rotational and vibrational state in the C state of N_2 is completely equilibrium with that in the X state(ground state).

In order to resolve this discrepancy, the measurement of the rotational and vibrational temperature in the X state of N_2 is inevitable. For this measurement, the Coherent Anti Stokes Raman Spectroscopy (CARS) method, is expected to have an important role [6]. Recently, Toyoda, et.al., tried to measure the rotational and vibrational temperature at X state of N_2 behind shock waves with hypervelocity(around 5 km/s) by using the CARS method [7,8]. According to the result shown in Refs. [7] and [8], even in CARS method, the low rotational temperature in the region just behind the shock front was observed. However, there were scattering of the temperatures in the equilibrium region. Therefore, before discussion of the behavior of the rotational temperature, the critical assessment of the CARS result itself in Refs. [7] and [8] is inevitable. For instance, by comparing the temperature obtained from the CARS method with that from the $N_2(2+)$ spectra at the region far from the shock wave where the thermochemical state is expected to be equilibrium, it can be verified whether the temperature determination method in the CARS has adequate accuracy.

Even though there exist some experimental results of rotational and vibrational temperatures derived from $N_2(2+)$ emission in the velocity range of 5 km/s [9, 10], the experimental condition is not completely consistent to that of Refs. [7] and [8]. For the direct comparison between CARS method and $N_2(2+)$ emission, the experimental result from $N_2(2+)$ in the consistent condition of Refs. [7] and [8] is inevitable.

Therefore, the purpose of this paper is to assess the temperature obtained from the CARS method by comparing the temperature obtained from the $N_2(2+)$ spectra. For this purpose, we tried to determine the temperatures by measuring the $N_2(2+)$ emission in the equivalent experimental condition of Refs. [7] and [8].

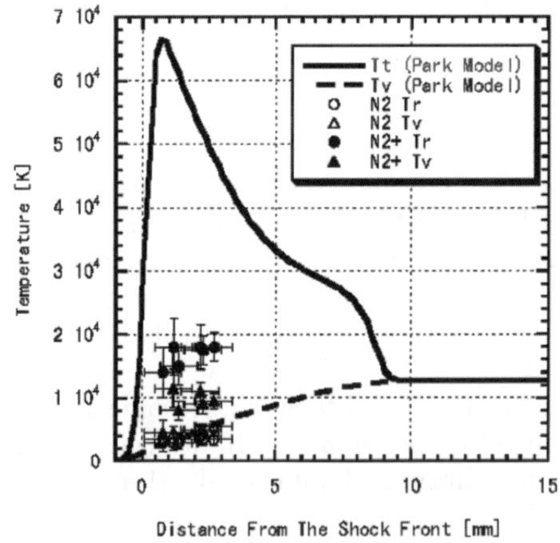

Fig. 1. Rotational and vibrational temperature distribution at 12 km/s derived from $N_2(2+)$ emission.

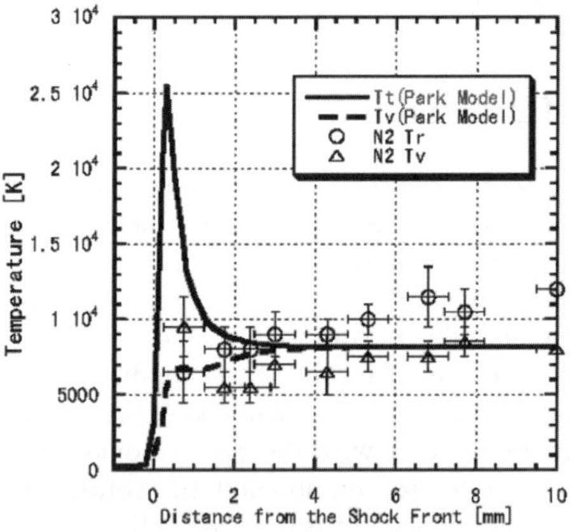

Fig. 2. Rotational and vibrational temperature distribution at 8 km/s derived from $N_2(2+)$ emission.

2 Experimental Method

The experiment was conducted by using the double diaphragm free piston shock tube at ISAS. In order to compare the CARS result obtained at Chiba University [7, 8], the ambient pressure was set to 10 torr and the corresponding shock velocity was in the range between 5 and 6 km/s.

Figure 3 shows the optical setup at the measurement section of the shock tube. In this shock tube, the flow property behind the shock wave is considered to be almost one-dimensional. As a matter of the fact, since the shock front is almost perpendicular to the wall of the tube and the investigated region behind the shock front is comparatively near the shock front, we can assume that the non-uniformity for the flow behind the shock wave, which might be caused by the thick boundary layer, is small and the radiative characteristics is almost uniform in the direction parallel to the shock front. As shown in Fig.3, a radiation emitted from a location apart from the shock front by a specific distance is focused on the entrance slit of the spectrometer (ORIEL MS127i) by means of an ellipsoidal mirror and is decomposed into a spectrum by the spectrometer. The spatial resolution of the measurement in this system is less than 1.2 mm.

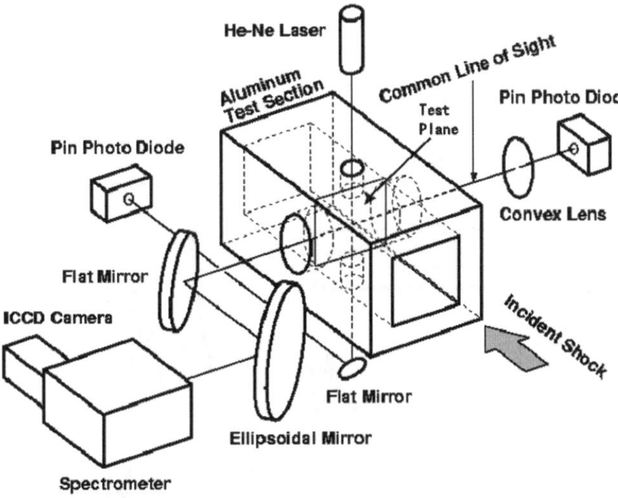

Fig. 3. Experimental setup of optical system

The spectrum image of the radiation decomposed by the spectrometer was recorded by means of an image-intensified CCD camera (ANDOR IN-STA SPECV IS510 ICCD). The ICCD camera was driven by the delayed signal generated by the pressure sensor mounted upstream of the measurement section. In each run of the shock tube, was recorded one ICCD image for the radiation spectrum emitted from a location with a selected distance apart from the shock front. Accumulating the radiation spectrum data for various locations, which was obtained by driving the ICCD camera with a various delayed time in repeated runs of the shock tube operations, an entire distribution of the radiation spectrum was obtained for the regions behind the shock front.

For this method to make sense, the reproducibility of each run of the shock tube is inevitable. To monitor the reproducibility, not only the pressure signals and the shock propagation velocity measured from them but also the distribution of the total radiation intensity distribution was monitored. For this purpose, the pin photodiode sensor was installed on the opposite side of the spectrometer. Throughout this experiment, the reproducibility was found to be fairly good.

The He-Ne laser in Fig.3 was used as a light source for the laser schlieren technique to detect the density rise of the shock front, which gives an origin of the distance of the location starting from the shock front.

3 Experimental Result and Discussion

From the emission spectrum of $N_2(2+)$, the physical quantities such as the rotational and vibrational temperatures are obtained from the spectrum fitting method. For the fitting method, the theoretical spectra were calculated by SPRADIAN [11,12]. Figure 4 shows the typical fitting result at the spatial position of 18 mm behind the shock wave. As shown, the calculated spectrum agrees with the experimental one. The rotational and vibrational temperatures derived from this fitting method are 7500 K and 5000 K, respectively.

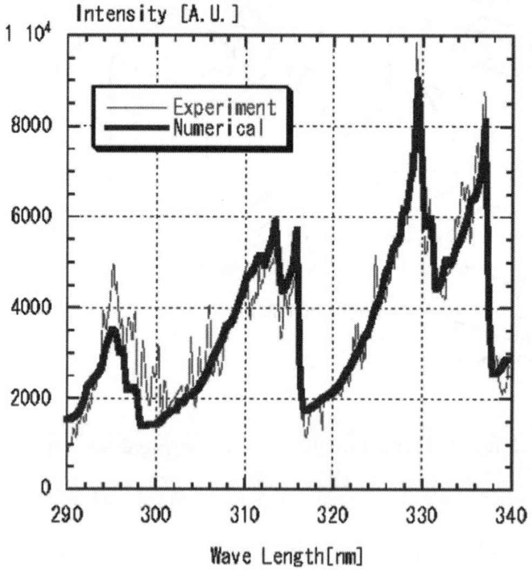

Fig. 4. Spectrum fitting for V_s=5.8 km/s, L=18 mm.

By applying this fitting method to the spectra data at several locations behind the shock front, the spatial distribution of the rotational and vibrational temperatures behind the shock front can be obtained. Figure 5 shows the temperature distribution based on the $N_2(2+)$ emission. Figure 5 also shows the numerically predicted translational and vibrational temperatures based on the Park model. According to the numerical result, the distance required to reach the equilibrium sate is as short as 1 mm. This is caused by the high initial ambient pressure of 10 torr. As for the experimental result, the temperature distributes, almost within ±2000 K from the temperature, at the equilibrium state suggested by the numerical calculation, even though there can be observed slight scattering. This experimental result shows that the thermochemical equilibrium state is attained in the region where the present experimental result was obtained.

Figure 6 shows the temperature distribution obtained from the CARS method which is cited from Refs. [7] and [8]. In the region beyond 15 mm from the shock front, the vibrational temperature distributes in the interval between 10^4 K and 1.5 $\times 10^4$ K. This temperature behavior is unreasonable. According to the recent accuracy estimation, this curious tendency is caused by the large error due to the low S/N ratio of the CARS spectra. Especially, the curiously high temperature region (far from the shock front, beyond 15mm) contains the large error of the temperature determination. Therefore, the temperature obtained from the CARS method is reasonable only in the region less than 15 mm from the shock front. In the region within 15 mm from the shock front, the temperature distributes in the wide extent(i.e., the large scattering of the temperature), though the temperature obtained from the CARS method distributes around the equilibrium temperature which is suggested from the Park model as shown in Fig.5.

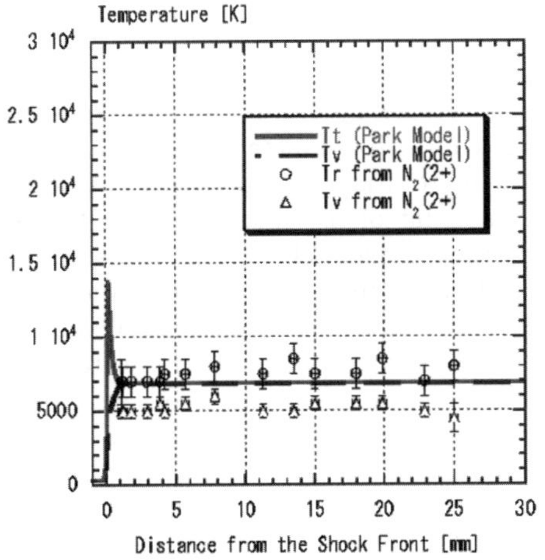

Fig. 5. Temperature distribution derived from $N_2(2+)$ emission at V_s=5.8±0.2 km/s, P_0= 10 torr.

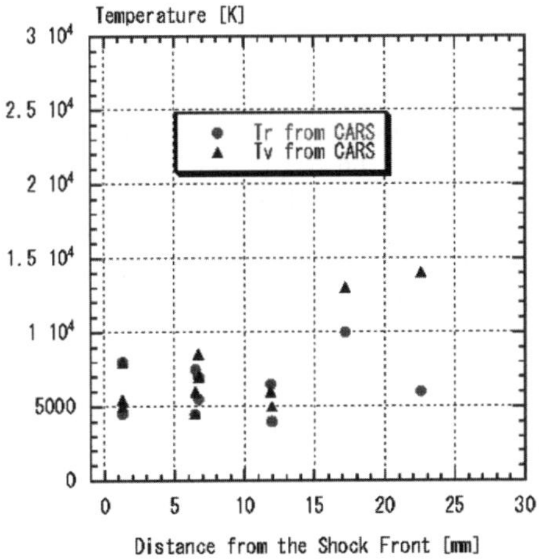

Fig. 6. Temepratures distribution derived from CARS method cited from Ref. [7].

From the comparison between Figs.5 and 6, the scattering in the temperature distribution is much larger for the CARS method than that for the $N_2(2+)$ emission spectroscopy. As for the temperature behavior, the rotational temperature is larger than the vibrational temperature in the $N_2(2+)$ emission spectroscopy as shown in Fig.5. On the contrary, the vibrational temperature is larger than the rotational temperature in the CARS method as shown in Fig.6. In other words, the temperature behavior in the CARS method is quite opposite trend compared with the emission spectroscopy. From the view point of the energy transfer mechanism by the shock wave, the temperature distribution obtained from the emission spectroscopy is quite reasonable, since in general the translational mode is firstly excited by the shock wave, and then, the energy transfer from the translational mode to the rotational, vibrational and electron excitation mode is occurred in this order. Therefore, there could be an error in the temperatures determined by the CARS method, and some improvement for the method to deduce the temperatures from the CARS spectra is required.

For the improvement of the temperature deduction from the CARS spectra, the detail model should be used for the calculation of the theoretical CARS spectra. For example, in present, for the calculation of the theoretical CARS spectra, the pressure effect is completely neglected. However, the pressure effect becomes significant in the condition that the pressure is larger than 1 atm [13]. In this experiment, since the initial ambient pressure is high (10 torr), the pressure in the region behind the shock wave with the velocity of 5 km/s is estimated in the order of 3 atm. Therefore, the pressure effect should be taken into consideration. Even though this improvement is inevitable for the further discussion of the rotational relaxation phenomena from the CARS result, unfortunately, due to the complexity of the precise calculation of the CARS spectrum, the calculation with the pressure effect and the temperature deduction based on the more precise theoretical spectra is remained for the future work.

4 Summary

For the assessment of the rotational and vibrational temperatures obtained from the CARS method, emission spectroscopy was conducted in the consistent experimental condition. From the $N_2(2+)$ emission spectroscopy, the rotational temperature was larger than the vibrational temperature. From the view point of the energy transfer by the shock wave, this trend in the temperature distribution is reasonable. On the other hand, from the CARS method, the quite opposite trend of the temperature distribution was observed. This was partly caused by the fact that the theoretical CARS spectrum was calculated based on the simple model in which the pressure effect was completely neglected. Since, in this experimental condition, the pressure effect can not be neglected. Therefore, for the improvement of the CARS method, the more precise model such as the pressure effect should be included for the future.

References

1. K., Fujita, et.al., Journal of Thermophysics and Heat Transfer, vol.16, pp77, 2002.
2. K., Fujita, et.al., Journal of Thermophysics and Heat Transfer, vol.17, pp210, 2003.
3. A., Matsuda, et.al., Journal of Thermophysics and Heat Transfer, vol.18, pp342, 2004.
4. A., Matsuda, et.al., Journal of Thermophysics and Heat Transfer, vol.19, pp294, 2005.
5. K., Fujita, et.al., AIAA Paper, AIAA 2002-3217, 2002.
6. Alan, C., Eckbreth, "Laser Diagnostics for Combustion Temperature and Species," Abacus Press, 1987.
7. K., Toyoda, H., Sato, H., Ando, K., Miyazaki, K., Takada, and K., Maeno, AIAA Paper, AIAA 2005-0832, 2005.
8. K. Miyazaki, H. Ando, K. Takada, K. Toyoda, and K. Maeno, Proceedings of the 25th International Symposium on Shock Waves, No.1262-1a, pp.877-882, Bangalore, India, 17-22 July, 2005.
9. Hayashi, K., et.al., Proceedings of 35th FDC(in Japanese), pp167-170, 2003.
10. Sharma, S.P., et.al., Journal of Thermophysics and Heat Transfer, vol.5, pp257, 1991.
11. Fujita, K., and Abe, T., ISAS Report., No.669., 1997.
12. Fujita, K., et.al., AIAA Paper, AIAA 97-2561, 1997.
13. Robert J. Hall, et.al., Optics Communications, vol.35, No.1, pp69-75, 1980.

Availability of the imploding technique as an igniter for large-scale natural-gas-engines

T. Tsuboi[1], S. Nakamura[1], K. Ishii[1], and M. Suzuki[2]

[1] *Department of Mechanical Engineering, Faculty of Engineering, Yokohama National University, 79-5 Tokiwadai, Hodogaya-ku, Yokohama 240-8501, Japan*
[2] *Faculty of Science, Toho University, 2-2-1 Miyama, Funabashi 274-8501, Japan*

Summary. For a large-scale Otto-engine we planned to produce a large-scale flame-jet using imploding detonation. Beside this, we will use the natural gas as a fuel. The detonability of methane is very weak. Therefore, we observed at first the DDT process for the mixture of methane + oxygen-enriched air in a one-dimensional detonation tube (Ex1) and secondly we measured the pressure profiles at imploding center in three-dimensional detonation chamber(Ex2). The results were that (1) one could make the DDT-distance shorter with increasing initial mixture-pressure, when the adequate Shchelkin spiral was selected, and the oxygen-enriched air was used. (2) Even in a small chamber one could produce the detonation by using the imploding chamber at high initial pressures $P>0.5$ MPa. (3) The flame-jet can be applied as an igniter in the large-scale Otto-engine.

1 Introduction

When one uses an internal combustion engine with a large diameter, he has to use a diesel engine in order to obtain the high rate of heat-release per crank-angle. When one uses the conventional Otto-engines, namely when one uses a spark igniter in a cylinder, the mixture cannot be burnt out in a short time in a large-scale combustion chamber, because the flame velocity is not high enough. One possible technique will be the combination of Otto and diesel engine-systems. Namely, the gas must be ignited in the wide place of combustion chamber. Beside this we are planning to use the natural gas as a fuel to reduce the generation of carbon-dioxide. At present one injects the heavy oil into the natural-gas-air mixture in the cylinder of diesel engine with low compression ratio. Since the ignition limit is lower than that of natural gas, one can use the heavy oil as an igniter. However, in this case there are several problems caused by the exhaust gas, such as soot, complex higher hydrocarbons, especially fuel NOx, SOx. One can not avoid the emission of these chemicals. When we consider the emission gas problem, we have to avoid the generation of such molecules.

A few groups ([1], [2]) have studied the methane-air detonation. However, the generation of self-sustained detonation in methane-air mixture in a small scale tube is usually very difficult. Therefore, in this study we tried at first to measure the DDT-process in the methane mixed with the oxygen-enriched air using various Shchelkin spirals (Ex1) in order to know whether the detonation or shock-induced ignition occurs in the methane-air mixture. In the second experiment we designed the chamber where the detonation converges three-dimensionally, and the high pressure is generated at the center. We measured the pressure profile at the converging center (Ex2) and confirmed that we can obtain the long flame-jet.

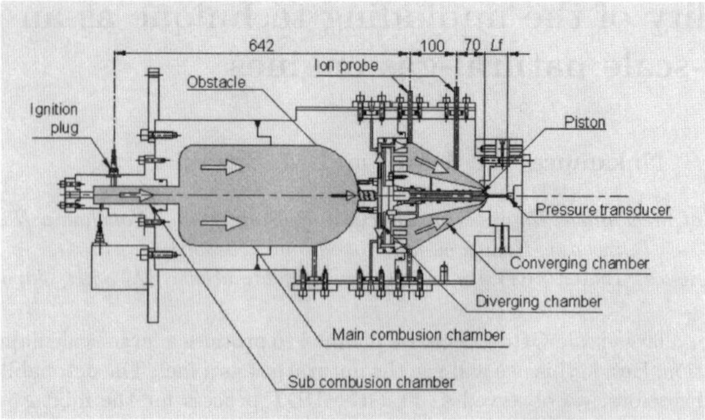

Fig. 1. Experimental apparatus for imploding detonation (Ex2):The mixture was ignited at the left side, propagated to the right side through the sub combustion chamber. The flame expanded from left side to right side in main combustion chamber and the unburned mixture was compressed in the converging chamber, together with the obstacle and diverging space. Due to this high initial pressure and obstacle, the flame grew up to the detonation. The very high pressure could be established at the center of converging room.

2 Experimental apparatus

In the one-dimensional detonation tube the DDT was observed for the mixture of methane and oxygen-enriched air. The tube had a 50 mm diameter, and was 5.8 m long with the Shchelkin spiral of l=600 and 1200 mm. The dimensions of the spiral proposed by [3] were selected. The ion probes were set at each 400 mm to measure the velocities of propagating flame in DDT process. A Kistler-type pressure transducer was set at the other end of the ignition plug.

The DDT in the initial pressures of 50, 100, and 150 MPa was observed in the mixture of methane with the normal air or the oxygen-enriched air with 26 % O_2.

Fig.1 shows the experimental setup for the imploding detonation.The mixture is ignited at the left side of the figure. Through the sub combustion chamber the flame went into the main combustion chamber. Here, the flame must propagate slowly to the right side, in order to compress the unburned mixture in the left side. At the center place the flame accelerated with an obstacle, which was carved like Shchelkin spiral, in order to make the DDT-distance short. The flame diverged behind the obstacle, and into the converging room. Here, the imploding detonation occurred. The pressure sensor was set at the converging center.

The air, which was enriched with 5 % oxygen, was used for the acceleration of DDT and the initial pressures from 0.5 to 0.8 MPa were selected. All experiments were performed for the stoichiometric mixture.

3 Results

3.1 DDT in a detonation tube

Effect of Shchelkin-spiral

Three types of Shchelkin-spiral were prepared for experients. The Dimensions are S01:l=600 mm, P=30 mm, d=38 mm, Br=0.42. S02:l=600 mm, P=45 mm, d=38 mm, Br=0.42. S03:l=1200 mm, P=45 mm, d=38 mm, Br=0.42. Here, l is the length, P is the pitch of the spiral, d is the inner diameter of the spiral, D is the outer diameter of the spiral. which was equal to the inner diameter of the detonation tube and the value Br is the blockage ratio, which is defined as 1-$(d/D)^2$. Firstly, the effect was observed by using the spiral S01 for the mixture of methane and normal-air under an initial pressure of 240 kPa. The Shchekin-spiral brought to the strong acceleration of the propagating flame. The high pressure-increase, whose pressure was measured at the opposite wall of spark plug, became higher from the beginning in the experiment. However, the DDT did not occur under this condition. As expected, the Shchelkin-spiral influenced strongly the acceleration of flame propagation.

We observed also the influence of the pitch and the length of spiral under our condition. There were no large influence on the pressure profiles and the velocities at pitches between 30 and 45 mm. However when we elongated the spiral length from 600 mm to 1200 mm, the propagation velocities became two times larger, and the pressure showed an increase of 3 times and the duration of high pressure at 1200 mm length decreased a half of the duration at 600 mm length. Fig.2 shows the results.

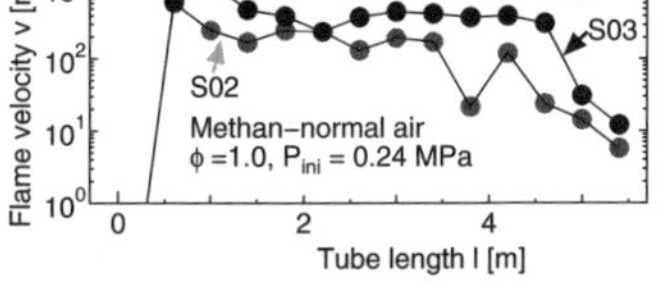

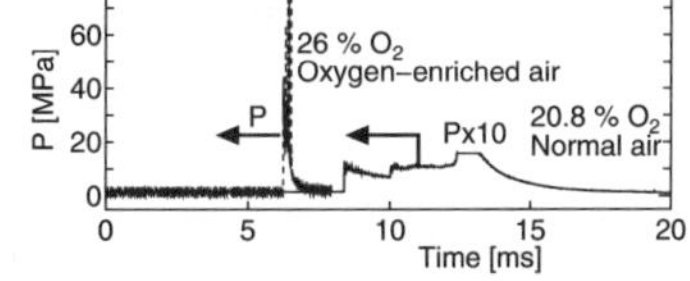

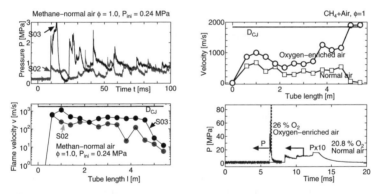

Fig. 2. (Left figure) Effect of the length of Shchelkin-spiral: The upper figure shows pressure profiles with Shchelkin spirals S02 and S03 at the tube end. The lower figure shows the relation between flame velocities and the distance from ignition plug. Conditions: Spiral S02(l=600 mm) and S03(l=1200 mm). Initial Pressure $P_{\rm ini}$=0.24 MPa, ϕ=1 of normal air.

Fig. 3. (Right figure) Velocities along the tube (upper figure) and pressure profiles at the tube end (l=1200 mm) (lower figure): Condition: Spiral S03. Initial Pressure $P_{\rm ini}$=0.15 MPa, ϕ=1 of oxygen-enriched air(26 % O_2)(open circles(upper) and solid line(lower)) and of normal air(20.8 % O_2)(open quadrangles(upper) and dashed line(lower)). The pressure in the case of the oxygen-enriched air reached to the pressure more than 50 MPa.

Influence of the initial gas condition

Influence of the oxygen-enriched air: We observed the pressure profiles at the tube end (l=1200 mm) and velocity profiles along the tube in the case of the oxygen-enriched air(26 %

O_2) and compared with those in the case of the normal air. Fig.3 shows the result: In the case of the oxygen-enriched air the detonation could be obtained even if the velocities once decreased after the acceleration of velocity due to Shchelkin spiral(l=1200 mm), i.e. the flame velocity reached the CJ-velocity near the end of the tube (upper figure) after the complex change of the velocity. The pressure profile at tube end showed that of detonation. On the contrary, in the case of the normal air no DDT process could be observed, though the reflected shock-pressure increased until more than 1 MPa from the initial pressure 0.15 MPa.

Initial pressure: We observed the velocity and the pressure for the case of the oxygen-enriched air mixture under the condition of the initial pressures 50, 100, 150 kPa. The flame velocity once decreased behind the Shchelkin spiral, and then it accelerated again after a certain period. The period became shorter in the tube with increasing initial pressure from 50 to 150 kPa. Fig.4 shows the pressure profiles at the tube end. The arrival time of the detonation became shorter, and the maximum pressure at the tube end was increased, when the initial pressure increased. However, if one observes the pressure-profiles precisely, the incident shock wave was observed in front of detonation (the pressure increase before the large pressure-increase at the initial pressure of 50 and 100 kPa in the upper figure of Fig.4). This means that the incident shock wave propagated before the detonation wave. This incident shock wave sometimes was not observed in the mixture of the initial pressure of 150 kPa. In that case the maximum pressure became small. The initial pressure of 150 kPa corresponded to the condition in our experimental system whether the incident shock wave influenced the pressure profile of the detonation. Therefore, the maximum pressure was sometimes high and sometimes low.

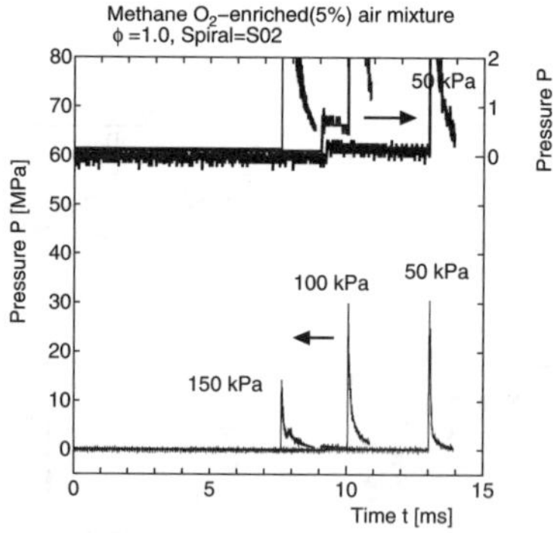

Fig. 4. Effect of initial pressure on pressure at the end wall: Condition: Spiral S02. Initial Pressure P_{ini}=0.05, 0.10, 015 MPa, ϕ=1 of oxygen-enriched air(26 % O_2). The upper pressure profiles(right axis) was an enlargement(10X) of the vertical axis of lower pressure profiles(left axis).

3.2 Imploding detonation

From the experimental results of the one-dimensional detonation tube we confirmed that it is possible to obtain the detonation-phenomena by using the Shchelkin spirals under the condition of oxygen-enriched air and high initial-mixture-pressure.

Therefore, we made the converging detonation chamber. In this chamber the pressure at converging center was measured in the mixture containing oxygen-enriched air of 26 % O_2 with the small obstacle similar to the Shchelkin spiral. The maximum pressure was observed not only at the first wave, but also at the second wave. For example, the pressure increased up to about 30 MPa due to converging shock wave and then the detonation arrived at the center. The pressure increased to 300 or sometimes to 500 MPa in the case mentioned above. The duration of the high pressure was very short. We could record only one or two high pressure-data in the computer because the sampling time was 1 μs in our recording system. Therefore the data themselves must have had a large experimental error. We changed the initial pressure from 0.5 to 0.8 MPa and we could obtain the maximum pressure of 500 MPa, as shown in Fig.5. In this Figure the first incident shock waves were also shown, as the open circles, which were called as "first wave". The length of the fire jet, which gushed out of the chamber, was photographed for the mixture of methane with 26 % oxygen in air and the maximum length was about 10 cm long

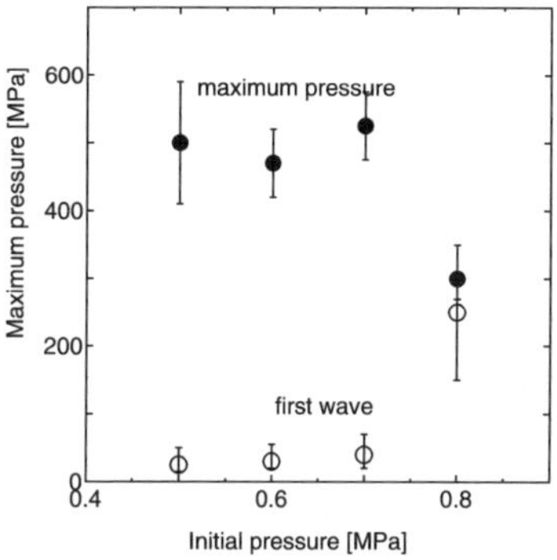

Fig. 5. The maximum pressure (solid circles) in converging detonation chamber and the first pressure(open circles) : ϕ=1, CH_4: O_2 : N_2= 13:26:74, L_f=10 mm

4 Discussion

When one converges the detonation, one can obtain the very high pressure and temperature. By using this high pressure the burned gas could be spout out into the main (i.e. large-scale) combustion chamber (the cylinder) of the internal engine. We expected that this high temperature gas can be applied as an igniter.

In the first experiment (Ex1) we could obtain the detonation (DDT) in methane-air mixture, and could confirm that the DDT-distance was made short, when (1) the appropriate Shchelkin was used, (2) the initial pressure was made increase, and (3) the oxygen-enriched air was used. Furthermore, we found that when we could obtain the unburned mixture compressed by the combustion before the flame, the DDT-distance could be made shorter and the maximum pressure of detonation became larger.

According to this results we designed the converging detonation chamber(Ex2). Fig.1 shows our chamber. When the mixture was ignited in the sub combustion chamber on the left side of the figure, the flame propagated in the main combustion chamber, and through the obstacles the flame velocity was accelerated. We planned that the detonation was to be produced until this moment, and the detonation to diverge in the diverging chamber. However, according to the measurement of the velocity in the converging chamber, the flame was still in DDT-process. The accelerated flame was converged to the center of the converging chamber. At this center, however, the very high pressure was observed, though the pressure perturbed in each experiment.

From these pressure profiles we could discuss as follows: The first shock-pressure increased at the center with increasing initial pressure. The sudden pressure increase was observed after the first wave, and the pressure showed the maximum value. Namely, the first wave must be shock-compressed air from P_1=0.5 MPa to P_2=30 MPa behind reflected wave at converging center, whose temperatures were T_1=293 K and T_2=880 K, by assuming the adiabatic compression by κ=1.35. Then, the gas was heated due to detonation. The calculated CJ-temperature was T_{CJ}=3588 K, P_{CJ}=216 MPa, D_{CJ}=1981 m/s under the initial condition of P_2=30 MPa and T_2=880 K. The measured pressure value 400 MPa must have been those of over-driven detonation. If we see the detonation from the initial condition 1, the pressure P_{CJ}=216 MPa is that of the over-driven detonation. The experimental apparatus "Ex2" was made for the first time in our laboratory. It was still not the best scale and form for obtaining the high and stable pressure. Especially the DDT-phenomena were very unstable. Sometimes the maximum pressure was 1/100 lower than the highest value, because the detonation could not be established at the center. The Figs.3.2 and 5 show only the results that the DDT was well established and the high pressure was obtained. Since the dimension of combustion chamber was not the best form for imploding phenomena, the measured maximum pressure fluctuated. We need further study on this technique. However, from the results we can say: The maximum pressures about 400 MPa could be obtained from the initial mixture pressure between 0.5 and 0.8 MPa. This pressure 400 MPa is high enough for the ignition source of Otto-engine, though these initial pressures (0.5 to 0.8 MPa) were still lower than those of an actual Otto-engine. We observed the 10 cm long flame jet of high temperature and 20 cm long jet of relatively high temperature at atmospheric pressure for the mixture of methane and oxygen-enriched air at the initial pressure of 0.5 MPa. The length of the flame jet also must be long enough as the ignition source of Otto-engine.

We confirmed from the above experiment and discussion that the maximum pressure at the converging center for the mixture of methane and oxygen-enriched air was higher than the CJ-pressure calculated from the initial mixture condition. This predicts that though the CJ-detonation is available, the converging detonation of methane is more suitable for an ignition source for large-scale Otto-engine, even if we use the natural gas as a fuel.

5 Conclusion

We observed the DDT process for mixture of methane + oxygen-enriched air in a one-dimensional detonation tube (Ex1) and the pressure profiles at imploding center in three-dimensional detonation chamber(Ex2). From these two types of experimental apparatus we found that (1) the DDT-distance was made shorter with increasing initial mixture-pressure, when the adequate type of Shchelkin spiral was selected, and the oxygen-enriched air was used, (2) Even in a small chamber one could produce the detonation by using the imploding chamber at high initial pressures P >0.5 MPa. (3) The length of flame-jet can be long enough as an igniter in the large-scale Otto-engine.

References

1. Wolanski, P., Kauffman, C.W., Sichel, M., Nicholls, J.A.: Detonation of methane-air mixtures. In:*Proc. 18th Symp. Int. Combust.* (The Combustion Institut, 1981) pp.1651-1660.
2. Kuznetsov,M., Ciccarelli, G., Dorofeev, S., Alekseev, V., Yankin, Yu, Kim, T.H.: Shock Waves **12** (2002) 215-220.
3. Lindstedt, R.P., Michels, H.J.: Combust. Flame **72** (1989) 169-181.

Experimental study of SiC-based ablation products in high-temperature plasma-jets

M. Funatsu[1] **and H. Shirai**[2]

[1] *Department of Mechanical Systems Engineering, Graduate School of Engineering, Gunma University, 1-5-1 Tenjin-cho, Kiryu, Gunma 376-8515, JAPAN*
[2] *Gunma University, 4-2 Aramaki-machi, Maebashi, Gunma 371-8510, JAPAN*

Summary. Ablation experiments of SiC-based materials have been conducted systematically in our laboratory. In the present study, physico-chemical phenomena of SiC-based ablation products in high-temperature and high-velocity micro-air plasma-jets at an atmospheric pressure are investigated experimentally. In the experiments, the SiC-based materials are inserted into the plasma-jets, the ablation products spouted out from the SiC-based materials are observed by a high-speed video camera. In addition, spectroscopic measurements are performed to investigate the radiative characteristics of the SiC-based ablation products in detail.

1 Introduction

When a spacecraft enters the Earth's or outer planet's atmosphere at a hypervelocity, an extremely strong shock wave is formed over the spacecraft. In general, the aerodynamic heating from such a flowfield to the spacecraft is caused by the transport phenomena. Usually, the heating is so severe that heat protection by an ablation method is required. For the ablation protection method, materials with low heat conductivity are used to the spacecraft surface as an ablation material. So, in the research fields, we have estimated numerically radiative and absorptive characteristics of ablation plasma generated from silicon carbide (SiC) ablation as exactly as possible [1]. The plasmas originated from SiC materials are expected to have many excellent advantages. In order to investigate the effectiveness of SiC ablation layers, the reduction of radiative heating from the plasmas to a body surface due to the layers has been investigated by using a simple fluid dynamical model for a stagnation shock layer with the effect of radiative reflection on the ablation surface. Then, the SiC heat shield has been estimated to be very effective in blocking the radiative heating for the Jovian entry conditions [1]. However, experimental or actually physical behaviors of the ablation materials in high-enthalpy flows as an arc-jet wind tunnel [2] and a micro-air plasma-jet [3] have not been clarified yet, because its ablation mechanisms include the complicated physico-chemical phenomena. So, in the previous study [4], in order to investigate experimentally the effectiveness as a heat shield of SiC-based and carbon-based materials, the basic heat-resistance experiments are performed in high-temperature and high-velocity micro-air plasma-jets.

In the present experiments, test pieces made of SiC-based materials as silicon carbide composite material (SiC/SiC) and SiC are inserted into the plasma-jets of about 10,000 K. In order to investigate the physical behaviors of the ablation materials such as spouted gases from the materials, recession processes, and interactions between the plasma-jets and the materials, ablating pieces are observed with a high-speed video camera. Moreover, radiation originated from the flows in the vicinity of the materials is measured spectroscopically.

2 Experimental setup

In order to investigate experimentally the physical behaviors of the ablation products such as spouted gases from the SiC-based materials, recession processes, and interactions between high-enthalpy flows and the materials, ablation experiments are performed in high-temperature and high-velocity micro-air plasma-jets.

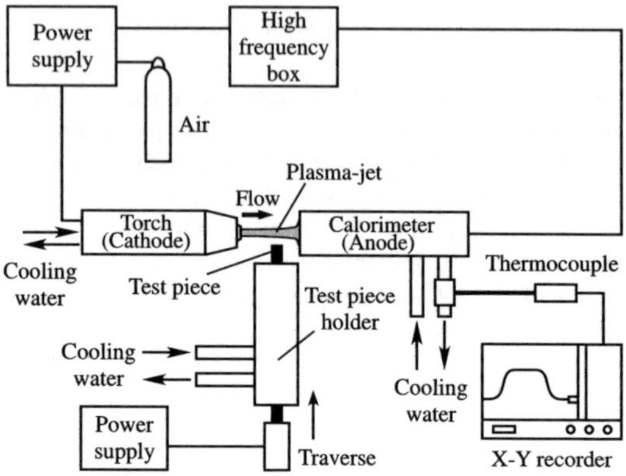

Fig. 1. Schematic of experimental setup

Figure 1 shows the schematic view of an experimental setup [4–6]. It consists of a plasma-jet generator, a gas supply assembly, a test-piece holder with the function as a calorimeter, and an automatic test-piece feed controlling device.

A plasma torch was cooled by water. Air (N_2:O_2=79:21 in % by volume) was used as a test gas, and introduced into the torch with a circulation and exhausted into a test section through a water-cooled small nozzle. The test gas, which was heated by an arc discharged between a cathode inside the nozzle and an external anode, became a micro-air plasma-jet. The distance between the nozzle and the anode was variable and was usually set 5.0 mm. The plasma-jets generated were very fine and stable.

In the ablation experiments, we used two small size nozzles of 0.4 and 0.7 mm in diameter. The test pieces were SiC-based materials as SiC and SiC/SiC, carbon-based materials as carbon and C/C. The pieces were cylindrical ones of 2 mm in diameter and 100 mm in length for SiC, and square-shaped ones of 2 mm in width and 100 mm in length for SiC/SiC, carbon, and C/C. The measured density was 3,250 kg/m^3 for SiC, 2,080 kg/m^3 for SiC/SiC, 1,830 kg/m^3 for carbon, and 1,740 kg/m^3 for C/C. In the experiments, the pieces were made to be the anode and were installed into a test-piece holder. In order to be positioned at a constant distance from the nozzle exit, the pieces were moved forward as the ablation progressed by the feed controlling device. In case of the experiments of SiC-based materials, however, the materials were perpendicularly inserted into the plasma-jets as shown in Fig. 1, because the electric conductivities of the materials were very low and the materials were unable to make to be the anode. The present experimental conditions were that discharge current and voltage for the nozzle of 0.4 mm in diameter were 10 A and 125-130 V, and those of 0.7 mm in diameter were 10 A and 150-155 V, respectively. In these conditions at an atmospheric pressure, we obtained the ablation properties of the materials, that is, weight loss, weight loss rate, and heat of ablation. The physical behaviors of the ablating pieces were simultaneously investigated by using the high-speed video camera with 10,000 frame per second (MEMRECAM fx K4, nac Image Technology Inc.).

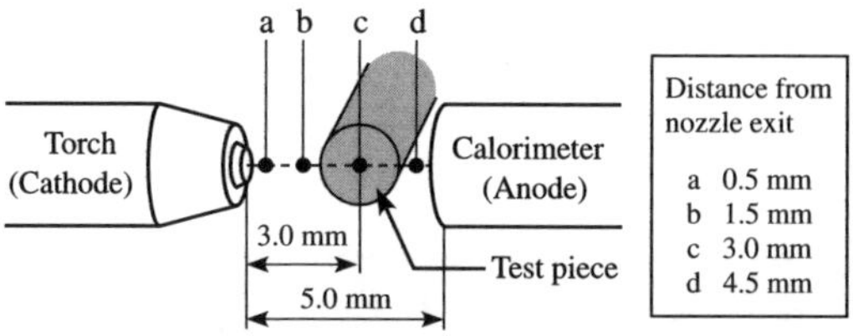

Fig. 2. Measuring points by spectrometer

Spectroscopic measurements were made by a spectrometer with a grating of 1,200 grooves/mm and a focal distance 250 mm (C5094, Hamamatsu Photonics K.K.). Figure 2 shows measuring points on a flow axis of the plasma-jets. The spectrometer was arranged on extension line of the material feed direction and located perpendicularly for the jets.

3 Results and discussion

3.1 Ablation properties

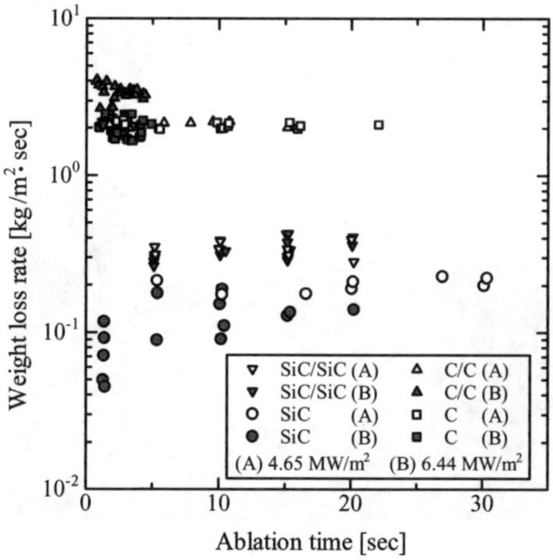

Fig. 3. Weight loss rate as a function of ablation time

Figure 3 shows the weight loss rate, which means weight loss per unit time and area, of the materials. In case of the condition (A), the heat flux q is 4.65 MW/m^2, the weight loss rates of SiC/SiC, SiC, C/C, and carbon are 0.34, 0.21, 2.11, and 2.08 $kg/m^2 \cdot$ sec, respectively. The weight loss rates of SiC-based materials are 1/6 to 1/10 of carbon-based materials. In case of the condition (B), q=6.44 MW/m^2, the weight loss rates of SiC/SiC, SiC, C/C, and carbon are 0.37, 0.14, 3.35, and 2.00 kg/m$^2\cdot$ sec, respectively. It is found that the weight loss rates of SiC-based materials are much slower than those of carbon-based materials for both conditions. Therefore, it is considered that the heat-resistance of SiC-based materials is much higher than that of carbon-based materials.

3.2 Observation of ablating materials with high-speed video camera

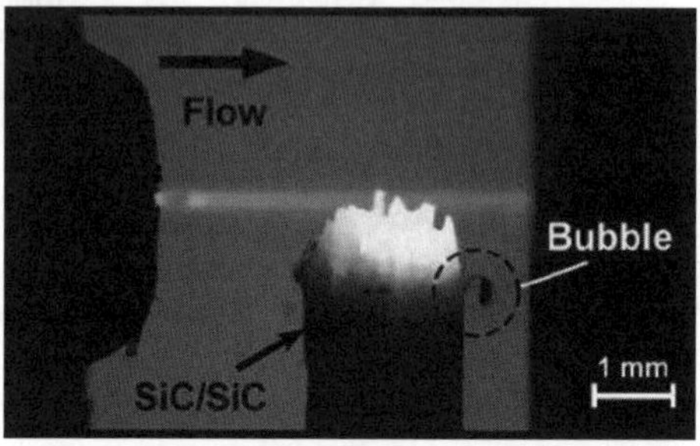

(a) SiC/SiC

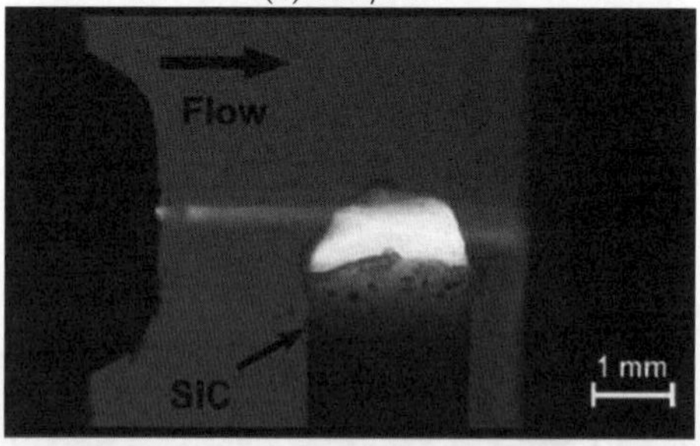

(b) SiC

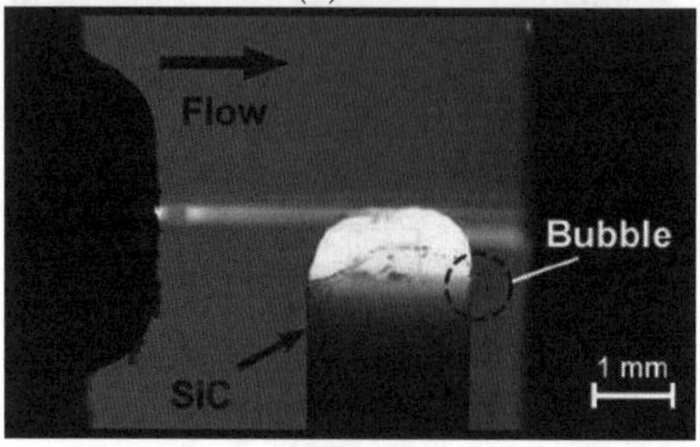

(c) SiC

Fig. 4. High-speed video camera images of ablating SiC-based materials

Figures 4 (a), (b), and (c) show shadowgraph images of the ablation experiments taken by the high-speed video camera. The video camera images of ablating behavior of SiC/SiC and SiC materials are obtained with 10,000 and 1,000 frame per second and

3 μsec and 6 μsec in shutter speed, respectively. The nozzle size is 0.4 mm in diameter. In the experiments, a shadowgraph method is used, which a test section is illuminated by a xenon lamp from an opposite side of the camera, because an extremely intense emission is originated from the ablating test piece. The test section is thus located between the camera and the lamp. In the figures, the plasma-jet flows from the nozzle at the left side to the test piece, which is installed into its holder, in the middle test section. It can be seen that the plasma-jet is very fine, a shock wave appears after the nozzle exit, and that the heating surfaces on the materials emit light intensely. In the actual moving images regarding the figures, spherically liquefied substances appear and disappear iteratively on the side-surfaces of the ablating materials. So, it is considered that these substances and lots of the dark dots on the side-surfaces of the ablating materials are like a bubble generated by thermal decomposition gases in the virgin SiC-based materials. Therefore, it can be said that the ablated SiC-based materials decompose thermally into C and Si. Then, the C sublimates, and the Si and its oxides vaporize via liquid phase on the ablated surface. The SiC-based materials, which have these mechanisms, could suppress the increase in the weight loss rate. However, we think that further investigation on microscopic measurement is needed in more detail.

3.3 Spectroscopic measurements

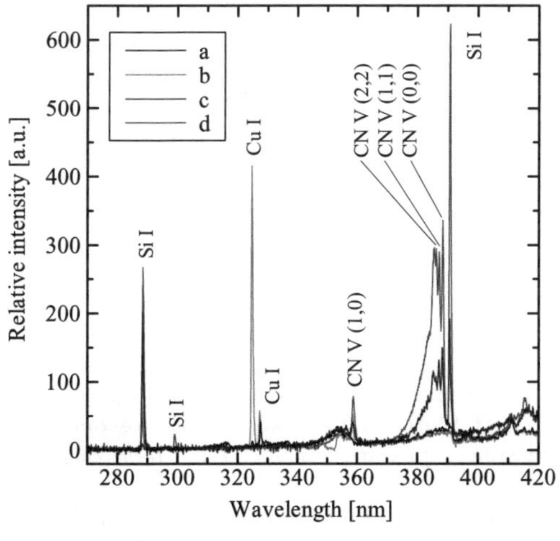

Fig. 5. Measured spectra of SiC ablation in wavelength region 270 to 420 nm

The emission spectra from the SiC ablation products are measured in the plasma-jets. Figure 5 shows the experimental spectra for all the measuring points (see Fig. 2) in the wavelength region of 270 to 420 nm. The spectra contained several molecular bands, namely intense CN Violet and weak N_2^+ First Negative (1–) bands, and atomic lines of copper, silicon, and so on. This reason for the Cu atomic lines generated is that the nozzle made of copper is partially melted and vaporized due to the heatings of the plasma-jets. The Si line nearby 290 nm is observed at the measuring points *b*, *c*, and *d*, and the Si line nearby 390 nm is done at the points *c* and *d*. The spectral intensities of Si atomic lines nearby 290 nm increase as the distance from the nozzle exit increases. From the result of detection of the Si lines at point *b*, it is considered that the Si gases might spout out from the SiC-based materials to the upstream region of the plasma-jets.

4 Conclusions

Physico-chemical phenomena of SiC-based ablation products in high-temperature micro-air plasma-jets were investigated experimentally. Main conclusions were as follows.

(1) The rates of weight loss of SiC-based materials were very slower than those of carbon-based materials. It was found that SiC-based materials might be effective ablation materials.
(2) From the high-speed video camera images of ablating SiC-based materials, the gases spouted out from side-wall of the materials.
(3) In the spectroscopic measurements of SiC ablation products, Si lines were measured in the upstream region between the nozzle exit and ablated SiC surface. Therefore, it was considered that Si gases might spout out to the upstream region from the material.

Acknowledgement. This research was supported by a Grant-in-Aid for Scientific Research, No.18760604 (Young-B), FY2006-2007, and was partially supported by a Grant for Joint Research Project as Development of Silicon-based Functional Materials, FY2005-2006, from the Ministry of Education, Culture, Sports, Science, and Technology of Japan. Furthermore, this research was supported by a Grant-in-Aid for Researchers Oversea Program, International Collaborations, FY2007, from the Japan Society for the Promotion of Science.

References

1. M. Funatsu and H. Shirai: "Reduction of radiative heating due to a SiC ablation layer," *Proc. 24th Int'l Sympo. Shock Waves* **1**, 221-226 (2004)
2. J. M. Donohue, D. G. Fletcher, and C. S. Park: "Emission spectral measurements in the plenum of an arc-jet wind tunnel," AIAA Paper 98-2946 (1998)
3. K. Kubota, M. Funatsu, H. Shirai, and K. Tabei: "Spectroscopic measurements of micro-air plasma-jets at an atmospheric pressure," *Trans. Jpn. Mech. Eng. B.*, (in Japanese), **71**-707, 1806-1812 (2005)
4. M. Funatsu, K. Ito, H. Shirai, M. Moteki, and F. Takakusagi: "Heat-resistance experiments of SiC-based ceramics in high-temperature air plasma-jets," *Proc. 25th Int'l Sympo. Space Technology and Science (Selected Papers)*, 750-755 (2006)
5. K. Ito, M. Funatsu, H. Shirai, M. Moteki, and F. Takakusagi: "Observation and spectroscopic measurement of SiC-based material ablation," *Proc. 38th Fluid Dynamics Conf.*, (in Japanese), 211-214 (2006)
6. M. Funatsu and H. Shirai: "Radiative properties of strong shock waves with SiC-based ablation products," *Abs. 5th Asian-Pacific Conf. Aerospace Technology and Science*, 63-64 (2006)

On pressure measurements in blast wave flow fields generated by milligram charges

S. Rahman[1], E. Timofeev[1], H. Kleine[2], and K. Takayama[3]

[1] *Department of Mechanical Engineering, McGill University, 817 Sherbrooke St. West, Montreal, Quebec H3A2K6, Canada*
[2] *School of Aerospace, Civil, and Mechanical Engineering, University of New South Wales/Australian Defence Force Academy, Canberra, ACT 2600, Australia*
[3] *TUBERO, Tohoku University, 2-2-1 Katahira, Aoba-ku, Sendai 980-8577, Japan*

1 Introduction

Pressure measurements are widely used in blast wave experiments as described by Cole [1] for underwater explosions and by Baker [2] for explosions in air. They are performed using various kinds of pressure transducers – small devices converting the pressure exerted on their sensitive surface into a recordable electrical signal. In so-called face-on measurements, the sensitive surface of a pressure transducer is positioned parallel to the blast wave front, so that the pressure behind the reflected wave is actually measured and the respective value for the incident blast is then deduced using the normal shock relations. In side-on measurements, the blast wave front is normal to the sensitive surface, and thus the pressure behind the blast wave is directly recorded.

An ideal pressure transducer would faithfully reproduce the pressure history at a given point in space. However, any real pressure transducer has a finite size, and therefore in a general case its sensitive membrane is subjected at any given instant to a non-uniform pressure distribution, which effectively results in spatial pressure averaging, i.e., a space-averaged rather than point-wise pressure value is recorded.

Furthermore, any actual transducer does not respond instantaneously to pressure variations in time. The so-called response time is an important characteristic of pressure transducers determined by their design. Any pressure variations with a characteristic time smaller than the response time are not present in the recorded signal. An effective time-averaging takes place.

The goal of the present study is to evaluate possible experimental errors due to the above spatial and temporal averaging effects for the pressure measurements in blast wave flow fields generated by milligram silver azide charges. The importance of this investigation stems, first of all, from the fact that laboratory experiments with milligram charges represent a very attractive alternative to field trials with large charge masses [3]. However, a number of shortcomings exist (see [3]), one of which is that pressure transducers cannot be scaled down as well and the above-mentioned averaging effects may become significant and result in sometimes unacceptable experimental errors. Secondly, it is common practice to use pressure measurement data for the calibration of numerical blast sources [4]. The accuracy of pressure measurements becomes then of paramount importance, since it is a cornerstone on which the validity of numerical simulations is based. It is known, for instance, that the location of the regular-to-Mach reflection transition for blast wave reflections is rather sensitive to the incident wave intensity. The calibration of the numerical blast source using erroneous pressure data may then lead to erroneous conclusions from the subsequent numerical simulations of the transition.

2 Numerical model of pressure transducers

In the present paper we carry out numerical simulations of the blast wave interaction with a pressure transducer to evaluate the influence of finite sensitive surface area and finite response time on the measured pressure values. The Euler equations in their integral form are taken as the governing equations for the flow fields. The gas (air) is assumed to be perfect with a constant specific heat ratio $\gamma = 1.4$. As a blast wave source, we use a hot, pressurized spherical balloon, which is a well-established approach when only the induced blast wave itself is of interest (see [4]). The common practice when using the blast source model is to calibrate its initial gasdynamic parameters (i.e., the initial gas pressure, temperature, and specific heat ratio) so that the resulting point-wise blast wave overpressures would correspond as closely as possible to the available experimental pressure records. In our case, precise initial calibration does not make sense because it is not known a priori how accurate the experimental pressure readings are and the very aim of our study is find out how much they differ from the actual pressure values in the flow field due to averaging effects. Thus, only a rough calibration has been performed for 10 mg silver azide charges which resulted in the following parameters of the numerical blast source: radius of the pressurized sphere: 1.771 mm; pressure ratio to ambient pressure: 2203; density ratio to ambient density: 330.5.

In blast wave pressure measurements, pressure transducers are usually flush-mounted on a flat plate to avoid any influence of blast wave diffraction over the pressure transducer body. In this case, both face-on and side-on measurement setups lead to the axisymmetrical problems schematically illustrated in Fig. 1.

In the face-on setup, the lower boundary of the computational domain represents the axis of symmetry, while the plane of symmetry forms its left boundary (as seen in Fig. 1a). The upper boundary of the computational domain is considered to be a wall placed far enough so that the reflected shock wave would not reach the pressure transducers during the observation time. The plates with pressure transducers are shown in Fig. 1a as vertical lines, such as A_1A_3. The segment A_1A_2 represents the sensitive surface while the rest (A_2A_3) is the transducer rim and the mounting plate. The pressure transducers themselves (PT_1, PT_2, PT_3) are schematically shown for illustration purposes only; they are not a part of the computational domain.

The pressure transducers are to be placed at different distances from the explosion center. To avoid multiple computer runs, we incorporate all transducers into a single computational setup (as an example, only three of them are shown in Fig. 1a). The

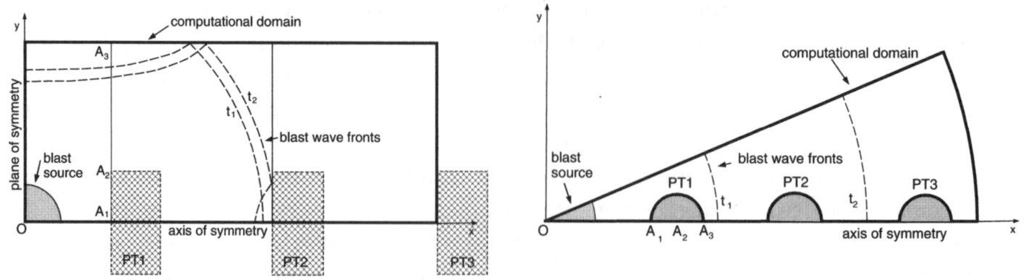

a b

Fig. 1. Numerical setup for the simulation of **a** face-on and **b** side-on pressure recordings.

computational experiment was performed as follows. At first, all the vertical boundaries corresponding to the locations of pressure transducers, such as A_1A_3 for transducer PT1 on Fig. 1a, were declared to be "invisible", i.e., the boundary conditions were implemented in such a way that the treatment of the boundary nodes was exactly the same as for the internal ones, so that the propagating blast wave does not "feel" the boundaries at all. The flow fields were saved as a data point array each time the primary blast wave approaches a pressure transducer location (for instance, at instant t_1 as shown in Fig. 1a). Then, the computations were resumed separately from each saved flow field while declaring the vertical boundary in front of the approaching blast wave to be a solid wall. The blast wave reflects from the wall as shown in Fig. 1a for an instant t_2. Two pressure values were recorded for each transducer as functions of time: the value at its central point (e.g. A_1) and the value averaged over the sensitive area, i.e., over the circle of radius A_1A_2. This approach models the spatial averaging effect of a pressure transducer.

A pie-shaped computational domain can be chosen for the simulation of side-on pressure measurements (Fig. 1b). The flow field at all times remains quasi-one-dimensional (spherical) and the axisymmetrical simulations were undertaken solely to provide for the possibility to average the pressure profiles over the sensitive area of the pressure transducers which are schematically shown in Fig. 1b. Two pressure values were recorded for each transducer as functions of time: the value at its central point (e.g. A_1) and the value averaged over the sensitive area, i.e., over the semi-circle of diameter A_1A_3.

The influence of finite response time of a pressure transducer can be evaluated using the time-averaging procedure described by the following equation:

$$\bar{p}(t) = \frac{\int_{t-T}^{t} p(t)\mathrm{d}t}{T} \quad . \tag{1}$$

In the above equation, $p(t)$ is the pressure history recorded in the course of blast wave propagation (either at the central point or averaged over the sensitive area); $\bar{p}(t)$ is the time averaged pressure response; T is the response time of a pressure transducer.

In the present paper the simulations are done for transducers with a circular sensitive area of 2.1 mm radius and 2 μs response time, which approximately corresponds to the parameters of a Kistler 603B transducer. A locally adaptive, unstructured, Euler solver [5] is employed.

3 Simultaneous side-on and face-on experimental measurements

In order to demonstrate and investigate the differences that the orientation of the gauge introduces in the recording of blast-induced overpressure, experiments with a setup indicated in Fig. 2 were also conducted. A charge (silver azide, AgN_3, with a nominal mass of 10 mg; for details on these charges and the ignition procedure see [6]) was suspended between two pressure transducers, one of which was mounted as a side-on gauge in the center of a thin sharp-edged so-called lollipop transducer, while the other sensor was part of a plate hit face-on by the blast wave. The charge was within experimental accuracy ($\pm$ 0.5 mm) exactly between the centers of both sensors and aligned with the upper surface of the lollipop gauge and the center of the face-on gauge. The distance between charge and sensor had to be chosen large enough to avoid the initial non-sphericity of the wave [6] and small enough to maintain sufficient wave strength and thus an acceptable signal-to-noise ratio. The available field of view of the visualization set-up (200 mm)

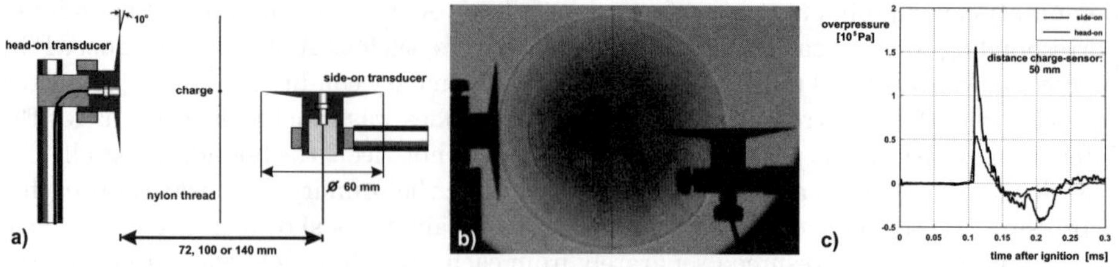

Fig. 2. Simultaneous side-on and face-on pressure measurements: **a** sketch of the experimental setup; **b** single frame of a time-resolved sequence (distance from charge 50 mm); **c** typical pressure traces (distance from charge 50 mm).

provided another upper limit. The distances of 36 mm, 50 mm and 70 mm respectively, were eventually settled on. In parallel, the process of ignition and blast wave formation and propagation was recorded in the form of time-resolved shadowgraphs, typically at recording frequencies of 500 000 frames per second (fps) with a frame exposure time of 250 ns (Fig. 2b). These tests were conducted with a Shimadzu HPV-1 high-speed video camera [7]. The optical records provided x-t data of the blast wave (scaled to 10 mg AgN_3 at standard ambient conditions, see [6]), from which the blast wave speed at the location of the gauges could be determined. Furthermore, the visualization allowed one to detect any possible irregularities during the interaction of the blast wave with the gauges and their holders. In some trials, the disk of the lollipop gauge was deliberately misaligned to see if there were visible changes in the reflection pattern. Each experiment yielded three simultaneously obtained sets of data: from the pressure transducers, two traces of side-on and face-on overpressure vs. time (Fig. 2c), as well as x-t data and thus shock speed vs. distance from the visualization. From face-on overpressure and shock speed the corresponding side-on overpressure was determined and compared to the value that the side-on gauge provided. The value based on shock speed was considered as the reference value as the optical records were the least dependent on the measurement device and therefore deemed the most accurate.

4 Results and discussion

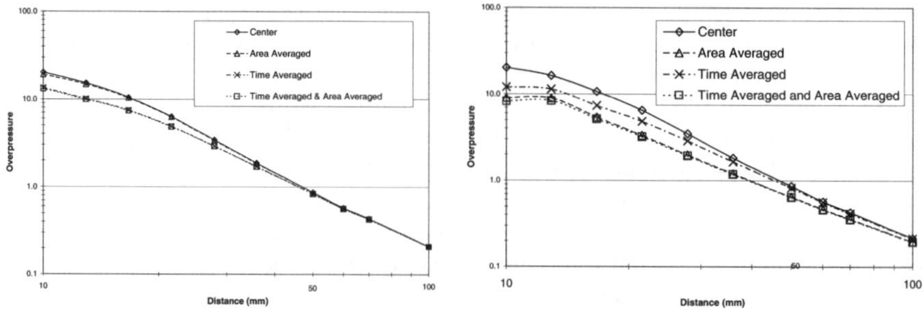

Fig. 3. Numerical simulation results for overpressure (in atm) vs. the distance from the charge center (in mm): **a** face-on measurements, **b** side-on measurements.

Numerical simulation results for face-on and side-on pressure measurements are shown in Fig. 3. In each case four curves are plotted: overpressure at the central point of the pressure transducer, overpressure averaged over the sensitive area, time-averaged overpressure at the central point, and time-averaged area-averaged overpressure. An implicit assumption is made that the combined time- and area-averaging effect can be evaluated by the consecutive application of the area averaging and time averaging procedures.

The maximum reduction of face-on pressure due to spatial averaging is about 5% and takes place when the pressure transducer is located very close to the charge, at 10 mm from the charge center. The error drops below 1% at distances higher than 20 mm. This is not surprising since at face-on measurements the pressure on the sensitive area is essentially non-uniform only when the blast wave radius is comparable with the size of the sensitive area (4.2 mm in diameter). As for time averaging at face-on measurements, its influence is much more significant. The error is as high as 32% at 10 mm and about 19% at 20 mm. It drops below 1% only beyond 60 mm from the charge center.

Unlike the face-on case, area averaging appears to be the most significant source of error for side-on pressure measurements: 53% at 10 mm, 11% at 50 mm. Even at 100 mm there is still a difference of 1.5% between area-averaged and central point values. This is because the pressure behind the blast wave front rapidly drops, resulting in a significant drop in the area-averaged pressure value when the front moves along the sensitive surface, as it is the case in side-on measurements. With increasing distance from the charge the characteristic distance for the pressure drop increases while the size of the pressure transducer remains the same, leading to the diminishing of the area-averaging effect. The influence of finite response time is less than that of spatial averaging but it is still quite significant, similar to the face-on case: 39% at 10 mm; above 10% up to a distance of 28 mm; below 1% only beyond 60 mm. Since spatial averaging significantly smoothes out the pressure profile, the subsequent application of the time-averaging procedure leads only to a minor additional reduction of pressure values.

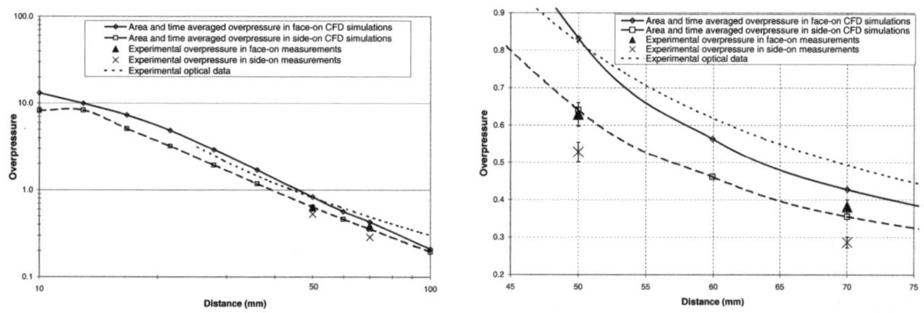

Fig. 4. Comparison of numerical and experimental results for face-on and side-on pressure measurements. Plot **b** enlarges the most important area of the logarithmic plot **a**.

The numerical and experimental results for face-on and side-on pressure measurements are compared in Fig. 4. The curve corresponding to the overpressure values obtained from optical records is also included. It should be noted again that the apparent disagreement between the numerical and experimental data is due to the fact that the numerical blast source was only approximately calibrated to produce the blast wave of roughly the same intensity as a 10 mg charge. More precise calibration does not make sense until the accuracy of experimental data is clarified. As seen from Fig. 4, the numerical simulations predict approximately the same difference between the face-on and side-on measurements as that observed in the simultaneous experimental measurements. This indicates that our simple model of spatial and temporal averaging due to finite size and finite response time of a pressure transducer works quite well.

Another observation is that even face-on experimental values at relatively large distances are noticeably below the values from optical records. This is partly because the face-on and side-on experimental values were obtained by extracting the maximum overpressure value, without resorting to pressure extrapolation procedures which are commonly used when processing experimental pressure signals [2]. Even though there is some justification behind the extrapolation procedures, they are of somewhat arbitrary nature and not unique. Their evaluation and comparison will be reported elsewhere. For the present study it suffices to say that pressure extrapolation would indeed bring the face-on results much closer to the optical values, at least at sufficiently large distances from the charge. However, the side-on results would still remain significantly lower.

5 Conclusions

The present study shows that the results of pressure measurements in experiments with small charges must be used with great caution. The effective spatial and temporal averaging of the pressure signal due to finite size and finite response time of pressure transducers may lead to a significant underestimation of blast wave intensities. The side-on setup is especially prone to this effect. Our numerical model of pressure transducers appears to provide very good qualitative agreement with experiment but it can hardly be used to deduce some kind of a universal correction factor. The face-on setup provides the results close to those from optical records if the pressure transducer is sufficiently remote from the charge, that is, if the characteristic time of pressure decay behind the blast front is much higher than the response time of the pressure transducer.

Acknowledgement. The study was partially supported by NSERC Discovery grant 298232-2004.

References

1. Cole RH: *Underwater explosions*, Princeton University Press (1948)
2. Baker WE: *Explosions in air*, University of Texas Press, Austin and London (1973)
3. Kleine H, Timofeev E, Takayama K: Shock Waves **14**(5/6):343 (2005)
4. Ritzel DV, Matthews K: Proc. 21st ISSW, Great Keppel, Australia, paper 6590 (1997)
5. Tahir R, Voinovich P, Timofeev E: `SolverII` – 2D, multi-block, multi-gas, adaptive, time-dependent, unstructured CFD application for MS Windows. Email: rabi.tahir@rogers.com, timofeev@sympatico.ca (1993-2007)
6. Kleine H, Dewey JM, Ohashi K, Mizukaki T, Takayama K: Shock Waves **13**(2):123 (2003)
7. Kleine H, Hiraki K, Maruyama H et al.: Shock Waves **14**(5/6):333 (2005)

Quantitative diagnostics of shock wave - boundary layer interaction by digital speckle photography

N. Fomin[1], E. Lavinskaya[1], P. Doerffer[2], J.-A. Szumski[2], R. Szwaba[2], and J. Telega[2]

[1] *Physical and Chemical Hydrodynamic Laboratory, Luikov Heat and Mass Transfer Institute, National Academy of Sciences of Belarus, ul. P.Brovki, 15, 220072, Minsk, (Belarus)*

[2] *Transonic Flows and Numerical Methods Department, The Szewalski Institute of Fluid-Flow Machinery, Polish Academy of Sciences, ul. Fiszera 14, 80-952 Gdan'sk (Poland)*

Summary. Digital laser speckle photography, an advanced technique of quantitative flow structure evaluation, has been applied for quantitative density field monitoring in transonic flow in the Shock Tunnel of the Institute of Fluid-Flow Machinery, Gdan'sk. Measurements of density distributions in the boundary layer of shock wave interacting with wall vortices as well as surface pressure were taken. Sub-pixel resolution software for the deflection angles of probing laser light evaluation using speckle technology is described and examples of flow fields are given.

1 Introduction

The research carried out concerns counteracting the shock induced separation by streamwise vortices. The streamwise vortex is generated in a boundary layer upstream of the shock. The main focus is to investigate the behaviour of the vortex interacting with the separation. The previous research proved that the vortex can cross the shock wave without dissipating, see Doerffer and Bohning, [1]. In case of separation, the oil visualization technique indicates the absence of vortex traces downstream of the separation. It is not clear whether the vortex is disintegrated by the reattachment, or whether it is lifted from the wall and remains further downstream the reattachment. Because the oil visualization is not of any further help in this case, we have turned to the quantitative flow visualization of the vortex core, which may deliver the answer to our question.

Traditionally, flow visualization techniques are used for qualitative flow analysis [2-5]. Recent progress in the field is connected with the use of coherent optical methods of flow diagnostics and is based on the rapid development of laser techniques and modern digital recording and acquisition systems, especially with the high resolution CCD matrices [6,7].

The line-of-sight flow visualization techniques like Schlieren, Schadow, Talbot or Holographic interferometry as well gain significantly in quantitative data treatment by using digital image acquisition. Attempts for 3D structure reconstruction in complex flowfields appeared due to the possibilities to accumulate a great amount of line-of-sight experimental information by tomographic flowfield reconstruction [8] or by comparison of experimental data with the results of numerical flow simulation [9,10].

One of the most attractive line-in-sight optical techniques in fluid mechanics is speckle photography. Digital versions of this technique have already been used successfully in a number of fluid mechanical applications, including quantitative measurements of turbulence statistics, combustion and shock wave phenomena [7]. The present paper deals with the application of digital laser speckle photography (DLSP) for quantitative monitoring of flowfield in transonic flow boundary layer.

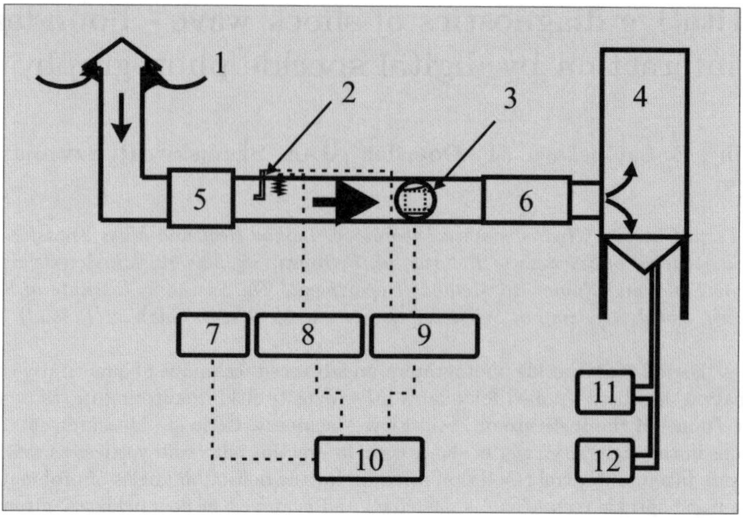

Fig. 1. Transonic wind tunnel scheme: 1, flow inlet; 2, Prandtl probe; 3, test section with optical windows; 4, vacuum tanks; 5, drying unit; 6, valve; 7, barometer; 7, thermometer; 8, pressure transducers; 10, PC; 11,12, pumps.

2 Experiment

2.1 Transonic tunnel

The experimental research was conducted in the transonic tunnel at The Institute of Fluid Flow Machinery of Polish Academy of Sciences in Gdan'sk, Poland [1]. The tunnel is depicted in the scheme on Fig. 1. The tunnel is an intermittent type, sucking the air from ambient conditions. The tunnel working scheme is as follows: two vacuum pumps (having power of 85kW together) evacuate two vacuum tanks (of volume $120m^3$) down to 95% vacuum. After reaching the desired pressure, the tunnel is ready for blow down. The test is realized by opening a valve (electrically controlled), thus connecting the tanks with the ambient air. The air, driven by the pressure difference, is being sucked from the atmosphere into the tanks flowing through the test section. Such operation secures constant conditions during the test. The air flowing into the tunnel is being driven through a layer of silica gel to extract humidity. With a throat of 100mm x 100mm the tunnel can run about 25-second blows every seven minutes. Apart form the optical instruments, described in a more detailed way later on, the measurements are supported by additional instruments (an intelligent pressure scanner PSI9010 for static pressure measurements, Kulite transducer of stagnation pressure, digital thermometer of stagnation temperature and a digital Druck Resonant Sensor Barometer, all coupled with a personal computer).

2.2 Digital speckle images recordings

Laser speckle interferometry enables the direct non-intrusive measurement of density gradients for optically transparent media with non-uniform refractive index distributions

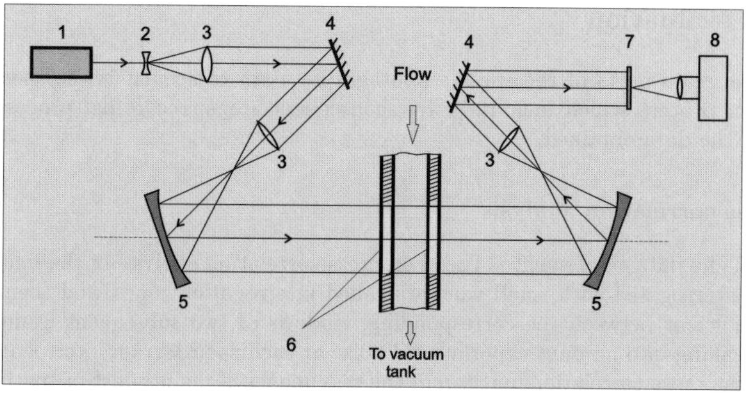

Fig. 2. Optical scheme of the speckle interferometer: 1, laser; 2,3, collimating lenses; 4, reflecting mirrors; 5, paraboloidal mirrors; 6, optical windows of transonic tunnel; 7, ground glass; 8 CCD camera.

[7]. An expanded parallel beam of laser light is transmitted through the test section. The wave front of the transmitted laser light is disturbed due to refraction of the beam on the density gradients. The test object is placed in front (on the left side) of the ground glass, and it is imaged by means of paraboloidal mirror 5 onto the plane of the ground glass, as shown in the Fig. 2. The ground glass works as a speckle-field generator and in the space behind it the laser light consists of the smallest granules of light - speckles. The speckles are recorded by a digital CCD camera of high resolution. The camera is focused onto a plane at distance ΔL from the ground glass. In the double exposure mode, (DEM), two speckle patterns are superimposed by recording two exposures on the same CCD matrix, like in PIV [6]. In contrast to PIV, no particles are injected into the flow, whereas the density gradients in the flow field studied provide the virtual particles (optical speckles) displacements.

By digital specklegram processing as described below, it is possible to determine two components of the speckle displacement at each specklegram interrogation point. These values can be easily converted into the components of the deflection angle of the light passing through the flow under study.

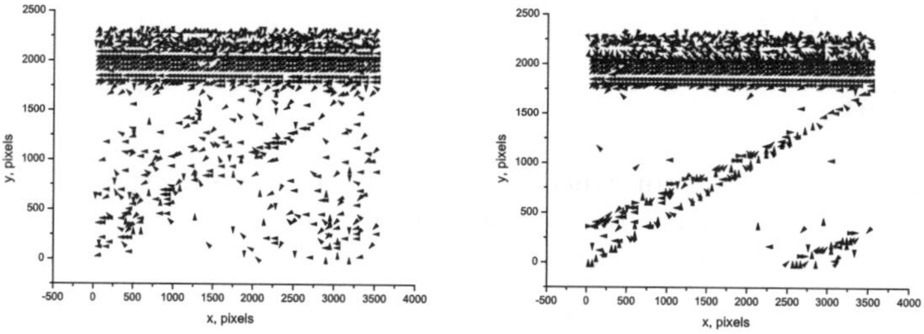

Fig. 3. Speckle displacement due to refraction of the probing laser light passed through the flow field.

3 Data evaluation

One of the main parts of the speckle photography data treatment is the speckle data acquisition process which includes a few consequent steps: recording, processing and analyzing the data obtained.

3.1 Cross-correlation analysis

For DEM, the data treatement is based on cross-correlation analysis of the images. The image is interrogated with small windows called interrogation zones, and then a correlation coefficient between the corresponding windows of two subsequent frames is calculated. Taking into account experimental noise at each specklegram interrogation zone $\tilde{\sigma}(m,n)$, the cross-correlation function of the two images is the product of spatial convolution of the images:

$$\Re_{1,2}(m,n) = I_1(m,n) \otimes I_2(m,n). \tag{1}$$

Corresponding Fourier spectra are:

$$F\{\Re_{1,2}\}(u,v) = F\{I_1\}(u,v) \bullet F\{I_2\}(u,v) + \sigma(u,v). \tag{2}$$

with σ - corresponding noise in the Fourier spectrum. The estimate of the sought cross-correlation function is

$$\tilde{\Re}_{1,2}(m,n) = F^{-1}\{F\{\tilde{I}_1\}(u,v) \bullet F\{\tilde{I}_2\}(u,v)\}, \tag{3}$$

with $\tilde{I}_1$, $\tilde{I}_2$ - filtered images.

3.2 Sub-pixel accuracy of evaluations

As the measured density gradients are essentially different in different flow regions, the sub-pixel resolution software for cross-correlation analysis of sequences of digital specklegrams is used here for the flowfield reconstruction. An advanced software has been designed and tested with noise filtration in Fourier plane and in combination with minimum quadratic difference method proposed by Merzkirch and Gui [11]. Statistical properties of density probability function of the probing laser intensity distribution have been determined and checked for Gaussian form. Fig. 3 contains quantitative map of the speckle displacement vectors, reconstructed with sub-pixel accuracy.

4 Discussion

For the complex flowfield reconstruction based on the results of speckle displacement measurements, numerical simulation of the light propagation through the field has been performed by solving the eikonal ray equation:

$$\frac{d}{ds}\left[n(\mathbf{r})\frac{d\mathbf{r}}{ds}\right] = \nabla n(\mathbf{r}), \tag{4}$$

where $\mathbf{r} = x \cdot \mathbf{i} + y \cdot \mathbf{j} + z \cdot \mathbf{k}$ is the position vector of a point on the ray and ds is an element of the arc length along the ray. The details of the reconstruction are given in [7,8] and illustrated on Figs. 4,5.

For the used here optical magnification $M = 1$, the dimensions of the reconstructed flowfield are $20 \times 30mm^2$. With the size of interrogation window being 32×32 pixels, the possible amount of instantaneously measured speckle displacement vectors is about 10000. It allows boundary layer monitoring with spatial resolution of about $30\mu m$.

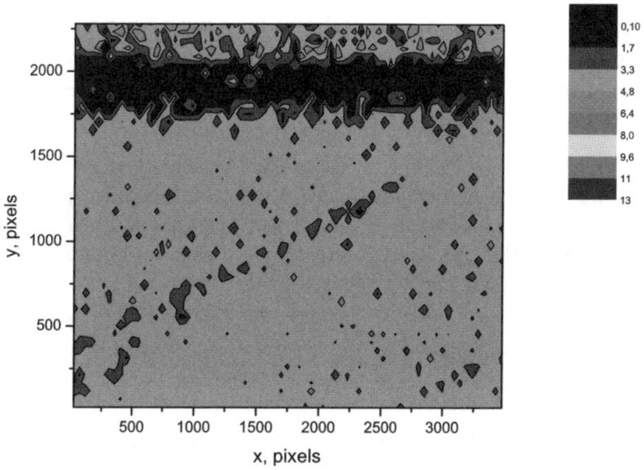

Fig. 4. Density gradients isoline in transonic flow.

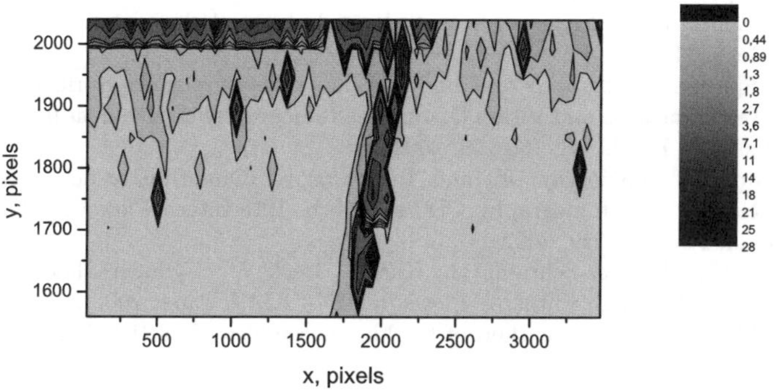

Fig. 5. Density gradients isoline in transonic flow (magnified).

Figs. 4,5 contain density gradient isolines in small fragment of the transonic flow, reconstructed with different magnification. The values are given in pixels, and the speckle displacement of one pixel corresponds to the density gradient of about $50kg \cdot m^{-4}$. We would like to emphasize, that it is quantitative data, and the technique doesn't need any calibration. The obtained values can be used for the validation of the CFD models and codes.

5 Conclusion

The line-in-sight DSP technique with sub-pixel resolution allows quantitative measurements of the density gradients in transonic boundary layer with sufficiently high spatial resolution. The technique is sensitive to fluctuations in refractive index and, unlike PIV, does not require seeding particles. The obtained quantitative values can be used for the validation of the CFD models and codes.

Acknowledgement. The research described in this publication has been performed in the frame of Agreement between Belarus and Polish Academies of Sciences and has been supported partly by INTAS (Innovation grant Ref. No 05-1000007-425), by Belarus Foundation for Fundamental Research (grants 07-070, T07-005) and by Belarus State Programs.

References

1. Doerffer P., Bohning R., Shock wave - boundary layer interacrion control by wall ventilation. Aerospace, Sciences, and Technology, paper 5208, 2003.
2. Merzkirch W., Flow Visualization, 2nd edition. Academic Press, Orlando, 1987.
3. Takayama K., Application of holographic interferometry to shock wave research. Proc. SPIE 398, 174- 180, 1983.
4. Ben-Dor G., Igra O., Elperin T., eds. Handbook of Shock Waves. Academic Press, New York, Vol. 1, 2001.
5. Settles G. S., Schlieren and Shadowgraph Techniques. Visualizing Phenomena in Transparent Media. Springer, New York, 2001.
6. Raffel M., Willert C., and Kompenhans J., Particle Image Velocimetry: A Practical Guide. Springer, Berlin, 1998.
7. Fomin N., Speckle Photography for Fluid Mechanics Measurements. Springer, Berlin, 1998.
8. Fomin N., Lavinskaya E., and Vitkin D., Speckle tomography of turbulent flows with density fluctuations. Exp. Fluids, 33, 160-169, 2002.
9. Abe A., Fomin N., Lavinskaya E., and Takayama K., Quantitative flow visualization by holographic and speckle tomography. CD Proc. of the 10th Intern. Flow Visualization Symp., Kyoto, Japan, Paper F0029, 2002.
10. Hannemann K., Martinez-Schramm J., Karl S., Beck W.: Cylinder Shock Layer Density Profiles Measured in High Enthalpy Flows in HEG. AIAA Paper 2002-2913, 2002
11. Gui L.C., Merzkirch W., A method of tracking ensembles of particle images. Exp. Fluids, 22, 465-468, 1996.

Part VI

Facilities

A simulation technique for radiating shock tube flows

R.J. Gollan[1], C.M. Jacobs[1], P.A. Jacobs[1], R.G. Morgan[1], T.J. McIntyre[1], M.N. Macrossan[1], D.R. Buttsworth[2], T.N. Eichmann[1], and D.F. Potter[1]

[1] *Centre for Hypersonics, The University of Queensland, Australia*
[2] *Faculty of Engineering and Surveying, The University of Southern Queensland, Australia*

Summary. We describe a numerical modelling technique used to simulate the gas flow in the complete X2 facility in non-reflected shock tube mode. The technique uses a one-dimensional model to simulate piston dynamics and diaphragm rupture and couples this to an axisymmetric simulation of the shock tube which captures viscous and finite-rate chemistry effects. This technique is used to simulate a nonequilibrium radiation condition relevant to a Titan atmospheric manoeuvre. The condition is a 7 km/s shock propagating into a N_2/CH_4 mixture at 80 Pa. The results show that the shock remains relatively planar at the exit of the shock tube such that there should be little difficulty for the optics. In terms of modelling, the finite-rate chemistry gas performs better than the equilibrium gas for these flows with regards to flow property estimates.

1 Introduction

At the University of Queensland, in the X2 facility, a series of experiments has been initiated to study a low pressure Titan-like atmosphere condition. In particular, a condition suggested at the Radiation of High-temperature Gases in Atmospheric Entry Workshop [1] is being investigated — TC2-T1. This condition is being investigated at other facilities also. Our motivation is to provide a complementary and independent set of data for this condition.

The focus of this paper is to describe a numerical simulation technique that is used to analyse the flows produced in our experimental facilities. We use this technique to simulate the current campaign of experiments at the TC2-T1 condition. Numerical simulations aid the wind tunnel testing by estimating a full set of flow properties that are not directly measurable. We aim to simulate the propagation of the shock into the test section in order to estimate the spatial variation of flow properties along the optical path used for the radiation measurements.

The preliminary part of the gas flow including, the piston dynamics, is treated by a quasi-one-dimensional model. The results of a one-dimensional simulation are used as input to a viscous axisymmetric simulation of the shock propagation through the shock tube and into the test section. This technique was first demonstrated by Jacobs et al. [2] in a simulation of an air condition at approximately 8 km/s in the X3 expansion tube facility. This technique was also used extensively in the thesis by Scott [3] in his study of air and Titan-like atmosphere gas conditions in the X2 facility. In this previous work, the simulation technique was applied to "normal" expansion tube operation. In the current work, the technique is applied to a non-reflected shock tube mode and finite-rate chemical effects for a Titan-like atmosphere gas are simulated in the axisymmetric part of the calculations.

2 Description of experiment

The X2 facility is nominally a free-piston driven expansion tube. In the experiments described here the facility is operated as a "classical" shock tube in a non-reflected mode of operation. A 35 kg piston travels through the driver tube of 4.5 m length with a bore of 0.257 m. The shock tube is 3.71 m and the acceleration tube is 5.70 m long. The bore for both the shock and acceleration tubes is 85 mm. In the shock tube arrangement (see Fig. 1(a)), there is no barrier for the flow between the shock tube and acceleration tube — these tubes are connected giving an effective shock tube length of 9.41 m.

The experiment aims to produce a shock speed of 7 km/s into low pressure gas at 80 Pa. The experimental shots for this condition are labelled as *x2s197* and *x2s198*. The operating conditions for these shots are shown in Table 1.

Table 1. Fill conditions in X2

Reservoir fill pressure	1.15 MPa gauge
Compression tube fill pressure	300 mbar (30000 Pa)
Test gas pressure	80 Pa
Primary diaphragm burst pressure	15.5 MPa
Reservoir gas	air
Driver gas	5.5% Ar, 94.5% He (by volume)
Test gas	95% N_2, 5% CH_4 (by volume)

The instrumentation for these shots included wall pressure sensors along the length of the tube as indicated in Fig. 1(a). A Pitot pressure sensor was located test section at approximately 4 cm downstream of the shock tube exit accounting for tunnel recoil and at a radial height of 28 mm above the centreline of the flow. Additionally radiation data were gathered with a photodiode and a photomultiplier which focussed on a region of the flow at the same axial location as the Pitot probe but 28 mm below the centreline. The experimental setup is shown in Fig. 1(b).

3 One-dimensional simulation

Quasi-one-dimensional simulations were completed using the `L1d2` code [4]. The code provides viscous simulations of the gas flow in a variable-area duct and models the piston dynamics. A number of gas slugs, pistons and diaphragms can be modelled in `L1d2` using a Lagrangian formulation for the gas slugs, with second order accuracy in both space and time. Flow in one dimension only is calculated and changes in duct area are assumed to be gradual. Boundary layers are approximated by the addition of wall shear stress to the momentum equation and heat transfer to the energy equation.

The `L1d2` computational results are used as both an aid in developing the experimental operating conditions and as input to the axisymmetric simulations. In the hybrid technique for complete simulation of the facility, the results of the one-dimensional simulation are used at the point in time of primary diaphragm rupture. The hot driver gas between the piston face and primary diaphragm is used as input initial conditions for the axisymmetric simulation.

For the one-dimensional simulations, four slugs of gas are used, as shown in Fig. 1(a). These slugs model the reservoir gas, the driver gas and the test gas (two slugs). The test

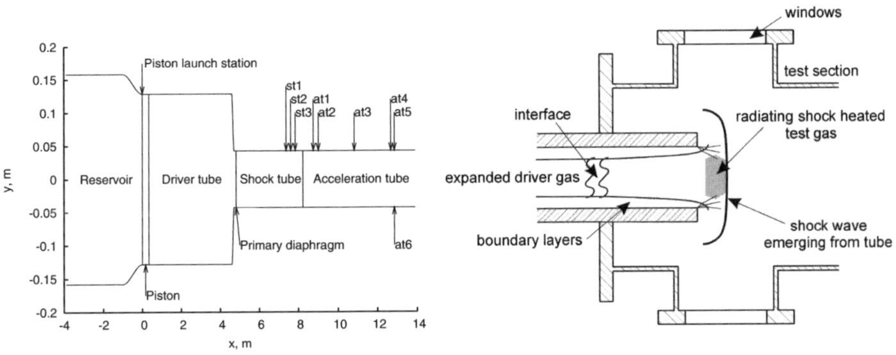

(a) (b)

Fig. 1. (a) X2 geometry for L1d simulations; (b) Pictorial view of the test-section in X2

gas is split into two slugs; the gas initially in the shock tube and the acceleration tube respectively, both at the same conditions. The piston is modelled initially as stationary in the firing position. A diaphragm is positioned at 4.81 m and given a hold time of 5 μs once the burst pressure is reached. A gas interface is placed between the slugs of test gas in the shock and acceleration tubes. At the reservoir end, the initial boundary is set to a stationary wall, and at the end of the acceleration tube the boundary is set to a free end. Two loss regions with a head loss coefficient of 0.35 are added at the changes in area. The initial temperature of all gas slugs is 296 K.

4 Axisymmetric simulation: method and results

In order to include the effects of boundary layer development and nonequilibrium chemistry, viscous axisymmetric simulations were performed with the code `mbcns2` [5]. This code integrates the finite-volume form of the compressible Navier-Stokes equations with an explicit time-stepping scheme. The Titan test gas is treated as a mixture of thermally perfect gases of which 13 component species are considered: N_2, CH_4, CH_3, CH_2, CH, C_2, H_2, CN, NH, HCN, N, C, H. An additional two species, Ar and He, are present in the simulation to represent the driver gas. Note that the argon is part of the driver gas only whereas the test gas is the mixture of 95% N_2/5% CH_4 reported in Table 1. The details of thermodynamic and transport properties calculations are given in [6]. The CH_4, CH_3, CH_2, CH and NH species are all present in mole fractions less than 1.0×10^{-3} and thus were ignored when computing mixture properties for viscosity and thermal conductivity. The flow solver is coupled to a finite-rate chemistry module [6] by using the operator-splitting approach. The reduced chemical kinetic scheme recommended by Gökçen [7] has been used but the ionization reactions have been ignored. The ionization reactions have little influence on the blunt body shock layer flow [8] and this should hold for the direct simulation of the propagating shock in the experimental flow.

The computational domain is shown in Fig. 2. The blocks shown in the tube and dump tank region were further subdivided giving a total of 130 blocks. The calculation which is performed in parallel on up to 65 processors. The centreline is treated as a symmetry condition. The remainder of the boundaries are specified as fixed temperature walls at 296 K with the exception of the right-hand end of the domain. Since only part

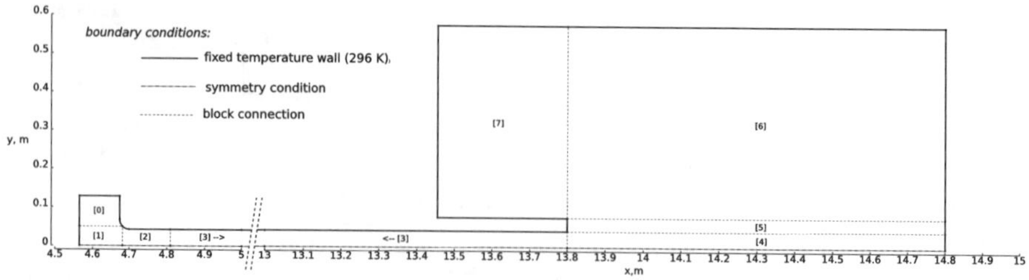

Fig. 2. Computational domain for axisymmetric calculations

of the dump tank is modelled, an outflow boundary condition is used there. In the simulations presented here, that boundary will have no influence as the simulation is terminated before any flow reaches the right end. Blocks 0, 1 and 2 are initially filled with high pressure, high temperature driver gas. The conditions for this gas are taken from the one-dimensional simulations computed using `L1d2`. The inital condition for the remaining blocks is the Titan-like atmosphere test gas at a pressure of 80 Pa and assumed temperature of 296 K. The discretisation of the domain consists of 3000 cells axially and 30 cells radially in the shock tube for the low resolution calculations. The discretistation of the remainder of the domain was chosen to give cells of approximately equal dimension to those in the shock tube.

Table 2 gives a comparison between the experimentally recorded shock speed and the numerical estimates. The comparison of shock speeds between experiment and simulation is one indicator of how well the simulations are approximating the experimental flow. The errors for the simulations given in Table 2 are quite high. Unfortunately the accuracy is limited by grid resolution and the modelling assumptions inherent in the `L1d2` calculation. In terms of grid resolution, the width of a finite-volume cell for the low resolution calculations is approximately 3 mm and thus the transducer can only be located to within ± 3 mm for the axisymmetric simulations. The `L1d2` calculations consistently estimate a higher driver gas temperature due to the model of heat transfer being essentially convective. The effect of a higher driver gas temperature is an increase the calculated shock speed in the numerical simulations. The axisymmetric, finite-rate calculations estimate a shock speed of 7200 ± 267 m/s which is within the uncertainty for the experimental values.

This calculation also shows good agreement with the pressure values recorded at the wall of the shock tube just before the exit into the test section. This is shown in Table 2. The noise present in the simulation would appear to be a numerical artefact. This noise is substantially diminished in the high resolution calculation (twice the number of cells axially) for the equilibrium gas when compared to the low resolution calculation. An interesting result is the poor performance of the equilibrium gas calculations in terms of shock speed estimate. This result suggests that the equilibrium gas assumption is inappropriate for these conditions and that a finite-rate chemistry gas model should be used to simulate these flows. Due to the size of the calculation, a high resolution calculation for the finite-rate gas has not been completed at the time of writing. However, the good agreement for the low resolution calculation means that we can apply these results to analyse the flow in test section.

One of the issues the numerical simulations aimed to address was the question of flow uniformity behind the shock as it enters the test section. Figure 3(a) shows contours of

Form of estimate	shock speed (m/s)
experiment	
x2s197	6948 ± 104
x2s198	7037 ± 105
one-dimensional modelling: `L1d2`	7347 ± 426
axisymmetric modelling: `mbcns2`	
finite-rate: low res.	7200 ± 267
equilibrium: low res.	8178 ± 302
equilibrium: high res.	8000 ± 255

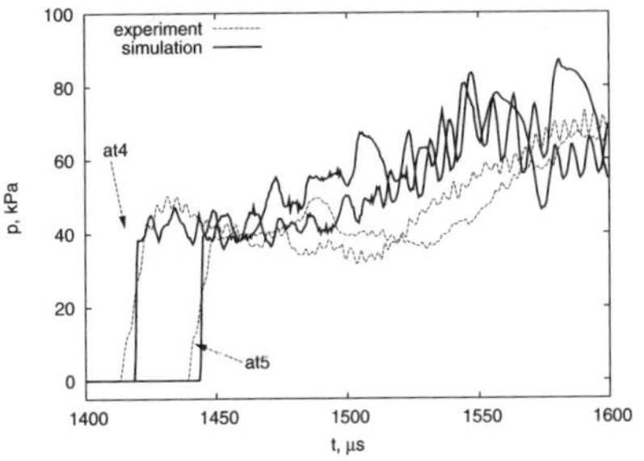

Table 2. Comparison of shock speed estimates in table. The error bounds are based on spatial and temporal sampling resolution. The graph shows a comparison of experimental and simulated pressure history for shot *x2s197*

temperature and density of cyanogen. The position of the shock is at the approximate location at which the optics is focussed during the experiment. Qualitatively, it appears that the shock is very planar and this is a desirable result for the experimentalists. It is estimated that the focussing region for the optics spans about 2 cm in the radial direction across the flow. In Fig. 3(b) the profiles for temperature and density of cyanogen at various radial locations are shown along with a calculation from a space-marching code, `poshax` [9]. In the immediate post-shock region, the profiles mostly coincide at various radial locations, with some discrepancy beginning at $r = 3.2\,\text{cm}$. Again this result is encouraging for the experimentalists as it identifies a region (radially) in the core flow with largely uniform conditions. The calculation from `poshax` is better able to capture the detailed post-shock relaxation as it does not suffer from the diffusion to due operator-splitting present in the `mbcns2` calculations. For future simulations, it would appear that a combination of one-dimensional modelling (`L1d2`) and steady flow space-marching analysis (`poshax`) is more useful than the large-scale axisymmetric simulations. However, the present axisymmetric simulations are still useful in terms of assessing the uniformity of the core flow.

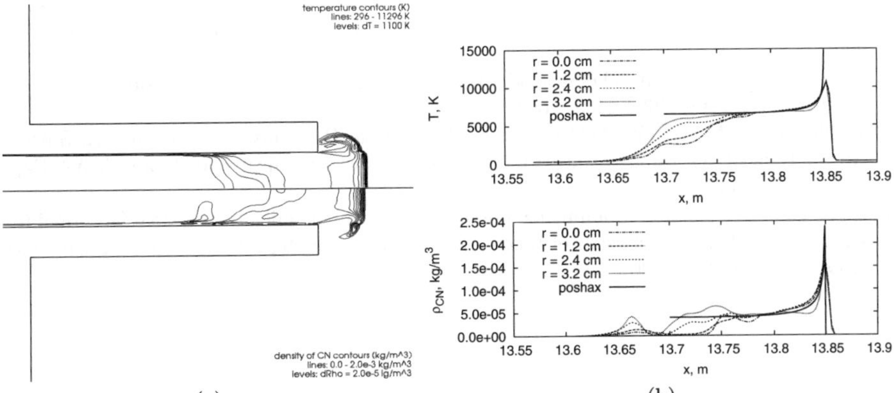

Fig. 3. (a) Contours of temperature and CN density at t = 1.37 ms; (b) Profile of temperature and CN density at t = 1.37 ms

5 Concluding remarks

In this work we have demonstrated a numerical simulation technique that models the non-reflected shock tube mode of the X2 facility. This technique uses a one-dimensional model to simulate the flow from the piston launch through to the rupture of the primary diaphragm. The hot driver gas, computed by the one-dimensionsal simulation, is used as input for a viscous axisymmetric calculation. We showed that the results are influenced by the type of gas model and that a chemical nonequilibrium gas model is most appropriate. This calculation is the first to simulate the flow of a reacting Titan-like atmosphere gas in a large-scale simulation of a shock tube facility.

There are a few issues related to the numerical modelling we hope to address in future work. These include a high resolution calculation for the finite-rate model and calibration of the one-dimensional model in terms of its convective heat transfer model.

Acknowledgements The large-scale parallel calculations reported in this work were performed on an Opteron cluster sponsored by Sun Microsystems and the Queensland State Government as part of a Smart State Research Facilities Fund grant.

References

1. A. Wilson (ed.) : The 2nd International Workshop on Radiation of High-Temperature Gases in Atmospheric Entry, European Space Agency, SP-629, November, 2006.
2. P. A. Jacobs, T. B. Silvester, R. G. Morgan, M. P. Scott, R. J. Gollan and T. J. McIntyre : Superorbital expansion tube operation: Estimates of flow conditions via numerical simulation, AIAA Paper 2005-694, 2005.
3. M. P. Scott : Development and Modelling of Expansion Tubes, PhD thesis, School of Engineering, The Unversity of Queensland, 2006.
4. P. A. Jacobs : Shock Tube Modelling with L1d, Department of Mechanical Engineering Report 13/98, The University of Queensland, November, 1998.
5. P. A. Jacobs : MB_CNS: A computer program for the simulation of transient compressible flows, Department of Mechanical Engineering Report 10/96, The University of Queensland, December, 1996.
6. R. J. Gollan : Yet another finite-rate chemistry module for compressible flow codes, Department of Mechanical Engineering Report 2003/09, The University of Queensland, July, 2003.
7. T. Gökçen : N_2-CH_4-Ar chemical kinetic model for simulations of atmospheric entry to Titan, AIAA Paper 2004-2469, 2004.
8. J. Olejniczak, M. Wright, D. Prabhu, N. Takashima, B. Hollis, E. Zoby, and K. Sutton : An analysis of the radiative heating environment for aerocapture at Titan, AIAA Paper 2003-4953, 2003.
9. R. J. Gollan : poshax: Computing post-shock thermochemically relaxing flow behind a steady normal shock, Division of Mechanical Engineering Report 2007/09, The University of Queensland, May, 2007.

Aerodynamic force measurement technique with accelerometers in the impulsive facility HIEST

H. Tanno, T. Komuro, K. Sato, and K. Itoh

Japan Aerospace Exploration Agency JAXA, Kakuda Space Center, 1-Koganesawa, Kimigaya, Kakuda, Miyagi, Japan

Summary. A feasibility study of a direct acceleration measurement technique was performed in the free-piston high enthalpy shock tunnel HIEST. The technique is simple and cost effective rather than aerodynamic force balance technique. With the present technique, aerodynamic force was obtained as products of measured acceleration and mass of the model, which model was weakly suspended with thin wires. Natural vibration of the test model, which often disturb precise measurement, can be removed through signal recovery process with a de-convolution calculation and with low-path filtering. Time response of the measurement technique guaranteed 0.1ms with a digital low-path filter, which cut-off frequency is 10kHz. For evaluation of the technique, wind tunnel test in HIEST was conducted to measure unsteady drag force of a 500mm length HB-2 standard model. The wind tunnel test results showed that the presenet technique can guaranteed time response as order of sub-ms without high-frequency messy noise. The comparison between the present measurement results and the measurement results in blow-down type wind tunnel showed that the difference is 5%.

1 Introduction

Free-piston shock tunnels [1] are the only ground test facilities to simulate high-temperature real gas flow, such as external flow around re-entry vehicles. The tunnels can produce high stagnation pressure and stagnation temperature, up to several hundred MPa and up to several thousand K, respectively. However, ms order test duration of the tunnels is extremely short comparing with the conventional wind tunnels. Hence, conventional force measurement technique, which has been generally used in long duration wind tunnels, can not be available because its time response is not sufficient.

For aerodynamic force measurement in such a short test duration, fast response force measurement technique is required. The direct acceleration measurement technique [2] was developed by CALSPAN in 1960's. In this technique, test models were weakly-restrained (suspended) with low stiffness support such as thin wires, so that effect of restorative force caused by model support can be neglected within short test period. Hence, aerodynamic force can be obtained simply as a product of measured acceleration and mass of a model. However, measured force shows messy oscillations, which were caused by mechanical vibrations coming from the insufficient rigidity of test model. Since these oscillations do not damp within short test time, they overlap with the relevant signal and disturb accurate measurement. With this oscillation, applicable model was restricted to small size (a few hundred mm). Duryea and Sheeran [3] reported that they restricted the test model size up to 300mm under the test flow of 10ms duration in their shock tunnel. If the effect of the oscillation can be removed, time resolution of the direct acceleration technique will be improved. The authors tried to apply a signal recovery process with deconvolution calculation to improve time response. With this trial, unsteady drag force of 80mm diameter sphere in a shock tube was successfully measured with time response of order of μs [4].

In this study, the signal recovery technique was tried to apply on aerodynamic force measurement in free-piston shock tunnel HIEST [5]. To evaluate a feasibility of the present technique, unsteady drag force measurement of HB-2 standard model of 500mm length in the free piston shock tunnel HIEST was performed. The measurement results was compared with the result obtained by aerodynamic balance in HIEST. The results also compared with the results obtained in blow-down type hypersonic wind tunnel HWT1.

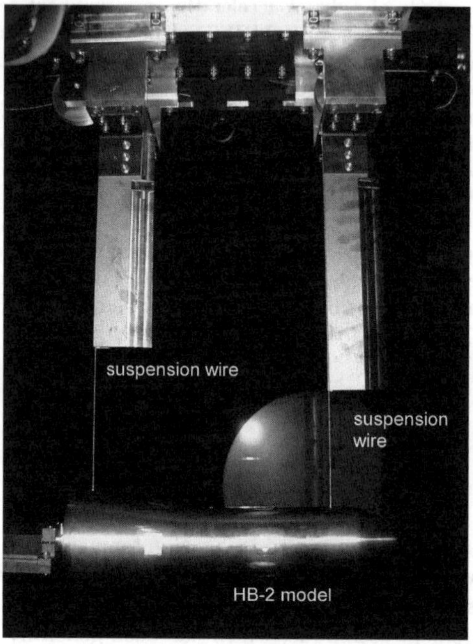

Fig. 1. HB-2 standard test model installed in the HIEST test section. The model was suspended with two thin wires.

2 HB-2 hypersonic standard test model

2.1 HB-2 standard test model

HB-2 model as shown in Fig.1 is a standard model used for evaluation of flow characteristics in various supersonic wind tunnels [6]. This model was made from aluminum alloy(A7075) of 500mm length and of 10.88kg weight. For the installation of piezoelectric type accelerometers (PCB 352C66: resonance frequency of 40kHz), the model has 50mm × 50mm square open-space in the vicinity of the model mass-center. In order to simplify the vibration characteristic of the model, configuration was designed as simple as possible.

The model was suspended with thin wires, which diameter was 0.5mm. Although high-stiffness large-diameter sting was not necessary for this measurement, a small diameter sting is required, which sting protects signal cables of accelerometers from high-enthalpy test flow. This sting was also used as a safety holder for the model, if the suspension wires accidentally broke.

3 Signal recovery method

To remove the oscillations caused by mechanical vibrations of the test model, a signal recovery process based on frequency domain de-convolution was applied as follows. If we assumed a measurement system as linear, the relation between output signal x, system function g and input force f can be related by the following convolution equation.

$$x(t) = \int_0^t g(t-\tau)f(\tau)d\tau \tag{1}$$

In this case, $x(t)$is the measured signal and $f(t)$ is the loaded force. To obtain the loaded force from measured signal, the integral must be inverted. By applying Fourier transform, Eq. (1) is converted to

$$X(\omega) = G(\omega)F(\omega) \tag{2}$$

where ω is the angular velocity. The capital letters represent the transformed function. Thus, $f(t)$ can be obtained through invert Fourier transform $\mathcal{F}^{-1}$ as follows.

$$\mathcal{F}^{-1}(F(\omega)) = \mathcal{F}^{-1}(X(\omega)/G(\omega)) \tag{3}$$

However, before this can be used, $G(\omega)$ has to be obtained with the calibration. In this study, impact test was performed as calibration. In the impact test $X_c(\omega)$as output from the accelerometer and $F_c(\omega)$ as output from the impact hammer can be measured, then $G(\omega)$can be determined from

$$G(\omega) = X_c(\omega)/F_c(\omega) \tag{4}$$

3.1 Dynamic characteristics of the model

To obtain the dynamic characteristics of the test model, impact test was performed. In the test, an impact hammer (PCB model 086C03) was used to initiate impulse response on the model. Fig.2a showed the force history input by the impact hammer and measured with an accelerometer mounted in the model. Fig.2b showed spectrum of vibration characteristics of the model. In the Fig.2b, there are some peaks, which represent resonance frequency of the mechanical vibrations; natural mode of vibration of the test model. The impact test showed that the the 1st vibration mode was 3.35kHz. The higher frequency modes were 5.6kHz and 8.2kHz.

Through the signal recovery process, frequency response function $G(\omega)$ was obtained. Since the impact hammer can not initiate the high frequency component, $F(\omega)$ decreases as the frequency gets higher. That means the high frequency component of $G(\omega)$ was amplified in Eq.(4) and that components will cause high frequency noise on de-convoluted signals. This high-frequency noise can be removed with low-path filter to cut-off higher

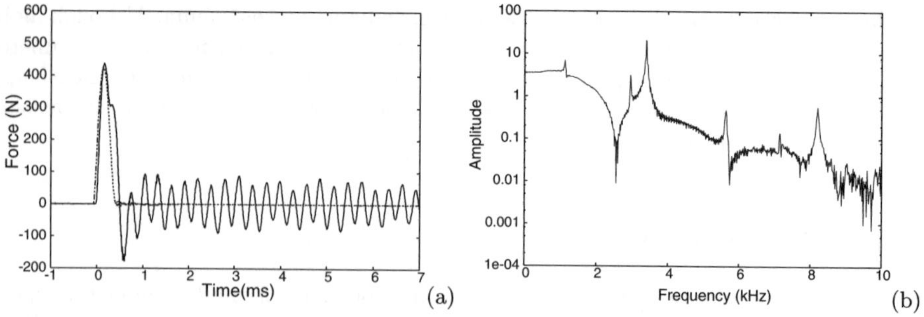

Fig. 2. Typical result of impact test was shown. a:Top figure shows axial force history of the HB-2 standard model and loaded force measured with a impact hammer. b:Bottom figure shows frequency characteristics of the HB-2 model

frequency component of $G(\omega)$. That means the time response was decided with the low-path filter characteristics. Since test free-stream velocity is at most 5km/s and model length is 500mm, required time for the test flow to pass over the model is 0.1ms or longer. The cut-off frequency of the low-path filter was hence decided to 10kHz.

4 Wind tunnel test results

Through the present test campaign, stagnation pressure was mostly held constant 20 MPa. On the other hand, stagnation enthalpy was varied from H_0=4MJ/kg to 12MJ/kg. The test flow conditions were calculated with an axis-symmetrical in-house nozzle flow code [7] as shown in Table 1. The test time in this study was specified as 3 to 5ms, when the free-stream Pitot pressure seems to be established.

Fig.3a showed an example of measured drag history under the condition of P_0=22MPa and H_0=7MJ/kg. Thin line and thick line shows the raw data, signal recovery data, respectively. The fluctuation caused by mechanical vibration of the model in the raw signal record can be removed with the signal recovery process as shown in thick line. The low-path filter used in the process was digital FFT filter, which is a function of the data-application software Origin 7.5 [8]. Since the frequency difference of the model vibration modes between the impact testing and wind tunnel testing, high frequency oscillation slightly remained. However this oscillation effect was negligible in the present measurement. It should be noted again that the cut-off frequency of the low-path filter applied in this study is 10kHz. Hence, the measurement guaranteed the analysis of the phenomena, which time-constant is less than 0.1ms. Fig.3b showed the comparison between the present measurement technique and the aerodynamic force balance technique, which was previously performed in HIEST [9]. The cut-off frequency of the low-path filter in the balance technique is almost 500Hz. The figure easily showed that the balance results had still heavy fluctuation, which was caused with irrelevant high frequency components. It should be noted that the present method can serve faster time response without high-frequency noise.

To evaluate the present measurement uncertainty, the comparison with other wind tunnel facilities was conducted as shown in Fig.4. In the comparison, reference data was used, which data was obtained in blow-down type hypersonic wind tunnel HWT1

Condition		A	B	C
Stagnation pressure	P_0 (MPa)	16	22	18
Stagnation enthalpy	H_0 (MJ/kg)	3.9	7	12
Free stream temperature	T_∞ (K)	350	710	1290
Free stream density	$\rho_\infty(10^{-3}kg/m^3)$	16	12	8
Free stream velocity	u_∞ (km/s)	2.7	3.5	4.5

Table 1. Free stream conditions obtained with the in-house nozzle flow code

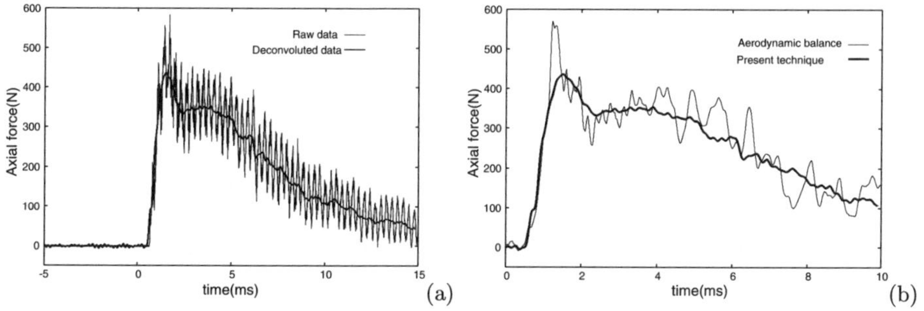

Fig. 3. Example of the measured axial force history with the HB-2 standard model in HIEST under condition H_0=7MJ/kg, P_0=22MPa. a:Thin line shows raw data from the accelerometer mounted in the model. Thick line shows signal recovery data through the de-convolution calculation. b: Thin line and thick line shows the balance measurement technique and present measurement technique.

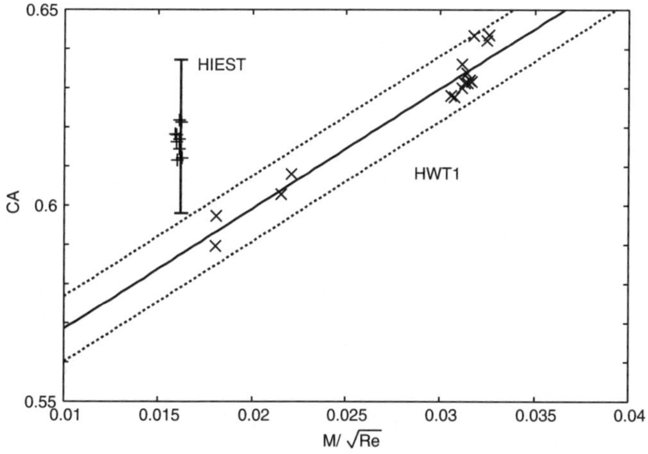

Fig. 4. Comparison of the CA with viscous interaction parameter $M_\infty/\sqrt{Re}$. Solid line shows the results obtained in HWT1(Hypersonic wind tunnel in JAXA Chofu). Dotted line shows 95% uncertainty. The open square shows the present results. Error bar shows 95% uncertainty.

located in JAXA Chofu [10]. Since Mach number and Reynold's number were different between each facilities, viscous interaction parameter $M_\infty/\sqrt{Re}$ was applied to compare each tunnel results. In HWT1, Measurement was performed with a HB-2 model of same configuration and same dimension. The uncertainty of the HWT1 results is less than 3%. On the other hand, the uncertainty of the HIEST is less than 6%. It should be noted that the number of measurement in HIEST was not enough to evaluate the uncertainty in detail. However, the figure showed that the results showed 5% higher than that of HWT1 measurements.

5 Conclusion

The new force measurement technique for impulsive facilities was evaluated in the free-piston shock tunnel HIEST. The wind tunnel test showed that the present technique has feasibility to measure aerodynamic force within short test duration as order of sub-ms. The technique can measure aerodynamic force with time response of 0.1ms without high-frequency noise. The uncertainty analysis revealed that the measured results were different from the blow-down wind tunnel results.

References

1. R. J. Stalker:Development of a hypervelocity wind tunnel. The Aeronautical Journal, 76(738):374-384 (1972)
2. L. Bernstein: Force measurement in short-duration hypersonic facilities, AGARDograph No.214, Department of Aeronautical Engineering, Queen Mary College, University of London (1975)
3. G. R. Duryea and W. J. Sheeran: Accelerometer force balance techniques. ICIASF'69 record, IEEE publication 69 C 19-AES (1969)
4. H. Tanno, K. Itoh, T. Saito, A. Abe, K. Takayama: Interaction of a shock with a sphere suspended in a vertical shock tube, Shock Waves, 13:191-200 (2003)
5. K. Itoh, S. Ueda, H. Tanno, T. Komuro, K. Sato: Hypersonic aerothermodynamics and scramjet research using high enthalpy shock tunnel, Shock Waves, vol. 12, pp. 93-98 (2002).
6. J.D. Gray: Summary report on aerodynamic characteristics of standard models HB-1 and HB-2, AEDC-TDR-64-137. (1964)
7. M. Takahashi, S. Ueda, T. Tomita, H. Tamura: Transient flow simulation of a compressed truncated perfect nozzle, AIAA Paper No.01-3681 (2001)
8. http://www.originlab.com/
9. K. Sato, T. Komuro, H. Tanno, S. Ueda, K. Itoh , S. Kuchi-ishi, S. Watanabe: An experimental study on aerodynamic characteristics of standard model HB-2 in high enthalpy shock tunnel HIEST. Prcoeedings of the 24th International symposium on Shock Waves, Beijing, China, July.11-16 (2004)
10. S. Kuchi-ishi, S. Watanabe, S. Nagai, S. Tsuda, T. Koyama, N. Hirabayashi, H. Sekine, and K. Hozumi: Comparative forceheat flux measurements between JAXA hypersonic test facilities using standard model HB-2 (Part 1: 1.27 m hypersonic wind tunnel results), JAXA-RR-04-035E (2005)

On the free-piston shock tunnel at UniBwM (HELM)

K. Schemperg and C. Mundt

Institut für Thermodynamik, Universität der Bundeswehr München (UniBwM), 85577 Neubiberg, Germany

Summary. The current status of the new free-piston shock tunnel HELM being built at UniBwM is described. The facility is nearing completion with all major components on site and assembled. HELM is a medium-sized facility to produce high enthalpy flow with total enthalpies up to 25MJ/kg with a typical total pressure of 100MPa. A special feature of the HELM shock tunnel is the integration of a plasma facility producing long duration flow at high enthalpy (up to 20MJ/kg) in the test section, e.g. for heating up model surfaces or investigations on flow and especially jet interactions. Described are the layout of the facility, calculations of the driving conditions and expectations for the shock tube flow.

1 Introduction

Free-piston shock tunnels [1] are used to investigate aerothermodynamic phenomena of high speed/high enthalpy flows, which is especially interesting for the development of capable reusable space transportation systems [2]. Considering the reentry into the Earth's atmosphere, vehicle velocities of approx. 8 km/s occur for low earth orbit and the high enthalpy of the flow causes the air molecules to dissociate and thus store thermal energy as chemical energy. The simulation of high enthalpy flows is an important method to gain further insight into re-entry phenomena, such as e.g. non-equilibrium gas dynamics, chemical reactions and radiation.

A shock tunnel facility simulates these conditions by generating high enthalpy, high density flows thus reproducing the free flight velocity in the range of 5-8 km/s and the binary scaling parameter (ρL). Currently a free-piston shock tunnel facility including a plasma burner named HELM (High Enthalpy Laboratory Munich) is built up at the University of the Armed Forces in Munich (UniBwM). Fig.1 presents a comparison of several ground based facilities to simulate typical reentry trajectories in terms of binary scaling parameter (ρL) and the flow velocity U, including the estimated values for the HELM facility.

2 Facility design

A schematic overview of HELM is also shown in Fig. 1. The arrangement consists, as for any free-piston shock tunnel, of the following main sections: the driver section consisting of an air buffer and the compression tube, the shock tube or driven section and the nozzle/test section. For operation, high pressure air in the air buffer is used to accelerate the piston along the compression tube. The driver gas is quasi-adiabatically heated and compressed and a diaphragm ruptures at the downstream end once sufficient pressure is obtained. A shock wave propagating along the shock tube is initiated. Reaching

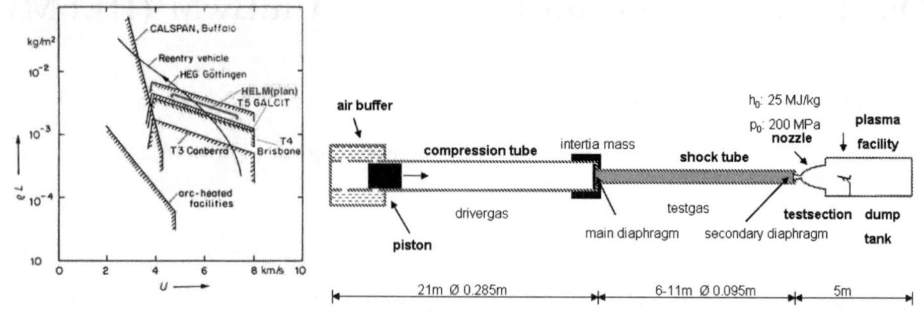

Fig. 1. Effects of various high enthalpy facilities [3] (left), schematic view and dimensions of the free piston shock tunnel HELM (right)

the downstream end, the shock wave is reflected and propagates back upstream leaving behind a region of uniform gas at high temperature and pressure.

This zone lasts for the order of milliseconds and is used as a gas reservoir for the nozzle. The facility is designed to achieve a total enthalpy h_0 in the range of 20-25MJ/kg with a total pressure p_0 of max. 200MPa. Thus flow velocities in the range of 5-7km/s can be produced. The overall compression tube length L is 21m (including piston launcher device) with an inner diameter D of 0,285m, the L/D-ratio results in 74. The shock tube consists of several tube sections, this allows different shock tube lengths l of 6, 8, 9 or 11m with an inner diameter d of 0,095m. It is planed to use several piston weights, the piston mass ranges from 30-120kg. The compression ratio calculated for HELM lies between 64 and 124. The nozzle exit diameter is max. 0,7m and the measuring time is expected to be 1-2ms.

Fig.2 shows details of the shock tunnel construction. The tunnel is mounted on a base frame with a total length of 42m. Movable carriages are installed to allow a displacement of the tube sections for operation and maintenance. The tubes are connected with a roller bearing mechanism to the carriages, so that the moving parts, the tube section and the nozzle can move freely during operation, but also to change configurations easily (e.g. shock tube length). The test section and the dumptank are in fixed position. Further

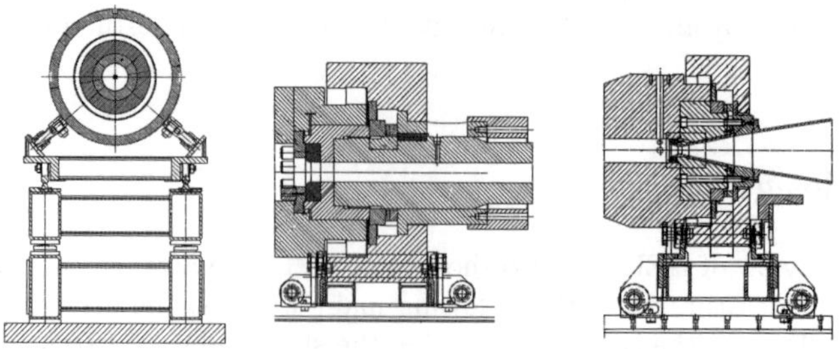

Fig. 2. Drawing of shock tunnel details: base frame and bearing of the tubes (left), main diaphragm (center), nozzle diaphragm (right)

Fig.2 shows the main diaphragm section with the connection of compression and shock tube and the nozzle diaphragm section at the shock tube end. At this location, three holes are positioned to enable optical access through sapphire-windows for temperature and species concentration measurements. Fig.3 gives an overview on the whole facility with the air buffer at the beginning of the compression tube, the inertia mass and a part of the nozzle, the total facility length is 50m and the approx. total weight 120t (moving and fixed parts). These dimensions represent the optimized and final configuration of the facility.

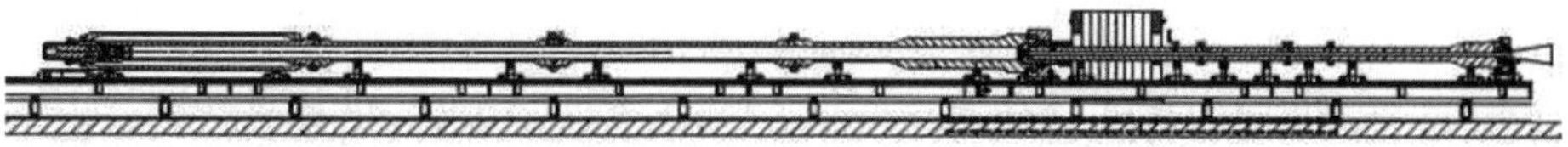

Fig. 3. Drawing of the free-piston shock tunnel HELM

3 Gasdynamic Layout

The gasdynamic layout and the calculation of the operating conditions of the facility were modeled by using the quasi-one-dimensional method L1d by P.A. Jacobs [4] with validation and parameters as in [5]. L1d is a computer code for the modeling of the gas-dynamic processes within transient-flow facilities, free-piston shock tunnels included. The numerical modeling is based on a quasi-one-dimensional Lagrangian description of the gas dynamics coupled with engineering correlations for viscous effects and point-mass dynamics for piston motion.

The calculations were used to optimize the facility geometry by parameter variation of compression and shock tube length, nozzle throat and shock tube diameter and air buffer volume. This results in an gasdynamically optimized geometry configuration, which was the basis for the mechanical engineering. L1d was also used to calculate expected operating conditions for the shock tunnel, especially the high pressure conditions. A medium enthalpy testcase was chosen for the optimization of test time, enthalpy and pressure.

The tube length used for the calculations corresponds to the effective gasdynamic tube length. Tuned piston and tailored tunnel operation is assumed. The piston motion of a tuned piston is such that the driver gas is continuously pressed into the shock tube and the pressure at the compression tube end is held almost constant. Tailored interface mode implies that the original values of the gas condition are chosen in such a way, that the reflected shock at the shock tube end brings the following contact surface to rest. These conditions result in a larger test time. Further investigations concerning the estimation of the operation modes tuned piston and tailored interface of the HELM shock tunnel were already done before [6].

Hornung [7] considered the length to diameter ratio for a shock tube to be around 90-100 for ideal operation. The shock tube length for HELM is variable from 6m, 8m, 9m to 11m. This results in a l/d-ratio of 70, 94, 106 and 129. The calculations showed, that for a shock tube length of less than 6m, significantly less test time is available. Simulations were done for shock tube length in the range of 7,90-10,90m, this results in an enthalpy variation of 12-12,5MPa and respectively pressure variation of 42-37MPa. Test time rises with increasing shock tube length, but at the same time max. enthalpy and shock speed is attenuated as the influence of the shock tube wall boundary layer increases. Using several shock tube segments and thus varying the shock tube length easily, further insight in this phenomena will be gained, once the shock tunnel starts operation. For a 50kg piston and a diaphragm burst pressure of 50MPa, the optimized shock tube length was chosen to be 8,90m with an inner diameter of 0,085m. With this configuration (compression tube length 19,90m) pressures from 40-78MPa and enthalpies from 12-15MJ/kg can be achieved (reservoir pressure 13MPa, Helium driver).

The compression tube length is limited by available space, but enthalpy and pressure rise with the tube length. The length was varied between 15,90-34,90m, resulting in a reservoir pressure range of 71,20-86,60MPa and an enthalpy range of 16,80-18,80MJ/kg. The optimized length for the compression tube is 20,90m with a diameter of 0,285m. With this configuration pressures up to 79MPa and enthalpies up to 18,20MJ/kg can be achieved.

With the optimized lengths of compression and shock tube, the nozzle throat diameter was varied from 0,015m to 0,040m, resulting in calculated test times in the range of 2,6-0,4ms. The test time is assumed to be the time between reflected shock and the interface reaching the nozzle throat. The optimized diameter is set to 0,025m. Enthalpies up to 17,80MJ/kg and pressures up to 94,20MPa can be reached having a test time of 1,2ms.

The air buffer reservoir volume was optimized to $1,2m^3$ (variation range $0,60 - 2,40m^3$) by considering a max. reservoir pressure of 20MPa and the heaviest piston with 120kg. The total moving mass of the shock tunnel is around 90t. The maximum recoil distance is set to 0,05m and thus the inertia mass will be in the range of 40-45t.

The diameter of the shock tube d was optimized to 0,095m to gain a suitable l/d-ratio of 94 with the fixed length of 8,90m. With this configuration, enthalpy and pressure range from 13,50-16,50MJ/kg and 65-105MPa having a test time of 1,7ms.

Fig.4 compares the previous ($helm169$) with the optimized facility geometry($helm144$) by means of achievable stagnation enthalpy and pressure in the nozzle reservoir. Calculated values by the ESTCj program (developed by P.A. Jacobs, 2002) are also included. The performance improvement is clearly visible. The calculation of the high pressure operation conditions was important for the mechanical engineering of the tunnel. Max. compression tube pressure is reached with max. air buffer pressure (20MPa). Max. shock speeds (and thus resulting max. shock tube pressures) can be reached with

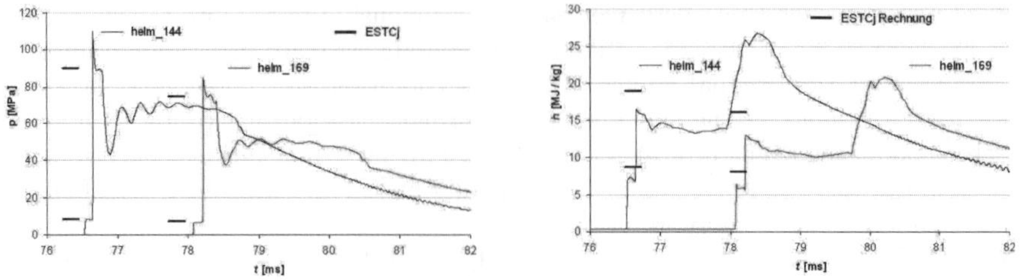

Fig. 4. Stagnation pressure p (left) and enthalpy h (right) in the nozzle reservoir - previous (*helm_169*) and optimized (*helm_144*) configuration (Helium driver) [8]

max. diaphragm burst pressure (100MPa). Calculations were accomplished for Helium, Argon and air drivers. Fig.5 shows results for the stagnation pressures in the nozzle reservoir and at the driver tube end for the high pressure condition.

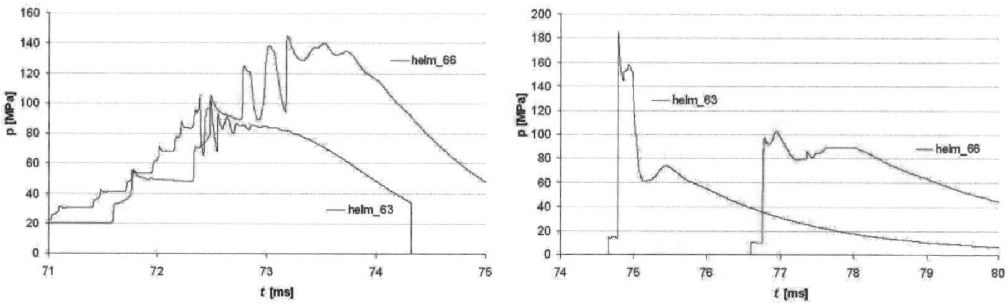

Fig. 5. High pressure condition: stagnation pressure p at the driver tube end (left) and in the nozzle reservoir (right) (*helm_63* : Helium driver, *helm_66* : Argon driver) [8]

4 Integration of plasma facility

A special feature of the HELM shock tunnel is the integration of a plasma facility producing long duration flow at high enthalpy in the test section, e.g. for heating up model surfaces [9] or investigations on flow interactions. The stream inflow is aligned vertically to the tunnel nozzle flow (see Fig.1). High temperature measurement methods and development, as well as calibration of sophisticated optical measurement techniques can be tested in a continuous stream before using them in short duration tests. The small size plasma facility achieves total enthalpies up to 20MJ/kg with Mach numbers < 3 and temperatures in the range of 300-5000K with a maximum electrical power of 250kW(DC)(present data for atmospheric operation [10]). The degree of efficiency was experimentally determined to 0,5. The air is heated by an electric arc which is generated by a Y-shaped assembly of electrodes, see Fig.6. To increase the durability of the electrodes up to 100 operating hours and to minimize the flow contamination, the root points of the electric arc are rotated by a rotating magnetic field during operation. The direction of rotation is compensated in the arc chamber. The gas flow reaches a settling chamber and then expands through a nozzle. The reservoir pressure achieves up to 40bar and the gas mass flow varies between 10-60g/s. Several modular nozzles with exit diameters of 10, 12 or 12.5mm can be installed and therefore free stream Mach numbers are tuneable [10]. Fig. 6 shows the plasma facility in operation and a detailed sketch of the plasma burner. The test duration can last up to 180 minutes. Fig.1 shows the simulation spectrum of the plasma facility in comparison with other ground based facilities.

5 Concluding Remarks

All major components of the shock tunnel are manufactured and the assembly is ongoing at present. Operation is planned to begin in 2007. A commissioning stage is planned where operation is to be proven followed by a calibration phase during which the free-stream

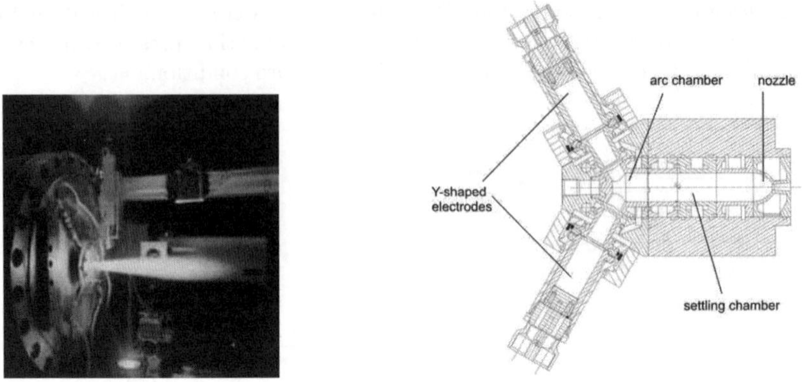

Fig. 6. Plasma facility in operation and drawing of plasma burner

conditions will be measured. During operation the following diagnostics will be available: strain, pressure and heat transfer gauges and a probe to record pitot pressure and stagnation heat transfer in the free stream flow of the test section which yields to the stagnation enthalpy.

The mechanical integration of the plasma facility will start soon after the commissioning stage of the shock tunnel. The test section is already prepared to flange on the plasma burner, the facility peripheral equipment, like power supply and water cooling system, must be installed yet. The first operation of the HELM laboratory with shock tunnel and plasma facility is currently planned for 2008/2009.

References

1. Stalker R.J.: *Development of a hypervelocity wind tunnel*, Aeronaut. J. 76, 1972.
2. Hornung H., Bélanger J.: *Role and Techniques of Ground Testing for Simulation of Flows up to Orbital Speed*, AIAA 90-1377, 1990.
3. Hornung H.: *28th Lanchester Memorial Lecture - experimental real-gas hypersonics*, Z. Fugwiss. Weltraumforsch. 12, 1988.
4. Jacobs P.A.: *Quasi-one-dimensional modelling of free-piston shock tunnels*, AIAA Paper 93-0352, 1993.
5. Mundt Ch., Boyce R., Jacobs P., Hannemann K.: *Validation study of numerical simulations by comparison to measurements in piston-driven shock tunnels*, accepted paper for publication, AST, 2007.
6. Schemperg K., Mundt Ch.: *Zur gasdynamischen Auslegung eines Stosswellenkanals*, GAMM Anual Meeting, 2004.
7. Hornung, H.: *Experimental Hypervelocity Flow Simulation, Needs, Achievements and Limitations*, GALCIT Report, 1997.
8. Kliche D.: *Numerische Untersuchungen zur Auslegung des Stosswellenkanals HELM*, Internal Report, Institut für Thermodynamik, Universität der Bundeswehr München, 2005.
9. Hirschel E.H.: *Thermal surface effects in aerothermodynamics*, Proceedings of the Third European Symposium on Aerothermodynamics for Space Vehicles, ESTEC, Noordwijk, ESA-SP-426, 1998.
10. Langkau R.U.: *Eine neue Forschungsanlage zur Untersuchung chemischer Reaktionen in Gasströmungen hoher Enthalpie*, PhD Thesis, Universität der Bundeswehr München, 1981.

Progress towards a microfabricated shock tube

G. Mirshekari and M. Brouillette

Department of Mechanical Engineering, Université de Sherbrooke, CANADA J1K 2R1

Summary. This paper presents the progress in the design, fabrication and testing of a microscale shock tube. A step-by-step procedure has been followed to develop the different components of the microscale shock tube separately and then combine them together to realize the final device. The paper reports on the progress in microfabrication of the microchannel and associated pressure sensor system.

1 Introduction

Shock tubes are common aerodynamics testing devices because of their relative simplicity and versatility. Recently there is a great interest in the development of Micro Electro Mechanical Systems (MEMS) in different fields of science and technology. In the shock waves research arena there is interest in theoretical and experimental observations of microscale impulsive phenomena in small devices [2]. There are also numerous macroscale applications of microscale shock flows, such as shock propagation in porous media, for example.

An ordinary shock tube consists of four basic components: a driver tube, a driven tube, a membrane and instrumentation. Our long term objectives is the fabrication and testing of a 10 μm hydraulic diameter fully instrumented shock tube. The present paper focuses on the fabrication of the tubes and pressure sensors.

2 Microchannel

Since the fabrication methods at the micron scales are different from those at macro scales, the geometry of a microscale shock tube has been selected to meet the possibility of microfabrication. Most of microfabrication procedures are borrowed from microelectronics. Two well known approachs are bulk micromachining and surface micromachining. Surface micromachining is based on the deposition and etching of different structural layers while bulk micromachining defines the structures by selectively etching inside the substrate. In both approaches a planar two-dimensional pattern is either etched or deposited on the substrate. This means that, for ease of fabrication, the common circular section tube has to be replaced with a rectangular section of dimensions a and b forming an aspect ratio a/b; the hydraulic diameter is then $D_h = 2ab/(a+b)$. For a rectangular section with a large aspect ratio, the hydraulic diameter is then approximately twice the smallest dimension.

A variety of microchannels have successfully been fabricated, with inlet and outlet plenums, as sketched in Fig. 1 . The fabrication procedure starts with 10μm Deep Reactive Ion Etching (DRIE) [1] of a silicon wafer to pattern the channels and inlet and outlet

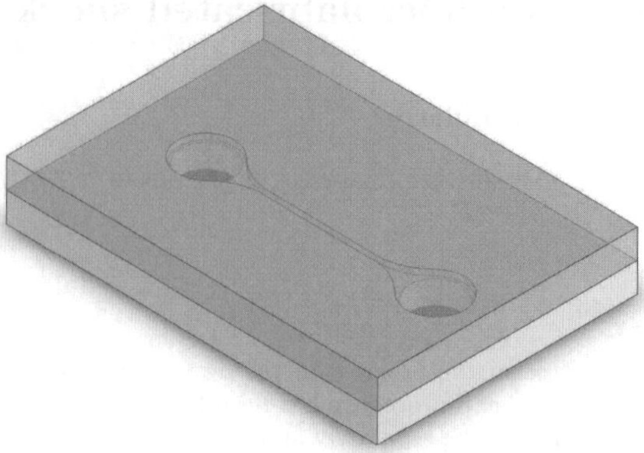

Fig. 1. The microchannel etched on a silicon wafer and covered with a Pyrex wafer

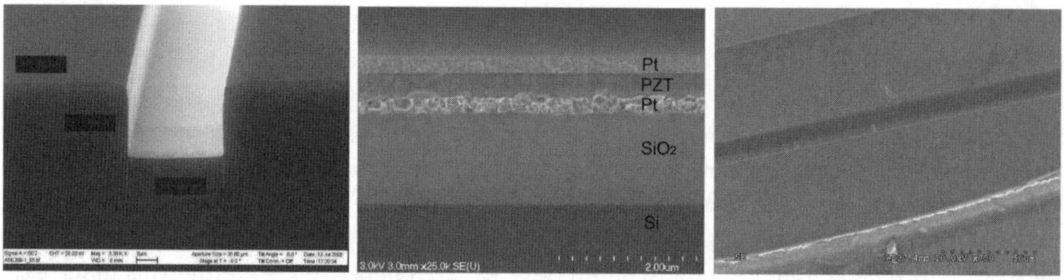

a b c

Fig. 2. SEM images of: (a) $6.6\mu m \times 9.4\mu m$ channel etched on the silicon substrate, (b) sensor cross-section, (c) success to create the entrance of $10\mu m$ channel in the side plan of the chip

plenums (see Fig. 2a). This is followed by $600\mu m$ etch through the wafer to create the inlet and outlet holes. The silicon wafer then covered with a Pyrex wafer using electrostatic (anodic) bonding [4] as shown in Fig. 1.

By themselves, these microchannels have been used to measure the pressure drop as a function of flow rate in a steady state compressible flow at microscales. Indeed, these measurements can be used to improve the numerical simulations of low Reynolds compressible flows [3]. Interestingly, in microchannel flow the gas is in thermal equilibrium with the surrounding solid walls. For a small channel the Mach number cannot be increased considerably by increasing the pressure ratio across the channel. To achieve higher Mach numbers, the value of $64L/Re_1D_h$ must be small, where L is the channel length and Re_1 the Reynolds number based on inlet conditions. For a fixed geometry this can be done by increasing the entrance Reynolds number. In the microchannel the value of $64L/Re_1D_h$ is roughly 30 and smaller values can be obtained by increasing the inlet pressure P_1. For example, it can be calculated that an entrance pressure of 100 atm is needed to produce a Mach number of about 0.55 for pressure ratio of 1.5; therefore, we conclude that it is not easy to experimentally achieve sonic flows in a microchannel.

In the study of these microchannel flows the continuum approximation is assumed to be valid at sufficient pressures. The Knudsen number $Kn \equiv \Lambda/D_h$, defined as the ratio of mean free path of the gas molecules (Λ) to the characteristic dimension of the device (D_h), is between 10^{-2} and 10^{-4} in the present experiments. This flow falls within the range of continuum flow, but which may be considered as a slip flow at low pressures and as a non-slip flow in high pressures. In the slip flow the velocity at the wall can no longer be considered to be zero as the length scale is commensurate with the mean free path of the gas; in this case the velocity near the wall is commonly described by the Maxwellian boundary condition.

The experimental setup consists of a chamber for installing the microchannels and establishing the inlet and outlet connections (Fig. 3). The measurement of flow rate is done by means of movement of a mercury bead inside a precision bore glass tube. This setup is entirely immersed in an ice-water tank to ensure a uniform and constant temperature. To examine a variety of pressure ratios (P_1/P_2) versus flow properties (Re, M) and geometrical properties (a/b, L/D_h), the experiment has been planned in a way that it can give the variation of pressure losses versus Mach number, Reynolds number, L/D and aspect ratio a/b.

The Reynolds number can be calculated from

$$Re = \frac{\rho_1 V_1 D_h}{\mu} = \frac{\dot{m} D_h}{\mu A} \tag{1}$$

where $A = a \cdot b$ is the section area of channel. The Mach number and static pressures at the entrance and the outlet of the channel can be calculated in terms of total pressures and the mass flow rate $\dot{m}$ from the following system of simultaneous equations

$$\begin{cases} P_{t1} A_{t1} - P_1 \alpha {D_h}^2 = \dot{m} V_1 \\ P_1 = \dfrac{\dot{m} P_{t1}}{\rho_{t1} \alpha {D_h}^2 V_1} \end{cases} \tag{2}$$

where subscript t denotes the total quantities and $\alpha = A/{D_h}^2$.

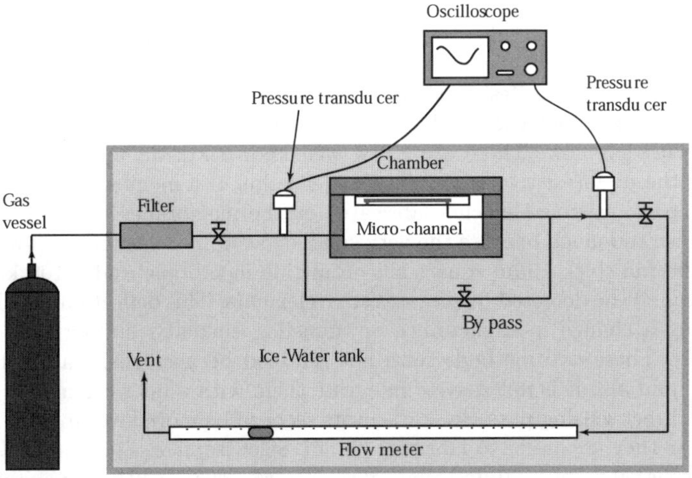

Fig. 3. Measurement setup established for precise volumetric flow rate measurement in microchannels

The results of experiments with microchannels confirm that the flow inside the microscale shock tube follows the laminar model over a wide range of Knudsen numbers (see Fig. 4).

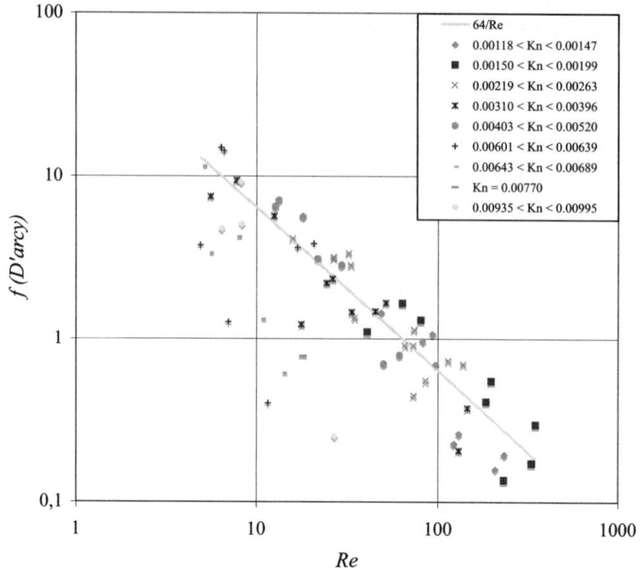

Fig. 4. The measured friction coefficient as a function of Reynolds number at different values of Knudsen numbers.

3 Micro-sensors

Like most of ordinary shock tubes, to acquire the practical information on the propagation of a shock wave in the microchannel, a number of pressure sensors needs to be installed along the wall to detect the desired change in pressures. In fact, in a microscale shock tube the size of sensors cannot exceed the size of channel and the sensor should be directly in contact with gas flow. There are a few well known MEMS approaches for sensing pressure. In the piezoresistive approach, the deflection of a membrane under pressure is detected by piezoresistive elements fabricated on the membrane. This approach cannot be scaled down too much because the very small deflection of membrane (for example 44 nm, for a 500 nm thick 10μm square silicon nitride membrane under 100 kPa pressure difference) cannot be detected by piezoresistive elements. This deflection may be detected by measuring a change in capacitance or tunneling current resulting from membrane deflection [7]. These two methods were not retained because their fabrication is very complicated and also it is not easy to integrate them with a microchannel.

Instead, direct sensing piezoelectric sensors seem to be more favorable for the present application as they are easier to fabricate [8], [5]. Sizewise, it was not expected to obtain a detectable signal from a 10μm square piezoelectric sensor as the capacitance of such sensor is too small. Therefore the sensors and channel are extended a few hundred times in the direction perpendicular to the shock wave propagation as shown in Fig. 2c. This allowed us to increase the surface of sensor (and the amount of produced electric charge) while keeping the hydraulic diameter around the desired value of 10μm.

In order to characterize the PZT sensors before integrating them with a channel, a variety of geometrically different sensors have been fabricated and tested. After packaging and wire bonding, the sensors were installed and tested at different shock Mach numbers in an ordinary shock tube. The results of testing at different Mach numbers are illustrated in Fig. 5. The sensors show a linear behavior in the range of experiment as expected. Also the output voltage of the sensors are almost the same for different geometries. Because of this it appears even smaller sensors could be possible. Figure 6b shows the oscilloscope trace of three sensors arranged along the shock wave propagation.

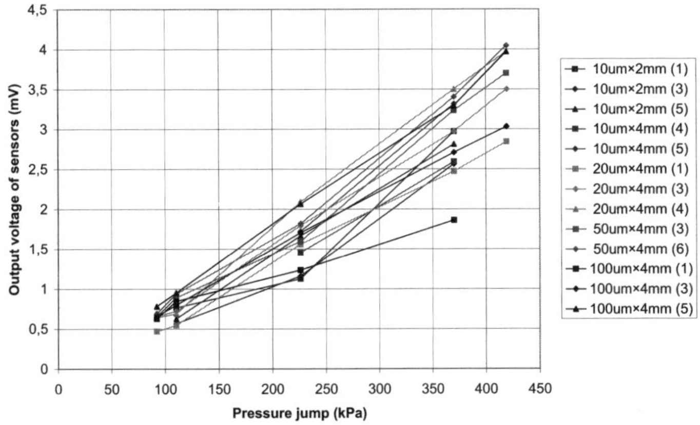

Fig. 5. Sensors output voltages as a function of shock pressure jump

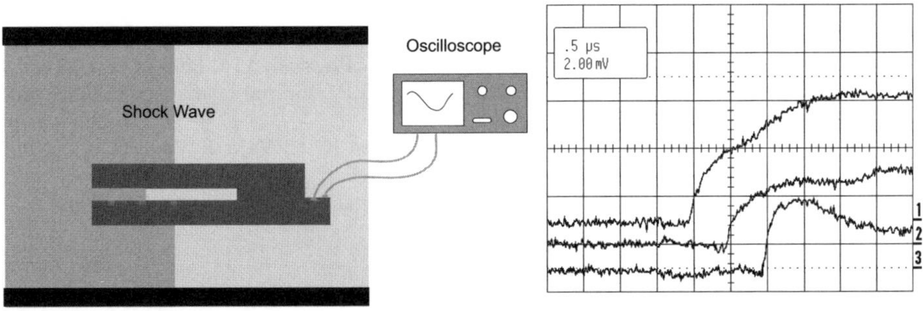

a b

Fig. 6. (a)The microfabricated shock tube (or sensors) inside an ordinary shock tube, (b) Oscilloscope trace of M=2.1 shock wave passing over three $10\mu m \times 2mm$ PZT sensors. The distances between sensors are $356\mu m$.

4 Micro shock tube

The next step is to integrate the microchannel with PZT sensors. Contrary to the previous channels, in this step the channel will not be fabricated by bulk etching of the silicon substrate. In this case the channel will be patterned on a $10\mu m$ thick SU8 layer developed on the top electrode of the sensors. Another silicon substrate covered with SU8 layer will be bonded to the sensor substrate by means of low temperature bonding process [6]. This bonding technique does not include high voltage and high temperature which may damage the sensors. Another challenge in this step is the creation of the entrance of channel as it is a feature on the side plan of the device. To achieve this a narrow trench etched on the silicon substrate acts as a guide for cleaving the wafer. The trench is then closed with a 8μm thick layer of PECVD silicon oxide to prevent surface defects during the next lithography steps. Fig. 2c shows the entrance of channel successfully created with this technique.

5 Conclusion

A microscale shock tube has been designed and is being fabricated stepwise. So far the channel and the sensors were fabricated and tested and we proceed to integrate them to realize a 10μm hydraulic diameter channel with a number wall pressure sensors. Flow tests with the microchannels showed that the flow in the channel follows a laminar behavior at different Knudsen numbers. The tests on the micro-sensors showed a linear behavior in the range of experiment.

References

1. Ayon A. A., Braff R., Lin C. C., Sawin H. H., Schmidt M. A.: Journal of Electrochemical society, **146**, (1999)
2. Brouillette M.: Shock waves, **13**, 12 (2003)
3. Gad-el-Hak M.: Mechanics and Industries, **2**, 44 (2001)
4. Knowels K.M., Van Helvoort A.T.J.: International Materials reviews, **51**, 5 (2006)
5. S.-H. Lee S. H., Esashi M.: Sensors and Actuators A: Physical, **114**, (2004)
6. Pan C. T., Yang H., Shen S.C., Chou M.C., Chou H.P.: Journal of micromechanicas and microengineering, **12**, (2002)
7. Stratton F. P., Kubena R. L., McNulty H. H., Joyce R. J., Vajo J.: Journal of vacuum science technology, **B16**, 4 (1998)
8. Zhang L., Ichiki M., Maeda R.: Journal of the European Ceramic Society, **24**, (2004)

Part VII

Flow Visualisation

A tool for the design of slit and cutoff in schlieren method

D. Kikuchi[1], **M. Anyoji**[1], **and M. Sun**[2]

[1] *Dept. of Aerospace engineering, Graduate School of engineering, Tohoku University, (Japan)*
[2] *Center for Interdisciplinary Research, Tohoku University, Aramaki aza aoba 6-3, Aoba-ku, Sendai, 980-8578, (Japan)*

Summary. We have developed a computer-aided analysis tool for shadowgraph and schlieren optical setups with a graphical user interface by handling both the optical setups and the computational fluid dynamics. It has been demonstrated that our simulation can display simple shadowgraph and schlieren images for two-dimensional flows. In this paper, the technique is further used to analyze the detailed effects of slits and cutoffs used in a typical schlieren optical setup for axisymmetric flows. The present study successfully reveals the influences of the two factors on the schlieren image of the supersonic flow over a sphere.

1 Introduction

In the past, shadowgraph and schlieren numerical images for two-dimensional flows and three-dimensional flows could be displayed by using formula of density gradients that are valid under straight-ray approximation in flowfield and for very simple optical setup [1]. And the method is not general enough for the purpose of the optimization of the optical setup. We attempt to display optical images by simulating the propagation of the rays in the flowfield only based on geometrical optics. It is necessary to simulate the inflection of the rays by the variation of the density and trace each ray in the test section. The flowfield in the test section is represented by the numerical result given by a solution-adaptive flow solver [2].

Up to now, the numerical shadowgraph and schlieren images for two-dimensional flows have been constructed [3, 4]. This paper describes the simulation of the numerical schlieren images for axisymmetric flows. Moreover, an detailed investigation of the effect of the slit and cutoff used in schlieren method is also reported.

2 Schlieren optical setup

To illustrate the possibility to design and optimize a schlieren optical setup for shocked flows, a typical schlieren setup as shown in Fig. 1 is considered. The entire setup consists of six optical elements, a light source on the leftmost end, a slit, two lenses (lens-1 and lens-2), a cutoff, and a recording plane on the rightmost end. The effect of the slit is considered by changing the shape and the position of the light source. The lens-1 and lens-2 are collimating lens, and defined by five parameters, its center coordinate (x, y, z), focus length, and diameter. Two cutoffs, circular bright-field cutoff and knife-edge, are tested in this paper. The circular bright-field cutoff is defined by four parameters, its center coordinate (x, y, z) and diameter. The knife-edge is defined by six parameters, its center coordinate (x, y, z), height, length, and rotation angle. The recording plane is the plane on which the image is projected, defined by eight parameters, its center coordinate

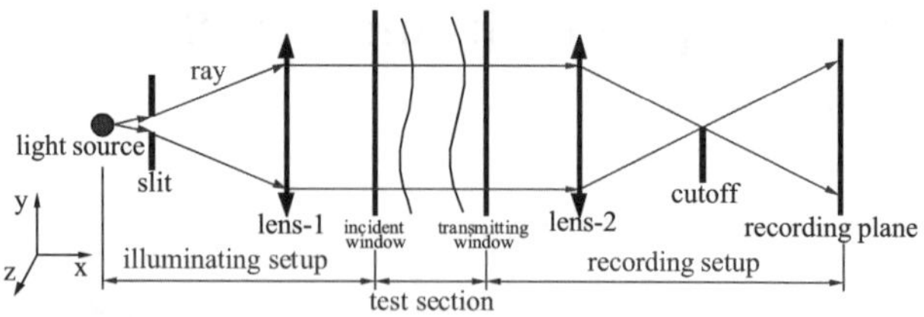

Fig. 1. Schlieren optical setup consisting of a light source, a slit, two lenses, a cutoff, and a recording plane.

(x, y, z), normal vector (x, y, z), height and length. In our simulation, the origin of the coordinates is set at the center of the test section. In order to evaluate and optimize the optical setup for a specified flowfield, it is necessary to construct the numerical image at the recording plane by simulating the propagation of the rays from the light source to the recording plane [3].

3 Simulation technique

3.1 Illuminating optical setup

In the simulation, a slit with a specific shape is regarded as a light source, and discretized as uniformly distributed point light sources. The shape of the slit is then represented by the distribution of these point light sources. The simulation starts from each point light source. The rays radiated from the point light sources located on the slit surface become almost parallel beams through the lens-1, and then enter the test section.

3.2 Test section

The axisymmetric flow considered is a supersonic flow over a sphere flying at $M = 3.0$ in air. First, we calculate the distribution of the density in the axisymmetric flowfield in test section. The distribution is given by the flow solver that is based on the solution-adaptive grid [2] as shown Fig. 2(a), denoted as the CFD-grid in this paper. The color lines are density contours.

Then, we construct the distribution of refractive index from that of the density according to Gladstone-Dale equation,

$$n = 1 + K\rho \tag{1}$$

where n, K and ρ denotes the refractive index, the Gladstone-Dale constant, and the density respectively.

We calculate the propagation of the rays from the distribution of refractive index. A ray is repeatedly traced from its starting point to its end point though the computational cells until it reaches the transmitting window as shown in Figs. 2(b) and (c). In each computational cell, the end point of the rays should be simulated from the coordinate of its starting point and direction vector.

Although a ray is assumed to propagate along a straight line within the computational cell in a three-dimensional space as shown in Fig. 2(b), its trajectory on the $z - r$

coordinates is generally curved as shown in Fig. 2(c). Therefore, the ray tracing for an axisymmetric flow is very different from that for a two-dimensional flow, and is more sophisticated. The end point is solved by combining two following relations.

The first relation comes from the straight-ray assumption in each computational cell, which can be expressed as:

$$(\mathbf{r} - \mathbf{r_0}) \times \mathbf{k} = \mathbf{0} \tag{2}$$

where $\mathbf{r} = (r_x, r_y, r_z)$ is the position vector of the end point, $\mathbf{r_0} = (r_{0x}, r_{0y}, r_{0z})$ is the position vector of the starting point, $\mathbf{k} = (k_x, k_y, k_z)$ is the direction vector of the ray in the computational grid. And the second relation is the three-dimenionsal surface equation defined by the edge in the $z - r$ coordinates as shown in Fig. 2(c),

$$r_{\mathrm{x}}^2 + r_{\mathrm{y}}^2 = R_i(r_z)^2 \tag{3}$$

where $R_i(r_z)$ is r-component of the CFD-grid, $i = 1, 2, 3, 4$ represents four side of the computational cell. And $R_i(r_z)$ must lie in the edge in the $z - r$ coordinates of the CFD-grid. By solving (2) and (3), eight general solutions can be obtained. Solutions lying beyond the edge are discarded. In the solutions left, the closest one to the starting point in x-direction is selected.

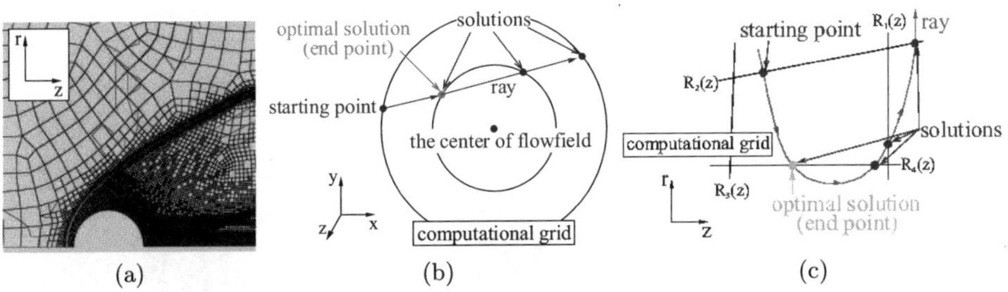

Fig. 2. (a) shows the CFD-grid used for simulating a supersonic flow over a sphere flying at $M = 3.0$ in air. The trajectory of a ray in axisymmetric flowfield: (b) in the $x - y - z$ coordinates; (c) in the $z - r$ coordinates, where r is the distance from the axis located at the center of the flowfield.

3.3 Recording optical setup

The rays radiated from the transmitting window of the test section are condensed on the region near the cutoff plane by the lens-2. Then, only the rays not blocked by the cutoff can reach the recording plane. At the recording plane, a numerical image is constructed according to the irradiance of all ray tubes. Initial ray tubes at the incident window are defined as a rectangular cell formed by every four rays and the ray tubes are deformed by the refraction of the rays in the test section. The ray tubes are generally deformed at the recording plane. The irradiance of the final ray tubes is obtained according to the conservation of light energy in the ray tubes,

$$I_1 S_1 = I_2 S_2 \tag{4}$$

where I_1, S_1 are the irradiance and the area of the initial ray tube, and I_2, S_2 are those of the final ray tube.

4 Results and discussion

4.1 Initial parameters of the optical elements

In our simulation, the center of the test section ($x = 0mm$) is set to be the origin of the x-axis, or the axis of the whole optical setup. The slit is initially located at $x = -1000mm$, which is exactly the left focus of the lens-1 ($f = 500mm$). The distance between the lens-1 and the center of the test section is $500mm$. The test section is $200mm$ wide, so that the incident and the transmiting window are located at $x = -100mm$ and $x = 100mm$ respectively. The lens-2 ($f = 200mm$) is set at $500mm$ behind the transmitting window, and the cutoff is inserted exactly at the location of the right focus of the lens-2, $x = 700mm$. The recording plane is set at $x = 900mm$, so that the focal plane of the recording system is located at the transmitting window. The flying sphere of $50mm$ in diameter is installed at the center of the test section. In this optical setup, if there is not density perturbation in the test section, the size of the image at the slit reduces to 2/5 at the cutoff.

4.2 Circular bright-field cutoff

Schlieren images for different sized circular bright-field cutoffs are shown in Fig. 3. For the purpose of comparison, the shadowgraph image that corresponds to the schlieren one using an infinitely large circular bright-field cutoff is shown in Fig. 3(f).

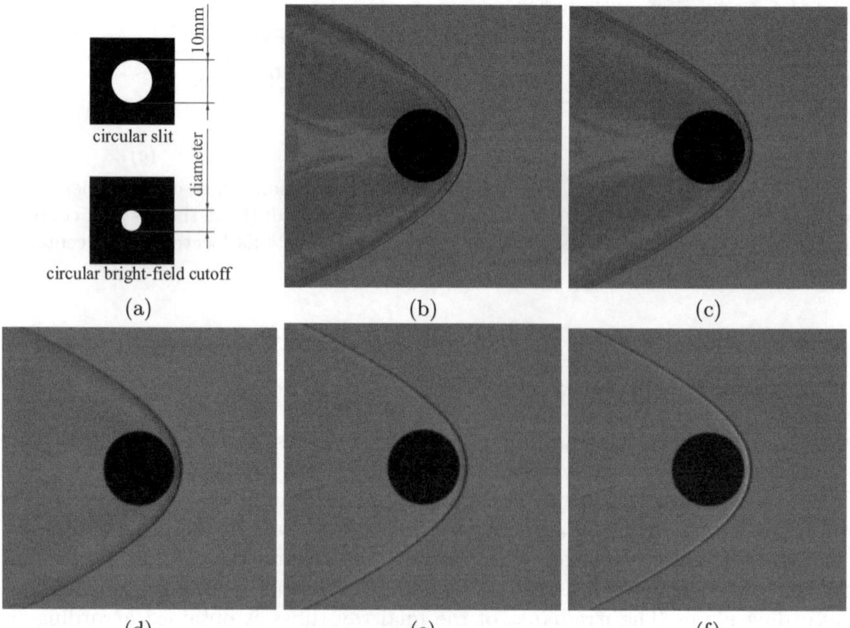

Fig. 3. Numerical schlieren images obtained by varying the diameter of a circular bright-field cutoff: (a) shows the circular slit of $10mm$ in diameter and the circular bright-field cutoff used; (b) $\phi = 2mm$ (with 75% cutoff of the $4mm$ high slit image); (c) $\phi = 3mm$ (44% cutoff); (d) $\phi = 4mm$ (0% cutoff); (e) $\phi = 5mm$ (roughly a shadowgraph image). (f) shows the corresponding shadowgraph image. It is seen that a small cutoff can reveal wave structure in wake better.

A circular slit of $10mm$ in diameter is used as the light source. In this case, the size of the slit image become $4mm$ at the cutoff. It is seen that the obtained schlieren image gradually approaches shadowgraph image by enlarging the diameter of circular bright-field cutoff.

4.3 Knife-edge

The effect of the location of a knife-edge on the image is investigated, and the numerical images shown in Fig. 4. A rectangular slit of $5mm$ in width and $2mm$ in height is used as the light source. In this case, the size of the slit image becomes $0.8mm$ high at the cutoff. The knife-edge is shifted from the bottom of the slit image to the top.

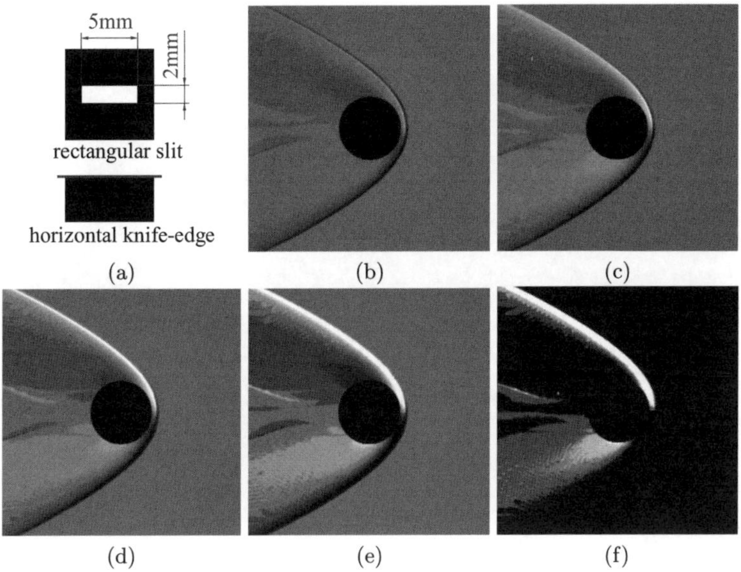

Fig. 4. Numerical schlieren images obtained by shifting a knife-edge from the bottom of the $0.8mm$ high slit image to the top: (a) shows the rectangular slit and the horizontal knife-edge used; (b) 0% cutoff (the tip of knife-edge is located at the bottom of the image); (c) 25% cutoff; (d) 50% cutoff (the center); (e) 75% cutoff (f) 100% cutoff (the top).

4.4 Concluding remarks

The effects of the size and location of the slit and the cutoff have been revealed. It is clear that the technique can be used to evaluate schlieren setups for not only two-dimensional flows but also axisymmetric flows. It can be an effective and useful tool for designing an optical setup especially for a large-scale facility.

However, the computing time required for constructing a high-resolution schlieren image using structured rays can be over dozens of minutes, so that practically it is not possible yet to get an immediate schlieren image after adjusting an optical element. There is room for future improvement. For this purpose, an adaptive ray tracing method is under development.

References

1. Yates L.A., *Images Constructed from Computed Flowfields*, AIAA J. 31: 1877-1884, 1993
2. Sun M. and Takayama K., *Conservative smoothing on an adaptive quadrilateral grid*, J. Comp. Phy. 150: 143-180, 1999
3. Sun M., *Computer Modeling of Shadowgraph Optical Setup*, the 27th International Congress on High-speed Photography and Photonics, 2006
4. Anyoji M. and Sun M., *Computer analysis of the schlieren optical setup*, the 27th International Congress on High-speed Photography and Photonics, 2006

Application of pressure-sensitive paints in high-speed flows

H. Zare-Behtash, N. Gongora, C. Lada, D. Kounadis, and K. Kontis

The University of Manchester, School of MACE
Sackville Street, Manchester M60 1QD, UK

1 Introduction

Pressure-sensitive paint (PSP) has become a useful tool to augment conventional pressure taps in measuring the surface pressure distribution of aerodynamic components (Mosharov et al. [1]). PSP offers the advantage of nonintrusive global mapping of the surface pressure.

During the last decades a number of research groups (McLachlan et al . [2] and Bjorge et al. [3]) have employed the PSP technique in high-speed flows. The present study involves the application of a Pressure-Sensitive Paint (Tris-Bathophenathroline Ruthenium Perchlorate), incorporated in a recipe developed in-house (Wong et al. [4]), onto two separate experimental setups, and aims to provide not only a quantitative but also a qualitative pressure map. The first case is to map the flowfield on the sidewall of a complex geometry (Fig. 1(a)). The model is divided into three main sections. A concave section, an area of constant cross-section and a region bounded by solid straight walls on the top and bottom surfaces, with the inlet pressure varrying in the range $1-3$ *bar*. The second case is to examine the control effectiveness of dimples on the glancing shock wave turbulent boundary layer interaction produced by a series of hemi - cylindrically blunted fins (Fig. 1(b)) at Mach number 0.8, performed in a transonic wind tunnel (Lada et al. [5]).

2 Theoretical Background

The PSP consists of a dispersion of luminescent probe molecules in an oxygen permeable binder layer. An excitation light source of wavelength λ_e and intensity I_e is used to promote molecules to an excited energy state (Carroll et al. [6] and Sakaue et al. [7]). For PSP there are two desirable mechanisms for the molecule to return to the ground state: luminescence at wavelength λ and intensity I or the transfer of energy by collision with an oxygen molecule, a process called dynamic or oxygen quenching. An increase in pressure P causes a corresponding increase in the partial pressure of oxygen P_{O_2} and an increase in the oxygen concentration within the binder layer n_{O_2}. This results in a larger level of oxygen quenching and lower luminescence intensity.

Because the amount of oxygen in the test gas can be related to static pressures, one can obtain pressure signals from the change in the luminescent intensity of PSP. Henry's law states that the concentration of oxygen in a liquid layer, n_{O_2}, is linearly proportional to the partial pressure of oxygen above the liquid, P_{O_2}, and is written as:

$$n_{O_2} = \sigma P_{0_2} \tag{1}$$

Henry's constant σ also displays a temperature dependence. The advantage of this technique over conventional pressure tap and surface transducer measurements is the ability to make global surface pressure measurements at a lower cost.

Tris-Bathophenanthroline Ruthenium is selected as the probe molecule. Sol-gel derived coating was used as binder for this paint. Sol-gel derived coatings are characterized by high thermal stability and, therefore, exhibit very low temperature-dependent viscosity changes.

A typical PSP system requires an illumination source for the PSP excitation, a photodetector for acquiring images and a data reduction tool to process the PSP images into pressure maps through the use of a calibration curve (Sakaue et al. [8]). For steady or time-averaged pressure measurements, the detector SNR can be improved by summation of a series of experiments (Sakamura et al. [9]). However, it is impossible to utilize such averaging techniques for truly unsteady pressure measurements.

The dark current is due to thermally generated electrons, hence these electrons are independent of the light falling on the CCD detector. However these electrons are counted as signal. The dark current signal of the image detector can be unstable, so it is desirable to acquire a dark image without any light in the test section before each series of experiments and to substract this dark image from all images of the series (Mosharov et al. [1]).

3 Experimental Setup

As it is mentioned in Sec. 1 two individual test cases were examined. For the first test section shown in Fig. 1(a) a series of transducers were mounted on the sidewall. The same wall was coated with the PSP paint with the front side covered with an optical grade perspex to allow optical access. For each inlet pressure examined a total of 25 images were captured with an exposure time of 200 ms.

For the second test case shown in Fig. 1(b) the models incorporated a hemi-cylindrical leading edge. Three different sweep angles were considered: 0°, 15°, 30°. A series of 2 mm

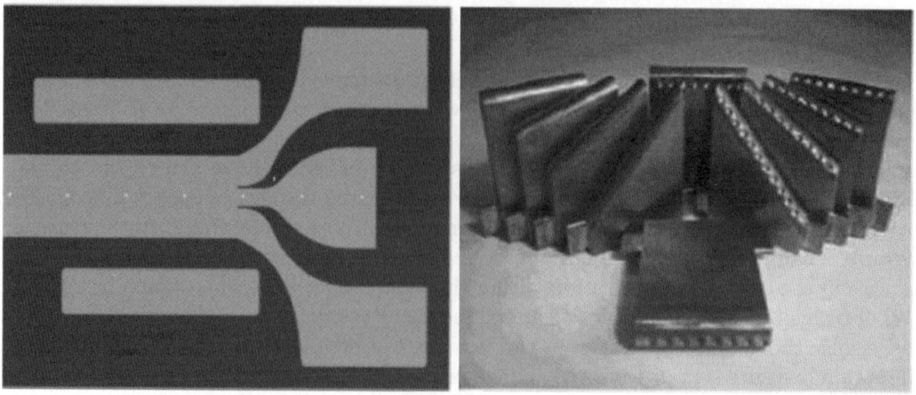

(a) (b)

Fig. 1. Cases under investigation. a) Convergent-divergent nozzle, b) Shock generator hemi-cylindrical blunt fins.

diameter, 1 mm deep and 3 mm spaced dimples were drilled across the hemi-cylindrical leading edge at angles 0°, 45° and 90° relative to the tip of the leading edge. The model blockage ratio was 8%. The side wall of the wind tunnel was coated with the PSP paint so that the effects of the various models would be depicted onto the images captured.

4 Results and Discussion

4.1 Calibration

For the wind tunnel case a-priori (static) calibration method was employed to determine the characteristics of the luminophore sensor. This took place in a pressure/temperature controlled chamber. The pressure varied in the range between 0-4 bar and the temperature could be controlled between -3° to $+60^oC$. The sample was initially illuminated by a pair of LED arrays with peak wavelength of 470 nm and then the luminescent emission was captured by a camera (LaVision Imager Intense) with a long-pass filter (610 nm) attached in front of the lens. The calibration was determined from the ratio of the wind-off and wind-on intensities. The Stern-Vomer coefficients were then used to obtain pressures from the intensity ratios.

Figure 2 shows the calibration lines obtained at different temperatures. The pressure sensitivity of the luminophore was found to increase with increasing temperature. In addition, it is clear the high temperature dependence of the paint, a factor that should be considered during the pressure measurements.

For the steady flow through the convergent divergent nozzle, in-situ calibration was applied. The reason why a-priori calibration was used rather than in-situ for the dimples is becasue of the inherently unsteady nature of the flow within the wind tunnel it would not be possible to obtain many pressure readings, and if one were to curve fit a third or even second order polynomial to these few points it would not provide a realistic relation between intensity ratio and pressure ratio.

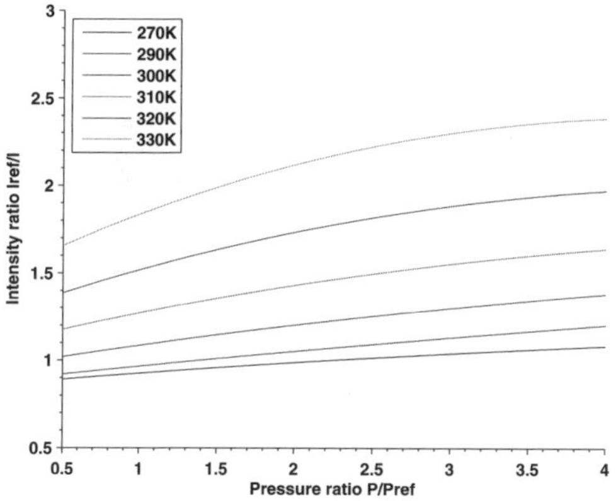

Fig. 2. PSP stern-volmer plot for various temperatures.

4.2 Convergent-divergent nozzle

Figure 3 shows the in-situ calibrated images acquired for two out of six inlet pressures for the convergent divergent nozzle. The profiles of the six inlet pressures along the central axis of the nozzle are plotted in Fig. 4.

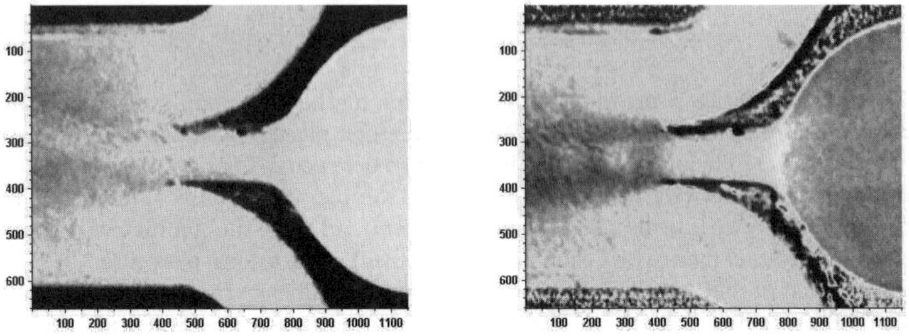

(a) (b)

Fig. 3. Pressure map for convergent-divergent nozzle. a) inlet pressure $P2 = 1.5\ bar$, b) inlet pressure $P6 = 2.5\ bar$.

One can clearly see that with increasing inlet pressure, the pressure within the converging concave section increases from 1.3 to 2.4 *bar*. But through the uniform area section the pressure appears rather uniform ($x = 400 - 500$ in Figs. 3 and 4) indicating that the nozzle is choked. At inlet pressures $P5 = 2.38\ bar$ and $P6 = 2.51\ bar$ we also start to notice the formation of the oblique shock structures and the resulting rise in pressure due to them, these are visible in Fig. 3(b) and also in the profiles in Fig. 4.

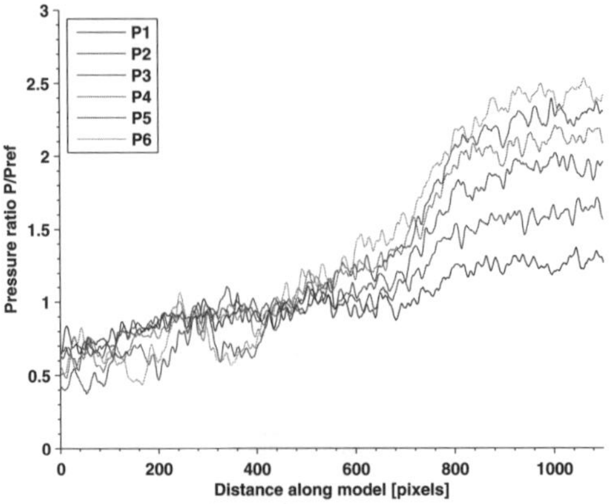

Fig. 4. Pressure profiles along the central axis of the convergent-divergent nozzle for various inlet pressures.

As for the accuracy of the PSP technique, Fig. 5 shows the pressure profile for the two extreme inlet pressures along with the presures obtained from the three transducers which are in the field of view of the images. The results show a promising correlation between the two techniques employed. What needs to be considered is that further research is currently being undertaken to improve this correlation further.

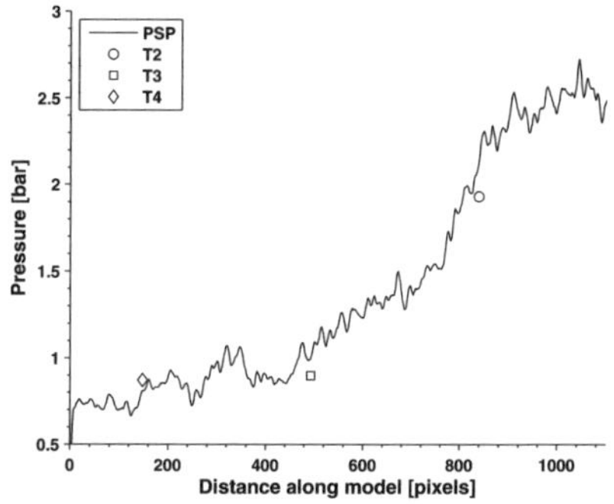

(a)

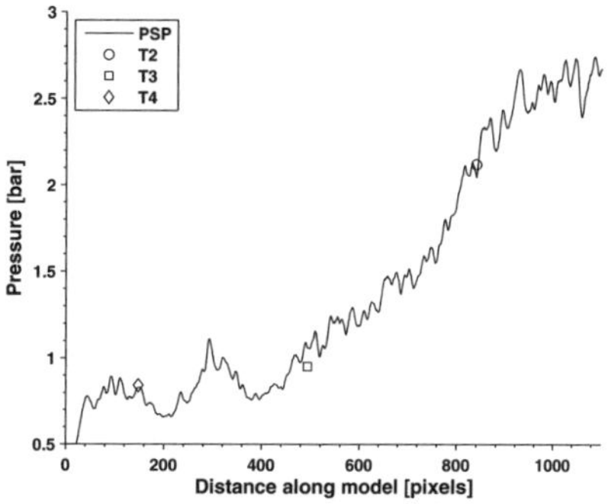

(b)

Fig. 5. Comparison between PSP and pressure tappings. a) inlet pressure $P5 = 2.38\ bar$, b) inlet pressure $P6 = 2.5\ bar$.

4.3 Transonic wind tunnel

Figure 6(b) shows the effect of dimples, located at 45° across the hemicylindrical leading edge, for the 0° angle of sweep at $M - 0.8$. For the no dimples case (Fig. 6(a)), the shock wave system is symmetrical. The dimples affect the local flow field around the leading edge. They induce adverse pressure gradients due to the local surface roughening causing the thickening of the boundary layer. As a result, the dimples alter the effective geometry at the leading edge region. However, due to their location, their effect is not symmetrical. On the upper side, a small in size normal shock wave is observed at approximately one-third of the chord-length of the fin. It is conjectured that the oncoming Mach number (on the upper side) drops quickly, at some distance from the surfcase, to a value where it is no longer possible to achieve the required pressure jump (flow deflection) across a weak oblique shock wave and the flow downstream turns subsonic. This means that it is not possible for a lambda-shock structure to exist. Presence of the dimples prevents the shock from moving too far downstream, which may lead to reduced shock motion and delay of buffet onset. On the lower section of the wall, an overall decrease of the pressure level is measured, occupying almost the full chord-length of the fin. On the upper section, a strong interaction region is mapped near the apex of the fin corresponding to the location of the normal shock wave.

4.4 Experimental uncertainties

The accuracy of the intensity-based PSP optical system is sensitive to drift and ageing of the light source and detector as well as paint in-homogeneities in dye concentration and

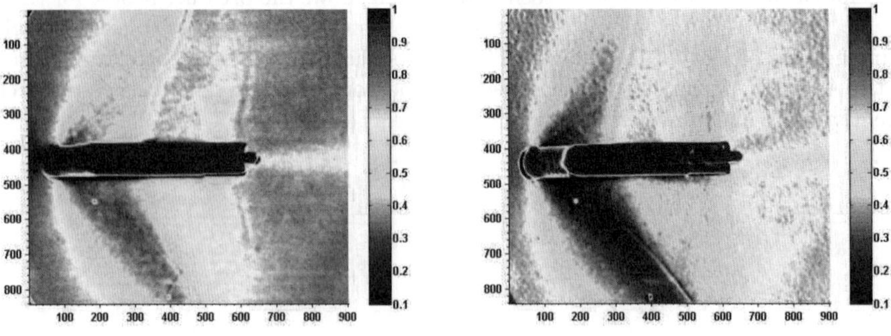

(a) (b)

Fig. 6. Calibrated pressure ratios for: a) no dimples, b) dimples at 45°.

film thickness. The ratio of wind-off image to wind-on image is processed to eliminate the effects of these factors. However, the effect of the variation of illumination is not cancelled. Aging is another potential source of error but all tests were completed within few hours. Because the models used were simple in shape and the aerodynamic loads were small, model deflection was negligible. The accuracy of the technique corrsponds to that of the thermocouples and pressure transducers used for calibration.

References

1. Mosharov V., Radchenko V., Fonov S., Luminescent Pressure Sensors in Aerodynamic Experiments. Central Aerodynamic Institute (TsAGI), (1998)
2. McLachlan B., Bell J., Park H., Kennedy A., Schreiner J., Smith S., Strong J., Gallery J., Gouterman M., Pressure-sensitive paint measurements on a supersonic high-sweep oblique wind model. Journal of Aircraft **32**, (1995)
3. Bjorge S., Reeder F., Subramanian C., Crafton J., Fonov S., Flow around an object projected into a supersonic freestream. AIAA Journal **43**, (2005)
4. Wong C., Amir M., Lada C., Kontis K., Molecular image sensing for pressure and temperature surface mapping of aerodynamic flows. 42nd AIAA Aerospace Sciences Meeting and Exhibit, (2004)
5. Lada C., Amir M., Wong C., Kontis K., Effect of Dimples on Glancing Shock Wave-Turbulent Boundary Layer Interactions. 42nd AIAA Aerospace Sciences Meeting and Exhibit, (2004)
6. Carroll B.F., Abbitt J. D., Lucas E. W., Morris M. J., Step Response of Pressure-Sensitive Paints. AIAA Journal **34**, (1996)
7. Sakaue H., Matsumara S., Schneider S. P., Sullivan J. P., Anodized Aluminium Pressure Sensitive Paint for Short Duration Testing. 22nd AIAA Ground Testing Conference, (2002)
8. Sakaue H., Gregory J. W., Sullivan J. P., Porous Pressure-Sensitive Paint for Characterizing Unsteady Flowfields. AIAA Journal **40**, (2000)
9. Sakamura Y., Matsumoto M., Suzuki T., High Frame-Rate Imaging of Surface Pressure Distribution using a porous Pressure-Sensitive Paint. Measurement Science and Technology **16**, (2005)

Doppler Picture Velocimetry (DPV) applied to hypersonics

A. Pichler, A. George, F. Seiler, J. Srulijes, and M. Havermann

French-German Research Institute of Saint-Louis (ISL), 5 rue du Général Cassagnou, F-68301 Saint-Louis, France, Tel.: +33 3 8969 5045, Fax: +33 3 8969 5048, e-mail: pichler@isl.tm.fr

Summary. The special wide-field Michelson interferometer designed at ISL transforms the Doppler frequency shift of light scattered by tracer particles crossing a light sheet into a shift of luminosity at the output of the Michelson interferometer, giving information on the particle velocity in the light sheet plane. To overcome former disadvantages of the Doppler Picture Velocimetry (DPV) the optical set-up as well as the Doppler picture processing algorithm were greatly improved. The new DPV algorithm now allows an automated calculation of the velocity profiles from the Doppler pictures without manual fringe tracing by the user as done in the past. The working principle of DPV and the new analysis algorithm are explained by means of experiments done at the ISL shock-tunnel facilities. The vertical velocity around a wedge and a sphere model in a Mach 6 flow has been measured. The results have been validated by comparison with Particle Image Velocimetry (PIV) at the same flow conditions.

1 The DPV technique

The Doppler Picture Velocimetry technique (DPV) uses a Michelson interferometer (MI) as described by Seiler et al. [1]. Concerning the optical set-up, the laser light is divided into two parts, see Figure 1, by a splitter cube. One part is focused onto dispersion plates (the use of two plates compared with only one (see [1] [2]), greatly reduces unwanted speckle noise in the reference image), the other part of the laser light illuminates a plane Σ which crosses the gas flow as a light sheet. In order to visualize the velocity distribution in this plane Σ, tracer particles are seeded into the flow and the light scattered from these particles is focused into the specially designed MI, which is used here as a wide field interferometer.

The light scattered from the light sheet plane Σ is, in the case of moving tracer particles, Doppler shifted. The frequency change of the scattered light, as seen by the MI, is transformed into changes $d(\Delta\varphi)$ of the phase difference $\Delta\varphi$ between the two legs of the Michelson interferometer. For the DPV measurements the MI is adjusted to interference fringes.

In order to test the DPV set-up, experiments with different bodies in a Mach 6 flow, have been done at the shock-tunnel facilities at ISL. Figure 2 shows a wedge in the the shock-tunnel.

The Doppler reference image (Figure 3, left) is taken simultaneously with the measured Doppler image (Figure 3, right) using CCD cameras (PCO Pixelfly).

The vertical velocity induced over the wedge surface causes a Doppler shift $d\nu$ of the frequency ν_L of the light scattered by the tracer particles, here titanium dioxide (TiO_2), resulting in a light intensity change dI across the wedge shock wave. This produces the fringe displacement in the Doppler image (figure 3, right) showing the velocity jump across the bow wave attached to the tip of the wedge.

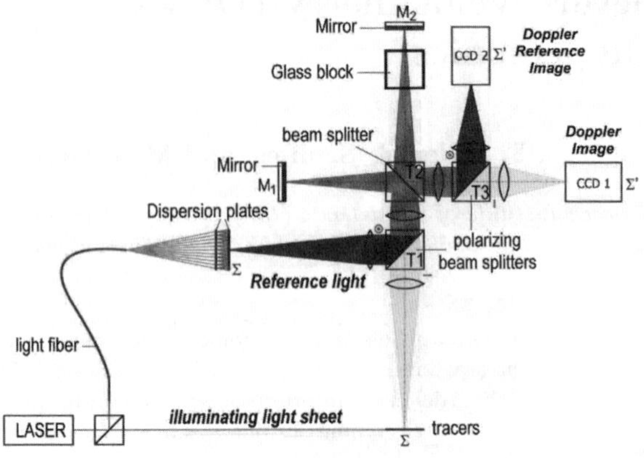

Fig. 1. Optical set-up of the Michelson interferometer

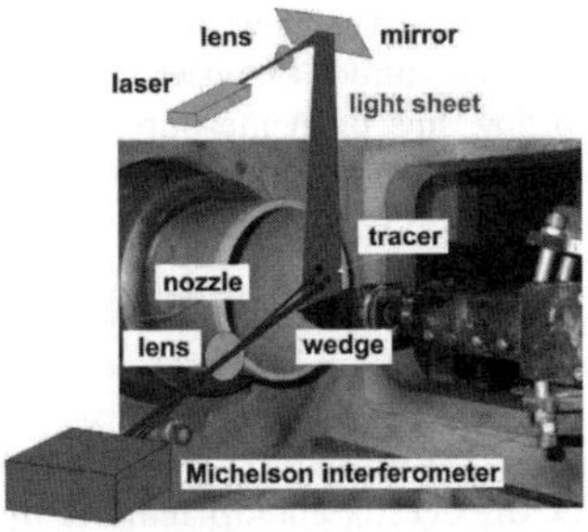

Fig. 2. Model arrangement for DPV in the shock-tunnel

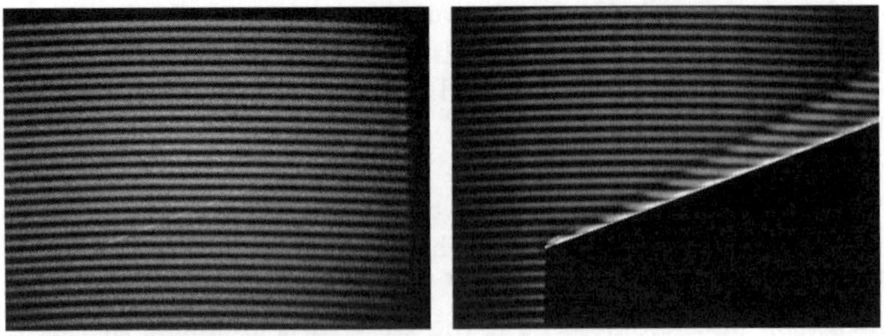

Fig. 3. Reference and Doppler image for a wedge in a Mach 6 flow

2 Automated DPV picture processing

In the captured Doppler images velocity variations are visible as Fizeau fringe displacements. However, the measured velocities cannot be well recognized by this method. A better form of visualization would be a pseudo-colour image of the velocity profile (see section 3 for examples).

All existing DPV picture analysis methods require manual or interactive fringe tracing by the operator. This procedure is quite time consuming[1], unavoidable input errors limit the precision of the results. To overcome these restrictions, Pichler et al. [2] developed a new automated DPV image processing software without need of fringe tracing.

The implemented analysis algorithm is based on a technique using Fast Fourier Transformations (FFT) to examine the Doppler images in the frequency domain, see Takeda et al. [1]. The software (written in Scilab 4.1) allows the automated calculation of the velocity profile from Doppler images (1000 x 1200 pixels) in about 90 seconds on a Standard PC[2].

A possible alternative to the FFT method described in this paper is the use of Least Square Estimators, under development by Pfaff [4].

2.1 Requirements

The DPV algorithm needs to measure the fringe displacements automatically. In order to do this correctly, the Doppler pictures must fulfill the following requirements:

1. The maximum fringe shift (measured in pixels) should not exceed half of the distance between the minima of two fringes, the fringe spacing. So the glass block used in the MI has to be adapted to the expected velocity range.
2. Reference and Doppler images need to be matched in size and rotation. Marker images should be taken before the experiment.

All Doppler image examples used to explain the algorithm steps applied fulfill the requirements discussed above.

2.2 Image filtering

The captured Doppler images contain the velocity variation information in form of fringe displacements. The Doppler images are filtered first to facilitate the automated detection of fringe shifts by the computer. All unwanted brightness variations (not belonging to the fringe system) and high-frequency background noise are removed from the pictures while preserving the original fringe displacements in all details. Figure 4 shows the filtered reference and Doppler pictures for the flow over the wedge from Figure 3.

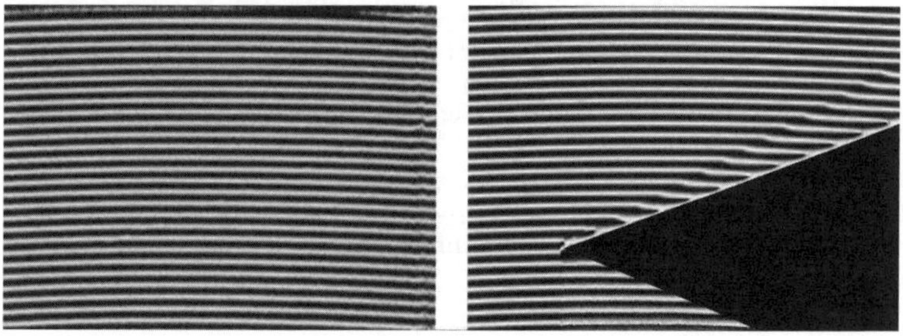

Fig. 4. Filtered Doppler pictures

[1] An experienced operator takes at least 30 minutes.

[2] Intel Pentium 4 (2 GHz) CPU, 1024 MByte RAM, Windows XP

2.3 Velocity profile calculation

The filtered images are processed by an automated DPV algorithm. Additionally, two black/white mask images are needed, one masking out non-analyzable areas (e.g. saturations, no tracers) and one image masking out the model contour to show its placement in the shock-tunnel and its dimensions.

Equation 1 describes the brightness information of each pixel of the filtered Doppler image as:

$$g(x,y) = c(x,y) \cdot e^{j \cdot 2\pi \cdot f_0 \cdot y} + c^*(x,y) \cdot e^{-j \cdot 2\pi \cdot f_0 \cdot y}. \quad (1)$$

A value of $g(x,y) = -1$ corresponds to the brightness minimum, a value of $g(x,y) = 1$ to the maximum, $j = \sqrt{-1}$. The phase shift can be found in $c(x,y)$ or its complex conjugate $c^*(x,y)$:

$$c(x,y) = \frac{1}{2} \cdot b(x,y) \cdot e^{j \cdot \Delta\varphi(x,y)}. \quad (2)$$

In order to extract the phase shift information in equation 2 each column of the Doppler image is transformed in the frequency domain using the FFT:

$$G(f,x) = C(f - f_0, x) + C^*(f + f_0, x). \quad (3)$$

Figure 5 visualizes the amplitude of the FFT for an example column of a typical Doppler image.

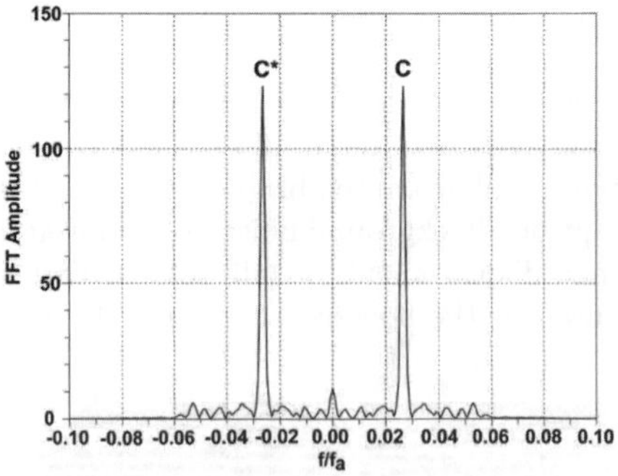

Fig. 5. FFT of an example column of a typical Doppler image

In the next step the sideband C is left shifted on the frequency axis by f_0 to get the spectrum C containing the phase difference information in the base band (demodulation).

To obtain the distribution of the phase difference $\Delta\varphi(x,y)$ the spectrum is transformed back from frequency domain $C(x,y)$ to the image domain $c(x,y)$ using the Inverse Fast Fourier Transformation (IFFT). Finally the complex logarithm of $c(x,y)$ is calculated as follows:

$$\log[c(x,y)] = \log[\frac{1}{2} b(x,y)] + j \cdot \Delta\varphi(x,y). \quad (4)$$

The phase difference $\Delta\varphi(x, y)$ can be found completely separated from any amplitude variation in the imaginary part of equation 4.

Having applied the operations described above to all image columns of both, reference and Doppler pictures, the variation of the phase differences between the Doppler and the reference images are calculated for each pixel:

$$d(\Delta\varphi)(x, y) = \Delta\varphi_D(x, y) - \Delta\varphi_L(x, y). \quad (5)$$

The phase difference variations are proportional to the velocity variation dv:

$$dv(x, y) = v(x, y) = \frac{A}{2\pi} \cdot d(\Delta\varphi)(x, y) \quad (6)$$

The value of the proportionality factor A in equation 6 depends on the length of the glass block used in the Michelson interferometer and the wavelength of the laser used. With the reference image being defined as situation of velocity zero the absolute velocity $v(x, y)$ is equal to the velocity variation $dv(x, y)$.

3 Velocity results

3.1 Wedge

The velocity picture of Figure 6 over a wedge surface shows in pseudo colours the vertical velocity component v present in both the free stream flow and behind the oblique attached shock wave. On the wedge, the measured velocity is of the order of $v \approx 550 \frac{m}{s}$, as visualized in orange in the velocity picture of Figure 6, being constant between bow wave and wedge surface.

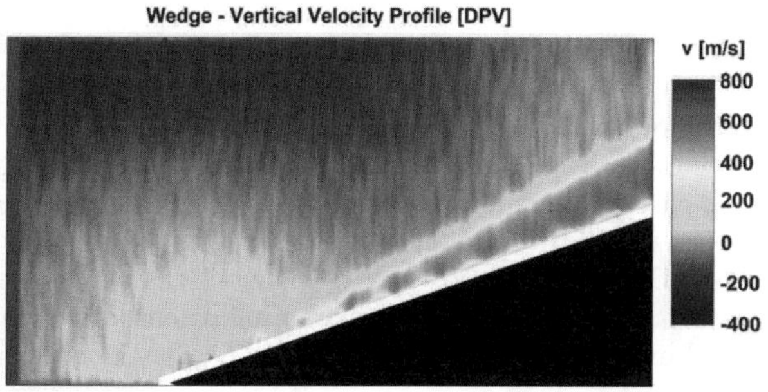

Fig. 6. DPV velocity distribution over the wedge visualized in pseudo colours

The DPV velocity results are in good agreement with Particle Image Velocimetry (PIV) measures at the same conditions [1], see figure 7, as well as theoretical estimates.

3.2 Sphere

The vertical velocity distribution for a sphere is represented in Figure 8. Over the sphere, the flow accelerates to slightly more than 500 $\frac{m}{s}$.

The sphere velocity measurements are in good agreement with PIV and theoretical calculations.

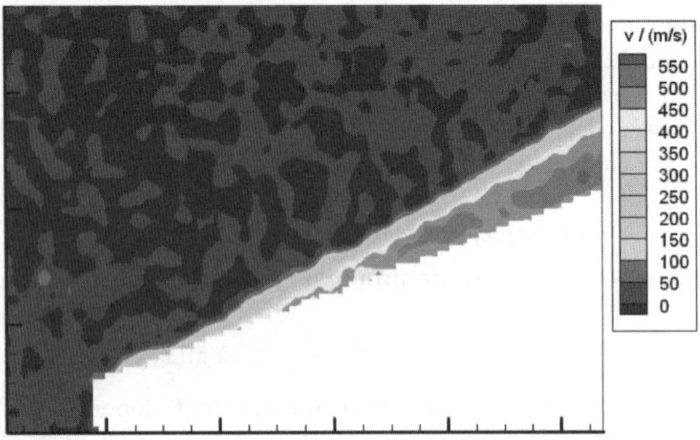

Fig. 7. PIV measurement of the velocity distribution over the wedge

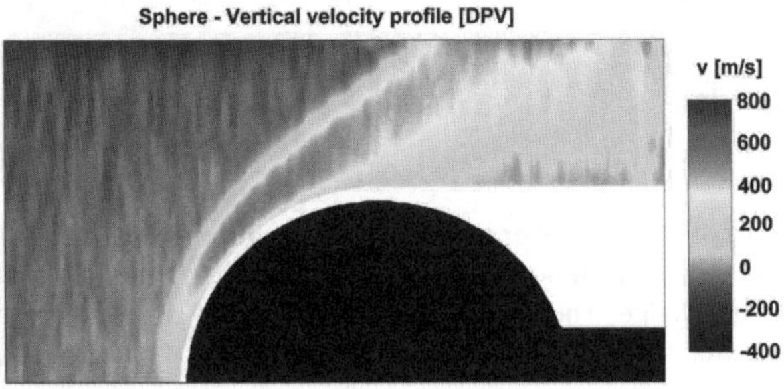

Fig. 8. DPV velocity distribution around a sphere

4 Conclusions

The MI-system used for taking DPV pictures was improved to further optimize the separation of the simultaneously illuminated reference and Doppler pictures. A new DPV algorithm has been developed to analyze the Doppler pictures automatically without user interaction. It has been tested by taking Doppler pictures of the flow around a wedge and a sphere. The measured and visualized velocity distributions agree very well with PIV measurements and theoretical estimations. Future applications of DPV are time-resolved and/or time-averaged measurement of velocity profiles in hypersonic flows.

References

1. Seiler F., George A., Srulijes J., Havermann M.: *Progress in Doppler Picture Velocimetry (DPV) Technique*, Proc. of the 12th International Symposium on Flow Visualization, Göttingen, Germany, 2006
2. Pichler A., George A.: *Das Dopplerbild-Verfahren (DPV). Automatische Auswertung nach der FFT-Methode*, ISL-Report R-109/2007, 2007
3. Takeda M., Ina H., Kobayashi S.: *Fourier-transform method of fringe-pattern analysis for computer-based topography and interferometry*, Optical Society of America, 1981
4. Pfaff R.: *Automated processing of the ISL Doppler images*, Bachelor thesis, FernUniversität Hagen, Febuary 2007

On the conservation laws for light rays across a shock wave: Toward computer design of an optical setup for flow visualization

M. Sun

Center for Interdisciplinary Research, Tohoku University, Sendai 980-8578, Japan *

1 Introduction

It has been our continuous attempt to develop a computer-aided technique for the design and optimization of an optical setup for compressible flow visualization. The basic idea is rather simple. By representing an experiment by numerical data, given an optical setup, one can get the corresponding numerical photo of the experiment in a computer, so that one can design and/or optimize the optical setup. Then one just assembles and allocates optical components in a real laboratory according to what he has decided in computer. The technique can significantly reduce time and labor required in conducting flow visualization especially for short-duration and costly experiments. It can also provide a tool for the investigation of any abnormality appeared in a real visualization result.

The idea relies on two solution techniques. One is how to get the proper numerical data to represent the experiment to be visualized. This can be achieved by using a reliable CFD solver. The costs in time and labor required to get the numerical data should be as low as possible, as least lower than those needed in conducting one or a few real shots of experiments. In addition, the solver should be easy-to-use for experimenters. This is being solved by constructing a virtual shock tube, with which one can get numerical data, as one operates in a real shock tube. For example, for the experiment of shock wave reflection, one just needs to specify the shock Mach number and the wedge angle, and then the virtual shock tube will provide related numerical data [6].

Another problem is how to construct a numerical photo from numerical data according to the optical principles for a given optical setup. The available methods to construct shadowgraph and schlieren images are based on numerous assumptions, such as parallel incident beams, and smooth variation of density field. The derivative formula originally derived from smooth flows has been used for granted for shock waves. The author suspected strongly the methodology, because it might violate a sort of conservation laws for light rays, and tried to find a *conservative* way to construct the numerical photo.

In the first attempt, a simple camera system, a portion of the shadowgraph setup, was tested. It contains only one design parameter, the location of the focal plane. The incident rays are assumed to be parallel and perpendicular to the test window, and to propagate along a straight path in flowfield. The angle of deviation of a ray is calculated by integrating the first order derivative of density over the path. The grayscale shadowgraph photo was successfully obtained by considering *the conservation of light energy* striking in a ray tube over a period of time [7].

In the second attempt, a complete shadowgraph setup consisting of a light source and two lenses was tested. Incident rays are allowed to incident at the test window at

* Also Interdisciplinary Shock Wave Laboratory, Transdisciplinary Fluid Integration Research Center, Institute of Fluid Science, Tohoku University, Sendai 980-8577, Japan

any angle, and to refract in the flowfield. The angle of deviation is solved according to the Snell's law, a solution technique satisfying *the conservation of the angle of deviation* across a shock wave. The possible total reflection is deliberately neglected. The strategy has been used to clarify the influences of the locations of both lenses and the light sources in the shadowgraph setup [8]. The early tests also indicated the possibility to evaluate the schlieren setup having additional slit and cutoff.

In this paper, the ray tracing method that satisfies the conservation laws is discussed in more detail and verified by two simple cases.

2 Optical setup

The optical setup to be tested in this paper was shown in [8]. The entire setup consists of five optical elements, a light source on the leftmost end, a slit, two lenses (lens-1 and lens-2), a cutoff, and a recording-plane on the rightmost end. The slit and the source light are treated as one light source with the shape of the slit. The lens-1 and lens-2 are collimating lens, and defined by five parameters, the center coordinates(x, y, z), the focus length, and the diameter. The optical axis is taken as the x-axis. The recording plane is the plane on which the image is projected, defined by eight parameters, the center coordinates, the normal vector, the height and the length. In our simulation, the origin of the coordinate is set at the center of the test section. In order to evaluate and optimize the optical setup for a specified flowfield, it is necessary to construct the numerical image at the recording plane by simulating the movements of rays from the light source to the recording plane.

3 Ray tracing method

The complete optical setup together with the flowfield is solved by using the ray tracing method, from the light source to the recording plane. Only geometrical optics is followed in this work. The technique has been commonly used in optical design. In one word, a ray goes straight, or gets refracted by an interface separating two media with different indices of refraction. The deviation of the refracted ray is determined from the Snell's law. The detailed procedure will not be repeated here, and the reader may refer to [3] for more information.

Tracing a ray through flow field was generally considered to be an integral of derivatives of density along the light path [2,4,9]. The method is valid for flows with smooth density variations. Strong refractions in condensed media [2], and their interfaces may be serious enough to invalidate this approach. In the present method, we determine the angle of refraction directly from the Snell's law, without recourse to density derivatives.

A ray moves straight within a grid cell, illustrated in Fig. 1. A ray enters the cell from point S, leaves from point E. A predictor-corrector approach is followed to calculate the location of end point E and the outgoing direction vector $\mathbf{K}_e$. In the predictor step, a ray travels in its initial direction $\mathbf{K}_s$, and the ending point E* can be obtained. Combining two refractive indices at the starting point S and the ending point E*, together with the orientation of the interface, one can obtain the predicted outgoing direction vector $\mathbf{K}^*$. The corrector steps follows the very similar procedure as the predictor step, the only difference is the initial direction vector $\mathbf{K}_s$ is replaced by $(\mathbf{K}_s+\mathbf{K}^*)/2$. The final outgoing

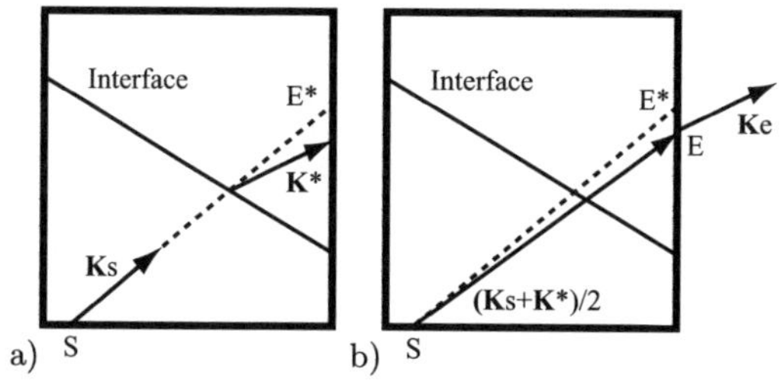

Fig. 1. Ray tracing in a grid cell: a) predictor step; b) corrector step

point E and the outgoing vector $\mathbf{K}_e$ are taken as the initial conditions for tracing the ray in the next cell. The procedure is repeated until the ray reaches the right window of the test section or the boundary of the numerical data.

The outgoing direction vector of the light ray crossing an interface in the grid cell is basically determined by the law of refraction, known as Snell's law, that states

$$n_1 sin(\theta_1) = n_2 sin(\theta_2), \tag{1}$$

where n_1, n_2 are the index of refraction at the starting and the end point respectively. The angle formed by the incident ray and the normal of the interface is denoted by θ_1, and the refracted ray θ_2, as sketched in Fig. 2a. We show here that if the deviation angle of a light ray is evaluated by (1), the total angle of deviation integrated for a ray traveling through a shock wave is unchanged, even for the numerically captured shock waves that are commonly represented by more than one grid cells. Suppose there are two additional layers n_1^*, n_2^* lying between n_1 and n_2 that denote the refractive indices on the both sides of a shock wave, as shown in Fig. 2b, applying (1) for the three interfaces, one gets

$$n_1 sin(\theta_1) = n_1^* sin(\theta_1^*) = n_2^* sin(\theta_2^*) = n_2 sin(\theta_2), \tag{2}$$

It is clear that the final angle of deviation depends only on the values of refractive indices at the starting and end point, no matter how many interfaces the ray has traveled.

However, the law of refraction gives no solution of θ_2, for $\frac{n_1}{n_2} sin(\theta_1) > 1$. In this case, the law of reflection is followed instead as sketched in Fig. 2c, $\theta_3 = \theta_1$. This phenomenon is known as total internal reflection. Notice that the law of reflection does not reflect

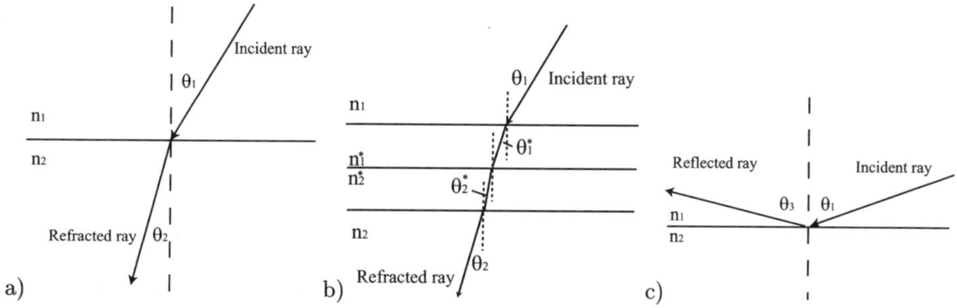

Fig. 2. a) The law of refraction; b) the conservation of the angle of deviation for a ray across a numerically captured shock wave; c) the law of reflection.

any variation of the refractive index. It suggests that under this situation the optical setup may fail to resolve flow variations. The internal reflection commonly appear in the visualization of a planar shock wave using the parallel incident beams, in which $\theta_1 \approx 1$.

Every three neighboring rays form a triangular cell that is defined as a ray tube. The initial ray tube is defined on the left window, and the irradiance of the tube can be calculated from the light source according to the geometrical optics. The final ray rube is located at the recording plane. It is generally deformed since a ray travels in its own path through the flowfield, and it may be affected by optical components differently. A deformed ray tube can possibly overlap with other ray tubes. The irradiance of the final rube satisfies

$$I_1 = S_0 I_0 / S_1,$$

where S_0 , I_0 are the area and irradiance of the initial ray tube, and S_1, I_1 are those of the final ray tube at the recording plane. An expanded ray tube therefore leads to a dark region, and a contracted ray tube creates a bright region. The irradiance at the overlapping regions is obtained by summating all ray tubes there, so that the overlapping regions often appear to be bright.

4 Tests

Two simple fields of the refractive index are tested in order to verify the solution strategy. In all tests, 40×40 structured rays are distributed over a $100mm \times 100mm$ test window, and the numerical data is represented by the same number of grid cells. The width of the test section is $200mm$. One test is a simple shock solution, the refractive indices on both sides satisfying

$$n^L = 1 + 12.9\kappa, n^R = 1 + 1.29\kappa,$$

where κ is the Gladstone-Dale constant, taken as 2.2×10^{-4} in this paper. The field corresponds to the refractive indices of 10 bar and 1 bar air filled on the two sides. The images obtained on the right window for a point light source is shown in Fig. 3. The light source is located at the left focal point of the lens-1 ($f = 500mm$), and its location is perturbed in z-direction. Shifting the light source in z-direction forms a tiny angle, ϵ, between the incident beams and the normal of the left window of test section, $\epsilon = z/f$, where z is the coordinate of the light source in z-direction. For $\epsilon = 0$ or $z = 0$, a plain image without any disturbances is generated, so it is not shown in the figure. For positive z values, the light rays come from the right to the left of the shock. A sharp dark band is followed by a bright band at the shock region. This observation agrees well with the second derivative of the refractive index. However, the images behave quite differently for negative z values, mainly due to the light rays experience an internal reflection at the shock front. A similar behavior is observed for a light source with a diameter of $5mm$, as shown in Fig. 4.

The second test case considered is a sine distribution,

$$n = 1 + 1.29\kappa[5.5 + 4.5sin(4\pi x)], \tag{3}$$

which contains the same maximum and minimum as the first test. The results are shown in Fig. 5. If the incident angle ϵ is large enough, the image appears as the sine function that roughly corresponds to the second derivative of the refractive index. For small incident angle, the portion of image appears to be plain, where the light rays undergo a total reflection.

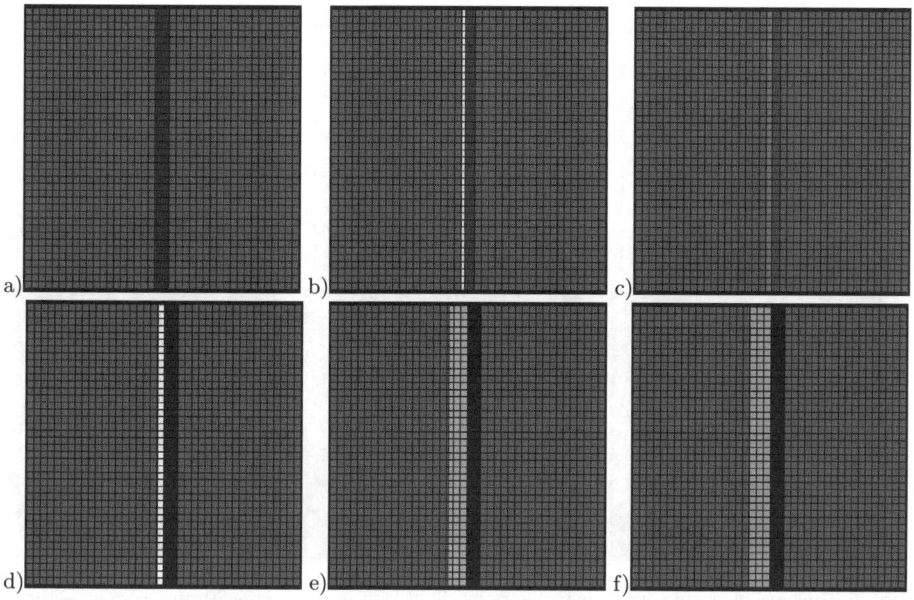

Fig. 3. Images of a shock wave using a point light source: a)z=-5mm; b)z=-3mm; c)z=-1mm; d)z=1mm; e)z=3mm; f)z=5mm;

Fig. 4. Images of a shock wave using a circular light source of $5mm$ in diameter: a)$z = -5mm$; b)$z = -3mm$; c)$z = -1mm$; d)$z = 1mm$; e)$z = 3mm$; f)$z = 5mm$.

Fig. 5. Images of a sine wave using a point light source: a)$z = -20mm$; b)$z = -10mm$; c)$z = -5mm$; d)$z = 5mm$; e)$z = 10mm$; f)$z = 20mm$.

5 Conclusion

A solution strategy for constructing a realistic image for the design and optimization of optical setup is summarized. The technique provides reasonable good results for two test problems. It is observed that the internal light reflection may affect the image as well.

References

1. Hecht E. (2002) Optics, 4th Ed., Addison Wesley
2. Merzkirch W., Flow Visualization, Academic Press, Inc., (1974); 2nd edition (1987).
3. Mouroulis P., Macdonald J. (1997) Geometrical Optics and Optical Design, Oxford Univ.
4. Settles G.S. (2006) Schlieren and Shadowgraph Techniques, Springer, 2nd printing.
5. Sun M., Takayama K. (1999) Conservative smoothing on an adaptive quadrilateral grid, J. of Computational Physics, V150: 143-180
6. Sun M (2007) A calculator for shock wave reflection phenomenon, ISSW26.
7. Sun M, Takayama K, Ikohagi K (2005) Computerized visualization of numerical data, ISSW25, Bangalore, India.
8. Sun M (2006) Computer modeling of shadowgraph optical setup, ICHPP27, Xi'an, China.
9. Yates L.A. (1993) Images constructed from computed flowfields, AIAA J. 31(10), 1877-1884.

Shock stand-off distance over spheres flying at transonic speed ranges in air

T. Kikuchi[1], D. Numata[1], K. Takayama[2], and M. Sun[3]

[1] *Graduate school of engineering, Tohoku University, 2-1-1, Katahira, Aoba, Sendai, Miyagi, 980-8577, Japan*
[2] *Tohoku University Biomedical Engineering Research Organization, Tohoku University,2-1-1, Katahira, Aoba, Sendai, Miyagi, 980-8577, Japan*
[3] *Center of Interdiciplinary Research, Tohoku University 6-3, Aoba, Aramaki, Aoba, Sendai, Miyagi, 980-8578, Japan*

1 Introduction

Shock stand-off distances over spheres vary very widely at transonic flow ranges from large distance to gradually that at supersonic flow ranges. In the past, the shock stand-off distance over blunt bodies were visualized intensively [1] [3] [2], however, the range of shock Mach number is close to 1.1 in air [4]. In the advent of the Apollo project, experimental and numerical studies of shock stand- off distance over spheres were performed at hypersonic speeds and recently to illustrate the effect of real gas on the shock stand-off distances, a series of ballistic range experiments was conducted at hypersonic flow ranges [5].

In 2001, the introduction of sonic cruisers operational at Ms ranging from 0.95 to 1.00 once revived the study of the drag forces in transonic flow range. Jameson reported recently new numerical result regarding the sonic cruiser [6]. The drag force reduction of supersonic biplane, which may reduce sonic boom at transonic and supersonic flow ranges, was reported .

In experimental determinations of shock stand-off distances over spheres at Ms = 1.01 to 1.1 in air, the wind tunnel experiments would never be useful but ballistic range tests with the combination of optical flow visualizations would give only a reliable way to collect data. Traditionally the shock stand-off distances at sonic speed ranges were tested in ballistic ranges, nevertheless it would be true that due to the inaccuracy existing in ballistic range data acquisitions, they were recognized not as major facilities for quantitative gas-dynamic studies. However, recently we became interested in the operation of two-stage gas guns convertible to high- speed gas dynamics study with combination of very elongated recovery tank, which enabled the two-stage gas gun to serve for ballistic ranges. Adjusting their operational method for a wider launch speed range, we succeeded to project a moderately large diameter sphere at high subsonic to low supersonic speeds without reducing the degree of reproducibility. We applied double exposure holographic interferometry to the ballistic range observations [7].

This paper reports a result of holographic interferometric observations on free flight spheres in atmospheric air at Ms = 1.01 to 1.1. We firstly describe the experimental setup and method of sphere speed measurements and secondly summarize the shock stand-off distances.

2 Experiment

2.1 Facility

We converted a two-stage light gas gun of the Shock Wave Research Center of the Institute of Fluid Science, Tohoku University to a single stage light gas gun operation. Figure 1 shows its high- pressure section, in which pressurized helium up to 5MPa of 45 litters in volume can be filled. A 51 mm diameter and 3.5m long acceleration tube is connected to the T shape section and opened to a 1.7 m diameter and 12 m long stainless steel recovery tank, along the side wall of which three pairs of 600 mm diameter observation ports are attached.

We use a diaphragm-less piston mechanism to hold the cylindrical sabot and sphere combination in front of the piston, which is backed up by auxiliary high-pressure helium slightly higher than the fill pressure in the T shaped section. Upon decreasing the auxiliary helium very quickly down to close to ambient pressure, the piston recedes resulting in the rush of pressurized helium onto the rear surface of the projectile. Hence the projectile is accelerated consistently if the fill pressures are precisely controlled. Eventually the scatter of projectiles muzzle speeds became less than 1%.

When the projectiles are ejected from the acceleration tube, a precursory shock wave exists always in front of a projectile, which may be strong enough to affect the process of sabot separation. To get rid of this effect, we connect a muzzle blast remover at the end of acceleration tube as shown in Fig.2. The muzzle blast should be attenuated quickly otherwise it remains strong so as to influence the entire flow fields around the sphere. The muzzle blast remover consists in principle of a co-axial perforated tube, along which the precursory blast wave attenuates through perforations.

Fig. 1. Ballistic range

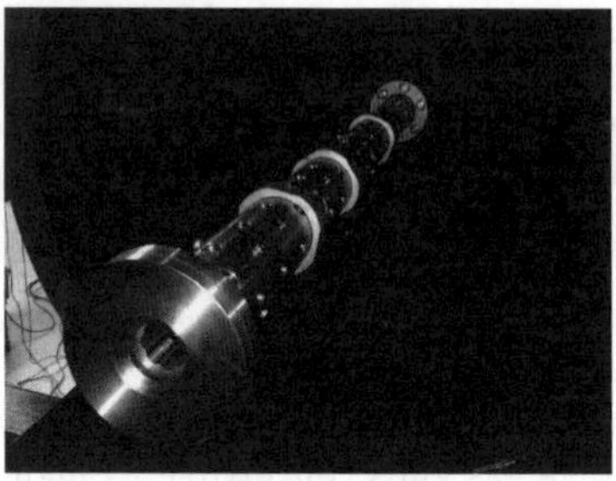

Fig. 2. Muzzle blast remover

2.2 Sabot separation

Removing the muzzle blast, we have to separate the sphere from the sabot. We accommodate a 40 mm diameter sphere made of polycarbonate in a 51 mm diameter sabot. To make the four piece sabot decomposed from the sabot-sphere combination without causing any extra force exerting on the sphere, we empirically learned the sabot shape. For higher speed sabot separation, it is relatively easy to split a sabot and an object body by taking advantage of so-called aerodynamic sabot separation. However, in transonic speed ranges, the difference of speeds of waves driven by sabot segments and that of the shock wave in front of the sphere is very small. Hence we have to take a special care in designing the experimental. A sphere model and a four-piece sabot are shown in Fig.3. The projectile is 40 mm diameter polycarbonate sphere of 40.3 gram in weight. Sabots are polycarbonate 4-parts type with hole of 28 mm in diameter on the bottom. Total mass of projectile and sabots is 101.5 gram.

Presumably this is one of the reasons why in transonic ballistic range experiments the total length of previous ballistic ranges were so extend as to be many hundred meters long. In this experiment, however, the laboratory space is so limited that we have to adopt a more direct control of suppressing waves and wavelets.

Although loosely separated by aerodynamic drag forces, the sphere model and four-piece sabot should be more distinctly separated from each other. We placed a thick steel plate with a relatively larger hole in its center so as for the sphere to pass through the center hole but sabot segments collide against the plate. The shock waves precursory to the sabot segments are attenuated by passing through the three-stages of baffle plates. As a result of this arrangement, just passing the visualization section, the sphere model is independent of any wavelets associated with blast waves.

Fig. 3. Sphere shaped projectile and 4-parts sabots

2.3 Optical arrangement

The optical arrangement of double exposure holographic interferometer is shown in Fig.4. Light source is a holographic double pulse ruby laser: the energy of approximate 1.0 J

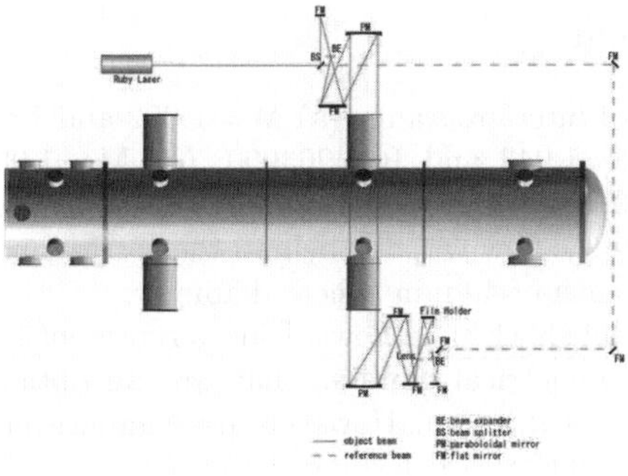

Fig. 4. Optical arrangement

and pulse duration of 25 ns. Source beam is split into 6:4 intensity ration, 60 % for object beam OB and 40 % for reference beam RB. OB is expanded and collimated between two 500 mm diameter paraboloidal schlieren mirrors passing through 600 mm diameter and 20 mm thick acrylic observation window. RB and OB take nearly identical light path and expose on a 100 mm x 125 mm holographic sheet film AGFA 10E75. The first exposure is performed before the shock arrival and the second one is synchronized by the arrival of the sphere at the test section.

2.4 Velocity measurement

We used at first VISAR (velocity interferometry from the surface of any reflector) to measure in-bore projectile acceleration. At the same time the projectile speed was measured by using a laser cutting method at the position putted between the sabot remover and visualization section. We then found that the terminal projectile speed measured by VISAR is that measured by the laser cut is not necessarily shows the identical value but the latter is always smaller, which implies that after the sabot separation spheres are slightly attenuated. We then placed two laser-cut devices in front of the observation section and assured the sphere speed. Experimental setup inside of recovery tank is shown in Fig.5.

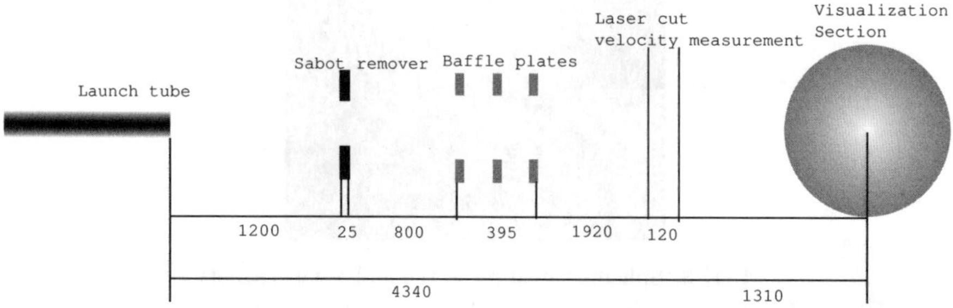

Fig. 5. Experimental set up

3 Results and discussion

3.1 Visualization results

Figures 6 show a series of interferograms: (a) Ms=1.002 and Re=925000; (b) Ms=1.025 and Re=949000; (c) Ms=1.043 and Re=963000; (d) Ms=1.066 and Re=983000; (e) Ms=1.080 and Re=995000; (f) Ms=1.104 and Re=110400, where Re is the Reynolds number and is defined by the flow variable behind the corresponding incident shock wave and the characteristic length of 40 mm sphere diameter.

Unlike two-dimensional shock tube flows, fringes represent the density integral along the OB path in axial symmetrical flows so that we can obtain corresponding density profiles from the fringe distribution. However, its comparison with numerical results would be made in a straightforward fashion if the numerical density can be presented to match with interferometric display. The present interferometric image clearly shows a detached shock wave over a sphere and its structure. Boundary layers developing along the sphere surface and the oscillatory wake structure are well recognized. The wake shows a wavy structure. This implies that although we assume the steady sphere motion, the entire flow over the sphere is subsonic and eventually unsteady.

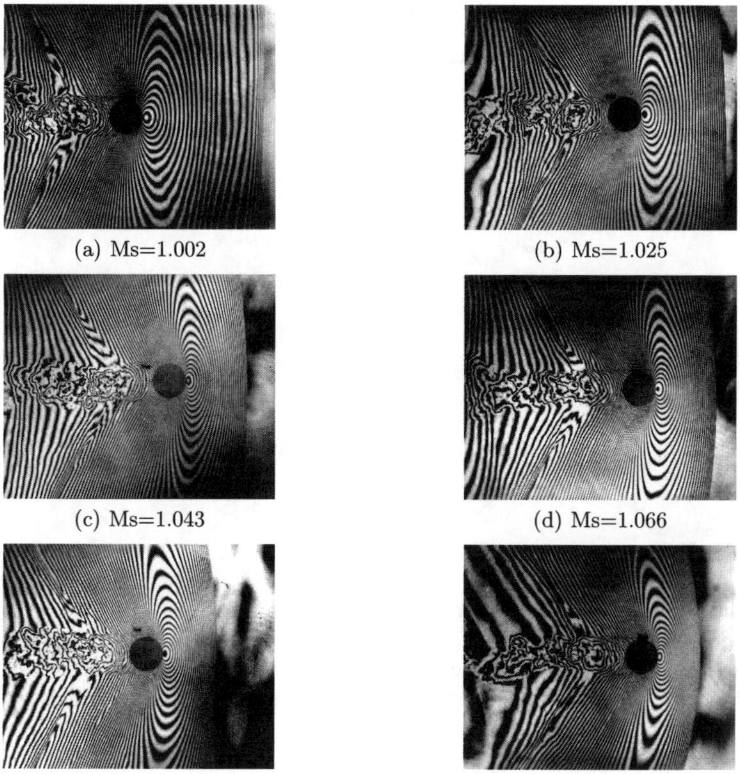

(a) Ms=1.002 (b) Ms=1.025

(c) Ms=1.043 (d) Ms=1.066

(e) Ms=1.080 (f) Ms=1.104

Fig. 6. Visualization results

3.2 Shock stand-off distance

We measured the shock stand-off distance directly on the interferograms and the result is summarized in Figs. 7, whereas the shock stand-off distance is normalized by the sphere diameter against the shock Mach number ranging from 1.0 to 12 in Fig.7(a) and 1.0 to 1.2 in Fig.7(b). The shock stand-off distance increases monotonously to infinite as the sphere speed approaches to Mach number unity. Even at high subsonic speed, detached shock waves exist in front of the sphere and propagate at the sonic speed. The shock stand-off distance increases with the elapse of time and hence the flow can no longer steady. Hence in that sense, the shock stand- off distance at Ms= 1.0 is infinite at longer time. The structure of detached shock waves would vary from partially dispersed one to fully dispersed one while Ms becomes smaller than unity. We noticed that our measured data points depart slightly from the prediction, which may be attributable to the error caused during velocity measurements.

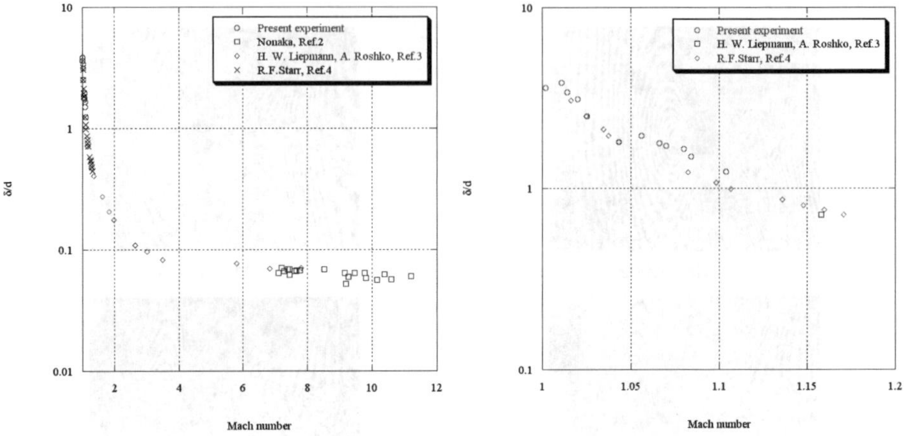

Fig. 7. Shock stand-off distance

4 Concluding remarks

We launched 40 mm diameter spheres in atmospheric air at Ms = 1.01 to 1.1 by using a gas gun. Holographic interferometric observations were performed and the shock stand-off distances in this speed range were estimated. We found on interferograms that the flow field cannot be steady even though the sphere speed is steady, which implies that numerical solution will never converge..

References

1. Juergen W. Heberle, George P. Wood, Paul B. Gooderum, "Data on shape and location of detached shock waves on cones and spheres", NACA technical note 2000 (1950)
2. Van Dyke,M.D.,"The Supersonic Blunt-Body Problem-Review and Extensions", J.Aeronaut.Sci, pp485-496(1958)
3. H. W. Liepmann, A. Roshko, Elements of gasdynamics, John Wiley & Sons, New York(1957)
4. R.F.Starr, A.B.Bailey, M.O.Varner,"Shock Detached Distance at Near Sonic Speeds" AIAA Journal technical note,Vol.14, pp537-539(1976)
5. S.Nonaka, H.Mizuno, K.Takayama, C. Park, "Measurement of Shock Standoff Distance for Sphere in Ballistic Range", Journal of Thermophysics and Heat Transfer Vol.14, pp225-229(2000)
6. Antony Jameson, "An Investigation of the Attainable Efficiency of Flight at Mach One or Just Beyond", 45th AIAA Aerospace Sciences Meeting and Exhibit, 37(2007)
7. K. Takayama, "Applications of holographic interferometry to shock wave research", SPIE 398:pp174-180(1983)

Shock tube study of the drag coefficient of a sphere

G. Jourdan[1], L. Houas[1], O. Igra[2], J.-L. Estivalezes[3], C. Devals[3], and E.E. Meshkov[4]

[1] *Polytech' Marseille, DME, IUSTI/UMR CNRS 6595, Université de Provence Technopôle de Château-Gombert, 13013 Marseille (France)*
[2] *The Pearlstone Center for Aeronautical Engineering Studies, Department of Mechanical Engineering, Ben Gurion University, Beer Sheva (Israel)*
[3] *ONERA CERT DMAE, 2 Av. E. Belin, B.P. 4025, 31055 Toulouse, Cedex 4 (France)*
[4] *Russian Federal Nuclear Center, Institute of Experimental Physics, 607190 Sarov (Russia)*

1 Introduction

For evaluating the motion of a solid object in a gaseous medium one has to know the drag coefficient of the considered object. It is therefore not surprising that significant research was done and published on this matter during the past century. In experimental investigations aimed at finding the drag coefficient of a solid sphere moving through a fluid various techniques were employed. These different techniques were used for evaluating the drag coefficient over a wide range of Reynolds and Mach numbers. Among the frequently used methods one should mention freely falling spheres in liquid or in air, spheres placed in wind tunnels, spheres mounted on flying aircraft, spheres towed in water channels and spheres flying in aero-ballistic range. In many of these experiments the relative velocity of the sphere was constant, or almost constant. Based on results from experiments conducted using the above mentioned techniques a 'standard drag coefficient' curve has been derived for a sphere. For low subsonic speeds one curve represents the spheres drag coefficient over a large range of Reynolds number; this curve is available in almost all fluid mechanics textbooks. This detailed and accurate knowledge available for the drag coefficient of a sphere is limited to the case where the sphere's relative velocity is constant, i.e., to a steady flow condition. However, in many engineering applications the sphere(s) motion through the fluid in which it is immersed is not steady; i.e., it experiences acceleration or deceleration. For example in nozzle flow of a solid propellant rocket, flows behind shock/blast waves propagating into a dusty gas, volcanic eruptions etc. Using the standard drag curve for such flows is, at least, questionable. For having a reliable simulation of a non-stationary two phase flow (solid particles immersed in a gaseous medium), an accurate knowledge of the particle drag coefficient is essential. Measurements of the sphere's drag coefficient in non-stationary flows were conducted during the past four decades. Among the published results one should mention the work of [1], [2], [3], [4], [5], [6], [7] and [8].

2 Experimental facilities

Experiments were conducted in the 80 by 80 mm cross-section, multi-phase variable inclination shock tube of the University of Provence. Details regarding the shock tube geometry and performances are available in [9]. Using this shock tube allows experiments with incident shock wave Mach number within the range $1.1 < M_{is} < 5$. This in turn allows the present investigations to cover both subsonic and supersonic post-shock flow cases. As was done in previous experiments [7], the sphere's drag coefficient, CD was deduced from its trajectory. In order to obtain an accurate trajectory of the investigated sphere the windows in the presently used test-section provided a field of view of 80 by 300 mm (300 mm in the flow direction). The spatial locations of the sphere during its motion, induced by the post shock flow, were recorded using shadowgraph technique coupled with high speed camera and a stroboscopic Nanolite flash lamp which was synchronized with the rotating-drum camera. This optical facility enables photographing every 70 μs during each experiment (i.e., having around 60 photos per run in the subsonic cases and about 35 photos per run in the supersonic cases), thereby ensuring accurate construction of the tested sphere's trajectory. PCB piezoelectric transducers were used for recording pressure histories needed for deducing the incident shock wave Mach number and for triggering optical and recording facilities. Recording the post shock pressure history is essential since we limited the construction of the sphere's trajectory to the duration in which a uniform post-shock flow prevails. During this period of time the post shock flow properties can easily be predicted using the Rankine-Hugoniot shock relations. Several type of spheres (solid or hollow sphere and expanded foam) composed of different material (Polystyrene, Polymer ATECA, Polypropylene and Nylon) were used in the present study, with a diameter ϕ and a material density ρ_p ranging from 500 μm to 6.5 mm and from 25 kg/m^3 to 1550 kg/m^3, respectively. This allows covering a large domain of Reynolds and Mach numbers, based on the relative velocity between the flow and spheres. A crucial point in the experiments was the problem how to keep the tested spheres away from the test section walls. After checking various options the following method was chosen. The tested sphere was suspended from the tube ceiling, close behind the entrance to the test section, on a wire taken from a spider-web. The spider-wire was strong enough to keep the small sphere suspended in the air until the arrival of the incident shock wave. Moreover, it quickly accelerates to the post-shock gas flow velocity before disintegrating. Fig. 1 contains three shadowgraphs taken 36 μs before the shock collision with the sphere (a), 34 μs after the incident shock wave hit the sphere (b) and 104 μs after its interaction

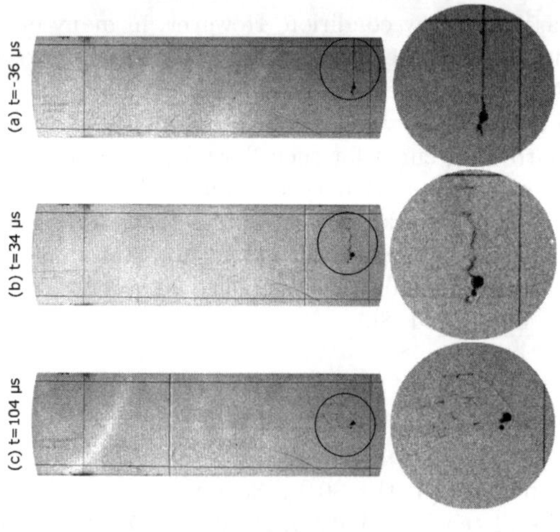

Fig. 1. Three spheres suspended on a spider wire, (a) before arrival of the incident shock wave, (b) just after and (c) shortly after the head-on collision between the spheres and the incident shock wave. In this case the flow behind the shock is supersonic and the incident shock wave is moving from right to left.

with the 1.92 mm diameter nylon sphere suspended on the spider-wire (c). As can be seen, the process of the spider wire acceleration and disintegration is completed very quickly after it collides, head-on with the incident shock wave. Disturbances produced by the spider-wire and its breaking were hardly noticeable. On the other hand, given that the post shock flow is supersonic, the detached shock wave from the sphere is easily noticed in Fig. 1c.

3 Theoretical background

When a solid particle is exposed to a gas flow, its response depends on the relative velocity that exists between the particle and the flow. For a low concentration of solid particles in a suspension, one can ignore interaction between solid particles and their contribution to the suspension pressure. In such a case the drag force acting on the solid particle(s) is the sole meaningful force that determines the particle motion. In experiments conducted in shock tubes with relatively small (and therefore light) particles, the particle experiences very large acceleration due to the very fast post-shock gas flow. Until the particle reaches the post-shock flow velocity, the relative velocity between the particle and the gas flow changes and the particle motion is non-stationary. Should the particle trajectory be recorded accurately, its drag coefficient could be evaluated from the particle's equation of motion (for details see [7], [9]).

4 Results and discussion

Unlike in the experimental work of [7] where many experiments were repeated in order to construct the sphere's trajectory, in the present case about 35 to 60 photos with time difference of 70 μs between successive frames were taken in a single test. Therefore, the obtained temporal locations of the flying sphere were more than needed for an accurate construction of the sphere trajectory. In the present results the accuracy in measuring the sphere location, from recorded shadowgraphs, is to within ± 1 mm. In some experiments a few particles were tested in the same run. In such cases the different spheres were suspended either along a line perpendicular to the flow direction for minimizing wake interference between tested spheres, or with significant longitudinal gap between them. In all experiments where the post-shock flow was subsonic, the flying sphere experiences a uniform post-shock flow until the arrival of the reflected shock wave from the tube end-wall. In all experiments where the post-shock flow was supersonic, the flying sphere experiences a uniform post-shock flow until the arrival of the contact surface. In evaluating the sphere drag coefficient only the part of the sphere's trajectory constructed inside the uniform post-shock flow zone was used. In this zone the post-shock flow conditions were derived using the Rankine-Hugoniot shock relations and the measured incident shock wave Mach number. In the following, the procedure used for deducing the drag coefficient of spheres exposed to the shock wave induced flow is outlined for only one case. A detailed description of additional investigated cases is available in [9]. Sample of obtained shadowgraphs for a subsonic case (run#166) is shown in Fig. 2, where two different spheres were employed. One, a nylon sphere (ϕ=1.96 mm and ρ_p=1204 kg/m^3) was kept suspended on a spider wire just at the entrance to the test section (t=0μs), the second sphere, made of polystyrene foam (ϕ=6.5 mm and ρ_p=25 kg/m^3), was placed far

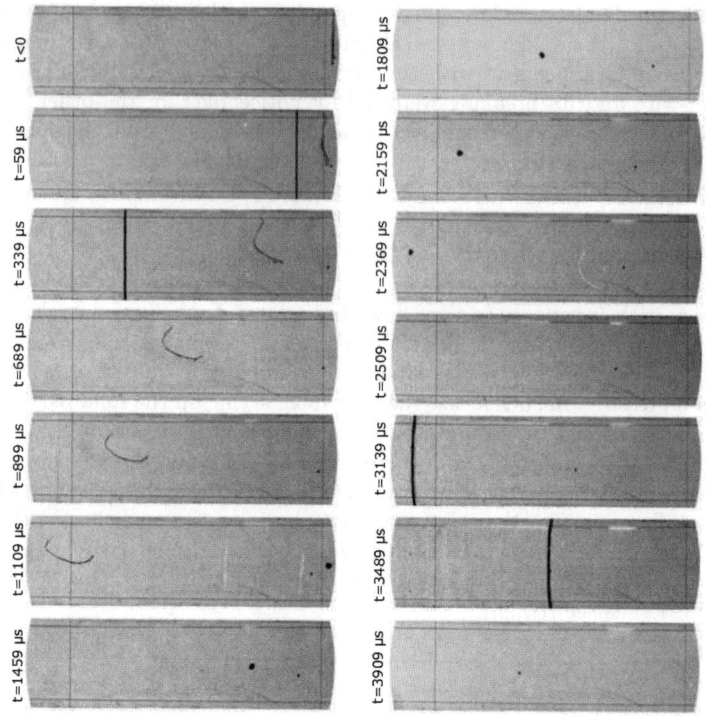

Fig. 2. Sample of shadowgraphs from run#166 in which the post shock flow is subsonic carrying a nylon ball (ρ_p=1204 kg/m^3) 1.96 mm in diameter and a foam ball (ρ_p=25 kg/m^3) 6.5 mm in diameter. The incident shock wave is moving from right to left.

upstream of the nylon sphere and therefore is not seen until t=1109 μs. It is apparent from these shadowgraphs that the spheres move along horizontal lines and therefore one can safely neglect gravity effect and a possible sphere's rotation. As expected, the sphere response to the shock induced flow is not instantaneous and for the first 300 μs after the collision with the incident shock wave the nylon ball hardly moves. The 6.5 mm polystyrene foam ball is significantly lighter than the 1.96 mm nylon ball. Although it was placed far upstream of the nylon ball it reaches the smaller ball at about t=1110 μs and soon thereafter passes it. Shortly before t=2500 μs the large light polystyrene foam ball reaches the end of the field of view and soon thereafter the reflected shock wave from the shock tube end wall enters the field of view. This significant difference in velocity between the two balls, will be seen clearly in the constructed spheres trajectories. Based on all recorded shadowgraphs, the sphere's trajectory was reconstructed and shown in Fig. 3a, where the shown straight lines represent the incident and the reflected shock waves, respectively. The sphere's velocity is presented in Fig. 3b; the straight solid lines indicate the gas velocity ahead and behind the incident shock wave. The sphere's velocity can be obtained by differentiating the previous curve of the recorded spheres' locations. Thus, the sphere's acceleration, needed for evaluating its drag coefficient, can be deduced via a second numerical differentiation. Since numerical differentiation yields a very noisy acceleration curve, we have to find a curve fit to the recorded spheres' locations. Here, a

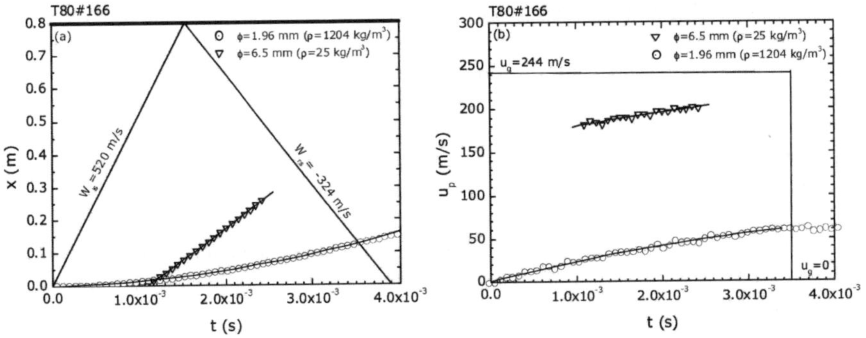

Fig. 3. Trajectories of spheres tested in run#166 (a) and sphere's velocities obtained by numerical differentiation of sphere's trajectories (b).

third order polynomial fit was used for constructing the sphere trajectory. The values obtained for the sphere's velocity and acceleration were then used for deducing the sphere's drag coefficient. Results obtained for the sphere drag coefficient CD versus Reynolds number Re_p, for the considered subsonic case (run#166), are shown in Fig. 4a. It is clear from these results that for the covered range of Re_p, the changes in the sphere's drag coefficient are moderate. Therefore, an average value of CD was deduced from results shown in Fig. 4a to be associated with the appropriate average value of Re_p. Summary of all results obtained in the present research is shown in Fig. 4b in the $CD - Re_p$ plane, where one point represents the average CD obtained in each experiment. The solid line

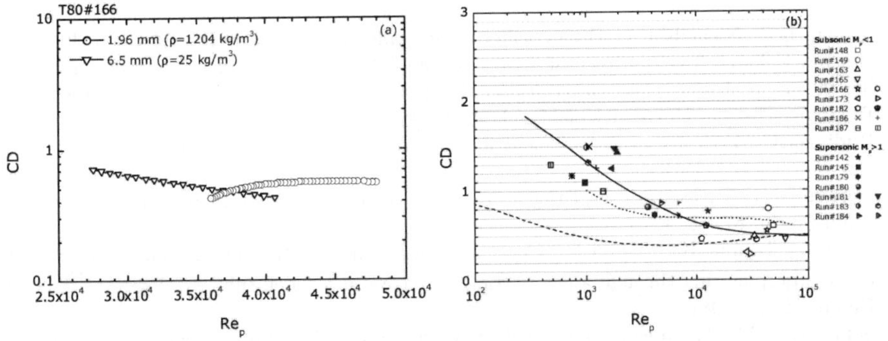

Fig. 4. The spheres' drag coefficient derived from results shown in Fig. 3(a) and summary of CD versus Re_p based on average values of CD (b).

appearing in this figure is a curve fit to the present results. The dotted line is the correlations proposed for CD by [7] and the dashed line is the standard drag curve proposed for a similar steady flow. The correlation describing present results is a third-order polynomial fit, similar to that proposed by [7] and given in the following equation.

$$log_{10}CD = -0.69555 + 1.2589\ log_{10}Re_p - 0.46458\ (log_{10}Re_p)^2 + 0.04532\ (log_{10}Re_p)^3$$

where Re_p is the Reynolds number based on the particle relative velocity. We can note a significant difference between the drag coefficient of a sphere in a steady flow (the standard drag curve) and that obtained in a non-stationary flow conditions which is significantly higher for Reynolds numbers within the range of $500 < Re_p < 10^4$. It was suggested by many authors [1], [4], [10], [11], that the flow unsteadiness contribution to the sphere's drag may be expressed in terms of a non-dimensional parameter $Ac = \frac{du_p}{dt}\phi/(u_g - u_p)^2$. In the present experiments this parameter is quite small, it is within the range $7\times10^{-5} < Ac < 4\times10^{-2}$. The fact that Ac is so small should not be surprising since in spite of the sphere's high initial acceleration its diameter is very small and its initial relative velocity is relatively high. Therefore, one could conclude that the parameter Ac is not a characteristic parameter in the considered flows.

5 Conclusions

The present results obtained for the sphere's drag coefficient, CD, strengthen past assessments that significant difference exists between values obtained for CD in steady and non-steady flows. Furthermore, throughout most of the investigated range of Reynolds numbers the obtained non-steady values are over 50% higher than those obtained in a similar steady flow case. The gap between the two increases with decreasing Reynolds numbers. The proposed correlations can be used safely in numerical simulations of dusty flows where the dust concentration is not high, and for Reynolds numbers within the range covered in the present investigation.

References

1. Selberg B.P., Nicholls J.A., Drag coefficient of small spherical particles, AIAA, 6, 401, 1968
2. Buckley F.T.Jr., An experimental investigation of the drag coefficient of particles accelerating within particulate clouds at transonic slip flow conditions by a light extinction technique, PhD Dissertation, University of Maryland, USA, 1968
3. Rudinger G., Effective drag coefficient for gas-particle flow in shock tubes, ASME J, Basic Eng. D., 92, 165, 1970
4. Karanfilian S.K., Kotas T.J., Drag on a sphere in unsteady motion in a liquid at rest, J. Fluid Mech., 87, 85, 1978
5. Quta E., Tajima K., Suzuki S., Cross sectional concentration of particles during shock process propagating through a gas-particles mixture in a shock tube, In: Proc. 13th ISSW, Buffalo, USA, 1981
6. Sommerfeld M., The unsteadiness of shock wave propagating through gas particle mixtures, Exp. in Fluid, 3, 197, 1985
7. Igra O., Takayama K., Shock tube study of the drag coefficient of a sphere in a non-stationary flow, Proc. R. Soc. Lond. A, 442, 231, 1993
8. Sun M., Saito T., Takayama K., Tanno H., Unsteady drag on a sphere by shock wave loading, Shock Waves J., 14, 3, 2004
9. Jourdan G., Houas L., Igra O., Estivalezes J.-L., Devals C., Meshkov E.E, Drag coefficient of a sphere in a non-stationary flow; new results, To be published, 2007
10. Crowe C.T., Nicholls J.A., Morison R.B., Drag coefficient of inert and burning particles accelerating in gas streams, In Proc. 9th Symp. on Combustion, 395, New York, USA, 1963
11. Temkin S., Metha H.K., Droplet drag in an accelerating and decelerating flow, J. Fluid Mech., 116, 297, 1982

Three-dimensional interferometric CT measurement of discharging shock/vortex flow around a cylindrical solid body

M. Ota[1], **T. Inage**[2], **and K. Maeno**[1]

[1] *Graduate School of Engineering, Chiba University*
1-33, Yayoi, Inage-ku, 263-8522 Chiba (Japan)
[2] *Graduate School of Science and Technology, Chiba University*
1-33, Yayoi, Inage-ku, 263-8522 Chiba (Japan)

Summary. To elucidate the dynamics of heat and fluid flows, three-dimensional aspects of the fields are of great significance. In the research of high-speed flow field including shock waves the common flow visualization methods are shadowgraph, color-schlieren, and interferometric methods. In these methods the three-dimensional (3-D) density information is integrated along an optical observation axis and go down to two-dimensional (2-D) in the represented image. The 3-D measurement of the complicated flow field including various phenomena has not been developed especially in high speed gasdynamic research. Experimental data are in the poor situation.

1 Introduction

The purpose of our investigation is to develop interferometric CT (Computed Tomography) technique to observe high-speed and three-dimensional (3-D) flow field that includes shock wave, and to clarify 3-D flow phenomena induced by shock waves. In our previous study, 3-D complex flow discharged from square nozzle and two parallel and cylindrical nozzles was measured by interferometric CT method [1]~ [4]. The shock Mach number at the exits of the nozzles were both higher and lower than 2.0. As a result, various phenomena of 3-D flow field were clarified by several imaging technique such as pseudo-color images, pseudo-schlieren images by pseudo-schlieren technique, 3-D isopycnic images, etc. We also applied a novel presentation method for flow field, which is named as "Distribution Combined Schlieren Image (DCSI) method [1], [4]" to our CT results to demonstrate the 3-D features of complex flow field. Filtered Back Projection (FBP) algorithm [5] has been used to reconstruct the 3-D density flow field in our previous study. In this paper ART algorithm is applied to the reconstruction from incomplete projection data caused by opaque object in a flow field.

2 Experimental Apparatus and Interferometric CT Measurement

We use a diaphragmless shock tube to produce a shock wave with good reproducibility. Figure 1 illustrates a schematic diagram of our experimental apparatus and our observation system. The observation system consists of a CCD camera, a Mach-Zehnder interferometer, a pulsed nitrogen laser, a delay/pulse generator, an oscilloscope, and a personal computer. The shock wave is generated by a diaphragmless shock tube driver in the low-pressure tube of 3.1 meters in length, and its inner cross section is 40 mm $\times$ 40 mm square. A rotating plug which has a cylindrical nozzle and a circular cylinder is installed at the end of the low-pressure tube. A cylindrical nozzle is open to the low-pressure test section. To obtain the 3-D image of flow field, we need the multidirectional projection data for a reproducible flow. A set of experiments has been performed for several rotation angles at the combination of fixed initial gas conditions for the high-pressure chamber and the low-pressure tube. Figure 2 shows coordinate system of rotating plug relative to the light pass s. We define x and y axes as shown in Fig. 2, where these axes rotate with rotating plug. The z is central axis of rotating plug and is perpendicular to x and y axes. Rotation angle θ can be controlled from outside the shock tube with introduced rotation driving equipment. The experiment is performed for 19 rotation angles between 0° and 90° at five-degree intervals while the light path s is fixed, taking benefit of the two-axis symmetrical characteristics of the flow field. The 3-D density distribution is reconstructed from a set of projection data for the same M_i and z_s. The Mach number of the incident shock wave M_i is calculated by pressure jump across the shock wave at the pressure transducer installed at 61mm ahead of the inlet of the rotating plug. In this paper M_i is fixed to 2.0.

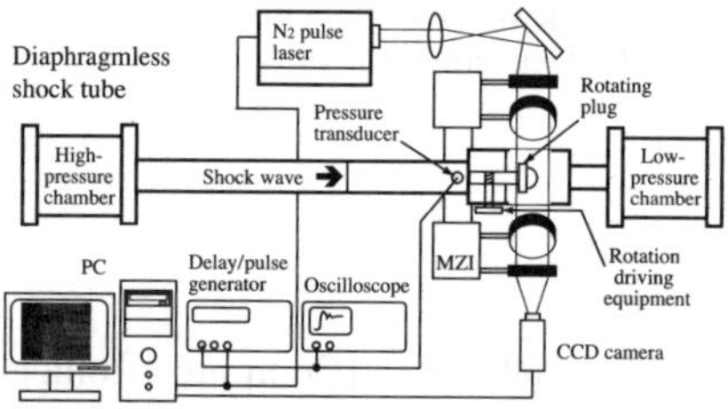

Fig. 1. Diaphragmless shock tube and experimental setup

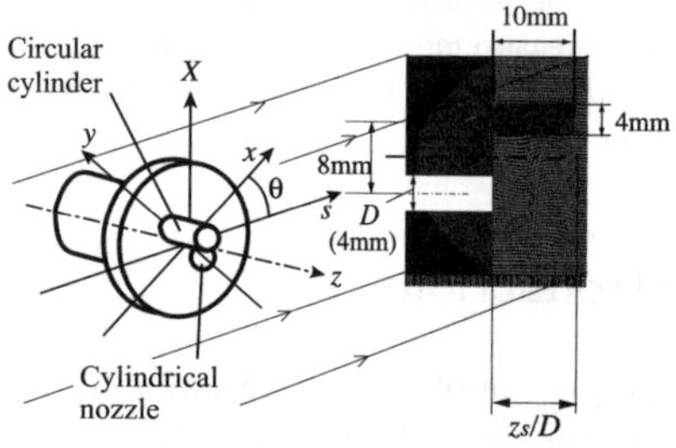

Fig. 2. Rotating plug with a circular cylinder and a cylindrical nozzle

The left image of Fig. 3 shows a finite-fringe interferogram at rotation angle $\theta = 90°$ taken by CCD camera with Mach-Zehnder. Thick blanked line indicates the central axis of a rotating plug, and thin blanked line indicates the central axis of a cylindrical nozzle

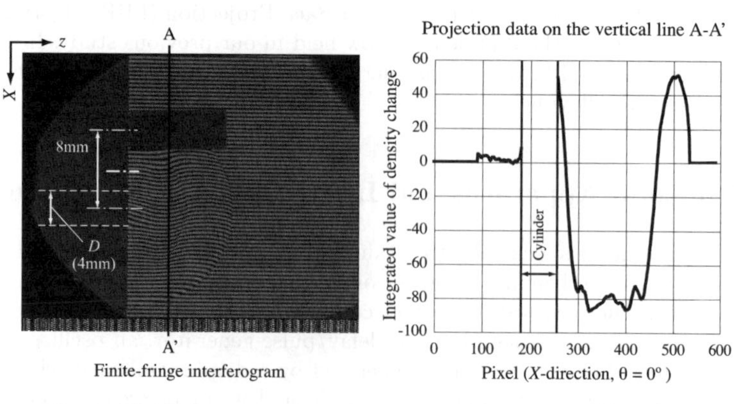

Fig. 3. Finite fringe interferogram and calculated projection data

and a circular cylinder. The distance between these two central axes is 4mm. In this figure z_s is a frontal position of the primary shock wave, D is a diameter of a cylindrical nozzle (4mm), and z_s/D is the normalized frontal position of the primary shock wave. In this paper we discuss the case data of $z_s/D = 2.74$.

In our interferometric CT measurement, three-dimensional density distribution is reconstructed from multidirectional projection data. Projection data are calculated from Eq. 1, where ΔH is displacement of fringe pattern, Δh is interval of fringe pattern, λ is wavelength of observation light, K is Gladstone-Dale index and d is the length of test section.

$$\int_0^d \{\rho(x,y,z) - \rho_0\} ds = \frac{\Delta H}{\Delta h}\frac{\lambda}{K} \tag{1}$$

Displacement of fringe ΔH and fringe interval Δh in the right hand side in Eq. 1 are calculated from finite-fringe interferogram in Fig. 3. Firstly, we calculate Δh at no flow area. Secondly, we calculate ΔH from displacement of fringe pattern comparing with no flow area. We obtain the integrated value of density change along the light pass from Eq. 1. The right plot of Fig. 3 shows the calculated projection data on the vertical line A-A' in the left image. The horizontal axis indicates the position X of the vertical line A-A', the origin of X and z coordinates is located in left top position of this image. The vertical axis of left plot indicates the calculated integrated value of density change and its value is normalized with the initial density (ρ_0) at no flow area. We repeat this process for all projection angles (from 0° to 90°) of one cross section to obtain the multidirectional projection data of one cross section. Then 2-D density distribution is reconstructed from these projection data with appropriate reconstruction algorithm. Finally, 3-D density distribution is obtained as collection of reconstructed 2-D density distribution.

3 ART reconstruction

The ART (Algebraic Reconstruction Technique) is one of the iterative reconstruction method and consists of assuming that the cross section consists of an array of unknowns, and then setting up algebraic equations for the unknowns in terms of the measured projection data [6]. ART is much simpler than FBP (Filtered Back Projection) method [5] which is the transform-based method and we have used in our previous study. For FBP a large number of projections is required for higher accuracy in reconstructed image, in the situation where it is not possible to obtain these projections the reconstructed image is suffer from many streaky noise. The reconstruction from incomplete projection data is more amenable to solution by ART.

The object in the observation area blocks off the observation light for interferometry, the calculated projection data contain the blank part which corresponds to the position of the circular cylinder as indicated in right plot of Fig. 3. Therefore the density distribution around the object has to be reconstructed from these lacked incomplete projection data. In this paper the density distribution of unsteady flow field around circular cylinder is reconstructed by ART.

4 Distribution Combined Schlieren Imaging - DCSI

We have succeeded to identify the various phenomena of high-speed flow discharged from nozzle by using interferometric CT measurement in our previous study. Pseudo-Schlieren technique is a useful tool for studying 3-D problems in shock dynamics. Recently we have also proposed a new method for visualization called Distribution Combined Schlieren imaging (DCSI) method [1], [4]. DCSI is a fusion of pseudo-schlieren image from density gradient and pseudo-color image of designated one image of flow properties. As pseudo-schlieren image only exhibits the steep density gradient line or surface, it is sometimes difficult to identify how the flow property changes across the steep gradient line or surface in the observed complex flow fields. In our CT measurement we can obtain the 3-D density distribution and then calculate density gradient data. Therefore, our experimental CT results can represent DCSI data of density distribution. In CFD simulation density and other properties of flow field can be represented, so a detailed discussion on the properties is also produced with DCSI method. In DCSI method we can duplicate the image of steep density gradient lines and another image of the flow property around these gradient lines simultaneously.

5 Results and Discussion

The resultant image in y-z cross section is illustrated in Fig. 4. The left is pseudo-color image of normalized density distribution, middle pseudo-schlieren image and right is DCSI image. The vertical thick line with white blank indicates rotating plug's wall and white blank indicates a cylindrical nozzle. The position of a circular cylinder is indicated with two horizontal blanked line. The vortex around discharging flow from a cylindrical nozzle is identified in pseudo-color image and primary shock wave (PSW), secondary shock wave (SSW), contact surface (CS) and reflected shock wave (RSW) from circular cylinder is clearly seen in pseudo-schlieren image. We can easily understand at sight that the discharging flow from a cylindrical nozzle is expanded and then compressed at SSW in DCSI image.

Figure 5 illustrates pseudo-color images of normalized density distribution in x-y cross section. Six cross sections (position A$\sim$F) that is parallel to rotating plug's wall are indicated. The normalized position of cross section (z/D) is shown at upper side of each image. The position of a circular cylinder is indicated with blanked circle. In position A$\sim$C, the cross sectional shape of vortex around the discharging flow from a cylindrical nozzle is captured. The reflected shock wave (RSW) from circular cylinder is seen in position D$\sim$F. Four slanting noise from a circular cylinder is appeared in position E and F, this is influence of reconstruction from incomplete projection data.

The DCSI image in x-y cross section is shown in Fig. 6. RSW around the circular cylinder is lightly indicated in the position A$\sim$C but the slanting noise is strong it is difficult to distinct the flow phenomena around the object. In position D$\sim$F the density gap across the RSW is large we can identify the cross sectional shape of RSW clearly. To clarify the complex 3-D phenomena with higher accuracy, reducing the noise that is contained in projection data obtained from finite-fringe interferogram is necessary.

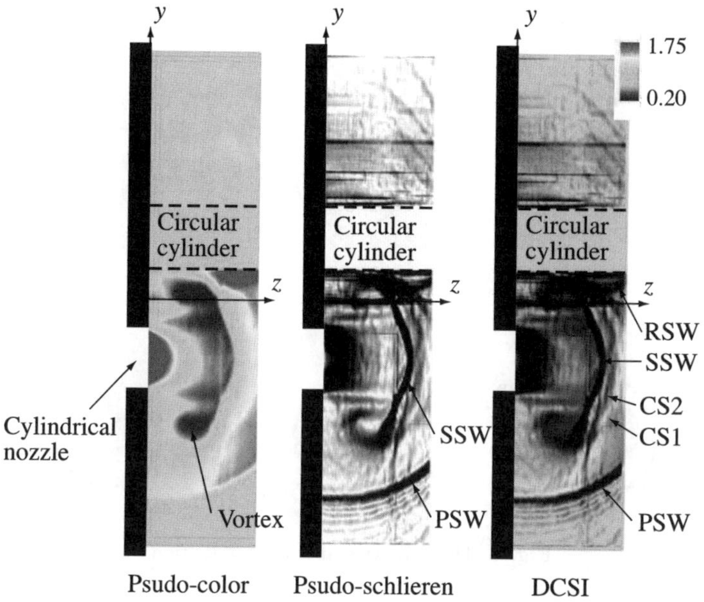

Fig. 4. Pseudo-color, pseudo-schlieren and DCSI image at y-z cross section

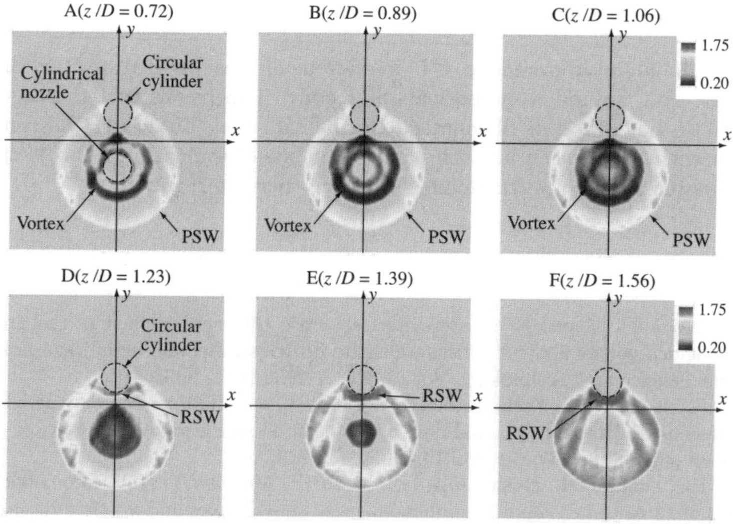

Fig. 5. Pseudo-color images of normalized density distribution at x-y cross senction

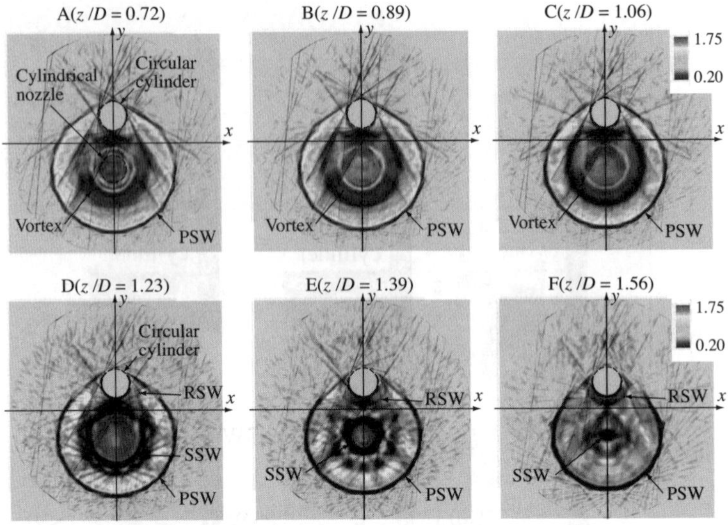

Fig. 6. DCSI images at x-y cross section

6 Conclusion

Three-dimensional interferometric CT measurement has applied to the discharging shock/vortex flow around a cylindrical solid body. The reconstructed image is influenced by noise but 3-D flow phenomena can be captured. To clarify the complex 3-D phenomena with higher accuracy some improvements of evaluation of projection data - image processing from finite-fringe interferogram - may be required.

References

1. Ota M., Koga T., Maeno K. : Laser interferometric CT measurement of the unsteady supersonic shock-vortex flow field discharging from two parallel and cylindrical nozzles, Measurement Science and Technology, Vol 17, No. 8, 2066-2071, 2006.
2. Maeno, K., Kaneta, T., Yoshimura, T., Morioka, T. and Honma, H., Pseudo-schlieren CT measurement of three-dimensional flow phenomena on shock waves and vortices discharged from open ends, Shock Waves, Vol.14, No.4, 239-249, 2005.
3. Honma H., Ishihara M., Yoshimura T., Maeno K., Morioka T., Interferometric CT measurment of three-dimensional flow phenomena of shock waves and vortices discharged from open ends, Shock Waves, Vol. 13, 179-190, 2003.
4. Ota M., Koga T., Maeno K. : Interferometric CT Measurement and Novel Expression Method of Discharged Flow Field with Unsteady Shock Waves, Japanese Journal of Applied Physics, Vol. 44, No. 42, L1293-1294, 2005.
5. Shepp L. A., Logan B. F. : The Fourier Reconstruction of a Head Section. IEEE Trans. Nucl. Sci. NS-21, 21-43, 1974.
6. Kak A. C., Slaney M. : Principles of Computerized Tomographic Imaging. IEEE, New York, 1988.

Vizualization of 3D non-stationary flow in shock tube using nanosecond volume discharge

I.A. Znamenskaya, I.V. Mursenkova, T.A. Kuli-Zade, and A.N. Kolycheva

Moscow State University, Faculty of Physics, Leninskie Gory, Moscow, 119992, Russia

Summary. Nanosecond-lasting homogeneous volume discharge is tested and used for visualization and investigation of stationary and non-stationary 2D and 3D flows in shock tube channel: shock wave diffraction on models, supersonic flow over models with bow shock; transonic separation zone. Pulse volume discharge with ultraviolet preionization by radiation from the sliding surface discharges was used. Different models were tested in supersonic and transonic flow. Total time of flow glow at discharge initiating in optical band (light-emitting image exposure) was shown to be less then 150-200 ns. Gas density non-homogeneity in transonic/supersonic flow when ionized by pulse volume discharge results in redistribution of discharge plasma flux. Integral plasma glow images of instant flow structure were recorded on colour film, black-and-white film, CCD and digital photo camera.

1 Introduction

3D high speed flows visualization is a special problem in experimental supersonic gas dynamics analysis. Shadow and interference methods are applied to 2D flows, and also some axisymmetrical flows. The main imperfection of optical methods is that, all density changes are summarized along a propagation direction of light beam and thus integral value of density change is recorded. Special methods should be used to obtain the images of 3D stationary and non-stationary shock configurations, separation zones, turbulent area. Some types of gas discharge are used for visualization and investigations of supersonic/hypersonic gas flows. Shock wave configurations [1], separation zone [2], vortexes [3], boundary layer positions [4], some density inhomogeneities were successfully visualized due to dependence of glow intensity gas local parameters. Various types of discharges of millisecond-lasting duration and quasy-stationary discharges were used. Stationary discharge current thermal heating of gas influences flow configuration and parameters including shock wave position [5]. High speed gas dynamics processes and microsecond-lasting interactions in shock tube channel can be visualized with pulse (nanosecond-lasting) volume discharge [6, 7].

2 Method of Visualization

Pulse volume discharge visualization method was used for obtaining non-stationary flow images. 3D flows in shock tube channel were visualized and compared to CFD images [6, 7]. There is no gas thermal heating in instantly ionized flow during the nano scale exposure time. Also no macro plasma instabilities arise in such a short time interval - that is why discharge gap area remains uniform in homogeneous flow. Discharge chamber was mounted in a shock tube channel (cross section $24mm \times 48mm$) as a special section.

Pulse volume discharge with ultraviolet preionization by radiation from the sliding surface discharges, which form plasma electrodes was used. Two sidewalls of the chamber are the quarts windows. Another two walls are plasma electrodes (or plasma sheet discharges, sliding on dielectric surfaces). Discharge area was 100 mm long (1 on fig. 1). The flow pressure was 8-60 kPa, initial plane shock waves Mach numbers were 2-5, gas flow behind incident shock wave Mach numbers $M = 0,9 - 1.7$. Models were mounted in shock tube of rectangular cross section in discharge gap area 10 cm long. Images of plane shock wave interaction with models were recorded from opposite windows as well as images of stable and non-stable transonic/supersonic flows over the models. Integral plasma glow images of instant flow structure were recorded with digital photo cameras (2 on fig. 1), on high sensitive film (color or black/white) and then scanned. Two glow images of ionized flow were taken from opposite windows of discharge chamber (fig. 1). Also CCD camera with nano-scale gate was used to investigate space glow distribution evolution in time (3 on fig. 1). Two volume discharge glow images in homogeneous gas are shown on fig. 2. Exposure (camera gate time) was 10 ns. Total volume discharge glow time is 160 - 200 ns depending on density and flow structure. It is in agreement with electric discharge current measurements. Non-stationary flow in shock tube does not change during this time. Gas density non-uniformity in transonic/supersonic flow when ionized by pulse volume discharge results in redistribution of a discharge plasma. Discharge glow intensity (electron concentration and electrical conductivity) depends on value of an ionization coefficient. Local ionization coefficient is a non-linear function of E/N. Thus glow intensity increases in areas of low density and high density gradients. There is no gas thermal heating in instantly ionized flow during the exposure time.

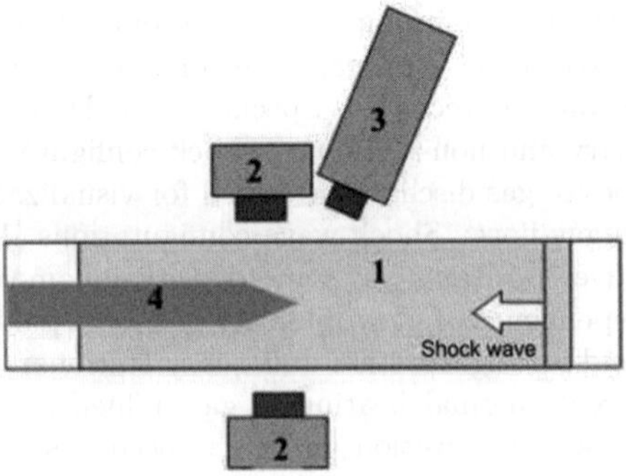

Fig. 1. Experimental arrangement

Local light flux intensity I and spectral characteristics of the spontaneous light emitted in volume discharge depends on the local parameters of the gas medium.

$$I = f((E/N), I_d, P, T, d) \tag{1}$$

E is the electrical field strength; N is the molecular number density, I_d is the discharge current, P is the pressure, T is temperature, d is the distance between plasma electrodes. Space factor of the electrodes can be neglected - initial electric field is homogeneous, transversal to flow direction.

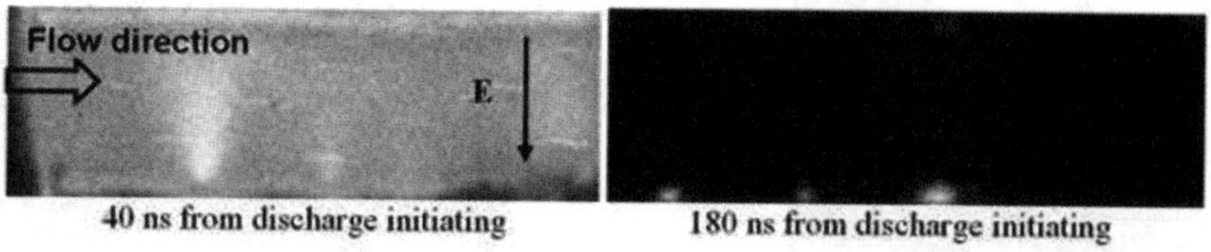

Fig. 2. Volume discharge glow image in homogeneous gas

Stationary and non-stationary flows over different axisymmetrical bodies were visualized with pulse volume discharge method. Images of 3D flows were computer processed and compared to consequent shadow images. Cones of different geometry, blunt cylinder and sphere were treated.

Different axisymmetrical models were tested in supersonic and transonic flow. Models of blunt cylinder (fig. 3), cone (fig. 4), cylinder, sphere (fig. 5) - were mounted in shock tube - in discharge gap area with symmetry axes parallel to the flow (4 on fig. 1).

Complex supersonic or transonic flow arises in channel after normal shock wave diffraction. Gas density in uniform flow behind the incident shock wave was calculated using Rankine - Hugoniot relations:

$$\rho_2/\rho_1 = (\gamma+1)M_0^2/[(\gamma-1)M_0^2+2] \tag{2}$$

In $30-40\mu s$ quasi-stationary flow over a model arises. Pulse volume discharge was initiated in different moments of gasdynamic process. Consequent images of plane shock wave diffraction on models were recorded as well as images of transonic/supersonic flow over the models.

3 Blunt cylinder

Instant discharge images of shock diffraction on blunt cylinder are presented at Fig. 3. While incident shock wave was spreading in discharge gap, most of plasma flux was in low pressure zone in non-disturbed gas in front of incident shock wave (Fig. 3 A,B). Non-stationary reflected shock configuration is not visualized. Fig. 3 presents image of stationary flow near cylinder, recorded when incident shock wave had passed off the discharge gap. The bow shock wave, line of flow separation, line of flow attachment, oblique shocks, oblique shocks reflected from the horizontal shock tube channel wall, are seen at the image. Shock layer is dark area (compressed zone). Shadow image of bow shock in stable flow near cylinder was compared to consequent pulse discharge image (Fig. 3 D).

4 3D non-stationary vortex flow

Fig. 4 presents images of discharge visualization of shock wave ($M = 3-4$) diffraction on cone, mounted in shock tube test chamber on a cylindrical holder. In cone rear side separation non-stationary turbulent zone of low density arises after shock diffraction. Thus 3D flow structures, vortexes in separation zones are visualized with pulse discharge plasma radiation. Successive images of vortexes (10-20 μs after shock contact with model)

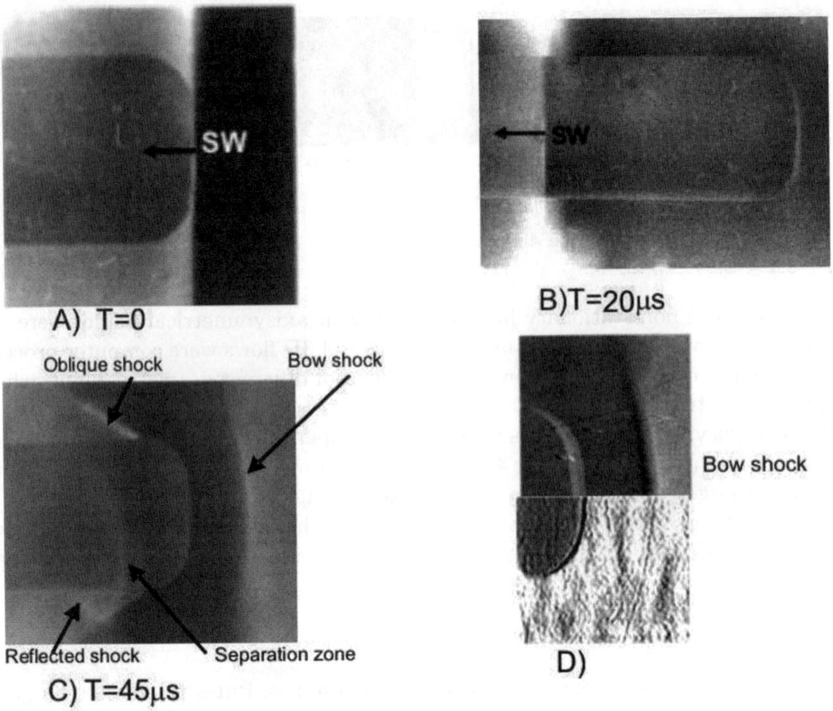

Fig. 3. Shock wave diffraction on blunt cylinder

illustrate non-stable vortex zone evolution. Pairs of images taken from opposite windows show vortex zone instant 3D configurations.

Fig. 5 presents discharge visualization of shock diffraction on sphere 12 mm in diameter. Main visualized flow structure elements are: shock wave moving to left (intersection with window glass), vortex separation zone.

5 Some flow analysis

Shock wave diffraction and stationary flow arising after shock wave diffraction on cone combined with cylinder 10 mm in diameter (fig. 6) were investigated. The aim was analysis of supersonic flow in shock tube with attached and detached conical bow shock. Evolution of reflected and then bow shock in time after diffraction was analyzed (angle to flow), also its dependence of shock configuration on flow Mach number, gas density. Fig. 7 shows experimental data obtained from processed discharge images: angle between reflected conical shock surface and symmetry axes. Solid line is computed value for stationary conical flow. Initial shock Mach number was 3,8; $t = 0$ - moment of shock contact with cone. Flow with stationary shock configuration arises at $40\mu s$ and later at $120\mu s$ angle increased slightly - perhaps due to tube walls influence.

Animations were created from series of processed discharge flow images.

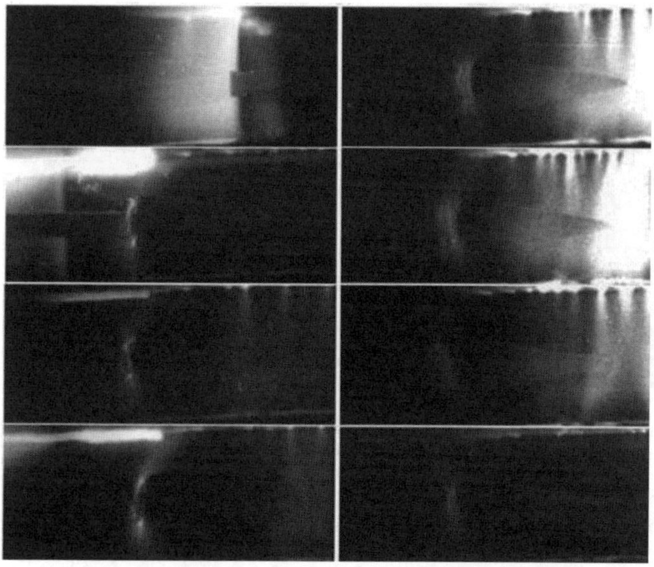

Fig. 4. Flow dynamics in cone rear side

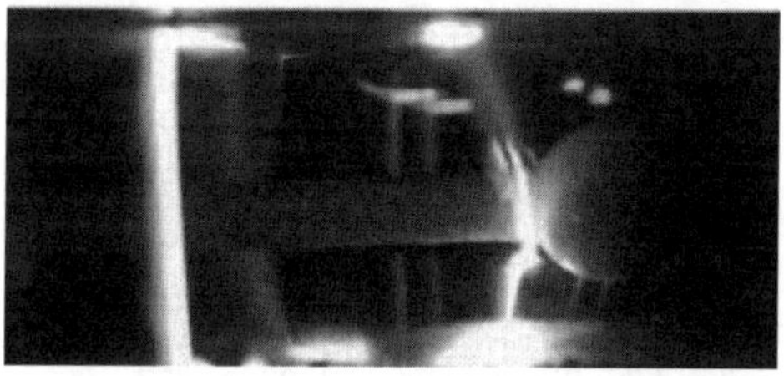

Fig. 5. Shock diffraction on sphere

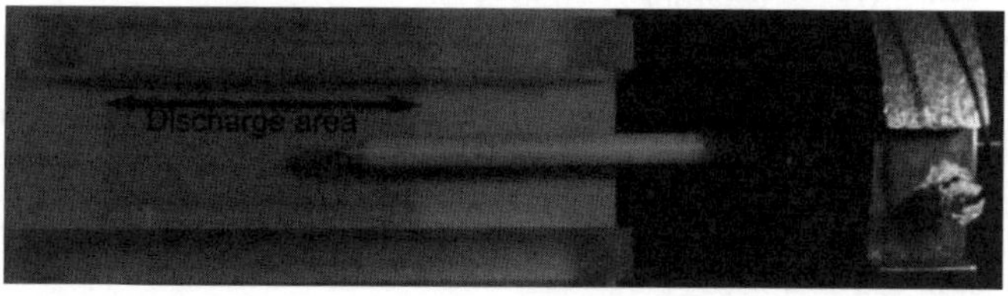

Fig. 6. View of cone model in open test section

6 Conclusion

Pulse volume discharge method images were obtained with exposure time 130-200 ns. Axisymmetrical and 3D supersonic gas flows over different bodies were tested. Flow images were computer processed and compared to consequent shadow and computed

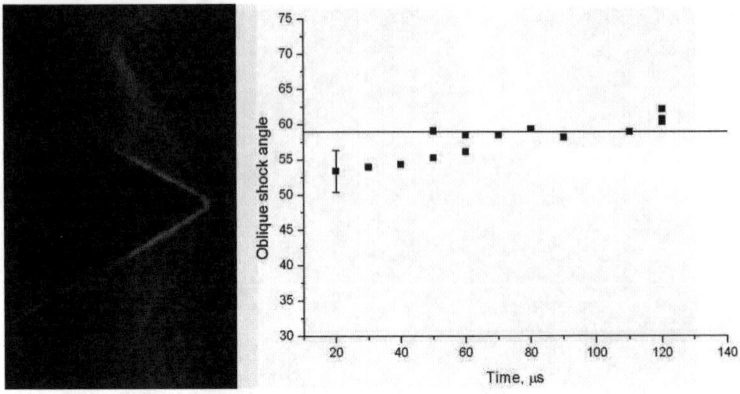

Fig. 7. Oblique shock angle dependence on time

data. Areas of low density are the zones of intense plasma fluxes. Instant flow images of tested flows present bow shock waves, lines of separation; line of apposition; vortexes in rear separation zone; oblique shocks. Non-stable 3D turbulent flows in rear models sides were visualized and analyzed.

Acknowledgement. Researches were supported by the Program No. 9 part 2 of the Russian Academy of Sciences and the Russian Foundation for Basic Research, grant No. 05-08-50247.

References

1. Alfyorov V. I. and Dmitriev L. M., Electric discharge in a gas flow in the presence of density gradients, Teplofizika Vysokikh Temperatur, vol. 23, p. 677, 1985
2. Jagadeesh G., Takayama K., Srinivasa Rao B.R., Nagashetty K., Reddy K.P.J., Separated Hypersonic Flow Visualization Using Electric Discharge, The 9th International Symposium on Flow Visualization, Edinburg, p. 151, 2000
3. Alfyorov V. I., On determining the flow density field in visualizing vortex cores by the high-voltage discharge method, Tr. TsAGI, no. 1421, p. 13, 1972
4. Nishio M., Methods for Visualizing Hypersonic Shock-Wave/Boundary- Layer Interaction Using Electric Discharge, AIAA Journal, vol.34, no.7, 1996
5. Meyer R., Palm P., Plonjes E., Rich J.W., and Adamovich I.V., The Effect of a Nonequilibrium RF Discharge Plasma on a Conical Shock Wave in a M=2.5 Flow, AIAA Journal, vol. 41, no. 5, p. 465-469, 2003
6. Znamenskaya I.A., Discharge imaging technique for shock tube studies,21 ISSW, Great Keppel, Australia, p. 2881, 1997
7. Dankov B.N., Kulickov V.N., Znamenskaya I.A., Visualization of Transient Separation Flow over 3D Model of Cone Shroud with Three Methods, Proc. of 10th International Symposium on Flow visualization, Kioto, Japan, ISBN4-906497-82-9, 2002

Part VIII

Hypersonic Flow

Assessment of the convective and radiative transfers to the surface of an orbiter entering a Mars-like atmosphere

N. Bédon, M.-C. Druguet, D. Zeitoun, and P. Boubert

Université de Provence, IUSTI-UMR CNRS 6595, Polytech'Marseille, Mécanique Energétique Technopôle de Château-Gombert, 5 rue Enrico Fermi, 13453 Marseille Cedex 13, France

Summary. This paper deals with the radiative transfers in the nonequilibrium gas flow that surrounds a space vehicle entering a Mars-like atmosphere. It presents the modeling and the numerical methodology recently developed at IUSTI to account for radiation of high-temperature gases and the coupling between the radiative transfers and the flowfield.

1 Introduction

When an orbiter enters a planetary atmosphere at hypersonic speed, the temperature raises and the molecular species dissociate in the shock layer surrounding the vehicle, leading to large energy tranfers to its surface. As these energy transfers might damage the vehicle, it is important to correctly predict them. If convective fluxes are well modeled and documented in the literature, the radiative transfers are not yet well understood. The need to understand and predict the radiative heating to the surface of a space vehicle entering a planetary atmosphere increases as the future space missions are getting more and more challenging. Preliminary computations regarding missions to Mars have revealed the importance of radiative heat transfers on both the afterbody – where the payload is located – and the forebody [1]. Back to Earth from such a space mission, the vehicles will be entering the Earth atmosphere at sufficiently high speed to face strong radiative heating.

In the framework of such future space missions, ESA and NASA have revived the study of radiation of nonequilibrium plasmas. ESA has recently organized a series of workshops to gather researchers from European countries to compete on specific test cases defined to study various aspects of radiation of high-temperature gases in planetary entry conditions. Among the many test cases proposed in the workshops [1], the test case 3 (TC3) is a generic axisymmetric model representative of a vehicle – called a Mars Sample Return Orbiter (MSRO) – entering a Mars-like atmosphere. It is designed to assess the radiative heating on the forebody and on the afterbody of the vehicle.

In the present work, we propose to estimate the convective and the radiative transfers in the shock layer enclosing the MSRO model and to its surface, when the orbiter enters a Mars-like atmosphere at high velocity. The assessment of these tranfers requires the estimate of the radiation of the chemical species of the Martian atmosphere at high temperature.

2 Description of the Test Case 3

The orbiter is composed of a sphere-cone – the forebody – attached to a cylindrical afterbody, as shown in Figure 1, where the dimensions are given in centimeters. It presents an axial symmetry, so only half of the vehicle is represented. For the purpose of this paper, we compute the flowfield and the radiative transfers on the front part of the orbiter only. The flow conditions correspond to a specific point of the trajectory of the orbiter during its entry into the Martian atmosphere. Those freestream conditions are : $U_\infty = 5223$ m/s, $T_\infty = 140$ K, $P_\infty = 7.87$ Pa and $\rho_\infty = 2.9933 \times 10^{-4}$ kg/m^3. The wall temperature is set at 1500 K on the forebody surface. The Mars-like atmosphere is here assumed to be composed of pure CO_2, that dissociates through the shock wave. A simple chemical kinetics model with the five following species CO_2, CO, O_2, C and O is used in our simulations. All the details about the physical models used in that test-case can be found in [2].

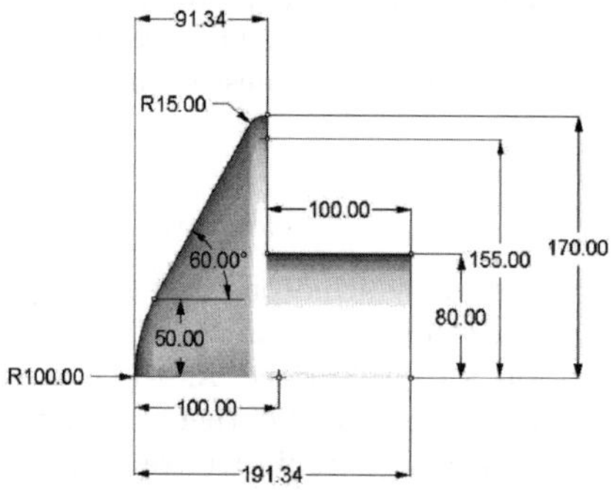

Fig. 1. Schematic of the Mars Sampler Return Orbiter.

3 Modeling of the radiation-flow interaction

We consider here that the global interaction between the radiative transfers and the flow past the MSRO can be described in four different steps. The first step consists in the determination of the flowfield around the orbiter, with the resolution of the Navier-Stokes equations. The temperature and mass fraction of each species being known everywhere, we may then calculate the emission and absorption coefficients of the gas mixture composing the Martian atmosphere at each point of the flowfield. It is then possible to compute the radiative fluxes within the flowfield and to surface of the body. Eventually, these radiatives fluxes are inserted into the conservative energy equation and the flow is calculated again. This iterative process is done until the solution has converged.

3.1 Computation of the flow past the MSRO

The flow is a nonequilibrium state and is governed by the compressible Navier-Stokes equations and the conservation equations for the species densities and the vibrational energies of the molecules. Their resolution is based on a finite volume formulation, implemented in the PINENS code (Parallel Implicit NonEquilibrium Navier-Stokes) [3]. An example of the flowfield computed on the forebody part of the orbiter is given in Figure 2, where the temperature contours in the shock layer are displayed. The resulting convective fluxes to the wall of the orbiter are plotted in Figure 3, where comparisons

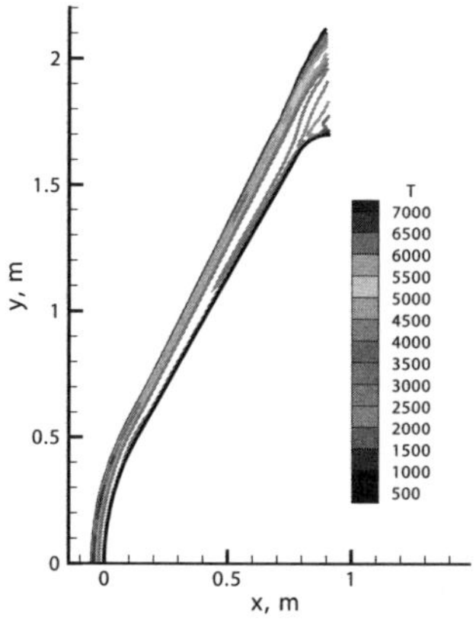

Fig. 2. Temperature contours on the orbiter forebody.

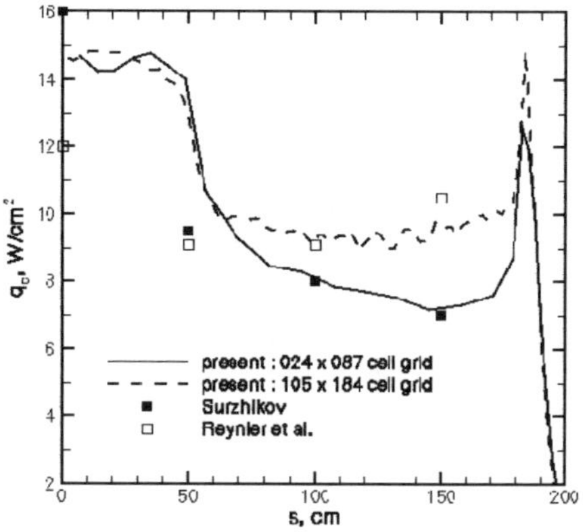

Fig. 3. Convective heat flux to the surface of the orbiter forebody.

between results computed on different grids (24×87 cells and 105×184 cells) show that the solution is still grid-dependent. A comparison with results of other authors [4,5] show that our results are in agreement with their results.

3.2 Emission and absorption coefficients

The determination of the emission and absorption spectra of the gas mixture is a difficult task. In reference [6], experiments have been carried out to determine the emission and absorption coefficients of different species, depending on the temperature. The method is used here to obtain the emission and absorption coefficients of the species CO, C and O. Note that the emission and absorption coefficients of the molecule CO_2 are not computed in the present work.

3.3 Radiative heat flux to the orbiter surface

The total radiative heat flux transfered to the wall of the orbiter is the sum over the spectrum of all spectral radiative heat fluxes, which have to be estimated in each computational cell. We consider here that the fluxes received by a given area of the wall come from the cells directly in front of it. It means we assume that each cell emits spectral radiative fluxes only in the direction normal to the surface of the orbiter. In other words, this method of computing the radiative flux is a one-dimensional approach. Thus, the spectral radiative heat flux $q^{\lambda}_{R_i}$ coming out of the cell i is composed of the spectral radiative heat flux $q^{\lambda}_{R_{i+1}}$ entering the cell i from the cell $i+1$, balanced by the transmissivity τ^{λ}_i of the cell i, and of the spectral radiative heat flux $dq^{\lambda}_{R_i}$ emitted by the different species present in the cell i. This is summarized in Figure 4. The spectral radiative flux $q^{\lambda}_{R_i}$ emitted by the cell i is then given by :

$$q^{\lambda}_{R_i}\, S_{out}(i) = \tau^{\lambda}_i\, q^{\lambda}_{R_{i+1}}\, S_{in}(i) + dq^{\lambda}_{R_i} \qquad (1)$$

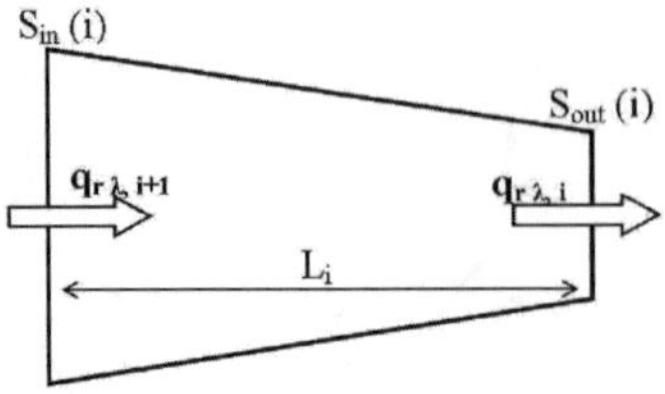

Fig. 4. Schematic view of the spectral radiative heat fluxes emitted by the cell i.

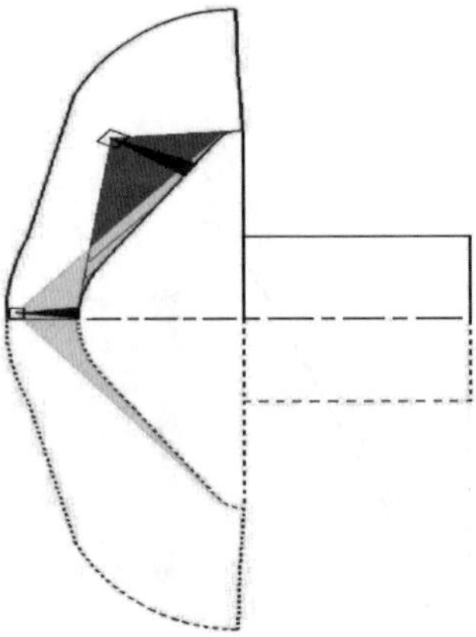

Fig. 5. Examples of the emission's angle θ from computational cells.

where $S_{in}(i)$ and $S_{out}(i)$ are respectively the surface of the "in" and "out" interfaces of the cell i, and where the transmitivity is given by

$$\tau_i^\lambda = e^{-K_i^\lambda \; L_i}. \tag{2}$$

K_i^λ is the spectral absorption coefficient, and L_i the length of the cell i. Note that all the radiative fluxes mentioned here can be written scalarly since we deal with a one-dimensional approach. The radiative flux $dq_{R_i}^\lambda$ emitted by the cell i within the solid angle Ω towards the orbiter surface is expressed as :

$$dq_{R_i}^\lambda = \Omega \; J_i^\lambda \; V_i, \tag{3}$$

where J_i^λ is the spectral emission coefficient and V_i the volume of the cell i. Since the whole problem is axisymmetric, we may suppose that the ratio of the solid angle Ω to the whole space 4π can be approximated by the ratio of the angle θ to the whole angle 2π, where θ is the angle in the 2D axisymmetric plane as shown in Figure 5. With those assumptions, the term $dq_{R_i}^\lambda$ can be written as follows :

$$dq_{R_i}^\lambda = \left(\frac{\theta}{2\pi}\right) 4\pi \; J_i^\lambda \; V_i. \tag{4}$$

In a quasi 1D approach, the angle θ in which the radiation propagates in the axisymmetric plane from one cell i of the flowfield towards the cell $i = 0$ on the orbiter surface may be defined as the blue angle in Figure 5. This angle is rather narrow when the i cell is far from the orbiter surface, but can reach the value of π when the cell i is very close to the orbiter surface. When this "narrow" angle is used in formula (4), the corresponding results are referred to as "angle : narrow". To take into account the fact that the radiation does not emit from a cell i solely into the direction normal to the orbiter surface, we propose to widen the angle θ and to compute it as the angle that includes all the rays that come out of the cell i and that can reach the surface of the orbiter (green angle in Figure 5). When this "large" angle is used in our simulations, the corresponding results are marked "angle : full". Finally, the maximum value that the angle θ can take is π. (The results are then labelled "angle : pi".)

Now that the spectral radiative flux is determined by the formulas (1–4), the total radiative flux to the orbiter surface can be computed as the sum over the spectrum of the spectral radiative fluxes on the cell near the wall ($i = 0$) :

$$q_R = \sum_{\lambda} q_{R_0}^{\lambda} \tag{5}$$

4 Radiative heat flux results

The approach described in the previous section has been applied to estimate the radiative transfers to the surface of the Mars Sampler Return Orbiter. Given the composition of the gas mixture (CO_2, CO, O_2, C and O), its thermodynamical state, and the model to determine the emission and absorption coefficients (that considers the radiation of the CO, C and O species only), we observe that the gas mixture is mainly radiating in a spectrum between 100 nm and 2000 nm. An example of the spectral radiative flux at the stagnation point of the flowfield is plotted in Figure 6. Three values of the angle θ in formula (4) have been used : the narrow angle (blue in Figure 4), the large angle (green in Figure 4), and the angle $= \pi$. The comparison of the spectra obtained at the stagnation point with those different angle shows the non negligible influence of that angle on the spectral radiative flux. Finally, we present in Figure 7 the total radiative

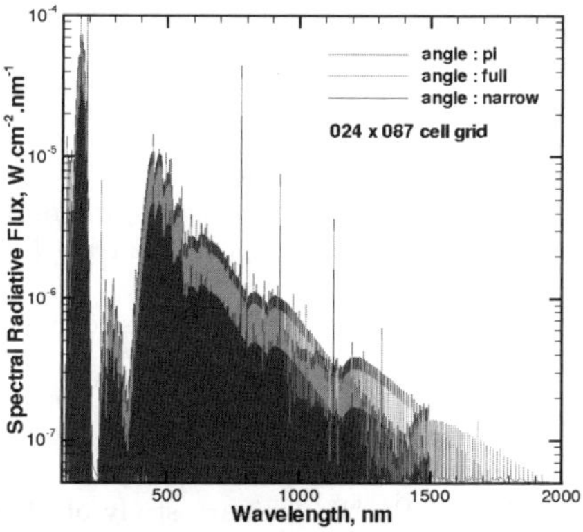

Fig. 6. Spectral radiative heat flux at the stagnation point.

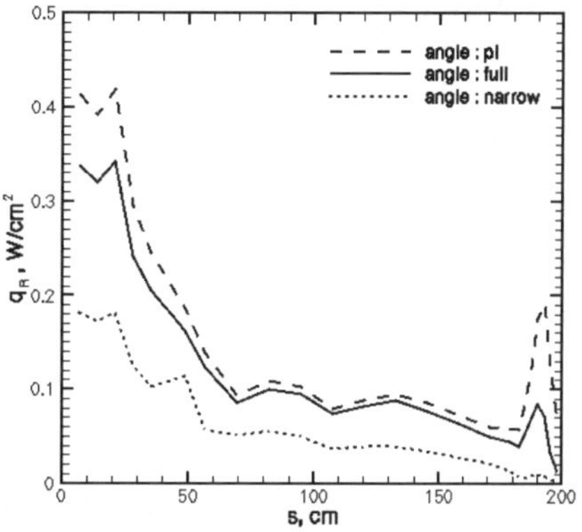

Fig. 7. Total radiative heat flux to the forebody surface.

flux to the surface of the orbiter forebody. Again, a comparison is made between the results obtained with different values of the angle θ in formula (4), and shows that the main difference is between the results obtained with the narrow and the large angle. The value of 0.34 W/cm^2 that we obtain for the radiative flux at the stagnation point ($s = 0$) is much lower than the value of 1.2 W/cm^2 obtained by S. Surzhikov [4] and by Y. Babou *et al.* [7]. This difference may come from the radiation of certain species (like CO_2) that is not taken into account in the present model.

5 Conclusion

We have proposed a model to predict the radiative transfers to the wall of a space vehicle entering the Martian atmosphere. The convective heat flux was obtained after simulating the flowfield with the resolution of the Navier-Stokes equations. The estimate of the radiative heat flux required the determination of the emission and absorption coefficients for the gas mixture. The spectral and total radiative fluxes shown in that paper are the first results we have obtained with the model recently developed to estimate the radiative transfers in nonequilibrium gas flows. Our approach needs to be improved in different ways. First we need to determine the emission and absorption coefficients for species like CO_2 that were not taken into account in the present simulations. Then we have to consider a fully two-dimensional approach to solve the radiative transfer equation and compare the results then obtained to those obtained with the quasi one-dimensional approach. Finally, we have to implement the full coupling between the radiative transfers and the flowfield by introducing the loss of energy due to radiation into the conservation equation of energy.

Acknowledgement. All the computational ressources used for the present simulations were provided by the Institut du Développement et des Ressources en Informatique Scientifique (IDRIS), 91403 Orsay, France.

References

1. ESA-SP-583. *Proceedings of the International Workshop on Radiation of High Temperature Gases in Atmospheric Entry – Part II*, Porquerolles (France), 2004
2. Charbonnier J.-M., Omaly P., "TC3 : Update of the axially symmetric testcase for high-temperature gas radiation prediction in Mars atmosphere entry", *Proceedings of the International Workshop on Radiation of High Temperature Gases in Atmospheric Entry – Part II*, Porquerolles (France), 2004
3. Druguet M.-C., Candler G.V., Nompelis I., "Effect of numerics on Navier-Stokes computations of hypersonic double-cone flows", *AIAA Journal*, **43**(3), pp. 616–623, 2005.
4. Surzhikov S., "TC3: Convective and radiative heating of MSRO for simplest kinetics model", *Proceedings of the International Workshop on Radiation of High Temperature Gases in Atmospheric Entry – Part II*, Porquerolles (France), 2004
5. Reynier Ph., Lino da Silva M., Marraffa L. and Mazoué F., "Preliminary study of Mars entry: application to the prediction of radiation", *Proceedings of the International Workshop on Radiation of High Temperature Gases in Atmospheric Entry – Part II*, Porquerolles (France), 2004
6. Rond C., "Etude expérimentale et numérique de la cinétique chimique et radiative hors d'équilibre à l'aval d'une onde de choc stationnaire", *Thèse de Doctorat*, Université de Provence (France), 2006
7. Babou Y., Rivière P., Perrin M., Soufiani A., "Prediction of radiative flux distribution over the front shield of a vehicle entering Martian atmosphere – Contribution to Test Case TC3", ESA-SP-629. *2nd International Workshop on Radiation of High Temperature Gases in Atmospheric Entry*, Rome (Italy), 2006

Base pressure and heat transfer on planetary entry type configurations

G. Park, S.L. Gai, A.J. Neely, and R. Hruschka

School of Aerospace, Civil, and Mechanical Engineering
UNSW@ADFA, Canberra, ACT 2600, Australia

Summary. The present paper discusses the base pressure and heat transfer of various blunt bodies at hypersonic speeds. It is shown that the base to stagnation pressure ratio of a blunt body is relatively independent of the Mach and Reynolds numbers in hypersonic flow. The base to stagnation heat transfer, however, shows a dependency on the latter, as it is found to increase with Reynolds numnber. Experimental results of surface heat transfer rate around a cylinder model are presented. These data were obtained in a free piston shock tunnel with total enthalpy of 3.98MJ/kg. The free-stream unit Reynolds number per metre and Mach number were 1.47×10^6 and 10.9 respectively. The fore-body heat transfer results show good agreement with the theories of Lees and Kemp et al. The base region heat transfer data showed a local minimum at around 130° from the stagnation point followed by a gradual recovery towards the rear stagnation region. The magnitudes of heat transfer ratio in the base region were representative of those in the base region of planetary entry type configurations.

1 Introduction

An understanding of the physics of base flow is necessary for the successful design of spacecraft and re-entry capsules. The existence of supersonic reverse flow as observed by Gnoffo [1] in his CFD calculations, for example, is one important aspect of the base flow phenomena at high Mach number. In the hypersonic regime, the density in the base region is relatively low and the recirculation vortex is energized by high rates of shear in the lower part of the shear layer due to high values of (M^*/M_e). Here M^* is Mach number on the dividing streamline and M_e is Mach number at the edge of shear layer. The high velocities in the recirculation region created by this vortex may give rise to supersonic reverse flow as a result of concentrated vorticity in the neck region. Skews and Kleine [2] recently postulated that the interaction of two shear layers from the opposing side towards the neck gives rise to a jet like flow. However, to date, there is no experimental verification of this supersonic reverse flow phenomenon in the base region.

Another important aspect regarding the base flow phenomena is a significantly high base pressure (P_b) with respect to the free-stream pressure (P_∞) in hypersonic flow. This is unlike what is observed in supersonic flow. Secondly, the base to stagnation pressure ratio of a blunt body (P_b/P_s) is found to be not only independent of Mach number at hypersonic speeds, but is also found to be relatively independent of the Reynolds number. While this has been shown to be true in the past for simple shapes like cylinders [3], [4], it is also seen to be true of blunt nosed conical shapes such as Apollo and Mercury capsule as will be shown in this paper. Although at present, there is sufficient information regarding the base pressure behind blunt bodies, that on base heat transfer (q_b) is sparse and what little there is, is limited to supersonic speeds. In recent years, however, some CFD data at high hypersonic speeds has become available [1].

As a first step, therefore, in the present paper, we describe some experimental results on heat transfer around a cylinder (q_θ) including the base region in hypersonic flow using a free piston shock tunnel. The experimental data are analysed and compared with the theories of Lees [5] as well as Kemp et al. [6]. The motivation to conduct these experiments using a two dimensional cylinder is that it is a generic blunt body on which a large body of information exists at least at subsonic and supersonic speeds. Secondly, it is an easier configuration to study such basic properties as base pressure, base heat transfer and separation at high Mach numbers. It is intended to extend the experiments to axisymmetric bodies such as re-entry type configurations.

2 Experimental details

The experiments were performed in the free piston shock tunnel T-ADFA located in the School of Aerospace, Civil, and Mechanical Engineering at UNSW@ADFA. Table 1. shows the reservoir and free-stream conditions used in the present experiments.

Table 1. Flow conditions

h_0 (MJ/kg)	P_0 (MPa)	P_∞ (Pa)	T_∞ (K)	ρ_∞ (kg/m^3)	U_∞ (m/s)	M_∞	Re_∞ (1/m)
3.98	12	251	142	0.0061	2513	10.9	1.47×10^6

The cylinder model used in the present experiments had an outer diameter of 32mm and the model aspect ratio was 4. The model was supported by side struts and chamfered end plates were attached to both sides to produce an acceptable two dimensional flow. The surface heat transfer rates were measured using coaxial thermocouples (K-type). The thermocouples were calibrated using a water dipping technique [7] and the thermal product was found to be 9690 ± 300 W$s^{1/2}/m^2K$. Fig.1. shows a schematic of the model installation and thermocouple location.

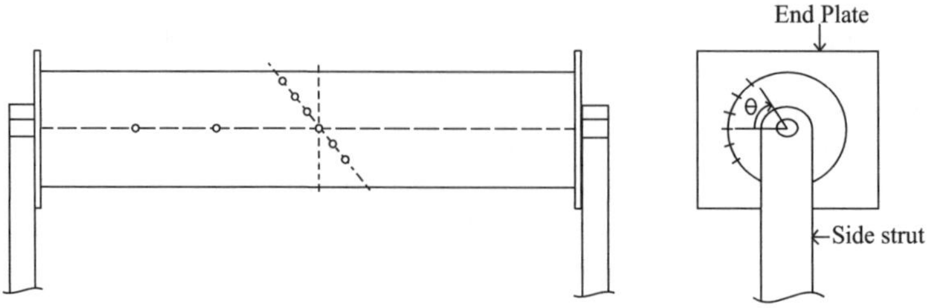

Fig. 1. Thermocouple location of the cylinder model

Each thermocouple is located every 20° apart and several thermocouples were located off center to check the two-dimensionality of the flow. Throughout the experiments, it was found that the off center measurements agreed well within ±10% of the centerline measurements, indicating a reasonably two-dimensional flow in the mid-span region.

3 Base pressure behind a blunt body at hypersonic Mach numbers

From the dimensional analysis, it can be shown that:

$$q_b/q_s \text{ or } P_b/P_s = f(M_\infty, Re_D, T_w/T_o, \gamma, Pr)$$

where (q_b/q_s) and (P_b/P_s) are the base heat transfer and base pressure ratios normalised by the respective stagnation values. The quantities on the right hand side of the equation are the Mach number, Reynolds number based on the model diameter, wall to stagnation temperature ratio, ratio of specific heats, and Prandtl number, respectively. This is true across a broad range of flow regimes [8]. However, we note that in hypersonic flow, the base to stagnation pressure ratio seems relatively independent of M_∞, Re_D, T_w/T_o, γ, and Pr where the P_b/P_s asymptotes to a constant value in accordance with the hypersonic Mach number independence principle. Fig.2.(a) illustrates this trend.

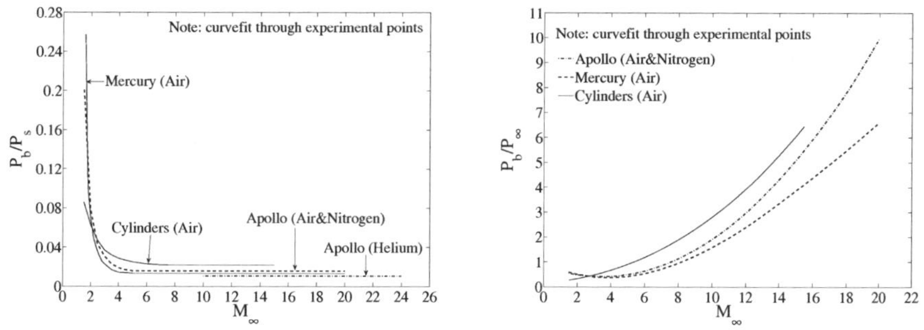

Fig. 2. (a) Base to stagnation pressure ratio (P_b/P_s) and (b) Base to free-stream pressure ratio (P_b/P_∞)

As can be seen from Fig.2.(a), the base to stagnation pressure ratio plateaus above $M_\infty \geq 6$ for all body types. Fig.2.(b) illustrates the variation of base to free-stream pressure ratio against the Mach number. The striking feature in Fig.2.(b) is the significant increase in the base pressure ratio at very high Mach numbers particularly beyond $M_\infty \geq 8$. For example, at a Mach number of 18, typical of reentry speeds, the base pressure is nearly 8 times the free-stream pressure for the Apollo Command module and is about 5.5 times the free-stream pressure for the Mercury Capsule. This is due to the entropy rise across the bow shock wave as discussed by Ferri and Pallone [9]. Another interesting feature from Fig.2. is that the base pressure ratio of the cylinder model is higher than the other two re-entry type configurations. This is thought to be due to the relieving effect of the axisymmetric configuration.

4 Blunt body heat transfer

The surface heat transfer measurements on the cylinder are shown in Fig.3. It can seen that the fore-body heat transfer distribution agrees well with the theories of Lees [5] and Kemp et al. [6] in Fig.3.(a). The theory of Lees assumes thermodynamic equilibrium and cold wall conditions whereas Kemp et al. assume thermodynamic equilibrium but varying properties (pressure gradient and enthalpy ratio) around the surface of the body. When obtaining the theoretical heat transfer estimates using the theories of Lees and Kemp et al., the pressure distribution around the surface was calculated using the expression given by Tewfik and Giedt [10] for a circular cylinder. The heat transfer results of Tewfik and Giedt in Fig.3. were obtained using their empirical expression [10]. The expression used by Lees [5] is:

$$q_w = 0.47\sigma^{-2/3}\sqrt{(\rho_e\mu_e)_0}\sqrt{u_\infty}h_{se}F(S)$$

where

$$F(S) = \frac{(1/\sqrt{2})(P/P_0)(w/w_0)(u_e/u_\infty)r_o^k}{\sqrt{\int_o^s (P/P_0)(u_e/u_\infty)(w/w_0)r_o^{2k}ds}}$$

where k=0 for a two-dimensional body and k=1 for an axisymmetric body, w=ratio of viscosity over the product of universal gas constant and temperature, σ=Prandlt number, s=distance along the surface from the stagnation point, r_0=body radius, h_{se}=enthalpy at stagnation point, and 0=conditions at stagnation point.
The expression used by Kemp et al. [6] is:

$$q/q_s = (r\rho_w\mu_w u_e/\sqrt{2\xi})(2\rho_{ws}\mu_{ws}(du_e/dx)_s)^{-1/2}(g_{\eta w}/g_{\eta ws})$$

where

$$[g_{\eta w}/(1-g_w)]/[g_{\eta ws}/(1-g_{ws})]$$

$$= (1+0.096\sqrt{\beta}) \div (1+0.096\sqrt{0.5}) = (1+0.096\sqrt{\beta})/1.068$$

and

$$\xi = 0.5x^2\rho_{ws}\mu_{ws}(du_e/dx)_s \text{ for a two-dimensional body}$$

where β=pressure gradient,η=distance normal to the surface, s=conditions at stagnation point, r=body radius, $(du_e/dx)_s$=velocity gradient at stagnation point, ρ_w=density at the wall, μ_w=viscosity at the wall, u_e=velocity at the edge of boundary layer, and g_w=wall to the edge of boundary layer enthalpy ratio

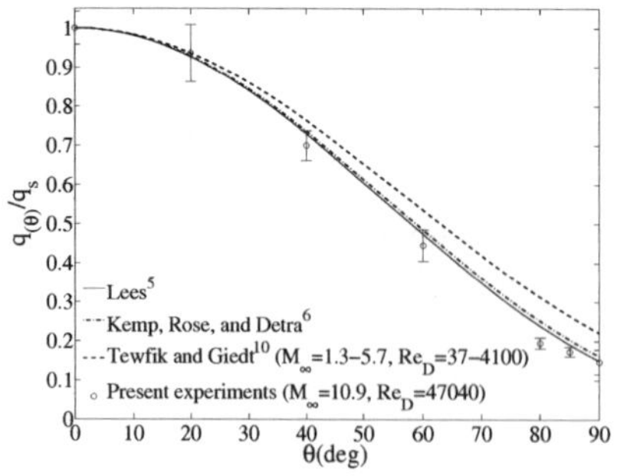

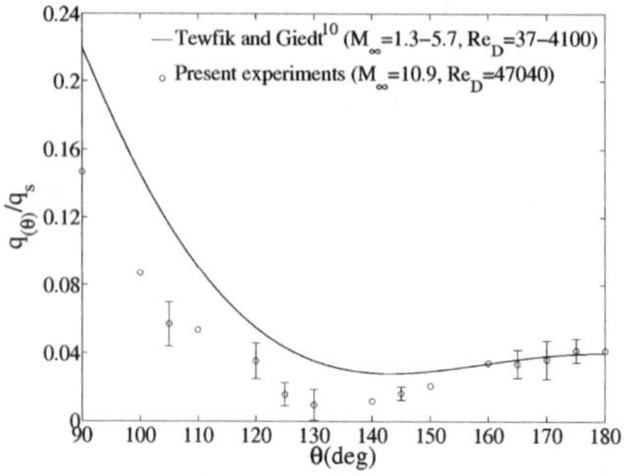

Fig. 3. (a) The fore-body heat transfer and (b) The base region heat transfer

The above equations use the local similarity assumption which seems a reasonable assumption for the present experiments. Fig.3.(b) shows the heat transfer in the base region of a cylinder. It is seen that there is a distinct local minimum at about 130° from the stagnation point after which it increases terminating in a near plateau between 160° and 180°. The heat transfer minimum seems to coincide with local pressure minimum defining a separation point as shown by Park et al. [8]. This recovery in heat transfer is a result of momentum of the re-circulating flow. Moreover, it can be seen that the heat transfer ratio of the present experiment is consistently lower than the results of Tewfik and Giedt [10] in supersonic flow. This is thought to be the result of the lower density in the base region, a consequence of the high hypersonic free-stream Mach number. The error bars in Fig.3. denote the shot-to-shot variations of the averaged steady heat transfer rates. In the present experiments, the steady flow was obtained approximately 700μs after the incident shock arrival and lasted approximately 300μs.

Fig.4. shows the base to stagnation heat transfer ratio of various blunt bodies against the Reynolds number based on the model diameter.

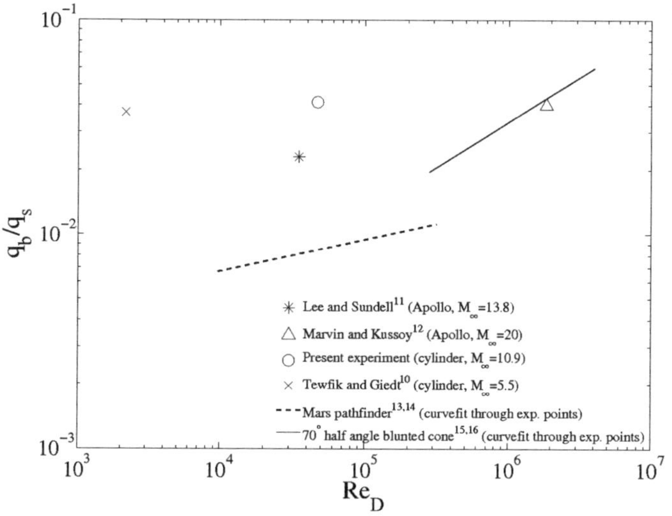

Fig. 4. The base heat transfer of various blunt bodies

From this limited data, it is seen that for axisymmetric configurations, the base heat transfer ratio seems to increase with Reynolds number. For the case of cylinder, however, such a trend is not clearly evident. It is also to be noted that, at a given Reynolds number, the cylinder base heat transfer is higher than the axisymmetric configurations. This is similar to what is seen in the case of the base pressure. It should be pointed out that all the data with axisymmetric configurations have been obtained with sting mounted models and may have been affected by interference effects.

5 Conclusions

It has been shown that while the base to stagnation pressure ratio is seen to be independent of both Mach and Reynolds number, the base to stagnation heat transfer ratio is found to be Reynolds number dependent. Experiments on a cylinder model in a shock tunnel showed that the data agree well with the theories of Lees as well as Kemp et al. Exploration of the base region showed a heat transfer minimum at the separation point followed by a gradual recovery towards the rear stagnation region. It was also found that the relative heat transfer rates in the base region were lower than those observed by Tewfik and Giedt in supersonic flow. This is due to lower density that exists in the base region in hypersonic flow as compared to supersonic flow.

References

1. Peter A. Gnoffo, Planetary-entry gas dynamics, Annu. Rev. Fluid. Mech., Vol.31, 1999, pp459-494
2. B. W. Skews and H. Kleine, Flow features resulting from shock wave impact on a cylindrical cavity, under publication in JFM, 2007
3. C. Forbes Dewey Jr., Near-wake of a blunt body at hypersonic speeds, AIAA journal, Vol.3, No.6, 1965, pp1001-1010
4. G. M. Gregorek and K. D. Korkan, An experimental observation of the Mach and Reynolds number independence of cylinders in hypersonic flow, AIAA journal, Vol.1, No.1, 1963, pp210-211
5. Lester Lees, Laminar heat transfer over blunt-nosed bodies at hypersonic flight speeds, Jet Propulsion, Vol. 26, No.4, 1956, pp259-269
6. Nelson H. Kemp, Peter H. Rose and Ralph W. Detra, Laminar heat transfer around blunt bodies in dissociated air, Journal of the aerospace sciences, July, 1959, pp421-430
7. S. G. Mallinson, Shock wave/Boundary layer interaction at a Compression corner in hyper-velocity flows, 1994, PhD Thesis, University of New South Wales
8. G. Park, S. L. Gai and A. J. Neely, A note on the base pressure of blunt bodies in hypersonic flow, AIAA paper, submitted, Mar, 2007
9. Antonio Ferri and Adrian Pallone, Note on the flow fields on the rear part of blunt bodies in hypersonic flow, WADC-TN-56-294, Aeronautical Research Laboratory, July, 1956
10. O. K. Tewfik and W. H. Giedt, Heat transfer, recovery factor, and pressure distributions around a circular cylinder normal to a supersonic rarefied-air stream, J. Aero. Sci., Vol.27, No.10, 1960, pp721-729
11. G. Lee and R. E. Sundell, Heat-transfer and pressure distributions on Apollo models at M=13.8 in an ARC-heated wind tunnel, NASA TM X-1069, Feb, 1965
12. J. G. Marvin and M. Kussoy, Experimental investigation of the flow field and heat transfer over the Apollo-capsule afterbody at a Mach number of 20, NASA TM X-1032, Feb, 1965
13. R. Bur, R. Benay, B. Chanetz, A. Galli, T. Pot, B. Hollis and J. Moss, Experimental and numerical study of the Mars Pathfinder vehicle, Aerospace Science and Technology 7 (2003), 510-516
14. R. Hollis and J. N. Perkins, Transition effects on heating in the wake of a blunt body, AIAA-97-2569, AIAA Thermodynamics conference, 32nd, Atlanta, GA, June 23-25, 1997
15. T. J. Horvath and C. B. McGinley, Blunt body near wake flow field at Mach 6, AIAA-96-1935, AIAA Fluid dynamics conference, 27th, New Orleans, LA, June 17-20, 1996
16. T. J. Horvath and Klaus Hannemann, Blunt body near-wake flow field at Mach 10, AIAA-1997-986, Aerospace Sciences Meeting and Exhibit, 35th, Reno, NV, Jan. 6-9, 1997

Combustion performance of a scramjet engine with inlet injection

S. Rowan, T. Komuro, K. Sato, and K. Itoh

Japan Aerospace Exploration Agency, Kakuda Space Center, 1 Koganesawa, Kimigaya, Kakuda-shi, Miyagi-ken, JAPAN 981-1525

Summary. The combustion performance of a two-dimensional scramjet engine model with inlet and tangential injection was experimentally investigated in the HIEST free-piston shock tunnel. Experiments were performed at a free-stream Mach number of 7, a stagnation enthalpy of 6 MJ/kg and a stagnation pressure of 30 MPa. The use of inlet injection or combined inlet and tangential injection resulted in reduced ignition times and improved combustion pressure levels when compared to tangential injection. In addition, a lower injection angle or downstream injection position for inlet injection has the potential to minimise inlet losses without significantly affecting engine pressure levels.

1 Introduction

One of the major obstacles to the development of scramjet-powered aerospace planes is the large amount of viscous drag they produce at hypersonic flight speeds. Depending on the vehicle configuration, the viscous drag can account for up to 40% of the overall drag [1]. The tangential injection of fuel is one of the most promising methods of viscous drag reduction, particularly in the presence of boundary-layer combustion [2]. However, tangential injection is not conducive to efficient mixing and combustion in the main-stream flow and, as a result, overall engine performance can be reduced. In this case, the use of a secondary injection scheme to promote mixing and combustion may lead to improved overall engine performance.

Previous studies have investigated the use of normal injection [3] and hyper-mixer injection [4] in conjunction with tangential injection to improve scramjet engine performance. While the secondary injection enhanced main-stream mixing and combustion as desired, it also adversely affected the reduction in the viscous drag produced by the tangential injection. One of the main reasons for this was the close proximity between the locations of the secondary injection and tangential injection. In the present study, the use of inlet injection to promote combustion while minimising interference with tangential injection is experimentally investigated. It is hoped that this will lead to an improvement in overall scramjet performance by maintaining efficient combustion when used in conjunction with tangential injection to reduce viscous drag.

2 Experimental Setup

2.1 Test Facility

Experiments were conducted in the High Enthalpy Shock Tunnel (HIEST) located at JAXA's Kakuda Space Center. HIEST is a free-piston driven shock tunnel capable of generating test flows with a maximum stagnation pressure of 150 MPa and a maximum stagnation enthalpy of 25 MJ/kg [5]. An axi-symmetric, contoured Mach 7 nozzle was used to produce a test flow with a stagnation enthalpy of 6 MJ/kg and a stagnation pressure of 30 MPa. The flow properties in the test section were determined using the numerical codes, ESTC [7] and NENZF [6]. Table 1 shows the average flow properties for each test condition. Air and nitrogen were used as the test gas. Gaseous hydrogen was used for fuel injection.

Table 1. Test conditions and calculated flow properties

Test Condition	7 MJ/kg	Uncertainty	Repeatability
Stagnation pressure (MPa)	30.2	-	-
Stagnation enthalpy (MJ/kg)	6.2	-	-
Pressure (kPa)	3.4	±15 %	±3 %
Temperature (K)	567	±12 %	±5 %
Density (kg/m^3)	0.021	±13 %	±3 %
Velocity (m/s)	3276	±5 %	±2 %
Mach number	6.9	±5 %	±1 %

2.2 Model

The scramjet engine model used in the experiments is shown in Fig. 1. The two-dimensional engine flowpath was designed to produce a static temperature of approx-

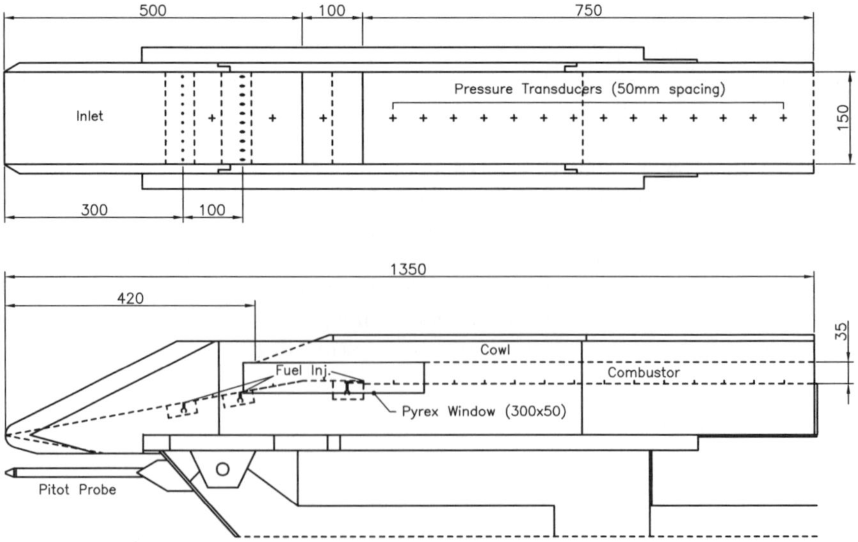

Fig. 1. Schematic of the scramjet engine model used in the experiments

imately 1500 K at the combustor entrance. The model has an overall length of 1350 mm and an overall width of 150 mm. The inlet, which has an angle of attack of 10 degrees, is 500 mm and is followed by a 100 mm-long isolator section. The combustor is a simple parallel duct that is 750 mm long and 35 mm high.

2.3 Fuel Injection

Four fuel injection configurations were used in the experiments - tangential injection, front inlet injection, rear inlet injection and combined inlet and tangential injection - as shown in Figure 2. Both inlet and tangential injection was accomplished using an array of ten 2 mm-diameter circular orifices equispaced at 15 mm intervals. Fuel was injected tangentially into the flow from from a 5 mm high backward-facing step at the combustor entrance (Fig. 2a). On the inlet, fuel was injected at an angle of 30 or 90 degrees relative to the free-stream. The inlet injection was conducted at one of two positions located 300 and 400 mm from the inlet leading edge (Fig. 2b,c). Tests were also conducted with a combination of tangential and front inlet fuel injection (Fig. 2d).

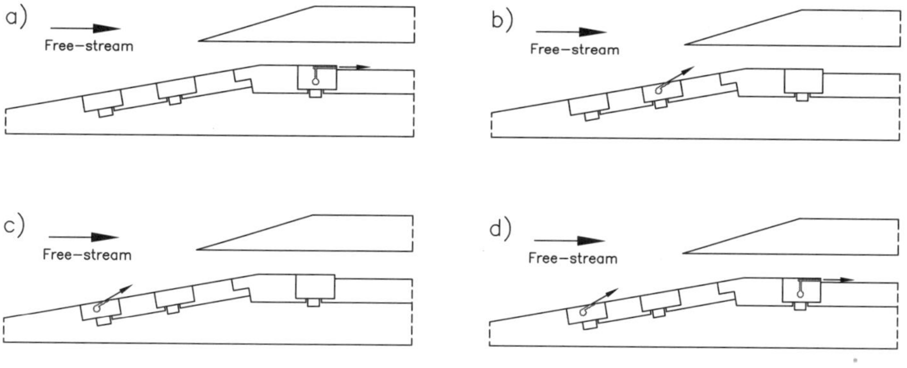

Fig. 2. Fuel injection configurations. **a)** Tangential injection **b)** Rear inlet injection **c)** Front inlet injection **d)** Combined inlet and tangential injection

2.4 Data Acquisition and Reduction

The scramjet engine model was instrumented with a total of 17 Kulite XCL-100 semi-conductor pressure transducers mounted along the engine centre-line. There were two transducers on the inlet, one transducer in the isolator and fourteen transducers in the combustor. The pressure transducers in the combustor were located at 50 mm intervals. The Pitot pressure and fuel system manifold pressure were also recorded for each test. The output of each transducer was averaged over the steady test time and normalised by the Pitot pressure of the flow to minimise any small variations in the free-stream flow conditions.

3 Results

3.1 Inlet Injection

Experiments were performed with inlet injection and compared to results obtained with tangential injection to investigate the effectiveness of the inlet injection. The results are shown in Figure 3. For the inlet injection tests, fuel was injected at 90 degrees from the front inlet position. All of the tests were conducted at a fuel equivalence ratio of 1. Air and nitrogen were used as the test gas to observe the effect of combustion.

As expected, the tangential injection does not induce strong fuel-air mixing and ignition does not occur until approximately 1100 mm from the leading edge. When inlet injection is used, the earlier injection of the fuel and the higher injection angle contribute to improved mixing and ignition occurs 100 mm earlier at approximately 1000 mm. While the earlier ignition of the fuel is advantageous as the required combustor length will be reduced, any drag increase produced by the use of inlet injection may negate this advantage. This issue will be the subject of future force measurement experiments.

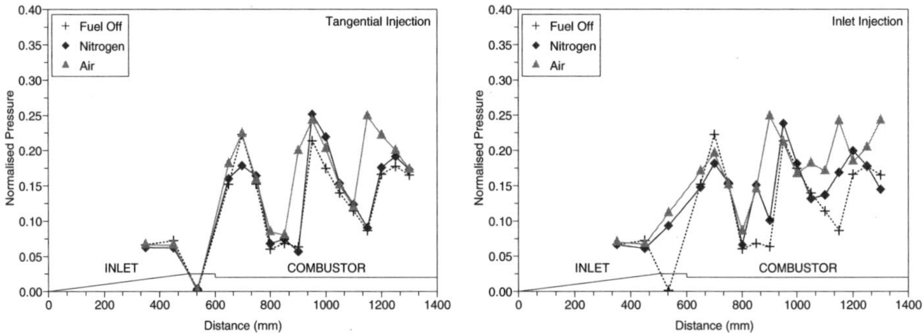

Fig. 3. Comparison of pressure distributions for tangential injection (left) and front inlet injection (right)

3.2 Injection Configuration

Each fuel injection configuration shown in Figure 2 was tested at a fuel equivalence ratio of 1. The average engine pressure distributions for front inlet injection, rear inlet injection and combined inlet/tangential injection are compared in Figure 4. An inlet injection angle of 90 degrees was used for all tests.

The pressure distributions obtained for rear and front inlet injection are shown on the left of Figure 4. While the use of front inlet injection generates a higher isolator pressure (at $x = 535$ mm) due to interference between the injection and leading edge shocks, both configurations produce similar combustion pressure levels towards the rear of the engine. As a result, rear inlet injection should lead to smaller drag losses and better overall engine performance.

Combined inlet and tangential injection is compared with front inlet injection on the right of Figure 4. In the combined injection case, the amount of fuel injected from the inlet is approximately halved and, as a result, the isolator pressure rise is significantly reduced. However, the observed pressure levels are similar to those obtained for inlet injection only. Therefore, the efficacy of the combined inlet and tangential injection will depend on how much the engine drag performance is improved by the use of tangential injection.

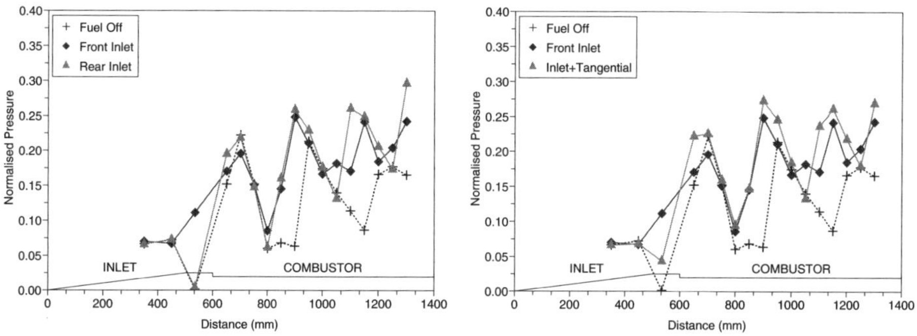

Fig. 4. Pressure distribution for front inlet injection compared with those obtained for rear inlet injection (left) and combined inlet and tangential injection (right)

3.3 Injection Angle

The effect of injection angle on the scramjet engine pressure distributions produced by front inlet injection is shown on the left of Figure 5. Results are shown for tests conducted with inlet injection angles of 30 and 90 degrees. While the isolator pressure is raised due to a stronger bow shock from injection, increasing the inlet injection angle does not result in any significant change in the distribution or magnitude of the pressure levels in the combustor section. Therefore, injection at 30 degrees can be used to minimise the bow shock losses associated with injection.

3.4 Equivalence Ratio

The effect of increasing the fuel equivalence ratio was also investigated in the case of front inlet injection. The results are shown on the right of Figure 5. Increasing the fuel equivalence ratio results in slightly earlier ignition at approximately x=950 mm and an increase in the combustor pressure level of approximately 15-20%. The amplitude of the combustor presure level oscillations is also smaller due to a weakening of the cowl shock. However, the isolator pressure is approximately an order of magnitude larger due to the increased strength of the injection bow shock. This is indicative of increased inlet losses and the overall performance of the engine is unlikely to be improved.

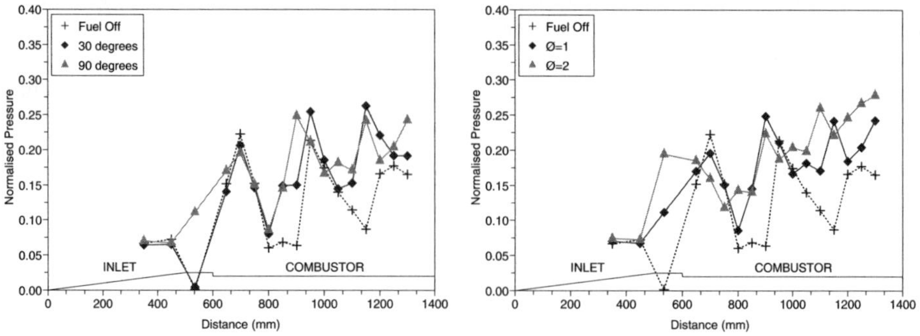

Fig. 5. Effect of injection angle (left) and equivalence ratio (right) on combustor pressure distribution with front inlet injection

4 Conclusions

A two-dimensional scramjet engine model equipped with inlet injection and tangential injection was tested in a free-piston shock tunnel with the aim of evaluating its combustion performance. The effects of changes in the injection configuration, injection angle and fuel mass flow rate were investigated.

Inlet injection led to earlier ignition and combustion in the engine with ignition occurring approximately 100 mm further upstream than tangential injection. The use of a lower injection angle or downstream injection position for inlet injection was found to have the potential to minimise inlet losses without significantly affecting engine pressure levels. These topics, along with the effect of combined inlet and tangential injection on overall engine performance, will be the subject of future studies.

References

1. D.P. Lewis, J.A. Schetz: Tangential Injection from Overlaid Slots into a Supersonic Stream. Journal of Propulsion and Power **13**, 1 (1997)
2. C.P. Goyne, R.J. Stalker, A. Paull, C.P. Brescianini: Hypervelocity Skin-Friction Reduction by Boundary-Layer Combustion of Hydrogen. Journal of Spacecraft and Rockets **37**, 6 (2000)
3. S. Rowan, A. Paull: Performance of a Scramjet Combustor with Combined Normal and Tangential Fuel Injection. AIAA Paper 2005-0615 (2005).
4. S. Rowan, M. Takahashi, T. Sunami, K. Itoh, T. Komuro, K. Sato: Viscous Drag Reduction in a Mach 12 Scramjet Engine, AIAA Paper 2005-3356 (2005)
5. K. Itoh, S. Ueda, T. Komuro, K. Sato, H. Tanno, M. Takahashi: Hypervelocity Aerothermodynamic and Propulsion Research Using a High Enthalpy Shock Tunnel HIEST. AIAA Paper 99-4960 (1999)
6. J.A. Lordi, R.E. Mates, J.R. Moselle: Computer program for the numerical simulation of non-equilibrium expansions of reacting gas mixtures. NASA Contractor Report CR-472 (1966)
7. M.K. McIntosh: Computer program for the numerical calculation of frozen and equilibrium conditions in shock tunnels. Technical Note CPD 169, Department of Supply, Australian Defence Scientific Service, Weapons Research Establishment (1970)

COMPARE, a combined sensor system for re-entry missions

A. Preci[1], G. Herdrich[2], M. Gräßlin[1], H.-P. Röser[1], and M. Auweter-Kurtz[3]

[1] *Institute of Space Systems, Universität Stuttgart, Pfaffenwaldring 31, 70569 Stuttgart, Germany*
[2] *Graduate School of Frontier Sciences, The University of Tokyo, Japan*
[3] *President of Universität Hamburg, Edmund-Siemers-Allee 1, 20146 Hamburg, Germany*

Summary. For the DLR mission SHEFEX II, the Institute of Space Systems (IRS) proposes a combined sensor system. This experiment combines both pyrometric and radiometric measurement during the re-entry phase. Additionally, total pressure can be measured via the optical path of the radiometer. The experiment will enable a separation of the radiative heat flux from the total heat flux and it will enable the specific enthalpy to be determined. By measuring at adequate wavelength ranges information on plasma composition can be gained.

1 Introduction

For the DLR mission Sharp Edge Flight Experiment SHEFEX II, the Institute of Space Systems (IRS) proposes a combined sensor system, which was developed as a concept earlier at the IRS. This experiment combines both pyrometric and radiometric measurement during the SHEFEX II mission. Additionally, total pressure can be measured via the optical path of the radiometer. The experiment will enable a separation of the radiative heat flux from the total heat flux and it will enable the specific enthalpy to be determined. Depending on the measured wavelength range of the radiometer the experiment might lead to the determination of the plasma composition. As an option, a spectrometer can be used instead of the radiometer to gain spectrally resolved data in the post-shock regime. The sensor system COMPARE (COMbined Planetary entry And trajectory Rebuilding Experiment) allows further in combination with numerical tools the reconstruction of crucial trajectory parameters such as velocity, Mach number, speed of sound and the ambient pressure and it enables also characterising of the TPS. In combination with a velocity measurement the atmospheric pressure can be determined as a function of time. The flight qualified pyrometric sensor system PYREX [1] will be used for the determination of the TPS temperature by measuring the rear side temperature of the TPS. PYREX was developed at the IRS and has already been flight qualified during the X38 campaign. For the measurement of the radiation heat flux a thermopile will be used. As the thermopile requires a duct through the heat shield material with optical access to the surrounding plasma, the total pressure can be measured via the optical path of the spectrometer using a piezo-resistive pressure sensor. The design of the duct geometry is based on experience gained during the Apollo 4 mission see [2], where the pollution of the optical window through ablating particles was negligible, and on experience gained during the design of the RESPECT sensor for EXPERT. The analysis of the measured data has to be performed using numerical tools, such as the nonequilibrium Navier-Stokes code URANUS, which was developed at the IRS. The axi-symmetric flow field solver URANUS considers an 11 components and 6 temperatures air model, implementing also different catalysis models for the different surface materials. The measured data will allow a validation of the numerical models used there or provide necessary information to improve those models.

2 The project SHEFEX II

SHEFEX II is a German Aerospace Center (DLR) project to be realised in cooperation with the industry, research centers and universities. It is the second mission of this kind, after SHEFEX I, which had a successful flight and data acquisition but couldn't be recovered after landing in the sea. The goal of SHEFEX II is to test new thermal protection concepts and investigate the applicability of a faceted geometry for the layout of re-entry vehicles. The big advantage of such simple shaped, faceted geometries is that they can be produced and mounted relatively easy compared to complex curved shapes. By designing the flat ceramic panels as serial parts and as expandable inserts further significant reductions in the costs of TPS manufacturing can be achieved. The project SHEFEX II will also provide the possibility of investigating aspects of the active aerodynamic control using canards.

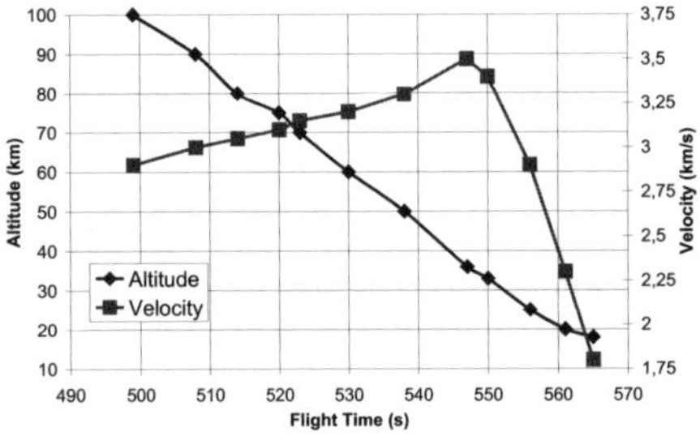

Fig. 1. Altitude and velocity of the SHEFEX II re-entry trajectory

It is planned to use a configuration of brasilian VS-40/44 sounding rockets as a launcher, which will bring the 350 kg heavy vehicle to an altitude of approx. $h = 200km$. The entry in Earths atmosphere will occur with an entry velocity of $v_e = 3km/s$ and an entry angle of $\gamma_e = -25°$. SHEFEX II reaches the maximum velocity of $v_{max} = 3.5km/s$ at an altitude of $h = 35km$. The peak mach number $Ma_{max} = 12.6$ will be reached at an altitude of $h = 80km$. The experiment time ends after 62 seconds at an altitude of $h = 20km$ and a mach number of $Ma = 7.7$.

2.1 Thermal loads during re-entry

No informations on the thermal loads during the SHEFEX II re-entry are issued by the DLR until now. Only data on the velocity and Mach number are available (see Fig. 1). Hence, an estimation of the thermal loads is necessary. The vehicle has a faceted front end with a very low nose radius of the cone end. For an estimation of the heat flux a Chapman approach is used:

$$\dot{q} = c\sqrt{\frac{\rho}{r_N}}v^x. \quad (1)$$

In this equation, ρ represents the density and r_N the nose radius. The coefficients c and x have the values $c = 5.214 \cdot 10^{-5}$ and $x = 3.15$. The density ρ is determined using an exponential approach for the atmosphere:

$$\rho(h) = \rho_0 \cdot e^{-\frac{h}{H}}. \quad (2)$$

In Eq. 2, $\rho_0 = 1.752kg/m^3$ represents the reference density, h is the altitude and $H = 6700m$ is a scaling factor.

As the nose radius is very low such that the Chapman approach (Eq. 1) can not be used properly, an effective nose radius has to be determined. In Fig. 2 the faceted shape of the SHEFEX II front end is approximated by an ellipse with the semimajor axis $a = 1500mm$ and the minor axis $2 \cdot b = 500mm$.

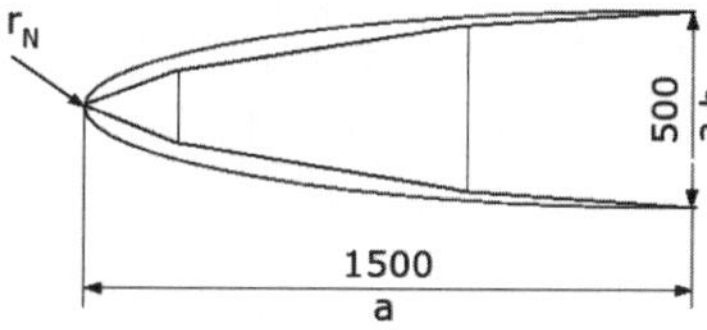

Fig. 2. SHEFEX II geometry and effective nose radius approximation

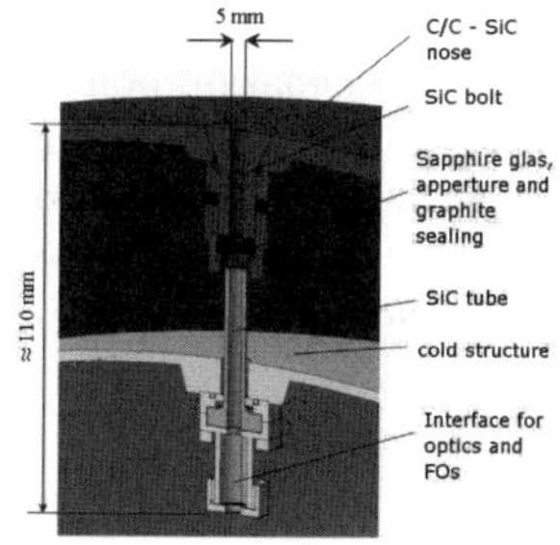

Fig. 3. RESPECT sensor mounting on the TPS

The radius of the curvature at the major axis is then used as the effective nose radius r_N. Eq. 3 yields an effective nose radius of $r_N = 41.7mm$.

$$r_N = \frac{b^2}{a}. \quad (3)$$

The temperature is determined using a radiation equilibrium approach at the surface. The emissivity of the surface material (C/C-SiC) is assumed to be $\varepsilon = 0.83$. Fig. 4 shows the estimated heat flux and temperature during the re-entry phase of SHEFEX II for the trajectory points in Fig. 1. The peak heat flux $\dot{q}_{max} = 4.2MW/m^2$ and peak temperature $T_{max} = 3077K$ are reached at approx. $h = 25km$, $Ma = 10$ and $\rho = 4.2 \cdot 10^{-2}kg/m^3$. It has to be stated out, that the stagnation point temperature will be considerably lower during flight, because of the planned active cooling of the nose tip. Consequentially, the loads shown in Fig. 4 represent a conservative case.

Preliminary project data show that the drag coefficient is $c_W = 0.2$. Considering a reference area of $A = 0.15m^2$ and the mass of the vehicle $m = 350kg$ the ballistic coefficient β can be calculated. In this case, the ballistic coefficient is $\beta = 11.7 \cdot 10^3kg/m^2$. The time gradient of the temperature can then be determined using Eq. 4 .

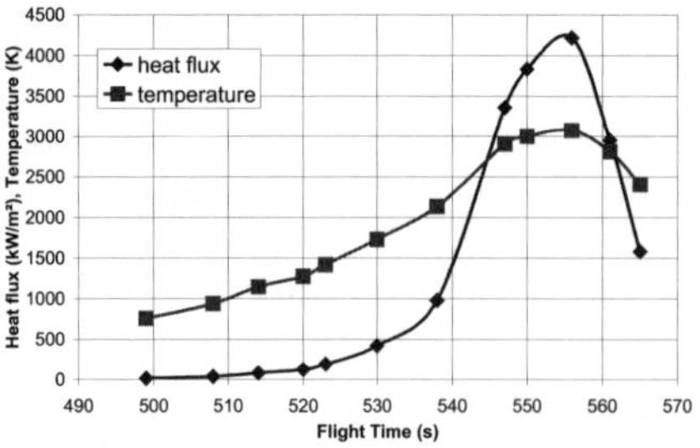

Fig. 4. Thermal loads during re-entry of SHEFEX II

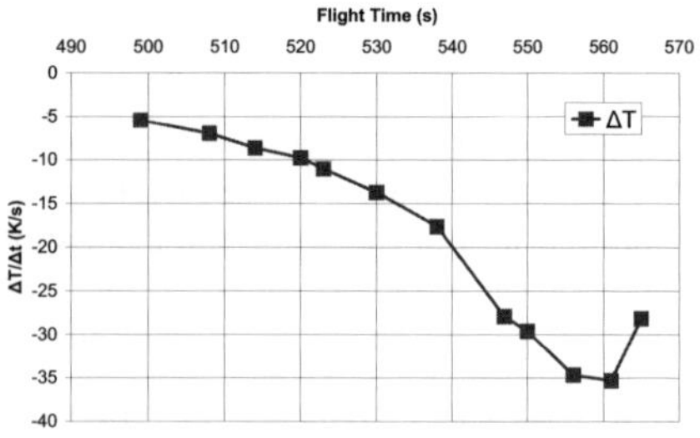

Fig. 5. Time gradient of the temperature during re-entry of SHEFEX II

$$\frac{\partial T}{\partial t} = -\frac{1}{8} T v \left[\frac{\sin(\gamma_e)}{H} + \frac{3.15\rho}{\beta} \right]. \tag{4}$$

Here is T the stagnation point temperature, v the velocity, ρ the density and β the ballistic coefficient. Fig. 5 shows that the highest time gradient is approx. 35 K/s. Using a margin of 2, the sampling rate of the pyrometer would have to be at least 70 Hz to guarantee a resolution of the measured temperature of 1 K. The pyrometers used for PYREX on X38 and on EXPERT achieve this requirement.

3 The Sensor System COMPARE

Pyrometer sensor

The measurement of the TPS rear side temperature will be realised using a pyrometer. Fig. 6 shows a scheme of the pyrometer sensor of COMPARE. For the pyrometer an InGaAs photodiode or a Si photodiode as already used in PYREX-KAT X38 can be applied. The InGaAs photodiode allows for the measurement of lower temperature as the linear sensitivity region is in the near infrared regime at wave lengths between $\lambda = 1.0\mu m$ and $\lambda = 1.6\mu m$. The emission of the TPS rear side is guided through a SiC tube mounted directly behind the TPS to the fibre optics. For the focusing of the emission on the fibre optics a flat-convex lens is used. The layout of the pyrometer sensor is mainly based on

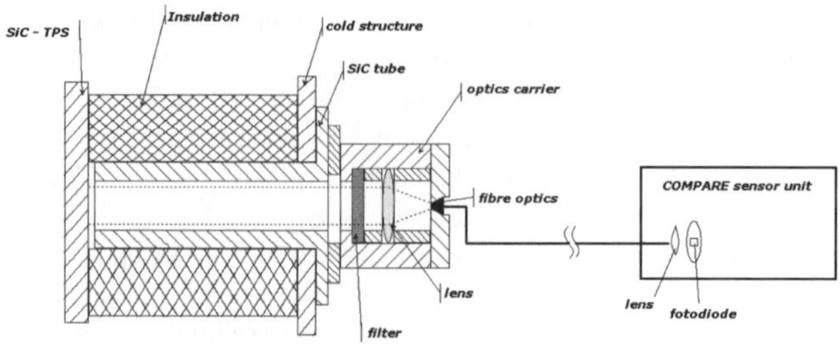

Fig. 6. Scheme of the pyrometer sensor of COMPARE

the already flight qualified PYREX-KAT X38 and on the PYREX [1] sensor currently being developed for the EXPERT mission. Hence, there is no need of a new development of the pyrometric sensor of COMPARE and of the electronics, as these can be adapted with minor modifications.

Radiometer sensor

A thermopile will be used as a radiometer detector on COMPARE. The advantage of thermopiles is their constant sensitivity over spectral range to be measured. This is absolute necessary if the intensity distribution of the radiation source is unknown. There is also a good experience made with thermopiles during the NASA Apollo 4 mission [2] and at the IRS during the characterisation of candidate TPS materials for the entry in the atmosphere of Titan [4]. The wavelength ranges for the measurement will be defined by the strongest radiation-bound during the re-entry.

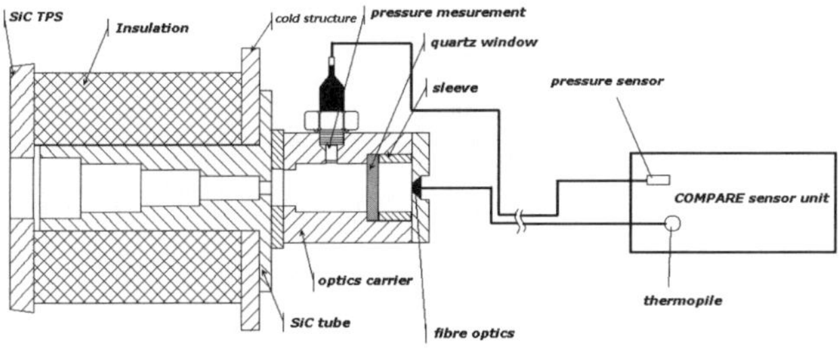

Fig. 7. Scheme of the combined radiometer and pressure sensor of COMPARE

Fig. 7 shows a scheme of the radiometer sensor of COMPARE. The layout of the SiC tube is mainly based on experience gained during the Apollo 4 mission. This design assures the highest gain of the radiation signal and the lowest possibility of pollution of the quartz window. The design of the interface between the SiC tube and the TPS is mainly based on the experience with the sensor system RESPECT on EXPERT. This interface shown in Fig. 3 was designed in a cooperation between the DLR Stuttgart and the IRS. It allows for the thermal decoupling of the hot structure from the cold structure.

Pressure sensor

To avoid additional interfaces and boreholes on the TPS the pressure sensor will be mounted on the optics carrier of the radiometer sensor (see Fig. 7). The total pressure will be measured in the chamber in front of the quartz window. A piezo-resistive sensor with a silicon chip membrane is foreseen to be used as the pressure sensor of COMPARE. Miniaturised pressure sensors are available on the market and their high frequency response allows for high resolution of the pressure measurement.

COMPARE Assembly

The sensor unit of COMPARE is mainly based on experience gained during the design and the up to come qualification of the IRS sensor systems PYREX, PHLUX and RESPECT [3]for the EXPERT mission. The total mass of the system COMPARE is approx.

$m = 2.5kg$, and the power requirement is 3 to 5 W. The dimensions of the sensor unit containing the sensors and the electronics has approx. the dimensions $140x140x90mm^3$. The data has to be stored externally and the power supply has to be provided by the vehicle. The transfer of the data will be realised using RS 422 interfaces with a 115.2 kBit/s transfer rate. To avoid the danger of wrongly connecting the data transfer and the power interface during mounting, male and female plugs are planned to be used for the connectors. A preliminary data word of COMPARE is shown in Tab. 1.

Table 1. Preliminary data word of COMPARE

Bytes	Data
1	Start byte
1	Temperature range
2	Temperature value
2	Value of the radiative heat flux
2	Pressure value
2	Internal time
1	Stop byte
11	Total

4 Summary

The combined sensor system COMPARE represents a good possibility of a combined measurement of temperature, radiative heat flux and total pressure during the re-entry phase of a vehicle such as SHEFEX II. Numerical tools allow for the reconstruction of important trajectory parameters (e.g. velocity, mach number, speed of sound, ambient pressure) out of the measured TPS rear side temperature, radiative heat flux and the total pressure. The sensor system COMPARE is particularly distinguished by the low mass, the low power requirement and the relatively low complexity of the system. The electronics of all components are placed in one sensor unit. The sensor system COMPARE is mainly based on experience gained in other projects such as X38 and EXPERT. Hence, most of the components of COMPARE have a high technology readiness level.

References

1. Auweter-Kurtz M., Fertig M., Herdrich G., Laux T., Schöttle U., Wegmann Th., Winter M. *Entry Equipments at IRS - In-flight Measurement during Atmospheric Entries.* Space Technology Journal, **Vol. 23**, Iss. 4, pp. 217-234, 2003
2. Park C. Stagnation-Point Radiation for Apollo 4 - A Review and Current Status. 35th AIAA Thermophysics Conference, Anaheim, CA, USA, 2001
3. Preci A., Lein S., Schüssler M., Auweter-Kurtz M., Fertig M., Herdrich G. and Winter M. Numerical Simulation and IRS Instrumentation Design for EXPERT. 25th AIAA Aerodynamic Measurement Technology and Ground Testing Conference, San Francisco, CA, USA, June 5-8, 2006
4. Röck W.: Simulation des Eintritts einer Sonde in die Atmosphäre des Saturnmondes Titan in einem Plasmawindkanal. Dissertation, Institute of Space Systems, Universität Stuttgart, Stuttgart (1999)

Drag reduction by a forward facing aerospike for a large angle blunt cone in high enthalpy flows

V. Kulkarni, P.S. Kulkarni, and K.P.J. Reddy

Department of Aerospace Engineering, Indian Institute of science, Bangalore, India

Summary. Drag reduction studies are conducted using a flat disc tipped aerospike for a 120-degree apex angle blunt cone model in high enthalpy flows. Accelerometer based force balance is used for the drag force measurement in the newly established free piston driven shock tunnel, HST3. Drag reduction upto about 58 percent has been achieved for Mach 8 flow of 5 MJ/kg specific enthalpy at zero degree angle of attack.

1 Introduction

Imposed bluntness at the nose of the hypersonic vehicle is necessary to alleviate the oncoming heat load. However, increased wave drag is the immediate consequence of the forced bluntness. Hence, research in the field of hypersonics is always centered on the reduction of wave drag encountered by the space vehicle during its ascent phase. Most of the drag reduction techniques are intended to modify the flowfield ahead of the stagnation point. In particular, deposition of energy [1], injection of supersonic jet [2], multi-stepped wake, [3] and retractable aerospike [4] are the different ways by which the possibility of drag reduction has been proved. However, the use of retractable spike is the simple and easy to implement technique. Viren et al [4] and Gnemmi et al [5] carried out shock tunnel testing to investigate possibility of drag reduction using different spike configurations and observed that the maximum reduction is possible with the disc tipped spike.

We have extended the drag reduction studies using aerospike for a 120-degree apex angle blunt cone to higher enthalpy flows in HST3, free piston driven hypersonic shock tunnel. Accelerometer force balance system is developed and used for the first time in the free piston driven shock tunnel during these studies. CFD analysis is also carried out for the drag reduction studies using the MBCNS code [6]. Details of the experimental facility, test model, force balance along with the experimental and CFD results are mentioned in the following sections.

2 Experimental Facility

Experimental results reported here are carried out in the newly established HST3, free piston tunnel. The HST3 tunnel, shown schematically in Fig. 1, is of moderate size having a piston weight of 20 kg. The tunnel consists of a 10m long 165mm internal diameter compression tube, 4.4 m long 39mm diameter shock tube, a convergent-divergent Mach 8 conical nozzle and 450mm long 300mmx300mm size test section connected to a dump tank. The piston is driven by nitrogen gas in 1m long 500mm diameter reservoir and the compression tube is filled with helium gas at 1 atm. pressure. The compression tube is provided with sensors at four locations to measure the acceleration and speed of the piston inside the compression tube during the run. The shock tube has two pressure sensors mounted known distance apart towards the end to monitor the shock speed and one pressure transducer at the end of the tube to measure the stagnation pressure at the entry of the nozzle. The tunnel has been calibrated for stagnation enthalpy of about 5MJ/kg. The flow quality and uniformity inside the test section is checked using the pitot rake on a traverse mechanism. Typical test time of the tunnel is $\sim$ 1.5ms. The performance of the tunnel is estimated using different numerical codes based on the measured pitot signals and typical tunnel operating parameters. The freestream conditions of the tunnel are listed in Table 1.

3 Test model and force balance

In the present studies we have chosen high drag producing 120-degree apex angle blunt cone model of base diameter 60 mm for demonstrating substantial drag reduction. We have selected flat aerodisc, which has diameter equal to $\frac{1}{4}$th the model base diameter and aerospike of length equal to the model base diameter, for producing the maximum change in the wave drag [4]. Hence, a forward facing 60mm long aerospike with a flat-faced aerodisc of 15 mm diameter is used. The model and spike are fabricated using an aluminum alloy. Weight of the model is 166 gram without the spike while it increases to 170gm with the addition of the spike.

The test model is provided with enough space from inside for the mounting of accelerometer force balance. Schematic of the model with the force balance is shown in Fig. 2. The balance system consists of a single rubber bush with a central sting attached to the model through a metallic ring. The model moves freely during the test time as the resistance offered by the rubber bush during the short run time is negligible [7]. The miniature accelerometer (PCB 303A 6403; 10mV/g sensitivity) is mounted inside the model along its axis to measure the acceleration experienced by the model during the hypersonic flow in the test section. The dynamic calibration of the force balance is carried out using the impulse hammer test. The impulse of the known value and the corresponding acceleration of the model are used to arrive at the proper system response function or transfer function.

4 Experimental studies

The above mentioned test model is mounted in HST3 at zero degree angle of incidence during the experiments for the calibrated freestream conditions mentioned in table [1]. Experiments carried out, for both with and without spike configurations, are seen to be consistent. However, particles from the paper diaphragm are found to cause sever hitting on the model surface. This particle hitting is observed to be harmful for the acceleration signal, since the acceleration signal is not able to remain steady for 1.5 ms, instead steady state of acceleration signal gets terminated after arrival of paper particles in test section, which is around 100-200 microseconds. Nevertheless, the acceleration signal is consistent with the rise time obtained from pitot followed by steady time of few hundred microseconds.

5 CFD simulations

The Multi block Compressible Navier Stokes code [6] is used for the computational analysis. Equilibrium chemistry is the important feature of this code for the simulation of real gas effects. The flow domain is divided into multiple blocks for the simulation and these blocks are further discretized using structured grid. The tunnel freestream conditions at the inlet, supersonic outlet, axisymetric boundary at the axis and usual wall boundary conditions are used for the simulations.

6 Results and discussion

Drag force is recovered from the experimentally acquired acceleration signal using the transfer function obtained from the dynamic calibration. Force signal, like acceleration signal, confirms the repeatability and the consistency of the measurement. It is seen that the time variation of the recovered drag signal for the blunt cone model is identical to the pitot signal. However, the effect of the particle hitting observed in the acceleration signal is also evident in the force signal. The drag coefficient calculated from these experiments is 1.25 for this model. The corresponding theoretical value estimated using the modified Newtonian theory [8] including the centrifugal force effects is 1.36. Over estimation of the drag force is seen in CFD analysis where the drag coefficient is predicted as 1.52.

The system response function obtained for the blunt cone without disc spike is used to recover the drag signal by deconvoluting with the experimental signal for the blunt cone model with the forward facing flat disc tipped aerospike. The typical drag signals with and without the disc spike are shown in Fig. 3. The reduction in the drag force due to the presence of the aerospike is very clear from this figure. The measured drag coefficient of 0.52 for the spiked model is found to be about 58 percent less than the drag coefficient of the model without aerospike. Drag reduction of 69 percent is predicted by the CFD simulation, which has overestimation of 19 percent in the percentage reduction. This discrimination in the estimation of percentage drag reduction between CFD and experiment may be due to assumptions involved in the simulations. The Mach number profile obtained from the CFD analysis with the spike configuration for the experimental conditions is shown in Fig. 4.

The accelerometer based force balance system is found to be capable of measuring the drag force for models in the free piston driven shock tunnel. The percentage drag reduction for the disc spike of similar configuration observed by Viren et al [6] is almost same at lower enthalpy as that observed during the experiments in free piston driven shock tunnel at higher enthalpy. Hence, the effectiveness of the disc spike (length equal to base diameter of the model and diameter equal to $\frac{1}{4}{}_t h$ of the diameter of test model) as a drag reducing agent is found to be consistent at higher enthalpy also.

7 Conclusion

Drag force measurement experiments are carried out on a 120-degree apex angle blunt cone model without and with flat disc tipped spike using the accelerometer force balance in the free piston driven shock tunnel. The experimentally obtained drag coefficient without the disc tipped spike has good agreement with the theoretical prediction while

CFD simulation over predicts the drag coefficient. Presence of the forward facing disc tipped aerospike at the nose of the 120-degree blunt cone model has reduced its total drag by 58 percent at the freestream stagnation enthalpy of 5 MJ/kg. CFD simulation has over predicted the drag reduction.

Acknowledgement. We thank DST, DRDL, AR&DB and The Director, IISc for the financial support. Also, we thank the HEA Lab team for the technical help during the experimental work.

References

1. K.Satheesh and G. Jagadeesh: *Effect of concentrated energy deposition on the aerodynamic drag of a blunt body in hypersonic flow*, Physics of fluids, 19, 031701 (2007)
2. Balla Venukumar, G. Jagadeesh and K.P.J. Reddy,: *Counterflow drag reduction by a supersonic jet for a blunt body in hypersonic flow*, Physics of fluids, 18, 118104 (2006)
3. V. Menezes, S. Kumar, Maruta, K. P. J. Reddy, and K. Takayama,: *Hypersonic flow over a multi-step afterbody*,Shock Waves Vol. 14, pp 421-424, (2005).
4. Viren M, Saravanan S, Reddy KPJ, : *Shock tunnel study of spiked aerodynamic bodies flying at hypersonic Mach number*, Shock Waves Vol 12, pp 197-204 (2002)
5. Gnemmi, P., Srulijes, J., Roussel, K., and Runne, K. : *Flow Field around Spike-Tipped Bodies*, AIAA Paper 2001-2464, June 2001.
6. Jacobs, P. A. : *MB_{CNS} : A computer program for the simulation of transient compressible flows*,Department of Mechanical Engineering Report 10/96 , The University of Queensland, December 1996.
7. Niranjan Sahoo, Kiran Suryavamshi, K. P. J. Reddy, and David J. Mee, : *Dynamic force balance for short duration hypersonic testing facilities*,Expts. Fluids, 38, 606 (2005)
8. Truitt RW: *Hypersonic Aerodynamics*, New York: The Ronald Press Company,(1959)

Table 1. Freestream conditions

Static pressure (kPa)	Static temperature (K)	Mach number	Stagnation enthalpy (MJ/kg)
0.284	316	~ 8	~ 5

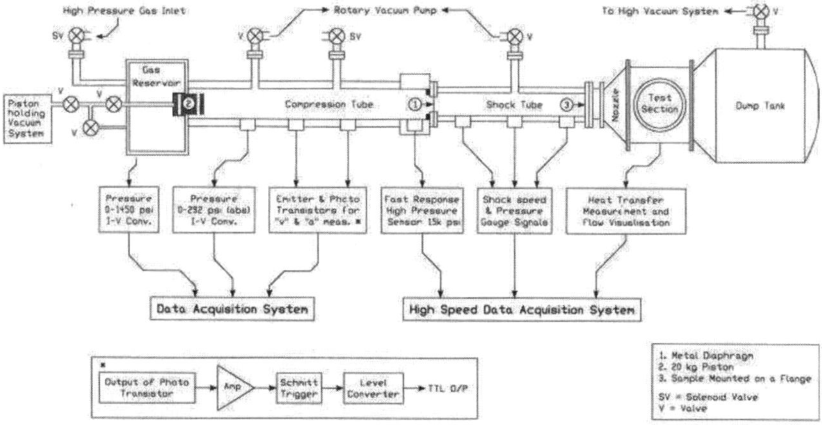

Fig. 1. Schematic diagram of free piston driven hypersonic shock tunnel HST3.

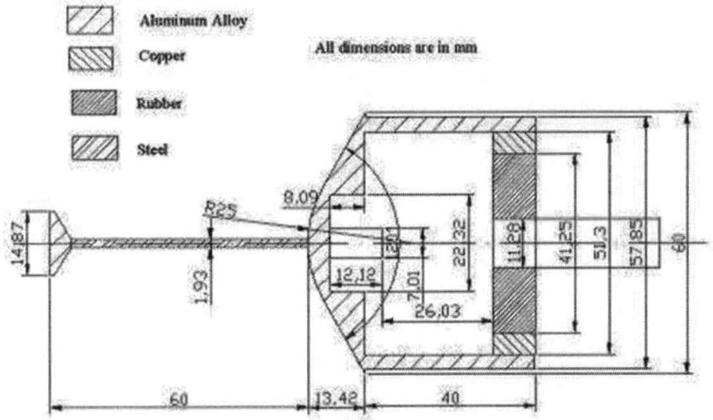

Fig. 2. Schematic diagram of the of the 120-degree apex angle blunt cone model with forward facing aerospike and accelerometer balance system.

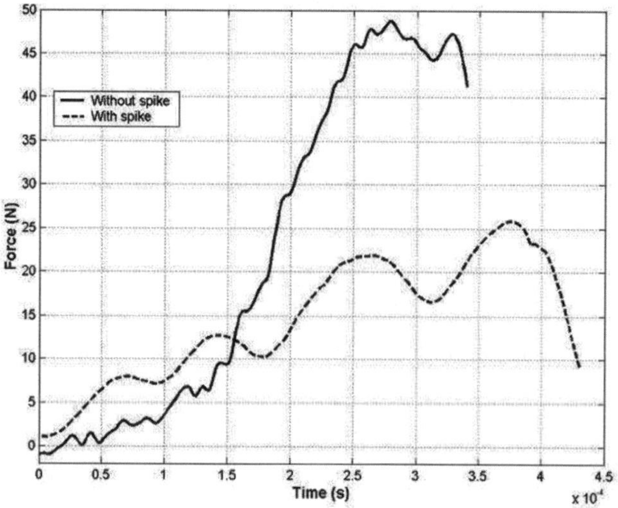

Fig. 3. Measured aerodynamic forces for the 120-degree apex angle blunt cone model with and without forward facing aerospike at flow enthalpy of ∼ 5 MJ/kg.

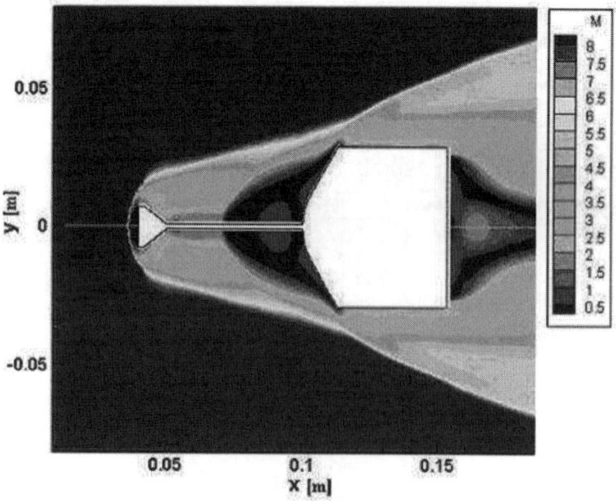

Fig. 4. Computed Mach number profiles around the blunt cone with forward facing disc tipped aerospike in Mach 8 flow.

Drag reduction by counterflow supersonic jet for a blunt cone in high enthalpy flows

V. Kulkarni and K.P.J. Reddy

Department of Aerospace Engineering, Indian Institute of science, Bangalore, India

Summary. Counterflow supersonic jet is used as a drag reduction device during the experiments in free piston driven shock tunnel, HST3. Accelerometer based force balance is employed to measure the drag force experienced by the 60-degree apex angle blunt cone model without and with the supersonic jet opposing the hypersonic flow. It is observed that the drag force decreases with increase in injection pressure ratio until the critical injection pressure is reached. Maximum reduction in drag force of 44 percent is recorded at the critical injection pressure ratio 22.36. Further increase in injection pressure ratio has reduced the percentage drag reduction. Change in nature of the flowfield around the model has also been observed across the critical injection pressure ratio.

1 Introduction

Penalty to the reduced heating rate due to higher bluntness is the increased wave drag experienced by the hypersonic vehicles. Requirement of the fuel goes up owing to increase in drag force. Therefore, research in the area of hypersonics is always oriented towards reduction of wave drag. Injection of supersonic jet from the stagnation point to alter the flowfield ahead of the nose is one of the interesting methods to reduce the wave drag. Finley [1] conducted experiments to study the effect of injection and reduction in pressure coefficient at Mach number 2.5. Modification in the flow field in the presence of a forward facing jet or plasma has been reported by Shang et al [2]. Shock tunnel studies of supersonic jet injection in hypersonic flow have also been conducted [3]. These counterflow drag reduction (CDR) studies are carried out either at lower enthalpies in shock tunnel or at low Mach numbers and low enthalpies in wind tunnels.

Establishment of free piston driven shock tunnel, HST3, has given the major impetus to conduct the CDR studies at higher Mach number and at higher enthalpy to capture the whole envelop of percentage drag reduction for various jet total pressures. Therefore, a 60-degree apex angle blunt cone model integrated with the accelerometer force balance is employed for CDR experiments in free piston driven shock tunnel. A solenoid-based injector is developed and used for injection of supersonic jet. Details of the experiments and results along with the test model and force balance are given in the following sections.

2 Experimental Facility

Experimental results reported here are carried out in the newly established HST3, free piston driven shock tunnel. The HST3 tunnel shown schematically in Fig. 1 is of moderate size having a piston weight of 20 kg. The tunnel consists of a 10m long 165mm internal diameter compression tube, 4.4 m long 39mm diameter shock tube, a convergent-divergent

Mach 8 conical nozzle and 2 m long 1 m diameter size test section cum dump tank. The piston is driven by nitrogen gas in 1m long 500mm diameter reservoir and the compression tube is filled with helium gas at 1 atm. pressure. The compression tube is provided with sensors at four locations to measure the acceleration and speed of the piston inside the compression tube during the run. The shock tube has two pressure sensors mounted known distance apart towards the end to monitor the shock speed and one pressure transducer at the end of the tube to measure the stagnation pressure at the entry of the nozzle. The tunnel has been calibrated for stagnation enthalpy of about 5MJ/kg. The flow quality and uniformity inside the test section is checked using the pitot rake on a traverse mechanism. Typical test time of the tunnel is $\sim$ 1.5ms. The performance of the tunnel is estimated using different numerical codes based on the measured pitot signals and typical tunnel operating parameters. The freestream conditions of the tunnel are listed in Table 1.

3 Test model and force balance

A 60-degree apex angle blunt cone model of base diameter 70 mm is used for the drag force measurement during the CDR studies. The 0.4 kg model is fabricated using an aluminum alloy. A 2mm diameter orifice is made at the stagnation point for the jet injection experiments. Connection of the orifice with the gas reservoir is made using the flexible tubing. Provision is also made inside the model for integration of the force balance. The schematic of the test model with the force balance is shown in Fig. 2. The balance system consists of two rubber bushes with a central sting attached to the model through two metallic rings. The model moves freely during the test time as the resistance offered by the rubber bush during the short run time is negligible [4]. The miniature accelerometer (PCB 303 A 20849; 10mV/g sensitivity) is mounted inside the model along its axis to measure the acceleration experienced by the model during the hypersonic flow in the test section. The dynamic calibration of the force balance has been accomplished using the impulse hammer test. The impulse of known value and the corresponding acceleration of the model are used to arrive at the proper system response function or transfer function.

4 Solenoid injector

A solenoid-based injector is fabricated and used to synchronize the injection of supersonic jet with the entry of hypersonic flow in the test section. A controller is designed and developed to supply the 24 V power to the solenoid valve during the experiment. Inlet of the injector is connected to the high-pressure cylinder while outlet is connected to orifice of the test model. High-pressure cylinder is provided with the regulator to alter the total pressure of supersonic jet to carry out injection experiments at various injection pressure ratios, $\frac{P_{0j}}{P_{02}}$, where P_{0j} is total pressure of jet and P_{02} is pitot pressure of hypersonic flow.

5 Experimental studies

The 60-degree apex angle blunt cone model is mounted at zero degree angle of incidence in the test section of HST3. Experiments without injection and with different injection pressure ratios are conducted for the calibrated freestream conditions. During the experiments with injection, trigger pulse to the solenoid controller is given from the pressure transducer at the end of compression tube. The trigger voltage is adjusted in such a way that the injection of supersonic jet synchronizes with the arrival of hypersonic flow in the test section. Experiments for different injection pressure ratios 7.45, 14.91, 22.36, 29.82 and 37.27 are carried out for the same freestream conditions by varying the total pressure of the supersonic jet. Acceleration of the model is recorded for all the test cases. Consistency in the acceleration signal, for each test case, has been observed.

6 Results and discussion

Experimentally obtained acceleration signal and the system response function are used to recover the drag force for the experiment without injection. Nature of thus obtained force signal is found similar to the pitot signal where the flow establishment time and steady test time are clearly seen. Drag coefficient is calculated from the recovered drag signal. Experimentally obtained drag coefficient is 0.61 while the corresponding theoretically [5] calculated drag coefficient is 0.63.

Drag force for injection experiments is recovered from the acceleration signal with the help of same transfer function used for experiments without injection. Reduction in the drag force is noticed for all the injection pressure ratios. The percentage reduction in drag coefficient for all the injection pressure ratios is shown in Fig. 3. Maximum reduction in the drag force corresponds to the injection pressure ratio 22.36. It is clear from the Fig. 3. that, drag force decreases with increase in injection pressure ratio. However, this decrease continues till the drag reduction reaches its maximum value at the injection pressure ratio 22.36. This pressure ratio corresponding to the maximum drag reduction, is termed as critical pressure ratio. Percentage reduction in drag coefficient decreases with further increase in injection pressure ratio after its critical value. This change in nature of the percentage of drag reduction across the critical injection pressure ratio is attributed to the change in nature of the flowfield around the model which has been observed during the force measurements. The drag force is unsteady in nature for the injection pressure ratios upto the critical value, which can be clearly seen in the force signal shown in Fig. 4. However, the force signal for the injection pressure ratio more than the critical pressure ratio shows similar steadiness observed in the force signal without injection (Fig. 5.).

7 Conclusion

Experiments for CDR are conducted with a 60-degree apex angle blunt cone model integrated with the accelerometer force balance. Experiments with injection are carried out for five injection pressure ratios by varying the total pressure of the jet. Maximum reduction in the drag force, 44 percent, is recorded for the critical injection pressure ratio of 22.36. Unsteadiness and steadiness in the force signal across the critical injection pressure are faithfully recorded using the accelerometer during the force measurement.

Acknowledgement. We thank DST, DRDL, AR&DB and The Director, IISc for the financial support. Also, we thank the HEA Lab team for the technical help during the experimental work.

References

1. Finley PJ: *The flow of a jet from a body opposing a supersonic free stream*, Journal of Fluid Mechanics 26 (Part 2): 337-368, (1966)
2. Shang J.S. J Hayes amd J. Menart: *Hypersonic flow over a blunt body with plasma injection*, Jl. Of space craft and rockets, Vol. 39. No. 3, 2002, pp 367-374 (a)
3. Balla Venukumar, G. Jagadeesh and K.P.J. Reddy,: *Counterflow drag reduction by a supersonic jet for a blunt body in hypersonic flow*, Physics of fluids, 18, 118104 (2006)
4. Niranjan Sahoo, Kiran Suryavamshi, K. P. J. Reddy, and David J. Mee, : *Dynamic force balance for short duration hypersonic testing facilities*, ,Expts. Fluids, 38, 606 (2005)
5. Truitt RW: *Hypersonic Aerodynamics*, New York: The Ronald Press Company,(1959)

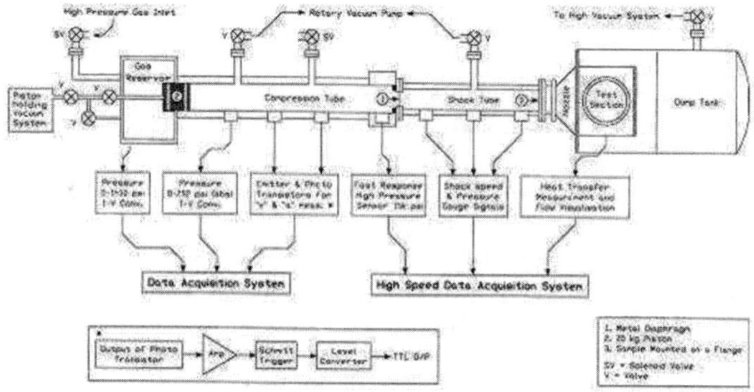

Fig. 1. Schematic diagram of the fully instrumented free piston driven hypersonic shock tunnel HST3.

Table 1. Freestream conditions

Static pressure (kPa)	Static temperature (K)	Mach number	Stagnation enthalpy (MJ/kg)
0.284	316	~ 8	~ 5

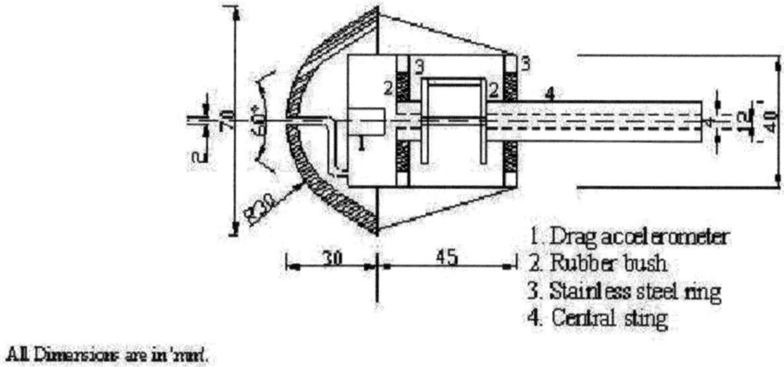

Fig. 2. Schematic diagram of the 60-degree apex angle blunt cone with the accelerometer based balance system and passage for supersonic jet injection at the stagnation point.

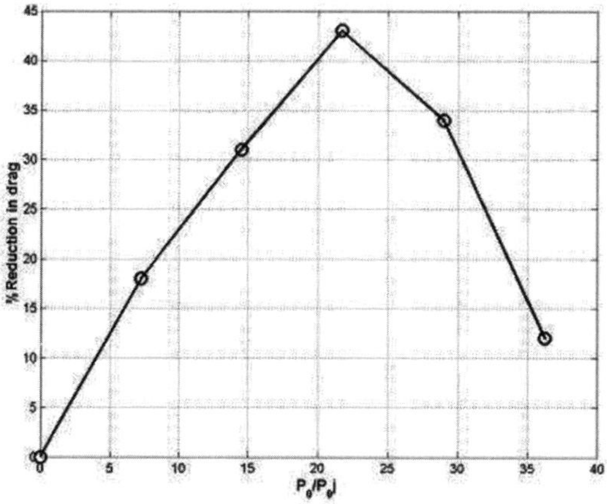

Fig. 3. Drag reduction corresponding to different injection pressure ratios.

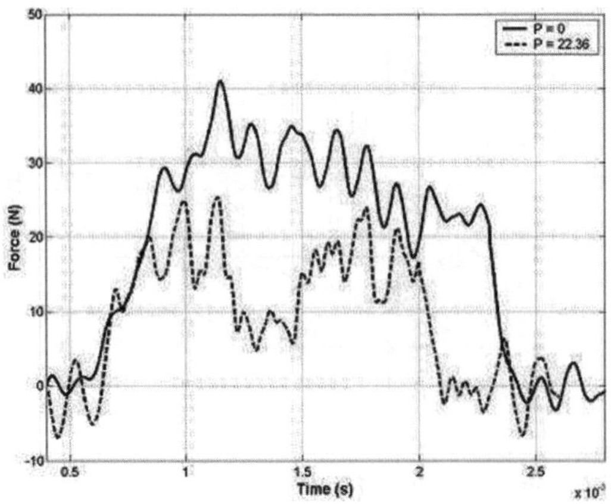

Fig. 4. Unsteady drag force signal recovered from the accelerometer signal for the injection pressure below the critical value.

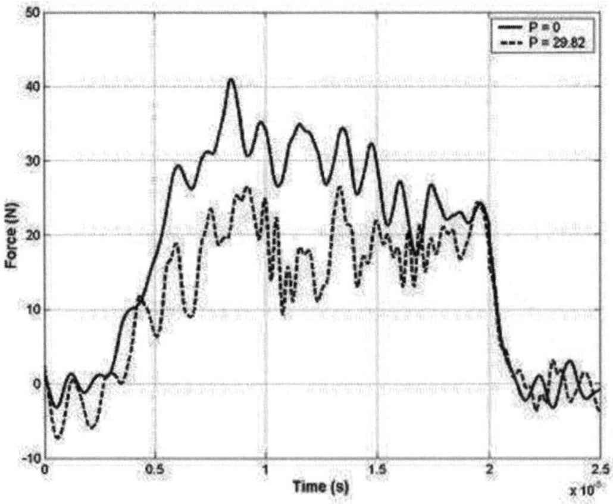

Fig. 5. Steady drag force signal recovered from the accelerometer signal for the injection pressure above the critical value

Effect of electric arc discharge on hypersonic blunt body drag

K. Satheesh and G. Jagadeesh

Dept. of Aerospace Engineering, Indian Institute of Science, Bangalore - 560 012,India

Summary. Experimental results on the effect of energy deposition using an electric arc discharge, upstream of a 60° half angle blunt cone configuration in a hypersonic flow is reported.Investigations involving drag measurements and high speed schlieren flow visualization have been carried out in hypersonic shock tunnel using air and argon as the test gases; and an unsteady drag reduction of about 50% (maximum reduction) has been observed in the energy deposition experiments done in argon environment. These studies also show that the effect of discharge on the flow field is more pronounced in argon environment as compared to air, which confirms that thermal effects are mainly responsible for flow alteration in presence of the discharge.

1 Introduction

When a body is exposed to hypersonic flow, it is subjected to extremely high wall heating rates owing to the conversion of the kinetic energy of the oncoming flow into heat through the formation of shock and viscous effects.This is one of the main problems that stands in the way of practically viable hypersonic transport. The conventional way of tackling this problem, is to configure the hypersonic vehicle so as to have a blunt fore-body, which helps minimise the heating rates by increasing the shock stand-off distance. But this also results in an increase in wave drag and puts the penalty of excessive load on the propulsion system.An alternative approach to solving this problem is to alter the flow field using external means without changing the shape of the body. The superiority of this method lies in the fact that the effective shape of the body can be altered to meet the requirements of low wave drag, without having to pay the penalty of increased wall heat transfer rate.

Various methods have been suggested in the literature for the alteration of the hypersonic flow field, one amongst which, uses localised energy deposition in the free stream to produce the required changes in the flow field. Theoretical and numerical studies reported widely in the literature have shown many potential uses for this method in improving the performance of hypersonic vehicles [1], [2].Various studies directed towards understanding the mechanism of flow alteration by energy deposition have also been reported, suggesting that the thermal effects play a vital role in the flow field alteration by energy deposition [3]. Experiments have also been reported on the use of microwave energy deposition to alter the flow field on a flat faced cylindrical model, and the results shows a decrease of stagnation pressure as a result of the interaction of the energy spot [4].It is also suggested in this paper that only 1% of the total energy goes into translational excitation and the rest is made use of for exciting the electronic and vibrational modes ans so the gas heating is strongly dependant on the relaxation of these modes.However, in spite of the considerable amount of computational work and preliminary experimental studies being done in this field, conclusive experimental demonstration of the use of energy deposition for flow field alteration at hypersonic Mach number is found lacking.In this backdrop, the main aim of the present study is to experimentally investigate the effect of energy deposition on the hypersonic wave drag on a 60° half angle blunt cone. In the following sections, we discuss the details of this experimental work, comprising of drag measurements and schlieren flow visualisation, directed at investigating the potential use of energy deposition by electric discharge as a drag reduction device in hypersonic flow.

2 The experimental facility

The experiments described here are conducted in the IISc hypersonic shock tunnel HST-2, which consists of a stainless steel shock tube of 50mm diameter connected to a convergent-divergent conical nozzle of 300mm exit diameter, with a 450mm long test section of 300mm x 300mm square cross section. Keeping in view the mechanism of flow alteration by energy deposition suggested in the literature, it was decided to conduct the experiments using two different test gases namely argon and air.In selecting argon as the test gas, it is expected that, being a mono-atomic gas, a larger percentage of the electrical energy deposited would be used in exciting the translational energy mode, thereby producing a more pronounced heating effect as compared to the electric discharge in air. The corresponding free stream conditions under which the experiments are carried out is listed in Table.1, and under these operating conditions, HST-2 generates a steady hypersonic flow for a period of almost $1.1msec$.These flow conditions are reproducible within ±5% accuracy.

Table 1. Typical free stream conditions in the shock tunnel

No.	M_∞	$P_\infty(pa)$	$\rho_\infty(kg/m^3)$	$T_\infty(k)$	$u_\infty(m/sec)$	Test gas
R1	6	1593	0.0352	157.7	1510.3	*Air*
R2	9.1	556	0.0357	75	1468	*Argon*

The experimental study has been carried out using a test model with a 60° half angle fore-body with a blunt nose as the test model,which represents a conventional re-entry capsule configuration. Considering the fact that that energy deposition is done using an electric discharge, it was fabricated out of an electrically insulating material(HYLEM), in order to avoid the possibility of an arc being established between the electrodes and the model.The model also houses the accelerometer based drag balance, the arrangement of which is shown in Fig.1. As can be observed from the figure, the model fitted with a drag accelerometer is supported by a rubber bush, which during the short testing time of the shock tunnel ensures unrestrained motion of the model subject to the aerodynamic forces. This drag balance was calibrated by applying a known impulse load using an impulse hammer (PCB GK291D01) and the experimental acceleration signals are then deconvolved with the system response obtained from the calibration, in order to recover the drag force history.

Since an electric discharge is used as the energy source in the experiments, the energy deposition apparatus consists of a high voltage power source and a pair of electrodes

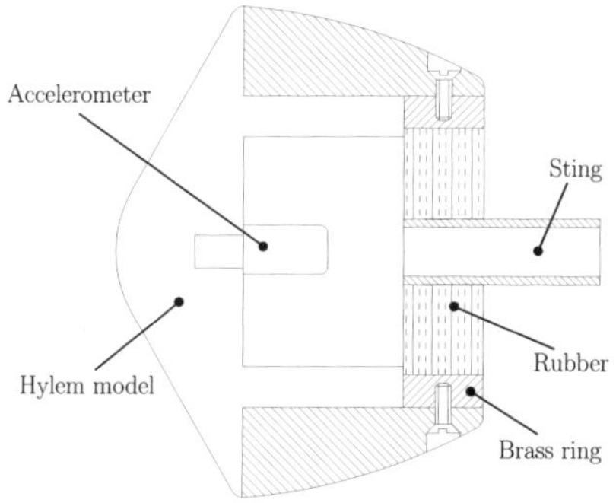

Fig. 1. Schematic diagram of the accelerometer based drag balance

placed in the flow field, between which the discharge is struck. To arrive at an optimum configuration of the electrode fixtures, the disturbance caused by these fixtures to the blunt body flow field was studied using a standard 'Z-type' schlieren arrangement.A high speed digital camera operating at a 10,000 fps is used for recording the schlieren images, and by studying these images for various electrode fixtures, an optimum arrangement with a minimal disturbance to the flow field was arrived at.Figure 2 shows a schematic of this arrangement along with the image of the blunt body flow field in presence of these fixtures, from a test conducted at condition R2 with the electrodes placed $70mm$ from the model. This arrangement has the electrodes fixed to the bottom corners of the test section, forming an angle of 62° with the flow direction and with their tips at a distance of $70mm$ from the model nose, and $4mm$ offset from the stagnation streamline. Even though the schlieren images shows that the blunt body flow field is affected by the

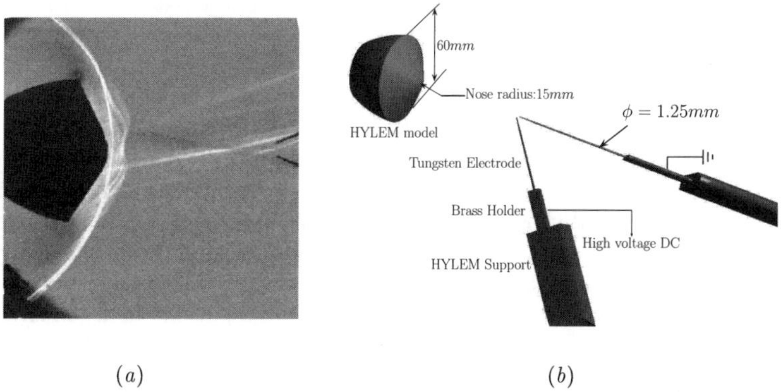

Fig. 2. The electrode fixtures used in the experiments. (**a**) Flow field. (**b**) Electrode configuration

presence of the electrode fixtures, it will be shown later that this has a negligible effect on its aerodynamic drag.

Electric discharge is produced between these two electrodes, one of which is at the ground potential, and the other electrode connected to a high voltage DC supply, the

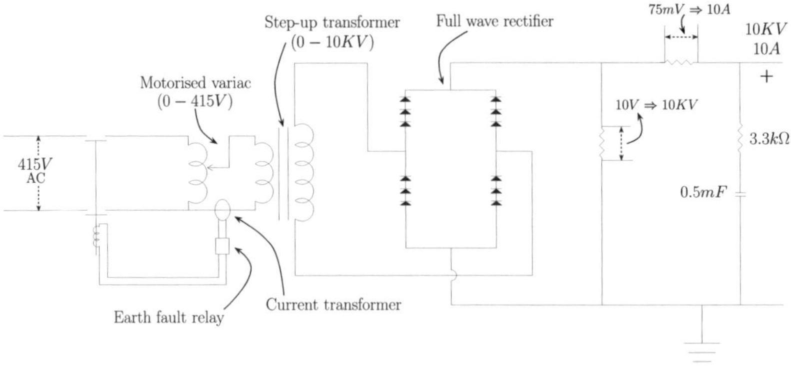

Fig. 3. Circuit diagram of the DC power source used for generating the electric arc

circuit diagram of which is shown in Fig.3. This power supply has a rated output voltage of 10KV with a 10A short circuit current. It consists of a step -up transformer for stepping up the line voltage (415V) to 10KV , which is then rectified using a full wave bridge rectifier along with a capacitor connected in parallel with the load to produce the required DC voltage.Provisions are also made in the circuit in order to monitor the voltage drop across the arc and the arc current to facilitate the estimation of the arc power.

3 Experimental results

Drag measurement along with simultaneous flow visualisation is carried out for the experimental conditions listed in table1.In these experiments, the power source is operated at the maximum power rating, with the electrodes placed at a distance of $70mm$ from the model.Even though the rated power output of the DC power source amounts to $100kw$,it is observed from the measured discharge voltage and current that only $0.3kw$ of energy goes into the discharge, with the remaining part of it being consumed probably by the impedance losses during the transient operation of the power source.

The sequential schlieren images of the blunt body flow field with energy deposition under the aforementioned experimental conditions is given in Fig.4. The time indicated in these images, recorded during a single experiment, is measured from the onset of hypersonic flow over the test model, and the alteration of the blunt body shock structure in presence of the discharge is clearly depicted in these images.These figures show a significant alteration in the flow field due to energy deposition, especially in argon environment,

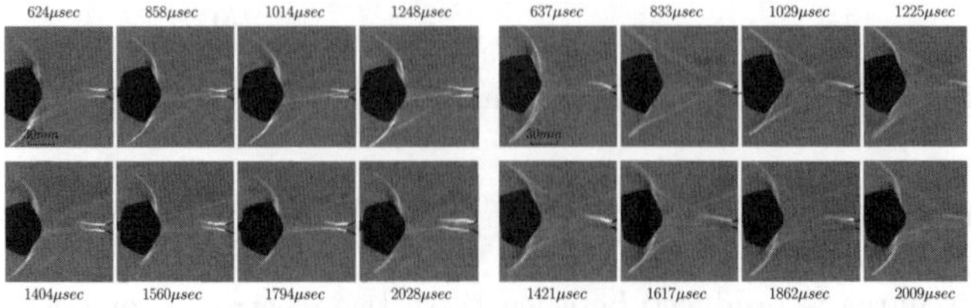

Fig. 4. Effect of energy deposition. **(a)** Condition R1. **(b)** Condition R2

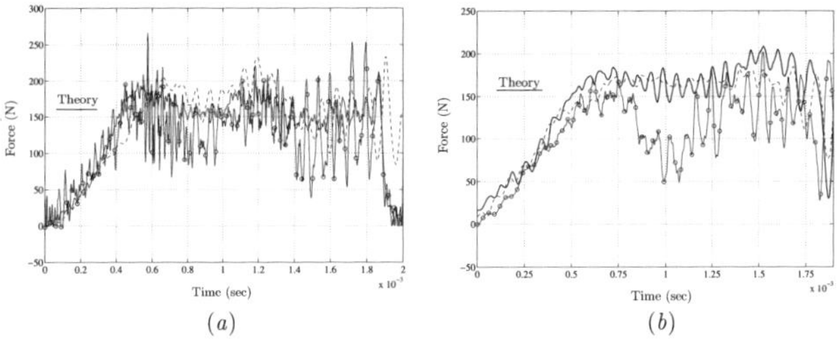

Fig. 5. Effect of energy deposition: drag measurements.——,Without electrodes; ·····,Without energy; —o—,With energy. (**a**) Condition R1. (**b**) Condition R2

with the entire model located within the aerodynamic shadow of the shock generated by the energy source.

Similar observations can also be made from the drag signals recorded during these experiments (Fig.5).The sudden drop in drag observed in Fig.5(b) can be correlated to the corresponding images in Fig.4(b), with the entire model engulfed within the separation region.These measurements show a maximum drag reduction of 50% in the experiments done in argon with an average reduction in drag of 27.4% and an average reduction in drag amounting to 18.5% in air. A more meaningful way for comparing efficiency of energy deposition under different test conditions could be by using the parameter E_{eff} defined as the ratio of the amount of power gained by drag reduction to the amount spent for reducing the drag, as follows [1].

$$E_{eff} = \frac{(D - D')V_\infty}{P_{el}}, \tag{1}$$

where $D - D'$ =Reduction in drag(N), and V_∞ =Free stream velocity(m/sec)
In the present experiments with argon as test gas, $E_{eff} = 227.6$ and similarly for the experiments with air as test gas $E_{eff} = 110.55$. So it is clear from both the schlieren images and drag measurements that the effectiveness of energy deposition is more in argon environment, thereby confirming that thermal effect plays a dominant role in the mechanism of flow alteration by energy deposition.

4 Conclusions

In conclusion, the concept of hypersonic blunt body drag reduction has been verified experimentally, and a confirmation of the importance of thermal effects on flow alteration using energy deposition is obtained. An unsteady drag reduction of about 50% is measured for a large angle blunt cone at Mach 9 by depositing the energy at about 1.17 body diameters ahead of the blunt cone in the IISc hypersonic shock tunnel using Argon as the test gas.

References

1. David Riggins,H.F.Nelson and Eric Johnson:AIAA Journal, **37**,4 (1999).
2. Machert.S.O, Shneider.M.N and Miles.R.B: Scramjet Inlet Control by Off-Body Energy Addition:A Virtual Cowl, AIAA Paper 2003-0032, 2003.

3. S.Merriman,Elke Ploenjes,Peter Palm and Igor.V.Adamovich: AIAA Journal, **39**,8 (2001).
4. Yu.F.Kolesnichenko, V.G.Brovkin, O.A.Azarova, V.G.Grudnitsky, V.A.Lashkov and I.Ch.Mashek: Microwave energy release regimes for drag reduction in superrsonic flows, AIAA 2002-0353,2002.

Effect of the nose bluntness on the electromagnetic flow control for reentry vehicles

H. Otsu[1], T. Matsumura[1], Y. Yamagiwa[1], M. Kawamura[2], H. Katsurayama[3], A. Matsuda[3], T. Abe[3], and D. Konigorski[4]

[1] *Shizuoka University, Johoku 3-5-1, Nishi-ku, Hamamatsu, Shizuoka, 432-8561, Japan*
[2] *University of Tokyo, Hongo 7-3-1, Bunkyo-ku, Tokyo, 113-8656, Japan*
[3] *ISAS / JAXA, Yoshinodai 3-1-1, Sagamihara, Kanagawa, 229-8510, Japan*
[4] *Astrium, Huenefeldstrasse 1-5, Bremen, 28199, Germany*

Summary. Effect of the nose bluntness on the magnetgic flow control system for reentry vehicles is investigated by the CFD analyses. The vehicle shape is assumed to be a reentry capsule EXPRESS with different nose configurations; spherical nose and flat head. CFD results suggest that the drag force of the vehicle with the spherical nose is enhanced more drastically than that with the flat head. This means that when the magnetic flow control system is applied to the vehicle with the spherical nose the drag force is enhanced efficiently. When the interaction parameter is sufficiently large, the enhanced drag force to the reentry vehicle does not depend on the nose bluntness. This suggests that the drag force can be controlled without considering the nose bluntness of the vehicle with the magnetic flow control system.

1 Introduction

The thermal protection system (TPS) for the reentry vehicles is one of the key technologies for development of the reentry vehicles. Most of TPS are designed to protect the vehicle from the severe reentry heating and are damaged during the reentry flight. Thus, current TPSs are needed to be replaced or refurbished for next flight. This means that the current TPS is not suitable for future reusable space transportation system.

As one of the candidates of TPS for the future reusable space transportation systmes, we are interested in the magnetic flow control system for mitigating the reentry heating. In this flow control system, the plasma flow behind the strong shock is controlled by the imposed magnetic field through the induced electric current and the Lorentz force as shown in Fig.1. As a result, the shock layer thickness and the boundary layer are expected to increase and the heat flux to the vehicle is mitigated. Additionally, the shock layer enhancement will enhance the drag force of the vehicle. Thus, this system is expected to enable us not only to control the flight performance of the vehicle but also to mitigate the aerodynamic heating on the vehicle.

This system had been cosidered to be promisible and investigated intensinvely 1960s. [1–3] But at that time, the sufficiently strong magnetic field is very difficult to create in the reentry vehicle. Nowadays various super-conducting magnet and coil are available to create the strong magnetic field. Thus, recently much work has been devoted to the study on the basic flow mechanism of the flow control and, as a result, its basic flow mechanism is being clarified not only numerically but also experimentally [4–6].

Most of the researches are focused on understanding the mechanism of the flow control system. When we consider this system as a future TPS, we need to investigate the feasibility of this system for reentry vehicles. To evaluate the feasibility of this system, we need to investigate the efficiency of this system along the reentry trajectory. Along

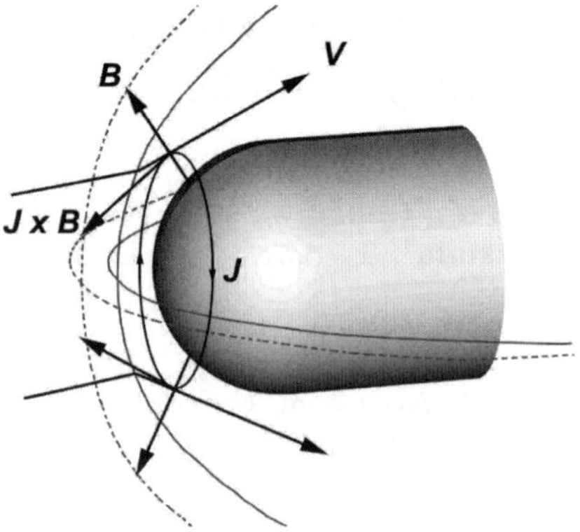

Fig. 1. Schematic view of the flow around the reentry vehicle with imposed magnetic field and the induced current

the reentry flight trajectory, the flow condition such as velocity and density changes drastically. The reentry trajectory mainly depends on the drag force to the vehicle. The drag force is expected to depend on both the vehicle shape and the magnetic flow control system.

In the present paper, in order to evaluate the feasibility of the magnetic flow control system, we estimated the effect of the shape of the reentry vehicle on the magnetic flow control system. Especially, we focused on the nose bluntness of the reentry vehicle, which is one of the mose important factor for designing the shape of the reentry vehicle. Because the nose bluntness of the vehicle is one of the key parameters which can change the aerodynamic and aerothermodynamic performance of the reentry vehicle.

2 Physical Models and Governing Equations

When the magnetic flow control system is applied to reentry vehicles, the shock layer in front of the vehicle is enhanced, which leads to the drag force enhancement. The change of the drag force will have a significant impact on the reentry trajectory. In order to estimate the drag force enhancement, we performed CFD analyses around the reentry vehicle including the electromagnetic effect. The govering equations are based on Navier-Stokes equations including the source term related to the electromagnetic force and the Joule heating. The governing equation is expressed as follows;

$$\frac{\partial \mathbf{U}}{\partial t} + \nabla \cdot \mathbf{F} = \begin{bmatrix} 0 \\ \mathbf{J} \times \mathbf{B} \\ \mathbf{J} \cdot \mathbf{E} \end{bmatrix}, \mathbf{U} = (\rho, \rho\mathbf{V}, E_t)^t, \tag{1}$$

where $\mathbf{F}$ are the flux vector including the viscous terms. And ρ, $\mathbf{V}$, E_t is the density, the velocity vector, and the total energy respectively.

Ohm's law is used to calculate the current density arranged in the the following manner,

$$\mathbf{J} = \sigma(\mathbf{E} + \mathbf{V} \times \mathbf{B}), \tag{2}$$

where **E**, **V**, **B** are the electric field vector, the velocity vector, magnetic field vector, respectively. In the present study, the vehicle shape is assumed to be an axis-symmetric. Thus, the electric field is not created around the vehicle. The magnetic field is defined by the magnetic dipole and the dipole axis is set to be parallel to the body axis. Thus, the magnetic field is also set to be symmetric to the axis of the vehicle.

The electric conductivity is assumed to be dependent on the temperature inside the shock layer only. In the present study, the conductivity is assumed to be the following relation;

$$\sigma = \sigma_0 \left(\frac{T}{T_{max}} \right)^2, \tag{3}$$

where σ_0 is the maximum value of the electric conductivity and T_{max} is the maximum temperature inside the shock layer.

The vehicle shape is based on the reentry capsule EXPRESS as shown in Fig.2. To investigate the effect of the nose shape on the efficiency of this system, we prepared two different body configurations; spherical nose and flat head. In the case of the flat head, the nose shape is assumed to be a elliptical shape with curvature radius of 2.0 [m]. The magnetic dipole is located at the center of the spherical nose as shown in Fig.2.

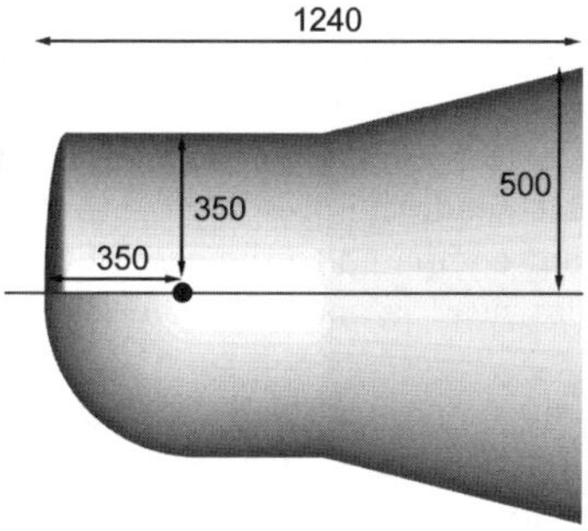

Fig. 2. Vehicle shape, EXPRESS reentry capsule with different nose bluntness; flat head (top) and spherical nose (bottom)

The governing equation is discretized by a finite volume method and the numerical flux is evaluated by AUSMDV scheme. [7] The time integration is performed by euler explicit method.

3 Numerical Results

The flight condition is that the reentry flight velocity is 7000 [m/s] at the altitude of 70 [km]. Under this condition, the maximum electric conductivity was estimated to be about 100 [S/m]. Thus, the maximum electric conducitivity, σ_0, is set to be 100 [S/m] The magnetic field strength is set to be 1.0 [T] at the stagnation point.

Figure 3 shows the pressure contours around the reentry vehicle. When the magnetic flow control system is applied to the vehicle with a spherical nose, the pressure contours

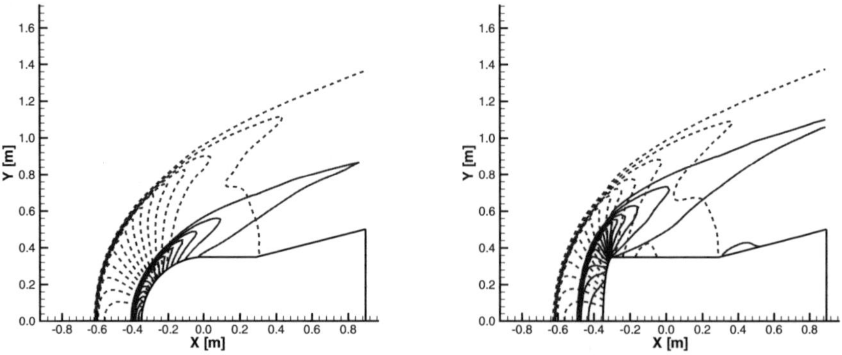

Fig. 3. Pressure contours around the reentry vehicle with different nose bluntness; spherical nose (left) and flat head (right). Solid line shows the pressure contours without imposed magnetic field. Dashed-line shows the pressure contours with imposed magnetic field.

changes drastically. Near the stagnation region, the magnetic flow control system enhanced the shock stand-off distance drastically. This result suggests that the drag force of the vehicle is enhanced drastically. Even when the magnetic flow control system is applied to the vehicle with a flat head, the pressure contours changes but the efficiency of this system is not so significant compared to the case of the spherical nose. This suggests that the drag force with a flat head is not enhanced significantly compared to the case of the spherical nose.

Next we focus on the pressure contours when the magnetic field is applied. As shown in Fig.3, the pressure contours for two cases are almost identical when the magnetic field is applied. Though this result sounds strange, this result can be understood as follows. When the effect of the induced electromagnetic force on the flow field is weak compared to the pressure inside the shock layer, the shock shape and the pressure contours mainly depend on the flow parameters such as Mach number and the vehicle shape. Thus, the pressure contours should be dependent on the vehicle shape. But as the induced electromagnetic force becomes strong, the shock shape and the pressure contours are mainly dependent on the parameters related to the electromagnetic effect such as the magnetic field strength and the electric conductivity inside the shock layer. The impact of the electromagnetic effect on the flow field can be estimated by the interaction parameter, Q, defined as follows,

$$Q = \frac{\sigma B^2 L}{\rho V} \tag{4}$$

where B, σ, L, ρ, V are the magnetic field strength, the electric conductivity, reference length, density and velocity, respectively. In the present study, the reference length is set to be a nose radius of the vehicle, 0.35 [m].

Figure 4 shows the pressure contours of the two different vehicle configurations for Q of 7.5, 30, and 120. In this figure, the magnetic field strength is set to be 0.5, 1.0, 2.0 [T] and the electric conductivity is fixed at 50 [S/m]. From this figure, we can observe that as the interaction parameter Q increases the shock stand-off distance increases and the

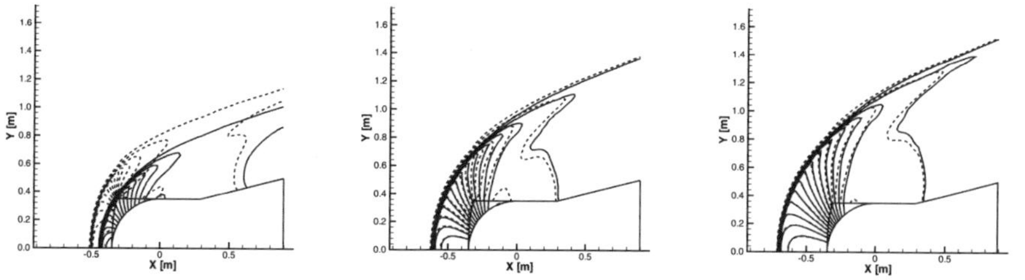

Fig. 4. Pressure contours around the reentry vehicles for Q of 7.5 (B =0.5 [T], left), 30 (B =1.0 [T], middle), 120 (B =2.0 [T], right). Solid and dashed lines show the pressure contours for the reentry vehicles with spherical nose and flat head, respectively.

difference of the pressure contours for reentry vehicles becomes negligible. This suggests that the nose bluntness has little impact on the pressure contours around the vehicle and the drag force of the vehicle when the interaction parameter is sufficiently large.

The drag force of the reentry vehicles with the imposed magnetic field is composed of the pressure drag, the friction drag, and the Lorentz force. The drag force related to the surface pressure and the skin friction is calculated by integration of the pressure and the skin friction along the vehicle surface. The drag force related to the Lorentz force is calculated as a reaction force to the super-conducting coil inside the vehicle.

The drag force of the reentry vehicles is summarized in Fig.5. The electric conductivity is set to be 50, 75, 100 [S/m] and the magnetic field strength is set to be 0.5, 1.0, 2.0 [T]. Thus, the interaction parameter Q increases up to about 240.

From this figure, we can observe that the drag coefficient for both configurations increase in proportion to the interaction paraneter, Q. For $Q \leq 30$, the drag coefficient in the case of the spherical nose increases more rapidly than that in the case of the flat head. This means that the drag coefficient of the spherical nose is more susceptible to the electromagnetic effect than that of the flat head. For $Q \geq 30$, the drag coefficients for both configurations become identical as Q increases. If the drag coefficients become identical, the reentry trajectories for both configurations will also become identical. This result suggests that from the viewpoint of the design of the vehicle configuration, the

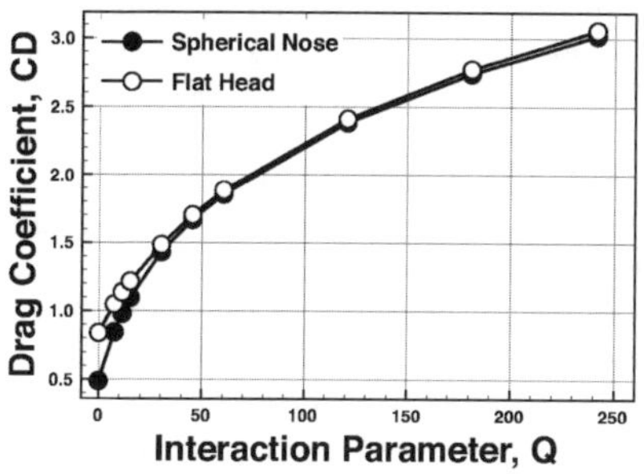

Fig. 5. Relationship between the drag coefficient and the interaction parameter, Q

drag force can be controlled without considering the nose bluntness of the reentry vehicle by the imposed magnetic field under the flight condition that the interaction parameter is sufficiently large.

For $Q = 240$, the drag coefficient for both configurations is estimated to be 3.0, which is much larger than that without imposed magnetic field. The interaction parameter of 240 is realized by the magnetic field of 2.0 [T] and the electric conducitivity of 100 [S/m] at the altitude of 70 [km] and filght velocity of 7.0 [km/s]. This means that by using the present system the drag coefficient of the reentry vehicle can be enhanced drastically and the reentry heating environment will be improved without considering the nose bluntness.

4 Conclusion

In the present study, we investigated the effect of the nose bluntness on the drag force enhancemnt of the reentry vehicle by the electromagnetic flow control system. For the test body configuration, we used the EXPRESS reentry capsule with different nose configuration; spheircal nose and flat head. Numerical results suggest that as the interaction parameter increases the drag force increases drastically.

When the interaction parameter is small, the magnetic flow control system can enhance the drag force of the the spherical nose more drastically than that of the flat head. But when the interaction parameter is sufficiently large, the nose bluntness has little impact on the drag force enhancement. This result suggests that the drag force can be controlled by the imposed magnetic field without considering the nose bluntness of the reentry vehicle. The drag coefficient of the reentry vehicle is estimated to be enhanced from 0.5 to 3.5 with the magnetic field strength of 2.0 [T], which shows that the magnetic flow control system can enable us to control the aerodynamic performance of the reentry vehicle efficiently.

Acknowledgement. This research was partly supported by the Ministry of Education, Science, Sports and Culture, Grant-in-Aid for Scientific Research (A) (2), 15206093, 2006.

References

1. A. R. Kantorwitz: "A survey of physical phenomena occurring in flight at extreme speeds," in Proceedings of the Conference on High-Speed Aeronatucs, edited by A. Ferri, N. J. Hoff, and P. A. Libby (Polytechnic Institute of Brooklyn, New York, 1955), pp. 335-339.
2. E. L. Resler and W.R. Sears: "The prospects of magneto-aerodynamics," J. Aeronatut. Sci. **25**, 235 (1958)
3. R.W. Ziemer and W.B. Bush: Phys. Rev. Letters, **1**, (1958)
4. J. Poggie and D. V. Gaitonde "Magnetic control of flow past a blunt body: Numerical validation and exploration," Phys. Fluids, **14**, (2002)
5. H. Otsu, T. Abe, D. Konigorski: "Influnce of the Hall effect on the electrodynamic heat shield system for reentry vehicles," AIAA 2005-5049, 2005.
6. Y. Takizawa, S. Sato, T. Abe and D. Konigorski: "Experiment on shock layer enhancement by electro-magnetic effect in reentry-related high-ehthalpy flow," AIAA 2005-4786, 2005.
7. Wada, Y. and Liou, M. S.: A Flux Splitting Scheme with High-Resolution and Robustness for Discontinuities, AIAA Paper 94-0083, January, 1994.

Enhanced design of a scramjet intake using two different RANS solvers

M. Krause and J. Ballmann

Lehr- und Forschungsgebiet für Mechanik, Templergraben 64, 52062 Aachen

Summary. A numerical and experimental analysis of a scramjet intake has been initiated at RWTH Aachen University. The paper presents an overview of the ongoing work on the numerical simulations using two different, well validated RANS solvers, namely FLOWer and QUADFLOW. Several geometry concepts, e.g. 2D intake, 3D intake with or without sidewall compression, using a single or double ramp configuration are investigated. The influence of different geometries on the flow, especially on the separation bubble in the isolator inlet as well as transition effects, and efficiency is discussed. Several numerical simulations have been performed applying a variety of turbulence models.

1 Introduction

The engine inlet of an air breathing hypersonic propulsion system mostly consists of several exterior compression ramps in front of by a subsequent interior isolator/diffusor assembly (see fig. 1 [2D intake], fig. 2 [3D intake]). Oblique shock waves without a final

Fig. 1. Hypersonic 2D inlet model without sidewall compression

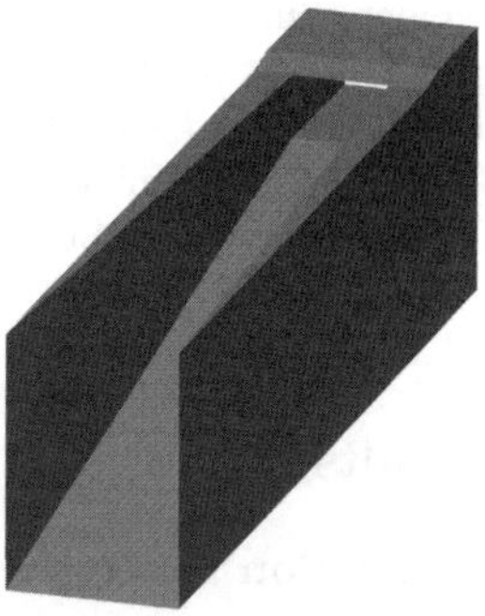

Fig. 2. Hypersonic 3D inlet model

normal shock are performing the compression of the incoming flow from hypersonic to supersonic speed. The interaction of shock waves with thick hypersonic boundary layers (BL) causes large separation zones that are responsible for a loss of mass flow and several other effects, like e.g. unsteady shock movement or high heat flux into the structure. Thereby the engine performance decreases. The high total enthalpy of the flow yields severe aerodynamic heating, further enhanced by turbulent heat flux. To realize hypersonic combustion a necessary amount of mass flow has to enter the combustion chamber. Additionally, a certain area ratio between the inflow capture cross-section and the isolator cross-section has to be preserved (mostly around 6). To get the required compression ratio and mass flow by simultaneously having moderate ramp angles the intake has to be very long. The combustion chamber performs probably best when it has an almost quadratic shape (width/height of max. 2), while the 2D intake performs best when its width to height width ratio is larger than 5. The shape of the combustion chamber cross-section is the crucial design element, so the intake has to have an appropriate shape and at the same time provide a good efficiency. To withstand aerodynamic heating during operation, ramp and cowl leading edges will have finite radii, which generate curved bow shocks on both edges producing large entropy layers. Thus an increase in aerodynamic wall heating is caused [1]. Furthermore, the shock detachment influences the captured mass flow and the flow conditions in the isolator intake. Remarkable differences occur by using a 3D geometry with converging sidewalls as discussed in chapter 3.

2 Numerical Method

Two RANS flow solvers have been used for the presented flow simulations, one is the structured explicit flow solver FLOWer and the other one is the unstructured, implicit, adaptive flow solver QUADFLOW. Both solvers employ finite volume methods containing flux formulations based on approximate Riemann solvers. FLOWer was developed under the lead of DLR and the QUADFLOW code is still being further developed. It uses an integrated concept of finite volume method, multiscale analysis and on this basis an adaptive grid generation applying B-Spline techniques. To simulate turbulent flow, a wide variety of low Reynolds number turbulence models for compressible fluid flow are implemented in both flow solvers. For example: Spalart - Allmaras (SA), Linear Explicit Algebraic Stress model (LEA), Local Linear Realizable (LLR), Shear Stress Transport (SST) and Wilcox $k-\omega$. The computations were performed on the SunFire SMP–cluster of RWTH Aachen University, the Jump cluster of the Research Centre Jülich and the NEC SX–8 cluster at Stuttgart University. Both Flow solvers are well validated and tested [2, 3].

3 Results

3.1 Validation for Hypersonic Flow Conditions

For describing the intake geometries we introduce a coordinate system as shown in fig. 3 where x points in freestream flow direction, z is defined along the leading edge of the ramp and y completes the right hand cartesian reference system. For Validation several calculations for the 2D intake SCR02 (fig. 3, 4 of the DLR Cologne have been done using nearly all common turbulence models like SA, LEA, LLR, SST, $k-\omega$ and the Speziale, Sakar and Gatski (SSG) Reynolds stress model. The flow conditions were as follows: M_∞= 6.0, $Re_l = 10.543 \cdot 10^6$ [1/m], T_∞=59 K, p_∞=747 Pa, $T_{wall} = 300K$. It was found that none of the applied two equation turbulence models yielded good results [4]. They overestimated the BL thickness and the separation within the isolator part of the scramjet. While comparing the measurements with the numerical simulations

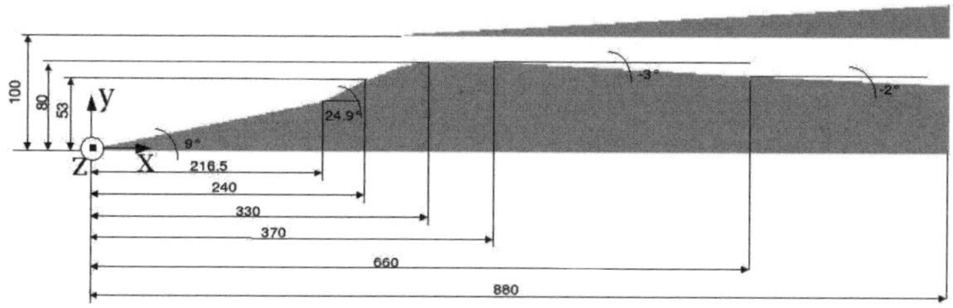

Fig. 3. Sketch of intake SCR02

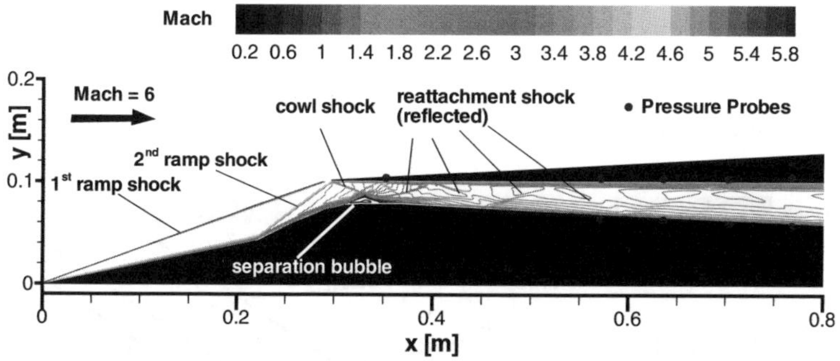

Fig. 4. Mach contours for scramjet intake SCR02, 2D simulation without suction

it was found that only the SA and the SSG model yielded appropriate results [5, 6]. One can see, that grid adaptation leaded to a better flow resolution in comparison to the computation on a sufficiently refined structured grid especially in the separation region of the inlet and at shocks. It can also be seen, that the strong cowl shock interacts on the lower isolator wall with the hypersonic BL and the expansion fan, which is generated by the convex ramp shoulder. A large separation bubble is produced which blocks half of the overall isolator height.

3.2 Comparison 2D vs. 3D Intake

Several computations were performed for purely 2D compression on a ramp as shown in fig. 1 with or without sidewall and with a convex curve or convex corner at the ramp shoulder for redirecting the flow by Prandtl-Meyer expansion into horizontal direction. Also several combinations of ramp angles for the double ramp configuration have been investigated. For the shown results the flow conditions were: $M_\infty = 8$, $Re_l = 2.945 \cdot 10^6$ [1/m], T_∞=226 K, p_∞=1172 Pa, $T_{wall} = 300K$. Fig. 7 shows results of a 3D simulation for a 2D intake with sidewalls. It is found that the sidewall has great influence on the flow field. The sidewall shocks are bending the ramp shocks in upstream direction yielding an increased loss in mass flow. The separation zone within the isolator is larger than in the 2D simulation, because of the sidewall BL. As expected, vortices are generated in the corners between sidewall and ramps. These vortices also lead to shock bending and very high heat loads on the surface. It was also worked out that the continuously curved ramp shoulder is favorable with respect to mass flow loss and to heat loads on the walls. It has

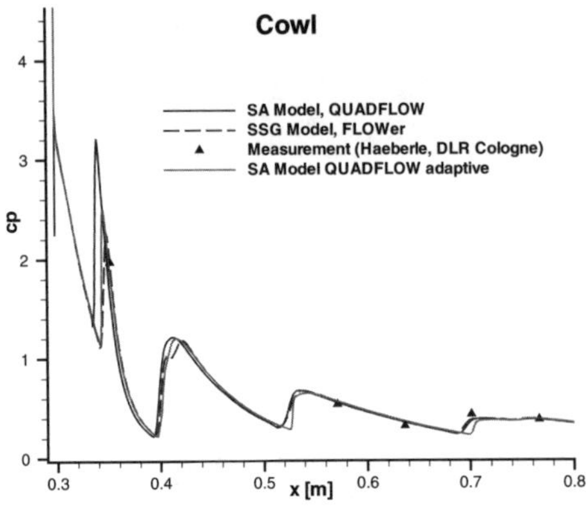

Fig. 5. Pressure distribution on cowl of SCR02, 2D simulation without suction

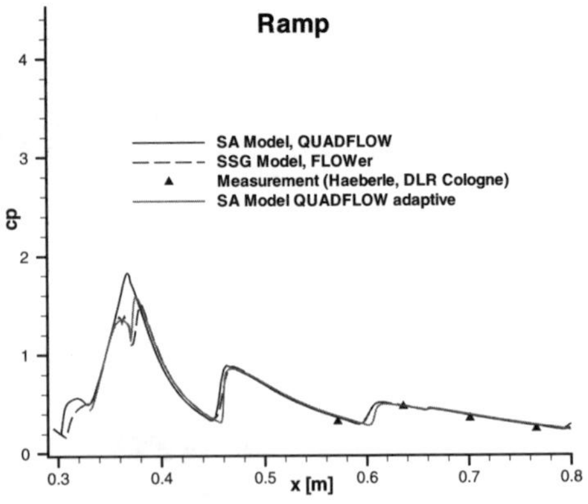

Fig. 6. Pressure distribution on ramp of SCR02, 2D simulation without suction

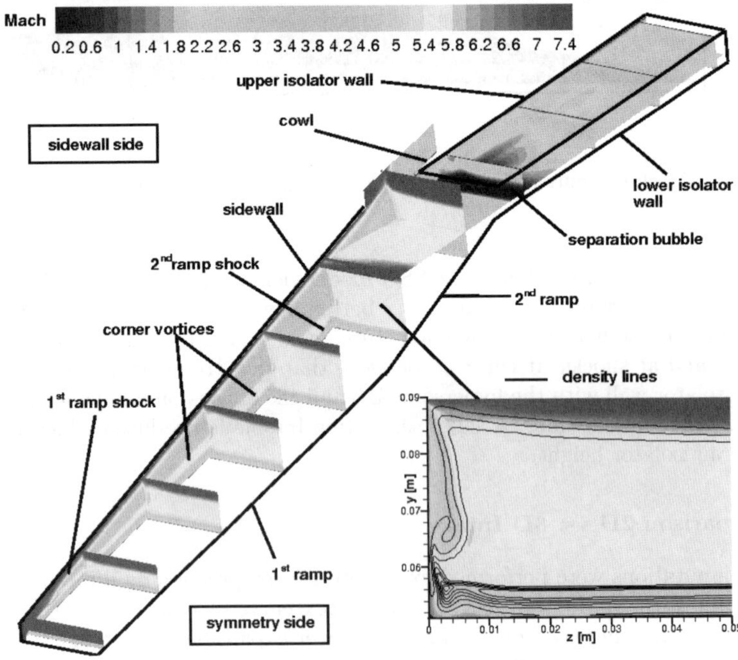

Fig. 7. Mach contours for scramjet intake investigated at Shock Wave Laboratory of RWTH

been proven that moderate ramp angles to compress and redirect the flow into the inlet lead to higher efficiency and more homogeneous isolator flow. As can be seen in fig. 7 the sidewall was minimized such that it just catches the first ramp shock. This resulted in smaller sidewall BL's, thus leaded to less shock bending and sidewall compression, and this finally resulted in less mass flow loss and higher intake efficiency.

To avoid the deficiencies of simply ramp compression(e.g. small compression ratio, long intake, high enthalpy and mass flow loss), studies of a 3D intake (like sketched in fig. 2) are conducted. Fig. 8 shows the computed Mach contours in the horizontal longitudinal mid-plane of the isolator. Here the sidewalls along the ramp converge with an angle of 2.8° and thus contribute to the compression of the flow, resulting in a much shorter and wider intake (intake length/width: 2D:L/W ≈ 14; 3D: L/W ≈ 4). The averaged Mach number and temperature at the isolator exit were approximately the same as computed for the 2D intake without sidewall compression. The sidewalls begin with the first ramp, and their leading edge is normal to the x-z plane and points in y direction. This leaded to thick BL's at the sidewalls and to big vortices generated in the corners. These vortices reduce the BL thickness in some regions, because slow wall material is transported away from the walls. At other locations they lead to separation zones at the sidewalls and the ramp. These additional separations produce additional shock - shock interactions. Therefore, different sidewall concepts will be studied.

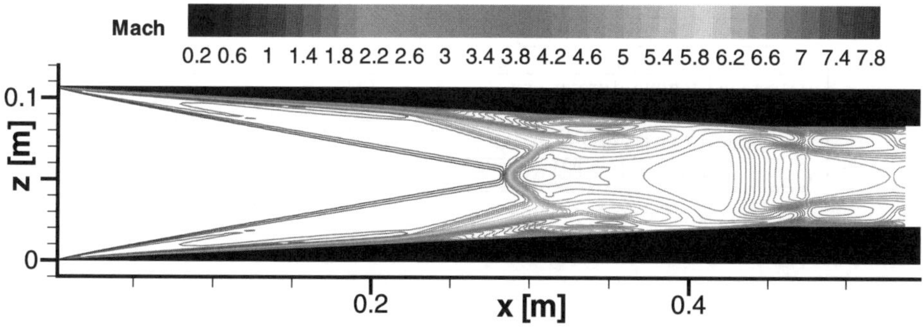

Fig. 8. Mach contours for 3D scramjet intake with sidewall compression

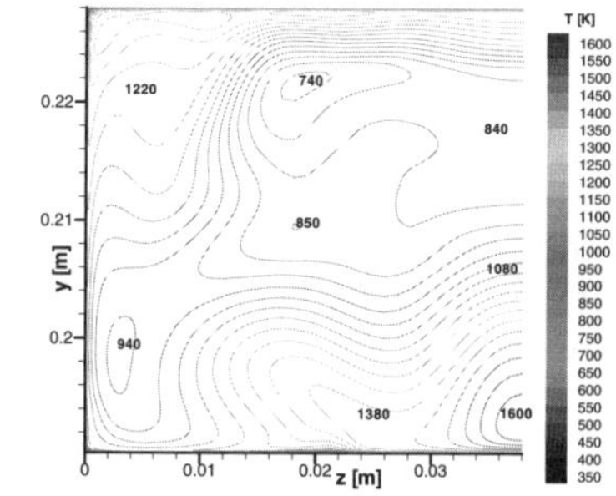

Fig. 9. Temperature distribution at isolator exit cross-section for 3D intake

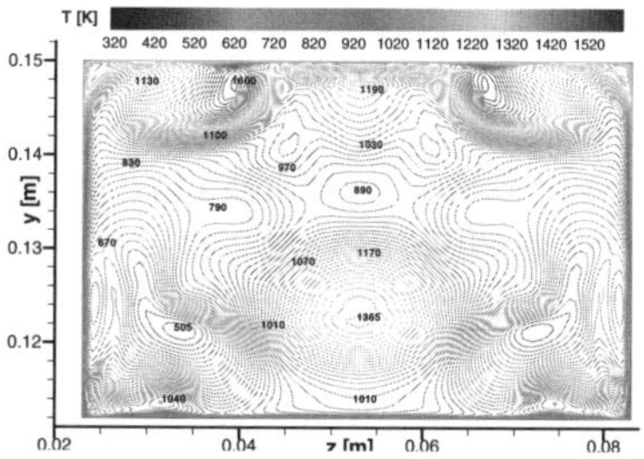

Fig. 10. Temperature distribution at isolator exit cross-section for 3D intake

The computations performed so far showed quite promising results for a 3D compression of the flow. Table 1 shows a comparison between the results (for same freestream flow conditions and isolator inlet cross-section) for a 2D intake (ramp angles of 7.5° and 19°, no sidewall compression, capture cross-section = 76 mm [width] x 228 mm [height]) and 3D (ramp angle 13.5°, 2.8° sidewall compression, capture cross-section = 106 mm [width] x 150 mm [height]) intake. Looking at these results the 3D intake seems to be superior at first glance, but there are several disadvantages. Due to the strong shock interactions and vortices the flow field is very inhomogeneous. There are large gradients of the flow quantities. It is also not clear how this complex flow system will react on disturbances in the free-stream conditions. Because of the inhomogeneous flow, there are great differences in the heat loads on the surface of the structure. At the exit of the isolator are zones with very high temperature (around 1600 K) and zones with very low temperature (around 550 K) generated next to each other. It will be to decide by the combustion chamber specialists if that is favourable or modifications of the compression flow is required. Further investigations have to be done for the 3D intake to get a broader basis for decision by combustion chamber arguments. Fig. 9 and 10 compare the temperature profiles at the isolator exit for the both intakes described in table 1.

values	2D with sidewall	3D 2.8° sidewall
mass flow possible [kg/s]	0.75355	0.69146
loss [%]	20.88	15.23
averaged temperature at exit[K]	1035	987
averaged contraction ratio at exit [bar]	28.5	42.7
averaged Mach number at exit	3.02	3.24
enthalpy loss [%]	8	4.8

Table 1. Comparison of 2D and 3D intake

4 Conclusions

For supersonic combustion with resulting positive engine thrust, the performance of the intake is essential. That is why a combined investigation based on numerical analysis and experimental examination of scramjet intake geometries is undertaken. The numerical investigation is of great importance, because one aim of the research is designing a well-suited hypersonic intake for the given flight trajectory. Both 2D and 3D compression seem to have advantages and disadvantages, so it will be the future work to elaborate all aspects of the different geometry concepts to find a compromise for a most appropriate hypersonic intake design. The experimental work will allow to validate the computational results, so that the numerical simulations can be used to investigate the flow behaviour more independently from experiments. To create a most efficient hypersonic intake, still a lot of further 3D numerical computations will be required.

References

1. J. Anderson: *Hypersonic and High Temperature Gas Dynamics*, MacGraw–Hill (1989)
2. B.U. Reinartz, J. van Keuk, T. Coratekin, and J. Ballmann: *Computation of Wall Heat Fluxes in Hypersonic Inlet Flows*, AIAA Paper 02-0506 (2002)
3. F.D. Bramkamp, Ph. Lamby, and S. Müller: *An adaptive multiscale finite volume solver for unsteady and steady flow computations*, Jounal of Computational Physics **197**, pp 460-490 (2004)
4. M. Krause, B. Reinartz, and J. Ballmann: *Numerical Computations for Designing a Scramjet Intake*, 25th ICAS Conference (2006)

Experimental and numerical investigation of film cooling in hypersonic flows

K.A. Heufer and H. Olivier

Shock Wave Laboratory, RWTH Aachen University, Germany

1 Introduction

At present reusable cooling techniques for hypersonic vehicles utilize a passive heat shield as it is applied for the Space Shuttle for example. This technique does not allow to adjust the heat protection performance during flight and therefore the peak heat loads are dimensioning for the entire flight path. An alternative to such a system can be the film cooling technique which is already state of art for cooling of turbine blades. Thereby mostly air is led through discrete holes in the blade to establish a cooling film on the surface that reduces the penetrating heat loads. Most of the investigations in the past were done for quasi two dimensional turbulent flow at low Mach numbers. An overview of more recent investigations about three dimensional film cooling is given in Ref. [1]. Beside of this the active cooling technique has also been investigated in some works concerning hypersonic flows [2]. Thereby porous materials were used to inject the cooling gas through the surface and therefore a cooling effect could already be achieved in the wall. This so called transpiration cooling technique is also applied for cooling of rocket combustion chambers [3]. All of these studies show that an active cooling system is an effective method to lower the heat loads into a structure. This gives the motivation to investigate the feasibility of an active cooling in laminar hypersonic flow as it mainly appear during the critical reentry phase.

2 General description of the flow problem

For first studies of film cooling in hypersonic flows a simple inclined flat plate model was chosen as reference case. Furthermore low enthalpy conditions were used to neglect any real gas effects, i.e. ideal gas behaviour can be assumed. For the investigations with film cooling transverse slots of sufficient span were used as blowing openings, so that the flow problem can be regarded as two-dimensional. Due to the injection of the cooling gas ideally a film should establish over the surface, which separates the hot flow from the wall. As a result of the added mass in the boundary layer its thickness increases. This leads to smaller gradients and therefore to lower heat fluxes and lower friction forces at the wall. This effect can additionally be influenced by the cooling gas properties like the temperature, the heat capacity or the thermal conductivity. Furthermore the different flow conditions between the cooling gas and the hot flow lead to gradients between these two layers. With an increasing distance from the blowing opening these gradients reduces, so that the temperature and velocity gradients at the wall increase again. At least a mixing between the two phases can be expected, so that due to this and the increasing wall gradients the cooling effect decreases the more downstream from the

blowing opening. In principle, an increasing cooling effect is expected with an increasing cooling gas mass flow.
For the later discussion of results some parameters have to be introduced. At first the cooling effectiveness is defined as follows:

$$\eta = 1 - \frac{\dot{q}_c}{\dot{q}_{nc}} \tag{1}$$

Here $\dot{q}$ indicates the heat flux to the wall and the indices c and nc the cases with and without cooling. This cooling effectiveness can also be correlated to the adiabatic wall temperatures for a stationary flight case [4]. One of the main influencing parameters of the cooling process is the blowing rate F, which is defined as the specific mass flow ratio of the cooling gas and the free stream:

$$F = \frac{\rho_c u_c}{\rho_\infty u_\infty} \tag{2}$$

Neglecting the influence of the cooling gas injection on the boundary layer thickness simple theoretical considerations analogous to [5] show that the cooling effectiveness for a laminar and incompressible flow depends on:

$$\eta = f\left(\frac{\sqrt{x'}}{F \cdot s} \cdot \frac{1}{\sqrt{Re_\infty}}\right) \tag{3}$$

The distance from the blowing opening $x' = x - x_{slot}$ is related to the centre of the blowing slot, s indicates the slot width and Re_∞ is the unit Reynolds number of the free stream.

3 Experimental Setup

As mentioned above a simple single ramp model consisting of an inclined flat plate at an angle of 30° with transverse slots is used for the fundamental research presented in this work. For heat flux measurements 30 coaxial thermocouples in combination with an infrared camera are used. The blowing slot is arranged at 45° facing downstream in respect of the surface. The distance of the slot from the sharp leading edge amounts to 55 mm and the opening has a width of 0.5 mm. As cooling gas nitrogen has been used. For the experiments the shock tunnel TH2 of the Shock Wave Laboratory is employed. The low enthalpy condition allows to calculate the flow parameters behind the front shock of the model under the assumption of ideal gas behaviour. These post shock values are used as free stream conditions on the flat plate for the numerical simulation [6].

4 Numerical Method

For the numerical simualtion an explicit CFD-scheme has been developed to solve the two-dimensional compressible Navier-Stokes equations for ideal gases [4]. The solution method is based on a Cartesian grid with non-equidistant grid cells. In two steps the solution for a new time step is achieved. First, the viscosity terms are neglected and the Euler solution is achieved by using the weighted average flux method. Thereby the fluxes at the cell boundaries are determined by an exact Riemann solver and oscillations of the solution in the regions of high gradients are damped by the limiter of Albada [7]. Second, a two step Runge-Kutta-scheme is applied to take the viscosity terms into account. The viscosity is determined by Sutherlands law and the thermal conductivity is achieved for a constant Prandtl number Pr = 0.72. For the calculations two different grids were used. The cell height of the first one amounts to $6.25 \cdot 10^{-6}$ m normal to the surface with an aspect ratio of 8. For more detailed investigations in the region of the blowing opening a grid of higher resolution is necessary with a cell height of $3.125 \cdot 10^{-6}$ m and an aspect ratio of 4. The blowing opening was simulated as a boundary condition with the assumption of a constant blowing ratio and cooling gas temperature along the slot.

5 Results

5.1 Influence of the cooling gas injection on the flow field

Regarding the principle of the film cooling process as described in chapter 2 mass is added to the boundary layer by the injection of the cooling gas. This leads to an increase of the boundary layer thickness which causes a deflection of the outer flow. The results of the numerical simulation presented in Fig. 1 for a blowing ratio of F = 0.13 show exactly these effects ($Re_\infty = 4.3 \cdot 10^6\ m^{-1}$, $Ma_\infty = 2.6$, $T_\infty = 488$ K). The oncoming boundary layer separates at about 4 mm in front of the blowing opening and as expected its thickness increases. Due to the separation of the flow a large vortex occurs which forms the separation bubble. Additionally two small vortices are visible at the edges of the blowing opening. These vortices are relatively small due to the low injection velocity.

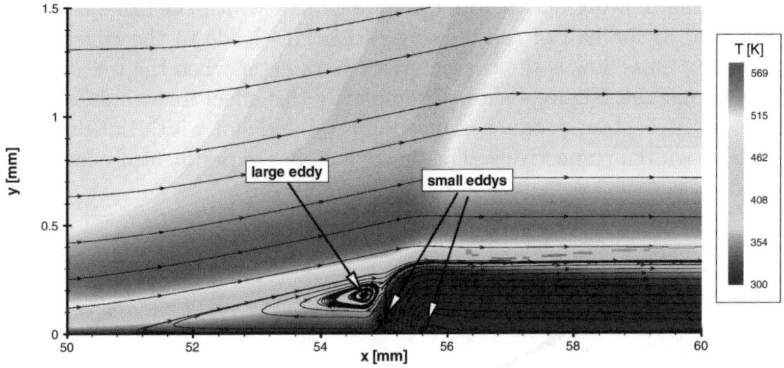

Fig. 1. Influence of the cooling gas injection on the flow field

5.2 Validation of the numerical scheme

In figure 2 experimental results in comparsion to equivalent CFD calculations are presented ($Re_\infty = 3.0 \cdot 10^6\ m^{-1}$, $Ma_\infty = 2.62$, $T_\infty = 635$ K). Heat flux distributions along the flat plate are shown for the cases without and with cooling (F = 0.096). Upstream the blowing position a good agreement of the experimental and numerical results with values from literature [8] is observed. At the slot position the heat flux to the wall reduces to its minimum and it increases again as expected with increasing distance from the slot. Plotting the cooling effectiveness versus the blowing ratio also the expected effect of increasing effectiveness with an increasing blowing ratio becomes obvious (Fig. 3). All of the CFD results fit satisfactory to the experimental values.

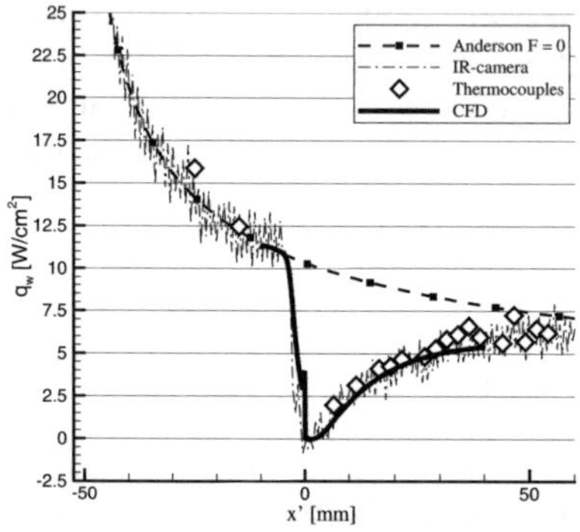

Fig. 2. Heat flux distribution in the case of cooling, F = 0.096

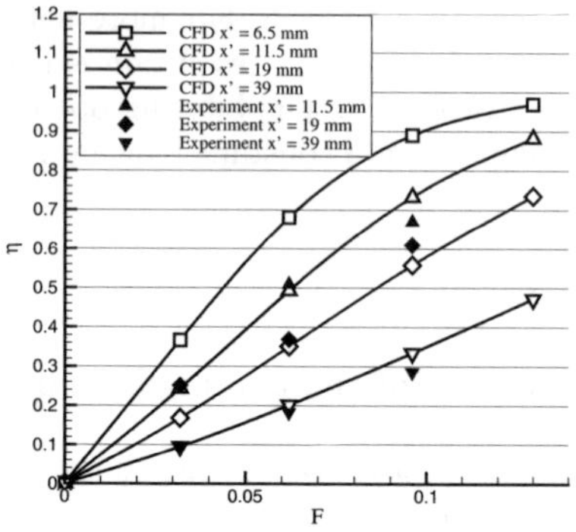

Fig. 3. Influence of the blowing ratio

5.3 Variation of the blowing geometry

For the study of the influence of the slot width this parameter has been varied from 0.25 mm to 2.0 mm at a constant blowing ratio. The results reveal that for a constant distance from the blowing opening and a doubled slot width the same cooling effectiveness is achieved as for half of the blowing ratio. This leads to the conclusion that the absolute cooling mass flow is the characteristic parameter when the free stream parameters are kept constant. Figure 4 shows exemplarily this effect for two different numerical simulations for a constant product of blowing ratio and slot width but different blowing ratios and slot widths respectively. This behaviour has experimentally been verified.

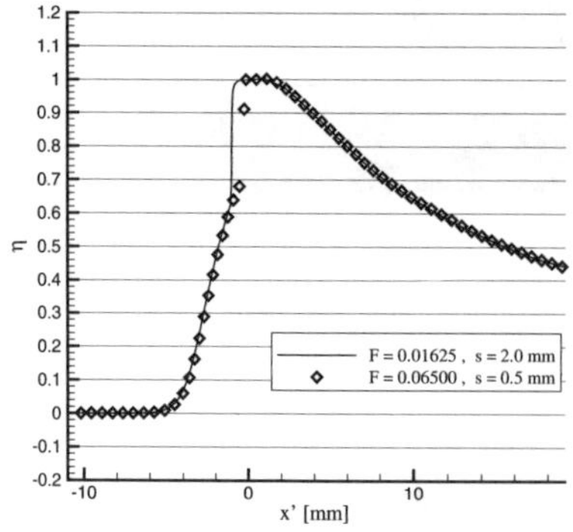

Fig. 4. Variation of the slot witdth

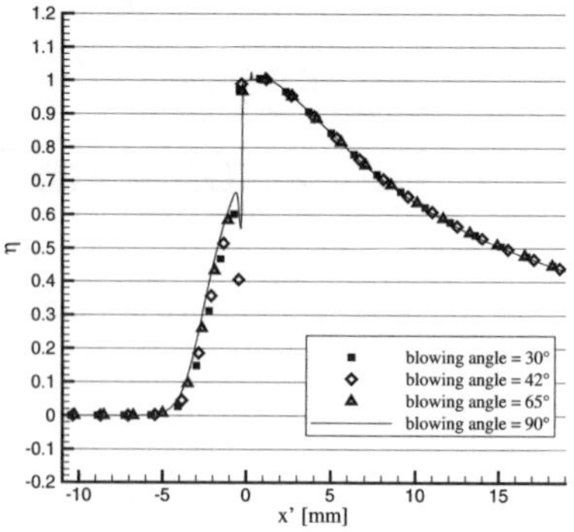

Fig. 5. Variation of the blowing angle

In literature for turbine blades an increase of the average cooling effectiveness with decreasing blowing angle is known for a cooling gas injection through discrete holes and for turbulent boundary layers. Due to the smaller injection angle the intensity of the vorticy structures around the injection holes is reduced by the lower blowing velocity normal to the surface. This leads to a lower disturbance of the main boundary layer and therefore to a less mixing between free stream and cooling gas and a higher cooling effectiveness respectively. In a pure laminar, two dimensional flow these effects do not occur so that the cooling effectiveness should not be affected by the blowing velocity normal to the surface. This is confirmed by the results of the slot width study where the blowing ratio and the velocity normal to the surface respectively have no influence on the cooling effectiveness for the same cooling mass flow. Figure 5 shows the expected results for different blowing angles from 30° to 90° for a constant blowing ratio of F = 0.065 and a constant slot width.

5.4 Variation of the characteristic free stream parameters

The Reynolds number effect has been investigated by varying the reference length, i.e. the distance from the leading edge to the blowing opening and keeping all other parameters constant. With an increasing distance from the leading edge the boundary layer thickness of the free stream increases which interacts with the cooling gas jet. For setting up a correlation which is capable to describe the cooling effectiveness for varying flow conditions two mechanism of the boundary layer thickening downstream of the injection slot have to be considered. These are given by the inital thickening due to the mass addition of the cooling gas at the injection slot and the known boundary layer growth which is proportional to the square root of the running length $\sqrt{x}$. To cover both effetc in eq. 3 an additional term of the form $\sqrt{x/x_{ref}}$ has been introduced, where the reference length x_{ref} has been empirically found to $x_{ref} = (x_{slot})^{1.16}$ with x_{ref} and x_{slot} in meter. The Mach number influence is described by the term $(\rho^* \cdot \mu^*)^{0.5} \cdot (\rho_\infty \cdot \mu_\infty)^{-0.5}$ where the values ρ^* and μ^* are determined by the reference temperature $T^* = T_\infty \cdot (1 + 0.72(T_r/T_\infty - 1))$ [5]. Figure 6 and 7 show the dependence of the cooling effectiveness on this correlation parameter for independent Mach and Reynolds number variation. It is obvious that this factor is able to correlate the results over a large range of flow conditons.

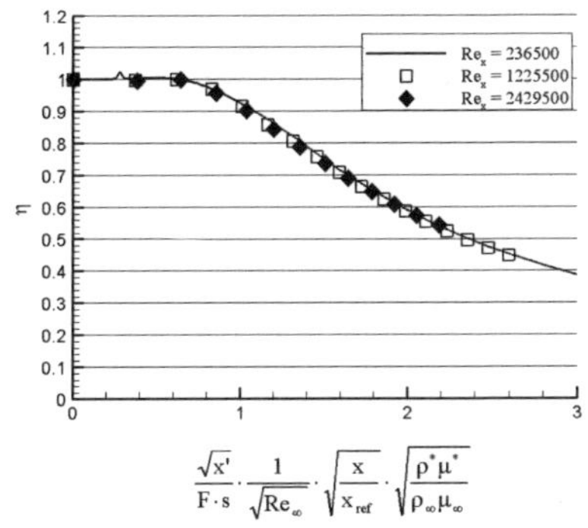

Fig. 6. Reynolds number variation

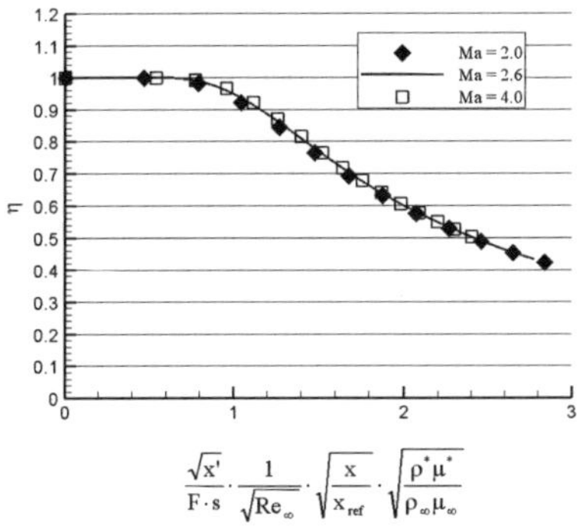

Fig. 7. Mach number variation

6 Conclusion

In the presented work film cooling for a two dimensional inclined flat plate configuration in hypersonic flow has been investigated. The geometry of the blowing opening as well as the free stream parameters have been varied. For the reference configuration CFD calculations show a good agreement to experimental results which validates the numerical code. The variations of the blowing slot width and the blowing angle are of no influence on the cooling effectiveness as long as the injected cooling mass flow is constant. Varying the free stream parameters leads to different interactions between the cooling gas flow and the oncoming boundary layer. For an increasing Reynolds number as well as for an increasing Mach number larger cooling effects have been observed. Their influence is described by a correlation factor for the cooling effectiveness. The blowing ratio and the cooling gas mass flow respectively show the expected effect that an increase of these parameters lead to higher cooling effectiveness.

Acknowledgement. This work has partially been funded by Helmholtz Gesellschaft (VH-VI-050).

References

1. Jung K.H.: *Mehrreihige Filmkühlung an gekrümmten Oberflächen*, PhD thesis, University Darmstadt, 2001
2. Cresci J.R., Libby P.A.: *The downstream influence of mass transfer at the nose of a slender cone*, Journal of the Aerospace Sciences 29: 815-826, 1962
3. Lezuo M.K.: *Wärmetransport in H2-transpirativ gekühlter Brennkammer*, PhD thesis, RWTH Aachen University, 1998
4. Heufer K.A., Olivier H.: *Film Cooling of an Inclined Flat Plate in Hypersonic Flow*, AIAA Paper 2006-8067, 14th AIAA/AHI Space Planes Conference and Hypersonic Systems Conference, Canberra, Australia, 2006
5. Goldstein R.J.: *Film Cooling*, In Advances in Heat Transfer. Irvine, T.F. and J.P. Hartnett Eds. Academic Press. New York 7: 321-379, 1971
6. Heufer K.A., Olivier H.: *Film Cooling for Hypersonic Flow Conditions*, 5th European Workshop on Thermal Protection Systems and Hot Structures, Estec, Noordwijk, Netherlands, 2006
7. Torro E.F.: *Riemann solver and numerical methods for fluid dynamics*, ISBN 3-540-65966-8, Springer Verlag, 1999
8. Anderson, J.D.: Hypersonic and high temperature gas dynamics. McGraw-Hill (1989)

Experimental and numerical investigation of jet injection in a wall bounded supersonic flow

J. Ratan and G. Jagadeesh

Department of Aerospace Engineering, Indian Institute of Science Bangalore-560012, India

Summary. Numerical and experimental studies of a supersonic jet (Helium) inclined at 45° to a oncoming Mach 2 flow have been carried out. The numerical study has been used to arrive at a geometry that could reduce an oncoming Mach 5.75 flow to Mach 2 flow and in determining the jet parameters. Experiments are carried out in the IISc. hypersonic shock tunnel HST2 at similar conditions obtained from numerical studies. Flow visualization studies carried out using Schlieren technique clearly show the presence of the bow shock in front of the jet exposed to supersonic cross flow. The jet Mach number is experimentally found to be ≈ 3. Visual observations show that the jet has penetrated up to 60% of the total height of the chamber.

1 Introduction

Recently there has been a resurgence of interest in scramjet (supersonic combustion ramjet) engines. People are trying different methods to achieve combustion at supersonic speeds in different parts of the world. The main hindrance behind combustion at supersonic speeds is supposed to be inadequate mixing of the fuel with oxidizer. Additionally total pressure loss and increase of wave-drag due to shocks arising as a consequence of injection of fuel into the oncoming supersonic stream sets a challenge in designing practical scramjet engines. In other words engine generated thrust should overcome the wave drag and frictional drag on the body efficiently.

Perhaps the simplest method of fuel injection is injection from the wall of the combustion chamber. In such a case the injected fuel interacts with the oncoming air and generates complex flow field structures. Boundary layer separation, separation shock, bow shock, reattachment shock, recirculation region, large scale structures are qualitative observations of the flow field around an injection port. Large scale structures originate due to engulfment of cross-flow into the injectant and merging of smaller eddies. Furthermore, jet free-stream shear layer, horseshoe vortex due to jet column, far-field counter rotating vortex pairs complicates the flow field.

The focus of most of the earlier numerical studies [1–4] has been more to validate the experimental findings rather than using it as a precursor tool for experiments. The earliest reported experiments in slot injection (injection through a slot across the span of a flat plate) is by Spaid and Zukoski [5]. The experiments were conducted for a range of supersonic speeds ($M_\infty = 2.61$, 3.50 and 4.54) with a very narrow slot width (0.2668 mm) on a flat plate. The experiments conducted by Aso *et. al* [6] provide wall pressure distributions and flow patterns on the wall through oil flow visualization. Qualitative studies of flow features using PLIF and Mie scattering by Gruber *et. al.* [7], quantitative velocity and turbulent field data using LDV by Santiago and Dutton [8] and wall pressure distribution using pressure sensitive paint by Everett *et. al.* [9] are also reported. With the advent of ultra fast cameras, it has been possible to study evolution of flow field structure with very small time interval. A. Ben-Yakar *et. al.* [10] have studied the evolution of vortical structure of Hydrogen and Ethylene transverse jets in a supersonic cross-flow experimentally. With a very high speed camera, they have been able to track the eddies at 1 microsecond interval. Mixing and jet penetration are found to be more for Ethylene than Hydrogen.

The focus of the present work is to study the penetration properties of a supersonic jet in a Mach 2 cross flow. The IISc. Hypersonic shock tunnel offers Mach 5.75 flow. So a numerical study (2D) has been done to arrive at a simple method to reduce the flow velocity to Mach 2 and also study the effect of jet injection into the oncoming flow. Experiments are carried out subsequently. The details of the study are explained in the subsequent sections.

2 Numerical studies

The Navier-Stokes equations written in vector integral conservation-law form can be expressed as

$$\frac{\partial}{\partial t}\int_V \mathbf{U}dv + \oint_S (\mathcal{H} - \mathcal{F}).\hat{n}ds = 0 \tag{1}$$

where $\mathbf{U}$ is a column vector, $\mathcal{H}$ and $\mathcal{F}$ are second-order flux tensor defined as

$$\mathbf{U} = \begin{bmatrix} \rho \\ \rho u \\ \rho v \\ E \end{bmatrix}, \quad \mathcal{H} = \begin{bmatrix} \rho u_\perp \\ \rho u u_\perp + p\hat{n}_x \\ \rho v u_\perp + p\hat{n}_y \\ (E+p)u_\perp \end{bmatrix}$$

$$\mathcal{F} = \begin{bmatrix} 0 \\ \tau_{xx}\hat{n}_x + \tau_{yx}\hat{n}_y \\ \tau_{xy}\hat{n}_x + \tau_{yy}\hat{n}_y \\ A \end{bmatrix}$$

Here ρ, u, v, p represent density, x, y component of velocity, pressure respectively, $u_\perp = u\hat{n}_x + v\hat{n}_y$, $\hat{n}_x, \hat{n}_y$ are the direction cosines of a cell normal, $E = \frac{p}{\gamma - 1} + \frac{1}{2}\rho(u^2 + v^2)$, $A = (u\tau_{xx} + v\tau_{yx} - q_x)\hat{n}_x + (u\tau_{xy} + v\tau_{yy} - q_y)\hat{n}_y$.

Numerical studies have been done before carrying out experiments in the IISc. hypersonic shock tunnel HST2. An explicit finite volume Navier-Stokes solver has been specifically developed and validated in the present study. Only 2D simulations are carried out. The inviscid fluxes are discretized using approximate Riemann solver of Roe [11]. To increase the spatial accuracy to third order, a MUSCL extrapolation strategy [12] is followed with Van Albada limiter [13] to suppress spurious oscillations near a strong discontinuity. Viscous fluxes are discretized by central differencing [14]. A multi step Runge-Kutta scheme is used for time integration [15]. Grid clustering(near the injection point) and stretching(near the wall) are avoided and a uniform grid is used throughout the domain.

The flow domain consists of the region between a 2D wedge and a flat plate. The region is shown later in a contour plot. After a couple of simulation the wedge angle is finalized to 30°, which reduces the Mach number from 5.75 to 2 near the middle of the parallel region. This is where the jet is injected at 45° to the plate.

All variables are specified at the inlet and extrapolated from interior at the outlet. The top boundary is a no slip wall except at the injection points where jet velocity, static pressure and temperature are specified. 12 injection points are used for injection, equal wall normal and downstream component of velocity are specified. The bottom boundary is a no slip wall. Boundary layer is allowed to develop for around 70000 time steps with CFL number as 0.1. After that the injection is started with a reduced CFL number (≈ 0.01). The Reynolds number based on length of the wedge and free-stream conditions of the experimental facility is 3.746×10^4.

Two cases are studied, one with the presence of top wall along the whole length and in the second half of the domain. For the first case a normal shock occurs at the end of the inclined portion which travels further upstream and reduces the mass flow rate into the parallel portion where the flow becomes subsonic. The normal shock disappears for the second case and reduces the Mach number to 2 near the middle of the parallel portion. Fig. 1 shows contour plot of Mach number in the presence of injection from the middle of the parallel region with superimposed streamlines. Wall is present only in the second half of the top boundary.

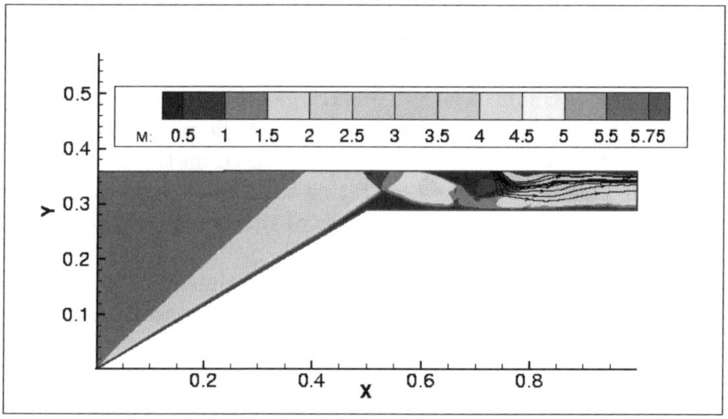

Fig. 1. Mach no. contour plot in the presence of injection with superimposed streamlines

3 Experimental set-up

3.1 IISc. Hypersonic Shock tunnel HST2

Experiments are carried out in IISc. hypersonic shock tunnel HST2. Details of free stream conditions that can be simulated in the IISc. shock tunnel are explained in [16] and [17]. Stagnation enthalpies up to 3 MJ/kg can be simulate for $\approx$ 1 ms in the facility. Free stream Mach numbers from 5.75-12 can be generated with a good repeatability ($\pm 5\%$). Free stream conditions for the present experiments are given in Table 1. Here Re_l is Reynolds number based on flat plate length.

Table 1. Free stream conditions for the present experiments

Driver gas	P_0, kPa	H_0, MJ/kg	M_∞	P_∞, kPa	T_∞, K	Re_l
N_2	290	0.8	5.75	0.24	100	1.873×10^4

3.2 Model geometry

The model has two parts. First part is a 210×120 mm^2 flat plate with a chamfered leading edge. This is to ensure that the shock due to finite thickness of the leading edge does not interfere with the flow. There is a through hole (5 mm diameter) on the plate inclined at an angle of 45° to the plate at a distance of 60 mm from the leading edge. The plate is held firmly with thin metallic strips from top wall of the test section during experiments. The other part is basically a wedge with inclined and horizontal portions with same width as that of the flat plate. This is held from back with a sting during experiments.

3.3 Injection system

Helium is injected using a fast acting valve (Parker, Kuroda, EJ-15 series). The response time of the valve is less that 1 millisecond. One end of the valve is connected to a high pressure bottle while the other end is connected to the flat plate through pipes. The injection port diameter is 5 mm. The valve is triggered using the signal from a pressure sensor mounted on the shock tube at 2 m upstream of the port. The synchronization of the starting point of injection and arrival of flow is based on the free stream conditions, shock Mach number and the distance separating the sensor and the port. The valve is mounted inside the test section to minimize loss of time taken by the injectant to reach the injection port from the exit of the valve.

To ascertain the jet jet Mach number during experiments, few initial experiments are carried out without cross flow with the pitot pressure sensor mounted 2 mm above the port. The pressure inside the test section is maintained equal to the static pressure of cross flow during experiments. The pitot pressure is recorded in and is denoted by (p_{02}). With the same setup, another run is taken with pressure inside the test section at atmospheric pressure. Again the pitot pressure is recorded (p_{01}). A representative profile of total jet exit pressure at atmospheric condition and at 3400 Pa are given in Figs. 2 and 3 respectively. By using the formula

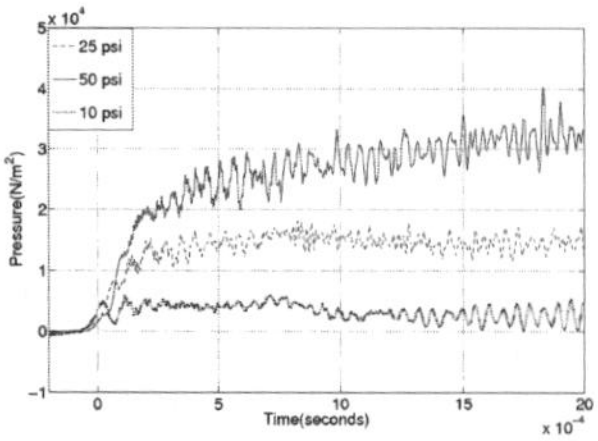

Fig. 2. Total jet pressure profiles at atmospheric condition.

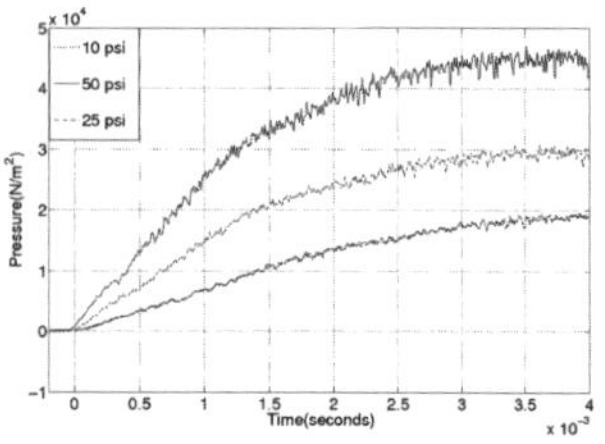

Fig. 3. Total jet pressure profiles at 3400 Pa.

$$\frac{p_{01}}{p_{02}} = \Big(\frac{2\gamma}{\gamma+1}M_1^2 - \frac{\gamma-1}{\gamma+1}\Big)^{\frac{1}{\gamma-1}} \Big(\frac{1+\frac{\gamma-1}{2}M_1^2}{\frac{\gamma+1}{2}M_1^2}\Big)^{\frac{\gamma}{\gamma-1}} \quad (2)$$

the exit Mach number of the jet is calculated. Once the Mach number is known, the jet static pressure is calculated using isentropic relation.

3.4 Flow visualization

A 'z-type' Schlieren set up has been used for visualizing the flow field. Details of the Schlieren set up can be found in [18]. Two concave mirrors of focal length 96 inches are used to obtain parallel beam of light. A 300 watts North Star lamp with C-Clamp base is used as a continuous light source. A high speed camera (Phantom 7.2, Ms Vision Research, USA) with a frame rate of 6688 s^{-1} at a full resolution of 800×600 is used to take sequential photos of jet interaction with the cross flow.

4 Experimental Results

Several experiments are carried out with and without injection. A pitot probe is used to monitor total pressure in the parallel portion of the model near the injection port for runs without injection. A static pressure probe has been also mounted flush with the parallel portion of the model to monitor the static pressure for runs without injection. From the two pressures, the Mach number of the flow near the injection point is calculated using Rayleigh supersonic pitot formula. It has come around Mach 1.9. The jet pressure, jet Mach number and jet to cross flow momentum flux ratio are shown in Table 2.

Table 2. Conditions for the jet

Run no.	p_j, Pa	M_j	$\frac{(\rho v^2)_j}{(\rho v^2)_{cf}}$
04	4456.08	2.86	2.9
05	2579.26	3.13	2
06	1300.79	3.54	1.3

Flow visualizations have been done for the total test time (≈ 800 microseconds) using the continuous light source. The frame rate and exposure of the camera are not sufficient to capture the time evolution of length scales as has been done in [10]. Therefore the focus is mainly to observe the jet penetration into the oncoming flow for inclined jets. Figs. 4 shows the photograph of the flow field for injection pressure of 50 psi.

5 Conclusion

An effort is made to determine a geometrical configuration that could reduce a hypersonic flow (Mach 5.75) to a supersonic flow (Mach 2) by quasi 2D boundaries. The flow field structure arising out of interaction between a supersonic jet and supersonic cross flow is also studied. In the beginning a 2D numerical simulation is carried out to determine the

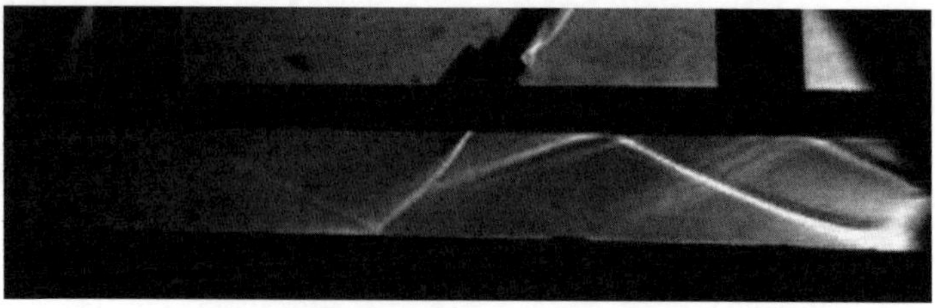

Fig. 4. Photographs of the flow field for injection pressure of 50 psi

geometry as well as jet parameters. Later experiments are done in the IISc. hypersonic shock tunnel HST2. Pitot and static pressure measurements confirmed near Mach 2 cross flow around the injection port. Total pressure measurements of the jet are used to find jet Mach number, jet to cross flow momentum flux ratio etc. The jet is injected at three different injection pressures. Flow visualizations have been done using Schlieren technique. Instantaneous photographs show increasing strength of the bow shock ahead of the injection port with jet increasing pressure. Also visual observations show jet penetration up to 60 percent of the height of the parallel region. Future studies would involve two jets from opposite side of the parallel region and detailed study of mixing of an injectant (Hydrogen) with the oncoming flow.

References

1. R. Dhinagaran and T. K. Bose: AIAA Journal **36**, 3 (1998)
2. Rizzetta D. P.: AIAA Journal **30**, 10 (1992)
3. F. Grasso and V. Magi: AIAA Journal **33**, 1 (1995)
4. P. Gerlinger, J. Algermissen and D. Bruggemann: AIAA Journal **34**, 1 (1996)
5. Spaid F. W. and Zukoski E. E.: AIAA Journal **6**, 2 (1968)
6. Aso S., Okuyama S, Kawai M., Ando Y.: AIAA Paper **93**, 0489 (1993)
7. Gruber M. R., Nejad A. S., Chen T. H., Dutton J. C.: Journal of Propulsion and Power **11**, 2 (1995)
8. Santiago J. G., Dutton J. C.: AIAA Journal **35**, 5 (1997)
9. Everett D. E., Woodmansee M. A., Dutton J. C., Morris M. J.: Journal of Propulsion and Power **14**, 6 (1998)
10. A. Ben-Yakar, M. G. Mungal and R. K. Hanson: Physics of Fluids **18**, 026101 (2006)
11. P. L. Roe: Journal of Computational Physics **43**, (1981)
12. B. Van Leer: Journal of Computational Physics **32**, (1979)
13. G. D. Van Albada, B. Van Leer, and W. W. Roberts: Astron. Astrophys. **108**, (1982)
14. Peyret R. and Taylor T. D.: In: *Computational Methods for Fluid Flow*, (Springer, Berlin, 1983) pp pages
15. K. G. Powell and B. Van Leer: *Tailoring explicit time marching schemes to improve convergence characteristics*, (Von Karman Institute Lecture series 90-03, 1990)
16. R. Joarder and G. Jagadeesh: Shock Waves **13**, 5, (2003)
17. Niranjan Sahoo: Simultaneous measurement of aerodynamic forces and convective surface heat transfer rates for large angle blunt cones in hypersonic shock tunnel. PhD thesis, Department of Aerospace Engineering, Indian Institute of Science, Bangalore (2003)
18. K. Satheesh, G. Jagadeesh and K. P. J. Reddy: Current Science **92**, 1, (2007)

Experimental investigation of cowl shape and location on inlet characteristics at hypersonic Mach number

D. Mahapatra and G. Jagadeesh

Aerospace Engineering Department, Indian Institute of Science, Bangalore, India

Summary. An experimental study is presented to show the effect of the cowl location and shape on the shock interaction phenomena in the inlet region for a 2D, planar scramjet inlet model. Investigations include schlieren visualization around the cowl region and heat transfer rate measurement inside the inlet chamber.Both regular and Mach reflections are observed when the forebody ramp shock reflects from the cowl plate. Mach stem heights of 3.3mm and 4.1mm are measured in 18.5mm and 22.7mm high inlet chambers respecively. Increased heat transfer rate is measured at the same location of chamber for cowls of longer lenghs is indicating additional mass flow recovery by the inlet.

1 Introduction

The success of a hypersonic vehicle, amongest many other factors, is dependent upon the successful inlet design. A good and reliable inlet will help in better performance of a supersonic combustion system that leads to a successful hypersonic flight mission. The inlet can have a variety of shapes to achieve the required compression. The optimum inlet design ensures the classical Shock-on-Lip(SOL) condition for least spillage (loss of mass flow to the combustor) for least loss in vehicle thrust. Since the SOL condition is satisfied at a single Mach number, the fixed- geometry inlets, though attractive for their simple and light-weight design, have to pay penalty in the whole spectrum of their fight domain and hence are limited to a small operational Mach number range On the other hand, variable inlet geometries are heavier and more complex in design than fixed-geometry inlets but operate over a wide range of Mach numbers and unlike fixed-geometry inlets they can be started at various flight conditions reducing the risk of engine unstart.

Both experimental and numerical studies have been reported in the litrature. Many types of high speed fixed-geometry [1] and variable geometry [2] inlet configurations have been studied. Some studies are reported where people have looked at role of cowl on inlet performance. Kanda et al [3] tested six inlet models in a Mach 4 wind tunnel where the parameters of studies were side plate sweep angle, contraction ratio and cowl geometry. Holland [4] examined both experimentally and computationally the effects of Reynolds number and cowl position on the internal shock structure and the resulting performance of a 3D sidewall compression scramjet inlet with leading edge sweep of 45 degrees at Mach 10. Boon and Hillier [5] performed numerical simulations for a dual mode axisymmetric hypersonic inlet for an inlet-cowl configuration at Mach 6 centrebody-cowl design condition and at two off-design Mach numbers 5 and 7.

Neverthless, experimental investigations on the cowl-inlet flow dynamics at hypersonic Mach number is still inconclusive. In this backdrop the objectives of the present investigations are :

- Carrying out schlieren flow visualization for a planar and single ramp scramjet inlet model to visualize the shock structures and observe the shock reflection pattern for two nominal Mach numbers 8 and 6 (design and off-design Mach number respectively) as a function of cowl position and shape.
- Determination of surface convective heat transfer in the rectangular duct to get an idea about the shock compression in the inlet after the flow crosses the cowl.
- Carrying out illustrative numerical simulations to compliment the experiments.

2 Model Design, Fabrication and Instrumentation

The model for the present experiments has been designed, fabricated and tested in the IISc hypersonic shock tunnel, HST2. It is a 2D, planar, single ramp scramjet inlet model designed for Mach number 8 and the Mach number inside the chamber after the shock compression is around 1.7. The main reason for adopting an inlet model of 2D planar configuration is to ensure that the model is as simple as possible,so that the flow complexity due to the shape of the model can be minimized. In addition, the single ramp compression is preferred to the multi-ramp compression mainly due to two reasons. First is, in case of multi-ramp configuration the overall length of the model would be too big to be accommodated in the facility. The second reason is to make sure we visualize the shock structure. Greater the turning angle of a single ramp stronger will be the shock coming from it and hence higher density gradient. The dimensions of the model are determined from the oblique shock theory so that the Shock-on-Lip (SOL) condition is satisfied for the design Mach number. A commercial 3D NS code CFX 5.7 was also used for preliminary parametric studies before arriving at the present configuration of the model. The final dimensions of the model are decided with a compromise between the theoretically obtained dimensions and the facility constraints (test section dimensions). The model has ramp angle 27^0, overall length 230mm and width 112mm. For the design condition the chamber has a dimension of 85mm X 100 mm X 10mm (lXwXh).

The model has been fabricated from commercially available hylam sheet (mechanical grade) of $\frac{1}{4}''$ thickness.The model parts are machined separately and then assembled together using mechanical fasteners. A set of cowls of different lengths and shapes and a set of chamber side plates of different height are machined separately. The cowl plates or chamber side plates can be replaced easily if required. Two slots are made, one on ramp and another inside the chamber to make provision for the heat transfer gauges and provision is also made for fixing the pressure transducers.

For the first set of experiments it was planned to measure the surface convective heat transfer to the model. The model is instrumented with twelve platinum thin film gauges, six on the ramp and six inside the chamber. The gauges are hand painted (Metello Organic Platinum ink, Engelhard,NJ) on a macor substrate. The distance between two consecutive gauges is 10mm.

3 Experiments in the shock tunnel

All the experiments are carried out in IISc, hypersonic shock tunnel (HST2) at the design Mach number 8.0 and an off-design Mach number 6.0 using air as the test gas. The typical free stream conditions during the present experiments is shown in table 1. The model is fixed to the top wall of the tunnel test section by means of a stainless steel sting.

Table 1. Typical free stream conditions during the present experiments

Expt. condition	M_∞	P_0 (bar)	H_0 (MJ/kg)	ρ_∞ (kg/m3)	p_∞ (pa)	Re_∞/m	T_∞ (K)
Design condn	7.9	22.87	2.02	0.0060	254.2	$1.1321e^6$	149.1
Off-design condn	5.14	10.45	1.623	0.0227	1680	$0.591e^6$	257.1

For schlieren flow visualization the camera was operated at 10,000 frames per second with a resolution of 450 X 450 pixels. A standard 300W North Star lamp with C-Clamp base is used as the continuous light source. Operation of the camera is synchronized with the shock tunnel flow using a trigger pulse generated by the pressure sensor located at the end of the tube. The light is switched on before the experiment and only the camera is triggered to synchronize with shock tunnel flow. Acquisition of data during experiment was done by using NI PXI-6115 DAS, (manufactured by National Instruments Pvt. Ltd) at 10MHz sampling rate.

4 Results and Discussions

Experiments are done using cowl plates of four different lengths and three chamber heights. Heat transfer measurements are performed along with schlieren flow vizualisation. Fig. 1 shows a comparison of shock structure in front of the cowl plate for two

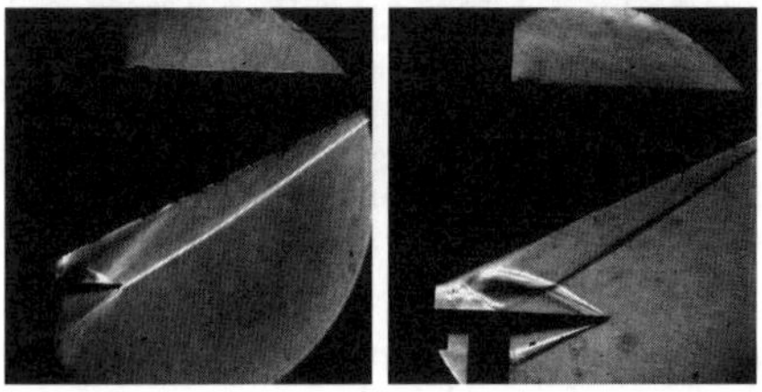

Fig. 1. Shock structures obtained from Schlieren in steady flow time(800μsec) of the tunnel for cowls of two different lengths

cowl lengths namely 111mm and 151mm with same chamber height 10mm recoreded during the steady flow time of the tunnel at free stream Mach number 8.0. In 111mm cowl case the SOL condition seems to be satisfied. A separation bubble which sits at the inlet is clearly visible. The inlet contraction ratio for this case is 7.4 which is more than the Kantrowitz limit [6] for self starting i.e 1.62. But contraction ratios more than Kantrowitz limit can be tolerated since it assumes a single normal shock at the inlet which is having higher total pressure loss than a series of oblique shocks. In contrast, for the 151mm cowl case there is no separation bubble present. A strong interaction between the forebody shock and cowl shock can be observed which seems to weaken the forebody shock. Fig. 2 gives a comparison of the shock structures for three heights of

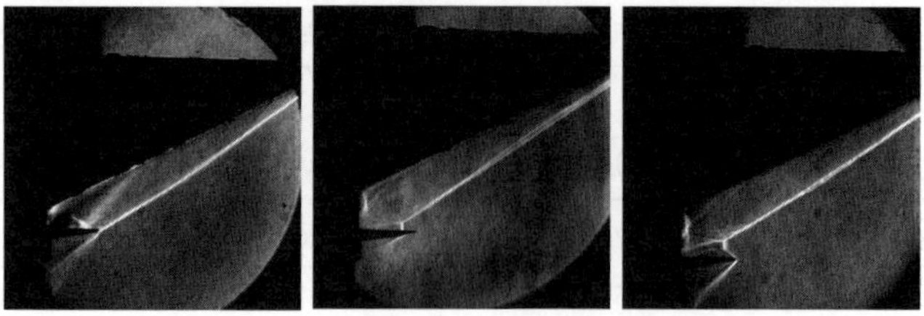

Fig. 2. Shock structures obtained from Schlieren in steady flow time(800μsec) of the tunnel for three different chamber heights

chamber namely 10mm,18.5mm and 22.7mm keeping the cowl length same i.e 111mm. The reflection pattern for the 111mm cowl case is regular where as in in 18.5mm and 22.7mm cases the Mach stems are clearly visible. In the 18.5mm case the Mach stem is touching the cowl plate forming a triple point whereas in 22.7mm case the Mach stem is above the cowlplate level forming a different pattrn of reflection. Mach stem length in both the cases are 3.3mm and 4.1mm respectively which are smaller in comparison to the height of the chamber.

Fig. 3 shows the shock reflection pattern between the cowls of same length but having sharp and blunt leading edge. In case of blunt leading edge cowl the forebody shock interacts with the cowl bow shock and enters inside the chamber instead of meeting at the shoulder as in sharp cowl case. No separation bubble could be seen in this case.

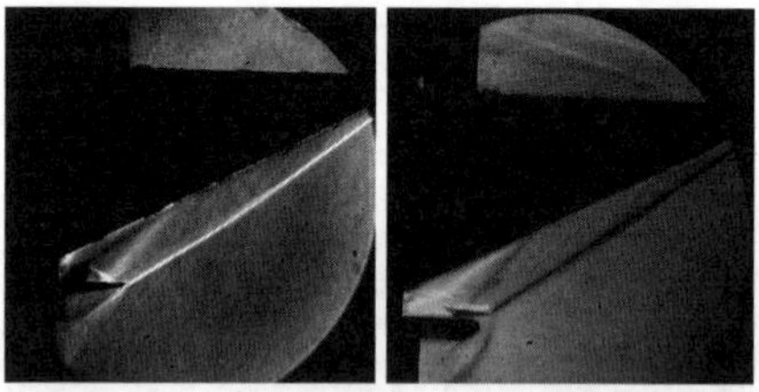

Fig. 3. Shock structures obtained from Schlieren in steady flow time(800μsec) of the tunnel for sharp and blunt leading edge cowl of same length

To have a better idea of the flow physics inside the inlet chamber, convective heat transfer rates are measured inside the chamber as well as on the ramp plate. Fig. 4(a) shows the heat transfer signals obtained at a fixed location inside the inlet chamber for three different cowl lengths namely 111mm, 131mm and 141mm at Mach 8. We can observe the highest heat transfer rate in case of 141mm cowl followed by 131mm and 111mm cowl case. This seems obvious since an increased length of cowl will increase the mass flow rate to the inlet resulting in more heat transfer at the same location provided all other parameters remain same. Fig. 4(b) shows the same plot for Mach 6.

Variation of heat transfer with chamber height is also studied. Fig. 5(a)shows the heat transfer variation at a fixed location(180mm from the forebody tip)inside the chamber at Mach 8. There is no sharp peak in any of the signals implies no reflected shock strikes that location. With 22.7mm chamber height the heat transfer rate value is highest followed

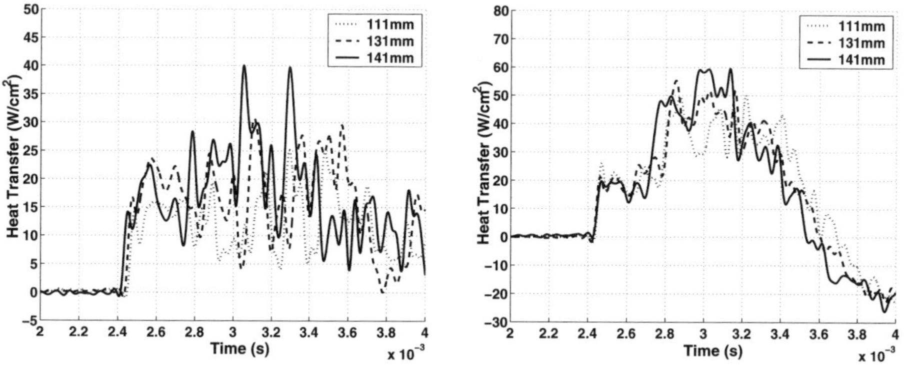

(a) (b)

Fig. 4. Surface convective heat transfer rates measured inside the inlet chamber for three different cowl lengths for (a) Mach 8 (b) Mach 6

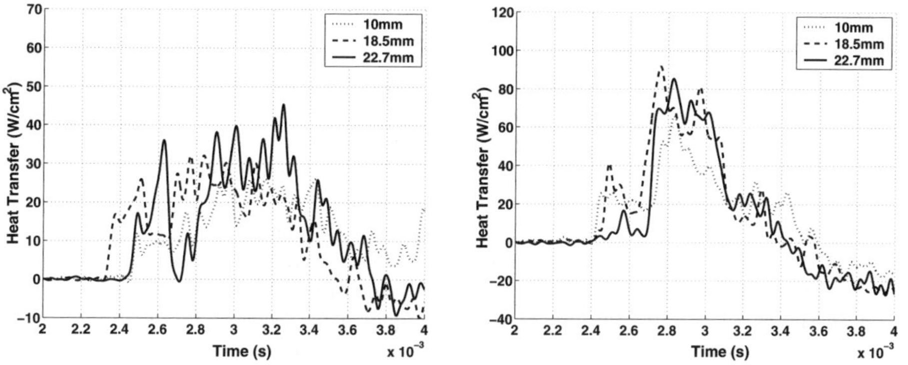

(a) (b)

Fig. 5. Surface convective heat transfer rate measured at a fixed location inside the inlet chamber for three chamber heights for (a) Mach 8 (b) mach 6

by 18.5mm and 10mm chamber heights which is due to increased mass flow rate with increase in chamber height. Similar trends in heat transfer rates can be observed for Mach 6 case (Fig. 5(b)).

In order to compliment the experiments exhaustive numerical simulations are carried out for all the experimental configurations of a model using a commercial 3D, compressible Navier-Stokes code, CFX5.7. Experimental free stream conditions are given as inputs to the code.

Fig. 6(a) shows the values of heat transfer rates along the length of chamber for four different lengths of cowl namely 111mm, 131mm, 141mm and 151mm at Mach 8. Values of heat transfer rates obtained from numerical simulations are also plotted. It can be observed that except for a few cases, they are matching well with-in $10 - 15\%$. Fig. 6(b) shows the values of heat transfer measured along the length of chamber for three different heights of the chamber, namely 10mm,18.5mm and 22.7mm at Mach 8. With exception of few cases they match well with the corresponding numerical values.

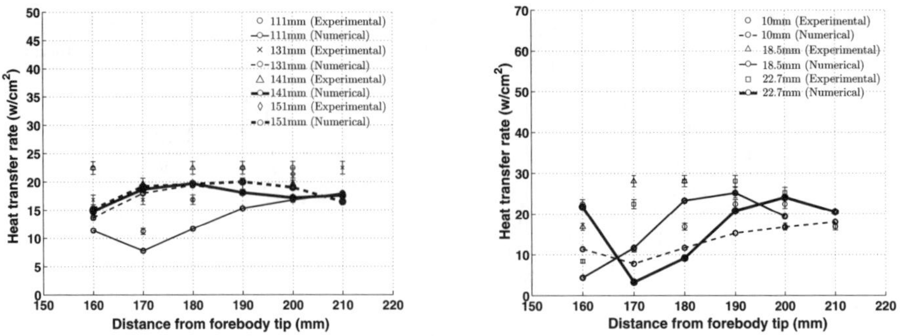

(a) (b)

Fig. 6. Variation of heat transfer rate along the length of chamber for (a) cowls of different lengths (b) chambers of different heights

5 Conclusion and future work

A planar, single ramp scramjet inlet model has been designed, fabricated and tested to investigate the shock interaction phenomena and resulting inlet performance. The experiments are carried out in IISc shock tunnel HST2 at two nominal Mach numbers 8.0 and 6.0. Schlieren flow visualization are carried out for different cowl lengths and shapes and different chamber heights to study the forebody and cowl shock interactions. With a 10mm chamber height the shock reflection is observed to be regular where as for the 18.5mm and 22.7mm chamber heights Mach reflections, having Mach stem heights of 3.3mm and 4.1mm (measured from schlieren Figs.)respectively, are seen. Heat transfer rate measurement inside the chamber for the above configurations matches reasonably well with the corresponding numerical values. Higher mass recovery (spillage reduction) with increase in cowl plate length for the given inlet chamber. Even results from CFD indicates $5 - 10\%$ mass flow rate increase for every 10mm increase of cowl plate length. But mass flow rate in the inlet chamber needs to be measured experimentally to draw such conclusions. Some more point measurements inside the chamber like static pressure measurement, needs to be carried out to have better idea about the flow dynamics inside the chamber.

References

1. Michael K. Smart and Carl A. Trexler, AIAA 2003-0012 (2003)
2. D.P. Mrozinski, J.R. Hayes, AiAA 99-0899 (1999)
3. Takeshi Kanda,Tomoyuki Komuro, Goro Masuya, Kenji Kudo, Atsuo Murakami, Kouichiro Tani, Yoshio Wakamatsu and Nobuo Chinzei: Journal of Propulsion and Power **7**, 2 (1991)
4. Scott D. Holland, AIAA 92-4026 (1992)
5. S. Boon and R. Hillier, AIAA 2006-12 (2006)
6. Kantrowitz, A and Donaldson, C, duP, *'Preliminary investigations of supersonic diffusers'*, NACA WR L 713, May 1945

Experimental investigation of heat transfer reduction using forward facing cavity for missile shaped bodies flying at hypersonic speed

S. Saravanan, K. Nagashetty, G. Jagadeesh, and K.P.J. Reddy

Department of Aerospace Engineering, Indian Institute of science, Bangalore, India

Summary. Forward facing circular nose cavity of 6 mm diameter in the nose portion of a generic missile shaped bodies is proposed to reduce the stagnation zone heat transfer. About 25% reduction in stagnation zone heat transfer is measured using platinum thin film sensors at Mach 8 in the IISc hypersonic shock tunnel. The presence of nose cavity does not alter the fundamental aerodynamic coefficients of the slender body. The experimental results along with the numerically predicted results is also discussed in this paper.

1 Introduction

The combination of low altitude and high flight velocity for a missile shaped body imposes severe heat transfer environments at the stagnation region of the nose tip. In the range of speeds of interest (2-4 km/s), the projectile nose is subjected to very high heating rates. As an example of the severity of the heating, the stagnation temperature at sea level and a velocity of about 2.6 km/s corresponds to the melting point of tungsten [1, 2]. Also, the shape change that occurs during ablation causes both aerodynamic as well as structural problems. Forward facing cavities have been of interest over the past 15 to 20 years as a possible way of reducing nose-tip heating of the vehicle. The forward facing cavity has the ability to modify the behavior of the bow-shock (i.e, to cause bow shock oscillation) and thereby pushing it away from the body [3,4,5]. Several cavity models have been used with varying geometry (i.e., diameter, L/D and lip radius) over the period and investigated both experimentally and numerically. A range of L/D (cavity length to diameter) with different shapes of lip cavities were investigated experimentally [6].Inspite of many reported studies of use of nose cavities at hypersonic speeds, so far there has been no published literature on the variation of heat transfer rate on a missile shaped body with nose cavity. Also, most of investigations have been computational in nature and very few experimental studies are reported in this area. In this backdrop, the objectives of the present study are (a)determination of surface convective heat transfer rates over a generic missile shaped body with and without forward facing cavity and (b) carrying out the numerical simulations to compliment the experimental results.

2 Experimental facility and instrumentation

In the present investigation, the shock tunnel (HST 2) is operated at a reservoir enthalpy of 2 MJ/kg with an effective test time of 1 ms, which results in stagnation temperature of 2000 K. All the shots are carried out in HST2 at free stream Mach number of 8 using helium and air as driver and driven gases, respectively. The typical test conditions of the experiment are shown in Table 1. The experimental data during the shock tunnel

Table 1. Typical free stream conditions during the present experiments

M_∞	P_0 (bar)	H_0 (MJ/kg)	ρ_∞ (kg/m3)	p_∞ (pa)	Re_∞/m	T_∞ (K)
8.0	19.4	1.96	0.0050	207	$0.98e^6$	143

testing has been acquired using NI PXI-6115 DAS (National Instruments Pvt. Ltd)data acquisition system. Operation of the NI data acquisition system is initiated with the shock tunnel flow using a trigger pulse generated by the output of the shock speed pressure sensor. The DAS settings were changed suitably depending on the signal levels.

3 Models and sensors

A blunted cone-cylinder-frustum with fins is used in the present study ((20.5°) cone half-angle and radius of 15 mm). The length of the cavity (L) was defined as the length from the nose lip to the cavity base. The photograph of the model with 6 mm cavity is shown in Fig.1. The entire model including the cavity is made up of an aluminum alloy. The test model has a provision for inserting detachable Macor substrates over the surface and platinum thin films are deposited on Macor. Platinum thin film gauges are deposited by using vacuum deposition. The typical response time of these sensors is $1 - 2\mu$ s and can conveniently be used in short duration test facilities. The newly fabricated circular cavity is attached to the cylinder-frustum portion of the model with triangular fins. Since, the focus of this work was on the stagnation region, an attempt has been made to get more number of thin films (seven gauges) on the nose cone with minimum spacing. In order to have a comparison of convective heat transfer rates, more or less same gauge location (platinum thin film sensors) have been chosen for the model without cavity. The remaining seven sensors are placed on the surface of the cylinder and frustum in order to obtain the complete heat transfer distribution over the model. All the gauges used in the experiments are calibrated and the value of α is found to be 0.0023 (per K).

4 Results and discussion

The heat transfer measurements are carried out only at zero degree angle of attack for a missile shaped body with and without forward facing cavity. The heat flux can be obtained from the voltage-time traces of the platinum thin film sensors and the typical surface temperature recorded by the gauge, located at axial distance of 18.9 mm is shown in Fig 2. It is evident from the figure that the amplitude of the voltage for the cavity configuration is just 50% of the model without cavity. A typical heat transfer signal obtained numerically, is shown in Fig 3 and the net reduction of 25% has been observed for cavity configurations near the stagnation zone. In order to compliment the experimental results, the numerical simulation is carried out for the above cavity model with a stagnation enthalpy of 2 MJ/kg and a free stream Mach number of 8. A commercially available 3-dimensional, compressible Navier-Stokes code namely CFX

5.7 has been used for the simulation. In simulation, the body geometry and freestream conditions correspond to experimental condition of the shock tunnel. The normalized surface heat transfer rates, over the surface of missile shaped model at Mach 8 along with CFD results is shown in Fig 4. The experimental results agree with computed results. The reduction in heat transfer rate is also plotted in Fig. 5 for stagnation enthalpy of 2 MJ/kg. Experimental results show that the temperatures (heat transfer rate) are much lower for a cavity configuration compared to the model without cavity, because of the recirculation region and cooling mechanism. Maximum reduction of 27% has been observed in the heat transfer rate at an axial location of 11.88 mm for 6 mm cavity. In addition, the reduction pattern in heating rate distribution is marginally affected in all gauge locations by the cavity.

5 Conclusion

The structure of the hypersonic flow over a cavity has been studied experimentally and numerically. Experimental data are seen to compare reasonably well with numerical predictions from a Navier-Stokes solver. A net reduction of 25% heat transfer rate is observed near the stagnation region for 6mm cavity configuration.

Acknowledgement. We thank HEA Lab team for their support during the experimental work and operation of HST2

References

1. Reinecke WG and Sherman M (1993) The survivability and performance on hypervelocity projectiles, In Proc. of the 14th Intl Symposium on Ballistics.
2. Reinecke WG and Guillot MJ (1995) Full scale ablation testing of candidate hypervelocity nose tip materials, In Proc. of the 15th Intl Symposium on Ballistics.
3. Johnson RH (1959), Instability in hypersonic flow about blunt bodies, Phys. of Fluid, 2, pp.526-532
4. Engblom WA, Goldstein DB, Ladoon D, Schneider SP (1997) Fluid dynamics of hypersonic forward-facing cavity flow. AIAA Journal of Spacecraft and Rockets, 33 (3), 353-359
5. Ladoon DW, Schneider SP, Schmisseur JD (1998) Physics of resonance in a supersonic forward-facing cavity. Journal of Spacecraft and Rockets, 33 (5), 626-632.
6. Silton SI, Goldstein DB (2005) Use of an axial nose-tip cavity for delaying ablation onset in hypersonic flow. Journal of Fluid Mechanics, 528, 297-321

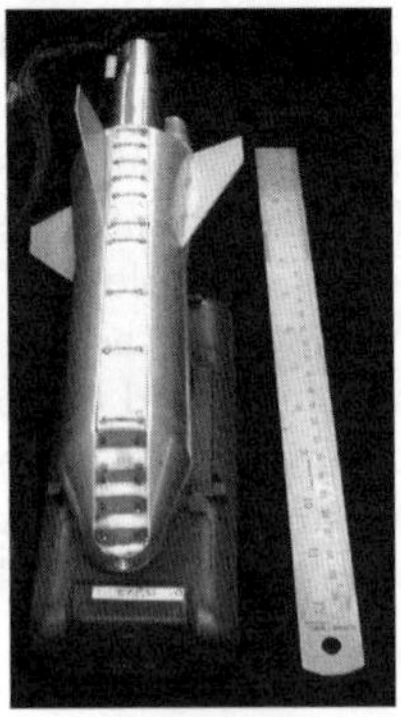

Fig. 1. Photograph of model with nose cavity fitted with platinum thin film sensors for measurement of convective heat transfer coefficients.

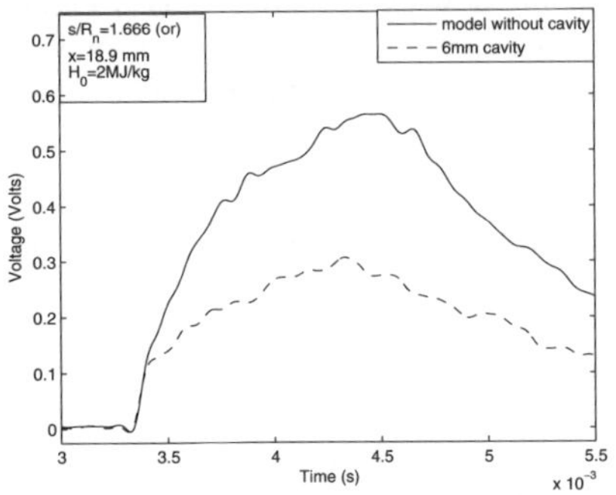

Fig. 2. Typical temperature-time history from platinum thin film sensors with and without forward facing cavity.

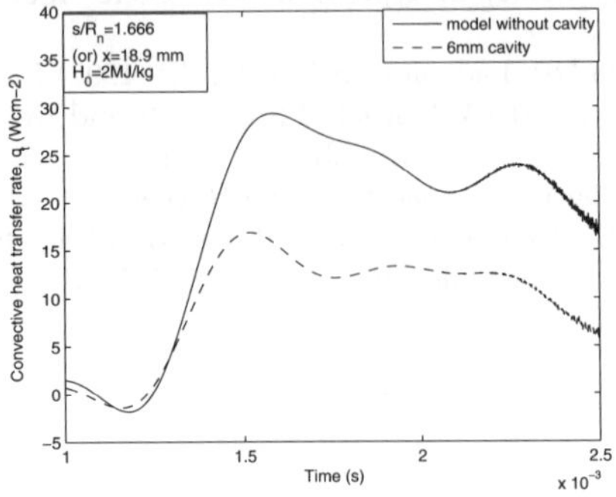

Fig. 3. Numerically integrated heat transfer signal for platinum thin film sensors with and without forward facing cavities.

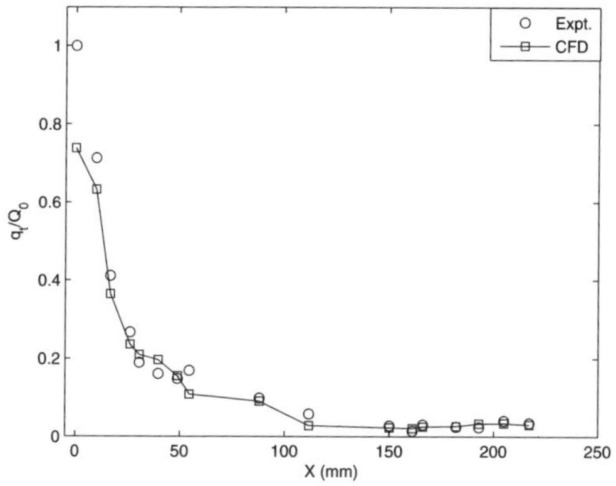

Fig. 4. Variation of normalized heat transfer rate over the surface of missile shaped body with forward facing cavity at 0° angle of attack.

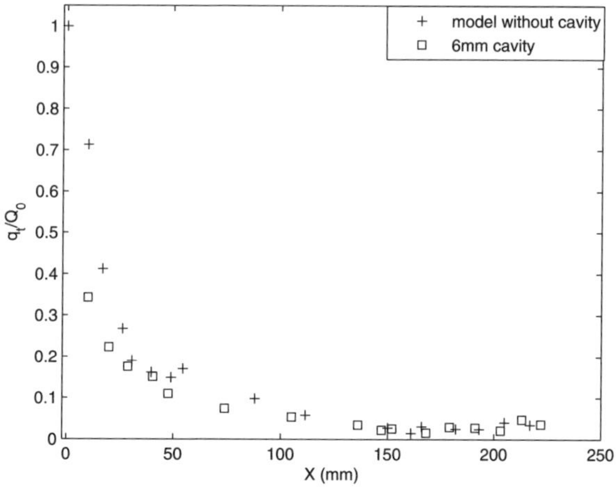

Fig. 5. Variation of convective surface heat transfer rates over the surface of missile shaped model flying at Mach 8 with various configurations

Extrapolation of a generic scramjet model to flight scale by experiments, flight data and CFD

A. Mack[1], **J. Steelant**[1], **and K. Hannemann**[2]

[1] *European Space Agency (ESA-ESTEC), Aerothermodynamics Section Keplerlaan 1, 2200 AG Noordwijk (The Netherlands)*
[2] *German Aerospace Center DLR, Spacecraft Section Bunsenstraße 10, 37073 Göttingen (Germany)*

Summary. Based on ground and flight testing, a generic full scale scramjet combustion chamber was investigated to extrapolate the results obtained at small scale experiments. Numerical CFD validation was done for the experimental data for a duct size of 9.8mm height and wall porthole injection. Taking previous results of mixing and combustion performance enhancement into account, the configuration was extrapolated to a real full scale vehicle with a duct size of 300mm height. The combustion process in the full scale combustion chamber showed comparable flow topology changes and pressure rise due to variation in equivalence ratio. The different working modes of the scramjet such as the pure supersonic mode at lower equivalence ratio, a mixed subsonic/supersonic mode with a locked reverse flow area behind the injector at moderate equivalence ratio and choked flow at high equivalence ratio could be reproduced for both small and large scale ducts.

1 Introduction

The development of future scramjet (supersonic combustion ramjet engines) is based on flight testing, CFD and ground testing. Ground testing concentrates on reproducing the key features of engine operation, without the cost and engineering challenges associated with flight testing. Experiments have shown that free piston driven shock tunnels are well suited for scramjet testing [1], however comparison with flight tests is necessary. In the High Enthalpy Shock Tunnel Göttingen (HEG) at DLR, the free stream conditions encountered during the HyShot flight test were reproduced [2]. The available flight data of the HyShot flight test program in 2002 were performed at Mach 7.8 from 37km to 23km altitude. A 1:1 duplication of the flight model was tested 2003 and 2007 in HEG. Good agreement between the numerical and experimental data could be achieved regarding surface pressure and heat fluxes [3] and flow topology changes at higher equivalence ratios [4]. In the past, extensive work was done on generic combustion chambers concerning the injection system [5], optimisation of the duct cross sections, the impact of the numerical modelling and the presence of turbulent boundary layers from e.g. intake ramps entering the ducts [6]. The present paper focusses on the extrapolation of the experimental and numerical data gained with the small scale combustion chambers of HyShot to the ones of full scale applications.

2 Numerical Modelling

The numerical investigations, computing the three dimensional flow were performed using the DLR TAU code on unstructured meshes. The DLR TAU-code has been validated in the past for different configurations at super- and hypersonic flow conditions [7] including

extensive studies of wall jets and combustion [8]. Different standard upwind solver such as Van-Leer, AUSM and AUSMDV are available. One and two equation turbulence models such as Spalart-Allmaras, k-ω and SST are implemented. The Evans/Schexnayder 7-species 8-reaction set [9] is taken as combustion modelling for the nonequilibrium flow calculations, the forward reaction rates are described by an Arrhenius law, whereas the backward rates are calculated by the equilibrium constants. The diffusion is modelled by Fick's law with a laminar and turbulent Schmidt-number of 0.7.
For the present investigations the duct flow has been treated as fully turbulent, however no interaction of the turbulence and the combustion process was taken into account. Previous investigations showed that the impact of the turbulence-combustion coupling modeled by an assumed pdf-approach is quite weak [10]. The ramp flow entering the combustion chamber after bleeding the boundary layer was treated as 2D in order to reduce the numerical effort. The 2D intake solution was extended in spanwise direction and applied by a Dirichlet boundary condition to the duct inflow plane. The wall temperature was assumed to be constant at 300K. The unstructured meshes were adapted and contained between 1.2 10^6 and 1.7 10^6 points.

3 Numerical Computation of the HyShot Configuration

In a first step, the experimental data of the HyShot configurations will be compared with numerical data to demonstrate the capability to model the driving flow and combustion phenomena in an adequate way. The unstructured, adapted grid contains 1.5 10^6 points, the combustion chamber containing approximately 1.2 10^6 points and the remaining being placed in the exhaust region. The typical flowfield topology is shown in Figure 1 (left). The bow shock of the injector can be seen clearly, also its reflections at the upper and lower wall and laterally at the two symmetry planes. Further downstream, the shocks, in addition with the pressure increase due to combustion, lead to local subsonic pockets and finally to subsonic flow close to the (injector) symmetry plane. Nevertheless, the major part of the combustion process takes place in supersonic regime. Due to the low equivalence ratio the penetration of the fuel is not high, it stays mainly in the lower part of the duct. Analysing the combustion process, no ignition delay can be observed; high temperature and pressure behind the detached bowshock of the injector ignite the mixture immediately.

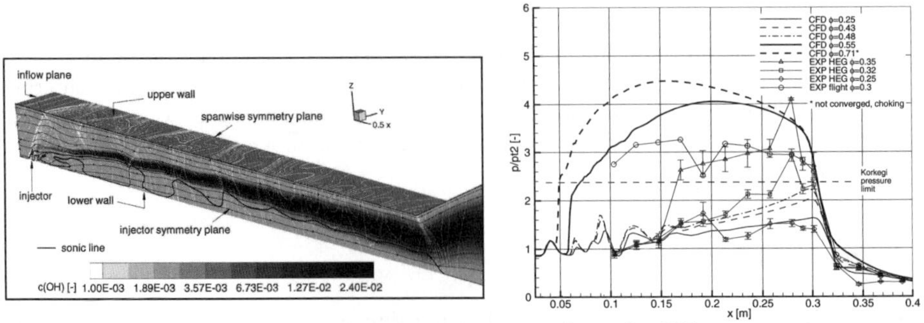

Fig. 1. HyShot: Flow topology and combustion process shown by OH concentration, pressure isolines and sonic line (left); surface pressure distribuations at lower wall (right)

Results of the surface pressure distribution are shown in Figure 1 (right) for fuel-off and fuel-on using different turbulence models. The two-equation k-ω model leads to a higher heat release and pressure rise due to an enhanced mixing and combustion process than the Spalart-Allmaras model. This is also a result of higher turbulent viscosity level in the developing boundary layers within the combustion chamber being enhanced by the shock reflections in the duct. The shock patterns are well resolved, the maximum dimensionless pressure reaches 1.8 (SA) and 2.02 (k-ω). The pressure peaks are shifted due to a slightly different flowfield for the different turbulence models. Since the shocks undergo multiple reflections in lateral direction and at upper/lower walls the shifts increase with increasing x. Analyzing the flowfield unveils that the flow is close to sonic speed with local subsonic pockets (close to the injector symmetry plane). Every strong shock impingement leads to a subsonic pocket. Due to the almost sonic flow over a large part of the combustor close to the injector plane the angles of the reflected shocks inside the duct are higher. Therefore, more but weaker pressure peaks occur in the combustor for the results based on the k-ω turbulence modelling. Comparison at the upper wall is also done in [3], revealing also good agreement with the experimental data. In addition, the heat fluxes calculated with the k-ω model reproduce the experimental heat fluxes with very good agreement. Due to some uncertainties of the injected hydrogen condition the equivalence ratio was varied for the numerical computations. Solutions from equivalence ratios of 0.25 to 0.71 (SA model) are compared in Figure 2 (left) with the results from HEG and flight experiment [2]. The solution is obtained on structured meshes, which do not fully resolve the local pressure pattern further downstream in the combustion chamber, whereas the averaged pressure increase due to combustion is prediced well [4]. The lower equivalence ratios show an almost linear increase in pressure due to combustion, which was identified as mainly supersonic combustion. The maximum pressure is below the Korkegi limit (p/pt2=2.4), which is an indicator for boundary layer separation. At higher equivalence ratio, the pressure suddenly increases behind the injector (Figure 2, right) due to a flow separation. This separation originates from the combustion chamber exit, where the critical pressure is reached first. From there, the separation moves upstream and gets locked behind the injector for an equivalence ratio of 0.55. A stable mode, with a large subsonic reverse flow area appears. Further downstream, the flow accelerates again to supersonic speed; but the combustion takes mainly place in the subsonic area. This behaviour can also be observed during freeflight (Figure 2, left), however

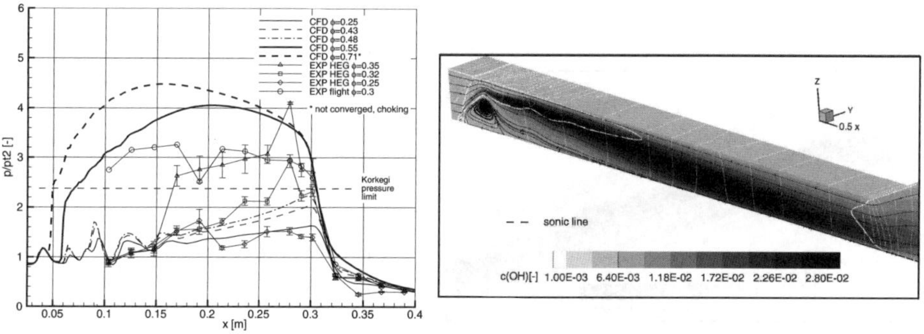

Fig. 2. Numerical and experimental surface pressure plots of HyShot, flight altitude 28km (left). Flow topology and combustion process at high equivalence ratio shown by OH concentration, pressure isolines and sonic line (right, $\phi = 0.55$)

at a lower equivalence ratio. The pressure increase due to heat release is so high that a normal shock occurs which travels even upstream the injector. No stable flowfield can be established in the computation, the flow chokes thermally.

4 Extrapolation to a Full Scale Combustion Chamber by CFD

A generic combustion chamber, which has the dimensions of a full scale hypersonic vehicle (LAPCAT M8 vehicle [11]), has been chosen in the past to investigate the mixing and combustion behaviour for different duct cross section aspect ratios (corrsponding to different injector spacings), injector types and flow conditions. The height is 0.3m. For the present cases, a generic boundary layer enters the duct. It has half the height of the computational domain, corresponding to 25% of the duct height. Although the intake boundary layer at HyShot was bled, the boundary layer thickness at the injector position (59mm downstream of bleed) is 16% of the duct height. The boundary layers entering the duct therefore are of the same order of magnitude.

Porthole injection on both the upper and lower combustion chamber wall is foreseen. Due to this, the upper boundary can be modelled as symmetry plane. The dimensionless duct cross section dimensions are 5x10 (HyShot approximately 5x5). The inflow Mach number is 3 (HyShot 2.5), static pressure 1 bar and 1300K static temperature. Comparable with the Hyshot results, mainly the same flow phenomena can be observed although the geometry and flow conditions are slightly different. Due to the higher Mach number in the generic duct, comparable flow topologies occur at slightly higher equivalence ratios.

For an equivalence ratio of 0.5, supersonic combustion is observed (Figure 3, left). The flow topology is comparable with Figure 2 (right) (HyShot). At an equivalence ratio of 0.7, a stable separation behind the injector occurs with a large subsonic reverse flow area comparable to the results with the small scale Hyshot experiment. At an equivalence ratio of 1, thermal choking takes place and no stable flowfield can be obtained. The surface pressure plot in a spanwise middle section (Figure 3, right) shows comparable results as for the small scale experiment. At low equivalence ratio (0.5), supersonic combustion takes place with a monotonic averaged pressure increase due to combustion. Local pressure peaks of the reflected bowshock appear. In the subsonic combustion areas for an equivalence ratio of 0.7 (Figure 4), the pressure increases in the injector vicinity rapidly due to the high heat release of the combustion process. The local pressures are

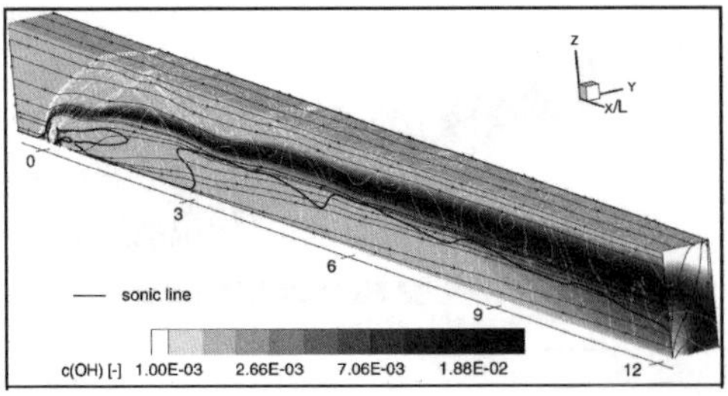

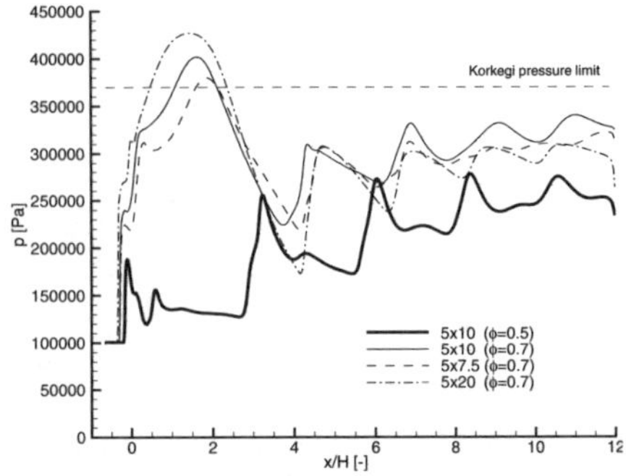

Fig. 3. Generic configuration: Flow topology and combustion process shown by OH concentration, pressure isolines and sonic line (left); surface pressure distribuations at lower wall (right)

above the Korkegi limit for flow separation. Further downstreams, the flow accelerates to supersonic speed and shows also a monotonic pressure rise.

For the full scale configuration, the target combustion chamber length is 2m. Due to computational efforts, the length is fixed to 1.8m, which results in a dimensionless length of the computational domain of L/H=12 (H=0.15m, half duct height, porthole injection from upper and lower wall); related to the HyShot model it is longer (L/H=30). In addition, the Mach number of the generic duct is higher compared to HyhShot, resulting in a lower heat release and pressure rise. Due to the larger duct length, the separation in the HyShot model begins at the end of the duct and travels against the flow direction until it is locked behind the injector. With the generic case, the reverse flow is established directly behind the injector. This is due to the smaller duct width, which results in combination with a high heat release to partial blocking of the duct flow. The different duct cross sections between HyShot (5x5) and the generic duct (5x10) leads to less vertical but more lateral injector bow shock reflections, where the vertical reflections are dominant. The reverse flow area of the generic configuration is trapped exactly between these reflections. As mentioned, the local pressures in the HyShot model are much higher than the Korkegi limit ($\phi = 0.55$, $\phi = 0.71$) whereas with the generic cases, the static pressures are only slightly above the Korkegi value ($\phi = 0.7$). This explains why the locked reverse flow area and the thermal choking for the generic duct appears at higher equivalence ratios. The Hyshot duct at $\phi = 0.55$ is much closer to choking than the generic model at $\phi = 0.7$ which can be seen at the sonic Mach line. Large part of the HyShot duct flow is subsonic across the whole duct section, therefore the pressure increases significantly in the whole chamber, compare Figure 2 (right).

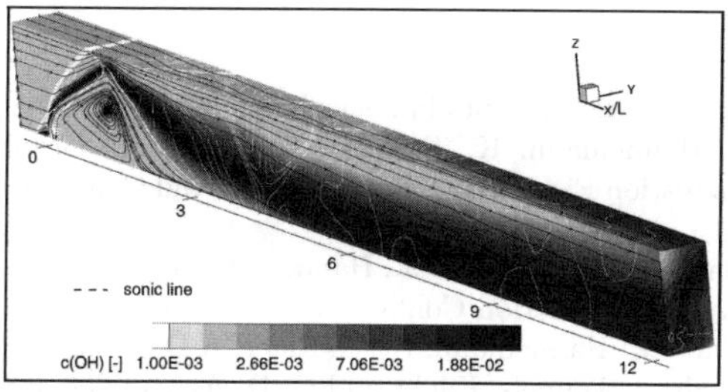

Fig. 4. Flow topology and combustion process of generic configuration at an equivalence ratio of 0.7 shown by OH concentration, pressure isolines and sonic line

Combustion efficiencies and total pressure loss of the generic configuration are shown in Figure 5 (left). The total pressure loss increases drastically when switching to subsonic combustion at $\phi = 0.7$ related to the enhanced combustion in the reverse flow area. The overall combustion efficiency is higher than at supersonic combustion, although more fuel is injected. The combustion process for the HyShot configuration in Figure 5 (right) shows a comparable behaviour; at mainly subsonic combustion the combustion efficiency is increased mainly in the reverse flow area, whereas the supersonic combustion process shows an almost linear increas with increasing x coordinate. The strong vortex being created by the injection process leads to a good mixing process of the hydrogen with the air also further downstream in the duct. In the expansion nozzle at $x > 300mm$ the combustion process still continues.

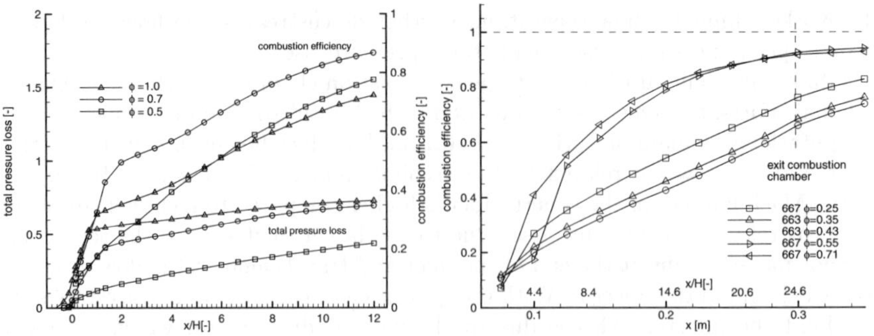

Fig. 5. Combustion efficiencies for different equivalence ratios of generic configuration (left) and HyShot (right)

5 Conclusion

The present comparison of the flow phenomena of a small scale combustion chamber with a numerical extrapolation to a large scale combustion chamber showed, that the main effects of the flow topology changes can be transferred to flight scale. Supersonic and subsonic combustion at lower and moderate equivalence ratios as well as choking at high equivalence ratios could be obtained at both scales, although these preliminary results need further investigation due to slight differences in geometry and flow condition.

References

1. Paull, A.: Scramjet measurements in a shock tunnel. AIAA 99-2450, 1999
2. Gardner, A.D., Hannemann, K., Steelant, S., Paull, A.: Ground Testing of the HyShot Supersonic Combustion Flight Experiment in HEG and Comparison with Flight Data. AIAA 2004-3345, 2004
3. Mack, A., Steelant, J., Schramm, J.M., Hannemann, K.: Sensitivity Analysis for the HyShot Generic Supersonic Combustion Configuration using CFD, ISABE-2007-1310, 2007
4. Mack, A., Steelant, J., Hannemann, K., Gardner, A.D.: Comparison of Supersonic Combustion Tests with Shock Tunnels, Flight and CFD, AIAA 2006-4684, 2006
5. Mack, A., Steelant, J.: Mixing Enhancement by Shock Impingement in a Generic Scramjet Combustion Chamber. European Conference on Computational Fluid Dynamics ECCOMAS CFD 2006, September 5-8 2006, Egmond aan Zee, The Netherlands, 2006
6. Mack, A., Steelant, J., Togiti, V., Longo, J.M.A.: Impact of boundary layer Turbulence on the Combustion Behaviour in a Scramjet. proceedings of EUCASS 2007, Paper 295, 2007
7. Mack, A., Hannemann, V., Validation of the Unstructured DLR-TAU-Code for Hypersonic Flows, AIAA 2002-3111
8. Karl, S., Hannemann, K., Application of the DLR Tau-code to the RCM-1 testcase: Penn State Preburner Combustor. Proceedings of the 3rd International Workshop on Rocket Combustion Modelling, 2006
9. Evans, J.S., Schexnayder Jr., C.J., Influence of Chemical Kinetics and Unmixedness on Burning in Supersonic Hydrogen Flames AIAA Journal, Vol. 18, No. 2, February 1980
10. Karl, S., Hannemann, K., Steelant, J., Mack, A.: CFD Analysis of the HyShot Supersonic Combustion Flight Experiment Configuration, AIAA 2006-8041, 2006
11. Sippel, M., Klevanski, J.: Preliminary Definition of the Supersonic and Hypersonic Airliner Configurations in LAPCAT. AIAA 2006-7984, 2006

Force measurements of blunt cone models in the HIEST high enthalpy shock tunnel

K. Sato, T. Komuro, M. Takahashi, T. Hashimoto, H. Tanno, and K. Itoh

Japan Aerospace Exploration Agency, Kakuda Space Center, 1 Koganesawa, Kimigaya, Kakuda, Miyagi, 981-1525, Japan

Summary. In the present paper, two blunt cone models with half-angles of 15 degrees were tested in a high enthalpy shock tunnel to investigate fundamental real gas effect on high enthalpy aerodynamic characteristics. The curvature radius of the nose of the models are 20 mm and 50 mm. The tests were carried out at 4 MJ/kg - 14 MJ/kg of stagnation enthalpy and the axial forces were measured by the force balance with acceleration compensation. The present results were compared with the results obtained by calculation of CFD. From the comparison between the experimental results and the results of CFD the real gas effect on axial force was discussed.

1 Introduction

Air stream around the vehicles which reenters into the atmosphere is compressed and heated by the very strong bow shock wave formed with a high-speed flight. High temperature in the shock layer carries out dissociation of the gas molecule. A part of the dissociated atom is re-combined while passing the surroundings of the body. The flow surrounding the body of the shock layer inner side is a chemical non-equilibrium flow, and the flow of such real gas influences the aerothermodynamics characteristics of the body. It is important in the design of the vehicle to study the chemical non-equilibrium flow around the body, and to evaluate the characteristics. The purpose of the present study is to investigate the influence on the aerodynamic characteristics due to the real gas effect to model shape. For this reason, the force measurement tests using blunt cone models was carried out in the high enthalpy shock tunnel HIEST at the Japan Aerospace Exploration Agency (JAXA) [1]. In the test, the axial force up to stagnation enthalpy 14 MJ/kg condition was measured and the influence of axial force was investigated to the two blunt cone models by which the curvature radii of blunt differ. Furthermore, the test results were compared with the calculation results of CFD.

2 Blunt cone test models and three component aerodynamic balance

The schematic view of two blunt cone models, an aerodynamic force balance, and the Sting's attachment is shown in Fig. 1 (a) and Fig. 1 (b). As shown in the figures, both models are axi-symmetrical with a half-angle of 15 degrees and a base diameter of 200 mm. The curvature radii of the nose of the model are 20 mm and 50 mm, respectively. The full length of the model with the curvature radius of 20 mm and the model with the curvature radius of 50 mm was 316 mm and 230 mm, respectively. The material of the models was made from the aluminum alloy (A7075). The mass of R20 model and R50 model including a model attachment adapter was 6.63 kg and 6.04 kg, respectively. In

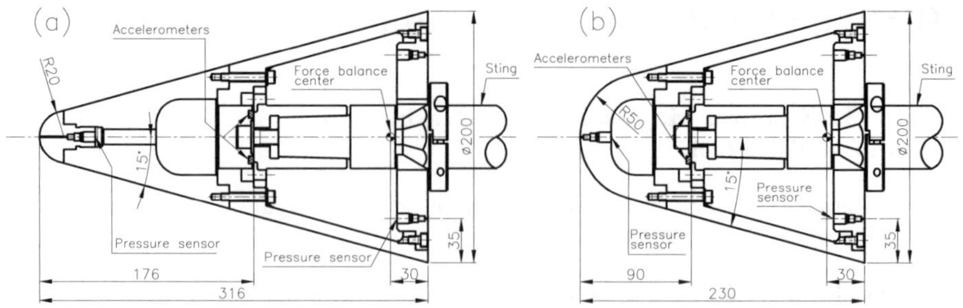

Fig. 1. (a) R20 blunt cone model, (b) R50 blunt cone model

order to measure the Pitot pressure in stream of air, the pressure sensor (Kulite, XCL-100-100A) was attached to the nose of a model. In the force measurement, the cross-beam type aerodynamic force balance of three-components [2] which has high stiffness was used. The capacity of axial force, normal force, and pitching moment of a balance is 980 N, 4900 N, and 147 Nm, respectively. The main coefficients of the axial force, normal force, and a pitching moment obtained from static load calibration of the balance were 11746 N/mV, 10803 N/mV, and 47.523 Nm/mV, respectively. The linearity on the axial force of a balance was within ±1.0% approximately over the static load of 250 N to 1250 N. In this measuring method, mechanical vibration of a model, a balance, and Sting is recorded with the output signal from the axial force kx of a balance because those mechanical vibration is not damped within the short test time of HIEST. Acceleration compensation was performed in order to reconstruct the axial force $f = kx+ma$ from this output signal included a mechanical vibration. For this reason, two acceleration sensors (ENDEVCO, 2250A-10) were attached to the front of a balance adapter. Where, m is the equivalent inertia mass of a model.

3 Dynamic calibration of the balance

The equivalent inertia mass m of the model for performing acceleration compensation was measured by both dynamic calibrations with a fracture stick method and an impact hammer method. The result of the dynamic calibration of R50 model to the acceleration sensors a_1 and a_2 was shown in Fig. 2. Where, the vertical axis of the figures are the deviations from the average value of the measured equivalent inertia mass. The natural vibration frequency of the balance and the acceleration sensors in the axial force direction was 1.13 kHz - 1.14 kHz. The average value of the equivalent inertia mass to the acceleration sensors a_1 and a_2 was 7.36 kg and 7.81 kg, respectively, and the deviations from the average value of an equivalent inertia mass was within ±2.0% and within ±1.5%, respectively. In the R20 model, natural vibration frequency was 1.10 kHz. The average value of the equivalent inertia mass to the acceleration sensors a_1 and a_2 was 7.88 kg and 8.32 kg, respectively, and the deviation was less than ±1.5% in both the acceleration sensor. The average value of the equivalent inertia mass in both models showed the tendency for the acceleration sensor a_2 side to be larger than a_1 side. For this difference,

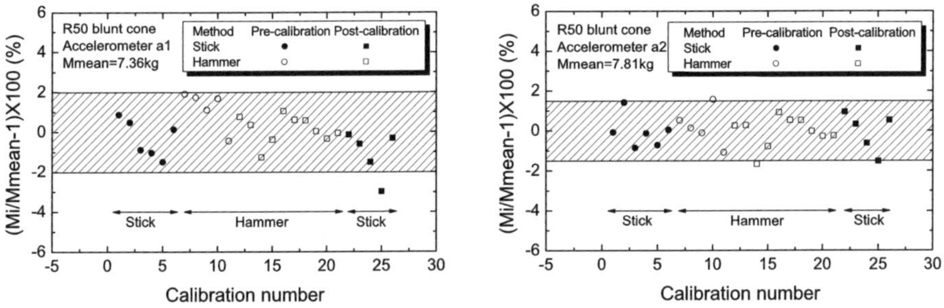

Fig. 2. Dynamic load calibration of axial force direction for R50 blunt cone model with acceleration sensor a_1 (left) and a_2 (right)

the inertia force ma in the wind tunnel tests used the average value of each the inertia force $(m_1a_1 + m_2a_2)/2$.

4 Test condition

In the test at HIEST, the contour nozzle was used, which throat diameter is 50 mm, exit diameter is 800 mm, the nozzle expansion area ratio is 256, and nozzle full length is about 2.8 m. The test condition using the nozzle is shown in Table 1. The test condition was estimated based on the calculation with non-equilibrium nozzle flow code NENZF [3]. The tests were performed at five different stagnation enthalpy conditions between 4 MJ/kg to 14 MJ/kg. On these test conditions, the uniform area of the flow obtained from the nozzle flow calibration test was less than about 150 mm in radius from a nozzle center [4]. The angle of attack of the models was 0 degree, and the center of the models was installed on the center axis of the nozzle. In the test, the models with a maximum diameter of 200 mm are put into the area of a uniform flow.

Table 1. Test conditions and flow conditions

Condition	Stagnation enthalpy H_0(MJ/kg)	Stagnation pressure P_0(MPa)	Static pressure P_1(kPa)	Static temperature T_1(K)	Velocity V(m/s)	Mach number M	Reynold's number Re(/m)
A	4	13	1.2	340	2730	7.4	1.7×10^6
B	7	18	2.1	650	3420	6.8	1.2×10^6
C	10	17	2.5	1060	4050	6.3	7.5×10^5
D	12	17	2.4	1180	4260	6.2	6.4×10^5
E	14	17	2.5	1340	4590	6.2	5.5×10^5

5 CFD

CFD analysis of the blunt cone models was conducted in order to compare with the test results in a high enthalpy flow. The computational area is the blunt portion and cone portion of a model, and the base portion was removed from the computation. The basic equations of CFD are axisymmetric Navier-Stokes equations. The flow was assumed to be laminar flow. The number of computational grids is 101 × 80 points. For the time integration, the lower-upper symmetric Gauss-Seidel (LUSGS) scheme was applied. Convective terms are modeled by the AUSMD/V scheme. Viscous terms are modeled by the central difference scheme. Chemically non-equilibrium phenomena in the high-temperature air was described by the Dunn-Kang model including 7 chemical species (O_2, N_2, O, N, NO, NO^+, e^-) while thermally equilibrium was assumed. The surface of the model was assumed to be fully catalytic and isothermal wall of 300 K.

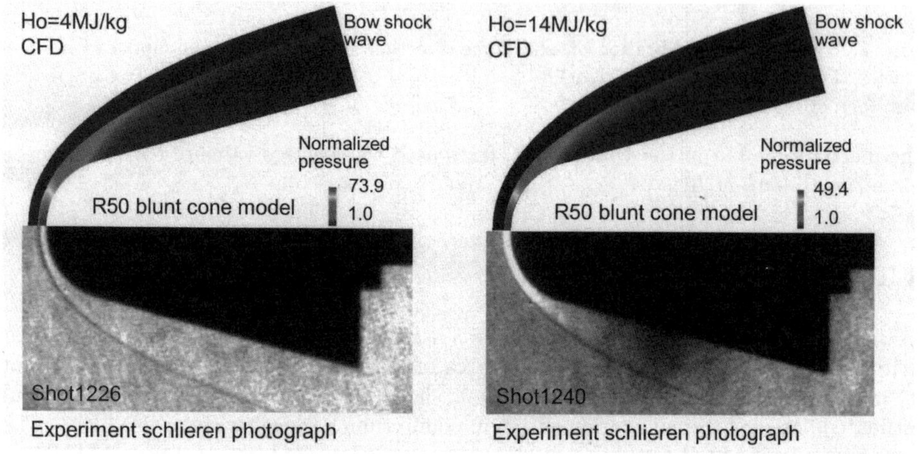

Fig. 3. Comparison of bow shock wave with CFD and Schlieren photograph at H_0=4 MJ/kg (left) and H_0=14 MJ/kg (right)

6 Result and discussion

For the R50 model, the pressure contour obtained by CFD and the experimental Schlieren pictures at test condition A and E are shown in left part and right part of Fig. 3, respectively. The Schlieren pictures were taken with the high speed camera (HADLAND, IMACON468). In the case of condition A, the shape of the shock wave in CFD and experiment agreed, and the shock stand-off distance at the nose of a model was about 6.0 mm. In the case of condition E, the shock stand-off distance in experiment was about 5.2 mm - 5.4 mm, and that in CFD was 5.0 mm. For the R20 model, the left part of fig. 4 shows the signal output of the axial force of a balance kx and the inertia force ma in the case of condition C and the reconstructed axial force f. The right part of fig. 4 shows the time history of an axial force coefficient C_A and the Pitot pressure measured at the model nose. The axial force of a balance kx shows the opposite phase to the inertia force ma. The reconstructed axial force f and Pitot pressure were settled down from about 2 ms. The test window for 2 ms was chosen from here. The result of CFD is also shown in these figures. The reconstructed axial force f is about 420 N at 3 ms in the test window, and this value agrees with the results of CFD. The Pitot pressure is slighty lower than the results of CFD. The axial force coefficient C_A was estimated from reconstructed axial force f, the dynamic pressure of the flow q and model base area A. In the test window, C_A of the test results agree with the results of CFD. For the R50 model, the test results and CFD results in the case of condition C are shown in Fig. 5. The reconstructed axial force f and Pitot pressure were settled down from about 2 ms like the case of R20 model. The reconstructed axial force f is about 670 N at 3 ms, and the result is increasing about 1.6 times to the force of R20 model on the same test condition. However, as compared with the results of CFD, it is low about 10%. On the other hand, the Pitot pressure which became settled from about 2 ms agree with the results of CFD. As a result, C_A of the test results in the test window is lower than the results of CFD. Over the test range, the test results of C_A for the R20 model and the R50 model are shown in Fig. 6 to viscous interaction parameter $C^{1/2}M/Re^{1/2}$, respectively with the results of CFD. Where, M, Re, and C are Mach number, the unit Reynold's number, and the constant, respectively. The experimental C_A was estimated by the value which averaged between test windows. The test window is 2 ms on condition A to condition D, and is 1 ms on condition E. The experimental C_A for the R20 model becomes high to the increase in the viscous interaction parameter, and the inclination is in the tendency which becomes small from H_0=10 MJ/kg. This tendency of the test results agreed also with the tendency in calculation of CFD. On the other hand, although the experimental C_A for the R50 model is agreed with the results of CFD only on condition A, experimental C_A on the

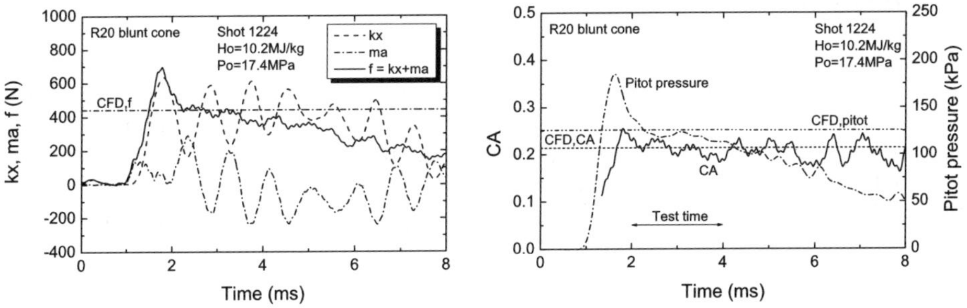

Fig. 4. Trace of axial force (left), Pitot pressure and C_A (right) for R20 blunt cone model at H_0=10 MJ/kg condition

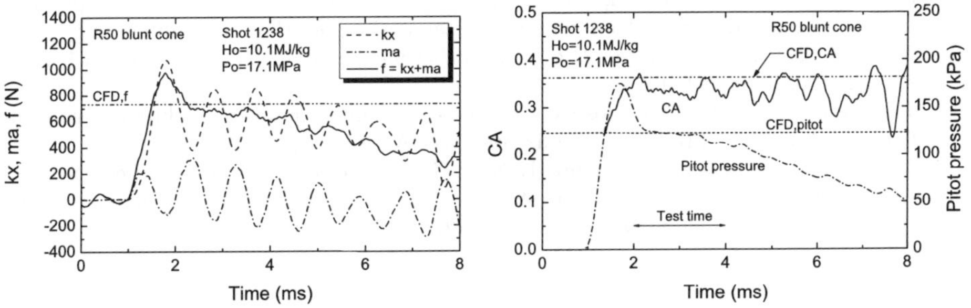

Fig. 5. Trace of axial force (left), Pitot pressure and C_A (right) for R50 blunt cone model at H_0=10 MJ/kg condition

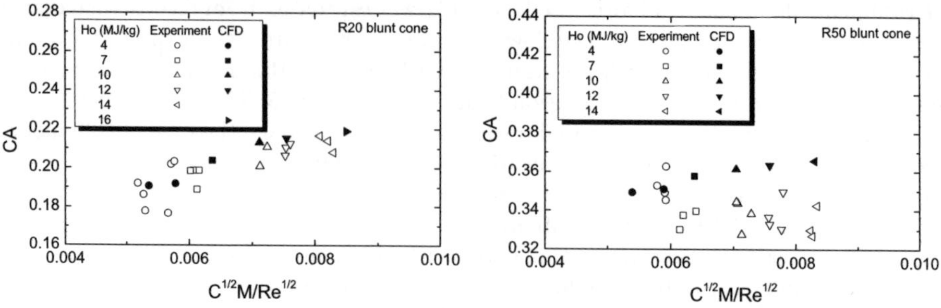

Fig. 6. Relation between axial force coefficient C_A and viscous interaction parameter $C^{1/2}M/Re^{1/2}$ for R20 blunt cone model (left) and R50 blunt cone model (right)

higher stagnation enthalpy beyond it are considerably smaller than the results of CFD. This result shows the tendency which shifts from an effect of viscous interaction with the increase in stagnation enthalpy conditions, and differs from the tendency of the test results and CFD results for R20 model. The difference of this result should be investigated further by doing the test which changed the curvature radius of blunt cone model.

7 Conclusion

The force measurement test of two blunt cone models was carried out in the high enthalpy shock tunnel HIEST. The result was compared with the result of CFD and discussed. The test results of C_A for R20 blunt cone model became high with the increase in a viscous interaction parameter, and agreed with the tendency of CFD. The test results of C_A for R50 blunt cone model shows the tendency which shifts from an effect of viscous interaction with the increase in stagnation enthalpy conditions, and were smaller than the results of CFD.

References

1. K. Itoh, S. Ueda, H. Tanno, T. Komuro, K. Sato: Hypersonic aerothermodynamic and scramjet research using high enthalpy shock tunnel. Shock Waves **12**, (2002) pp 93-98
2. K. Sato, T. Komuro, H. Tanno, S. Ueda, K. Itoh, S. Kuchiishi, S. Watanabe: An experimental study on aerodynamic characteristics of standard model HB-2 in high enthalpy shock tunnel HIEST. In: *Proc. of the 24th ISSW at Beijing, china July 11-16, 2004*, vol 1, ed by Z. Jiang (2004) pp 263-268
3. J.A. Lordi, R.E. Mates, J.R. Mossele: Computer program for numerical solution of nonequilibrium expansions of reacting gas mixtures. NASA CR-472 (1966)
4. T. Hashimoto, S. Rowan, T. Komuro, K. Sato, K. Itoh, M. Robinson, J. Martinez-Schramm, K. Hannemann: Application of HEG static pressure probe in HIEST. In: *Proc. of the 26th ISSW at Göttingen, Germany July 15-20, 2007*

Investigations of separated flow over backward facing steps in IISc hypersonic shock tunnel

P. Reddeppa, K. Nagashetty, and G. Jagadeesh

Department of Aerospace Engineering,
Indian Institute of Science, Bangalore-560012 (India)

Summary. Two backward facing step (2 mm and 3 mm step height) models are selected for surface heat transfer measurements. The platinum thin film gauges are deposited on the Macor inserts using both hand paint and vacuum sputtering technique. Using the Eckert reference temperature method the heating rates has been theoretically calculated along the flat plate portion of the model and the theoretical estimates are compared with experimentally determined surface heat transfer rate. Theoretical analysis of heat flux distribution down stream of the backward facing step model has been carried out using Gai's non-dimensional analysis. Based on the measured surface heating rates on the backward facing step, the reattachment distance is estimated for 2 and 3 mm step height at nominal Mach number of 7.6. It has been found from the present study that for 2 and 3 mm step height, it approximately takes about 10 and 8 step heights downstream of the model respectively for the flow to re-attach.

1 Introduction

Identification of the exact location of separation point, reattachment point and length of separation are very critical for characterizing the separated hypersonic flow around the bodies. Two-dimensional backward facing step model is a simple geometry that is useful to steady the separated flow features at hypersonic speeds. Some experimental studies carried out in different hypersonic ground test facilities using backward facing step model are reported in open literature [1,3,6].Correlations on heat flux distribution downstream of the backward facing step have been developed[2,3] by few research groups. The disagreement on the nature of heat transfer at reattachment of a backward facing step is still not clear. Moreover, there is no experimental data on heat transfer rate distribution in the separated flow around a step at enthalpies of 0.5 to 3.0 MJ/kg.

2 Objectives

The main focus of the present paper are the following,
1. To measure the surface convective heat transfer rates on a backward facing step model at Mach 7.6 and compare the experimentally measured data with the predicted values from available theories.
2. Identification of the reattachment point location on the backward facing step model from the measured heat transfer rates.
3. To determine suitable correlating parameters using the measured heat transfer rates and characterize the flow down stream of the backward facing step models at hypersonic speeds.

3 Test Models and Experimental Conditions

The schematic diagram and the photographs of the 2 mm and 3 mm step backward facing step models along with the platinum thin film gauges are shown in the Fig 1. The 3 mm step height model is 122 mm long with an upstream flat plate length of 72 mm and width of the model is 60 mm. Total of 23 platinum thin film heat transfer gauges are used for surface convective heat transfer measurements along the length of the model. The 2 mm step height model is 100 mm in length which is slightly smaller than the above configuration and at the same time the ratio of upstream flat plate length to step height has been maintained (24) for both the step models, with upstream flat plate portion is 48 mm long and width of the model is 85 mm. Total of 18 platinum thin film gauges are used for surface convective heat transfer measurements along the length of the model. The surface heating rate measurements are carried out in the IISc hypersonic shock tunnel (HST2) at a Mach number of 7.6 and enthalpy of ~2.3 MJ/kg and test time in the tunnel is ~1 ms. The experimental free stream conditions used in the study are given in the Table 1.

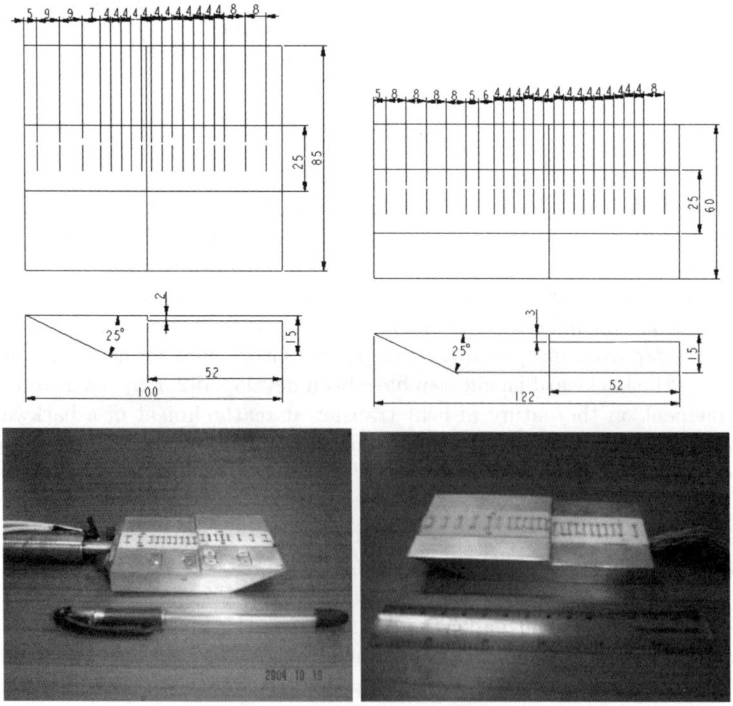

Fig. 1. Schematic diagrams and photographs of the 2 and 3 mm (step height)backward facing step models, used in the present study.

Table 1. Typical experimental conditions in the IISc hypersonic shock tunnel HST2

$H_0\left[\frac{MJ}{kg}\right]$	T_0 [K]	$\rho_\infty\left[\frac{kg}{m^3}\right]$	p_∞[Pa]	T_∞[K]	P_0[MPa]	Ma_∞	$u_\infty\left[\frac{km}{s}\right]$	$Re_\infty\left[\frac{1}{m}\right]$
2.4	2367	0.00473	257	189	1.76	7.6	2.1	7.764x10^5

4 Results and Discussion

4.1 Surface heating rates on backward facing step model at hypersonic speeds

On an average 3 to 5 experiments were carried for a given test condition and model configuration to evaluate the repeatability of the measurements. Out of them, the results from three good runs are taken for the analysis and where the deviations are found to be small, the average of 3 runs is considered in the analysis. The measured variation of the Stanton number along the length of the backward facing step model of 2 and 3 mm step heights, are shown in Figs 2 and 3 respectively.

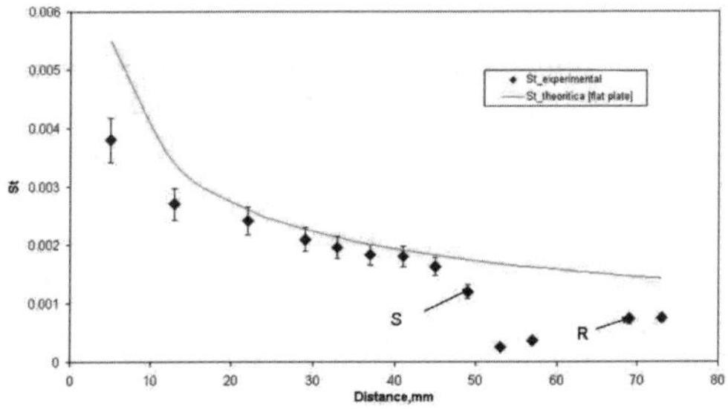

Fig. 2. Comparison of experimental Stanton number with theoretical estimate of flat plate Stanton number on backward facing step (2mm step height) at Mach 7.6.

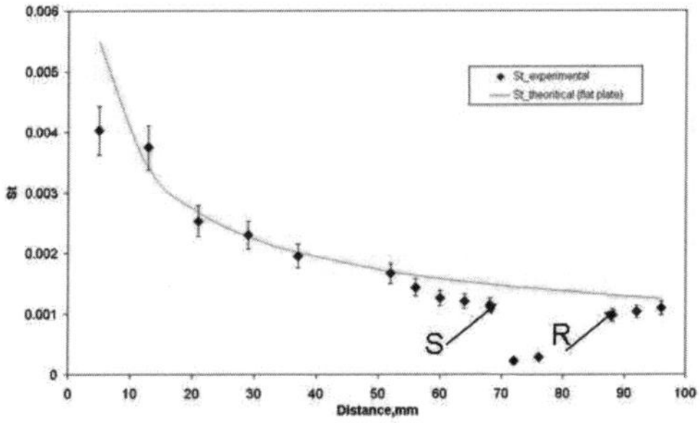

Fig. 3. Comparison of experimental Stanton number with theoretical estimate of flat plate Stanton number on backward facing step (3mm step height) at Mach 7.6.

Theoretical heat transfer rate is calculated using the Eckert [4] reference temperature method assuming the flat plate throughout the length of the model. In the down stream of step locations, we could not record the output from all the gauges because of the very small convective heat transfer rates in the separation bubble. The flow separates from corner of the step and reattaches in the down stream of the step. The separation and reattachment points are also shown in the same diagram.

4.2 Theoretical considerations for flow down stream of the step

Gai [3] has developed the heat flux functional relation based on Chapman's[2] separated flow analogy to determine the functional dependence of the heat flux down stream of the rearward facing step model using dimensional analysis. To determine what parameters must be matched for similarity in Stanton number distributions down stream of a backward facing step model and assuming that step is orthogonal to the flat plate surface we can write

$$St(x'/h) = f[M_\infty, Pr, \gamma, \alpha, Re_l, T_W/T_0, \delta_l/h, \delta_l/r_d] \tag{1}$$

This functional relation ship down stream of the step can be simplified as,

$$St(x'/h) = f[V_\infty^*/M_\infty \tau] \tag{2}$$

Where,the viscous interaction parameter and the well known hypersonic small disturbance parameter are given by $V_\infty^* = (M_\infty^* \sqrt{C^*})/Re$ and $M_\infty\ \tau$ (where, $\tau = h/L$)respectively. Therefore, for hypervelocity flows, the heat transfer behind a small step is dependent on the viscous interaction parameter and the hypersonic small disturbance parameter. Under these assumptions, the heat transfer behind a rearward facing step (where, $\tau << 1$) in hypervelocity flow is a function of $[V_\infty^*/M_\infty\ \tau]$for a low to moderate Reynolds numbers. Note that this is identical to Chapman parameter in moderate to high supersonic Mach number flows. Gai et.al. [3] also found that, if the viscous interaction parameter is matched between different data sets, then the type of distribution and the level of heat transfer should also match.

The non-dimensional experimental Stanton number data are plotted in Fig.4 along with important experimental results of Gai et.al [5] [and Wada and Inoue [6].The similarity values of viscous interaction parameter and hypersonic small disturbance parameters are tabulated in Table 2.

Table 2. Viscous interaction parameter and small disturbance theory similarity parameters obtained by other researchers along with the values for the present shock tunnel experiments for backward facing step model

Test	$V_{\infty,L}^*$	M_∞	$V_{\infty,L}^*/M_\infty$
Gai et.al	0.019	0.918	0.02609
Wada and Inoue	0.016	0.867	0.01854
Present expts.	0.026	0.316	0.0824

All the data of measured heat transfer rates on the backward facing step follows the same trend even though they are generated in different experimental facilities. It is quite clear that at these moderate enthalpy levels after separation the flow essentially goes

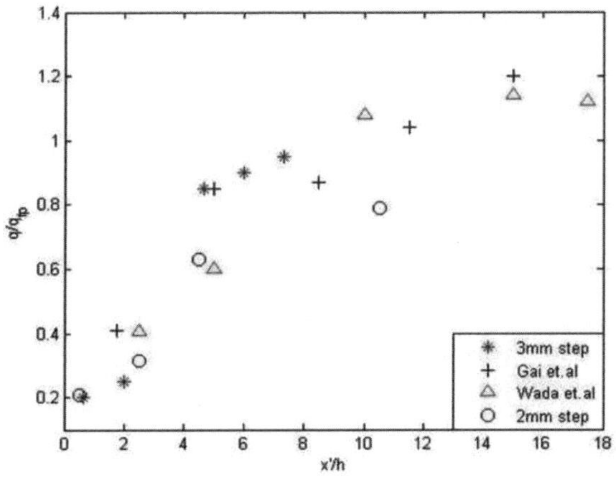

Fig. 4. Comparison of Gai, Wada & Inoue and present experiments (backward facing step) for similar values of the parameters and (Present Expts: M=7.6, hypersonic shock tunnel for 2 mm and 3 mm step height; Gai et al M=10, expansion facility; Wada & Inoue: M=10, Gun tunnel).

back to flat plate regime in a linear fashion and there are no sudden non-linear jumps. It implies that for the present experimental conditions at Mach 7.6 whenever the flow encounters small changes in the geometry it re-attaches in a linear fashion (i.e., without any spikes) in the measured values of surface convective heating rates. The results indicate that Gai's non-dimensional parameter group (viscous interaction and hypersonic small disturbance) can be correlated to backward facing step in the region of separated flows. The reattachment process is shown to take place over a length of 6 step heights after the initial rise in heat flux. The start of initial rise in heat fluxes itself takes ~4 step heights.

5 Conclusions

From the experimentally measured heat transfer rates we can clearly identify both the flow separation and re-attachment points in the hypersonic flow field around backward facing step models. Measured surface heat flux down stream of step (in the separated flow region) is compared with Gai's non-dimensional analysis. The heat flux values smoothly reach the flat plate value similar to Gai's prediction. It is shown in these experiments for the first time that Gai's theoretical similarity analysis for backward facing step geometry is also valid for moderate to low stagnation enthalpy flows. The flow takes nearly 10 and 8 step heights to reattach in case of 2 and 3 mm step height backward facing step models respectively.

References

1. Rom, J. and Seginer, A., [1964], "Laminar heat transfer to a two-dimensional backward facing step from the high-enthalpy supersonic flow in the shock tube, AIAA J, Vol.2, pp. 251-255.

2. Chapman, D. R., "A theoretical analysis of heat transfer in regions of separated flow", Tech rep., NACA TN 3792, 1956.
3. Hayne, M. J., Mee, D. J., Morgan, R. G., Gai, S. L., and Mcintyre, "Heat transfer and flow behind a step in high enthalpy superorbital flow", The Aeronautical Journal, pp. 435-442, July2003.
4. Eckert, E., "Engineering relations for friction and heat transfer to surfaces in high velocity flow", Journal of Aeronautical Sciences, Vol. 22, No. 8, pp. 587-587, 1955.
5. Gai S. L., Reynolds N. T., Ross C. and Baird J. P., [1989] " Measurements of heat transfer in separated high-enthalpy dissociated laminar hypersonic flow behind a step", J. Fluid Mech., Vol. 199, pp. 541-564.
6. Wada, I., and Inoue, Y., [1973], "Heat transfer behind the backward facing step in the hypersonic flow", Proceedings of the tenth International Symposium on Space Technology and Science, pp. 425-432.

Measurement of aerodynamic forces for missile shaped body in hypersonic shock tunnel using 6-component accelerometer based balance system

S. Saravanan, G. Jagadeesh, and K.P.J. Reddy

Department of Aerospace Engineering, Indian Institute of science, Bangalore, India

Summary. This paper reports the basic design of a new six component force balance system using miniature piezoelectric accelerometers to measure all aerodynamic forces and moments for a test model in hypersonic shock tunnel (HST2). Since the flow duration in a hypersonic shock tunnel is of the order of 1 ms, the balance system [1] uses fast response accelerometers (PCB Piezotronics; frequency range of 1-10 kHz) for obtaining the aerodynamic data. The balance system has been used to measure the basic aerodynamic forces and moments on a missile shaped body at Mach 8 in the IISc hypersonic shock tunnel. The experimentally measured values match well with theoretical predictions.

1 Introduction

In planning missile mission, it is necessary to make accurate prediction (measurements) of the aerodynamic forces acting on a missile shaped model both during its ascent and descent stages. The aerodynamic force measurements are necessary for determining many flight vehicle parameters that are important for performance and stability of the vehicle. Various techniques have been used by the researchers to measure the force components in impulse facilities which include stress wave balance, semiconductor strain gauges, piezoelctric pressure transducers, pressure sensitive paints, free flying technique and accelerometer balance system. In a stress wave force balance system [2] the aerodynamic loading may be determined by interpretation of the stress waves which travel through the supports. Although this technique is successful in impulse facilities, it needs a relatively large model size so that the reflected stress waves from the model end do not interact with the measured signal. Moreover, exhaustive static as well as dynamic calibrations needs to be done prior to actual model testing in impulse facilities. Alternatively, the individual values of the pressure signals obtained from piezo electric pressure transducers [3] can also be resolved in particular direction and then integrated over the model surface to get the total aerodynamic forces on the body. The spatial resolution of the transducers on the model surface and the process of resolving the pressure and integrating it all over the model surface may introduce errors into the calculation of aerodynamic forces. Joshi and Reddy [4], used an accelerometer balance system (three-component) rigidly attached to the inside wall of hollow model and this model in turn is attached to the central sting by 2 rubber springs. During the tunnel run, the hollow model along with the 3 accelerometers will move and the resulting accelerations in the vertical and axial directions are given by the accelerometers outputs. However the balance is limited to measure only the fundamental aerodynamic coefficients. Hence, an attempt has been made in high enthalpy laboratory to initiate the design and development of six component accelerometer balance system and also to evaluate the performance of the balance for a known model configuration, like, missile shaped body. In this back drop, six component accelerometer balance system has been designed and fabricated for measuring aerodynamic forces on a missile shaped body, in this study.

2 Experimental facility and instrumentation

Experiments were conducted in IISc hypersonic shock tunnel. The tunnel was operated with reflected mode at stagnation pressure and temperature of 2 MPa and 2000 K, respectively. Unit Reynolds number of 1.0 Million per meter was achieved for the above reservoir condition. The test section has a square chamber having dimensions of 300mmX300mmX450mm length with circular optical quality glass windows on either side of it for flow visualization. Very high speed DAS, namely NI PXI-6115 has been used in order to acquire adequate data within the available run time. It has 24 independent data recording analogue channels with each channel having a speed of 10 MHz sampling rate and 64MB onboard memory and interfaced with a computer.

3 Accelerometer balance fabrication and model configuration

In the present study an internally mounted six component accelerometer force balance [1] was designed and fabricated to measure all the six fundamental forces and moments experienced by the missile shaped body flying at hypersonic Mach number of 8 (Fig.1). The newly designed 6-component balance system consists of 6 accelerometers (PCB Piezotronics, Models 352C67 and 303A, frequency range of 10 kHz)in which four are mounted on the balance system while two are mounted on the body (one is mounted along the axis of the model and another one in the base of the circumference of the model). The system involves measuring simultaneously the acceleration of the model in 3 normal directions in addition to rotational acceleration of the model. The fundamental equations have been derived [1] for computing aerodynamic forces and coefficients from the measured accelerations. The test model is a combination of cone, cylinder, frustum and triangular fins geometry and is shown in Fig.1. Duralumin has been used as the material for the fabrication of model. The model is a blunted cone, which has a total length of 40.8 mm with vertex angle of 41°, after body with a length of 111 mm, frustum with 75 mm length and vertex angle of 10° and four triangular fins. The blunt nose has a radius of 15 mm. The base diameter and overall length of the model are 51 and 227 mm, respectively. The model weighs around 0.552 kg and has its center of gravity located along the axis at a distance of 75 mm from the base.

4 Accelerometer balance theory

The basic principle of the accelerometer force balance system is based upon the behavior of spring mass system with single degree of freedom. Assuming that all the forces are varying with time and axial force acts through C.G. of the model. In addition, springs are linear and offer no resistance and no damping force in the system. The main principle of this design is that the model along with the internally mountable accelerometer balance system are supported by rubber bushes [5] and there-by ensuring unrestrained free floating conditions of the model in the test section during the flow duration and the resulting accelerations are measured from the outputs of the corresponding accelerometers. Thus, the aerodynamic coefficients are computed using the following relations and from the accelerations measured by the axial force, front lift, aft lift, side force 1, side force 2 and rolling moment accelerometers.

$$N(t) = m\left(\frac{b\xi_1 + a\xi_2}{a+b}\right) \tag{1}$$

$$C(t) = m * \xi_3 \tag{2}$$

$$e_N = \left(\frac{I}{m}\right)\left(\frac{\xi_1 - \xi_2}{b\xi_1 + a\xi_2}\right) \tag{3}$$

$$C_m = \left(\frac{N(t)}{q_\infty S}\right)\left(\frac{(x_{cg})_{base} + e_N}{D_B}\right) \tag{4}$$

$$Y(t) = m\frac{b\xi_4 + a\xi_5}{a+b} \tag{5}$$

$$e_Y = \left(\frac{I}{m}\right)\left(\frac{\xi_4 - \xi_5}{b\xi_4 + a\xi_5}\right) \tag{6}$$

$$C_Y = \left(\frac{Y(t)}{q_\infty S}\right)\left(\frac{(x_{cg})_{base} + e_Y}{D_B}\right) \tag{7}$$

$$C_{R,M} = \left(\frac{m\xi_6}{2}\right)\left(\frac{1}{q_\infty S}\right) \tag{8}$$

Where, ξ_1 (front lift), ξ_2 (aft lift), ξ_3 (drag), ξ_4 (side force 1), ξ_5 (side force 2)and ξ_6 (rolling moment) are the measured accelerations at the indicated location in the balance system and in the model, a is the distance between centre of gravity of the model and c.g of front lift and side force 1 accelerometers, b is the distance between centre of gravity of the model and c.g of aft lift and side force 2 accelerometers. Here, m and I are the mass and moment of inertia of the model, respectively. q_∞ is the free stream dynamic pressure and S is the reference area.

5 Results and discussion

For the present investigation, helium is used as the driver gas to generate required enthalpy flow and air is used as the test gas. The free stream conditions in the test section during the experiments is shown in Table 1. Two to three experiments at each condition are carried out to check the repeatability of the signals. All the experimental data are deduced by averaging results during the steady flow duration. The typical acceleration response obtained from the accelerometers mounted in the six-component balance system and in the model are shown in Fig.2. Variation of fundamental aerodynamic coefficients are plotted for a missile shaped body with angle of attack and is shown in Fig.3. The other three components, namely side force, yawing and rolling moment are also presented in the same Fig.3. The theoretical values of aerodynamic coefficients namely lift, drag and pitching moment for the missile model at different angles of attack are estimated by using the modified Newtonian theory [6], taking into account of centrifugal force effects over the hemispherical nose portion of the blunt cone of a missile. In order to compliment the experimental results, the commercially available CFX-Ansys 5.7 3-d compressible Navier stokes code has been extensively used to obtain the flow field features around a 41° apex angle missile shaped model with varying angle of attack at hypersonic Mach number of 8. The boundary conditions used in this simulation are based on the experimental free stream conditions in the test section of the shock tunnel. Numerically computed aerodynamic force coefficients over the test model with various angles of attack are calculated by integrating the surface pressure over the missile shaped body. From Figure 4, it is clearly seen that, the measured lift and pitching moment coefficients matches fairly well with Newtonian theory at all angles of attack compared to CFD prediction. In the case of drag component, the experimental values are low when compared with theory whereas at higher angle of attack, it matches with theory. For the symmetric configuration of chosen test model and with β and α equal to 0°, the lateral coefficients were essentially small and the measured data showed large side-force, yawing-moment and rolling moment coefficients and the results are plotted in the same figure. It is quite surprising to see that both side force and yawing moment coefficients are independent of angle of attack and the scatter in rolling moment coefficient is not clearly understood at this time.

6 Conclusion

The designed six-component accelerometer balance system is used to measure all aerodynamic force coefficients for missile shaped model, flying at hypersonic Mach number of 8. Fundamental aerodynamic results generated using this accelerometer balance system are in good agreement with predictions from modified Newtonian theory and CFD. Future research effects will be directed towards further improving the performance of the balance of the system.

References

1. Saravanan S, Jagadeesh G, Reddy KPJ (2004) A new 6-component accelerometer force balance for short duration ground testing facilities. 24th International Symposium on Shock Waves, Beijing, China, July 20-25.
2. Mee DJ, Daniel WJT, Simmons JM (1996) Three-component force balance for flows of millisecond duration. AIAA Journal, 34 (1), 590-595
3. Srulijes J, Gnemmi P, Runne K, Seiler F (2002) High-pressure shock tunnel experiments and CFD calculations on spike-tipped blunt bodies. AIAA Paper 2002-2918
4. Joshi MV, Reddy NM (1986) Aerodynamic force measurements over missile configurations in IISc shock tunnel at M = 5.5. Experiments in Fluids 4: 338-340
5. Sahoo N, Mahapatra DR, Jagadeesh G, Gopalkrishnan S, Reddy KPJ (2003) An accelerometer balance system for measurement of aerodynamic force coefficients over blunt bodies in a hypersonic shock tunnel. Measurement Science and Technology, 14: 260-272
6. Truitt RW (1959) Hypersonic Aerodynamics. The Ronald Press Company, New York

Table 1. Typical free stream conditions during the present experiments

M_∞	P_0 (bar)	H_0 (MJ/kg)	ρ_∞ (kg/m3)	p_∞ (pa)	Re_∞/m	T_∞ (K)
8.0	20.1	2.01	0.0052	212	$1.05e^6$	149

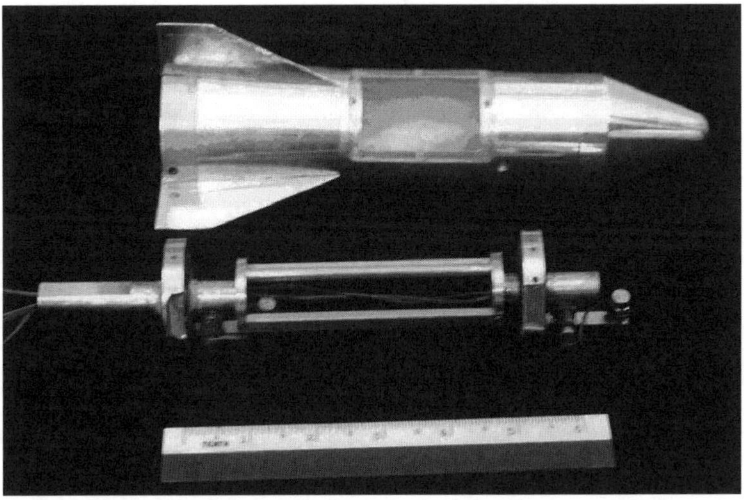

Fig. 1. Photograph of a test model with 5o flare and fins along with the six-component accelerometer force balance system

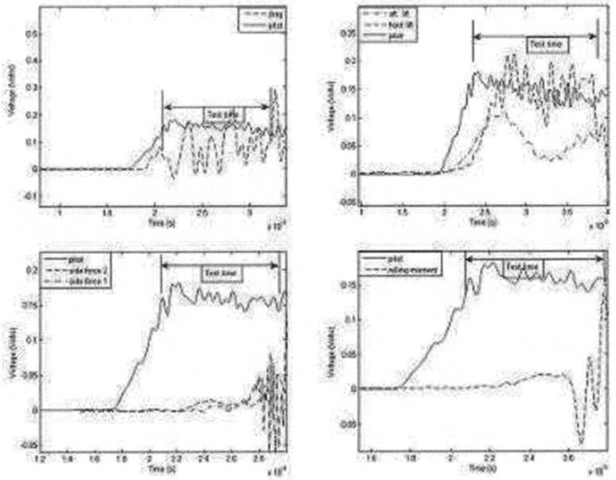

Fig. 2. Typical signals from various accelerometers mounted in the model and accelerometer balance system for the missile shaped body flying at Mach 8

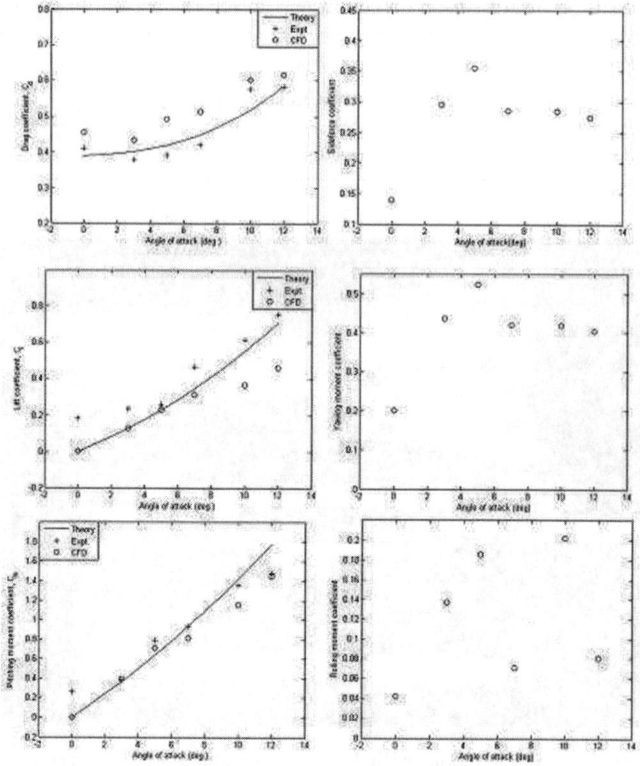

Fig. 3. Variation of aerodynamic coefficients for the 41 deg apex angle of missile model with blunt cone at Mach 8 for various angles of attack.

Measurement of shock stand-off distance on a 120° blunt cone model at hypersonic Mach number in Argon

K. Satheesh and G. Jagadeesh

Dept. of Aerospace Engineering, Indian Institute of Science, Bangalore - 560 012,India

Summary. The flow around a 120° blunt cone model with a base radius of $60mm$ has been visualised at Mach 14.8 and 9.1 using argon as the test gas, at the newly established high speed schlieren facility in the IISc hypersonic shock tunnel HST2.The experimental shock stand off distance around the blunt cone is compared with that obtained using a commercial CFD package.The computed values of shock stand off distance of the blunt cone is found to agree reasonably well with the experimental data.

1 Introduction

Location of the bow shock ahead of the blunt body dictates the near flow field features around bodies in hypersonic flow that are usually dominated by viscous and entropy swallowing effects.Shock stand off distance is a very convenient parameter for validating any analytical or numerical model before using them for predicting the other flow field features.In recent times new hypersonic drag reduction techniques such as upstream energy deposition is being proposed.It is desirable to conduct the preliminary investigations of these concepts using mono-atomic gases like argon, since only the excitation of the translational energy mode plays a dominant role in these processes.But in the open literature there is very little information on shock stand off distances in argon especially at high Mach numbers(8 and above).In the present study, the hypersonic flow around a blunt body has been visualised in a shock tunnel using a high speed camera.The visualised shock stand-off distance around the blunt cone is compared with the results from a computational fluid dynamic study.

2 The experimental facility

The experiments described here are conducted in the IISc hypersonic shock tunnel HST-2, which consists of a stainless steel shock tube of 50mm diameter connected to a convergent-divergent conical nozzle of 300mm exit diameter, with a 450mm long test section of 300mm x 300mm square cross section [1]. The free stream conditions under which the experiments are carried out is listed in Table.1, and under these operating conditions, HST-2 generates a steady hypersonic flow for a period of almost $1.1msec$.These flow conditions are reproducible within $\pm 5\%$ accuracy.

The experimental study has been carried out using a test model with a 60° half angle fore-body with a blunt nose as the test model,which represents a conventional re-entry capsule configuration.

Table 1. Typical free stream conditions in the shock tunnel

No.	M_∞	$P_\infty(pa)$	$\rho_\infty(kg/m^3)$	$T_\infty(k)$	$u_\infty(m/sec)$	Test gas
R1	14.8	27	0.005	26	1432	*Argon*
R2	9.1	556	0.0357	75	1468	*Argon*

2.1 The schlieren set-up

The schlieren imaging technique is a widely used method for visualising inhomogeneities in transparent gaseous medium. This technique works by making use of the refraction of light due to the spatial variations in the refractive index of the media under study.A parallel beam of light is allowed to pass through the test area, after which it is focused on a knife edge.Due to the inhomogeneities in the media, the light from different parts of the test area will be refracted by different amounts and thus will not focus at the same point. The knife edge then selectively blocks the light rays coming from different parts of the test area; depending on its orientation and the amount of refraction undergone by the light rays.As a result of this selective blockage of light, the inhomogeneities(density changes) in the test area shows up as intensity gradients in the resultant image; the contrast (the fractional change in intensity with respect to the background intensity)of which at a point being directly proportional to the density gradient at the corresponding point in the test flow [2]. A 'z-type' schlieren set-up (Fig.1) was used for the visualisation of the hypersonic flow field; wherein a pair of concave mirrors of focal length of $96''$ are used to obtain the parallel beam of light and then for focusing it on the knife edge.

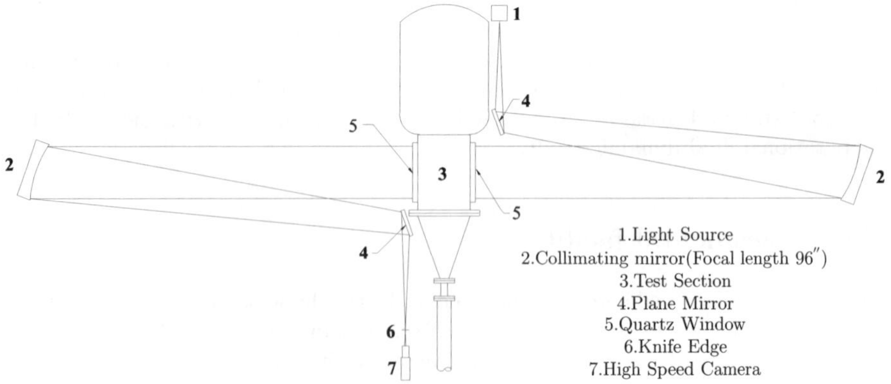

Fig. 1. Schematic diagram of the high speed schlieren visualisation set-up used in the present experiments

In the present experiments a high speed camera (Ms Vision Research USA) that is capable of recording 6,688 frames per second at full resolution of 800 x 600 pixel using a SR-CMOS imaging array is used to record the images. As we keep increasing the recoding speed the pixel resolution reduces and hence for the given condition we have to optimise the operational conditions of the camera. This camera has the continuous adjustable resolution feature that enables us to enhancing the speed of the camera with adjustable 16 x 8 pixel increments. In the present experiments the schlieren images have been recorded using by operating the camera at 10000 frames per second speed with a resolution of 450×450 pixels . A 300 Watts halogen lamp is used as the continuous light source.Operation of the camera is synchronised with the shock tunnel flow using a trigger pulse generated by the pressure sensor located at the end of the shock tube.The light is switched on before the experiments and the camera is triggered to synchronise with shock tunnel experiments.

3 Results and Discussions

The schlieren images recorded by the high speed camera during the experiment reveals the various features of the shock tunnel flow.Figure. 2 is a set of sequential photographs recorded during a typical shock tunnel experiment. The shock stand-off distance

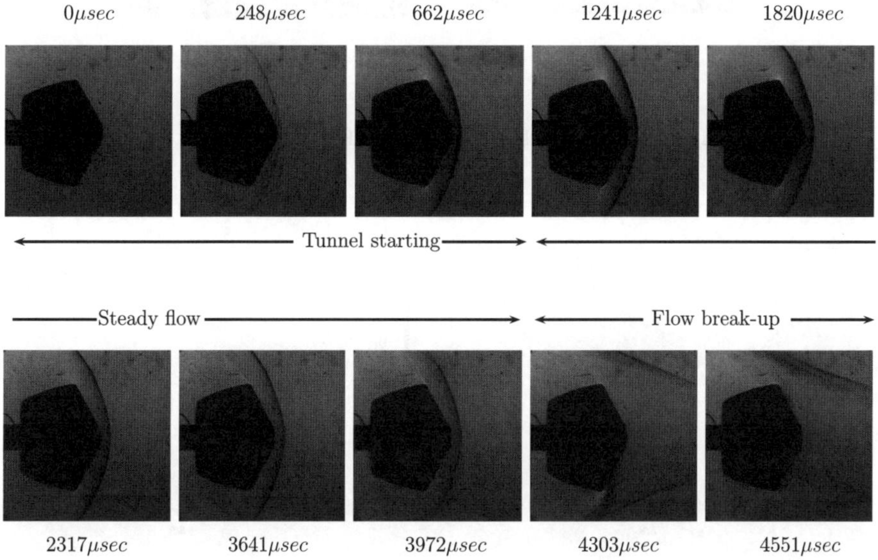

Fig. 2. The dynamics of shock tunnel flow evolution corresponding to condition R1

especially along the stagnation streamline is a unique signature of the body geometry and the flow Mach number [3]. In fact most of the blunt body hypersonic flow fields are characterized by the shock stand-off distances ahead of the blunt body. In order to quantify the results from the high speed visualisation studies the shock stand-off distance along the length of the blunt cone was measured from the schlieren images. In order to complement the experiments the hypersonic flow around the blunt cone was also simulated by solving the Navier-Stokes equation in the axi-symmetric formulation using the commercial CFD (Computational Fluid Dynamics) package CFX 5.7. The experimentally visualised bow shock wave in front 120° blunt cone model along with the results from the computational study is shown in figure(Fig.3). The variation of the shock stand-off distance along the surface of the blunt cone obtained from both experiments and CFD studies are shown in Fig 4.A reasonably good agreement can be observed between the experimental and computed values.

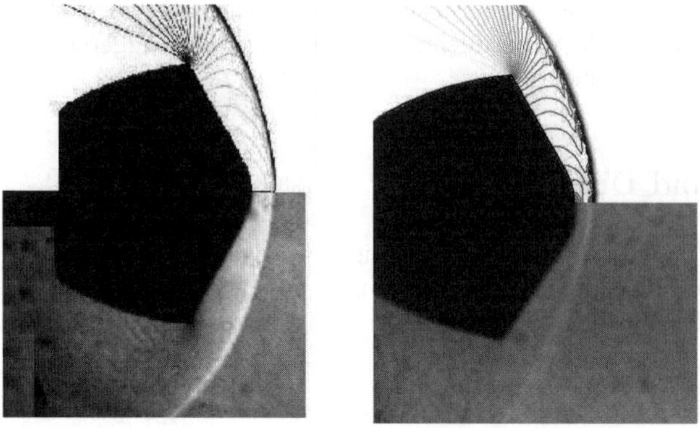

(*a*) (*b*)

Fig. 3. The flow field. (**a**) Condition R2. (**b**) Condition R1

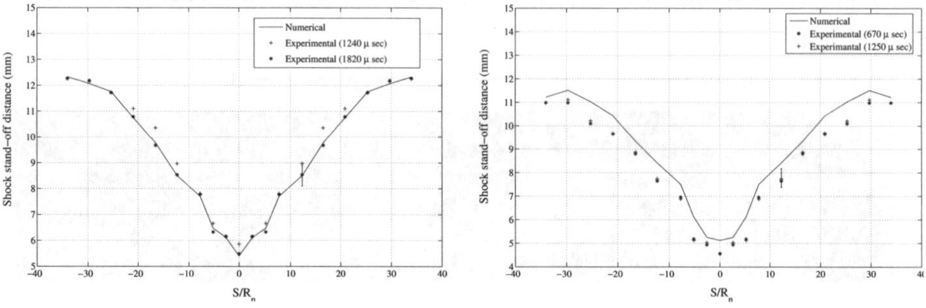

(*a*) (*b*)

Fig. 4. The shock stand-off distance. (**a**) Condition R2. (**b**) Condition R1

References

1. Reddy, N.M, Jagadeesh G, K. Nagashetty and Reddy. K.P.J:Journal of Indian Academy of Sciences, **21**,6 (1996).
2. Settles.G.S:*Schlieren and shadowgraph techniques*(Springer,2001)
3. Olivier.H:Journal of Fluid Mechanics, **413** (2000).

Model for shock interaction with sharp area reduction

J. Falcovitz[1] **and O. Igra**[2]

[1] *Institute of Mathematics, The Hebrew University of Jerusalem, Israel*
[2] *Department of Mechanical Engineering, Ben-Gurion University of the Negev, Beer-Sheva, Israel*

Summary. We model analytically the interaction of a planar shock wave reflected from a sharp reduction in a conduit cross-section area, with very large area ratio. Previously published studies for shock interaction with an area change assumed a short smooth connecting nozzle, where at large times a steady flow evolves. We assume that the flow converges through the sharp area transition via a smoothly-formed quasi-one-dimensional streamtube, reaching sonic (choked) or subsonic (unchoked) velocity. In the chocked case a CRW is required to match pressures between the sonic flow and the transmitted shock state. The model leads to a pair of equations for velocity and pressure at a contact discontinuity, analogous to the wave interaction curves that resolve a Riemann problem. We find that the overpressure amplification ratio of the transmitted shock is in the range of $1.5 - -2.0$ for very strong and very weak incident shock, respectively. A good agreement is obtained between the model predictions and two-dimensional GRP simulations.

1 Introduction

Shock wave propagation in conduits having varying cross-section is frequently encountered in engineering design. For example in an accidental explosion in coal mines. The flow field resulting from the interaction of an incident shock wave with a segment of area contraction or enlargement in a conduit was studied by [3]. A similar study was performed for the interaction of a centered rarefaction wave (CRW) with an area reduction segment in ducts [2]. In both cases two long uniform area conduits were joined by a segment of smoothly varying cross-section area. The analytical model for predicting the (large-time) reflected/transmitted waves was based on the key assumptions of a *steady quasi-one-dimensional flow in the connecting segment*, and that all waves involved were *simple waves.*

Restricting the incident wave to either a shock or a just-initiated CRW, produces a flow characterized by a vanishing length-scale. Therefore, the only length-scale related to the wave interaction with the area-change duct segment is its length L_d. Hence, when the the distance between the transmitted and reflected waves increases relative to L_d, they approach the simple waves envisioned by the aforementioned theory.

Evidently, an abrupt (discontinuous) area change cannot be treated within this framework. In the absence of a smooth connecting duct, how can an alternate characteristic length be associated with the flow through the sharp entrance to the narrow duct? Having inspected 2-D (planar or axisymmetric) simulations of this flow process, we observed that at large times the flow immediately downstream the entrance was approximately steady. Thus, in the absence of a smooth converging duct segment, the entrance flow streamtubes conform to a virtual area-change segment whose length L_v is proportional to the diameter of the narrow conduit D_2. The ratio L_v/D_2 depends on the incident shock intensity, and was found to be of $O(1)$.

The existence of such "virtual duct flow" implies that the present model can be formulated, in analogy with the previously mentioned wave interaction theory, as a flow field comprising a reflected shock and a self-similar transmitted wave, separated by a segment of steady flow. Referring to the schematic diagrams in Fig. 1, the expanding space between the reflected shock front and the endwall (marking the sharp area contraction) is regarded as a virtual plenum chamber consisting of quiescent fluid. This large volume of high-pressure fluid feeds a steady converging flow into the narrow conduit. As a simplifying assumption, the narrowest streamtube in that flow is taken as the narrow conduit itself (ignoring any further contraction resulting from flow separation immediately downstream of the sharp entry).

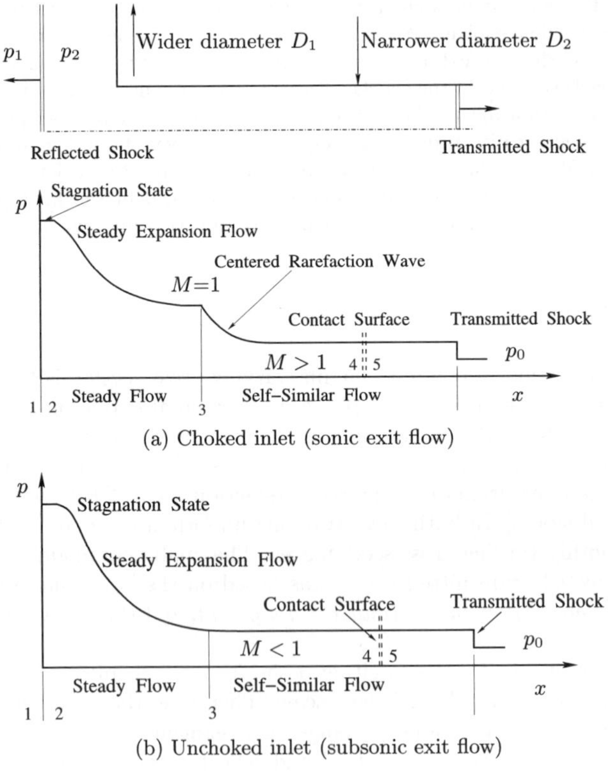

Fig. 1. Flow model downstream of duct entrance: (a) Sonic case; (b) Subsonic case

2 Outline of the flow model

The flow evolving at the entry to the narrow duct and downstream is depicted schematically in Fig. 1. There are two cases: in case (a) the flow at the interface between the steady and self-similar flow regimes is choked (sonic); in case (b) the flow through the steady/self-similar interface is subsonic. Referring to the index marking flow states on the model diagram, the flow process is described as follows.

- The incident shock produces a jump transition from the ambient state $(\cdot)_0$, to the elevated-pressure state $(\cdot)_1$.
- The reflected shock, approximated as a head-on reflection from a rigid wall, produces the higher-pressure quiescent state $(\cdot)_2$, which serves as an effective plenum chamber driving the flow into the narrow duct.
- A steady subsonic expansion flow evolves between the virtual stagnation state $(\cdot)_2$ $(M_2 = 0)$ and the interface state $(\cdot)_3$ $(M_3 \leq 1)$.
- In the sonic case $(M_3 = 1)$ a centered-rarefaction-wave (CRW) evolves to bridge the pressure gap $(p_3 > p_4)$ between the interface and the contact surface states.
- In the subsonic case $(M_3 < 1)$ there is no pressure gap $(p_3 = p_4)$ and no CRW arises.
- A contact discontinuity separating states $(\cdot)_4$ and $(\cdot)_5$ is always formed.
- State $(\cdot)_5$ is produced by the transmitted shock propagating into ambient state $(\cdot)_0$.

Some remarks concerning this flow model are due. As in [4], where a CRW propagating through a smooth converging segment into a narrow conduit was studied, a stationary interface separating a (quasi-1D) steady flow segment from an expanding self-similar flow segment arises (here it is point 3). The significance of the contact discontinuity, not required in the case of an isentropic CRW [4], is that the state $(\cdot)_4$ fluid is produced by isentropic expansion flow from the "plenum chamber" $(\cdot)_2$, while state $(\cdot)_5$ is produced by the transmitted shock propagating into the ambient state $(\cdot)_0$. The entropy difference between these two states (separated by the contact discontinuity) is thus analogous to the entropy difference across the contact surface in a shock tube problem.

The analytic model is presented in some detail below. From the model equations it was found out that for a perfect gas having $\gamma = 1.40$ the sonic case requires incident shocks of $M_{s,i} > 1.7183$ (for weaker shocks the subsonic case prevails).

3 Flow model equations

The state behind the incident shock is obtained from the perfect gas Rankine-Hugoniot relations

$$p_1 = p_0 \left[1 + \tfrac{2\gamma}{\gamma+1}\left(M_{s,i}^2 - 1\right)\right], \qquad \rho_1 = \rho_0 \tfrac{\mu^2 p_0 + p_1}{p_0 + \mu^2 p_1}, \qquad u_1 = c_0 \tfrac{p_0 + p_1}{\gamma p_0 M_{s,i}}, \tag{1}$$
$$\mu^2 = \tfrac{\gamma-1}{\gamma+1;}, \qquad c_0 = \sqrt{\tfrac{\gamma p_0}{\rho_0}}\,.$$

The head-on reflected incident shock approximately produces state $(\cdot)_2$ as follows

$$p_2 = p_1 \tfrac{(2\mu^2+1)p_1 - \mu^2 p_0}{p_0 + \mu^2 p_1}, \qquad \rho_2 = \rho_1 \tfrac{\mu^2 p_1 + p_2}{p_1 + \mu^2 p_2}, \qquad u_2 = 0\,. \tag{2}$$

The steady isentropic expansion flow between stagnation state $(\cdot)_2$ and interface state $(\cdot)_3$ is

$$p_3 = p_2 \left[1 + \tfrac{\gamma-1}{2} M^3\right]^{-\frac{\gamma}{\gamma-1}}, \qquad \rho_3 = \rho_2 \left[1 + \tfrac{\gamma-1}{2} M^3\right]^{-\frac{1}{\gamma-1}}, \tag{3}$$
$$c_0 = \sqrt{\tfrac{\gamma p_3}{\rho_3}}, \qquad u_3 = M_3 c_3\,,$$

where in the sonic case $M_3 = 1$, but in the subsonic case the value of $M_3 < 1$ is unknown. Assuming the sonic case, a CRW separating states $(\cdot)_3$ and $(\cdot)_4$ is governed by the Riemann-Invariants relations

$$u_4 + \tfrac{2}{\gamma-1} c_4 = u_3 + \tfrac{2}{\gamma-1} c_3 \,, \qquad c_4 = \sqrt{\tfrac{\gamma p_4}{\rho_4}} \,, \qquad \tfrac{p_4}{\rho_4^{\gamma}} = \tfrac{p_3}{\rho_3^{\gamma}} \,. \tag{4}$$

Across the contact we have continuity of pressure and velocity

$$u_5 = u_4 \,, \qquad p_5 = p_4 \,, \tag{5}$$

and the relations across the transmitted shock are

$$M_{s,t} = \left(\tfrac{\gamma+1}{4}\right)\left(\tfrac{u_5}{c_0}\right) + \left[1 + \left(\tfrac{\gamma+1}{4}\right)^2 \left(\tfrac{u_5}{c_0}\right)^2\right]^{\frac{1}{2}} \,, \tag{6}$$

$$p_5 = p_0 \left[1 + \tfrac{2\gamma}{\gamma+1} \left(M_{s,t}^2 - 1\right)\right] \,, \qquad \rho_5 = \rho_0 \tfrac{\mu^2 p_0 + p_5}{p_0 + \mu^2 p_5} \,,$$

where by ρ, p, c, γ, u, M, $M_{s,i}$, $M_{s,t}$ we denote, respectively, the fluid density, pressure, speed of sound, polytropic index, flow velocity and Mach number, incident and transmitted shock wave Mach number.

In the sonic case ($M_3 = 1$) the system 2–7 is interpreted as follows. The expansion flow is readily evaluated from Eqs. 2–4 as function of the incident shock Mach number $M_{s,i}$. The remaining equations constitute a Riemann problem resolved by a right-facing shock, a left-facing CRW, the two waves being separated by a contact discontinuity. Having evaluated state $(\cdot)_3$, the sub-system 4–7 is solved by calculating the intersection of the left and right u, p interaction curves, thus producing the contact velocity and pressure.

In the subsonic case the CRW vanishes and states $(\cdot)_3$,$(\cdot)_4$ are identical. However, the contact discontinuity $(\cdot)_4$–$(\cdot)_5$ remains, and the pressure and velocity continuity relations 5 remain as well. The system 2–7 reduces to a kind of Riemann problem, where the right wave is the transmitted shock and the left wave is replaced by the steady expansion flow $(\cdot)_1$–$(\cdot)_3$. The state $(\cdot)_4 = (\cdot)_3$ is evaluated as function of M_3 using Eqs. 2–4. The post-shock state $(\cdot)_5$ is evaluated, as before, by Eq. 7. Here again, the solution is determined by solving the velocity and pressure continuity at the contact surface $(\cdot)_4$–$(\cdot)_5$.

4 Transmitted shock results

Let the intensity of the transmitted shock be represented by the pressure amplification factor

$$A_t = \frac{(p_5 - p_0)}{(p_1 - p_0)} \,. \tag{7}$$

The main physical results of our interaction model are displayed in Fig. 2 as the transmitted pressure amplification factor for a range of incident shock intensity encompassing both the sonic and subsonic cases. The calculations and data processing were performed by a simple code written specifically for this study, which is available at `http://www.bgu.ac.il/:80/me/laboratories/shockwave/` (look up Calculations Page). The code shows how the model equations were solved iteratively. For a strong incident shock the model predicts a lower bound of $A_t \approx 1.5$, while for a weak shock the asymptotic value

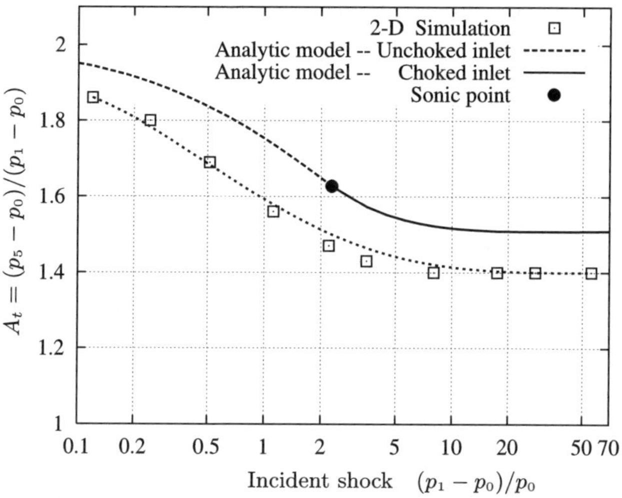

Fig. 2. Transmitted shock amplification factor as function of incident shock intensity

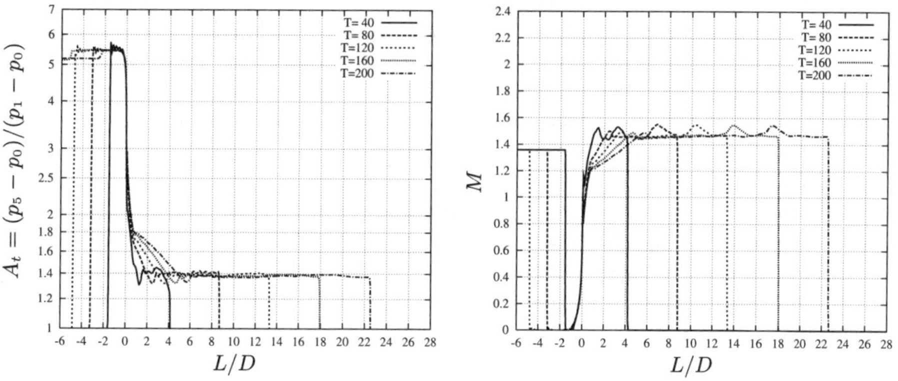

Fig. 3. Pressure amplification and flow Mach number profiles for incident shock $M_{s,i} = 3$.

is $A_t \approx 2$. This trend conforms to the classical feature that a weak incident shock (with nearly isentropic compression) is reflected with overpressure amplification factor of 2. On the other hand, a very strong incident shock is reflected head-on with an asymptotic overpressure ratio of 8, and the post-shock state is characterized by high entropy. Hence, the "plenum chamber" state $(\cdot)_2$ produces a transmitted shock having a relatively low overpressure ratio of 1.5.

For testing the model we turned to numerical simulation, producing the A_t points in Fig. 2, that show the same trend as the model, with about 7% lower values. The 2D numerical simulations were performed by the GRP scheme [1], on an axisymmetric configuration with narrow tube of $D_2/2 = 5\,[mm]$ and wider tube of $D_1/2 = 50\,[mm]$. The 50×340 computational rectangle was divided into 100×680 square cells, and the initial conditions were stationary ambient air ($\rho_0 = 1.29\,[kg/m^3]$, $p_0 = 0.1\,[MPa]$). In Fig. 3 we show the pressure amplification ratio and (axial) flow Mach number profiles at several times $T\,[\mu s]$, where L is the distance from narrow tube entrance and $D = D_2 = 10\,[mm]$ is its diameter. The transmitted shock amplification was read from the leading shock clearly visible in Fig. 3.

Concerning the agreement between the simulation and the model, we first observe that the transmitted wave is of constant magnitude as in the theory. A fan-like CRW pattern visible in both the pressure and Mach number profiles, corresponds to the CRW in the model, although the M-lines appear to fan out not exactly from the sonic point but from a point of slightly higher Mach number. This demonstrates that a stationary sonic point is approximately obtained in the simulation, and that it separates the steady flow regime (left) from the time-dependent (presumable self-similar) flow regime (right), as in the present model. Summarizing our results, it is apparent from Fig. 2 that the analytical model provides a reasonable estimate for the transmitted shock intensity. Throughout the investigated range, the model overestimates the transmitted shock overpressure amplification ratio by 7% to 9%.

5 Concluding remarks

The present model predicts that at an abrupt large area contraction the transmitted shock overpressure magnification factor is in the range of $2 \sim 1.5$. The first value agrees with the acoustic approximation to a head-on shock reflection. The second is obtained asymptotically for strong shocks, and its low value is probably due to the large jump in entropy accompanying the nearly head-on reflection of the incident shock. This asymptotic value indicates that while a head-on reflection in air can produce an 8-fold amplification of the incident shock overpressure, the shock transmitted through a narrow tube is amplified by only 1.5.

Another interesting observation is concerning the issue of quasi-one-dimensional modeling as compared to fully 2D-axisymmetric flow. The reflected and transmitted shocks obtained in the 2D computations were found to be nearly planar, and in that sense there is a good agreement with the 1D model. However, as was demonstrated in [4], when the incident wave is a CRW the flow evolving about the area contraction segment is genuinely two-dimensional, so that a 1D approximation is in qualitative disagreement with the actual flow.

References

1. Ben-Artzi M., Falcovitz J. *Generalized Riemann problems in computational fluid dynamics*, Cambridge University Press, London, 2003.
2. Gottlieb J.J., Igra O. *Interaction of rarefaction waves with area reduction in ducts*, JFM 137:285–305, 1983.
3. Greatrix D.R., Gottlieb J.J. *An analytical and numerical study of a shock wave interaction with an area change*, University of Toronto, Institute for Aerospace Studies UTIAS Rep. No. 268, 1982.
4. Igra O., Wang L., Falcovitz J. *Non-stationary compressible flow in ducts with varying cross-section*, Proc. Instn. Mech. Engrs., Part G, 212:225–243, 1998.

Modelling dissociation in hypersonic blunt body and nozzle flows in thermochemical nonequilibrium

E. Josyula[1] **and W.F. Bailey**[2]

[1] *Air Force Research Laboratory, Wright-Patterson Air Force Base, Ohio 45433 (USA)*
[2] *Air Force Institute of Technology, Wright-Patterson Air Force Base, Ohio 45433 (USA)*

Summary. The generalized depletion equations, considering state-to-state kinetics of dissociating nitrogen, are solved to predict the extent of vibrational depletion in the temperature range of 3000-10000 K for hypersonic blunt body flows, where the translational temperature is higher than vibrational temperature ($T > T_v$) in the shock layer. The prediction of nonequilibrium dissociation rates using the vibration-dissociation coupling model are significantly lower than the Park's two-temperature model rates and helps explain the restricted success of the Park's dissociation model in certain temperature ranges of hypersonic flow past a blunt body. Unlike the blunt body flows, energy transfers in expanding nozzle flows occurs for a vibrational temperature much higher than the translational temperature ($T_v > T$) and the vibrational population experiences an enhancement.

1 Introduction

The proper treatment of energy transfer between nonequilibrium molecular energy modes and dissociation has important implications in the accurate prediction of the aerothermodynamics of gas systems. At high temperatures molecular collisions result in the exchange of the translational, rotational, vibrational, and electronic energies of the collision partners. Nonequilibrium vibrational energy distributions are required for prediction of dissociation rates, interpretation of radiation experiments, and interpretation of ionic recombination rates [1,2].

The nonequilibrium conditions behind the shock wave of a blunt body exemplify the case where the translational temperature is greater than the vibrational temperature ($T > T_v$). The higher rate of dissociation versus recombination in blunt body flows makes the vibrational population depletion significant [3]. In expanding nozzle flows, however, the vibrational temperature freezes near the throat and the translation temperature rapidly falls as a result of the expansion process [4]. As the translational temperature drops in the axial direction, the dissociation rate drops significantly. This recombination-dominant flow leads to a population enhancement in the vibrational manifold under the non-equilibrium reactive conditions. A vibration-dissociation coupling model, thus, has to account for this enhancement to accurately predict the nonequilibrium chemistry [5]. The paper demonstrates a dissociating blunt body flow with population depletion in the shock layer and a dissociating expanding nozzle flow with population enhancement at the nozzle exit.

2 Governing Equations

The global conservation equations in mass-averaged velocity form to simulate the flowfield are presented in this section.

$$\frac{\partial}{\partial t}(\rho_n) + \nabla \cdot (\rho_n \mathbf{u}) = \dot{\omega}_n \ n{=}0,1,... \tag{1}$$

$$\frac{\partial}{\partial t}(\rho \mathbf{u}) + \nabla \cdot (\rho \mathbf{u}\mathbf{u} - p\bar{\bar{\delta}}) = 0 \tag{2}$$

$$\frac{\partial}{\partial t}(\rho e) + \nabla \cdot [\rho(e + p/\rho)\mathbf{u}] = 0 \tag{3}$$

Eqns. 1 to 3 describe the conservation of mass, momentum and energy in the flowfields of interest. Eqn. 2 and 3 represent the conservation of total momentum and energy, respectively. The conservation Eqn. 1 is written for the mass density in quantum level n. The source term $\dot{\omega}_n$ derived from the vibrational master equations is made up of the relevant energy exchange processes consisting of the V-T, V-V and dissociation processes.

The rate equation for change in vibrational population accounting for V-T and V-V transfers with dissociation is given by,

$$\dot{\rho}_j = \gamma_j v(1+\varphi_j) = k_{j+1,j}\rho_{j+1}\rho + k_{j-1,j}\rho_{j-1}\rho - k_{j,j-1}\rho_j\rho - k_{j,j+1}\rho_j\rho + \tag{4}$$
$$\sum_{n=0}^{n_0-1}\{r_{j+1,j}^{n,n+1}\rho_{j+1}\rho_n + r_{n+1,n}^{j-1,j}\rho_{j-1}\rho_{n+1} - (r_{j,j-1}^{n,n+1}\rho_j\rho_n + r_{n+1,n}^{j,j+1}\rho_j\rho_{n+1})\} - k_{d_j}\rho_j\rho$$

where, $\dot{\rho}_j \equiv \frac{d\rho_j}{dt} = \frac{d\rho_j^o}{dt}\{1+\varphi_j(t)\} + \rho_j(t)\dot{\varphi}_j(t) \approx \gamma_j v(1+\varphi_j)$ Following Osipov and Stupochenko [6], $\frac{d\rho}{dt} \equiv v = -\sum_{i=0}^{n_0}(k_{d_i}\rho_i\rho)$. Here ρ_j is the mass density associated with the vibrational population in state j. The superscript "o" denotes equilibrium conditions. In Eqn. 4, $k_{j+1,j}$ is the V-T rate coefficient for vibrational transitions to occur from $j+1 \rightarrow j$. The variables $r_{j+1,j}^{n,n+1}$ and $r_{n+1,n}^{j-1,j}$ are the V-V rate coefficients for vibrational exchanges between two molecules from/with $j+1 \rightarrow j$, $n \rightarrow n+1$. The rate constant for dissociation from vibrational state j is k_{d_j}. Note that $\sum_{n=0}^{n_0}\rho_n = \rho$.

The V-T and V-V terms may be re-grouped and the populations written in terms of the equilibrium population and state depletion factor, φ_j, giving rise to a set of cascade equations. These cascade equations were simplified by starting with the ground state (0) equation and writing the sum of the V-T and V-V entries in terms of the sum of the dissociation terms on the left-hand-side and the last term on the right-hand-side. This expression is then substituted for the (1,0) V-T and V-V exchanges in the equation for level 1. The remaining V-T and V-V terms involving level 2 in the level 1 equation are then expressed in terms of the sum of the dissociation terms from levels 0 and 1. This leads to the general cascade equation (Eqn.5), where we have solved the j_{th} rate equation for φ_{j+1} and accumulate exchanges with lower levels and dissociation in the term D_j. :

$$\varphi_{j+1}(1 + \sum_{n=0}^{n_0-1}\{\frac{r_{j+1,j}^{n,n+1}\rho_{j+1}^o\rho_n^o(1+\varphi_n)}{k_{j+1,j}\rho_{j+1}^o\rho}\}) = \varphi_j + \tag{5}$$
$$\sum_{n=0}^{n_0-1}\{\frac{r_{j+1,j}^{n,n+1}\rho_{j+1}^o\rho_n^o}{k_{j+1,j}\rho_{j+1}^o\rho}(\varphi_{n+1}-\varphi_n)\} + \varphi_j\sum_{n=0}^{n_0-1}\{\frac{r_{j+1,j}^{n,n+1}\rho_{j+1}^o\rho_n^o(1+\varphi_{n+1})}{k_{j+1,j}\rho_{j+1}^o\rho}\} + \frac{D_j}{k_{j+1,j}\rho_{j+1}^o\rho}$$

where we have used: $v = -\sum_{i=0}^{n_0} k_{d_i}\gamma_i(1+\varphi_i)$ and defined
$D_j = -\sum_{i=0}^{j}[\{\sum_{m=0}^{n_0} k_{d_m}\gamma_m(1+\varphi_m)\} - k_{d_i}\gamma_i](1+\varphi_i)$.

The coupled rate equations, Eqn. 5 was linearized and an iterative solution achieved. The solution was validated against a solution of the Master Equations. Deviations of the two solutions of less than 2% were achieved over the temperature range of 3000 to 10000 K. The factor φ, which denotes the deviation of the quasi-steady state distribution from the equilibrium could then be used to determine the nonequilibrium dissociation rate from an equilibrium dissociation rate, presented in the following section.

3 Population Depletion in Nonequilibrium Dissociating Shock Wave Flows

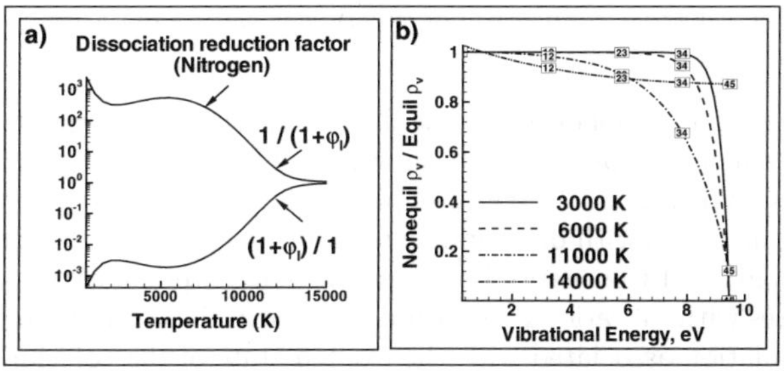

Fig. 1. (a) Equilibrium and Nonequilibrium population ratio and Reduction factor for Park's Dissociation, (b) Ratio of population distribution showing depletion in quantum levels

Figure 1a shows the term $1 + \varphi_l$ and its inverse which gives the dissociation reduction factor from equilibrium rates plotted as a function of temperature. Note that when $\varphi_l < 0$, the effect is to deplete the population in the vibrational states. The reduction factor, $\frac{1}{1+\varphi_l}$, is the ratio of the level population distribution at equilibrium to the nonequilibrium state. The reduction factor which accounts for the deviation from the quasi-steady state distribution can range from 1 to 3 orders of magnitude between the temperatures of 5,000 K to 10,000 K at which there is considerable nitrogen dissociation, Figure 1a. This factor applied as correction to the equilibrium dissociation rates to account for the depletion effects in the vibrational levels helps in explaining the validity of Park's two-temperature model only in certain temperature ranges.

The depleted vibrational state populations in N_2 resulting from the ladder climbing dissociation model, dissociation restricted to last level, are presented in Figure 1b. The population density, shown as a factor of the equilibrium population undergoes a reduction in the highly excited vibrational states. The loss is significant in the upper levels shown for the temperatures between 3,000 K and 6,000 K. At higher temperatures, the depletion extends to lower levels also as can be seen for 11,000 K. However, at temperatures above 11,000 K, the reduction in the vibrational level population diminishes and at the highest temperature of 14,000 K, the reduction is small.

The extent of this departure from equilibrium is determined by a complex interplay of the relative values of the energy transfer rates, V-V and V-T, the dissociation rate and vibrational bias of the dissociation reaction. Here, using the generalized depletion analysis, we analyzed the role of V-T transfer and V-V exchange in nonequilibrium dissociation of Nitrogen using the standard "ladder model" of dissociation. The rate (cm^3/s) of the V-V and V-T rates are presented in Figs. 2b at 4000 K. The V-T rate corresponds

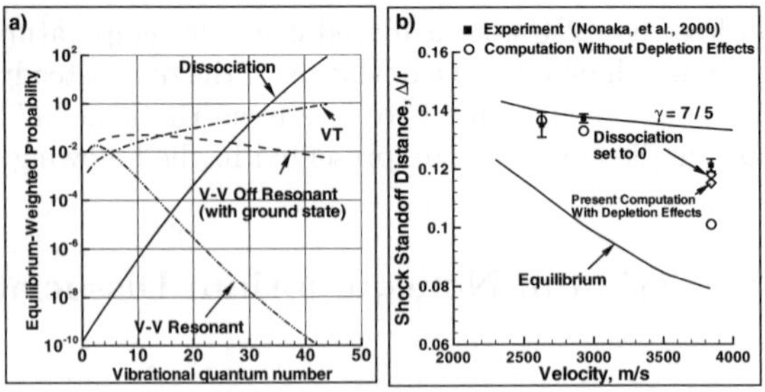

Fig. 2. (a) Transition probabilities of energy exchanges in N_2 dissociation at 4000 K, (b)Shock-standoff distance vs. velocity - comparison of experiment and computation

to $j+1 \rightarrow j$ and increases monotonically with the quantum level vibrational energy. Two relevant V-V exchanges are presented, both weighted by the relative equilibrium population of the vibrational state of the collision partner: (1) the weighted V-V exchange with the ground state increases with quantum level vibrational energy and then decreases due to increasing disparity of the energy being exchanged, (2) population-weighted resonant exchange increases with vibrational quantum number initially and then decreases due to the reduced population associated with the excited state of the collision partner.

Also, shown is the equilibrium dissociation rate. This is the value required to support the equilibrium dissociation rate from a single, given vibrational state. In the "ladder model" the state of interest is the last bound vibrational state. For a given set of V-V and V-T transition rates, the relative kinetic importance of the V-V terms, $r_{j+1,j}^{n,n+1}$, the V-T term, $k_{j+1,j}$, and the dissociation term, D_j appearing in Eqn. 5 can be assessed. It can be concluded that for Nitrogen at these temperatures, the dissociation probability for the last state is high relative to both the weighted V-V and V-T rates and that V-T is more important than V-V. Regarding depletion of the dissociating state, one would anticipate that depletion will be significant because the dissociation rate exceeds the V-T rate considerably. Since the V-T probabilities are higher than those of weighted V-V energy transfers, the effect of V-V on the population depletion is negligible.

Mach No.	T_∞ (K)	$T_{v\infty}$ (K)	p_∞ (Pa)	Radius (m)	Medium	Validation
11.18	293	293	1200	0.007	Air	Exp [7]

Table 1. Freestream conditions of flow past blunt body

The vibration-dissociation coupling model was tested on a Mach 11.18 air flow past a blunt body, the case chosen due to the inability of the Park model to predict the shock-standoff distance correctly [3]. A shock-standoff distance comparison of the experimental data and computational predictions is shown in Fig. 2. The shock-standoff distances at velocities less than 3000 m/s are close to frozen flow conditions and the predictions using existing vibration-dissociation models *without* depletion effects are close to those *with* depletion effects. However at higher flow velocities, the departure of the computation from data is 17% due to faster dissociation rates predicted by the Park model. The present computation which includes the depletion effects predicts a shock-standoff distance very close to data, Fig. 2b.

4 Population Enhancement in Nonequilibrium Dissociating Expanding Nozzle Flows

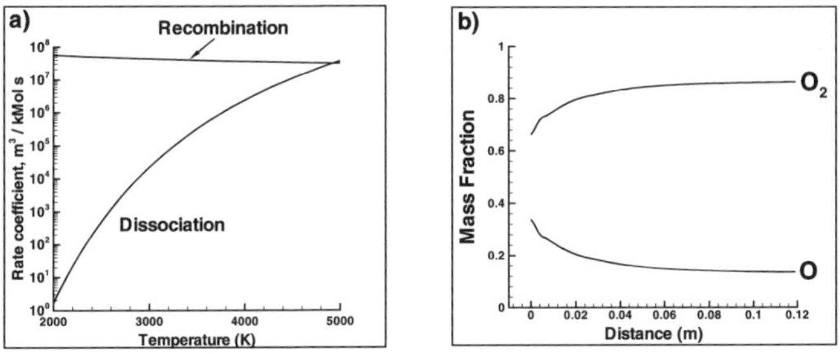

Fig. 3. (a) Dissociation and Recombination Rate for the $O_2 + O_2 = 2O + O_2$ reaction, (b) mass fraction along length of nozzle

As the temperature decreases downstream, the effect of the population enhancement is to adjust the rate of dissociation and recombination by lowering the mass fraction of O_2 and increasing that of O along the nozzle. From Fig. 3a it is seen that when the temperature drops from 5000 K to 3000 K, the recombination rate coefficient is nearly the same. For the same temperature drop, the fall in the dissociation rate coefficient is significant and decreases by a factor of about 1700. Also, the recombination rate coefficient is higher than the dissociation rate downstream of the throat. For this recombination-dominant flow, the calculations of the vibrational population enhancement demonstrated in the present study provide a useful relationship to calculate the effective nonequilibrium dis/recomb rates from equilibrium rates.

An expanding nozzle flow of oxygen was modeled to study the effect of vibrational and reactive kinetics on the flowfield [4]. The nozzle with a throat radius of 3 mm has an area ratio of 50. A throat temperature of 5000 K with 0.3376 oxygen atom mass fraction and $p_{throat} = 9479kPa$ was specified. The static pressure at throat is 9479 kPa. Fig. 3a shows

Location	T (K)	T_v (K)	O_2 mass fraction	Density kg/m^3	Enhancement factor from Boltzmann in last level
at exit	2548	2761	0.862	0.0622	120

Table 2. Nozzle locations and flow conditions for which population enhancement demonstrated

the dissociation (k_d) and the recombination coefficient (k_r), with the following definitions. For the dissociation reaction, $k_f = C_f T^{\eta} \exp(-\Theta_d/T) exp(-\Theta_d/T)$, where C_f, and η are from experiments and Θ_d is the characteristic temperature of dissociation. This effective temperature, T in the above equation is commonly calculated in today's hypersonic CFD codes by the empirical two-temperature dissociation model attributed to Park. The present numerical study was conducted to perform a state-to-state kinetic modeling of the nonequilibrium flow physics leading to physically-meaningful vibration-dissociation coupling models for a better estimation of the effective temperature of dissociation. The atom recombination rate is obtained from $k_r = k_f/K_{eq}$, where K_{eq} is the equilibrium constant determined from curve fits of experimental data. The T and T_v along the length of the nozzle are shown in Fig. 4a. Since T decreases very rapidly in an expanding nozzle, the dissociation rates fall sharply and recombination of atoms increases along the length of the nozzle. Resulting mass fraction of the O_2 and O species along the length of the nozzle is shown in Fig. 3b.

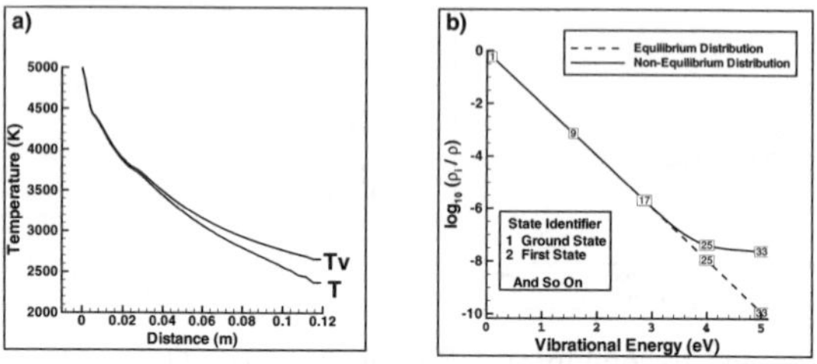

Fig. 4. (a) Translational and vibrational temperatures along length of nozzle, $T_{throat} = 5000\ K$, $p_{throat} = 9479\ kPa$, Medium=O_2, $\alpha_O = 0.3376$, (b)Population enhancement at the nozzle exit, relative to equilibrium population distribution

The vibrational population enhancement in nonequilibrium nozzle flow is demonstrated in Fig. 4b for nozzle exit. The population distributions corresponding to the Boltzmann are compared to the nonequilibrium populations obtained from solving the master equations accounting for vibrational and dissociation kinetics. (Refer to Table 2 for nozzle flow details.) The population enhancement at the exit is a factor of 120. The effect of population enhancement is to lower the effective recombination rate. Even without the consideration of the effects of the population enhancement on the dissociation rates, the temperature drop along the nozzle has the effect of reducing the dissociation rates. However, the effect of population enhancement is to lower this reduction.

References

1. Macheret, S.O., Adamovich, I.V. Semiclassical Modeling of State-Specific Dissociation Rates in Diatomic Gases. Journal of Chemical Physics, vol. 113, no. 17, 2000
2. Josyula, E., Bailey, W.F., Gudimetla, V.S. Rao: Modeling of Thermal Dissociation in Nonequilibrium Hypersonic Flows. AIAA Paper 2006-3421, 2006
3. Josyula E., Bailey W.F., Vibration-Dissociation Coupling Model for Hypersonic Blunt-Body Flow: AIAA Journal, vol. 41, no. 8, 2003
4. Shizgal B.D., Vibrational Nonequilibrium in a Supersonic Expansion with Reaction: Application $O_2 - O$. Journal of Chemical Physics, vol. 104, no. 10, pp. 3579-3597, 1996.
5. Josyula E., Bailey W.F., Vibrational Population Enhancement in Nonequilibrium Dissociating Hypersonic Nozzle Flows: Journal of Thermophysics and Heat Transfer, vol. 18, no. 4, 2004
6. Osipov, A.I., Stupochenko, E., Kinetics of the Thermal Dissociation of Diatomic Molecules I: Small Impurity of Diatomic Molecules in a Monatomic Gas: Combustion, Explosion and Shock Waves, vol. 10, no. 10, 1974
7. Nonaka S., Mizuno, H., Takayama, K., Park, C., Measurement of Shock-Standoff Distance for Sphere in Ballistic Range: Journal of Thermophysics and Heat Transfer, vol. 14, no. 2, 2000

Numerical and experimental investigation of viscous shock layer receptivity and instability

A. Kudryavtsev, S. Mironov, T. Poplavskaya, and I. Tsyryulnikov

Khristianovich Institute of Theoretical and Applied Mechanics, Russian Academy of Sciences, Siberian Division, 4/1 Institutskaya St., 630090 Novosibirsk (Russia)

Summary. A hypersonic viscous shock layer over a flat plate at the free-stream Mach number $M_\infty = 21$ is investigated both numerically and experimentally. The shock layer is excited by either external acoustic field or blowing/suction on the plate surface. The interaction of disturbances with the leading edge shock wave and their development in the boundary layer are simulated by solving the Navier-Stokes equations with a high-order shock-capturing scheme. Data of numerical simulations are compared very favorably with measurements conducted in a hypersonic nitrogen wind tunnel using the electron-beam fluorescence techniques. Numerical simulations show that the interference between natural and artificially excited disturbances can be used for active control of the instability development. This conclusion is confirmed experimentally.

1 Introduction

A better understanding of the mechanisms governing emergence and evolution of disturbances in a viscous shock layer (VSL) is a necessary condition for the development of efficient methods for controlling the laminar-turbulent transition in the flow around hypersonic flying vehicles. These mechanisms may differ substantially from those investigated in supersonic flows at lower Mach numbers. In recent years at ITAM (Novosibirsk) a number of experiments were performed on the receptivity and instability of the flat plate VSL at a very high Mach number, $M_\infty = 21$ [1,2]. However, possibilities of experimental modeling of receptivity and evolution of disturbances in hypersonic wind tunnels are rather limited. As concerns numerical simulations, mostly they have been performed for lower Mach numbers [3,4].

The present paper describes the results of numerical and and experimental investigations of receptivity and evolution of perturbations in a VSL on a flat plate for $M_\infty = 21$ and a moderate Reynolds number ($Re_L = 1.44 \cdot 10^5$). The problem of interaction of the VSL with free-stream acoustic disturbances of slow and fast modes as well as with perturbations introduced in the vicinity of the plate leading edge by means of periodic blowing and suction is studied numerically by solving the 2D Navier-Stokes equations. The computational results are compared with the characteristics of density fluctuations measured in experiments performed at the same flow conditions.

2 Experimental approach and numerical model

2.1 Experimental facility and techniques

The experiments are performed in the T-327A nitrogen hypersonic wind tunnel. The T-327A is a continuous blowdown facility with gas exhaust into a 100 m^3 vacuum chamber.

The running time is 30 s. The pressure in a plenum chamber is 8 MPa, the stagnation temperature is 1200 K. The exit diameter of a conical water-cooled nozzle is 220 mm, the nozzle cone angle is 8°, and the uniform flow core diameter is 100 mm. The streamwise Mach number gradient is 3 m^{-1}. The test model is a trapezium, the length is $L = 240$ mm ($Re_L = 1.44 \cdot 10^5$), the leading edge width is 100 mm, the trailing edge width is 80 mm. The wedge angle at the leading edge is 7°, the radius of edge bluntness is 0.05 mm.

A Pitot probe with internal diameter 0.5 mm is used to measure the total pressure distribution and obtain the Mach number distribution. The method of electron-beam fluorescence is used to measure the mean density distributions and the density fluctuations. The electron energy is 14 keV, the current is 0.5 mA, the beam diameter in vacuum is 1 mm. The optical system includes a fast lens, a filter with a bandwidth 360-500 nm and diaphragms. The light filter accepts the radiation of the first negative and second positive gaps of VRT of nitrogen molecules and molecular ions. The range of disturbance frequencies measured is f = 1-50 kHz.

Two different kinds of flow disturbances are studied: 1) natural disturbances, i. e. broadband noise consisting of external acoustic waves of slow mode emitted by the turbulent boundary layer near the nozzle throat and 2) artificial disturbances — mass flow rate fluctuations introduced by an oblique aerodynamic whistle (6mm of diameter, the frequency of fundamental harmonics is 20 kHz). The whistle is mounted just below the plate leading edge and generates perturbations which are equivalent to a periodical blowing and suction.

2.2 Numerical formulation

The 2D Navier-Stokes equations along with the equation of state for a perfect gas are solved. The convective terms are calculated with the MP5 (monotonicity-preserving, 5th order) scheme developed in [5]. This scheme has a built-in analyzer, which can distinguish between solution discontinuities and smooth extremum points. The diffusive terms are approximated by the 4th order central finite differences. Time stepping is performed by the 3rd order Runge-Kutta method.

A part of the lower side of the rectangular computational domain coincides with the plate surface. The plate temperature is $T_w = 300$ K. As rarefaction effects in the problem considered are fairly significant, the boundary conditions on the plate surface take into account the velocity slip and temperature jump (velocity slip is approximately 17% of the free-stream value at $x/L = 0.1$ and about 7% near the trailing edge). The inflow boundary is located at a distance of a few computational cells upstream from its leading edge. The height of the computational domain is chosen such that the bow shock wave (SW) emanating from the leading edge does not interact with the top boundary. The outflow boundary is situated downstream from the plate trailing edge so that the flow in the exit cross section is fully supersonic. A uniform computational grid consists of 1050 cells in the streamwise direction and 240 cells in the crossflow direction.

First, the steady flow is computed. The computed mean flow has been compared with experimental data and excellent agreement of the total pressure, density, and Mach number distributions has been observed.

In solving the problem of interaction of the viscous shock layer with natural disturbances, the latter are introduced by prescribing the variables on the inflow boundary as a superposition of the steady basic flow and a planar monochromatic acoustic wave. The boundary conditions on the plate surface are the same as those used to find the steady solution, except for zero temperature perturbations (because of considerable thermal inertia of the plate). After introduction of perturbations, the Navier-Stokes equations are integrated in time until the solution reaches a quasi-periodic regime.

Blowing-suction type disturbances are simulated by imposing the boundary condition for the transverse mass flow rate on the portion of the plate surface between $x_1 = 0.065L$ and $x_2 = 0.08L$.

3 Results

Figure 1 shows the mean density flowfield (Fig. 1a) and the instantaneous field of density fluctuations in the case of excitation of the VSL by external acoustic disturbances (Fig. 1b). The frequency of acoustic disturbances is $f = 19.2$ kHz. The amplitude of the density perturbations (normalized with the free-stream density) is $A = 0.0286$. The solid and dashed isolines in Fig. 1b show the positive and negative fluctuations of density, respectively. It is seen from the figure that there are two regions with the maximum fluctuations of density: on the shock wave itself and on the upper edge of the boundary layer where the mean density changes rapidly because of an increase in static temperature. The fluctuations on the SW and on the boundary-layer edge are opposite in phase. Following to the linear theory of interaction of disturbances with a SW [6], it can be demonstrated that under the given flow conditions, the external acoustic disturbances passing through the SW can generate only waves of the entropy-vortex mode, no transmitted acoustic wave exists.

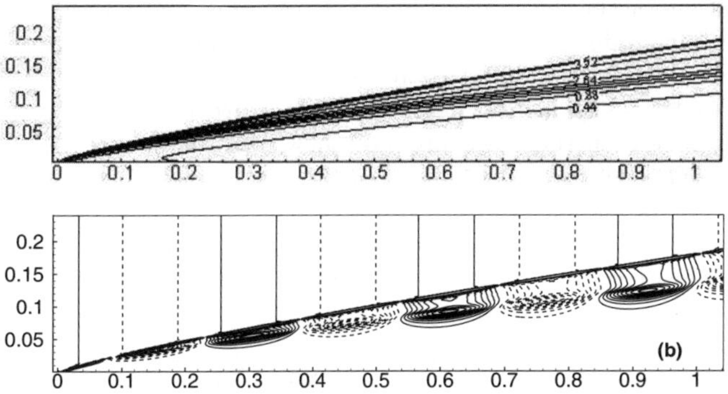

Fig. 1. Isolines of mean density (a) and density fluctuations at $A = 0.0286$, $f = 19.2$ kHz

It seems of interest to compare the computed growth rates of disturbances on the SW with the data of the linear stability theory. The computation by the locally parallel linear stability theory has been performed taking into account the influence of a closely located SW [7]. Figure 2 shows the growth rates of disturbances. The solutions obtained from the locally parallel linear stability theory (dashed lines) are seen to be in close agreement with the results of direct numerical simulations of the VSL excited by slow or fast external acoustic waves (solid lines) as well as with the experimental data for the slow acoustic disturbances (triangles).

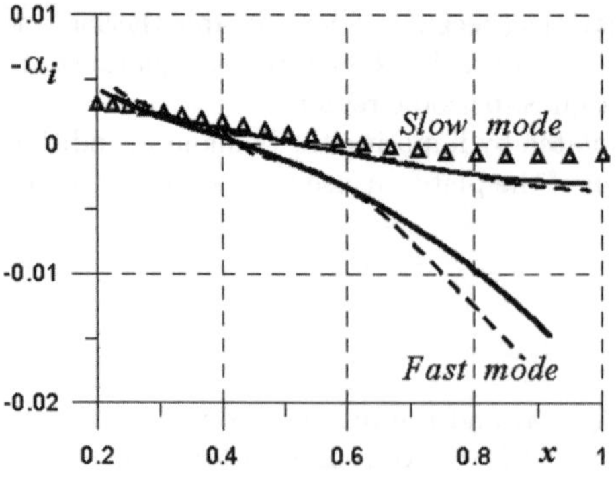

Fig. 2. Theoretical, computational, and experimental growth rates of disturbances measured on the SW

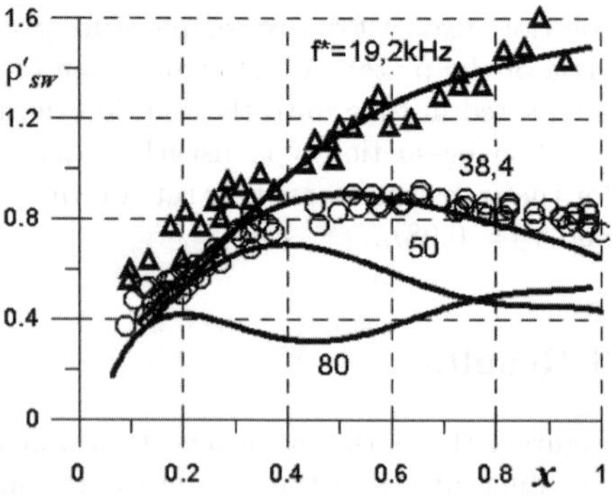

Fig. 3. Maximum amplitudes of density fluctuations excited by slow acoustic waves of different frequencies

The solid lines in Fig. 3 show the variation along the plate of the computed maximum amplitudes of density fluctuations (on the SW) for slow external acoustic disturbances of different frequencies $f = 19.2$ kHz, 38.4 kHz, 50 kHz, and 80 kHz. Symbols in Fig. 3 show the experimentally measured amplitudes at the same frequencies. The computational and experimental data are in good agreement. It can also be seen from the figure that an increase in frequency yields a non-monotonic variation of the maximum amplitude of density fluctuations along the streamwise coordinate. This phenomenon can be probably explained by the interaction of external flow perturbations and vortex disturbances in the VSL, which have different streamwise wavenumbers and propagate with different velocities. This interaction is manifested on the SW as amplitude beatings.

Numerical simulations show that if the VSL is perturbed by a blowing and suction near the leading edge then the flowfields (not shown) in many aspects resemble those generated by external acoustic disturbances. In Fig. 4 the results obtained from numerical simulation of the development of disturbances generated by the blowing/suction are compared with experimental data. As can be seen, the calculated dependence of amplitude on the streamwise coordinate for the fundamental harmonics with $f = 20$ kHz (1)

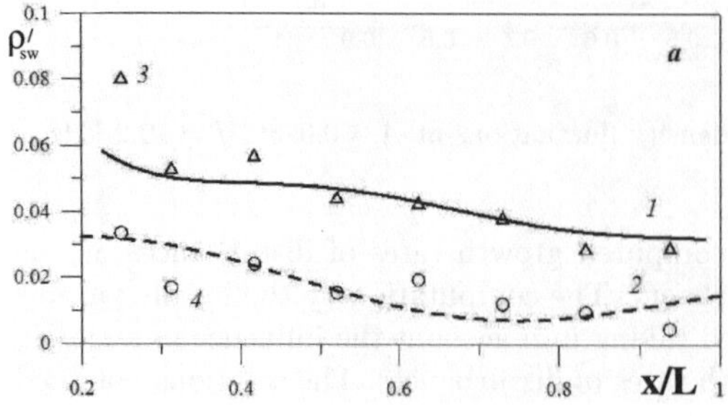

Fig. 4. Maximum amplitudes of density fluctuations excited by by blowing/suction with amplitude $A = 0.286$.

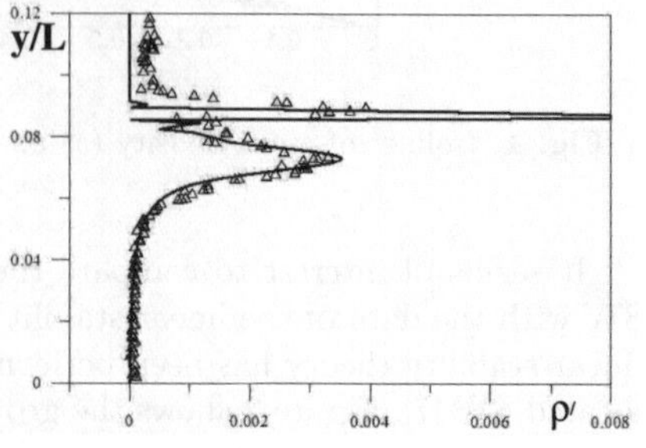

Fig. 5. Distribution of amplitude of density fluctuations at the cross-section $x/L = 0.42$. $f = 20$ kHz, $A = 0.286$

and the first higher harmonics with $f = 40$ kHz (2) is in close agreement with experimental measurements (3, 4). The distribution of disturbance amplitude across the boundary layer is also successfully predicted by numerical simulation. In Fig. 5 the amplitude of density fluctuations at the cross-section $x/L = 0.42$ is shown as an example. The greater peak corresponds to the position where the SW is situated whereas the position of the second maximum coincides with the outer edge of the boundary layer.

The similarity of the flow disturbances generated in the VSL by external acoustic waves and a periodical blowing/suction provides a possibility of active control of the evolution of disturbances. It has been shown with the help of numerical simulations that the introduction of periodic perturbations of the blowing-suction type on the flat-plate surface in the opposite phase to external acoustic disturbances leads to a significant suppression of the instability wave propagating in the boundary layer. This phenomenon is illustrated by Fig. 6.

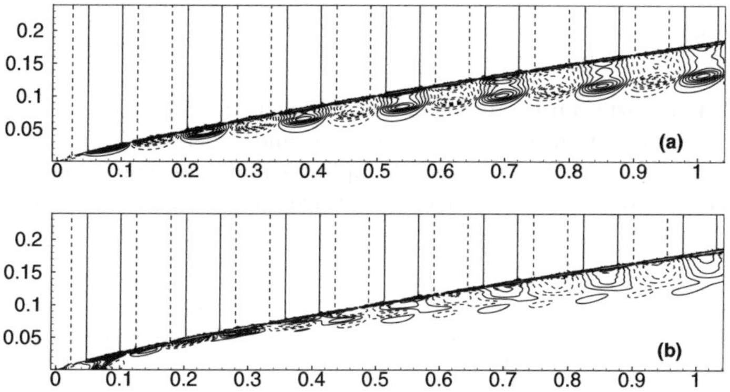

Fig. 6. Isolines of density fluctuations excited by a slow acoustic wave with $f = 38.4$ kHz and $A = 0.001$ (a) and by the same acoustic wave in superposition with an opposite-phase blowing/suction with $A = 0.06$

This numerical prediction has been confirmed experimentally. For this purpose, controlled acoustic disturbances are introduced in the wind tunnel plenum chamber. Their phase is synchronized with the blowing/suction introduced with an oblique aerodynamic whistle mounted on the test model. Variation of the phase difference leads to beatings of the disturbance amplitude. Figure 7 shows the relative amplitude of disturbances measured on the boundary layer edge at $x/L = 0.63$ for different phase shifts (circles) along with the results of numerical simulations (triangles). The simulations have been performed for two cases, namely when the phases of two types of disturbances are opposite or, on the contrary, coincide.

4 Conclusion

Numerical simulations and experimental investigations of a hypersonic viscous shock layer excited by external acoustic waves have shown that the intensity of vortical disturbances generated behind the shock wave is maximum near the shock wave and on the boundary

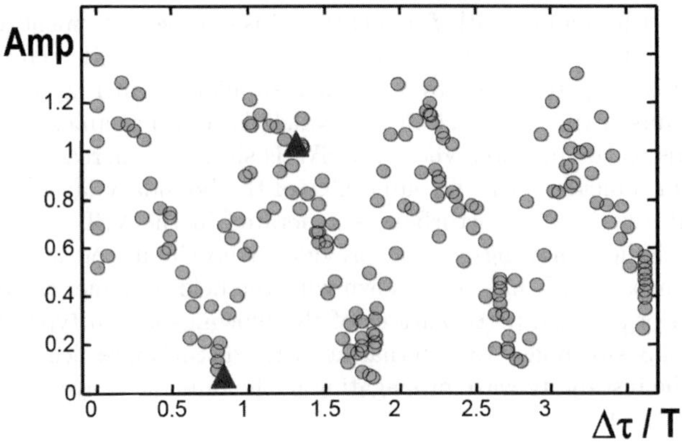

Fig. 7. Interference of disturbances generated by external acoustic wave and blowing/suction on the flate surface.

layer edge. Similar disturbances are also observed when the shock layer is excited by a periodic blowing and suction. Both the simulations and experiments give evidence that the blowing/suction perturbation of a tuned amplitude and phase can be used to suppress the disturbances generated by external acoustic waves and delay the development of flow instability.

Acknowledgement. This work was supported by Russian Foundation for Basic Research (Grant No. 05-08-33436). Computer resources were provided by Siberian Supercomputer Center (Novosibirsk) and Joint Supercomputer Center (Moscow).

References

1. Mironov S.G., Maslov A.A.: An experimental study of density waves in hypersonic shock layer on a flat plate. Phys. Fluids, vol.12, no.6, 2000
2. Mironov S.G., Maslov A.A.: Experimental study of secondary instability in a hypersonic shock layer on a flat plate. J. Fluid Mech., vol.412, 2000
3. Ma Y., Zhong X.: Receptivity of a supersonic boundary layer over a flat plate. Part 2. Receptivity to free-stream sound. J. Fluid Mech., vol. 488, 2003
4. Egorov I.V., Sudakov V.G., Fedorov, A.V.: Numerical simulation of supersonic boundary-layer receptivity to acoustic disturbances. Fluid Dynamics, no. 2, 2006
5. Suresh A., Huynh S.T.: Accurate monotonicity-preserving schemes with Runge-Kutta stepping. J. Comput. Phys., vol. 136, 1997
6. McKenzie J.F., Westphal K.O.: Interaction of linear waves with oblique shock waves. Phys. Fluids, vol. 11, 1968
7. Maslov A.A., Poplavskaya T.V., Smorodsky B.V.: Stability of a hypersonic shock layer on a flat plate. Comptes Rendus Mecanique, vol. 332, 2004

Numerical rebuilding of the flow in a valve-controlled Ludwieg tube

T. Wolf, M. Estorf, and R. Radespiel

Institute of Fluid Mechanics, TU Braunschweig, Bienroder Weg 3, 38106 Braunschweig, Germany

Summary. The Hypersonic Ludwieg Tube Braunschweig (HLB) is a valve-controlled wind tunnel that has been designed for a Mach number of $Ma = 5.9$ and a Reynolds number range from $2.5 \cdot 10^6 \mathrm{m}^{-1}$ up to $2.5 \cdot 10^7 \mathrm{m}^{-1}$. The intermittent working principle implies an unsteady onset of flow which leads to a delay of the time frame suitable for measurements, which is amplified due to the usage of a valve. This work numerically simulates the starting process. The flow field in the HLB is numerically rebuilt for two operating points including valve opening. The results are used to distinguish the starting-process from the time frame with steady flow conditions.

1 Introduction

Aerothermal design of reusable reentry vehicles requires good knowledge of aerodynamic and thermal loads on all structural components. Although numerical methods for hypersonic flow simulations are already well developed the experimental proof is still essential. Because of their low operational cost and good flow quality Ludwieg-tube type cold blow-down tunnels are of interest. They need no pressure control device or large settling chamber as conventional blow-down facilities. A further reduction of operational effort is possible by utilisation of a fast-acting valve. According to this approach, documented by Koppenwallner et al. [1], a hypersonic testing facility has been established at the Technical University at Braunschweig in 2002. Although there have been extensive experimental and numerical investigations to assess the flow quality during measurement time, i.e. with steady flow conditions the fundamental understanding of the instationary onset of flow and its impact on results gathered during measurement time is still incomplete.

The inflow in the test section of the HLB during a wind tunnel run is different from the conditions in free atmosphere. These deviations are partly due to the design of the wind tunnel, namely variations transverse to flow direction in the wake downstream of the valve body and diverging streamlines in the test section. The remaining deviations stem from the intermittent working principle of a Ludwieg-tube. Obviously, the time frame applicable for measurements is delayed by the amount of time it takes the flow to start up, fully develop and settle down. In valve-controlled wind tunnels the delay is related to the opening velocity of the valve and therefore is usually larger compared to facilities using burst diaphragms. During the onset of flow a model in the test section is exposed to aerothermodynamic loads which strongly differ from the conditions during measurement time. While the latter is investigated in Wolf et. al. [2] using a coupled approach, this work focuses on the onset of flow in the HLB and the differences with respect to conventional diaphragm controlled facilities.

2 Description of the facility

A schematic drawing of the HLB is given in Fig. 1. The assembly is divided into a high pressure and a low pressure section, which are separated by a pneumatically driven, fast-acting valve. The high pressure section consisting of the 17 m long storage tube with a 3 m long heated segment can be pressurised to up to 30 bar. The length of the heated segment corresponds to the amount of air which escapes from the storage tube during a wind tunnel run. The low pressure section comprising the hypersonic nozzle, the test section, the diffuser and a 6 m^3 dump tank is evacuated to about 1 mbar before each run. The heating to up to 500 K prevents condensation effects of the expanding gas in the nozzle. The valve body contains a cylinder with a pneumatically driven piston which fits into the nozzle throat with its conical end. The valve is closed after 100 ms to prevent a pressure equalisation and thus allows for a short idle period of the facility between two successive runs. The nozzle has been designed for a Mach number of $Ma = 5.9$ in the test section elaborately (cf. references in [2] for details on the design and flow quality in the HLB). The length of the storage tube limits the time frame suitable for measurements with steady flow conditions to approximately 80 ms.

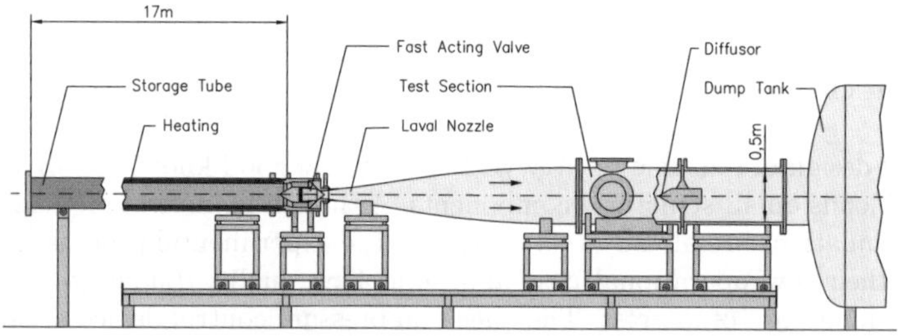

Fig. 1. Sketch of the hypersonic testing facility HLB.

The opening of the valve piston has a strong influence on the onset of flow in the wind tunnel. Consequently, the piston speed variation in time is an important input parameter for the moving boundaries in the numerical simulations. The experimental setup for measuring the piston velocity is shown in the left image of Fig. 2 and consists of a photodiode attached to the front of the valve body and a line pattern on the side of the piston. The alternating voltage signal of the photoelectric relay is recorded by an oscilloscope during opening and closure. The resulting piston path and its velocity is shown in the right graph of Fig. 2. The piston is accelerated to a maximum velocity of approximately 6 $m\,s^{-1}$, until its direction is temporarily reversed which results from the limited ventilation of the cylinder. The end position is reached after approximately 23 ms.

3 Numerical methodology

The TAU-Code developed at DLR (cf. Gerhold [3])is used to solve the mass-averaged Navier-Stokes equations. For the time-dependent simulations in this work a dual time-stepping approach with the implicit LU-SGS relaxation solver for the inner iterations is chosen (cf. Dwight [4]). Air is treated as an ideal gas, using Sutherland's law for the whole

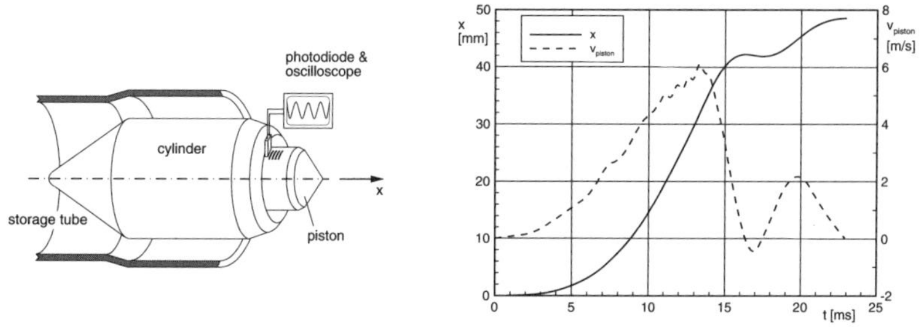

Fig. 2. Schematic experimental setup for measuring the piston velocity (left). Measured piston path and velocity during opening (right).

temperature range. With respect to the flux scheme, several test cases were investigated, comprising a strong normal shock wave, a single mach reflection (SMR), Sod's shock tube and a simplified Ludwieg tube. The AUSMPW+ scheme as described in Kim. et. al. [5] showed the best overall performance and is used in this work. The Menter SST turbulence model is employed for the fully turbulent flow in the HLB. The computational grid is created in two steps. First, the domain bound by the HLB axisymmetry axis, the valve and the outer contour of the wind tunnel is meshed. Next, the 2D grid is converted into an axisymmetric grid with an opening angle of 3°. This allows for faster solver operation, increased accuracy near the symmetry axis and less grid points compared to a full 3D computation. Of course this approach neglects effects due to temperature stratification as well as non-symmetric flow separation. Yet, the numerical effort of these simulations, esp. for wind tunnel flows with extreme variations of time scales and length scales, up to now forbids a calculation of the full three-dimensional flow field.

The valve opening is incorporated by means of grid deformation. As shown in Fig. 2, the piston's traverse path is 48.5 mm. This distance cannot be covered on a single grid as the cells in the vicinity of the piston would become too skewed, leading to an increased discretisation error. Therefore, the traverse path is divided into segments. The first grid is created with an initial gap of 1.1 mm, as the solver cannot handle touching boundaries. This leads to a reduction of the simulation time to approximately 20 ms. Before the calculation of each physical time step the boundary of the piston is shifted, according to the measurements. A maximum deformation distance of 5 mm within a single traverse segment is regarded as acceptable, leading to 10 segments. At each transition from a segment to its successor the actual solution is transferred. An interpolation routine is used here, based on the inverse distance weighting method. The error in the density distribution after two successive interpolations is less than 0.1% in smooth regions and below 5% at discontinuities. The geometric conservation law (GCL) is satisfied.

The time-dependent flow in the HLB is numerically rebuilt for the stagnation pressures $p_0 = 3.25$ bar and $p_0 = 8.75$ bar. The state indicated by "0" refers to the conditions in the storage tube before a run. The conditions downstream of the expansion wave with a Mach number $Ma \approx 0.067$ in the storage tube can be calculated using textbook methods, leading to an intermediate state. A further isentropic expansion to a nozzle Mach number $Ma = 5.9$ yields the reference conditions in the test section given in Table 1. In order to resolve the boundary layer properly, the distance of the first point layer next to the wall is chosen such that $y^+ \leq 1$ for almost the whole domain, except for the vicinity of the nozzle throat where values of $y^+ > 1$ could not be avoided. The boundaries are

Table 1. Reference flow conditions in the test section assuming isentropic expansion.

p_0	3.25 bar	8.75 bar
Mach number	5.9	
Temperature	61.189 K	
Density	$11.865 \cdot 10^{-3}$ kg m^{-3}	$31.945 \cdot 10^{-3}$ kg m^{-3}
Pressure	2.083 mbar	5.610 mbar
Unit Reynolds number	$2.699 \cdot 10^6$m^{-1}	$7.267 \cdot 10^6$m^{-1}

treated as isothermal walls with a temperature of 500 K in the storage tube, along the valve body and the piston, while the temperature of the nozzle and test section walls is 295 K. A linear temperature drop is imposed on the part of the contour where the piston touches the nozzle throat in closed position. The length of the computational domain is approximately 6.2 m, including the 3 m long heated section of the storage tube. Despite a coarse grid in the storage tube the number of grid points varies from 280 000 to 330 000 depending on the piston position. As a result of the high velocity (esp. during the beginning of the start up) and the small grid cells the size of the physical time step is limited to $\Delta t = 5 \cdot 10^{-8}$ s for the first 1.2 ms and then can be increased to $\Delta t = 5 \cdot 10^{-7}$ s which leads to a large number of time steps for the simulated run time of 20 ms. The number of sub-iterations per physical time step is chosen to 100, ensuring a reduction of the residual by at least four orders of magnitude while the flow is highly unsteady and convergence to machine accuracy for the rest of the simulation. This computational effort and I/O operations during each unsteady time step lead to a duration of approximately six weeks on 32 cpus of a p690-series compute server per simulation.

4 Results

The variation of the static pressure and the density along the HLB axisymmetry axis in time is shown in Fig. 3 with the nozzle throat located at $x = 0$ mm. In the left figure the typical pattern of primary and secondary shock wave with the shear layer in between is clearly visible. This structure is followed by an additional shock while the discontinuity at $x = 340$ mm is a reflected shock, which depends on the piston position and moves upstream with increasing valve gap. Compared to the starting process in a conventional shock tube as described in [6], these additional shocks stem from the turning of the flow along the valve piston as well as the varying area between the piston and the nozzle contour. The right figure illustrates the movement of the shocks with the increasing valve gap. As a part of the detailed analysis of the flow in the HLB, an exemplary density gradient distribution is shown in Fig. 4. After 1.5 ms run time the aforementioned configuration with three shocks traveling in downstream direction and the upstream moving shock is clearly visible. The variation of the Mach number at the center of the entry plane of the test section in time is shown in the left image of Fig. 5. Depending on the stagnation pressure the initial shock arrives after 3.15 ms or 3.65 ms run time. The flow topology with the gas front and the two consecutive shocks can be clearly identified. After the second shock at $t \approx 8$ ms the results are almost identical. Oscillations are present until $t \approx 15$ ms while the deviations at $t = 12$ ms are the remnants of the relocation of the sonic line which takes place when the area of the valve gap becomes larger than the nozzle throat. As long as the transition to supersonic speed occurs across

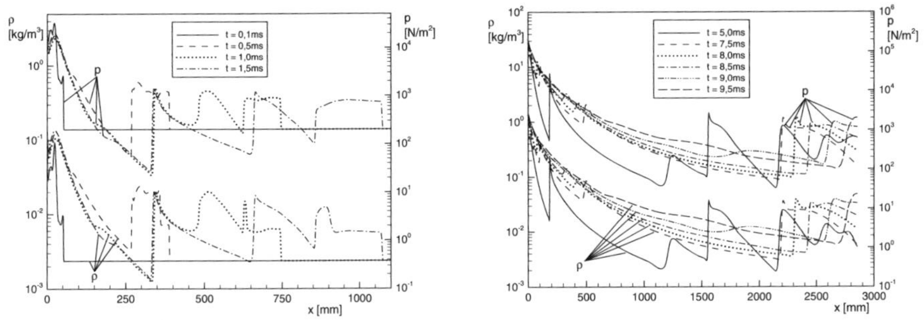

Fig. 3. Variation in time of density and pressure along the axisymmetry axis of the HLB for $p_0 = 3.25$ bar.

the valve gap, the streamlines separate from the nozzle contour downstream of the nozzle throat. This region is preceded by an oblique separation shock. Throughout the relocation of the sonic line, this shock moves upstream and travels as a planar shock wave into the storage tube at sonic speed.

This shock is also visible in the right image of Fig. 5 which shows the variation in time of the calculated and measured static pressure divided by p_0. The pressure transducer is located at the outer HLB contour near the valve body. While the valve gap becomes wider and the pressure decreases, the results show a good agreement. As soon as the shock mentioned above passes the transducer as a result of the sonic transition, the pressure increases. After that the flow is not stationary but exhibits oscillations which are different in the experiment and the simulation. In the experiment low-frequency oscillations with a large amplitude are visible. They might be identified as oscillations normal to the tube axis as described by Hartig and Swanson [7]. They are possibly excited by the non-axisymmetric propagation of an acoustic wave (expansion wave at the beginning or compression wave after sonic transition). From the experimental results in the beginning a frequency of 1060 Hz is extracted, which goes down to 730 Hz when information from the wider, non-heated section of the storage tube reaches the sensor (not shown in the graph). These values conform with the respective frequencies of 1280 Hz and 755 Hz for the first mode of a non-axisymmetric transverse oscillation, which is the result of the theoretical approach of Hartig and Swanson [7]. Due to the restrictions in the simulations, only axisymmetric oscillations may appear. An analysis of the simulated results yields a frequency slightly below 5000 Hz, while theory predicts 4855 Hz (second mode without axial periodicity or first mode with axial periodicity), resulting in an overall

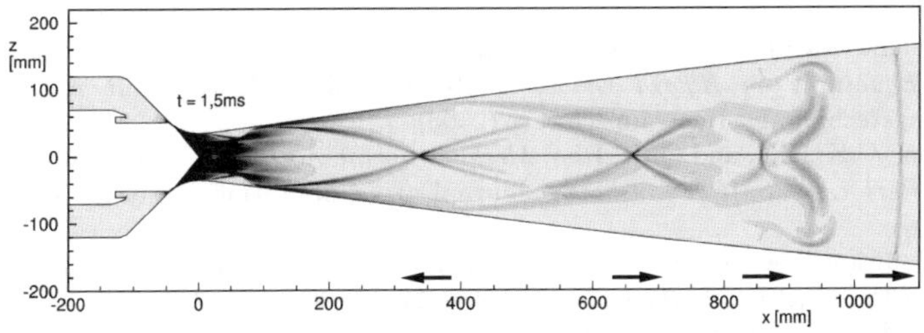

Fig. 4. Calculated density gradient distribution in the HLB for $p_0 = 3.25$ bar after 1.5 ms run time.

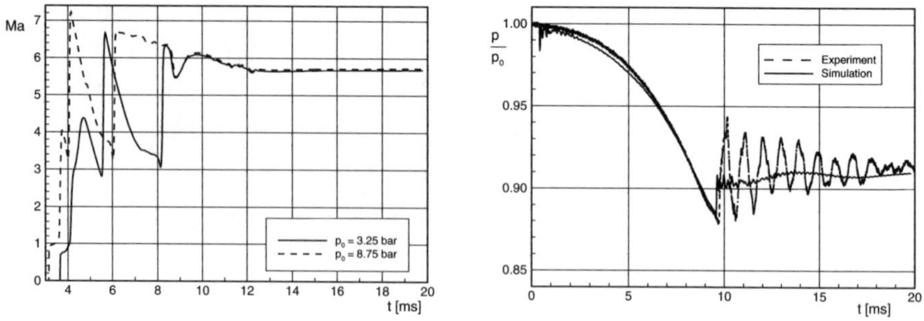

Fig. 5. History of Mach number at the center of the entry plane of the test section (left); Calculated and measured pressure variation in the HLB storage tube (right).

good accordance between theory and measurement as well as simulation. Apart from the oscillations the agreement of the calculated and measured mean values is very good.

5 Conclusion

The flow in the hypersonic Ludwieg-tube HLB during the first 20 ms of run time is numerically rebuilt for two stagnation pressures, including the measured valve opening but neglecting non-axisymmetric effects. Compared to wind tunnels using burst-diaphragms, the results show the typical onset of flow with additional shocks due to the fast-acting valve. Perturbations are induced until the relocation of the sonic line, which occurs when the valve has almost opened completely. Consequently, the duration until the flow conditions in the test section are steady is about 15 ms. Pressure oscillations in the storage tube visible in the numerical and experimental results agree well with theory.

References

1. G. Koppenwallner, R. Müller-Eigner, and H. Friehmelt: *HHK Hochschul-Hyperschall-Kanal: Ein 'Low-Cost' Windkanal für Forschung und Ausbildung*, DGLR-Jahrbuch **2**, pp 887–896 (1993)
2. T. Wolf, M. Estorf, and R. Radespiel: *Investigation of the Starting-Process in a Ludwieg-Tube*, Journal of Theoretical and Computational Fluid Dynamics **21**, 2, pp 81–98 (2007)
3. T. Gerhold: Overview of the Hybrid RANS Code TAU. In: *MEGAFLOW - Numerical Flow Simulation for Aircraft Design*, Notes on Numerical Fluid Mechanics and Multidisciplinary Design **89**, pp 81–92, ed by N. Kroll and J.K. Fassbender, Springer, Berlin Heidelberg (2005)
4. R.P. Dwight: *An Implicit LU-SGS Scheme for Finite-Volume Discretizations of the Navier-Stokes Equations on Hybrid Grids*, DLR-FB 2005-05, Institut für Aerodynamik und Strömungstechnik, DLR, Braunschweig (2005)
5. K.H. Kim, C.K. Kim, and O.-H. Rho: *Methods for the Accurate Computations of Hypersonic Flows, Part I: AUSMPW+ scheme*, Journal of Computational Physics **174**, 1, pp 38-80 (2001)
6. A.-S. Mouronval, A. Hadjadj, A.N. Kudryavtsev, and D. Vandromme: *Numerical investigation of transient nozzle flow*, Shock Waves **12**, pp 403–411 (2002)
7. H.E. Hartig, C.E. Swanson: *'Transverse' Acoustic Waves in Rigid Tubes*, Physical Review **54**, pp 618–626 (1938)

Numerical study of shock interactions in viscous, hypersonic flows over double-wedge geometries

Z.M. Hu[1], **R.S. Myong**[1,2], **and T.H. Cho**[1,2]

[1] *Research Center for Aircraft Parts Technology, Geyongsang National University, Jinju 660-701, South Korea*
[2] *School of Mechanical and Aerospace Engineering, Geyongsang National University, Jinju 660-701, South Korea*

Summary. The characteristics of the boundary layer in a supersonic flow can be drastically altered by a shock wave. The interaction between shock wave and boundary layer plays an important role in the design and operation of high-speed vehicles and propulsion systems. In this paper, the shock/shock interaction and the shock/boundary interaction over a compression ramp and a double-wedge structure in supersonic/hypersonic flows were studied numerically by solving the Favre-averaged Navier-Stokes equations. The numerical simulations show that the viscosity effect results in different interaction transition point, and induce more complex flow patterns, such as the separation shock, shocklets, shear unsteadiness, and complex interactions among them. With the presence of viscous effects the pressure load imposed on the wedge surface can be changed a great deal comparing to the inviscid flow simulations.

1 Introduction

Shock/shock, shock/boundary-layer interactions in the supersonic/hypersonic flows have a dramatic effect on heat transfer, skin fraction, and separation. The characteristics of the boundary layer in a supersonic flow can be drastically altered by a shock wave. For example, shock interaction caused the airframe damage to X-15. During a flight in which a ramjet test model was attached to the ventral fin of the X-15A. An unforeseen shock interaction produced high heating loads which burn through portion of the fin and there were substantial damage to the substructure and to subsystems enclosed in the ventral fin. There is also interest in shock interaction heating which has been found to result in extremely high heat loads at the engine inlet cowl on the high speed vehicle proposed by NASP. The interactions of shock and shock, shock and boundary layer are also important and fundamental issues occur in the inlet flows of a hypersonic air-breath Scramjet which has a multi-ramp shock-compression-driven intake. Such kind of damage to the ventral fin on the X-15 caused by shock/shock interaction phenomena was defined as one of unknown unknowns by Bertin and Cummings in [1,2] for the develop of hypersonic vehicles before the X-15A accident. Unknown unknowns are normally discovered during flight tests and could bring on drastic consequences to the survival of the vehicle or of the crew and lead to unacceptable increases in the costs to develop the vehicles [1].

Edney [3] used shock polar diagrams to classify the interactions of oblique shock waves and bow shocks on a cylinder firstly. Through his experimental researches, it was realized that anomalously high heating and pressure loads can be induced by shock interactions on supersonic or hypersonic vehicles and small changes in the geometry can lead to large changes in the overall flow structure. Olejniczak et al [4] numerically studied inviscid, perfect gas shock interactions on double-wedge like geometries. It was found that there is a critical Mach number, above which underexpanded jet flow always occurs along the wedge surface. The interaction of jet with the adjacent subsonic region results in large-amplitude pressure variations on the wedge surface. Ben-Dor et al [5] revealed that there is a hysteresis and self-induced oscillations in the shock flow pattern for various angles of inclination of the second wedge. Again, this study only solved the inviscid perfect gas equations for simplification. A large number of papers have been published on this topic. However, there are still some questions unanswered up to now. For example, what are the effects of the viscosity and finite-rate chemical reactions which are coupled to the fluid motion? Does the shock/boundary-layer interaction cause substantial change to the flow structures of shock/shock interaction?

In this paper, Favre-averaged Navier-Stokes equations are solved to study on the detailed and individual flow structures of the interaction among shock waves and boundary-layers. For consideration of the numerical technologies, it is a great challenge to capture the flow features in such thin shock layers induced by a compression ramp or a double-wedge like geometry in the supersonic/hypersonic flow, including various types of shock interactions, strong shear layers, separations, and shock/boundary-layer interactions, etc. In the following sections, after a brief description of the governing equations and turbulence models, shock-wave/boundary-layer interactions are simulated and compared with available experimental data to calibrate the code and the turbulence models. Then, for double-wedge supersonic flows, only laminar and inviscid flows of interest are studied and discussed in present study.

2 Numerical algorithms

For the high supersonic and hypersonic turbulent flows, the governing equations shall be Favre-averaged or mass-averaged Navier-Stokes equations. The equations for multi-component system in two dimensional Cartesian coordinates were used to govern the FANS mean flow. For the turbulent flow, several kind of two-equation $k-\omega$ models were used, such as Wilcox's (1988, 1998) models [6,7] and Menter's (1994) BSL and SST models [8].

For the flows of interest, the flowfields generated by the interaction of shock wave with boundary layer are inherent unsteady. However, turbulence models used in RANS/FANS methods do not account for the unsteady motion of the shock [9]. This is one of the main reasons for their poor performance in strong inviscid-viscous interactions. For example, in compression ramp flows, conventional RANS models have failed to accurately predict the location of the separation shock, size of separation bubble at the ramp corner, mean velocity profiles downstream of the interaction and heat transfer [10,11]. Shock-unsteadiness correction [12] was applied to turbulence models for the simulation of supersonic compression ramp flows. But the modification delays reattachment and leads to slow recovery of the boundary layer on the ramp [9] and other problems while perform better for capturing the separation spot than the original models. Two possible mechanisms have been proposed to explain the shock system unsteadiness: the first one is the turbulence of the incoming boundary layer, and the second is that the separation shear layer amplifies the low frequencies of the contracting/expanding bubble motion, which are felt upstream through the large subsonic region of the separated zone [13].

For numerical algorithms, the dissipative terms were discretized by a fourth order centered difference scheme, while the convective terms were discretized using the second order DCD (dispersion controlled dissipative) scheme proposed by Jiang et al for shock wave capturing [14,15].

3 Compression ramp flows and code calibration

The principal elements of the flowfield structure include the amplification of the turbulence by the shock waves in the boundary-layer and external flow, formation of a new layer in the near wall part of the attaching flow, formation of Taylor-Göttler vortices, and reverse transition in the separation region due to the favorable pressure gradient and decreasing of the local Reynolds number in the reverse flow caused by the decease of velocity in the direction of the separation point. Generally, the flow characteristics are the boundary layer, the separation bubble, the separation shock, and the reattachment shock. One test case of supersonic flow within a compression ramp (computational domain and the structured grid are shown in Fig. 1) is now calculated for code calibration and model effect comparison. The flow condition is defined as follows: $p_\infty = 483kPa, T_\infty = 303K, M_\infty = 2.94, \delta = 0.00827m, Re_\delta = 2.8 \times 10^5, \alpha = 20^o$.

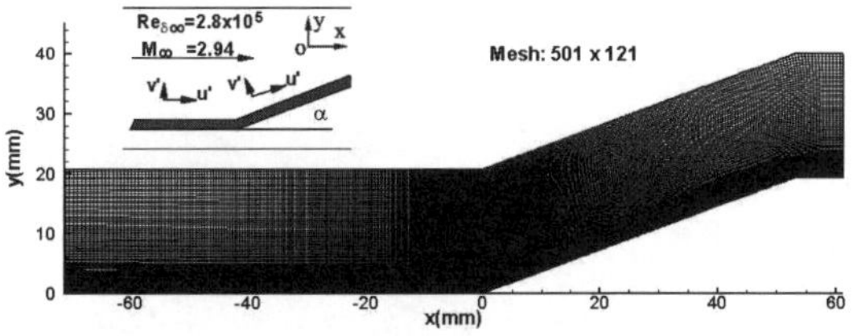

Fig. 1. Computational domain and the structured grid.

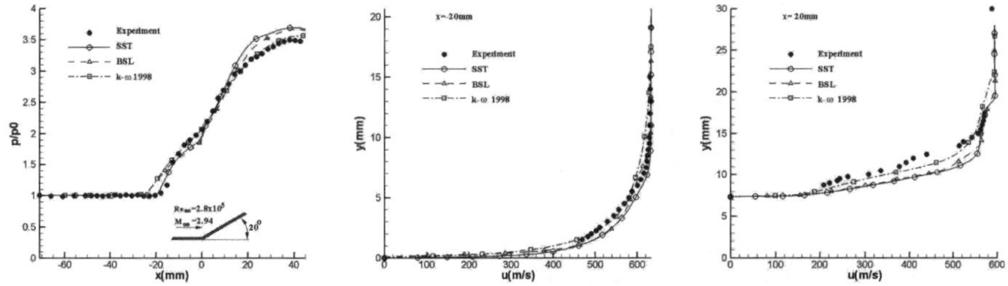

Fig. 2. Comparison of normalized surface pressure and velocity components with experimental measurements [16].

For the comparison of the effects of turbulence models on the shock/boundary interaction, the Wilcox's 1998 two-equation model [7] and Menter's SST and BSL models [8] are used for the following simulation. The experimental measurements [16] of wall pressure and velocity component, as well as the numerical results are shown in Fig. 2. Either in the separation region or in the reattachment region, the Wilcox's 1998 model performances better than the other two models according to the wall pressure. However, upstream of the separation bubble, the best one for capturing the velocity profile is the Menter's SST model, and then the BSL model and the Wilcox's 1998 model in sequence. Downstream the reattachment region, none of the turbulence models can predict the velocity profile well. Free-stream conditions, separation unsteadiness, and three-dimensional effects shall be considered for a better prediction of the inviscid/viscous interaction in future studies. The numerical experience here proved that the resulted flowfield suffers some from the inflow boundary conditions. It needs further analysis and discussion.

4 Double-wedge flows

In this section we consider a hypersonic flow over a double-wedge geometry as shown by Fig. 3. The oblique shock waves induced by the two wedges interact with each other, and the shock waves interact with the boundary layer and separation flow over the wedge corner. The interactions result in the formations of complex shock wave, shear layer, and recirculation flow patterns. For the following simulations, the Mach number of the free-stream flow is set to be 7, and the length ratio $L_2 : L_1 = 5cm : 10cm, \theta_1 = 15^o$, while change θ_2 for different cases. Single-block structured girds with 701×201 nodes were used to discretize the computational domains for all the double-wedge geometries.

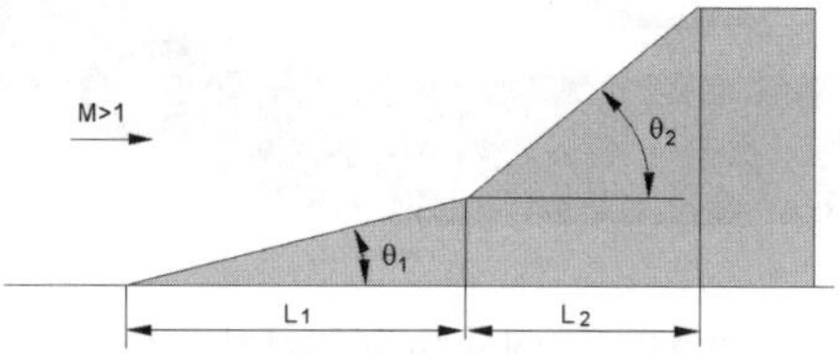

Fig. 3. Double-wedge geometry.

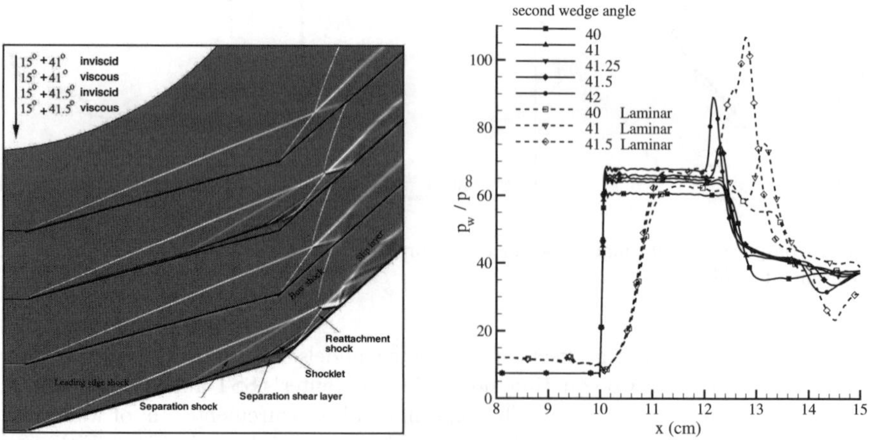

Fig. 4. Viscosity effect on the type VI and V shock interaction transition (filled symbols-inviscid cases, open symbols-viscous cases).

The viscous effect on the shock interactions was studied using laminar simulation. As illustrated by the profiles with open symbols in Fig. 4, the interaction type VI completely changes into type V while boundary layer induces new flow structures in the case $\theta_2 = 41^o$. Unexpectedly, the maximum pressure load on the wedge surface with $\theta_2 = 41.5^o$ increases about one third when viscosity is considered. As shown in Fig. 4, the separation flow near the wedge corner triggers off a separation shock, and the unsteady shear layer in the separation flow brings on several shock-lets. The separation shock strengthens the leading edge shock, and the shock-lets and the separation bubble affect the reattachment shock. All these flow characteristics change the flow conditions around the shock interaction spot, and alter the shock interaction patterns. With the presence of shock-lets and shear flow unsteadiness in the separation region, the effects of the separation region can no longer be modelled simply as an extra deflection in an inviscid flow.

Fig. 5 shows the comparison of numerical(left top) and experimental(right top) [17,18] schlieren for type V interaction when $\theta_2 = 45^o$. The leading shock, the bow shock, the separation shock, and the reattachment shock were well captured. However, the shear-layer unsteadiness within the separation bubble was over predicted since only laminar simulation was used in this case, and the over-predicted unsteadiness induced strong shock-lets structures. As a consequence, the shock interaction structure appears some difference from the experimental measurement due the the effect of the over-predicted shock-lets on the shock/shock interaction structure. The over-prediction indicates that an advanced turbulence model or LES and DNS should be used to appropriately model the separation flow over the wedge corner. Unsteady flow features associated with the movements of the separation shock and the curved shock were also illustrated in [19]. The experimental results show that the movements of the separation shock and curved shock don't seem to be connected to each other [19]. Fig. 5 (lower part) shows the self-oscillation of the shock/shock interaction structure which was reported [5] to be triggered by the out-phase of the propagations of the triple points within the shock/shock interaction frame. To properly model the unsteadiness induced by shock-lets and shock/boundary interaction is a great challenge to any existing turbulence models and need further studies.

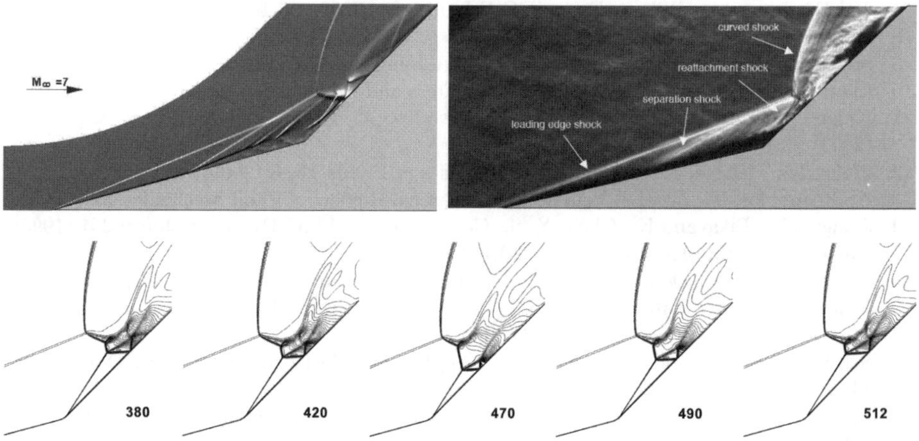

Fig. 5. Type V interaction and the shock/boundary layer interaction structures. Top left: numerical schlieren; top right: experimental schlieren [17,18] (with permission); low: sequential isopycnics showing the oscillation of the interaction structures(inviscid solutions).

5 Conclusion

In this paper, the shock wave/boundary layer interactions occurring in the supersonic/hypersonic flows induced by the compression ramps and double-wedge structures were numerically studied. Some turbulence models were applied to the compression ramp flows. It concludes that the viscosity and boundary layer alter the transition point, and induce more complex flow patterns, such as the separation shock, the shock-lets, and the shear unsteadiness. The pressure load imposed on the wedge surface with presence of viscosity effect can be changed a great deal comparing to the inviscid flow simulations. Further study will be needed since the fact that the shear-layer unsteadiness within the separation bubble was over predicted without turbulence consideration in the current simulation.

Acknowledgement. This work was supported by Korea Research Foundation Grant No. KRF-2005-005-J09901 and the Second Stage Brain Korea 21. The authors would like to thank Dr. Schrijer, F.F.J. at Delft University of Technology for his courtesy and Schlieren figure.

References

1. Bertin, J.J., Cummings, R.M.: Progress in Aerospace Sciences **39**, (2003)
2. Bertin, J.J., Cummings, R.M.: Annual Review of Fluid Mechanics **38**, (2006)
3. Edney, B.: Anomalous heat transfer and pressure distributions on blunt bodie at hypersonic speeds in the presence of an impinging shock. Rep.115, The Aerospace Research Institute of Sweden, Stockholm, Swede (1968)
4. Olejniczak, J., Wright, W.J. and Candler, G.V.: Journal of fluid Mechnics **352**, (1997)
5. Ben-Dor, G., Vasilev, E.I., Elperin, T. and Zenovich, A.V.: Physics of Fluids **15**,12 (2003)
6. Wilcox, D.C.: AIAA Journal **26**,11(1988)
7. Wilcox, D.C.: One-equation and two-equation Models. In: *Turbulence Modelling for CFD*, 2nd Edition, ed by Wilcox, D.C., (DCW Industries, 2000) pp 103-218
8. Menter, F.R.: AIAA Journal **32**,8 (1994)
9. Sinha, K., Mahesh, K., Candler, G.V.: AIAA Journal **43**, 3 (2005)
10. Liou, W.W., Huang, G., Shih, T.H.: Computers and Fluids **29**, 3 (2000)
11. Knight, D. Yan, H., Panaras, A.G., Zheltovodov, A.: Progress in Aerospace Sciences **39** (2003)
12. Sinha, K., Mahesh, K., Candler, G.V.: Physics of Fluids **15**, 8 (2003)
13. Andreopoulos, Y., Agui, J.,Briassulis, G.: Annual Review of Fluid Mechanics **32** (2000)
14. Jiang, Z.L., Takayam, K., Chen, Y. S.: Computational Fluid Dynamics Journal **4** (1995)
15. Jiang, Z.L.: Acta Mechanica Sinica **20**,1 (2004)
16. Settles, G. S. and Dodson L. J.: Hypersonic shock/boundary-layer interaction Database: New and corrected data. NASA-CR-177638, Department of Mechanical Engineering, Penn State University, University Park, PA 16802, (1994)
17. Schrijer, F.F.J., Scarano, F., and et. al.: Application of PIV in a hypersonic Double-ramp Flow. AIAA Paper 2005-3331 (2005)
18. Schrijer, F.F.J., Scarano, F., and van oudheusden, B.W.: Expeiments in Fluids **41**,2 (2006)
19. Schrijer, F.F.J., van oudheusden, B.W., and et. al.: Quantitative visualization of a hypersonic double-ramp flow using PIV and schlieren. 12th International Symposiumon Flow Visualization, Götthinggen, Germany, Sep. 2006

Numerical study of thermochemical relaxation phenomena in high-temperature nonequilibrium flows

S. Kumar[1], H. Olivier[1], and J. Ballmann[2]

[1] *Shock Wave Laboratory, RWTH Aachen University, Germany*
[2] *Mechanics Laboratory, RWTH Aachen University, Germany*

Summary. To study the atmospheric reentry phase of a space vehicle, it is necessary to correctly understand the thermochemical nonequilibrium processes coupled to the aerodynamic aspects of this critical phase. In a typical hypersonic flow about a blunt body, the region between the body surface and the shock is the site of intensive thermochemical processes. The different internal energy modes of the molecules are far from their equilibrium state. The energy exchanges between these different modes occur according to the relaxation time associated to each process. Detailed physico-chemical models for air in chemical and thermal nonequilibrium are needed for a realistic prediction of hypersonic flow fields. In the framework of the present work, an adaptive CFD code *QUADFLOW* has been extended to take into account real gas effects for a five components air model. Different thermochemical models have been implemented in the code. The uncertainties associated with the physico-chemical modelling and their influence on the flow field are discussed with the help of computational results.

1 Introduction

The flow field around a blunt body in hypersonic flow is characterized by the presence of a strong bow shock wave in front of the body. Under certain high enthalpy free stream conditions, the temperature behind the bow shock wave will become so high that the excitation of internal energy modes and/or dissociation of the molecules can occur. These high temperature effects are initiated and proceed through collisions of the molecules. To reach a full excitation of e.g. the vibrational energy mode, a molecule needs to collide a number of times with other molecules. The same is true for a chemically reacting medium to reach its equilibrium composition at a given pressure and temperature. In general, these processes have characteristic time scales for reaching equilibrium. In hypersonic flow fields, the flow velocity can be very high and the density very low. The characteristic time needed for the equilibrium thermodynamic state and equilibrium chemical composition can lag those at the local pressure and temperature significantly. The flow is then said to be in chemical and thermal nonequilibrium. Some practical consequences of the high temperature phenomena in hypersonic (so-called real gas effects) are significant changes in shock stand-off distances, peaks in thermal loads, skin friction drag, forces and moments on the vehicle. It is important to note that not only heat transfer and skin friction but even a pressure dependent aerodynamic quantity such as the pitching moment, can be affected by high temperature chemistry. Thus, an improved understanding of various thermochemical processes is required in order to accurately predict the flow characteristics. In this study, an adaptive CFD code *QUADFLOW*, being developed at RWTH Aachen, was used for the computations. The influence of thermochemical models on the relaxation processes are discussed with the help of computational results.

2 Description of *QUADFLOW*

Reference [1] describes the detailed features of the adaptive CFD code *QUADFLOW*. The method consists of an integrated framework, including a novel grid generation technique using B-splines, advanced adaptation criteria based on multiscale analysis and a flow solver, which is capable to operate on arbitrary unstructured meshes. The discretization scheme has to meet the requirements of the adaptation concept and has to fit well with the mesh generation. This requires the development of a finite volume scheme for fairly general cell partitions that is surface based and can cope, in particular, with hanging nodes and possible unstructured parts in complicated regions of the flow domain. For this purpose, the locally adapted grid is treated as a fully unstructured mesh with arbitrary polygonal/polyhedral control volumes in two and three space dimensions, respectively. Several approximate Riemann solvers and flux-vector splitting schemes have been incorporated. A linear, multidimensional reconstruction of the conservative variables is applied to increase the spatial accuracy. In order to avoid oscillations in the vicinity of local extrema and discontinuities, a limiter function with TVD property is used. In its basic form, the flow solver assumes a thermally and calorically perfect gas model. In the context of real gas simulations, upwind schemes as extended for chemically reactive gas mixtures have been implemented. Various physico-chemical modelling to take into account relaxation processes have been incorporated. Further details on the implementaion of the physico-chemical models in *QUADFLOW* can be found in Ref. [6].

3 Results and discussions

3.1 Inviscid air flow over a circular cylinder

The hypersonic two-dimensional inviscid air flow over a circular cylinder was investigated, with the free stream conditions as follows: $\rho_\infty = 0.0016\ kg/m^3$, $T_\infty = 196\ K$, $M_\infty = 12.7$, $X_{N_2} = 0.75$, $X_{O_2} = 0.25$. The cylinder has a diameter of 0.0406 m. The computation was made considering calorically perfect, chemical nonequilibrium [8] and thermochemical equilibrium gas models. Figure 1 shows the iso-density plot for the three different flows. The comparison of temperature distribution along the stagnation line is shown in Fig. 2. The shock stand-off distance becomes smaller with the internal energy excitations and/or chemically reactions. This is evident from both the figures shown, where it is largest in the case of calorically perfect gas and smallest for thermochemical equilibrium flow. The analytical stagnation point temperature as calculated using $\left(T_S = T_\infty \left(1 + 0.5(\gamma - 1)M_\infty^2\right)\right)$ is 6519 K and the computed value is 6614 K. Thus an excellent agreement between analytical and computational result has been achieved. The analytical stagnation point temperature for the present test case in thermodynamic equilibrium, as determined from jump conditions across the shock is 3342 K, which is in quite good agreement with the computed value of 3430 K. The agreement regarding the shock stand-off distance between the present computation and as calculated using the empirical correlation of Billig [2] is quite good, as shown in Fig. 2.

3.2 Relaxation behind normal shocks

The rates at which the relaxation processes occur behind strong shock waves have an important effect on the flow field around high-speed missiles and re-entering vehicles. The stationary, inviscid one-dimensional flow of a five component air in thermal and chemical nonequilibrium behind a normal shock is investigated. The free stream conditions are: $V_\infty = 6000\ m/s$, $p_\infty = 100\ Pa$, $T_\infty = 300\ K$ and the post normal shock conditions are: $V_s = 945\ m/s$, $p_s = 35070\ Pa$,$T_s = 16630\ K$. Figures 3 to 5 show the plots of mass fraction of molecular components and temperatures for the chemical reaction rate models of Park 87 [9], Park 90 [10] and Gupta [3]. All the results are obtained using the nonpreferential dissociation model of Treanor and Marrone [11], [7] and Klomfaß's vibration-translation and vibration-vibration models [5]. In the plots, the temperatures are normalised with the value of the translational temperature directly behind the normal shock. At high temperatures, say, around 10000 K and above, chemical and vibrational relaxation times are of the same order. Therefore, the chemical reactions run widely under vibrational nonequilibrium, which is evident from the results. The relaxation length depends on the rates of different chemical reactions. The reaction rate sets of Park 90 and Gupta model are found to be fastest and slowest, respectively. Therefore, the relaxation length corresponding to Park 90 and Gupta model are largest and smallest, respectively. Even though the relaxation curves for N_2, O_2 and NO differ in the Figs. 3 to 5, the same final equilibrium condition is arrived. This is consistent with the fact that the final equilibrium condition for a given free stream condition is independent of a chemical rate set, since it corresponds to infinite reaction rates.

3.3 Inviscid nitrogen flow over a circular cylinder

In this section, results of a similar study as that of the one-dimensional flow for a two-dimensional inviscid nitrogen flow over a circular cylinder are presented. The free stream conditions taken from one of the experiments of Hornung [4] are as follows: $V_\infty = 5594\ m/s$, $\rho_\infty = 0.00498\ kg/m^3$, $T_\infty = 1833\ K$, $X_N = 0.073$. The cylinder has a radius of 0.0127 m. It is assumed that the vibrational energy of N_2 in the free-stream is in equilibrium with other modes of energy and thus $T_{N_2}^{(vib)} = 1833\ K$.

Figure 6 shows the comparison of experimental and computational interferograms of the flow field for the chemical kinetic models of Park 85, Park 90 and Gupta. For all the three cases, the vibration-dissociation coupling of Treanor and Marrone [7], [11] with nonpreferential dissociation was considered. For the vibration relaxation time required in the vibration-translation relaxation model, the model of Klomfaß [5] was used. The agreement regarding the shock stand-off distance is quite satisfying but the differences in the flow field are clearly seen. Thus it is inferred that the shock stand-off distance is not as sufficient parameter as the density field to validate the physiochemical modelling. It is observed from the figures that the Park 90 rate set produces the flow field which is in quite good agreement with the experimentally obtained flow field. Figure 7 (left side) shows the temperature distribution along the stagnation line for these three chemical kinetic models. The shock stand-off distance is largest with the Park 85 model and smallest with the Gupta model. This is consistent with the fact that the rate constants for N_2 dissociation for the Park 85 model is smaller than that for the Gupta model and that the Park 90 rate set lies in between.

In case of the *nonpreferential* dissociation model, dissociation occurs with equal probability from any vibrational level in any collision that has sufficient translational energy to effect the dissociation. In case of the *preferential* model, the dissociation probability is higher for higher vibrational levels. The molecules have to first ladder climb to come to a higher excited level before dissociation takes place. Thus a lag in vibrational excitation has the effect of inhibiting the dissociation process because of the lack of highly excited (and thus easily dissociable) molecules. Thus the rate of dissociation is smaller in case of the preferential dissociation model compared to the nonpreferential model. The slower dissociation corresponds to a lower density rise and thus to a larger shock stand-off distance, as is evident in Fig. 7 (right side). The shock stand-off distance can differ as much as by 10 % and the flow field can significantly differ from the experimentally observed flowfield. Reference [6] provides further detailed studies of this test case with various models and their influence on the flow field.

4 Conclusions

The rates at which the relaxation processes occur behind strong shock waves have an important effect on the flow field around hypersonic vehicles. Using one- and two-dimensional computational results, we have shown the sensitivity of flow field with respect to the physico-chemical modelling. Different thermochemical models produce different flow fields. A practical quantity like the shock stand-off distance can be changed by more than 10% by changing the thermochemical models for the same free stream conditions. Thus, the reliable application of CFD codes for hypersonic flow simulations requires an improved input by way of physico-chemical models and associated data.

References

1. Bramkamp F., Lamby Ph., Müller S., An Adaptive Multiscale Finite Volume Solver for Unsteady and Steady Flow Computations, Journal of Computational Physics, vol. 197, no. 2, pages 460-490, 2004
2. Billig F.S., Shock-Wave Shapes around Spherical- and Cylinderical-Nosed Bodies, Journal of Spacecraft and Rockets, vol. 4, no. 6, pages 822-823, 1967
3. Gupta R. N., Yoss J. M., Thompson R. A., Lee K.P., A review of reaction rates and thermodynamic transport properties for an 11-species air model for chemical and thermal nonequilibrium calculations to 30000K, NASA-RP-1232, 1990
4. Hornung H.G., Nonequilibrium Dissociating Nitrogen Flow over Spheres and Cylinders, Journal of Fluid Mechanics, vol. 53, pages 149-176, 1972
5. Klomfaß A., Hyperschallströmungen im thermodynamischen Nichtgleichgewicht, Ph.D thesis, RWTH Aachen University, Germany (ISBN 3-8265-0887-4), 1995
6. Kumar S., Numerical Simulation of Chemically Reactive Hypersonic Flows, Ph.D thesis, RWTH Aachen University, Germany (ISBN 3-8322-5065-4), 2005
7. Marrone P.V., Treanor C.E., Chemical Relaxation with Preferential Dissociation from Excited Vibrational Levels, The Physics of Fluids, vol. 6, no. 9, pages 1215-1221, 1963
8. Park C., On the Convergence of Computation of Chemically Reacting Flows, AIAA paper 85-0247, 1985
9. Park C., Assessment of Two-Temperature Kinetic Model for Ionizing Air, AIAA paper 87-1574, 1987
10. Park C., Nonequilibrium Hypersonic Aerothermodynamics, Wiley Interscience, 1990
11. Treanor C.E., Marrone P.V., Effect of Dissociation on the Rate of Vibrational Relaxation, The Physics of Fluids, vol. 5, no. 9, pages 1022-1026, 1962

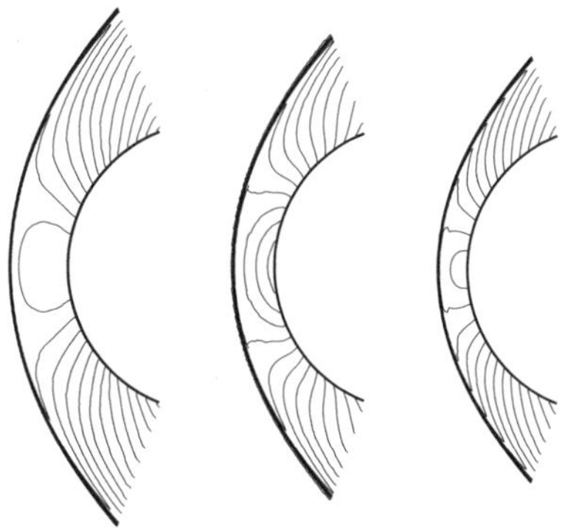

Fig. 1. Iso-density plots (left: calorically perfect, center: chemical nonequilibrium, right: thermochemical equilibrium)

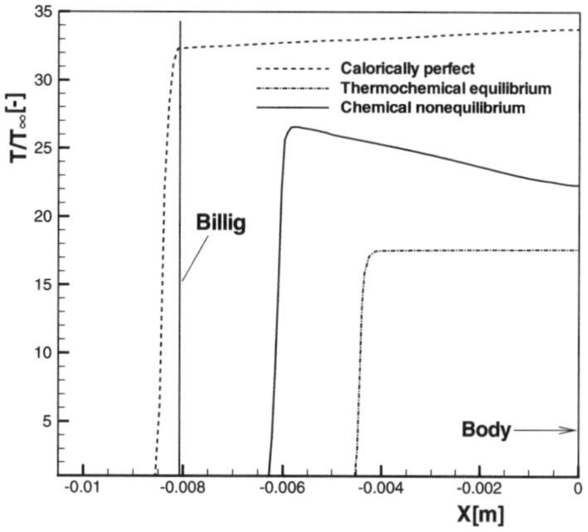

Fig. 2. Temperature distribution along the stagnation line

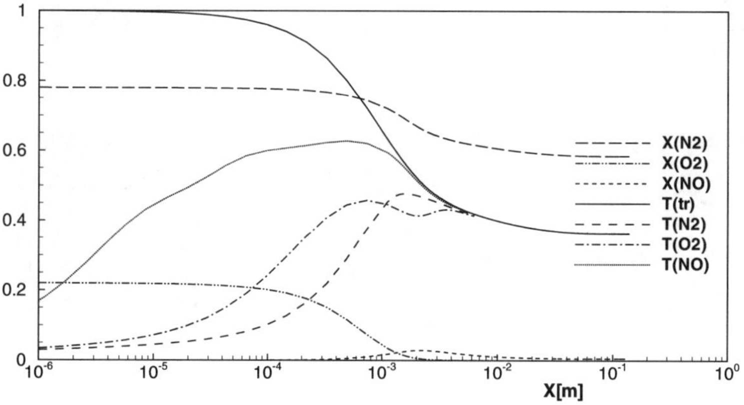

Fig. 3. Nondimensional profiles of X_i and $T^{(i)}$, Park 87 rate constants

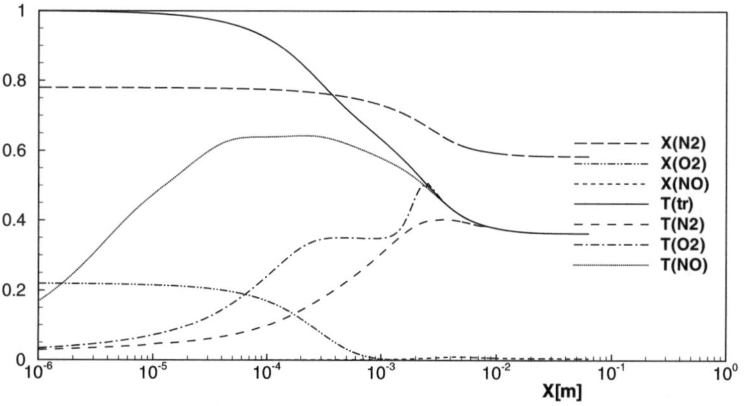

Fig. 4. Nondimensional profiles of X_i and $T^{(i)}$, Park 90 rate constants

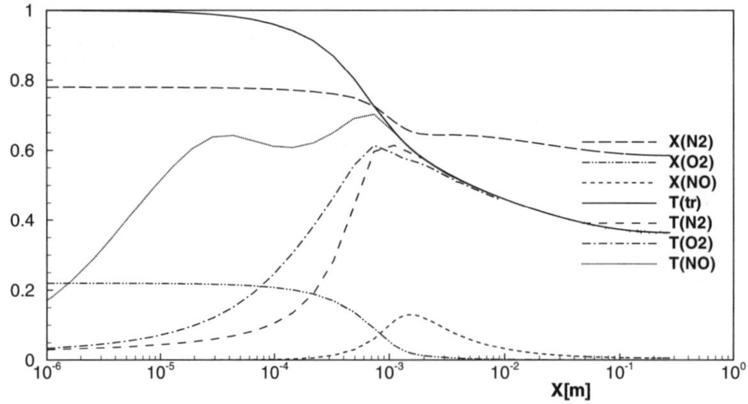

Fig. 5. Nondimensional profiles of X_i and $T^{(i)}$, Gupta rate constants

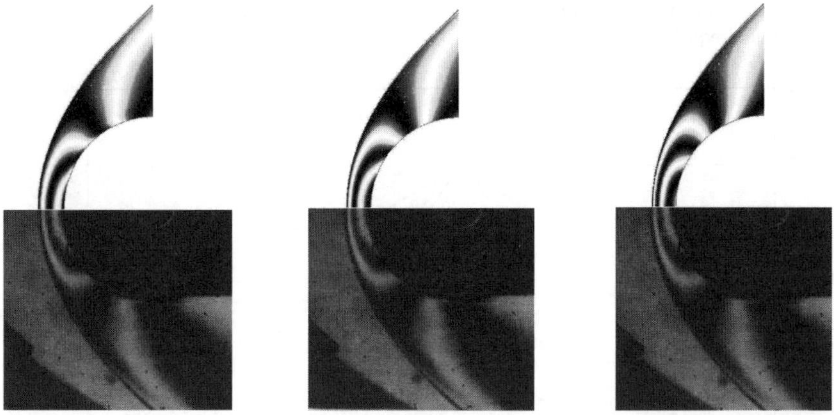

Fig. 6. Comparison of interferograms (bottom: experimental interferogram, top: computed interferograms (left: Park 85, center: Park 90, right: Gupta model))

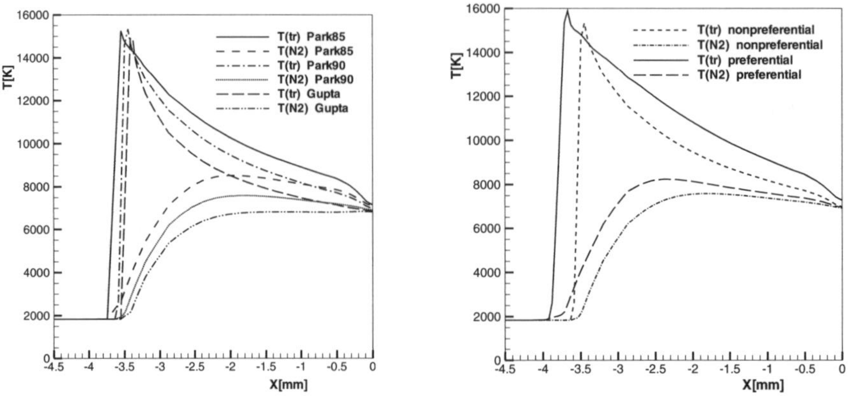

Fig. 7. Temperature distribution along the stagnation line

Numerical study of wall temperature and entropy layer effects on transitional double wedge shock wave/boundary layer interactions

T. Neuenhahn and H. Olivier

Shock Wave Laboratory, RWTH Aachen University, Germany

1 Introduction

Recent and upcoming hypersonic flight tests and upcoming future hypersonic applications require some leading edge bluntness to reduce the high heat loads in these stagnation regions. On the other hand blunt leading edges induce additional total pressure loss due to the strong bow shock in the vicinity of the blunt leading edge. Thus a small but finite leading edge bluntness is required. Moreover, the entropy layer generated due to the bluntness has significant influence on the flow field downstream as e.g. a shock wave/boundary layer interaction (SWBLI). This has been investigated for a compression corner with a blunt leading edge by various authors [1], [2], [3], [4]. Holden [1] observed that the separation bubble first grows and than decreases with increased leading edge radius. He identified that the reversal point is defined by a value between 0.5 and 0.7 of the combined bluntness-viscous interaction parameter [5] which rates the influence on the leading edge shock due to the leading edge bluntness and due to the displacement of the developing boundary layer. Additional to the effect of leading edge bluntness the effect of the elevated wall temperature on the shock wave/boundary layer interaction of a compression corner has been investigated [9], [10] showing that the separation bubble increases with elevated wall temperature. The combined influence of leading edge bluntness and elevated wall temperature on a SWBLI of a compression corner has been studied quiet recently [4] being the issue of this paper. The experiments agree with the findings by Holden [1] for the leading edge bluntness for constant wall temperature and also with the observations for the elevated wall temperature [9], [10]. Finally, the heat transfer measurements indicated a transitional behaviour of the shock wave/boundary layer interaction thus the commercial CFD code CFX is employed for this numerical study because it features the Menter/Langtry transition model.

2 Numerical method

The commercial code CFX [6] can solve the steady or unsteady Navier-Stokes equations implicitly using a finite volume method and a second order advection sheme. The employed mesh generator is ICEM Hexa 11.0 allowes the generation of high quality hexahedral grids. For all computated cases a grid convergence study has been performed showing the grid independence. At the wall $y^+ < 1$ was ensured.

2.1 Validation for hypersonic flow

Bluntness and viscous effects in the leading edge region

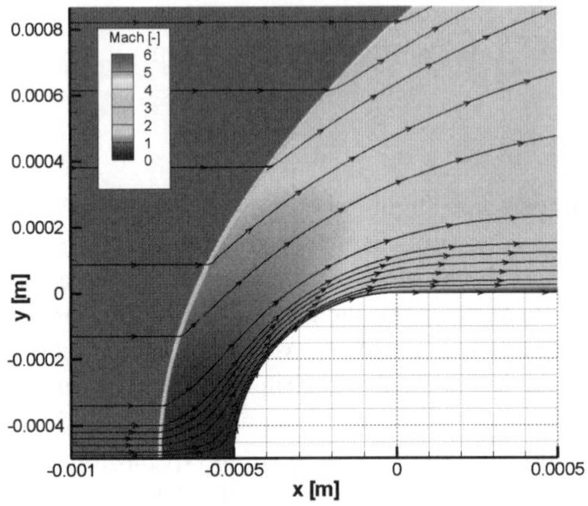

Fig. 1. Mach number distribution of leading edge region

To validate the commercial code for the hypersonic regime the first test case is a flat plate with leading edge radii between 0 to 1 mm and Mach numbers between 2 and 7.5 (Fig. 1). For bluntness dominated flows the computations agree with the empirical formulation for shock stand-off by Billig [7] (Fig. 2). The Blast Wave Theory (BWT) [7] describing the pressure distribution downstream of a blunt leading edge is employed for bluntness dominated flows. The aggrement in the area of validity [8] is good, see e.g. the 1.0 mm blunt leading edge computation (Fig. 3). Moreover, the comparison of the computation for 1.0 mm leading edge radius for inviscid and viscous flow show the same wall pressure distribution thus indicating a complete bluntness dominated leading edge region. As is well known the viscous effects downstream of a sharp leading edge are described by the strong and weak viscous interactions theory (VIT). For the sharp leading edge the pressure distribution determined by the CFD computation and by the VIT fits well (Fig. 3).

Laminar Shock Wave/Boundary Layer Interaction

The next validation case is a hypersonic laminar SWBLI of a flat plate/ramp model investigated by Lewis et al. [9]. The flat plate had a length of 63.5 mm followed by a 10.5° inlcined ramp. The experiments have been performed in GALCIT Mach 6 wind tunnel which provided a free-stream Mach number of 6.06 and unit Reynolds numbers between $0.15 \cdot 10^4$ and $5 \cdot 10^4$ 1/m. The variation of the free-stream Reynolds nummber allowed to show the known behaviour of laminar SWBLI which have an increased separation region with increased Reynolds number (Fig. 4). The position of the separation in the

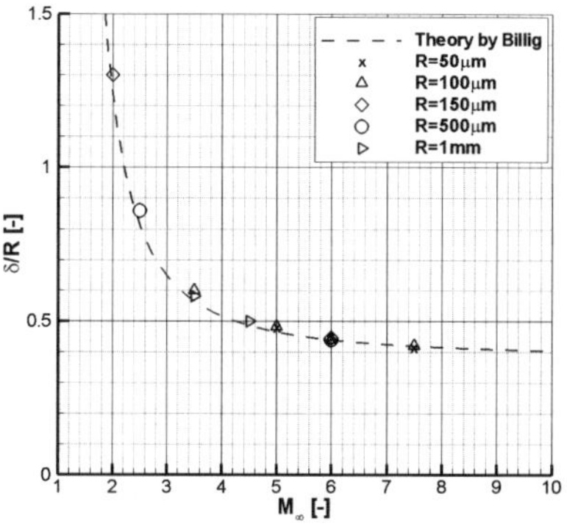

Fig. 2. Shock stand-off for different leading edge radii and different Mach numbers, $Re_\infty = 2.40 \cdot 10^6 - 7.19 \cdot 10^6 1/m$

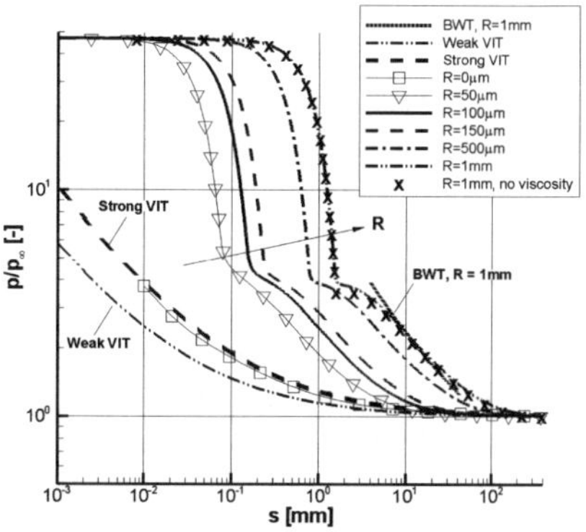

Fig. 3. Pressure distribution for different leading edge radii at Ma = 6, $Re_\infty = 7.19 \cdot 10^6 1/m$

experiment and the numerical solution does not fit perfectly. It is observed that for the numerical solution the separation occurs more downstream. In the previous test cases the model was cooled having a wall-to-free-stream-temperature ratio of 1.70 thus allowing to investigate the wall temperature effect with the model run under adiabatic conditions. These experiments showed the increased separation length with increased wall temperature as observed by other researchers [10], [4] (Fig. 5). The offset in the pressure values might be due to measurement uncertanties of the free-stream pressure because the code shows its capabilities to handle hypersonic flat plate flows.

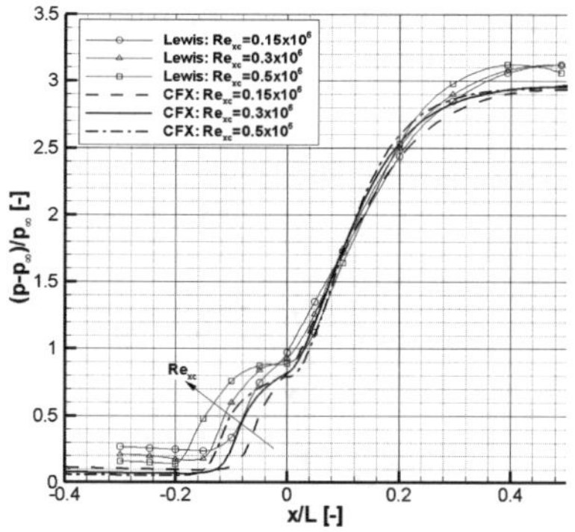

Fig. 4. Reynolds number effect

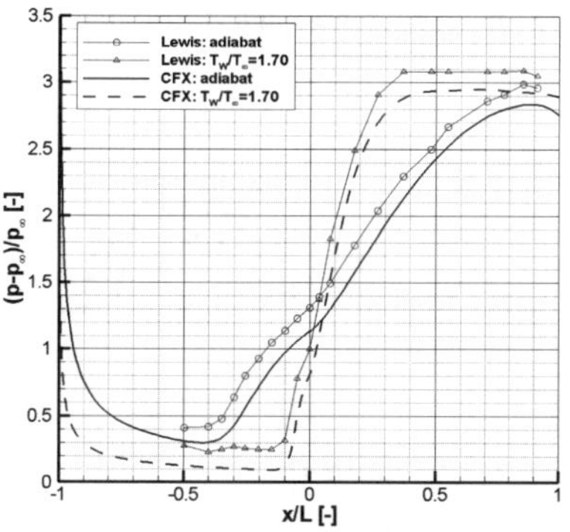

Fig. 5. Wall temperature effect

3 Heated double wedge configuration with different leading edge radii

The double ramp configuration has been investigated in the hypersonic shock tunnel TH2 of the Shock Wave Laboratory at RWTH Aachen University. The simulated free-stream conditions of tunnel condition I are a Mach number of 8.1, unit Reynolds number $3.8 \cdot 10^6$ 1/m, stagnation enthalpy 1.54 MJ/kg and static pressure of 520 Pa. The two ramps have a length of 0.18 m and 0.255 m, and are inclined to the free-stream 9° and 20.5°, respectivley. The model has a width of 0.27 m thus fullfilling the criteria for two-dimensional flow given by [9]. Different wall temperatures of 300 K, 600 K and 760 K have been realized by an electric heating. A complete description of the experiments is given in [4]. Figures 6 and 7 show a comparison of schlieren images of the experiment and the computation for a leading edge radius of 0.5 mm and a wall temperature of 760 K. Moreover, Figure 6 indicates the separation and reattachement positions marked in the following pressure and Stanton number distributions showing a good aggreement.

Figures 8 and 9 show the pressure coefficient and the Stanton number for the sharp leading edge (R = 0 mm) and a wall temperature of 300 K for laminar, transitional and turbulent flows. The computation of laminar flow overpredicts the separation size and

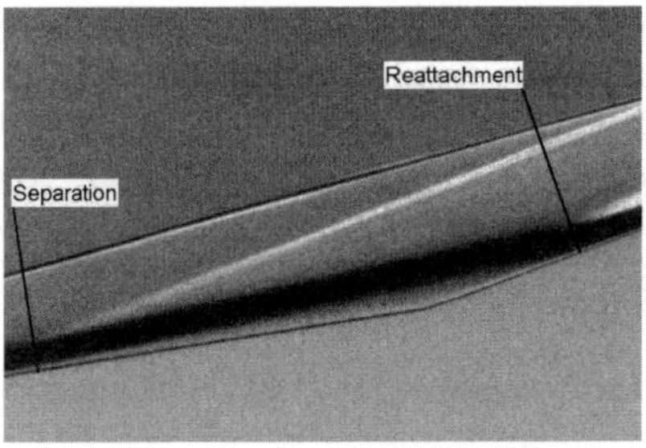

Fig. 6. Schlieren Image: Experiment

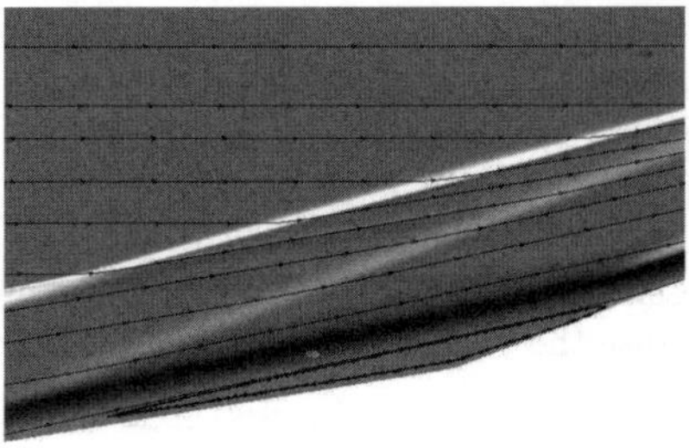

Fig. 7. Schlieren Image: CFX

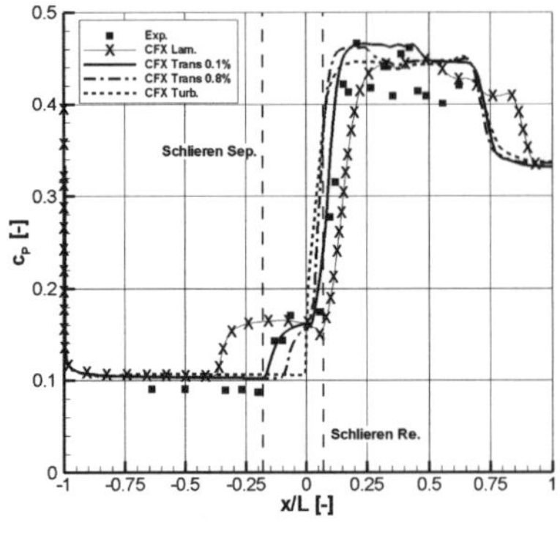

Fig. 8. Pressure coefficient for a sharp leading edge at wall temperature of 300 K

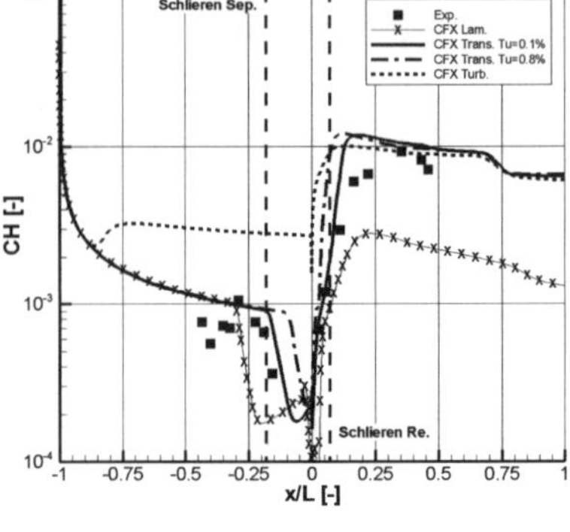

Fig. 9. Stanton number for a sharp leading edge at wall temperature of 300 K

underpredicts the reattachment heat flux. The turbulent case shows no separation and overpredicts the heat flux upstream of the kink. The transitional flow aggrees with the heat fluxes of the laminar boundary layer upstream and the turbulent boundary layer downstream of the interaction thus transition occurs in the separation bubble. The size of the separation length increases with the prescribed free-stream turbulence level. The best aggrement is obtained for an assumed free-stream turbulence level of 0.1% being used for all further computations. The results for different leading edges at wall temperatures of 300 K are given in Figures 10 and 11. The computations show the general trend of reduced reattachment pressure and heat flux with increased leading edge radius, but have only a fair agreement with the measurements. The increase in the separation length is captured but do not show the reversal trend in separation length proposed by Holden [1].

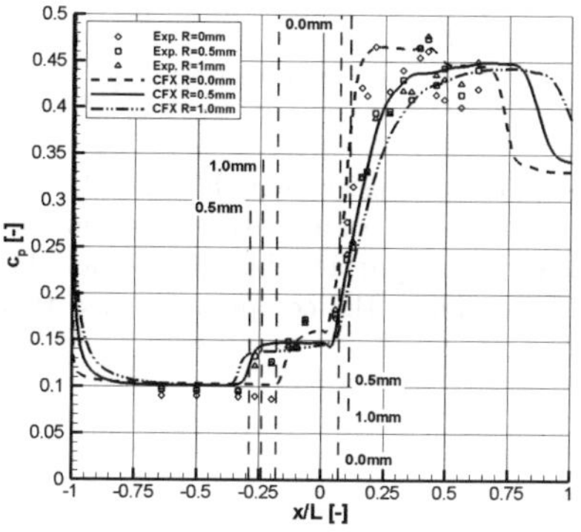

Fig. 10. Pressure coefficient for different leading edge radii for a wall temperature of 300 K

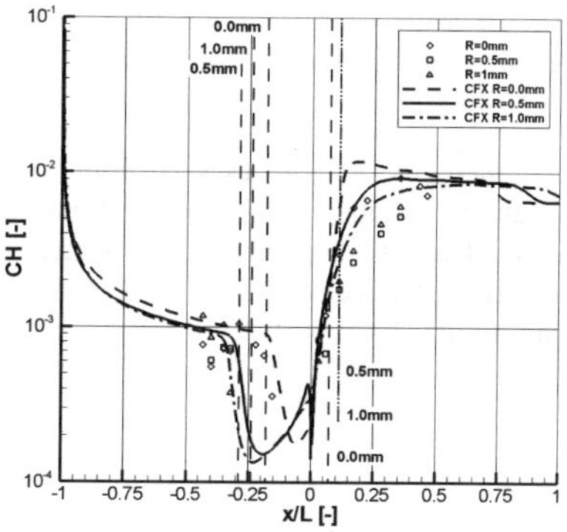

Fig. 11. Stanton number for different leading edge radii for a wall temperature of 300 K

Figures 12 and 13 show the wall temperature influence for the double wedge configuration with a sharp leading edge. The increase in separation size is well captured, but in the reattachment region the reduced pressure and heat flux coefficients are slightly overpredicted. This reduction observed in experiment and numerics is due to the interaction of the SWBLI of the corner and the leading edge shock. This behaviour for the elevated wall temperature is in opposite to the known trend of the SWBLI of a flat plate/ramp configuration with increasing reattachment pressure and heat flux [10].

4 Conclusion and Outlook

The capability of the commercial code CFX for simulating the hypersonic, bluntness and viscous dominated flow along flat plates has been shown. The effects of Reynolds number and the wall temperature on a laminar SWBLI have been determined satisfactorly. The computation of transtional SWBLI is still subject of ongoing research. The employed transition model by Menter/Langtry gives a fair aggreement with the experiment, but does not capture the reversal trend in the separation length.

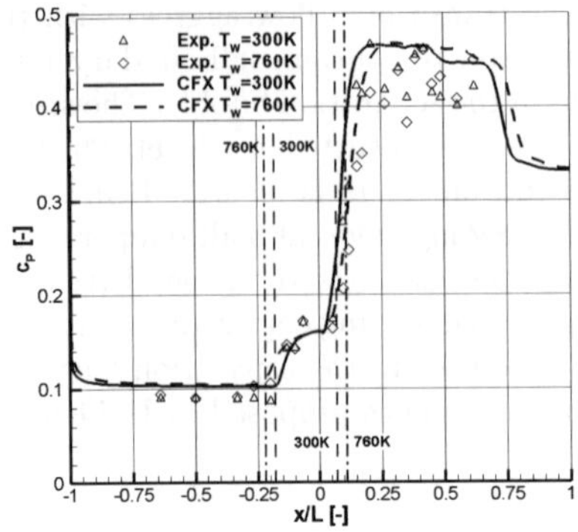

Fig. 12. Pressure coefficient for different wall temperatures and a sharp leading edge

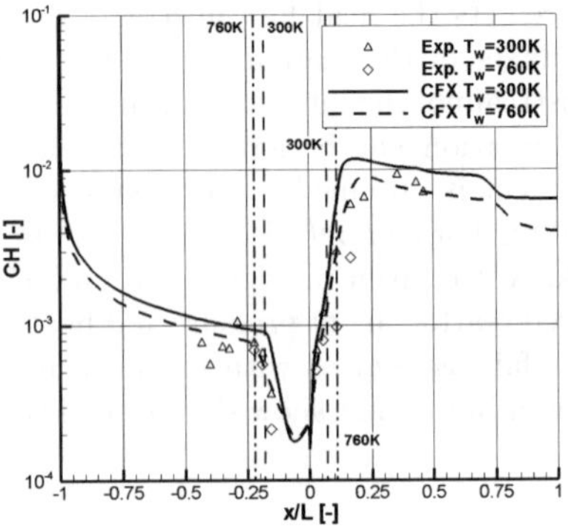

Fig. 13. Stanton number for different wall temperatures and a sharp leading edge

Acknowledgement. This work was funded by the Deutsche Forschungsgemeinschaft DFG under the Graduier-tenkolleg "Aerothermodynamische Auslegung eines Scramjet-Antriebssytems für zukünf-tige Raumtransportsysteme" which is gratefully acknowledged. Further, the authors like to thank B. Akih Kumgeh and A. Peters for their contribution to the computation and their scientific input.

References

1. Holden, M.S.: *Boundary-layer displacement and leading-edge bluntness effects on attached and separated laminar boundary layers in a compression corner. Part II: Experimental study.*, AIAA Journal **Vol. 9**, No. 1 (1971) pp. 84-93.
2. Mallinson, S.G., Gai, S.L., Mudford, N.R.: *Leading-edge bluntness effects in high enthalpy, hypersonic compression corner flow*, AIAA Journal **Vol. 34**, No. 11 (1996) pp. 2284-2290
3. Reinartz, B., Ballmann, J.: *Numerical simulation of wall temperature and entropy layer effects on double wedge shock/boundary layer interactions*, AIAA Paper 2006-8137, 14th AIAA/AHI Space Planes Conference and Hypersonic Systems Conference, Canberra, Australia (2006)
4. Neuenhahn, T., Olivier, H.: *Influence of the wall temperature and the entropy layer effects on double wedge shock boundary layer interactions.*, AIAA Paper 2006-8136, 14th AIAA/AHI Space Planes Conference and Hypersonic Systems Conference, Canberra, Australia (2006)
5. Cheng, H.K. et al.: *Boundary layer displacement and leading-edge bluntness effects in high-temperature hypersonic flow*, Journal of Aerospace Science **Vol. 28**, No. 5 (1961) pp. 353-381
6. Menter,F.R. et al.: *CFD simulations of aerodynamic flows with a pressure-based method.*, 24TH International Congress of the Aeronautical Sciences, Yokohama, Japan (2004)
7. Anderson, J.D.: Hypersonic and high temperature gas dynamics. McGraw-Hill (1989)
8. Lukasiewitz, J.: *Hypersonic flow – Blast analogy*, AEDC TR-61-4, (1961)
9. Lewis, J.E., Kubota, T., Lees, L.: *Experimental investigation of supersonic laminar, two-dimensional boundary-layer separation in a compression corner with and without cooling.*, AIAA Journal **Vol. 6**, No.1 (1968)
10. Bleilebens, M., Olivier, H.: *On the influence of elevated surface temperatures on hypersonic shock wave/boundary layer interaction at a heated ramp model.*, Shock Waves, **Vol. 15**, No. 5, (2006) pp. 301-312

Similarity laws of re-entry aerodynamics - analysis of reverse flow shock and wake flow thermal inversion phenomena

S. Balage, R. Boyce, N. Mudford, H. Ranadive, and S. Gai

University of New South Wales at Australian Defense Force Academy, Canberra, ACT 2600, Australia

Summary. A computational fluid dynamics (CFD) based dimensional analysis of the flow in the base region of a planetary re-entry configuration is presented. Reynolds number and free stream Mach number are found to have the dominant influence on the Mach number of the trapped recirculating flow in the wake. Prandtl number is found have the strongest influence on the temperature of the wake recirculating flow. Two associated flow phenomena, wake reverse flow shock (WRFS) and wake flow thermal inversion (WFTI) are introduced. The governing role of the Prandtl number on the wake flow energy budget and thus the base region heating is discussed.

1 Introduction

The present work uses computational fluid dynamics (CFD) to investigate the steady-state, perfect gas fluid dynamics of a planetary re-entry capsule at zero angle of attack with the emphasis on the flow physics of the base region. The study employs non-dimensional similarity parameters to generalize its findings. The fluid dynamics phenomena of the fore-body flow have been well documented for many years. Knowledge of the base flow aerodynamics, however, remains sparse. The present work is motivated by the need to understand the physics of the base flow to provide knowledge for re-entry spacecraft design optimisation and risk reduction processes.

Dimensional analyses of the Navier-Stokes equations show that the Reynolds number (Re), Prandtl number (Pr), ratio of heat capacity (γ) and Mach number (M) form a set of pi groups suitable for the study of the re-entry aerodynamics with adiabatic walls. With the aid of these similarity parameters, this document investigates two phenomena associated with the flow in the base region, the wake reverse flow shock (WRFS) and wake flow thermal inversion (WFTI). The former has been observed before in numerical studies (for example, [2]) but never in experiments. The latter is, to our knowledge, a new phenomenon. A CFD based investigation of the dependence of the maximum reverse flow Mach number on the Reynolds number and the free stream Mach number is presented. The effect of the Reynolds number on the maximum reverse flow Mach number and recovered base stagnation temperature is found to be qualitatively similar to that of the free stream Mach number. The energy budget for the trapped wake recirculation flow is discussed. The Prandtl number is proposed as the primary determining parameter between the balance of energy of work done on the trapped recirculation region via viscous interactions and heat transfer out of it by thermal diffusion. The dependence of the base stagnation point temperature on the Prandtl number is documented.

The prototype re-entry configuration considered is that of the Beagle II mission to Mars, the demise of which provided the motivation for the present work. The approach adopted in this study is to perturb the similarity parameters around a selected datum point to reveal the flow physics. The datum planetary atmosphere is modelled as an ideal gas with properties corresponding to CO_2. The datum free stream conditions are those of a point at $26km$ altitude on a typical entry trajectory through the Martian atmosphere. The static temperature used is $192K$, the static pressure is $54Pa$ and the velocity is $4000ms^{-1}$. The Mach number is 18.19 for γ 1.333 and the free stream Re_D is 544270.

The CFD code used here is the commercial code CFD^{++} [3]. CFD^{++} can solve both the steady or unsteady compressible Navier-Stokes equations, including multi-species and finite-rate chemistry modelling. For the present work, the calculations performed are double precision and of second-order accuracy in both time and space. Concerning the spatial discretization, total variation diminishing (TVD) polynomial interpolation with MinMod limiting is used, while an implicit Runge-Kutta method is used for the time integration. A structured 2 dimensional mesh is used in this study. The mesh blocking topology is an O-grid adjacent to the body geometry surrounded by a C-grid, which captures the wake flow. The steady-state, ideal gas CFD computations are performed on a 2-dimensional axi-symmetric model of the external shape of the Beagle II's thermal protection system with minor modifications. The CFD did not model the backward facing step cut at the aft side of the shoulder nor the launch clamp ring at the base [4].

2 Flow Conditions

One of the purposes of the present work is to investigate the effects of similarity parameters on wake flows of different scales. The CFD simulations are conducted using dimensional values for the gas properties with modifications to adjust the relevant non-dimensional quantities. Therefore, the far-field or inlet boundary conditions for the CFD simulations are required in the form of velocity (U), temperature (T) and pressure (P). The present study employs the total enthalpy and Mach number to specify the-far field fluid dynamics at the chosen trajectory point, thus requiring a transformation between the two sets of quantities. The enthalpy equation and the definition of the Mach number provide two linear equations in U^2 and T with a unique solution for velocity and temperature. Given free stream Mach number M_∞, total enthalpy and γ, the free stream velocity and temperature can be obtained by solving these two equations. The free stream temperature serves as a constraint to the relationship between the free stream pressure and density via the ideal gas law. Given a prescribed inlet temperature, a desired free stream density can then be obtained, to achieve to a desired Reynolds number, by adjusting the inlet pressure. We shall call this method of changing the free stream Reynolds number to obtain similar solutions the *density adjustment method.* Scaling the reference viscosity μ_0 and conductivity k_0 of the Sutherland law allows matching of the Reynolds and Prandtl numbers for flows with different body length scales. We shall call this method of obtaining similar solutions *viscosity adjustment method.*

3 Similarity

Laminar, ideal gas (CO_2), steady-state CFD simulations are used here in the investigation of similarity parameters. A study of generating similar solutions with density method is first conducted. Meshes for both full scale Beagle II and 1/10 scale version are constructed and CFD simulations are carried out with the free stream pressure and density ten times that of the datum values for the 1/10 case. The other gas parameters and the free stream conditions retain their datum point values. The converged solutions show agreements typically within 0.2% between the two length scales. The solution Mach number, temperature and flow speed are compared in figures 1. and 2. The body length scales are matched during the post processing for the purpose of comparison in the diagrams below. Other parameter fields examined in the study include total temperature, total pressure, viscosity, thermal conductivity and Prandtl number fields.

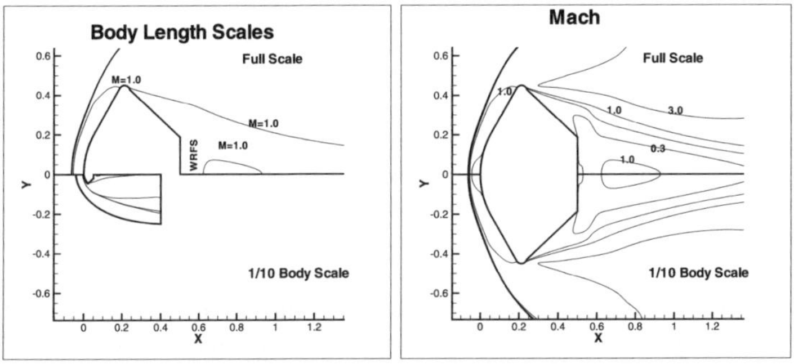

Fig. 1. Left- The 1/10 scale solution (bottom half) made similar to full scale model (top half)using density adjustment method in true spatial scale. Right- The 1/10 model expanded by x 10 during post processing. Mach number solution fields are plotted. Note the reverse flow supersonic region. WRFS occures as flow leaves $M = 1$ region towards the base of the probe.

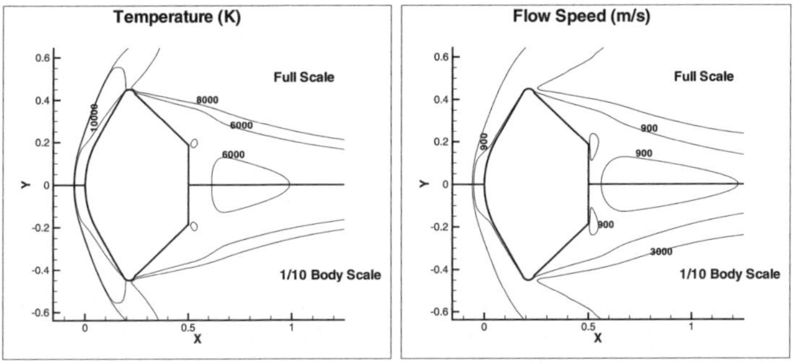

Fig. 2. Left- The temperature fields. Right-The velocity fields. Top half is full scale model and the bottom half is 1/10 scale similar solution matched with density adjustment method

The 1/10 scale solutions are also shown to be similar to the full scale solution via the viscosity adjustment method which necessitates decreasing μ_0 by ten times. Here k_0 is decreased 10 times in order to match the Prandtl number. Similar agreements as for the density adjustment method are observed.

By using both the density and viscosity adjustment methods, several selected values of γ and Mach number solutions have been matched for 1/10th body length scale models with that of the full-scale model. The solution at our selected datum point shows a

Wake Reverse Flow Shock. In the solutions of the 1/10 model with the datum far-field conditions, without any matching of the similarity parameters, the WRFS vanishes. The overall velocities in the latter case are lower and the wake flow never reaches the supersonic values. The other solution fields such as temperature, normalized pressure, etc. are distictly different from that of the full scale model. The above observations confirms that similar solutions are obtained by matching M_∞, γ, Re_∞ and Pr, thus verifying the findings of dimensional analysis.

4 Wake Flow Shock and Wake Flow Thermal Inversion

The fluid dynamic phenomena associated with a given flow geometry can be viewed as a consequence of the relative strengths of the similarity parameters. The flow dynamic effects of the Beagle II geometry are investigated first by changing the free stream Reynolds, Mach and Prandtl numbers away from the datum point while keeping the remainder of the similarity parameters and total enthalpy the same. The results of these investigations are presented in figure 3 where the results are plotted in the form of the maximum reverse flow Mach number and the base stagnation temperature normalized by total free stream temperature. This format allows us to see the maximum values recovered by the thermal and kinetic components by the wake flow. We call this diagram *wake energy recovery plot* (WERP). Two turbulent models and a reacting flow simulation are also studied. Finally, the total specific enthalpy of the flow is perturbed from the datum value while keeping the same M, γ and free stream density. Figure 3 shows that Re_∞ and M_∞ data lies on one class of curve and Pr data lies in a different class of a curve.

The Prandtl numbers of the simulations are changed from the datum by changing the thermal conductivity k_0 while other similarity parameters are kept constant. The Prandtl number therefore is altered throughout the entire domain. The total temperature of the base flow (thus the base stagnation temperature) approaches the free stream total temperature as the base Prandtl number approach unity. Note that the base Prandtl number is taken to be the Prandtl number in the vicinity of the dividing streamline in the wake region and is not a precise value. For base Prandtl numbers higher than approximately unity, the wake total temperature exceeds that of the free stream. We label this flow condition *wake flow thermal inversion* (WFTI). The Prandtl curve of figure 3 has a negative slope. This is due to the increased temperature of the base flow reducing the maximum base Mach number, M_{bmax}.

We speculate that the Reynolds curve of Figure 3 indicates that the base flow responds to increased availability of mechanical energy of the free stream, due to an increase in Reynolds or Mach free stream values, by receiving more work done on it via viscous forces. This is reflected in the higher value of the maximum base flow Mach number, M_{bmax}, and the temperature ratio of the base centre temperature to the total free stream temperature, T_{base}/T_{tot}, recovered at the centre of the base. As Reynolds or Mach numbers are increased beyond a critical point, M_{bmax} would exceed unity leading to the *Wake Reverse Flow Shock* (WRFS) as the flow transit back to subsonic speeds near the base stagnation point.

The effect of perturbing the free stream total enthalpy is also studied. Figure 3 shows that a 50% increase in total enthalpy would shift the datum point considered to the left and a decrease of 50% would shift it to the right. The effect on T_{base}/T_{tot} is not as great as on M_{bmax} and the two points lie close to the Reynolds (and Mach) curve. This

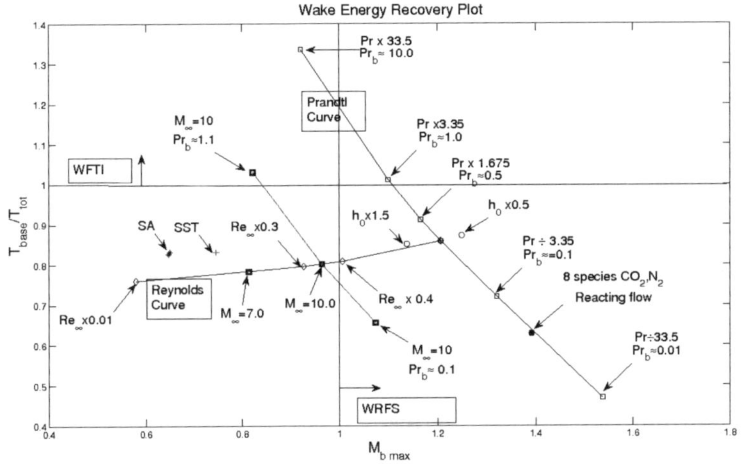

Fig. 3. The wake energy recovery plot (WERP)-The plot of wake flow maximum Mach number vs the ratio of base stagnation temperature to total temperature for changes of Re_∞,M_∞ and Pr. The point of intercept of the Reynolds and Prandtl curves is the datum re-entry condition described in the section 1.

observation is consistant with the resulting changes in Re_∞ and M_∞ due to the change in the free stream enthalpy.

Based on the above observations we argue that the base flow dynamics associated with the Prandtl number are different from that of the Mach and Reynolds numbers. We propose that while the Mach and Reynolds numbers controls the availability of the kinetic energy of the free stream for the viscous forces to work on the trapped base flow, the Prandtl number governs the balance of the energy in the base flow. At steady-state, in laminar flow, the energy balance of the wake flow is between the work done on it due to viscous interactions and heat transferred out of it due to thermal conductivity. The Prandl number forms the natural non-dimensional parameter for this interaction indicating where the balance lies. This role of the Prandtl number is further supported by the observation that WFTI occurs at base Prandtl number equal to unity.

The effect of turbulence models is investigated for both the Spallart-Almaras model and SST model. In both cases the turbulent viscosity is increased in the base region, mostly near the wake neck region, decreasing the effective Reynolds number of the base region flow. This is also evident through the reduction of the wake length. Both turbulent solutions occupy points on Figure 3 which correspond to lower Reynolds number and higher Prandtl number to the datum point. This agrees with the concept of increased viscosity due to the addition of turbulent viscosity overpowering the increase of effective thermal conductivity due to eddy diffusion. The WRFS does not appear in these turbulent models due to reduced Re. An eight species reaction model for $97\% CO_2$ and $3\% N_2$ by mass fraction [1] [4] has also been computed and occupies a position on figure 3 that correspond to a low Prandtl number. This can be explained by the reduction of the temperature due to the predominance of endothermic reactions in the base flow as well as on the reduction of the effective Prandtl number due to the increase of *reaction conductivity*, a mode of heat transfer due to mass diffusion.

5 On the uses of the Wake Energy Recovery Plot

Representations similar to WERP in figure 3 could be used by space probe design teams in the analysis of error in CFD computations as they describe the interaction of the *flow parameters* (T_{base}/T_{tot}, M_{bmax}), the *similarity parameters* (Re, Pr) and *flow phenomena* (WRFS, WFTI). For example, one could determine from figure 3 that the uncertainty in Prandtl number would lead to a greater effect in base heating than say, Reynolds number. Similarly, the effects of numerical artifacts, such as artificial viscosity, once estimated, could be projected on to the quantities of interest. Mesh convergence could also be observed for multiple quantities of interest simultaneously and an appropriate mesh could be made at critical values of the flow phenomena such as the WRFS to capture the shock.

6 Conclusions and Future Work

The present work establishes the WRFS and WFTI phenomena as a theoretical possibility for steady-state, laminar, compressible flow. Their association with Reynolds number and Prandtl number is established. While the actual wake reverse flow shock and the thermal inversion may not be relevant to most of the re-entry flow, the nature of the Reynolds and Prandtl number effect on the wake flow energy distribution and thus the base heating remains directly relevant. The intriguing possibility of a part of a gas flow attaining temperatures higher than its free stream total temperature is to be investigated further to find out if any real gas would support the requirements. The narrowing down of the thermal and kinetic energy maxima to two classes of curves on WERP significantly constrains the flow dynamics possibilities of a given wake flow. The dependence of the curves on the shape of the re-entry configuration is being investigated. The WERP would have consequences in development of empirical formulae for base heating. This work is supported by ARC DP0666941.

References

1. Chen Y.K., Henline W.D., Stewart D.A. and Candler G.V., Navier Stokes Solutions with Surface Catalysis for Martian Atmospheric Entry. Journal of Spacecraft and Rockets **30**, No 1. (1993)
2. Gnoffo, P. A. Planetary-Entry Gas Dynamics. Annu. Rev. Fluid Mech **31**,459-94 (1999)
3. Goldberg, U., Batten P., Palaniswamy, S. Hypersonic flow Predictions uing Linear and Nonlinear Turbulence Closures. AIAA J. Aircraft **37**,671-675 (2000)
4. Liever, P.A. Habchi, S.D., Burnell S.J., Lingard, J.S. Computational Fluid Dynamics Predictions of the Beagle 2 Aerodynamic Database. Journal of Spacecrft and Rockets **40**, No.5 632-638 (2003)

Simultanous measurements of 2-D total radiation and CARS data from hypervelocity flow behind strong shock waves

K. Maeno[1], M. Ota[1], A. Matsuda[2], B. Suhe[3], and K. Arimura[3]

[1] *Graduate School of Engineering, Chiba University, 1-33 Yayoi, Inage, Chiba 263-8522, Japan*
[2] *Institute of Space and Astronautical Science / JAXA, Yoshinodai 3-1-1, Sagamihara, Kanagawa, Japan*
[3] *graduate student, Graduate School of Science and Technology, Chiba University*

Summary. For the heatproof design of a re-entry space vehicle, the radiative and non-equilibrium heating from shocked air ahead of the vehicle plays an important role on heat flux to the wall surface as well as convective heating. So far, various researches have been conducted. In our research Coherent Anti-stokes Raman Spectroscopy (CARS) method has been applied for measuring rotational and vibrational temperatures behind hypervelocity shock wave. These temperatures are obtained from fitting program, which is fitting experimental spectra into theoretical spectra. The purpose of this paper is to show the results of CARS-2D simultanous measurement. In addition, inspectation of the past temperature results and the factor in causing difference on the fitting has been investigated.

1 Introduction

During space vehicle reentries into the atmosphere, the heating problem of space vehicles becomes remarkable. So far, great efforts have been done to understand radiating high enthalpy flows. When the reentry velocity is over 10km/s, a radiative heating from the shocked air ahead of the vehicle plays an important role on the heat flux to the wall surface of the structure as well as convective heating. Many experimental studies have been carried out using shock tube in order to clarify the radiative features of shocked air. In our previous works, a CCD camera system has been introduced for spontaneous radiation observation of strong shock waves in low-density air. The system consists of an imaging spectrograph, a streak camera, a gated image-intensified CCD camera and a personal computer. Some interesting features have been obtained independently for the total radiation and for the spectral radiation behind strong shock waves with over 10 km/s in air. In pur previous study a spectroscopic measurement together with total radiation observation by two CCD cameras, was also carried out. This measurement was focused on the equilibrium radiation region followed the nonequilibrium radiation peak just behind the shock front with 10 km/s. The Boltzmann plot method, which was based on the second positive band of nitrogen molecules, was used in order to evaluate its vibrational temperature from the spectroscopic data. The method, however, has some assumptions and time averaging. From this reason, our Boltzmann plot data have relatively large errors. Recently, Coherent Anti-Stokes Raman Spectroscopy (CARS) has been developed for temperature and concentration measurement for mainly combustion research. CARS is a nonlinear, four-wave optical mixing technique for fast flow, combustion, and so on. The CARS can produce a strong and coherent signal. This strong signal is detected even in backgroud luminosity. The temperatures are determined by fitting computer-generated theoretical spectra to the observed CARS spectrum. In this study CARS is applied as more precise technique for vibrational and rotational temperatures of nitrogen in hypervelocity flow behind the strong shock wave. The CARS signal should be acquired in the stable radiation area behind the shock wave. Therefore, to grasp where the stable radiation area is, we constructed the CARS-2D total radiation measurement system. This system enables us to get CARS signal and total radiation image simultaneously. In this paper the measurement of vibrational/rotational temperatures of hyper-velocity flow behind shock waves by CARS has been established over 5 km/s shock velocity.

2 CARS method and spectral fitting

The theory of the CARS effect has been developed extensively. Incident laser beams at frequencies ω_1 (pump beam) and ω_2 (Stokes beam) interact through the third order nonlinear electric susceptibility $\chi^{(3)}_{CARS}$ to produce coherent radiation $I_3(\omega_3)$ at $\omega_3 = 2\omega_1 - \omega_2$ as follows

$$I_3 = \frac{\omega_3^2}{n_1^2 n_2 n_3 c^4 \epsilon_0^2} I_1^2 I_2 |\chi^{(3)}_{CARS}|^2 l^2 \left(\frac{sin\frac{\Delta kl}{2}}{\frac{\Delta kl}{2}} \right)^2 \tag{1}$$

$$\Delta k = 2k_1 - k_2 - k_3 \tag{2}$$

where n_1, n_2, and n_3 are the refractive index at ω_1, ω_2, ω_3, respectivley; c is the velocity of light; l is interaction length; k_1, k_2, and k_3 are the wave vectors of the pump, Stokes and CARS beams, respectively; ϵ_0 is the permittivity of free space. The CARS signal is enhanced on the condition of phase matching, $\Delta k = 0$. Then four wave vectors form phase-matching diagram (BOXCARS) shown in Fig.1. The incident laser beams, i.e, two pump beams (k_1) and a Stokes beam (k_2), are aligned in order to satisfy vector relation. $\chi^{(3)}_{CARS}$ is shown by

$$\chi^{(3)}_{CARS} = \sum_j K_j \frac{\Gamma_j}{2\Delta\omega_j - i\Gamma_j} + \chi_{nr} \tag{3}$$

where the j summation is over vibration-rotation transitions in the vicinity of $\omega_1 - \omega_2$; Γ_j is the Raman line width (FWHM); χ_{nr} is a background contribution due to electrons and remote resonances; and $\Delta\omega_j$ is detuning $\omega_j - (\omega_1 - \omega_2)$. ω_j is Raman resonance. The module K_j is shown to be related to the Raman cross-section by

$$K_j = \frac{2n_1 c^4}{n_2 \hbar \omega_2^4} N \Delta_j \frac{d\sigma}{d\Omega}\bigg|_j \Gamma_j^{-1} \tag{4}$$

where N is the number density of the Raman active molecule; Δ_j is the population difference between the upper and lower vibration-rotation states; and $(d\sigma/d\Omega)\big|_j$ is the cross-section for spontaneous Raman scattering.

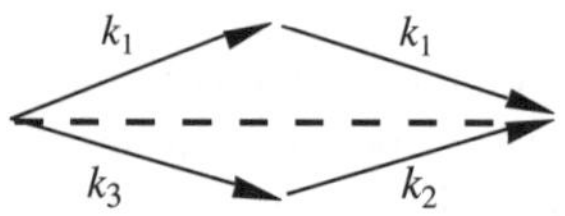

Fig. 1. Phase-matching diagram of BOXCARS.

On the assumption that molecules have Boltzmann distributions based on the rotational (T_r) and vibrational (T_v) temperatures, Δ_j can be expressed as

$$\Delta_j = \frac{(2J+1)g_I}{Q_r Q_v}\left[exp(\frac{-F_{v,J}hc}{kT_r})exp(\frac{-G_v hc}{kT_v}) - exp(\frac{-F_{v+1,J}hc}{kT_r})exp(\frac{-G_{v+1}hc}{kT_v})\right] \quad (5)$$

where $F_{v,J}$ and G_J are the rotational and vibrational energy terms, respectively; g_I is the spin degeneracy, 6 for even-J rotational levels and 3 for odd-J levels in N_2; and Q_r and Q_v are the rotational and vibrational partition functions, respectively.

From equations (1),(3) and (4), CARS signal is approximately proportional to the square of number density of molecule. Therefore, it is not easy to detect the CARS signal from low-pressure and radiating fast gas. As written in the following section, the measured CARS spectra are fitted with calculated CARS signals by treating T_v, T_r, Γ as free paratmeters to decide the temperatures.

3 Experimental apparatus and diagnostic system

A free-piston, double-diaphragm shock tube has been used to generate strong shock waves in low-density gas. The cross section of the low-pressure tube is 40 mm × 40 mm square. The observation window of the test section is mounted near a focal lens with some distance from the sidewall of the shock tube. This distance prevents high power laser from destroying the window. The side wall of shock tube has two small halls along the optical path of laser beam. The shock velocity is measured by using ion probes mounted on the sidewall of the shock tube test-section. Figure 2 shows a layout of diagnostic system

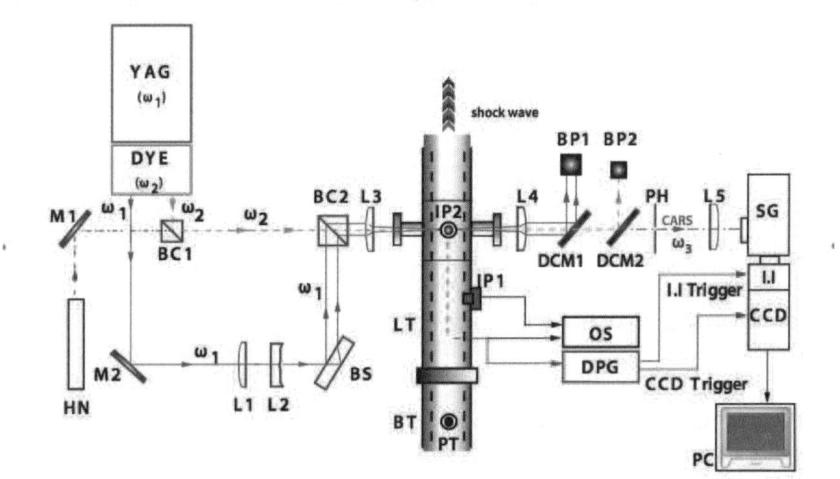

Fig. 2. Layout of CARS measurement system. Code: HN, He-Ne laser; M, mirror; W, window; BC, beam combiner; L, lens; BS, beam splitter; DCM, dichroic mirrors; BP, beam pocket; PH, pine hole; IP, ion probe; PT, pressure transducer; DPG, delay pulse generator; SG, spectrograph; OS, oscilloscope; LT, low-pressure tube; BT, buffer tube.

for the CARS measurement in shock waves. This system consists of second harmonics of Nd:YAG laser (ω_1), a dye laser (ω_2), a spectrograph, and an ICCD. The laser beam (ω_1) is divided in two by a beam splitter (BS). These beams are combined to the laser beam (ω_2) by a beam combiner (BC2). Then three laser beams are focused with desirable angles in the shock tube. The CARS spectra is detected by the ICCD. The entrance slit width is set to 100 μm throughout the observation. The ICCD is mounted on the focal exit of the spectrograph. The Nd:YAG laser (Continuum PL8010) is used for the pump beam (ω_1). At a second harmonics beam (532 nm), the line width is 0.5 cm^{-1} FWHM, the pulse duration is 10 ns. The power of Stokes beam is about 80 mJ at both narrow-band and broadband oscillation. The optical path length of pump beam is adjusted to synchronize the arrival time of two beams at a measurement point. Two pump beams and a Stokes beam are focused into shock tube through a focal lens. The laser incidence route is shown in Fig. 3. The plane containing both ω_2 and ω_3 beams is perpendicular to the plane in two ω_1 beams. In this case, phase-matching condition has to be satisfied as well as planer BOXCARS. This method is called as folded BOXCARS, especially, multiplex folded BOXCARS using broadband Stokes beam at single shot. This method has advantage in separating CARS signal from pump- and Stokes beams.

The three beams (ω_1 and ω_2) must be irradiated on time when the shock wave running at up to 10 km/s just arrives at the observation window. Above CARS system, there is a 2D total radiation measurement system (2D system). The 2D system consists of a mirror and another ICCD camera. Both systems are structured so as not to disturb mutually, and then enable us to have simultanous measurement of CARS signal and 2D total radiation image.

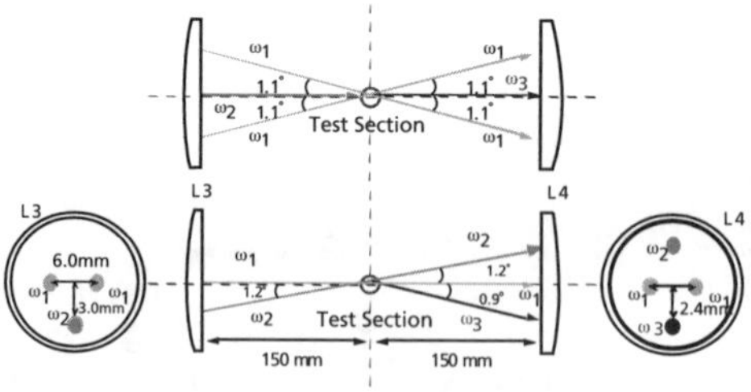

Fig. 3. Laser incidence route (top view and side view).

4 Experimental results and discussion

4.1 CARS spectra of shocked air with broadband Stokes beams

The measured spectrum data with a theoretical spectrum are shown in Fig. 4. The theoretical spectrum has been calculated with shifting two parameters, T_v, T_r. Test gas is room air. The measurement position is 6.7mm behind the shock front. Velocity of the shock wave is 5.4km/s, Mach number of 15.63. Initial pressure is 1333Pa. After the fitting, the vibrational and rotational temperatures are estimated as 8500K, 7500K, respectively. From this fitting, we can obtain the vibrational and rotational temperatures. However, the disagreement of the theoretical spectrum from measured data is still seen in the high wavenumber region.

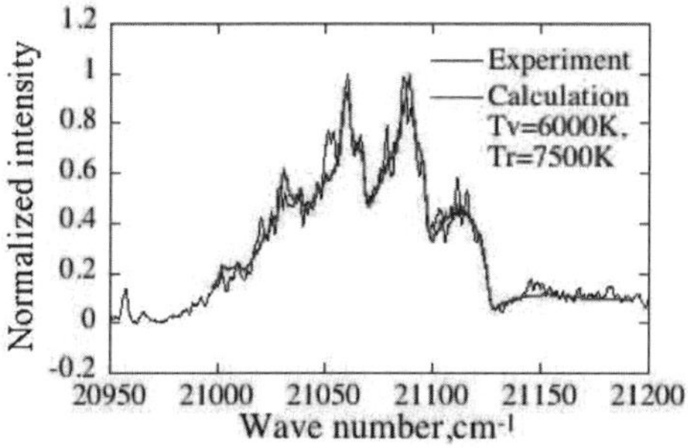

Fig. 4. CARS spectra of shocked air with broadband Stokes beams.

4.2 2D total radiation measurement

We also performed the simultaneous detecting CARS signal and acquiring 2D total radiation. The purpose of this experiment is to improve accuracy of the data obtained from CARS measurement and to inspect whether the confusion is seen at the shock front. The result of the simultanous measurements is presented in Fig. 5. The figure. 5 shows the image of two dimensional total radiation (2D image) and the image of CARS signal. Test gas is room air and the measurement position is 1.3mm behind the shock front. Velocity of shock wave is 5.38km/s. Mach number of 15.54. Initial pressure is 1333Pa. In this figure shock wave flashes by from right to left in the round observation window.

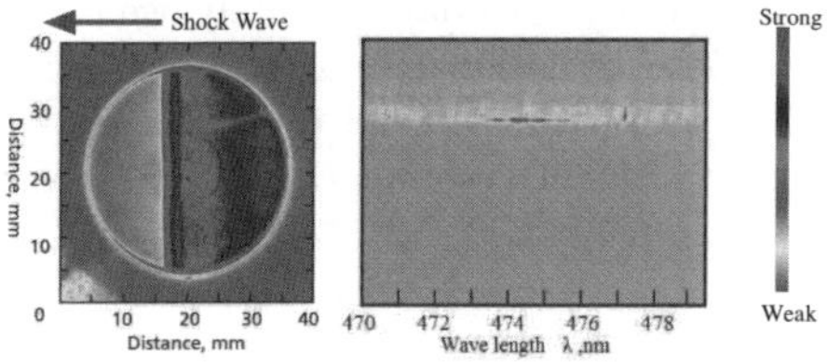

Fig. 5. The reasult of simultaneous measurement of CARS and 2D total radiation, Left; 2D image of shocked flow, Right; CARS signal image

4.3 Accuracy evaluation of the measurement

In our study rotational and vibrational temperatures have been determined by spectral fitting. Even though the agreement of fitting has been improved, an error range of this measurement method and the causes of fitting error should be investigated. Figure 6 shows a fitting error graph between experimental CARS spectrum and the theoretical spectrum. Vertical axis is normalized error value and horizontal one is temperature. From this graph, error range has been calculated about 2000K. Taking into account our fitting order is every 1000K, this graph shows good accuracy of our fitting method.

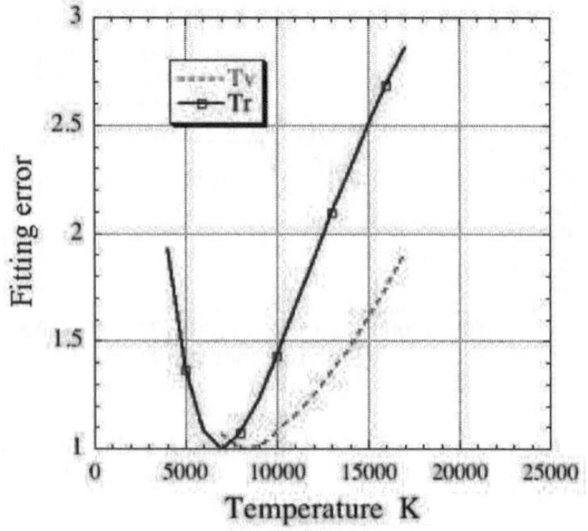

Fig. 6. CARS error value graph

5 Summary

The CARS measurement system has been established for temperature measurement of unsteady and hypervelocity flow behind the strong shock waves. Measured shapes of CARS spectra agree well with theoretical fitting curves. Then, vibrational and rotational temperatures are measured by fitting theoretical spectrum to the experimental one. This procedure has been proved by the accuracy evaluation about error range between either spectrum. In order to obtain the CARS data precisely, we have to be careful to measure them from the location where radiation homogeneity is established behind the shock front. We need more data to investigate the temperatures of shock waves using this CARS-2D simultanous measurement system.

References

1. J. Koreeda, Y.Ohama. and H.Honma. Proceedings of 20th International Symposium on Shock Waves, 1181-1186, Pasadena, CA, USA, 1995.
2. K. Miyazaki et.al., Proceedings of 25th International Symposium on Shock Waves, 877-882, Bangalore, India, 2005.

Shock tunnel testing of a Mach 6 hypersonic waverider

K. Hemanth, G. Jagadeesh, S. Saravanan, K. Nagashetty, and K.P.J. Reddy

Department of Aerospace Engineering, Indian Institute of Science, C. V. Raman Avenue, Bangalore, 560012, India

Summary. A waverider is a lifting body configuration whose upper surface is parallel to the free stream, and the lower surface aerodynamically so designed, that the resulting shock at the design Mach number, is always attached with the leading edge of the vehicle. This prevents spillage from high pressure (lower) surface to the low pressure (upper) surface.In the present study a conical waverider has been designed, fabricated and tested at Mach 6 in the IISc hypersonic shock tunnel HST2. The measurements show that the waverider has a lift to drag ratio of 4.28 at the designed Mach number. Exhaustive FEM and CFD studies are also carried out to complement the force measurements in the tunnel.

1 Introduction

The maneuverability requirements of future hypersonic vehicles require aerodynamic configuration which is capable of efficiently integrating non-circular airframes, lifting capabilities and propulsion components in such a fashion so as to minimize the aerodynamic heating and radar detectability. In this context the concept of waveriders cruising at hypersonic speeds [1, 1] is very promising especially in the backdrop of availability of next generation structural materials that can withstand very high temperatures.

A simple way to understand the concept of a waverider is to dichotomise its name-WAVE + RIDER i.e. something that "rides" on a "wave". To be more specific, a waverider can be understood as a lifting body configuration comprising of two surfaces-upper and lower. The upper surface is parallel to free stream and offers no obstruction to the flow and hence the pressure on the upper surface is close to the free stream pressure. On the other hand, the lower surface obstructs the flow and results in a shockwave. But the lower surface is aerodynamically tailored, such that at the design Mach number, the resulting shock coincides with the leading edge of the body (the edge formed by joining the upper and lower surfaces) all along. In other words, the shock is attached to the leading edge and the vehicle seems to be "riding its own shock wave". This attached shock prevents the spillage of flow from the high pressure side (lower surface) to the low pressure side (upper surface) thereby resulting in very high Lift/Drag ratios which are necessary for sustained hypersonic flight.

While considerable amount of research has gone into evolving techniques to design waveriders, precious little information is available on their performance under hypersonic speeds. Not many attempts measure the basic aerodynamic forces acting on the waverider at hypersonic speeds and evaluate them. The objectives of the present study were as follows: 1)To aerodynamically design a conically derived waverider for a nominal Mach number of 6 from basic hypersonic conical flow principles. 2) To design, fabricate and calibrate an internally mountable rubber based 3-component accelerometer force balance and then obtain the performance characteristic of waverider in the IISc hypersonic shock tunnel (HST2) both at design (Mach 6) conditions. 3)To carry out FEM studies on the model and the force balance by taking inputs from CFD to complement the force measurements. 4)To carry out illustrative CFD studies on the designed waverider to complement the experiments.The details of the experimental and theoretical studies are described in the subsequent sections.

1.1 Experimental study

Based on the methodology suggested by Rasmussen [3] in the present study, a conical waverider has been designed assuming inviscid flow past circular and elliptic cones using analytical expressions to describe and characterize such flows. The conical waverider (Fig. 2) has been designed for optimum performance at a free stream Mach number of 6. Considering the advantages of cone-derived waveriders as listed in Broadway and Rasmussen [4], it was decided to design a cone-derived waverider. The general sequence followed in obtaining the aerodynamic shape of a waverider from a known flow field involves the following steps: 1) Establish the flow fields i.e. flow past the cone at design conditions; 2) Choose a stream surface in the flow; Step 3) Generate the surface which is the intersection of the stream surface with the conical shock. Replace this portion of stream surface by a solid surface which will be the lower surface of the vehicle; 4) Once the lower surface is established, from the leading edge of the lower surface generate a surface which is parallel to the free stream. This free stream surface will be the upper surface of the vehicle. The logic of following the above sequence is that in any flow field, a stream surface can be replaced by a solid surface. So when the chosen portion of the stream surface is replaced by a solid surface, that surface will generate the exactly same shock pattern which will then be coinciding with the leading edge of the surface all along.

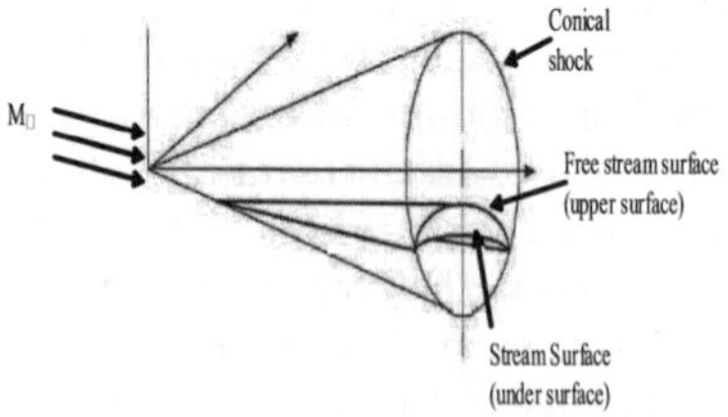

Fig. 1. Schematic of the shock cone and stream surface (the flow generating cone is not shown)

Based on the aerodynamic design the waverider model (Fig. 2) is manufactured using rapid prototyping and numerical machining techniques.The waverider model is about 180 mm long, 80 mm wide and the thickness of the model towards the base is 18 mm. The model is attached to sting with a neoprene rubber bush (10 mm thick) to ensure that the model will experience virtually unrestrained movement during shock tunnel testing.Once we measure the model accelerations during hypersonic flow, and since we know the mass of the model (89 gm) from Newtons law we can easily deduce the aerodynamic coefficients. Three accelerometers are mounted on the waverider to measure the drag, frontlift and aftlift experienced by the model during hypersonic flow. Since the thickness of the

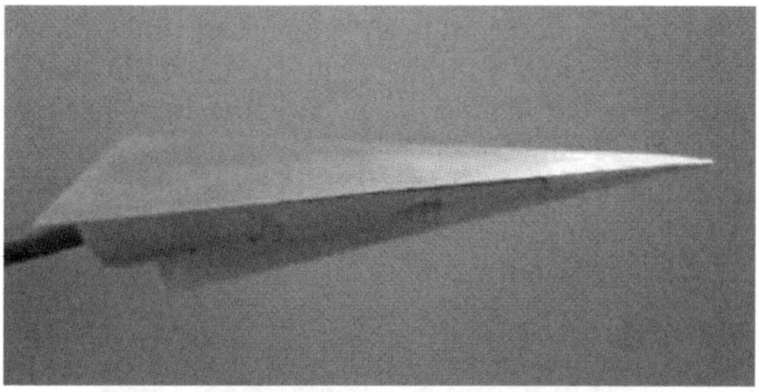

Fig. 2. Isometric view of the hypersonic waverider model used in the present study

Table 1. Nominal Free stream conditions in HST2

Driver Gas	Helium
Shock Mach number, M_s	3.78
Stagnation pressure, P_o(kPa)	1250
Stagnation Temperature, T_o(K)	1628
Stagnation enthalpy, H_o(MJ/kg)	1.65
Freestream Mach number, M_∞	5.6
Freestream static pressure, P_∞(kPa)	1.17
Freestream static temperature, $T\infty$ (K)	231
Free Stream Density, ρ_∞ (kg/m^3)	0.019
Freestream unit Reynolds number (per meter)	2.2x10^6

waverider model is only about 18 mm and the minimum height of the accelerometers (PCB-303A,PCB Piezotronics;10 mV/g)is about 12 mm it is really a challenge to mount the accelerometers properly within the waverider model. Both CFD and FEM tools have also been exhaustively used to design a rubber based accelerometer force balance system used in the present study. The HST2 shock tunnel is capable of simulating Mach numbers ranging from 3.5 to 13 and stagnation enthalpies ranging from 0.7 MJ/kg to 5 MJ/kg.The typical experimental conditions in the shock tunnel used for aerodynamic force measurement of waverider are shown in Table 1.The measured acceleration signals are deconvolved to obtain the basic aerodynamic coefficients. Further finite element modelling of the force balance system is carried out to complement the measurements.

1.2 FEM and CFD studies of the waverider

From the basic operation of the IISc shock tunnel, it is known that the conditions in the test section start from the initial conditions and rise for about a period of 0.4 ms, after which the conditions remain steady for a period of about 0.6 ms which is the window of the test time. A similar transient loading is applied in the FEM analysis where the loads are zero initially and rise to the steady state values over a period of 0.4 ms, after which the loads remain constant for a period of 0.6 ms, the value of the loads being equal to the aerodynamic loads as obtained from the steady state CFD analysis.The aerodynamic loads on the waverider are chiefly the pressure loads. Though viscous effects result in shear stresses, the high L/D of the waverider is primarily because of the pressure distribution on the upper and lower surfaces. Accordingly, the average values of pressure on the upper and lower surfaces were calculated from the results of the CFD analysis. For the present study, the FEM analysis was carried out using the commercial package MSC.NASTRAN 2004 (MSC Software). The geometry of the model including the rubber bush was built in the CAD package AUTOCAD MECHANICAL DESKTOP 6. The geometry was then imported into the pre-processing package MSC.PATRAN 2004 and the 3-d finite element model of the waverider along with the rubber bush used in the analysis is shown in Fig. 3.

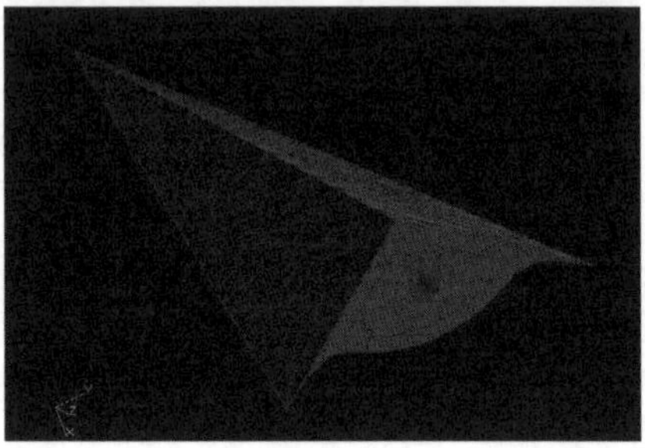

Fig. 3. The typical mesh used in the finite element modelling of hypersonic waverider

The waverider part was assigned the properties of Duralumin and the rubber bush was specified as an isotropic material with E = 3 MPa [5]. Four-node tetrahedral elements were used to build the finite element model. The details are as follows: 1. Number of nodes = 67222. Number of elements = 3395 Two types of loading were applied on the finite element model: 1) Time varying pressure applied uniformly over both the upper and lower surfaces 2) Displacement constraints on the hole in the rubber bush. Since the rubber bush is rigidly bonded to the supporting sting which is held firmly, the displacements are constrained to be zero in all the three directions at the central hole of the rubber bush. The magnitudes of the pressures applied were: 1) 3045.2 Pa on the lower surface and 2) 1200.6 Pa on the upper surface. The nature of time variation of the applied pressures was similar to the time history of the experimental conditions in the test section which were determined from the pitot pressure history in the test section.

The typical experimental and FEM signals from the lift accelerometer along with the acceleration history from the FEM studies is shown in Fig. 4.The typical signals obtained from the drag accelerometer and the FEM results is shown in Fig. 5. Based on the results from the force measurements in shock tunnel at designed conditions the waverider shows a coefficient of drag of 0.0091 while the corresponding coefficients of lift and pitching moments are 0.039 and 0.031. This shows the present waverider has a lift to drag ratio of about 4.28.

Numerical Study

Illustrative numerical simulations are also carried out to complement the experiments using CFX- Ansys software.The geometry of the model and the grid for the analysis were generated using the commercial grid generation package ICEM-CFD (Ansys International, USA). The boundary conditions were accordingly set as symmetric conditions on this plane of symmetry. 3-D Navier Stokes equations were solved in the finite volume

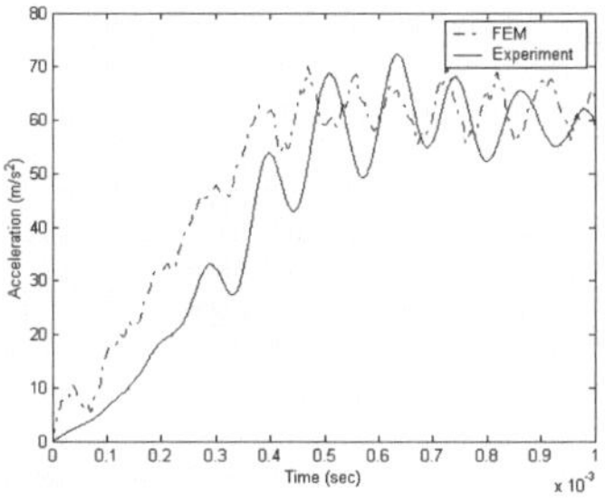

Fig. 4. Typical signal recorded by the front lift accelerometer along with the simulated signal from FEM studies

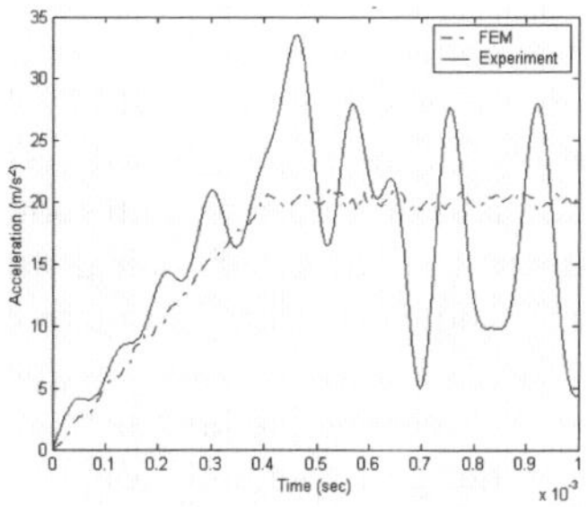

Fig. 5. Typical signal recorded by the drag accelerometer along with the simulated signal from FEM studies

formulation for the steady flow past the waverider. The density distribution along the leading edge of the waverider obtained from CFD shown in Fig. 6 clearly shows that the shock wave is attached all along the leading edge for the shock tunnel test flow conditions at Mach 5.6 and this is consistent with the concept of waverider.The shock pattern around the vehicle indeed confirms that the vehicle is a "Waverider". This observation is quite significant in the light of the fact that though the waverider was designed from analytical solutions for inviscid conical flow, the behavior is not very different for flow with viscosity.The basic aerodynamic coefficients obtained from the experiments along with the results from the CFD studies are shown in Table 2.

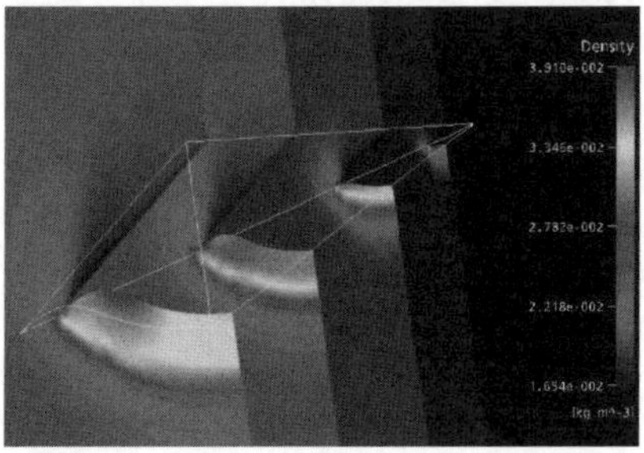

Fig. 6. The density distribution on the IISc hypersonic waverider simulated from CFD studies

Table 2. Aerodynamic coefficients from Experiments and CFD analysis

Aerodynamic coefficient	Experiment	CFD
C_l	0.039	0.032
C_d	0.0091	0.0077
C_m	0.031	-
L/D	4.28	4.15

2 Conclusions

A Mach 6 hypersonic conical waverider has been designed from basic conical flows, fabricated and tested in the IISc hypersonic shock tunnel HST2. The basic aerodynamic forces acting on the waverider is measured using rubber based accelerometer force balance. The IISc waverider designed for nominal Mach number of 6 shows a lift to drag ratio of about 4.28. This is in agreement with the L/D ratios of conical waveriders reported in the open literature. Both FEM and CFD studies are carried out to complement the experiments. The agreement between experiments and CFD results is good. Future experimental campaign aims at measuring the surface convective heating rates and the skin friction on the waverider at both design and off-design hypersonic Mach numbers.

Acknowledgement. The authors would like to place on record sincere thanks to Aeronautical Research and Development Board, New Delhi for sponsoring the research project on waveriders.We also thank all the staff of HEA laboratory for the help during the experiments.

References

1. Kuchemann, D., and Weber, J: Progress in Aeronautical Sciences **9**, (1968)
2. Townend, L., H.: Progress in Aeronautical Sciences **18**, (1979)
3. Rasmussen, M.L.: Journal of Spacecraft and Rockets **17**, 6(1980)
4. Broadaway, R., Rasmussen, M.L.: Aerodynamics of a simple Cone Derived Waverider In: *AIAA Paper 84-0085*, AIAA 22nd Aerospace Sciences Meeting, Nevada, Jan.1984.
5. Sahoo, N., Mahapatra, D., R., Jagadeesh, G., Gopalakrishnan, S., and Reddy, K., P., J: Measurement Science Technology **14**,(2003).

Supersonic flow over axisymmetric cavities

K. Mohri and R. Hillier

Department of Aeronautics, Imperial College London

Summary. Laminar and turbulent computations of rectangular cavities in Mach 2.2 flow are presented. Unsteady open cavities are shown in all laminar cases (L/D = 1.33, 10.33, 11.33 and 12.33) whilst turbulent simulations exhibit closed cavity flows for L/D of 11.33 and 12.33. The turbulent computations are supported by experimental schlieren images using a spark light source.

1 Introduction

The focus of this research is the computational and experimental study of Mach 2.2 flow over annular cavities, of rectangular section, on a body of revolution as shown in Figs 1 and 2. Rectangular cavity flows have been classified into three main groups [1, 2]. In an *open cavity* the separated shear layer reattaches onto the rear corner and in a *closed cavity* it reattaches on the cavity floor. The intermediary state between the two types where flow may also switch from one to the other is termed as *transitional.* Factors that affect the type of cavity flow are the nature and thickness of the approach boundary layer, Mach number, Reynolds number and cavity geometry.

Open cavities (that is the shorter length-to-depth ratios) are often unsteady. The semi-empirical model proposed by Rossiter [3] describes a feedback loop between vortex shedding and acoustic radiation from the cavity front and rear respectively and a modified version is provided by Heller and Bliss [4]. Although successful in some cases, the validity of these formulae for high supersonic flows has been challenged [5, 6].

Knowledge of cavity flow physics is critical as we try to control or make use of cavities in various applications. Examples include reduction of noise and structural loading and enhanced fuel and air mixing in hypersonic propulsion systems.

2 Cavity Model and Test Conditions

The cavity model is shown in Fig. 1. The cavity depth (D = 3 mm) and cone apex angle are fixed and L may be varied between $3-37$ mm. The free stream test conditions are: Mach number (M_∞ = 2.2), static pressure (P_∞ = 25.25 KPa), static temperature (T_∞ = 146.34 K) and unit Reynolds number ($R_e = 32\text{x}10^6\ m^{-1}$).

The supersonic tunnel provides a run time of 20 to 30 seconds at the nominal Mach number of 2.2. It has a blow-down configuration with supply total pressure of $270\text{x}10^3$ Nm $^{-2}$. The model sits co-axially in the working section. An 8.0 Mega pixel digital camera and spark light source (spark time of order 0.1 μs) were used for image capturing.

3 Numerical Procedure

The CFD code used is second order accurate, in both space and time, and has been successfully used in various compressible flow studies, including cavities [7–9]. The laminar and turbulent (using the one-equation Menter turbulence model) simulations presented here are for cavities with L/D ratio of 1.33, 10.33, 11.33 and 12.33. Transition is set at the cavity front corner for the turbulent cases.

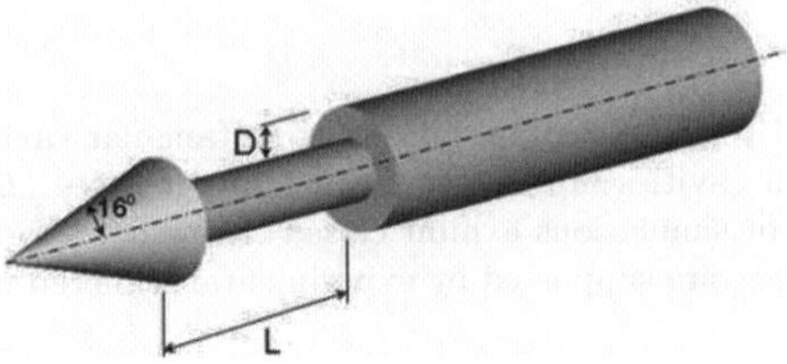

Fig. 1. Cavity model geometry

3.1 Computational Mesh Domain and Boundary Conditions

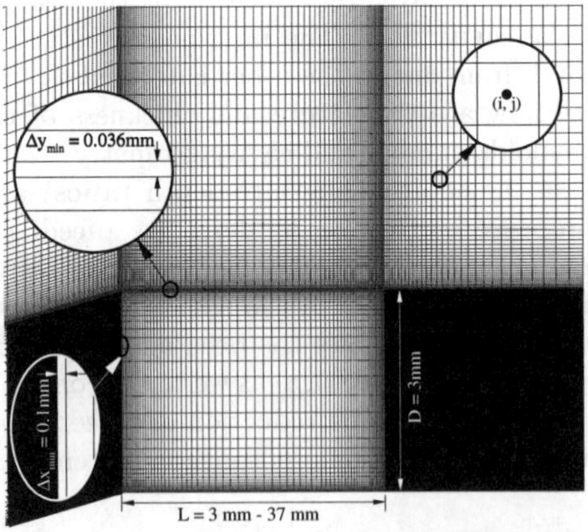

Fig. 2. Details of mesh grid used for simulations. Flow is left to right

A typical mesh is shown in Fig. 2. The coarsest mesh comprises $10L$ (L, the cavity length in mm) cells along the cavity length, and 50 cells along the cavity depth. Mesh refinement was achieved by doubling the number of cells in the i and j body directions. Comparison of time-averaged data for surface pressure and skin friction indicated that the finer mesh results could be regarded as mesh independent.

The inflow boundary cell values are kept fixed at the flow test conditions. The outflow boundary (downstream of the cavity) has a continuation condition whereby cells take the flow values of the adjacent upstream flow field cell. Walls are specified as isothermal at 300 K.

4 Results

Laminar simulations are presented in Fig 3. The flow is open and unsteady. Vortices convect downstream and stretch and distort when negotiating the rear lip of the cavity and emerging onto the afterbody. With increasing L/D, more violent and larger structures are seen towards the rear wall.

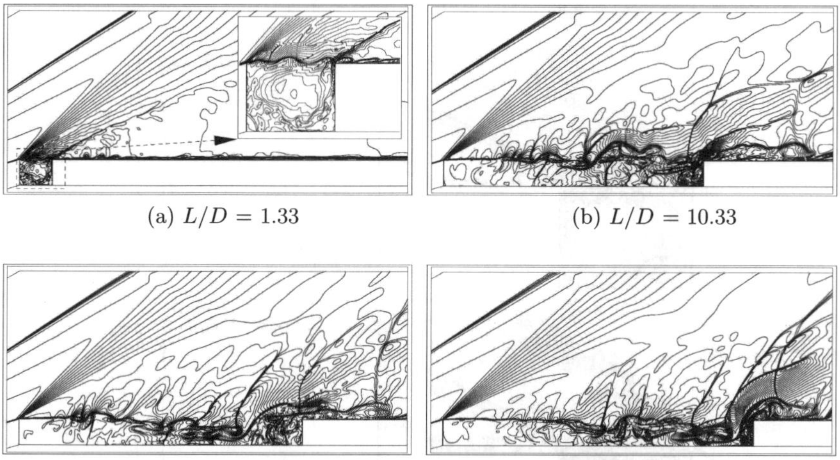

(a) $L/D = 1.33$ (b) $L/D = 10.33$

(c) $L/D = 11.33$ (d) $L/D = 12.33$

Fig. 3. Instantaneous density contours for laminar simulations. ρ/ρ_∞ range is $0 - 2.5$ for all, except for (a)$L/D = 1.33$ which shows $0 - 1.45$, at 5% increments in ρ/ρ_∞ for all. Flow is left to right

Instantaneous density contours and corresponding surface pressures for $L/D = 1.33$ are shown in Fig. 4. s/D represents the wetted distance along the cavity surface from the front upper corner to the rear upper corner ($s/D = 0 - 3.3$). The time interval $= tU_\infty/D$ between each picture is 0.44. The average convective velocity of a vortex midway along the cavity length is $U_{vortex}/U_\infty \approx 0.5$. It must be noted that this velocity does depend on its streamwise position within the shear layer. Picture 1 shows the shear layer is lifted from the cavity rear corner corresponding to relatively low pressure at this point. A vortex (V1) is approaching the rear corner. An upstream travelling wave reflected from the rear wall has travelled approximately $\frac{1}{3}L$ in the upstream direction. As V1 approaches the rear corner it causes a pressure rise at this point. It can be seen that by picture 3, maximum pressure ratio is reached (2.56) at the rear corner. By this point V1, upon reaching the rear corner, has been distorted whilst it emerges onto the afterbody. Three further vortices follow behind. As V1 leaves the cavity, the shear layer lifts off the rear corner and relieves the pressure. By picture 5, V1 is still visible on the afterbody whilst it travels downstream. The two vortices following (V2 and V3) appear to be swept downwards into the cavity and are distorted by the wave system and viscous interaction with the cavity rear wall to such an extent that they are no longer visible. Vortex V4 seems to assume the same position as V1 and the shear layer looks similar to that in picture 1. The cycle shown in Fig. 4 illustrates that for every vortex emerging onto the afterbody, two vortices following behind enter the cavity instead.

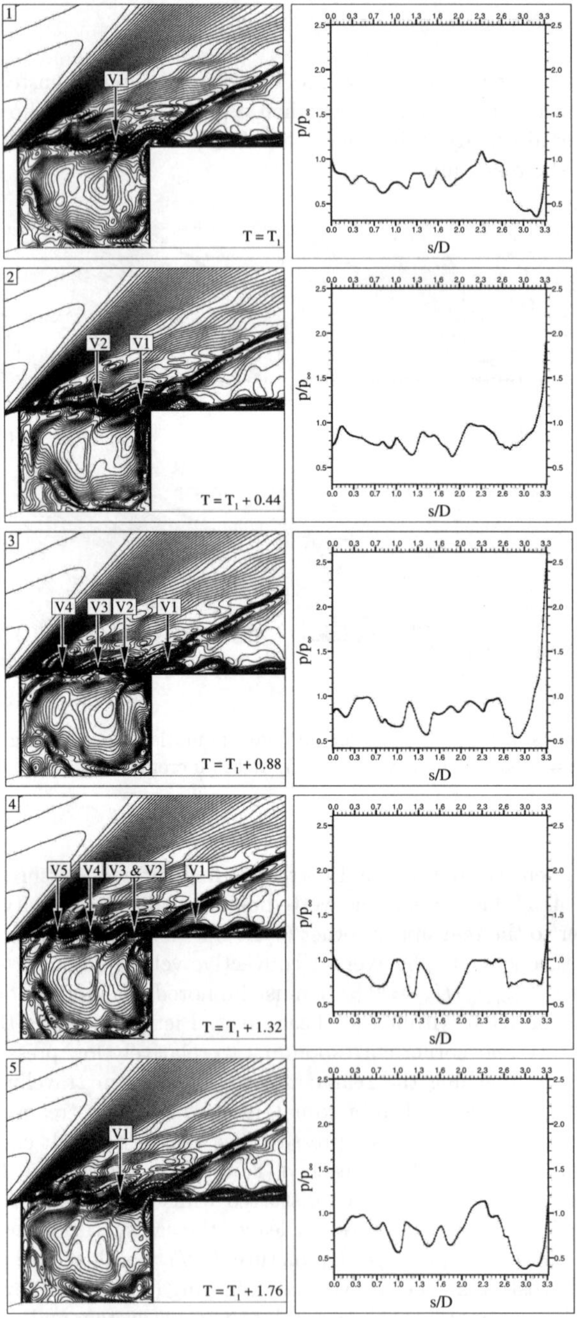

Fig. 4. Instantaneous density contours from laminar simulation of $L/D = 1.33$ (left) and their corresponding surface pressure (right). ρ/ρ_∞ range is $0 - 1.5$ at 2% increments in ρ/ρ_∞. Flow is left to right

The pressure time-history at the cavity front corner is given in Fig. 5. The oscillations are periodic and multiple modes are demonstrated. Low frequency oscillations may be seen on the left image and the zoomed in image on the right shows the high frequency oscillations. By inspection of the computed time-history, the basic high frequency of fluctuations is found to be $fL/U_\infty = 2.06$.

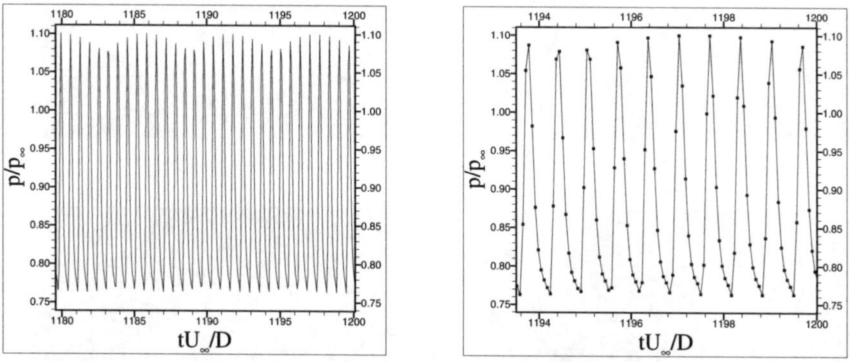

Fig. 5. Laminar CFD Pressure time-history for $L/D = 1.33$ on the cavity front corner

In turbulent simulations the forebody flow is laminar and transition is forced at the beginning of the free shear layer. The results show open cavities for $L/D = 1.33$ and 10.33, and closed for $L/D = 11.33$ and 12.33. This result is supported by the experimental schlieren images of Fig. 6, obtained both with the bare model and with a tripping device attached at the front corner. The same cavities give open and closed flow in both cases. Thus, it can be concluded that the free shear layer is turbulent. The schlieren images also exhibit the axisymmetry of the large scale flow features. In Fig. 7, the numerical density gradient in the normal direction is compared to the experimental schlieren for $L/D = 12.33$ case, showing close agreement between CFD and experiment.

Acknowledgement. This research is funded by EPSRC.

References

1. P. J. Dissimile et al: Journal of Fluids Engineering **122**, 32 (2000)
2. R. L. Stallings and L. J. Wilcox: NASA Technical Report **2683** (1987)
3. J. E. Rossiter: Aeronautical Research Council Technical Report **3438** (1964)
4. H. H. Heller and D. B. Bliss: AIAA Paper 75-491 (1975)
5. C. K. W. Tam and P. J. W. Block: Journal of Fluid Mechanics **83**:373-399 (1978)
6. O. H. Unalmis et al: AIAA **42** 10:2035-2041 (2004)
7. A. P. Jackson et al: Journal of Fluid Mechanics **427**:329-358 (2001)
8. R. Hillier et al: Shock Waves **13**:375-384 (2003)
9. S. Creighton: Hypersonic Flow Over Non-Rectangular Cavities. PhD thesis, Imperial College, London, UK (2003)

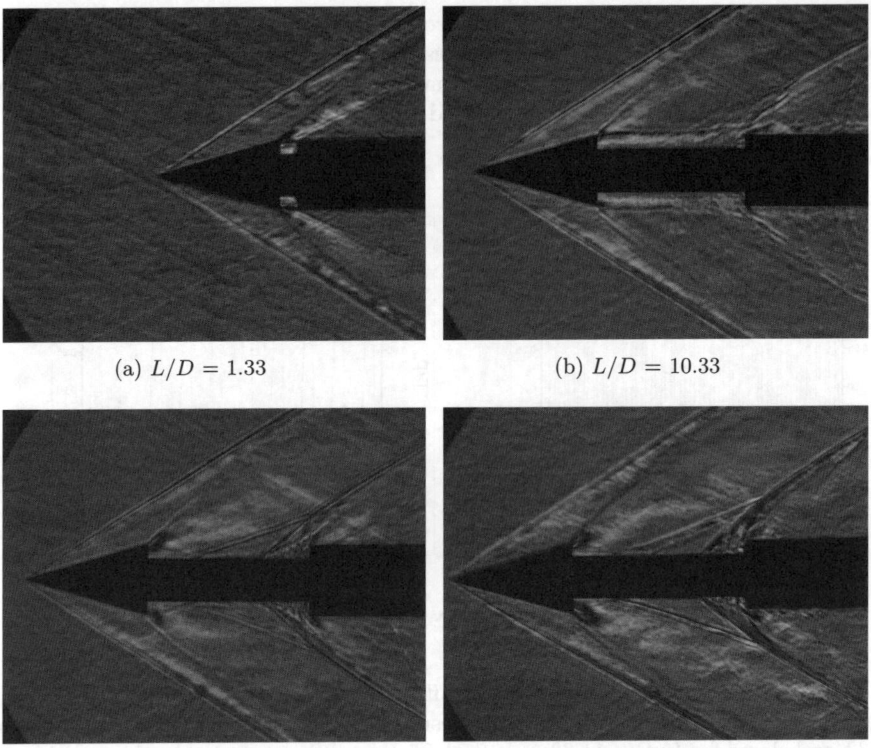

(a) $L/D = 1.33$ (b) $L/D = 10.33$

(c) $L/D = 11.33$ (d) $L/D = 12.33$

Fig. 6. Schlieren visualisation images of cavities

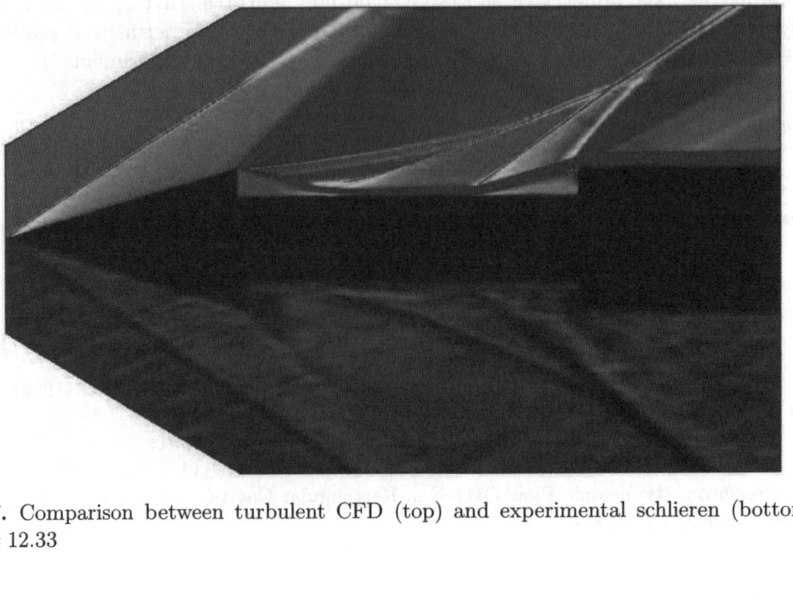

Fig. 7. Comparison between turbulent CFD (top) and experimental schlieren (bottom) for $L/D = 12.33$

Tandem spheres in hypersonic flow

S.J. Laurence[1], R. Deiterding[2], and H.G. Hornung[1]

[1] *Graduate Aeronautical Laboratories, California Institute of Technology, Pasadena, CA 91125, USA*
[2] *Oak Ridge National Laboratory, P.O. Box 2008 MS6367, Oak Ridge, TN 37831, USA*

1 Introduction

The problem of determining the forces acting on a secondary body when it is travelling at some point within the shocked region created by a hypersonic primary body is of interest in such situations as store or stage separation, re-entry of multiple vehicles, and atmospheric meteoroid fragmentation. The current work is concerned with a special case of this problem, namely that in which both bodies are spheres and are stationary with respect to one another. We first present an approximate analytical model of the problem; subsequently, numerical simulations are described and results are compared with those from the analytical model. Finally, results are presented from a series of experiments in the T5 hypervelocity shock tunnel in which a newly-developed force-measurement technique was employed.

2 Analytical modelling

First we describe an analytical model used to predict the forces acting on the secondary body for configurations in which it lies entirely within the primary shocked region. For simplicity, a uniform freestream consisting of a perfect, inviscid gas is assumed.

The situation under consideration is shown in Fig. 1 and the problem parameters are indicated. We wish to determine the axial (drag) and lateral (lift) force components on the secondary sphere, focussing in particular on the behaviour of the lift component.

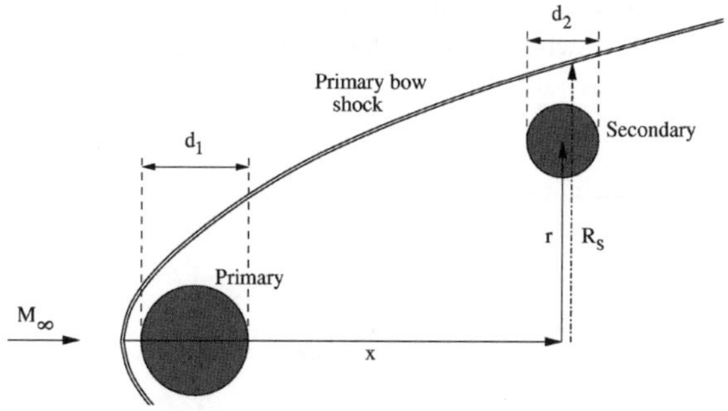

Fig. 1. Representative physical situation for which the analytical model is developed.

The assumption of a large freestream Mach number allows us to make use of the axisymmetric blast wave analogy to approximate the flow conditions inside the primary bow shock. Given knowledge of these conditions, we seek an approximation to the surface pressure distribution on the secondary, as, in hypersonic blunt-body flows, the viscous force contributions are often negligible in comparison to the pressure contributions.

Consider first how the pressure distribution may differ from that if the secondary were in the uniform freestream, where the flow symmetry would result in a zero-lift configuration. First, the flow direction is no longer aligned with that of the freestream, but is deflected by the primary bow shock, shifting the stagnation point to the inner side of the body. This will provide a repulsive contribution to the lateral force, which we will equate with a positive lift contribution. Second, as the dynamic pressure increases sharply in the direction of the primary bow shock, the effective dynamic pressure will be higher on the outer side of the secondary than on the inner side, from which will arise an attractive contribution to the lateral force. Note that, to a first approximation, only this latter effect will depend on the secondary body size, so we might expect the lift coefficient to become increasingly negative as the diameter ratio, d_2/d_1, is increased.

These effects may be quantified by introducing a reference surface pressure distribution and modifying it appropriately: here we use the modified Newtonian description first proposed by Lees [1]. The first effect noted above is easily accounted for, given the rotational symmetry of the sphere. The second effect may be approximated by Taylor-series expanding the stagnation point pressure that appears in the modified Newtonian description to the linear terms in x and r on the surface of the sphere. The force coefficients may then be evaluated by integrating the pressure with the appropriate normal component over the body surface. Using Euler angles, one may then show, for example,

$$\begin{aligned} C_L = \frac{1}{2}\sin\delta(p_0' - p_1') - \frac{1}{15}\frac{d_2}{d_1}\sin 2\delta\frac{\partial(p_0' - p_1')}{\partial(x/d_1)} \\ - \frac{1}{15}\frac{d_2}{d_1}(2 - \cos 2\delta)\frac{\partial(p_0' - p_1')}{\partial(r/d_1)} - \frac{2}{3}\frac{d_2}{d_1}\frac{\partial p_1'}{\partial(r/d_1)}, \end{aligned} \quad (1)$$

where the variables p_1 and δ are the static pressure and flow angle, respectively, as given by the blast wave analogy, and p_0 is the stagnation point pressure. The reference point for these variables is taken to be the centre of the secondary sphere. For convenience we have introduced the notation $p_i' = p_i/\frac{1}{2}\rho_\infty V_\infty^2$ and have also assumed that isosurfaces in the blast wave solution are locally flat over the volume occupied by the secondary.

Figure 2 shows the lift coefficients that result from this expression for various diameter ratios at two values of the downstream displacement. The ratio of specific heats is γ=1.4.

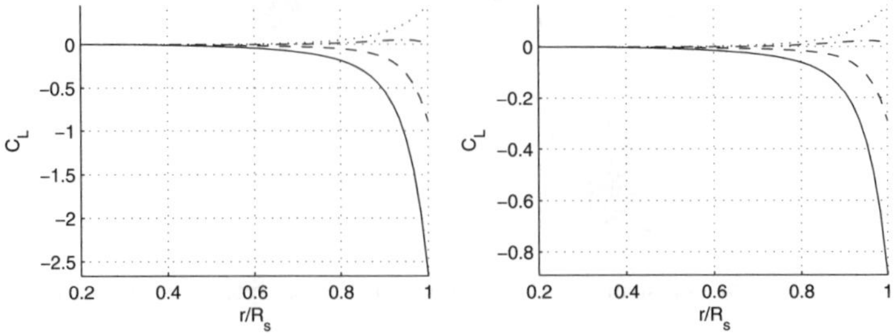

Fig. 2. Theoretical secondary lift coefficients for x/d_1=4 (left) and x/d_1=32 (right): ——, $d_2/d_1 = 1/2$; - - -, $d_2/d_1 = 1/4$; – · –, $d_2/d_1 = 1/8$; · · ·, $d_2/d_1 = 1/16$.

Note that the lift coefficient becomes increasingly negative as the secondary body size is increased, as predicted above. The lift value magnitude decreases as the downstream displacement is increased, but the qualitative behaviour remains unchanged.

3 Computational modelling

Computational modelling of the hypersonic tandem spheres problem has been carried out using the AMROC (Adaptive Mesh Refinement in Object-oriented C++) software developed by R. Deiterding. Further details of this software may be found in [2].

Simulations were carried on the IBM Power4+ machine DataStar at the San Diego Supercomputing Center. The Euler equations were solved for a perfect gas using a time-explicit finite volume approach together with a level-set-based ghost fluid method for the embedded boundaries. The numerical flux was evaluated by employing the flux-vector splitting scheme after Van Leer within the second-order accurate MUSCL-Hancock slope limiting technique. Godunov splitting was used for the multi-dimensional extension.

The problem parameters chosen were a diameter ratio of 1/2, downstream displacements of 1.5 and $4d_1$ (centre-to-centre), and freestream Mach numbers of 10 and 50. For each combination of parameters, a series of simulations was performed in which the secondary lateral position was varied incrementally from immediately behind the primary body to outside the primary bow shock. For all simulations, three additional refinement levels were used, each with an isotropic refinement factor of 2. The physical domain was either $3.9{\times}3.9{\times}3.1d_1$ or $6.4{\times}5.5{\times}3.1d_1$, with one primary diameter corresponding to 82 cells at the highest level of refinement. The CFL number in all cases was 0.8, with a computational cost per simulation of typically 1000 h CPU using 48 processors. A computational schlieren image from one of these simulations is shown in Fig. 3.

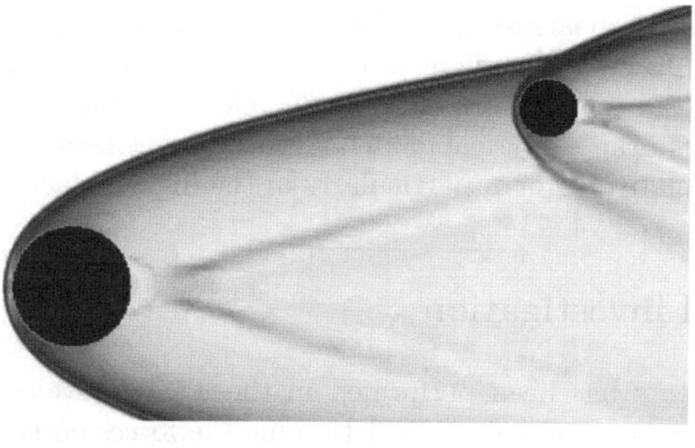

Fig. 3. Computational schlieren image: $x/d_1{=}4$ (centre-to-centre), $M_\infty{=}50$.

The flow was generated in a given run by ramping up the velocity at the inlet boundary to the appropriate steady value. The lift and drag values on each body were calculated on-the-fly by evaluating numerically the surface integral of hydrodynamic pressure forces over the body and average values were calculated over an appropriate time period.

In Fig. 4, results from the analytical model and computations are compared. To provide a meaningful comparison, the lateral displacements have been normalized by the primary bow shock radius in each case. This is necessitated by the under-prediction of the shock radius by the blast wave analogy.

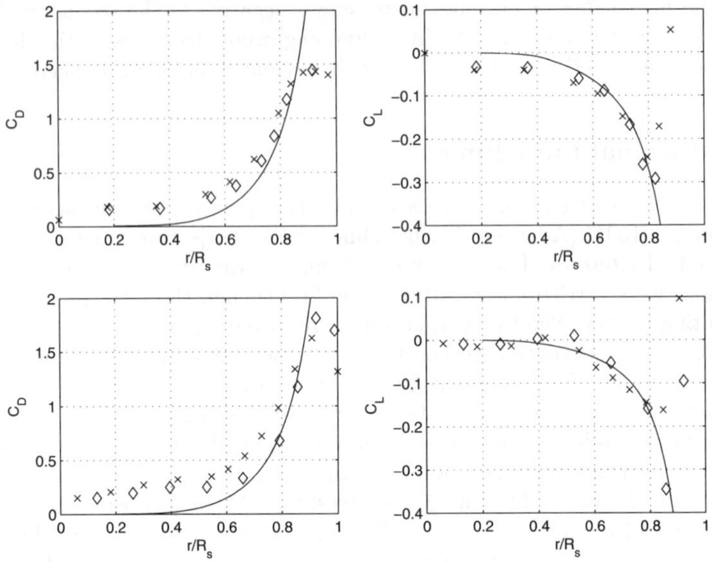

Fig. 4. Comparison of theoretical model with computations for a diameter ratio of $d_2/d_1=1/2$ and downstream displacements of 1.5d$_1$ (top) and 4d$_1$ (bottom): —, theoretical; ×, computational, $M_\infty=10$; ⋄, computational, $M_\infty=50$. The ratio of specific heats in all cases is $\gamma=1.4$.

Substantial disagreement between theoretical and computational values is observed as r/R_s approaches 1, as impingement of the primary bow shock on the secondary results in the breakdown of the analytical model. The presence of the strong entropy wake in the blast wave solution also appears to lead to the analytical drag profile decaying too quickly as r/R_s is decreased to zero (we might also expect the absence of a wake region in the analytical model to lead to disagreement for r/R_s close to zero). Given the assumptions that have been made in the analytical model, however, agreement wih the computational values is otherwise relatively good, particularly in the higher Mach number case.

4 Experimental investigation

A series of experiments has been performed in the T5 hypervelocity shock tunnel at the California Institute of Technology to determine the forces on the secondary body in several tandem spheres configurations. The test gas in all experiments was carbon dioxide. For each shot, reservoir conditions were calculated from measured pressures and the incident shock speed using ESTC (Equilibrium Shock Tube Calculation), due to McIntosh [3]. Freestream conditions were then calculated at the relevant point downstream with NENZF (Non-Equilibrium NoZzle Flow), due to Lordi, Mates & Moselle [4]. All shots were carried out at nominally the same condition: representative values for the reservoir pressure and enthalpy were 17 MPa and 9.4 MJ/kg, while the freestream velocity, density, and Mach number were typically 3100 m/s, 0.035 kg/m^3 and 4.5, respectively.

A schematic of the apparatus for this series of experiments is shown in Fig. 5. The primary sphere, of diameter 64 mm, was mounted rigidly to the test section by means of a sting and supporting plates. The secondary sphere, of diameter 32 mm, was suspended from the roof of the test section by two cotton threads in a V-arrangement. These threads were destroyed by the onset of the flow, allowing for free-floating model behaviour during the test time. A padded catcher mechanism, mounted directly behind the secondary sphere, served to terminate the motion after the conclusion of the steady flow period.

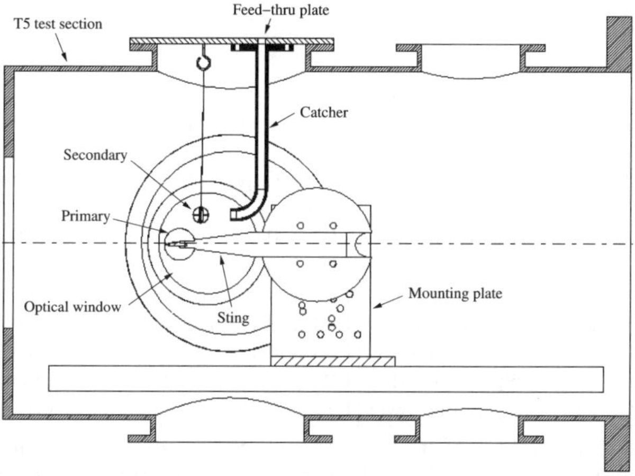

Fig. 5. Arrangement of apparatus in the T5 test section (to scale)

Two means of force-measurement were employed in this series of experiments. The first made use of a sequence of high-speed images obtained from the T5 optical system, consisting here of a conventional Z-arrangement schlieren with a Vision Research Phantom v5 digital camera. Images were recorded at either 25,000 or 38,000 frames per second and were then processed by an image-tracking algorithm to determine the x and y coordinates of the secondary body in each frame, as well as the scaling factor that transformed from image to physical dimensions. Assuming both the lift and the drag to be constant over the test time, this allowed quadratic polynomials to be fitted to each of the x and y displacement profiles as functions of time, with the acceleration in each case being given by twice the quadratic coefficient, and the forces following trivially.

The second means of measuring the drag force on the secondary sphere was through direct measurements of the acceleration by an internally-mounted uniaxial accelerometer. This method was not used in all of the experiments in the current series, however, due to constraints that the accelerometer cabling placed on the model geometry.

The experimental force coefficients are tabulated in Table 1 and are compared with coefficients obtained from further numerical simulations in AMROC, carried out at appropriate freestream conditions and including the effects of the conical T5 nozzle. Figure 6 shows experimental and computational schlieren images from one such experimental condition. As may be seen, the shock impingement points on the secondary body are relatively close in the two images, although the experimental shock radius appears slightly smaller. Impingement occurred in the majority of experiments: in such situtations, the lift value especially is sensitive to the exact location of the impingement point. Despite this fact, agreement between the force coefficients obtained from the two experimental techniques and with those from computations is generally good, in most cases lying

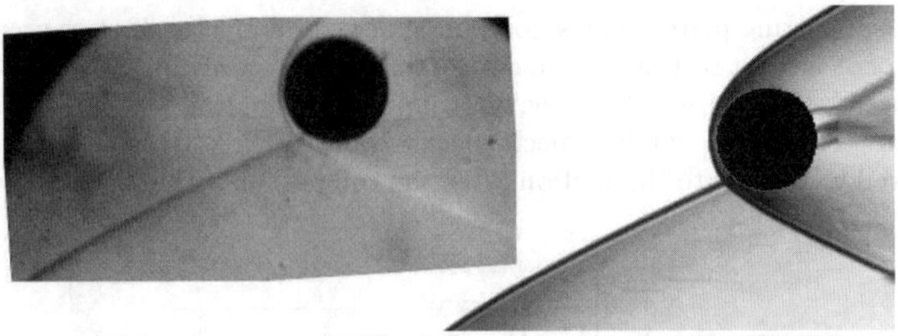

Fig. 6. Experimental and computational schlieren images of T5 shot 2330. The experimental image has been rotated to correct for the rotation in the T5 optical system.

Shot number	Axial disp.	Lateral disp.	Shock position	Experimental C_D	Experimental C_L	Computational C_D	Computational C_L
2322	-	-	Sphere outside	0.97 ± 0.07	(accelerometer)	0.956	(perfect gas)
				0.95 ± 0.07	(images)	0.995	(real gas)
2326	1.07	1.21	Impinging upper	1.44 ± 0.09	0.02 ± 0.11	1.40	0.23
2327	1.25	1.18	Just impinging	1.43 ± 0.11	(accelerometer)	1.34	
2328	1.25	1.18	Just impinging	1.35 ± 0.11	0.07 ± 0.11	1.34	0.01
2329	1.50	1.21	Sphere inside	1.08 ± 0.15	-0.13 ± 0.09	1.05	-0.05
2330	1.50	1.67	Impinging lower	1.11 ± 0.08	0.29 ± 0.05	1.01	0.28

Table 1. Force coefficients obtained from experiments and corresponding numerical simulations. Displacements are given in primary body diameters and are centre-to-centre values.

within the experimental error. Note, however, that the computational drag values lie consistently slightly lower than the experimental values: this may be explained by the absence of viscous forces in the numerical simulations.

5 Conclusion

The problem of determining the forces acting on the secondary body in a hypersonic tandem spheres configuration has been explored analytically, computationally, and experimentally. An analytical model based on the blast wave analogy was developed, and the resulting secondary force coefficients showed reasonable agreement with results from perfect-gas numerical simulations. A series of experiments in the T5 hypervelocity shock tunnel, employing a newly-developed force measurement technique, was also carried out, with results again showing reasonable agreement with those from numerical simulations.

References

1. Lees. L: Hypersonic Flow, Proc. 5th Int. Conf. Los Angeles, 1955, pp 241–276
2. Deiterding R.: Detonation structure simulation with AMROC. *High Performance Computing and Communications 2005, Lecture Notes in Computer Science 3726*, 2005, pp 916–927
3. McIntosh M.: Computer Program for the Numerical Calculation of Frozen and Equilibrium Conditions in Shock Tunnels. Australian National University, Canberra A.C.T., 1968
4. Lordi J. A., Mates R. E., Moselle, J. R.: Computer program for the numerical solution of nonequilibrium expansions of reacting gas mixtures. NASA CR-474, 1966

Three dimensional experimental investigation of a hypersonic double-ramp flow

F.F.J. Schrijer, R. Caljouw, F. Scarano, and B.W. van Oudheusden

Delft University of Technology, Faculty of Aerospace Engineering
Kluyverweg 1, 2629 HS Delft (The Netherlands)

Summary. The flow over a 15°-45° double compression ramp was studied at Mach 7.5. CFD computations are compared to 2 component PIV (particle image velocimetry) measurements. Furthermore stereoscopic PIV was used to measure the three component velocity vector, enabling to perform a 3D flow survey. The overall flow topology is assessed and special attention is devoted to the separated region. Finally the effect of a sharp leading edge on the separation region is investigated.

1 Introduction

The flow over a double compression ramp is studied experimentally in a short duration Ludwieg-tube facility at Mach 7.5 [5]. The large second ramp angle introduces a shock detachment accompanied by an Edney type V interaction. A schlieren image of the flow is given in figure 1. From previous studies using high-speed schlieren imaging it was found that the flow field shows unsteady behavior [6] which is typical for these type of flows [8]. Movement of the separation shock was detected as well as the movement of the curved shock. The unsteady phenomena occurring at separation were ascribed to transition in the separated shear layer. The movement of the curved shock is believed to be caused by a shock hysteresis phenomena as described by Ben-dor et al. [2]. This in combination with the relatively large separated region and the possible occurrence of transition in the shear layer are the reason that flow simulation by means of CFD is non-trivial [3]. In the current study the flow is investigated more in detail. Two component velocity measurements at the center of the model described in detail in [6] are compared to a turbulent CFD computation. Following the results from this comparison, a three component flow survey is performed using stereoscopic PIV.

2 Experimental setup

The model used in the experiments is a double compression ramp featuring a 15° first and 45° second ramp angle. The total model length is 15 cm where the first ramp is 10 cm length and the model width is 11 cm. The setup used for the 2C PIV measurements can be found in Schrijer et al. [6]. The model leading edge radius was increased to 1 mm by adding a piece of tape to the model nose to obtain a uniform geometry.

To measure the three component velocity field, stereo PIV was applied. The illumination was performed by a Quantel laser having 200 mJ per pulse. The typical time separation between the pulses was 1 μs. The laser sheet had a thickness of 1.5 mm. The particle images were recorded using two PCO sensicam QE frame-straddling CCD cameras having a resolution of 1376 × 1040 pixels corresponding to a field of view of 6 × 4.5

cm^2. TiO_2 was used as flow tracer which has a nominal particle size of approximately 500 nm. The typical window size used for image interrogation is 32×32 pixels^2, this corresponds to a measurement volume of 1.4×1.4 mm^2. Because of inertial forces the particle traces will slip in regions with high accelerations (shocks), it was experimentally determined that the particle relaxation time over a shock is $\tau = 2.5$ μs [5]. The actual slip length depends linearly on the particle velocity, the maximum slip length is 2.5 mm, however it is smaller in regions with lower velocities.

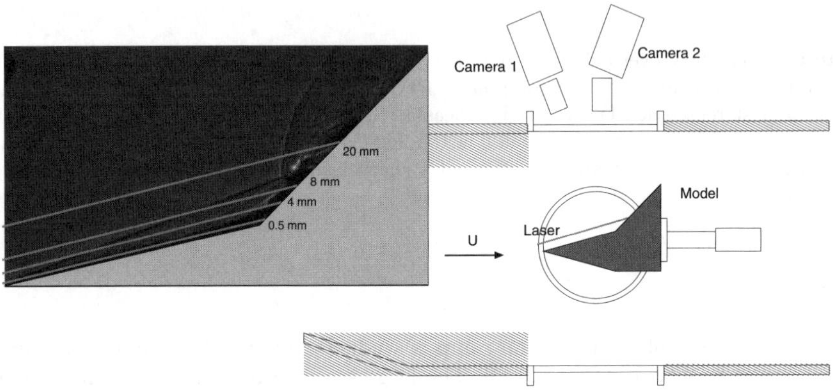

Fig. 1. Stereo PIV setup and measurement location

The 4 planes that are investigated are oriented parallel to the first ramp, they are located at 0.5, 4, 8 and 20 mm from the model surface. In figure 1 the schematic overview of the setup is given including an insert of a schlieren image showing the exact location of the laser sheet with respect to the model and flow features. The measured velocity components are linked to the orientation of the laser sheet; U, V and W are respectively the in-plane horizontal, spanwise and out-of-plane components. The obtained velocity fields are averages of typical 10 recordings.

3 CFD comparison to 2C-PIV study

A two dimensional CFD computation was performed using a Navier Stokes second order finite volume flow solver (LORE) [7]. The inflow conditions were equal to the wind tunnel free stream conditions. The computational mesh featured 2.1 million cells. Only a fully turbulent calculation converged to a stable solution where Menter's shear stress transport model was used as turbulence model. In figure 2 the computational results are shown in combination with experimental results which are discussed in detail in Schrijer et al. [6]. It was found that the flow field showed good qualitative agreement. However the separation region in the CFD computations was considerably larger compared to the experiments causing the shock interaction to occur further away from the model surface. This was primarily ascribed to the three dimensionality of the flow field that empties the separation bubble thus making it smaller. To assess the amount of three dimensionality of the flow, it was investigated using stereo PIV.

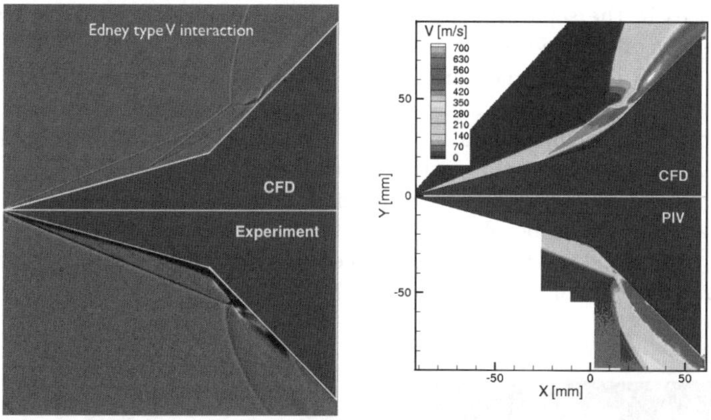

Fig. 2. Comparison between CFD simulation and experimental results; synthetic schlieren versus experiments (left) and computed vertical flow component versus PIV vertical flow component (right)

4 Flow field overview

First the overall flow field will be discussed using the results obtained from the 4, 8 and 20 mm planes. In figure 3 the particle image recording is shown in combination with the velocity field for the plane at 4 mm from the surface.

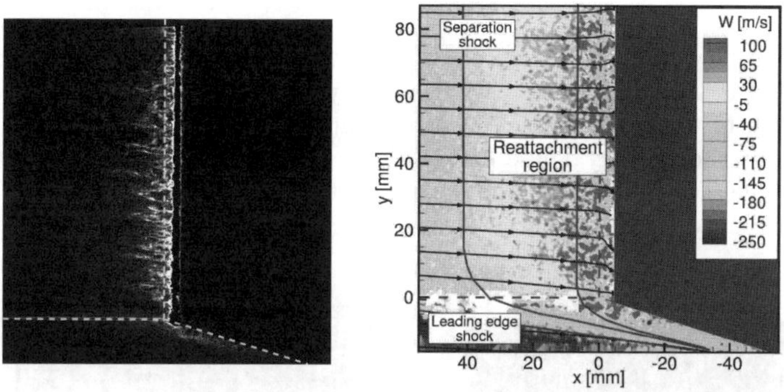

Fig. 3. Results for the plane at 4 mm, particle image recording (left) and flow field (right), contour represents out-of-plane velocity component

Progressing downstream toward the second ramp, an increase is found in the particle density caused by the thermodynamic density increase across the separation shock. This is also observed from the velocity field where the vertical flow component increases when separation occurs. Further downstream the plane crosses the reattachment region (not yet a shock) which is again is associated to an increase in particle density and out-of-plane velocity component. Finally approaching the model surface, streaks are visible in the reattachment region. These streaks are conceived to be caused by the presence of

Görtler vortices in the separated shear layer. These vortices are common to occur in separated regions of double-ramp flows [4].

Figure 4 shows the flow measured in the plane at 8 mm from the model surface. The overall flow topology is the same as the 4 mm plane. Again an increase in particle density is observed for the separation shock with an increase of out-of-plane velocity component. At reattachment the particle density increases again due to an increase of thermodynamic density. Here the reattachment region is coalesced into the reattachment shock. Furthermore it can be observed that the particle density is less uniform in the small region behind the reattachment shock compared to the free stream. This is caused by the turbulent nature of the flow at reattachment.

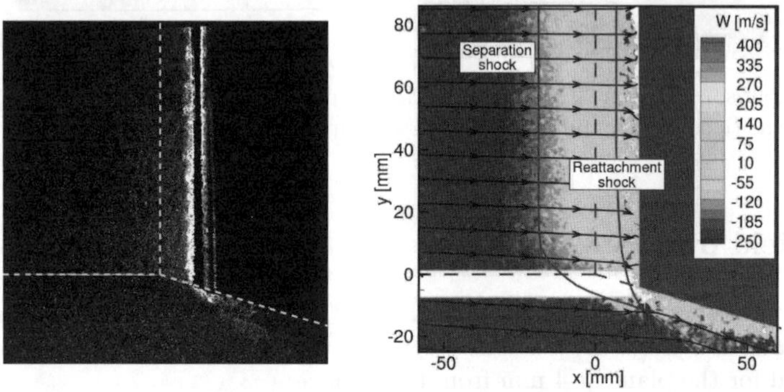

Fig. 4. Results for the plane at 8 mm, particle image recording (left) and flow field (right).

Finally the flow field at 20 mm above the first ramp surface is shown in figure 5. Clearly visible is the increase in particle density due to the curved shock. After the uniform region of increased particle density, empty blobs are observed that mark the presence of the shear layer. These empty blobs were also observed in the 2C particle image recordings, see [6]. Crossing the shear layer the presence of the wall jet is visualized by an increase of non-uniform particle seeding density. The same is visualized from the

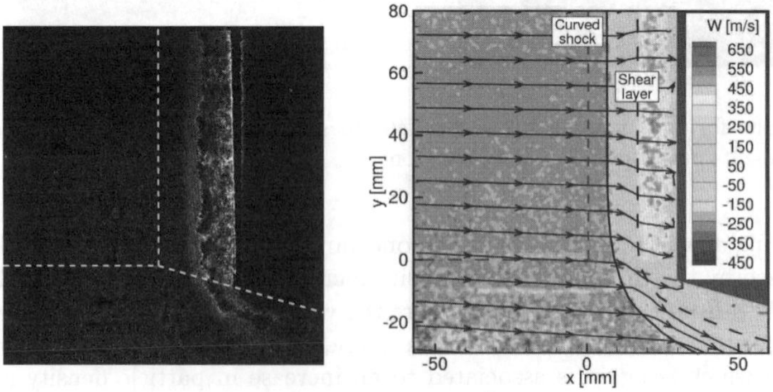

Fig. 5. Results for the plane at 20 mm, particle image recording (left) and flow field (right).

velocity contours, clearly showing the increase in out-of-plane component when crossing the curved shock and an even further increase when entering the wall jet region.

Considering the macroscopic shape of the shock wave pattern established over the model it can be concluded that wave are essentially 2D over a relatively large portion of the center of the model.

5 Surface flow

To investigate the flow in the separated region, a plane at 0.5 mm from the model surface was investigated. Here the largest 3D effects are expected to occur. Again the particle image recording and velocity field are depicted in figure 6. As can be seen from the recorded images, again the particle density increase is observed when crossing the separation shock. Subsequently in the separated region traces of longitudinal streaks can be observed, these are believed to be caused by the emergence of the Görtler vortices, similar to the ones observed for the plane at 4 mm. The velocity fields show a large span-wise velocity gradient in the separated region causing an emptying of the separation bubble. Because of this the separation bubble is reduced in size compared to a two dimensional bubble.

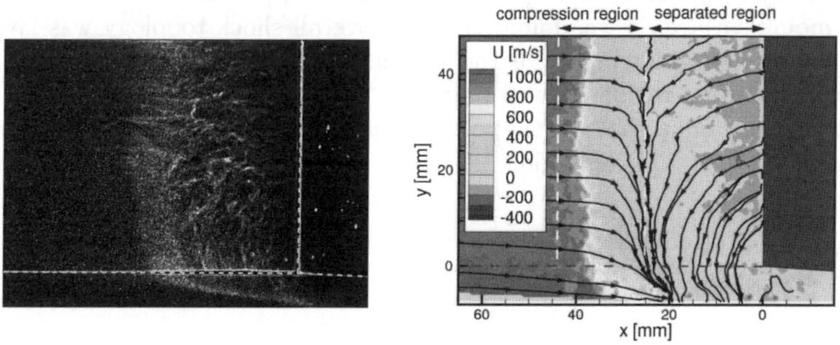

Fig. 6. Results for the plane at 0.5 mm, particle image recording (left) and flow field (right).

6 Influence of leading edge shape

Additionally the separation region was also investigated for a sharp leading edge where the nose radius was 0.1 mm. The results are shown in figure 7, comparing this to the results for the rounded leading edge (figure 6) it can be seen that the compression and separation region is dramatically altered. The extent of the compression and separated region is reduced and at some location the separation region is completely absent. Looking at the particle image recording it can be seen that at this location streaks can be observed emanating from the leading edge which destroy the separated region. From literature it is known that the leading edge has a big influence on the wavelength and presence of Görtler vortices [1] however it appears that small irregularities can also have a significant influence on the extent of the separated region.

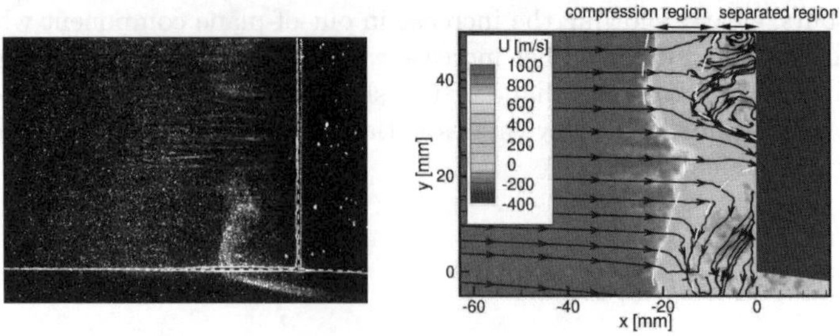

Fig. 7. Results for the plane at 0.5 mm for the model with sharp leading edge, particle image recording (left) and flow field (right).

7 Conclusions

The flow over 15°-45° ramp was studied at Mach 7.5. A turbulent CFD simulation on a two dimensional computational grid was made. Comparison with experimental results obtained using 2C PIV and schlieren showed a qualitatively good agreement. However the separation region was found to be considerably larger in the computations. This was ascribed to the three dimensionality of the flow field. Stereo PIV measurements yielding all three velocity components were performed at several planes at 0.5, 4, 8, and 20 mm above the model surface. It was found that the overall shock topology was largely two dimensional over the greater part of the model. Focussing the attention to the plane crossing the separation region showed that here the flow was highly three dimensional. Furthermore it was found that the shape of the model leading edge has a dramatic effect on the extent of the separation region.

Acknowledgement. The authors would like to acknowledge Dr.L.Walpot and D.Sileri for providing the CFD data.

References

1. Aymer de la Chevalerie D, Fonteneau A, De Luca L, Cardone G (1997) Görtler type vortices in hypersonic flows: the ramp problem, Experimental Thermal and Fluid Science, Vol. 15
2. Ben-Dor G, Vasilev EI, Elperin T, Zenovich AV (2003) Self-induced oscillations in the shock wave flow pattern formed in a stationary supersonic flow over a double wedge, Physics of Fluids, Vol. 15, No. 12
3. Druguet M, Chandler GV, Nompelis I (2005) Effect of numerics on Navier-Stokes computations of hypersonic double-cone flows, AIAA Journal, Vol. 43, No. 3
4. Navarro-Martinez S, Tutty OR (2005) Numerical simulation of Görtler vortices in hypersonic compression ramps, Computers & Fluids, Vol. 34
5. Schrijer FFJ, Scarano F, van Oudheusden BW (2006) Application of PIV in a Mach 7 double-ramp flow, Experiments in Fluids, Vol. 41
6. Schrijer FFJ, van Oudheusden BW, Dierksheide U, Scarano F (2006) Quantitative visualization of a hypersonic double-ramp flow using piv and schlieren, 12th international symposium on flow visualization, Göttingen, Germany
7. Walpot LMGFM (2002), Development and application of a hypersonic flow solver, PhD thesis, Delft University of Technology
8. Wright MJ, Sinha K, Olejniczak J, Candler GV, Magruder TD, Smits AJ (2000) Numerical and experimental investigation of double-cone shock interactions, AIAA J., Vol. 38, No. 12

Triple point shear layers in hypervelocity flow

M. Sharma[1], **L. Massa**[2], **and J.M. Austin**[1]

[1] *Department of Aerospace Engineering, University of Illinois at Urbana-Champaign, IL 61801*
[2] *Department of Mechanical Engineering, Clemson University, SC 29634*

1 Introduction

Thermochemical processes such as dissociation and vibrational excitation can have a substantial impact on the gas dynamics of planetary entry. A critical question that confronts vehicle designers is the role of such molecular effects on transition and turbulence. In high stagnation enthalpy flows, thermochemical processes have been observed to affect transition to turbulence in boundary layers through modifications to the mean flow profile as well as to flow stability [1–3].

In the present work, we consider a free shear layer as a model problem in which thermochemical nonequilibrium effects on hydrodynamic stability can be investigated. We report on an experimental and analytical study examining free shear layers generated by a shock triple point. Initial experimental investigations of a triple point free shear layer in a hypervelocity flow and a spatial linear stability analysis with detailed thermochemical modeling are presented.

In order to produce well-characterized experimental inflow conditions and to avoid boundary layer complications associated with splitter plate geometries, we consider a free shear layer created via a Mach reflection. Two opposing wedges are mounted perpendicular to the free stream, shown schematically in Figure 1. The shear layers separate a supersonic, relatively cold, gas stream and a subsonic, relatively hot gas stream. Pressure-flow deflection polars are used to produce initial estimates of the flow properties across the contact surfaces and to identify appropriate experimental run conditions. Sample shear layer properties are summarized in Table 1.

2 Experimental Approach

Experiments are carried out in the newly-constructed 152 mm diameter hypervelocity expansion tube (HET) facility at the University of Illinois [4]. By varying the initial pressure and composition of the gases in the driver, driven, and expansion sections, the facility can access a range of test flow conditions with stagnation enthalpies (approximately 8-10 MJ/kg) in which real gas effects can be studied. In air, the facility is designed to achieve Mach numbers from 3.0 to 7.1. In the broader scope of this study, we will consider three test gases: air, carbon dioxide and argon. The flexibile operation of the facility permits the facility run conditions to be chosen to match the free stream conditions as closely as possible for each test gas. Calculated free stream conditions and wedge angles for each gas are shown in Table 2.

The results presented in this paper focus on air as the test gas. Table 3 presents the comparision of experimental data with theoretical predictions for the air run condition

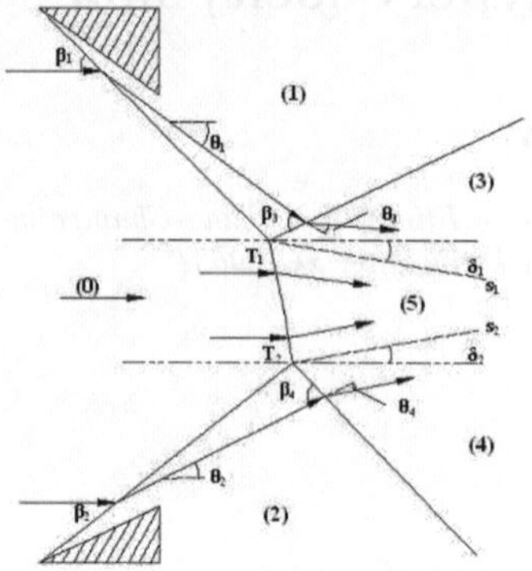

Fig. 1. Schematic of an asymmetric Mach reflection.

	u m/s	ρ kg/m^3	T K	p kPa
State 0	3980	0.005	740	0.974
State 2	3380	0.021	2985	17.92
State 4	2910	0.048	4462	61.36
State 5	720	0.025	8368	60.97

Table 1. Properties at triple point T2 closest to the 25° wedge for nominal Mach 7.29 condition in air.

	Air	Ar	CO_2
Mach number	7.29	7.28	7.21
T, K	740	820	1080
p_{pitot}, kPa	67	90	82
p, kPa	0.974	1.15	1.28
u, m/s	3980	3875	3720
ρ, kg/m^3	0.005	0.007	0.006
Test time, μs	160	160	170
θ_1, degrees	35	25	45
θ_2, degrees	25	25	15
Initial Pressures			
Driver, kPa	4400	4400	4400
Driven, kPa	1.5	1.55	1.1
Expansion, mTorr	200	250	300

Table 2. Theoretical freestream conditions.

	Theory	Exp.
Mach number	7.29	**7.1**
p_{pitot}, kPa	67	**64**
U_s, m/s	2126	**2069 $\pm$43**
Test time, μs	158	**100**
β_1, degrees	44.69	**44.4 $\pm$0.3**
β_2, degrees	32.03	**32.5 $\pm$0.3**

Table 3. Comparison between theoretical and experimentally measured flow parameters.

described in Table 2. The primary shock velocity U_s was measured using time-of-arrival data from wall-mounted pressure transducers. Transducers mounted normal to the flow on the sting within the test section were used to measure pitot pressure and provide the trigger signal for flow visualization purposes. A sample pitot pressure survey 54 mm downstream of the tube exit for the air test condition is shown in Fig. 2a.

The opposing wedge configuration used to create a Mach reflection, Figure 2b, consists of two wedges mounted via a backing plate to the sting with a 25.4 mm separation distance between each wedge tip. A pitot probe is located 31.75 mm below the test section centerline, which is within the core flow diameter as verified by previous experiments [4]. In order to minimize three-dimensional phenonema such as wedge edge reflection [5], the wedges were designed with inlet aspect ratios of 1.25 and wedge aspect ratios of 5.

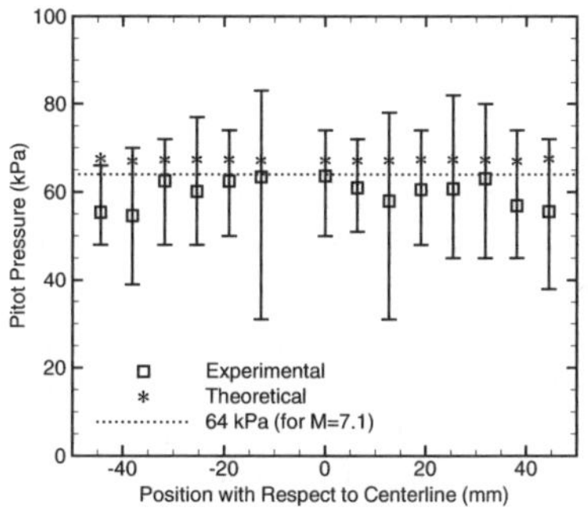

(a)

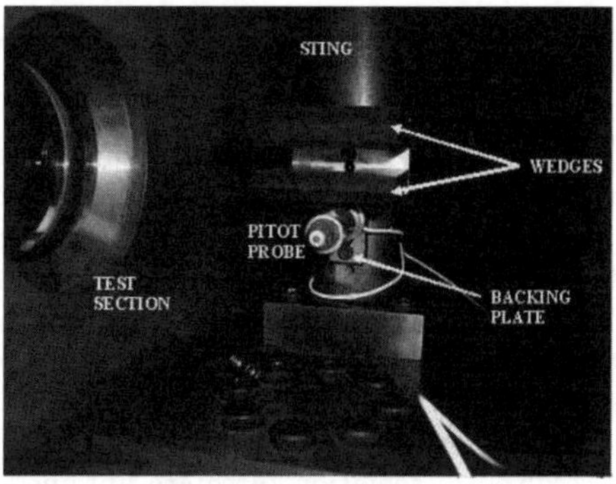

(b)

Fig. 2. a) Pitot survey at 54 mm downstream of tube exit. Test gas is air, nominal Mach number is 7.29. Probe was located on vertical tube axis. b) Sting with backing plate, off-vertical axis pitot pressure transducer and wedges in HET test section.

A schlieren visualization of a Mach relection created in a Mach 7.1 free stream is shown in Figure 3. The Mach reflection features such as the two triple points, the Mach stem, the incident and reflected shocks, and the two shear layers can be observed.

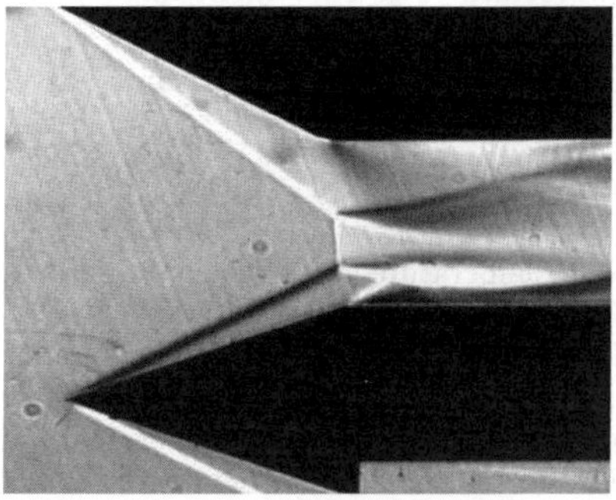

Fig. 3. Mach reflection created over two oppposing wedges with $\theta_1 = 35°$ and $\theta_2 = 25°$ in a Mach 7.1 free stream in air. The wedge aspect ratio is 5 and the inlet aspect ratio is 1.25. Field of view is 32 mm.

3 Spatial linear stability analysis

We carry out a spatial linear stability analysis of hypersonic free shear layers to examine the effect of thermochemical nonequilibrium on associated growth rates. The shear-layer profiles are obtained by solving the 2D Navier-Stokes equations in the boundary-layer form. The parabolic form of the equation are transformed into the shear layer variables ξ, η using fourth order central stencils in the η direction and solved with a spatial marching scheme starting from the experimental initial conditions. The computational grid includes 500 points in the η direction and the spatial marching integrator relative tolerance has been set to $\times 10^{-7}$. A total of 131 unknowns are solved for each grid point; they include 129 species plus velocity and enthalpy. The present analysis focuses on the effect of the thermochemical state and the convective Mach number M_c on the growth rate eigenvalues.

A six species model for air (O_2, N_2, NO, O, N, and Ar) is assumed. We analyze and compare three cases: frozen, nonequilibrium and equilibrium flows. Nonequilibrium air is considered to be in vibrational and chemical nonequilibrium, but in rotational equilibrium (rigid rotator model) at the translation temperature. State resolved energy transfer in adiabatic inelastic molecule collisions is of key importance in understanding nonequilibrium processes. Non-reactive transition rates are modeled using the semi-classical 3D FHO-FR collision model described in [6–8]. The analytical formulas are extended to high collision energies by a reformulation of the steepest descent integration procedure. The present approach requires the solution of a non-linear equation for each rate evaluation. State resolved dissociation rates are obtained by summing bound-free transition probabilities over a set of transition levels exceeding the maximum bounded energy by 20 levels. For reactive collisions, state-resolved reaction rates are taken from the work of Bose and Candler [9,10].

The initial convective Mach number is defined based upon the frozen speed of sound at the inlet plane; the value corresponding to the experiment is $M_{c*} = 0.683$. A parametric study is performed with M_c in $[\frac{M_{c*}}{2}, 2.5M_{c*}]$, by multiplying the velocity on either side of the shear layer by a factor, while keeping the enthalpy and pressure constant. Results are presented at a distance of 3 cm from the inlet plane. The thermal distribution is shown in Figure 4. The equivalent vibrational temperature T_{ve} for a diatomic molecule in vibrational nonequilibrium is defined as the translational temperature that would support the given vibrational energy at equilibrium. The equivalent dissociation temperature T_{de} is the equilibrium translation temperature that yields a degree of dissociation equal to the nonequilibrium case. Results show that at nominal and halved

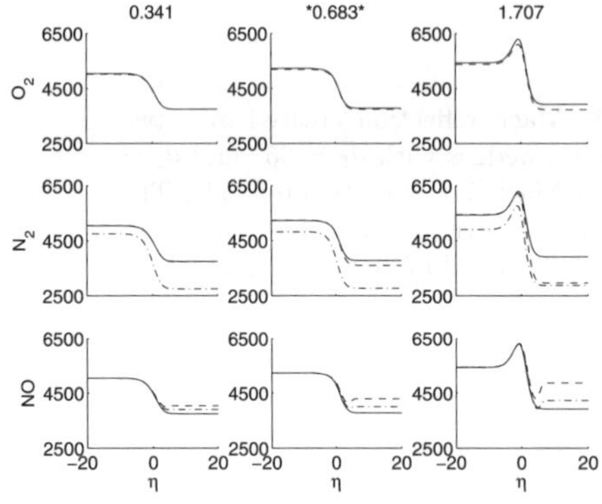

Fig. 4. Temperature distribution in the shear-layer at three convective Mach numbers. In the first two rows: the solid line represents the translational temperature, the dashed line is the equivalent vibrational temperature and the dash-dotted line is the equivalent dissociation temperature. In the third row the dissociation temperature is substituted with the first vibrational temperature, defined as $\frac{e_{v,1}-e_{v,0}}{k\log(N_0/N_1)}$ where N are mole numbers in the subscripted vibrational levels.

inlet speed the dissociation is primarily responsible for departure from equilibrium. At an inflow velocity equal to 2.5 times the nominal value, T_{ve} and T_{de} are equally distant from the nonequilibrium translational temperature. The oxygen and nitrogen energy level distributions are approximately log-linear.

The 2D eigenvalues are presented in Figure 5a as a function of the frequency for different values of M_c. The change between a single subsonic eigenmode and a double supersonic eigenmode spectrum occurs for the frozen case at $M_c \approx 1$, in agreement with Jackson and Grosch's analysis [11]. Supersonic eigenmodes appear at $M_c < 1$ for the equilibrium case, first panel second row. This behavior occurs because M_c is evaluated on the basis of the frozen post shock conditions and does not accurately represent the equilibrium model. Jackson and Grosch [11] identified a threshold condition for the existence of supersonic unstable waves which for an equilibrium flow translates to the requirement $M_{c,e} > 1$, where the convective Mach number is now based upon the equilibrium speed of sound, For all cases discussed in Figure 5a the ratio between the equilibrium convective Mach number and the frozen convective Mach number is equal to 1.345, so that for the first panel of the second row, $M_{c,e} = 1.147$, a value greater than 1. The nonequilibrium flow transitions between subsonic instability and supersonic instability at values of M_c that are in between those corresponding to the frozen and equilibrium models. Three-dimensional growth rates are reported in Figure 5b. The equilibrium shear-layer is more unstable than the frozen case. The maximum growth rates of the nonequilibrium case fall between the two. For large convective Mach numbers, the nonequilibrium case has a frequency range narrower than both equilibrium and frozen cases. The maximum growth rates diminish with an increase in convective Mach number, as expected. This consideration along with the condition $M_{ce} > M_c$ and the fact that the equilibrium growth rate exceeds the frozen one, implies that is not possible to collapse the α_i vs. Mach number relation onto a single curve by simply replacing the Mach number with the appropriate convective counterpart, as was done for frozen cases of different composition by Jackson and Grosch [12].

The eigenfunctions for the most amplified 3D rate are reported in Figure 6a. The difference between the flow model solutions is marked, with the nonequilibrium solution in between the frozen and the equilibrium cases. In general, for all cases analyzed in this research, nonequilibrium eigenfunctions are closer to the equilibrium than to their frozen counterparts.

Translational and vibrational temperature eigenfunctions are reported for three nonequilibrium cases in Figure 6b. The marked difference between eigenfunctions corresponding to T_{v1} and T_{ve} points to the unsuitability of double temperature nonequilibrium

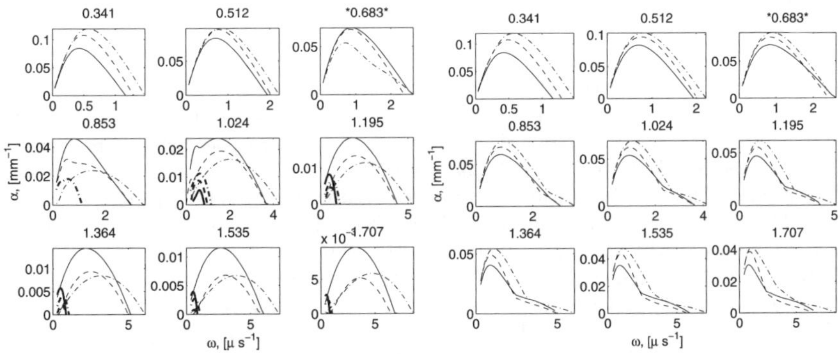

(a) (b)

Fig. 5. a) 2D spatial growth rates and b) 3D spatial growth rates as a function of the frequency ω. In the 3D case, the maximum growth rate over β is reported for each frequency, where β is the spanwise wave number. Solid line: frozen flow, dashed line: nonequilibrium flow, dash-dot line: equilibrium flow. The thick lines represent the second eigenmode appearing at $M_c \approx 1$.

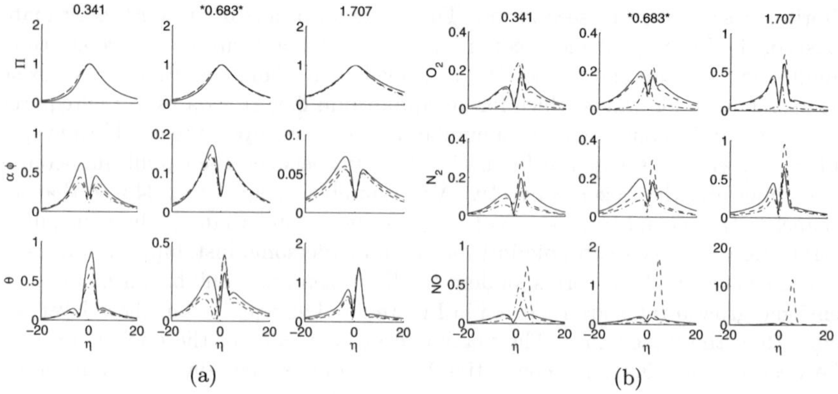

Fig. 6. a) Modulus of the eigenfunctions at the most amplified (ω, β) at three convective Mach numbers. Lines as in Figure 5. b) Modulus of the temperature eigenfunctions with the most amplified growth over (ω, β) at three convective Mach numbers. Solid line: T, dashed line: T_{v1}, dash-dot line: T_{ve}.

approximations for modeling instability and transition in high speed flows. The large peak in the T_{v1} eigenfunction for NO is a consequence of the spectrum of nascent states assigned by the second Zeldovich reaction to products. The presence of this feature demonstrates the importance of accurately determining the state-to-state transfer rates.

Acknowledgement. This work was supported in part by the Air Force Office of Scientific Research, with Dr. John Schmisseur as Technical Monitor.

References

1. M.R. Malik and E.C. Anderson: Physics of Fluids A **3**, 5, pp 803–821 (1991)
2. G. Stuckert and H.L. Reed: AIAA Journal **32**, 7, pp 1384-1393 (1994)
3. M.L. Hudson, N. Chokani, and G.V. Candler: AIAA Journal **35**, 6, pp 958–964 (1997)
4. A.T. Dufrene, M. Sharma, and J.M. Austin: 45th AIAA Aerospace Sciences meeting (2006)
5. B.W. Skews: Shock Waves **7**, pp 373–383 (1997)
6. I.V. Adamovich and J.W. Rich: J. Chem. Phys. **109**, 18, pp. 7711-7724 (1998)
7. S.O. Macheret and I.V. Adamovich: J. Chem. Phys. **113**, 17, pp 7351-7361 (2000)
8. I.V. Adamovich: AIAA Journal **39**, 10, pp 1916–1925 (2001)
9. D.Bose and G.V. Candler: J. Chem. Phys. **104**, 8, pp 2825–2833 (1996)
10. D.Bose and G.V. Candler: J. Chem. Phys. **107**, 16, pp 6136–6145 (1997)
11. T.L. Jackson and C.E. Grosch: J. Fluid Mech. **208**, pp 609–637 (1989)
12. T.L. Jackson and C.E. Grosch: Phys. Fluids **2**, 6, pp 949–954 (1990)

Part IX

Ignition

Auto-ignition of hydrogen-air mixture at elevated pressures

A.N. Derevyago, O.G. Penyazkov, K.A. Ragotner, and K.L. Sevruk

Physical and Chemical Hydrodynamic Laboratory, Luikov Heat and Mass Transfer Institute, National Academy of Sciences of Belarus, ul. P.Brovki, 15, 220072, Minsk, (Belarus')

Summary. Ignition times and auto-ignition modes of hydrogen-air mixtures have been studied behind reflected shock waves. Experiments were performed at temperatures 830 – 1450 K, pressures $2-21$ atm, and equivalence ratio of $\phi = 1.0$. Ignition delay times were determined using OH emission profiles, pressure and ion current measurements. Different auto-ignition modes of the mixture (strong, transient, and weak) were identified by comparing velocities of reflected shock wave at different distances from the reflecting wall. The influence of the interaction of the reflected shock wave with the dynamic boundary layer on the ignition time measurements in a shock tube has been examined.

1 Introduction

The critical conditions and mechanisms of hydrogen oxidation are explained in the theory of chain reactions by the competition of the reactions of branching and termination of chains [1-4]. One of the most important consequences of this theory is the existence of hydrogen -oxygen explosion limits in a temperature-pressure plane [2] and the complex functional dependence of the ignition delay time on the pressure and temperature of the reaction mixture [5].

The reaction rates of hydrogen oxidation ($T > 800$ K) at low post-shock pressures (> 0.2 MPa) has been thoroughly studied and validated with the use of a large amount of shock-tube experiments behind incident and reflected shock waves in mixtures highly diluted with argon. At these conditions, the ignition delay usually is inversely proportional to the oxygen partial pressure and fast chain branching reactions [1-4] are dominating in reaction mechanism of hydrogen oxidation.

The growing of initial pressures (> 0.2 MPa) increases the role of the chain branching reactions with the participation of HO_2 molecules formed due to trimolecular reaction $H+O_2+M = HO_2+M$, which favors the removal of active hydrogen atoms from the fast chain branching cycle and increase the characteristic time of hydrogen oxidation. The functional behavior of induction time on temperature and pressure at elevated pressures is usually established by using detailed numerical simulations [6, 7], or from empirical approximations based on low-pressure measurements. Moreover, existing reaction schemes of hydrogen oxidation are not validated extensively by experimental measurements at high pressures. Only few works [8-11] were devoted to studies of high-temperature hydrogen auto-ignitions at elevated pressures of $2-9$ atm. At the same time, the prospects of hydrogen applications in power engineering and industry call for a knowledge of hydrogen auto-ignitions at high pressures.

The present work gives results of experimental studies on auto-ignition of stoichiometric hydrogen-air mixtures at pressures of up to 21 atm and temperatures of $830 - 1450$ K.

2 Experimental setup

The experimental configuration and cross-section of the test section is illustrated in Fig. 1. A stainless steel shock tube of 50 mm diameter and 8.5 m length was used in the experiments. The runs were performed in stoichiometric hydrogen-air mixtures at mean post-reflected shock pressures of $2-21$ atm. The ranges of post-shock conditions studied in this work are summarized in Table 1. Commercial grade hydrogen of 99.9% purity and compressed air were used for mixture preparations. An electronic pressure meter controlled the initial pressure in the shock tube with accuracy of ± 0.2 $mm\ Hg$. Gas parameters behind incident and reflected shock waves were computed by using the shock-adiabatic curve assuming frozen chemistry, known temperature dependence of heat capacity, and shock wave velocity measurements with accuracy $\pm 0.5\%$ at different locations along the tube.

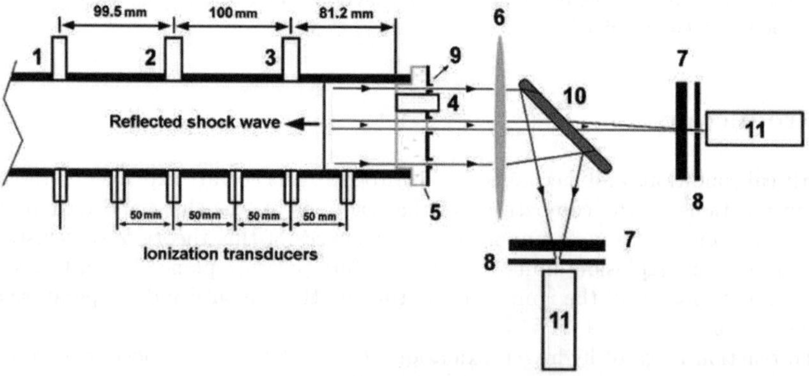

Fig. 1. Test section of the shock tube: 1-4 - high-frequency pressure transducers; 5 - quartz reflecting wall; 6 - lens (f = 40 cm); 7 - interference filters; 8 - diaphragms; 9 - ring diaphragm; 10 - beam-splitter; 11 - photomultipliers.

Mixture	P, atm.	T, K
1	2.1-3.86	890-1433
2	5.1-7.72	989-1385
3	5.1-7.72	989-1385
4	5.1-7.72	989-1385
5	5.1-7.72	989-1385
6	5.1-7.72	989-1385

Table 1. The experimental conditions of hydrogen-air mixtures

Pressure variations at different cross-sections of the shock tube were measured by high-frequency *PCB* pressure sensors, Model 113*A*24, with response time less than 1.5 μs and with a 1.5 mm spatial resolution (Fig.1). Light emission measurements were made by imaging the gas column (ϕ 5 mm) along the tube axis onto the first photomultiplier detector. The 0.5 mm aperture ensured the angle selection of transmitted radiation and passed only light beams propagating along the tube axis. To identify the influence of the shock wave bifurcation on the hydrogen auto-ignition in the boundary layer, the emission from an annular gas volume of outside diameter 44 mm and inside diameter 38 mm selected by means of the annular diaphragm was imaged onto the second photomultiplier (Fig. 1). The beam splitter divided the output spectrum in two optical paths to provide the simultaneous observations of the gas luminosity in selected directions. The luminosity from OH ($\lambda = 306.2$ nm; $\Delta = 2.4$ nm) radicals was used to measure autoignition of the mixture. Ignition-delay times were compared with the pressure measurements at reflecting wall. The ignition time was defined as the time difference between shock arrival at the end wall and the onset of emission at required intensity levels from the selected gas columns along the tube axis. The optical setup was calibrated for ignition-delay time measurements in both spectral paths.

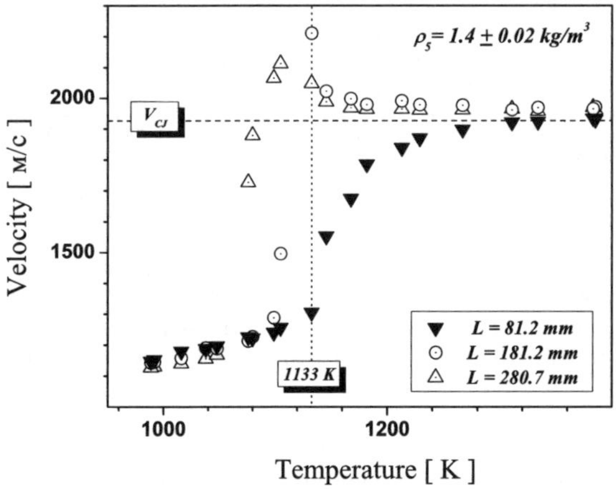

Fig. 2. Velocities of reflected shock wave at 81.2, 181.2 and 155 mm from reflecting wall vs. post-shock temperature in stoichiometric hydrogen-air mixtures and corresponding position of the strong ignition limit. Post-shock density is 1.4 ± 0.02 kg/m^3.

Three distinct autoignition modes of the gas mixture namely strong, transient, and weak [12-14] were identified based on velocity measurements of the reflected shock wave relative to the gas flow behind the incident shock at different locations from the reflecting wall [15]. Visible speed was calculated by processing shock-arrival times at pressure sensors along the tube. Figure 2 illustrates the typical dependence of reflected shock-wave velocity on post-shock temperature. On the basis of pressure and emission observations the inflection point of velocity curve at low temperatures for distances of 81.2 mm was used for determining the position of the strong explosion limit.

3 Result

The results of induction time measurements for two selected post-shock densities are presented in Fig. 3. At high temperatures of $T > 1100\ K$, the strong ignition of hydrogen-air mixtures was realized in the experiments. Because the detonation onset occurs near the reflecting wall the arrival time of reaction front to ion sensor located at a distance of 31.2 mm from is almost independent on the temperature. For strong auto-ignitions, the induction time measurements along the tube axis and boundary layer gives the same results within the experimental error, which indicates that the gas ignition is quite homogeneous.

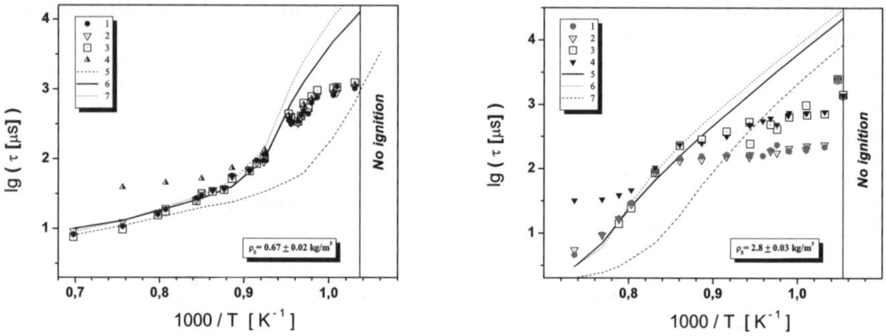

Fig. 3. Induction time vs. the reciprocal temperature in a stoichiometric hydrogen-air mixture for post-shock density of 0.67±0.02 kg/m^3 (a) and 2.8±0.03 kg/m^3 (b): 1, 2 - OH emissions along the tube axis and in a boundary layer; 3 - pressure measurements; 4 - ion current measurements; 5 - reaction mechanism [16]; 6 - reaction mechanism [17]; 7 - reaction mechanism [15].

The predictions of detailed reaction mechanisms [16, 18] correlate well with experiments at temperatures of $T > 1100\ K$. The best agreement is observed for low pressures mixtures ($2.1 - 3.86\ atm$). At temperatures of $T > 1250\ K$ and pressures of $7 - 15\ atm$, the reaction mechanisms [16, 18] give $1.2 - 1.5$ times shorter induction times. In this case, the mechanism [18] exhibits better agreement with the experimental data (Fig. 3). Thus, predictions of the detailed hydrogen reaction mechanisms at elevated pressures $P > 7$ atm and temperatures of $T > 1250\ K$ are not sufficiently adequate.

A significant discrepancy between the measurements and calculations is observed at temperatures of $T < 1100\ K$. Experiments show that at high pressures and low temperatures the induction time remains practically independent on the temperature. As the temperature decreased, the strong regime of hydrogen ignition changed to a mild one [13-15]. Under these conditions, the pressure recording becomes insensitive to the instant of the onset of ignition and gives overestimated values for induction times (Fig. 3). The emission measurements have shown that both the ignition mode and the temperature dependence of the ignition time are changed at $T < 1200\ K$. Emissions of OH radicals in the periphery of the tube start to be earlier than at the tube axis. Thus, when the regime changes the hydrogen auto-ignition becomes inhomogeneous. This tendency becomes the most pronounced in high-density mixtures (Fig. 3).

An analysis shows that at elevated pressures and low temperatures ($T < 1200\ K$), the interaction of reflected shock-wave with a dynamic boundary layer affect strongly on the nonuniformity of the flow and the autoignition of hydrogen-air mixture. Under these conditions, comparisons of experimental and numerical results become difficult and require knowledge of both the reaction kinetics and the local flow parameters behind incident and reflected shock waves. Domains of different autoignition modes in stoichiometric hydrogen-air mixtures are presented in Fig. 4. As seen in the figure, weak ignitions have been observed only in a narrow range of post-shock conditions at low temperatures. Strong initiations and transient regimes, resulting to deflagration to detonation transition upstream the reflecting surface, occupy the major part of the diagrams.

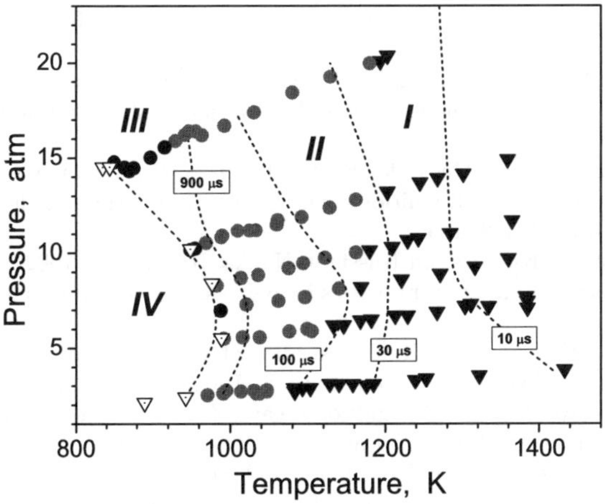

Fig. 4. Autoignition domain in stoichiometric hydrogen-air mixture in a P-T plane: 1 - strong ignition; II - transient ignition; III - weak ignition; IV - no ignition.

4 Conclusion

The autoignition of a stoichiometric hydrogen-air mixture at pressures of $2-21\ atm$ and temperatures of $840-1430\ K$ has been investigated numerically and experimentally.

It has been established that the reaction mechanism of A. A. Konnov (version 4.0) [16] and the GRIMECH 3.0 [18] provide a good qualitative agreement with experimental data at temperatures of $T > 1200\ K$; however, they should be somewhat refined to improve the quantitative agreement with the experimental data at elevated pressures.

It was shown that the interaction of the reflected shock wave with the boundary layer significantly influences on the results of induction time measurements and hydrogen autoignitions at temperatures of $T < 1100\ K$.

The data obtained can be used for analysis of combustion in engines and gas-turbines, improvement of the reaction mechanism of hydrogen oxidation, and predictions of the detonability limits and explosion safety of hydrogen-air mixtures.

Acknowledgement. This work was sponsored by State Research Programs of the Republic of Belarus "Hydrogen 18, "Hydrogen 36" and "Thermal Processes 56".

References

1. Semenov N. N., Some Problems of Chemical Kinetics and Reactivity. [in Russian], Izd. AN SSSR, Moscow, 1958.
2. Lewis D., von Elbe G., Combustion, Flames and Explosions in Gases. Academic Press, New York, 1961.
3. Kondrat'ev V. N., Nikitin E. E., Chemical Processes in Gases.[in Russian], Nauka, Moscow, 1981.
4. Ivanova A. N., Andrianova Z. S., and V. V. Azatyan. Application of the general approach to obtaining of ignition limits in the reaction of hydrogen oxidation. Khim. Fiz., 17(8), 91-100, 1998.
5. Gel'fand B. E., Popov O. E., Medvedev S. P., Khomik S. V., Agafonov G. L., and Kusharin A. Yu., Special features of self-ignition of hydrogen-air mixtures at high pressure. Dokl. Akad. Nauk, 330(4), 457-459, 1993.
6. Dimitrov V. I., Simple Kinetics. [in Russian], Nauka, Novosibirsk 1982.
7. Maas U., and Warnatz J., Ignition processes in hydrogen-oxygen mixtures. Combust. Flame, 74(1), 53-61 1988.
8. Schott G. L., and Kinsey J. L., Kinetic studies of hydroxyl radicals in shock waves II: Induction times in the hydrogen-oxygen reaction. J. Chem. Phys., 29, 1177-1182, 1958.
9. Skinner G. B., and Ringrose G. H., Ignition delays of a hydrogen-oxygen-argon mixture at relatively low temperature. J. Chem. Phys., 42, 2190-2192, 1965.
10. Snyder A. D., Robertson J., Zanders D. L., and Skinner G. B., Shock Tube Studies of Fuel-Air Ignition Characteristics. Report AFAPL-TR, 65-93 1965.
11. Belford R. L., and Strehlow R. A., Shock tube technique in chemical kinetics. Ann. Rev. Phys. Chem., 20, 247- 272, 1969.
12. Saytzev S. G., and Soloukhin R. I., On autoignition of the adiabatically heated gas mixture. Proc. Combust. Inst., 8, 344-347, 1962.
13. Voevodsky V. V., and Soloukhin R. I., On the mechanism and explosion limits of hydrogen-oxygen chain self-ignition in shock waves. Proc. Combust. Inst., 10, 279-283, 1965.
14. Meyer J. W., and Oppenheim A. K., On the shock induced ignition of explosive gases. Proc. Combust. Inst., 13, 1153-1164, 1971.
15. Penyazkov O. G., Ragotner K. A., Dean A. J., Varatharajan B., Autoignition of propane-air behind reflected shock waves. Proc. Combust. Inst., 30, 1941-1947, 2005.
16. Konnov A. A., Detailed reaction mechanism for small hydrocarbons combustion. Release 0.4, 1998, http://homepages.vub.ac.be/ akonnov.
17. Konnov A. A., Detailed reaction mechanism for small hydrocarbons combustion. Release 0.5, 1998, http://homepages.vub.ac.be/ akonnov.

Discrepancies between shock-tube and rapid compression machine ignition at low temperatures and high pressures

E.L. Petersen[1], M. Lamnaouer[1], J. de Vries[1], H. Curran[2], J. Simmie[2], M. Fikri[3], C. Schulz[3], and G. Bourque[4]

[1] *Mechanical, Materials and Aerospace Engineering, University of Central Florida, Orlando, FL, 32816, USA*
[2] *Chemistry Department, National University of Ireland, Galway, Ireland*
[3] *IVG, Universitaet Duisburg-Essen, Duisburg, Germany*
[4] *Rolls-Royce Canada, Montreal, Canada*

1 Introduction

Ignition delay time data at elevated pressures and low-to-intermediate temperatures continue to be of interest because of their importance for the validation of chemical kinetics models at practical engine conditions. However, it has been noticed recently that shock-tube ignition data can disagree significantly with rapid compression machine (RCM) data at higher pressures and temperatures below about 1100 K. The basic trend is that the ignition delay times obtained from shock-tube experiments are faster than model predictions and data obtained in RCM's over this lower-temperature region, sometimes by an order of magnitude or more.

These discrepancies are exemplified by propane-air ignition, and this case is utilized to highlight the salient issues that comprise the topic of the present paper. Figure 1 presents a comparison of shock-tube, RCM, and model ignition data for propane-air at an equivalence ratio (ϕ) of 0.5 and an average pressure of 30 atm (details on the facilities and model are provided below). The shock-tube data consist of the results from Herzler *et al.* [1,2], Cadman and Thomas [3], and new data by the primary authors at overlapping conditions; the RCM data are recent results from the NUI Galway authors. At the higher temperatures above about 1100 K, the shock-tube data from all three groups and the model are in excellent agreement. For temperatures below about 1100 K, where RCM data are available, there is a dramatic deviation between the available shock-tube data and the model/RCM data.

Some important observations can be drawn from this propane example: 1) shock-tube data from different facilities and groups tend to agree amongst themselves, even in reproducing the faster ignition at the lower temperatures; 2) RCM ignition data have longer apparent times than the shock-tube data; and, 3) RCM data tend to agree more with what is predicted by existing chemical kinetics models. The present paper brings attention to the observed similarities and discrepancies. Provided below are brief descriptions of the experimental facilities and some additional examples showing similar trends.

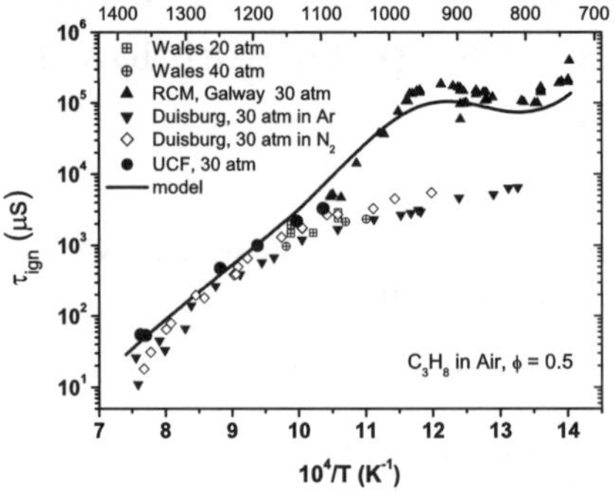

Fig. 1. Fuel-lean ($\phi = 0.5$) propane-air ignition results from several experiments and the model. Wales data are from [3]; Duisburg data are from [1,2]; UCF data from present study

2 Facilities Summary

2.1 Shock Tubes

Table 1 summarizes some details of the high-pressure shock tubes utilized for the bulk of the experiments mentioned in this paper (this is not meant to be an exhaustive list).

Table 1. Shock-tube specifications for selected high-pressure ignition experiments. UCF is from [4]; BC is from [5]; Wales is from [3]; Duis is that of [1]; Stan is from [6]

Specifications	UCF	BC	Wales	Duis	Stan
Driver Dia, mm	76	59	64	90	75
Driver Length, m	3.5	3.2	3.0	6.1	3.0
Driven Dia, mm	162	59	64	90	50
Driven Length, m	10.7	4.3	3.8	6.4	5.0
Diaphragm	AL	AL	Mylar	AL	AL
Max Pressure, atm	100	200	NA	700	600
Pressure Diagnostic	side, end	end	end	side	side
Emission (OH*, CH*)	side, end	end	side	side	side
Tailoring Gas	He/CO_2	He/air	He/Ar	He/Ar	NA
Max Test Time, ms	14	5	6	15	NA

In undiluted fuel-air mixtures at elevated pressures, the ignition event is often coincident with large pressure increases and strong gas dynamic coupling, usually leading to detonation-like wave patterns. Some representative pressure and CH* traces as seen from endwall measurements are provided in Fig. 2. The left figure is from one of the UCF propane-air ignition points from the present study, the ignition times of which are seen in Fig. 1. Note that the ignition time is clearly defined by the strong ignition features in both the pressure and emission traces. In the right plot, a typical methane-based ignition experiment (50/50 CH_4/C_2H_6) is shown for 827 K, 19.7 atm. This particular test exhibited a rather large pressure increase beginning at time τ_1 prior to the main ignition event at τ_{ign}. For this example, the pressure increased by as much as a factor of 2.5 prior to the strong ignition. As shown by Fieweger *et al.* [8], the strong ignition feature corresponds to a nearly homogeneous ignition event and approximates the 0-D model; the pre-ignition rise in pressure (and CH* emission) is representative of a mild ignition event originating in localized zones.

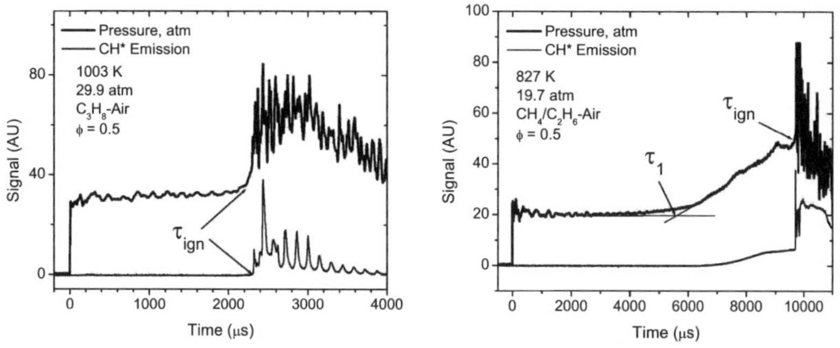

Fig. 2. Sample pressure and emission data. Left figure: C_3H_8 experiment from the present study. Right figure: 50/50 CH_4/C_2H_6 in Argon-based "air", taken from the set of experiments presented by de Vries and Petersen [7]

2.2 Rapid Compression Machine

A rapid compression machine simulates the compression stroke of a single engine cycle and so it allows autoignition phenomena to be studied in a more ideal, constant, and controllable environment than a reciprocating engine. The fundamental objective of an RCM is to heat the test gas as rapidly as possible to a high temperature and pressure with minimal heat losses. The NUI Galway RCM is different from most other RCMs in that it has a twin-opposed piston configuration described previously [9], resulting in a fast compression time of a little more than 16 ms. In addition, creviced piston heads are used to improve the post-compression temperature distribution in the combustion chamber [10, 11]. This particular design was adopted following studies at MIT [12–14], that found the creviced design resulted in an almost homogeneous temperature field in the post-compression period.

The adiabatic compressed gas temperature, T_C, and pressure, p_C, are calculated using the initial temperature, T_i, initial pressure, p_i, and the experimentally measured compressed gas pressure, p_C, employing Gaseq [15] which uses the isentropic relation between T and p. Typically, the time for compression is very rapid, that is 16.6 ms, therefore heat losses during this period are very low. If the ignition delay times are short, below about 100 ms, heat losses have only a small effect on the unfolding chemistry. However, as the ignition delay times increase, the heat losses to the walls have a greater impact on the experimental results. Even though ignition delays were observed up to 400 ms following compression, ignition delay times greater than 100 ms were found to be less consistent.

3 Kinetic Model

The chemical kinetic mechanism was developed and simulations performed using the HCT (Hydrodynamics, Chemistry and Transport) program [16]. The detailed chemical kinetic reaction mechanism used in these calculations was based on the hierarchical nature [17] of reacting systems. The hydrogen submechanism is based on that which has recently been validated by O'Conaire *et al.* [18] in the temperature range 298–2700 K, at pressures from 0.05 to 87 atm, and equivalence ratios from 0.2 to 6.0. The kinetic mechanism employed for the methane/ethane system is based on that published by Fischer *et al.* [19, 20] in their dimethyl ether study.

The C_3 sub-mechanism is based on the modeling work of Curran *et al.* [21,22] using the thermochemical parameters and rate constant rules described in their work on iso-octane oxidation. This mechanism has recently been used successfully to simulate the oxidation of methane/propane mixtures at high pressures [23]. The complete kinetic mechanism consists of 118 different chemical species and 663 elementary reactions and is available by contacting (henry.curran@nuigalway.ie).

4 Ignition Time Comparisons

Figure 3 shows further comparisons between shock tubes and the kinetics model at conditions of elevated pressure (20, 40 atm) and lower temperatures. In the left plot, the pure methane ignition times agree with the model at the higher temperatures but begin to deviate substantially for temperatures below about 1200 K. The CH_4 data of de Vries and Petersen [7] and Petersen *et al.* [24] agree well with the Duisburg data from the present study, including the faster ignition at the lower temperatures. The Wales data [25] show a similar trend but are faster by a factor of about two from the UCF and Duisburg data. This difference may be due to the use of τ_1 by the Wales group to define ignition time rather than the main strong ignition event. Also, for the lowest temperatures, the CH_4 ignition tends to not show any strong ignition event within the available test time, so even the ignition times from UCF and Duisburg are defined for the CH_4-only results near the onset of the mild ignition event.

Also shown in the left plot of Fig. 3 are data and modeling for CH_4/C_2H_6 blends containing 25-50% ethane. Again, there is good agreement between model and data at higher temperatures but not at the lower temperatures. Note that the data from both UCF [7, 24] and Duisburg (this study) agree well. In all cases for the CH_4/C_2H_6 blends, the ignition is defined by the strong ignition event. However, at the lower temperatures, there were in some cases large pre-ignition pressure rises (one of which is shown in Fig. 2). As mentioned by Fieweger *et al.* [8], when such a pressure increase occurs, the τ_{ign} data can be made to agree with the zero-dimensional kinetics by taking into account the corresponding temperature rise due to the compression. When such an adjustment is made, the corrected data show very good agreement with the kinetics model. The adjusted temperatures for the two points shown were derived assuming isentropic compression and taking an average temperature.

The right plot in Fig. 3 shows the results of recent data taken at UCF to correspond with one of the natural gas mixtures tested by Huang and Bushe [5] in air at $\phi = 1.0$. At lower temperatures, the new data are faster than the model, and they also show pre-ignition pressure increases prior to the main τ_{ign} event. When these data are corrected for the corresponding temperature increase (red symbols), they too tend to show better agreement with the model predictions.

Ideally, the adjustment calculation should be performed with the chemical kinetics in conjunction with an imposed T, p increase to account for the accelerated ignition chemistry due to the pressure increase. Nonetheless, the simple corrections seen in Fig. 3 are good evidence that the apparent accelerated ignition may be due to the pre-ignition pressure increase resulting from the presence of deflagration kernels in the test area. This adjustment technique requires the original pressure traces for each experiment, which are not usually provided in the open literature.

Also noticeable in the right plot of Fig. 3 is the apparent disagreement between the new UCF data and those of [5]. This discrepancy may be due to differences in the definition of ignition delay time between the two data sets. For example, if the τ_1 time were plotted for the UCF data, also shown in the right plot of Fig. 3, there is much better agreement between the two shock tubes.

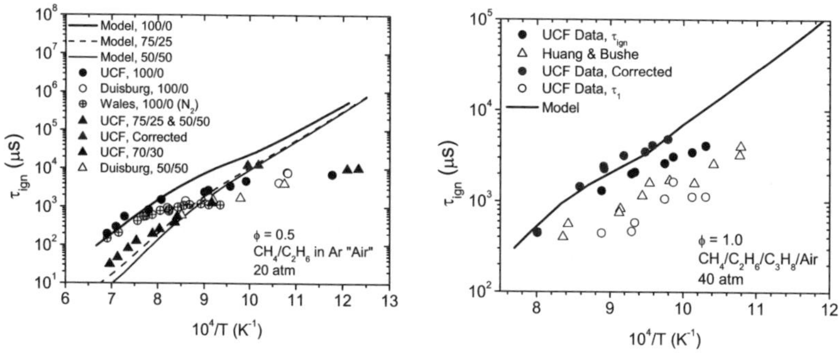

Fig. 3. Ignition for methane fuel blends. Left plot: CH_4 and CH_4/C_2H_6 fuels. Right plot: 0.0867 CH_4 + 0.0034 C_2H_6 + 0.0011 C_3H_8 in air at $\phi = 1.0$

Figure 4 shows another comparison but this time for a heavier fuel, iso-Octane, at $\phi = 1.0$, including data from several different RCMs. There is good agreement between the shock-tube data of Fieweger *et al.* [8], the RCM data, and the model. There is also fair

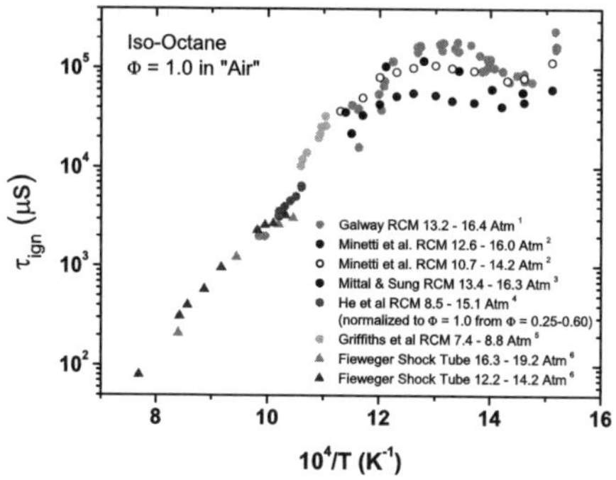

Fig. 4. iso-Octane ignition delay times from several sources. 1-recent Galway data; 2-ref. [26]; 3-ref. [27]; 4-ref. [28]; 5- [29]; 6-ref. [8]

agreement amongst the different RCM data. Further comparisons between shock tube, RCM, and kinetics modeling are underway. Although there is an apparent, considerable discrepancy between model/RCM and shock-tube experiments, some of the discrepancy can be explained by the effects of pressure increases in the shock-tube data, as shown herein. However, it is still not clear what causes the earlier, repeatable ignition events in the shock-tube experiments that lead to the this pre-ignition pressure increase.

References

1. Herzler J., Jerig L., Roth P.: Combust. Sci. Tech. **176**, 1627 (2004)
2. Herzler J., Jerig L., Roth P., Schulz C.: *Shock Tube Study of the Ignition of Propane at Intermediate Temperatures and High Pressures*, Proc. Euro. Combust. Mtg. (2005)
3. Cadman P., Thomas G.O.: J. Phys. Chem. Chem. Phys. **2**, 5411 (2000)
4. Petersen E.L., Rickard M.J.A., Crofton M.W., Abbey E.D., Traum M.J., Kalitan D.M.: Meas. Sci. Tech. **16**, 1716 (2005)
5. Huang J., Bushe W.K.: Combust. Flame **144**, (2006)
6. Davidson D.F., Hanson R.K.: Int. J. Chem. Kinet. **36**, 510 (2004)
7. de Vries J., Petersen E.L.: Proc. Combust. Inst. **31**, 3163 (2007)
8. Fieweger K., Blumenthal R., Adomeit G.: Combust. Flame **109**, 599 (1997)
9. Affleck W.S., Thomas A.: Proc. Inst. Mech. Engrs. **183**, 365 (1968)
10. Brett L.: Ph.D. thesis, National Univ. of Ireland, Galway (1999)
11. Brett L., MacNamara J., Musch P., Simmie J.M.: Combust. Flame **124**, 326 (2001)
12. Lee D.: Ph.D. thesis, Massachussetts Institute of Technology, Cambridge, MA, USA, (1997)
13. Lee D., Hochgreb S.: Combust. Flame **114**, 531 (1998)
14. Lee D., Hochgreb S.: Int. J. Chem. Kinet. **30**, 385 (1998)
15. Morley C.: Available from http://www.c.morley.ukgateway.net/gseqrite.htm
16. Lund C.M., Chase L.: LLNL Report UCRL-52504, revised (1995)
17. Westbrook C.K., Dryer F.L.: Prog. Energy Combust. Sci. **10**, 1 (1984)
18. Ó Conaire M., Curran H.J., Simmie J.M., Pitz W.J., Westbrook C.K.: Int. J. Chem. Kinet. **36**, 603 (2004)
19. Fischer S.L., Dryer F.L., Curran H.J.: Int. J. Chem. Kinet. **32**, 713 (2000)
20. Curran H.J., Fischer S.L., Dryer F.L.: Int. J. Chem. Kinet. **32**, 741 (2000)
21. Curran H.J., Gaffuri P., Pitz W.J., Westbrook C.K.: Combust. Flame **114**, 149 (1998)
22. Curran H.J., Gaffuri P., Pitz W.J., Westbrook C.K.: Combust. Flame **129**, 253 (2002)
23. Petersen E.L., Kalitan D.M., Simmons S., Bourque G., Curran H.J., Simmie J.M.: Proc. Combust. Inst. **31**, 447 (2007)
24. Petersen E.L., Smith S.D., Hall J.M., de Vries J., Amadio A., Crofton M.W.: ASME Paper GT2005-68517 (2005)
25. Goy C.J., Moran A.J., Thomas G.O.: ASME Paper GT2001-0051 (2001)
26. Minetti R., Carlier M., Ribaucour M., Therssen E., Sochet L.R.: Proc. Combust. Inst. **26**, 747 (1996)
27. Mittal G., Sung C.J.: Combust. Sci. Tech., in press
28. He X., Donovan M.T., Zigler B.T., Palmer T.R., Walton S.M., Wooldridge M.S., Atreya A.: Combust. Flame **142**, 266 (2005)
29. Griffiths J.F, Halford-Maw P.A., Mohamed C.: Combust. Flame **111**, 327 (1997)

Ignition delay studies on hydrocarbon fuel with and without additives

M. Nagaboopathy[1], G. Hegde[1], K.P.J. Reddy[1], C. Vijayanand[2], M. Agarwal[2], D.S.S. Hembram[2], D. Bilehal[2], and E. Arunan[2]

[1] *Department of Aerospace Engineering, Indian Institute of Science, C. V. Raman Avenue, Bangalore, 560012, India*

[2] *Inorganic and Physical Chemistry Department, Indian Institute of Science, C. V. Raman Avenue, Bangalore, 560012, India*

Summary. Single pulse shock tube facility has been developed in the High Temperature Chemical Kinetics Lab, Aerospace Engineering Department, to carry out ignition delay studies and spectroscopic investigations of hydrocarbon fuels. Our main emphasis is on measuring ignition delay through pressure rise and by monitoring CH emission for various jet fuels and finding suitable additives for reducing the delay. Initially the shock tube was tested and calibrated by measuring the ignition delay of $C_2H_6 - O_2$ mixture. The results are in good agreement with earlier published works. Ignition times of exo-tetrahdyrodicyclopentadiene ($C_{10}H_{16}$), which is a leading candidate fuel for scramjet propulsion has been studied in the reflected shock region in the temperature range 1250 - 1750 K with and without adding Triethylamine (TEA). Addition of TEA results in substantial reduction of ignition delay of $C_{10}H_{16}$.

1 Introduction

Our laboratory has been doing thermal decomposition studies using single pulse shock tube for molecules of interest to atmospheric chemistry [1,2]. Recently the shock tube has been modified to study ignition delay of various hydrocarbon fuels and to find out suitable additives for enhancing their ignition. There are several reports available on various hydrocarbon fuel ignition delay studies. Generally ignition delay of hydrocarbons increases with increasing number of carbon atom in the structure. In alkanes ethane has the lowest ignition delay and methane has the longest delay, branched hydrocarbons have longer ignition delay time than their linear chain counterparts [3–5]. Ignition delay of ringed hydrocarbons does not follow any order as found in chained hydrocarbons, but it mainly depends on the structure of the molecule [6]. Exo-tetrahydrodicyclopentadiene is a large ringed structure hydrocarbon molecule, produced by hydrogenation of dicyclopentadiene and having molecular formula $C_{10}H_{16}$. It has high volumetric energy density and relative stability than other cyclic compounds and this is proposed as a leading candidate fuel for supersonic combustion ramjets (scramjet) propulsion. Knowledge about the ignition behaviour of $C_{10}H_{16}$ at a given condition will help to predict the design of combustion system for aerodynamic vehicle. Shock tube is an effective tool for understanding the combustion process of $C_{10}H_{16}$, but low vapour pressure of $C_{10}H_{16}$ at room temperature makes it difficult for shock tube study in gaseous state. Some investigations on ignition delay times have already been done using shock tube [7–11]. The present study addresses the ignition delay measurements of $C_{10}H_{16}$ with and without adding Triethylamine (TEA), through pressure rise in the reflected shock condition and recording CH radical emission due to the ignition. Initially the shock tube was calibrated with $C_2H_6 - O_2$ mixture, ignition studies then continued with $C_{10}H_{16}$ mixtures.

2 Experimental Method

2.1 Experimental setup

Experiments were performed in high purity helium driven stainless steel shock tube CST2, which is of 39mm diameter, having driver section of 1.97 m length and driven section of 4.2 m, separated by an aluminum diaphragm. A ball valve is introduced at 1.5 m from the end flange of the driven section for uniform heating of the test sample. Two optical view ports have been provided near to the end flange of driven section to study real time absorption and emission spectra during combustion. One of the view ports is connected with a vacuum monochromator coupled with photomultiplier by an optical fiber bundle to carry out emission spectroscopic studies. The schematic diagram of the CST2 assembly is given in the Fig. 1. For every experiment, the shock tube was pumped down to a pressure less than 10^{-4} torr. Initial pressure (P_1) of the sample section was measured using an IRA pressure transducer, and shock velocities were measured through three PCB pressure transducers (PT), which are connected over the last 1.5 m of the driven section. Reflected shock wave parameters are calculated using standard normal-shock relations.

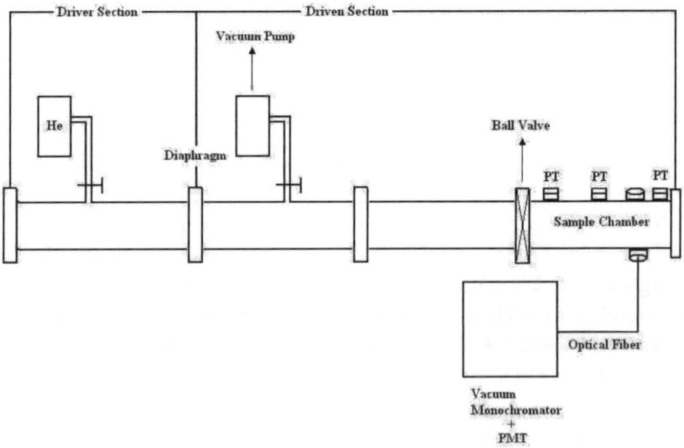

Fig. 1. Schematic sketch of CST2

2.2 Ignition studies

Initial study of ignition delay was carried out with $C_2H_6 - O_2$ mixture diluted in argon and these results have been used to calibrate the shock tube. Experiments were performed with ethane mixture of equivalent ratio (ϕ) 1. The measured ignition delay times from S-curve pressure rise were in good agreement with the published results. The efforts of studying ignition delay by optical diagnostics method have been developed successfully and tested with CH radical emission with the monochromator - PMT centred at 431.5

nm. Here, the ignition delay time is defined as the time between the arrival of reflected shock and the onset of CH emission. Investigation of ignition times of $C_{10}H_{16} - O_2$ without adding TEA are done in the reflected shock region in the temperature range of 1250 - 1750 K. A mixture of 0.2% concentration of $C_{10}H_{16}$ added with oxygen in the equivalent ratio 1 and diluted in argon has been taken for the study. The test sample has been mixed uniformly in a separate stainless steel chamber for a period of one hour using a circulation pump.

Further investigations on addition of TEA with $C_{10}H_{16}$ are carried out to explore the ignition behaviour. In this study $C_{10}H_{16}$ and TEA have been taken in the proportion of 0.9% and 0.1%, mixed with oxygen in the equivalent ratio of 1 and diluted with argon. The mixture is circulated for one hour to yield proper mixing of the compounds. Experiments are performed with the mixture under the same conditions as those done without addition of TEA and a substantial reduction in the ignition delay time is observed.

3 Results & Discussion

3.1 Ignition Time Data

Ignition times are obtained in several experiments (some of the results are shown in the Table-1) for $C_{10}H_{16}$ with and without adding TEA.Typical pressure rise and CH emission due to ignition of $C_{10}H_{16}$ at 1397 K is shown in the Fig. 2. The ignition delay time reffered here is the measure of the time delay between the pressure rise due to the arrival of the reflected shock and that due to the onset of ignition. In this case ignition delay was observed as 340 μs. Fig. 3 shows the pressure and CH emission signal for $C_{10}H_{16}$ with TEA mixture at 1281 K. In all the cases the ignition time depends on reflected shock temperature - pressure, equivalent ratio of the mixture and the concentration of the sample loaded. It is observed that an increase in the value of any of these parameters leads to a decrease in the ignition delay time. The calibration efforts of shock tube with $C_2H_6 - O_2$ mixture at various temperatures showed ignition delay between 92 μs - 1.44 *ms*. Experiments carried out on $C_{10}H_{16}$ without and with addition of TEA results in ignition delay of 50 - 900 μs and 70 - 690 μs respectively. The log τ vs 1/T plot for these experiments with and without addition of TEA are shown in Fig. 4 and 5. It is clearly evident from the two plots that the ignition delay time of $C_{10}H_{16}$ reduced with addition

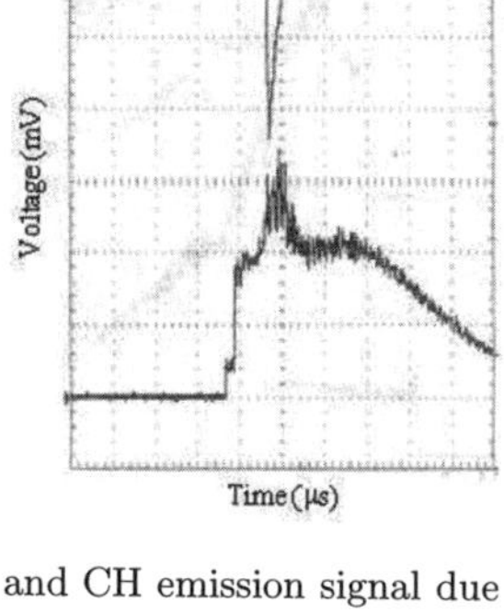

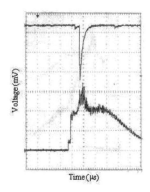

Fig. 2. Pressure rise and CH emission signal due to ignition of $C_{10}H_{16}$

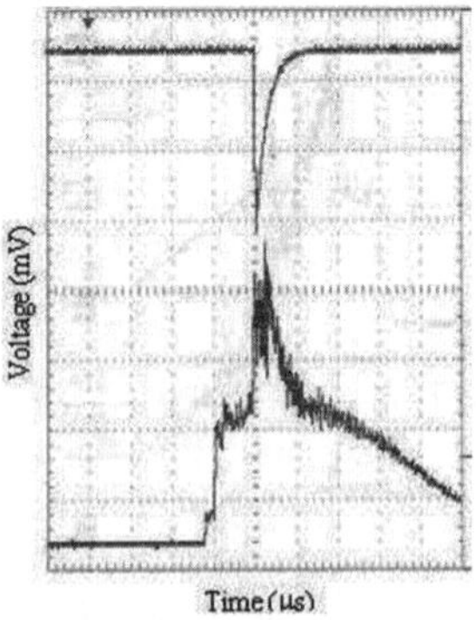

Fig. 3. Pressure rise and CH emission signal due to ignition of $C_{10}H_{16}$ + TEA

of TEA as compared to the ignition delay of pure $C_{10}H_{16}$. These plots have been used to determine the Arrhenius parameters of the reaction. Table-2 has a comparison of our ignition delay measurements with earlier reported results. It shows that the activation energy is significantly reduced in the presence of TEA.

Table 1. Some experimental results on ignition delay of $C_{10}H_{16}$ with and without TEA addition

Ignition delay of $C_{10}H_{16}$			Ignition delay of $C_{10}H_{16}$ with TEA		
T_5 (K)	P_5 (atm)	Delay (μs)	T_5 (K)	P_5 (atm)	Delay (μs)
1433	16.43	900	1375	14.55	530
1391	15.30	780	1384	14.71	690
1385	15.56	800	1403	15.22	130
1583	18.84	160	1422	15.36	390
1516	18.86	310	1460	15.99	560
1517	18.80	240	1474	16.66	450
1438	15.44	280	1476	17.06	400
1439	16.93	520	1486	17.48	330
1465	16.32	190	1491	16.54	250
1471	16.56	120	1506	16.79	350
1540	17.61	50	1512	18.60	340
1575	20.31	70	1546	18.46	80
1675	19.02	110	1563	16.90	310
1438	15.44	280	1574	17.99	110
1439	16.93	520	1575	18.91	110
1661	20.40	70	1581	16.97	220
1516	18.22	280	1593	18.31	150
1516	17.83	300	1627	17.97	70
1458	16.49	420	1690	19.27	110
1496	17.76	550	1721	19.54	70

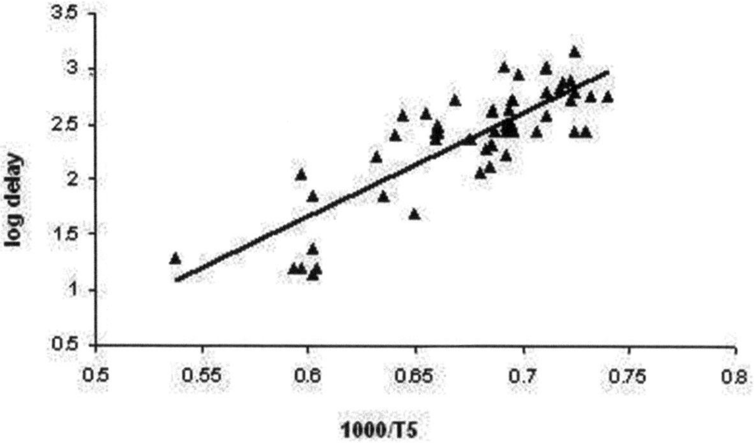

Fig. 4. Arrhenius plot of results from ignition delay experiments on $C_{10}H_{16}$

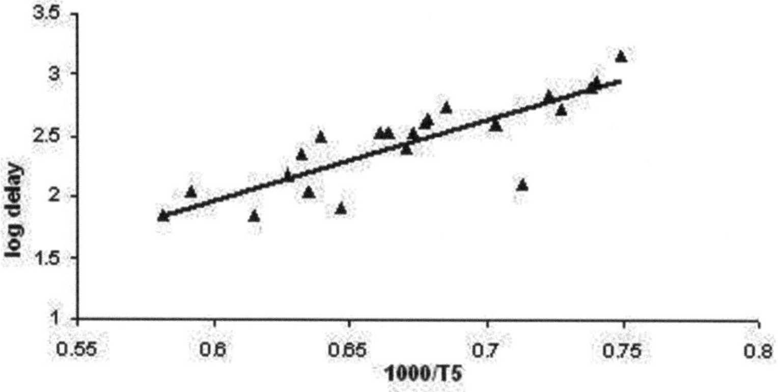

Fig. 5. Arrhenius plot of results from ignition delay experiments on $C_{10}H_{16}$ + TEA

Table 2. Arrhenius parameters for the ignition delay data on $C_{10}H_{16}$ with and without TEA

Reference	T5 (K)	Pressure (atm)	Log A	E_a (Kcal/mole)
[7]	1350 – 1550	1.2	-	53.7
[8]	1150 – 1500	3 – 8	-	43.1
[9]	1149 – 1688	1.7 – 9.3	-	34.8
$C_{10}H_{16}$ (this work)	1267 – 1686	13.5 – 20.4	-4.00	43.2 (± 4.1)
$C_{10}H_{16}$ + TEA (this work)	1335 – 1721	13.9 – 19.5	-2.15	30.7 (± 4.3)

4 Conclusion

The chemical shock tube facility developed at the high temperature chemical kinetics laboratory has been modified for ignition delay measurements. A database of ignition delay measurement through S–curve pressure rise and CH emission are compiled for a large ringed structure hydrocarbon fuel $C_{10}H_{16}$. With the addition of TEA, an appreciable reduction in the ignition delay times of $C_{10}H_{16}$ is noticed. TEA addition also reduces the activation energy of the fuel by more than 20%. More experiments are to be carried out in the near future with different additives. Fourier Transform Infrared spectroscopic investigations and Gas Chromatograph analysis of the initial and final products are to be done for the mixtures. This will help in understanding the kinetics and the combustion mechanism of $C_{10}H_{16}$.

Acknowledgement. The chemical shock tube facility has been established at the High Enthalpy Aerodynamics Laboratory by the active collaboration between IPC and AE departments. This effort has been supported by funds from ISRO, DRDL, DST-FIST and IISc.

References

1. B. Rajkumar, K.P.J. Reddy, E. Arunan: J. Phys. Chem A **106**, 8366 (2002)
2. B. Rajkumar, K.P.J.Reddy, E. Arunan: J. Phys. Chem A **107**, 9782 (2003)
3. N. Lamoureux, C.-E. Paillard, V. Vaslier: Shock Waves **11**, 309 (2002)
4. J.M. Simmie: Prog. Energy and Comb. Science **29**, 599 (2003)
5. A. Toland, J.M. Simmie: Combustion and Flame **132**, 556 (2003)
6. V.G. Slutsky, O.D. Kazakov, E.S. Severin, E.V. Bespalov, S.A. Tsyganov: Combustion and Flame **94**, 108 (1993)
7. D.F. Davidson, D.C. Horning, R.K. Hanson: AIAA **99**, 2216 (1999)
8. M.B. Colket, L.J. Spaddaccini: J.Propulsion and Power **17**, 315 (2001)
9. E. Olchansky, A. Burcat: ISSW24 5962
10. D.W. Mikolaitis, C. Segalm, A. Chandy: J.Propulsion and Power **19**, (2003)
11. S.C. Li, B. Varatharajan, F.A. Williams: AIAA **39**, (2001)

Ignition of hydrocarbon–containing mixtures by nanosecond discharge: experiment and numerical modelling

I.N. Kosarev, S.V. Kindusheva, N.L. Aleksandrov, S.M. Starikovskaia, and A.Y. Starikovskii

Physics of Nonequilibrium Systems Lab, Moscow Institute of Physics and Technology

Summary. The efficiency of nonequilibrium plasma of pulsed discharge as an igniter of combustible mixtures at elevated temperatures was investigated via shock tube technique. Experiments were carried out behind a reflected shock wave. The experiments were carried out with a set of stoichiometric mixtures C_nH_{2n+2}:O_2 (10%) diluted by Ar (90%) for hydrocarbons from CH_4 to C_5H_{12}. The temperature behind the reflected shock wave varied from 950 to 2000 K, and the pressure was 0.2 to 2.0 atm. Numerical modelling has been performed to compare the autoignition and ignition by a nanosecond discharge for CH_4 – C_5H_{12} containing mixtures. The results of the calculations demonstrate reasonable correlation with the experimental data.

1 Introduction

Ignition of fuel–containing mixtures is an important problem in fundamental and applied combustion research. Different mechanisms are considered to be responsible for artificial ignition under the action of low–temperature nonequilibrium plasma: vibrational excitation or excitation of lower electronic states at low electric fields [1], dissociation by electron impact via high-energy electronic levels at high electric fields, including dissociation at interaction of molecules with electronically excited species [2], ion chemistry [3] and so on.

To reveal this or that mechanism, two approaches are typically used: numerical modelling and quantitative experimental measurements combined with numerical analysis. It is quite typical for the first approach to numerically analyze the discharge effect on a combustible mixture by artificially injecting different radicals at the initial moment of calculations. So, paper [4] investigated numerically the kinetic paths to radical-induced ignition of a methane-air mixture. They were able to obtain the ignition starting even from 300 K at 1 atm mixture pressure injecting CH_3, CH_2, CH and C. Discussing the second approach, it is worth mentioning the papers [5], [6], where the authors compare quantitative experimental data on plasma - assisted ignition for RF and nanosecond pulsed plasma with numerical modelling and discuss possible mechanisms of combustion initiation by nonequilibrium plasma.

Indeed, the comparison of the experiments based on simultaneous control of both plasma and combustion measurements give a unique opportunity to understand the mechanism of plasma assisted ignition (PAI) and combustion (PAC) in a wide range of initial parameters. The present paper is aimed at developing a numerical model for the description of PAC under the action of nanosecond pulsed discharge and testing the results of the calculations on the experimental data.

2 Experiments

A typical experimental setup is described elsewhere [6]. In each experiment the delay time for autoignition was compared with the ignition delay time for ignition by a nanosecond discharge developed in the form of a fast ionization wave (FIW) in the end–plate section of the shock tube. This type of discharge produces uniform in space low temperature nonequilibrium plasma. Typically, discharge front propagates with high velocity (a few cm per nanosecond) from the high–voltage electrode to the low–voltage one. The section was made of dielectric material and the discharge developed between the end plate of the shock tube and stainless steel part 20 cm apart. The shock tube had a rectangular cross-section of 25x25 mm.

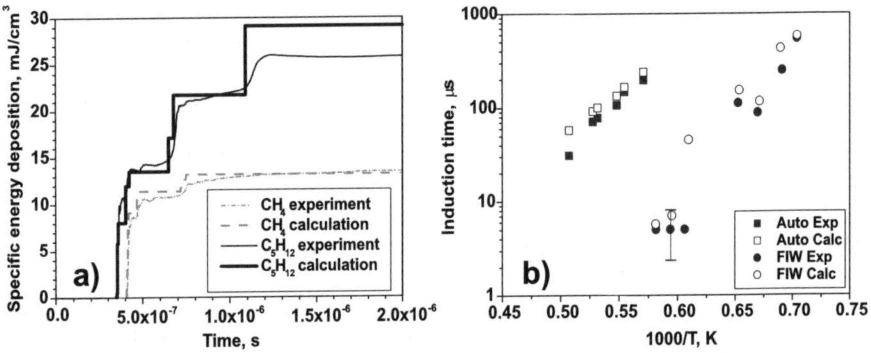

Fig. 1. (a) – Example of experimentally measured and calculated energy input for CH_4 and C_5H_{12} – containing mixtures; (b) – Comparison of experimentally obtained and calculated ignition delay times for CH_4:O_2:Ar mixture. “Auto” means autoignition, “FIW”- ignition by nanosecond discharge

The ignition delay time (or, in other words, the induction time for ignition) was determined from a sharp increase of OH (306 nm) or CH (431 nm) emission. The system for monitoring shock wave parameters included a system for measuring the velocities of the shock waves by means of the laser schlieren technique and a system for controlling initial pressure. The gas density (ρ_5), pressure (P_5), and temperature (T_5) behind the reflected shock wave were determined from the known initial gas mixture composition, the initial pressure, and the velocity of the incident shock wave. The nanosecond discharge was initiated at the instant at which the reflected shock wave arrived at the observation point. High–voltage pulses were produced with a ten–stage GIN–9 Marx high-voltage generator, which provided voltage amplitude in the range 60–160 kV at the high–voltage electrode. The duration of the pulse was about 40 ns, and 1-2 re-reflections were observed during 0.2 μs depending upon the parameters of the experiment.

The experiments were carried out with a set of stoichiometric mixtures C_nH_{2n+2}:O_2 (10%) diluted by Ar (90%) for hydrocarbons from CH_4 to C_5H_{12}. The temperature behind the reflected shock wave varied from 950 to 2000 K, and the pressure was 0.2 to 2.0 atm.

3 Numerical modelling

Numerical simulation of production of active particles was reduced (i) to the simulation of production of active particles (electrons, ions, excited particles, atoms and radicals) in a high electric field in the discharge and (ii) to the simulation of conversion of the active particles in the afterglow. Active particles under consideration were excited Ar atoms, O, H atoms, radicals of hydrocarbon molecules, electrons and positive ions. Production of negative ions and complex positive ions (Ar_2^+, O_4^+, etc.) was neglected because of high (about 1000 K) gas temperatures. At the end of plasma decay, only atoms and radicals were assumed to dominate the composition of active particles, whereas the presence of long–lived excited states was neglected. In the discharge front there is generally a very short and high overshoot of the applied voltage that cannot be experimentally resolved. To take into account the production of electrons and ions in this, poorly known, phase, we considered the initial electron density in the beginning of the main pulses of applied voltage to be an adjusted parameter. This parameter was determined when calculating the evolution in time of the discharge current or energy input and adjusting it to corresponding measured data (see, for example, Figure 1,a). In the experiments under consideration, the main pulse of applied voltage was generally followed by several re–reflections, and we have taken this into account.

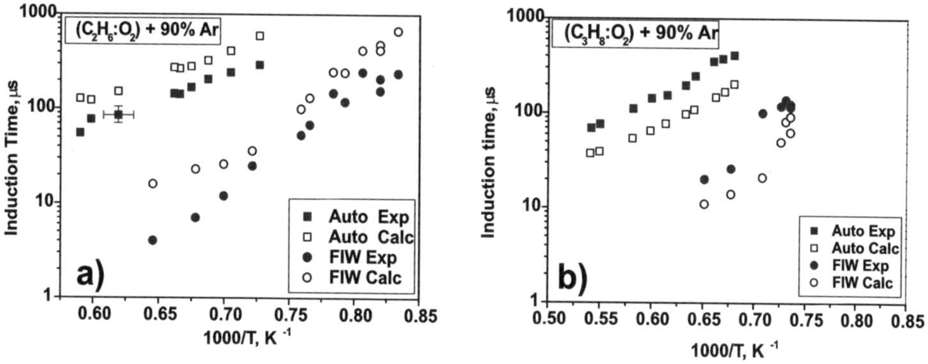

Fig. 2. Comparison of experimentally obtained and calculated ignition delay times for C_2H_6:O_2:Ar (a) and C_3H_8:O_2:Ar (b) mixtures. "Auto" means autoignition, "FIW" — ignition by nanosecond discharge

The evolution of the densities of active particles between and behind voltage pulses was simulated using a numerical solution of the balance equations. In this phase, in a zero electric field charged particles were removed due to dissociative electron–ion recombination producing an additional number of atoms and radicals. The last particles were also produced due to charge exchange of some positive ions on hydrocarbon molecules and due to quenching of excited Ar atoms by oxygen and hydrocarbon molecules. In our kinetic model, in the end charged and excited particles were excluded from the consideration and the gas mixture consisted of initial neutral species and atoms and radicals produced in the discharge and in its afterglow. The densities of atoms and radicals were used further as input parameters for a computer program to simulate the ignition. Under the conditions considered, O–atoms are the dominant active particles. The main mechanism for production of O atoms is electron impact dissociation of O_2 during the discharge. The production of atoms and radicals due to electron impact dissociation of hydrocarbon molecules is less important. De–excitation of Ar atoms increases density of H–atoms and radicals like CH_3 and C_5H_{11} in early afterglow of the discharge.

It is expected that the uncertainty in the cross section for electron impact dissociation of C_3H_8, C_4H_{10} and C_5H_{12} does not affect strongly the results of our calculations. To verify this, additional calculations were carried out with the cross sections of dissociation of these molecules increased by an order of magnitude. This led to an increase in the densities of H atoms and radicals of hydrocarbon molecules by a factor of 6–7, whereas the density of O atoms decreased by tens percent. As a result, the total density of atoms and radicals increased in the mixtures with C_3H_8, C_4H_{10} and C_5H_{12} only by 10%. This does not affect noticeably the induction time in the mixtures considered; our calculations showed that this time is sensitive to the total density of atoms and radicals produced by the discharge and less sensitive to the composition of active particles.

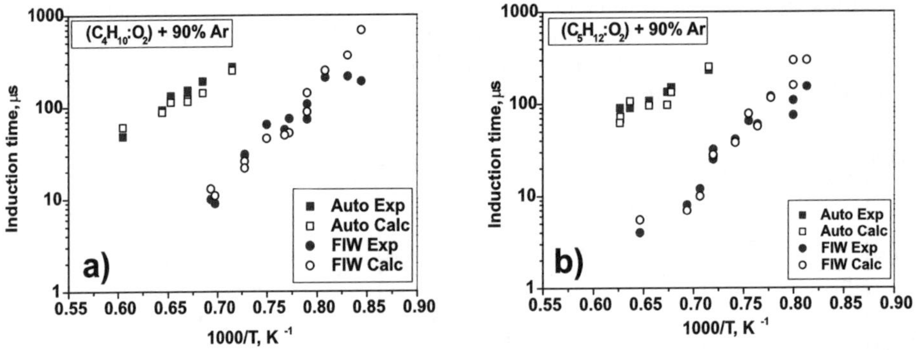

Fig. 3. Comparison of experimentally obtained and calculated ignition delay times for C_4H_{10}:O_2:Ar (a) and C_5H_{12}:O_2:Ar (b) mixtures. "Auto" means autoignition, "FIW" — ignition by nanosecond discharge

A zero–dimensional simulation of autoignition and artificial ignition by the discharge were performed at constant pressure using CHEMKIN code [7]. Autoignition was modelled using mechanisms proposed by Tan [8] and Konnov [9] for C_2H_6–C_3H_8 containing mixtures, and mechanism used by Zhukov and Starikovskii [10] for CH_4, C_4H_{10} and C_5H_{12} containing mixtures. We assumed that the nonequilibrium plasma of a pulsed gas discharge may be considered as an instantaneous injection of dissociated species and radicals, as described above.

4 Results and discussion

Methane containing mixture was the only one where we were not able to keep the same temperatures and pressures for the autoignition experiments and experiments with the discharge. Temperature range is clear from the Figure 1,b. Pressure was within the range 1.9–2.1 atm for the autoignition and 0.6–1.1 atm for the ignition by the discharge. It is clearly seen that these regimes are adequately described by kinetic scheme supplemented with the "discharge + afterglow" calculations described in previous section. Let us note that the mechanism [10] has been developed for a wide pressure ranges (up to hundreds of atm) and under our conditions it practically coincides with the well–known and widely tested GRI2.11 mechanism [11] for CH_4–O_2 combustion.

The results of the calculations for ethane–propane containing mixtures are given in Figure 2. Here it is worth mentioning that two different schemes for calculations of these mixtures were used in the present paper. The reason of using Tan scheme [8] for C_2H_6–C_3H_8 containing mixtures was dictated by the fact that this mechanism gives the same relative position of autoignition curves for C_2–C_3, as we have under our experimental conditions. Scheme proposed by Konnov [9] also gives quite reasonable agreement for experimental and calculated data for autoignition of ethane and propane – containing mixtures. Both schemes give correct value of the ignition delay shift under the action of the discharge. Finally, Figure 3 gives the comparison between experiments and calculations for C_4H_{10} and C_5H_{12}-containing mixtures. Here we have quite reasonable agreement both for autoignition and for the ignition by nanosecond discharge, and the shift of the ignition delay time is mainly determined by the production of oxygen atoms in the discharge and in the nearest afterglow.

5 Conclusions and Acknowledgements

So, we investigated ignition of hydrocarbon-containing mixtures under the action of pulsed nanosecond discharge. Comparison of the results of experiments and numerical modelling give a reasonable agreement and allow to conclude that, at our experimental conditions, the most important channel of the energy input is electron impact dissociation of molecular oxygen in the discharge. This study was partly supported by EOARD/ CRDF (grants 1349 and 1513) and Russian Foundation for Basic Research (grants 05-02-17323-a and 05-03-32975-a).

References

1. Lukhovitskii B I, Starik A M and Titova N S 2005 *Combustion, Explosion, and Shock Waves* **41** 386—94
2. Kossyi I A, Kostinsky A Yu, Matveyev A A and Silakov V P 1992 *Plasma Sources Sci. Technol.* **1** 207—20
3. Williams S, Popovic S, Vuskovic L, Carter C, Jacobson L, Kuo S, Bivolaru D, Corera S, Kahandawala M and Sidhu S *42nd AIAA Aerospace Sciences Meeting and Exhibit* (Reno, Nevada, USA, 5–8 January 2004) AIAA—2004—1012
4. Campbell C S and Egolfopoulos F N 2005 *Comb. Sci. Technol.* **177** 2275—98
5. Chintala N, Bao A, Lou G and Adamovich I V 2006 *Comb. and Flame* **144** 744—56.
6. Bozhenkov S A, Starikovskaia S M, Starikovskii A Yu 2003 *Comb. and Flame* **133** 133–146
7. Kee R J, Miller J A and Jefferson T H, Report No. SAND 80-8003, Sandia National Laboratory, Livermore, CA (1980).
8. Tan Y, Dagaut P, Cathonnet M, Boetther J C 1994 *Combust. Sci. Technol.* **103** 133.
9. Konnov A, Detailed reaction mechanism for small hydrocarbons combustion, http://homepages. vub.ac.be/akonnov/
10. Zhukov V P, "Autoignition of Saturated Hydrocarbons at High Pressures and Initiation of Detonation by Nanosecond Discharge", PhD Thesis, Moscow Institute of Physics and Technology (2004).
11. http://www.me.berkeley.edu/gri-mech/ releases.html

Laser-based ignition of hydrogen-oxygen mixture

Y.V. Tunik[1], **O. Haidn**[2], **and O.P. Shatalov**[1]

[1] *Institute for Mechanics of Moscow State University, 1, Michurinskij prospect, 119192 Moscow, Russia*
[2] *DLR, Heat Technology Institute of Space Propulsion, 74239 Lampoldshausen, Germany*

Summary. Two possible results of an impulse energy injection into the gas mixture are investigated. In the first case it is supposed, that in the gas an equilibrium spherical plasma spot is formed and its disintegration leads to ignition of a surrounding mixture. In the second case, injected energy excites internal degrees of freedom of molecules and atoms and after that converts into translation and rotation energy of the gas, increases the gas temperature and initiates the mixture combustion. In both cases the ignition is accompanied by combustion propagation and gas dynamic perturbances, which are studied on the basis of the one-dimensional equations of dynamics of viscous and heat-conducting gas taking into account the detailed kinetics of chemical transformations and global relaxation of injected energy.

1 Introduction

Now powerful laser-based discharges are investigated for creation of non-uniformities in gas flows near bodies (see, for example, [1]). The present paper is devoted to other applications of a laser impact on the gas mixtures. Here are considered two models of a laser-based ignition of a combustible gas. Classic ignition theory studies two main problems: an ignition of a gas from a spot source and an ignition of homogeneously heated gas. In addition, the thermal explosion theory gives only a qualitative insight about gas ignition with heat losses. Classical approach demands information about heat release rate or about a "global reaction" describing chemical transformations as a single whole. Other possibility of theoretical description of gas ignition and combustion is connected to development and use of detailed kinetics of appropriate chemical reactions.

2 Kinetic model of hydrogen ignition and combustion

There are some kinetic models of chemical transformations in mixtures of hydrogen and oxygen. In the present work the kinetic scheme offered in [2] is used. The model takes into account 10 components $M^{(i)}$: H_2, O_2, O_3, OH, HO_2, H_2O, H_2O_2, H, O, Ar, which participate in 9 reactions of dissociation and in 26 exchange reactions. Thus, in view of distinction of reactions of dissociation in collisions with different partners, the kinetic scheme contains 116 reactions $\sum_i \nu_{ir} M^{(i)} \leftrightarrow \sum_i \bar{\nu}_{ir} M^{(i)}$ Here ν_{ir} and $\bar{\nu}_{ir}$ are stoichiometric factors of direct and reverse reactions. Results of various testing of kinetic model [2] are presented in [3]. On the basis of this model the ignition of homogeneously heated H2-O2 mixture was described in adiabatic conditions at constant pressure or constant volume and also the gas ignition behind a shock wave in a shock tube. Numerical results are in a good agreement with theoretical and experimental data from [4]. These calculations describe both the induction period and the heat release zone caused by chemical transformations. This heat release in stoichiometric mixtures of hydrogen and oxygen diluted by argon has explosive character. Increase of argon concentration leads to growth of ignition delay time of these mixtures.

3 Disintegration of an equilibrium plasma spot located in the centre of the spherical chamber

In this work the problem of ignition of stoichiometric hydrogen-oxygen mixtures diluted by argon after impulse injection of laser energy E_R into spherical volume located in the centre of spherical chamber is solved. Radius of the spherical chamber R_s equals 25cm, radius of the energy injection cavity r_s equals 0.5cm. If typical time of exchange and relaxation processes smaller than characteristic gas dynamic time $\tau_g = r_s/a_0$ (a_0 is a sound velocity in the rest gas), then the state of mixture in spherical cavity r_s can be considered as equilibrium one in initial moment of ignition process. In this case the problem of gas ignition is the problem of equilibrium plasma spot disintegration.

It is assumed that plasma is weakly ionized and consists of free electrons and positive and negative ions of molecules, atoms and radicals listed above. Their equilibrium concentrations Y_i (number of moles in a mass unit of the mixture) are calculated from the law of conservation of elementary particles forming all components of the plasma: electrons, atoms of hydrogen, oxygen and argon. Equilibrium temperature is determined from the first law of thermodynamics, which in our case of constant gas density is equivalent to the condition of conservation of thermal energy of a mass unit of the mixture: $e(Y_i, T_{eq}) = E_0 + E_R$. Here E_0 is the thermal energy of the rest gas mixture: $E_0 = e_0(Y_{i0}, T_0)$. At initial instant the equilibrium values of temperature, pressure and component concentrations are set inside the spherical cavity r_s of laser energy injection. Fig. 1 presents the dependence of equilibrium plasma temperature on injected energy E_R at constant gas density in mixtures with 10, 40 and 70% of argon. Presence of free electrons is observed at equilibrium temperature more than 9000^0K (fig.2). Outside the energy deposition spot the gas mixture is under initial conditions at the temperature $T_0 = 300^0$K and normal pressure 0.101325MPa.

3.1 Equations

Disintegration of the equilibrium plasma spot is investigated on the basis of the one-dimension equations of dynamics of multi-component, heat conductive and viscous gas taking into account above-mentioned detailed kinetics of chemical transformations:

$$\frac{\partial \rho r^\nu}{\partial t} + \frac{\partial \rho u r^\nu}{\partial r} = 0, \frac{\partial \rho Y_i r^\nu}{\partial t} + \frac{\partial \rho Y_i u r^\nu}{\partial r} = \rho \omega_i r^\nu$$

$$\frac{\partial \rho u r^\nu}{\partial t} + \frac{\partial \rho u^2 r^\nu}{\partial r} + r^\nu \frac{\partial (p - \bar{\mu}_s \partial u/\partial r)}{\partial r} = 0, \tag{1}$$

$$\frac{\partial \rho e r^\nu}{\partial t} + \frac{\partial \rho u (h + u^2/2) r^\nu}{\partial r} + r^\nu \frac{\partial (u \bar{\mu}_s \partial u/\partial r + \lambda \partial T/\partial r)}{\partial r} = 0$$

Here conventional for gas-dynamic symbols are used, in the studied spherical case $\nu = 2$, rates of component concentration changes are calculated according to the law of mass action $\rho\omega_i = \sum_r (\bar{\nu}_{ir} - \nu_{ir}) \left(k_{fr} \prod_j (c_j)^{\nu_{jr}} - k_{br} k_{br} \prod_j (c_j)^{\bar{\nu}_{jr}} \right)$, c_i is the component mole number in a volume unity ($c_i = \rho Y_i$), k_{fr} and k_{br} are rate constants of direct and reverse reactions. Considered gas mixture is assumed to be the perfect gas with state equations $h = \sum_i Y_i H_i$ and $p = \rho T R_0/\mu$, μ is molecular weight ($\mu^{-1} = \sum_i Y_i$), enthalpy of one mole of a substance is determined by Gibbs function $\Phi_i(T)$ [5]: $H_i(T) = T^2 \partial \Phi_i(T)/\partial T + H_i(0)$. Viscosity μ_s and heat conductivity λ coefficients of the mixture are determined by Wilke's approximations [6] ($\bar{\mu}_s = \frac{4}{3}\mu_s$).

3.2 Dynamics of the plasma spot disintegration

The problem is solved numerically. Godunov's secondary order difference method with a moving mesh distinguishing of head shock wave and contact break is used as a base. Below at figures $r_0 = 10$cm.

Disintegration of equilibrium plasma spot in the mixture with 94% of argon leads to slow combustion (fig.3: the gas temperature spatial distributions at successive time moments, $E_R = 50E_0$.) Combustion propagates in practically the rest gas and at almost constant pressure. The flame speed equals approximately to 6.5 cm/s at the distance of 2.2 cm from the center (fig.4, curve 1: the flame trajectory). It is the same if the injected energy is two times more (curve 2).

In the mixture with 70% of argon the arising blast wave quickly weakens and doesn't ignite the mixture at low value of injected energy. Combustion propagates due to the gas heat conductivity. Combustion product expansion changes the gas parameters in front of the combustion zone and accelerates the flame propagation (fig.5, curve 1: the flame trajectory at $E_R = 50E_0$). At 10 cm from the vessel centre the flame speed equals to 55 m/s. If dimensionless value of injected energy is equal to 100, the gas ignition occurs just behind the arising shock wave. But in the course of time the flame lags from the head shock wave (fig.5, curve 2). Flame acceleration gives rise to a volume explosion near the vessel wall. At that maximum wall pressure is about 30bars (fig.6)

When argon concentration constitutes 10%, the radiation impulse generates detonation. In the case of the high value of injected energy this detonation is over driven with Mach number more than 6.7 (fig.7, curve 2: the flame trajectory at $E_R = 100E_0$). It is formed practically straight off the plasma spot disintegration. Maximum wall pressure equals approximately to 80bars. At the low value of injected energy ($E_R = 50E_0$) arising blast wave doesn't ignite the mixture. High temperature is only in expanded plasma spot during any time. No combustion propagation is obtained numerically, if heat conductivity isn't taken into account. Combustion propagates after heating up of the gas surrounding the plasma spot. Explosive heat release of chemical reactions in this mixture generates the stream towards the centre of symmetry. This forms reflected shock wave, which quickly transforms to C.-J. detonation (fig.7, curve 1). Wall pressure doesn't exceed 20bars.

4 Excitation of internal degrees of freedom of atoms and molecules and mixture ignition

Theoretically it can be considered the case when impulsive laser impact results only in excitation of vibrational and electronic degrees of freedom of molecules and atoms. In this case the gas ignition is caused by non-equilibrium transition of the injected energy into translational and rotational gas energy. Internal energy of a mass unit of such mixture e is the sum of translational-rotational energy e^{tr} and nonequilibrium energy $\tilde{e}$ of vibrational and electronic degrees of freedom, i.e. $e = e^{tr} + \tilde{e} = \sum_i Y_i \left(\varepsilon_i^{tr} + \tilde{\varepsilon}_i\right)$. Here ε_i^{tr} and $\tilde{\varepsilon}_i$ are respective energies of one mole of a substance. Because redistribution of injected energy on internal degrees of freedom does not change its value, change of nonequilibrium energy $\tilde{e}$ is caused by this energy relaxation $\tilde{\varepsilon}_i$ and chemical reactions

$$d\tilde{e}/dt = \sum_i \left(Y_i \left(\bar{\varepsilon}_i - \tilde{\varepsilon}_i\right)/\tau_i^R + \tilde{\varepsilon}_i dY_i/dt\right). \tag{2}$$

Here $\bar{\varepsilon}_i$ is an equilibrium internal energy of one mole of a substance, τ_i^R is relaxation time of nonequilibrium energy $\tilde{\varepsilon}_i$. The change of mixture composition influences weakly on the transition of injected energy into translation and rotation energy of gas molecules and atoms at low temperature during induction period in initial stage of relaxation process. Therefore it is assumed that in equation (2)

$$\sum_i \tilde{\varepsilon}_i dY_i/dt \approx 0 \tag{3}$$

and a relaxation of non-equilibrium internal energy can be described by the one simple equation

$$d\tilde{e}/dt = \left(\bar{e} - \tilde{e}\right)/\tau^R. \tag{4}$$

Here the equilibrium internal energy $\bar{e}$ can be determined as a difference between equilibrium thermal energy and energy of translational and rotational degrees of freedom at temperature T, i.e. $\bar{e} = e^{(T)} - e^{tr}$, $e^{(T)} = \sum_i Y_i H_i(T) - p/\rho$, $e^{tr} = \sum_i Y_i \varepsilon_i^{tr}$, $\varepsilon_i^{tr} = 0.5 J_i R_0 T + q_i$, J_i is the number of translational and rotational degrees of freedom, q_i is the formation energy of a substance. The equality $\sum_i Y_i \left(\bar{\varepsilon}_i - \tilde{\varepsilon}_i\right)/\tau_i^R = \left(\bar{e} - \tilde{e}\right)/\tau^R$, which follows from equations (2-4), can be considered as a definition of the effective time of global relaxation of non-equilibrium internal energy. In this work the typical time of this transition process is calculated like the characteristic time of vibration energy relaxation: $\tau^R = \bar{\tau}/p$, $\lg\bar{\tau} = AT^{-1/3} + B$, $\lg\bar{\tau} = AT^{-1/3} + B$.

The equation (4) for non-equilibrium energy $\tilde{e}$ is solved together with the equations (1), where the mixture enthalpy must be redefined as $h = \sum_i Y_i \varepsilon_i^{tr} + \tilde{e} + p/\rho$.

In initial moment in the spherical cavity r_s non-equilibrium energy $\tilde{e} = E_R$, but gas temperature and pressure are not changed in the whole volume of the spherical chamber: $T = T_0$,$p = p_0 = p_{AT}$. In this case the gas ignition depends on relations between relaxation, chemical and gas dynamic typical times.

In the mixture with 10% of argon the relaxation of the injected energy increases time of detonation formation at low value of injected energy. At high energy detonation initiation depends on the relaxation rate. If A= 50 and B=-14.6 ($\bar{\tau}$ (atm s)) (fig.8: the flame trajectory) the C.-J. detonation (curve 2) is obtained instead of over driven detonation (curve 1). If relaxation rate is ten times less, the detonation formation looks like in the case of equilibrium plasma spot disintegration at low value of injected energy (curve 3).

Figures 9 and 10 (flame trajectory) correspond to the mixture with 70% of argon at E_R/E_0 =50 and 100, correspondingly. Equilibrium plasma spot disintegration (curves 1) is compared with relaxation model (curves 2 at A= 50 and B=-12.3). The finite relaxation rate increases the gas burning time.

5 Conclusions

1. Proposed models allow describe different combustion regimes caused by laser-based ignition of hydrogen-oxygen mixtures diluted by argon in confined spherical vessel. They are detonation, slow combustion and accelerating flame.

2. At considered values of injected energy the combustion regime is determined by the mixture composition. Injected energy and its relaxation rate influence only on the time of this regime formation.

3. Heat conductivity is an important factor for theoretical solution of the question about minimum energy of detonation initiation in hydrogen-oxygen mixtures.

4. Equilibrium plasma spot is the reliable source of ignition of the hydrogen-oxygen mixtures.

Acknowledgement. This work is supported by the grant N 07-01-00281-a of Russian Fund of Basis Research and grant of President of Russian Federation N SS-6791.2006.1

References

1. P.K. Tretiakov, A.F. Garanin, G.I. Grachev et.al: Doklady Ak. Nauk **351**, 3, 1996
2. L.B. Ibraguimova, G.D. Smekhov, O.P. Shatalov: Electronic Journal "Physico-Chemical Kinetics in Gas Dynamics", http://WWW. Chemphys.edu.ru/, 2003
3. V.A. Pavlov, O.P. Shatalov, Yu.V. Tunik: Numerical and experimental validation of a database of H2 +O2 mixture combustion. In: *Nonequilibrium processes. Combustion and detonation*, vol 1, ed by G.D. Roy, S.M. Frolov, A.M. Starik (Torus Press, Moscow 2005) pp 3-12
4. E. Schultz, J. Shepherd: *Validation of detailed reaction mechanisms for detonation simulation*, (Graduate Aeronautical Laboratories, California Institute of Technology, Pasadena, CA 91125), Explosion Dynamics Laboratory Report FM99-5.
5. *Thermo-dynamic properties of individual substances.* Reference edition. V.1, book 1. Edited by L.V. Gurvich et al. Moscow, Nauka, 1978, 496pp.
6. C.R. Wilke: J. Chem. Phys. **18**, 4, 1950

6 Figures

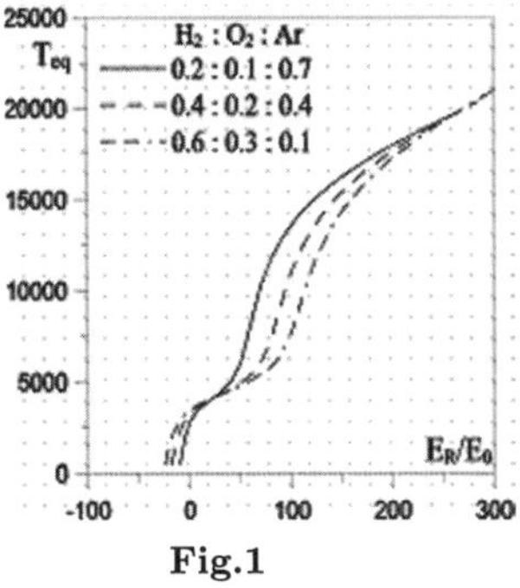

Fig.1

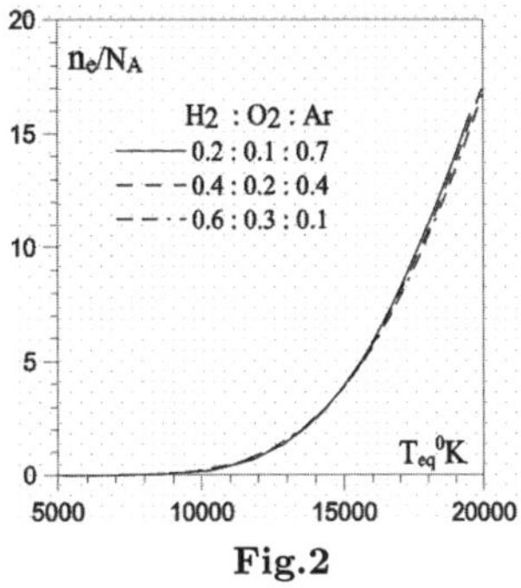

Fig.2

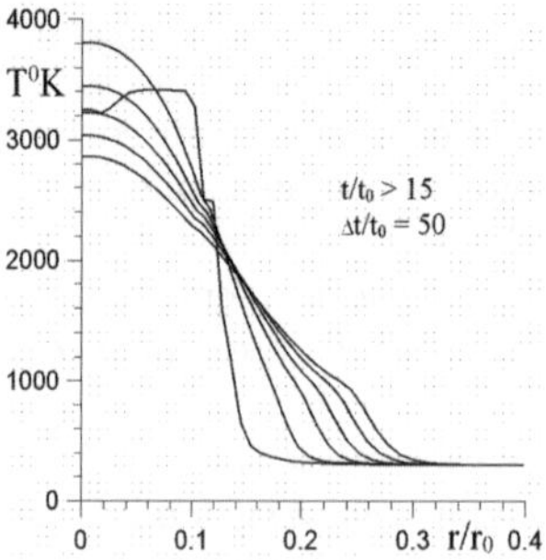

Fig.3

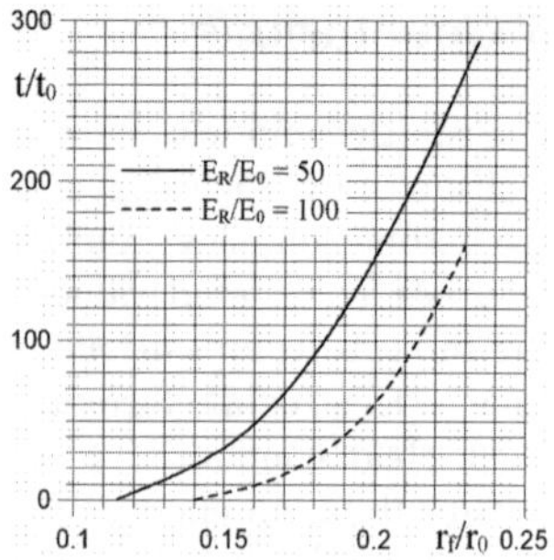

Fig.4

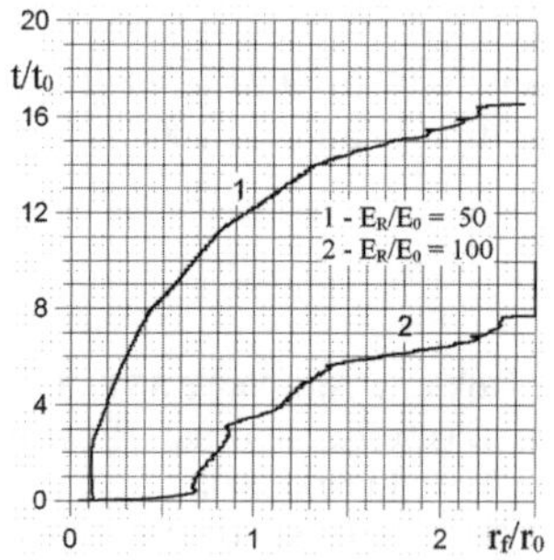

Fig.5

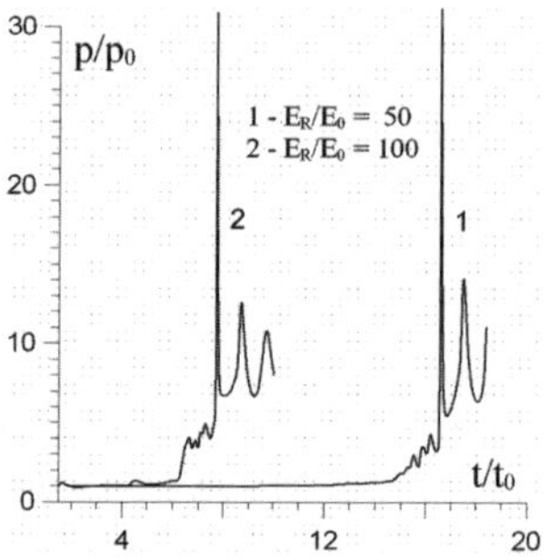

Fig.6

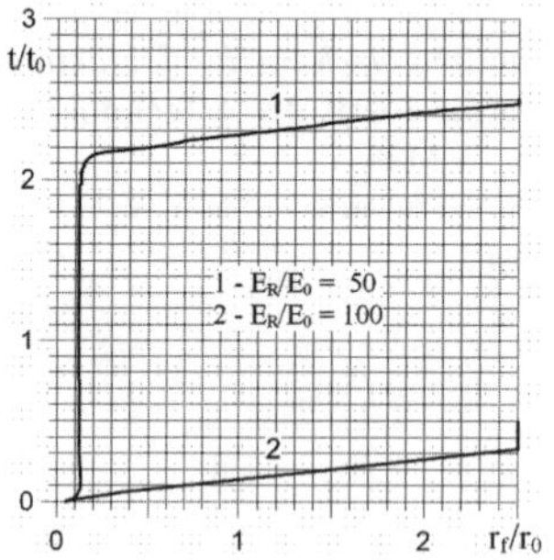

Fig.7

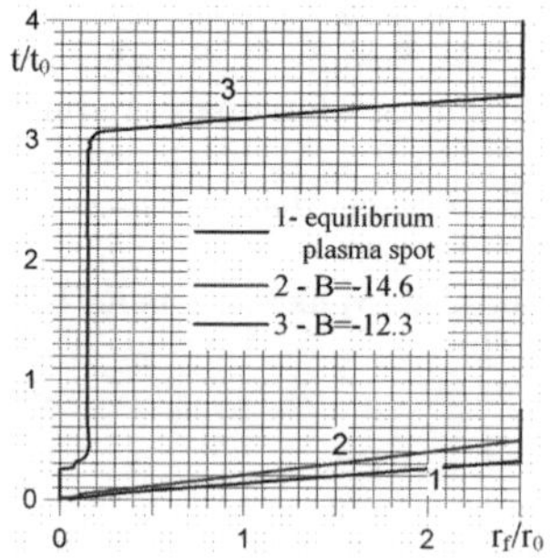

Fig.8

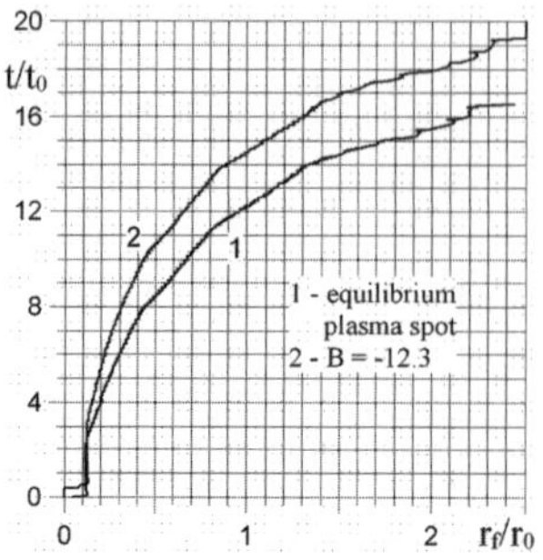

Fig.9

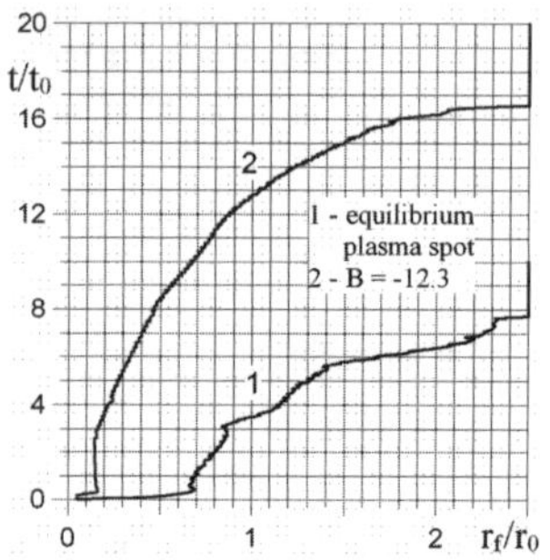

Fig.10

Measurements of ignition delay times and OH species concentrations in DME/O_2/Ar mixtures

R.D. Cook, D.F. Davidson, and R.K. Hanson

Department of Mechanical Engineering, Stanford University, Stanford, CA, USA

1 Introduction

Oxygenated fuels are becoming increasingly important in combustion applications. In diesel engines, oxygenate additives have demonstrated some potential to reduce soot emissions, which are a major barrier to expanding the use of diesel engines [1]. In addition, biofuels contain many oxygenated species, further motivating the study of detailed oxygenate chemistry. Dimethyl ether (DME) is an attractive fuel to study because it is a simple oxygenate with a good experimental and modeling database [2–7]. Two types of measurements are presented in this study: ignition delay times, and OH time-histories. Ignition delay times provide useful targets for evaluating the overall performance of detailed oxidation mechanisms and providing direct comparisons with past work. In very rich mixtures, however, there is no distinct ignition event, so OH profiles can be used instead. OH profiles are also useful because they can often provide information and constraints on a much smaller set of reactions than ignition delay times.

There have been two previous studies of DME ignition delay times. A study by Pfahl, et al. examined ignition delay times at 13 and 40 atm and from 650 to 1250 K in stoichiometric mixtures [3]. This study is particularly interesting because it provided ignition delay time data at conditions relevant to diesel engine conditions and demonstrated significant NTC (negative temperature coefficient) behavior. An additional study by Dagaut, et al. was performed to provide a set of high temperature ignition delay time data at 3.5 atm from 1200 to 1600 K [4]. Their study also varied equivalence ratio, $0.5 \leq \phi \leq 2$. The two previous studies provide a broad and complementary set of data. However, they do not provide enough information to fully correlate the effects of pressure and equivalence ratio.

DME oxidation models have been previously developed based on the ignition delay time data and other reactor studies. A model by Dagaut, et al. was based on their high temperature ignition delay time measurements as well as both high and low temperature jet-stirred reactor (JSR) measurements [4]. Lawrence Livermore National Labs (LLNL) have also developed a model which has considered the previous ignition delay time and JSR data [5], as well as variable pressure flow reactor (VPFR) measurements [6,7]. The most recent LLNL model agrees well with the previous reactor studies, and even does a good job of modeling ignition delay time in the NTC region. Unfortunately, previous studies did not measure species concentrations of important combustion radicals. The current study provides the first measurements of the important combustion radical OH in the DME oxidation system.

2 Experimental Setup

The experimental device used was a 15 cm diameter stainless steel shock tube. All measurements were performed in the reflected shock region, at a measurement location 2 cm from the endwall of the tube. Mixtures were prepared manometrically in a high-purity mixing assembly. Research grade Ar (99.999%) and O_2 (99.999%) were supplied by Praxair Inc., and DME (99+%) was supplied by Sigma-Aldrich Co. Between experiments, the shock tube and mixing manifold were routinely pumped down to ~3 μtorr and ~1 mtorr, respectively, to ensure the purity of the test mixture.

Ignition delay times were measured by monitoring OH* emission near 306 nm. Emission from the test section was focused through a window onto a Thorlabs PDA55 photodetector modified with a UV-enhanced Si photodiode (Hamamatsu S1722-02). A vertical slit constrained the maximum axial dimension of the measurement volume to ~3 mm and a Schott glass UG-5 filter was used to block unwanted visible light, while passing the desired OH* emission near 306 nm. OH species concentrations were measured using the frequency-doubled output of a narrow-linewidth ring dye laser, tuned to the peak of the R1(5) absorption line in the OH A-X (0, 0) band near 306.7 nm. Visible light near 613.4 nm was generated by pumping Rhodamine 6G dye in a Spectra Physics 380 laser cavity with the 5 W, continuous wave, output of a Coherent Verdi laser at 532 nm. The visible radiation was intracavity frequency-doubled using a temperature-tuned AD*A nonlinear crystal. This generates approximately 1 mW of light near 306.7 nm. Using a common-mode rejection detection setup, a minimum absorbance detection limit of $< 0.1\%$ could be achieved, resulting in a minimum OH detection sensitivity of ~0.3 ppm at 1500 K and 1.3 atm. Further details of the OH detection setup are described elsewhere [8,9].

3 Measurements

Ignition delay times have been measured for mixtures at 1.8 and 3.3 atm and $\phi = 0.5$, 1, and 2 in order to explore the dependence of ignition delay time on equivalence ratio and evaluate model performance at two different pressures. Additional ignition delay time measurements were performed at $\phi = 1$ and pressures from 1.6–6.6 atm to examine the pressure dependence of DME ignition delay times at high temperatures. The model used for comparison in this work is the most recent LLNL model, since it accurately describes a broad range of previous experimental work [6, 7]. Ignition delay time results at 1.8 and 3.3 atm from the current study are plotted in Figure 1 along with a comparison to the LLNL model, and the previous study of Dagaut, et al. Figure 2 shows the pressure dependence of ignition delay times, along with a comparison to the correlation derived in the current study (Equation 1).

$$\tau_{ign} = 4.21 \times 10^{-5} * P^{-0.66} * \phi^{0.72} * exp(22690/T), \tag{1}$$

where τ_{ign} is in sec, P is in atm, and T is in K.

OH time-histories have been measured in rich DME/O_2/Ar mixtures, $\phi \sim 3$, P $\sim$ 1.3 atm, and temperatures from 1400–1750 K. Some initial measurements were performed in mixtures containing 1.5% DME, but the concentration was later lowered to 0.15% for most experiments in order to minimize the effects of energy release and isolate a smaller group of sensitive reactions. Example OH profiles are shown in Figure 3.

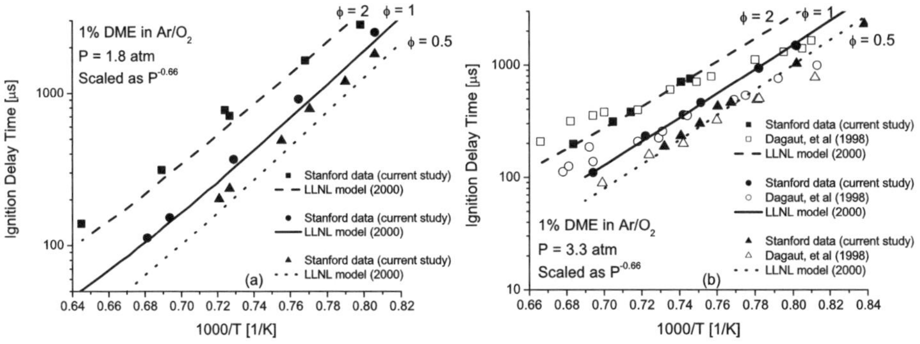

Fig. 1. Comparison of ignition delay times measured in current study to previous work and LLNL model at $\phi = 0.5$, 1, 2 and pressure equal to (a) 1.8 atm and (b) 3.3 atm

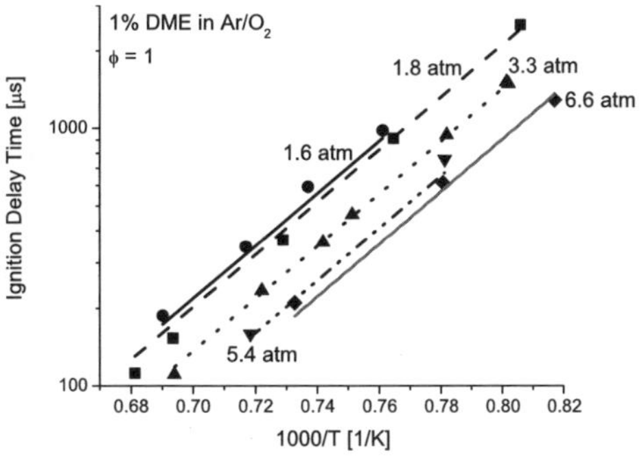

Fig. 2. Ignition delay times measured at pressures from 1.6–6.6 atm (symbols) and comparison to correlation derived from current study (lines)

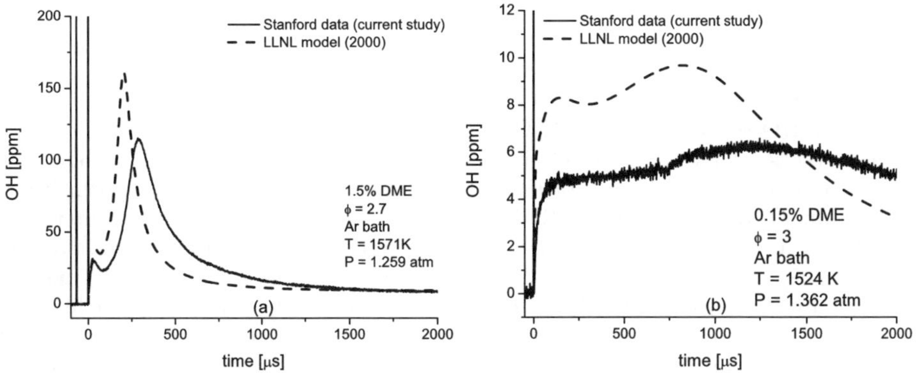

Fig. 3. OH profiles in rich DME oxidation and comparison to model at two different DME concentrations, (a) 1.5% and (b) 0.15%

4 Comparison to Previous Work

As shown in Figure 1, the DME ignition delay times agree quite well with the LLNL model at 1.8 and 3.3 atm. The measured activation energy is just slightly ($\sim 6\%$) lower than the model prediction and the measured dependence on equivalence ratio is also accurately predicted by the model. However, the current study suggests a pressure scaling of $P^{-0.66}$, compared to the model's pressure scaling of approximately $P^{-0.39}$. Since the previous high-temperature ignition delay time study only covered a single pressure, the current work provides the first measurement of the pressure dependence of DME ignition delay time at high temperatures. The overall magnitudes of the ignition delay times measured in this work also agree well with the previous high-temperature study of Dagaut, et al., although there is a noticeable difference in the measured activation energies. The activation energy measured here is considerably higher and in much better agreement with the LLNL model prediction.

Although the LLNL model does very well predicting ignition delay times, it uniformly overpredicts the measured OH profiles. Some features, like the presence of two local OH peaks in some cases, are predicted by the model, but the magnitudes of the peaks are quite different from the measurements. In mixtures with 1.5% DME, some of the discrepancy could come from uncertainties due to energy release and the accompanying rise in temperature. The reflected shock region is modeled in this study as a constant volume and constant internal energy reactor, but that approximation is not entirely accurate. More accurate results can be obtained by using more dilute mixtures since there is less energy release. However, the dilute mixtures containing only 0.15% DME still produce much less OH than is predicted by the LLNL model. OH sensitivity was calculated based on the model for both the concentrated and dilute mixtures, but even the dilute mixtures were sensitive to a very large number of reactions. At early times ($t < 50\ \mu s$), though, a reasonably small set of sensitive reactions could be identified (Figure 4). Adjusting the rates relevant at early times can also effectively lower the amount of OH predicted by the model throughout the experiment. Therefore, close examination of the following set of reaction rates may significantly improve the overall performance of the LLNL model with respect to OH:

(R1) $H + O_2 \Leftrightarrow OH + O$
(R2) $CH_3OCH_3 \Leftrightarrow CH_3 + CH_3O$
(R3) $CH_3OCH_3 + H \Leftrightarrow CH_3OCH_2 + H_2$
(R4) $CH_3OCH_3 + OH \Leftrightarrow CH_3OCH_2 + H_2O$
(R5) $CH_2O + OH \Leftrightarrow HCO + H_2O$

The reaction R1 is an extremely important reaction in nearly all hydrocarbon combustion systems and has been extensively studied previously, hence its rate coefficient is well-known. Of the five reactions, R5 is the least sensitive, and previous measurements, including a recent direct, high-temperature measurement by Vasudevan, et al., are very consistent [9]. Therefore, the other three reactions, DME decomposition (R2) and abstraction by H (R3) and OH (R4), are the ones for which the rates can most likely be adjusted to improve agreement between experiment and model. In fact, a recent direct measurement of R3 indicates a rate almost double the one used in the current LLNL model [10]. It is also interesting to note that the recently measured rate of R3 is in excellent agreement with the value used in a previous version of the model [5]. Adjusting this rate alone greatly improves agreement between the current study and the model. However, the model still significantly overpredicts OH, suggesting that the rates of reactions R2 and R4, which are relatively uncertain at the conditions of the current study, may

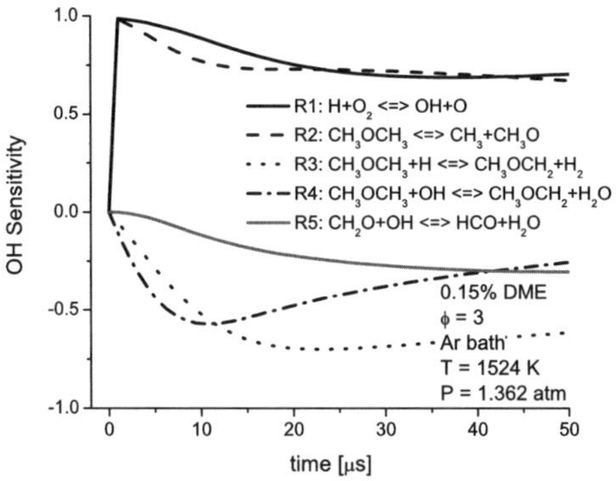

Fig. 4. OH radical sensitivity analysis of 0.15% DME mixture

still need to be adjusted. Direct measurements of these rate coefficients would of course be preferable.

5 Conclusions

The chemistry of DME oxidation was studied in the reflected shock region of a shock tube. Ignition delay times were measured using OH* emission over a broad range of conditions. This new study also covers a range of pressures (1.6–6.6 atm), which demonstrates the pressure dependence of DME ignition delay times at high temperatures. The current study agrees well with the previous high-temperature study [4] as well as the LLNL model [6, 7], but predicts a larger pressure dependence than the model. OH profiles were measured in very rich DME mixtures ($\phi \sim 3$) and compared to the model. Although the model is able to capture some features, it predicts more than the measured amount of OH. Detailed sensitivity analysis was then performed to identify the rates that are most likely responsible for the discrepancy, and that may be targets for future studies.

Acknowledgement. This work was supported by the Global Climate and Energy Project (GCEP).

References

1. S.C. Sorenson and S.E. Mikkelsen, SAE Paper 950064 (1995)
2. U. Pfahl, K. Fieweger, G. Adomeit: Proc. Combust. Inst. **26**, 781 (1996)
3. P. Dagaut, J.-C. Boettner, M. Cathonnet: Proc. Combust. Inst. **26**, 627 (1996)
4. P. Dagaut, C. Daly, J.M. Simmie, M. Cathonnet: Proc. Combust. Inst. **27**, 361 (1998)
5. H.J. Curran, W.J. Pitz, C.K. Westbrook, P. Dagaut, J.-C. Boettner, M. Cathonnet: Int. J. Chem. Kinet. **30**, 229 (1998)
6. S.L. Fischer, F.L. Dryer, H.J. Curran: Int. J. Chem. Kinet. **32**, 713 (2000)
7. H.J. Curran, S.L. Fischer, F.L. Dryer: Int. J. Chem. Kinet. **32**, 741 (2000)
8. J.T. Herbon, R.K. Hanson, D.M. Golden, C.T. Bowman: Proc. Combust. Inst. **29**, 1201 (2002)
9. V. Vasudevan, D.F. Davidson, R.K. Hanson: Int. J. Chem. Kinet. **37**, 98 (2005)
10. K. Takahashi, O. Yamamoto, T. Inomata, M. Kogoma: Int. J. Chem. Kinet. **39**, 97 (2007)

Shock tube study of artificial ignition of N_2O:O_2:H_2:Ar mixtures

I.N. Kosarev, S.M. Starikovskaia, and A.Y. Starikovskii

Physics of Nonequilibrium Systems Lab, Moscow Institute of Physics and Technology

Summary. Problem of ignition acceleration is of crucial importance from both scientific and technological standpoints. In order to investigate detailed kinetics of ignition and ways of its acceleration experiments the were carried out behind the reflected shock wave for the temperature and pressure ranges 850 – 2100 K and 0.1 – 1.0 atm respectively. We examined ignition of N_2O:H_2:Ar = 1:1:8 and O_2:N_2O:H_2:Ar = 0.3:1:3:5 mixtures. The autoignition, ignition by low temperature nonequilibrium plasma of nanosecond high-voltage discharge and flash-photolysis aided ignition have been compared. Numerical scheme to describe the autoignition in the system has been proposed and tested.

1 Introduction

The kinetics of N_2O–containing mixtures itself is of great interest, because nitrous oxide is a good source of atomic oxygen in both the ground ($O(^3P)$) and excited ($O(^1D)$) states. As a source of atomic oxygen, N2O is widely used in kinetic experiments. The question of the efficiency of using nonequilibrium plasmas for initiation of combustion still remains open. The discharge produces a lot of species: electronically excited atoms and molecules, ionized particles, vibrationally excited components. The relative role of different species is under discussion now. To check the role of excited and dissociated species we carried out experiments with excitation by laser flash–photolysis with dissociation of molecules of the mixture at wavelength of 193 nm. The idea to use laser flash–photolysis or focused laser radiation for initiation of chemical reactions is not new and have been widely used by researches ([1]- [5]), while the comparison of autoignition, ignition by nonequilibrium plasma and laser flash–photolysis is given for the first time.

2 Experiments

The ignition delay time in N_2O–containing mixtures was measured in a square shock tube (ST). The working channel of the tube was 1.6 m long. The experiments were performed behind a reflected shock wave at a distance of 5 mm from the end plate. The thermodynamic parameters of the gas behind the reflected shock wave were determined on the basis of the ideal ST theory. During the experiments the quality of the ST channel was under special consideration. The maximal size of the steps and clearances did not exceed 0.05 mm when mounting windows and assembling ST sections. Optical diagnostics of the shock wave was performed using 3 pairs of quartz windows of diameter 20 mm. Furtheron, the cross section closest to the ST end plate will be referred to as a diagnostic cross section. The incident shock wave velocity was determined using three laser schlieren gages located in series at three points spaced at 226 and 250 mm. The deceleration of the shock wave was taken into account. The mixtures to be investigated were prepared using high-purity gases O_2(99.99%), H_2(99.999%), Ar(99.998%) and N_2O of 98% purity. The accuracy of determination of gas mole fractions was at least 0.0025 and was mainly defined by the purity of N_2O and by the inleakage to the low–pressure chamber. Compressed air was used as a driver gas. The working chamber and the high–pressure chamber were separated by a metallized lavsan diaphragm 12 μm thick. In order to measure the ignition delay time the emission was measured in the diagnostic cross section at a wavelength of 306.4 nm that corresponds to transition of the excited OH radical to the ground state ($A^2\Sigma \rightarrow X^2\Pi$). The delay time was determined as the time between the schlieren signal which corresponds to the coming of reflected shock wave to the diagnostic cross section and the point of intersection of the time axis with the leading front of emission. The point of intersection was determined tangentially to the curve constructed at the point of the maximum of signal derivative. The optical system included a quartz condenser which focused the image onto the slit of MDR–23 monochromator. The emission was detected by a FEU–100 photomultiplier. A vertical 2x20 mm slit was installed on the window. The signals from the schlieren sensors and photomultiplier were registered by TDS 3014 Tektronix oscillograph (150 MHz bandwidth).

The experiments were repeated with the same mixture for autoignition, discharge initiation of ignition and ArF laser flash–photolysis. Ignition delay times were compared for all cases. The diagnostic system used for ignition by the discharge, described in details elsewhere [6] was modified to control ignition by laser flash-photolysis. A flash of UV–radiation of ArF excimer laser ("Center of Physical Devices" production, Troitsk,) was initiated behind the reflected shock wave instead of the nanosecond discharge. Laser output reached 0.2 J. The laser radiation was supplied to the dielectric section of the shock tube through the optical window perpendicularly to the shock tube axis in the same cross–section were we performed measurements in a case of the ignition by nanosecond discharge. Laser spot had approximately rectangular shape with 5x21 mm dimensions and was controlled by special sensible paper. A piroelectric detector (PEM21) with 21 mm diameter of receiving site was used to determine energy input from laser radiation into the gas. Signal from PEM21 was controlled every time before the experiment, when the system was pumped up and then during the experiment. Knowing spectral transmission of MgF_2 windows in this spectral range (were controlled by Varian Cary 50 Spectrophotometer) we calculated energy input into a gas. We organized experiment so, that the energy input in the discharge and the energy input from a laser were within one order of magnitude. Space uniformity of the excitation by the discharge and by laser radiation were controlled additionally.

3 Experimental Results

Table 1 gives results of the experiments on autoignition, ignition by the discharge and ignition by laser flash–photolysis. Cross–section of measurements is situated at the distance 55 mm from the end plate of the shock tube. In the Table, τ_{auto} is a delay time for autoignition, τ_{flash} is a delay time in the case of ignition by nanosecond flash laser pulse, τ_{FIW} is a delay time in the case of ignition by nanosecond high–voltage pulse of negative polarity with the amplitude 60 kV and duration 40 ns. Temperature, pressure, and gas number density behind the reflected shock wave (T_5, P_5 and n_5 respectively) are given in the Table together with the volumetric energy input.

Table 1. Experimental data on autoignition, ignition by laser flash–photolysis and ignition by pulsed high–voltage nanosecond discharge. Distance from the end plate of the shock tube is equal to 55 mm, N_2O:H_2:O_2:Ar = 1:3:0.3:5 mixture.

Mixture	ε mJ·cm^{-3}	T_5, K	P_5, atm	$n_5, 10^{18}$ cm^{-3}	τ_0, μs	τ_{fl}, μs	τ_d, μs
N_2O-H_2-O_2-Ar	-	998	0.44	3.25	463	-	-
N_2O-H_2-O_2-Ar	-	1027	0.45	3.21	368	-	-
N_2O-H_2-O_2-Ar	-	1168	0.24	1.50	175	-	-
N_2O-H_2-O_2-Ar	-	1091	0.28	1.88	259	-	-
N_2O-H_2-O_2-Ar	-	1125	0.26	1.69	218	-	-
N_2O-H_2-O_2-Ar	-	1071	0.31	2.13	300	-	-
N_2O-H_2-O_2-Ar	-	1064	0.35	2.44	306	-	-
N_2O-H_2-O_2-Ar	-	1027	0.39	2.81	378	-	-
N_2O-H_2-O_2-Ar	-	1132	0.26	1.70	219	-	-
N_2O-H_2-O_2-Ar	2.4	965	0.44	3.36	-	242	-
N_2O-H_2-O_2-Ar	2.5	971	0.44	3.37	-	278	-
N_2O-H_2-O_2-Ar	3.9	1021	0.46	3.34	-	235	-
N_2O-H_2-O_2-Ar	3.1	1046	0.45	3.17	-	215	-
N_2O-H_2-O_2-Ar	3.4	1077	0.41	2.79	-	190	-
N_2O-H_2-O_2-Ar	2.5	1118	0.37	2.41	-	147	-
N_2O-H_2-O_2-Ar	1.8	1176	0.32	2.02	-	126	-
N_2O-H_2-O_2-Ar	3.0	955	0.49	3.74	-	288	-
N_2O-H_2-O_2-Ar	2.0	924	0.51	4.02	-	309	-
N_2O-H_2-O_2-Ar	2.4	909	0.54	4.39	-	367	-
N_2O-H_2-O_2-Ar	3.1	905	0.59	4.78	-	344	-
N_2O-H_2-O_2-Ar	3.5	890	0.62	5.09	-	416	-
N_2O-H_2-O_2-Ar	3.4	1004	0.43	3.14	-	-	101
N_2O-H_2-O_2-Ar	3.7	1040	0.39	2.80	-	-	92
N_2O-H_2-O_2-Ar	4.4	1064	0.35	2.40	-	-	64
N_2O-H_2-O_2-Ar	4.6	1091	0.31	2.07	-	-	54
N_2O-H_2-O_2-Ar	3.4	976	0.51	3.82	-	-	177
N_2O-H_2-O_2-Ar	1.5	929	0.53	4.23	-	-	246
N_2O-H_2-O_2-Ar	1.7	909	0.56	4.53	-	-	261
N_2O-H_2-O_2-Ar	1.3	900	0.60	4.91	-	-	296
N_2O-H_2-O_2-Ar	2.9	970	0.45	3.37	-	-	139

Inaccuracy in the temperature measurements due to the data scattering does not exceed 20 K. Another source of the inaccuracy lies in the non–ideality of the processes in the shock tube. Indeed, we use 1D shock tube theory and data from three schlieren–systems to determine T_5 and P_5 on the basis of known velocity of the shock wave, and initial gas pressure and temperature, not taking into account possible influence of boundary layers on the parameters of the system. We use weakly diluted mixture and the cross-section of measurements is situated far from the end plate of the shock tube.

To approach to the conditions of ideal shock tube theory, we have performed an additional set of experiments, were we compared delay times for the autoignition and ignition by unfocused radiation of ArF laser in a case when cross–section of measurements is situated 5 mm apart the end plate of the shock tube. At these conditions, the experiments on the ignition by pulsed discharge have not been carried out because of the discharge peculiarities: typically, to perform exact control of the discharge parameters, we have to perform measurements at the distance equal to 1–2 diameters of the electrode. Additional experiments have been performed in gas mixture without O_2 additions, this allowed us to validate constructed numerical scheme.

Data obtained in the cross-section situated 5 mm apart the end plate of the shock tube are given in the Table 2. In addition to gas parameters (T_5, P_5, ρ_5), volumetric energy input and ignition delay time, we give here $O(^1D)$ atoms density calculated from the energy input. It was assumed that the quantum efficiency of the process of photon absorption by N_2O molecule is equal to unity.

Table 2. Experimental data on autoignition and ignition by laser flash–photolysis. Distance from the end plate of the shock tube is equal to 5 mm, N_2O:H_2:O_2:Ar = 1:3:0.3:5 mixture.

Mixture	$([O(^1D)]/n_5)$	T_5, K	P_5, atm	$n_5, 10^{18}$ cm^{-3}	$^*\tau_0$, μs	τ_{fl}, μs	τ_d, μs
N_2O-H_2-O_2-Ar	-	979	0.45	3.40	134	-	-
N_2O-H_2-O_2-Ar	-	952	0.48	3.72	155	-	-
N_2O-H_2-O_2-Ar	-	931	0.51	4.04	177	-	-
N_2O-H_2-O_2-Ar	-	916	0.55	4.37	199	-	-
N_2O-H_2-O_2-Ar	-	902	0.58	4.69	233	-	-
N_2O-H_2-O_2-Ar	-	883	0.60	4.97	291	-	-
N_2O-H_2-O_2-Ar	-	865	0.62	5.23	480	-	-
N_2O-H_2-O_2-Ar	-	870	0.63	5.26	459	-	-
N_2O-H_2-O_2-Ar	-	865	0.66	5.60	355	-	-
N_2O-H_2-O_2-Ar	-	852	0.69	5.88	451	-	-
N_2O-H_2-O_2-Ar	$0.7 \cdot 10^{-3}$	1157	0.26	1.64	-	77	-
N_2O-H_2-O_2-Ar	$1.6 \cdot 10^{-3}$	1227	0.19	1.15	-	86	-
N_2O-H_2-O_2-Ar	$0.7 \cdot 10^{-3}$	1180	0.21	1.31	-	101	-
N_2O-H_2-O_2-Ar	$1.1 \cdot 10^{-3}$	1107	0.29	1.91	-	105	-
N_2O-H_2-O_2-Ar	$0.7 \cdot 10^{-3}$	1080	0.32	2.15	-	110	-
N_2O-H_2-O_2-Ar	$1.6 \cdot 10^{-3}$	1055	0.33	2.28	-	95	-
N_2O-H_2-O_2-Ar	$1.9 \cdot 10^{-3}$	1068	0.38	2.59	-	91	-
N_2O-H_2-O_2-Ar	$1.6 \cdot 10^{-3}$	1030	0.38	2.68	-	113	-
N_2O-H_2-O_2-Ar	$1.4 \cdot 10^{-3}$	1018	0.40	2.87	-	126	-
N_2O-H_2-O_2-Ar	$1.3 \cdot 10^{-3}$	1006	0.42	3.06	-	127	-
N_2O-H_2-O_2-Ar	$1.5 \cdot 10^{-3}$	1024	0.47	3.33	-	110	-
N_2O-H_2-O_2-Ar	$1.2 \cdot 10^{-3}$	1001	0.45	3.26	-	139	-
N_2O-H_2-O_2-Ar	$3.4 \cdot 10^{-3}$	1165	0.22	3.14	-	273	-
N_2O-H_2-O_2-Ar	$3.7 \cdot 10^{-3}$	1244	0.17	2.80	-	228	-
N_2O-H_2-O_2-Ar	$4.4 \cdot 10^{-3}$	1195	0.21	2.40	-	222	-
N_2O-H_2-O_2-Ar	$4.6 \cdot 10^{-3}$	1195	0.17	2.07	-	235	-
N_2O-H_2-O_2-Ar	$3.4 \cdot 10^{-3}$	1172	0.20	3.82	-	254	-
N_2O-H_2-O_2-Ar	$1.5 \cdot 10^{-3}$	1352	0.10	4.23	-	149	-
N_2O-H_2-O_2-Ar	$1.7 \cdot 10^{-3}$	1278	0.14	4.53	-	190	-
N_2O-H_2-O_2-Ar	$1.3 \cdot 10^{-3}$	1150	0.23	4.91	-	261	-
N_2O-H_2-O_2-Ar	$2.9 \cdot 10^{-3}$	1121	0.26	3.37	-	288	-
N_2O-H_2-O_2-Ar	$1.5 \cdot 10^{-3}$	1094	0.28	4.23	-	305	-
N_2O-H_2-O_2-Ar	$1.7 \cdot 10^{-3}$	1068	0.30	4.53	-	366	-
N_2O-H_2-O_2-Ar	$1.3 \cdot 10^{-3}$	1107	0.36	4.91	-	268	-
N_2O-H_2-O_2-Ar	$2.9 \cdot 10^{-3}$	1061	0.37	3.37	-	365	-

Induction delay times are given in Figure 1, a. The shift of the ignition delay is clearly seen for the action of a laser flash–photolysis and for the action of the discharge. Absorption of laser radiation shifts the ignition threshold by 100–150 K, while the nanosecond discharge gives a value of 150–200 K. Volumetric energy input for both laser flash-photolysis and for the discharge is given in Fig. 1, b. Energy absorbed under the action of laser radiation increases with gas number density. As for the discharge, at the given conditions the energy input decreases with gas density [7]. This peculiarity is explained by the fact that the discharge development depends significantly upon gas density. With pressure increase at fixed gas temperature, discharge front decelerates and, finally, at high pressures, a uniform nanosecond breakdown transforms to a nanosecond pulsed corona from high-voltage electrode. In spite of the opposite trend, the energy input for both series of experiments is within one order of magnitude.

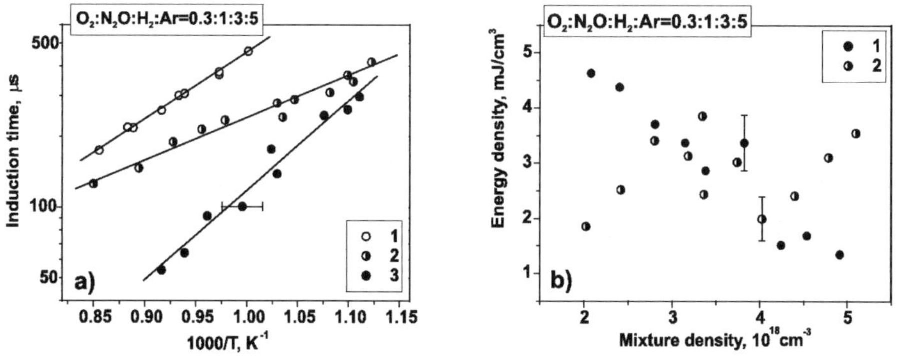

Fig. 1. a) Ignition delay time vs temperature. Comparison of autoignition (1), ignition by laser flash–photolysis (2) and by the discharge (3); b) Energy input in experiments with the discharge (1) and with flash–photolysis (2).

Zero–dimensional simulation of ignition was performed using CHEMKIN computer codes [8]. The calculations were performed at constant pressure. This mechanism was supplemented with the reactions of formation and quenching of OH radical in the $A^2\Sigma$ state for comparison of the numerical calculation results. The scheme was validated using experimental results on ignition [9] and unimolecular decomposition of N_2O [10]. For a case of modelling of experiments with laser flash-photolysis, reactions with $O(^1D)$ become to play an important role:

$$O(^1D) + H_2 = H + OH \tag{1}$$

$$O(^1D) + N_2O = N_2 + O_2 \tag{2}$$

$$O(^1D) + N_2O = NO + NO \tag{3}$$

It should to be noted that last two reactions decrease an efficiency of laser flash-photolysis as an initiator of ignition, leading to products which are "useless" in our experimental conditions. Calculations demonstrated rather good correlation with the experimental data for autoignition and significant discrepancy for ignition by laser flash-photolysis. To get coincidence between experimental and numerical data, we have to multiply $O(^1D)$ density obtained in experiments by factor of 0.3, otherwise the discrepancy between experiment and modeling is rather noticeable, but qualitatively we can see the same tendency as in experiments. The mentioned reactions are relatively well–known and their variation within reasonable limits lead to only insignificant change in ignition delay time. So, the numerical scheme and the results of experiments need additional analysis to obtain quantitative agreement.

4 Conclusions

So, autoignition, ignition by nanosecond discharge and ignition by laser flash–photolysis were investigated at the same experimental conditions for N_2O–containing mixtures using shock tube technique. Shift of the ignition by nanosecond discharge and laser flash–photolysis has been obtained. Numerical scheme to describe the autoignition has been proposed. Qualitative numerical analysis of ignition flash-photolysis has been performed.

Acknowledgement. This study was partly supported by EOARD/ CRDF (grants 1349 and 1513), Russian Foundation for Basic Research (grants 05-02-17323-a and 05-03-32975-a), and CRDF (Award MO-011-0). The authors are grateful to Dr. Nikolai Popov from Skobeltsyn Institute for Nuclear Physics (Moscow State University) for numerous fruitful discussions concerning the kinetic scheme.

References

1. Alden M, Edner H, Grafstroem P and Svanberg S 1982 Two-photon excitation of atomic oxygen in a flame *Opt Comm* **42(4)** 244-6
2. Forch B E and Miziolek A W 1986 Oxygen-atom two=photon resonance effects in multi-photon photochemical ignition of premixed H_2/O_2 flows *Optics Letters* **11(3)** 129-31
3. Lavid M and Stevens J G 1985 Photochemical ignition of premixed hydrogen/oxidizer mixtures with excimer lasers *Comb. and Flame* **60** 195-202
4. Chou M–S and Zukowski T J 1991 Ignition of $H_2/O_2/NH_3$, H_2/air/NH_3 and $CH_4/O_2/NH_3$ mixtures by excimer–laser photolysis of NH_3 *Comb. and Flame* **87** 191-202
5. Lavid M, Nachshon Y, Gulati S K and Stevens J G 1994 Photochemical ignition of premixed hydrogen/oxygen mixtures with ArF laser *Comb. Sci. and Techn.* **96** 231-45
6. Bozhenkov S A, Starikovskaia S M, Starikovskii A Yu 2003 Nanosecond gas discharge ignition of H2 – and CH4 – containing mixtures. *Comb. and Flame* **133** 133-146
7. Starikovskaia S M, Kukaev E N, Kuksin A Yu, Nudnova M M and Starikovskii A Yu 2004 Analysis of the spatial uniformity of the combustion of a gaseous mixture initiated by a nanosecond discharge. *Comb. and Flame* **139** 177–87
8. Kee R J, Miller J A and Jefferson T H 1980 *Report No. SAND 80-8003, Sandia National Laboratory, Livermore, CA*
9. Hidaka Y, Takuma H and Suga M 1985 *J. Phys. Chem.* **89** 4093–5
10. Zuev A P and Starikovskii A Yu 1992 *Soviet Journal of Chemical Physics. Gordon and Breach Science Publishers* **10(3)** 520—40

Shock tube study of kerosene ignition delay

S. Wang, B.C. Fan, Y.Z. He, and J.P. Cui

Key Laboratory of High Temperature Gas Dynamics, Institute of Mechanics, Chinese Academy of Sciences, Beijing 100080, China

1 Introduction

The ignition delay time of a fuel is usually concerned as a characteristic time to scale the duration of the gas flow passing through the combustion chamber, which is a criterion to measure the capability of the ignition and sustaining of the flame in the engine [1]. When using the Computational Fluid Dynamics (CFD) to simulate the real process in a combustion chamber, a simplified chemical reaction model is necessary. One of the criteria of the validation of the model is that it should correctly reflect the behavior of the ignition delay time [2,3]. The ignition delay time is also a mark to indicate what an additive is more effective to enhance the ignition process for hydrocarbon fuels [4,5].

Kerosene is an important hydrocarbon fuel being used for supersonic combustion. There appears to be very little ignition delay time data available for kerosene. Mullins' data have been used to characterize the ignition behaviors for kerosene [6]. However, Mullins' data have been obtained in a combustion rig fed with hot vitiated air [7]. In addition to chemical characteristics, these data include the effects of many physical factors, such as fuel atomization, and air vitiation. It should be noted that Mullins' data showed fuel atomization has a small effect on the ignition delay for kerosene, but the atomization is known to be important to liquid fuel ignition. From a purely academic viewpoint, ignition delay measurements are necessary, which include only the chemical component of the ignition process and exclude the vitiation effect, because these data are more fundamental and in any case of more widespread application to science generally.

The shock tube is widely used in the study of the ignition delay time of hydrocarbon fuels. Davidson et al. [8] have noted that the uncertainty in the concentration of the fuel vapor due to the adsorption of the fuel vapor on the shock tube wall is one of the error sources in ignition time measurements of liquid fuels. To account for this effect, they put forward a technique to measure the in situ vapor concentration of JP-10 using laser absorption at 3.39μm [9]. However, kerosene is a complex mixture of many hydrocarbon components rather than a single component fuel. To assess the degree of fuel vapor adsorption and enable a more accurate measurement of the fuel vapor concentration in gas phase may be important criteria in determining the reliable ignition delay time. In this study, the gaseous mean concentration of kerosene was determined by directly measuring the vapor pressure of kerosene using a high-precision vacuum gauge, in conjunction with gas chromatography to assess the adsorption contents for the different components of kerosene. Then, the ignition delay times for kerosene diluted in argon were measured under wider conditions. Furthermore, many existing measurements in shock tube have been carried out at the temperature range higher than 1300K, which is not enough for application in the scramjet engine working at the lower temperature part. The current study is concerned with the ignition delay time for kerosene over a wider temperature range, especially extending the lower temperature bound to 1000K.

2 Experimental

The experiments were carried out in a shock tube at the Institute of Mechanics, Chinese Academy of Sciences. The driver section has a length of 2.0m and the driven section is 1.8m in length. Both have circular cross sections with an inner diameter of 44mm. The driven section was evacuated down to the ultimate pressure of 1×10^{-2}Pa by a turbomolecular pump. The outgassing rate was smaller than 5×10^{-3}Pa/min. The incident shock speeds were measured by two piezoelectric transducers mounted on the shock tube sidewall. The conditions behind the reflected shock were calculated from the incident shock speed using the one-dimensional shock relations, and the contributions of oxidizer and fuel to the specific heat ratio and the sound speed were taken into account. A piezoelectric transducer mounted on the endwall of the driven section was used to monitor the pressure evolution in the reflected shock region. A quartz window was installed on the sidewall very close to the endplate of the driven section to monitor the emission from the ignition process in the reflected shock region. The emission focused through a lens was detected by using a photomultiplier after passing through a monochromator centered at the emission line of OH radical. The pressure and emission signals were recorded finally by a transient A/D transducer. A schematic of the facility is shown in Fig. 1.

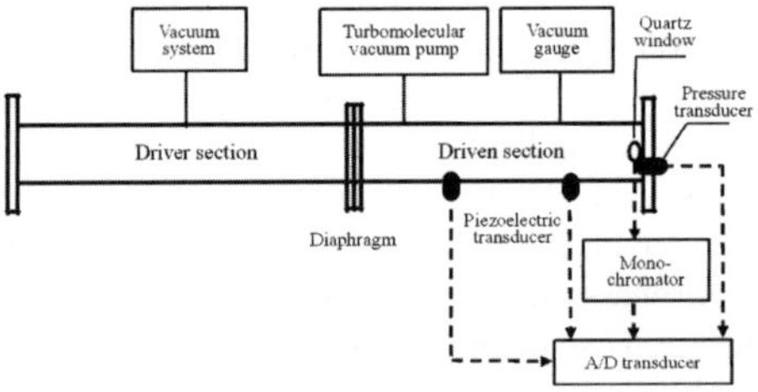

Fig. 1. Schematic of shock tube facility

A longer observation time is required as ignition times increase at the lower temperature region. For the purpose of the present effort the shock tube was run under conditions for a tailored interface, resulting in an observation time of about 6ms. Therefore, instead of pure He, a mixture of N_2 and He with the specific ratio was used as the driver gas to decrease the sound speed of the region 3 of shock tube and achieve conditions, in which the interaction of the reflected shock with the contact surface produces no secondary reflected wave. Under conditions for a tailored interface, the lower temperature bound of experiments was extended to 1000K in the current study.

Kerosene is a complex mixture of many heavy hydrocarbon components. The vapor saturation pressures of heavy hydrocarbons at room temperature are low. There exists a severe adsorption on the wall in shock tube experiments for kerosene, resulting in the uncertainty in determination of the composition of the test gas mixture. The very low vapor pressure of kerosene also limits the experimental concentration range. To minimize the degree of adsorption and increase the test vapor pressure, the shock tube and the mixing tank for the mixture of O_2 and Ar were preheated and maintained at 343K throughout the experiments. The further increase in temperature of preheating shock tube has not been adopted, because of the limitation of the working condition for measurement transducers. Since kerosene is a complex mixture of many heavy hydrocarbon components, the adsorption content of different component differs, leading to the gas composition is different from the liquid composition. A gas chromatograph with a flame ionization detector was used for the measurements of the adsorption content for kerosene. Fig. 2 are the chromatograms, and the left was obtained by directly injecting liquid kerosene sample into the gas chromatograph. After injecting a liquid kerosene sample into the shock tube and mixing fully with Ar, then a gas sample from the shock tube was introduced into the gas chromatograph, the right chromatogram in Fig. 2 was obtained, showing the decrease for peaks of retention times longer than 22min. Fig. 2 indicates clearly the difference in the adsorption content of different component for kerosene on the shock tube wall. The components of longer retention times have the higher boiling points and adsorb more easily on the shock tube wall.

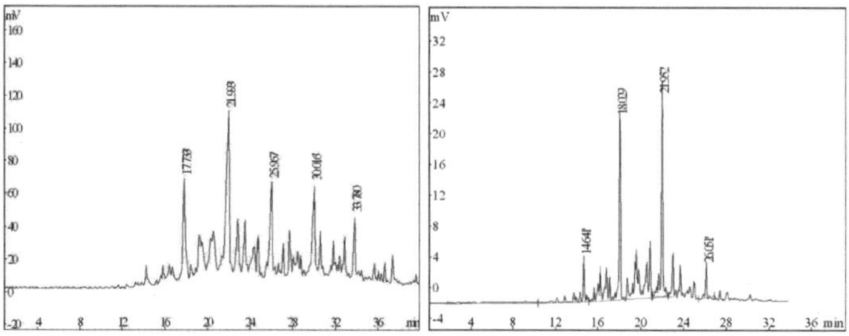

Fig. 2. Chromatograms for kerosene obtained both by directly injecting a liquid sample into the gas chromatograph(left) and when a gas sample from the shock tube was introduced into the gas chromatograph(right)

In the present study, a simulant modified fuel for kerosene was prepared by adding some heavy hydrocarbon components into the original kerosene in proportion to the adsorption content to compensate the loss in the gas phase through the adsorption on the wall. Thus the gas composition of the modified fuel was almost same as the original kerosene. In the current study, the ignition delay measurements were made for this modified kerosene, and the mean gas concentration for kerosene was determined by directly measuring the kerosene vapor pressure and using one-formula surrogate $C_{10}H_{22}$ to represent kerosene. After the shock tube was filled with kerosene, the kerosene vapor pressures in adsorption equilibrium were measured directly by using a high-precision vacuum gauge. A membrane vacuum gauge (Beijing Vacuum Instrument Factory, Model ZDM-1) was selected to measure the kerosene pressure, with a resolution of 0.1Pa when the working pressure is under 1000Pa. The time interval from injecting kerosene to the diaphragm rupture was about 10min. To estimate the adsorption extent, the adsorption curves of kerosene in the driven section maintained at 343K were measured during 10min after injecting 10μL, 20μL and 30μL of liquid kerosene, respectively, the results are shown in Fig. 3. These curves show the adsorption equilibriums are achieved almost in 5min and completely in 10min.

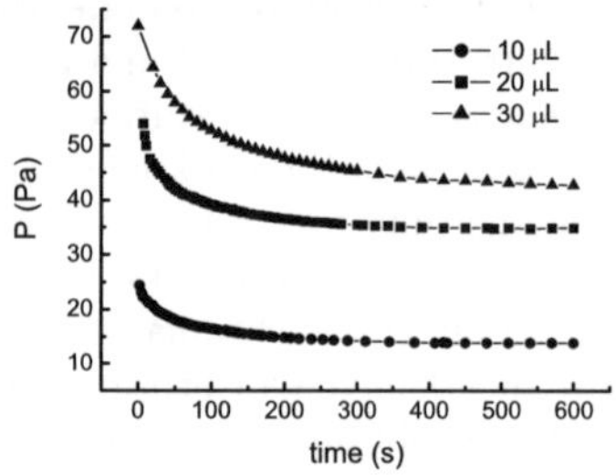

Fig. 3. Adsorption curves of the modified kerosene

After the driven section was evacuated to the ultimate pressure, a quantitative liquid kerosene was injected into the driven section, then the pressure was measured 5 min after the fuel evaporated and adsorbed. Then, the mixture of O_2 (99.995% pure) and Ar (99.99% pure) as diluent was added into the driven section, and an additional time of 5 min was allowed for the gases to mix fully, and the final total pressure was measured 10min after injecting kerosene. Our experience shows that 5min is sufficient to mix the gases fully. Thus the kerosene concentrations could be determined by using the direct pressure measurements based on the adsorption curves. The time interval of 10min from injecting fuel to the diaphragm rupture was specified to ensure both achievement of kerosene adsorption equilibrium and homogeneity of the test mixture. The driver section was simultaneously filled with N_2 and He at a required ratio as the tailored interface operation is achieved.

Experiments were performed over the pressure range of 1.8-5.0atm, the temperature range of 1030-1860K, and fuel concentrations of 0.1-0.33% mole fraction. The total pressure and the emission of OH radical at 309nm in the reflected shock region were recorded after the bursting of the diaphragm. The radical emission was used as the marker to identify the ignition time. An example of pressure and OH emission data is shown in Fig. 4. P_5 indicates the passing of the reflected shock and the pressure evolution, and

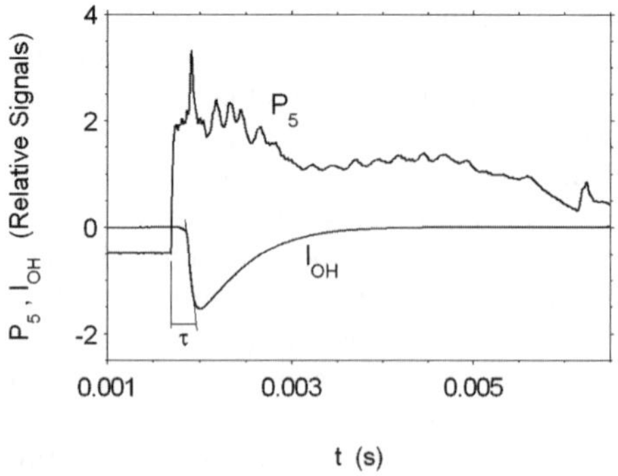

Fig. 4. Example of OH emission and pressure data. Reflected shock conditions: 1458K, 0.12% kerosene, 6.5%O_2, and 93.4%Ar

I_{OH} indicates the OH emission evolution in the reflected shock region. It can be seen that the abrupt increases in pressure and emission occur at the same time, and the onset of ignition is unequivocal, especially compared to the more conventional determination from the abrupt increase in pressure [10]. The ignition delay time is defined as the time interval between the passing of the reflected shock and the onset of ignition, which is usually determined by fitting a straight line to the initial rapid rising portion of the emission and extrapolating until it intersects with zero, as shown in Fig. 4.

3 Results

Ignition delay times were measured for four different fuel/O_2/Ar mixtures with stoichiometric ratios of ϕ=2.0, 1.0, 0.5 and 0.25, respectively. The initial compositions of the diluent gases and the injection volumes of kerosene are given in Table 1.

Table 1. Initial composition

No.	Composition of diluent gas	Injection volume of kerosene	ϕ
1	2.5%O_2+97.5%Ar	30 μL	2.0
2	5.0%O_2+95.0%Ar	30 μL	1.0
3	8.0%O_2+92.0%Ar	20 μL	0.5
4	6.5%O_2+93.5%Ar	10 μL	0.25

The kerosene ignition delay time data are shown in a logarithmic plot of the ignition time vs the reciprocal temperature in Fig. 5. The slopes of the ignition curves are almost the same with increasing ϕ, suggesting that the ignition activation energy varies only slightly. A least square analysis was performed to correlate the kerosene ignition time data with temperature and concentrations of kerosene and oxygen. For the temperature range of 1030-1870K, the correlation was obtained with an R^2 correlation coefficient of 0.990 in the following form

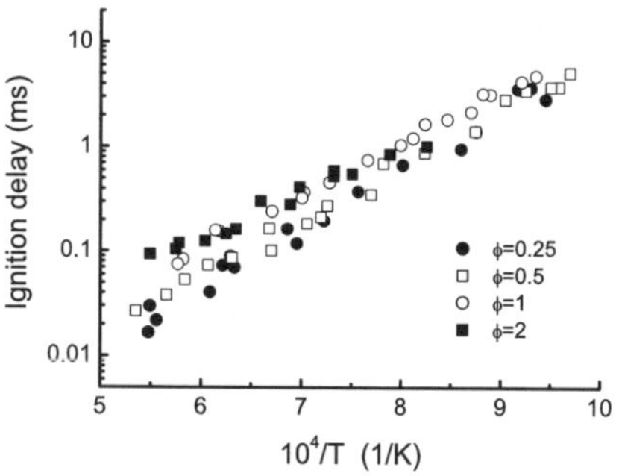

Fig. 5. Ignition delay times for kerosene

$$\tau = 2.58\times10^{-7}[Kerosene]^{0.42}[O_2]^{-0.40}exp(\frac{100663}{RT}) \quad (1)$$

where τ is in seconds, ϕ is the stoichiometric ratio, $[Kerosene]$ and $[O_2]$ are in mol/cm^3, and the activation energy is in J/mol.

The ignition delay time correlation for kerosene shows a relatively weak power dependency (-0.4) for oxygen, but the kerosene concentration power dependency (0.42) is the same as other single-component heavy hydrocarbon fuels that have a strong oxygen power dependency of about -1.0 [11].

Acknowledgement. This work was supported by the National Natural Science Foundation of China (Grant No. 90305021).

References

1. Tishkoff J. M., Drummond J. P., Edwards T., Nejad A. S., Future direction of supersonic combustion research: Air Force/NASA workshop on supersonic combustion, AIAA Paper 97-1017(1997)
2. Held T. J., Marchese A. J., Dryer F. L., A semi-empirical reaction mechanism for n-heptane oxidation and pyrolysis, Combust. Sci. and Technol., **123**, 107(1997)
3. Li S. C., Varatharajan B., Williams F. A., Chemistry of JP-10 ignition, AIAA J., **39**(12), 2351(2001)
4. Davidson D. F., Horning D. C., Hanson R. K. Hitch B., Shock tube ignition time measurements for n-Heptane/O_2/Ar mixtures with and without additives, *Proceedings of the 22nd International Symposium on Shock Waves*, Imperial College, London, UK, 191(1999)
5. Sidhu S. S., Graham J. L., Kirk D. C., Maurice L. Q., Investigation of effect of additives on ignition characteristics of jet fuels:JP-7 and JP-8, *Proceedings of the 22nd International Symposium on Shock Waves*, Imperial College, London, UK, 285(1999)
6. Veretennicov V. G., Autoignition study on kerosene in supersonic flow, Moscow Aviation Institute, SPC 96-4089, Moscow, Russia, 1997
7. Mullins B. P., Studies on the spontaneous ignition of fuels injected into a hot air-stream part 2: Effect of physical factors upon the ignition delay of kerosene-air mixtures, Fuel, **32**, 234(1953)
8. Davidson D. F., Horning D. C., Hanson R. K., Shock tube ignition time measurements for n-Heptane/O_2/Ar and JP-10/O_2/Ar mixtures, AIAA Paper 99-2216(1999)
9. Davidsion D. F., Horning D. C., Herbon J. T., Hanson R. K., Shock tube measurements of JP-10 ignition, Proc. Combust. Inst., **28**, 1687(2000)
10. Brown C. J., Thomas G. O., Experimental studies of shock-induced ignition and transition to detonation in ethylene and propane mixtures, Combust. Flame,**117**, 861(1999)
11. Colket M. B., Spadaccini L. J., Scramjet fuels autoignition study, J. Prop. Power, **17**(2), 315(2001)

Shock-tube study of the ignition delay time of tetraethoxysilane (TEOS)

A. Abdali[1], M. Fikri[1], H. Wiggers[1,2], and C. Schulz[1,2]

[1] *IVG, Universität Duisburg-Essen, Lotharstr. 1, 47057 Duisburg, Germany*
[2] *CeNIDE, Center for Nanointegration Duisburg-Essen, Duisburg, Germany*

Summary. Ignition delay times of tetraethoxysilane (TEOS) mixtures diluted with dry and humid synthetic air were measured behind reflected shock waves by monitoring the chemiluminescence emission of the CH* radical near 431 nm. The measurements were carried out for temperatures ranging from 1150 to 1350 K and an average pressure of 8 bar. The ignition-delay times of TEOS were determined as a function of water concentration. Additionally, TEOS stability in presence of humid air was investigated by means of Fourier transform infrared spectroscopy (FTIR) to insure that the mixture does not react prior to shock-wave heating. The results show that the stability of TEOS is not affected under the experimental conditions but that the presence of moisture in the TEOS/air mixture increases the ignition-delay times systematically.

1 Introduction

Silica nanoparticles find a large variety of practical applications. They are widely used as filler in plastics and coatings to improve material properties like hardness, tensile strength, abrasion resistance, and thermal stability. Especially, applications with a demand of high transparency require small silica particles with specific size, morphology, and surface coating [1]. Therefore, the synthesis of particles with highly-defined particle-size distributions is desired. Additionally, the degree of agglomeration is an important quality index in a number of applications. For example, agglomerated, nanostructured particles are needed in manufacturing of fillers and catalysts. In contrast, non-agglomerated nanoparticles are needed in ceramics, composites and electronics [2]. Non-agglomerated particles are usually produced by wet chemistry but these processes tend to be costly and may result in powders of limited consistency during large scale manufacture as the synthesis involves many process steps [3]. In this context, particle generation from the gas phase is of high interest because of its high selectivity, reduced number of process steps, continuous processing and, therefore, reduced manufacturing costs.

Typical gas-phase processes for the production of silica nanoparticles use hot-wall reactors, flame reactors and spray pyrolysis. Tetraethoxysilane (TEOS) as a halide-free and inexpensive precursor material is of growing interest for SiO_2 particle synthesis from the gas-phase. Many processes require the mixing of TEOS with air before the synthesis step. This paper reports on the kinetics of tetraethoxysilane decomposition by ignition of TEOS in air at high temperatures. These experiments provide valuable information for a controlled formation of highly-defined silica nanoparticles for specific applications. Because gas phase processes for nanoparticle formation are often based on flame-synthesis, not only ignition of TEOS in dry air but also the ignition of TEOS in presence of humidity has been investigated.

The thermal decomposition of TEOS under inert conditions has been investigated by Herzler et al. using a single-pulse shock tube [4]. The distribution of gaseous products was monitored by gaschromatography, and the main stable products were found to be ethylene and ethanol. From the product distribution and yield, a mechanism was derived for the initial thermal decomposition of TEOS and the subsequent reactions. Chu et al. [5] also studied the thermal decomposition of TEOS in a static system with species detection by Fourier-transform infrared spectroscopy (FTIR) in the temperature range between 700 and 820 K. The main product they found was ethanol and the decomposition could be described assuming a six-ring elimination mechanism. The thermal decomposition of TEOS was also studied in a low-pressure CVD reactor at 13 Pa and 950 K. Reaction products as well as the rate of film growth were determined and fitted with a new model. This model quantitatively accounts in the first step for the ethylene formation and the condensation of silanol with TEOS, which finally leads to the formation of ethanol as a stable product [6].

2 Experimental set-up

The experiments were carried out in a shock tube to study the ignition-delay time of the precursor (TEOS) as a function of temperature. The shock tube consists of a stainless steel tube with an internal diameter of 8 cm. It is divided by a diaphragm into driver and driven (test) section with a length of 3.8 m and 7.2 m, respectively. The diaphragms consist of 300-μm thick aluminum foil and were chosen to achieve the desired temperature behind the reflected shock. The driven section can be baked out and pumped down to pressures below 10^{-6} mbar by a turbo molecular pump and the driver section to 10^{-2} mbar by a rotary pump. Gas mixtures were prepared in a stainless-steel UHV mixing vessel, which can also be evacuated by a combination of a rotary pump and a turbo-molecular drag pump. Homogeneous mixtures were achieved by injecting liquid TEOS into the evacuated stainless-steel mixing vessel with subsequent addition of dry synthetic air. The ratio of TEOS in the gas mixture was always 1% by mass. For the experiments in humid air the desired amount of water was injected directly into the driven section of the shock tube, and then mixed with the TEOS/air mixture from the vessel. The investigations were carried out in dry as well as in humid air with a moisture level of 5% H_2O by mass. In order to investigate the influence of water concentration on the ignition-delay time of TEOS, the water concentration was varied from 0 to 10% H_2O by mass.

The ignition was observed by measuring the onset of CH*-chemiluminescence emission at 431 nm. The output light was selected by a band-pass filter and detected with a photomultiplier located downstream on the measurement plane. All ignition-delay times shown in this work were determined by extrapolating the steepest increase of the emission signal to its zero level on the time axis.

The gases and liquids used were of highest commercial purity: Ar $>$ 99.9999%, TEOS $>$ 99% and synthetic air $>$ 99.9999%. The reflected-shock temperatures are obtained from measurements of the incident shock-wave velocity assuming a one-dimensional behavior. A series of experiments tested the stability of TEOS in presence moist air at room temperature. The mixture was analyzed by FTIR using a commercial spectrometer (Bruker) with a resolution of 0.25 cm^{-1}.

3 Results and discussion

The stability of TEOS/air/water mixtures was investigated by FTIR spectroscopy. The composition of the fresh mixture (3.5 mbar TEOS, 1.2 mbar H_2O diluted in 570 mbar He) as well as the mixture after waiting for 30, 75 and 135 min was investigated. The results are illustrated in the figure . After 135 min the spectrum shows a minor absorption band near 1000 cm^{-1} (Q-branch) which is assigned to ethylene. Ethanol was not detected under our experimental conditions. These minor product formation implies the full stability of tetraethoyxsilane on the time scale of interest. The band at 2250 cm^{-1} (CO_2) is due to the impurity in the cell. In all shock-tube experiments the onset of

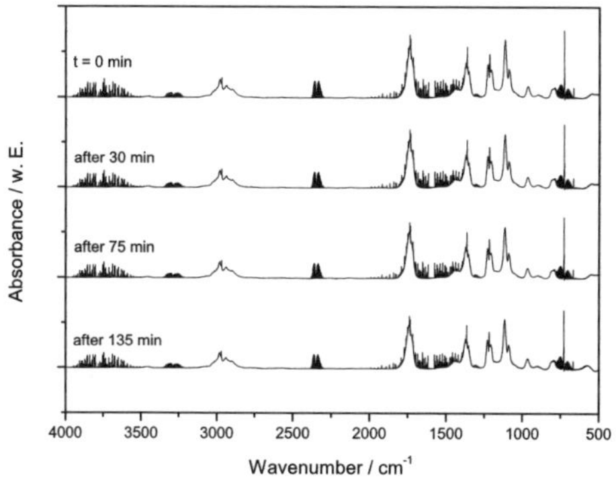

Fig. 1. FTIR spectrum of TEOS/H_2O/He mixture obtained immediately after mixing and at different residence times at room temperature.

CH* chemiluminescence and the pressure increase were used to deduce the ignition-delay times. A typical measurement illustrating the ignition-delay time is given in fig. 2. To neglect the influence of the pressure on the experimental results, all experiments were carried out at $p = 8$ bar.

The results of the ignition-delay times of TEOS in dry and humid air measured from the chemiluminescence onset are depicted as an Arrhenius-like representation (fig. 3). It shows that the ignition-delay time depends very strongly on the temperature and can be expressed approximately by straight line. It shows also that the presence of moisture in the mixture increases the ignition-delay times of TEOS. The influence of moisture on the ignition-delay time is not yet understood. The ignition-delay time of TEOS in dry air is expressed as a function of temperature:

$$\tau = 5.86 \times 10^{-9} \exp(28800 \text{ K}/T) \qquad (1)$$

and the ignition-delay time of TEOS in humid air was determined as:

$$\tau = 7.76 \times 10^{-6} \exp(21400 \text{ K}/T) \qquad (2)$$

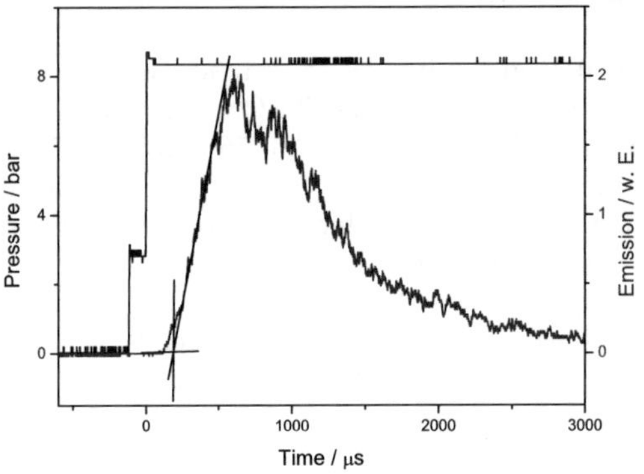

Fig. 2. CH* chemiluminescence and pressure as a function of time after the arrival of the reflected shock at $T_5 = 1190$ K. The observed ignition delay time at this conditions is $\tau = 190$ μs.

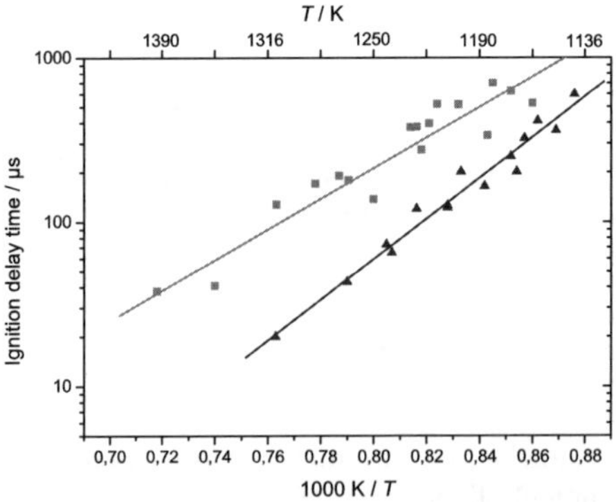

Fig. 3. Ignition-delay times of TEOS in dry air (red triangles) and in presence of moisture $[H_2O] = 5$ % per mass (green squares).

To clarify the influence of water on the ignition-delay time measurements were performed at various temperatures and different amounts of water vapor. Therefore, the ignition-delay times of TEOS/air mixtures were investigated as a function of water/TEOS mass-ratio and temperature (fig. 4). It shows that the experimental results obtained at different temperature and water/TEOS ratios from 0 to 10 by mass. It can be clearly seen that even traces of water in the gas mixture increase the ignition delay times significantly up to an order of magnitude. Moreover, at water/TEOS ratios above 2 no further increase of ignition delay times was found.

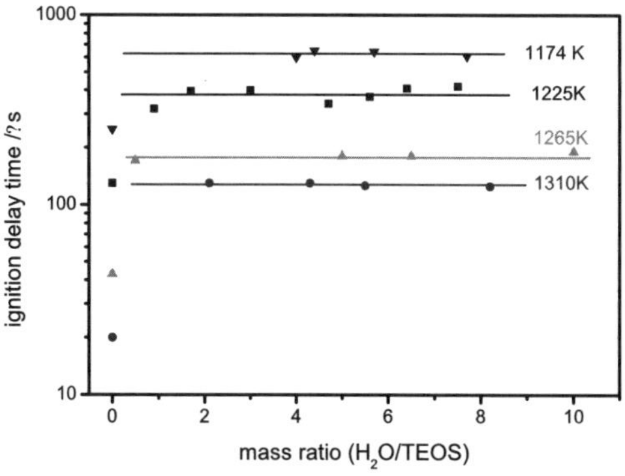

Fig. 4. Ignition-delay times of tetraethoxysilane as a function of water concentration at various temperatures: The lines show the average values of the measurements in the regime with H_2O.

4 Conclusions

The ignition-delay times of TEOS in dry and humid synthetic air was measured in shock-tube experiments as a function of temperature and air humidity. Arrhenius-type expressions for the temperature dependence of the ignition-delay time are given for dry and humid air. The presence of water significantly increases the ignition-delay time. Investigations with variable water/TEOS mass-ratios show that the ignition-delay times increase initially even in the presence of traces of water and remain constant at water/TEOS ratios above 2 by mass. The stability of TEOS vapor at room temperature in the presence of humidity was investigated by means of FTIR spectroscopy and shows that TEOS does not react with water at room temperature on the time scale of the mixing prior to the shock-wave experiment.

References

1. A. Gutsch, H. Mühlenweg, et al. Tailor-made nanoparticles via gas-phase synthesis. Small, 1(1): 30-46, 2005
2. R. Mueller, H. K. Kammler, S. E. Pratsinis, A. Vital, G. Beaucage, P. Burtscher. Non-agglomerated dry silica nanoparticles. Powder Technol. 140: 40-48, 2004.
3. S. E. Pratsinis. Material synthesis in aerosol reactors. Chem. Eng. Prog. 85(5): 62-66, 1989.
4. J. Herzler, J. A. Manion, et al. Single-Pulse Shock Tube Study of the Decomposition of Tetraethoxysilane and Related Compounds. J. Phys. Chem. A 101(30): 5500 -5508, 1997.
5. J. C. S. Chu, R. S., M. C. Lin. Thermal Decomposition of Tetraethyl Orthosilicate in the Gas Phase: An Experimental and Theoretical Study of the Initiation Process. Journal of Physical Chemistry 99: 663-672, 1995.
6. H. Takeuchi, H. Izami, and A. Kawasaki. Decomposition study on TEOS. Mater. Res. Soc. Symp. Proc. 334: 45, 1994

Author Index

Keyword Index

Printing: Krips bv, Meppel, The Netherlands
Binding: Stürtz, Würzburg, Germany